FUNDAMENTALS OF DRAFTING TECHNOLOGY

FUNDAMENTALS OF DRAFTING TECHNOLOGY

David A. Madsen

Department Chairman Drafting Technology
Authorized Autodesk Training Center for AutoCAD
Authorized SmartCAM Training Center
Member, Board of Directors, American Design Drafting Association
Authorized Softdesk Training Center for Civil Engineering Authorized AutoCAD Learning Center
Clackamas Community College, Oregon City, OR

Terence M. Shumaker (Chapters 1 CADD, 2, and 14)

Drafting/CADD Instructor
Manager
Authorized Autodesk Training Center for AutoCAD
Clackamas Community College, Oregon City, OR

J. Lee Turpin (Chapter 6)

Former Drafting Instructor/Department Chairman
Vocational Counselor
Clackamas Community College, Oregon City, OR

Delmar Publishers Inc.

I(T)P™

NOTICE TO THE READER

Publisher does not warrant or guarantee any of the products described herein or perform any independent analysis in connection with any of the product information contained herein. Publisher does not assume, and expressly disclaims, any obligation to obtain and include information other than that provided to it by the manufacturer.

The reader is expressly warned to consider and adopt all safety precautions that might be indicated by the activities described herein and to avoid all potential hazards. By following the instructions contained herein, the reader willingly assumes all risks in connection with such instructions.

The publisher makes no representations or warranties of any kind, including but not limited to, the warranties of fitness for particular purpose or merchantability, nor are any such representations implied with respect to the material set forth herein, and the publisher takes no responsibility with respect to such material. The publisher shall not be liable for any special, consequential, or exemplary damages resulting, in whole or in part, from the readers' use of, or reliance upon, this material.

Cover Design: design M design W

Delmar Staff
Executive Editor: Michael McDermott
Development Editor: Sheila Davitt
Project Editor: Elena Mauceri, Barbara Riedell
Editorial Assistant: Gwen Ceruti-Vincent
Production Supervisor: Larry Main
Art/Design Supervisor: Judi Orozco
Art Coordinator: Cheri Plasse
Design Coordinator: Lisa Bower

For information address Delmar Publishers Inc.
3 Columbia Circle Drive, Box 15–015
Albany, New York 12212–5015

Printed in the United States of America
Published simultaneously in Canada
by Nelson Canada,
a division of The Thomson Corporation

10 9 8 7 6 5 4 3 2 1 XX 99 98 97 96 95 94

Library of Congress Cataloging-in-Publication Data

Madsen, David A.
Fundamentals of drafting technology/David A. Madsen [with] Terence M. Schumaker, J. Lee Turpin.
p. cm.
Rev. ed. of: Engineering drawing and design. c1991.
Includes index.
ISBN 0-8273-5238-7
1. Mechanical drawing. I. Schumaker, Terence M. II. Turpin, J. Lee. III. Madsen, David A. Engineering drawing and design. IV. Title.
T353.E617 1994
604.2—dc20 93-10536
CIP

Contents

PREFACE viii

ACKNOWLEDGMENTS xi

THE SHORT HISTORY OF MECHANICAL DRAWING xii

INTRODUCTION 1

Drafting Technology 1
Drafting Occupations 1
Professional Organization 9
How to Become a Drafter 9
Drafting Occupational Levels 9
Drafting Job Opportunities 9
Drafting Salaries and Working Conditions 10
Computers in Drafting 10

CHAPTER 1 DRAFTING EQUIPMENT, MEDIA, AND REPRODUCTION METHODS 13

Drafting Equipment 13
Drafting Furniture 14
Drafting Pencils, Leads, and Sharpeners 14
Technical Pens and Accessories 16
Erasers and Accessories 17
Drafting Instruments 18
Drafting Machines 25
Scales 31
Cleaning Drafting Equipment 34
Drafting Media 34
Papers and Films 34
Sheet Sizes, Title Blocks, and Borders 36
Diazo Reproduction 41
Safety Precautions for Diazo Printers 42
Photocopy Reproduction 43
Microfilm 43
Professional Perspective 44
Drafting Equipment Problems 45

CHAPTER 2 INTRODUCTION TO COMPUTER-AIDED DESIGN AND DRAFTING (CADD) 47

What Is Computer Graphics? 47
Developing New Skills 48
The New Drafting Environment 49
Computers in Drafting 50
Industry and CADD 53
The CADD Job Market 54
Educational Preparation for CADD 55
The Language of the Computer 56
Drawing with the Computer 58
Computer Drafting Equipment 62
Future CADD Hardware and Software 73
CADD Materials 74
Getting Started with CADD 78
Creating and Using Symbols 82
Storing a Drawing 83
Professional Perspective 84
Engineering Problem 84
Introduction to CADD Problems 85

CHAPTER 3 SKETCHING, LETTERING, AND LINES 87

Sketching 87
Tools and Materials 87
Straight Lines 88
Circular Lines 88
Measurement Lines and Proportions 90
Procedures in Sketching 91
Multiview Sketches 92
Isometric Sketches 93
Lettering 97
Single-Stroke Gothic Lettering 97
Microfont Lettering 97
Other Lettering Styles 98
Lettering Legibility 100
Lettering in Industry 100
Vertical Freehand Lettering 100
Lettering Techniques 102
Composition 102
Making Guidelines 102
Other Lettering Aids and Guideline Methods 104
Points to Remember 104
Lettering Guide Templates 105
Mechanical Lettering Equipment 105
Using Lettering Equipment 106
Machine Lettering 106
Typewriters 107
Transfer Lettering 107
Lines 109
Kinds of Lines 110
Pencil and Ink Line Techniques 117
Professional Perspective 121
Engineering Problem 121
Sketching Problems 123
Lettering Problems 123
Line Problems 124
Drawing with the Computer Problems 127

CHAPTER 4 GEOMETRIC CONSTRUCTION 129

Line Characteristics 129
Geometric Shapes 129
Common Geometric Constructions 132
Constructing Polygons 135
Constructing Circles and Tangencies 141
Constructing an Ellipse 145
Professional Perspective 148
Engineering Problem 148
Geometric Construction Problems 150

CHAPTER 5 MULTIVIEWS AND AUXILIARY VIEWS 153

Multiviews 153
View Selection 156
Projection of Circles and Arcs 159
Third-angle Projection 162
First-angle Projection 163
Layout Techniques 164
Auxiliary Views 168
Plotting Curves in Auxiliary Views 172
Enlargements 172
Secondary Auxiliary Views 172
Professional Perspective 175
Engineering Problem 175
Multiviews and Auxiliary View Problems 176

CHAPTER 6 DESCRIPTIVE GEOMETRY 191

Standards of Projection 192
Projection of a Point 193
Lines in Space 195
Classification of Lines 195
Planes 200
Professional Perspective 204
Engineering Problem 204
Descriptive Geometry Problems 206

CHAPTER 7 MANUFACTURING PROCESSES 211

Manufacturing Materials 211
Manufacturing Processes 216
Machine Processes 221
Tool Design 235
Statistical Process Control (SPC) 236
Professional Perspective 238
Engineering Problem 238
Manufacturing Processes Problems 239

CHAPTER 8 DIMENSIONING 240

Dimensioning Systems 240
Basic Dimensioning Concepts 245
Notes for Size Features 255
Location Dimensions 259
Dimension Origin 261
Dimensioning Auxiliary Views 261
General Notes 262
Tolerancing 264
Dimensions Applied to Platings and Coatings 268
Maximum and Minimum Dimensions 269
Machined Surfaces 273
Design and Drafting of Machined Features 278
Symbols 279
Professional Perspective 279
Engineering Problem 282
Dimensioning Problems 286

CHAPTER 9 FASTENERS AND SPRINGS 300

Screw-thread Fasteners 300
Thread-cutting Tools 302
Thread Forms 303
Thread Representations 305
Thread Notes 309
Measuring Screw Threads 311
Threaded Fasteners 312
Washers 316
Dowel Pins 317
Taper and Other Pins 317
Retaining Rings 318
Keys, Keyways, and Keyseats 318
Rivets 318
Springs 319
Professional Perspective 323
Engineering Problem 324
Fasteners and Springs Problems 324

CHAPTER 10 SECTIONS, REVOLUTIONS, AND CONVENTIONAL BREAKS 332

Sectioning 332
Cutting-plane Lines 332
Section Lines 334
Full Sections 335
Half Sections 336
Offset Sections 336
Aligned Sections 337
Unsectioned Features 337
Conventional Revolutions 339
Broken-out Sections 339
Auxiliary Sections 340
Conventional Breaks 340
Revolved Sections 341
Removed Sections 342
Professional Perspective 343
Engineering Problem 344
Sectioning Problems 344

CHAPTER 11 GEOMETRIC TOLERANCING 367

General Tolerancing 367
Symbology 368
Datum Feature Symbols 368
Datum Reference Frame 370
Datum Features 370
Datum Target Symbols 374
Datum Axis 374
Feature Control Frame 377
Basic Dimensions 378
Geometric Tolerances 378
Material Condition Symbols 393
Location Tolerance 396
Datum Precedence and Material Condition 400

Position of Multiple Features 400
Composite Positional Tolerancing 402
Projected Tolerance Zone 403
Virtual Condition 407
Combination Controls 408
Professional Perspective 408
Engineering Problem 410
Geometric Tolerancing Problems 412

CHAPTER 12 CAMS, GEARS, AND BEARINGS 425

Mechanisms 425
Cams 425
Cam Displacement Diagrams 426
Construction of an In-line Follower Plate Cam Profile 430
Construction of an Offset Follower Plate Cam Profile 432
Drum Cam Drawing 432
Gears 434
Gear Types 434
Spur Gear Design 436
Designing and Drawing Spur Gears 439
Designing Spur Gear Trains 439
Designing and Drawing the Rack and Pinion 441
Designing and Drawing Bevel Gears 443
Designing and Drawing Worm Gears 443
Bearings 443
Drawing Bearing Symbols 446
Bearing Codes 447
Bearing Selection 448
Gear and Bearing Assemblies 452
Professional Perspective 454
Cam, Gear, and Bearing Problems 455

CHAPTER 13 WORKING DRAWINGS 460

Detail Drawings 460
Assembly Drawings 462
Types of Assembly Drawings 465
Identification Numbers 470
Parts Lists 472
Purchase Parts 474
Engineering Changes 474
Professional Perspective 476
Engineering Problem 479
Working Drawings Problems 479

CHAPTER 14 PICTORIAL DRAWINGS 503

Pictorial Drawing 503
Isometric Projection 505
Types of Isometric Drawings 506
Isometric Construction Techniques 506
Dimetric Pictorial Representation 511
Trimetric Pictorial Representation 512
Exploded Pictorial Drawing 513
Oblique Drawing 513
Perspective Drawing 515
One-point Perspective 516
Two-point Perspective 517
Three-point Perspective 518
Circles and Curves in Perspective 519
Basic Shading Techniques 521
Layout Techniques 522
Professional Perspective 522
Engineering Problem 522
Pictorial Drawing Problems 523

CHAPTER 15 ENGINEERING CHARTS AND GRAPHS 529

Rectilinear Charts 530
Surface Charts 537
Column Charts 537
Pie Charts 539
Polar Charts 540
Nomographs 540
Trilinear Charts 542
Flowcharts 542
Distribution Charts 543
Pictorial Charts 543
Professional Perspective 544
Engineering Problem 544
Engineering Charts and Graphs Problems 547

APPENDICES 551

Appendix A Abbreviations 552
Appendix B Tables 556
Appendix C American National Standards of Interest to Designers, Architects, and Drafters 583
Appendix D Unified Screw Thread Variations 584
Appendix E Metric Screw Thread Variations 586
Appendix F Geometric Tolerancing Symbol Trees 587

GLOSSARY 589

INDEX 595

Preface

Fundamentals of Drafting Technology is a practical, basic textbook that is easy to use and understand. The content may be used as presented, or the chapters may be rearranged to accommodate alternate formats for traditional or individualized instruction. *Fundamentals of Drafting Technology* may be used for the following courses:

- COMPUTER-AIDED DESIGN DRAFTING (CADD)
  Hardware, software, and applications
- MECHANICAL DRAFTING
  Sketching, lettering, lines, geometric constructions, multiviews and auxiliary views, dimensioning, fasteners, sectioning, working drawings, details, assemblies, parts lists, engineering changes, and geometric tolerancing
- DESCRIPTIVE GEOMETRY
- MANUFACTURING PROCESSES
- GEOMETRIC TOLERANCING
- MECHANISMS
  Gears, cams, bearings, and seals
- PICTORIAL DRAWINGS
- ENGINEERING CHARTS AND GRAPHS

MAJOR FEATURES

Fundamentals of Drafting Technology has these important features:

- CADD throughout
- ANSI and related standards emphasized
- Professional perspectives
- Step-by-step layout methods
- Engineering layout techniques
- CADD techniques

Fundamentals of Drafting Technology provides a practical approach to drafting as related to the American National Standards Institute (ANSI) standards. Also presented, when appropriate, are standards and codes related to specific engineering fields. One excellent and necessary foundation to engineering drawing and the implementation of a common approach to graphics nationwide is the emphasis of standardization in all levels of drawing instruction. When students become professionals, this text will go along as a valuable desk reference.

Each chapter provides realistic examples, illustrations, and problems. The examples illustrate recommended design presentation based on ANSI standards and other related national standards with actual industrial drawings used for reinforcement. The correlated text explains drawing techniques and provides professional tips for skill development. Step-by-step layout methods provide a logical approach to setting up and completing the drawing problems.

INDUSTRIAL APPROACH TO PROBLEM SOLVING

The drafter's responsibility is to convert the engineering sketch or instructions to formal drawings. The text explains how to prepare drawings from engineering sketches by providing the learner with the basic guides for layout and arrangement in a knowledge-building format. One concept is learned before the next is introduced. Problem assignments are presented in order of difficulty within each chapter and throughout the text. The concepts and skills learned in one chapter are used in subsequent chapters so that by the end of the text the student has the ability to solve problems using a multitude of previously learned activities. The problems are presented as engineering sketches or actual industrial layouts in a manner that is consistent with the engineering environment. Early problems provide suggested layout sketches. It is not enough for students to duplicate drawings from given assignments; they must be able to think through the process of drawing development. The goals and objectives of each problem assignment are consistent with recommended evaluation criteria based on the progression of learning activities.

COMPUTER-AIDED DESIGN AND DRAFTING

Computer-aided design and drafting (CADD) is presented throughout the text. CADD topics include:

- CADD hardware
- CADD software
- CADD material requirements
- Specific CADD applications
- CADD template menus and symbol libraries for specific engineering drafting applications
- Increased productivity with CADD
- Parametrics in CADD applications

COURSE PLAN

The introduction to *Fundamentals of Drafting Technology* provides a detailed look at drafting as a profession and includes occupations, professional organizations, occupational levels, opportunities, and computers.

Drafting Equipment, Media, and Reproduction Methods

The study of engineering drawing in this text begins with equipment, materials, and reproduction for manual drafting with specific instruction on how to use tools and equipment.

Introduction to CADD

Next you receive an in-depth introduction to CADD, including developing CADD skills, computer languages, drafting with the computer, using menus, CADD in industry, the CADD environment, the job market, computer drafting equipment, CADD drawing materials, and future CADD hardware and software.

Sketching, Lettering, Lines, and Geometric Construction

You then learn how sketching, lettering, lines, and geometric constructions may be properly used in both manual and computer-aided design and drafting in accordance with ANSI standards.

Multiviews and Auxiliary Views

A complete study of multiviews and auxiliary views, in accordance with ANSI standards, provides accurate and detailed instruction on topics such as view selection and placement, first- and third-angle projection, and viewing techniques. You are provided with a step-by-step example on how to lay out a multiview drawing.

Descriptive Geometry

This text provides basics of the descriptive geometry coverage that students need, including discussion of lines, planes, angles, slope, bearing, and intersecting lines and planes. Problems are presented as actual engineering situations.

Manufacturing Processes

This is a complete introduction to manufacturing processes, including product development, manufacturing materials, material numbering systems, hardness and testing, casting and forging methods, design and drafting, complete machining processes, computer numerical control, computer-integrated manufacturing, machine features and drawing representations, surface texture, design of machine features, tool design, and statistical process control.

Dimensioning

This chapter is in accordance with ANSI Y14.5M and provides complete coverage on dimensioning systems, rules, specific and general notes, tolerances, symbols, and dimensioning for CAD/CAM. You are provided with a step-by-step example showing how to lay out a fully dimensioned multiview drawing.

Fasteners and Springs

This chapter covers the complete range of fastening devices available, including screw threads, thread cutting, thread forms, thread representations and notes, washers, dowels, pins, rings, keys and keyseats, rivets, and springs.

Sections, Revolutions, and Conventional Breaks

This chapter explains in detail every type of sectioning practice available to the mechanical engineering drafter. Additionally, the chapter includes treatment of unsectioned features, conventional revolutions, and conventional breaks.

Geometric Tolerancing

This chapter provides a complete introduction to the interpretation and use of geometric tolerancing symbols and terms as presented in ANSI Y14.5M. It features excellent coverage on datums, feature control, basic dimensions, geometric tolerances, material condition, position tolerance, virtual condition, geometric tolerancing, and CADD.

Cams, Gears, and Bearings

This chapter provides you with extensive coverage of gears and cams. You are given detailed information on the selection of bearings and lubricants. The use of vendors' catalogue information is stressed. Actual mechanical engineering design problems are provided for gear train and cam plate design.

Working Drawings

Working drawings—including details, assemblies, and parts lists. The most extensive discussion on engineering changes available.

Pictorial Drawings

This chapter is a complete review of 3-D drafting techniques used in manual and computer-aided drafting. The content includes extensive discussions in isometric, dimetric, trimetric, perspective, exploded drawing, and shading methods.

Engineering Charts and Graphs

This chapter offers complete coverage on rectilinear, log and semilog scales, surface, column, pie, concurrency, alignment, organizational, statistical process control (SPC), distribution, and pictorial charts. The chapter includes an in-depth analysis of chart design.

SPECIAL FEATURES

The text contains CADD discussion and examples, engineering layout techniques, working from engineer's sketches, professional practices, and actual industrial examples. The problem assignments are based on actual real world engineering sketches or layouts.

SECTION LENGTH

Chapters are presented in individual learning segments that begin with elementary concepts and build until each chapter provides complete coverage of each topic.

APPLICATIONS

Special emphasis has been placed on providing realistic problems. Problems are presented as engineering sketches in a manner that is consistent with industry practices. Most of the problems have been supplied by the industry. Each problem solution is based on the step-by-step layout procedures provided in the chapter discussions. Problems are given in order of complexity so students may be exposed to a variety of engineering experiences. Early problems recommend the layout to help students save time. Advanced problems require the students to go through the same thinking process that a professional faces daily, including drawing scale, paper size selection, view layout, dimension placement, sectioning placement, and many other activities. Problems may be solved using manual or computer techniques as determined by the individual course guidelines. Chapter tests are provided in the Instructor's Guide for complete coverage of each chapter and may be used for student evaluation or as study questions.

Acknowledgments

I would like to give special thanks and acknowledgment to the many professionals who helped formulate this work, and to those who reviewed the manuscript in an effort to help me publish the best possible text:

Bruce Bainbridge
Clinton Community College

Pat Courington
Mid Florida Technical Institute
Valencia Community College

James A. Hardell
Virginia Polytechnic Institute

Tesfahun Berhe
University of Minnesota

I would also like to thank Mark and Wayne Neimeyer for preparing professional AutoCAD illustrations for this text.

The quality of this text is also enhanced by the support and contributions from industry and vendors. The list of contributors is extensive, and acknowledgment is given at each figure illustration. The following individuals and companies gave an extraordinary amount of support:

American National Standards Institute
New York, New York

Rachel Howard
American Design and Drafting Association
Rockville, Maryland

Wallace N. Burkey
Aerojet TechSystems Company
Sacramento, California

Bill Curtis
Curtis Associates
Portland, Oregon

Robert DeWeese
Consul and Mutoh Ltd.
Anaheim, California

Robert F. Franciose
ANSI Y14 Committee
San Jose, California

Jane Frarey
Accugraph Corporation
Ontario, Canada

Wayne Hodgins
Autodesk, Inc.

Steven F. Horton
San Jacinto College

Chester Jensen
Fox Valley Technical Institute

Howard Kaufman
John Doleva
Chartpak
Northampton, Massachusetts

Stanley Kresses
Trott Vocational Technical School

Jim Mackay
Berol USA RapiDesign
Burbank, California

Jim Melloy
Autodesk, Inc.

Tom Charland
Cynthia Ann Murphy
Daniel Partner
Computervision Corporation
Bedford, Massachusetts

Jan Murphy
T & W Systems
Huntington Beach, California

Tom Pearce
Stanley Hydraulic Tools, A Division of The Stanley Works
Portland, Oregon

Renee Randall
CALCOMP
Anaheim, California

Mike Stegmann
Hyster Company
Portland, Oregon

Mark Struges
Autodesk, Inc.

The Short History of Mechanical Drawing

J. H. Oakey

We are all aware of the marvelous drawings and the inventive genius of Leonardo da Vinci. It is naturally assumed that he was the originator of drafting, and that after his work, all inventors and engineers carefully duplicated his designs in some form in their drawings. This is not so. History is not well documented in the realm of inventors and engineers. The only source for the detailed history of Leonardo's work is his own careful representation. His drawings were those of an artist. They were 3-D and they generally were without dimensional notations. Craftsmen worked from the 3-D representation, and each machine or device was one off; parts were not interchangeable. In fact, interchangeability was not realized except in special demonstrations until the development of the micrometer in the late 1800s, and even then it was not easily achieved.

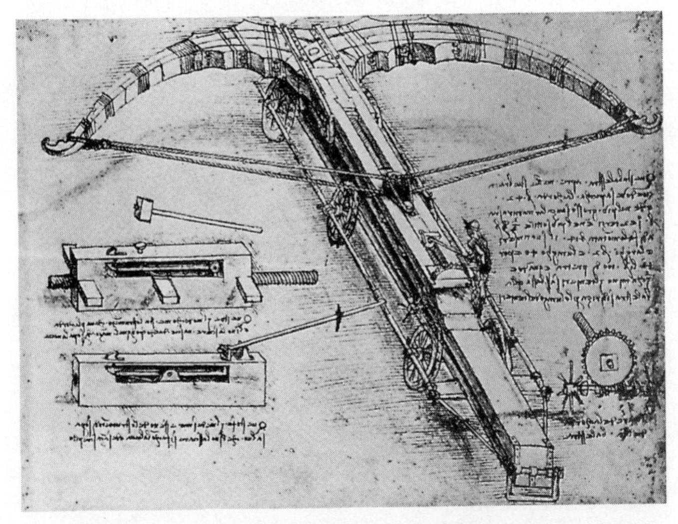

Without the concept of interchangeability, accurate drawings were not necessary. Inventors, engineers, and builders worked on each product on a one-off basis, and parts were manufactured from hand sketches or hand drawings on blackboards. Coleman Sellers, in the manufacture of fire engines, had blackboards with full-size drawings of parts. Blacksmiths formed parts and compared them to the shapes on the blackboards. Coleman Sellers's son, George, recalls lying on his belly using his arm as a radius for curves as his father stood over him directing changes in the sketches until the drawings were satisfactory. Most designs used through the 1800s were accomplished by first completing a hand sketch of the object to be built. These were then converted into wooden models (3-D modeling) from which patterns were constructed. This practice was followed well into this century by some. Most of us are familiar with the stories of Henry Ford and his famous blackboards. What is news is that these were also the Henry Ford "drafting tables." Henry would sketch cars and parts three dimensionally and have patternmakers construct full-size models in wood.

An effort to create a program to standardize drawing came when the Franklin Institute was founded in Philadelphia in 1824. It was founded "to advance the general interests" of mechanicians and entrepreneurs "by extending a knowledge of mechanical science." One of the goals of the Institute was to establish a mechanical drawing school. Unfortunately, the academic program was never realized because two factions argued over whether to base it on classical academics (Latin and Greek) or on science and practical courses. Eventually the founders abandoned the academic purposes of the Institute.

Also in 1824, in upstate New York, Amos Eaton founded a school "for the purpose of instructing persons . . . in the application of science to the common purposes of life." This school has grown to be known today as Rensselaer Polytechnic Institute.

In 1862, with the coming of federal government assistance in the form of the Morrill Act, more technical schools emerged, and we can assume that mechanical drawing *slowly* became a part of the intellectual tool kits of trained mechanics, engineers, and inventors.

This can, in part, put to rest the common thought (as we look at pictures of pyramids, steam engines, and other engineering examples) that "it all started with a drawing." Certainly there were freehand sketches, cartoons, and other forms of graphic models. It is probably accurate to say that before most things were built, they were tested with a 3-D model. Just as mechanical drafting slowly took the place of hand sketches and wooden models, today computer design and drawing and 3-D computer modeling are likely to rapidly replace board drafting. The eventual utility and order of computer modeling and design remain to be established as part of the developing graphic process.

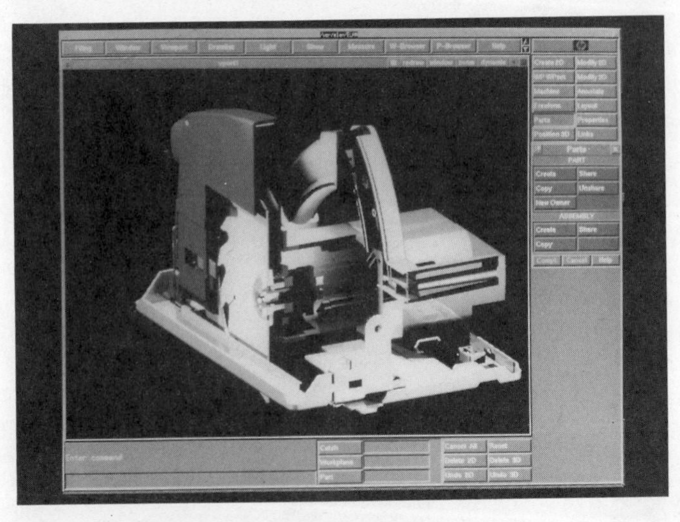

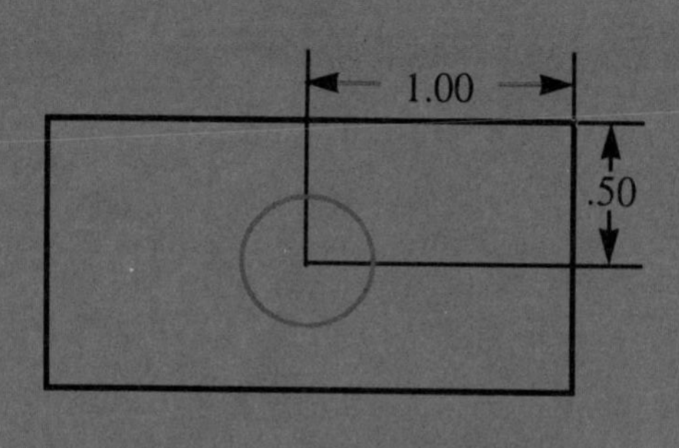

Introduction

DRAFTING TECHNOLOGY

According to the *Dictionary of Occupational Titles*, published by the U.S. Department of Labor, drafting is grouped with professional, technical, and managerial occupations. This category includes occupations concerned with the theoretical and practical aspects of such fields of human endeavor as architecture; engineering; mathematics; physical sciences; social sciences; medicine and health; education; museum, library, and archival sciences; law; theology; the arts; recreation; administrative specialties; and management. Also included are occupations in support of scientists and engineers and other specialized activities such as piloting aircraft, operating radios, and directing the course of ships. Most of these occupations require substantial educational preparation, usually at the college, junior college, or technical institute level.

Men and women employed in the drafting profession are often referred to as drafters. A general definition of drafter, as prepared by the Career Information System at the University of Oregon, is as follows:

> Drafters translate ideas and sketches of engineers, architects, and scientists into detailed drawings which are used in manufacturing and construction. Their duties may include interpreting directions given to them, making sketches, preparing drawings to scale, and specifying details. Drafters may also calculate the strength, quality, quantity, and cost of materials. They utilize various drafting tools and computer equipment, engineering practices, and math to complete drawings.

DRAFTING OCCUPATIONS

There are several types of drafting technology occupations. While drafting in general has one basic description, specific drafting areas have unique conceptual and skill characteristics. The types of drafting occupations fall into three general professional areas; architecture, engineering, and surveying. The following are specific drafting areas as defined by the *Dictionary of Occupational Titles.*

Architectural Drafter

Draws artistic architectural and structural features of any class of buildings and like structures. Draws designs and details, using drawing instruments. Confirms compliance with building codes. May specialize in planning architectural details according to structural materials used. (See Figures I–1[a] and I–1[b].)

Landscape Drafter

Prepares detailed scale drawings from rough sketches or other data provided by Landscape Architect. May prepare separate detailed site plan, grading and drainage

ENTRY PORCH ℄ HT. EL.=114.0
TYP. GABLE ℄ HT. EL.=111.0
STAND. ℄ HT. EL.=109.0
MAIN LEVEL EL.=100.0
LIVING RM. EL.=98.75

(a)

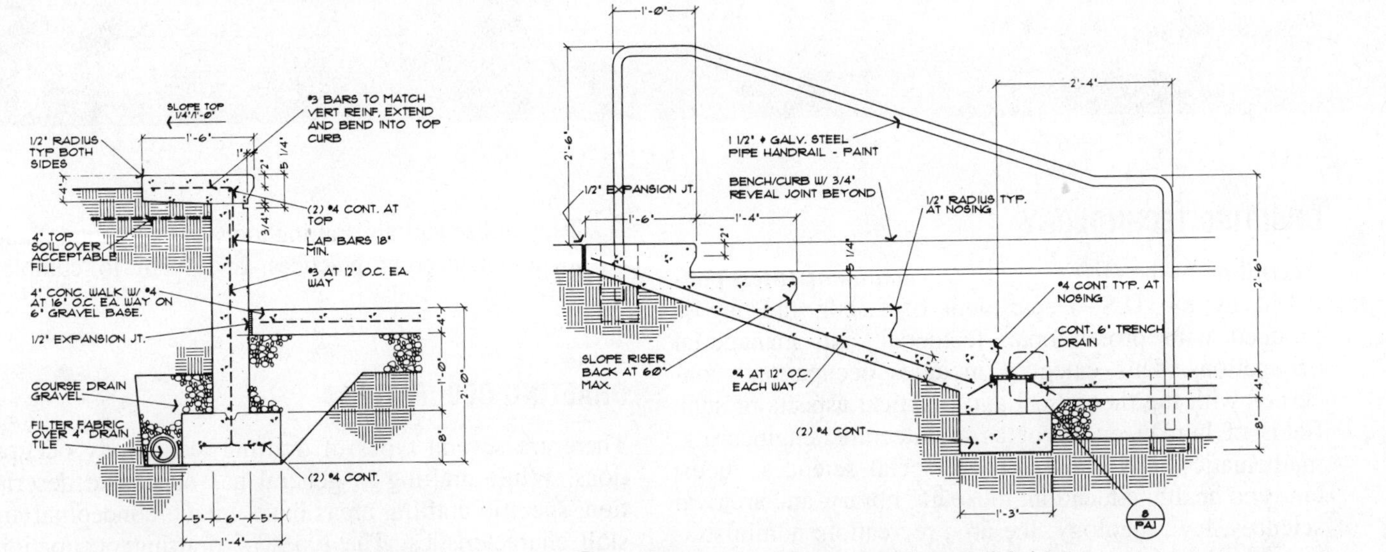

SLOPE TOP 1/4"/1'-0"
1/2" RADIUS TYP BOTH SIDES
#3 BARS TO MATCH VERT REINF. EXTEND AND BEND INTO TOP CURB
1'-6"
5 1/4"
(2) #4 CONT. AT TOP
4" TOP SOIL OVER ACCEPTABLE FILL
LAP BARS 18" MIN.
#3 AT 12" O.C. EA. WAY
4" CONC. WALK W/ #4 AT 16" O.C. EA. WAY ON 6" GRAVEL BASE.
1/2" EXPANSION JT.
COURSE DRAIN GRAVEL
FILTER FABRIC OVER 4" DRAIN TILE
(2) #4 CONT.
5" 6" 5"
1'-4"
2'-0"

6 BENCH/CURB DETAIL
PA.1 1" = 1'-0"

1'-0"
1 1/2" ⌀ GALV. STEEL PIPE HANDRAIL - PAINT
2'-4"
1/2" EXPANSION JT.
BENCH/CURB W/ 3/4" REVEAL JOINT BEYOND
1/2" RADIUS TYP. AT NOSING
1'-6"
1'-4"
#4 CONT TYP. AT NOSING
CONT. 6" TRENCH DRAIN
SLOPE RISER BACK AT 60° MAX.
#4 AT 12" O.C. EACH WAY
(2) #4 CONT
1'-3"
2'-6"

7 SECTION AT EXTERIOR STAIR
PA.1 1" = 1'-0"

FACE OF EXISTING CONC. WALL W/ PLASTER FINISH
PREMOLDED EXPANSION JOINT FILLER
CONT. 3/8" SEALANT OVER FILLER ROD
SYNTHETIC PLASTER FINISH OVER 2" EXPANDED POLYSTYRENE INSULATION BOARD
1 1/2" METAL FURRING CHANNELS AT 24" O.C.
GYPSUM VENEER PLASTER ON 5/8" GYP LATH
CONT. 3/8" SEALANT
5/8" P.C. PLASTER ON SELF-FURRING LATH OVER 8" CMU

9 BLD'G EXPANSION JOINT DETAIL
PA.1 3" = 1'-0"

SYNTHETIC PLASTER OVER 2" EXPANDED POLYSTYRENE INSULATION BOARD
CONT. 3/8" SEALANT OVER COMPRESSABLE FILLER ROD.
CARRY EXTRUDED POLYSTYRENE INSULATION BOARD TO TOP OF FTG.
CARRY PLASTER 4" MIN BELOW FIN. GRADE OR CONC. WALK.
5/8" P.C. PLASTER ON SELF FURRING LATH SHOWN DASHED SEE FLOOR PLANS FOR LOCATI
3/8" PREMOLDED EXPANSION JOINT FILLER
SYNTHETIC PLASTER OVER 2" EXTRUDED POLYSTYRENE INSULATION BOARD.
8" CMU WALL SEE PS.1 FOR TYPICAL REINF.
FINISHED FLOOR LINE
5 1/4"

10 CONSTRUCTION JOINT DETAIL
PA.1 3" = 1'-0"

(b)

Figure I–1 (a) Computer-generated elevations. *Courtesy Piercy & Barclay Designers, Inc.*; (b) computer-generated architectural details. *Courtesy Soderstrom Architects PC.*

plan, lighting plan, paving plan, irrigation plan, planting plan, and drawings and detail of garden structures. May build models of proposed landscape construction and prepare colored drawings for presentation to client. (See Figure I–2.)

Electrical Drafter

Prepares electrical-equipment working drawings and wiring diagrams used by construction crews and repairmen who erect, install, and repair electrical equipment and wiring in communications centers, power plants, industrial establishments, commercial or domestic buildings, or electrical distribution systems, performing duties described under Drafter. (See Figure I–3.)

Aeronautical Drafter

Specializes in preparing engineering drawings of developmental or production airplanes and missiles and ancillary equipment, including launch mechanisms and scale models of prototype aircraft, as planned by Aeronautical Engineer.

Electronic Drafter

Drafts wiring diagrams, schematics, and layout drawings used in manufacture, assembly, installation, and repair of electronic equipment, such as television cameras, radio transmitters and receivers, audioamplifiers, computers, and radiation detectors, performing duties as described under Drafter. Drafts layout and detail drawings of racks, panels, and enclosures. May conduct service and interference studies and prepare maps and charts related to radio and television surveys. May be designated according to equipment drafted. (See Figure I–4.)

Civil Drafter

This category is also known by the following titles: Drafter, Civil Engineering; Drafter, Construction; and Drafter, Engineering. Prepares detailed construction drawings, topographic profiles, and related maps and specification sheets used in planning and construction of highways, river and harbor improvements, flood control, drainage, and other civil engineering projects, preforming duties as described under Drafter. Plots maps and charts showing

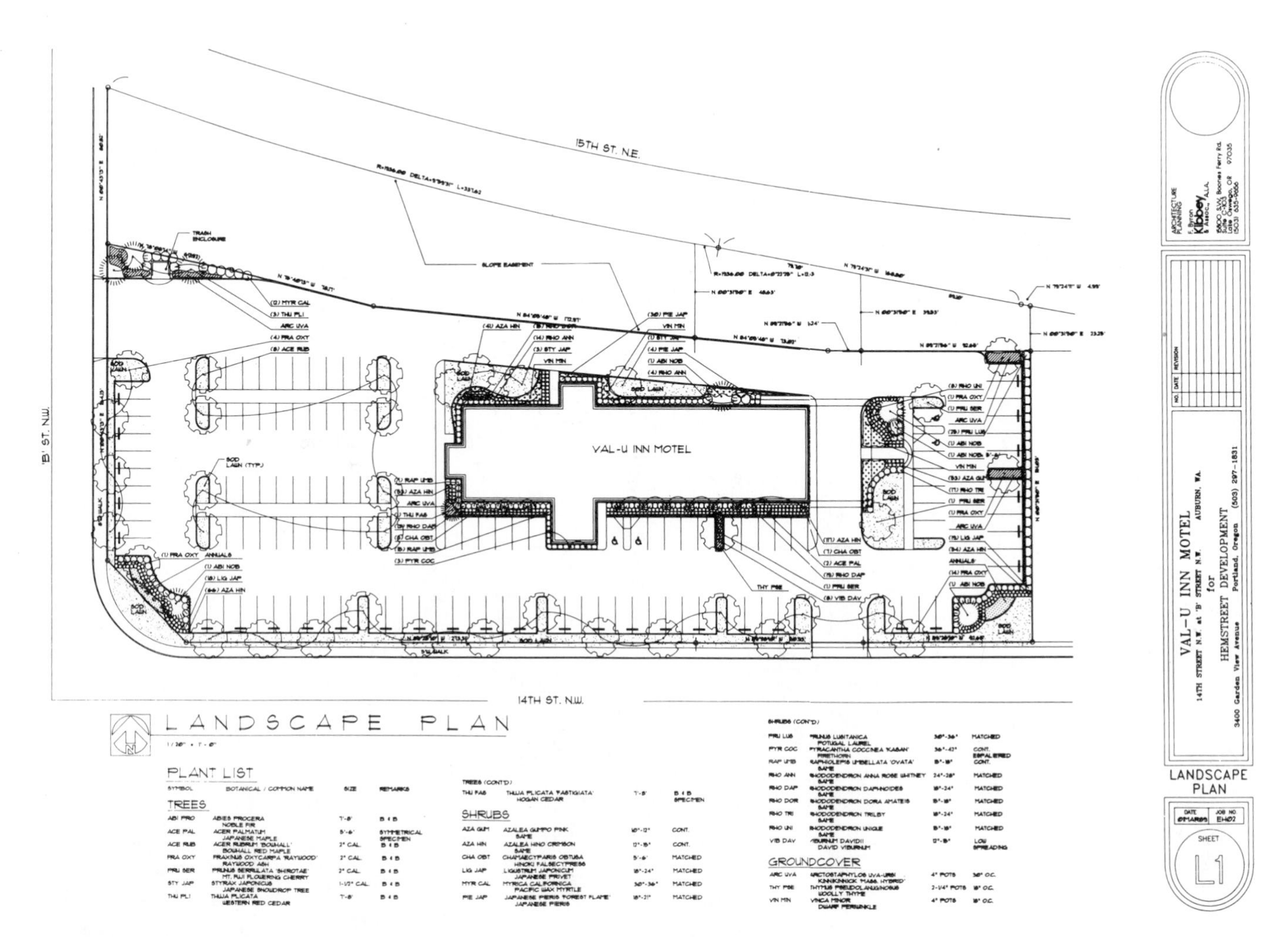

Figure I–2 Computer-generated landscape plan. *Courtesy OTAK, Inc., for landscaping, and Kibbey & Associates, for the site plan.*

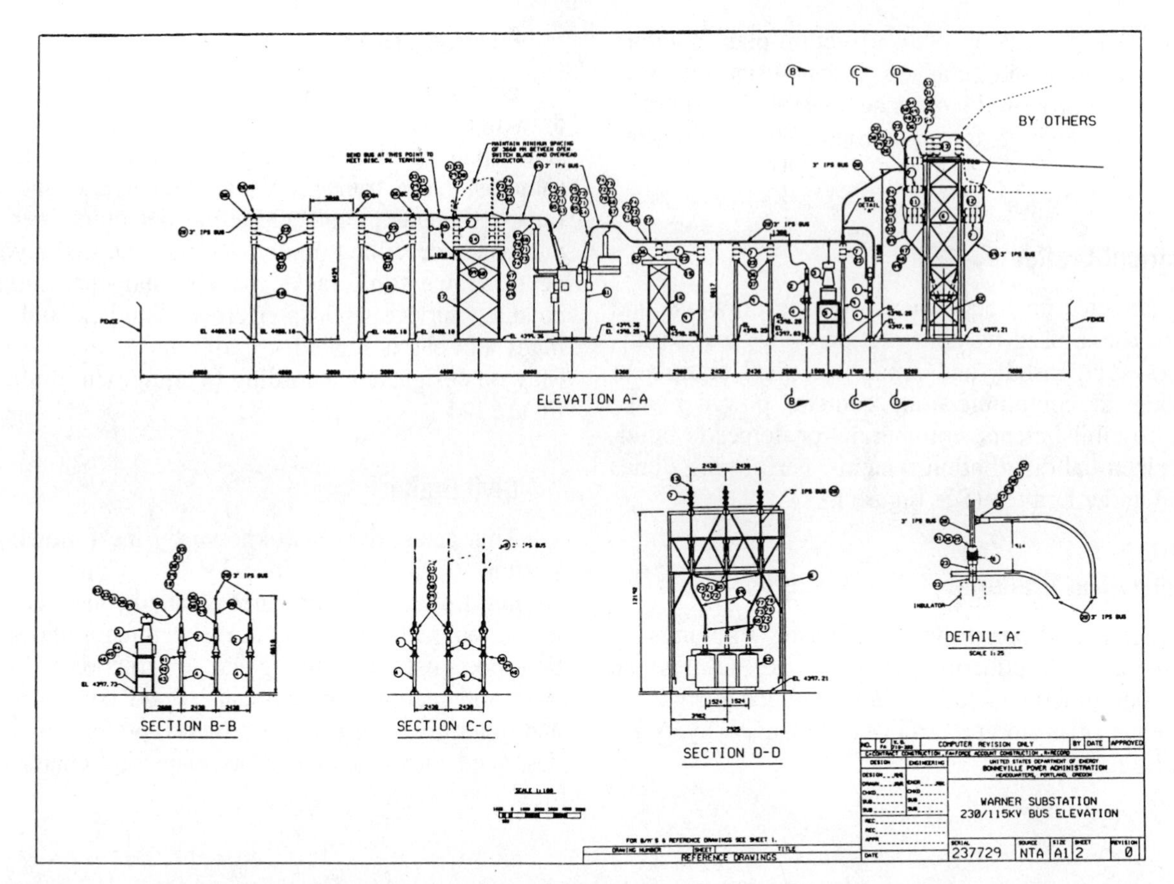

Figure I–3 Computer-generated electrical drafting plan and elevation. *Courtesy Bonneville Power Administration.*

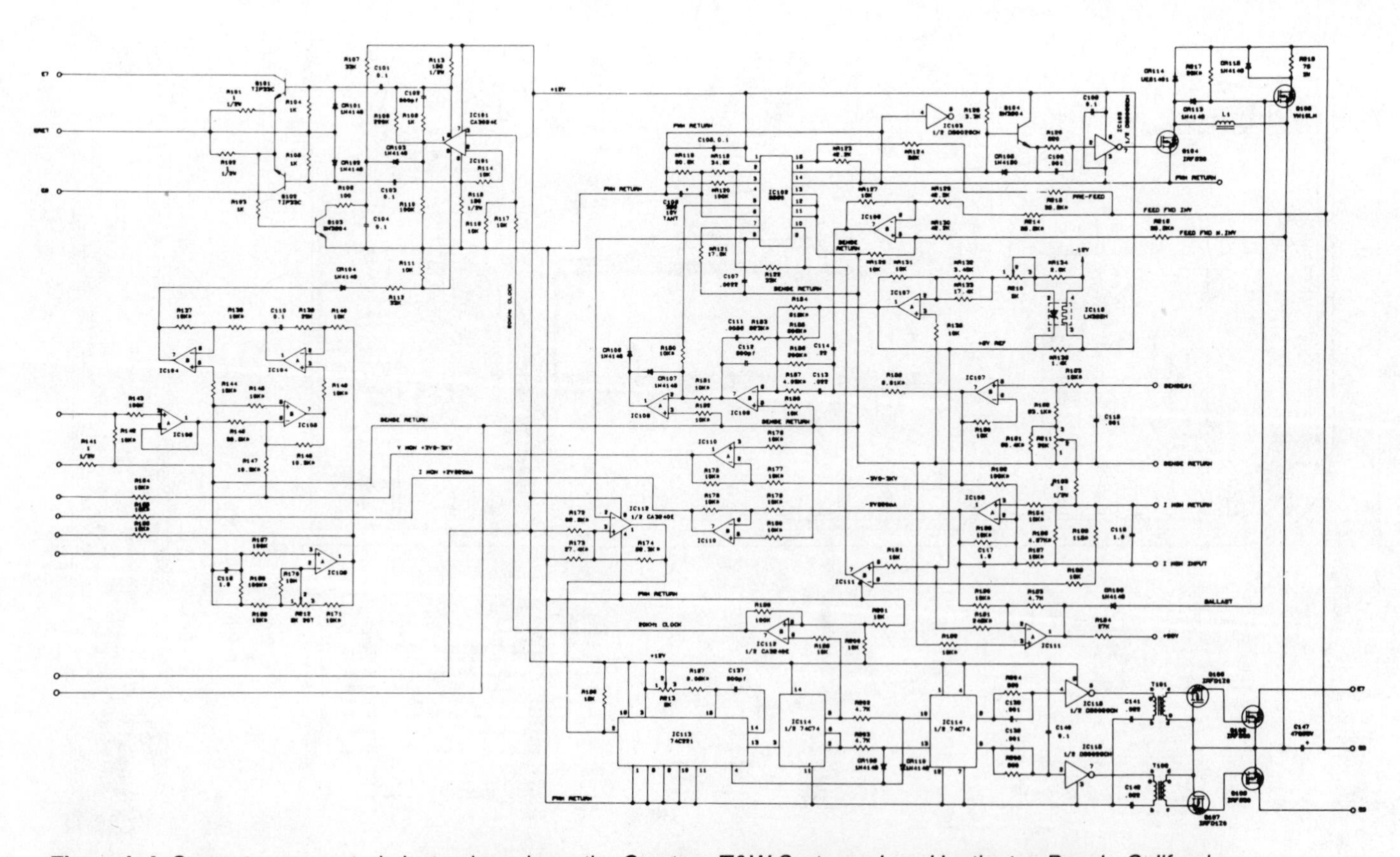

Figure I–4 Computer-generated electronics schematic. *Courtesy T&W Systems, Inc., Huntington Beach, California.*

profiles and cross sections indicating relation of topographical contours and elevations to buildings, retaining walls, tunnels, overhead power lines, and other structures. Drafts detailed drawings of structures and installations such as roads, culverts, fresh water supply and sewage disposal systems, dikes, wharfs, and breakwaters. Computes volume of excavations and fills and prepares graphs and hauling diagrams used in earthmoving operations. May accompany survey crew in field to locate grading markers or to collect data required for revision of construction drawings. May be designated according to type of construction. (See Figure I–5.)

Structural Drafter

Performs duties of Drafter by drawing plans and details for structures employing structural reinforcing steel, concrete masonry, wood, and other structural materials. Produces plans and details of foundations, building frame, floor and roof framing, and other structural elements. (See Figure I–6.)

Castings Drafter

Drafts detailed drawings for castings, which require special knowledge and attention to shrinkage allowances and such factors as minimum radii of fillets and rounds.

Patent Drafter

Drafts clear and accurate drawings of varied sorts of mechanical devices for use by Patent Lawyer in obtaining patent rights.

Tool Design Drafter

Drafts detailed drawing plans for manufacture of tools, usually following designs and specifications indicated by Tool Designer.

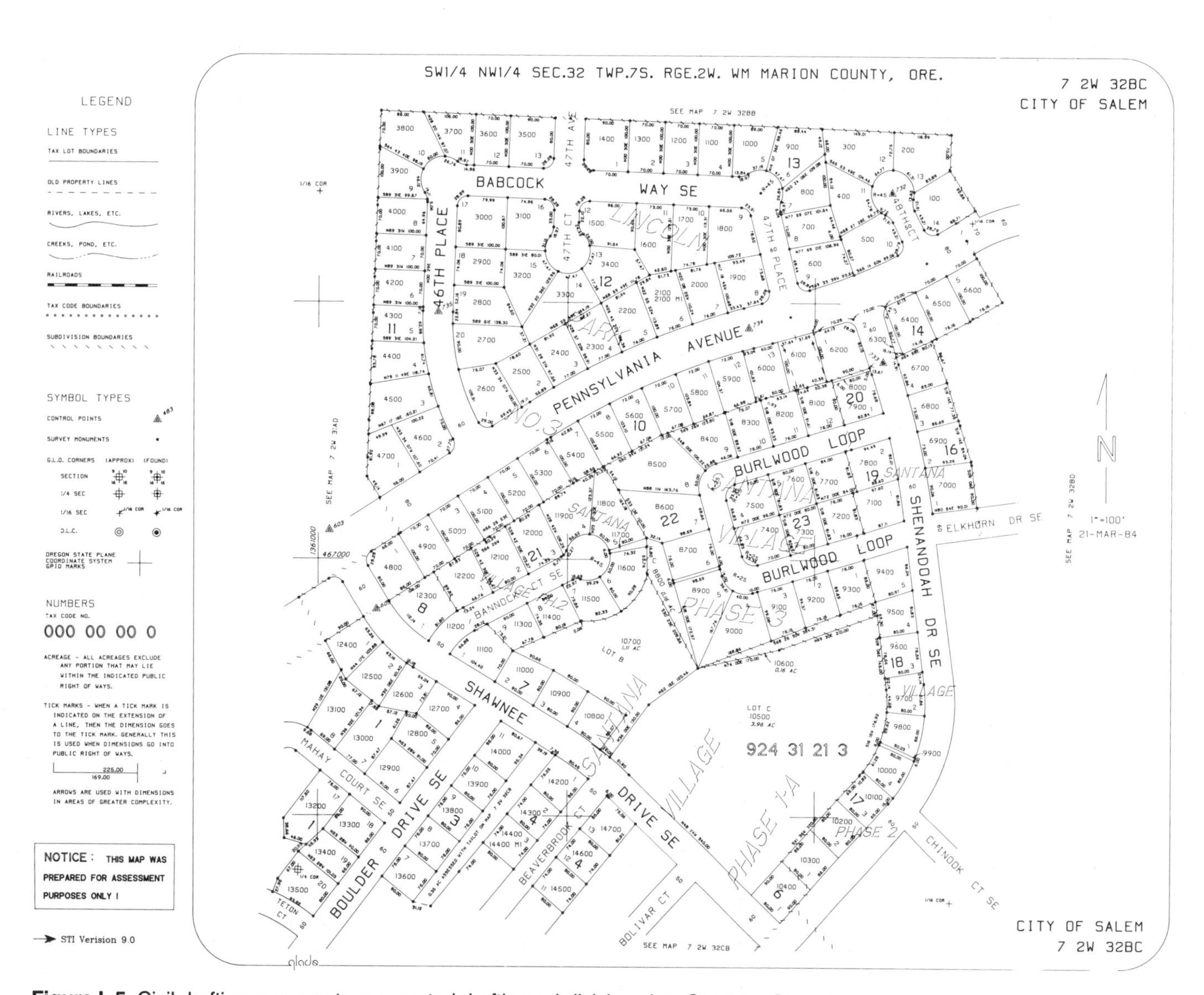

Figure I–5 Civil drafting, a computer-generated drafting subdivision plat. *Courtesy Glads Program.*

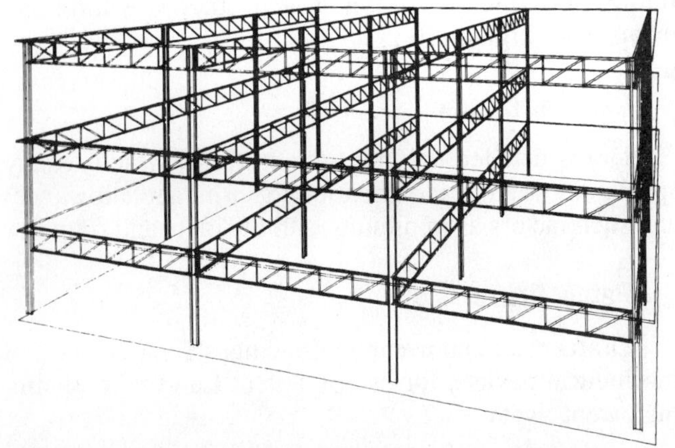

Figure I–6 Computer-generated perspective. *Courtesy Computervision Corporation.*

Mechanical Drafter

Drafts detailed working drawings of machinery and mechanical devices indicating dimensions and tolerances, fasteners and joining requirements, and other engineering data. Drafts multiple-view assembly and subassembly drawings as required for manufacture and repair of mechanisms. Performs other duties as described under Drafter. Mechanical drafting, in general, is the core of the engineering drafting industry. (See Figure I–7.)

Directional Survey Drafter

Plots oil- or gas-well boreholes from photographic subsurface survey recordings and other data. Computes and represents diameter, depth degree, and direction of inclination, location of equipment, and other dimensions and characteristics of borehole.

Geological Drafter

Draws maps, diagrams, profiles, cross sections, directional surveys, and subsurface formations to represent geological or geophysical stratigraphy and locations of gas and oil deposits. Performs duties described under Drafter. Correlates and interprets data obtained from topographical surveys, well logs, or geophysical prospecting reports, utilizing special symbols to denote geological and geophysical formations or oilfield installations. May finish drawings in mediums and according to specifications required for reproduction by blueprinting, photographing, or other duplication methods.

Geophysical Drafter

Draws subsurface contours in rock formations from data obtained by geophysical prospecting party. Plots maps and diagrams from computations based on recordings of seismograph gravity meter, magnetometer, and other petroleum prospecting instruments and from prospecting and surveying field notes.

Heating and Ventilating Drafter

Also known as Heating Ventilating and Air Conditioning (HVAC) Drafter. Specializes in drawing plans for installation of heating, air-conditioning, and ventilating equipment. May calculate heat loss and heat gain for

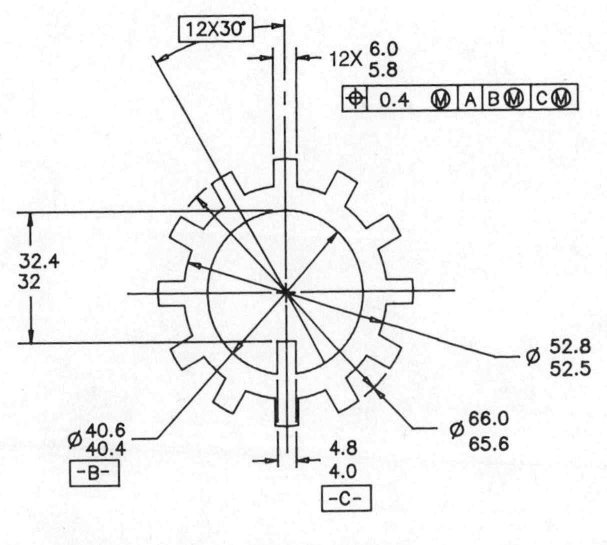

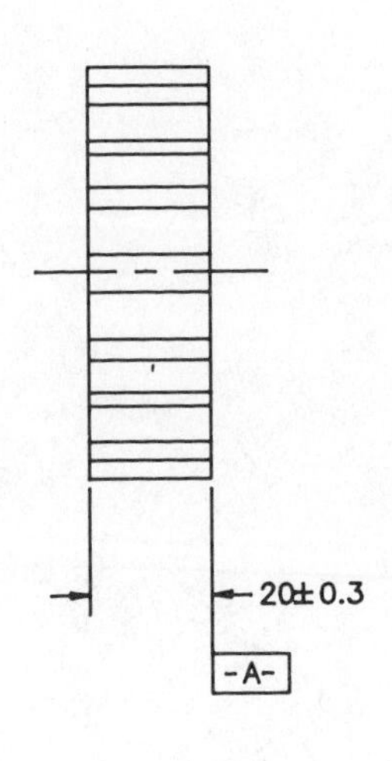

3. FINISH ALL OVER 1.6 μM

2. REMOVE ALL BURRS AND SHARP EDGES.

1. INTERPRET ALL DIMENSIONS AND TOLERANCES PER ANSI Y14.5–1982

NOTES:

Figure I–7 Computer-generated mechanical drafting.

buildings for use in determining equipment specifications, following standardized procedures. May specialize in drawing plans for installation of refrigeration equipment. (See Figure I–8.)

Plumbing/Industrial Pipe Drafter

Also known as Piping Drafter. Specializes in drafting plans for installation of plumbing and piping equipment for residential, commercial, and industrial installations. (See Figure I–9.)

Automotive Design Drafter

Designs and drafts working layouts and master drawings of automotive vehicle components, assemblies, and systems from specifications, sketches, models, prototype and/or verbal instructions, applying knowledge of automotive vehicle design, engineering principles, manufacturing processes and limitations, and drafting techniques and procedures, using drafting instruments and work aids. Analyzes specifications, sketches, engineering drawings, ideas, and related design data to determine critical factors affecting design of components based on knowledge of previous designs and manufacturing processes and limitations. Draws rough sketches and performs mathematical computations to develop design and work out detailed specifications of components. Applies knowledge of mathematical formulas and physical laws. Performs preliminary and advanced work in development of working layouts and final master drawings adequate for detailing parts and units of design. Makes revisions to size, shape, and arrangement of parts to create practical design. Confers with Automotive Engineer and others on staff to resolve design problems. Specializes in design of specific type of body or chassis components, assemblies or systems such as door panels, chassis frame and supports, or braking system.

Oil and Gas Drafter

Drafts plans and drawings for layout, construction, and operation of oil fields, refineries, and pipeline systems from field notes, rough or detailed sketches, and specifications. Develops detail drawings for construction of equipment and structures, such as drilling derricks, compressor stations, gasoline plants, frame, steel, and masonry buildings, piping manifolds and pipeline systems, and for manufacture, fabrication, and assembly of machines and machine parts. (See Figure I–10.)

Technical Illustrator

Lays out and draws illustrations for reproduction in reference works, brochures, and technical manuals dealing with assembly, installation, operation, maintenance, and repair of machines, tools, and equipment. Prepares drawings from blueprints, designs, mockups, and photographs by methods and techniques suited to specified reproduction process or final use, such as diazo, photo-offset, and projection transparencies, using drafting and optical equipment. Lays out and draws schematic, perspective, orthographic, or oblique-angle views to depict function, relationship, and assembly sequence of parts and assemblies, such as gears, engines, and instruments. Shades or colors drawing to emphasize details or to

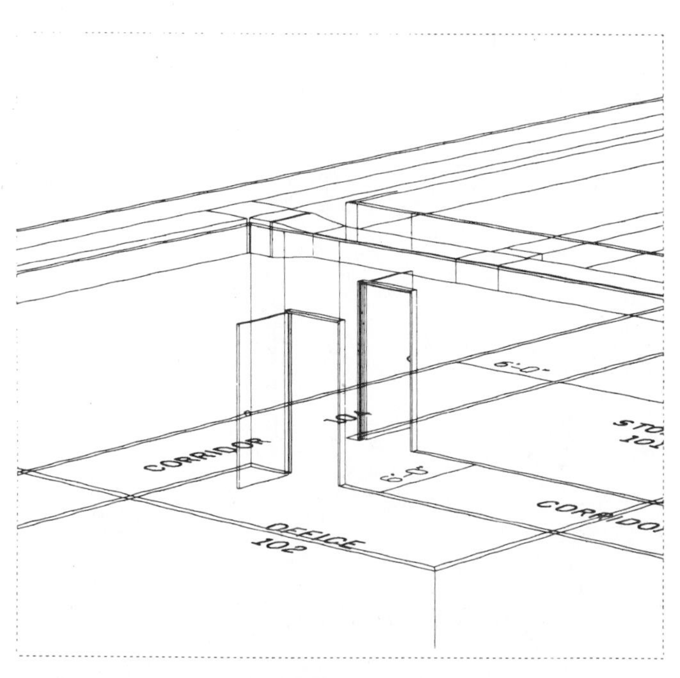

Figure I–8 Computer-generated HVAC pictorial. *Courtesy Computervision Corporation.*

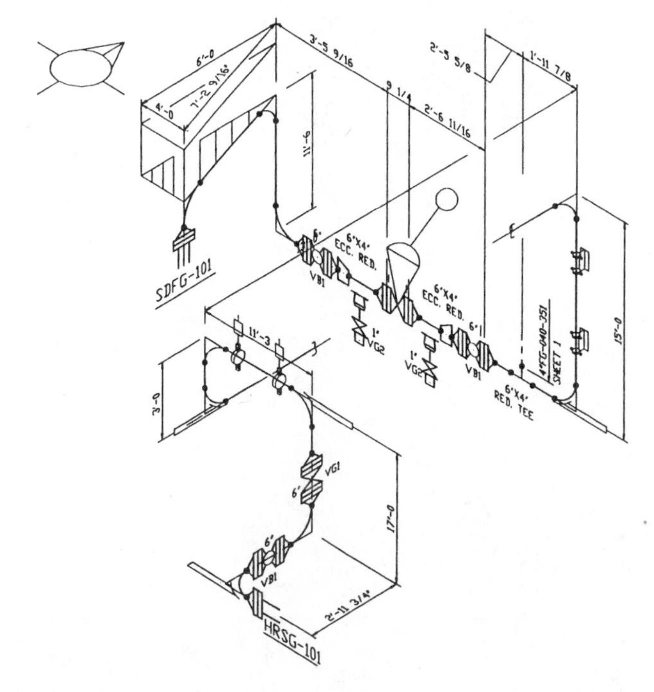

Figure I–9 Piping CADD isometric layout. *Courtesy Autodesk.*

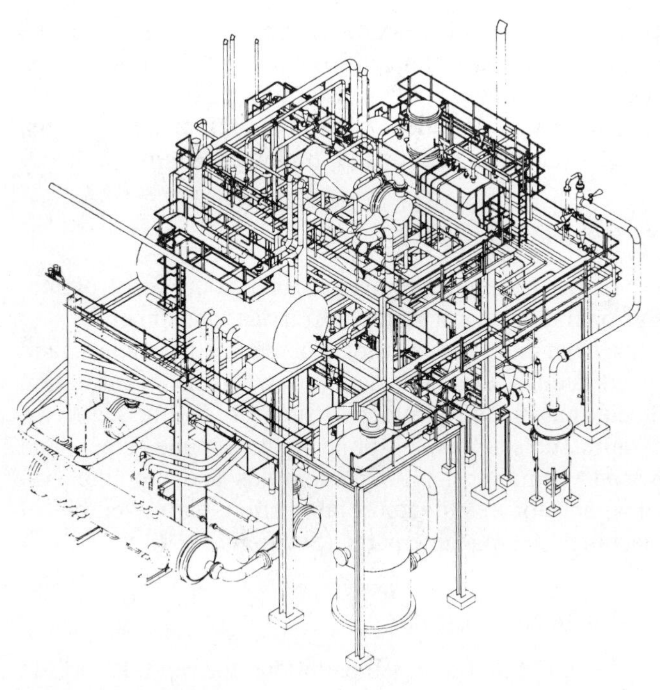

Figure I–10 Piping CADD pictorial. *Courtesy Computervision Corporation.*

eliminate undesired background, using ink, crayon, airbrush, and overlays. Pastes instructions and comments in position on drawing. May draw cartoons and caricatures to illustrate operation, maintenance, and safety manuals and posters. (See Figure I–11.)

Cartographic Drafter

Draws maps of geographical areas to show natural and construction features, political boundaries, and other features. Analyzes survey data, survey maps and photographs, computer- or automated-mapping products, and other records to determine location and names of features. Studies records to establish boundaries of properties and local, national, and international areas of political, economic, social, or other significance. Geological and topographical maps are drawn by Cartographic Drafter.

Photogrammetrist

Analyzes source data and prepares mosaic prints, contour-map profile sheets, and related cartographic materials requiring technical mastery of photogrammet-

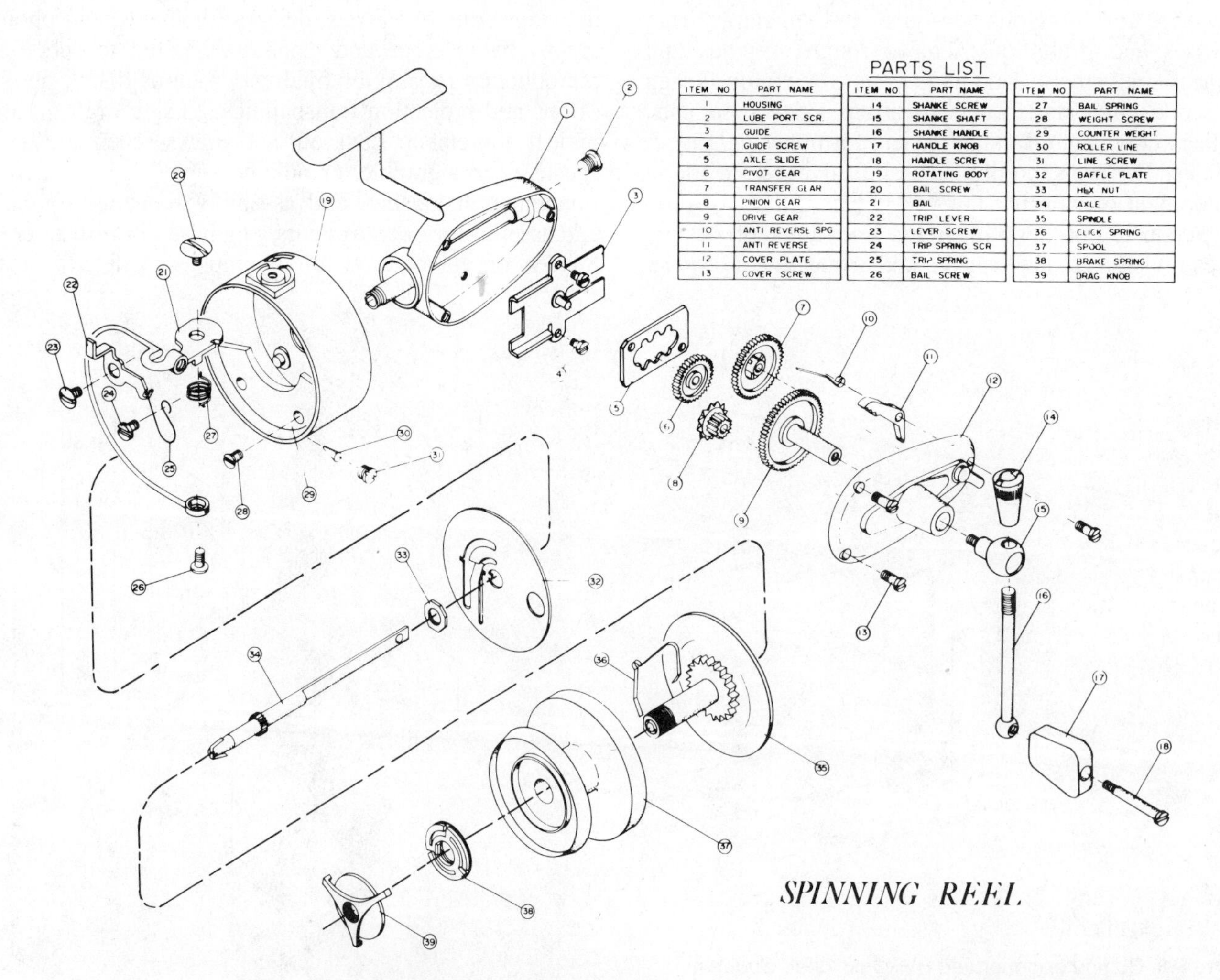

PARTS LIST

ITEM NO	PART NAME	ITEM NO	PART NAME	ITEM NO	PART NAME
1	HOUSING	14	SHANKE SCREW	27	BAIL SPRING
2	LUBE PORT SCR.	15	SHANKE SHAFT	28	WEIGHT SCREW
3	GUIDE	16	SHANKE HANDLE	29	COUNTER WEIGHT
4	GUIDE SCREW	17	HANDLE KNOB	30	ROLLER LINE
5	AXLE SLIDE	18	HANDLE SCREW	31	LINE SCREW
6	PIVOT GEAR	19	ROTATING BODY	32	BAFFLE PLATE
7	TRANSFER GEAR	20	BAIL SCREW	33	HEX NUT
8	PINION GEAR	21	BAIL	34	AXLE
9	DRIVE GEAR	22	TRIP LEVER	35	SPINDLE
10	ANTI REVERSE SPG	23	LEVER SCREW	36	CLICK SPRING
11	ANTI REVERSE	24	TRIP SPRING SCR	37	SPOOL
12	COVER PLATE	25	TRIP SPRING	38	BRAKE SPRING
13	COVER SCREW	26	BAIL SCREW	39	DRAG KNOB

Figure I–11 Technical illustration, exploded isometric assembly.

ric techniques and principles. Prepares original maps, charts, and drawings from aerial photographs and survey data and applies standard mathematical formulas and photogrammetric techniques to identify, scale, and orient geodetic points, estimations, and other planimetric or topographic features and cartographic detail. Graphically represents aerial photographic detail, such as contour points, hydrography, topography, and cultural features, using precision stereoplotting apparatus or drafting instruments. Revises existing maps and charts and corrects maps in various stages of compilation. Prepares rubber, plastic, or plaster three-dimensional relief models.

PROFESSIONAL ORGANIZATION

The American Design Drafting Association is a nonprofit, professional organization dedicated to the advancement of design and drafting.

The basic knowledge and skill required to do board drafting is also helpful in preparing input data for or operating computerized drafting systems, such as digitizers, plotters, and photo composition devices.

Whatever means are used to generate drawings, persons skilled in drafting techniques will be in demand. Therefore, drafting can be a rewarding and challenging career to those who can visualize and portray ideas graphically.

HOW TO BECOME A DRAFTER

Persons considering drafting as a career should be able to visualize what is to be drawn and be mechanically minded. Potential drafters should be willing to work hard and learn new things, and be able to adjust to varying working conditions. Above all, they should be neat and accurate in their work.

Today, most industries, to fill their drafting job openings, are employing only graduates who have received specialized training in design and drafting from technical institutes, junior or community colleges, and vocational schools. The specialized training includes basic and advanced drawing, mathematics (algebra, geometry, and trigonometry), physics or chemistry, English, humanities, technology courses, and courses in specific fields such as aeronautical, architectural, automotive, electrical, electronics, illustrative, mapping, mechanical, piping, structural, and sheet metal drafting. Computer-aided drafting is essential training for today's employment market.

DRAFTING OCCUPATIONAL LEVELS

There are several levels of advancement for drafters, which are generally based on educational background and practical experience. For advancement, most companies require career employees to hold a two-year college or trade-school degree; however, persons with an equivalent or larger amount of practical experience in a specific job-related area can obtain advanced occupational levels also.

Occupational levels differ slightly from one industry to another although the general classifications are generally accurate nationwide. Suggested educational levels have been added for reference.

DRAFTING JOB OPPORTUNITIES

Drafting job opportunities, which include all possible drafting employers, fluctuate with national and local economies. This is no different from most other employment in industry or construction. Drafting is tied closely to manufacturing and construction so that a slowdown or speedup in these industries nationally affects the number of drafting jobs available. This same effect upon drafting opportunities may be experienced at the local level or with specific industries. For example, construction may be strong in one part of the country, and slow in another, so the demand for drafters in those localities is strong or slow accordingly. Other examples are demonstrated when automobile manufacturers experience poor sales, and drafting opportunities decline, or when hi-tech industries expand, and more drafters are needed. Public and private indicators suggest that the demand for drafters will continue to be strong for graduates of two-year post-secondary drafting curriculums. The emphasis should be placed on drafting skills, computer-aided drafting, math, English, and oral- and written-communication skills.

The types of drafting jobs available are also controlled by local demands. A predominance of mechanical-drafting jobs are found in metropolitan areas where manufacturing is strong, while in outlying areas there may be more civil or structural drafting jobs. Each local area has a need for more of one type of drafting skill than another. Also, drafting curriculums in different geographical areas generally specialize in the fields of drafting that help fill local employment needs. Some drafting programs offer a broad-based education so that graduates may have versatile employment opportunities. When selecting a school look into curriculum, placement potential and local demand. Talk to representatives of local industries for an evaluation of the drafting curriculum.

Opportunities for advancement for drafters are excellent, although dependent upon the advancement possibilities of the specific employer. Advancement also depends upon an individual's initiative, ability, product knowledge, and willingness to continue to be educated. Additional education for advancement usually includes increased levels of mathematics, pre-engineering, engineering, computers, and advanced drafting. Drafting has traditionally been an excellent stepping-stone to designing, engineering, and management.

DRAFTING SALARIES AND WORKING CONDITIONS

Salaries in drafting professions are comparable to salaries of other professions with equal educational requirements. Employment benefits vary according to each individual employer. However, most employers offer vacation and health insurance coverage, while others include dental, life, and disability insurance.

COMPUTERS IN DRAFTING

The industrialized world is wrestling with the dramatic and constant changes of the electronic revolution and its child, the computer. The speed-of-light capabilities of the computer have shaped the gathering, processing, storage, and retrieval of information into a major industry. Every day our lives are touched by the effects of computers. Contrary to science-fiction works of years past, computers have not grown to occupy the core of the earth, but have shrunk to microscopic proportions, and now occupy tiny spaces inside our machines and tools.

The tools of the drafting trade have also felt the electronic touch of computers. The silicon chip, that maze of delicate circuitry, has found its way into drafting tools. The heart of modern drafting tools is now the computer. The core of a manual drafting workstation, parallel bars, arm- and track-drafting machines, are rapidly becoming obsolete. The tables they are attached to are being used as reference space, and handy places for books, papers, and vendor catalogs.

Now the drafter, or operator, uses electronic hardware to construct drawings, not on paper, but on what looks like a television screen. The screen is called a monitor, video display screen, or video display terminal. The drafting process is now called *computer-aided design drafting*, or CADD. Most people say "computer-aided drafting" and use the acronym of CAD. Actually, CAD means computer-aided design.

ONCE IS ALWAYS ENOUGH WITH CADD

by Karen Miller
Graphic Specialist with Tektronix, Inc.

Reusability is one of the important advantages of CADD. With CADD it is never necessary to draw anything more than once. This advantage is even enhanced by developing a CADD symbols library. Building a parts library for reusability has increased productivity, decreased development costs, and set the highest standards for quality at the Test and Measurement Documentation Group at Tektronix, Inc.

The parts library began by reusing 2D isometric parts created by the AutoCAD illustrator and saved as blocks (symbols) in a Parts Library Directory. By pathing the named blocks back to this library directory, each block becomes accessible to any directory and drawing file. This allows the CADD illustrator to insert library symbols into any drawing simply by typing the block name.

The illustrator adds new parts to the library as a product is disassembled and illustrated. Each part is given the next available number as its block name in the library, as shown in Figure I–12.

Originally, the AutoCAD drawings were combined with text (written using Microsoft Word) in Ventura Publisher to create technical publications. Now, the AutoCAD drawings are added to document files in Interleaf technical publishing software. The entire parts library is available to both AutoCAD and Interleaf users (Figure I–13).

From the point of view of cost management, the parts library has saved hundreds of hours of rework. From the illustrator's view, the parts library helps improve productivity and frees time for new or complex projects.

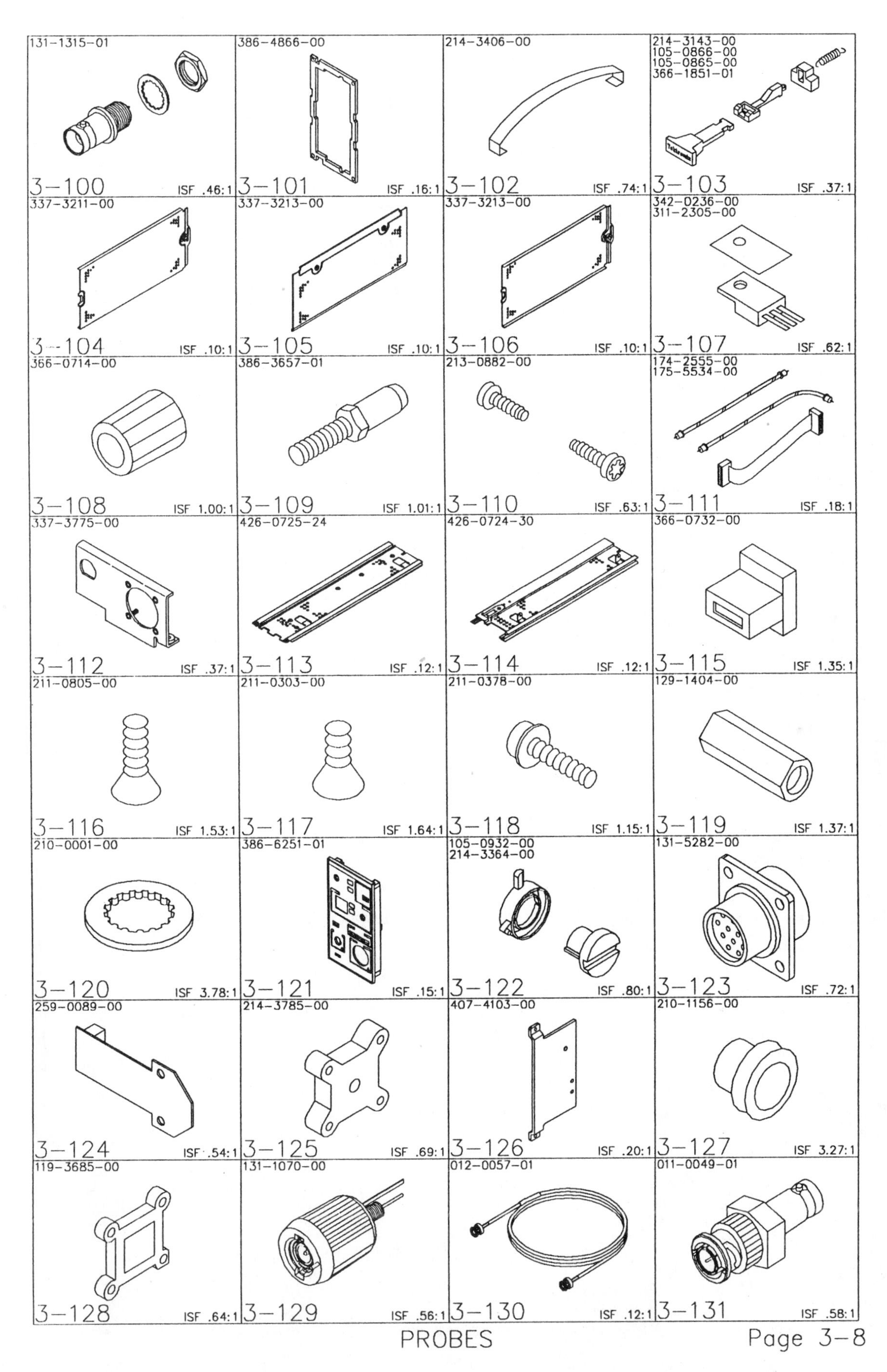

Figure I–12 *Courtesy Karen Miller, Technical Illustrator, Tektronix Inc. Beaverton, OR.*

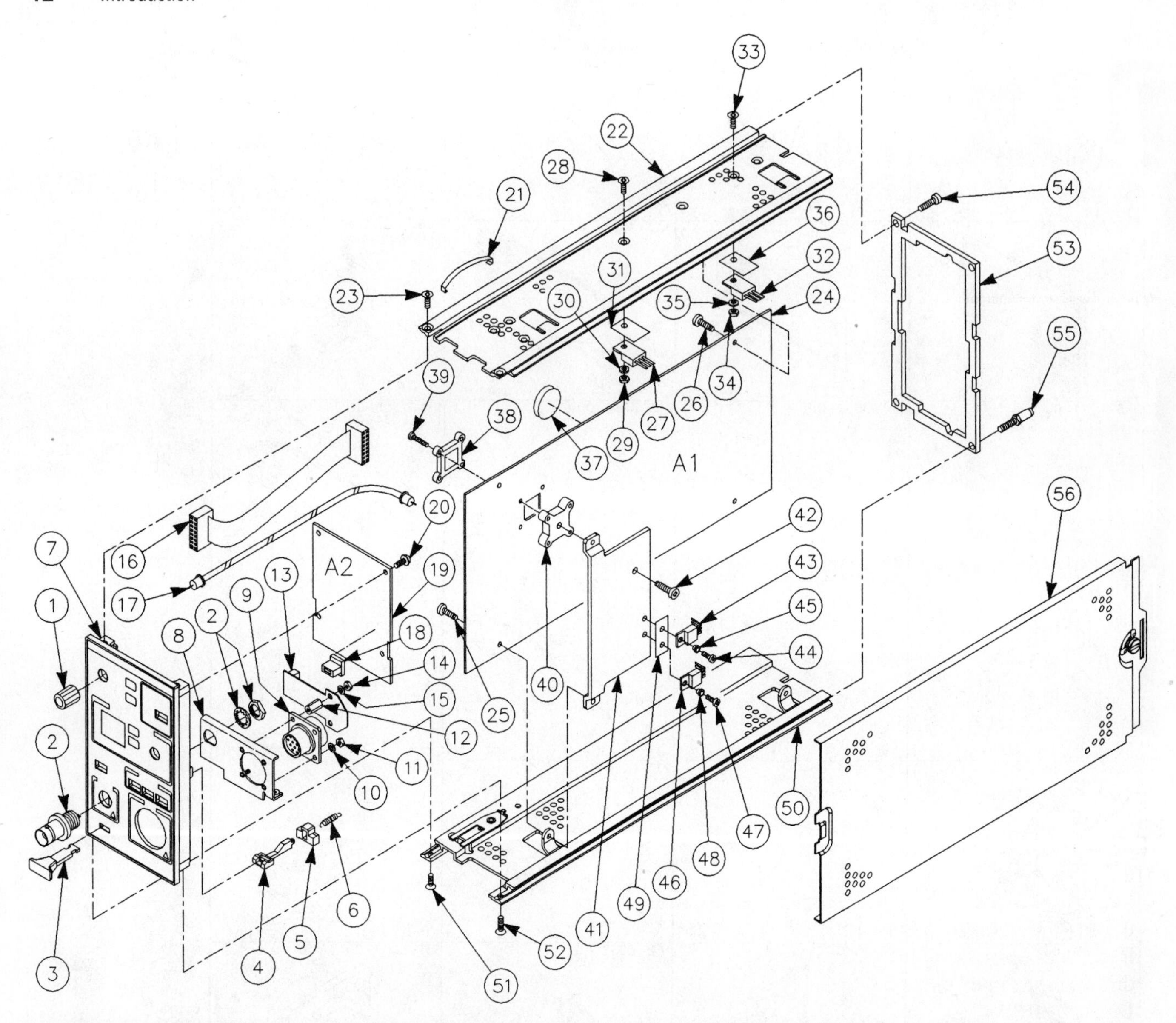

Figure I–13 Tektronix current probe amplifier. *Courtesy Karen Miller, Technical Illustrator, Tektronix Inc. Beaverton, OR.*

CHAPTER 1

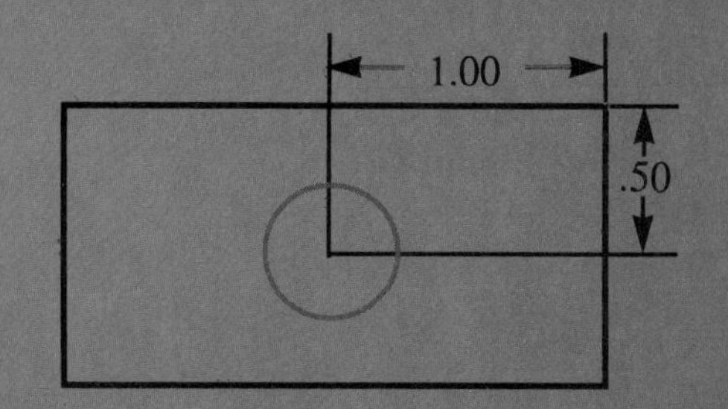

Drafting Equipment, Media, and Reproduction Methods

LEARNING OBJECTIVES

After completing this chapter you will:

- Describe and demonstrate the use of various manual drafting tools and equipment.
- Read engineer's, architect's, and metric scales, and drafting machine verniers.
- Discuss and use drafting media, sheet sizes, and title block information.
- Explain common reproduction methods.

DRAFTING EQUIPMENT

Drafting tools and equipment are available from a number of vendors that sell professional drafting supplies. For accuracy and long life, always purchase high-quality equipment. Local vendors can be found by looking in the yellow pages of your telephone book under headings such as: Drafting Room Equipment & Supplies, Blueprinting, Architects' Supplies, Engineering Equipment & Supplies, or Artists' Materials & Supplies.

Drafting supplies and equipment can be purchased in a kit or items can be bought individually. Many vendors have produced drafting kits, which are often available for economical prices. Whether equipment is purchased in a kit or by the individual tool, the items that are normally needed include the following:

- One mechanical lead holder with 4H, 2H, H, and F grade leads, or the automatic pencils discussed next. This pencil and lead assortment allows for individual flexibility of line and lettering control, or use in sketching.
- One 0.3-mm automatic drafting pencil with 4H, 2H, and H leads.
- One 0.5-mm automatic drafting pencil with 4H, 2H, H, and F leads.
- One 0.7-mm automatic drafting pencil with 2H, H, and F leads.
- One 0.9-mm automatic drafting pencil with H, F, and HB leads. Drafters may elect to purchase two or more pencils and use a different grade of lead in each. Doing so reduces the need to constantly change leads. Some drafters use a light-blue lead for layout work.
- Lead sharpener.
- 6-in. bow compass.
- Dividers.
- Eraser. Select an eraser that is recommended for drafting with pencil on paper.
- Erasing shield.
- 8-in. 30°–60° triangle.
- 8-in. 45° triangle.
- Irregular curve.
- Scales:
  1. Triangular architect's scale.
  2. Triangular civil engineer's scale.
  3. Triangular metric scale.
- Drafting tape.

- Circle template (small holes).
- Lettering guide.
- Arrowhead template.
- Sandpaper sharpening pad.
- Dusting brush.

DRAFTING FURNITURE

Tables

There is a large variety of drafting tables available, ranging from economical models to complete professional workstations. Drafting tables are generally sized by the dimensions of their tops. Standard tabletop sizes range from 24 × 36 in. to 42 × 84 in.

The features to look for in a good-quality, professional table include:

- One-hand tilt control.
- One-hand or foot height control.
- The ability to position the board vertically.
- An electrical outlet.
- A drawer for tools and/or drawings.

Some manufacturers ship tables with tops that are ready to draw on. Others ship tables with tops of steel or basswood that need to be covered. Basswood can be a drawing surface; however, most offices commonly cover drafting-table tops with smooth, specially designed surfaces. The material, usually vinyl, provides the proper density for effective use under normal drafting conditions. After compass or divider points have pierced the surface, the small holes close and so provide a smooth surface for continued use. Drafting tape is commonly used to adhere drawings to the tabletop, although some drafting tables have magnetized tops and use magnetic strips to attach drawings.

Chairs

Better drafting chairs have the following characteristics:

- Padded or contoured seat design.
- Height adjustment.
- Foot rest.
- Fabric that allows air to circulate.
- Sturdy construction.

DRAFTING PENCILS, LEADS, AND SHARPENERS

Mechanical Pencils

The term *mechanical pencil* is applied to a pencil that requires a piece of lead to be manually inserted; by some physical action such as twist, push, or pull, the lead is mechanically, or semiautomatically, advanced to the tip. A common lead holder, shown in Figure 1–1, is a mechanical pencil. Pressing the thumb button on the end of this model opens the chuck (clamping jaws) and releases the lead. For proper use, the lead should extend about 1/4 in. beyond the end of the chuck. Mechanical pencils vary in price. Some are made of plastic while others are made of metal. Get one that is comfortable and well made. Some inexpensive lead holders wear out quickly and often the chuck at the tip fails to hold the lead tightly during use.

Leads are bought separately, although one lead usually comes with the holder. You need to purchase other leads to correctly make a variety of lines and letters. Leads are graded by hardness and designated by a number and letter, or one or two letters. The designation is usually found along or on one end of the lead. Always sharpen the end opposite the designation.

Automatic Pencils

The term *automatic pencil* refers to a pencil with a lead chamber that, at the push of a button or tab, advances the lead from the chamber to the writing tip and, when a new piece of lead is needed, advances the new piece to the tip. Automatic pencils are designed to hold leads of one width so you do not need to sharpen the lead. These pencils are available in several different lead sizes. Drafters have several automatic or mechanical pencils. Each pencil has a different grade of lead hardness and is used for a specific technique. (See Figure 1–2.)

Lead Grades

The leads that you select for line work depend on the amount of pressure you apply and other technique factors. Experiment until you identify the leads that give the best line quality. Leads commonly used for thick lines range from 2H to F, while leads for thin lines range from 4H to H, depending upon individual preference. Construction lines for layout and guidelines are very lightly drawn with a 6H or 4H lead. Figure 1–3 shows the different lead grades.

Figure 1–1 Mechanical pencil, also known as lead holder. *Courtesy Koh-I-Noor Rapidograph, Inc.*

Figure 1–2 An automatic pencil; lead widths are 0.3, 0.5, 0.7, or 0.9 mm. *Courtesy Koh-I-Noor Rapidograph, Inc.*

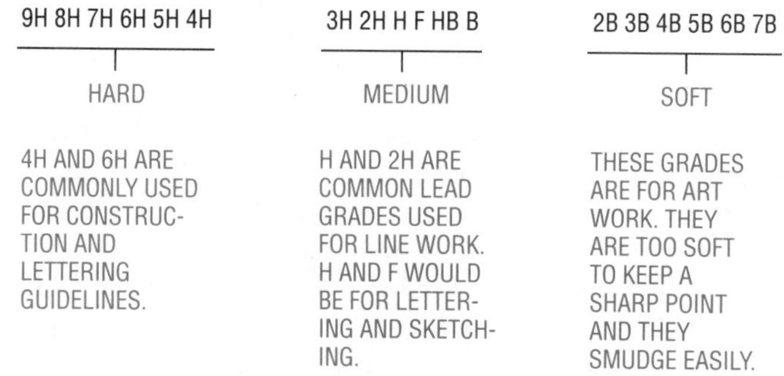

Figure 1–3 The range of lead grades.

Polyester and Special Leads

Polyester leads, also known as plastic leads, are for drawing on *polyester drafting film*, often called by its trade name, *Mylar®*. Plastic leads come in grades equivalent to F, 2H, 4H, and 5H and are usually labeled with a prefix and number. Pentel™, for example, uses P1 and P2. Some companies make a combination lead for use on both vellum and polyester film.

Colored leads have special uses. Red lead is commonly used for corrections. Red prints as a black line. Blue leads can be used on the original drawing by the supervisor to tell the drafter the corrections that need to be made. Blue lead will not show on the print when the original drawing with blue lines on it is run through a diazo machine. Some drafters use light-blue lead for all layout work and guidelines because it will not reproduce.

PROFESSIONAL HINT

Keep your pencil straight from side to side and tilted about 60° with the direction of travel. Try not to tilt the lead under or away from the straightedge. Rotate a lead holder between your fingers as you draw a line so that the tip stays uniformly conical. Automatic pencils do not require such rotation, although some drafters feel more comfortable rotating any pencil. The automatic pencil should be held near vertical, because if it is held at 60°, the lead breaks easily. Also, in a vertical position, the full surface of the lead is used. Provide enough pressure and go over each line enough times to make a line dark and crisp. Take care not to make it too thick. Figure 1–4 shows some basic pencil motions.

Students must experiment with different lead points and grades of lead in mechanical pencils to determine their best individual results. A comparison should also be made between the mechanical and automatic pencils to determine best results.

Sanding Block

Especially useful for sharpening compass lead, this is one of the simplest devices used for sharpening. Sandpaper stapled to a wooden paddle is called a sandpaper, or sanding block. Plastic lead fills sandpaper rapidly so several sheets are needed as compared to graphite lead. To avoid smudging your drawing, use the sanding block away from your drawing table and dispose of the graphite carefully.

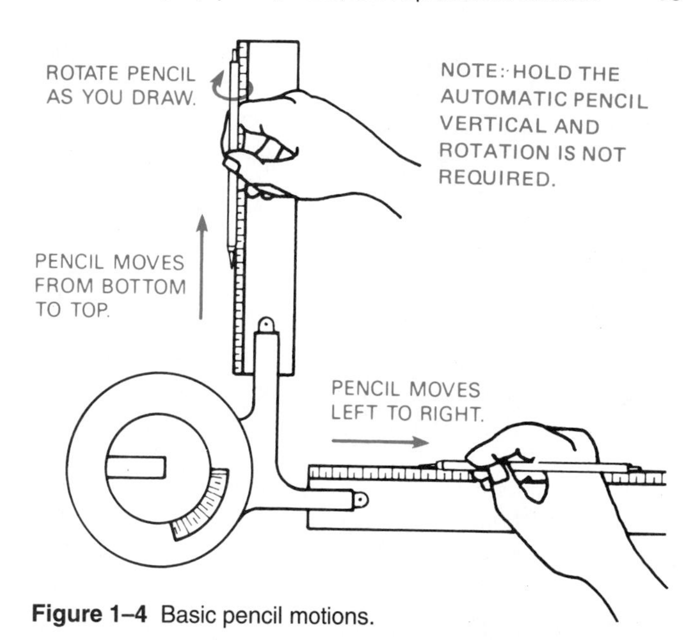

Figure 1–4 Basic pencil motions.

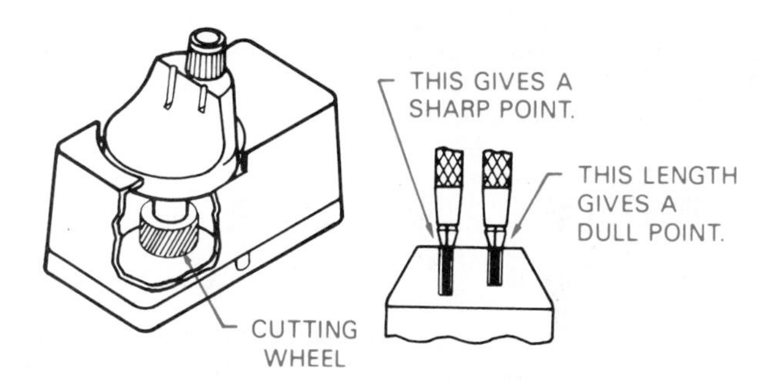

Figure 1–5 Cutting wheel lead pointer.

Pocket Pointer

A portable sharpener for mechanical pencils is the pocket pointer. This pointer contains blades that sharpen lead to a conical point. To sharpen a point, you either hold the pencil still with one hand and rotate the sharpener with the other hand, or vice versa. The cutting blades on this sharpener wear out with use and must be replaced periodically. The pocket pointer works with either graphite or plastic lead.

Cutting-wheel Pointer

The best type of conical-point sharpener for mechanical pencil holders is a mechanical lead pointer with a tool-steel cutting wheel. This unit must be clamped to the drawing board. The basic principle of this pointer is shown in Figure 1–5. Use the slots provided in the top to expose the right length of lead to get a sharp or slightly dull point. The slightly dull point is used for lettering. This sharpener works on graphite or polyester leads equally well. The cutting wheel lasts a long time and the unit is easy to clean by washing it in a sink. An occasional drop of light machine oil on the shaft keeps the cutting wheel working properly.

Electric Lead Pointer

The electric lead pointer is an excellent device for quick pencil sharpening. Most units have gages for proper lead extension and sharpen points to a perfect taper in about half a second. Their heavy-duty motors are quiet and maintenance free. Usually this machine has a storage area for lead holders. Lead pointer attachments are also available that turn an electric eraser into a pencil pointer.

PROFESSIONAL HINT

When you have sharpened the tip of plastic lead, push the lead back in the holder as far as practical because plastic lead is weaker than graphite and breaks easily. Pushing the lead back is a good practice with graphite lead also.

TECHNICAL PENS AND ACCESSORIES

Technical Pens

Also known as technical fountain pens, technical pens have improved in quality and ability to produce excellent inked lines. These pens function on a capillary action where a needle acts as a valve to allow ink to flow from a storage cylinder through a small tube, which is designed to meter the ink so a specific line width is created. A technical pen is shown in Figure 1–6. Technical pens may be purchased individually or in sets. The different tip sizes used to make various line widths range from a narrow number 6 × 0 (.005 in./0.13 mm) to a wide number 7 (.079 in./2 mm). Figure 1–7 shows a comparison of some of the different line widths available with technical pens.

Figure 1–6 Technical pen. *Courtesy Koh-I-Noor Rapidograph, Inc.*

6x0	4x0	3x0	00	0	1	2	2½	3	3½	4	6	7
.13	.18	.25	.30	.35	.50	.60	.70	.80	1.00	1.20	1.40	2.00
005 in	007 in	010 in	012 in	014 in	020 in	024 in	028 in	031 in	039 in	047 in	055 in	079 in
13 mm	18 mm	25 mm	30 mm	35 mm	50 mm	60 mm	70 mm	80 mm	1 00 mm	1 20 mm	1 40 mm	2 00 mm

Figure 1–7 Technical pen line widths. *Courtesy Koh-I-Noor Rapidograph, Inc.*

Technical pens are available in different price ranges as determined by the kind of material used to make the point. Each kind of point has a recommended use. Stainless steel points are made for use on vellum. These points wear out very rapidly when used on polyester film. Tungsten carbide points are recommended for polyester film and can be used on vellum. Jewel points provide the longest life and can be used on vellum or polyester film with the best results. They also provide the smoothest ink flow on polyester film.

In addition to having the advantage of a constant line width, technical pens have a reservoir that allows the drafter to make inked lines for a long period of time before ink must be added. Technical pens may be used with templates to make circles, arcs, and symbols. Compass adapters to hold technical pens are also available. Technical pen tips are designed to fit into scribers for use with lettering guides. This concept is discussed further in Chapter 3.

Most pen manufacturers have designed pen holders or caps that help keep pen points moist and ready for use. When pens are used daily, cleaning may be necessary only periodically. Some symptoms to look for when pens need cleaning include the following:

- Ink constantly creates a drop at the pen tip.
- Ink tends to flow out around the tip holder.
- The point plunger does not activate properly. If you do not hear and feel the plunger when you shake the pen, then it is probably clogged with thick or dried ink.
- When drawing a line, ink does not flow freely.
- Ink flow starts with difficulty.

The ink level in the reservoir should be kept between one-quarter to three-quarters full. If the cylinder is too full, the ink may not flow well; if too little ink is present, then the pen may skip when in use.

Pen Cleaning

Read the cleaning instructions that come with the brand of pen that you purchase. Some pens require disassembly for cleaning while others should not be taken apart. Some manufacturers strongly suggest that pen tips remain assembled. The main parts of a technical pen include the cap, nib, pen body, ink cartridge (or reservoir), and pen holder as seen in Figure 1–8. The nib contains a cleansing wire with drop weight and a safety

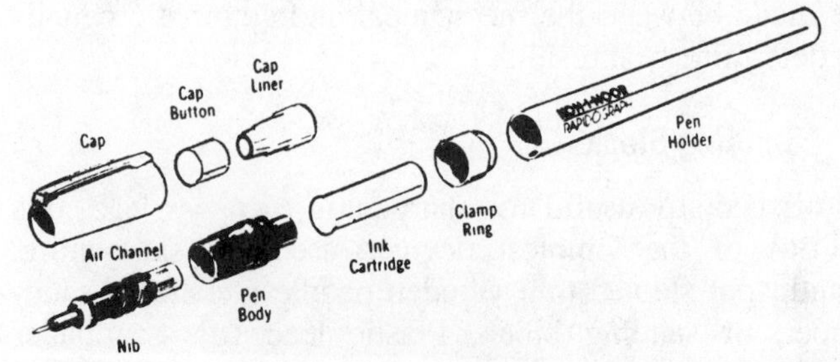

Figure 1–8 The parts of a technical pen. *Courtesy Koh-I-Noor Rapidograph, Inc.*

plug. Most manufacturers recommend that the nib remain assembled during cleaning as the cleansing wire is easily damaged. To fill the technical pen, unscrew the cap, holder, and clamp ring from the pen body. Remove the cartridge and fill it with ink to within 1/4 of the top. Slowly replace the cartridge and assemble the parts. Pens should be cleaned before each filling or before being stored for a long period of time. Clean the technical pen nib, cartridge, and body separately in lukewarm water or special cleaning solution.

Ultrasonic pen cleaners are available to clean points. Pens are placed in a tank where millions of energized microscopic bubbles, generated by ultrasonic action, carry cleaning solution into the smallest openings of the drawing point to scrub the tube inside and out.

Syringe pressure pen cleaners and point starters are also available for cleaning pens. These cleaners use pressure and suction for the cleaning action. These units are provided with a connector for use as a pen starter. Specially formulated pen cleaner should be used for best results in either the ultrasonic or syringe units. Pen cleaner can also be used to soak pen points for cleaning by hand.

Ink

Drafting inks should be opaque, or have a matte or semi-flat black finish that will not reflect light. The ink should reproduce without hot spots or line variation. Drafting ink should have excellent adhesion properties for use on paper or film. Certain inks are recommended for use on film in order to avoid peeling, chipping, or cracking. Inks recommended for use in technical pens also have nonclogging characteristics. This property is especially important for use in high-speed computer-graphics plotters.

When selecting an ink, be sure to purchase one for the job you want done. First, determine how the ink will be applied, that is, from a technical pen, computer plotter, air brush, fountain pen, calligraphy pen, or with a brush. Second, determine the surface the ink will be used on, such as paper (vellum or bond), polyester film, or acetate. Third, determine if your use requires the ink to be opaque, fast drying, waterproof, or erasable.

ERASERS AND ACCESSORIES

The common shapes of erasers are rectangular and stick. The stick eraser works best in small areas. There are three basic types of erasers: pencil, ink, and plastic. The plastic eraser is used for plastic lead or ink on polyester film. These erasers are identified by their white or translucent color. Select an eraser that is recommended for the particular material used. When used with ink, apply very light pressure, and take the utmost care because the friction developed by the speed of erasure can easily damage the drafting film surface and prevent the adhesion of ink when redrawing over the erased area. Moistening the eraser during use helps to reduce any damage to the drawing surface.

Erasing Tips

When erasing, the idea is to remove an unwanted line or letter. You do not want to remove the surface of the paper or polyester film. Erase only hard enough to remove the unwanted line. However, you must bear down hard enough to eliminate the line completely. If all the line does not disappear, ghosting results. A ghost is a line that seems to have been eliminated but still shows on a print. Lines that have been drawn so hard as to make a groove in the drawing sheet can cause a ghost too. To remove ink from vellum, use a pink or green eraser, or an electric eraser. Work the area slowly. Do not apply too much pressure or erase in one spot too long or you will go through the paper. On polyester film, use a vinyl eraser and/or a moist cotton swab. The inked line usually comes off easily, but use caution. If you destroy the matte surface of either a vellum or polyester sheet you will not be able to redraw over the erased area.

Lead may be picked up from the drawing board surface and transmitted to the back of the drawing surface. When this happens, the drawing must be turned over and the graphite removed from the back surface of the drawing.

Electric Erasers

Professional drafters use electric erasers. Those with cords that plug in are best, but cordless, rechargeable units are also available. When working with an electric eraser, you do not need to use very much pressure because the eraser operates at high speed. The purpose of the electric eraser is to remove unwanted lines quickly. Use caution, these erasers can also remove paper quickly! Figure 1–9 shows an electric eraser and its optional lead pointer attachment.

Erasing Shield

Erasing shields are thin metal or plastic sheets with a number of differently shaped holes. (See Figure 1–10.) They are used to erase small, unwanted lines or areas. For example, if you have a corner overrun, place one of the slots of the erasing shield over the area to be removed while covering the good area.

Eradicating Fluid

Eradicating fluid is primarily used with ink on film or for removal of lines from sepia (brown) prints. Sepia prints are often used to make corrections rather than correcting the original drawing. The eradicating fluid is most often applied with a brush or cotton swab. If in doubt about its application, follow the manufacturer's

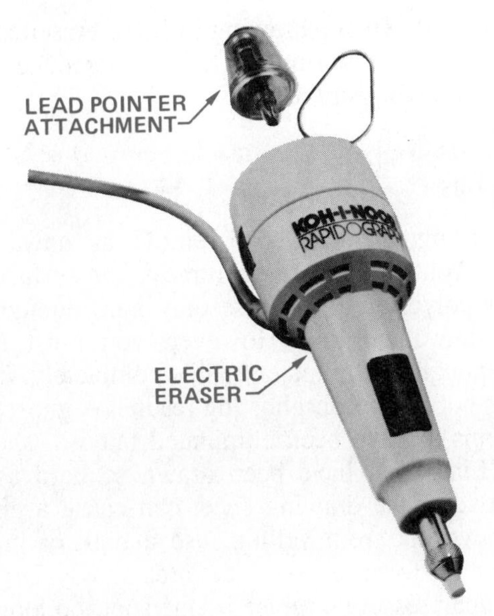

Figure 1–9 Electric eraser with optional lead pointer attachment. *Courtesy Koh-I-Noor Rapidograph, Inc.*

Figure 1–10 Erasing shield. *Courtesy Koh-I-Noor Rapidograph, Inc.*

instructions; however, application is usually done by lightly moistening the area to be corrected. Then the solution is wiped with a tissue, being careful to remove all residue. This process is continued until all residue is gone. Eradicating fluid is especially effective for removing aged ink lines and for erasure of large areas.

Cleaning Agents

Special eraser particles are available in a shaker-top can or a pad. Both types are used to sprinkle the eraser particles on a drawing to help reduce smudging and to keep the drawing and your equipment clean. The particles also help float triangles, straightedges, and other drafting equipment to reduce line smudging. Cleaning powders are designed to prepare the drafting surface prior to working on it. Dry cleaning pads are fabric covered and contain fine, graded art gum that sifts through the fabric weave when the pad is tapped on the drawing surface. Use this material sparingly since too much of it can cause your lines to become fuzzy. Cleaning powders are not recommended for use on ink drawings, nor on polyester film.

Dusting Brush

Use a dusting brush to remove eraser particles from your drawing. Doing so helps reduce the possibility of smudges. Avoid using your hand to brush away eraser particles because the hand tends to cause smudges which reduces drawing neatness. A clean, dry cloth works better than your hand for removing eraser particles, but a brush is preferred.

After about a month of use, the brush picks up graphite particles and casts a slight film over the drawing when used. For this reason, the brush should be cleaned regularly with soap and water.

DRAFTING INSTRUMENTS

Kinds of Compasses

Compasses are used to draw circles and arcs. However, using a compass can be time consuming. Use a template, whenever possible, to make circles or arcs more quickly. A compass is especially useful for large circles.

There are several basic types of compasses:

1. Friction-head compass—not as accurate or stable as other types.
2. Drop-bow compass—mostly used for drawing small circles. The center rod contains the needle point and remains stationary while the pencil or pen leg revolves around it. (See Figure 1–11.)
3. Circuit-scribing instrument—a modified drop-bow compass used to cut terminal pads, prepare printed circuit layouts on scribe-coat film.
4. Bow compass (center wheel)—operates on the screwjack principle by turning the large knurled center wheel. The bow compass, shown in Figure 1–12, is used for most drawings. This is the type of compass commonly used by professionals.
5. Beam compass—a bar with an adjustable needle, and a pencil or pen attachment for swinging large arcs or circles. Also available is a beam that is adaptable to the bow compass. Such an adapter works only on bow compasses that have a removable break point, not on the fixed point models.

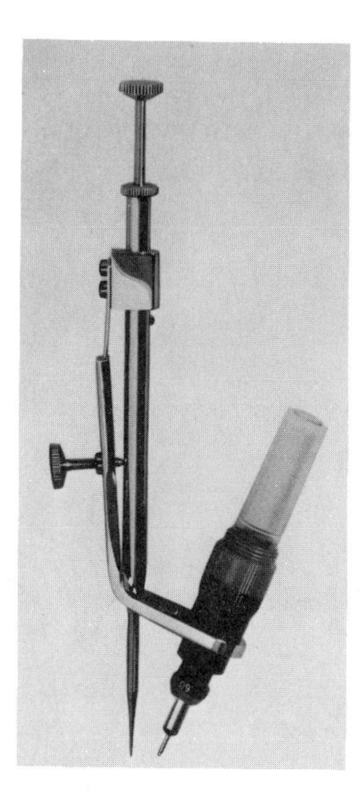

Figure 1–11 Drop-bow compass, technical pen model. *Courtesy J. S. Staedtler, Inc.*

Compass Use

Keep both the compass needle point and lead point sharp. The points are removable for easy replacement. The better compass needle points have a shoulder on them. The shoulder helps keep the point from penetrating the paper more than necessary. Compare the needle points in Figure 1–13.

The compass lead should, in most cases, be one grade softer than the lead you use for straight lines because less pressure is used on a compass than a pencil. Keep the compass lead sharp. An elliptical point is commonly used with the bevel side away from the needle leg. Keep the lead and the point equal in length. Figure 1–14 shows properly aligned and sharpened points on a compass.

Use a sandpaper block to sharpen the elliptical point. Be careful to keep the graphite residue away from your drawing and off your hands. Remove excess graphite from the point with a tissue or cloth after sharpening. Sharpen the lead often.

Some drafters prefer to use a conical point in their compass. This is the same point used in a lead holder. If you want to try this point, sharpen a piece of lead in your lead holder, then transfer it to your compass, as shown in Figure 1–15.

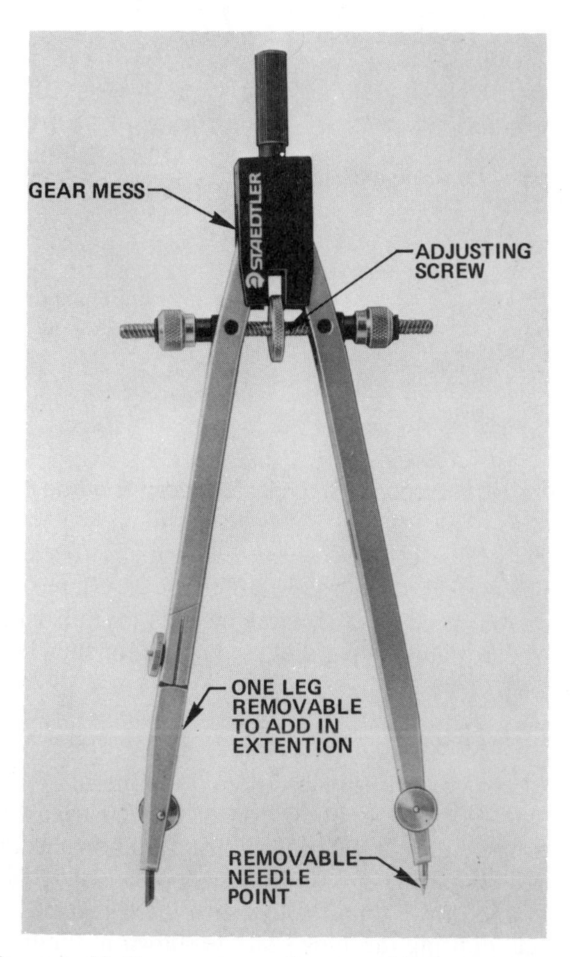

Figure 1–12 Bow compass. *Courtesy J.S. Staedtler, Inc.*

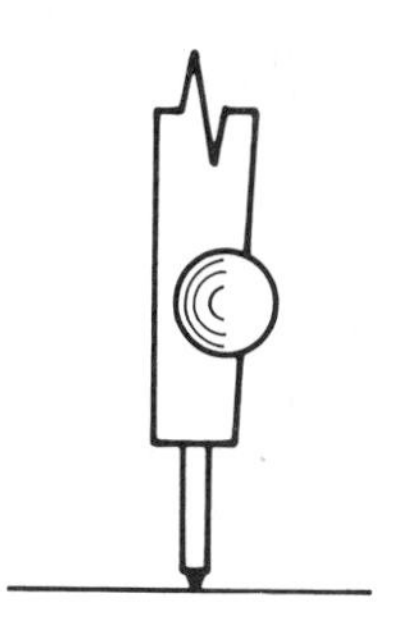

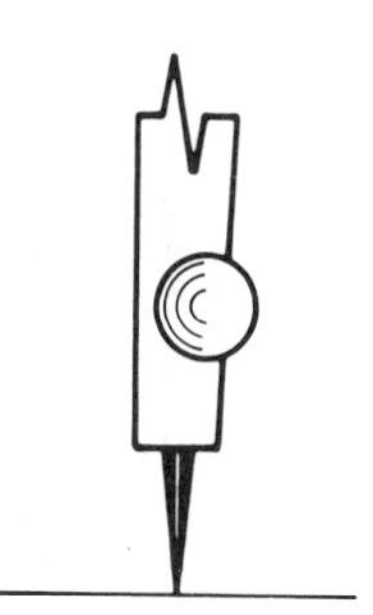

Figure 1–13 Compass points.

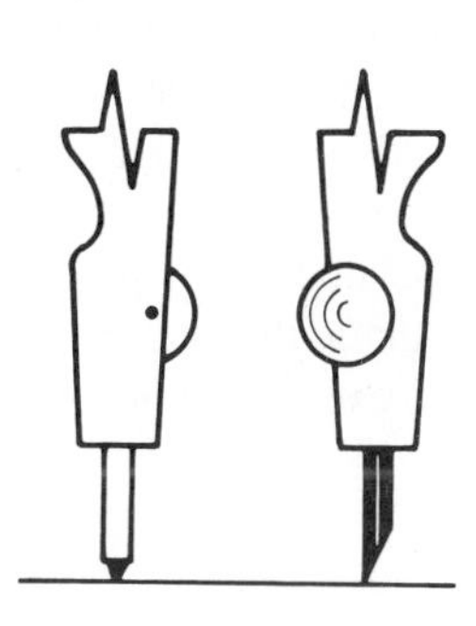

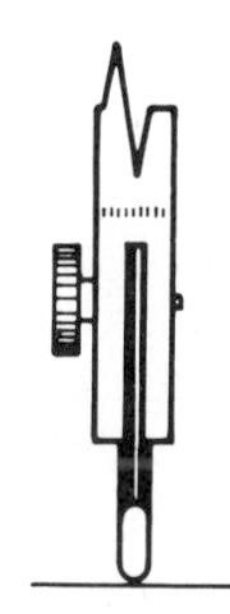

Figure 1–14 Properly sharpened and aligned elliptical compass point.

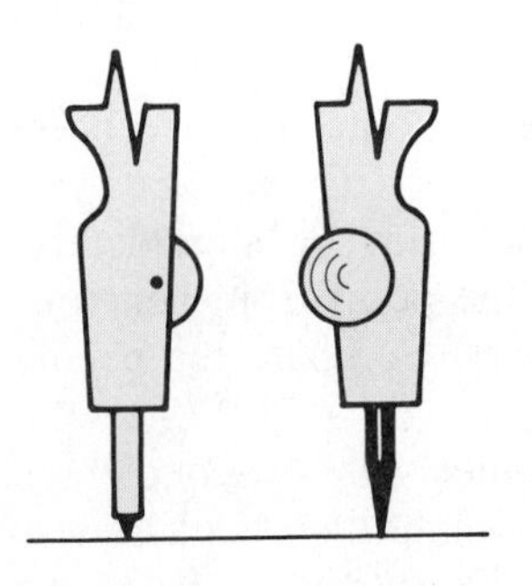
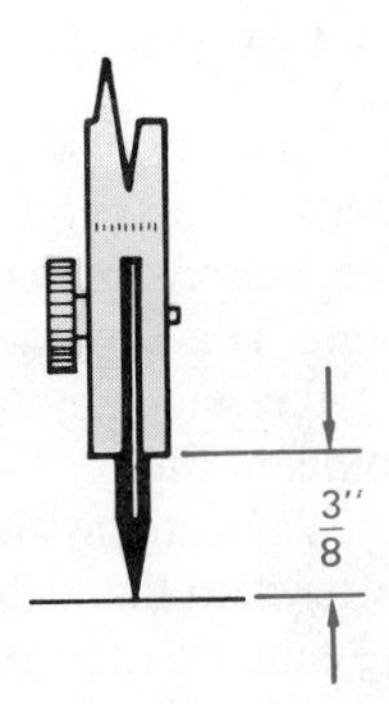

Figure 1–15 Conical compass point.

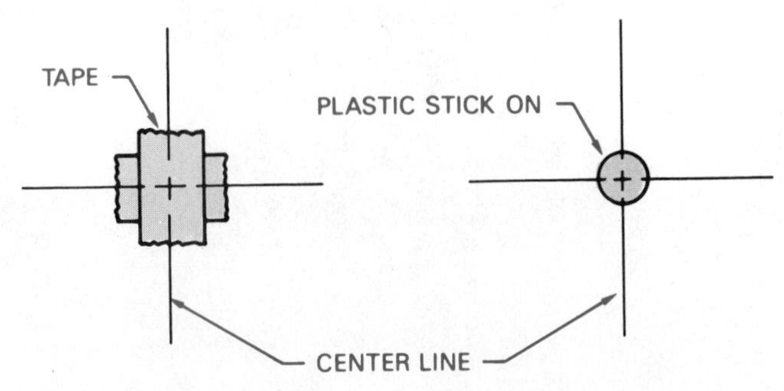

Figure 1–16 Drawing sheet protection from compass point.

If you are drawing a number of circles from the same center, you will find that the compass point causes an ugly hole in your drawing sheet. Reduce the chance of making such a hole by placing a couple of pieces of drafting tape at the center point for protection. There are small plastic circles available for just this purpose. Place one at the center point, then pierce the plastic with your compass. (See Figure 1–16.)

Dividers

Dividers are used to transfer dimensions or to divide a distance into a number of equal parts.

Note: Do not try to use dividers as a compass.

Some drafters prefer to use bow dividers because the center wheel provides the ability to make fine adjustments easily. Also, the setting remains more stable than with standard friction dividers.

A good divider should not be too loose or tight. It should be easily adjustable with one hand. In fact, you should control a divider with one hand as you lay out equal increments or transfer dimensions from one feature to another. Figure 1–17 shows how the divider should be handled when used.

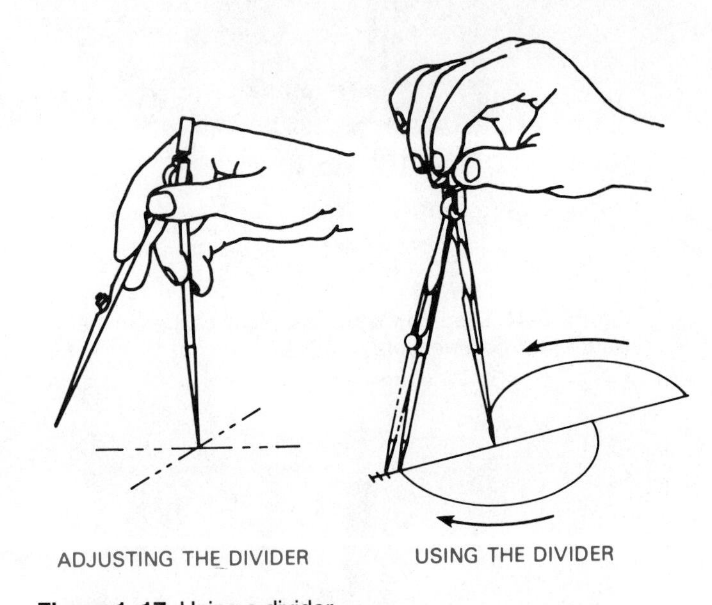

Figure 1–17 Using a divider.

Proportional Dividers

Proportional dividers are used to reduce or enlarge an object without the need of mathematical calculations or scale manipulations. The center point of the divider is set at the correct point for the proportion you want. Then you measure the original size line with one side of the proportional divider and the other side automatically determines the new reduced or enlarged size. Figure 1–18 shows a proportional divider.

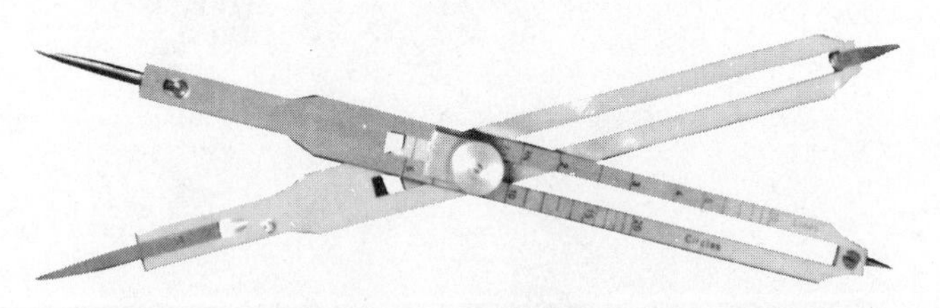

Figure 1–18 Proportional divider. *Courtesy Teledyne Post.*

Parallel Bar

The parallel bar slides up and down the board to allow you to draw horizontal lines. (See Figure 1–19.) Vertical lines and angles are made with triangles in conjunction with the parallel bar. The parallel bar is commonly found in architectural drafting offices because architectural drawings are frequently very large. Architects often need to draw straight lines the full length of their boards and the parallel bar is ideal for such lines.

Triangles

There are two standard triangles. One has angles of 30°–60°–90° and is known as the 30–60 triangle. The other has angles of 45°–45°–90° and is known as the 45° triangle. Figure 1–20 shows these popular triangles.

Some drafters prefer to use triangles in place of a vertical drafting machine scale as shown in Figure 1–21. The machine protractor or the triangle can be used to

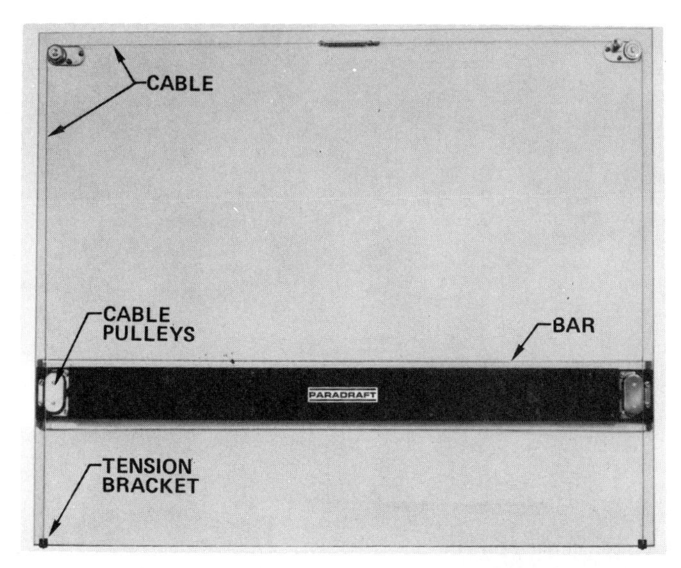

Figure 1–19 Parallel bar. *Courtesy Charvoz-Carsen Corporation.*

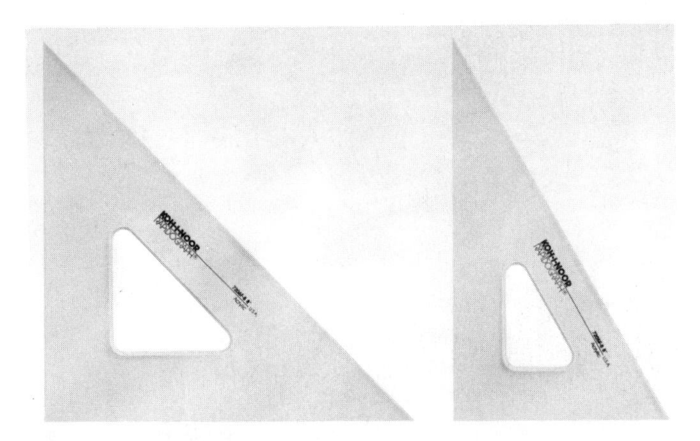

Figure 1–20 45° and 30°–60° triangles. *Courtesy Koh-I-Noor Rapidograph, Inc.*

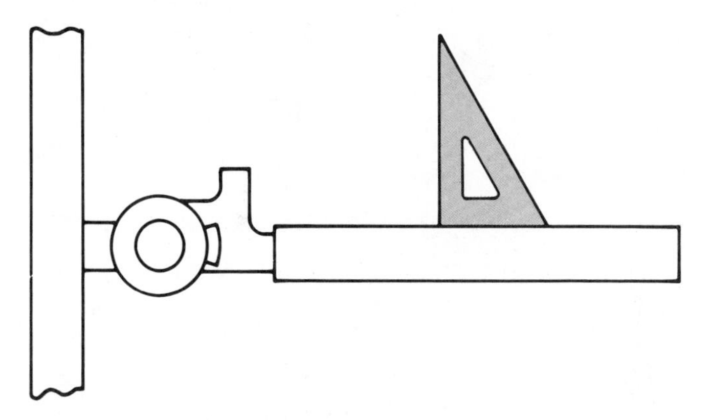

Figure 1–21 Using a triangle with a drafting machine.

make angled lines. Drafters who use parallel bars rather than drafting machines also use triangles to make vertical and angled lines.

Triangles may also be used as straightedges to connect points for drawing lines without the aid of a parallel bar or machine scale. Triangles are used individually or in combination to draw angled lines in 15° increments. (See Figure 1–22.) Also available are adjustable triangles with built-in protractors that are used to make angles of any degree up to a 45° angle. (See Figure 1–23.)

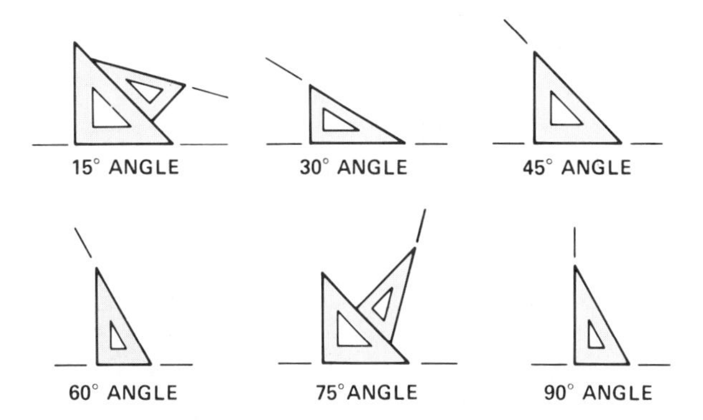

Figure 1–22 Angles that may be made with the 30°–60° and 45° triangles individually or in combination.

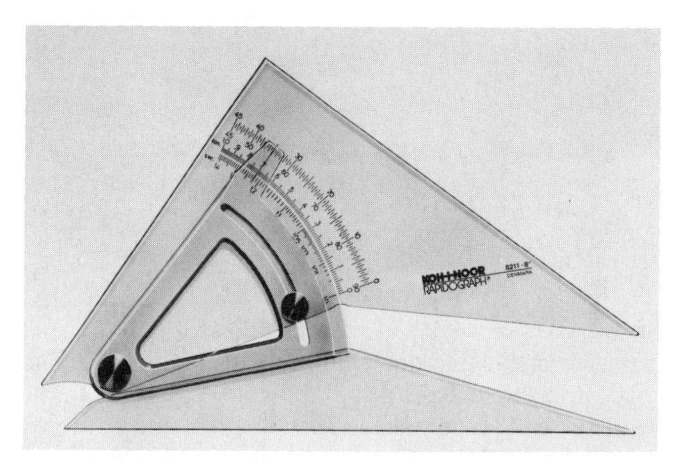

Figure 1–23 Adjustable triangle. *Courtesy Koh-I-Noor Rapidograph, Inc.*

Templates

Circle Templates. Circle templates are available with circles in a range of sizes beginning with 1/16 in. The circles on the template are marked with their diameters and are available in fractions, decimals, or millimeters. Sample circle templates are shown in Figure 1–24. Circle templates are calibrated in decimal, fraction, and metric increments. A popular template is one that has circles, hexagons, squares, and triangles.

Always use a circle template rather than a compass. Circle templates save time and are very accurate. For best results when making circles, try to keep your pencil or pen perpendicular to the paper. To obtain proper wide lines with a pencil, use a slightly rounded, soft-point mechanical pencil or a 0.9-mm automatic pencil. Another way to keep a proper thickness for wide lines is to use two template sizes that are very close together. For example, to draw a .75 in. diameter circle, use a .75 in.

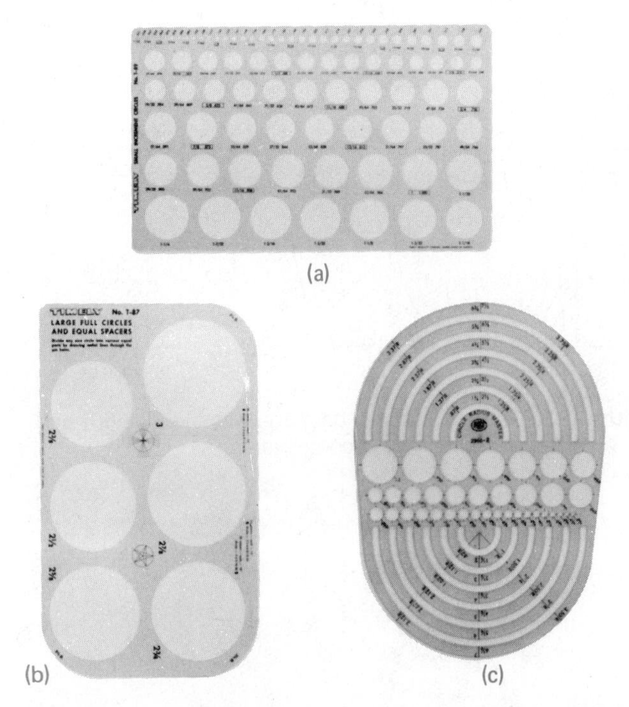

Figure 1–24 (a) Small circles; (b) large full circles; (c) large half-circles. *Courtesy Teledyne Post.*

diameter template to draw the circle, then draw a second concentric circle with a .7812 diameter template. The two lines will blend together, creating one line of proper line thickness. Be careful that the two lines do blend and do not appear as two separate lines. This technique does not apply to drawing thin-line circles or arcs.

To use a circle template properly, first draw the centerlines of your circle. Then exactly align the dashes on the template with the centerlines as shown in Figure 1–25. Proceed to trace the outline of the circle.

To draw arcs with a circle template, use one of two methods. One method is to draw the centerlines of the arc, align the template with the proper diameter, and draw the arc. Keep in mind that the template circles are marked in diameter while the size of an arc is given in radius. Remember to divide the template size in half to find the proper arc radius. The other method of drawing arcs is to lightly draw outside construction lines, then fill in the arc to the points of tangency as shown in Figure 1–26. Be sure the connection between the arc and the straight line is smooth.

When using a circle template and a technical pen, keep the pen perpendicular to the paper. Some templates have risers built in to keep the template above the drawing sheet. Without this feature there is a risk of ink running under a template that is flat against the drawing. If your template does not have risers, purchase and add template lifters, use a few layers of tape placed on the underside of the template (although tape does not always work well), or place a second template with a larger circle under the template you are using. (See Figure 1–27.)

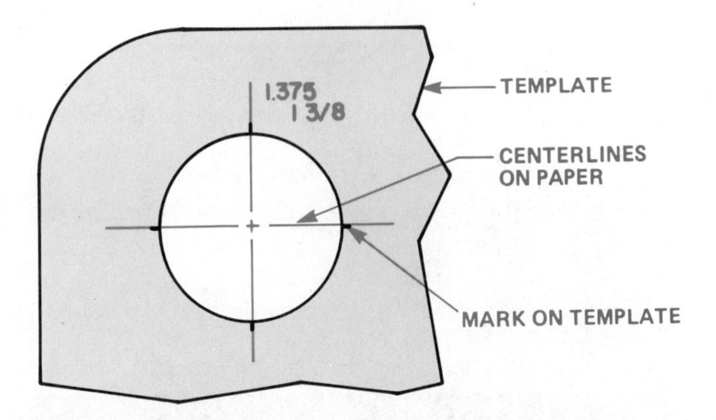

Figure 1–25 Using a circle template to draw a circle.

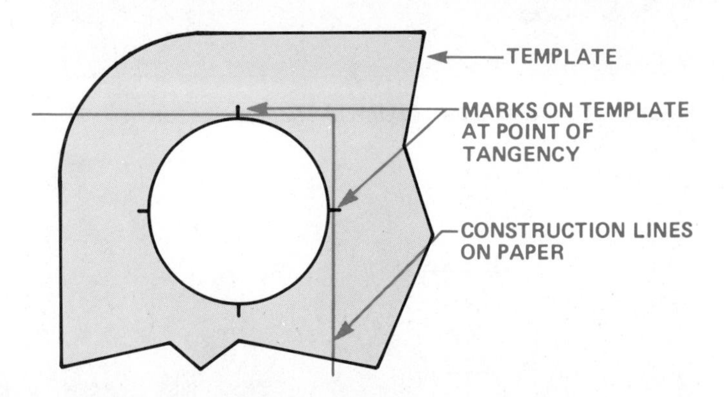

Figure 1–26 Using a circle template to draw an arc.

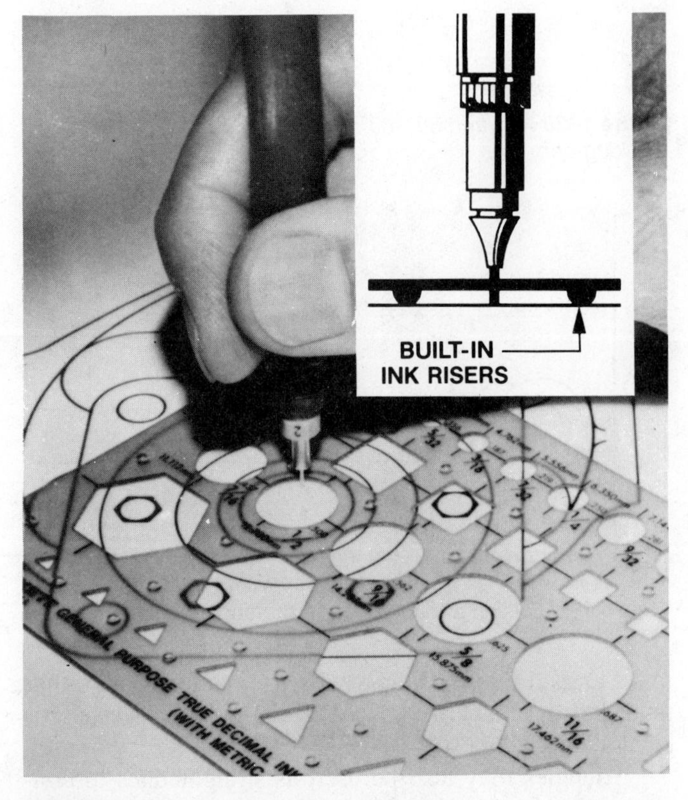

Figure 1–27 A template with built-in risers for inking. *Courtesy Chartpak.*

Ellipse Templates. Ellipses are circles seen at an angle. The angular relationship of an ellipse to a circle is shown in Figure 1–28. The parts of a circle are shown in Figure 1–29, while the parts of an ellipse are shown in Figure 1–30.

The kind of pictorial drawing known as isometric projects the sides of objects at a 30° angle in each direction away from the horizontal. These projections establish two of the three isometric axes. The third isometric axis is vertical. (See Figure 1–31.) Isometric circles are ellipses aligned with the horizontal, right, or left planes of an isometric box. These ellipses have centerlines that are parallel to two of the three major axes on an isometric drawing. When drawing circles in isometric you need to consider the plane in which the ellipse is to be drawn. Isometric ellipse templates automatically position the ellipse at the proper angle of 35° 16' as shown in Figure 1–32.

If, when drawing an isometric ellipse, you position the template so that the alignment marks are at a 30° angle from horizontal, as shown in Figure 1–33, then that ellipse goes on or parallel to the top surface of the isometric cube. The isometric ellipse is then in the hori-

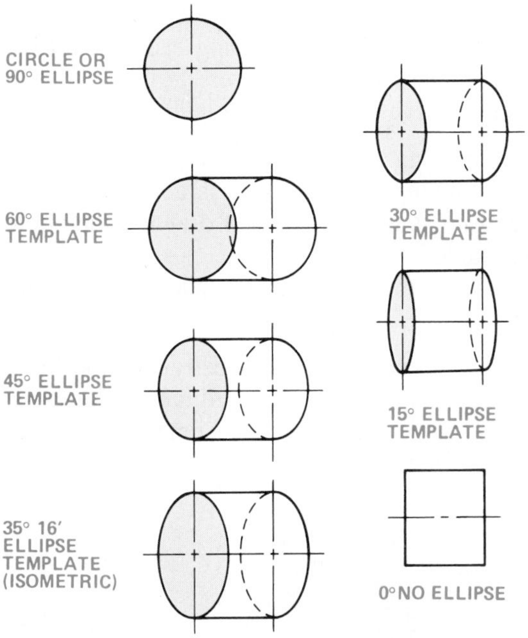

CIRCLE OR 90° ELLIPSE
60° ELLIPSE TEMPLATE
45° ELLIPSE TEMPLATE
35° 16' ELLIPSE TEMPLATE (ISOMETRIC)
30° ELLIPSE TEMPLATE
15° ELLIPSE TEMPLATE
0° NO ELLIPSE

Figure 1–28 Ellipses are established by their relationship to a circle turned at various angles.

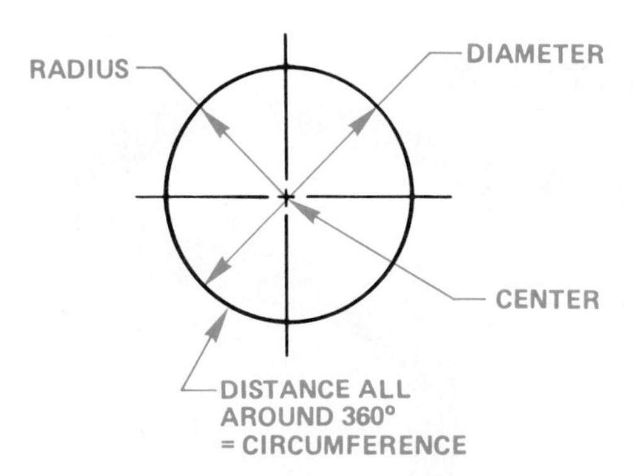

Figure 1–29 Parts of a circle.

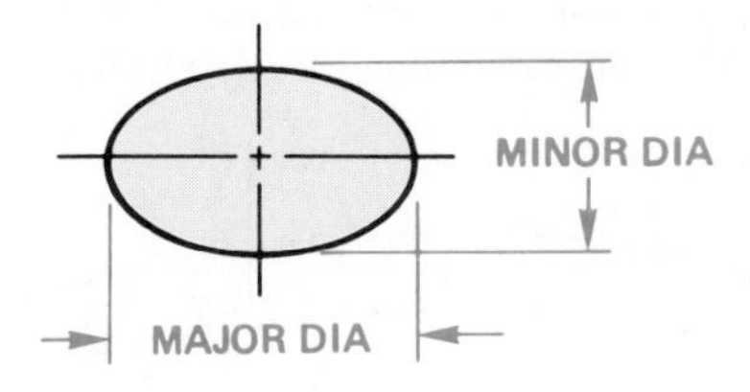

Figure 1–30 Parts of an ellipse.

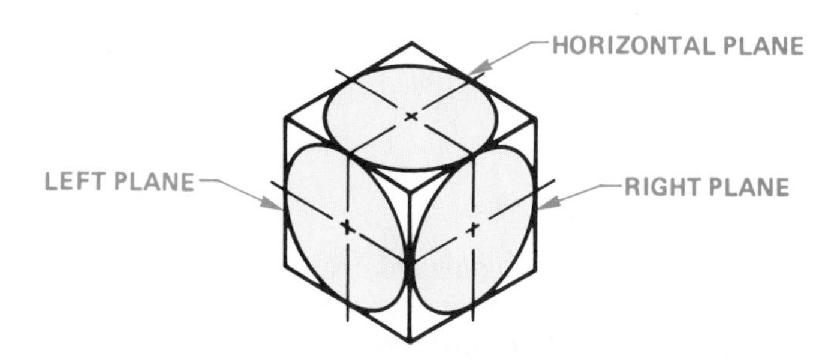

Figure 1–31 Ellipses in isometric planes.

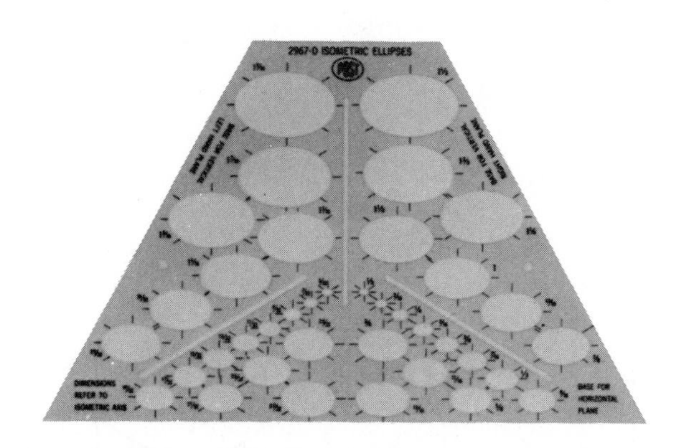

Figure 1–32 Isometric ellipse template. *Courtesy Teledyne Post.*

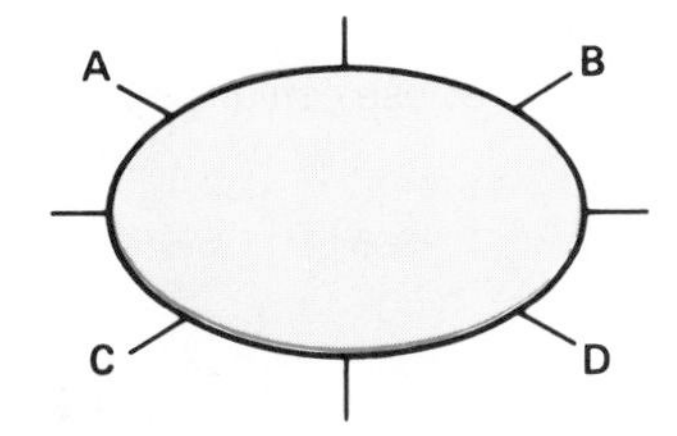

Figure 1–33 Positioning the isometric ellipse in the horizontal plane.

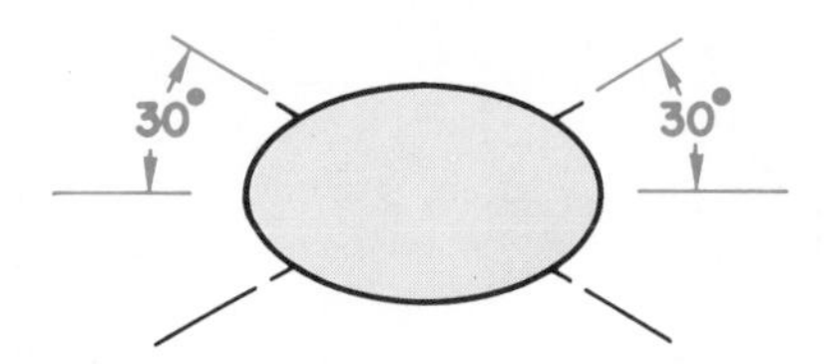

Figure 1–34 Isometric ellipse in the horizontal plane.

zontal plane as seen in Figure 1–34. The centerlines of the ellipse touch the midpoints of each edge of the isometric square, as shown in Figure 1–35. To position the ellipse on the other isometric axes, use the same four alignment marks. Place two of them vertically and the remaining two align on the 30° right or left axis, as shown in Figure 1–36.

Isometric drawing is discussed in detail in Chapter 14.

PROFESSIONAL HINT

Never use a compass if you can use a template. Templates increase drafting speed and are very accurate.

Irregular Curves

Irregular curves are commonly called French curves. These curves have no constant radii. An irregular curve is shown in Figure 1–37(a) while a flexible curve is shown in Figure 1–37(b). Figure 1–38 shows a radius curve, composed of a radius and tangent. The radius on these curves is constant and their sizes range from 3 ft to 200 ft, which are commonly used in highway drafting. In addition to these two kinds of curves there are available ship's curves. The curves in a set of ship's curves become progressively larger and, like French curves, have no constant radii. They are used for layout and development of ships' hulls.

In order to draw an irregular curve, points on the curve are developed or plotted. The points are then connected with a light line. Then a flexible curve is bent to fit, or a portion of an irregular curve is matched to the light line to include at least three plotted points. Care must be taken to make the curve flow smoothly. This smooth flow of the line is accomplished by never drawing the full length of the curve, but by overlapping each successive setting of the irregular curve. (See Figure 1–39.)

Use a French curve under the following conditions:

1. When drawing an irregular curve.
2. When drawing a portion of a large diameter circle and no standard means of drawing the circle is available.
3. When blending tangents, as shown in Figure 1–40.

Do not use a French curve to specify a shape to be manufactured unless X and Y coordinates along the curve can be established and dimensioned. Whenever possible, draw a line such as the one shown in Figure 1–40 with radii.

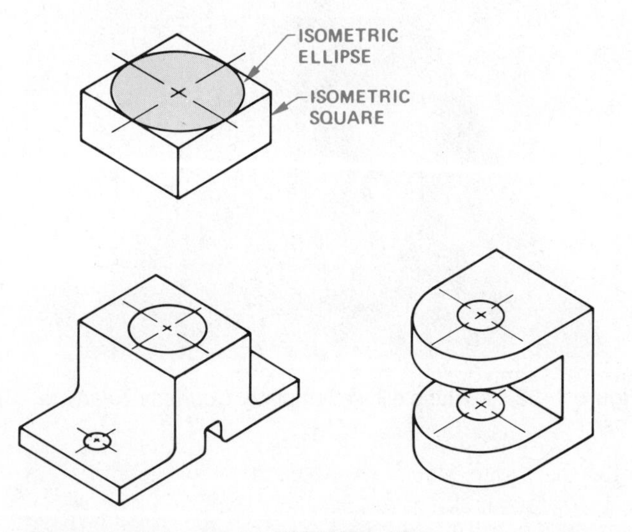

Figure 1–35 Isometric ellipse centerlines.

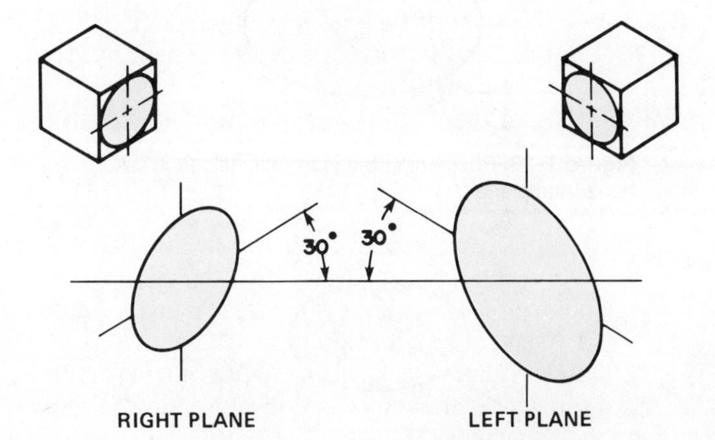

Figure 1–36 Right and left isometric ellipse planes.

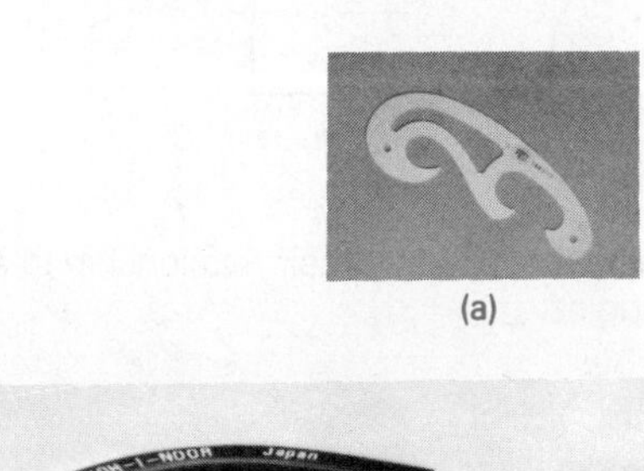

(a)

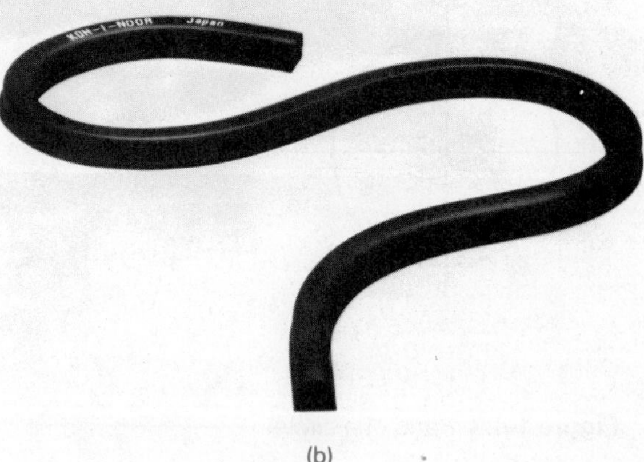

(b)

Figure 1–37 (a) Irregular or French curve. *Courtesy Teledyne Post*; (b) flexible curve. *Courtesy Koh-I-Noor Rapidograph, Inc.*

Figure 1–38 Radius curve.

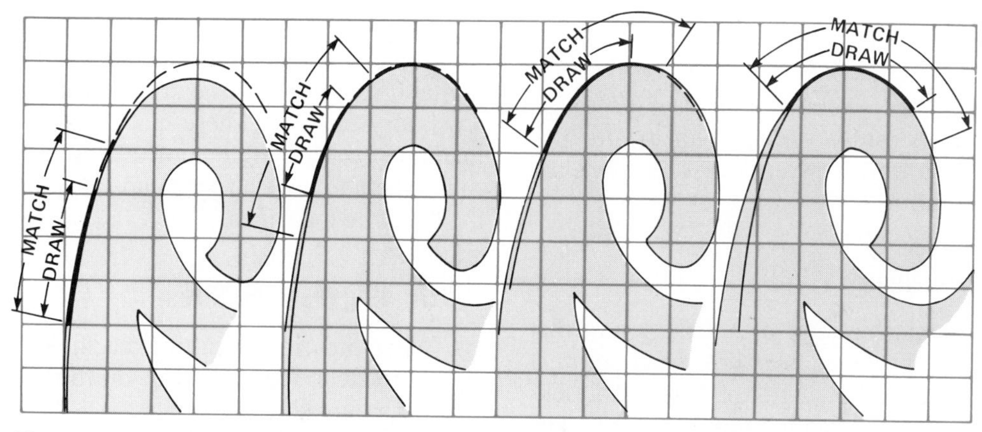

Figure 1–39 Irregular curve development.

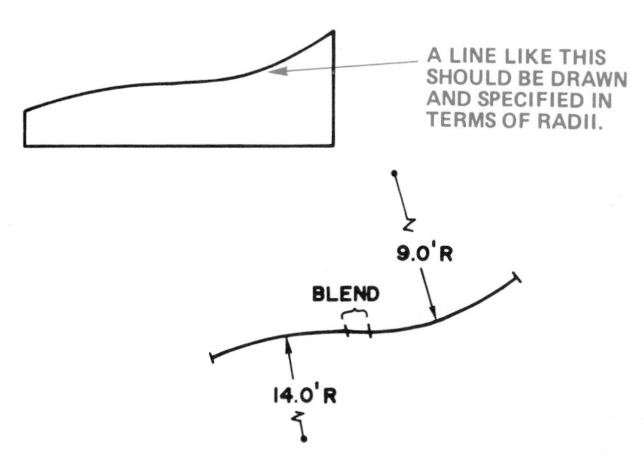

Figure 1–40 Drawing arcs as radii.

DRAFTING MACHINES

Drafting machines, for the most part, take the place of triangles and parallel bars. The drafting machine maintains a horizontal and vertical relationship between scales, which also serve as straightedges, by way of a protractor component. The protractor allows the scales to be set quickly at any angle. There are two types of drafting machines, arm and track. Although both types are excellent tools, the track machine has generally replaced the arm machine in industry. A major advantage of the track machine is to allow the drafter to work with a board in the vertical position. A vertical drafting surface position is generally more comfortable to use than a horizontal table. For school or home use, the arm machine may be a good economical choice.

Arm Drafting Machine

The arm drafting machine is compact and less expensive than a track machine. The arm machine clamps to a table and through an elbow-like arrangement of supports allows the drafter to position the protractor head and scales anywhere on the board.

The components of an arm drafting machine, shown in Figure 1–41, are as follows in numbered order:

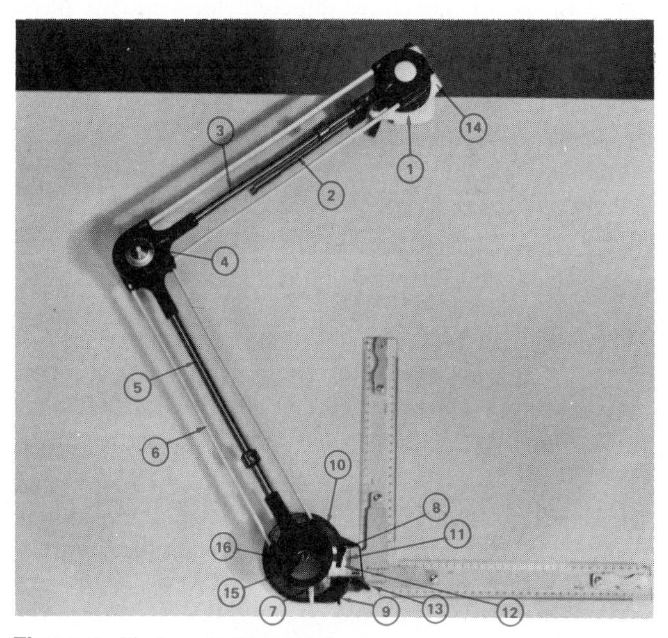

Figure 1–41 Arm drafting machine.

1. Screw board clamp.
2. Cartridge stabilizer—balances the machine on boards inclined to 25°.
3. Single tension bar—permits band adjustment.
4. Adjustable disc brake—hand tightens to increase machine stability on boards inclined to 35°.
5. Tension bars and bands—assure dependability and accuracy.
6. Semi-rigid band covers—permit an instant check on band tension.
7. 360° baseline setting—with positive lock baseline clamp. Allows for a fast alignment of the scale to any reference line on a drawing.
8. Dual action index control—press the control for automatic 15° settings; press and lift to release the index for free rotation to any angle.
9. Vernier clamp—locks the protractor at any angle.
10. Protractor scale—engine divided to 1° to 180° in both directions.
11. Vernier scale—engine divided to 5' (minutes).

12. Magnifier—on some models.
13. Micro-adjuster—assures fine scale alignment to the baseline.
14. Swing-free hinge—permits the machine to be raised to a rest position vertical to the board.
15. Translucent protractor cover—on some models.
16. Protractor head.

Track Drafting Machine

A track drafting machine has a traversing arm that moves left and right across the table and a head unit that moves up and down the traversing arm. There is a locking device for both the head and the traversing arm. The shape and placement of the controls of a track machine vary with the manufacturer although most brands have the same operating features and procedures. Figure 1–42 shows the component parts of a track drafting machine. The track drafting machine has the following advantages over the arm drafting machine:

1. The track machine is more stable and in some cases more accurate than an arm machine.
2. You can draw with your table inclined at a steep angle and the head will not slide down the table as it will with an arm machine.
3. Both the head and traversing arm may be locked in any position. This feature is important when using a lettering guide or other equipment that requires a stationary position.

As with the arm machines, track drafting machines have a vernier head that allows the user to measure angles accurately to 5'. A track drafting machine head protractor and vernier scale are shown in Figure 1–43. An optional dial head, available on some brands, allows for quick and easy angular measurements without the use of a vernier. A dial head is shown in Figure 1–44.

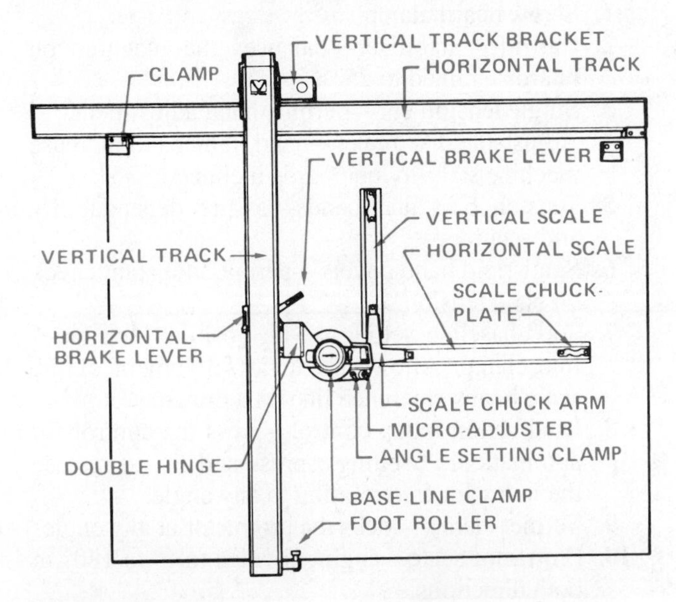

Figure 1–42 Track drafting machine and its parts. *Courtesy Consul & Mutoh, Ltd.*

Digital Display Machine

Some track drafting machines are available with a digital display of angles and X–Y coordinates along with a memory function. These machines are more expensive than other track drafters; however, some companies have found them to have a speed and accuracy advantage over the old-style machines.

Drafting Machine Sizes

When ordering a drafting machine, the specifications should relate to the size of the drafting board on which it will be mounted. For example, a 37½ × 60 in. machine would properly fit a table of the same size.

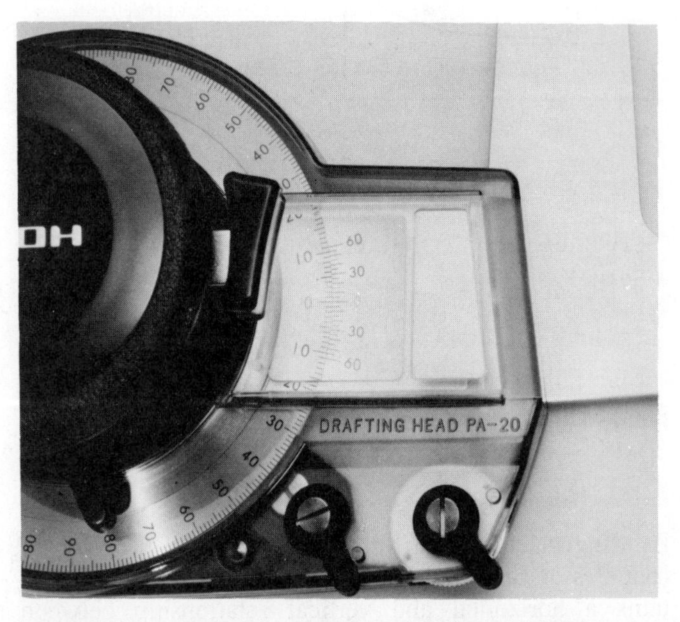

Figure 1–43 Drafting machine head protractor and vernier. *Courtesy Consul & Mutoh, Ltd.*

Figure 1–44 Drafting machine dial head. *Courtesy Consul & Mutoh, Ltd.*

Coordinate Reading and Processing Drafting Machine

There is a drafting machine available that offers a digital reading of angles, X and Y, and polar coordinates and has a resolution of 5'. A computer may be added to this machine that will greatly speed the time-consuming calculations that are done during layout and after a drawing is complete. The head and display of this machine are seen in Figure 1–45. A coordinate reading and processing machine will help perform the following tasks:

1. Provide a digital display of X and Y or polar coordinates.
2. Determine length measurements of straight or curved lines.
3. Determine angular measurement.
4. Provide a full range of scales in inches or millimeters.
5. Calculate an area or circumference.
6. Simultaneously calculate areas and find centers.

Controls and Machine Head Operation

Drafting machine heads contain the controls for horizontal, vertical, and angular movement. Although each brand of machine contains similar features, controls may be found in different places on different brands. (See Figure 1–46.) Most machines have the following controls:

1. Baseline adjustment—this releases the scales so they can move but the protractor will not be affected.
2. Index control—permits automatic stops every 15°. It can also be pushed in and locked to enable you to adjust the machine to any angle.
3. Indexing clamp—used to lock the protractor at intermediate angles (those other than 15° increments) so you can draw an accurate line without the protractor moving.

To operate the drafting machine protractor head, place your hand on the handle and using your thumb, depress the index thumbpiece. Doing so allows the head to rotate. Each increment marked on the protractor is one degree with a label every 10°. As the vernier plate (the small scale numbered from 0 to 60) moves past the protractor, the zero on the vernier aligns with the angle that you wish to read. For example, Figure 1–47 shows a reading of 10°. As you rotate the handle, notice that the

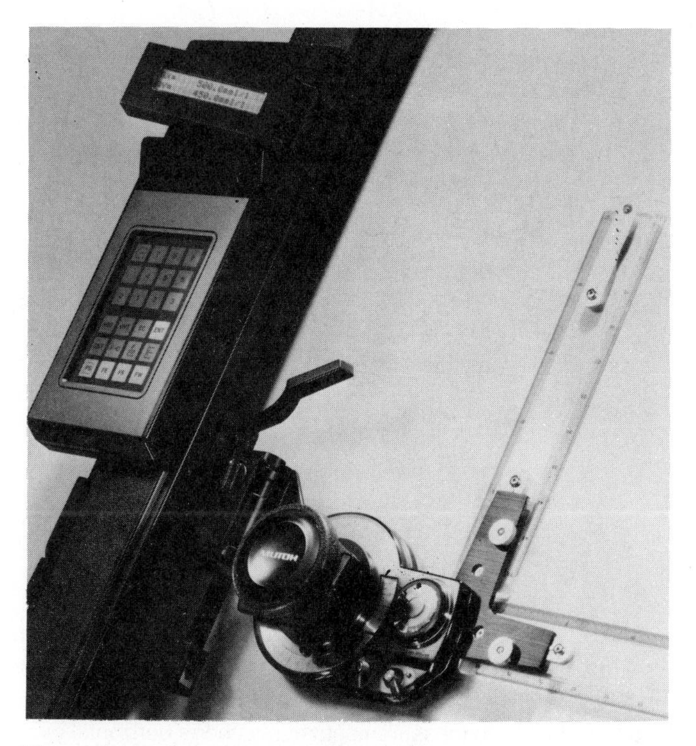

Figure 1–45 Coordinate reading and processing drafting machine. *Courtesy Consul & Mutoh, Ltd.*

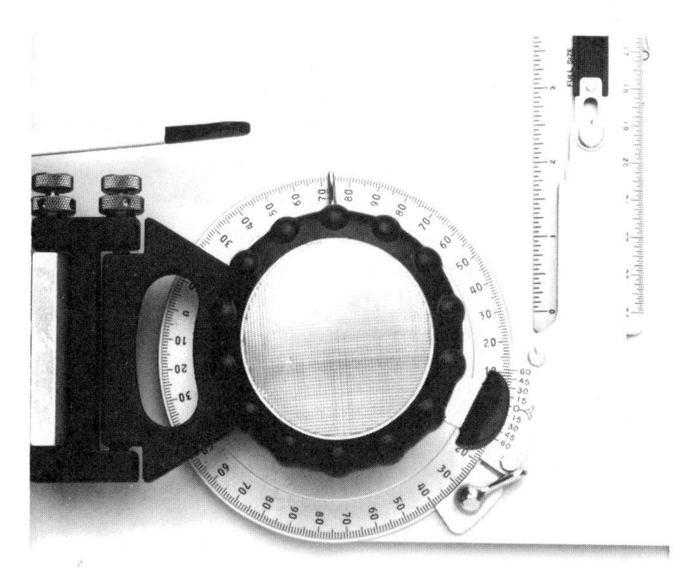

Figure 1–46 Drafting machine head controls and parts. *Courtesy VEMCO.*

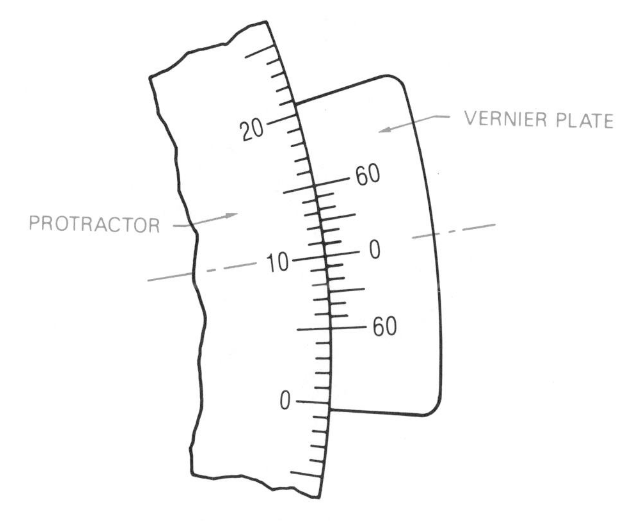

Figure 1–47 Vernier plate and protractor showing a reading of 10°.

head automatically locks every 15°. To move the protractor past the 15° increment, you must again depress the index thumbpiece.

Having rotated the protractor head 40° clockwise, the machine is in the position shown in Figure 1–48. The vernier plate at the protractor reads 40°, which means that both the horizontal and vertical scale have moved 40° from their original position at 0° and 90° respectively. The horizontal scale reads directly from the protractor starting from 0°. The vertical scale reading begins from the 90° position. The key to measuring angles is to determine if the angle is to be measured from the horizontal or vertical starting point. See the examples in Figure 1–49.

Measuring full degree increments is easy since you simply match the zero mark on the vernier plate with a full degree mark on the protractor. See the reading of 12° in Figure 1–50. The vernier scale allows you to measure angles as accurately as 5' (minutes). Remember, 1 degree equals 60 minutes (1° = 60') and 1 minute equals 60 seconds (1' = 60"). Some machines are accurate to 1'.

Reading and Setting Angles with the Vernier. To read an angle other than a full degree we will assume the vernier scale is set at a *positive angle* as shown in Figure 1–51. Each mark on the vernier scale represents 5'. First see that the angle to be read is between 7° and 8°. Then

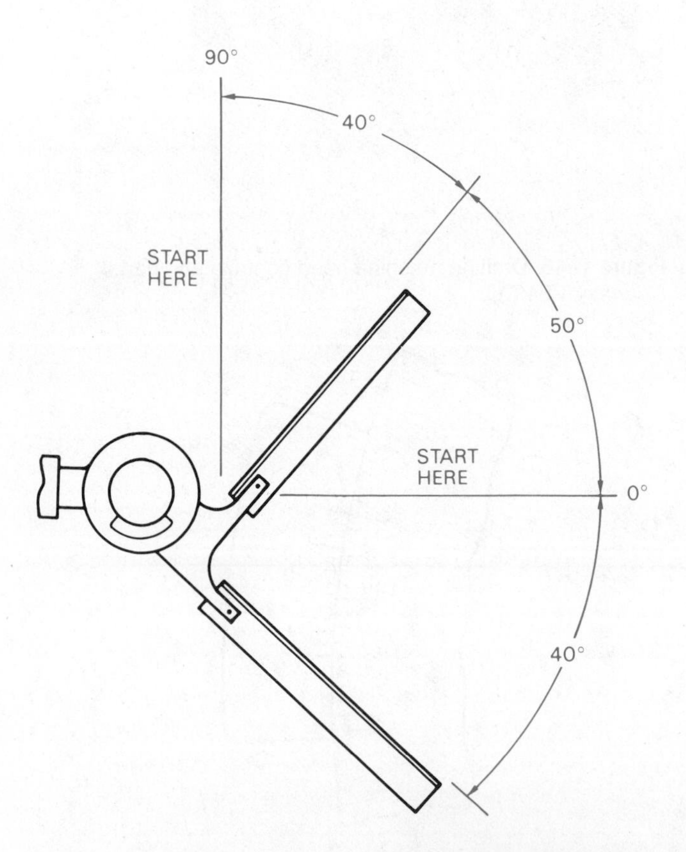

Figure 1–48 Angle measurement with the drafting machine.

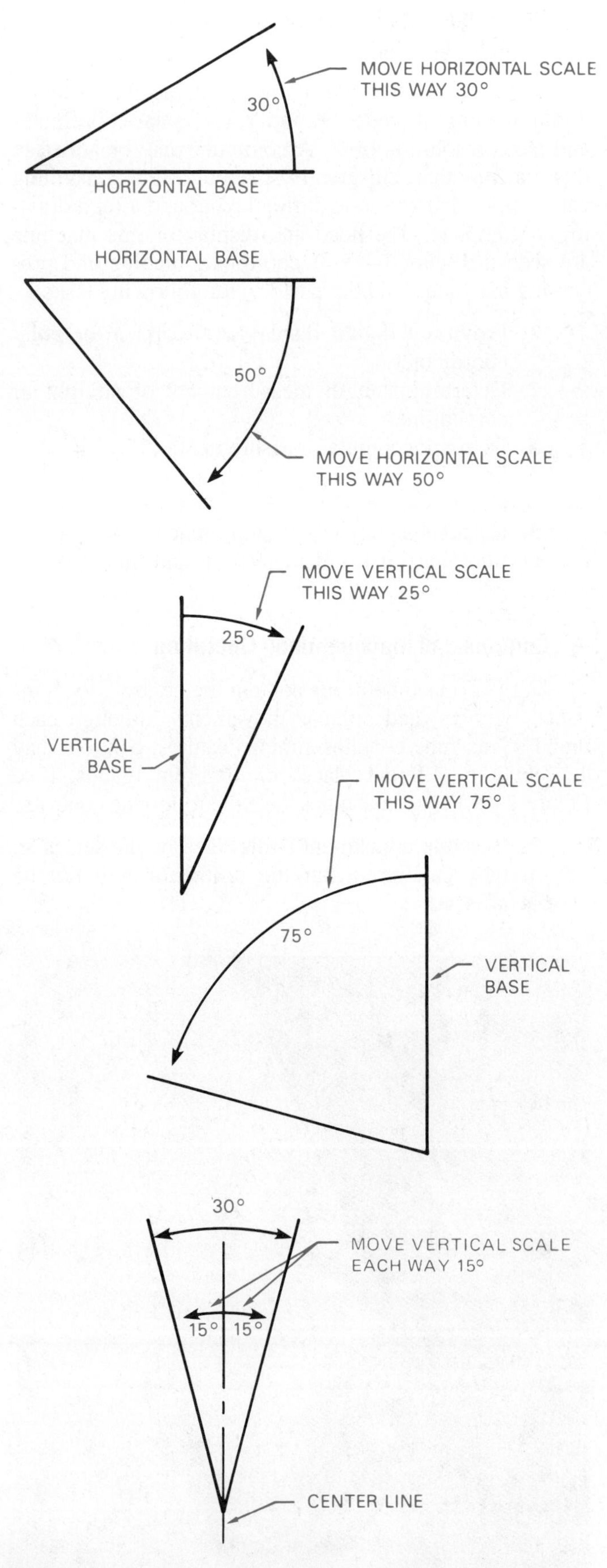

Figure 1–49 Angle measurements from either a horizontal or vertical reference line.

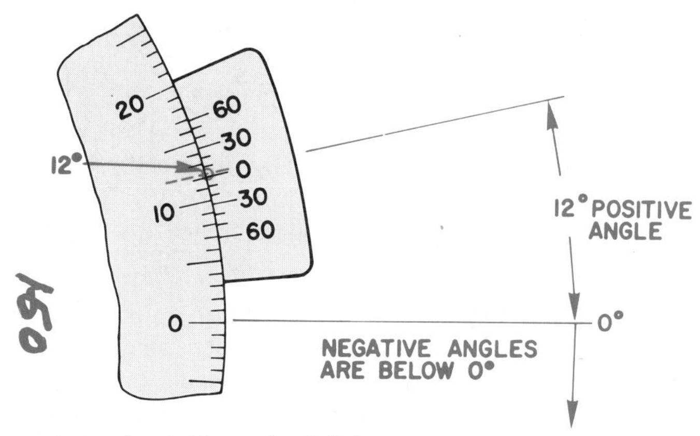

Figure 1–50 Measuring full degrees.

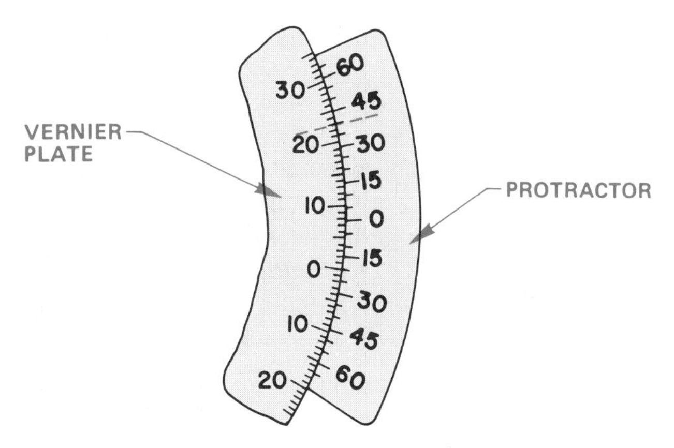

Figure 1–51 Reading positive angles with the vernier.

find the 5' mark on the *upper* half of the vernier (the direction in which the scale has been turned) that is most closely aligned with a full degree on the protractor. In this example, it is the 40' mark. Add the minutes to the degree just passed. The correct reading then, is 7°40'. The procedure for reading *negative* angles is the same, except to read the minute marks on the *lower* half of the vernier. The example shown in Figure 1–52 reads 4°25'.

Suppose you wished to set the angle 7°40' as shown in Figure 1–51. First release the protractor brake and disengage the indexing mechanism with the thumb control. Rotate the protractor arm counterclockwise until the zero of the vernier is at 7°. Then slowly continue the rotation until the 40' mark on the upper half of the vernier aligns with the nearest degree mark on the protractor. Lock the protractor brake and draw the line. The procedure for setting negative angles is essentially the same except for turning the protractor head in a clockwise motion. For example, to set the reading of Figure 1–52, rotate the protractor arm clockwise until the zero aligns with the 4° mark, then slowly continue the rotation until the 25' mark on the lower half of the vernier aligns with the nearest degree mark on the protractor.

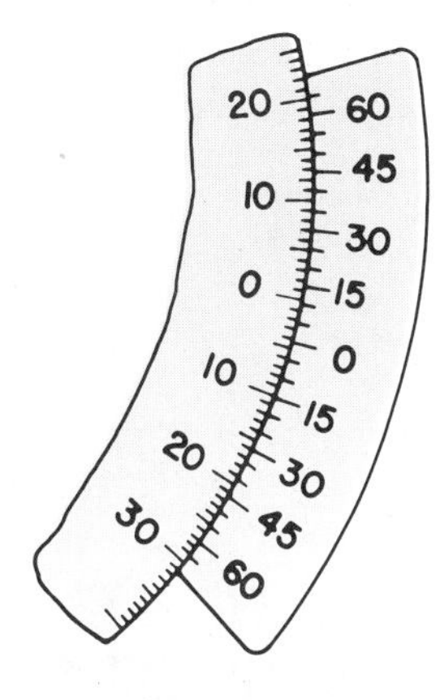

Figure 1–52 Reading negative angles with the vernier.

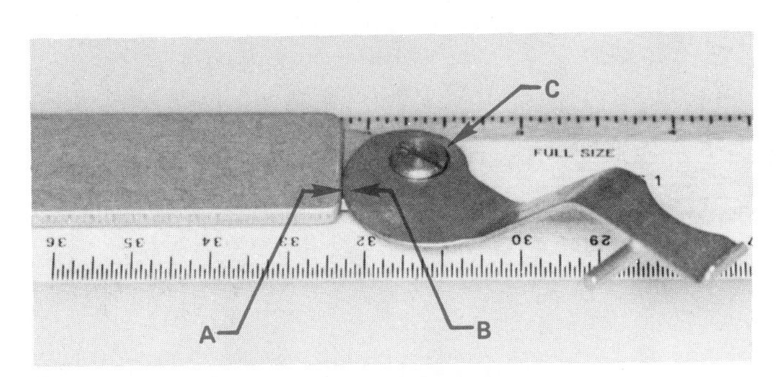

Figure 1–53 Scale removal. *Courtesy VEMCO.*

Machine Setup

To insert a scale in the baseplate chuck, place the scale flat on the board and align the scale chuckplate with the baseplate chuck on the protractor head. Firmly press, but do not drive the scale chuckplate into the baseplate chuck. To remove a scale, use the scale wrench as shown in Figure 1–53. Slip the wrench over screw C and turn clockwise, thus pressing curved section B strongly against section A of the baseplate chuck. Removing a scale by hand without the aid of a key could result in damage to the scale and/or machine.

Scale Alignment. Before drawing with any drafting machine, the scales should be checked for alignment and, if needed, be adjusted at right angles to each other. For best results with a track drafting machine, the scales should also be aligned with respect to the horizontal track. Both operations can be accomplished through the following procedure:

Step 1 Tighten the flat-head screw nearest the end of the scale on each scale chuckplate. Insert the scales in the baseplate chuck and press them firmly into place. Release the inner scale chuckplate lock-screw on the horizontal scale. Set the scale near the center of its angular range of adjustment and tighten the lock-screw.

Step 2 Draw a reference line parallel to the horizontal track by:

a. Locking the vertical brake and releasing the horizontal brake.

b. Placing a pencil point at zero on the horizontal scale and moving both the pencil and protractor head together laterally along the board. (See Figure 1–54.) *Caution:* Merely drawing the pencil along the scale will not assure the line being parallel to the horizontal track.

Step 3 Release the lock nut on the micrometer baseline screw and turn this screw until the scale is brought parallel to the reference line. (See Figure 1–55.) Tighten the lock nut firmly. Some machines have a baseline wing-nut. If yours does, release the baseline wing-nut and bring the scale parallel to the reference line. Tighten the baseline wing-nut. For those machines that have a baseline zero on the protractor (usually found to the left of the handle), first loosen the baseline wing-nut and align the arrow to 0°. Then lock the baseline wing-nut. Finally, loosen the horizontal scale chuckplate lock-screw and adjust the horizontal scale to the reference line, then retighten the screw.

Step 4 Remove the horizontal scale, turn it 180°, and replace it. Loosen the scale chuckplate lock-screw, adjust the scale parallel to the reference line, and tighten the lock screw. Now the horizontal scale is properly aligned when inserted from either end.

Step 5 Move the head 90° clockwise and adjust both ends of the vertical scale in the same manner as the horizontal scale and along the same reference line drawn in step 2b. (See Figure 1–56.)

By following this procedure, you have established a reference line setting that is parallel to the horizontal track and adjusted the scales so that they are parallel to the track and perpendicular to each other. For satisfactory results when drawing, the screws on the scales must be tight and the scale chuckplates pressed firmly into the chucks. Use good judgment when tightening any mechanism as too much force can cause components to break or wear out rapidly.

The alignment procedure of steps 1 and 2 should be checked periodically, even daily. Doing so may seem like a lot of work and trouble, but once you have gone through the procedure several times it becomes routine. By checking and adjusting the scale alignment often, you are sure that your drawings are accurate. One of a drafter's great frustrations is to prepare a layout and then discover that the machine is not properly aligned.

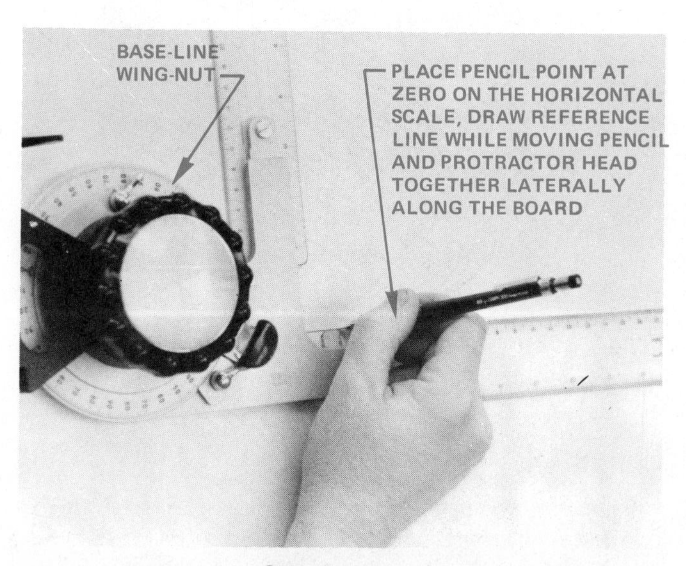

Step 2a

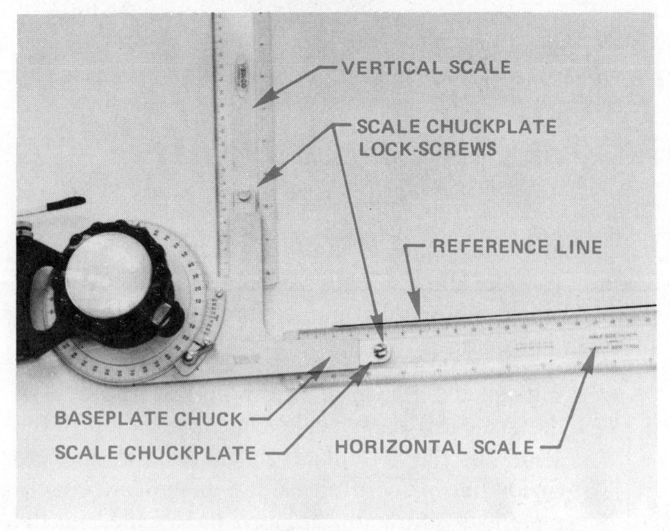

Step 2b

Figure 1–54 Scale alignment — steps 1 and 2.

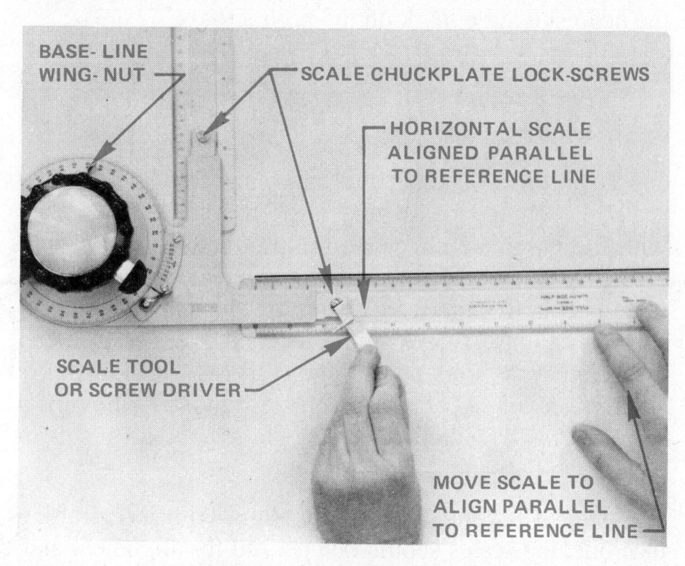

Figure 1–55 Scale alignment — steps 3 and 4.

SCALES

Scale Shapes

There are four basic scale shapes as shown in Figure 1–57. The two-bevel scales are also available with chuckplates for use with standard arm or track drafting machines. These machine scales have typical calibrations and some have no scale reading for use as a straightedge alone. Drafting machine scales are purchased by designating the length needed, 12, 18, or 24 in., and the scale calibration such as metric, engineer's full scale in 10ths and half scale in 20ths, or architect's scale 1/4" = 1'–0" and 1/2" = 1'–0". Many other scales are available.

Scale Notation

The scale of a drawing is usually noted in the title block or below the view of an object that differs in scale to that given in the title block. Drawings are scaled so that the object represented can be illustrated clearly on standard sizes of paper. It would be difficult, for example, to make a full size drawing of a Boeing 747, thus a scale that reduces the size of such a large object must be used. Machine parts are often drawn full size or even twice, four, or ten times larger than full size, depending upon the actual size of the part.

The scale selected, then, depends upon:

- The actual size of the part.
- The amount of detail to be shown.
- The paper size selected.
- The amount of dimensioning and notes required on the part.

The following scales and their notation are frequently used on mechanical drawings.

Full scale	= FULL or 1:1
Half scale	= HALF or 1:2
Quarter scale	= QUARTER or 1:4
Twice scale	= DOUBLE or 2:1
Four times scale	= 4:1
Ten times scale	= 10:1

Some scales used on architectural drawings are noted as follows:

1/8" = 1'–0"	1" = 1'–0"
1/4" = 1'–0"	1 1/2" = 1'–0"
1/2" = 1'–0"	3" = 1'–0"

Some scales used in civil drafting are noted as follows:

1" = 10'	1" = 50'
1" = 20'	1" = 60'
1" = 30'	1" = 100'

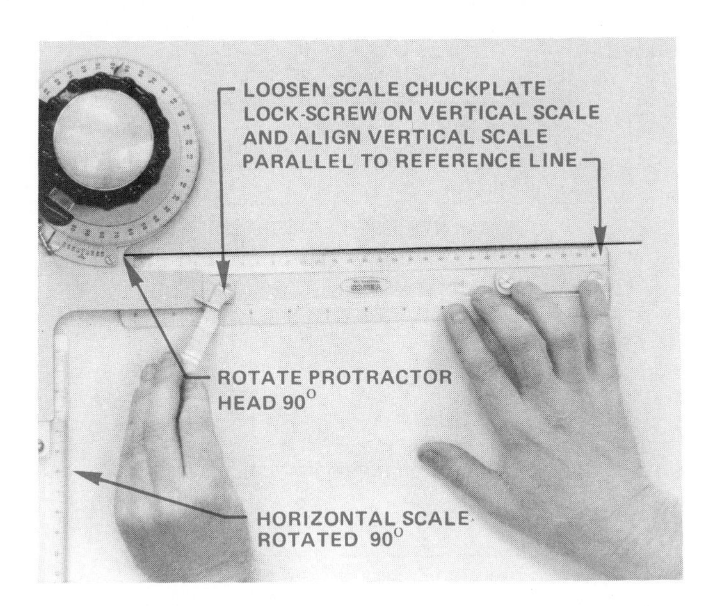

Figure 1–56 Scale assignment — step 5.

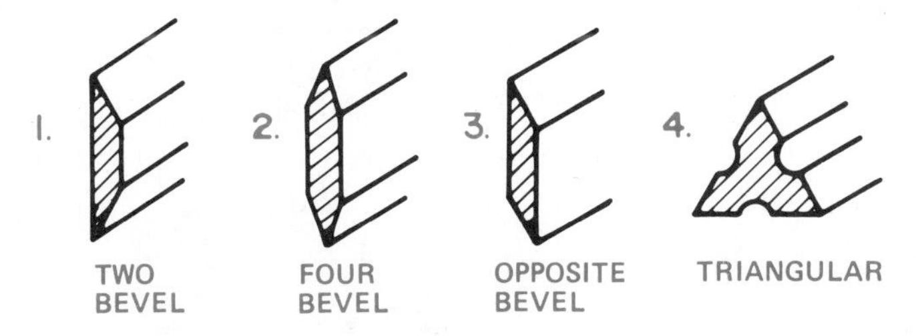

Figure 1–57 Scale shapes.

Metric Scale

ANSI According to the American National Standards Institute:

> The commonly used SI (International System of Units) linear unit used on engineering drawings is the millimeter. On drawings where all dimensions are either in inches or millimeters, individual identification of units is not required. However, the drawing shall contain a note stating: UNLESS OTHERWISE SPECIFIED, ALL DIMENSIONS ARE IN INCHES (or MILLIMETERS as applicable). Where some millimeters are shown on an inch-dimensioned drawing, the millimeter value should be followed by the symbol, mm. Where some inches are shown on a millimeter-dimensioned drawing, the inch value should be followed by the abbreviation, IN.
>
> Metric symbols are as follows:
>
> millimeter = mm
> centimeter = cm
> decimeter = dm
> meter = m
> dekameter = dam
> hectometer = hm
> kilometer = km

Some metric-to-metric equivalents are the following:

10 millimeters = 1 centimeter
10 centimeters = 1 decimeter
10 decimeters = 1 meter
10 meters = 1 dekameter
10 dekameters = 1 kilometer

Some metric-to-U.S. customary equivalents are the following:

1 millimeter = .03937 inch
1 centimeter = .3937 inch
1 meter = 39.37 inch
1 kilometer = .6214 miles

Some U.S. customary-to-metric equivalents are the following:

1 mile = 1.6093 kilometers = 1609.3 meters
1 yard = 914.4 millimeters = .9144 meters
1 foot = 304.8 millimeters = .3048 meters
1 inch = 25.4 millimeters = .0254 meters

To convert inches to millimeters, multiply inches by 25.4 mm.

Figure 1–58 shows the common scale calibrations found on the triangular metric scale. One advantage of the metric scales is that any scale is a multiple of 10; therefore, any reductions or enlargements are easily performed. In most cases, no mathematical calculations should be required when using a metric scale. Whenever possible, select a direct reading scale. If no direct reading scale is available you may use multiples of other scales. For example, Figure 1–59 shows how the 1:1 scale can be interpreted as a 2:1 scale if necessary. Double size (2:1) also can be read directly on the 1:5 scale by counting each 100 mm unit division as 10 mm. To avoid the possibility of error, avoid multiplying or dividing metric scales by anything but multiples of ten.

Civil Engineer's Scale

The triangular civil engineer's scale contains six scales, one on each of its sides. The civil engineer's

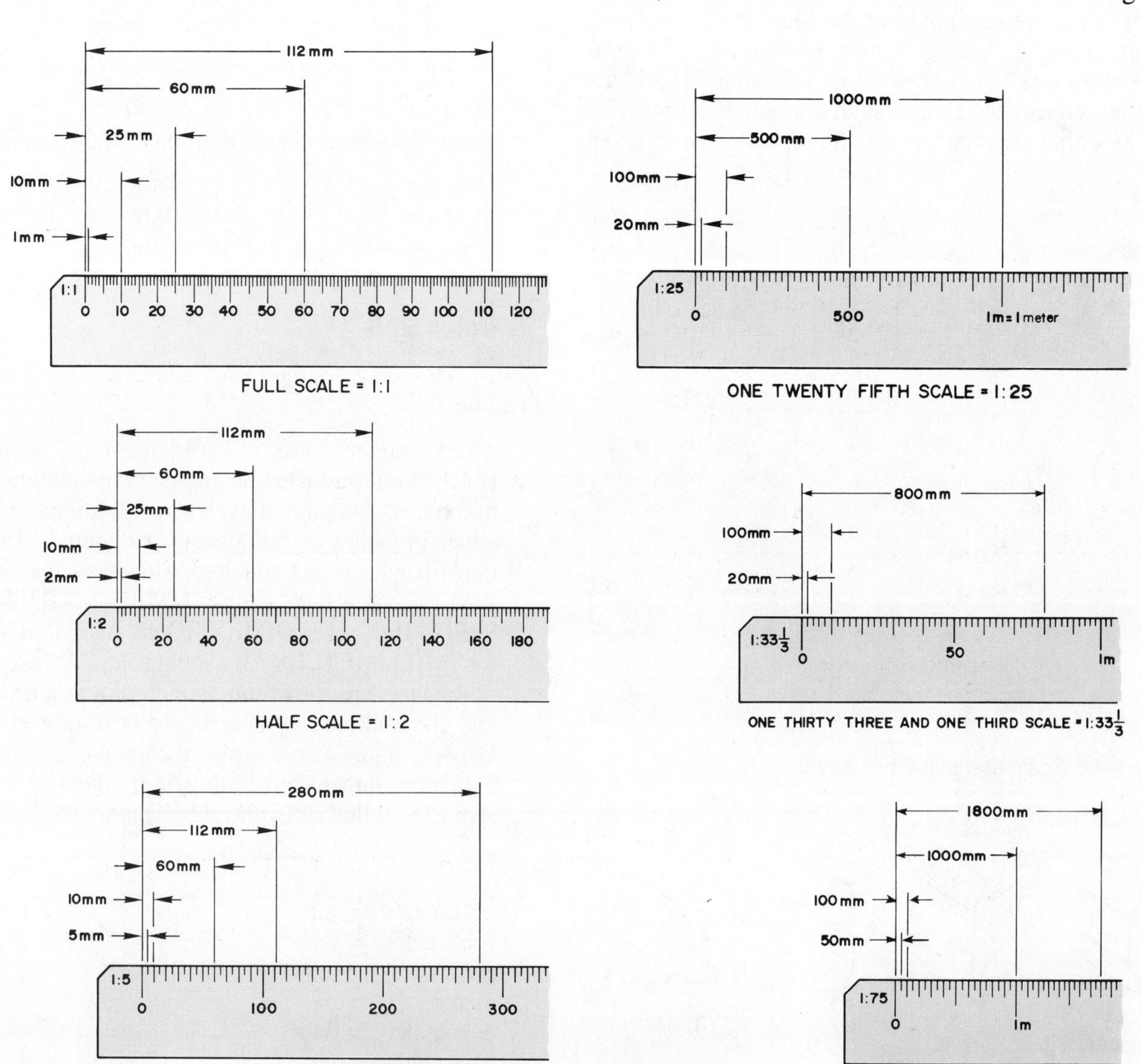

Figure 1–58 Metric scale calibrations.

scales are calibrated in multiples of ten. The scale margin displays the scale represented on a particular edge. The following table shows some of the many scale options available when using the civil engineer's scale. Keep in mind that any multiple of ten is available with this scale.

CIVIL ENGINEER'S SCALE					
Divisions	Ratio	Scales Used With This Division			
10	1:1	1"=1"	1"=1"	1"=10'	1"=100'
20	1:2	1"=2"		1"=20'	1"=200'
30	1:3	1"=3"		1"=30'	1"=300'
40	1:4	1"=4"		1"=40'	1"=400'
50	1:5	1"=5"		1"=50'	1"=500'
60	1:6	1"=6"		1"=60'	1"=600'

The 10 scale is often used in mechanical drafting as a full, decimal-inch scale, shown in Figure 1–60. Increments of 1/10 (.1) in. can easily be read on the 10 scale. Readings of less than .1 in. require the drafter to approximate the desired amount, as shown in Figure 1–60. Some scales are available that refine the increments to 1/50th of an inch. The 10 scale is also used in civil drafting for scales of 1"=10' or 1"=100' and so on. (See Figure 1–61.)

The 20 scale is commonly used in mechanical drawing to represent dimensions on a drawing at half scale (1:2). Figure 1–62 shows examples of half-scale decimal dimensions. The 20 scale is also used for scales of 1"=2', 1"=20', and 1"=200', as shown in Figure 1–63.

The remaining scales on the engineer's scale may be used in a similar fashion. For example, 1"=5', 1"=50', and so on. The 50 scale is popular in civil drafting for drawing plats of subdivisions.

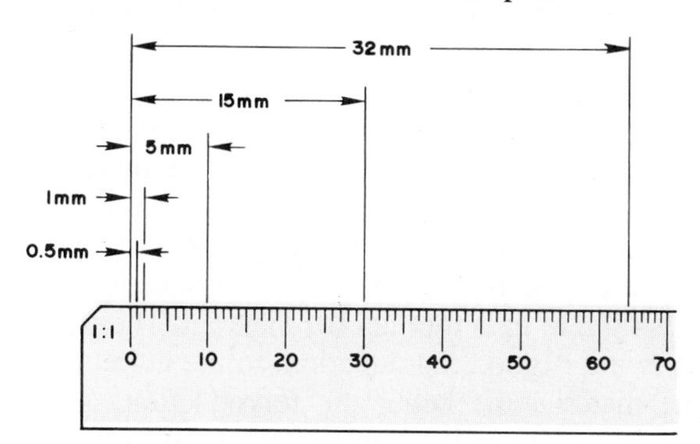

Figure 1–59 Metric double scale (2:1).

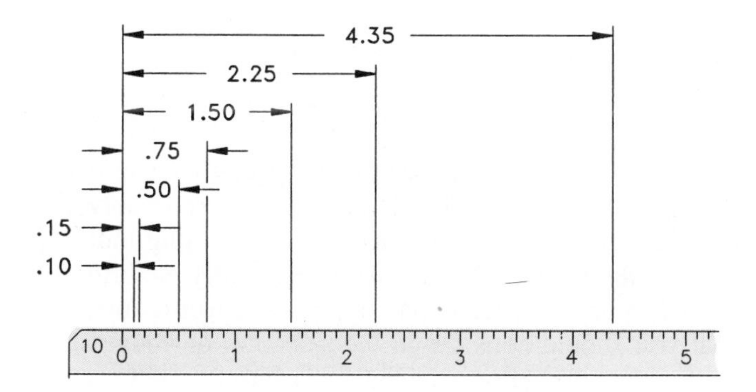

Figure 1–60 Full engineer's decimal scale (1:1).

Architect's Scale

The triangular architect's scale contains 11 different scales. On ten of them each inch represents a foot and is subdivided into multiples of 12 parts to represent inches and fractions of an inch. The eleventh scale is the full scale with a 16 in the margin. The 16 denotes that each inch is divided into 16 parts and each part is equal to 1/16th of an inch. Look at Figure 1–64 for a comparison between the 10 engineer's scale and the 16 architect's scale. Figure 1–65 shows an example of the full architect's scale, while Figure 1–66 shows the fraction calibrations.

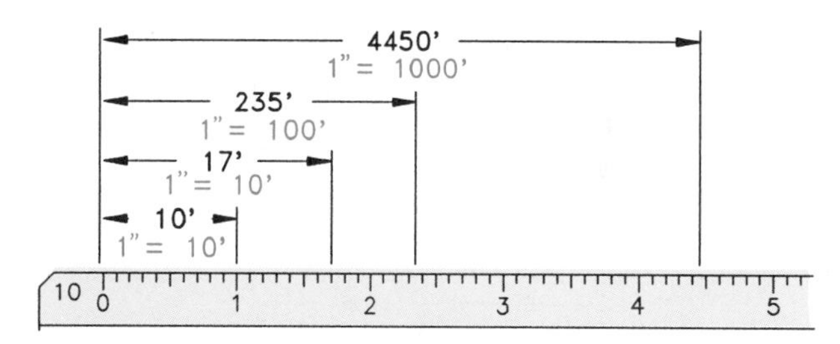

Figure 1–61 Civil engineer's scale, units of 10.

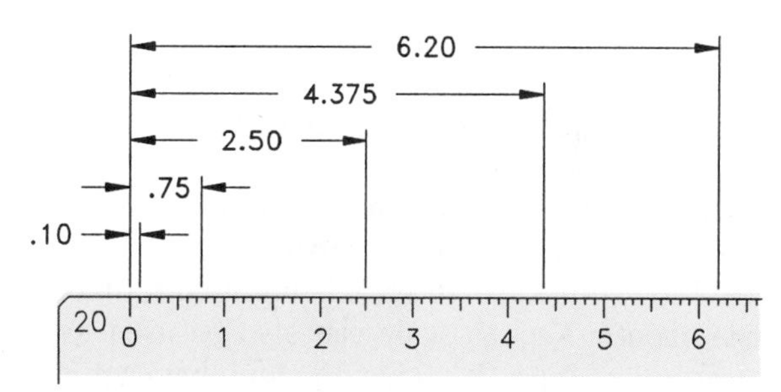

Figure 1–62 Half scale on the engineer's scale (1:2).

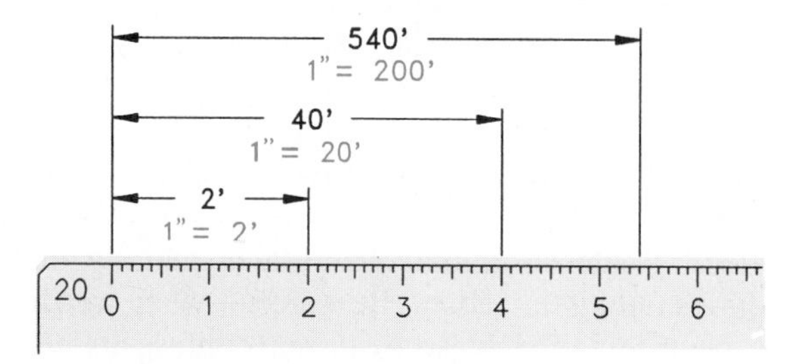

Figure 1–63 Civil engineer's scale, units of 20.

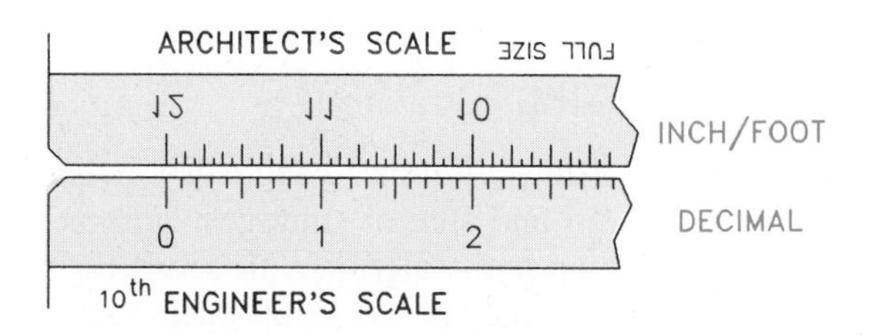

Figure 1–64 Comparison of full engineer's scale (10) and architect's scale (16).

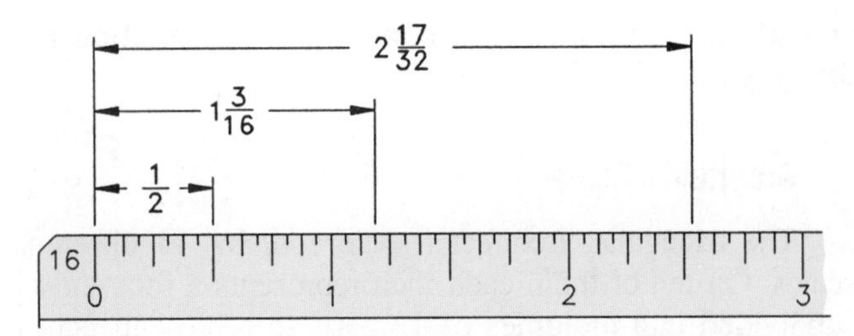

Figure 1–65 Full (1:1), or 12" = 1'–0", architect's scale.

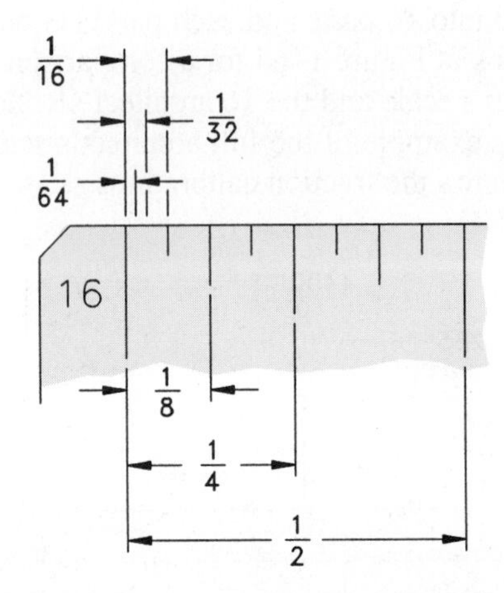

Figure 1–66 Enlarged view of architect's (16) scale.

The 16 scale can be used to illustrate a half-scale drawing. At half scale the 1/16 in. graduations equal 1/8 in. Be cautious when you use the full scale as a half scale because it is easy to make errors in the calculations. Figure 1–67 shows the 16 scale being used for half-scale measurements. The 16 scale can also be used to make twice-size drawings, 2:1 scale, by doubling each dimension. For example, 2"=1".

Look at the architect's scale examples in Figure 1–68. Note the form in which scales are expressed on a drawing. The scale notation may take the form of a word; full, half, double, or a ratio, 1:1, 1:2, 2:1; or an equation of the drawing size in inches or fractions of an inch to one foot, 1"=1'–0", 3"=1'–0", 1/4"=1'–0". The architect's scale commonly has scales running in both directions along an edge. Be careful when reading a scale from left to right so as to not confuse its calibrations with the scale that reads from right to left.

CLEANING DRAFTING EQUIPMENT

Cleaning the drafting equipment each day helps keep your drawing clean and free of smudges. This is easily done with mild soap and water or with a soft rag or tissue. Avoid using harsh cleansers or products that are not recommended for use on plastic. It is also a good practice to clean your hands periodically to remove graphite and oil.

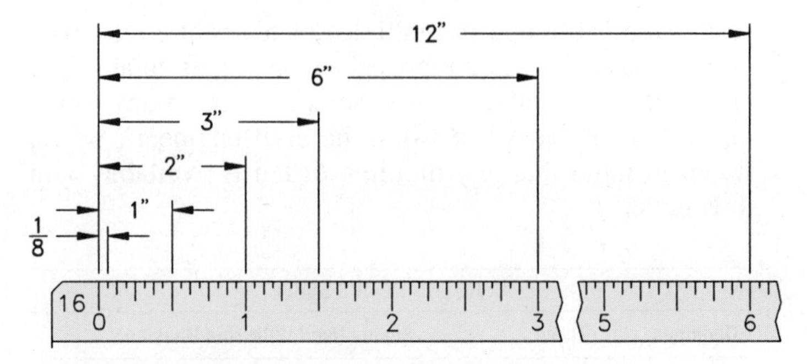

Figure 1–67 Half (1:2) architect's scale.

DRAFTING MEDIA

Several factors other than cost should influence the purchase and use of drafting media. The considerations include durability, smoothness, erasability, dimensional stability, and transparency.

Durability should be considered if the original drawing will have a great deal of use. Originals may tear or wrinkle and the images become difficult to see if the drawings are used often.

Smoothness relates to how the medium accepts line work and lettering. The material should be easy to draw on so that the image is dark and sharp without a great deal of effort on the part of the drafter.

Erasability is important because errors need to be corrected and changes frequently made. When images are erased, ghosting should be kept to a minimum. *Ghosting* is the residue that remains when lines are difficult to remove. These unsightly ghost images are reproduced in a print. Materials that have good erasability are easy to clean up.

Dimensional stability is the quality of the media to not alter size due to the effects of atmospheric conditions such as heat and cold. Some materials are more dimensionally stable than others.

Transparency is one of the most important characteristics of drawing media. The diazo reproduction method requires light to pass through the material. The final goal of a drawing is good reproduction, so the more transparent the material the better the reproduction, assuming that the image drawn is professional quality.

PAPERS AND FILMS

Vellum

Vellum is drafting paper that is specially designed to accept pencil or ink. Lead on vellum is probably the most common combination used in the drafting industry today. Many vendors manufacture quality vellum for drafting purposes. Each claims to have specific qualities that you should consider in the selection of your paper. Vellum is the least expensive material having good smoothness and transparency. Use vellum originals with

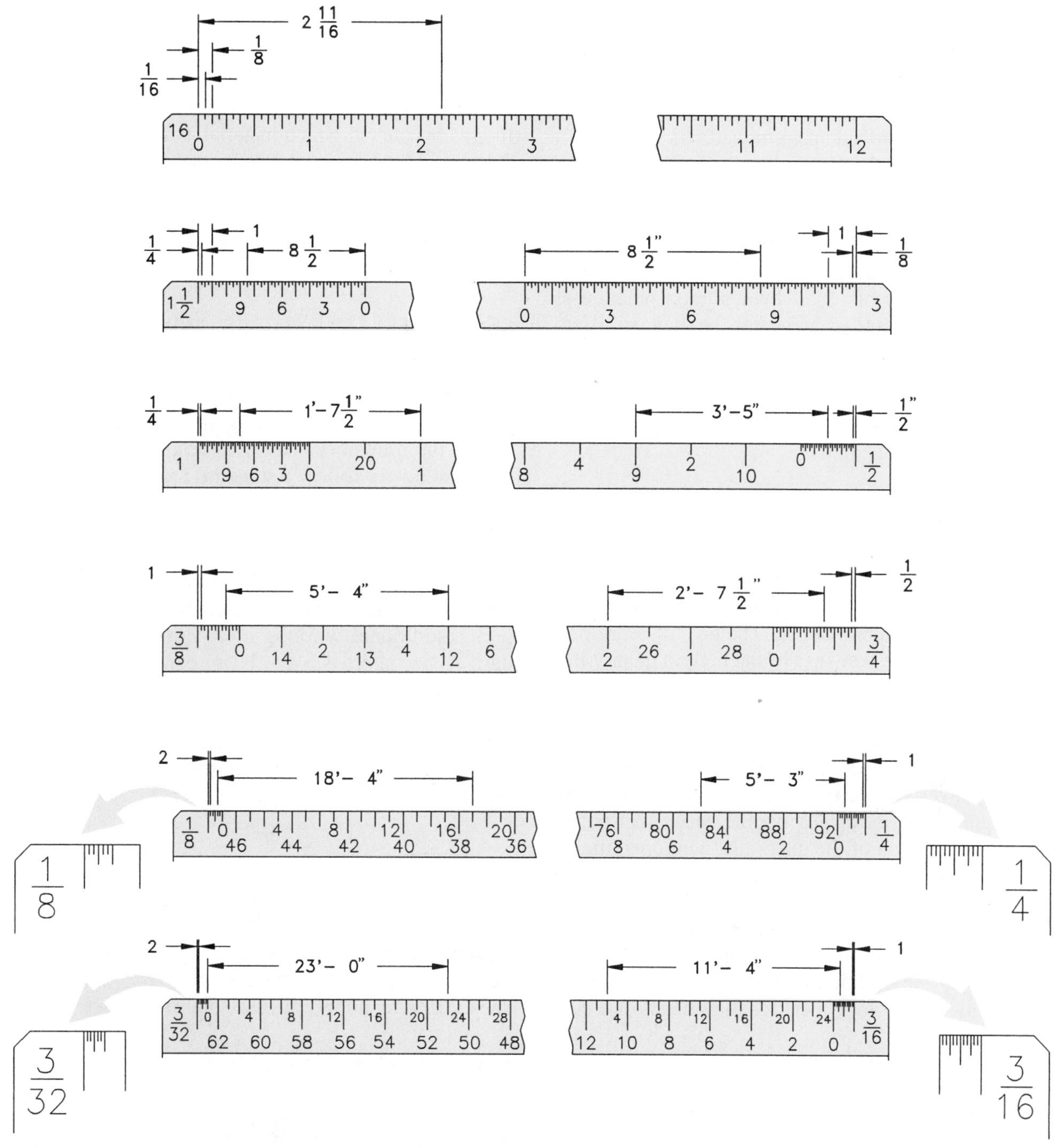

Figure 1–68 Architect's scale examples.

care. Drawings made on vellum that require a great deal of use could deteriorate, as vellum is not as durable a material as others. Also, some brands are better erased than others. Affected by humidity and other atmospheric conditions, vellum generally is not as dimensionally stable as other materials.

Polyester Film

Polyester film, also known by its brand name, Mylar®, is a plastic "stable base" material that offers excellent dimensional stability, erasability, transparency, and durability. Drawing on Mylar® is best accomplished using ink or special polyester leads. Do not use regular graphite leads as they smear easily. Drawing techniques that drafters use with polyester leads are similar to graphite leads except that in general the polyester leads are softer and feel like a crayon when used.

Mylar® is available with a single or double matte surface. Matte is surface texture. The double matte film has texture on both sides so that drawing can be done on either side if necessary. Single matte film is the most common in use with the nondrawing side having a slick surface. When using Mylar® you must be very careful not to damage the matte by erasing. Erase at right angles to the direction of your lines and do not use too much pressure. Doing so will help minimize damage to the matte surface. Once the matte is destroyed and removed, the surface will not accept ink or pencil. Also, be cautious about getting moisture on the Mylar® surface. Oil from your hands can cause your pen to skip across the material.

Normal handling of drawing film is bound to soil it. Inked lines applied over soiled areas do not adhere well and in time will chip off. It is always good practice to keep the film clean. Soiled areas can be cleaned effectively with special film cleaner.

Mylar® is much more expensive than vellum; however, it should be considered where excellent reproductions, durability, dimensional stability, and erasability are required of original drawings.

Reproduction

The one thing most designers, engineers, architects, and drafters have in common is that their finished drawings are intended for reproduction. The goal of every professional is to produce drawings of the highest quality that give the best possible prints when they are reproduced.

We have discussed many of the factors that influence the selection of media for drafting; however, the most important factor is reproduction. The primary combination that achieves the best reproduction is the blackest, most opaque lines or images on the most transparent base or material. Each of the materials mentioned makes good prints if the drawing is well done. If the only concern is the quality of the reproduction, ink on Mylar® is the best choice. Some products have better characteristics than others. Some individuals prefer certain products. It is up to the individual or company to determine the combination that works best for their needs and budget. The question of reproduction is especially important when sepias must be made. (Sepias are second- or third-generation originals. See Sepias in this chapter.)

Look at Figure 1–69 for a magnified view of graphite on vellum, plastic lead on Mylar®, and ink on Mylar®. Judge for yourself which material and application provides the best reproduction. As you can see from Figure 1–69, the best reproduction is achieved with a crisp, opaque image on transparent material. If your original drawing is not good quality, it will not get better on the print.

SHEET SIZES, TITLE BLOCKS, AND BORDERS

All professional drawings have title blocks. Standards have been developed for the information put into the title block and on the surrounding sheet adjacent to the border so the drawing is easier to read and file than drawings that do not follow a standard format.

Sheet Sizes

Standard sheet sizes are shown in Figure 1–70. Alternate sheet sizes are available for companies that prefer additional working area. For example: The standard A-size sheet is 8½ × 11 in. or 9 × 12 in. Additional optional sizes include B size, 11 × 17 in. or 12 × 18 in.; C size, 17 × 22 in. or 18 × 24 in.; and D size, 22 × 34 in. or 24 × 36 in. Also available is roll stock in standard widths for industries that require continuous lengths for an extra-long drawing surface.

Zoning

Some companies use a system of numbers along the top and bottom margins and letters along the left and right margins called zoning. Notice in Figure 1–70 that numbered and lettered zoning begins on C-size drawing sheets. Zoning allows the drawing to read like a road map.

Figure 1–69 A magnified comparison of graphite on vellum, plastic lead on Mylar®, and ink on Mylar®. *Courtesy Koh-I-Noor Rapidograph, Inc.*

For example, the reader can refer to the location of a specific item as D–4, which means that the item can be found at or near the intersection of D across and 4 up or down.

Title Blocks

Companies generally have title blocks and borders preprinted on drawing sheets to reduce drafting time and cost. Some companies use an adhesive title block so that one standard title block can be attached to any size drawing sheet. These title blocks may also be a cost-saving practice as preprinted blocks and borders are a little more expensive. Drawing sheet sizes and sheet format items such as borders, title blocks, zoning, revision columns, and general note locations have been standardized so that the same general relationship exists between engineering drawings from companies across the country. Each company may use a slightly different design, although the following basic information is located in approximately the same place on most engineering drawings:

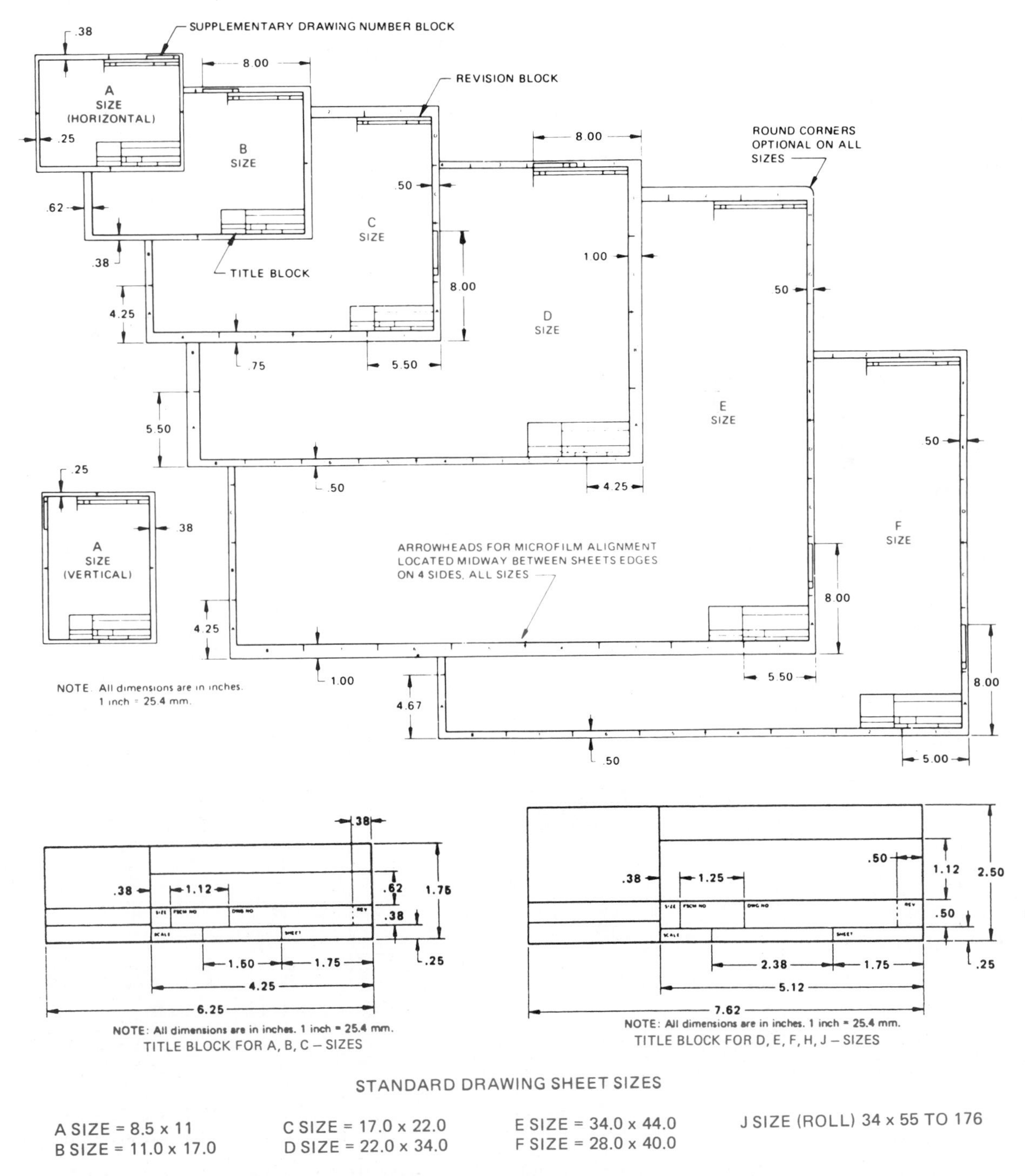

Figure 1–70 Standard drawing sheets sizes. *Reprinted with permission by ASME—ANSI Y14.1—1980.*

1. Title block. Lower right corner.
   a. Company name.
   b. Confidential statement.
   c. Unspecified dimensions and tolerances.
   d. Sheet size.
   e. Drawing number.
   f. Part name.
   g. Material.
   h. Scale.
   i. Drafter signature.
   j. Checker signature.
   k. Engineer signature.
2. Revision Column. Upper right corner, over or adjacent to the title block.
   a. Revision symbol, number or letter.
   b. Description.
   c. Drafter.
   d. Date.
3. Border Line.
   a. With or without zoning.

Figure 1–71(a), (b), (c), and (d) shows title blocks from several different industries. Notice the similarities among these title blocks. Some components in each title block are the same. The location of specific items may differ slightly and some companies require more detailed information than others but, in general, the title block format is the same.

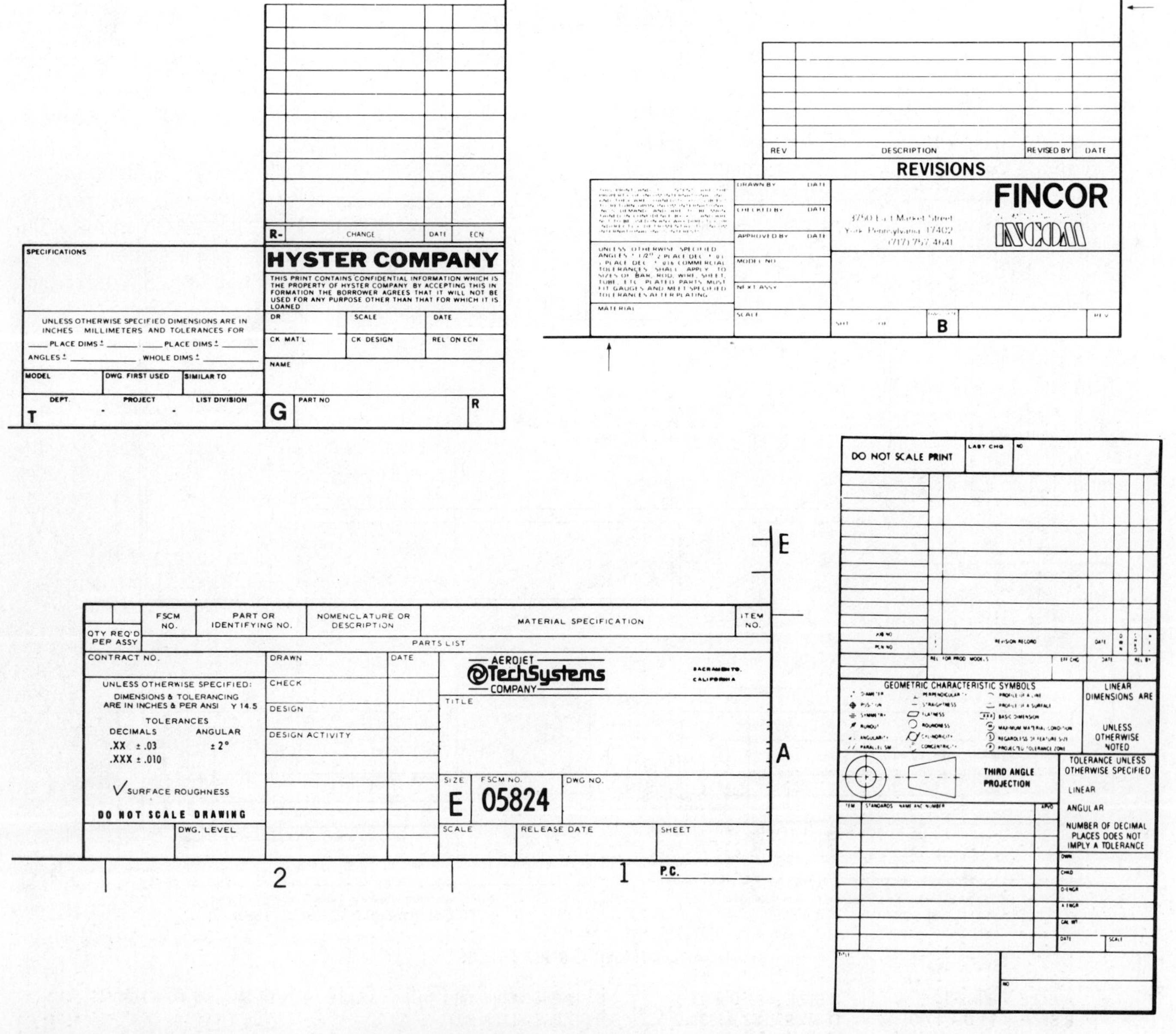

Figure 1–71 Sample title blocks. (a) *Courtesy Hyster Company;* (b) *courtesy Aerojet Techsystems Company;* (c) *courtesy FINCOR-INCOM International, Inc.;* (d) *courtesy Chrysler Corporation.*

Title Block Definitions

- *Tolerances* are discussed in detail in Chapter 8, although it is important to know that a tolerance is a given amount of acceptable variation in a size or location dimension. All dimensions have a tolerance.
- *Millimeters and inches.* All dimensions are in millimeters (mm) or inches (in.) unless otherwise specified.
- *Unless otherwise specified* means that, in general, all of the features or dimensions on a drawing have the relationship or specifications given in the title block unless a specific note or dimensional tolerance is provided at a particular location in the drawing.
- *Unspecified tolerances* refers to any dimension on the drawing that does not have a tolerance specified. This is when the dimensional tolerance required is the same as the general tolerance shown in the title block.
- *Revisions.* When parts are redesigned or altered for any reason the drawing will be changed. All drawing changes are commonly documented and filed for future reference. When this happens the documentation should be referenced on the drawing so that users can identify that a change has been made. Before any revision can be made, the drawing must be released for manufacturing.

Title and Revision Block Instructions

The title block displayed in Figure 1–72(a) shows most of the common elements found in industrial title blocks. The revision block, Figure 1–72(b), found in this format is located in the upper right corner of the drawing sheet.

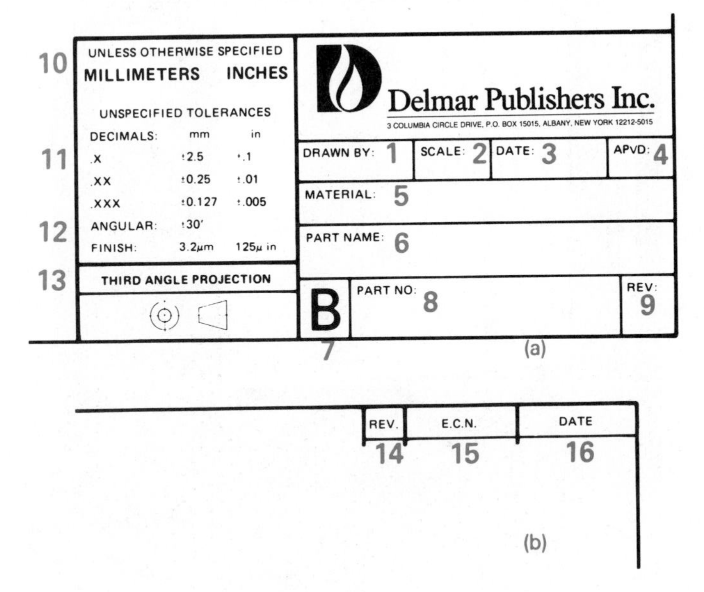

Figure 1–72 (a) Title block elements; (b) revision block elements.

The large numbers shown on Figure 1–72 refer to the following instructions for completing the title and revision blocks. All lettering should be .125 or .188 in. high as indicated in the instructions. The lettering style should be vertical upper-case Gothic freehand, or mechanical lettering as specified by the instructor or company standards. Computer graphic systems are to use vertical upper-case letters unless otherwise specified.

1. DRAWN BY: .125 in. high lettering. Identify yourself by using all your initials such as DRC, DAM, JLT, unless otherwise specified.
2. SCALE: .125 in. high lettering. Examples of scales to fill in this block are FULL or 1:1, HALF or 1:2, DBL or 2:1, QTR or 1:4, NONE.
3. DATE: .125 in. high lettering. Fill in this block using the following order: day, month, and year such as 18 NOV 93.
4. APVD: .125 in. high lettering. This block is to be used by the instructor or checker who initials his or her approval of the drawing.
5. MATERIAL: .125 in. high lettering. Describe here the material used to make the part, for example, BRONZE, CAST IRON, SAE 4320. (The material is given in each problem.)
6. PART NAME: .188 in. high lettering. Insert here the name of the part, such as COVER, HOUSING. (The part name is given in each problem.)
7. B: .188 in. high if unprinted. Denotes B-size drawing. Other options in this location refer to sheet sizes such as A, C, D, E, or F. When preprinted borders are used, this sheet size designation is usually printed. When adhesive title blocks are used, this area is often left blank to be filled in by the drafter.
8. PART NO: .188 in. high lettering. Fill in the drawing number of the part. Most companies have their own part numbering system. While numbering systems differ, they are often keyed to categories such as the disposition of the drawing (casting, machining, assembly), materials used, related department within the company, or a numerical classification of the part. (Part numbers for drawing problems will be given.)
9. REV: .188 in. high lettering. Fill in the revision letter or number of the part or drawing. A new or original drawing is 0 (zero). The first time a drawing is revised the 0 changes to a 1, for the second drawing change a 2 is placed here, and so on. A letter (A, B, C, etc.) or symbol may also be used.
10. UNLESS OTHERWISE SPECIFIED: Unless otherwise specified, drawing dimensions will be given in millimeters or inches. When the predominent dimensions are in MILLIMETERS

PRODUCING A CADD DRAWING

The following discussion gives you an overview of how drawings are produced on a CADD system.

The term *producing* means to display the drawing in some fashion on one of several different pieces of hardware. The most obvious is to display it on the screen as seen in Figure 1–73. This is accomplished with the REDRAW command. A quick print of the screen display is available to those with a system that has a hardcopy unit. A *hardcopy*, by the way, is a paper copy, while *softcopy* is video or audio format.

A second type of print hardcopy is produced by the printer. A PRINT command handles this job. A printed copy is useful for checking purposes much the same as a hardcopy unit print. The size is limited with both types of prints and therefore most drawings produced in this manner are not to scale.

The highest quality reproduction of a CADD drawing is produced by a pen plotter using wet-ink pens as shown in Figure 1–74. Quick, true-scale plots of good quality are produced by electrostatic plotters. Plotted prints are started with the PLOT command at the computer. The operator can request a plot of a drawing presently in the computer memory, or of any drawing file that is on a disk. Such a request can be made from the workstation, or from other terminals in the office. Some companies that use large, wet-ink pen plotters, will leave the plotting of drawings to the night shift. Then only one shift has to load and clean ink pens and make sure the plot is running properly. Other companies use plotters that accept roll paper. A list of drawings to be plotted can be given to the computer just before the day shift is over, and the plotter will produce a continuous stream of drawings until the entire list is plotted. A take-up reel behind the plotter rolls up all the drawings, which are then cut and trimmed the next day.

Drafters with small firms usually request a plotted copy at any time they need one. Many plotters use felt-tip pens of varying widths and colors that can be left in their carousels and receptacles without fear of drying out. Using felt-tip pens makes it easy to get a plot whenever needed because the hassle of working with wet-ink pens is eliminated.

Figure 1–73 Drawings can be displayed quickly on the screen. *Courtesy Hewlett-Packard Company.*

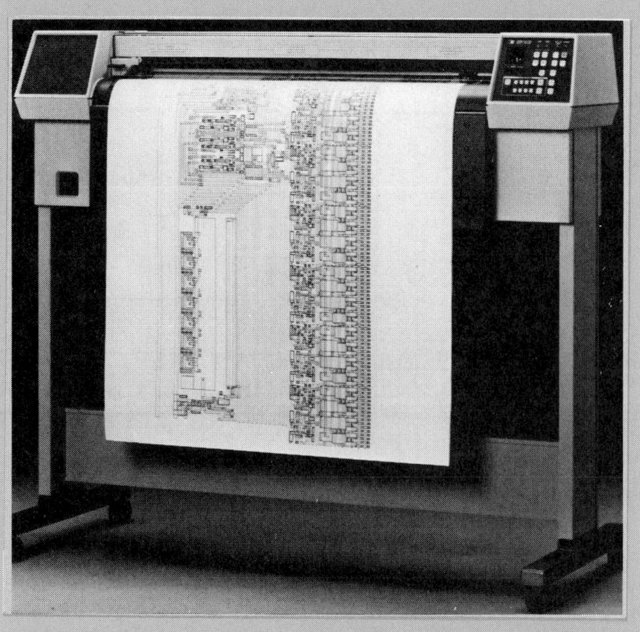

Figure 1–74 An E-size pen plotter. *Courtesy Hewlett-Packard Company.*

neatly blacken out INCHES; when in INCHES, blacken out MILLIMETERS. If dual dimensioning is used, place brackets around the [INCHES], or [MILLIMETERS] dimension that follows the primary dimension.

11. UNSPECIFIED TOLERANCES: One, two, and three place decimals are established as MILLIMETERS or INCHES in a manner the same as item 10. For example, if the drawing dimensions are predominately MILLIMETERS then INCH related tolerances are blackened out. If INCH dimensions are in brackets then each INCH tolerance is placed in brackets, [ ]. Angular tolerances for unspecified angular dimensions are ± 30'.

12. FINISH: In this black fill in the unspecified surface finish, for surfaces that are identified for finishing without a specific callout.

13. THIRD-ANGLE PROJECTION: The drawing assignments in this text are done using third-angle projection unless otherwise specified. A complete discussion of this topic is found in Chapter 5.

14. REV: Fill in the revision letter or number in this block, such as A, B, C, etc. Succeeding letters are to be used for each Engineering Change Notice (ECN) or group of related ECNs regardless of quantity of changes on an ECN. *Note:* the Rev: block in the title block must be changed to agree with the last REV. letter in the revision block.

15. E.C.N.: In this block, fill in the Engineering Change Notice (ECN) letter covering the Engineering Change Request (ECR), or group of ECRs that require the drawing to be revised.

16. DATE: Fill in the day, month, and year on which the ECN package is ready for release to production, such as 6 APR 93, 19 JUN 93, etc.

Drafting revisions are completed in chronological order by adding horizontal columns and extending the vertical column lines.

DIAZO REPRODUCTION

Diazo prints, also known as ozalid dry prints, or blue-line prints, are made with a printing process that uses an ultraviolet light that passes through a translucent original drawing to expose a chemically coated paper or print material underneath. The light does not go through the dense, black lines on the original drawing. Thus the chemical coating on the paper beneath the lines is not exposed. The print material is then exposed to ammonia vapor which activates the remaining chemical coating to produce blue, black, or brown lines on a white or clear background. The print that results is a diazo, or blue-line print, not a blueprint. The term *blueprint* is now a generic term that is used to refer to diazo prints even though they are not true blueprints.

Diazo is a direct print process and the print remains relatively dry. There are other processes available that require a moist development. Blueprinting, for example, is an older method of making prints, which uses a light to expose sensitized paper placed under an original drawing. The prints are developed in a water wash that turns the background dark blue. The lines from the original are not fixed by the light and wash out leaving the white paper to show. So a true blueprint has a dark blue background with white lines.

The diazo process is less expensive and less time-consuming than most other reproductive methods. The diazo printer is designed to make quality prints from all types of translucent paper, film, or cloth originals. Diazo prints may be made on coated roll-stock or cut sheets. The operation of all diazo printers is similar, but be sure to read the instruction manual or check with someone familiar with a particular machine if you are using a machine that is new to you. A diazo printing machine is shown in Figure 1–75.

Making a Diazo Print

Before you run a print using a complete sheet, run a test strip to determine the quality of the print you will get. Cut a sheet of diazo material into strips, then use a strip, not a whole sheet, to obtain the proper speed setting.

To make a diazo print, place the diazo material on the feedboard, coated (yellow) side up. Position the original drawing, image side up, on top of the diazo material, making sure to align the leading edges. Use diazo material that corresponds in size with your original.

Using fingertip pressure from both hands, push the original and diazo material into the machine. The light passes through the original and exposes the sensitized diazo material except where images (lines and lettering) exist on the original. When the exposed diazo material and the original drawing emerge from the printer section, remove the original and carefully feed the diazo material into the developer section, as shown in Figure 1–76.

In the developer section the sensitized material that remains on the diazo paper activates with ammonia vapor and blue lines form, although some diazo materials make black or brown lines. Special materials are also available to make other colored lines, or to make transfer sheets and transparencies.

Figure 1–75 Diazo printer. *Courtesy Ozalid Corporation.*

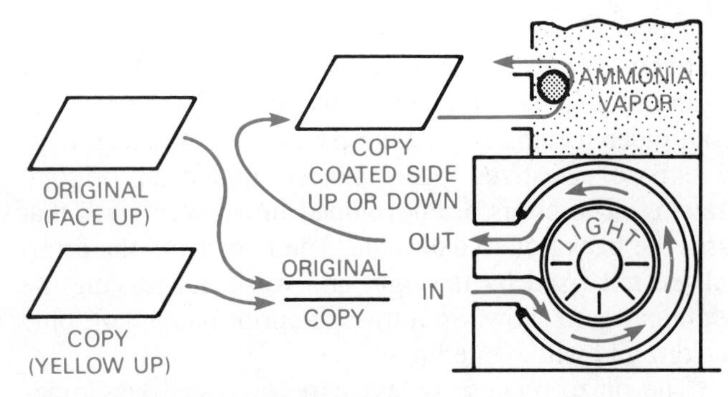

Figure 1–76 Diazo print process.

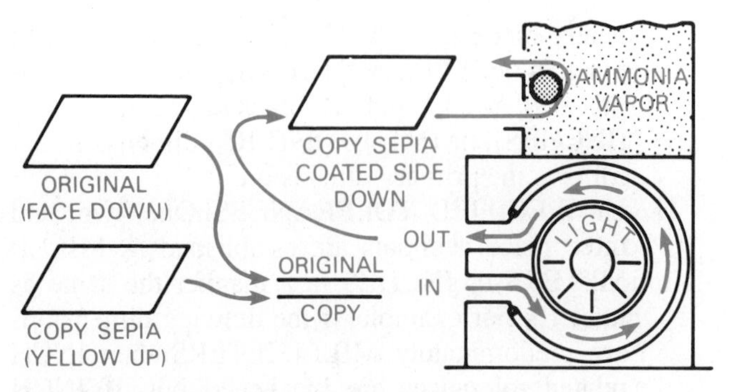

Figure 1–77 Reverse sepia print process.

Storage of Diazo Materials

Diazo materials are light sensitive so they should always be kept in a dark place until ready for use. Long periods of exposure to room light cause the diazo chemicals to deteriorate and reduce the quality of your print. If you notice brown or blue edges around the unexposed diazo material, it is getting old. Always keep it tightly stored in the shipping package and in a dark place such as a drawer. Some people prefer to keep all diazo material in a small, dark refrigerator. Doing so preserves the chemical for a long time.

Sepias

Sepias are diazo materials that are used to make secondary originals. A secondary or second-generation original is actually a print of an original drawing that can be used as an original. Changes can be made on the secondary original while the original drawing remains intact. The diazo process is performed with material called sepia. Sepia prints normally form dark brown lines on a clear (translucent) background; therefore, they are sometimes called brown lines. In addition to being used as an original to make alterations to a drawing without changing the original, sepias are also used when originals are required at more than one company location.

Sepias are normally made with the original face down on the sensitized sepia material. The balance of the print process is the same as just described except that the sepia should be fed into the developer coated-side down. Doing so allows the coating on the sepia material to come into direct exposure with the ammonia in the development chamber. Experiment both ways and observe the quality of your reproduction. Standard diazo copies may reproduce better with the coated-side down; it depends on your machine and materials.

Sepias are made in reverse (reverse sepias) so that corrections or changes can be made on the matte side (face side) and erasures can be made on the sensitized side. Sepia materials are available in paper or polyester. The resulting paper sepia is similar to vellum and the polyester sepia is Mylar®. Drawing changes can be made on sepia paper with pencil or ink, and on sepia Mylar® with polyester lead or ink. Figure 1–76 shows an illustration of the diazo print process, and Figure 1–77 shows the sepia print process.

SAFETY PRECAUTIONS FOR DIAZO PRINTERS

Ammonia

Ammonia has a strong, unmistakable odor. It may be detected, at times, while operating your diazo printer. The levels encountered are extremely low and harmless. Diazo printers are designed and built to provide safe operation from ammonia exposure. Some machines have ammonia filters while others require outside exhaust fans. When print quality begins to deteriorate and it has been determined that the print paper is in good condition, the ammonia may be old. Ammonia bottles should be changed periodically, as determined by the quality of prints. When the ammonia bottle is changed it is generally time to change the filter also.

When handling bottles of ammonia or when replacing a bottle supplying a diazo printer, the following precautions are required.

Eye Protection. Extreme care should be taken to avoid direct contact of ammonia with your eyes. Always wear safety goggles, or other equivalent eye protection, when handling ammonia containers directly or handling ammonia supply systems.

Avoid Contact with Ammonia and the Ammonia Filter. Avoid contact of ammonia and the machine's ammonia filter with your skin or clothing. Ammonia and ammonia residue on the filter can cause uncomfortable irritation and burns when it touches your skin.

Avoid Inhaling Strong Concentrations of Ammonia Fumes. The disagreeable odor of ammonia is usually sufficient to prevent breathing harmful concentrations of ammonia vapors. Avoid prolonged periods of inhalation close to open containers of ammonia or where strong, pungent odors are present. The care, handling, and storage of ammonia containers should be in accordance with the suppliers' instructions and all applicable regulations.

First Aid. If ammonia is spilled on your skin, promptly wash with plenty of water, removing your clothing if necessary, to flush affected areas adequately. If your eyes are affected, as quickly as possible irrigate with water for at least 15 minutes.

Anyone overcome by ammonia fumes should be removed to an area of fresh air at once. Apply artificial respiration, preferably with the aid of oxygen if breathing is labored or has stopped. Obtain medical attention at once in event of eye contact, burns to the nose or throat, or if the person is unconscious.

Ultraviolet Light Exposure. Under the prescribed operating instructions, there is no exposure to the ultraviolet rays emitted from the illuminated printing lamps. However, to avoid possible eye damage, under no circumstances should anyone attempt to look directly at the illuminated lamps.

PHOTOCOPY REPRODUCTION

Photocopy printers, also known as engineering copiers, make prints up to 24 in. wide, and up to 25 ft. long from originals up to 36 in. wide by 25 ft. long. Prints can be made on bond paper, vellum, polyester film, colored paper, or other translucent materials. The reproduction capabilities also include instant print sizes ranging from 45 to 141 percent of the original size. Larger or smaller sizes are possible by enlarging or reducing in two or more steps, that is by making a second print from a print.

Almost any large original can be converted into a smaller sized reproducible print, and then the secondary original can be used to generate diazo or photocopy prints for distribution, inclusion in manuals, or for more convenient handling. Also, a random collection of mixed-scale drawings can be enlarged or reduced and converted to one standard scale and format. Reproduction clarity is so good that halftone illustrations (photographs) and solid or fine line work have excellent resolution and density.

Engineering copiers are rapidly replacing diazo print machines in the engineering and drafting world because of their versatility and competitive cost. Also, engineering copiers do not use ammonia, which is an important consideration when purchasing a print machine.

MICROFILM

Microfilm is photographic reproduction on film of a drawing or other document which is highly reduced for ease in storage and sending from one place to another. When needed, equipment is available for enlargement of the microfilm to a printed copy. Special care must be taken to make the original drawing of the best possible quality. The reason for this is that during each "generation," the process of "blowback" makes the lines narrower in width and less opaque than the original. The term *generation* refers to the number of times a copy of an original drawing is reproduced and used to make other copies. For example, if an original drawing is reproduced on sepia or microfilm, and the sepia or microfilm is used to make other copies, this is referred to as a *second generation* or *secondary original.* When this process has been done four times, the drawing is called a fourth generation. The term *blowback* means bringing the drawing from the microfilm back to a printed copy. A true test of the original drawing's quality is the ability to maintain good reproductions through the fourth generation of reproduction.

In many companies original drawings are filed in drawers by drawing number. When a drawing is needed, the drafter finds the original, removes it, and makes a copy. This process works well although, depending upon the company's size or the number of drawings generated, drawing storage often becomes a problem. Sometimes an entire room is needed for drawing storage cabinets. Another problem occurs when originals are used over and over. They often become worn and damaged, and old vellum becomes yellowed and brittle. Also, in case of a fire or other kind of destruction, originals may be lost and endless hours of drafting vanish. For these and other rea-

CAD/CAM

The optimum efficiency of design and manufacturing methods exists without reproducing a single copy of a drawing of a part. Computer graphic systems may now directly link the engineering department by way of computer-aided design (CAD) to the manufacturing department machines operated by computer-aided manufacturing (CAM). Through the use of numerical control dimensioning and layout practices, the drafter or designer can prepare an engineering drawing on a computer graphics system and then have the computer convert the drawing's coordinate data to a numerical control computer tape. The computer software for this translation is known as a post processor and is the same program that reproduces data for manufacturing with a numerically-controlled machine tool. When the electronics are connected directly from engineering to manufacturing, there is a direct link between CAD and CAM. A copy of the actual drawing is never generated, unless someone needs a hard copy (print) for review away from the computer terminal.

sons, many companies are using microfilm for storage and reproduction of original drawings. Microfilm is used by industry for photographic reproduction of original drawings into a film-negative format. Cameras used for microfilming are usually one of three sizes: 16, 35, or 70 mm. Microfilming processor cameras are now available that prepare film ready for use in about 20 seconds.

Aperture Card and Roll Film

The microfilm is generally prepared as one frame or drawing attached to an aperture card, or as many frames in succession on a roll of film. The aperture card, shown in Figure 1–78, becomes the engineering document and replaces the original drawing. A 35-mm film negative is mounted in an aperture card, which is a standard computer card. Some companies file and retrieve these cards manually, while other companies punch the cards for computerized filing and retrieval. The aperture card, or data card as it is often called, is convenient to use. All drawings are converted to one size, regardless of the original dimensions. All engineering documents are filed together, instead of having originals ranging in size from 8½ × 11 in. to 34 × 44 in. Then for easy retrieval each card is numbered with a sequential engineering drawing and/or project number. Although the aperture card is more popular, some microfilm users prefer roll film, as seen in Figure 1–79. One disadvantage of roll film is that the reader has to look through a roll of film images to find the one drawing needed. This system is used in some government agencies for storage and retrieval of local maps and plats.

When a drawing is prepared on microfilm, generally two negatives are made. One negative is placed in an active file and the other is placed in safe storage in case the first is damaged or if an alternate is needed to make drawing changes and revisions. In many cases the original drawings are either stored out of the way or destroyed.

Microfilm Enlarger Reader-printer

Microfilm from aperture cards or rolls can be displayed on a screen for review, or enlarged and reproduced as a copy on plain bond paper, vellum or offset paper-plates. Most microfilm reader-printers can enlarge and reproduce prints ranging from A to C size. The prints are then available for disbursement to manufacturing, sales, or engineering department personnel.

Computerized Microfilm Filing and Retrieval

The next step in the automation of engineering-drawing filing systems is the computerized filing and retrieval of microfilm aperture cards. Implementation of this kind of system remains very expensive. Large companies and independent microfilming agencies are able to take advantage of this technology. Small companies can send their engineering documents to local microfilm companies, known as job-shops, where microfilm can be prepared and stored. Companies that use computerized microfilm systems can have a completed drawing microfilmed, stored in a computer, and available on-line in the manufacturing department in a few minutes.

PROFESSIONAL PERSPECTIVE

For centuries drafting was done in the form of artistic sketches. Then in the early 20th century, drafters began using tools such as straightedges and compasses. More recently, drafting has been done with drafting machines. Drafting machines have become increasingly accurate, with some manufacturers providing dial, digital, and coordinate reading and processing machine heads to help

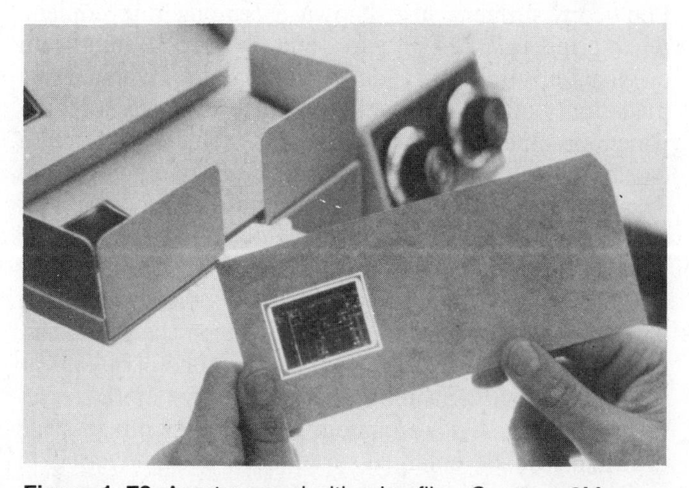

Figure 1–78 Aperture card with microfilm. *Courtesy 3M Company.*

Figure 1–79 Roll microfilm. *Courtesy 3M Company.*

increase productivity and accuracy. Wooden pencil and even mechanical lead holders have given way to the automatic pencil, which no longer requires sharpening. The drafter has available several line widths at his or her fingertips. Yet manual drafting remains a very technical skill, with individuals who possess artistic talents taking to the profession more rapidly and easily than others.

Some ask the questions, Will manual drafting be around in the future or will computer-aided design drafting (CADD) take its place? If CADD does take the place of manual drafting, will the drafter of tomorrow need the skill and artistic talents once needed? The answer to the second question may be no. The CADD drafter has different skill requirements. CADD accuracy is almost perfect and the final product is flawless in appearance. The computer as a drafting tool is demanding to the patience and concentration of the user. However, the new breed of drafter often takes to this electronic environment very well. Some professionals say that it takes a manual drafter about 100 hours of training and CADD use to equal manual drafting performance. After this initial break-in period, the productivity potential is dependent on the project complexity and the drafter's creativity. Some estimates show a 10 or 20 to one increase in productivity over manual drafting. Manual drafting, however, has not disappeared yet, and the professional drafter should have a good balance between manual and CADD abilities

CADD VS. MICROFILM

Microfilm has become an industry standard for storage and access of drawings. Large international companies especially rely on the microfilm network to insure that all worldwide subsidiaries have the ability to reproduce needed drawings and related documents. One big advantage of microfilm is the ability to archive drawings. Archive means to store something permanently for safekeeping. The use of CADD in the engineering and construction industries has made it possible to create and store drawings on flexible disk, tape, or optical disk. This has made it possible to quickly retrieve drawings that have been stored. A big advantage of this advancement involves local use of these drawings. When the CADD-generated drawings are retrieved, they are of the same quality as when they were originally drawn. The CADD drawings can be used to make multiple copies, or to redesign the product. Yet, microfilm remains an important part of the engineering environment, because microfilmed drawings last forever and there is currently a broad base of microfilm use worldwide. The optical disk comes closest to achieving the archival quality of microfilm, and perhaps one day a satellite network will exist that allow drawings on disk to be transmitted anywhere instantly. Until storage on disk becomes archival quality, microfilm will remain as a long-term method of saving drawings.

DRAFTING EQUIPMENT PROBLEMS

PART 1. READING SCALES AND DRAFTING MACHINE VERNIERS

1. Given the following engineer's scale, determine the readings at A, B, C, D, and E.

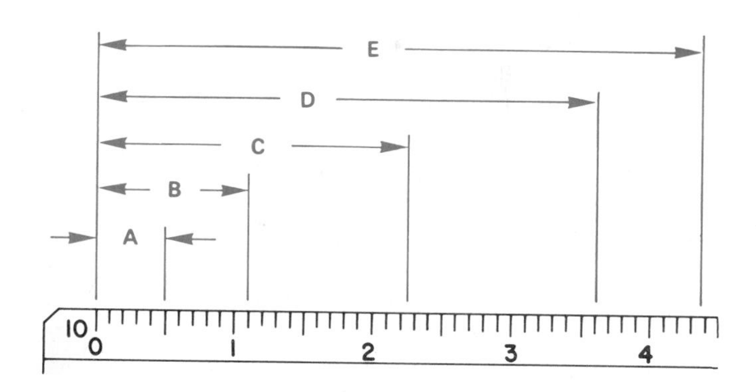

FULL SCALE = 1:1

2. Given the following engineer's scale, determine the readings at A, B, C, and D.

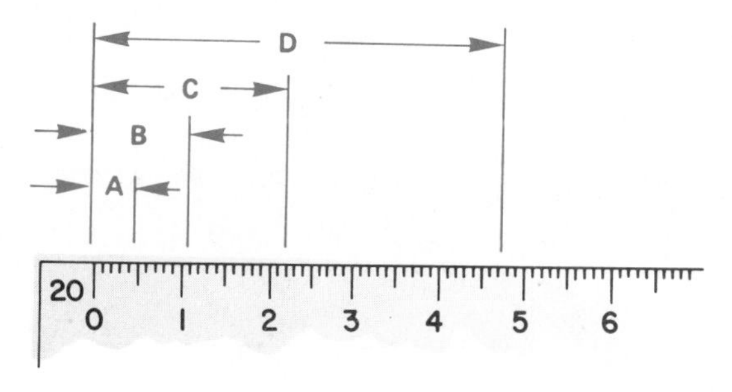

HALF SCALE = 1:2

3. Given the following architect's scale, determine the readings at A, B, C, D, E, and F.

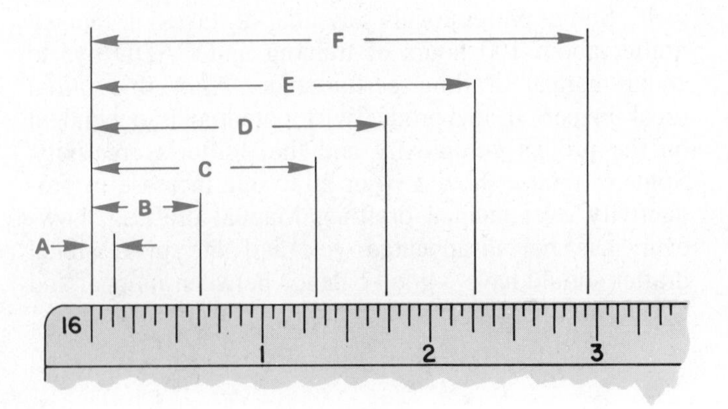

FULL SCALE = 1:1

4. Given the following metric scales, determine the readings at A, B, C, D, and E.

FULL SCALE = 1:1

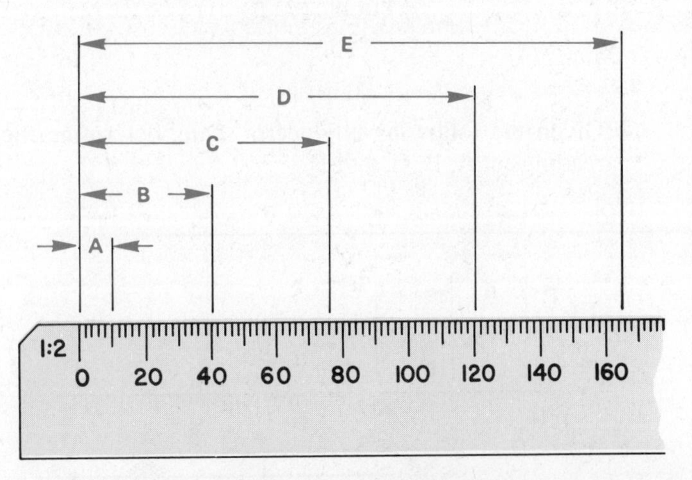

HALF SCALE = 1:2

5. Given the following drafting machine protractors, determine the angular readings.

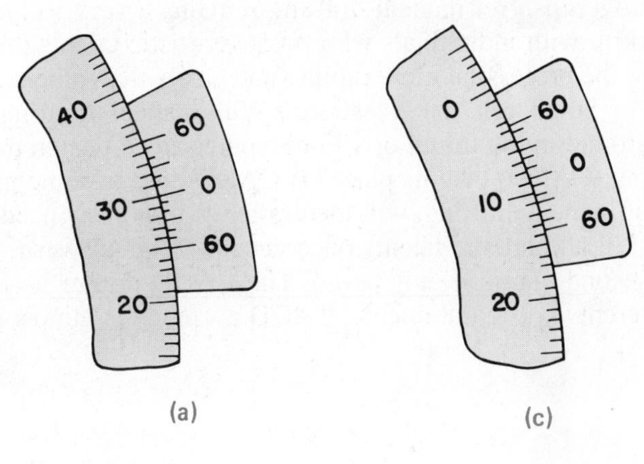

(a) (c)

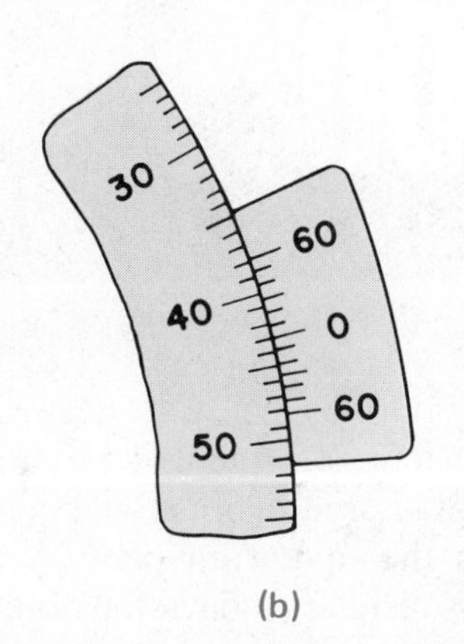

(b) (d)

PART 2. READING A TITLE BLOCK

6. Given the following title block, describe the numbered elements 1 through 13.

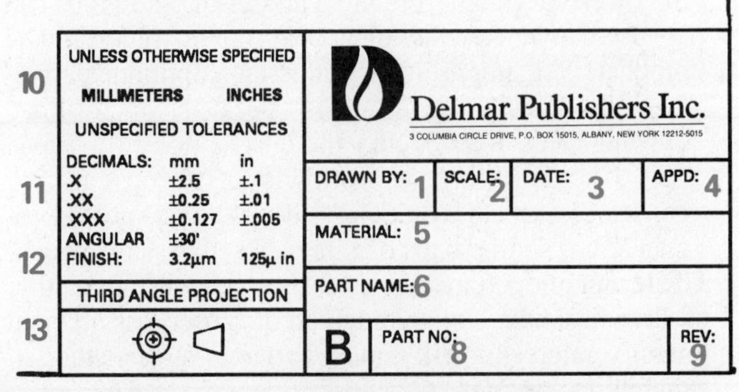

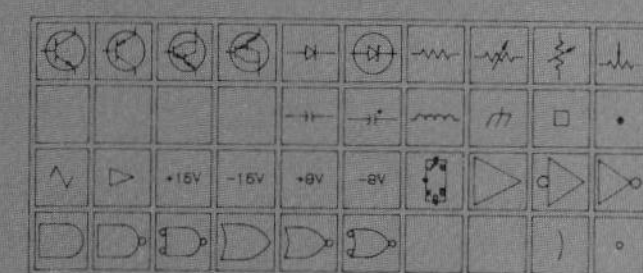

CHAPTER 2

Introduction to Computer-aided Design and Drafting (CADD)

LEARNING OBJECTIVES

After completing this chapter you will:

- Describe the new motor and mental skills needed to be a CADD drafter.
- Explain the CADD environment.
- Discuss computers in drafting and how the industry uses CADD.
- Evaluate CADD training and the job market.
- Discuss and evaluate CADD equipment, including monitors, computers, input devices, output devices, storage, and software.
- Explain the use of different CADD materials, supplies, and media.
- Get started with CADD by exploring the use of the menu, using pointing devices, manipulating the drawing, drawing symbols, creating a symbol library, and storing a drawing.
- Solve an engineering problem dealing with CADD.

WHAT IS COMPUTER GRAPHICS?

The term *computer graphics* refers to the entire spectrum of drawing with the aid of a computer, from straight lines to color animation. An immense range of artistic capabilities resides under the heading of computer graphics, and drafting is just one of them. Digital terrain modeling and analysis is one of the specialized areas that have developed as a result of computer graphics. (See Figure 2–1.)

Companies directly involved in electronics design, engineering, and manufacturing were the first to experiment and work with computer graphics. As they tested equipment and software, new demands arose that required changes and advances in both hardware and software. Equipment and programs got better, faster, more powerful, less expensive, colorful, and fun to operate.

As engineers and drafters learned to work with the new tools in a technically demanding atmosphere, artists were using the new tools in an atmosphere of unlimited structure. Hence the hastened development of three-dimensional displays, solids modeling, palettes of thousands of colors, and animation. (See Figure 2–2.) We see the structure and freedom brought by engineers and artists to computer graphics as we watch animated television network or station logos. Complex mathematical calculations are combined with the creative eye of the artist to achieve these effects.

The world of computer graphics is dynamic. The developments in technique and equipment have advanced rapidly, and applications continue to spread. The drafting applications of CADD are unlimited. This advance in technology changed the way we manipulate certain new tools and the ways we think when dealing with them. In this chapter we'll take a look at a few of these tools and see how to deal with the electronic workstation.

Figure 2–1 Example of digital terrain model. *Courtesy SysScan a.s. Norway.*

Figure 2–2 The screen display of this piping system was generated with a 3–D modeling software package. *Courtesy Applications Development, Inc.*

DEVELOPING NEW SKILLS

Drafting requires both manual and mental dexterity. Using tools involves motor skills. The drafter must also interpret, visualize, and achieve aesthetic layouts which are accomplished by using mental skills. The use of these two types of skills has not changed with the advent of computer graphics, but the tools and techniques associated with them have. Let's examine these two types of skills a little closer as they apply to computer graphics.

Motor Skills

Typing and digitizing are the principal motor skills you will be using while working with a CADD system.

Most commercial CADD systems require data entry with both digitizer and keyboard. The digitizer, or menu tablet, may contain all of the commands used with the system plus a number pad and letters of the alphabet (alphanumeric key pads). It is possible to enter all data

Figure 2–3 A joystick is located at the right of this keyboard. *Courtesy Megatek.*

by way of the menu tablet, but doing so would be slow and tedious. Many commands and parameters are entered more quickly using the keyboard. Therefore, we find that the two skills of typing and digitizing are often used in roughly equal amounts depending on the configuration of the equipment.

A CADD operator using a digitizer is required to be able to pinpoint certain spots on a drawing, often looking through a set of crosshairs in a puck, and then pressing a button to *digitize* the point. Many systems use the pen-shaped stylus that is held in the hand like a pencil. The operator points to a spot and it is digitized.

Another form of digitizing requiring fine motor skills is the use of a joystick, a device similar to those found in some airplanes. The stick is mounted in a small box or at one side of the keyboard and can be moved in any direction. (See Figure 2–3.) A set of crosshairs on the screen is moved when the stick is moved. A light touch is often needed to accurately operate a joystick.

Operators using large systems that operate with tape and disk packs may seldom be required to load or change disks or tapes. Systems using flexible disks may require the operator to load disks into the disk drives as shown in Figure 2–4. This operation requires a steady hand and light touch as the disks may be damaged if inserted into the drive too hard or too fast.

Although some systems demand accuracy in locating points with the stylus or puck, many allow for sloppy pointing. If you are close when pointing to a specific item, the computer looks for it, or *snaps* to the nearest line, feature, or coordinate point. Such a feature is an advantage because it allows an operator to develop speed without concentrating on the exact location of points.

Figure 2–4 Operator is loading a floppy disk into a disk drive.

Manual drafting is actually much more demanding in the area of motor skills than is CADD. This fact may open drafting careers to people who do not possess certain motor functions. CADD systems using speech activated commands will open the field of drafting and design even further for persons with limited motor skills. Many people possess unseen talents in design, drafting, and art that are hidden behind a physical handicap. The removal of certain barriers by the computer will enable many of these people to nurture and develop those talents despite certain physical limitations.

Mental Skills

Your brain is divided into two parts called the left and right hemispheres. Research has shown that each side can interpret its environment independently. The left side analyzes and categorizes data, then makes a logical decision based on the facts. The right side dispenses with logic and operates on gut feelings, intuition, and impulse. Both sides are at work when you are operating a computer graphics system. Let's refer to these two aspects of mental skills as analytical and synthetical, the former meaning to analyze or take something apart in a logical manner, and the latter being to combine certain aspects to form a result that could not normally be anticipated.

As a computer drafter you need to approach problems and drawings in a systematic manner. Planning a drawing is especially important because your time at the computer may be limited. Much of the data you enter is based on the Cartesian coordinate system; therefore a working knowledge of these coordinates is imperative.

Analytical skills also involve the basic function of knowing which button to push and when. This is related to the ability to give verbal instructions. Remember, you are issuing commands to the computer and they must be given in the correct sequence. If they are not, you may encounter difficulty.

The intuitive side of the brain, the right side, plays an important part in the computer drafter's job. A manual drafter must make decisions that often follow no logical scheme, such as style and layout, or scale and positioning. These same decisions are at play as you create drawings on the computer. How should I rotate this object for the best visual effect? What does this object look like when viewed from this direction? Will labeling look best here or there? How balanced is the drawing when notes are placed here? These are questions often answered by the gut feelings coming from the right side of the brain.

The workings and responsibilities of the two sides of the brain is a subject covered by many books and deserves our attention. Be constantly aware that it takes both sides of the brain to be creative.

THE NEW DRAFTING ENVIRONMENT

Throughout the course of this book you will be introduced to many of the techniques and concepts, equipment, and materials that are a part of computer graphics. The intent is to provide you with a basic understanding of these concepts and tools, and give you some insight into the world of CADD. It should be every drafting student's responsibility to keep abreast of the developments that are continually changing the face of design, drafting, and engineering. The drafting trade has experienced its greatest structural change, and even greater changes may be on the way. Read, study, visit companies, go to conferences, trade shows, and conventions. Know what is happening in your field and if you do, the next big changes may not be a surprise.

The Electronic Atmosphere

The environment that computers create is substantially different from a manual drafting atmosphere. The CADD office environment is electronic, therefore the atmosphere is more highly charged. Operators tend to think faster and work faster because the computer asks for information and never rests, whereas paper drawings on a board just lie there and do not talk back. (See Figure 2–5.)

An air of productivity often permeates an office equipped with CADD, either because the new tools enable drafters to work faster with greater accuracy, or because management expects greater productivity. In either case, CADD drafters have stated in surveys that they feel less inclined to stop and visit with their neighbor than they did when they worked at a drafting board.

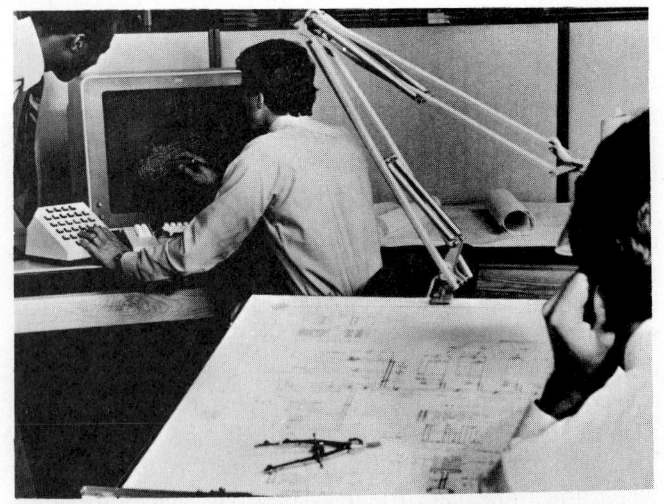

Figure 2–5 The tools of the trade have changed, and so has the intersection between drafter and drawing. *Courtesy International Business Machines Corporation.*

Greater anxiety and stress may be a part of the CADD environment, especially for those in the early days of training. Many drafters complain of eyestrain after long hours staring at a monitor. But this can be solved by fewer continuous hours at the workstation. There is a lot to learn when training on a new system, and tension can develop when the drafter realizes that he or she is supposed to eventually get up to a productive speed. Contributing to the anxiety are operator errors, hardware problems, and software bugs. Feelings of intense frustration abound when the computer crashes, or encounters a fatal error that instantaneously destroys a drawing. But errors can be corrected, hardware can be repaired, and bugs can be eliminated. With continued training and building of confidence and capabilities, the CADD drafter soon learns to deal with the tools and work around any shortcomings they may have.

Benefits and Capabilities

Each of the problems mentioned is, for the most part, an exception. The benefits of creating drawings on the computer far outweigh any negative aspects. Proficient CADD operators like the speed and accuracy that is inherent in the computer. The ability to draw something once and then use it any number of times simply by positioning it on a drawing is one of the greatest assets of a CADD system. The powerful editing capabilities of the computer can lead to tremendous time savings and have relieved many drafters of the fear and hatred of revisions and changes.

The need for hand lettering has been virtually eliminated by CADD. A variety of lettering fonts are available, and some even resemble the freehand styles used by architects. The lettering consistency offered by CADD means that a single drawing can be worked on by several people, and the finished plot will appear consistent and neat.

Practice should be a priority for new CADD drafters. There is much to learn with a computer-aided drafting system. The software is extensive and filled with many possibilities. Therefore, it will behoove the student and new employee to practice as much as possible on the CADD system and try to discover its many features. Most systems have a much greater capability than they are ever used for, and if you do some exploring, perhaps you can tap some of those hidden capabilities.

As you continue your study of drafting and design, keep in mind that creativity has not been replaced by the computer. Only the tools have changed, allowing entirely new horizons to be opened for exploration. With the new tools at your fingertips consider the limitations of your brain as the only obstacles to exploration.

COMPUTERS IN DRAFTING

The industrialized world is wrestling with the dramatic and constant changes of the electronic revolution and its child, the computer. The speed-of-light capabilities of the computer have shaped the gathering, processing, storage, and retrieval of information into a major industry. Every day our lives are touched by the effects of computers. Contrary to science-fiction works of years past, computers have not grown to occupy the core of the earth, but have shrunk to microscopic proportions, and now occupy tiny spaces inside our machines and tools.

The tools of the drafting trade have also felt the electronic touch of computers. The silicon chip has found its way into drafting tools. (See Figure 2–6 (a) and (b).) The heart of modern drafting tools is now the computer.

The drafter, or CADD operator, uses electronic hardware to construct drawings, not on paper, but on what looks like a television screen as seen in Figure 2–7. The screen is called a *monitor*, *video display screen*, or *video display terminal* (VDT). The drafting process is now called computer-aided design drafting, or CADD. Most people say "computer-aided drafting" and use the acronym CAD. Actually, CAD means computer-aided design.

The Reasons for Using a Computer

The first commercially produced computer drafting system was introduced by IBM in 1964. Computer drafting systems began to mature and prove themselves in the late 1970s. Sales of computer graphics terminals reached approximately 22,000 in 1979. Since that time there has been an exponential increase in the sales of computer graphics systems.

A visit to your local industries and engineering firms will make you aware of the popularity of computer-aided

Figure 2–6 (a) Pencil shows relative size of integrated circuit microprocessor chip. *Courtesy Digital Equipment Corporation.*

Figure 2–6 (b) Silicon chips mounted on an integrated circuit board control the workings of computers. *Courtesy National Instruments.*

Figure 2–7 The CADD drafter constructs drawings on a video display terminal (VDT), commonly referred to as a monitor. Written information appears on the small screen and the drawing on the monitor. *Courtesy Summagraphics Corporation.*

drafting. But why is it so popular? Why were computers introduced as the tool to replace the drafting machine and pencil? Why are companies scrambling to convert to computer drafting?

Computers are extremely fast. The contents of an entire encyclopedia can be transmitted in a fraction of a second. Computers are accurate. For example, computer-controlled robots can perform repetitive welds on new cars without the slightest variation in tolerance. (See Figure 2–8.) Large numbers of functions can be performed simultaneously by a single computer. Several operators can use the same computer at the same time via separate workstations, and each one can get instantaneous response from the computer. (See Figure 2–9.)

Each of these characteristics helps people increase productivity, which is the key to the success of using computer graphics. Without an increase in productivity it is doubtful that computer graphics would have ever gained as much popularity as they have.

But a fast machine cannot account for the productivity gain by itself; it needs competent human operators. The computer may be fast, but as yet, it cannot think, make judgments, or reason. Humans, on the other hand, may be slow, but they are intelligent and can use their powers of reasoning to solve problems and overcome obstacles. Together, the speed of the computer and the intelligence of the human form a productive combination.

Figure 2–8 Computer-controlled welding robots can perform repetitive welds on automobile assemblies with little variation in quality or dimensional tolerances. *Courtesy Cincinnati Milacron.*

(a)

Figure 2–9 Several CADD operators at different workstations can be served by one network computer. *Courtesy Schlumberger Technologies CAD/CAM Division.*

(b)

Figure 2–10 (a) The principal drafting tool was once the arm drafting machine as seen in this old photo of an engineering department; (b) the track drafting machine added greater speed and productivity to the drafting operation. *Courtesy Bruning.*

The Nature of the Change to Computers

Drafters have entered an era of dramatic change that does not happen very often. This radical change in drafters' tools has brought about a change in the way they must think and work. It has been devastating to many drafters who have worked at a trade in essentially the same way all their lives. Long-time drafters have seen gradual changes come along to make their jobs easier. The improvement on the T-square was the parallel bar. Then came the drafting arm, followed by the track drafting machine. Each was a major step for the drafting trade. Templates and technical pens increased productivity. But those were all minor changes for the field of drafting. (See Figures 2–10(a) and(b).)

Dr. Richard Byrne, an expert in communications and the computer revolution, discussed the nature of change at a convention of the American Institute of Design and Drafting in 1984. He suggested that change is manifested in two basic forms. The first occurs in cycles, and is called cyclical change. We see this type of change in clothing, hairstyles, color trends in decorating, fads, and the like. The styles may be altered over the years, but they often return to a previous form. Cyclical change isn't that traumatic. Most of us can handle it.

The second type of change is much more dramatic, and traumatic. It is referred to by Dr. Byrne as structural. And when structural change occurs, we don't go back. The advent of the automobile, air travel, and television are all examples of structural change. It is doubtful that we would give up automobiles, air travel, and television unless we absolutely had to.

The drafting, design, and engineering fields have

Figure 2–11 High-voltage electrostatic reciprocators (robot painters) apply a base coat to new cars at a Chrysler assembly plant. *Courtesy Chrysler Corporation.*

Figure 2–12 The reduction in size of CADD equipment has enabled small engineering and architectural firms to purchase systems. *Courtesy Hewlett-Packard Company.*

been rocked by a major structural change. Many of those manually operated tools that evolved and slowly developed over the years have been replaced by computers and computer-driven equipment, such as the robot automobile painters in Figure 2–11. Now that these innovations have been introduced into industry, it is unlikely that we will return to previous methods of manufacture. And there is no turning back.

INDUSTRY AND CADD

Who Uses CADD?

Practically everyone, since there are no longer companies that are too small to use CADD. The size and price of computer drafting systems have decreased to the point that even the one-person architectural company can afford one. Computer drafting systems are small enough to occupy just a few square feet of desk space. CADD systems are now available for everyone whereas once only large companies with large amounts of money to spend could afford one. Now individual contractors, designers, and architects can have one in their home or office. (See Figure 2–12.)

When CADD was first introduced, its principal thrust was in electronics design and drafting. But then, as CADD's popularity rose, the industry applied it to other areas. Process piping, HVAC, structural, architectural, and civil drafting, cartography, and sheetmetal drafting soon received the attention of CADD program developers. Now CADD has found its way into every corner of the drafting and design industry and has even become firmly ensconced in art. (See Figure 2–13.)

Figure 2–13 CADD is just a part of the entire computer graphics field, which also includes art. *Courtesy Numonics.*

Computers have found their way into the manufacturing process, too. Computerized welding machines, machining centers, punch-press machines and milling machines are commonplace. Many firms are engaged in the *CAD/CAM* process. CAD/CAM stands for computer-aided design/computer-aided manufacturing. In a CAD/CAM system, a part is designed on the computer and transmitted directly to the computer-driven machine tools that manufacture the part as seen in Figure 2–14. Within that process there are other computerized steps along the way.

The combination of the entire design, material handling, manufacturing, and packaging process is referred to as CIM, or computer integrated manufacturing. Within this process, the computer and its software controls, if not

Figure 2–14 Numerical control (NC) information required by the machining center in the background can be generated on the CAD/CAM system in the foreground. *Courtesy Computervision Corporation.*

all, then major portions of the manufacturing. A basic CIM system may include transporting the stock material from a holding area to the machining center at which several machining functions are performed. From there the part may be moved automatically to another station at which additional pieces are attached, then on to an inspection station, and from there to shipping or packaging.

Designers and engineers are also using CADD. For example, a part can be designed as a model on the screen of a computer in a process called *solids modeling*. The model can first be drawn as a *wire form*, or outline, in which you can see all edges. (A wire form object is shown in Figure 2–15(a).) Then the model outline can be filled in, or *rendered*, with various colors and shades on the monitor. In addition, the model may be rotated to any viewing angle desired, appear to be cut into and broken apart, then put through a series of simulated tests, and analyzed for strengths and defects in design. (See Figure 2–15[a] and [b].) This process can be referred to as *CAE*, or computer-aided engineering. (See Figure 2–16.)

THE CADD JOB MARKET

Job openings for drafters will continue to fluctuate in response to the health of the economy, but the position of drafter will be around for years to come. The job title is changing though. Most often, a CADD drafter is termed a *CADD operator*. But *operator* can mean almost anyone who uses a CADD system. Hence, we may begin to see the use of *CADD drafter*, *CADD designer*, or *CADD design-drafter*.

As a drafter gains experience, he or she may move into a design position or that of *technician*, depending on the company, the nature of the work, and the job titles the company uses. Drafters have always had the opportunity to advance into engineering, sales, estimating, contracting, managerial positions, and even company ownership. (See Figure 2–17.) After gaining experience with a specific system, a drafter may move into programming and customizing the company's software (computer programs) and symbol libraries (files of regularly used symbols). This could even lead to positions of systems analyst, computer sales, CADD

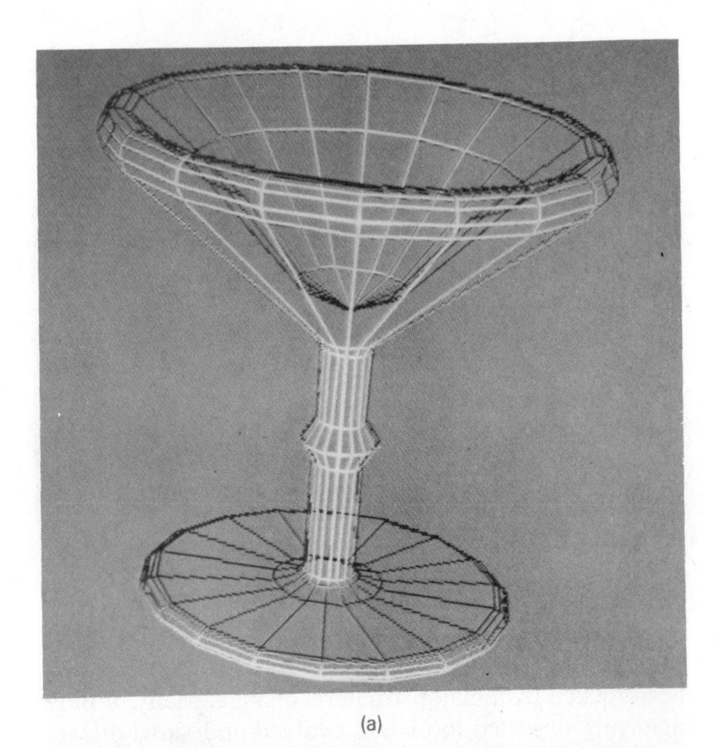

(a)

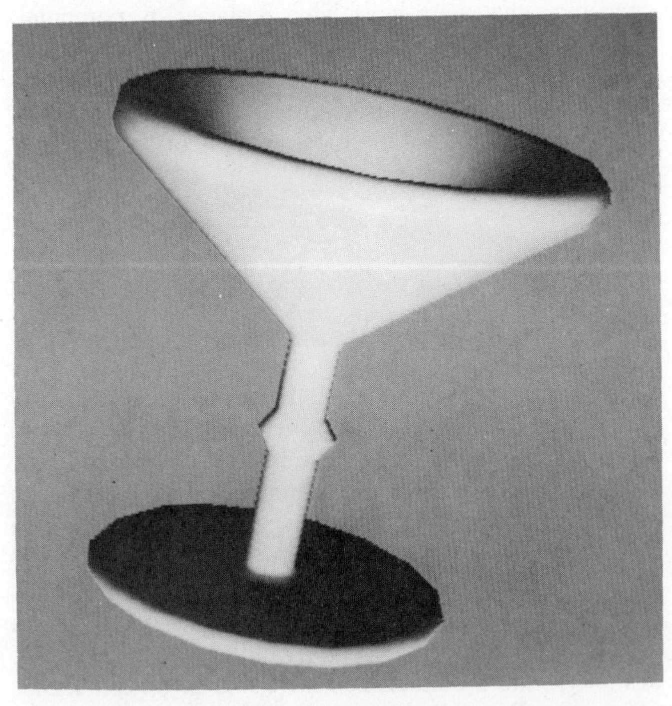

(b)

Figure 2–15 (a) All the points and vectors of an object are given to the computer to create a three-dimensional wire form; (b) a solid form can be created by filling in, or painting and shading, a wire form. *Courtesy Megatek.*

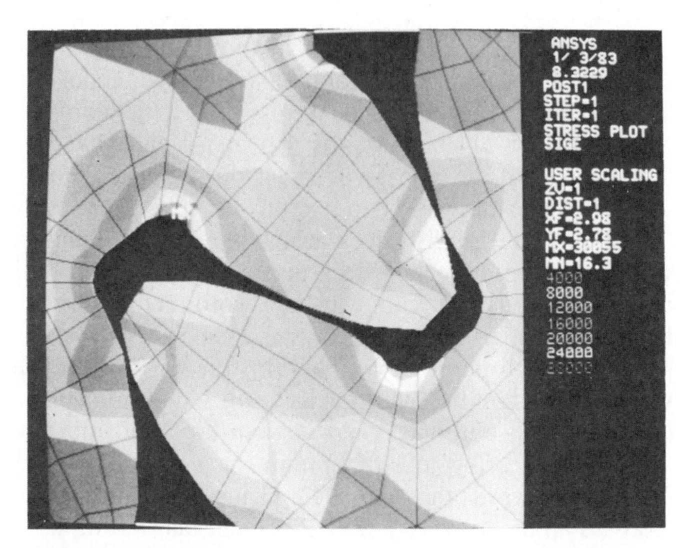

Figure 2–16 Objects such as these gear teeth can be subjected to simulated tests and stress analysis on a computer screen. *Courtesy Swanson Systems, Inc.*

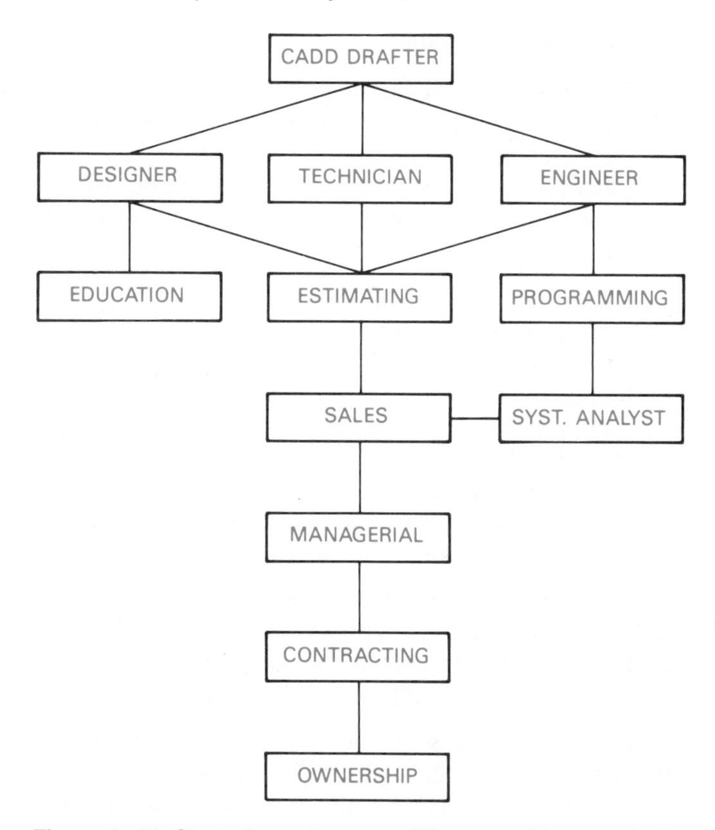

Figure 2–17 Chart shows the many different positions to which an experienced CADD drafter can advance.

training, or CADD instructor in public and private schools. (See Figure 2–18.)

The present need for CADD operators is great. Most companies that purchase a new CADD system train their existing employees to operate it. Also, they may hire drafters with CADD experience to help implement the new system. The demand for experienced operators and students with CADD training will continue to be strong as industry converts to CADD.

Figure 2–18 Drafters with CADD experience can become instructors in public schools or vendors of CADD systems. *Courtesy Houston Instruments.*

EDUCATIONAL PREPARATION FOR CADD

How much does a drafting or engineering student need to know about the operations of a computer-aided drafting system in order to become a good CADD operator? Are skills learned on one CADD system transferable to another? Will the skills change in a few years? Such questions continue to arise and probably will be debated for years.

In the meantime, you, the drafting student, need to keep a few things in mind. The best CADD operators are those people trained in the time-proven techniques and concepts of creating a nice-looking drawing. Every CADD operator needs a strong drafting and design education spiced liberally with mathematics, problem solving, and analytical skills. Employers are looking for people with a solid, basic education, which means possessing the ability to read, write, and spell correctly, think logically, communicate effectively, deal with people rationally, and accept responsibility.

A knowledge of computer programming and programming languages is not necessary to become a CADD operator. What is most important is to gain some familiarity with the basic operations of a computer and the other pieces of equipment at a CADD workstation. After all, almost everyone learns how to drive a car without learning to be a mechanic. There is always a period of training during which time the new employee learns the company standards and how to apply them to its CADD system. Once you learn the basics of operating a computer, you can operate any of them with a minimum amount of training. The greatest amount of time required by someone new to a system is spent learning how to run the computer software (programs).

CADD Training in School

The pressure is high for educators to develop computer-aided drafting classes and purchase CADD systems. This is not always possible or feasible. Even if your school drafting department does not have CADD, it should not deter you from studying the technology. Try to attend demonstrations, cooperative work exchanges, conventions, conferences, workshops, and seminars where CADD equipment is found. Taking field trips to industrial sites that use CADD equipment can develop and strengthen one's knowledge of computer-aided drafting. Remember, what is important is not whether your school has a computer drafting system, but whether you study computer-aided drafting.

The purpose of this textbook is not to instruct the reader in the workings of any one specific brand of computer-aided drafting system. Such instruction is covered in detail in other textbooks and manuals. The references in this text to CADD equipment, skills, techniques, and concepts used in the industry are intended to be general. All of the assignments in this book can be done manually or on any CADD system. From this book you will learn the techniques of drafting, while the specifics of operating your CADD system will be learned from the manufacturer's instruction manual and hands-on experience.

THE LANGUAGE OF THE COMPUTER

Probably one of the most difficult aspects of communicating with computers is just that — communicating. Computers were taught to communicate by people who developed computer programming languages, and unfortunately those languages instruct computers to use words and terms unfamiliar to most people. Anyone interested in computers has been forced to learn to communicate with them on their level. But we are beginning to see a reversal and computers are beginning to communicate with people on the human level. Until such communication becomes universal, some of us will learn programming languages and the rest of us will learn the vocabulary and jargon that enable humans and computers to get along. You won't be studying programming in this text, but we will take a brief look at the words and languages used to communicate with computers.

Vocabulary and Jargon

Any subject requires a knowledge of certain terms. Most jobs or hobbies involve the use of specific items or concepts that have names and meanings. The more involved we get in our pursuits the more knowledge we acquire about the equipment, techniques, and concepts used in that pursuit. We are often unknowingly learning a specialized vocabulary, and to speak to anyone about an aspect of our interest demands the use of specific words.

The world of computers is full of jargon, which is the specialized, technical vocabulary associated with a subject. All of the words and acronyms have meanings, but often plain English does a better job of explaining what we mean, especially to a person not skilled in the field. An acronym, by the way, is a word that is formed from the initial letters of a name, such as BASIC (*B*eginner's *A*ll-purpose *S*ymbolic *I*nstruction *C*ode). As you become familiar with CADD you will learn many terms, some useful and others seldom used but impressive sounding. Try to sift through the vocabulary and retain those words that you need. Using vocabulary correctly is an asset for any skilled person, but tossing jargon around indiscriminately may only serve to confuse and irritate those persons unfamiliar with the field.

Computer Languages

Computers receive immediate instructions from information that the user provides. This immediate instruction activates a block of detailed instructions inside the computer program. The program is written in one of several different programming languages, a few of which are listed here.

BASIC (Beginner's All-purpose Symbolic Instruction Code). BASIC is an algebraic programming language that is easy to learn and easy to use. BASIC has a relatively small number of commands and statement formats. There are many versions of BASIC developed to operate on specific brands of computers. BASIC is not a structured programming language and can be difficult to interpret if the program was not logically designed. BASIC was developed at Dartmouth College in the mid-1960s.

FORTRAN (Formula Translator or Translation). FORTRAN is a high-level programming language that is used for scientific, mathematical, and engineering applications. It was developed by the IBM Company in 1954.

Pascal. Pascal is another high-level language that was developed in the early 1970s. It is noted for its structured design and simplicity. It was named after Blaise Pascal, the seventeenth century French mathematician who developed the first desk calculator.

C. C is a high-level structured programming language designed for use on microcomputers. It was developed by Bell Laboratories. C is a popular language for use in programming CADD software.

LISP (LISt Processing). LISP is a high-level programming language used in artificial intelligence (AI). Used to customize and enhance such software as AutoCAD, which uses its own dialect of AutoLISP.

The operator of a CADD system does not need to know computer programming techniques. In fact the operator does not need to know anything about programming to create drawings with a CADD system, just as a person who drives a car doesn't have to be an automobile mechanic. But a CADD drafter's potential is greatly increased if he or she possesses some knowledge of programming. Drafters with programming ability may be able to customize the software to accommodate company standards and requirements. Drafters with such skills may even be able to move into positions of programmer, systems analyst, or troubleshooter for the system.

As a drafting student you will find it in your best interest to take computer programming courses while in school. Start with a course in BASIC, and if it seems like something you'd like to continue, move on to one of the higher level languages. The avenues of job advancement will be enhanced with programming abilities.

Commands and Prompts

Computers follow instructions well, but not blindly. They often ask questions in response to instructions issued to them. Instructions and questions are two important forms of computer communication to which you are exposed as soon as you turn on the machine. When you wish to get a program running in a computer you must first load the program into the computer. Type the word LOAD, and you have issued a command to the computer. (See Figure 2–19.) A command is a specific instruction, which, when entered by the user, causes a certain action to occur, such as loading the program, or drawing a line. You may then type RUN, which commands the computer to run the program. Some systems execute the load and run procedure automatically.

Once the program is up and running (known as having been "booted") the user will issue many commands in the drawing process. For instance, you may tell the computer to draw a polygon. The computer needs additional information to perform the task. The computer may prompt you, that is, ask you a question such as, "How

Figure 2–19 Typing is a skill used often in CADD. *Courtesy Calcomp.*

many sides?" You must respond with a number. What would happen if you typed a letter by mistake? You would then get an error message, or prompt, on the screen, such as "incorrect response," or "please enter a number." The prompts vary from one program to the next. Some programs are claimed to be *user friendly* and others are noticeably *user hostile*. Such terms refer to the ability of the computer to relate in human terms, and so reflect the ability of the programmer to think in such a manner.

The command and prompt form of communication will become familiar to you after just a few minutes of working at the computer. This form of communication is displayed on the monitor for your reference. Some displays begin at the top of the screen and move (scroll) down, while others begin at the bottom and scroll up. As you continue to work with the computer you will eventually encounter audible prompts called beeps. These often indicate an error on the part of the operator but can also be an acknowledgment. Responding to a computer's audible output is not critical to operating a CADD system; therefore hearing is not a criterion for becoming an operator. Most systems have more than one tone of beep, indicating different things which you will quickly learn.

The Keyboard

The layout of the computer keyboard varies between different brands of CADD systems. One keyboard is shown in Figure 2–20. The standard QWERTY typewriter keys are arranged the same, and the additional keys can be found along the top, left, and right sides of the standard keyboard. Consider the following descriptions of some of the most common keys found on a CADD system keyboard.

Return Key. The name of this key varies, but it is most commonly labeled RETURN, or ENTER, and is usually located at the right side of the keyboard. It is similar to the carriage return key on a typewriter. The computer responds to a line of data entry after this key is pressed.

Escape Key. This key allows you to escape from the current command or activity that you're engaged in.

Cursor Keys. The function of these keys is to move the screen cursor. They are arranged in a group and appear as four arrows pointing in the four compass directions. They are often called arrow keys.

Home Key. This key is used in conjunction with the cursor keys. It moves the cursor to either the top left corner of the screen or to the beginning of the line. It can be found in the center of the cluster of cursor keys or at the upper left of the keyboard.

Backspace Key. If you make a mistake in typing, press this key and you can delete what you just typed. This key can also serve to move the screen cursor back along the line just typed.

(a)

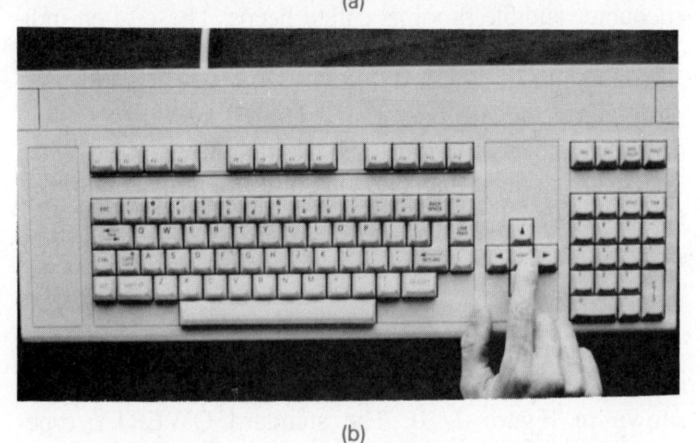

(b)

Figure 2–20 (a) Standard typewriter keyboard. *Courtesy Sharp.* (b) Typical alphanumeric computer keyboard. *Courtesy Texas Instruments, Inc.*

Control Key. This key is similar to a typewriter shift key because it is usually pressed along with another key (or keys) to initiate a function or command within the program. The control key is used extensively in word processing but its operation is often replaced with function keys in CADD systems.

Function Keys. These are additional keys located just about anywhere on the keyboard. (See Figure 2–20(b).) A popular location is along the top of the keyboard. The number and grouping pattern varies by manufacturer. They are used by the program to activate specific functions or commands. Their use may change in the program, or from one program to the next, hence the need for function key masks or overlays. The overlay is usually a plastic template with written labels that fits around the keys. Some computers and programs display the labels of the function keys at the bottom of the screen, which eliminates the need for key overlays.

Calculator Keypad. This cluster of keys is usually referred to as the keypad and contains the numbers 0–9, a decimal point, and arithmetic function symbols.

Become intimately familiar with the keyboard you will be working with, and you will be rewarded with less wasted time searching for keys and smoother-running sessions at the computer overall.

DRAWING WITH THE COMPUTER

Creating good drawings with a CADD system hinges on just a few things. One of the most important is your ability to learn and remember. Another is your hand-to-eye coordination between the digitizing surface and the screen. A third is your ability to understand and visualize coordinate systems as they apply to shapes on different planes and at a variety of rotation angles. The fourth is your typing ability, although this varies among CADD systems.

In this section we will look at how you use the above-mentioned skills to create drawings. The intention here is not to discuss the techniques used with a particular brand of CADD system, but to describe in general the concepts, techniques, and commands used when working with any computer drafting system. A familiarity with the basics of CADD will allow you to more readily grasp the particulars of any one system.

Cartesian Coordinate System

Absolute Coordinates. The *Cartesian coordinate system* is a rectangular coordinate system that locates a point by its distances from intersecting, perpendicular planes. The word Cartesian comes from Cartesius, the Latinized form of Descartes, a seventeenth-century French mathematician and scientist. Drawings done with a computer are based on the Cartesian (rectangular) coordinate system, so it is imperative that you develop a thorough understanding of it.

The two-dimensional rectangular coordinate system is illustrated in Figure 2–21. The point of intersection of the dark vertical and horizontal lines (planes) is called the *origin*. The origin has a value of zero. Values along the horizontal increase as you move to the right of the thick vertical line. This is called the *X axis*. The *Y axis* is the horizontal line and values increase as you move upward from the origin. The values just mentioned are positive. Note in Figure 2–21 that each axis can also have negative values. Negative X values are to the left of the vertical axis, and negative Y values are below the horizontal axis. Study this figure carefully before continuing.

The best way to understand the rectangular coordinate system is to work with a piece of 10 lines to the inch graph paper. Locate the intersection of any two heavy perpendicular grid lines on your graph paper. Mark this point as the origin. Both X and Y have values of zero at the origin. At each inch mark on the horizontal line to the right of the origin, write the values 1, 2, 3, 4, etc. Do the same for the inch lines above the origin along the vertical axis line. (See Figure 2–22.) Now locate the point of X=2, Y=2. First count two inches to the right on the X axis, then count two inches straight up on the Y axis. This point has the value of X=2, Y=2. Only one point in this coordinate system can have that value.

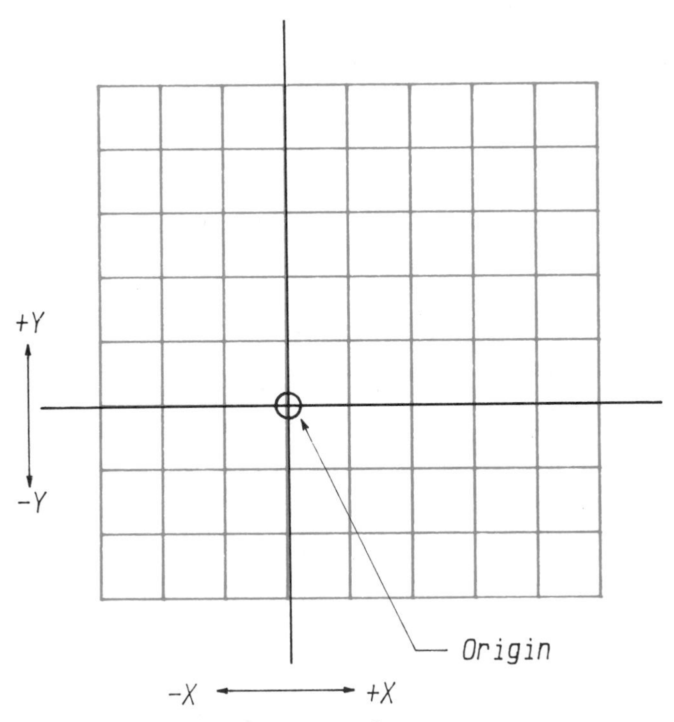

Figure 2–21 Two-dimensional Cartesian (rectangular) coordinate system.

Not all points will have even inch values. Locate the point X=4.5, Y=2.9 on your grid paper. This point is located directly from the origin and not from any other point. Coordinate values are always written with the X value given first. Therefore, this point could be written 4.5, 2.9 and you would know what it means. Count the grid squares along the X and Y axes to find this point, or use a scale and measure. Remember that every point must have at least two coordinate values. A point of just 4.5 could be any location 4.5 in. from the vertical line.

Relative or Incremental Coordinates. The Cartesian coordinate system uses absolute coordinates. This means that each point is referenced to the origin of the coordinate system and not the values of the points around it. The origin is fixed and does not move. Objects can also be drawn using a moving origin. In this method, each point becomes the origin and the next point is located in relation (relative) to the previous point. This method is called *relative* or *incremental* coordinates. If each point becomes the origin for the next point, then numbers to the left and below the origin are negative. See Figure 2–23 for an example of this method.

The incremental coordinate method may be useful when using the dimensions of each line to lay out a part.

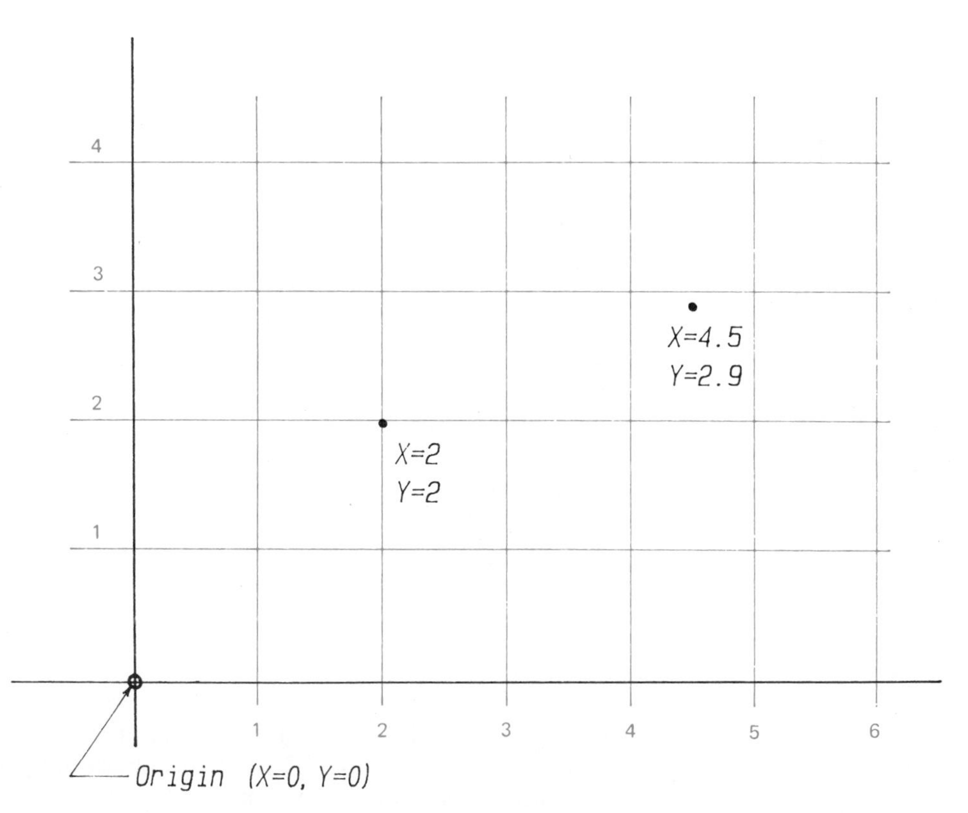

Figure 2–22 Points located on a rectangular coordinate grid.

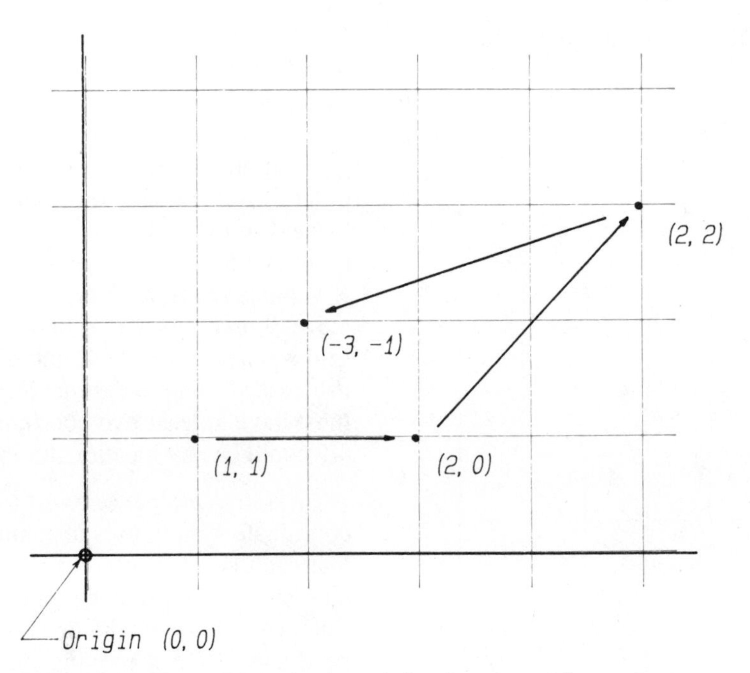

Figure 2–23 Points located using relative (incremental) coordinates.

It is important that you keep in mind the positive and negative aspects of this method, because each point becomes the origin of a coordinate system. For example, moving to the left of a point gives the next point a negative X value. The object in Figure 2–24 shows the relationship of points on an object constructed with relative coordinates.

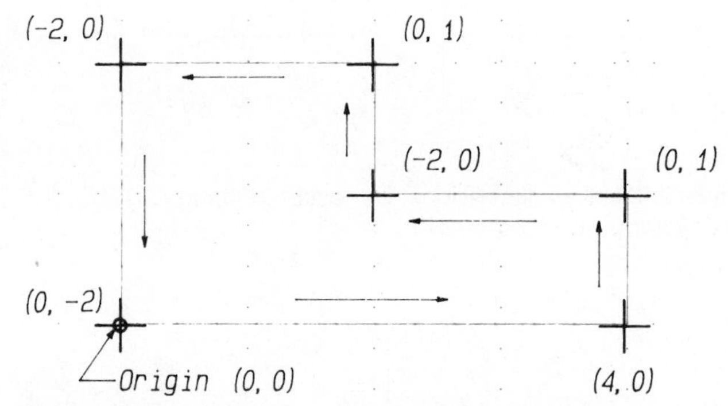

Figure 2–24 Object drawn using relative coordinates.

Polar Coordinates. Not all points can be located properly with a rectangular coordinate system. For example, a point may be at a 45° angle from horizontal at a radius distance of 1.5 in. from a reference point. Such a location defines a point by *polar* coordinates. Note in Figure 2–25 that the horizontal line has a value of 0°. Angles are measured in a counterclockwise direction from horizontal. The radius distance is measured from a specified reference, or center, point.

Polar coordinates are relative and can be used to create a drawing if all the angles of the part are known. Figure 2–26 is the same object drawn in Figure 2–24, but the values given are polar coordinates. The first number at each corner of the object is the angle from the previous point (origin), and the second number is the radius from the origin. Study this object until you are familiar with the concept of polar coordinates.

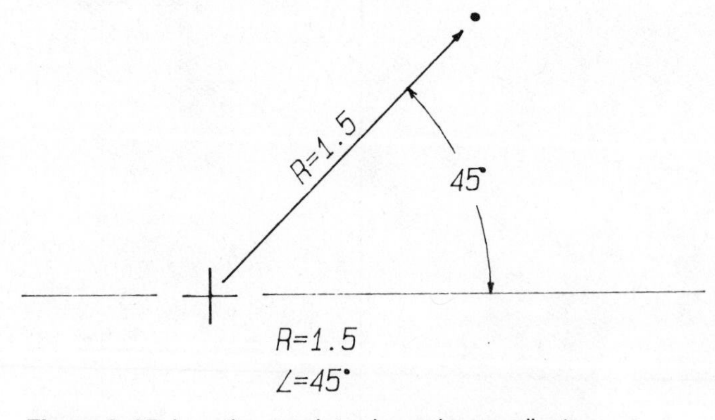

Figure 2–25 Locating a point using polar coordinates.

Three Dimensional Coordinates. To construct a three-dimensional coordinate system, all we need do is add a third dimension, the Z *axis*. The third axis, Z, projects out toward the viewer, Figure 2–27. The intersection of these three axes forms the origin, which has the numerical value of X=0, Y=0, and Z=0. Any point that is behind the origin on the Z axis has a negative value.

The object shown in Figure 2–28 is shown in wire form, and the X, Y, Z coordinate values of each point are indicated. The first number of each set is X, the second Y, and the third Z. Study the object in Figure 2–28 until you understand the technique to plot points in three dimensions (3–D).

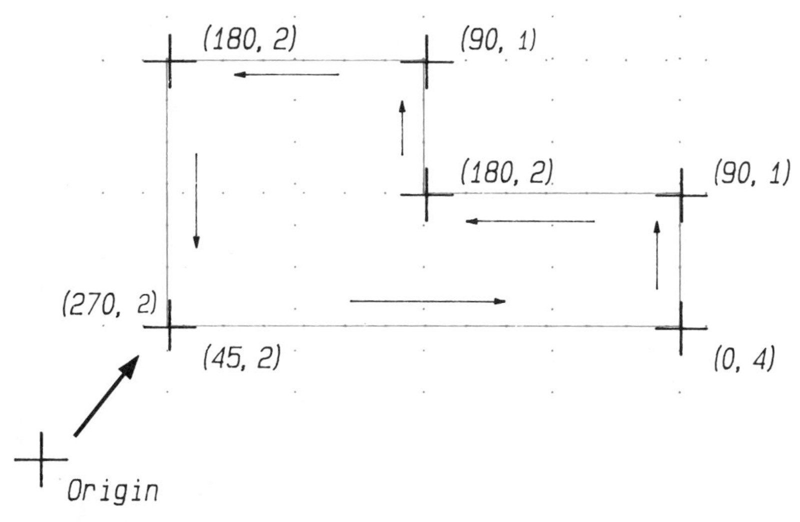

Figure 2–26 Object drawn using polar coordinates.

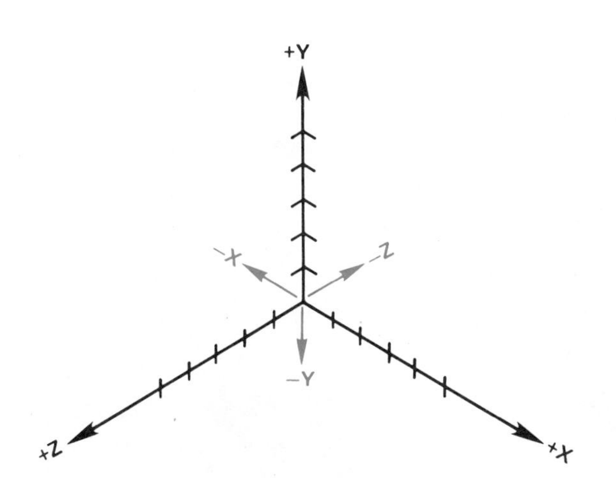

Figure 2–27 Three-dimensional coordinate system.

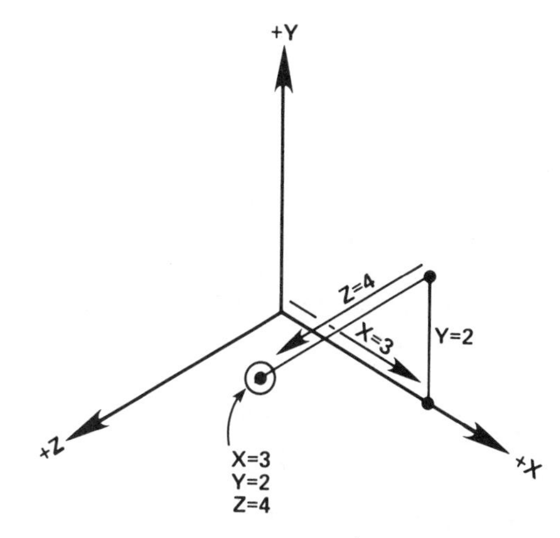

Figure 2–29 Locating a point using 3–D coordinates.

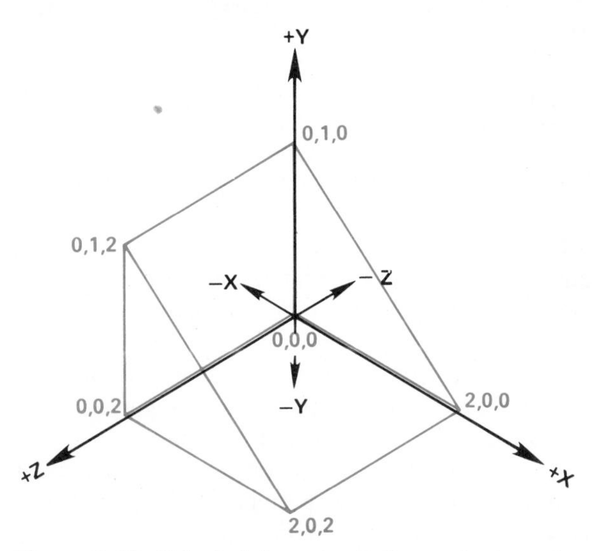

Figure 2–28 Object plotter using 3–D coordinate system.

EXERCISE 2–1

Now try your hand with the object given in the table on page 62. The only information you are provided is the coordinate values for each point on the object. Locate each point given on the chart and draw a line as you go. Lay out a three-dimensional axis like the one shown in Figure 2–27. Use a scale to measure the coordinate values given and plot them as shown in the example in Figure 2–29. The solution to this exercise is found at the end of the chapter.

Drawing in 3–D. The addition of the *Z axis* adds depth to the object and allows the CADD drafter to create a model form. A popular misconception about computer graphics is that the entire object can be shown in three dimensions (3–D) just by entering coordinates for a couple of views. This is not always true. The important thing to realize is that every point on the drawing must be defined by numerical locations, or coordinate points, for the computer to generate a 3–D form.

THREE-DIMENSIONAL COORDINATE EXERCISE

Point	X	Y	Z
1	0	0	0
2	3	0	0
3	3	1	0
4	1	3	0
5	0	3	0
6	0	3	4
7	1	3	4
8	3	1	4
9	3	0	4
10	0	0	4
11	0	3	4
12	1	3	4
13	1	3	0
14	3	1	0
15	3	1	4
16	3	0	4
17	3	0	0
18	0	0	0
19	0	0	4

The process of entering three-dimensional data into a computer graphics system is relatively simple. The principal component is the drafter's ability to visualize three dimensions and distinguish and separate any number of planes in one view. If data is entered in 3–D, an origin point is established. See Figure 2–27. Axis lines can be placed on the screen, and points can then be plotted to create a shape. The 3–D coordinate system axes can be rotated to suit the operator.

Note in Figure 2–28 that the coordinate system can accommodate negative coordinates. An object can be drawn in any or all of the four quadrants. An object can just as easily be drawn using all negative coordinates as positive ones. Most drafters work with the positive side of the coordinate system unless it is more convenient to locate the origin in the center of the part or a feature.

You will soon realize that a CADD system becomes amazingly versatile when all of the vertices of an object have been entered. Once the computer knows the location of all of the points of an object it can turn and rotate the object to any angle, as with the robot arms shown in Figure 2–30. And that is one of the features of CADD that makes it so appealing in engineering, design, testing, and manufacturing.

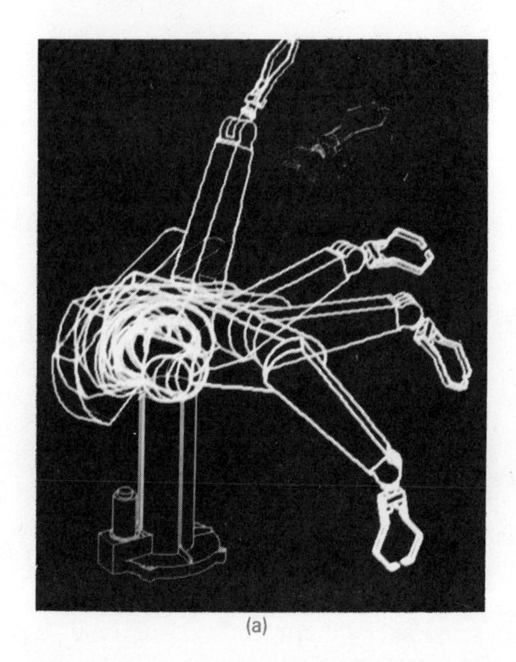

(a)

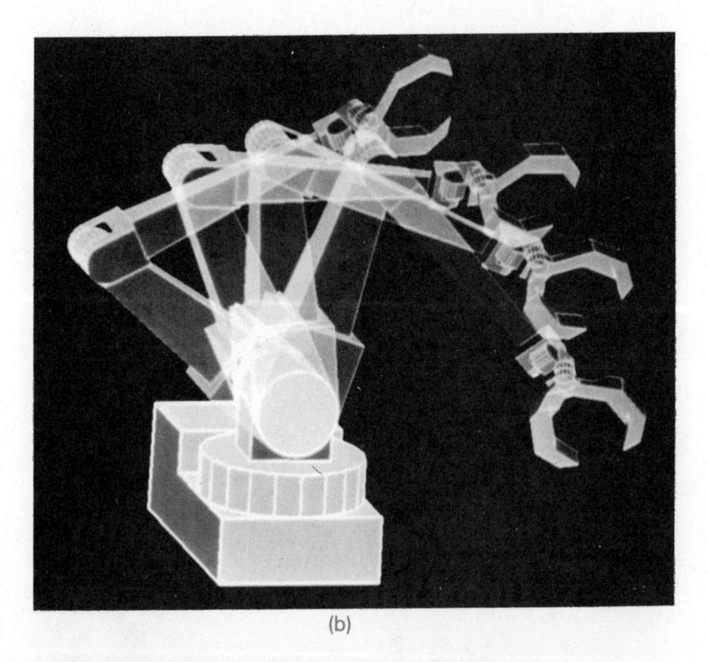

(b)

Figure 2–30 (a) These robot arms can be rotated in any direction once all of the points have been entered into the computer. *Courtesy Computervision Corporation*; (b) another example of robot arm movement. *Courtesy Spectragraphics Corporation.*

COMPUTER DRAFTING EQUIPMENT

The tools now used by the drafter are electronic machines collectively referred to as hardware. An individual CADD workstation relies on a computer for data processing, calculations, and communications with other pieces of peripheral equipment. Peripheral equipment is any additional hardware item that is needed and run by the computer and which performs functions that the computer cannot handle. A typical CADD workstation is shown in Figure 2–31.

The additional functions and services provided by the peripheral equipment fall into three categories: input, output, and storage. Input means to put information inside the computer to be acted on in some way. Input can come from the keyboard, a function board, or a digitizer. Output refers to information that is sent from the computer to a receiving point. The receiving point may

Figure 2–31 A typical CADD workstation with dual display screens and peripheral equipment. *Courtesy Calcomp.*

be a display on a video monitor, an inked drawing done by a plotter, a quickly reproduced image of the screen display generated by a hardcopy unit, or printed images and data provided by a printer. Storage refers to disk and tape drives which allow the operator to store programs, drawing files, symbols, and data.

We will examine all of the hardware mentioned here, and look closely at additional equipment in each of the categories described. All of the equipment is rendered functional by computer programs, or software, which is also discussed.

Monitors

The most obvious component of the CADD workstation is the video display screen, or monitor. We are all familiar with this piece of equipment because it looks much like a TV screen. The most common forms of monitors are vector and raster displays. Figure 2–32 is an example of a monitor. The monitor contains an electron gun that sits behind the screen and shoots electrons, which strike the screen causing it to glow. Screens may show a single color (monochrome) or a full color display. The monitor comes in a variety of forms which we examine here.

Vector. This type of monitor is also known as stroke writing, vector writing, line drawing, cursive, or calligraphic writing. The beam of the electron gun is aimed at X and Y coordinate points (called pixels) and connects these points with a line, much like the way you draw a line with a pencil. Each line and feature on the screen is rewritten frequently to allow the operator to view the image as constant. This is called a refresh system because the image is refreshed many times per second.

If large amounts of data are displayed on a refresh screen, it is possible to get a flicker because it takes longer for the electron gun to return to an individual pixel to refresh it. During that lapse of time the pixel may begin to fade. When the electron beam strikes the fading pixel again, it brightens, which is seen as flicker. All of the data is stored as coordinates (vectors); therefore curved lines may appear to have a pointed shape because curves are drawn with short straight line segments connecting many vectors. Since any element on the screen can be deleted and redrawn without repainting the entire screen, the vector terminal can be used to animate images.

Figure 2–32 The monitor resembles a television set and can display data in full color or monochrome. *Courtesy NEC Technologies, Inc.*

Raster. A raster display is similar to a television display. The screen is composed of a gridwork of tiny squares which appear as dots and are known as pixels. The word *pixel* is an acronym for picture (pix) element (el). The resolution of the screen, or the crispness of the image, is determined by the number of pixels. Each pixel is actually a *bit*, as shown in Figure 2–33. A bit is a binary digit and can be either on or off. A bit that is receiving a signal is on, and is lit. The absence of a signal means the bit is off. A raster display is often called a bit map because it is a map composed of pixels. Low-resolution raster screens may produce lines that have jaggies, or a jagged look. This is especially true for angled lines and curved features. The lines in Figure 2–34(a) and 2–34(b) show how an image would appear on raster screens of different resolution if magnified. Individual lines and features can be deleted without repainting the screen because the image is constantly refreshed. The brightness of each individual pixel can be adjusted to show highlights and shading.

Color and Monochrome. The advantages of color images on a video display screen are becoming well documented in industry. The color monitor is similar to a color TV set. A pixel on a color screen is actually composed of three dots, red, green, and blue (RGB). Three electron

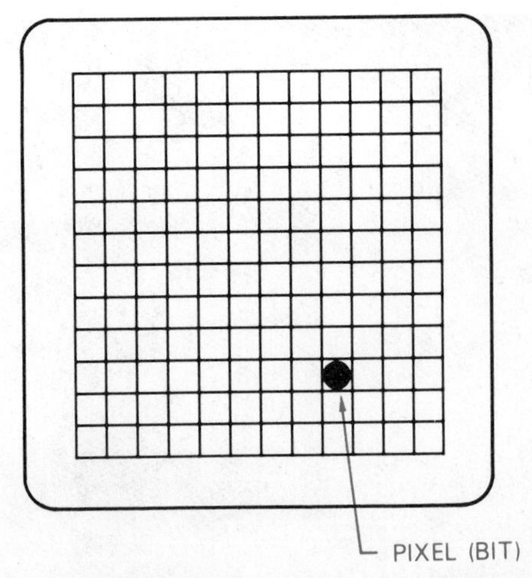

Figure 2–33 Video display screens are composed of a grid, or framework of pixels (bits), or dots, that can be turned on to display graphics or alphanumeric characters.

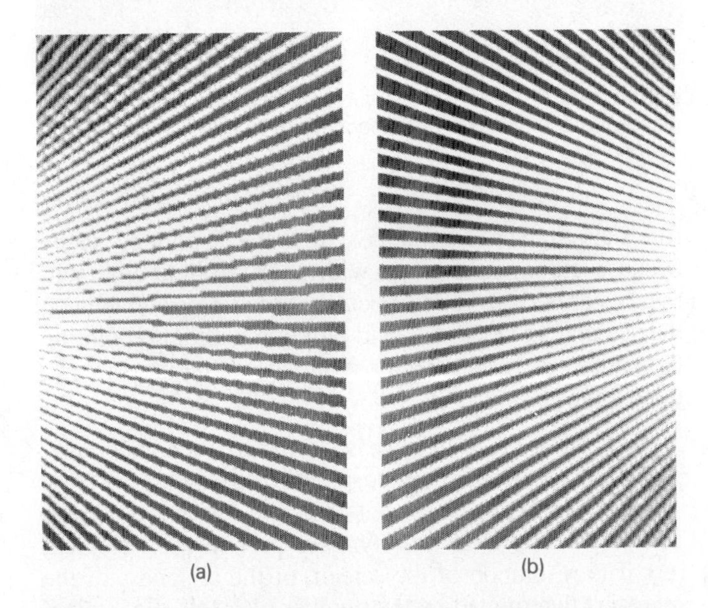

Figure 2–34 (a) A raster display produces jagged lines called jaggies; (b) but the jaggies can be reduced by high-resolution screens. *Courtesy Megatek.*

guns supply these three primary colors. Monochrome screens can be purchased with a variety of single display colors such as green, amber, white, black on white, blue, and red.

Matrix. There are ways of producing a display other than shooting electrons at a screen. The general term, *matrix display*, refers to rows and columns of elements or components that can be turned on or off, much the same as pixels. Some forms of matrix displays are liquid crystal display (LCD), light-emitting diode (LED), electroluminescent panels, and plasma display.

A plasma, or gas-discharge display, is shown in Figure 2–35. It is composed of two flat panels enclosing a layer of inert, ionized gas. One panel contains X axis wires and the other Y axis wires. When an X, Y coordinate is selected, the X and Y wires at that location are charged, causing the gas to glow at that point. This is similar to a glowing pixel in a raster display. The electroluminescent panel uses a thin-film phosphor in place of ionized gas.

Figure 2–35 Portable microcomputers with a flat-panel display. *Courtesy Toshiba America Information Systems, Inc.*

These matrix screen displays are also called *flat-panel displays.* They do not require the deep cabinet needed by the monitor to house the electron gun. Present research and development is pointed toward flat-panel displays because of the reduction in physical size and the lower power requirements.

The Heart of a Computer

The processing function of the CADD workstation is handled by the central processing unit (CPU). The CPU is normally an integrated circuit (IC) chip, dedicated to a specific function. Additional chips contain specific instructions that allow a certain component of the CADD system to operate. These instructions can only be read by the computer, and then acted on. They are called *read only memory*, or ROM. This is the memory that the computer never forgets. Other chips within the computer are empty. Information can be put into these chips and taken out again at any time. They are called *random access memory* (RAM).

The size of a computer's RAM is specified by the number of characters (letters or numbers) that it can hold. A character is formed by a group of eight bits (binary digits). When these eight bits are acted on as a whole unit it is called a *byte*. Thus the capacity of a computer's RAM is referred to by its number of bytes of memory, usually expressed in thousands or millions. One thousand bytes is called a K (kilo) and a million bytes is a "mb" (megabyte). A 640K computer has approximately 640,000 bytes of RAM.

It is common to hear someone mention that a computer is a 32-bit machine. That means the computer can process 32 bits of information simultaneously. How many bytes is that? Older computers in use are 16- or 20-bit machines.

The Mainframe Computer

The large computer that occupies its own climactically controlled room, such as the one shown in Figure 2–36, is commonly known as a mainframe computer. The term *mainframe* at one time referred to the size of the cabinet holding the computer. Most early CADD systems relied on a mainframe computer, and some still do. The mainframe computer may possess from 32- to 100-million bytes of memory. Mainframe computers often handle many other functions in addition to the graphics requirements of the CADD workstations. They are often referred to as host computers because they serve so many different terminals. They can support up to several thousand terminals on-line at one time.

The Minicomputer

The minicomputer represents a break from the restrictions of the centrally located mainframe computer. The minicomputer is smaller physically and in memory than a mainframe, but it can accomplish the same tasks. A plus for the minicomputer is its ability to support up to several hundred user terminals simultaneously (depending on the requirements of the users). The abilities of the minicomputer and its cost relative to the mainframe make it highly compatible with the requirements of many companies. CADD systems operating on a minicomputer are shown in Figure 2–37.

The Microcomputer

The microcomputer is also known as a desktop computer or a personal computer. (See Figure 2–38.) It possesses at least one microprocessor chip and normally between 640K and 8MB of RAM. Many micros now provide for memory expansion up to 16 megabytes. Most micros serve one terminal, but the trend is turning toward networking, which is connecting several terminals to share resources such as a large hard disk. Networking gives a single microcomputer some of the same capabilities as the minicomputer.

The appearance of the microcomputer can take many forms. Figure 2–39(a), (b), and (c) show different types of microcomputers. The computer itself does not have to be combined with the video display screen. The keyboard is often separate from the computer. Some micros have disk drives built in, and some even have printers built in. A group of microcomputers, called *portables*, can be carried by a handle, plugged in anywhere, and operated on the spot. They have detachable keyboards and built in disk drives.

Figure 2–36 The mainframe computer can support many workstations and occupies its own climactically controlled room. *Courtesy Intergraph Corporation.*

Figure 2–37 A minicomputer, shown in the background, can serve several workstations. *Courtesy Computervision Corporation.*

Figure 2–38 A microcomputer and peripherals are used in several CADD systems.

The microcomputer is the machine that allows CADD to penetrate markets that were not feasible with high-priced mainframes and minicomputer systems. Many schools now provide CADD training with microcomputer-based systems such as the one in Figure 2–40. Small companies can afford CADD systems, not only because of their low price, but because they do not take up a large amount of physical space.

(a)

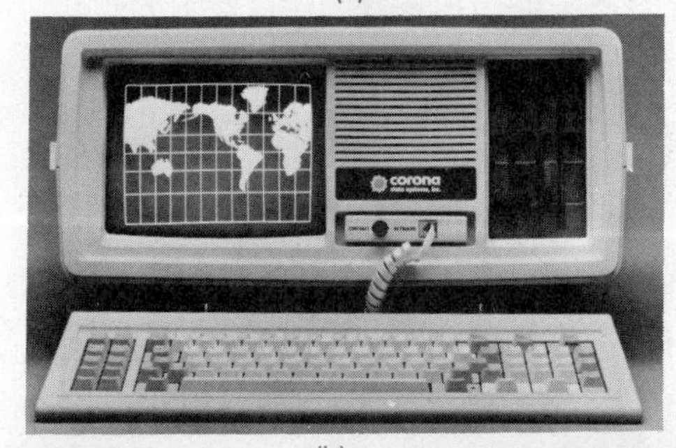

(b)

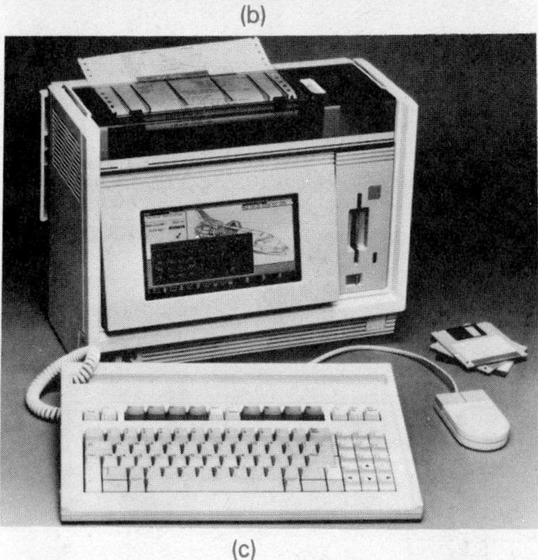

(c)

Figure 2–39 (a) A microcomputer with dual disk drives and detachable keyboard. *Courtesy NEC Information Systems, Inc.*; (b) transportable microcomputer with dual disk drives and detachable keyboard. *Courtesy Corona Data Systems, Inc.*; (c) transportable microcomputer with a flat-panel display, built-in printer, disk drive, mouse, and detachable keyboard. *Courtesy Hewlett-Packard Company.*

Input Devices

Keyboard. There are several methods of putting information into a computer. The most obvious is with the familiar alphanumeric keyboard (alpha-letters, numeric-numbers). (See Figure 2–41.) Written instructions can be given to the computer from the keyboard. Graphic instructions can also be entered from the keyboard in the form of X, Y, Z coordinates, circle diameters, number of polygon sides and curve radii. Using a keyboard could be a tedious method if all the drawing data were given to the computer in this fashion, but it isn't. Only specific instructions are issued to the computer from the keyboard.

Figure 2–40 Schools can provide CADD training with microcomputer-based systems.

Figure 2–41 This computer keyboard resembles that of a typewriter with the addition of several keys and a calculator keypad. *Courtesy of Capital Spectrum, Inc. for Dell Computers.*

At first glance, most keyboards look much the same, but closer inspection reveals their differences. In addition to the standard letters and numbers, a keyboard may contain a variety of special control keys, cursor keys, and calculator keypad. The cursor is a short underline or small box that shows your position on the screen. The cursor keys have arrows on them pointing in four different directions. Pressing these keys moves the cursor in the direction shown by the arrows.

Function Board. The keyboard may contain additional keys that are called function keys, or soft keys. They can be programmed to perform any function that the programmer desires. (See Figure 2–42.) The number of function keys on a keyboard can vary from four to twelve. The SHIFT, CTRL, and ALT keys can triple the number of functions available.

In addition to the function keys, some CADD systems make use of a function board. (See Figure 2–43.) This board contains several keys or buttons that can be programmed to activate specific commands or symbols. The same board can be changed as the program progresses. The operator just attaches a new mask or overlay to identify the keys.

Similar to the function board is the menu pad, menu tablet, or command board shown in Figure 2–44. Commands are displayed on the menu tablet as symbols or words. These are often printed on a menu overlay that is attached to the tablet or function board. A menu is a selection of commands, often related in some way. Some boards have pressure sensitive elements that allow the individual commands to be selected by the press of a finger. Others may use a stylus, or puck. A command issued to the computer can change the menu on the board and the operator must attach a new menu overlay. In this manner the drafter can use any number of symbols and commands.

Digitizer. The second method of entering drawing data into the computer is by pointing to it. This is done using an input device called a digitizer. The digitizer is often called a graphics tablet, or tablet for short. It is a low-profile piece of hardware that has a flat, smooth surface covering a gridwork of thin wires. Figure 2–45(a) and (b).) A pointing cursor is attached to the digitizer by a wire.

The pointing cursor used with the digitizer can take two standard forms, stylus or puck. The stylus looks like a ballpoint pen with a wire coming out of the top. (See

Figure 2–42 The dark-colored function keys along the top of the keyboard allow the operator to select specific functions or commands. *Courtesy SysScan a.s. Norway.*

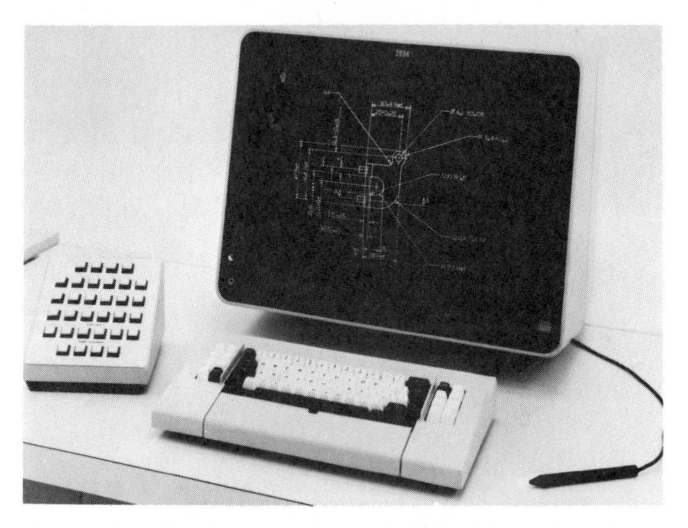

Figure 2–43 The function board to the left of the keyboard can be programmed to activate specific commands or symbols. *Courtesy International Business Machines Corporation.*

Figure 2–44 The operator can select a command or symbol from a menu tablet. *Courtesy Calcomp.*

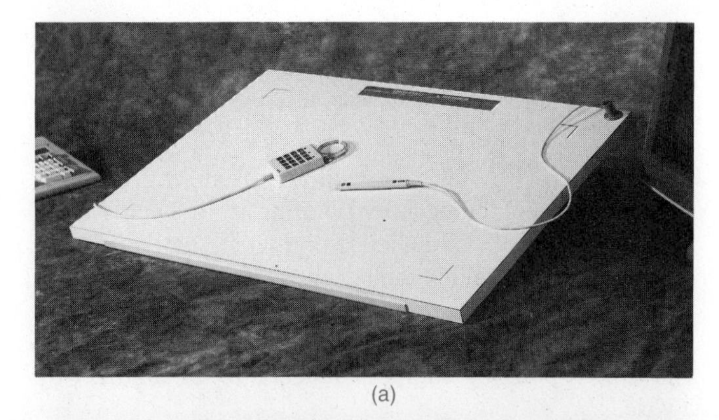

(a)

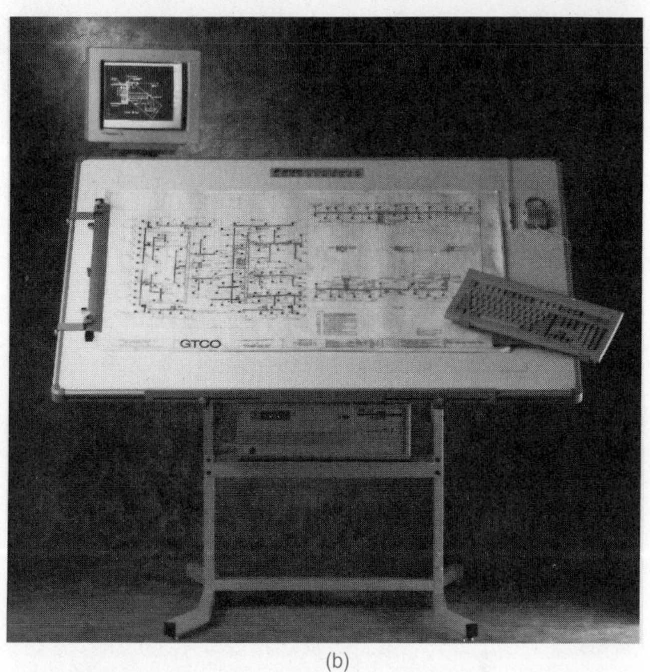

(b)

Figure 2–45 (a) Drawing data can be entered into the computer by using a digitizer; (b) a table-model digitizer — note both stylus and puck. *Courtesy GTCO Corporation.*

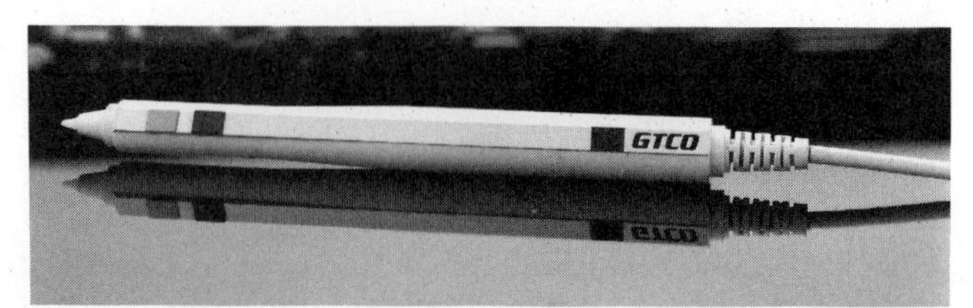

Figure 2–46 A hand-held stylus resembles a ballpoint with a wire coming out of the top. *Courtesy GTCO Corporation.*

Figure 2–46.) The puck does resemble a hockey puck, but it is usually rectangular in shape and may have several buttons on it. It too is attached to the digitizer by a wire. To locate points with the puck you must look through a small lens containing crosshairs. Two different types of pucks are shown in Figure 2–47(a) and (b).

When the stylus or puck is brought near the active surface of a digitizer tablet, its position is determined by the electronics inside the tablet. As the stylus is moved across the tablet, crosshairs on the video display screen track its movement. By pressing the stylus to the tablet at the beginning of a line, the coordinates for that point are quickly transmitted to the computer graphics system. Move the stylus to the end of the line and touch the tablet at that point. The final coordinates are then transmitted. A digitizer tablet operator can reduce an entire drawing to digits without ever stopping to calculate a single number.

Digitizer tablets are also used to give instructions to a graphics system. This input method uses a menu having several graphics commands printed on it. A command is a specific instruction that is issued to a computer. The user simply presses the stylus on the desired command and the computer system accepts it. (See Figure 2–48.)

Several other input devices exist that perform the same function as the stylus and puck, but do not have to be used with a digitizer. The most well-known is familiar to anyone who has ever played a video game. It is a joystick, as seen in Figure 2–49. This little spring-loaded stick allows you to move the crosshairs on the screen by tilting the joystick in any direction. The stick always returns to the center of its receptacle, but the crosshairs remain in the last position they were placed.

The trackball is a ball mounted inside a box as seen in Figure 2–50. The operator simply pushes on the ball to make it rotate and the crosshairs on the screen move in the direction of rotation.

Turn the trackball over so that you are holding onto the box and rolling the ball on the table, and you have the principle of the mouse. The mouse may have a small roller ball on its underside, or two wheels mounted at 90° to each other. The mouse, shown in Figure 2–51, is moved about on a flat surface in order to move the crosshairs on the screen. The optical mouse requires a special board on which to roll because it operates without moving parts. (See Figure 2–52.) The mouse and trackball are also attached to the terminal with a wire.

Output Devices

The primary output device is the video display screen which was discussed earlier. But there are several other methods of producing a drawing. Sending it to the plotter is a popular option, be it pen or electrostatic plotter. A quick print can be transmitted to the hardcopy unit, which performs a function similar to a photostatic copier. The printer can be used to get a paper copy of drawing data, program listings, or even a print of the screen display. Not all CADD systems use each of these pieces of equipment.

(a)

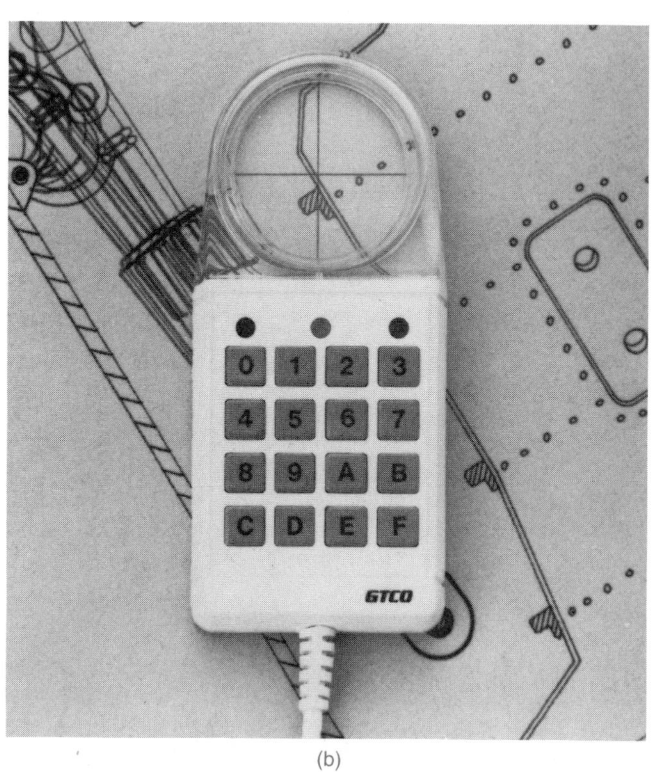

(b)

Figure 2–47 A four-button puck; (b) a 16-button puck. *Courtesy GTCO Corporation.*

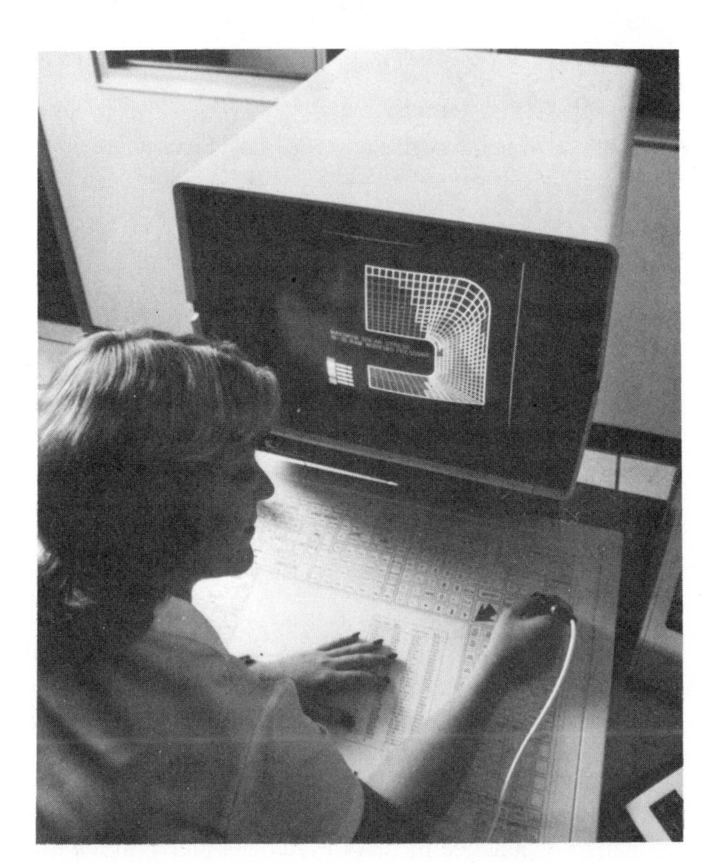

Figure 2–48 Using a stylus to select a command from the digitizer menu. *Courtesy Computervision Corporation.*

Figure 2–49 A joystick allows the operator to move the screen cursor. *Courtesy Chessell-Robocom Corporation.*

Figure 2–50 The operator's left hand is on a trackball that moves the screen cursor. *Courtesy COMTAL Image Processing Systems.*

Figure 2–51 The mouse to the right of the keyboard is moved on a flat surface to control the screen cursor. *Courtesy Capital Spectrum, Inc. for Dell Computers.*

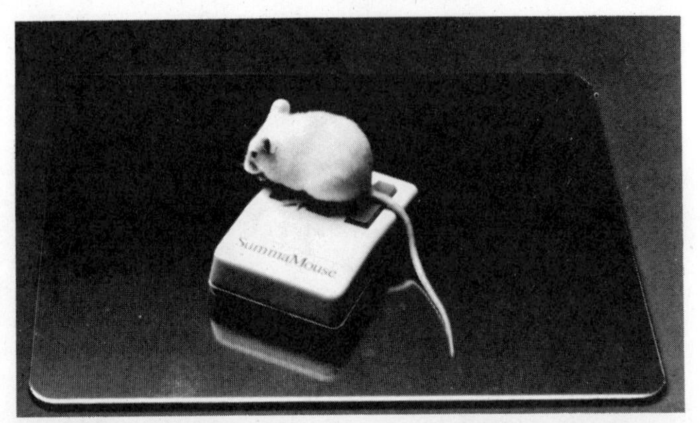

Figure 2–52 The optical mouse uses a reflective surface to control the movement of the screen cursor. *Courtesy Summagraphics, Inc.*

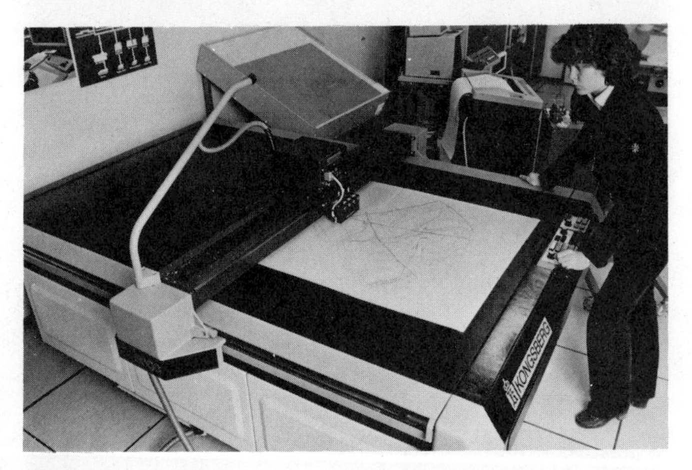

Figure 2–53 A large flatbed plotter. *Courtesy SysScan a.s. Norway.*

Flatbed Plotter. On a vector graphics display screen, lines and features are drawn using X and Y coordinates. This information is stored in the computer as a display list. The display list contains all the information that the plotter needs to function. The information stored on a raster screen (bit mapped) must first be converted to a display list before it is sent to the plotter because the plotter operates in two directions only. Movement of the pen is along the X and Y axes. Some plotters allow the pen to move in both directions, while others move the paper in one direction and the pen in the other.

The flatbed plotter uses a pen that moves in both directions. The paper size that can be used on this plotter is determined by the width and length of the bed. A vacuum or an electrostatic charge can be used to hold the paper firmly to the bed. The pen is held by an arm as seen in Figure 2–53.

Drum Plotter. A drum plotter is shown in Figure 2–54. A drum plotter may take on one of two different forms. The true drum plotter accepts roll paper and has tractor-feed holes on the sides. The drawing can be any length, but the width is determined by the size of the plotter. A variation of this is the drum plotter that has a permanently attached piece of Mylar® mounted on tractor or pin wheels. The drawing paper or film is then taped to the Mylar. The drum plotter does have a space-saving advantage over the larger flatbed plotter because it

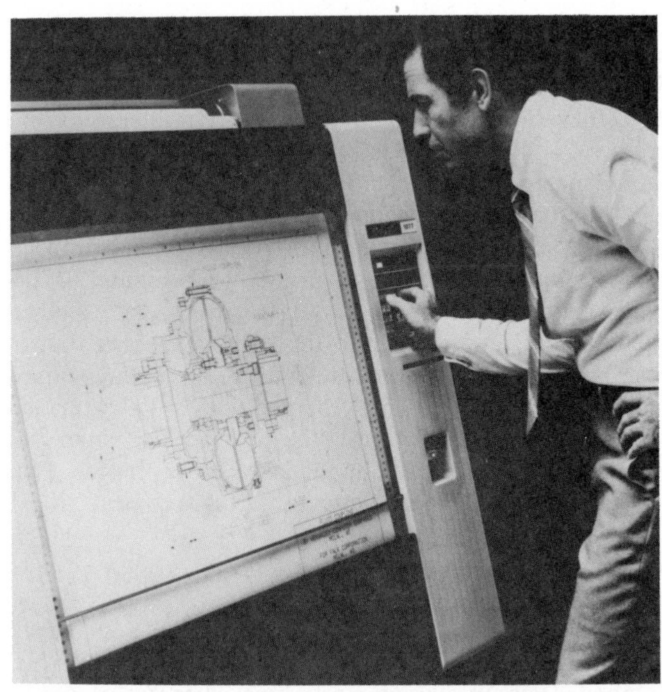
Figure 2–54 This drum plotter can take up to E-size drawings. *Courtesy Calcomp.*

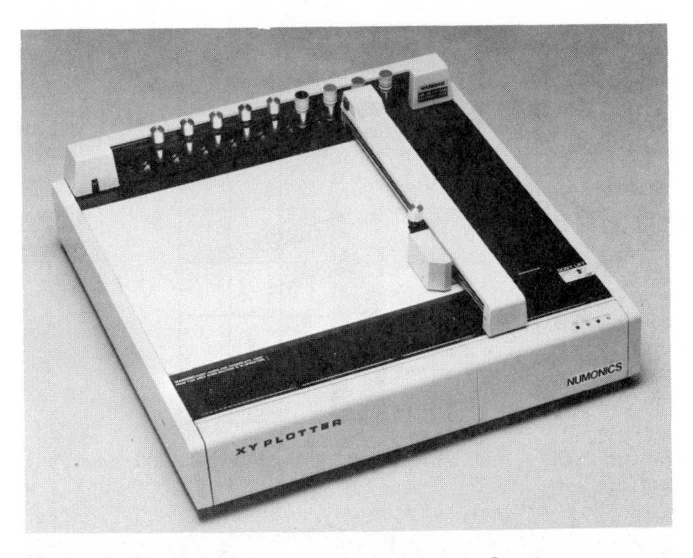

Figure 2–55 A small ten pen, A-size plotter. *Courtesy Numonics.*

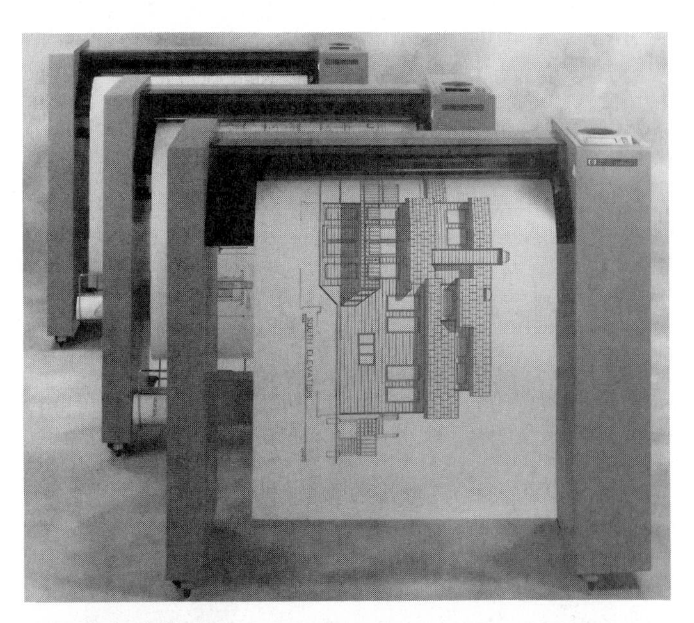
Figure 2–56 These plotters are mobile and compact. *Courtesy Hewlett-Packard Company.*

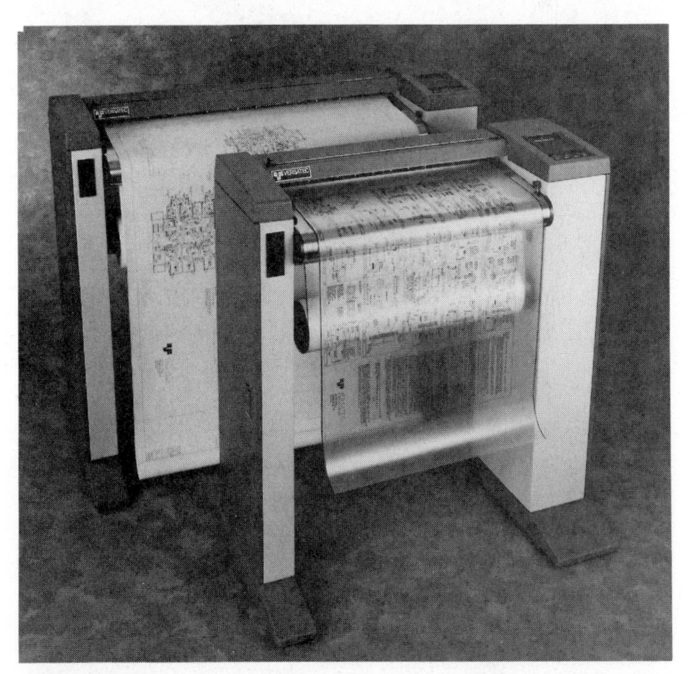
Figure 2–57 An electrostatic plotter. *Courtesy The Xerox Engineering Systems Company.*

stands upright. The mobility of the plotter was enhanced with small versions as shown in Figure 2–55 and models that roll where they are needed. (See Figure 2–56.) The type of drum plotter shown is called a microgrip plotter. It uses grit-wheel paper-grippers. Some people refer to it as a pinch-wheel plotter. This plotter moves the pen in one direction and the paper in the other.

Electrostatic Plotter. The high-speed electrostatic plotter is often referred to as a printer because of the nature of its process. As with drum plotters, the paper moves in one direction on an electrostatic plotter. A line of closely spaced wire nibs is charged to produce dots on coated paper. Pressure or heat is then used to permanently attach ink that adheres to the charged dots. These plotters are approximately ten times faster than pen plotters. Electrostatic plotters are finding their way into more CADD applications because of their speed, resolution, and ability to print in full color. (See Figure 2–57.)

Impact Printer. A variety of printers fall into the category of impact printer. The action of this kind of printer is to strike a ribbon and paper to produce a character or image. Two well-known members of this group are the daisy-wheel and dot-matrix printers. The daisy wheel produces characters similar to those of a typewriter and is often called a letter-quality printer (LQP). (See Figure 2–58.) The dot-matrix printer, seen in Figure 2–59, hammers tiny wires in specified locations to create a character composed of dots. A matrix is just an arrangement of rows and columns (Figure 2–60). Scoreboards in stadiums around the country flash scores using a matrix of lights.

Nonimpact Printer. Impact printers must use an inked ribbon, much like a typewriter does, but the nonimpact printer may be found in several different forms. The thermal printer uses tiny heat elements to burn dot-matrix characters into treated paper. Blue or black images can be produced. The electrostatic printer has been described separately. The ink-jet printer, seen in Figure 2–61, squirts miniscule droplets of ink at the paper to produce a dot-matrix character. Some of these printers spray a continuous stream of ink at the paper.

The best quality image produced by a nonimpact printer comes from the laser printer. (See Figure 2–62.) The laser printer is more closely related to the electrostatic copier process than to the striking, poking, burning, and spraying printing methods. A laser draws lines on a revolving plate that is charged with a high voltage. The light of the laser causes the plate to discharge. An ink/toner will then adhere to the laser drawn images. The ink is then bonded to the paper by pressure or heat.

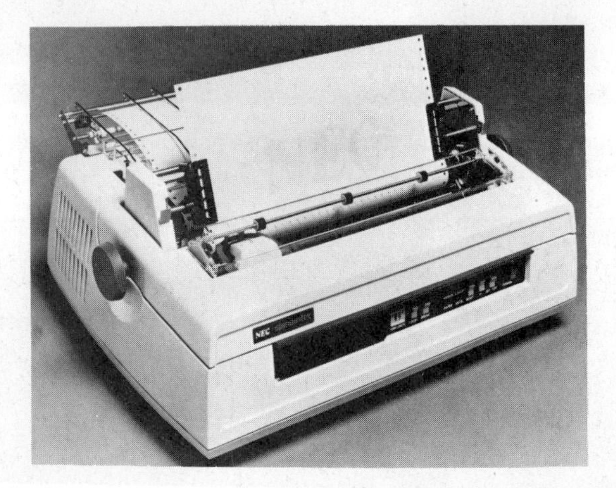

Figure 2–58 A letter-quality printer. *Courtesy NEC Information Systems, Inc.*

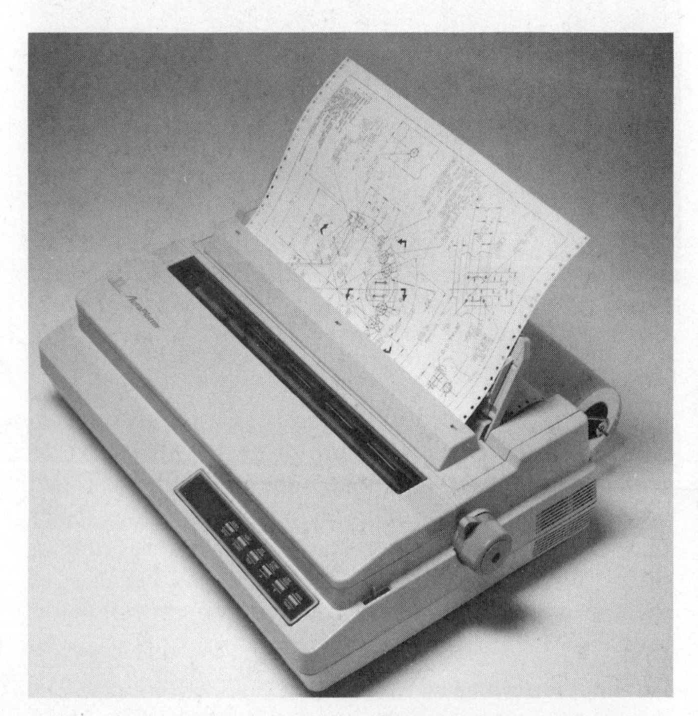

Figure 2–59 A dot-matrix printer. *Courtesy JDL, Japan Digital Laboratory.*

Storage

Storage can be either temporary or permanent. Temporary storage is often referred to as memory. This integrated circuit storage works fine when the power is on. But when the power goes off, the contents of temporary storage dissipate into the atmosphere. The temporary

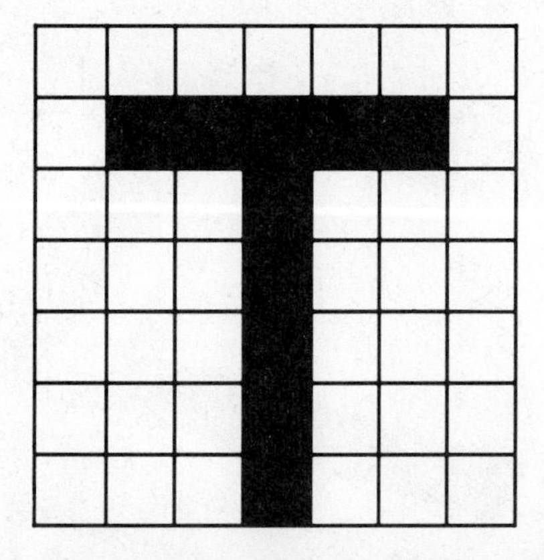

Figure 2–60 A matrix is an arrangement of rows and columns.

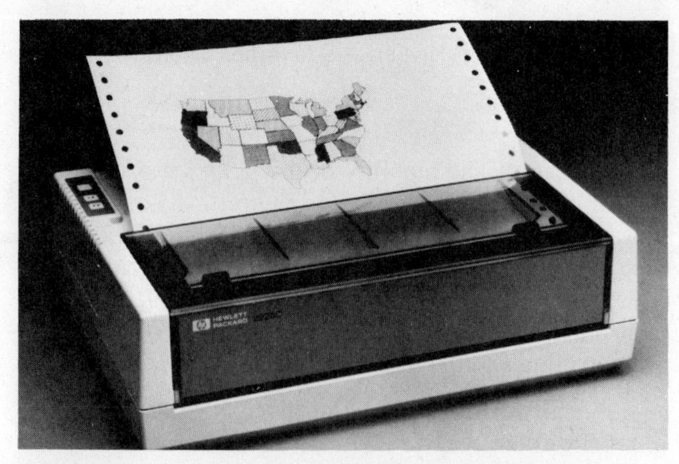

Figure 2–61 An ink-jet printer. *Courtesy Hewlett-Packard Company.*

storage that is used to handle drawing data is random access memory (RAM). Drawings need greater protection than is available with RAM, so they are transferred to magnetic tape or disk. This transfer of data is handled by a tape drive, seen in Figure 2–63, or disk drive, shown in Figure 2–64. Once the drawing files have been secured on tapes or disks, they are in permanent storage. They may be filed in racks in the computer room, or sent to a vault as backup copies.

When both tape and disk drive units are used within a company, the current drawings are stored on magnetic disks. At the end of the working day, or whenever policy dictates, the contents of the disks are dumped (transferred) to the magnetic tape drives and saved on tape. This cleans off the disks for reuse. Copies can then be made of the tapes for backups.

Disk and tape drives are often referred to as mass storage devices because they are devices that store large amounts of data. Disks and tapes come in a variety of sizes and are discussed in greater detail later in this chapter.

Software

A computer program that provides specific instructions to enable a computer to do a certain task is called software. When CADD software is purchased from a manufacturer it is known as a systems program. This specialized software provides a service for a specific endeavor such as integrated circuit (IC) layout, process piping design, solids modeling, HVAC, structural, and architectural. The company that purchases a software package may eventually begin to modify the contents. It may even create some programs of its own. This can be referred to as applications software that the customers can apply to their own needs.

Figure 2–62 A laser printer. *Courtesy NEC Technologies, Inc.*

Figure 2–63 This tape-drive unit accepts a tape cartridge. *Courtesy Danford Corporation.*

FUTURE CADD HARDWARE AND SOFTWARE

As the density of integrated circuits increases and memory capacities rise, larger programs will be able to operate faster with less power. Microcomputers are taking on the characteristics of mini and mainframe systems. Smaller CADD systems are able to store and handle more data because of their greater computing power.

The trend is toward total automation of the design, testing, and manufacturing process, and the integration of the disciplines of drafting and design. Eventually these tasks will be accomplished by a single computer system.

Figure 2–64 A disk pack is used in the disk drive. *Courtesy Computervision Corporation.*

These systems will have diagnostic abilities to pinpoint problems within the equipment, and redundant or auxiliary components that become active when another component fails.

Flat-panel display screens will continue to increase in popularity and may eventually overshadow video display technology. But video display is still viable and continued research will produce screens of superior resolution. Raster screens are available with resolutions of approximately 150 × 200 pixels, up to 1000 × 1000 pixels. In the near future we will see displays of up to 4000 lines.

Color monitors will become standard as will color hardcopy and color plotters. The increased use of three-dimensional (3–D) modeling will make color indispensable. Touch-sensitive screens may also be used more. Such screens are already on the market. (See Figure 2–65.)

Figure 2–65 A microcomputer with a touch-sensitive screen. *Courtesy Hewlett-Packard Company.*

A variety of input methods may be developed but the voice method may soon be preferred. Systems already exist that accept voice commands. Voice synthesizer chips are being used commercially in a variety of applications to speak certain phrases or instructions. But CADD computers will recognize a certain number of spoken words (commands) and translate them into action. Another possible method of input may be thought. The operator may one day wear a kind of sweatband around the head that contains tiny electrodes. The electrodes may sense thought patterns related to an object or shape and translate the patterns to lines and features on a large, color video screen. The part is then beamed to manufacturing and made by numerically controlled (NC) machines and robots.

Solids modeling and realistic shaded color models of parts and systems are becoming more commonplace as computer power increases. The use of 3–D may even evolve into the realm of holography. Lasers can be used to project a three-dimensional image in space. Designers and engineers could walk around the object and examine it from all sides. Animated screen displays enable us to view an object from all angles now, but the use of holography may expand those abilities. Projectors could be mounted around an open lot and the image of a new building could be projected onto the site at full scale. It could then be examined, tested, and changed on the very spot it is to be built.

The manner in which we work with computers and models may be dramatically changed by the technology called *virtual reality* (VR). The user wears a helmet composed of two small flat-panel screens, one for each eye. On one hand is worn a *data glove* containing fiber optic sensors for each finger. The virtual reality user is actually inside the model and can see his or her hand in the screen. Not only can the user move around inside the model, but can interact with it. The implications for this technology are tremendous, and the only limitation for its uses is our imaginations.

The progress of computer technology may branch down a path not mentioned here, but even so, the future will see smaller, more powerful systems that are easy to use, and there will be many more of them. It is up to today's drafting student to keep abreast of the rapidly changing scene of CADD, and try to remain open to the major innovations that are sure to come. Education in your specific career field should be thought of as a continuing process wherein you constantly study the techniques and tools used within your trade.

CADD MATERIALS

Drawing Media

Drafting tools may have changed radically but the finished product is still rendered on paper (vellum) or drafting film (Mylar®). The same paper sizes that are used in manual drafting are used in computer drafting. Those sizes are discussed in Chapter 1. The size of the plotter will, of course, determine the size of the paper used. Drafters occasionally must create drawings that are larger than standard paper sizes. This is accomplished by using roll paper or Mylar® that is available in standard widths. Figure 2–66 shows a plotter using roll paper.

Other papers used in computer drafting are printer papers and hardcopy unit papers. Printer paper used with dot-matrix and daisy-wheel printers is basically a bond paper available in several weights (thicknesses) and sizes, and may have tractor-feed holes along the sides. Treated papers that are sensitive to heat are used in thermal printers. Hardcopy units can use a variety of mediums, depending on the process, such as electrostatic, ink jet, or photo. A dry silver paper is used with the photo plotter type of hardcopy unit.

Drawing Pens

Computer graphics technology has succeeded in eliminating one traditional drawing instrument from the drafting repertoire: the pencil. Plotters use a variety of pens and pencils.

The types of pens used in plotters are liquid ink, wet ink, liquid roller, felt tip, ceramic tip, and ballpoint. Some examples are shown in Figure 2–67. The best quality lines are obtained with the wet-ink pen. This type of pen is seen in Figure 2–68, and is basically an adaptation of the technical pen. Wet-ink pens have a variety of tip widths and a reservoir for ink. Most plotters can hold several pens and automatically switch from one pen to a different pen to change line width or color. The main problem with using multiple wet-ink pens is preventing the tips from drying between use. Since the use of wet-ink pens can use time and be costly, most preliminary plots are done with felt-tip or ballpoint pens. Either kind of pen is less expensive, and several widths and colors can be used at the same time without the worry of wet-ink pens. A plotter using felt-tip pens is seen in Figure 2–69.

Some felt-tip pens do not draw well on vellum or Mylar®, so a special high-quality paper must be used. Adapters are available that allow wet-ink pens to be used in plotters that are made for felt-tip pens. Some schools prefer to use only felt-tip pens because they are clean and do not dry out fast. Felt-tip pens do dry out if left uncapped for a long period, so you should get in the habit of removing and capping the pens in your plotter before you leave the workstation.

Figure 2–66 An E-size drum plotter that uses roll paper. *Courtesy Hewlett-Packard Company.*

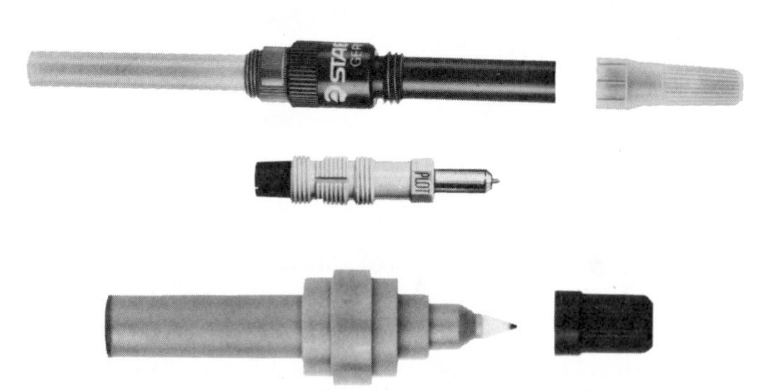

Figure 2–68 Wet-ink plotter pens. *Courtesy Mars Plot, J.S. Staedtler, Inc.*

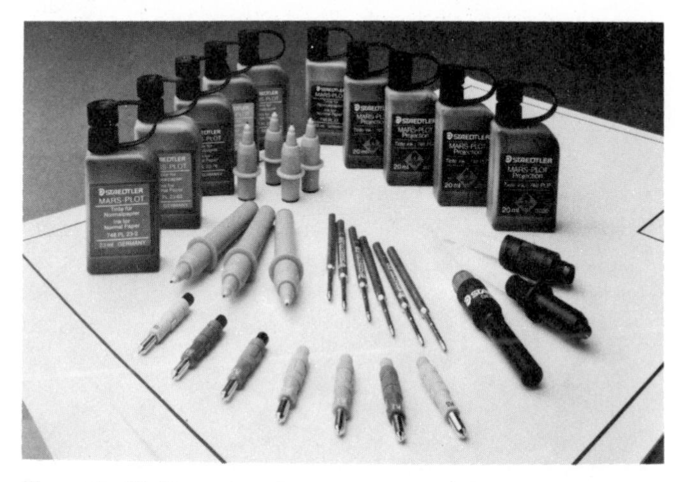

Figure 2–67 Examples of various types of plotter pens. *Courtesy Mars Plot, J.S. Staedtler, Inc.*

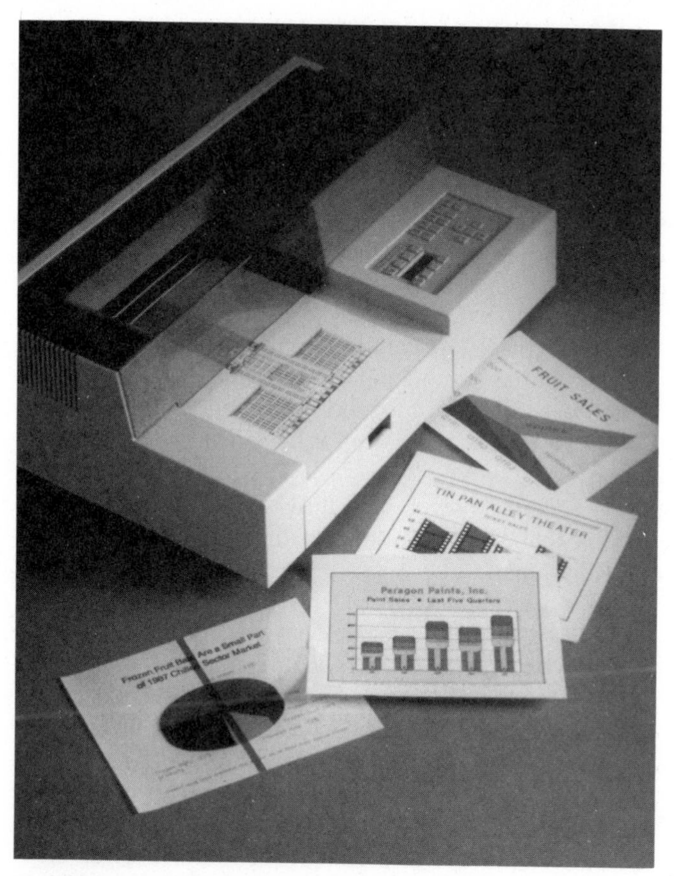

Figure 2–69 Felt-tip pens are being used in this A-size plotter. *Courtesy Hewlett-Packard Company.*

Ballpoint pens and pencils offer some relief from the drying problem, but they do not provide the high-quality lines of wet-ink pens. For this reason ballpoint pens and pencils are used most often for preliminary drawings and check prints.

Storage Media

The storage file for the computer drafting system is the magnetic tape or disk. The tape or disk drive holds the drawing file as magnetic impulses called *bits*. The drive can also read the data from the tape or disk and send it to the computer. The tape or disk drive unit is often referred to as the mass storage device. The term *mass storage* technically means a large-capacity storage device but it is commonly understood to mean any external storage beyond that inside the computer.

Magnetic Disks

The most popular form of storage media is the magnetic disk. The disk itself is made of either a rigid or flexible material, usually plastic, which is coated with a magnetic oxide. The disk resembles a brown phonograph record without grooves. The read/write heads of the disk drive skim just above the surface of the disk and deposit or remove magnetic impulses. Disks come in a variety of sizes and forms.

Mainframe and minicomputer systems often use the removable hard-disk module or disk pack. The disk pack contains several rigid disk platters 14 in. in diameter connected with a spindle. (See Figure 2–70.) When the disk pack is not being used (off line) it is stored in a plastic case with a handle on top. The bottom of the plastic case is removed before the pack is inserted into the disk drive. Once in the drive the plastic cover is removed to provide room for the access arms to get between the disks. The access arms are the read/write heads. When not inside the disk drive, the disk packs must be securely attached to the bottom and cover of the plastic case. This protects them from dust and other contamination. Care should be taken not to drop or otherwise mishandle the disk pack.

A second type of hard disk is the fixed hard disk, which cannot be removed from the disk drive. This is a rigid disk made of plastic or metal that is permanently sealed inside the disk drive. A fixed hard disk is faster and holds more data than its flexible cousin, and is usually offered with small computers such as micros and minis.

The most well-known of the magnetic disks is the flexible disk. It is also called a floppy disk or diskette. The floppy disk is much thinner than a hard disk and thus is floppy. It is encased in a paper cover and comes with a dust jacket that must be removed before the disk is inserted into the drive. Floppies come in three standard sizes. The regular diskette has a diameter of 8 in., the diameter of the mini floppy is 5¼ in., and the micro floppy is 3½ in. The micro floppy is encased in rigid plastic and the surface of the disk is not exposed. When the disk is inserted into the disk drive, a shutter opens to expose the disk to the read/write head. Examples of the various forms of disks are shown in Figure 2–71.

Disks and disk jackets should never be written on with pencil or ballpoint pen. Labels should be written first, then attached to the disk. Disks are magnetic and can be ruined if placed too near magnetic materials, so keep magnets out of the computer workstation area. Any magnetic field can rearrange the information stored on a disk, so you must be on the lookout for more than just

Figure 2–70 Disk pack. *Courtesy Uarco Incorporated.*

Figure 2–71 Examples of various types of flexible disks. *Courtesy Devoke Data Products.*

horseshoe-shaped magnets. Motors, television sets, loudspeakers, and ringing telephones all have magnetic fields that can damage the information stored on nearby floppy disks. Temperature extremes can damage disks, so try to store them in a room with a fairly comfortable temperature range. Dust, hair, and assorted atmospheric debris can contaminate a disk and render it worthless. See that your computer room is clean and that disks do not lie around outside their jackets and storage boxes.

Floppy disks must be handled carefully. Never touch the exposed areas of the disk. When inserting a floppy into the disk drive, do so gently, in the manner shown in Figure 2–72. If a disk is inserted into the disk drive too hard, it may bend or crease. An important consideration in classrooms is to keep disk drives and disks away from chalkboards. Chalk dust is quite unfriendly to disks and could provide untold headaches.

File drawers must be labeled and arranged in some sort of logical fashion to facilitate easy storing of drawings. So too must floppy disks. New disks must be initialized or formatted. The INITIALIZE or FORMAT command is issued to the computer after inserting a new disk into the drive. The disk drive then proceeds to divide the disk into sectors and tracks to accommodate the storage of files or records. An area for the disk directory is also established by the drive. The sectors of a disk resemble slices of a pie, and the tracks are like the grooves on a phonograph record. (See Figure 2–73.)

Figure 2–72 The proper way to insert a flexible disk into a disk drive.

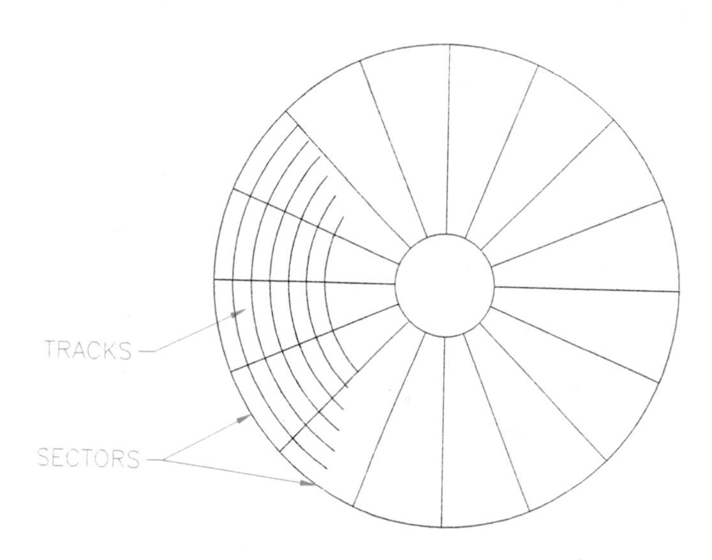

Figure 2–73 A disk is divided into pie-shaped sectors, and tracks that circle the disk-like grooves on a phonograph record.

Magnetic Tape

A less expensive type of magnetic storage media is magnetic tape. It is used primarily for off-line storage of drawings and data. As was mentioned earlier in this chapter, companies using mainframes and minicomputers with tape drives and disk packs usually use disks for daily work, and then dump, or transfer, the drawings to tape as backup storage on a regular basis. Disks are used during working hours because they are faster and hold more data.

Floppy disks are faster than tape because they are direct-access media. The disk is divided into sectors, tracks, and bytes. A specific location on a disk is much like the numbers of the houses on a street or apartments in a building. In fact a specific location on a disk is also called an address. When the disk drive discovers the address, it can go directly to that location. Hence the term direct access.

Magnetic tape, on the other hand, is said to have sequential access. The information on tape is recorded on parallel tracks. There are eight tracks, each containing one bit. Therefore the width of a tape can hold one byte and the bytes are recorded in blocks. A ninth track on the tape is used for automatic checking purposes. (See Figure 2–74.) Because tape is a long, thin strip, it must be wound and rewound to find the correct address, a process that takes time. The location of your drawing may be at one end of the tape. Once the tape drive learns the address, it must then wind the tape until it finds the location. All the addresses are located in sequence and must be accessed sequentially.

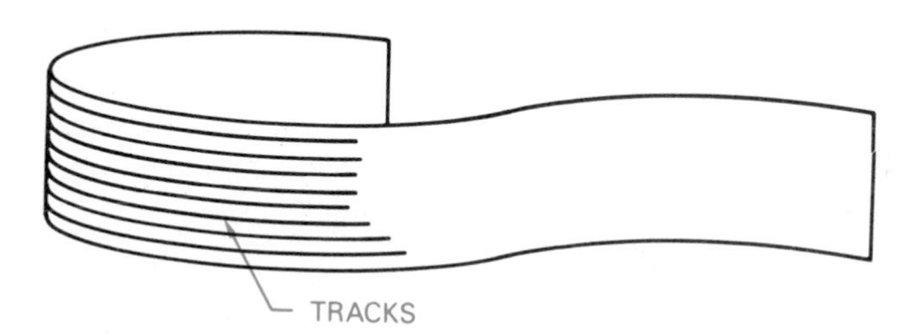

Figure 2–74 Magnetic tape stores data on parallel tracks.

Magnetic tape also comes in a variety of forms. The reel-to-reel tape that is used in mainframes and some minis looks quite similar to the reel-to-reel tape that is used on tape decks. The standard size is ½ in. wide. (See

Figure 2–75.) As was mentioned previously, the reel-to-reel tape is most often used for off-line storage of drawings and data because of its relatively low cost and smaller bulk when compared to disk packs. Tapes used for storage must be recopied periodically. If they are not recopied, the coils of tape may contaminate each other. Tape is tightly coiled and if left that way for several years, the magnetism of the individual bits may begin to merge with others or adjacent parts of the tape, thus creating an unreadable tape. This problem is not encountered with disks.

Magnetic tape is also available in cartridge and cassette form. These come in various lengths, much the same as audio tape. All tape is coated on one side, whereas disks can be coated on one or both sides and be designated as single sided or double sided. The same precautions hold for magnetic tape as were mentioned for disks; always keep them inside their storage cases when not in use, and never touch the exposed tape.

Figure 2–75 The reel-to-reel tape drive unit on the right can send stored drawings to the drum plotter on the left. *Courtesy Gerber Scientific Instrument Company.*

Optical Disks

Optical disks offer an economical, high-speed method of storing computer-generated drawings. Some advantages of optical storage disks over magnetic media include increased storage volume, transfer of information to a standard output device or directly to microfilm. One single optical disk can store a trillion bytes of information, which is accessible in only a few seconds. Optical disks are available in 5¼-, 12-, and 14-in. sizes.

CD ROMs

The fastest growing storage technology is the *compact digital read only memory*, or CD ROM. This is the same technology that is used for audio CD ROMs. the thin plastic platter is treated on one side with tiny pits and a reflective surface. Digital impulses are read by a laser and transformed into either audio or digital information. Many companies are providing parts catalogs and drawing libraries of components such as doors, windows, fasteners, etc., on CD ROM. CD ROM recorders will enable the disks to be used much as floppy disks are used now, for temporary and permanent storage of data.

GETTING STARTED WITH CADD

Most network CADD systems require that the user *log on* to begin the session. Logging on can be as easy as sitting down and pressing one key, or as involved as entering sets of letters or numbers, or typing a secret password.

A common requirement for logging on is for the user to enter the name or number of the workstation or terminal, a password (often the user's initials) that has been assigned to that specific operator, and then the name and number of the part or drawing that is to be worked on. If a new drawing is to be generated then it must be given a name or number. The system may take a few seconds to search for the required drawing before it is ready.

If the operator is beginning a new drawing, there may be some initial parameters that need to be established. These are the variables that you initially set up. Consider what you do when you begin a new drawing manually. You must determine the paper size, the scale, the type of drawing, and drawing format (title block, bill of material, revision block, etc.). Such decisions must also be made before beginning your drawing on a CADD system. Manually you must decide where to begin each view, which in CADD is the origin. You must also determine your text size, although this may change during the course of your drawing.

Another important initializing function that is inherent with some CADD systems is the mode in which the drafter will be working. The model mode and the drawing mode are two that are common to some systems. The *model mode* enables the drafter to generate a part in three dimensions. There is no specific layout involved because the information you are entering is creating a model in space, built around a coordinate system. The *drawing mode* will allow the operator to extract information (views) from the model to create a drawing that is then plotted on paper.

Using the Menu

The *menu* is a list of items (commands, numbers, symbols, etc.) from which you can select, similar to a menu in a restaurant. An example of a typical menu is seen in Figure 2–76. You select the options on the menu by using some sort of pointing device. The menu is usually a sheet of paper or a plastic overlay that is printed with all of the commands used in the particular CADD system. The menu is attached to an electronically sensitive surface called a tablet, or digitizer. A menu can also appear on the monitor. Commands may be selected using function keys, cursor keys, a stylus, mouse, or by touch. The screen-menu commands may appear anywhere around the periphery of the screen, depending on the CADD system.

Menu Layout. An important aspect of the menu is the manner in which it is arranged. There is no ideal arrangement for a menu, but the words, commands, symbols, and functions should be grouped in related clusters and positioned to minimize the hand movement needed to select items. All menus are arranged in such a manner, but some are better than others. That is why the menus that come with commercial CADD systems are often customized by the user company to suit its needs.

The arrangement and shape of a menu depends on the digitizing tablet or board to which it is attached. Some menus are located along the bottom of the digitizer while others may be positioned to one side. Another arrangement positions the menu items around the blank area used for digitizing points on a drawing. The most important aspect of the menu is the grouping of items in related clusters. The item groups are defined by borders, space, or colors, and are usually identified by a heading that defines their relationship. Some common headings are keyboard, keypad, construction, text, edit, dimension, geometry, symbol library, shapes, lines, measure, display, and plot. This list is by no means complete but should give the reader an idea of the types of categories found on menus. A good example of a typical menu layout is seen in Figure 2–77.

One of the most valuable skills that the CADD drafter can develop when learning to use a menu tablet is memory. Not computer memory, but operator memory. The sooner you remember where certain commands are physically located on the menu, the quicker you will be able to select them in order to construct a drawing. A good method to use when learning the menu is to begin with general items and work toward specific items. First learn what the command groups are, then where the command groups are located. Then as you begin to use the system, try to remember which commands are in each group. You will do better by first learning the commands in the groups you use most often, then adding those used the least.

Pointing Devices. The most popular method of selecting items on a menu tablet is with a pointing device, which is a general term for a variety of instruments that are attached to the digitizer or computer terminal by a wire. Of course the most well-known pointing device is the finger, which is used with systems that have function

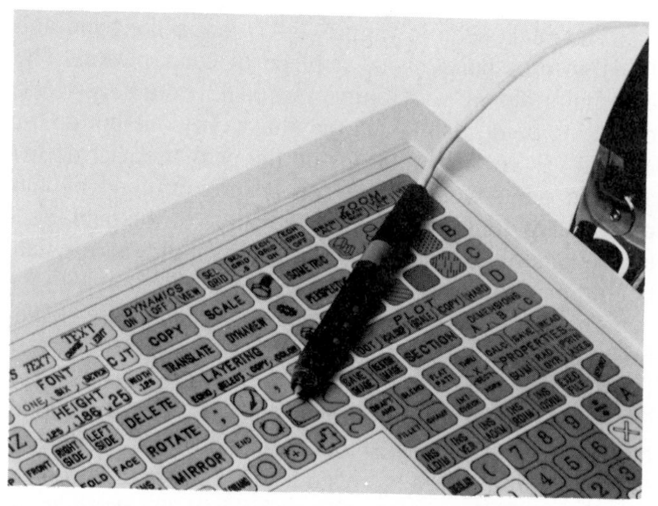

Figure 2–76 A typical tablet menu showing commands and functions. *Courtesy Computervision Corporation.*

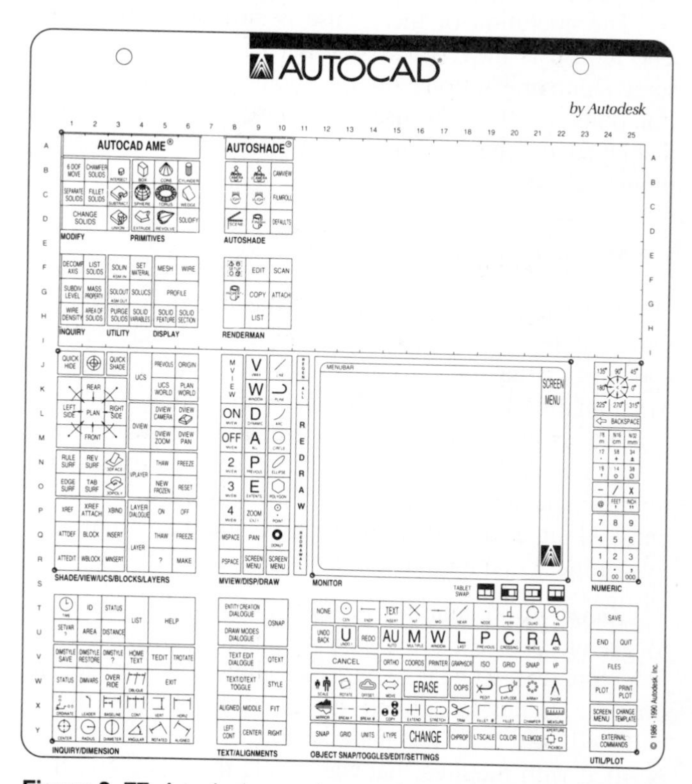

Figure 2–77 A typical menu layout. *Courtesy Autodesk, Inc.*

boards, buttons, or sensitized menu boards. Touch-sensitive computer screens are also in use. The most common pointing devices are hardware items such as the cursor, mouse, and light pen. These pointing devices were discussed earlier in this chapter, but we will review them again before we see how a drawing is constructed.

The digitizer cursor is the most common pointing device and should not be confused with the screen cursor. The digitizer cursor is an input device that is held in the hand and the screen cursor is a small box or line on the video display screen that indicates current position. The puck and stylus are two forms of digitizer cursors.

The puck, seen in Figure 2–47, fits in the hand and has thin crosshairs inside a piece of clear plastic. The crosshairs aid in positioning the puck accurately when digitizing points. The puck may have only one button, or as many as 16. Most buttons on the puck are user definable. In other words, they can be programmed by the user for any function. The identification of the puck buttons is one of the first things drafters should learn when beginning to operate a CADD system.

The stylus requires no memorization because it has no buttons. It resembles a ballpoint pen and is attached to the digitizer or computer by a wire. (See Figure 2–78.) Commands and functions are selected from the menu tablet using the stylus, which responds with a click or tone when a selection has been made. You should become familiar with the feel of the stylus and the amount of pressure required to activate it because, like the puck, a missed click may lead to errors and wasted time.

The operation of the mouse is similar to the puck. The mouse is moved across a flat surface to move the screen cursor. Buttons on the mouse activate specific functions or allow the user to choose from menu items displayed on the screen. Figure 2–79 shows a typical mouse.

Figure 2–78 This operator is using a stylus to draw contour lines on the screen. *Courtesy SysScan a.s. Norway.*

Figure 2–79 A three-button mouse. *Courtesy Summagraphics Corporation.*

Manipulating the Drawing

The subject of manipulating the drawing covers a wide range of capabilities within the CADD system, capabilities that may vary from one system to another. Manipulating the drawing begins with editing functions and moves into other aspects of creating a drawing, such as layout, design, additions, initial set up, examining microscopic detail, standing far away from the object, and looking at the object from any angle. Let's look at a few of these manipulative aspects.

Zoom. A popular command that enables the operator to get close to the work is a command called ZOOM. Selecting the ZOOM command allows the drafter to draw a rectangle around an area that needs to be enlarged. After the rectangle is drawn, the entire screen is filled with the area that was inside the rectangle. It is as if a window were placed around an area, and then the viewer walked closer to the window so that the area inside the frame was all that could be seen. This is a great command to use when working in a detailed area of a drawing. Once inside a window, the command can be used again and again to get even closer to the object or feature. Figure 2–80 illustrates how the ZOOM command works.

Mirror. The MIRROR command has freed the drafter and designer from tedious and redundant activities that limit the amount of creative time spent on a project. Commands such as MIRROR and REFLECT can give reverse copies of an element or symbol by simply specifying an axis on which to reflect the object. The image can be duplicated at any distance from the original, or appear to be attached to it, such as the object in Figure 2–81.

Copy. Once a feature or symbol has been drawn or located on a drawing, it can be copied once with the COPY command, or it can be copied at regular intervals any number of times. With the COPY command you need only identify the item to be copied, then indicate the position where it is to be copied. A similar command, called ARRAY, enables you to copy an object in a row and column arrangement, or in a circular pattern copied about a center point (polar array).

Area Fill. By selecting a command called AREA FILL, or HATCH, the drafter can choose a sectioning pattern, point to an enclosed polygon or circular shape, and automatically the area inside the polygon is filed with the pattern.

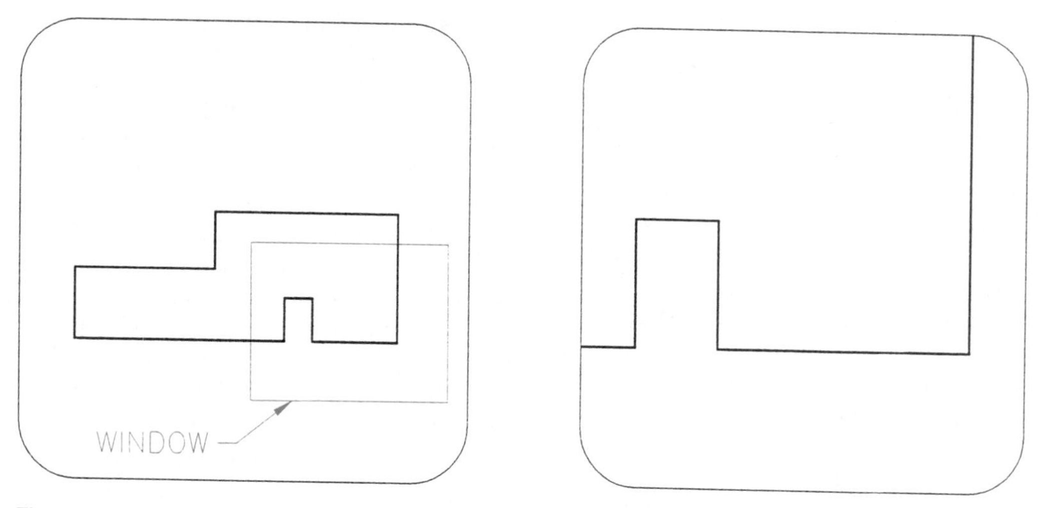

Figure 2–80 The ZOOM command brings the object, or a feature of the object, closer.

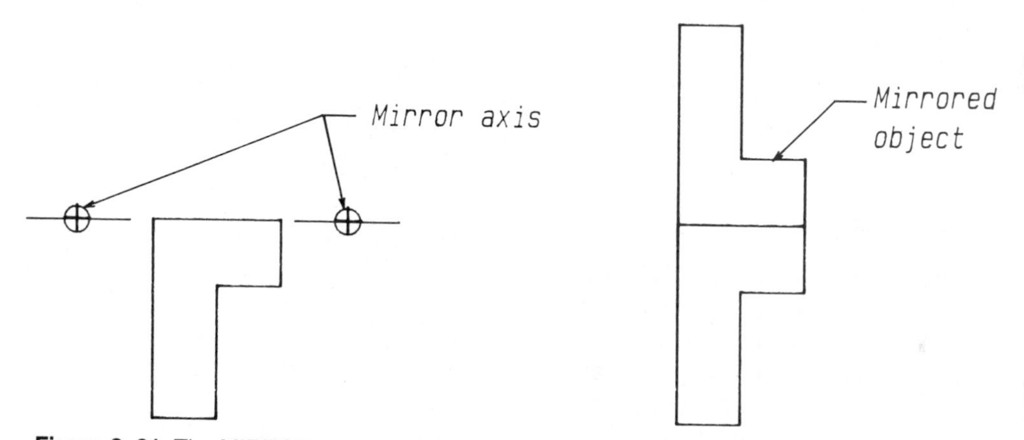

Figure 2–81 The MIRROR command can be used to reflect a feature.

Grid. The GRID command can be used to do a couple of things. The first is to paint a grid of dots, or tick marks, on the screen in any increment needed, be it millimeters, decimal inches, or feet. A grid snap or grid roundoff can also be set which snaps any digitized point, line, or feature to a grid intersection.

Image that you have grid lines on the screen that are 2 in. apart. Any point that you pick on the tablet will be moved to the nearest 2-in. grid intersection. Figure 2–82 illustrates this process. Try to digitize a point along a horizontal line from point A. Purposely move the pointer up somewhat from the intended line to point B1. If your new point is less than 1 in. vertically from the previous point it will be snapped to the previous Y coordinate, point B. If your second point, B2, was greater than an inch above the previous Y coordinate, the new point will be snapped to the next grid intersection, point C. This is a great aid in drawing straight lines and aligning features.

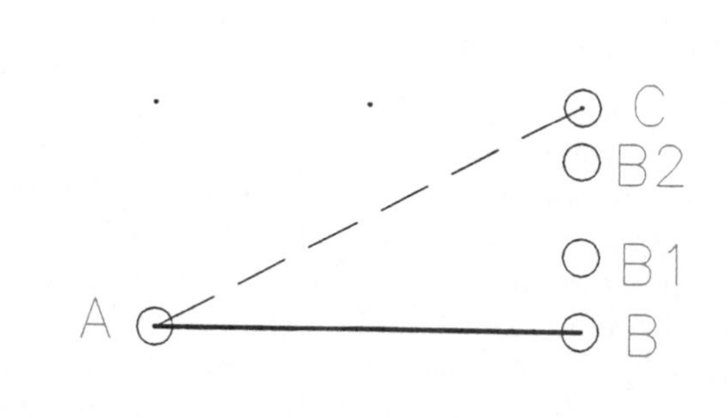

Figure 2–82 Example of a grid snap.

Viewpoint. After an object has been drawn, we may want to stand back and view it from a different angle. The view may be either orthographic or three-dimensional. Of course, you remember that a 3–D view can only be displayed if all of the coordinate points of the object have been entered into the computer. Systems that have the ability to rotate objects in 3–D open an exciting world of possibilities for the CADD drafter. It's not only exciting, but fun to watch an object rotate on the screen in front of you. Some systems enable the operator to rotate an object with a group of arrows on the menu that point in different directions and represent the different axes of rotation. Pressing one of these boxes a click at a time moves the object on the screen in the indicated direction. Other systems may employ a type of function box with knobs that can be turned in the desired direction of rotation. Other knobs on this box can be used to zoom in on and away from the object.

These are but a few of the specialized commands and functions found in different CADD systems. Each system will have commands similar to the ones discussed. As you learn the capabilities of your CADD system you will probably discover uses for these commands that are not mentioned here or in the documentation for the system itself. That is one of the exciting aspects of computer aided drafting: exploring the creative possibilities locked within those microscopic chips.

CREATING AND USING SYMBOLS

One of the great time-saving features of CADD technology is the ability to draw a feature once, save it, and then use it over and over without ever having to draw it again. That is why one of the first things a company does after purchasing a CADD system is to begin developing a symbol library, cell library, or symbol directory. This is a file of symbols that can be called up, located on the menu tablet, and used on a drawing. The area for a symbol library is shown on the menu in Figure 2–83.

Drawing a Symbol

The process of creating a symbol is a simple one. Begin a drawing with the name of the symbol as the file name. When you start the actual drawing of the symbol, begin by placing the origin of your coordinate system at a spot on the symbol that would be convenient to use as a point of reference or location on a drawing. Note the symbols in Figure 2–84. They all have small circles at their location points. This is the spot that you point to on your drawing to locate the symbol.

When the symbol has been drawn it must then be stored, but in a slightly different manner from a regular drawing. It is a symbol and not a drawing, and must be stored as such. Commands such as BLOCK and CREATE SYMBOL enable the drafter to save the drawing as a symbol. This allows the symbol to be used in ways that a regular drawing cannot be used.

Creating a Symbol Library

After all of the required symbols have been drawn they can be used in a variety of ways. It is important that all the symbols or blocks be located in a common area, such as a floppy disk, or a directory on the hard disk drive. The simplest is to just insert or merge them with an existing drawing, calling the symbol up by its file name. If you have the ability to customize the software's menus, the symbols can be added to new screen or pull-down menus for easy access. In addition, they can be added to an area of the tablet menu as discussed previously. These methods of locating the symbols in menus allow them to be picked and then placed in the drawing, thus eliminating the need to type file names.

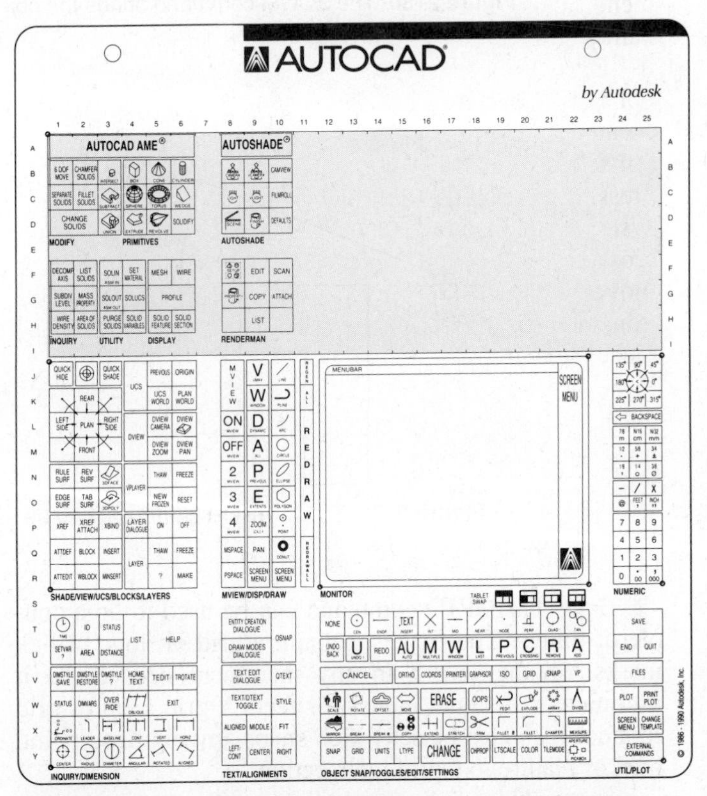

Figure 2–83 The upper portion of this menu overlay houses the symbol library. *Courtesy Autodesk, Inc.*

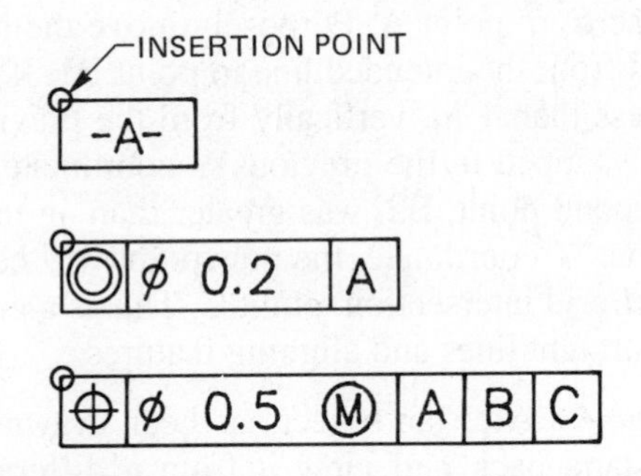

Figure 2–84 Location or insertion points on symbols.

If symbols are placed on a digitizer or tablet menu, the final task would be to have all the symbols on one template plotted in a pattern that matches their layout on the template. This can then be used as an overlay, or mask on the menu, which would look similar to those shown in Figure 2–85.

Using Symbols

Most of the editing features that are used to revise drawings can be used to alter symbols. In addition, there are several commands that enable the drafter to work with the symbols. The entire symbol template can be selected for use by choosing a command such as MENU. Another method of choosing a specific symbol template is to have all of the template names displayed on the menu tablet. Then it is just a matter of pointing to the name of the template that is needed.

One of the first things that can be done with a symbol is to select it. This is easily done by pressing the pointer inside the particular symbol box. With the symbol selected, it can be located on the drawing by simply pressing the pointer to the spot where it is desired. Some systems may automatically display the symbol at the crosshairs location. As the puck is moved, the symbol moves with it. When the symbol is in the desired location, simply press a button on the puck and the symbol is placed in that spot.

SYMBOL LIBRARY: PIPING

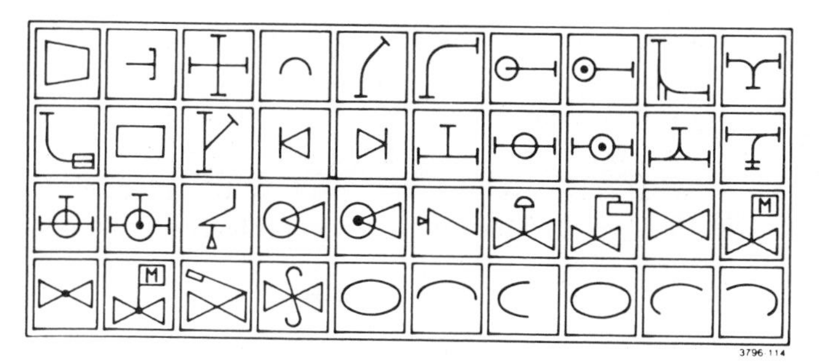

SYMBOL LIBRARY: SCHEMATIC

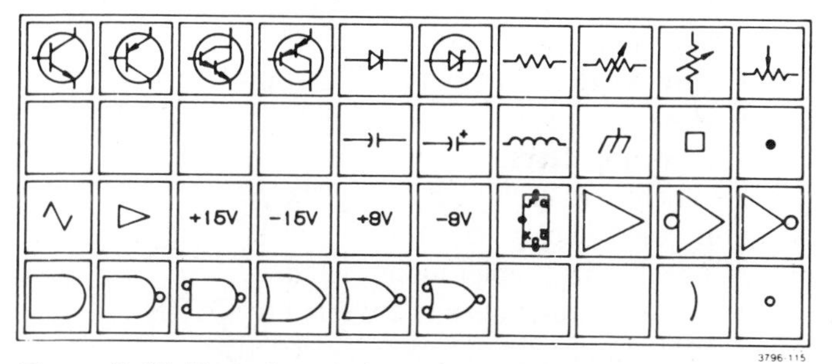

Figure 2–85 Typical symbol template overlays.

Symbols are fluid just like the rest of the drawing, and can be changed, moved, renamed, and copied. Commands such as ROTATE, COPY, MIRROR, RENAME, DELETE, and MOVE can all be used to work with symbols. These commands may be located in a group on the menu near the symbol template. It is possible that the symbol was drawn at a certain size, but once placed on the drawing it is too small. That is no problem because the size of the symbol can be changed using a SCALE, or SYMBOL SCALE command.

STORING A DRAWING

All drafters need a place to store their drawings. Most companies have large flat files in which drawings are stored. When a print is needed, the drawing is removed and run through the diazo machine and then returned to the file. Should you wish to make changes to your original you remove it from the file, take it to your drafting board, make the corrections, then make a print, and refile the original drawing. The storage file for the computer drafting system is the disk, or flexible disk. For permanent storage the magnetic tape is the storage file. The disk drive shown in Figure 2–86 does the actual storing of your drawing on the disk. It can also read the data from the disk and send it to the computer.

The act of storing a drawing that is on the screen is handled by menu commands such as SAVE, and END. After selecting the proper command you may be given a chance to change the file name of the drawing. After that, a press of the RETURN, or ENTER, key sends the drawing on its way to the disk. Some systems eliminate the need for the operator to store the drawing, because they are constantly sending the drawing data to the disk. This is a great feature because it removes the consequences of a computer crash, or malfunction. Should the

Figure 2–86 A disk pack is being inserted into a disk drive.

computer go down (malfunction) during the course of work on a drawing, nothing would be lost except the command that was in progress.

How often should a CADD operator store a drawing? This could be a matter of personal preference or company policy. Practice varies, but it is a good habit to save your work every 10 to 15 minutes, just in case something happens to the computer.

PROFESSIONAL PERSPECTIVE

Working with a CADD program can be exciting, frustrating, and fun. The frustration may come in the early stages of learning on a CADD system. It results from looking at the screen too long, trying to remember new commands and techniques, and working at the computer too many hours because it is so fascinating. If you have never worked with a computer before to create drawings, you are in for a lot of fun.

Be prepared to be challenged by the complexity of CADD programs, the speed of the computer, and the program's desire for information. You will be communicating with the computer. It will ask you questions and prompt you for information. You will need to think quickly and know your next move. Thorough planning should be a part of drawing with the computer. This means even drawing a freehand sketch before you sit down at the computer. You can meet the challenge by organizing your work and using school or company standards.

Keep in mind that when you work with a CADD system, you are working with a machine that is highly predictable, but occasionally frustrating. The computer will talk back to you and demand information. It can occasionally do things you don't expect, but seldom does it do anything that you did not tell it to do. It allows you to be extremely productive and creative. And it can eat your drawing in a split second. So save your work regularly!

There are abundant jobs for qualified CADD operators. The best operators are those who have good organizational and planning skills, good drawing skills, and know the capabilities of the CADD system with which they were trained. The best jobs will be obtained by those students who can use the computer's commands effectively and who know where and how to find the information they need quickly. They must develop an intimate knowledge of the reference manuals and on-screen help available with the CADD program. In addition, those students who possess the skills needed to customize the CADD software to suit their specific needs will be in high demand by employers. Your portfolio should contain diskettes of your best drawings in all specialty fields of drafting and engineering. An excellent presentation for these drawings is in the form of a continuously running "slide show" that automatically displays your drawings.

ENGINEERING PROBLEM

Suppose you were asked to: a) Evaluate CADD software packages to determine the best choices for your company; and b) Evaluate microcomputers and peripherals to run the CADD software. What would you do? There are no "right" choices, only a lot of different ones, all of which could do the job.

One solution to this problem is to set out with an organized plan, and most important, stick to it. The following lists contain items that you should keep in mind when evaluating and recommending CADD software and hardware.

Software

- What kinds of drawings does the company create?
- Are the drawings the company's product?
- Are the drawings used in the manufacture of a product?
- How complex are the drawings?
- Is 3–D required for a majority of the drawings?
- Are client drawings used and edited?
- Do the drawings have to be sent to a client or contractor?
- If the previous two items were answered "yes," what kind of system is used by the clients?
- How much money is available to spend?
- How many packages must be purchased?
- If more than ten packages must be purchased, does the software manufacturer provide site licenses?
- How good is the software documentation?
- Is quality training available? If so, is it local, or must you travel?
- How often are software upgrades issued?
- What is the average cost of software upgrades?
- How good is the software support? Is there a hotline number?

Hardware

- How much money is available to spend?
- How many workstations does the company want?
- How many workstations does the company need?
- How large are the typical drawings?
- How much space is available for hardware?
- How important is speed in the computer?
- Where will archive drawings be stored?
- Is color display required?
- Are digitizer tablet menus needed?
- Will old drawings be converted to CADD?
- What kind of paper copy is needed?
- Will paper copies be reproduced? If so, how?

- What size of flexible disk is needed?
- If client drawings are used, how are they received — by disk or by phone modem?
- How many people will be using the system?
- Do people in the office work with overlay drafting now?
- Do several people use the same drawing, or copies of the same base?
- Is interoffice communication important?
- Is it important that several employees be made aware of any updates to project drawings?
- What are the hardware requirements of the CADD software?

These lists include a sampling of the many things to be considered and decisions that must be made before a CADD system is purchased. Unfortunately, many companies rush into the purchase of a system without answering even a few of these questions. This eventually leads to headaches that could have been avoided. Remember, the more planning that is involved at the beginning of any project, the less time will be required to execute the project.

INTRODUCTION TO CADD PROBLEMS

Problem 2–1 This problem involves writing, not drawing. Choose one of the topics listed below and write a report on that subject. Should you choose to write a report on the topic 1. CADD system, select one manufacturer and focus your report on that specific brand of system. Be specific, especially if the manufacturer makes more than one model of CADD system. Check with your instructor about any specifics relating to your report. Reference materials can be trade magazines and journals, textbooks, company brochures and instruction manuals, telephone interviews, personal interviews, field trips and attendance at seminars, workshops, conferences and conventions.

1. CADD system (Choose a specific brand and model)
2. Computer: mainframe, mini, or micro
3. Input devices (hand-held)
4. Digitizer, graphics tablet
5. Output devices: plotter, printer, display screen (monitor or terminal)
6. Storage devices: disk drive, tape drive
7. Drawing media: paper, film, pens
8. Storage media: disks, tape

Problem 2–2 Write a short report that discusses the differences among pen plotters, electrostatic plotters, dot-matrix printers, and laser printers. You may use any sources that can provide the needed information, such as textbooks, magazines, advertising brochures, demonstrations, interviews, and personal experiences. Be sure to discuss the following points in your report:

- Type of media required (paper, pens, etc.)
- Media size
- Price
- Quality of final product
- Usefulness in your application
- Warranties and service contracts
- Durability
- Reputation
- Length of time on the market

Problem 2–3 Create a specification sheet for purchasing a single CADD workstation and the software to run it. You can do this as a report, a form, or a list. The assignment should include the following pieces of equipment and software:

- IBM compatible microcomputer — 486 — 33Mhz — 200Mb hard disk — 2 1.2Mb flexible disks — VGA color graphics card — 2 serial ports
- 13" color monitor
- 11"x11" digitizer tablet with a 12-button cursor
- B-size, 8-pen plotter
- MS-DOS software — latest version
- CADD software (AutoCAD, VersaCAD, etc.) — latest version

Keep the following things in mind when creating your specification list:

- Contact at least three computer hardware vendors for prices.
- Provide each vendor with the same specifications.
- Do not specify brand names of equipment unless instructed to do so.
- Determine the brand of CADD software you want before obtaining prices.
- You may have to go to different vendors for the CADD software.

Problem 2–4 Using existing equipment, or vendor's catalogs, measure the equipment required for a microcomputer CADD workstation (computer, monitor, keyboard, and pointing device), and design a workstation table. Sketch the table first, then draw it manually or on your CADD system.

After drawing the table, create a computer lab arrangement of from 12 to 20 CADD workstations using your table design. Consider the following when designing the lab:

- Chalkboard or marker board location
- Glare from windows
- Aisles and access
- Power connections
- Location of pen plotters or printers
- Storage for paper, plotter pens, and supplies
- Instructor podium or workstation
- Projection system for instructor workstation (use overhead projector)

Problem 2–5 Construct a comparison sheet on three pen plotters, each made by a different manufacturer, that are all in the same price range. The plotters should have the following features:

- Hold up to D-size paper
- Hold 8 pens, either felt tip or wet ink. Interchangeable pen carousels are acceptable.
- Portable (mounted on rollers)
- Allow manual pen control
- Allow for viewing the drawing in the middle of a plot (pause)
- Cost less than $6000

Check with local computer vendors for brochures and specifications. Read computer magazines and journals for plotter advertisements and articles that compare and rate plotters. Arrange for demonstrations by vendors or companies that use pen plotters. Plotters can have features additional to those mentioned above, or you can delete and add features to the list as desired or instructed.

SOLUTION TO EXERCISE 2–1 (PAGE 61.)

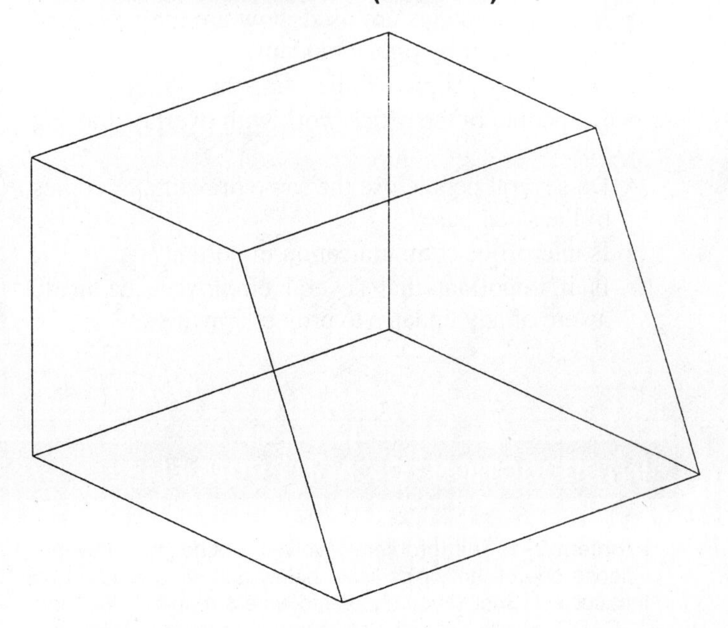

CHAPTER 3

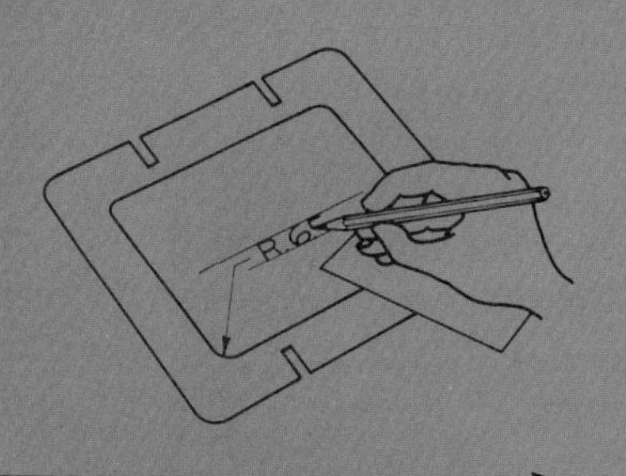

Sketching, Lettering, and Lines

LEARNING OBJECTIVES

After completing this chapter you will:

- Sketch lines, circles, arcs and multiviews.
- Do freehand lettering, and use mechanical lettering equipment.
- Use a CADD system to prepare text.
- Draw ANSI standard lines using manual drafting and computer-aided drafting.
- Solve an engineering problem using manual and computer-aided drafting.

SKETCHING

Sketching is freehand drawing, that is, drawing without the aid of drafting equipment. Sketching is convenient since all that is needed is paper, pencil, and an eraser. There are a number of advantages and uses for freehand sketching. Sketching is fast visual communication. The ability to make an accurate sketch quickly can often be an asset when communicating with people at work or at home. Especially when technical concepts are the topic of discussion, a sketch may be the best form of communication. Most drafters prepare a preliminary sketch to help organize thoughts and minimize errors on the final drawing. The computer operator usually prepares a sketch on graph paper to help establish the coordinates for drawing components. Some drafters use sketches to help record the stages of progress when designing until a final design is ready for implementation into formal drawings. A sketch can be a useful form of illustration in technical reports. Sketching is also used in job shops where one-of-a-kind products are made. In the job shop, the sketch is often used as a formal production drawing. When the drafter's assignment is to prepare working drawings for existing parts or products, the best method to gather shape and size description about the project is to make a sketch. The sketch can be used to quickly lay out dimensions of features for later transfer to a formal drawing.

TOOLS AND MATERIALS

Sketching equipment is not very elaborate. As mentioned, all you need is paper, pencil, and an eraser. The pencil should have a soft lead; a common number 2 pencil works fine. A mechanical pencil with an F or HB lead is also good, as is an automatic 7- or 9-mm pencil with F or HB lead. The pencil lead should not be sharp. A dull, slightly rounded pencil point is best. Different thicknesses of line, if needed, can be drawn by changing the amount of pressure you apply to the pencil. The quality of the paper is not too critical either. A good sketching paper is newsprint, although almost any kind works. Actually, paper with a surface that is not too smooth is best. Many engineering designs have been created on a napkin around a lunch table. Sketching paper should not be taped down to the table, so there is no need for tape. The best sketches are made when you are able to move the paper to the most comfortable drawing position. Some people make horizontal lines better than vertical lines. If this is your situation, then move the paper so that vertical lines become horizontal. Such movement of the paper may not always be possible, so it does not hurt to keep practicing all forms of lines for best results.

STRAIGHT LINES

Lines should be sketched in short, light, connected segments as shown in Figure 3–1. If you sketch one long stroke in one continuous movement, your arm tends to make the line curved rather than straight, as shown in Figure 3–2. Also, if you make a dark line, you may have to erase if you make an error, whereas if you draw a light line there often is no need to erase.

Following is the procedure used to sketch a horizontal straight line with the dot-to-dot method:

Step 1 Mark the starting and ending positions, as in Figure 3–3. The letters A and B are only for instruction. All you need are the points.

Step 2 Without actually touching the paper with the pencil point, make a few trail motions between the marked points to adjust the eye and hand to the anticipated line.

Step 3 Sketch very light lines between the points by stroking in short light strokes (2–3 in. long). Keep one eye directed toward the end point while keeping the other eye directed on the pencil point. With each stroke, an attempt should be made to correct the most obvious defects of the preceding stroke so that the finished light lines are relatively straight. (See Figure 3–4.)

Step 4 Darken the finished line with a dark, distinct, uniform line directly on top of the light line. Usually the darkness can be obtained by pressing on the pencil. (See Figure 3–5.)

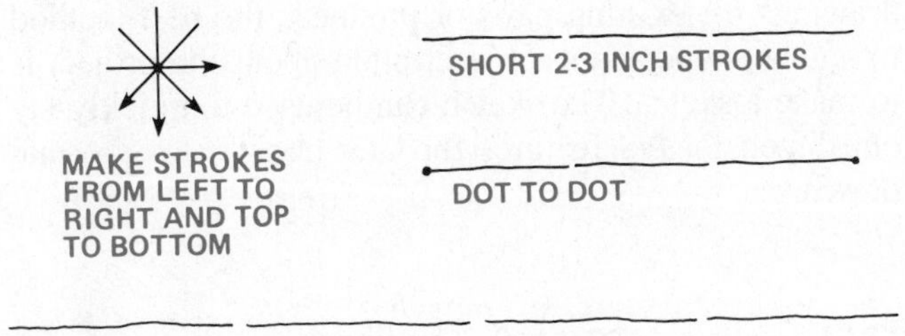

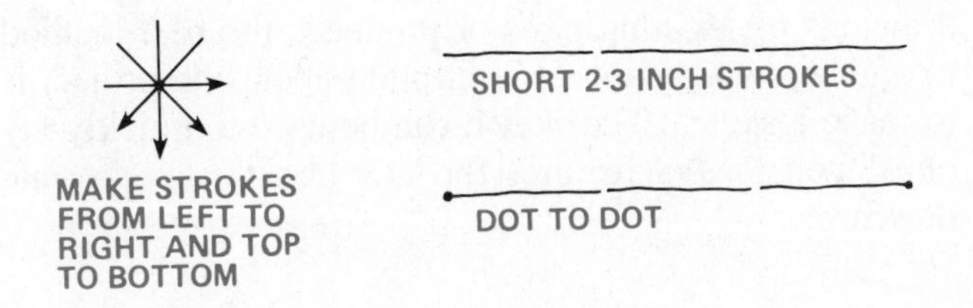

Figure 3–1 Sketching short line segments.

Figure 3–2 Long movements tend to cause a line to curve.

A • • B

Figure 3–3 Use dots to identify both ends of a line — step 1.

A • • B

Figure 3–4 Use short, light strokes — step 3.

PROFESSIONAL HINT

1. Start with light lines so corrections can be made without erasing.
2. Do not tape down your paper so you can turn it to different angles for easier sketching.

CIRCULAR LINES

Figure 3–6 shows the parts of a circle. There are several sketching techniques to use when making a circle; this text explains the trammel and hand-compass methods.

Trammel Method

Step 1 Make a trammel to sketch a 6-in. diameter circle. Cut or tear a strip of paper approximately 1 in. wide and longer than the radius, 3 in. On the strip of paper, mark an approximate 3-in. radius with tick marks such as A and B in Figure 3–7.

Step 2 Sketch a straight line representing the circle radius at the place where the circle is to be located. On the sketched line, locate with a dot the center of the circle to be sketched. Use the marks on the trammel to mark the other end of the radius line as shown in Figure 3–8. With the

Figure 3–5 Darken to finish the line — step 4.

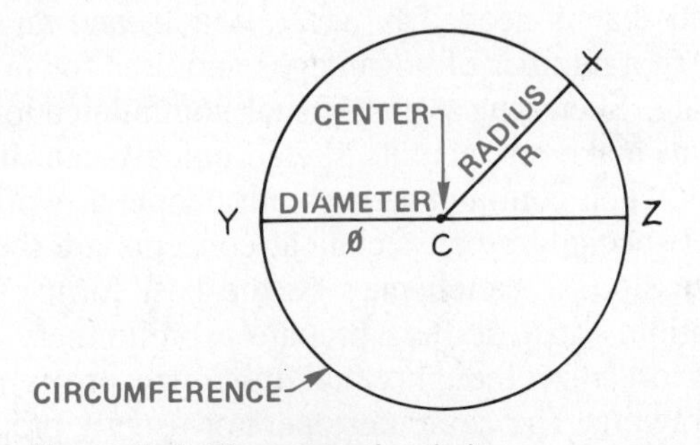

Figure 3–6 The parts of a circle.

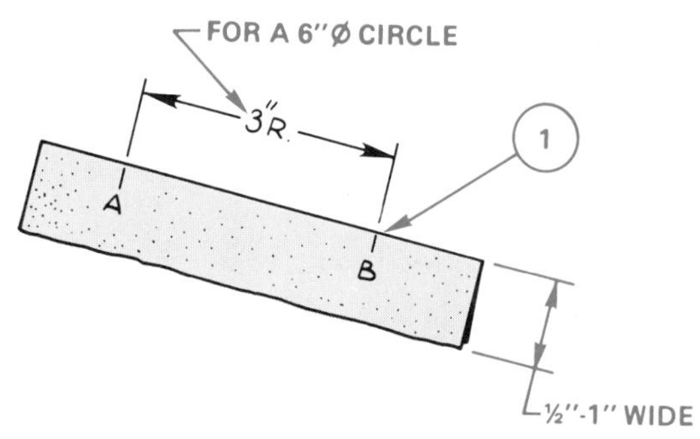

Figure 3–7 Make a trammel — step 1.

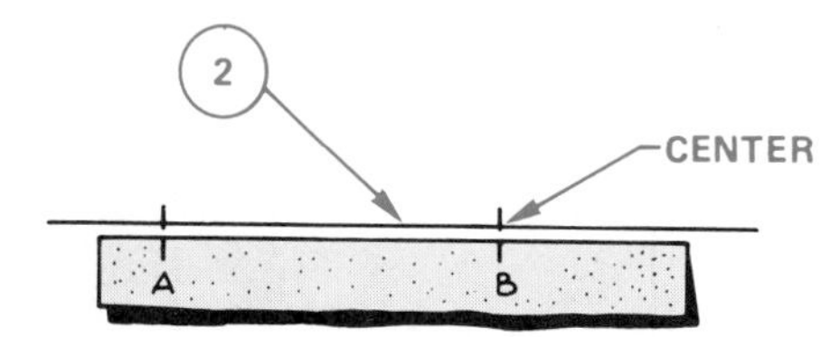

Figure 3–8 Locate the center of the circle — step 2.

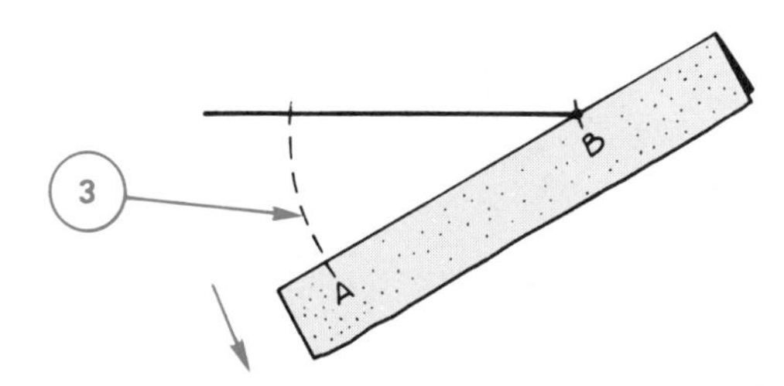

Figure 3–9 Begin the circle construction — step 3.

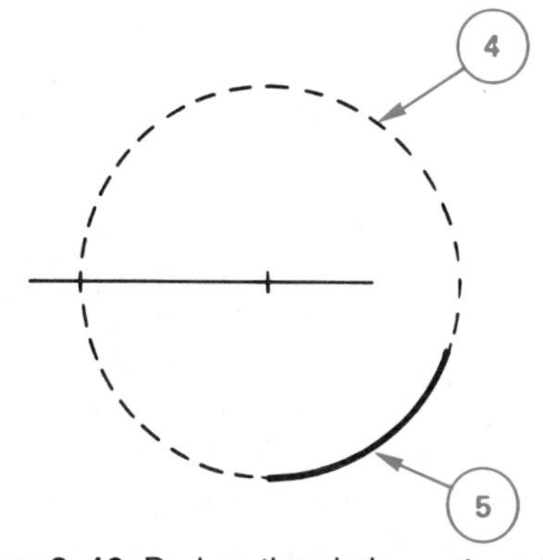

Figure 3–10 Darken the circle — steps 4 and 5.

trammel next to the sketched line, be sure point B on the trammel is aligned with the center of the circle you are about to sketch.

Step 3 Pivot the trammel at point B, making tick marks at point A as you go, as shown in Figure 3–9, until the circle is complete.

Step 4 Lightly sketch the circumference over the tick marks to complete the circle, then darken it (step 5) as shown in Figure 3–10.

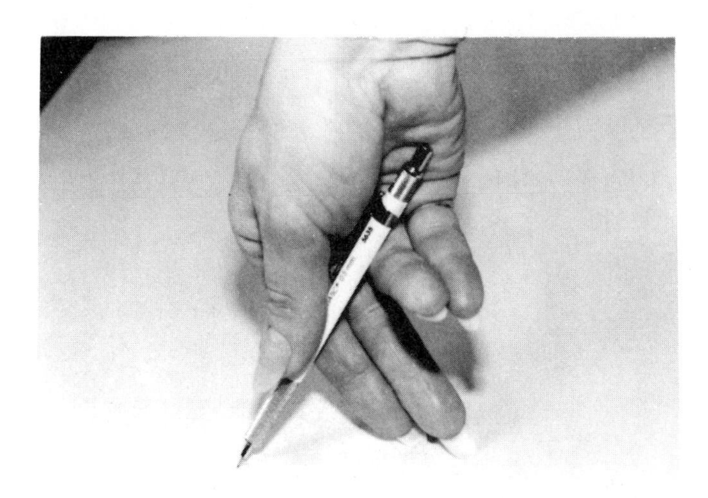

Figure 3–11 Holding the pencil in the hand compass — step 2.

Hand-Compass Method

The hand-compass method is a quick and fairly accurate method of sketching circles, although it is a method that takes some practice.

Step 1 Be sure that your paper is free to rotate completely around 360°. Remove anything from the table that might stop such a rotation.

Step 2 To use your hand and a pencil as a compass, place the pencil in your hand between your thumb and the upper part of your index finger so your index finger becomes the *compass point* and the pencil becomes the *compass lead.* The other end of the pencil rests in your palm as shown in Figure 3–11.

Step 3 Determine the circle radius by adjusting the distance between your index finger and the pencil point. Now, with the desired approximate radius established, place your index finger on the paper at the proposed center of the circle.

Step 4 With your desired radius established, keep your hand and pencil point in one place while rotating the paper with your other hand. Try to keep the radius steady as you rotate the paper. (See Figure 3–12.)

Step 5 You can perform step 4 very lightly and then go back and darken the circle or, if you have had a lot of practice, you may be able to draw a dark circle as you go.

Another method, generally used to sketch large circles, is to tie a string between a pencil and a pin. The distance between the pencil and pin is the radius of the circle. Use this method when a large circle is to be sketched, since the other methods may not work as well.

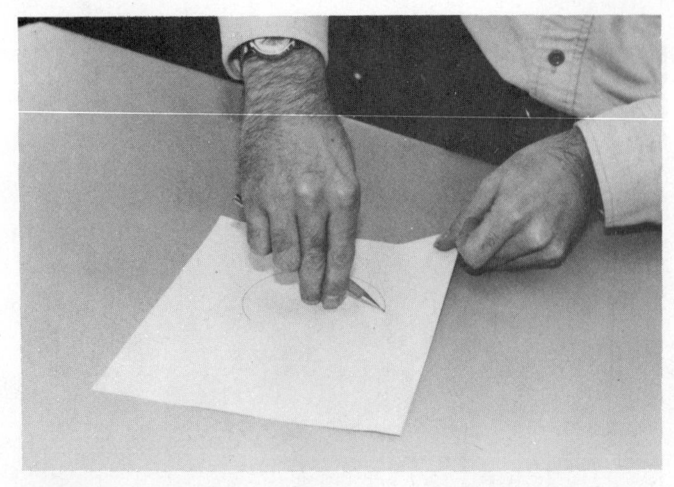

Figure 3–12 Rotate the paper under your finger center point — step 4.

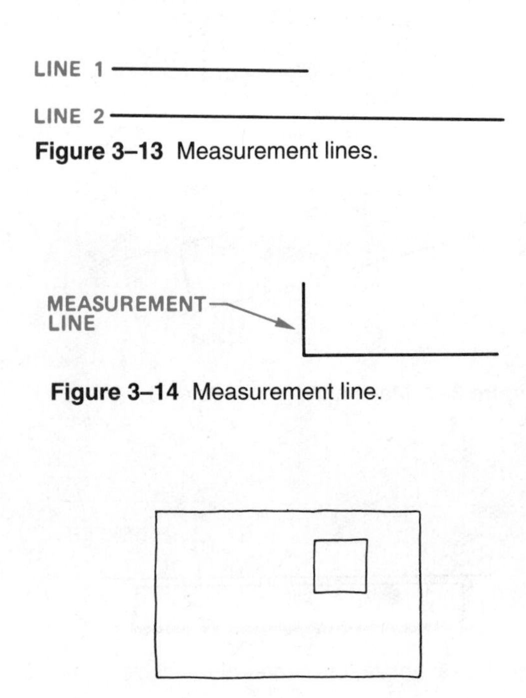

Figure 3–13 Measurement lines.

Figure 3–14 Measurement line.

Figure 3–15 Space proportions.

MEASUREMENT LINES AND PROPORTIONS

When sketching objects, all the lines that make up the object are related to each other by size and direction. In order for a sketch to communicate accurately and completely, it must be drawn in the same proportion as the object. The actual size of the sketch depends on the paper size and how large you want the sketch to be. The sketch should be large enough to be clear, but the proportions of the features are more important than the size of the sketch.

Look at the lines in Figure 3–13. How long is line 1? How long is line 2? Answer these questions without measuring either line, but instead relate each line to the other. For example, line 1 could be stated as being half as long as line 2, or line 2 called twice as long as line 1. Now you know how long each line is in relationship to the other (proportion), but we do not know how long either line is in relationship to a measured scale. No scale is used for sketching, so this is not a concern. So, whatever line we decide to sketch first will determine the scale of the drawing. This first line sketched is called the *measurement line*. Relate all the other lines in the sketch to that first line. This is one of the secrets of making a sketch look like the object being sketched.

The second thing you must know about the relationship of the two lines in the above example is their direction and position relative to each other. For example, do they touch each other, are they parallel, perpendicular, or at some angle to each other? When you look at a line ask yourself the following questions (for this example use the two lines given in Figure 3–14):

1. How long is the second line?
   a. same length as the first line?
   b. shorter than the first line? How much shorter?
   c. longer than the first line? How much longer?
2. In what direction and position is the second line related to the first line?

A typical answer to these questions for the lines in Figure 3–14 would be as follows:

1. The second line is about three times as long as the first line.
2. Line two touches the lower end of the first line with about a 90° angle between them.

Carrying this concept a step further, a third line can relate to the first line or the second line and so forth. Again, the first line drawn (measurement line) sets the scale for the whole sketch.

This idea of relationship can also apply to spaces. In Figure 3–15, the location of the hole can be determined by space proportions. A typical verbal location for the hole in this block might be as follows: the hole is located about one-half hole space from the top of the object or about two hole spaces from the bottom, and about one hole space from the right side or about three hole spaces from the left side of the object. All the parts must be related to the whole object.

Block Technique

Any illustration of an object can be surrounded with some sort of a rectangle overall, as shown in Figure 3–16. Before starting a sketch, see the object to be sketched inside a rectangle in your mind's eye. Then use the measurement-line technique with the rectangle, or block, to help you determine the shape and proportion of your sketch.

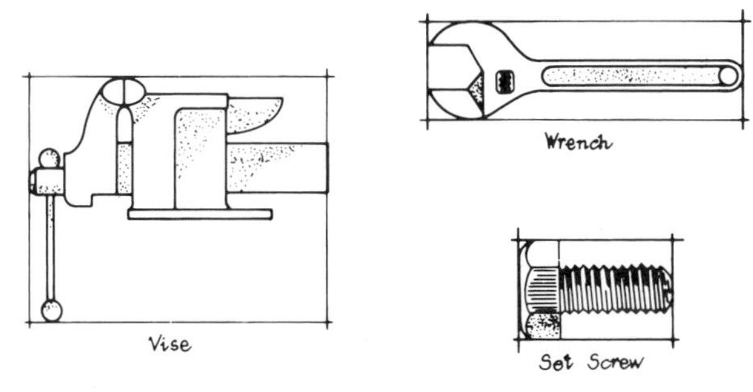
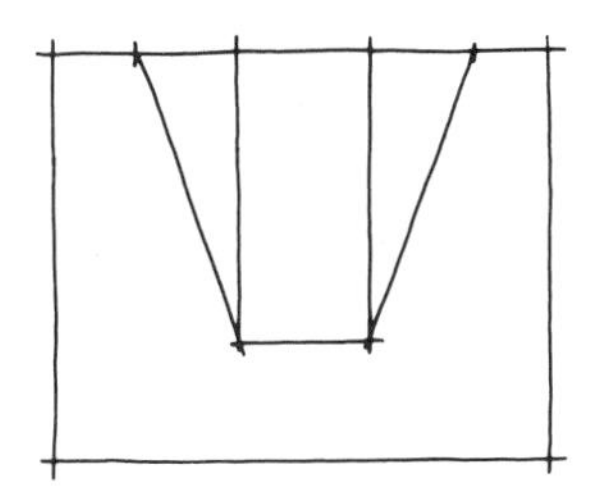

Figure 3–16 Block technique.

Figure 3–17 Outline the drawing area with a block — step 1.

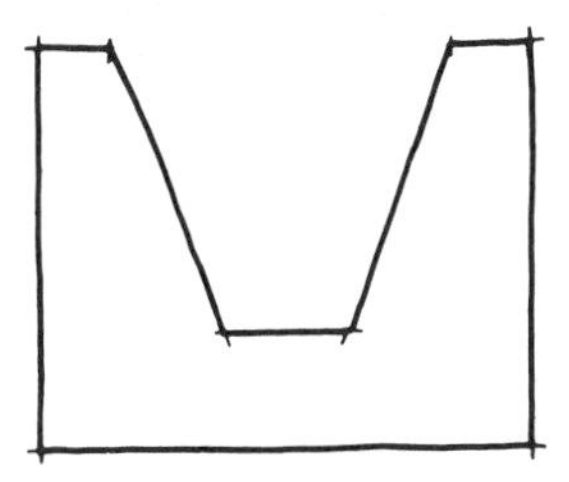

Figure 3–18 Draw features to proper proportions — step 2.

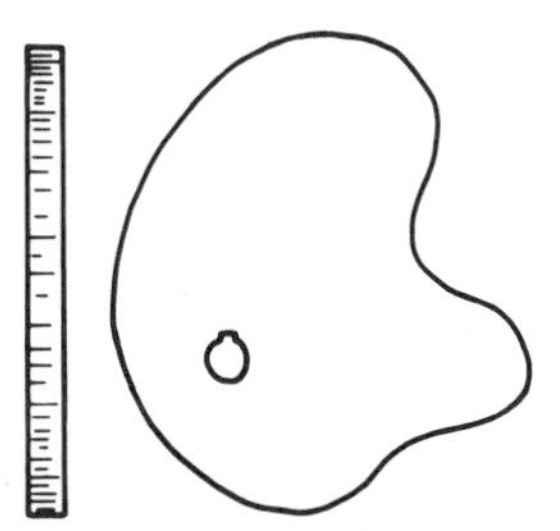

Figure 3–19 Darken the object lines — step 3.

PROCEDURES IN SKETCHING

Step 1 When starting to sketch an object, visualize the object surrounded with a rectangle overall. Sketch this rectangle first with very light lines. Sketch the proper proportion with the measurement-line technique, as shown in Figure 3–17.

Step 2 Cut sections out or away using proper proportions as measured by eye, using light lines, as in Figure 3–18.

Step 3 Finish the sketch by darkening in the desired outlines for the finished sketch. (See Figure 3–19.)

Irregular Shapes

By using a frame of reference or an extension of the block method, irregular shapes can be sketched easily to their correct proportions. Follow these steps to sketch the cam shown in Figure 3–20.

Step 1 Place the object in a lightly constructed box. (See Figure 3–21.)

Step 2 Draw several equally spaced or random horizontal, vertical, or diagonal lines that will cross or touch features of the object such as corners, and edges, as in Figure 3–22(a). Many drafters prefer the evenly spaced grid method shown in Figure 3–22(b). If you are sketching an object already drawn, just draw your reference lines on top of the object's lines to establish a frame of reference. If you are sketching an object directly, you have to visualize these reference lines on the object you sketch.

Figure 3–20 Cam.

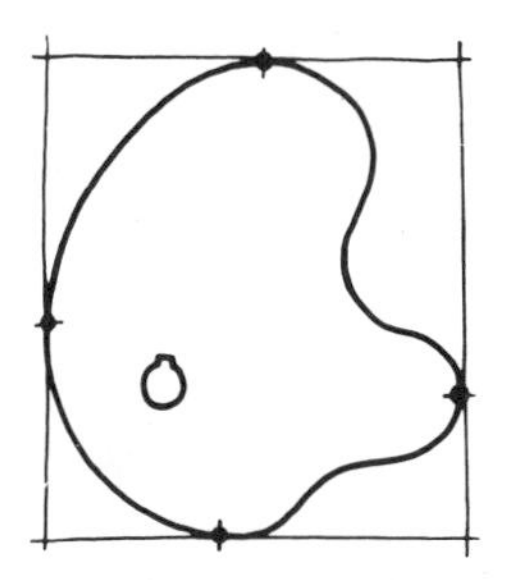

Figure 3–21 Box the object.

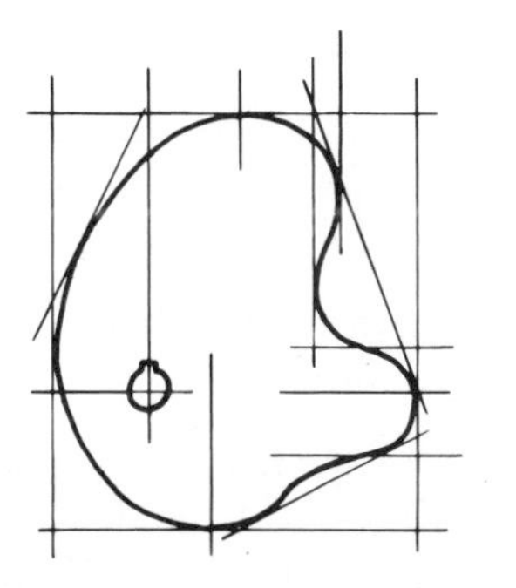
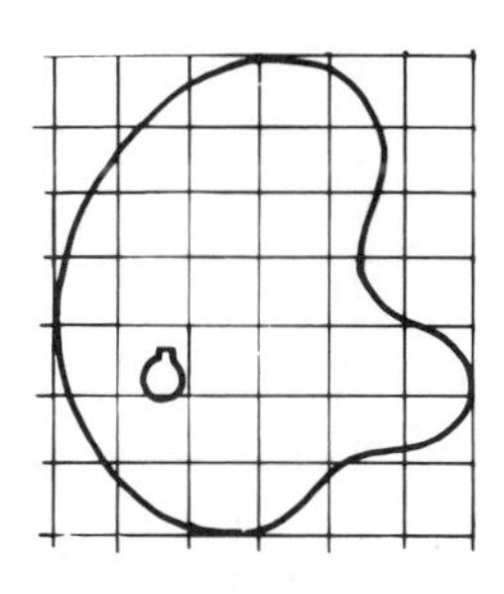

Figure 3–22 (a) Random-line grid; (b) evenly spaced grid.

Step 3 On your sketch, correctly locate a proportioned box similar to the one established on the original drawing or object, as shown in Figure 3–23.

Step 4 Using the drawn box as a frame of reference, include the grid or random lines in correct proportion, as seen in Figure 3–24(a) or (b).

Step 5 Then, using the grid, sketch the small irregular arcs and lines that match the lines of the original, as in Figure 3–25(a) or (b).

Step 6 Darken the outline for a complete proportioned sketch, as shown in Figure 3–26.

MULTIVIEW SKETCHES

Multiview projection is also known as orthographic projection. Multiviews are two-dimensional views of an object that are established by a line of sight that is perpendicular (90°) to the surface of the object. When making multiview sketches, a systematic order should be followed. Most drawings are in the multiview form. Learning to sketch multiview drawings will save you time when making a formal drawing.

Multiview Alignment

To keep your drawing in a standard form, sketch the front view in the lower left portion of the paper, the top view directly above the front view, and the right-side view to the right side of the front view. (See Figure 3–27. The views needed may differ depending upon the object. Multiview arrangement is explained in detail in Chapter 5.

Multiview Sketching Technique

Steps in sketching:

Step 1 Sketch and align the proportional rectangles for the front, top, and right side of the object given in Figure 3–28. Sketch a 45° line to help transfer width dimensions. The 45° line is established by projecting the width from the top view across and the width from the right-side view up until the lines intersect as shown in Figure 3–29.

Step 2 Complete the shapes by cutting out the rectangles, as shown in Figure 3–30.

Step 3 Darken the lines of the object as in Figure 3–31. Remember, keep the views aligned for ease of sketching and understanding.

Step 4 In the views where some of the features would be hidden, show those features with hidden lines, which are dashed lines as shown in Figure 3–32.

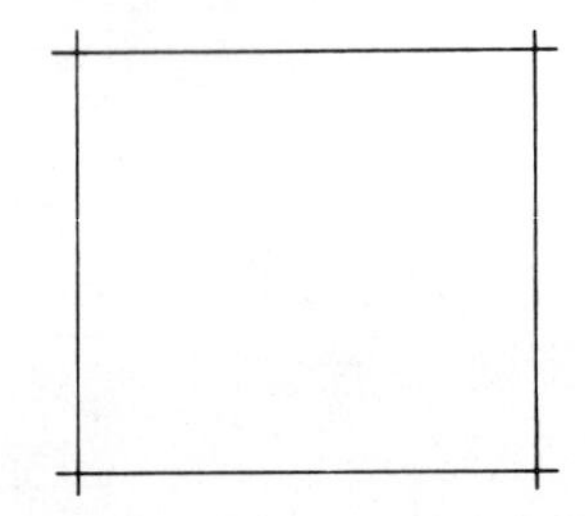

Figure 3–23 Proportioned box.

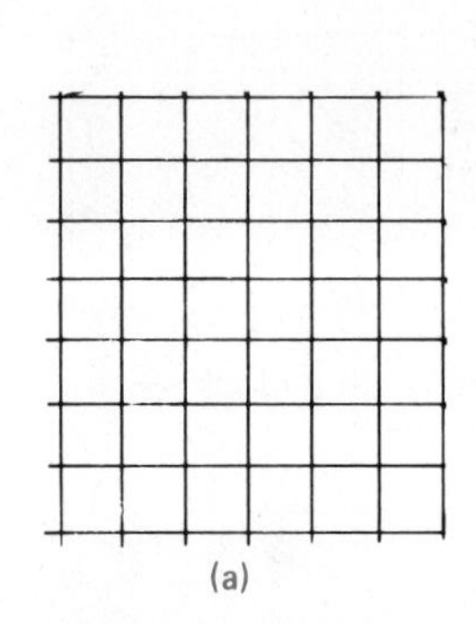

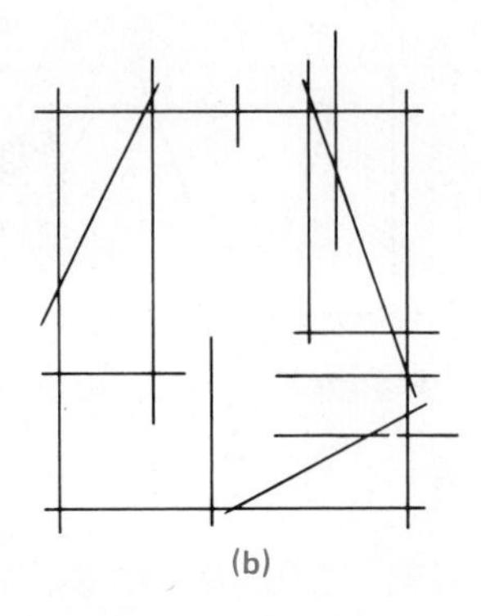

Figure 3–24 (a) Regular grid; (b) random grid.

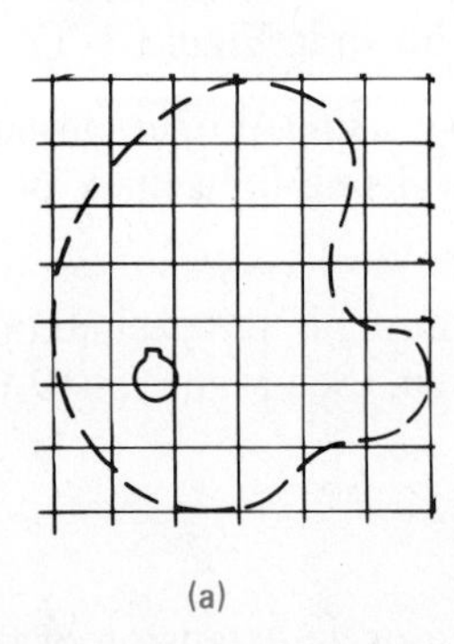

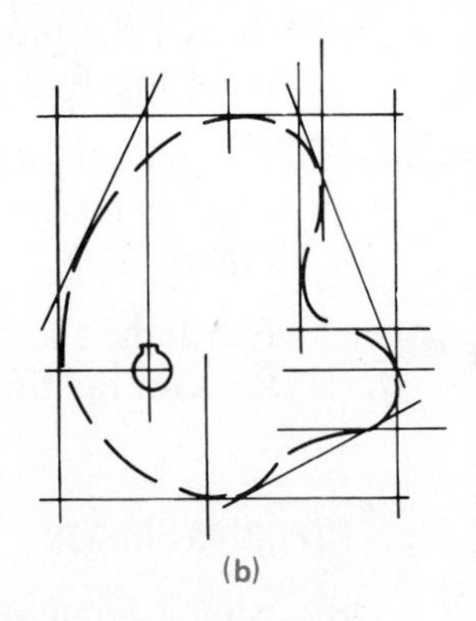

Figure 3–25 (a) Sketched shape using the regular grid; (b) sketched shape using a random grid.

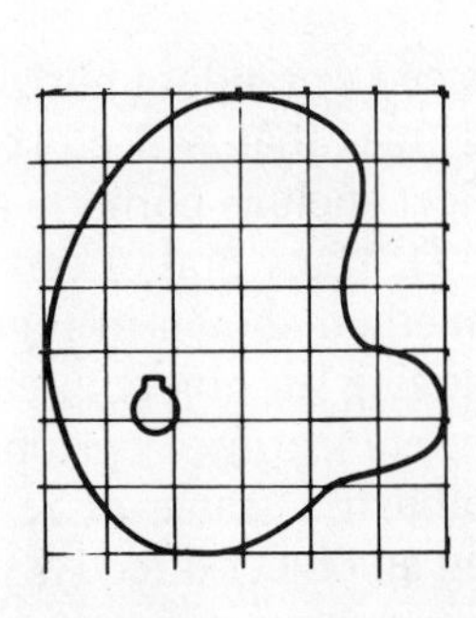

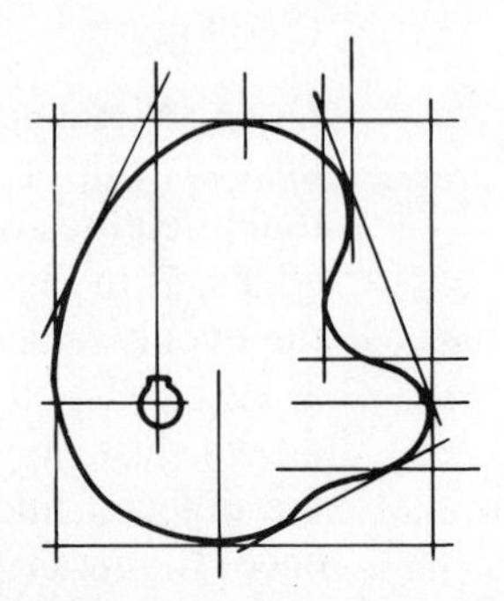

Figure 3–26 Completely darken the outline of the object.

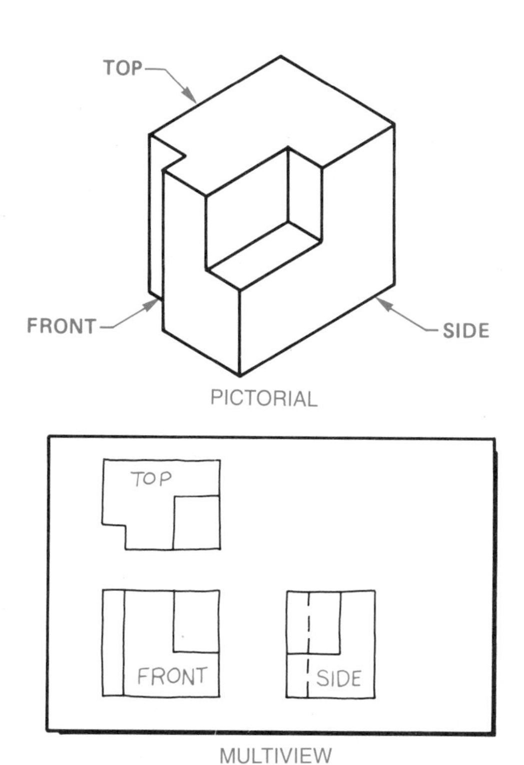

Figure 3–27 Views of the object shown in multiview and pictorial.

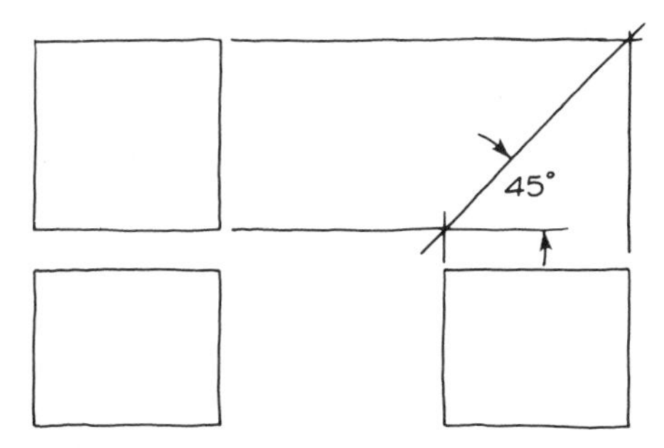

Figure 3–29 Block out views — step 1.

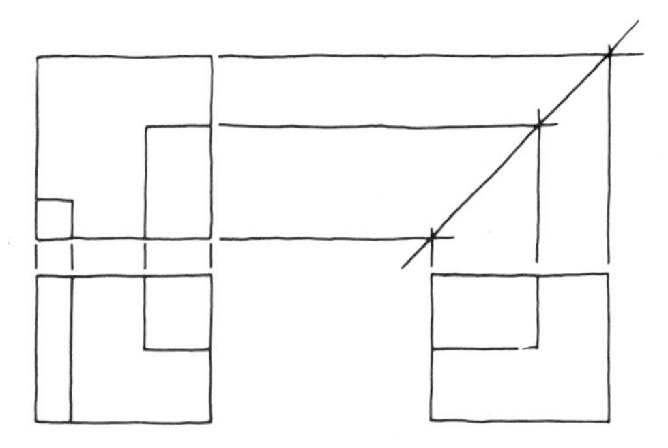

Figure 3–30 Block out shapes — step 2.

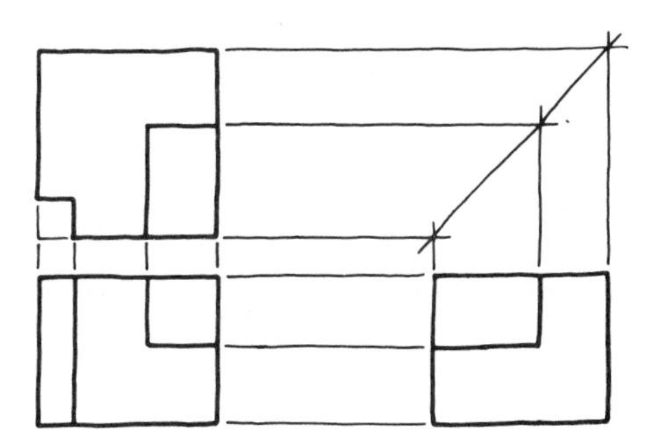

Figure 3–31 Darken all object lines — step 3.

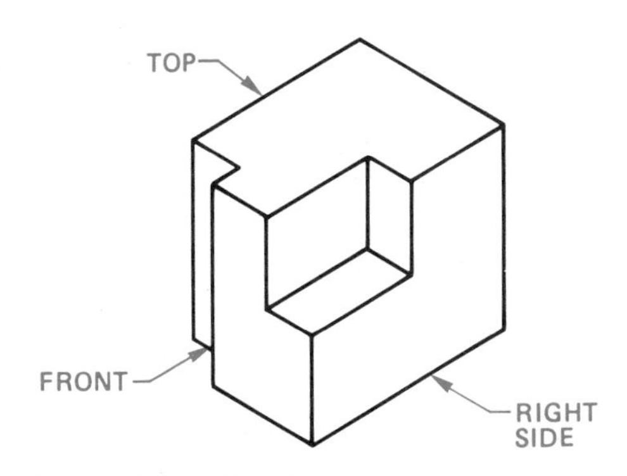

Figure 3–28 Pictorial view.

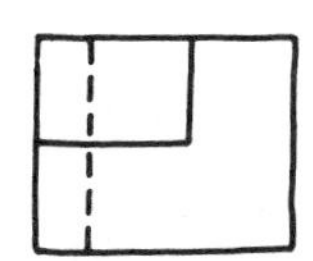

Figure 3–32 Draw hidden features — step 4.

ISOMETRIC SKETCHES

Isometric drawings provide a three-dimensional pictorial representation of an object. Isometric drawings are easy to draw and make a very realistic exhibit of the object. The surface features or the axes of the objects are drawn at equal angles from horizontal. Isometric sketches tend to represent the objects as they appear to the eye. Isometric sketches help in the the visualization of an object because three sides of the object are sketched in a single three-dimensional view. Chapter 14 covers isometric drawings in detail.

Establishing Isometric Axes

In setting up an isometric axis, you need four beginning lines: a horizontal reference line, two 30° angular lines, and one vertical line. Draw them as very light construction lines. (See Figure 3–33.)

Step 1 Sketch a horizontal reference line (consider this the ground-level line.)

Step 2 Sketch a vertical line perpendicular to the ground line and somewhere near its center. The vertical line will be used to measure height.

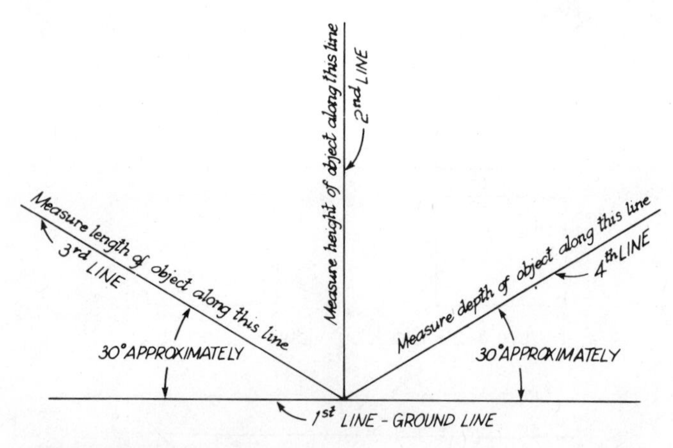

Figure 3–33 Isometric axis.

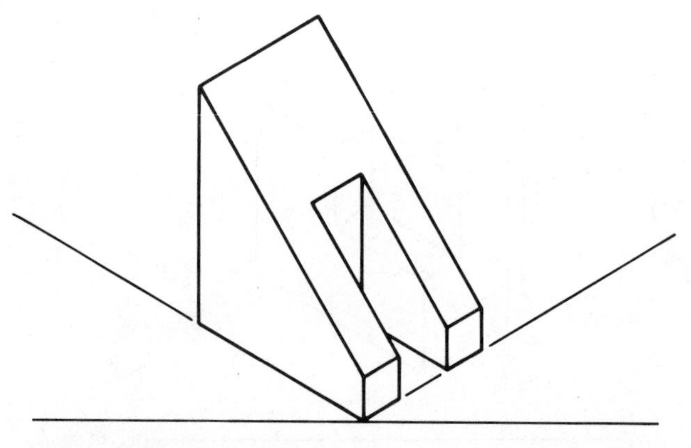

Figure 3–35 Given object.

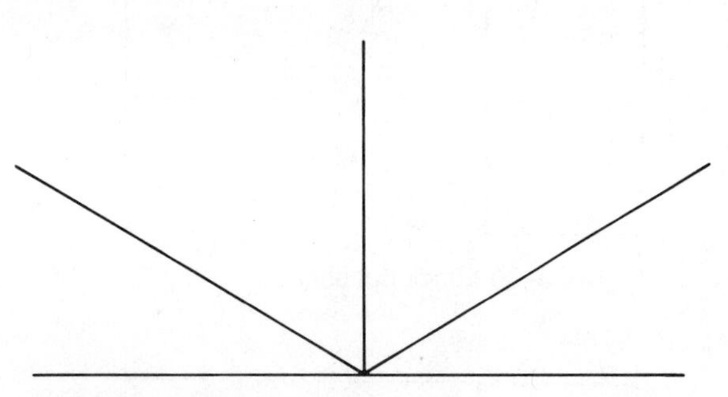

Figure 3–34 Sketch the isometric axis — step 3.

Figure 3–36 Imagine a box around the object — step 4.

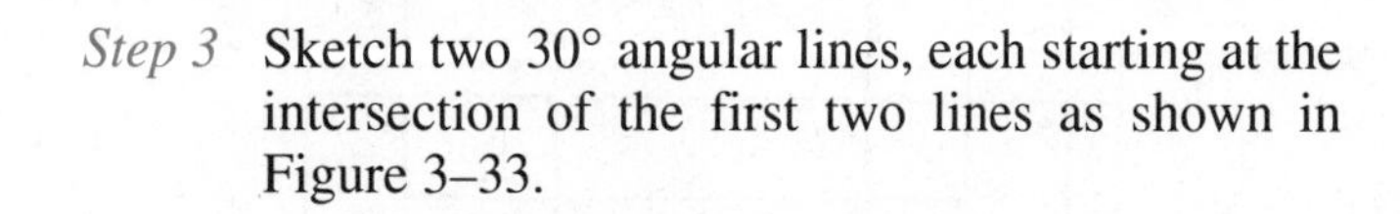

Step 3 Sketch two 30° angular lines, each starting at the intersection of the first two lines as shown in Figure 3–33.

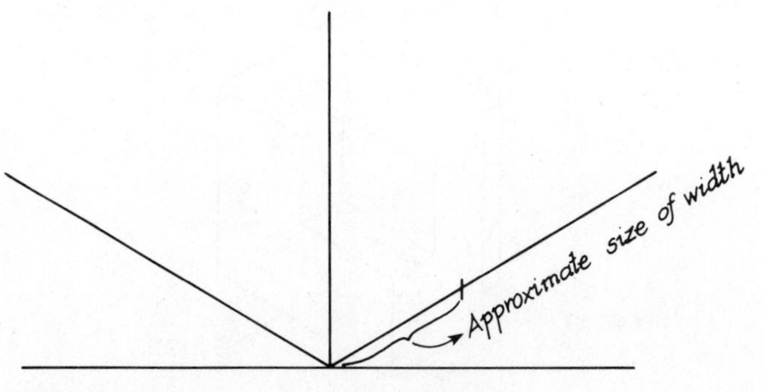

Figure 3–37 Lay out the width — step 4.

Making an Isometric Sketch

The steps in making an isometric sketch are as follows:

Step 1 Select an appropriate view of the object.

Step 2 Determine the best position in which to show the object.

Step 3 Begin your sketch by setting up the isometric axes. (See Figure 3–34.)

Step 4 By using the measurement-line technique, draw, using correct proportion, a rectangular box which could surround the object to be drawn. Use the object shown in Figure 3–35 for this explanation. Imagine the rectangular box in your mind, as in Figure 3–36. Begin to sketch the box by marking off the width at any convenient length (measurement line), as in Figure 3–37. Next estimate and mark the length and height as related to the measurement line. (See Figure 3–38.) Sketch the three-dimensional box by using lines parallel to the original axis lines. (Figure 3–39.) Sketching the box is the most critical part of the construction. It must be done correctly; otherwise your sketch will be out of proportion. All lines drawn in the same direction must be parallel.

Step 5 Lightly sketch in the slots, holes, insets, and other features, which define the details of the object. By estimating distances on the rectangular box, the features of the object are easier to sketch in correct proportion than trying to draw them without the box. (See Figure 3–40.)

Step 6 To finish the sketch, darken all the object lines (outlines), as in Figure 3–41. For clarity, do not show any hidden lines.

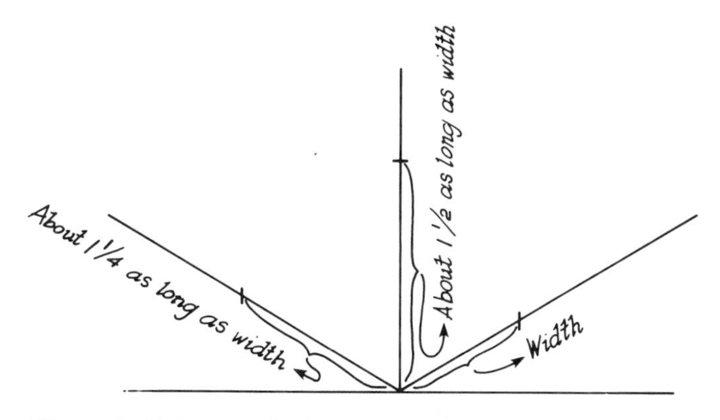

Figure 3–38 Lay out the length and height — step 4.

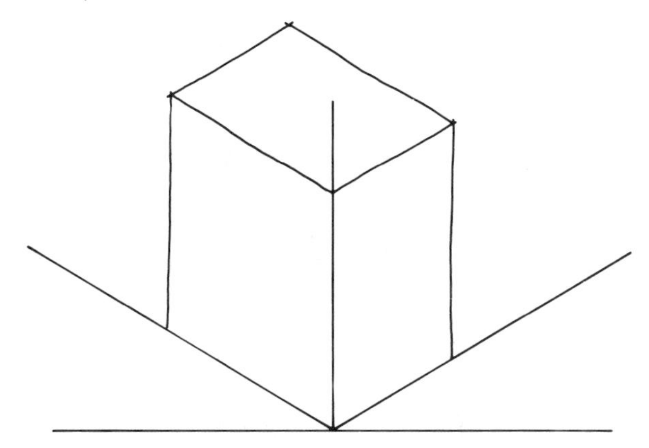

Figure 3–39 Sketch out the three-dimensional box — step 4.

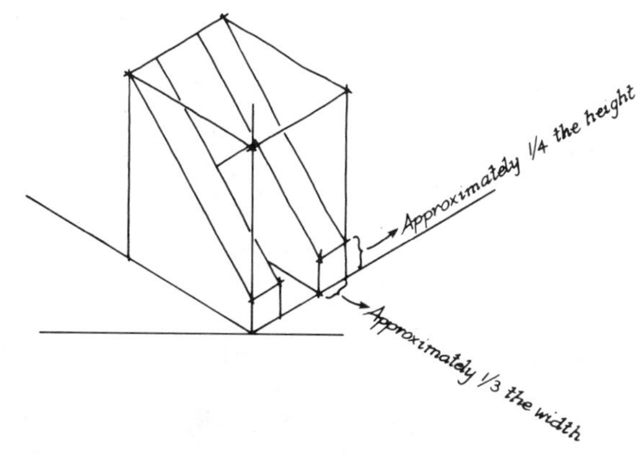

Figure 3–40 Sketch out the features — step 5.

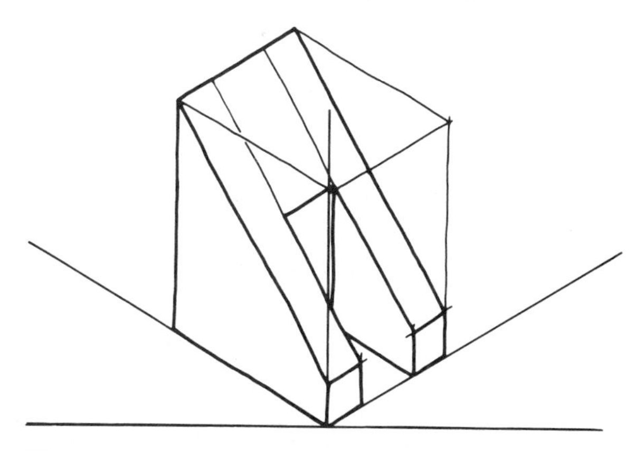

Figure 3–41 Darken the outline — step 6.

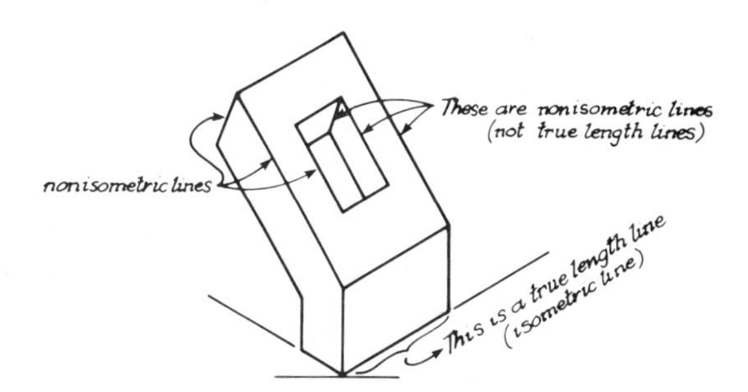

Figure 3–42 Nonisometric lines.

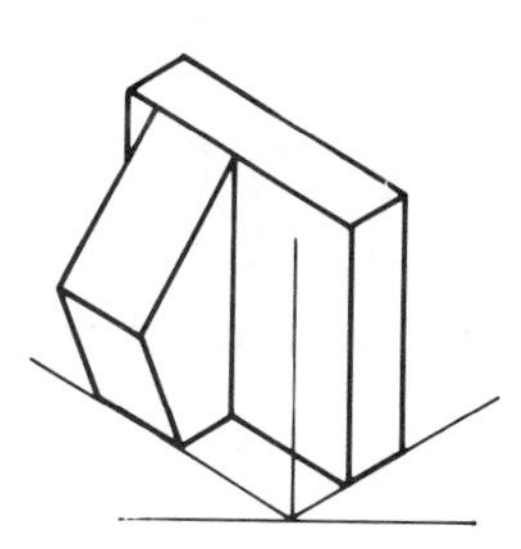

Figure 3–43 Guide.

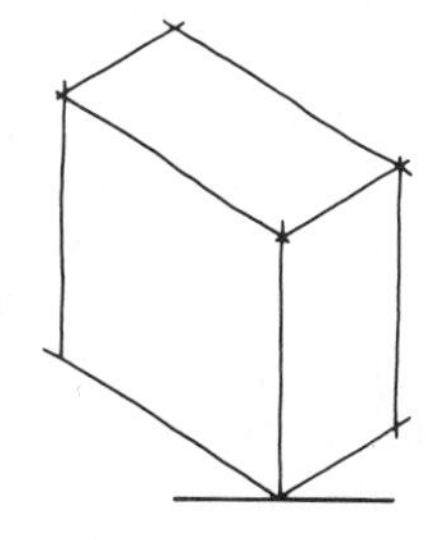

Figure 3–44 Sketch the box — step 1.

Nonisometric Lines

Isometric lines are lines that are on or parallel to the three original isometric axes lines. All other lines are nonisometric lines. Isometric lines can be measured in true length. *Nonisometric lines* appear either longer or shorter than they actually are. (See Figure 3–42.) You can measure and draw nonisometric lines by connecting their end points. You can find the end points of the nonisometric lines by measuring along isometric lines. To locate where nonisometric lines should be placed, you have to relate to an isometric line. Follow through these steps, using the object in Figure 3–43 as an example.

Step 1 Develop a proportional box, as in Figure 3–44.

Step 2 Sketch in all isometric lines, as shown in Figure 3–45.

Step 3 Locate the starting and end points for the nonisometric lines. (See Figure 3–46.)

Step 4 Sketch the nonisometric lines, as shown in Figure 3–47, by connecting the points established in step 3. Also darken all outlines.

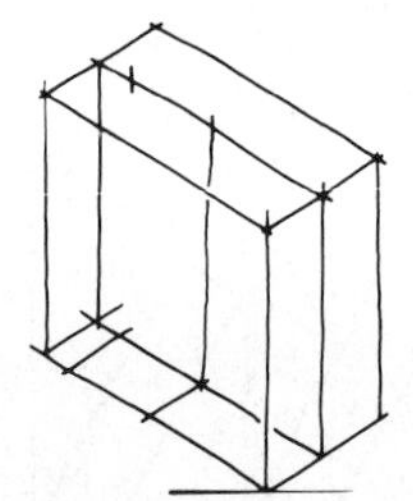

Figure 3–45 Sketch isometric lines — step 2.

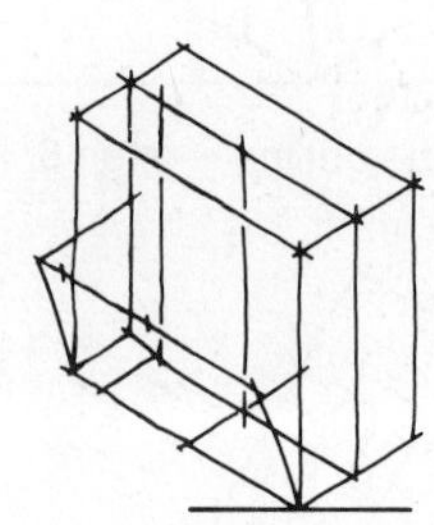

Figure 3–46 Locate nonisometric line end points — step 3.

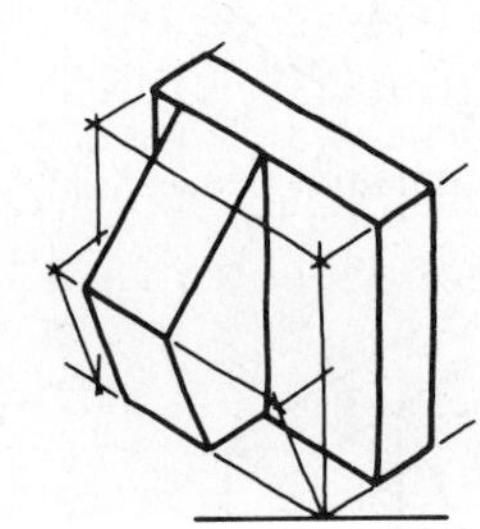

Figure 3–47 Complete the sketch and darken all outlines — step 4.

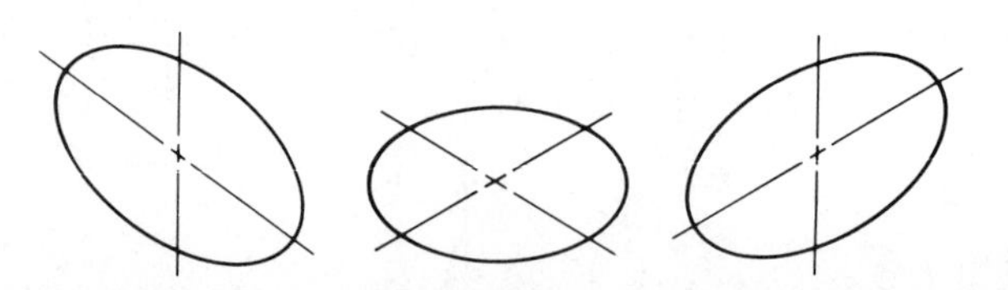

Figure 3–48 Isometric circles.

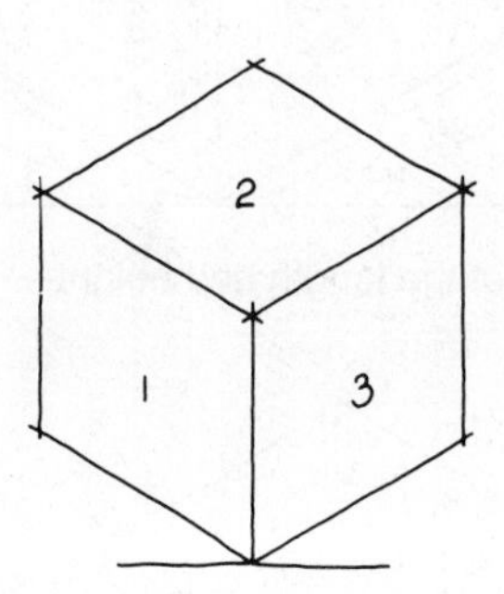

Figure 3–49 Draw an isometric cube — step 1.

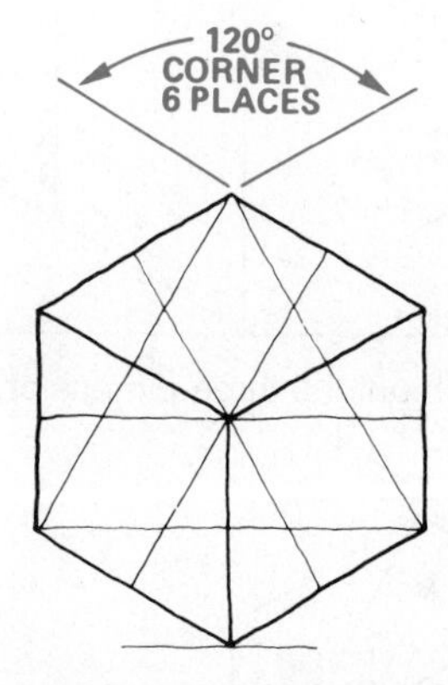

Figure 3–50 Four-center isometric ellipse construction — step 2.

Sketching Isometric Circles

Circles and arcs appear as ellipses in isometric views. To sketch isometric circles and arcs correctly, you need to know the relationship between circles and the faces, or planes, of an isometric cube. Depending on which face the circle is to appear, isometric circles look like one of the ellipses shown in Figure 3–48. The angle the ellipse (isometric circle) slants is determined by the surface on which the circle is to be sketched.

To practice sketching isometric circles you need an isometric surface to put them on. The surfaces can be found by first sketching a cube in isometric. A *cube* is a box with six equal sides. Notice, as shown in Figure 3–49, that only three of the sides can be seen in an isometric drawing.

Four-center Method. The four-center method of sketching an isometric ellipse is easier to perform, but care must be taken to form the ellipse arcs properly so the ellipse does not look distorted.

Step 1 Draw an isometric cube similar to Figure 3–49.

Step 2 On each surface of the cube, draw line segments that connect the 120° corners to the centers of the opposite sides. (See Figure 3–50.)

Step 3 With points 1 and 2 as the centers, sketch arcs that begin and end at the centers of the opposite sides on each isometric surface. (See Figure 3–51.)

Step 4 On each isometric surface, with points 3 and 4 as the centers, complete the isometric ellipses by sketching arcs that meet the arcs sketched in step 3. (See Figure 3–52.)

The four-center method is fast, but care should be taken to keep the isometric ellipses from looking like the examples in Figure 3–53.

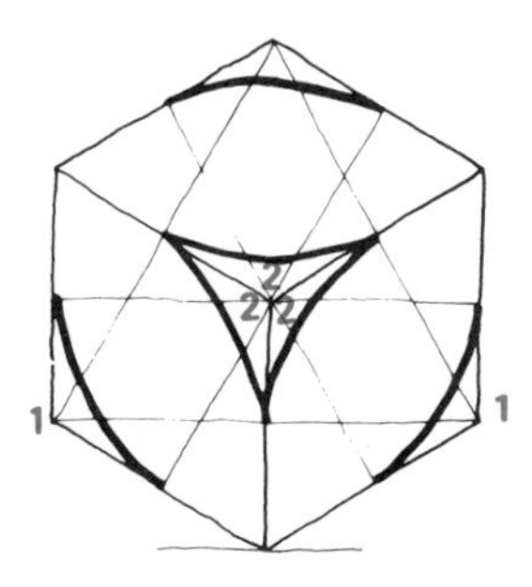

Figure 3–51 Sketch arcs from points 1 and 2 as centers — step 3.

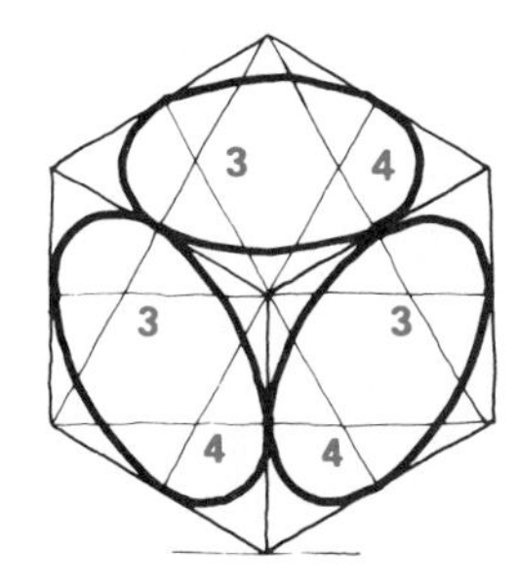

Figure 3–52 Sketch arcs from points 3 and 4 as centers — step 4.

Figure 3–53 Poorly sketched isometric ellipses.

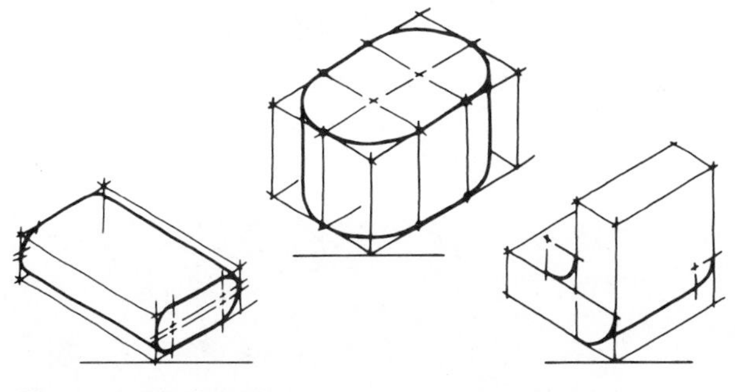

Figure 3–54 Sketching isometric arcs.

Sketching Isometric Arcs

Sketching isometric arcs is similar to sketching isometric circles. First, block out the overall configuration of the object, then establish the centers of the arcs. Finally, sketch the arc shapes as shown in Figure 3–54. Remember that isometric arcs, just as isometric circles, must lie in the proper plane and have the correct shape.

LETTERING

Information on drawings that cannot be represented graphically by lines may be presented by lettered dimensions, notes, and titles. It is extremely important that these lettered items be exact, reliable, and entirely legible in order for the reader to have confidence in them and never have any hesitation as to their meaning. This is especially important when using reproduction techniques that require a drawing to be reduced in size such as with a photocopy or microfilm, or when secondary originals are made from the original. Secondary originals are copies that may be used as originals to make changes and other copies. The quality of each generation of secondary original is reduced, thus requiring the highest quality from the original drawing. Poor lettering will ruin an otherwise good drawing.

SINGLE-STROKE GOTHIC LETTERING

ANSI The standard for lettering was established in 1935 by the American National Standards Institute (ANSI). This standard is now conveyed by the document ASME Y14.2M–1992, *Line Conventions and Lettering*.

The standardized lettering format was developed as a modified form of the Gothic letter font. The term *font* refers to a complete assortment of any one size and style of letters. The simplification of the Gothic letters resulted in elements for each letter that became known as single-stroke Gothic lettering. The name sounds complex but it is not. The term *single stroke* comes from the fact that each letter is made up of a single straight or curved line element that makes it easy to draw and clear to read. The reason that industry has accepted this style of letter is because the letters can be drawn very quickly while maintaining legibility. There are upper- and lower-case, vertical, and inclined Gothic letters, but industry has become accustomed to using vertical upper-case letters as standard. (See Figure 3–55.)

MICROFONT LETTERING

When adequately sized standard single-stroke Gothic lettering is created with clear open features there is generally no problem with microfilm or photocopy reductions, and many companies that use microfilm continue to use standard lettering. However, an alternate lettering design has been developed by the National Micrographics Association that is intended to provide greater legibility when reduced for microfilm applications. This lettering, known as Microfont, is a style that provides a more open font for better reproduction capabilities. These letters may be made freehand, with lettering templates and guides, on typewriters, or with computer graphics. (See Figure 3–56.)

ABCDEFGHIJKLMNOP
QRSTUVWXYZ&
1234567890

Figure 3–55 Vertical upper-case, single-stroke Gothic letters and numbers.

ABCDEFGHIJKLMNO
PQRSTUVWXYZ
1234567890

Figure 3–56 Microfont alphabet and numerals.

OTHER LETTERING STYLES

Inclined Lettering

Some companies prefer inclined lettering. The general slant of inclined letters is 68°. One edge of the Ames Lettering Guide has a 68° slant, which may be used to help maintain the proper angle. Structural drafting is one field where slanted lettering may be commonly found. Figure 3–57 shows slanted upper-case letters.

Lower-case Lettering

Occasionally, lower-case letters are used; however, they are very uncommon in mechanical drafting. Civil or map drafters use lower-case lettering for some practices. Figure 3–58 shows lower-case lettering styles.

Architectural Styles

Architectural lettering is much more varied in style than mechanical lettering; however, neatness and readability are essential. (See Figure 3–59.)

Figure 3–57 Upper-case inclined letters and numbers.

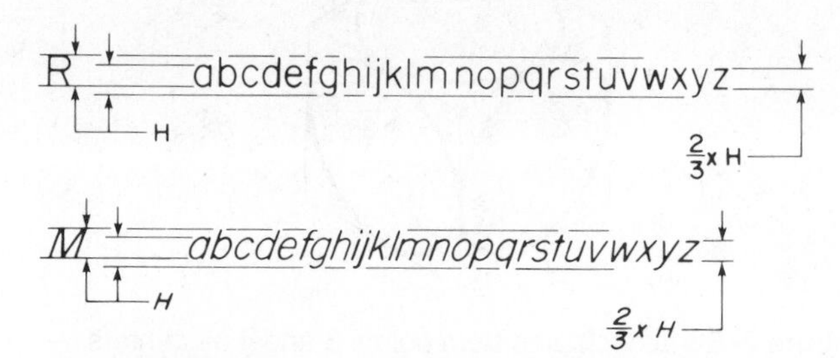

Figure 3–58 Lower-case lettering.

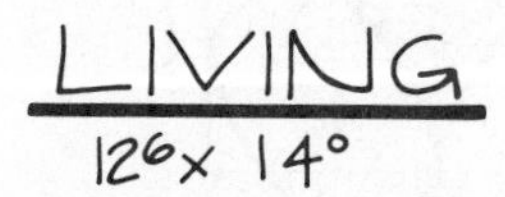

ABCDEFGHIJKLM
NOPQRSTUVWXYZ
1 2 3 4 5 6 7 8 9 10

Figure 3–59 Architectural lettering.

LETTERING

Lettering with a CADD system is one of the easiest tasks associated with computer-aided drafting. It is just a matter of deciding on the style or font of lettering to use, and then locating the text where it is needed. Lettering is called text when using CADD.

The CADD drafter often rejoices when the time comes to place text and notes on the drawing because no freehand lettering is involved. The computer places text of a consistent shape and size on a drawing in any number of styles, or fonts. The FONT or TEXT command is one of several that can be found in a section of the menu-labeled text, or text attributes. The drafter is also able to specify the height, width, and slant angle of characters (letters and numbers). Most systems maintain a certain size of text called the *default* size that is used if the operator does not specify one. The term *default* refers to any value that is maintained by the computer for a command or function that has variable parameters. The default text height may be ⅛ in., but the user can change it if need be.

Lettering Styles

Many CADD systems possess a variety of lettering styles, or fonts. The drafter can select the style to use simply by picking a menu command, symbol, or by typing a command at the keyboard. Figure 3–60 shows some of the styles and sizes of characters that can be used in CADD. The size and style of characters used is dictated by the nature of the drawing.

Locating Text

The process of locating text has not changed. The drafter still needs to decide where to locate dimensions, notes, and parts lists, but with CADD the process of placing notes is just a bit more technical. Most CADD systems provide for keyboard entry of location and size coordinates, thus allowing the drafter to accurately locate text. But using the keyboard to input text location coordinates is a tedious process. Often key strokes are combined with digitizer stylus commands to generate text.

Text is located by pointing to one of several places on the lettering. Figure 3–61 shows an example of several points on the text that can be used for location purposes. Not all CADD systems use all the points shown, but most systems have a command known as *text* that is used for placing written information on the drawing. Some systems may allow the operator to locate text between two points. The computer calculates the size of each letter so the text fits in the desired space.

The first decision regarding text is to determine its height, width, and slant angle. Most CADD systems maintain a default text size that is used by the computer if the operator decides not to change it. The angle of rotation, direction, or text path is also determined by the drafter. The text path is the angle from horizontal that the text will lie on. An example of text path is shown in Figure 3–62. Text located on a horizontal line has a direction of 0°, and text that reads from the bottom up vertically has a direction of 90°.

With the size and slant angle decided, you then need to locate it on the drawing and type the text at the keyboard. This process can occur in a couple of ways. You could be asked to first digitize the text location using a pointing input device and then type the text. The second method is the reverse of the first. Type the text and it appears on the screen with the crosshairs at the point of location that you previously specified. Then *drag* the text to the proper location, press a button on the puck or keyboard, and it is in place. The nice thing about locating note and labels on a drawing with a CADD system is the ability to move them around instantaneously, as often as you want, without erasing holes in your drawing.

ABCDEFGHIJKLMNOPQRSTUVWXYZ
acbcdefghijklmnopqrstuvwxyz
1234567890

ABCDEFGHIJKLMNOPQRSTUVWXYZ
acbcdefghijklmnopqrstuvwxyz
1234567890

ABCDEFGHIJKLMNOPQRSTUVWXYZ
acbcdefghijklmnopqrstuvwxyz
1234567890

Figure 3–60 A sample of character font styles.

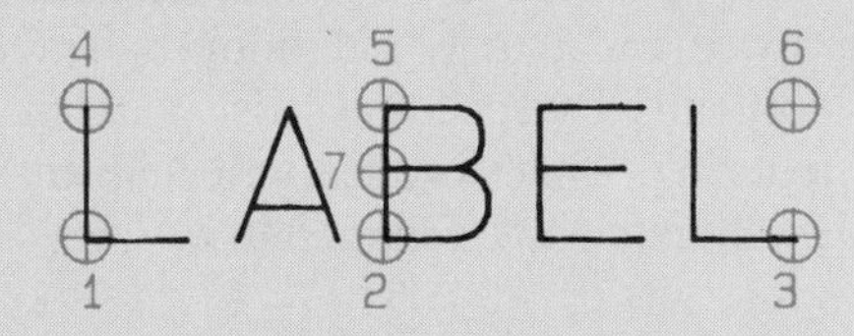

Figure 3–61 Text can be placed on a drawing by choosing one of several location points shown here.

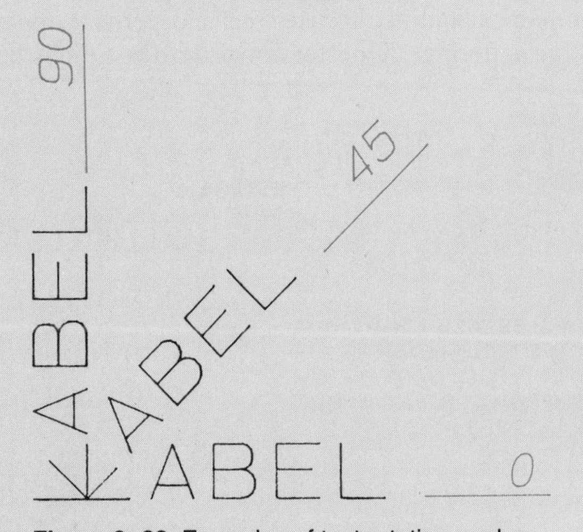

Figure 3–62 Examples of text rotation angles.

LETTERING LEGIBILITY

ANSI The minimum recommended lettering size on engineering drawings is .125 in. (3 mm). All dimension numerals, notes, and other lettered information should be the same height except for titles, drawing numbers, and other captions. Titles and subtitles, for example, may be a minimum of .25 in. (6 mm) high.

Either vertical or inclined lettering may be used on a drawing depending on company preference. However, only one style of lettering should be used on a drawing. Lettering must be dark, crisp, and opaque for the best possible reproducibility.

The composition or spacing of letters in words and between words in sentences should be such that the individual letters are uniformly spaced with approximately equal background areas. This usually requires that letters such as I, N, or S be spaced slightly farther from their adjacent letters than L, A, or W. A minimum recommended space between letters in words is approximately .06 in. (1.5 mm). The space between words in a note or sentence should be about the same as the height of the letters. The horizontal space between sentences in a note or paragraph should be equal to twice the height of lettering.

Notes should be lettered horizontally on the sheet. When lettering notes, sentences, or dimensions requires more than one line, the vertical space between the lines should be a minimum of one-half the height of letters or .06 in. (1.5 mm). The maximum recommended space between lines of lettering is equal to the height of the letters. Some companies prefer to use the minimum space to help conserve space while other companies prefer the maximum space for clarity. Individual notes should be separated vertically by spaces at least equal to twice the height of lettering. This distance will help maintain the identity of each note. (See Figure 3–63.)

LETTERING IN INDUSTRY

Companies continue to use freehand vertical upper-case Gothic lettering. A professional drafter should be able to letter rapidly with clarity and neatness. While freehand lettering should conform closely to the single-stroke Gothic style, each drafter may have a slightly different representation. Most industries prefer that drafters produce lettering that looks the same from one drawing to the next. Also, when drawing changes are made, the individual making the change should attempt to match the lettering on the original drawing. The entry-level drafter may be assigned the task of drawing changes and so should have a flexible lettering style. Lettering templates and guides are available in both standard vertical Gothic letters and in Microfont letters. When companies are working on government projects, lettering guides may be a requirement. Computer graphics is here to stay and ultimately there will be little, if any, need for freehand lettering. There is no doubt that computer-generated lettering is fast, easy, and very legible, although drafters, as professionals, will probably attempt to keep their lettering skills for at least one generation after a complete change to computer graphics has been made.

VERTICAL FREEHAND LETTERING

Vertical freehand lettering is the standard for mechanical drafting. The ability to perform good-quality lettering quickly is important. A common comment among employers hiring entry-level drafters is the ability to do quality lettering and line work. Although standard, not all companies require freehand lettering. Some companies allow drafters the flexibility of freehand lettering or using a template. Many companies are now changing to computer-aided drafting and traditional lettering skills may become obsolete.

Always use lightly drawn horizontal guidelines that are spaced equal to the height of the letters. Some people need vertical guidelines to help keep their letters vertical. The ability to perform quality freehand lettering requires a great deal of practice for most people.

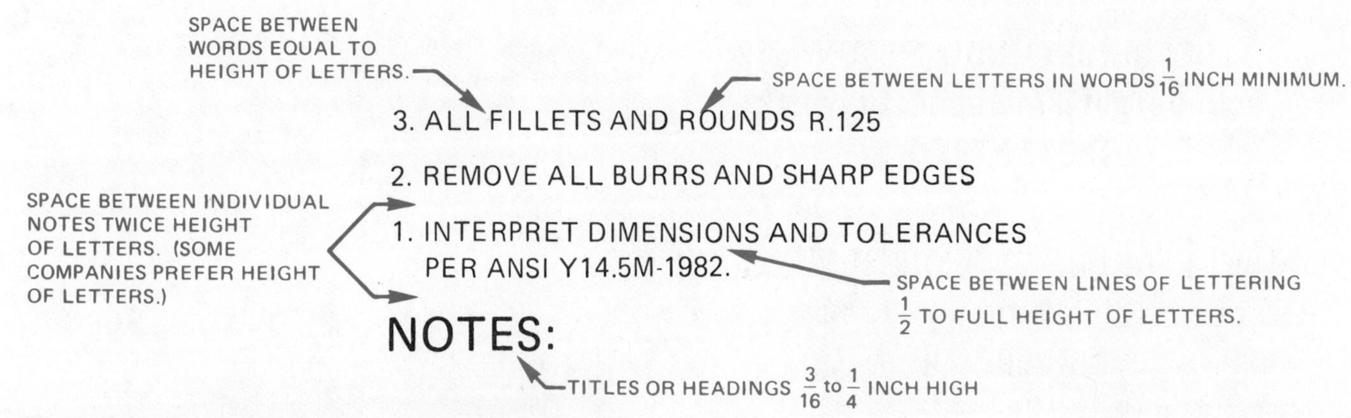

Figure 3–63 Spacing of letters, words, and notes.

PROFESSIONAL HINT

Use 2H, H, or F pencils for lettering. Try them all, but use the one that gives you the best results of dark smooth letters.

Vertical Capital Letters

Straight Elements. Your pencil point should be sharpened and slightly rounded to produce black, bold lines for large letters. Many drafters prefer using a 0.5 mm automatic pencil for lettering. This kind of pencil does not need sharpening and H, F, or HB leads are usually easy to control for effective lettering. Remember, you need to experiment with pencils and leads to determine which gives you the best results. Letters should be dark and crisp. Fuzzy letters print as fuzzy lines.

Letters composed of straight lines are shown in Figure 3–64. You should become familiar with these letter forms and the strokes needed to make them. The arrows in Figure 3–64 indicate the direction of the stroke used to form the letters. However, if these strokes are uncomfortable, you should develop your own procedure. Try the recommended strokes first. Because these letters are made of single-stroke elements, try not to let the strokes combine as the result may be curves where straight elements should be.

Horizontal guidelines should be used for all lettering at all times. Use vertical guidelines if you have difficulty keeping your letters vertical. Guidelines should be very light 6H or 4H pencil lines. Some drafters prefer to use a light-blue lead rather than a graphite lead for guidelines. Light blue will not reproduce in the diazo or photocopy processes. When lettering, protect your drawing by resting your hand on a clean protective sheet placed over your drawing. This prevents smearing and smudging, as shown in Figure 3–65.

Curved Elements. Letters that contain arcs are shown in Figure 3–66. Notice the difference in the sizes of arcs and circles used for different letters. The recommended letter elements, as with straight elements, are made up of a series of suggested strokes. These strokes when used as shown provide the best lettering results. Vertical guidelines are often an asset for the best lettering results.

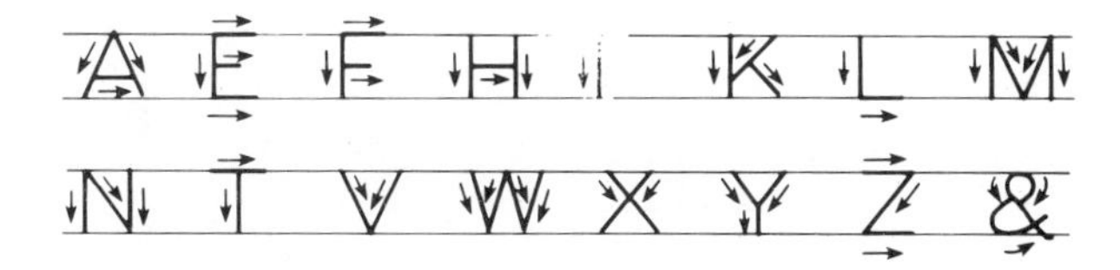

Figure 3–64 Recommended strokes for vertical uppercase Gothic letters with straight elements.

Vertical Numerals and Fractions

Vertical numerals, as seen in Figure 3–67, are also made up of recommended strokes. Numerals are the same height as capital letters. Generally all lettering on a drawing should be the same height, except for special notes and titles which may be larger.

Fractions. Fractions are not as commonly used on engineering drawings as decimal inches or millimeters. Fractional dimensions generally denote a larger tolerance than decimal numerals. When fractions are used on a drawing, the fraction numerals should be the same size as other numerals on the drawing. The fraction bar should be drawn in line with the direction of the dimension. For example, if all dimensions read horizontally, all fraction bars are horizontal. The fractions numerals should not touch the fraction bar. A space of 1.5 mm or

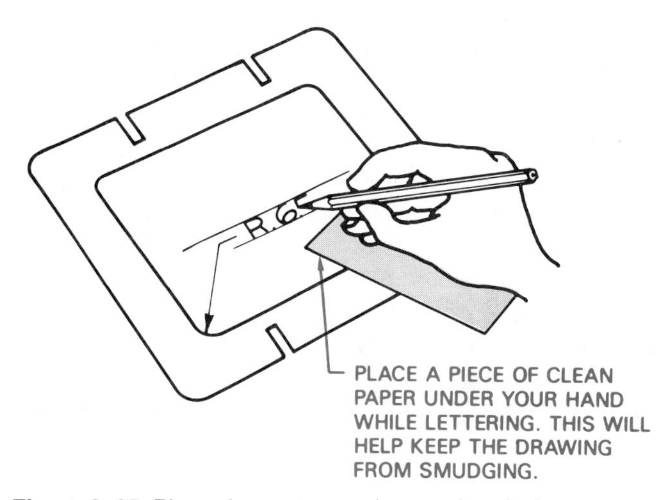

Figure 3–65 Place clean paper under your hand when lettering.

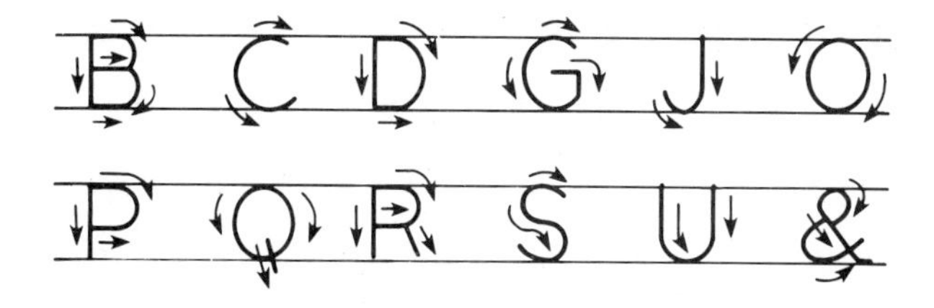

Figure 3–66 Recommended strokes for vertical uppercase Gothic letters with curved strokes.

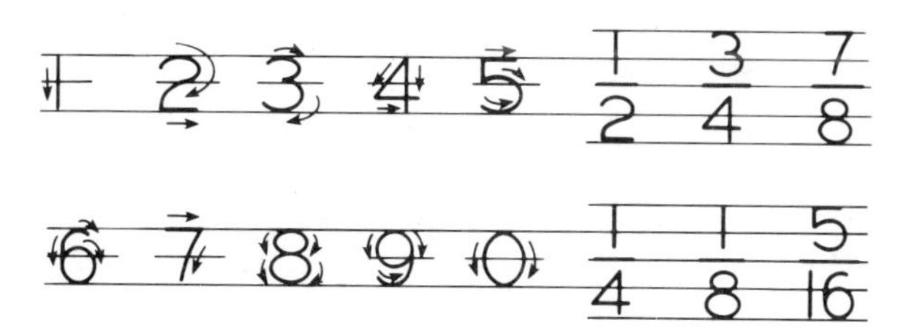

Figure 3–67 Vertical numerals and fractions.

.06 in. between the numeral and bar is suggested. The fraction bar may be diagonal in certain situations such as when used in a general note or in a drawing title. (See Figure 3–68.)

Decimal Points. The placement of the decimal point in a decimal dimension is critical. If the decimal point is crowded or drawn too lightly, it may be lost and the result is an unclear dimension. Always make the decimal point dark and bold. Also space the numerals far enough to clearly provide room for the decimal point. Two-thirds the height of letters is recommended. (See Figure 3–69.)

LETTERING TECHNIQUES

Always use guidelines. Straight, even letters of consistent height look better than letters of varying heights. Even when using guidelines be sure to extend each letter directly to the guidelines. Letters that periodically extend beyond or fall short of the guidelines tend to make the words or notes irregular.

Try an H or F pencil for lettering if you have a light touch or try a 2H pencil if your touch is heavy. Use a slightly rounded, conical point on your mechanical pencil or a 0.5 mm automatic pencil. Lines should be black, crisp, and sharp. All vertical lines are made perpendicular, starting at the top for each stroke; all horizontal lines are made from left to right. Balance angles for the letters A, V, W, X, and Y about a vertical guideline. Use a round form for curved letters. Be careful to not allow any space to show between the letter and the guideline.

COMPOSITION

As a rule of thumb, curved letters can be placed close together; straight letters should be placed further apart. See Figure 3–70. Good lettering composition is evident when all letters in a word look as if they have the same amount of space between them. Practice dividing the open space between letters. The open space between letters in each word should *appear* to be equal. To achieve this appearance it is often necessary to shorten the horizontal strokes of open letters such as L and J. When strokes are parallel and next to each other as the WA in WALL and both Ns and I in PLANNING should be placed a little farther apart.

If your letters are wiggly or if you are nervous, try pressing hard to make your lines straighter. If you are pressing too hard, try to relax the pressure. Also try making (drawing) each letter as rapidly as possible. This tends to eliminate wiggly letters. If your lead is too hard, wiggly letters could result; try a softer lead.

MAKING GUIDELINES

Guidelines are lightly drawn lines equal to the height of letters in distance apart. As previously mentioned, some drafters prefer to use a light-blue lead so guidelines will not reproduce. A rule of thumb is that guidelines should be drawn so lightly that when the drawing is at arm's length the lines should be very difficult to see.

Ames Lettering Guide

Probably the most commonly used device for making guidelines is the versatile Ames Lettering Guide. With an Ames Lettering Guide it is possible to draw guidelines and sloped lines for lettering from $1/16$ to 2 in. in height. Disk numbers from 10 to 2 give the height of letters in thirty-seconds of an inch. If $1/4$ in.-high letters are required, rotate the disk so the 8 ($8/32'' = 1/4''$) is at the index mark on the bottom of the frame. (See Figure 3–71.)

Use of the Column with Equally Spaced Holes. The column of equally spaced holes (second from right on the disk) is perhaps the most versatile column of holes since it results in uniformly spaced lines. Every other hole in this column is equivalent to the disk number setting. When set at 8, the distance from the first to third hole is $1/4$ in. The middle hole in each set of three is for a guideline that assists in forming upper-case letters such as A, B, E, F, H, P, and R. (See Figure 3–72.)

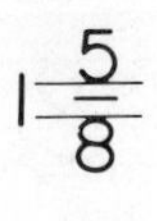

1-5/8   15/8

Figure 3–68 Fractions.

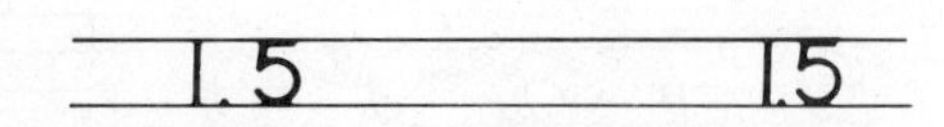

Figure 3–69 Spacing of decimal point in numerals.

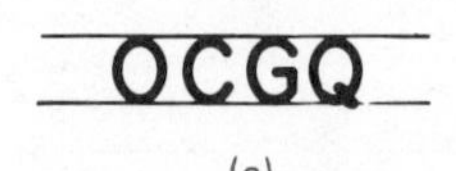

(a)

MNIHLTAW

(b)

Figure 3–70 (a) Curved letter composition; (b) straight letter composition.

Use a sharpened 4H or harder mechanical pencil or a 0.3-mm or 0.5-mm automatic pencil and place the guide, readable side up, against the top of the straightedge. Place the pencil in the top hole. Keep the pencil perpendicular to the paper and slightly inclined in the direction of travel. With the pencil in the top hole, slide the guide to the right, lightly drawing a line. Keep the base of the guide in contact with the straightedge until the end of the line has been reached. To make the remaining guidelines, move the pencil down hole by hole, and alternately slide the guide to the left and then to the right. You now have one set of three guidelines 1/4 in. high. To draw two more sets, repeat the procedure using the remaining holes in this column.

If no middle guideline is desired, or for 1/8 in. lettering, use the same 8 setting. If 3/32- or 1/16-in. heights are desired with no middle guideline, use the 6 or 4 setting respectively at the bottom of the disk, or use the index mark on the disk (near the 2/3 fraction) and set the disk at the desired mark on the frame.

Cross-sectioning. For parallel lines, needed for such purposes as section lines, including brick, tile, and concrete block, or a music staff, set the index mark on the disk (near the 2/3 fraction) at the desired mark on the frame (1/8, 3/32 or 1/16). Set your straightedge parallel to the desired lines and draw lines alternately to the right and left.

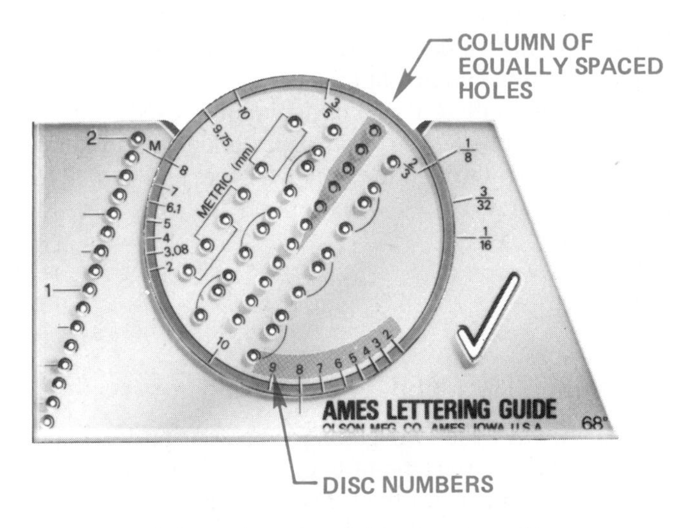

Figure 3–71 Ames Lettering Guide. *Courtesy Olson Manufacturing Company, Inc.*

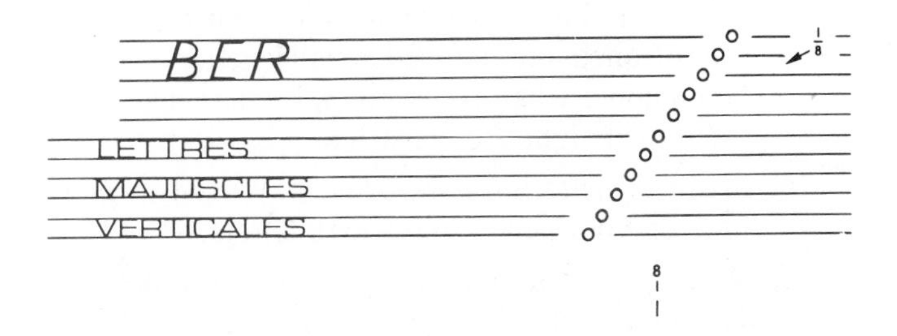

Figure 3–72 Making .125- and .25-in. guidelines. *Courtesy Olson Manufacturing Company, Inc.*

Four Guidelines. To aid in forming lower-case letters such as g, j, p, q, and y, use a fourth guideline. Use the top, 2nd, 4th and 5th holes. (See Figure 3–73.) If two sets of guidelines are required, use the 7th, 8th, 10th, and bottom holes.

Slanted or Vertical Guidelines. Slanted or vertical guidelines can be drawn easily to help keep your letters vertical for mechanical drafting or slanted for structural drafting. (See Figure 3–74.) Always remember, draw your slope lines lightly and use only enough to maintain slope uniformity.

Metric Guidelines. The numbers and set of six holes to the left of the disk relate to metric heights for guidelines. (See Figure 3–75.) This column of six holes offers the drafter the option of spacing guidelines equally (right brackets) or at half space (left brackets). The 3.08-, 6.1- and 9.75-mm calibrations are for standard letter heights used in United States drawings. By setting 3.08 opposite the left frame index *M*, guideline spacing will be .12 in. Similarly, setting 6.1 and 9.75 opposite *M* will give .24

Figure 3–73 Four guidelines for lower-case letters. *Courtesy Olson Manufacturing Company, Inc.*

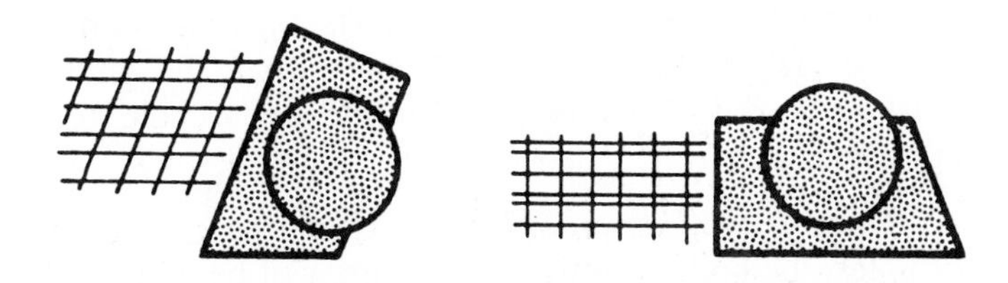

Figure 3–74 Inclined or vertical guidelines. *Courtesy Olson Manufacturing Company, Inc.*

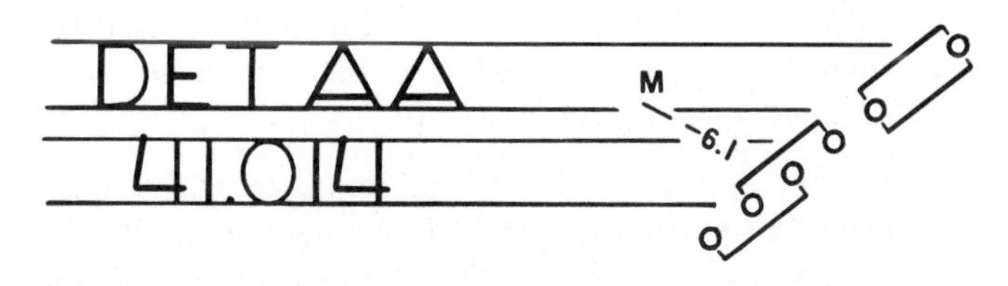

Figure 3–75 Metric guidelines. *Courtesy Olson Manufacturing Company, Inc.*

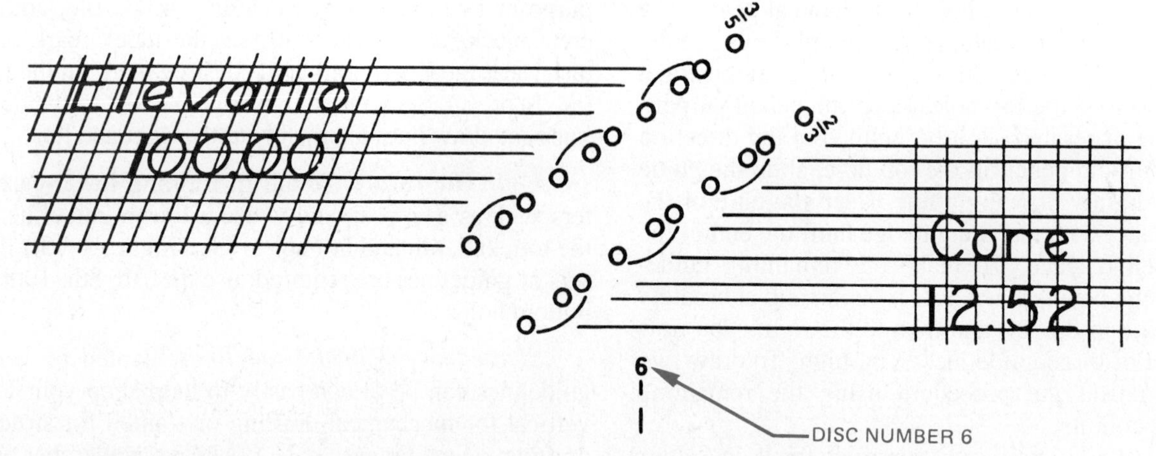

Figure 3–76 Two-thirds and three-fifths columns. *Courtesy Olson Manufacturing Company, Inc.*

and .38 in. high guidelines respectively. Other metric heights are calibrated on the disk for use if desired.

Two-thirds Ratio Column of Holes. Using the column of holes marked 2/3 (see Figure 3–76), the body of lower-case letters will be 2/3 that of capitals. To draw guidelines 3/16 in. high, set the disk at 6 (6/32" = 3/16") opposite the bottom index mark on the frame. Begin by placing the pencil point in the *second* hole from the top. Draw guidelines as described before by moving the guide alternately right to left and back, and moving the pencil down hole by hole. You now have a set of 3 guidelines 3/16 in. high. To draw two more sets, repeat the procedure using the rest of the holes in the 2/3 column. Note that sets of three holes are grouped by elliptical lines.

To draw more than three sets of guidelines, move the straightedge and guide down until the top hole in the 2/3 column coincides with the bottom line of the last set of guidelines. *Do not use this top hole.* The top hole is only to give proper spacing between lines when the guide must be moved. Place the pencil in the second hole and draw lines as before.

Three-fifths Column of Holes. Using the 3/5 column of holes, the body of lower-case letters will be 3/5 that of capitals. (See the left side of Figure 3–76.) This ratio is usually used by civil engineers. To use this column, proceed as described above for the 2/3 column of holes.

Column of Holes on the Frame. The 1/8 in. spaced holes on the left-hand side of the frame may be used for title blocks, grid lines, section lining, dimension line spacing, and spacing from 0 to 2 in.

Symbol Template. The symbol template on the right-hand side of the frame is used for making the ANSI control surface finish mark, new style finish mark, and the short leader and arrow line for the welding symbol. Refer to Figure 3–71.

OTHER LETTERING AIDS AND GUIDELINE METHODS

Other guideline lettering aids for equidistant spacing of lines have parallel slots ranging in width from 1/16 to 1/4 in. These lettering guideline aids are not as complex as the Ames Lettering Guide, but they are also not as flexible.

Another method of making guidelines used by a few drafters is to place 1/8-in. grid paper under the drawing. The lines show through the drawing sheet and guidelines need not be drawn.

POINTS TO REMEMBER

1. Always use two guidelines as the professionals do. Three are helpful for beginners. The center guideline is for the letter crossbars.
2. To adjust eye and hand coordination, first letter lightly, then darken your lettering. Doing so allows you to correct mistakes. This technique should not be necessary after you get some lettering experience.
3. Make all letters and numbers on the drawing at least 1/8 in. high, none smaller. If your drawing is microfilmed and reduced, smaller letters will not be readable when enlarged. All letters should be the same height. Be consistent.
4. Make all lettering dark. You may have to press hard on your pencil to get dark letters.
5. When practicing lettering, practice no more than 15 minutes per day. Otherwise your hand may cramp and any further practice may be of less value. Attempt to gain speed and neatness as you practice.

Berol. RapiDesign. R-960 VERTICAL LETTERING GUIDE WITH AUTOMATIC EQUAL SPACING GUIDE LINES

(a)

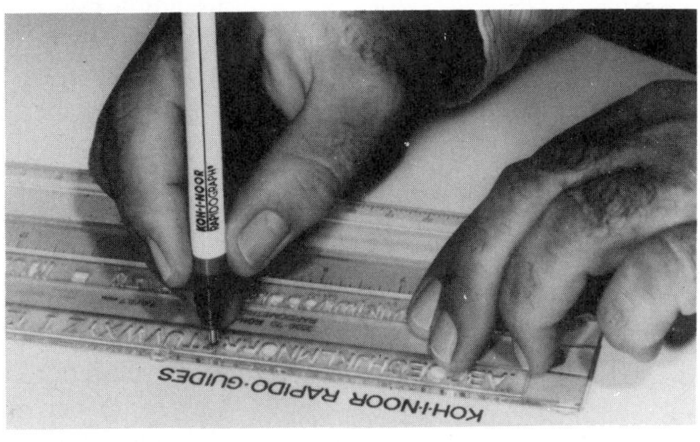

(b)

Figure 3–77 (a) Lettering guide template. *Courtesy Berol USA RapiDesign;* (b) lettering guide template in use. *Courtesy Koh-I-Noor Rapidograph, Inc.*

LETTERING GUIDE TEMPLATES

Vertical upper-case Gothic lettering is standard for mechanical drafting. Although professional drafters are excellent at lettering, each individual's lettering is slightly different. Some companies prefer that drafters use lettering guides so uniformity is maintained. Standard lettering guide templates are available with vertical Gothic letters and numerals ranging in height from 3/32 to 3/8 in. (See Figure 3–77.) Lettering guides are also available in many other lettering styles, including slanted Gothic, block, Futura, Old English, and Microfont letters in either upper- or lower-case.

MECHANICAL LETTERING EQUIPMENT

Mechanical lettering equipment is available in kits with templates for letters and numerals in a wide range of sizes. A complete lettering equipment kit includes a scriber, templates, tracing pins, and lettering pens. Figure 3–78 shows the component parts of a lettering equipment set.

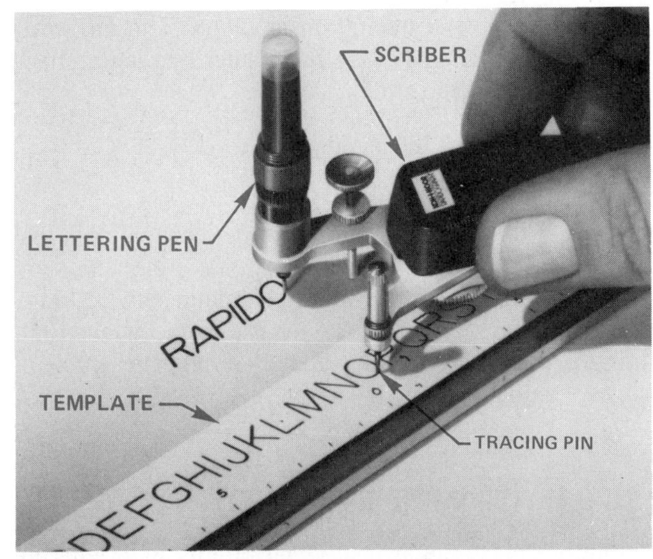

Figure 3–78 Components of a lettering equipment set, and using the mechanical lettering equipment — steps 1 and 2. *Courtesy Koh-I-Noor Rapidograph, Inc.*

Templates

Most templates have a convenient scale on the bottom edge for quick centering and letter spacing. Also, the distance between the bottom of the letters and the bottom of each template is uniform, so templates can be exchanged without changing the position of the straightedge.

Scriber

The scriber is used with templates to 500 (.5 in.) and with pens from number 000004 to 8. The scriber is adjustable from vertical to 22° forward-slanted letters by opening and closing the scriber arm. It is not necessary to move the templates or change position when adjusting to different slants. (See Figure 3–79.)

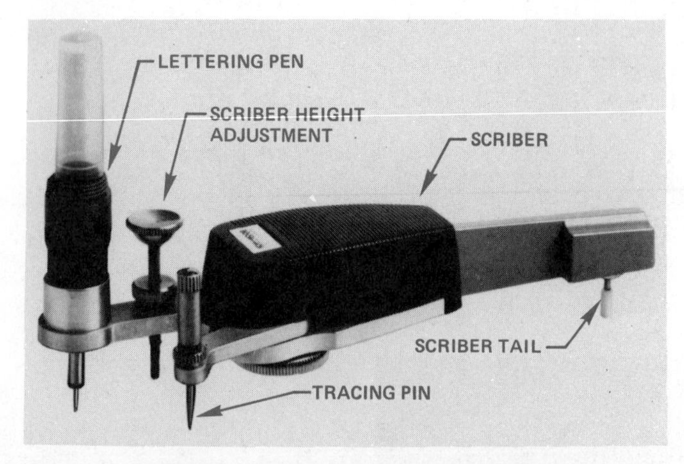

Figure 3–79 Scriber. *Courtesy Koh-I-Noor Rapidograph, Inc.*

Tracing Pins

There are three kinds of tracing pins. The proper pin to use depends on the template being used. Small letter templates require a thin tracing pin, while large templates require a thick tracing pin.

- The tracing pin indicated by a red-colored ring is especially for templates 60 (.06 in.) and 80 (.08 in.)
- The tracing pin with no special indication is suitable for all other templates.
- A double-end tracing pin is included in most sets and is applicable for all templates by rotating from one end to the other.

Lettering Pens

The lettering pen is similar to the technical pen discussed in Chapter 2. The scriber holds either the standard lettering pen or the technical fountain pen. Many drafters prefer technical pens for use in mechanical lettering equipment since they contain a larger reservoir of ink for continued use.

USING LETTERING EQUIPMENT

Skilled users can letter neatly and rapidly with lettering equipment. Expect to practice awhile before you are able to letter well. Most lettering is done with ink, but pencil adaptors are available.

Step 1 Choose the template best suited for the letter size and style required.

Step 2 Place the template at an appropriate distance from the lettering position. A straightedge should be placed below the template to allow for smooth sliding. The track drafting machine scales work well when locked in the horizontal position for horizontal lettering or in the vertical position for vertical lettering. (See Figure 3–78.)

Step 3 Choose the proper pen for your work from the large variety available. Notice that specific pen tip sizes are recommended for certain templates. If too large a pen tip is used with a very small template, the letters will fill in completely and be illegible.

Step 4 Loosen the scriber base screw. After the scriber tracing arm has been set for vertical or slant letters, tighten the base screw. Insert the standard lettering pen or a filled technical fountain pen into the scriber arm until the shoulder on the point is perfectly seated, then tighten the side screw.

Step 5 Place a scrap piece of drawing paper or Mylar® on the board for this step. Place the tail pin of the scriber into the guide channel on the bottom of the template. Next place the tracing pin into the engraved letter channel on the template. Gently lower the pen point until it reaches the scrap drawing paper. Adjust the pen height-adjusting screw so the point slightly touches the paper. A dot of ink should flow onto the paper or Mylar®. You may occasionally need to tap the pen tip lightly on scrap paper or a napkin to help keep the ink at the tip fluid. It is frustrating to get ready to draw letters when the ink is not ready.

Step 6 Now, while you keep the tail pin of the scriber in the guide channel on the bottom of the template, manipulate the tracing pin through the letter channels and the pen point will automatically form the letters on the paper. (See Figure 3–78.)

Important Things to Remember

- Always lay the template along a straightedge. The locked blade of the track machine is best.
- Insert the pen into the scriber until its shoulder is perfectly seated, then tighten the side screw.
- Use the proper tracing pin.
- Rest the scriber on the scriber stand when not in use.
- Clean pens after use.
- Practice.

MACHINE LETTERING

Lettering machines are available in a variety of fonts, styles, and letter sizes to prepare drawing titles, labels, or special headings. These types of lettering features are especially useful for making display letters or cover sheets, or rendering drawing titles. Figure 3–80 shows a lettering machine that has over 300 different lettering styles and sizes available on disks. This particular machine uses a typewriter keyboard for quick preparation of lettering. A personal computer may also interface with the lettering machine to increase speed and provide additional flexibility. Lettering machines prepare strips of lettering on clear tape with an adhesive back for placement on drawing originals. The tape is also available in a variety of colors for special displays and presentation drawings.

Some lettering machines perform directly on the drawing, and may provide the ability to use application cassettes to expand the set of characters to include special symbols. The character spacing may be automatically adjusted to fit the desired letters or symbols within a specified limited space. Dimension lines and numerals may be plotted simultaneously, and dimensions may also be scaled for plotting. Characters, sentences, phrases, or notes which are often repeated may be stored in memory and easily recalled when needed.

TYPEWRITERS

Specially designed typewriters are used to type certain information on engineering drawings. Some companies use such a machine to help save time on items that would require many hours of freehand or template lettering. The information that may be typed on drawings includes general notes, parts lists, or other lengthy documentation. (See Figure 3–81.)

Figure 3–80 Lettering machine. *Courtesy AED Corporation*

TRANSFER LETTERING

A large variety of transfer lettering fonts, styles, and sizes are available on sheets. These transfer letters may be used in any combination to prepare drawing titles, labels, or special headings. They may be used to improve the quality of a presentation drawing, or simply for all drawing titles. Figure 3–82 shows a few of the variety of letters available.

Transfer letters may be purchased as vinyl letter sheets where individual letters are removed from a sheet and placed on the drawing in the desired location as shown in Figure 3–83. Transfer letters are also available on sheets where letters are placed on the drawing by rubbing with a burnishing tool as shown in Figure 3–84. Rub-on letters, as they are often called, offer a high-quality lettering that is excellent for drawing titles and special displays. Drawing aids, in the form of transfer tapes, may be used to prepare borders, line drawings, or special symbols as shown in Figure 3–85.

Figure 3–81 Typewriter used in industry. *Courtesy Hyster Corporation.*

Figure 3–82 Variety of transfer lettering fonts. *Courtesy Chartpak.*

1. Peel away excess material from sheet. Letters and numbers will remain attached, ready for use.

2. Remove desired letters from sheet. Be sure to plan ahead by counting the number of characters to be used.

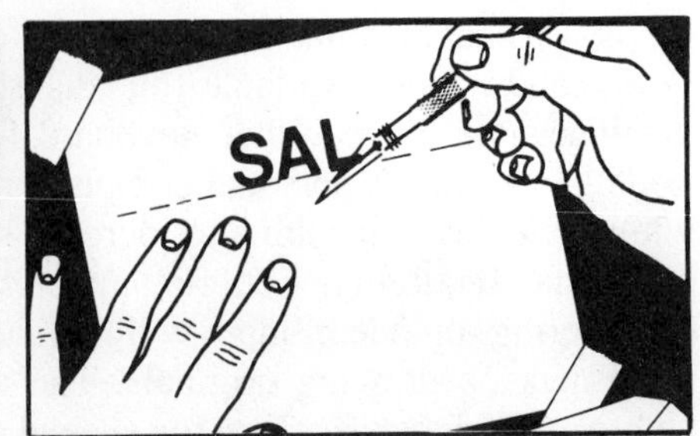

3. After drawing guidelines on object to be lettered, place letters carefully on surface and rub firmly in place. Vinyl lettering will adhere to almost any outdoor or indoor surface.

Figure 3–83 Using vinyl transfer letters. *Courtesy Chartpak.*

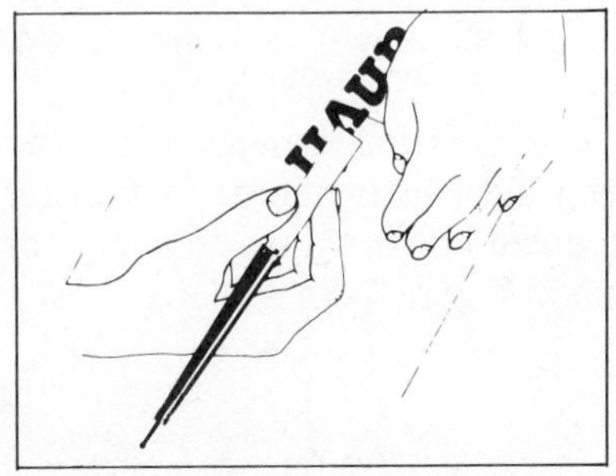

1. POSITION THE LETTER.

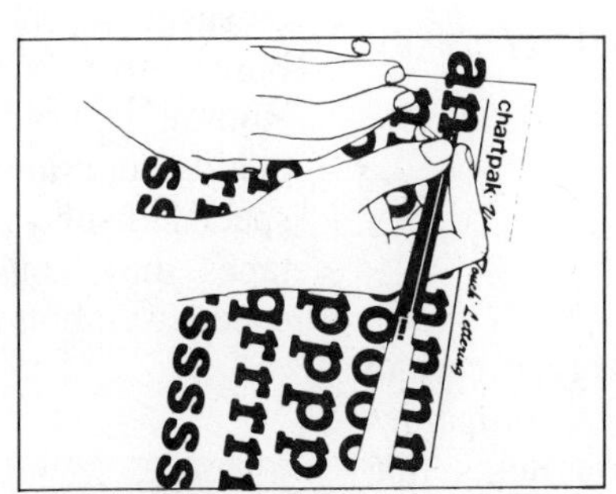

2. RUB OVER THE LETTER WITH A BURNISHER. BE SURE TO COVER ALL AREAS AND FINE LINES.

3. REMOVE THE SHEET BY CAREFULLY LIFTING IT FROM THE CORNER. FOR MAXIMUM ADHESION, PLACE THE BACKING SHEET OVER LETTER AND RUB AGAIN WITH THE BURNISHER.

Figure 3–84 Using rub-on letters. *Courtesy Chartpak.*

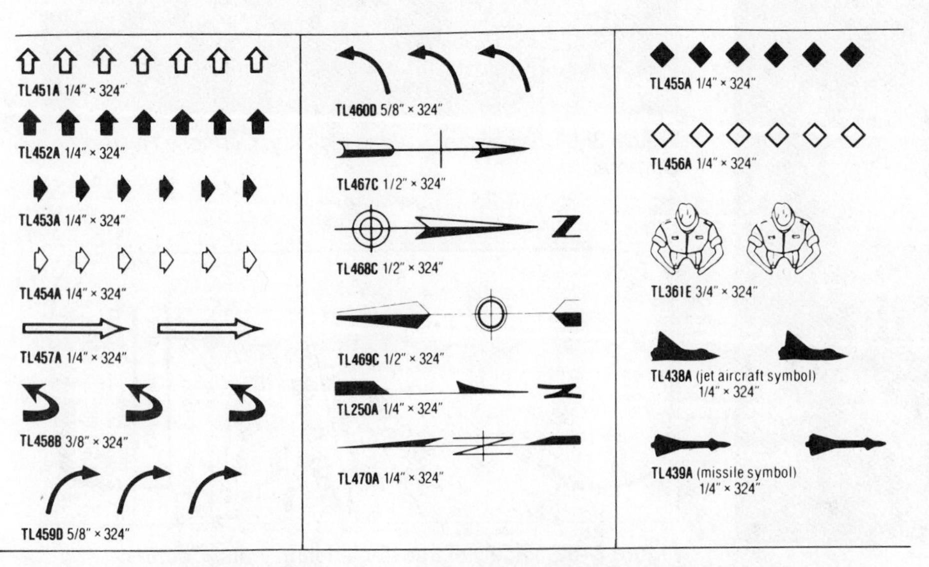

HOW TO MITER CORNERS

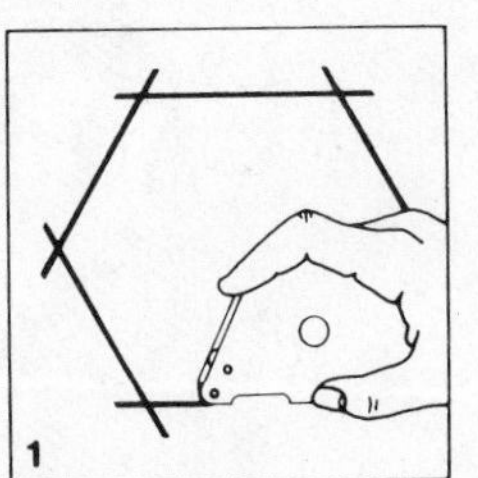

1. To miter corners overlap tape at corners, using tape pen for ease in handling.

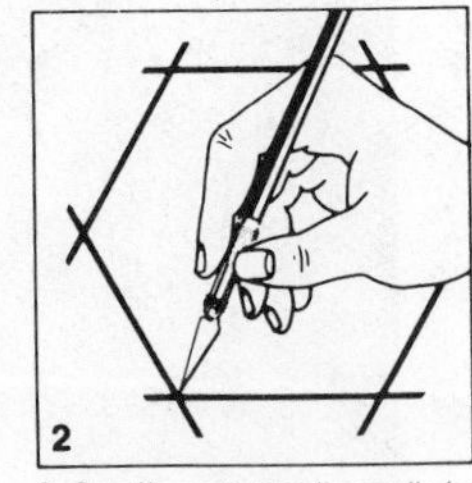

2. Cut off excess tape diagonally from outside to inside corner. Press firmly to cut through both layers of tape.

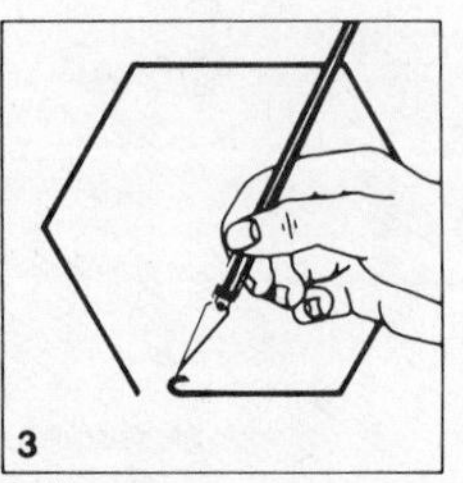

3. Lift tape with art knife and remove excess.

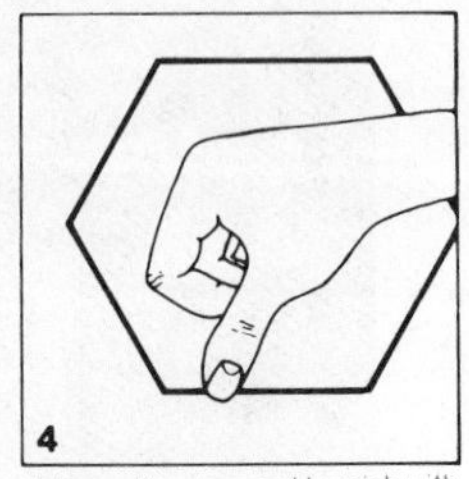

4. Reposition tape and burnish with fingers to secure corner.

Figure 3–85 Drafting aids: borders, lines, and special symbols. *Courtesy Chartpak.*

LINES

Drafting is a graphic language using lines, symbols, and notes to describe objects to be manufactured or built. Lines on drawings must be of a quality that will easily reproduce. All lines are dark, crisp, sharp, and of the correct thickness when properly drawn. There is no variation in darkness, only a variation in thickness, known as line contrast. Certain lines are drawn thick so that they will stand out clearly from other information on the drawing. Other lines are drawn thin. Thin lines are not necessarily less important than thick lines, but they are subordinate for identification purposes.

ANSI The American National Standards Institute (ANSI) recommends two line thicknesses with bold lines twice as thick as thin lines. This line standard relates to both manual and computer-aided drafting. Standard line thicknesses are 0.7 mm or .032 in. for thick lines, and 0.35 mm or .016 in. for thin lines. The actual width of lines may be more or less than the recommended thickness depending upon the size of the drawing or the size of the final reproduction. Drawings that meet military documentation standards require three thicknesses of lines: thick, medium, and thin. Figure 3–86 shows widths and types of lines as taken from ANSI standard, *Line Conventions and Lettering*, ANSI Y14.2M–1979. Figure 3–87 shows a sample drawing using the various kinds of lines.

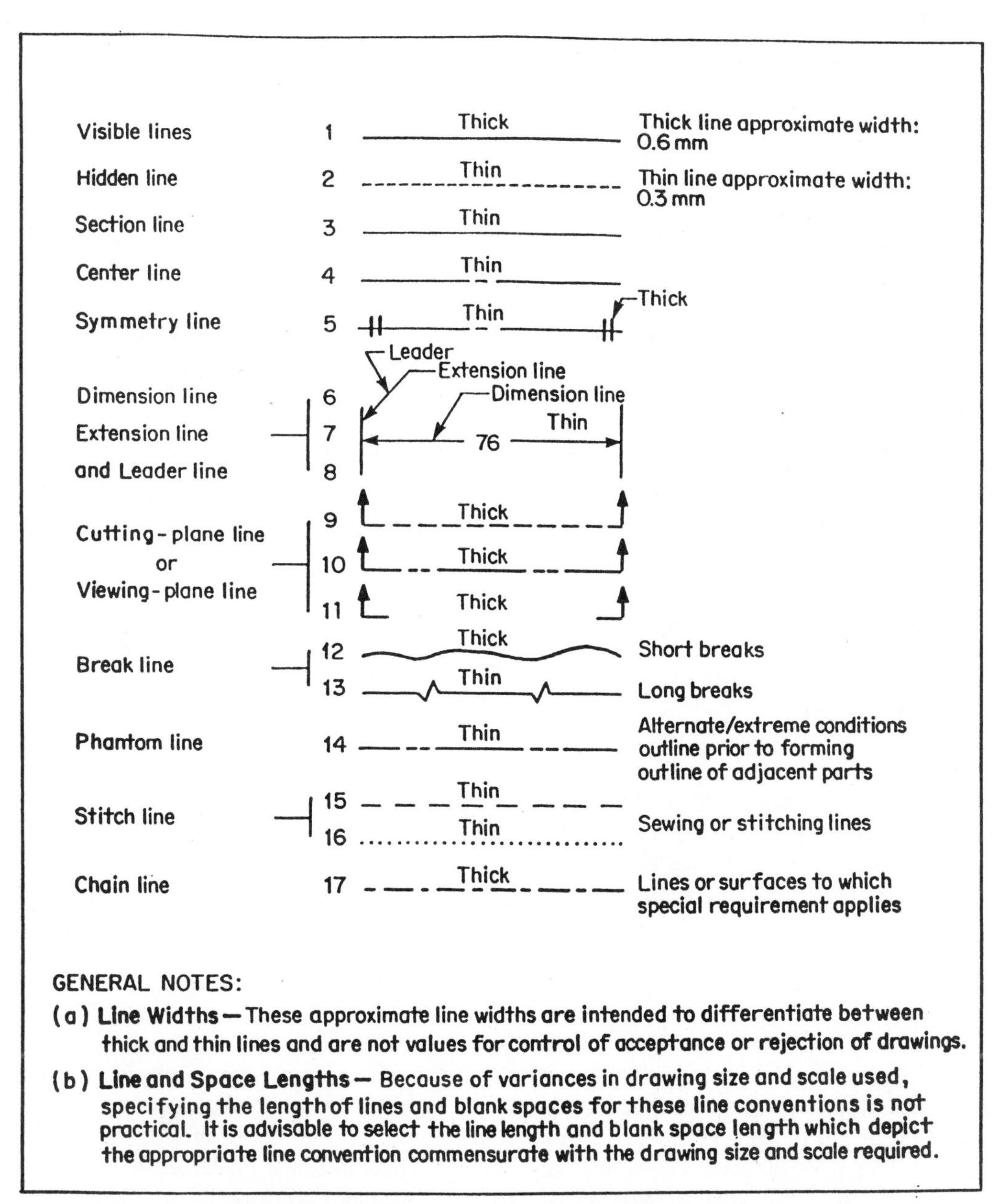

Figure 3–86 Line conventions, width and type of lines. *Reprinted with permission from ANSI — ASME Y14.2M–1992.*

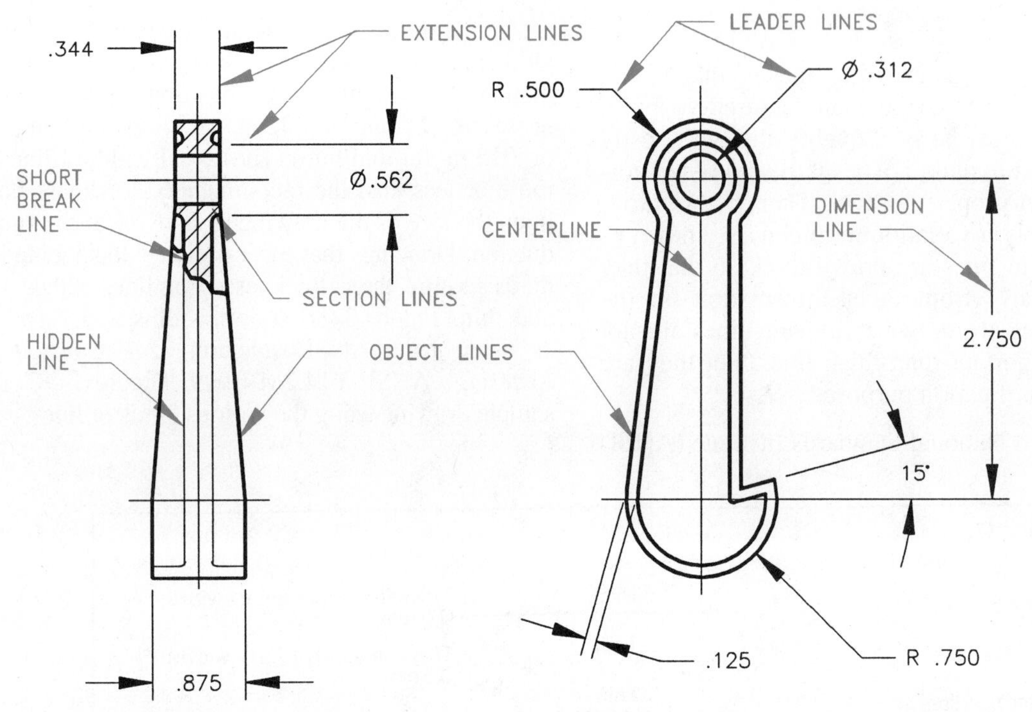

Figure 3–87 Sample drawing.

KINDS OF LINES

Construction and Guidelines

Construction lines are used for laying out a drawing. Construction lines are drawn very lightly so that they will not reproduce and will not be mistaken for any other line on the drawing. Construction lines are drawn with a 4H to 6H pencil and, if drawn properly, will not need to be erased. Use construction lines for all preliminary work.

Guidelines are drawn the same as construction lines and will not reproduce when properly drawn. Guidelines are used to keep lines of lettering in perfect alignment and of a constant height. For example, if lettering on a drawing is to be .125 in. high, then guidelines are drawn very lightly .125 in. apart. Guidelines must always be used for lettering. Some drafters prefer to use a light-blue lead for all guidelines and layout work. Light-blue lead will not reproduce and may be cleaner than graphite lead.

Object Lines

Object lines, also called visible lines, or outlines, describe the visible surface or edge of the object. They are drawn as thick lines as shown in Figure 3–88. Thick lines, remember, are usually drawn at 0.6 mm or .03 in. wide. Drafters usually use a soft lead, H or F, to draw object lines properly. You need to experiment to see which lead gives you the best line quality. For inked drawings, object lines can be made with a number 2.5 (0.70 mm) or a number 3 (0.80 mm) pen.

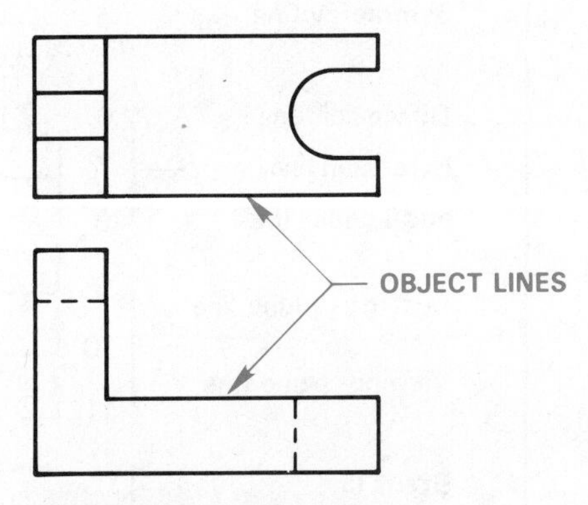

Figure 3–88 Object lines.

PROFESSIONAL HINT

1. To determine if your object line is dark and crisp enough, turn the paper over and hold it up to the light. The line should have a dark, consistent density. A light table, if available, also works very well to check darkness.
2. A conical, slightly rounded mechanical pencil point is most commonly used for drawing thick lines. A dull lead point will give you a wide, fuzzy line, but usually not a dark line. Rotate the pencil and go over each line enough times for proper darkness.
3. Most manual drafters use automatic pencils with a 0.7 mm lead for object lines.

DRAWING LINES

The basics of drawing lines with a CADD system are applicable to all systems, whether they use a puck, stylus, light pen, thumbwheels, joystick, or mouse. The term *pointers* refers to the pointing device used to select commands and digitize points. The term *digitize* means to press the stylus to the tablet or press the proper puck or keyboard button to digitize, or draw, a point. Several different commands will be given for one function in the following discussions, to give the reader an idea of the terms used in different systems.

Once you are in the digitize mode, your location on the screen is indicated by crosshairs. When you move the pointer, the crosshairs move. Unless you press a button or click the pointer, nothing is drawn on the screen. Some systems may continually display the coordinates of the crosshair location on the screen. These values change as you have the pointer. When constructing your drawing it is best to look at the video display as you draw. You will then be able to see if your lines are not straight, or if the drawing does not look right.

Begin drawing a line by first selecting the proper command such as POINT, INSERT POINT, PLACE LINE, or LINE. This command is used to locate corners of the object and connect them with straight lines. Digitize a point on the tablet and the coordinates of that point are displayed on the screen. Select another point and a straight line is drawn between the two. This shows as line 1 in Figure 3–89(a). Each selection of the pointer after that draws a line. The steps required to draw an object are shown in Figure 3–89(a)–(d). There is no need to select the INSERT POINT command between each digitized point. If you wish to break the line and move to another feature, select the INSERT POINT command again. This allows you to locate a new coordinate point without drawing an unwanted line.

Some systems use what is called a *rubber band line*. Once you have digitized one point, a line is attached to that point and to the crosshairs and moves with the crosshairs. Digitizing another point freezes the first line and another is now attached to the crosshairs.

Different types of lines can be indicated either before or after digitizing the points. Specifying the type of line before digitizing points is done by selecting a command such as LINE TYPE, LINE, or DASH. This command may then display several prompts asking for dash length, spacing, frequency, and the like. Any line drawn after you have entered a type will have the characteristics that you have chosen, and the type of line will change only after you select a new type of line. Should you wish to change an existing line to a different type, you can select the LINE TYPE command, set the new parameters, then digitize a point on the line. Some systems may require you to first activate an edit command such as MODIFY because you are changing something that has already been drawn.

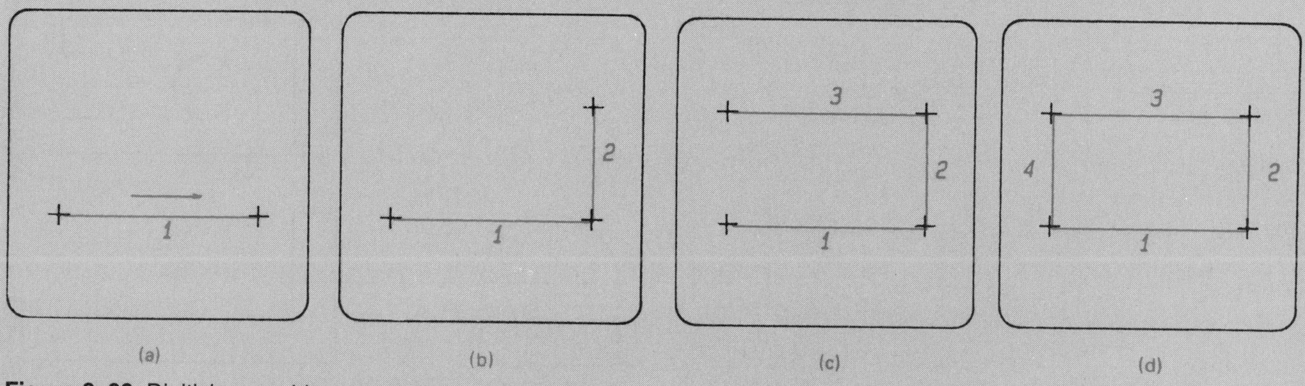

Figure 3–89 Digitizing an object one point at a time.

Hidden Lines

A *hidden line* represents an invisible edge on an object. (See Figure 3–90.) Hidden lines are thin lines, half as thick as object lines for contrast. Figure 3–91 shows hidden lines properly drawn with .125 in. dashes spaced .06 in. apart. Hidden lines, as all thin lines, can be drawn effectively with a sharp 2H or H mechanical pencil, a 0.3 mm to 0.5 mm automatic drafting pencil, or a number 0 (0.35 mm) technical pen tip when inking. The best way to draw hidden lines is to draw dash lengths and spaces by eye. This takes some practice, but is the fastest way.

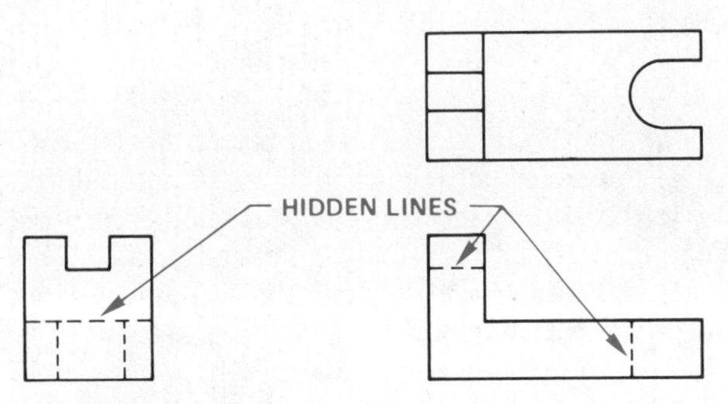

Figure 3–90 Hidden lines.

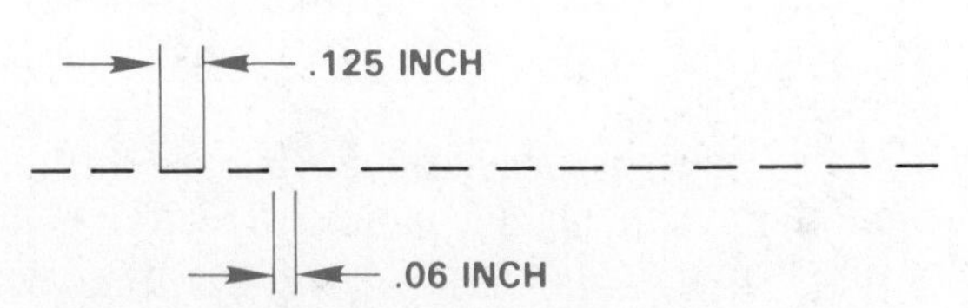

Figure 3–91 Hidden line representation.

Hidden Line Rules

The drawings in Figure 3–92 show situations where hidden lines meet or cross object lines and other hidden lines. These situations represent rules that should be followed when possible.

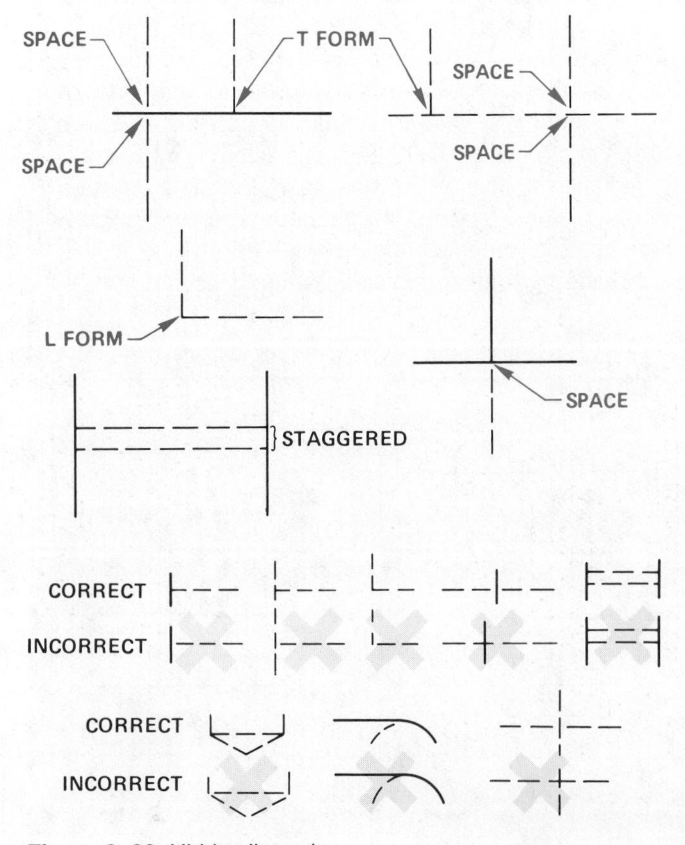

Figure 3–92 Hidden line rules.

Centerlines

Centerlines are used to show and locate the centers of circles and arcs, and are used to represent the center axis of a circular or symmetrical form. Centerlines are thin lines on a drawing. They should be about half as thick as an object line. The long dash is about .75 to 1.50 in. The spaces between dashes are about .062 in. and the short dash about .125 in. long. The length of long lines varies for the situation and size of the drawing. Try to keep the long lengths uniform throughout the centerline. (See Figure 3–93.) Small centerline dashes should cross only at the center of a circle or arc. (See Figure 3–94.) Small circles should have centerlines as shown in Figure 3–95.

Centerlines for holes in a bolt circle may be drawn either of two ways, depending upon how the holes will

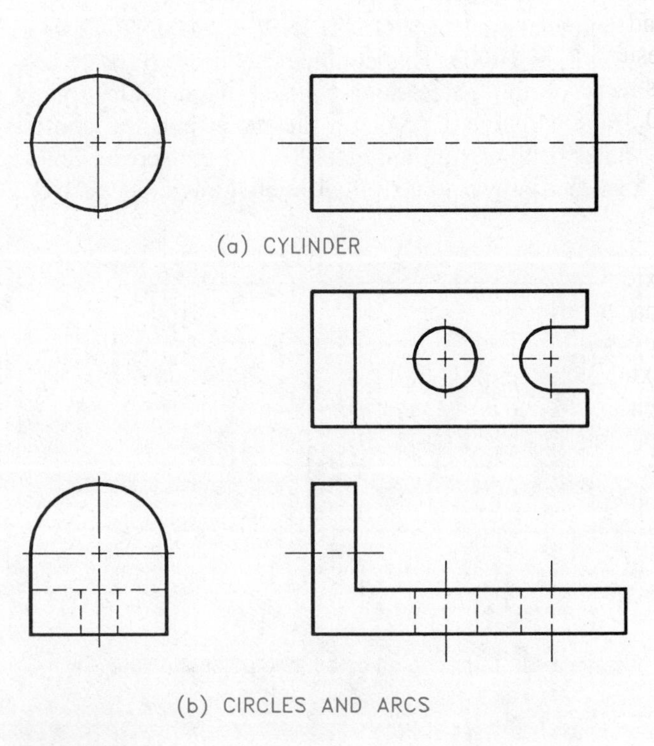

Figure 3–93 Centerline representation and examples.

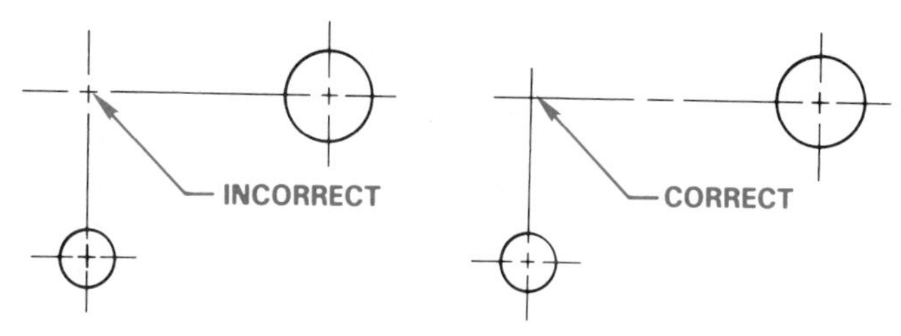

Figure 3–94 Centerline rules.

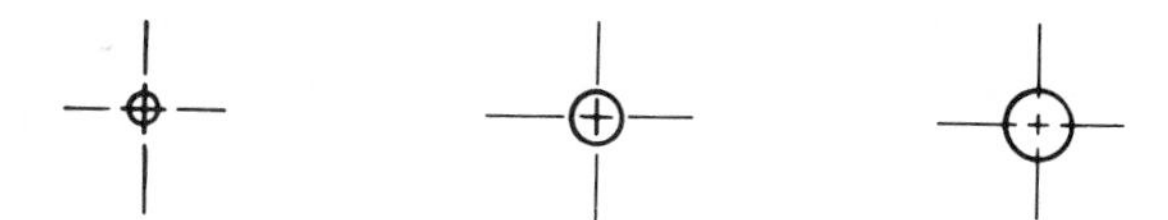

Figure 3–95 Centerlines for small circles.

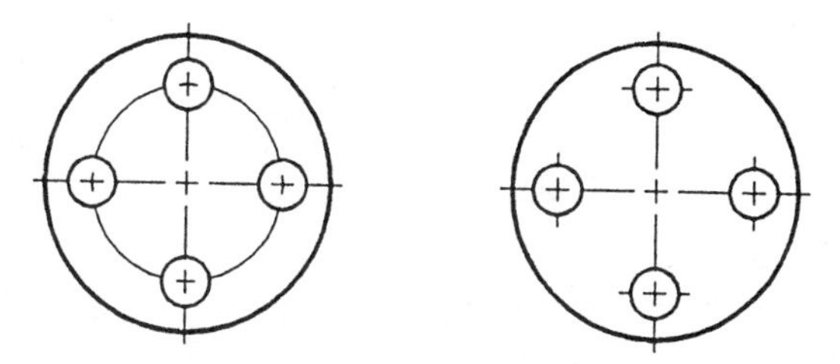

Figure 3–96 Bolt circle centerline options.

be located, as shown in Figure 3–96. A bolt circle is a pattern of holes arranged in a circle.

When a centerline represents symmetry, as in the centerplane of an object, the symmetry symbol, shown in Figure 3–97, may be used if needed for clarity.

The centerline is commonly drawn with a sharp 2H or H lead, or a 0.3- to 0.5-mm automatic pencil. A 0.3-mm pencil may provide best results with a soft lead and a 0.5-mm pencil with a harder lead. Remember the results should be dark, crisp, and sharp lines that are half as thick as object lines. When inking, a number 0 (0.35-mm) technical pen gives best results.

Extension Lines

Extension lines are thin lines used to establish the extent of a dimension. Extension lines begin with a short space from the object and extend to about .125 in. beyond the last dimension, as shown in Figure 3–98. Extension lines may cross object lines, centerlines, hidden lines, and other extension lines, but they may not cross dimension lines. Circular features, such as holes, are located by their centers in the view where they appear as circles. In this practice, centerlines become extension lines as shown in Figure 3–99.

Dimension Lines and Leader Lines

Dimension lines are thin lines capped on the ends with arrowheads and broken along their length to provide a space for the dimension numeral. *Dimension lines* indicate the length of the dimension. (See Figure 3–100.)

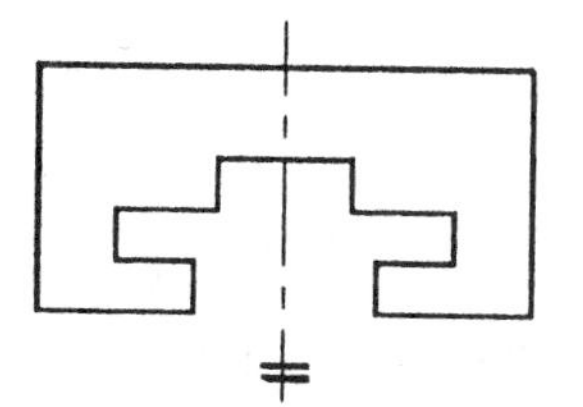

Figure 3–97 Symmetry symbol, thick.

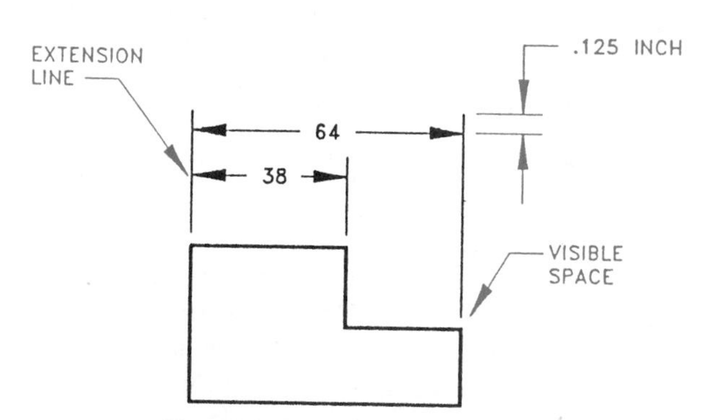

Figure 3–98 Extension lines.

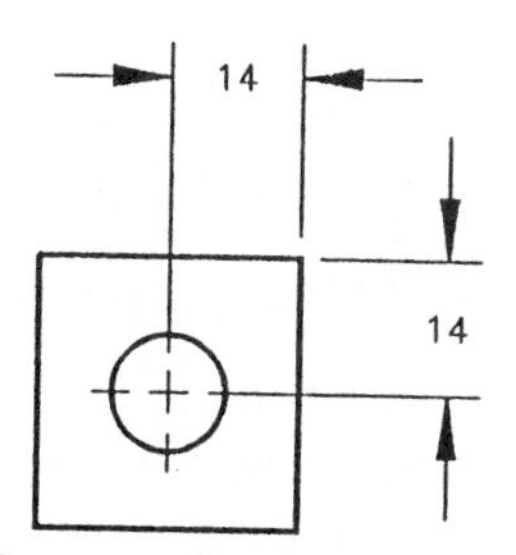

Figure 3–99 The centerline becomes an extension line.

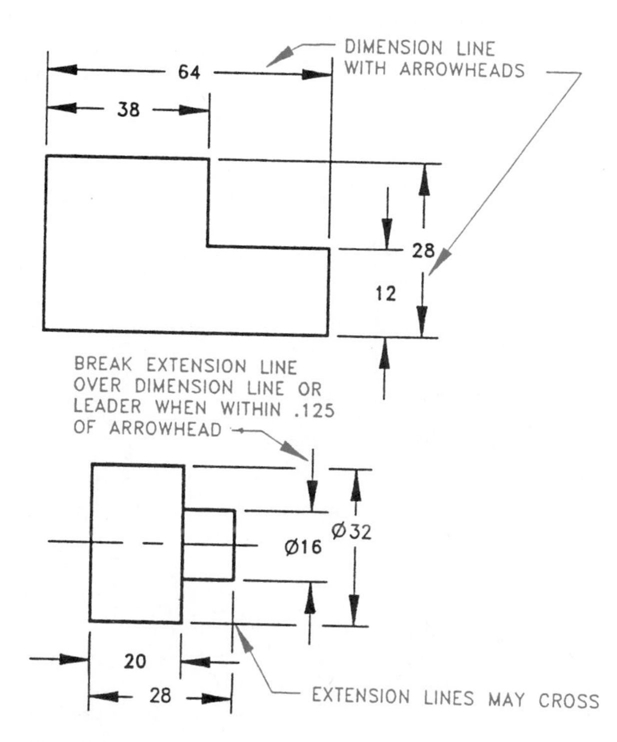

Figure 3–100 Dimension lines.

Leaders, or leader lines, are thin lines used to connect a specific note to a feature as shown in Figure 3–101. Leaders may be drawn at any angle, but 45°, 30°, or 60° lines are most common. Slopes greater than 75° or less than 15° from horizontal should be avoided. The leader has a .25-in. shoulder at one end that begins at the center of the vertical height of the lettering and an arrowhead at the other end pointing to the feature. If the leader in Figure 3–101 were to continue from the point where the arrowhead touches the circle, it would intersect the center. (See Figure 3–102.)

Arrowheads

Arrowheads are used to terminate dimension lines and leaders. Properly drawn arrowheads should be three times as long as they are wide. All arrowheads on a drawing should be the same size. Do not use small arrowheads in small spaces. (Limited space dimensioning is covered in Chapter 8.) Some companies require that arrowheads be drawn with an arrow template, while others accept properly drawn freehand arrowheads. Individual company preference dictates whether arrowheads are filled in or left open as shown. Microfilming requires that arrowheads be filled in for clarity. There are four optional arrowhead styles shown in order of preference in Figure 3–103.

Cutting-plane and Viewing-plane Lines

Cutting-plane lines are thick lines used to identify where a sectional view is taken. *Viewing-plane lines* are also thick and are used to identify where a view is taken for view enlargements or for partial views. Cutting-plane and viewing-plane lines are properly drawn in either of the two ways shown in Figure 3–104. The approximate dash and space sizes are given in the figure indicated. The cutting-plane line takes precedence over the centerline when used in the place of a centerline.

The scale of the view may be increased or remain the same as the view from the viewing plane, depending on the clarity of information presented. When the location of the cutting plane or viewing plane is easily understood or if the clarity of the drawing is improved, the

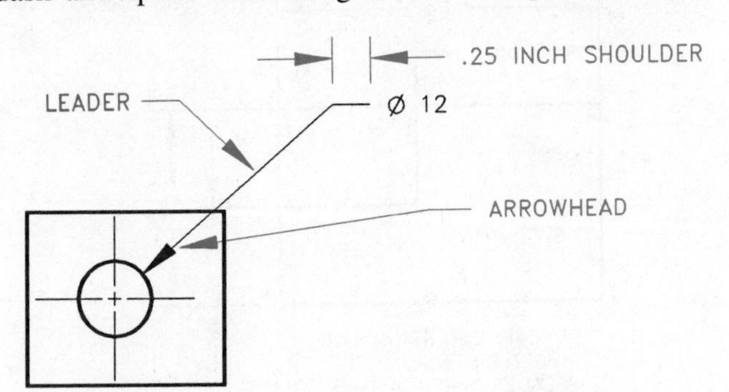

Figure 3–101 Leader line.

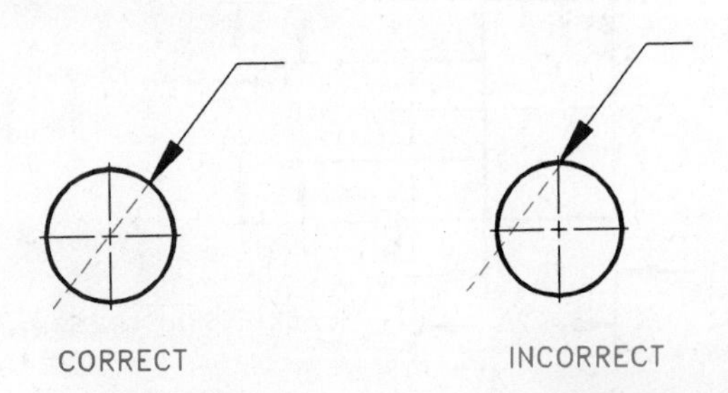

Figure 3–102 Circle-to-leader line relationship.

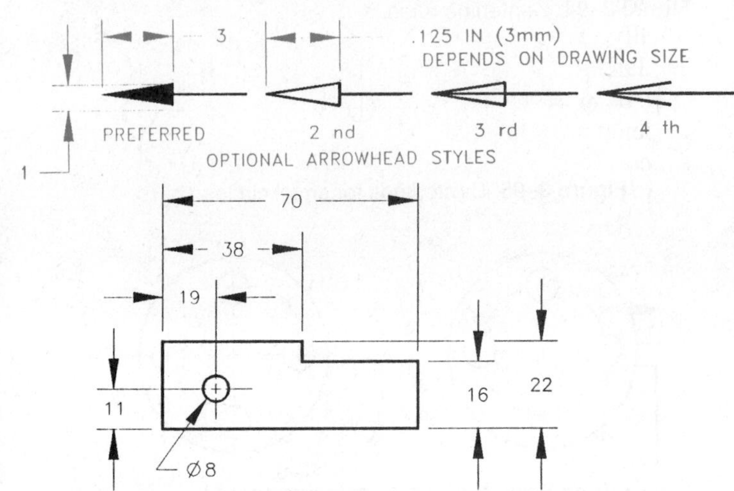

Figure 3–103 Arrowheads.

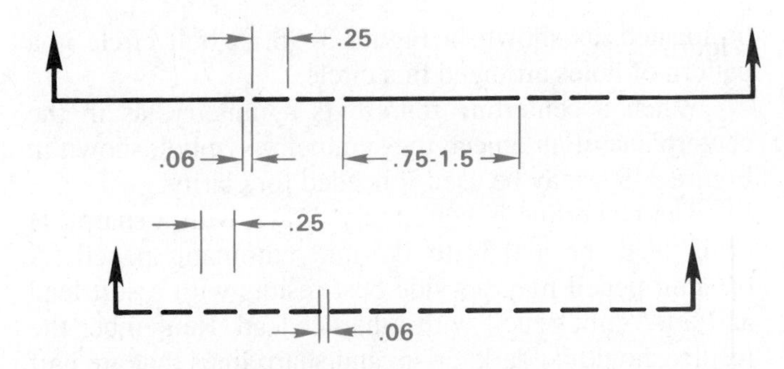

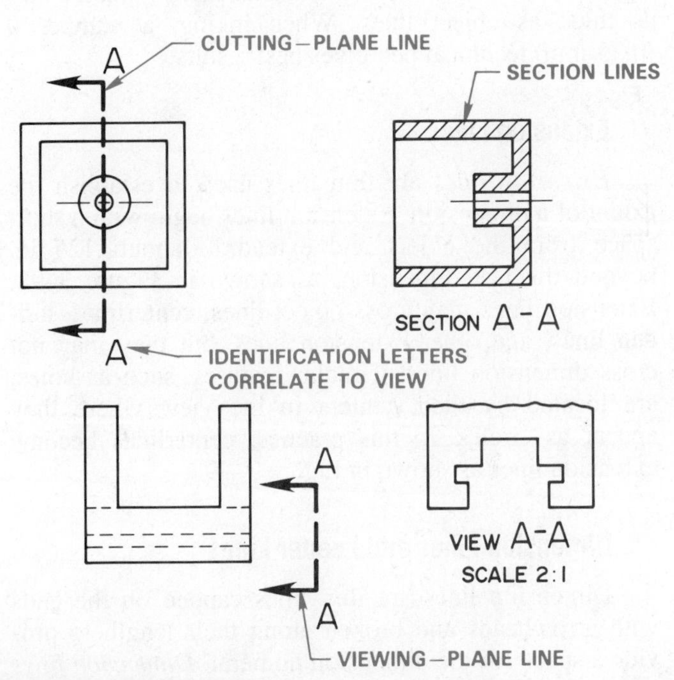

Figure 3–104 Cutting- and viewing-plane lines.

portion of the line between the arrowheads may be omitted as shown in Figure 3–105.

Section Lines

Section lines are thin lines used in the view of a section to show where the cutting-plane line has cut through material. (See Figure 3–106.) Section lines are drawn equally spaced at 45°, but may not be parallel or perpendicular to any line of the object. Any convenient angle may be used to avoid placing section lines parallel or perpendicular to other lines of the object; 30° and 60° are common. Section lines that are more than 75° or less than 15° from horizontal should be avoided. Section lines should be drawn in opposite directions on adjacent parts. (See Figure 3–112.) For additional adjacent parts any suitable angle may be used to make the parts appear clearly separate. The space between section lines may vary depending on the size of the object. (See Figure 3–107.) Figure 3–108 shows correct and incorrect applications of section lines. When a very large area requires section lining, you may elect to use outline section lining, as shown in Figure 3–109.

The section lines shown in Figures 3–106 through 3–109 were all drawn as general section-line symbols. General section lines can be used for any material and are specifically used for cast or malleable iron. Coded section-line symbols, as shown in Figure 3–110, are not commonly used on detail drawings, as they are more difficult to draw and the drawing title block usually identi-

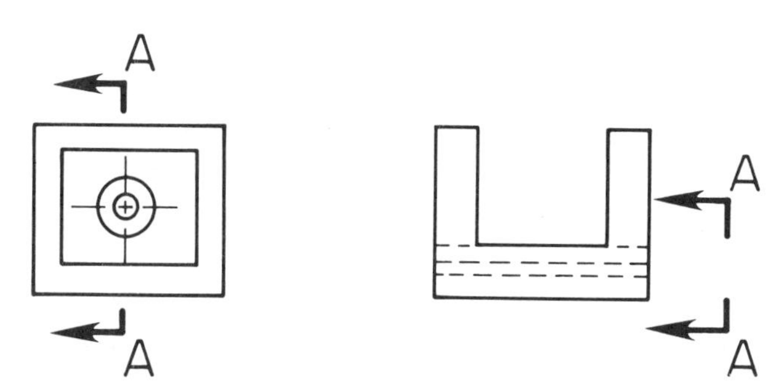

Figure 3–105 Simplified cutting- and viewing-plane lines.

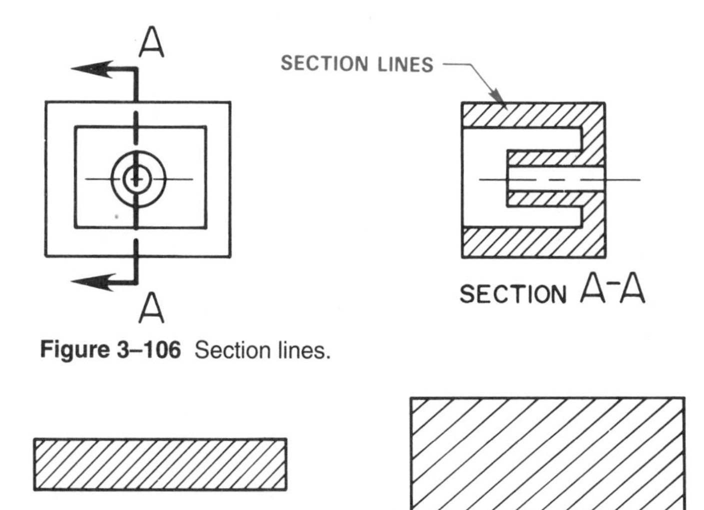

Figure 3–106 Section lines.

Figure 3–107 Space between section lines.

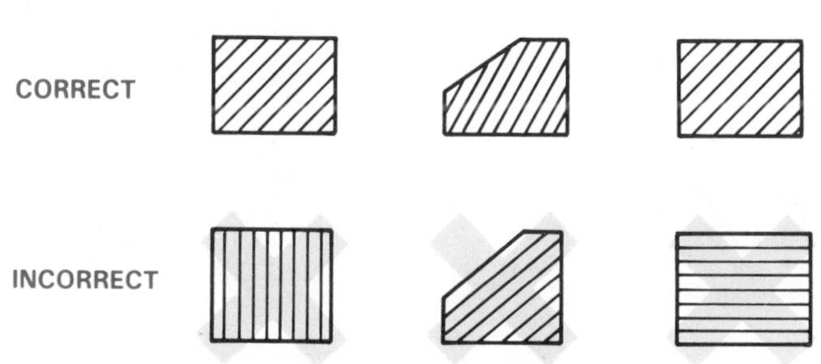

Figure 3–108 Correct and incorrect section lines.

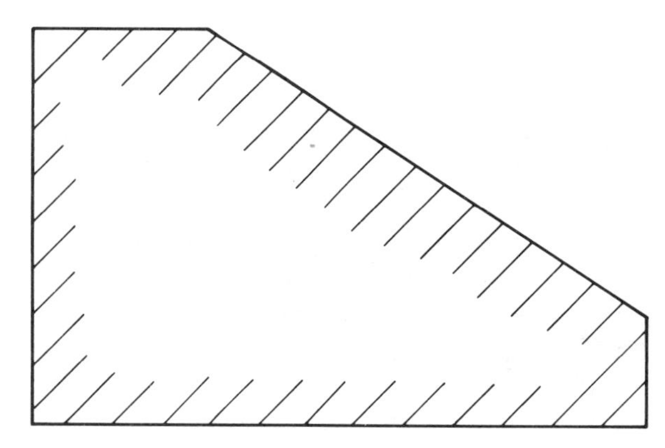

Figure 3–109 Outline section lines.

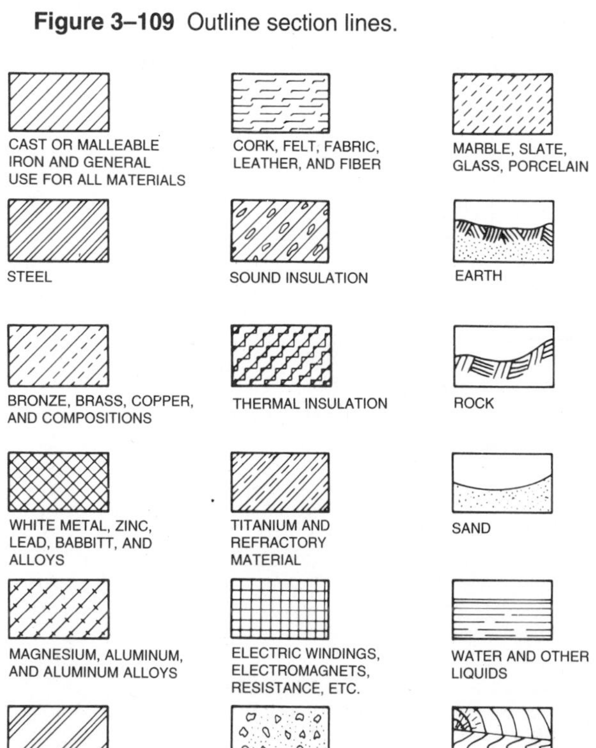

Figure 3–110 Coded section lines.

fies the type of material in the part. Coded section lines may be used when the material must be clearly delineated, as in the section through an assembly of parts made of different materials. (See Figure 3–111.) Very thin parts may be completely blackened rather than section lined. This option is often used for a gasket as shown in Figure 3–112.

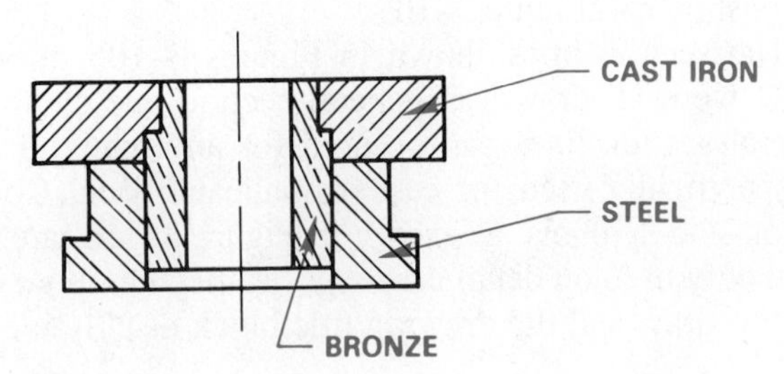

Figure 3–111 Coded section lines in assembly.

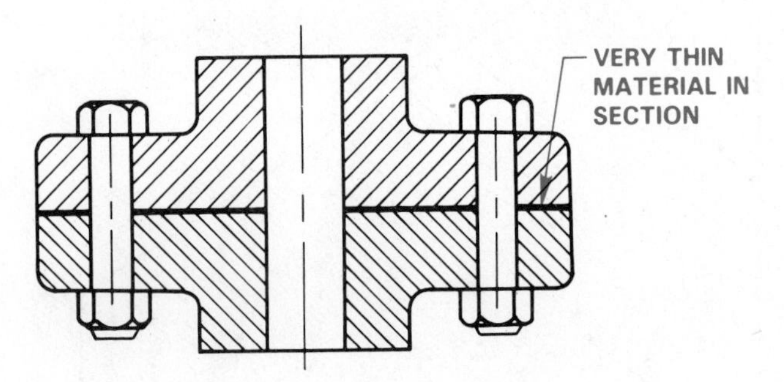

Figure 3–112 Very thin material in section.

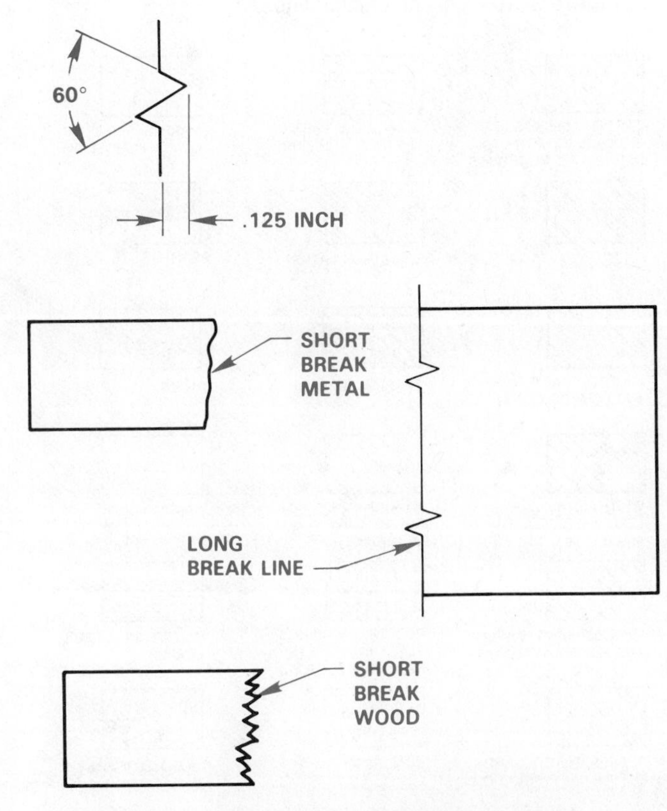

Figure 3–113 Long and short break lines.

Break Lines

There are two types of *break lines*: the short break and long break line. The thick, short break is very common on detail drawings, although the thin, long break may be used for breaks of long distances at the drafter's discretion. (See Figure 3–113.) Other conventional breaks may be used for cylindrical features as in Figure 3–114.

Phantom Lines

Phantom lines are thin lines made of one long and two short dashes alternately spaced as shown in Figure 3–115. Phantom lines are used to identify alternate positions of moving parts, adjacent positions of related parts, or repetitive details. (See Figure 3–116.)

Chain Lines

Chain lines are thick lines of alternately spaced long

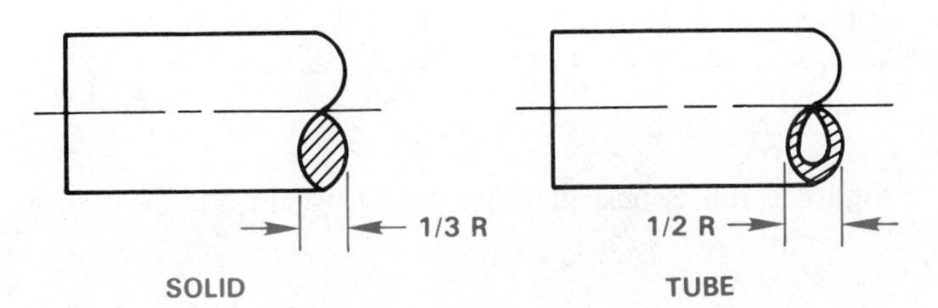

Figure 3–114 Cylindrical conventional breaks.

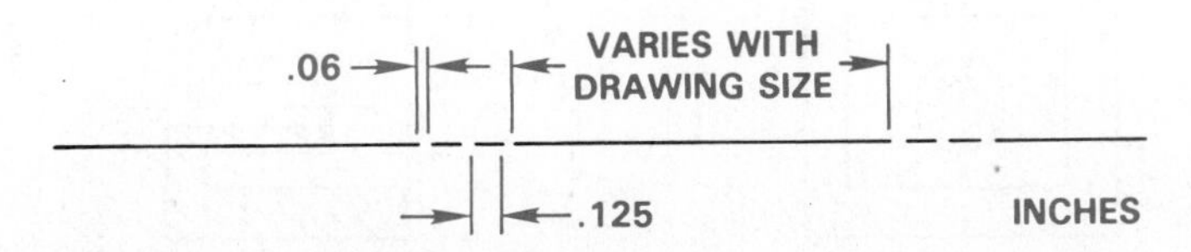

Figure 3–115 Phantom line.

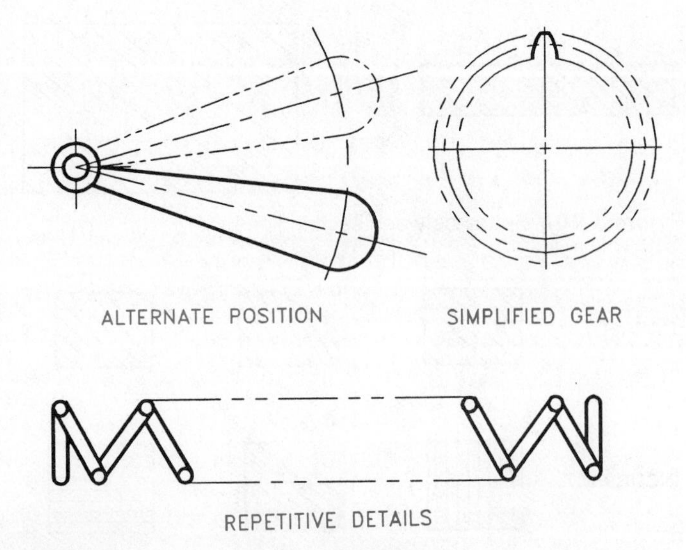

Figure 3–116 Phantom line examples.

and short dashes used to indicate that the portion of the surface adjacent to the chain line receives some specified treatment. (See Figure 3–117.)

Stitch Lines

There are two types of acceptable *stitch lines*. One is drawn as thin, short dashes, the other as .016-in. diameter dots spaced .12 in. apart. They are used to indicate the location of a stitching or sewing process as shown in Figure 3–118.

PENCIL AND INK LINE TECHNIQUES

Pencil Techniques

There are four basic properties of correctly drawn lines: uniformity, contrast, darkness, and sharpness. Line uniformity means that all lines are drawn their proper thickness without variation. For example, all thick lines (object, cutting-plane) are the same degree of thickness. All thin lines (center, dimension, extension) are the same thinness. Line contrast is the variation that exists between different types of lines; thick and thin.

ANSI According to ANSI standards, thick and thin are the two line thicknesses used. Military (MIL) standards recommend three line thicknesses: thick (cutting and viewing plane, short break, and object); medium (hidden and phantom); and thin (center, dimension, extension, leader, long break, and section).

All lines should be the same darkness with no variation. Lines must be drawn opaque so that light will not pass through. Lines are sharp when edges are clear and crisp. Fuzzy edges on lines result from the texture of the paper, or a dull or too-soft pencil point. After sharpening your mechanical pencil, round the point slightly in order to keep the tip from breaking. If you use an automatic pencil, do not let the lead out too far. (See Figure 3–119.)

Wash your hands and equipment frequently to keep your drawings clean. Try not to handle your pencil leads.

Lay out your entire drawing with construction lines before darkening any lines. A combination of pressure and pencil lead hardness makes the best lines. Draw with your pencil angled about 45° in the direction of travel and 90° with the scale or triangle. If you are right handed, it is generally best to draw horizontal lines from left to right and vertical lines from bottom to top. The opposite may be true for left-handed drafters. (See Figure 3–120.) Do not use excessive pressure. You do not have to dig trenches in the drawing board. The more pressure you use, the more difficult it is to erase. If your hand cramps while drawing or lettering, you are forcing yourself into an unnatural stroking technique. Relax! Try

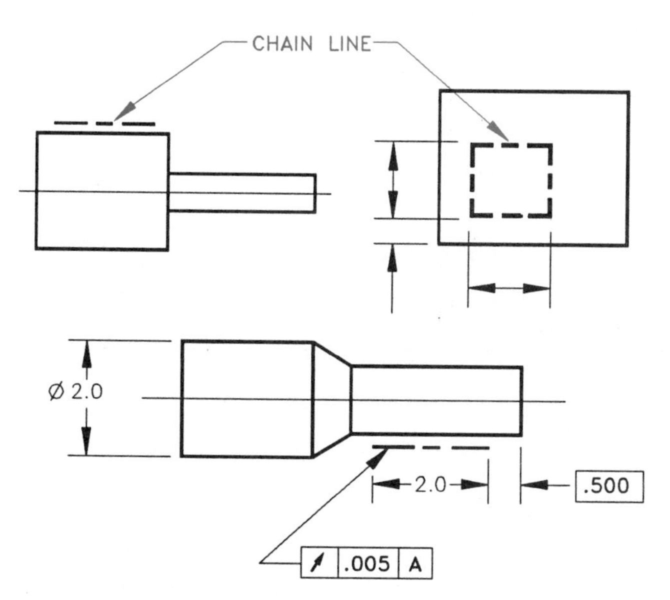

Figure 3–117 Chain lines.

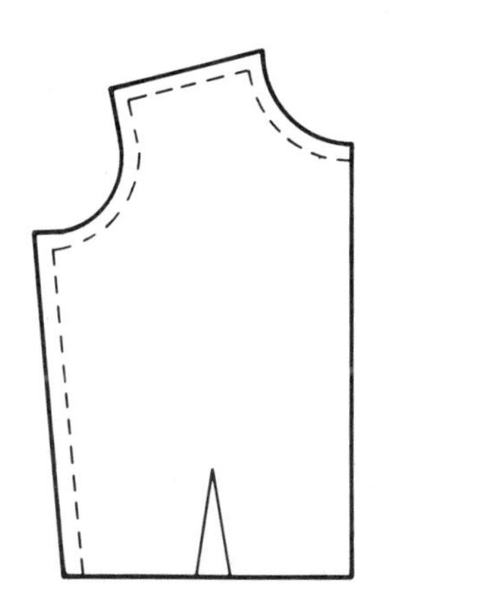

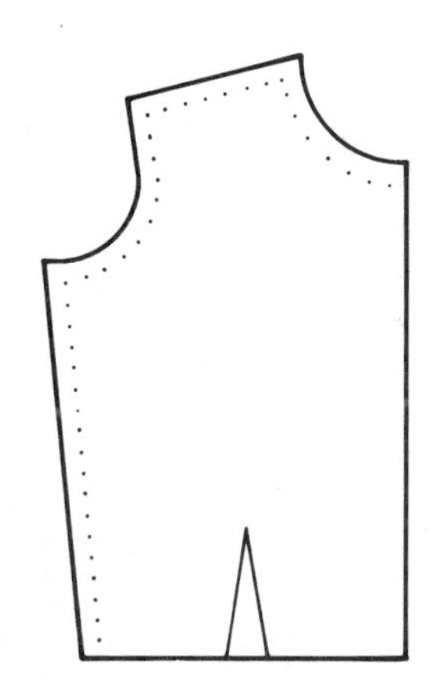

Figure 3–118 Stitch lines.

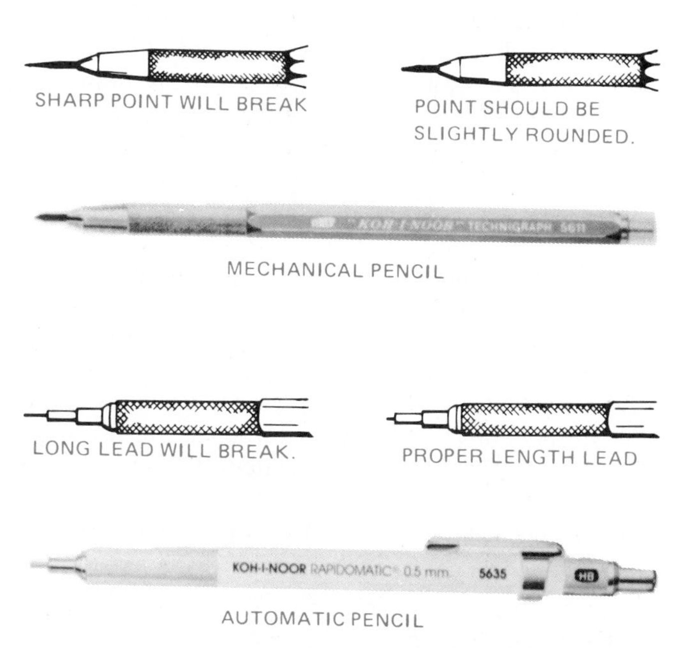

Figure 3–119 Pencil lead lengths, pencils. *Courtesy Koh-I-Noor Rapidograph, Inc.*

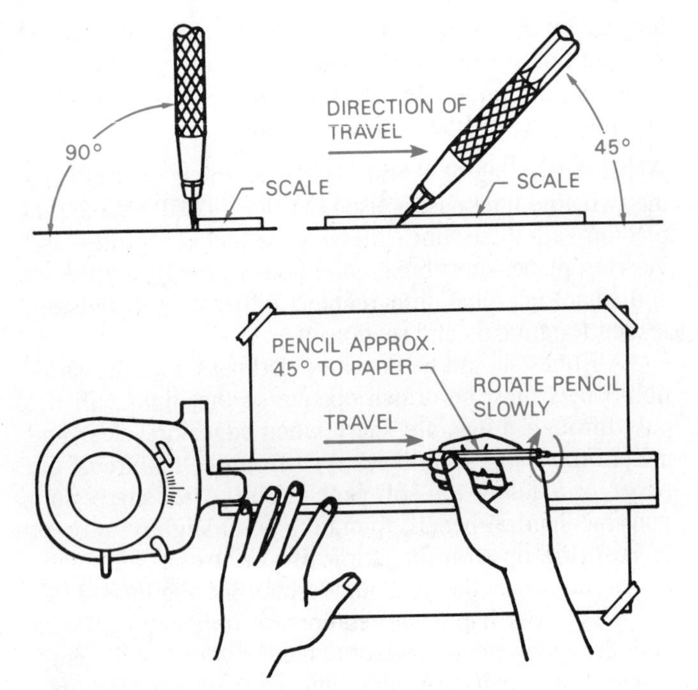

Figure 3–120 Pencil use.

different methods until you find a suitable system of strokes.

Make thick lines with the mechanical pencil using a slightly rounded point and several passes over each line. Use a 0.7- or 0.9-mm automatic pencil held vertically to establish full contact between the lead and the paper. Rotate your pencil as you draw a line to keep the line width constant. In order to obtain lines that are dark, crisp, and sharp, you may need to go over each line several times. If light passes through your lines, they are drawn too lightly. Hold your drawing up to a light or place it on a light table to see if light passes through.

Always draw radii first, then connect the straight lines. This is also a good technique for ink drawings. (See Figure 3–121.) If you make the straight lines first, you may have trouble aligning the arcs.

Line Layout

Draw thin lines first. Draw thin horizontal lines beginning from the top of the sheet and work downward. Draw vertical thin lines next, starting on the left side of the sheet and working toward the right side for right-handed drafters. Next draw all circles and arcs. Last, draw all horizontal and vertical thick lines in the same direction as thin lines. Do all lettering when the line work is complete. Try to avoid going back over lines or lettering that has already been drawn in order to help avoid smudging.

Inking Techniques

Ink or graphite only adheres cleanly to one side of vellum. If the ink beads or skips while making a line, you are working on the wrong side of the paper. If your vellum is in a pad, the top side is the good side. Mark that side when you remove a sheet from the pad. Paper with preprinted borders and title blocks have the correct side predetermined. Some manufacturers have a water mark that can be seen when the paper is held up to the light. When the water mark is readable, the paper is right side up. For Mylar®, there is either a single or double matte surface. Matte refers to the textured surface as opposed to a glossy surface. Double matte surface may be drawn on either side. Single matte surface requires you to draw on the matte side, not the shiny, or glossy, side.

Use template lifters or templates and other equipment designed for inking that keep the edge of the tool slightly off the paper. Template lifters are plastic shapes that can be attached to the back of a template or other instrument to help keep the instrument off the drafting surface, as seen in Figure 3–122. Another option is to place another template under the one being used. (See Figure 3–123.) There are also template risers available

Figure 3–121 Draw circles and arcs first.

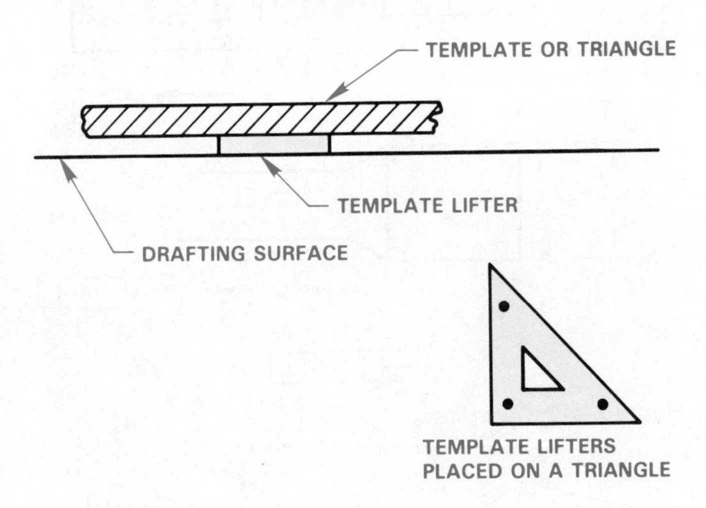

Figure 3–122 Template lifters.

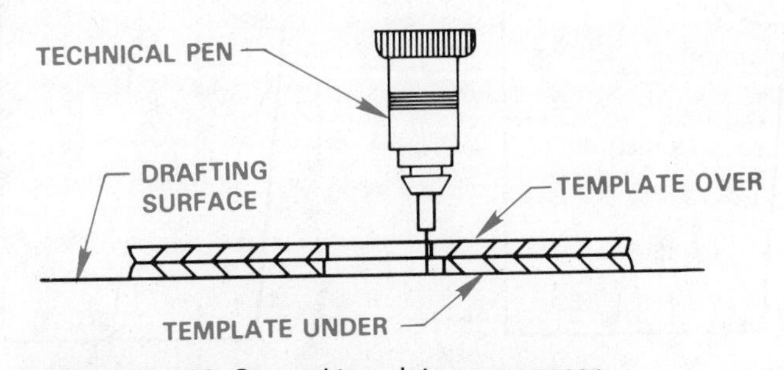

Figure 3–123 Second template as a spacer.

that are long plastic strips that fit on the edges of the template to help keep it off of the drawing surface. Figure 3–124 shows how the template riser functions when attached to a template. Drafting machine scales have long been manufactured with edge relief for inking. Now, templates, triangles, and other equipment are being made with ink risers built in, as shown in Figure 3–125.

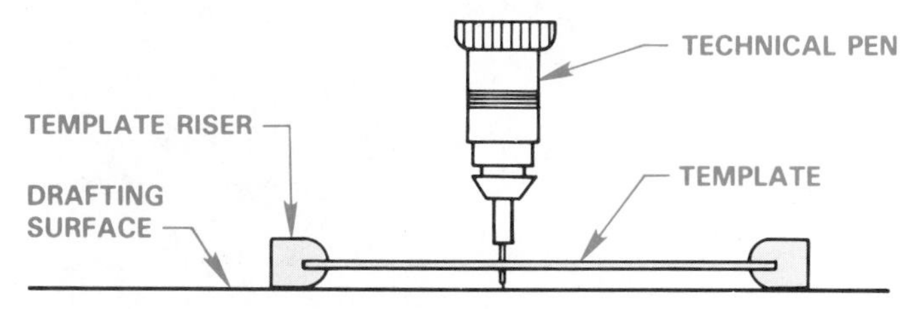

Figure 3–124 Template risers.

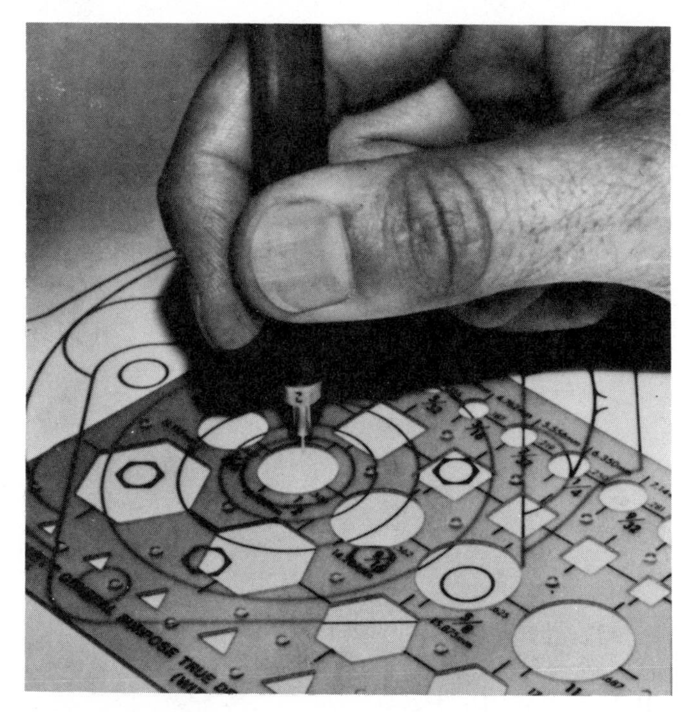

Courtesy Chartpak.

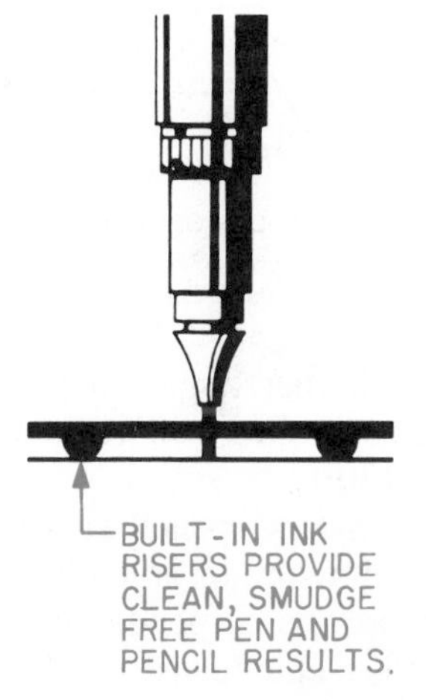

Figure 3–125 Built-in ink risers. *Courtesy Chartpak.*

If you draw with ink without taking these precautions, the ink could easily flow under the template and cause a mess as shown in Figure 3–126.

Periodically check your technical fountain pens for leaks around the tip to prevent your hands from smearing your drawing. Also check the tip for a drop of ink before you begin a line. (See Figure 3–127.) Have a piece of cloth or a tissue available to help keep the tip free of ink drops. Keep your pens clean and loaded with fresh ink to help keep proper ink flow for trouble free use. Also, keep a piece of paper handy for scratching your pen to start the ink flowing.

Ink your lines beginning with thin horizontal lines and work from top to bottom and from left to right if right handed. The technique of drawing horizontal lines top to bottom and vertical lines left to right doesn't eliminate graphite smudges as with lead, but allows inked lines to dry as you go along and saves valuable drafting time. Inked lines are dry when they do not appear glossy. Glossy lines are wet; keep equipment away from them. Next, ink all circles and arcs. Ink all thick horizontal and vertical lines in the same manner as thin horizontal and vertical lines. Finally, do all lettering.

When using technical pens, hold the pen perpendicular (90°) to the paper for the best results. Do not apply any pressure to the pen. Allow the pen to flow easily over the vellum or Mylar®. Figure 3–128 shows proper technical pen use. If the technical pen is not held 90° to the surface, the resulting line may be fuzzy or rough on the edge. Move your technical pen at a constant speed. Do not go too fast and do not slow at the ends of lines.

Figure 3–126 Ink smear.

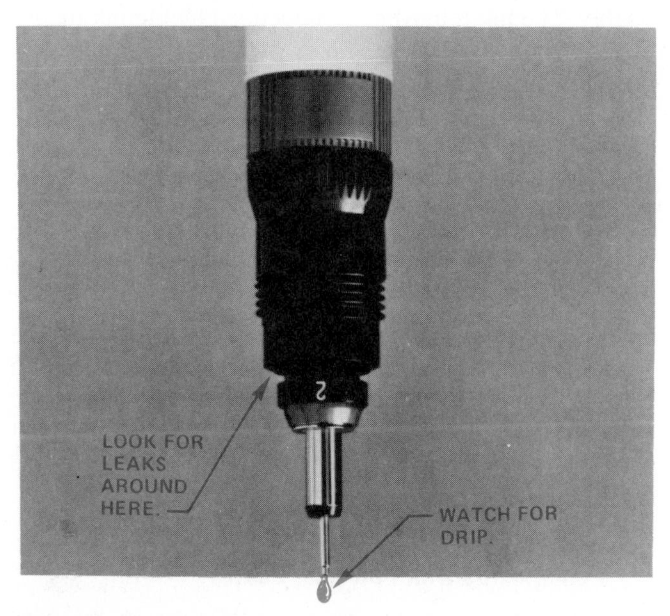

Figure 3–127 Clean ink drip from pen. *Courtesy Koh-I-Noor Rapidograph, Inc.*

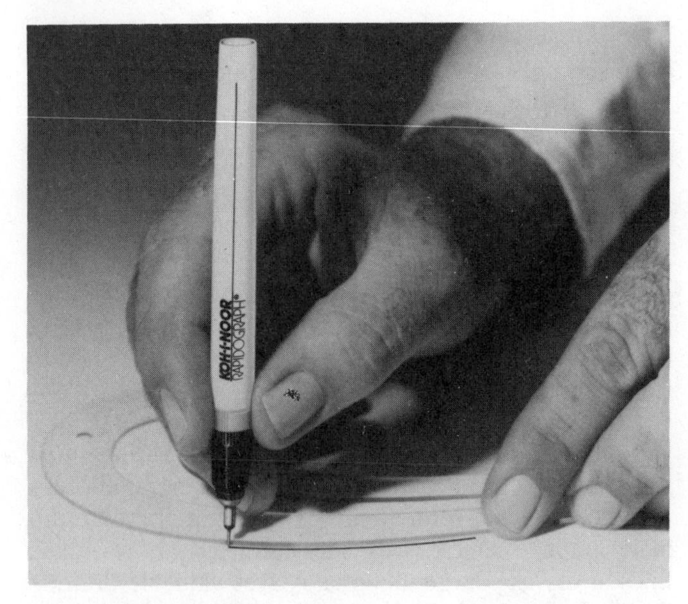

Figure 3–128 Proper technical pen use. *Courtesy Koh-I-Noor Rapidograph, Inc.*

Figure 3–129 Ink removal from Mylar® with plastic eraser. *Courtesy J.S. Staedtler, Inc.*

Following these hints will help keep your line consistent in width and the pen point from skipping.

Erasing Ink

Erasing ink from vellum is possible in small areas but doing so may destroy its surface. A smooth surface cannot be inked over again satisfactorily. An electric eraser would be best used here, but overenthusiasm will burn a hole through the paper. Electric erasers are excellent tools, but be careful to touch the paper surface lightly.

To erase on polyester film, it is best to apply a little water with a felt tip or a clean cotton swab. Remove any excess water with a blotter. Allow the area to dry thoroughly. The ink remover recommended by the ink manufacturer can also be used. A polyester eraser also removes ink. (See Figure 3–129.) Use an eraser with care so you do not destroy the matte surface, which makes further drafting in that area difficult. Some polyester erasers have erasing fluid inside to make ink removal easier, or use an eraser with a small amount of water.

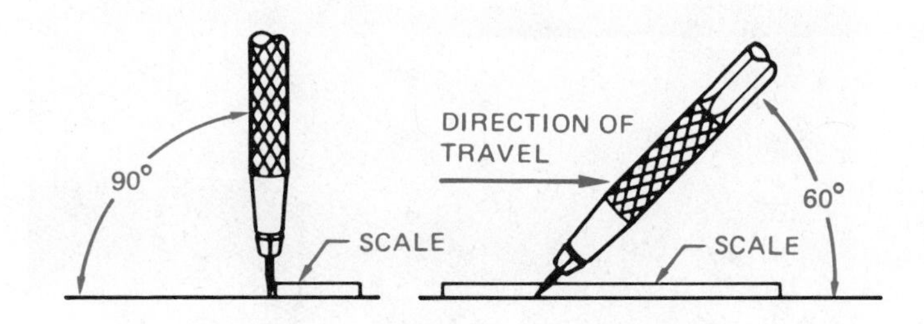

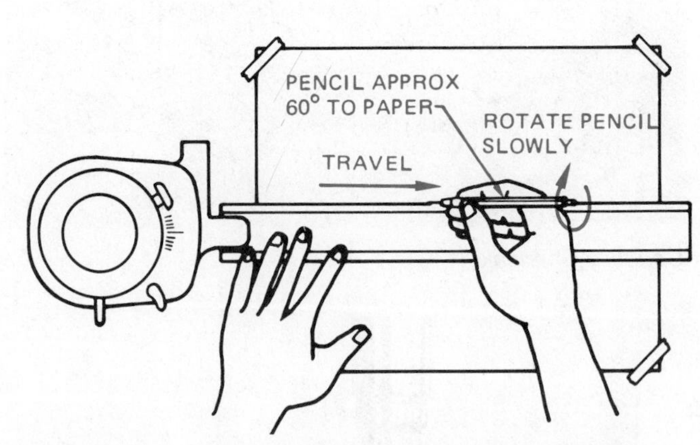

Figure 3–130 Line technique, polyester lead on polyester film (Mylar®).

Polyester Lead on Film

When pencils are used on Mylar® rather than ink, the recommended lead is made of polyester. The quality of a line on polyester film depends on how well the matte finish is maintained. If you use single matte film, do not try to draw on the glossy side; use polyester lead on the matte side.

Pencil Skills with Polyester Lead

- Reduce the angle of the pencil on film to between 55° and 65°. Sharper angles may penetrate the matte and lesser angles will reduce line quality. (See Figure 3–130.)
- Maintain a constant pencil and wrist angle throughout the drawing of an entire line. The tendency to increase the perpendicularity toward the end of the line is a chief cause of film *gouging*.

- Draw with a single line, in one direction. Tracing a line in both directions deposits a double line, and causes smearing and damage to the matte.
- Draft with a light touch. Drafting films require up to 40 percent less pressure than other media. Smearing and embossing can be reduced with less pressure. It will take practice to use polyester lead after drawing with graphite lead.
- Draw with a slightly rounded point when using a mechanical lead holder, or use an automatic lead holder of proper thickness.
- As you move the pencil, rotate it. This will help to keep the lines uniform in width and reduce point wear.
- The surface of your drafting board is also a factor in line quality. Use a recommended backing material on your table. Check with your local vendor.
- Erase with a vinyl eraser. If an electric eraser is used, be very careful not to destroy the matte.

PROFESSIONAL PERSPECTIVE

Sketching

Freehand sketching is an important skill if you are a manual or a CADD drafter. Preparing a sketch before you begin a formal drawing may save many hours of work. The sketch assists you in the layout process in that it allows you to:

- Decide how the drawing should appear when finished.
- Decide how big to make the drawing.
- Determine the sheet size for manual drafting or the screen limits for CADD.
- Establish the coordinate points for the computer drawing.

A little time spent sketching and planning your work saves a lot of time in the final drafting process. Sketches are also a quick form of communication in any professional environment. You can often get your point across or communicate more effectively with a sketch.

Lettering

The appearance of manual drawings may be greatly enhanced by quality freehand lettering. An otherwise good drawing may look unprofessional with poor lettering. However, good freehand lettering does not come easily for some people; it takes a lot of practice. The only substitute for practice is an inherent talent. For some people, lettering comes naturally.

CADD lettering is a different story. Lettering on a computer is as easy as typing, and the lettering comes out fast and is perfect every time. Another exciting aspect of CADD lettering is the many styles available. There are even lettering styles available that duplicate the artistic appearance of the best freehand architectural lettering. Freehand lettering may not be important when the computer age totally takes over the drafting industry. For now, however, lettering on manual drawings is very important.

Lines

Quality line work is dependent on the opaqueness of the lines. The diazo reproduction process creates an excellent print if the light is unable to pass through the lines. If lines appear fuzzy, take a magnifying glass and look at the line density. If the line does not make an opaque image, you will not get a good print. Do not blame it on the print machine. You cannot make poor lines look good if they are not opaque. The job is difficult when drawing with pencil on vellum. You need the right combination of pressure, lead hardness, and skill to produce properly executed lines. This combination is different for each individual. Some people draw acceptable lines with an F lead while others are successful with a 2H lead. Also, one lead may not work well for all lines. You may need to have a few automatic pencils with different lead grades. If you are inking on polyester film, you will get good-quality lines, but you may have trouble getting used to working with the pitfalls of ink, such as smearing and error cleanup. Either manual technique takes a lot of practice.

Drafting with the computer is a different story, because all you need to do is make lines on the screen and then it is up to the plotting or printing process to reproduce a quality drawing on paper. You can choose a check print on a dot matrix printer for a quick copy where quality is unimportant, or you can ink plot on vellum or polyester film for excellent plots. You can even plot different line thicknesses by using thick or thin pens as appropriate. You can also plot in different colors if you want to emphasize a part of the drawing.

ENGINEERING PROBLEM

The engineering drafter often works from sketches or written information provided by the engineer. When you receive an engineer's sketch and are asked to prepare a formal drawing, you should follow these steps:

Step 1 Whether you are using manual or CADD drafting, prepare your own sketch the way you think it should look on the final drawing, taking into account correct drafting standards.

Step 2 Evaluate the size of the object so you can determine the scale and sheet size for your final manual drawing or the screen limits for your final computer-aided drawing.

Step 3 Lay out the drawing either very lightly using construction lines or light blue lead. The construction lines may be easily erased if you make a mistake. If you are using a computer, begin the layout on the screen and if you make an error, just erase it and try again. Editing your CADD work is fast and easy.

Step 4 Complete the final manual drawing by darkening all construction lines to proper ANSI standard line weights and run a copy for checking. After completing the computer drawing, a check plot may be made on the plotter or printer.

The example in Figure 3–131 shows a comparison between the engineer's rough sketch and the finished drawing.

A CADD drafter can easily work from points established on a Cartesian coordinate system as discussed in this chapter. When this is required, you need to first determine the type of coordinate system being used: absolute, incremental, or polar. When a situation of this kind occurs you may be given the X and Y values for each point in a chart similar to Figure 3–132. To proceed with a problem of this type, you can either locate the points on the screen and then connect them with lines or other shapes, or you can draw lines between the Cartesian coordinates and complete the object as you go. The solution to the coordinates given in this demonstration is shown in Figure 3–133.

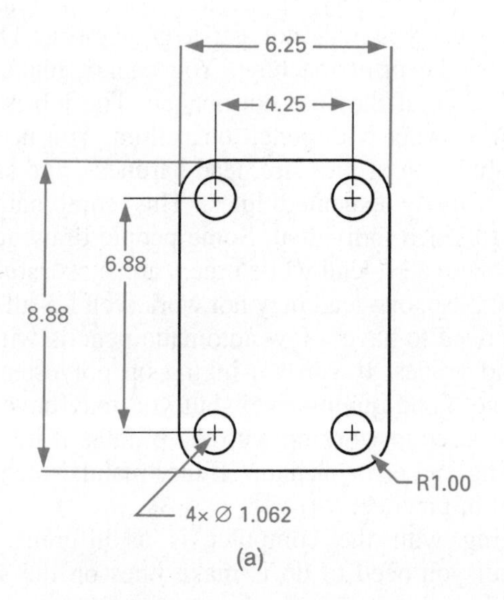

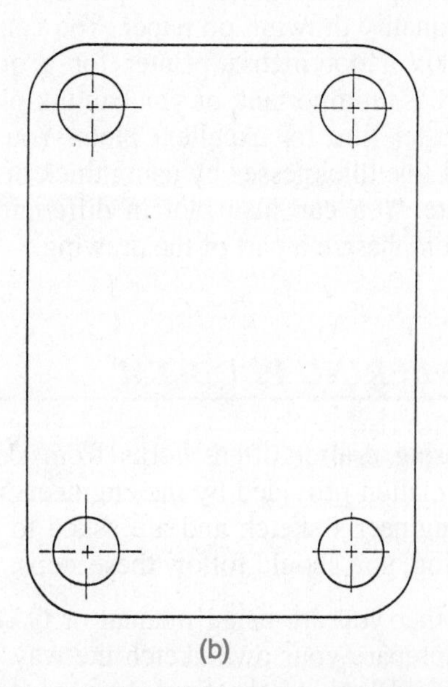

Figure 3–131 A comparison between an engineer's rough sketch (a) and the finished drawing (b).

POINT	X	Y
1.	2.8	2.3
2.	7.7	2.3
3.	7.7	5.6
4.	5.2	5.6
5.	4.1	4.1
6.	2.8	4.1
7.	2.8	2.3

Figure 3–132 X and Y point coordinates.

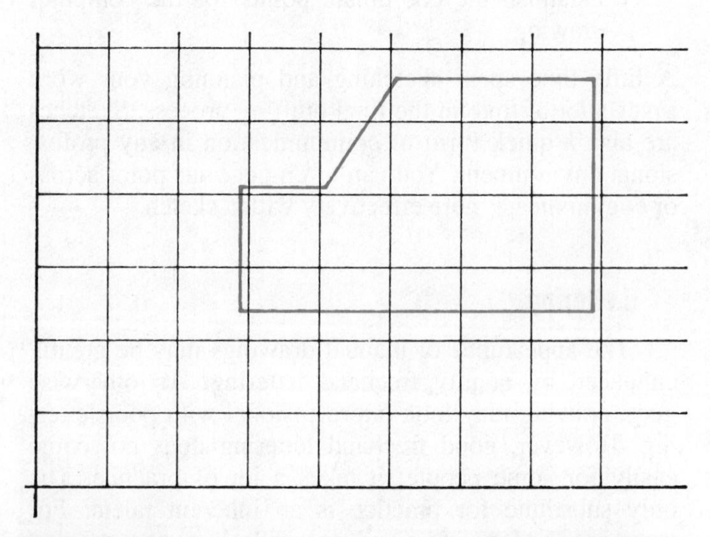

Figure 3–133 The solution to a Cartesian coordinate system drawing problem.

PART 1. SKETCHING PROBLEMS

DIRECTIONS

Use proper sketching materials and techniques to solve the following sketching problems, on 8½ × 11 in. bond paper or newsprint. Use very lightly sketched construction lines for all layout work. Darken the finished lines, but do not erase the layout lines.

Problem 3–1 List on a separate sheet of paper the length, direction, and position of each line shown in the drawing below. Remember, do not measure the lines with a scale. Example: Line 2 is the same length as line 1, and touches the top of line 1 at a 90° angle.

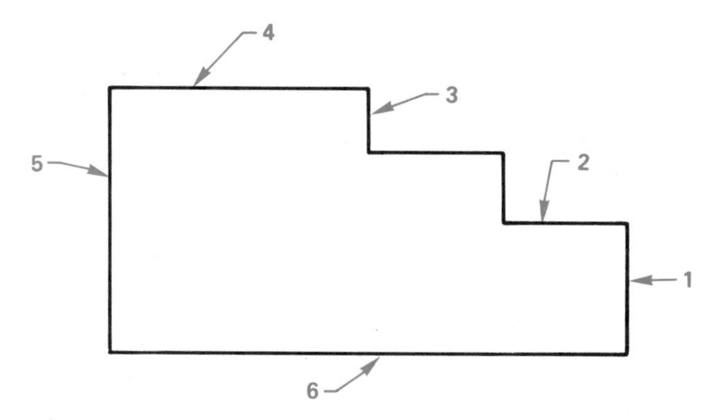

Problem 3–2 Use the box method to sketch a circle with approximately a 4-in. diameter. Sketch the same circle using the trammel and hand-compass methods.

Problem 3–3 Find a stapler, tape dispenser, or coffee cup and sketch a two-dimensional frontal view using the block technique. Do not measure the object. Use the measurement-line method to approximate proper proportions.

Problem 3–4 Find an object with an irregular shape such as a French curve and sketch a two-dimensional view using either the regular grid or random grid method. Sketch the correct proportions of the object without measuring.

Problem 3–5 Given the three objects in the accompanying sketch, without measuring, the front, top, and side views of each. Use the multiview-alignment technique discussed in this chapter.

Problem 3–6 Using the same object selected for problem 3–3, sketch an isometric representation. Do not measure the object but use the measurement-line technique to approximate proportions.

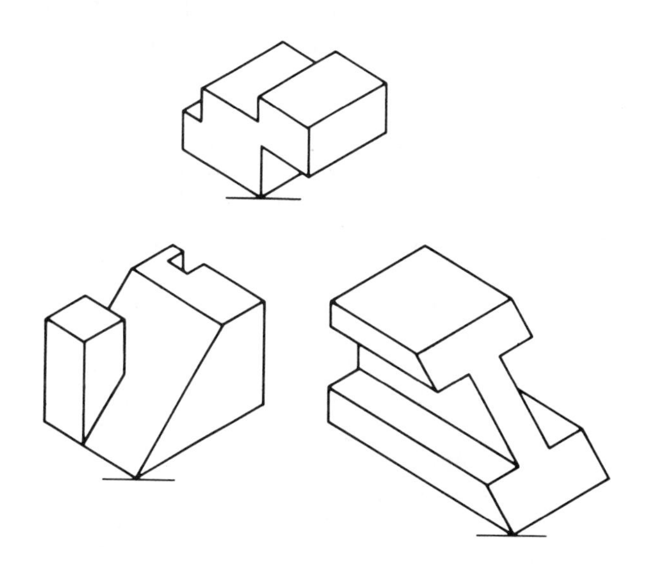

PART 2. LETTERING PROBLEMS

DIRECTIONS

Use vertical freehand Gothic lettering, or computer-generated lettering as required by the specific instructions for each assignment. Do all freehand lettering on an A-size drawing sheet with .125-in. letters. Space lines of lettering .125 in. apart. Use guidelines for all freehand lettering.

Problem 3–7 Use vertical Gothic freehand lettering to letter the following statement:

THE QUALITY OF THE FREEHAND LETTERING GREATLY AFFECTS THE APPEARANCE OF THE ENTIRE DRAWING. DRAFTERS LETTER IN PENCIL OR INK. PROPER FREEHAND LETTERING IS DONE WITH A SOFT, SLIGHTLY ROUNDED POINT IN A MECHANICAL DRAFTING PENCIL, OR A 0.5 MM AUTOMATIC PENCIL WITH 2H, H, OR F GRADE LEAD DEPENDING UPON INDIVIDUAL PRESSURE. LETTERING IS DONE BETWEEN VERY LIGHTLY DRAWN GUIDELINES. THESE GUIDELINES ARE DRAWN PARALLEL AND SPACED EQUAL TO THE HEIGHTS OF THE LETTERS. GUIDELINES HELP TO KEEP YOUR LETTERING UNIFORM IN HEIGHT. LETTERING STYLES MAY VARY BETWEEN COMPANIES. SOME COMPANIES REQUIRE THE USE OF LETTERING DEVICES.

Problem 3–8 Use vertical Gothic freehand lettering to letter the following statement:

MOST MECHANICAL DRAFTING THAT DOES NOT USE CAD DOES USE VERTICAL FREEHAND LETTERING. THE QUALITY OF THE FREEHAND LETTERING GREATLY AFFECTS THE APPEARANCE OF THE ENTIRE DRAWING. MANY MECHANICAL DRAFTING TECHNICIANS USE FREEHAND LETTERING WITH PENCIL ON VELLUM OR WITH POLYESTER LEAD ON MYLAR®. LETTERING IS COMMONLY DONE WITH A SOFT,

SLIGHTLY ROUNDED LEAD IN A MECHANICAL PENCIL OR A 0.5-MM LEAD IN AN AUTOMATIC PENCIL. LETTERING IS DONE BETWEEN VERY LIGHTLY DRAWN GUIDELINES. GUIDELINES ARE SPACED PARALLEL AT A DISTANCE EQUAL TO THE HEIGHT OF THE LETTERS. GUIDELINES ARE REQUIRED TO HELP KEEP ALL LETTERS THE SAME UNIFORM HEIGHT.

Problem 3–9 Use vertical Gothic freehand lettering and a CADD system to letter the following notes:

1. INTERPRET DIMENSIONS AND TOLERANCES PER ANSI Y 14.5M–1982.
2. UNLESS OTHERWISE SPECIFIED, ALL DIMENSIONS ARE IN MILLIMETERS.
3. REMOVE ALL BURRS AND SHARP EDGES.
4. ALL FILLETS AND ROUNDS R 6.
5. CASE HARDEN 62 ROCKWELL C SCALE.
6. AREAS WHERE MATERIAL HAS BEEN REMOVED SHALL HAVE SMOOTH TRANSITIONS AND BE FREE OF SCRATCHES, GRIND MARKS, AND BURNS.
7. FINISH BLACK OXIDE.
8. PART TO BE CLEAN AND FREE OF FOREIGN DEBRIS.

Compare the difference between freehand and CADD lettering as to speed and appearance.

PART 3. LINE PROBLEMS

DIRECTIONS

1. Using the selected engineer's layout as a guide only, make an original drawing using manual or CADD as required by your course objectives. Select an appropriate scale. Draw *only* the object lines and centerlines as appropriate for each problem. Keep in mind that the engineer's sketches are rough and not meant for tracing.
2. Use construction lines (very lightly drawn) to prepare the entire drawing. When satisfied with the product, darken the drawing using proper line technique.
3. Complete the title block.
   a. The title of the drawing is given.
   b. The material the part is made of is given.
   c. The drawing number is the same as the problem number.
   d. Use FULL scale.

Problem 3–10 Object lines (in.)

Part Name: Plate
Material: .25-in.-Thick Mild Steel

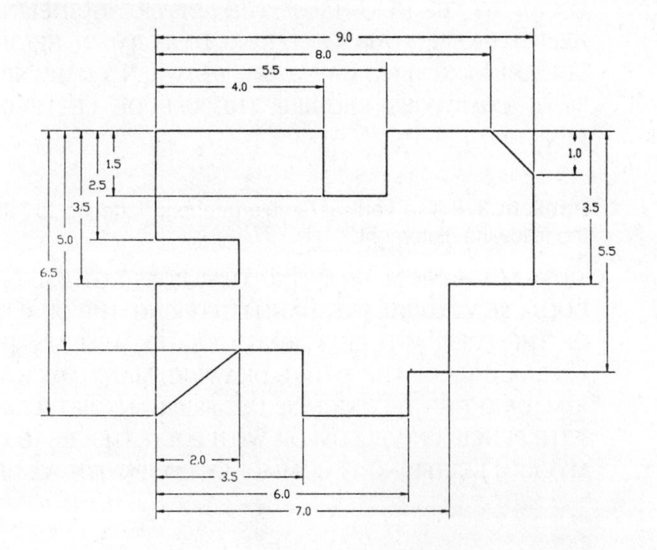

Problem 3–11 Straight object lines only (in.)

Part Name: Milk Stencil
Material: .015-in.-thick wax-coated cardboard
Used as a stencil to spray paint identification on crates of milk.

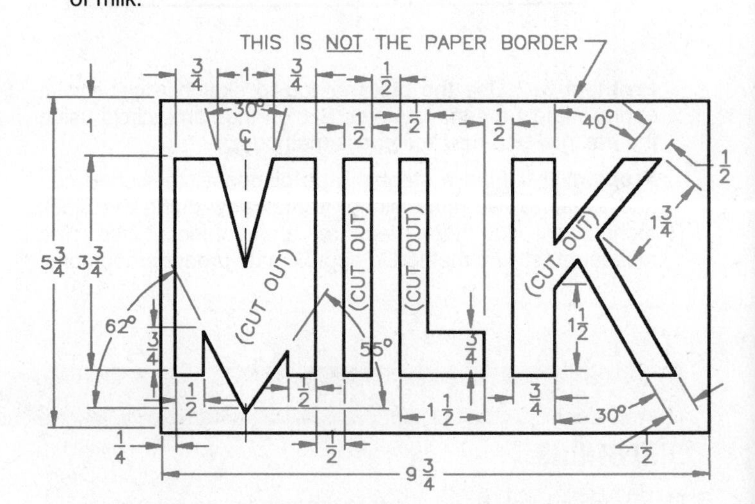

Problem 3–12 Circle and arc object lines and centerlines (in.)

Part Name: Connector
Material: .25-in.-Thick Steel
Courtesy Allied Systems Company.

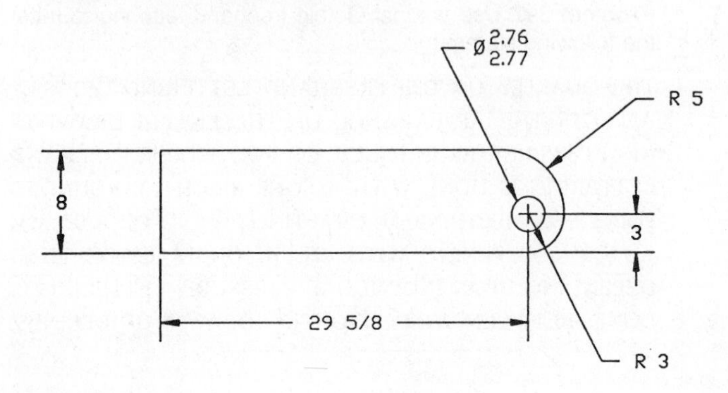

Problem 3–13 Arcs, object lines, and centerlines

Part Name: Latch
Material: .25-in.-Thick Mild Steel
Courtesy Allied Systems Company.

Problem 3–14 Straight line and arc object lines and centerlines (metric)

Part Name: Plate
Material: 10-mm-Thick HC–112
Used as a spacer to separate electronic components in a computer chassis.

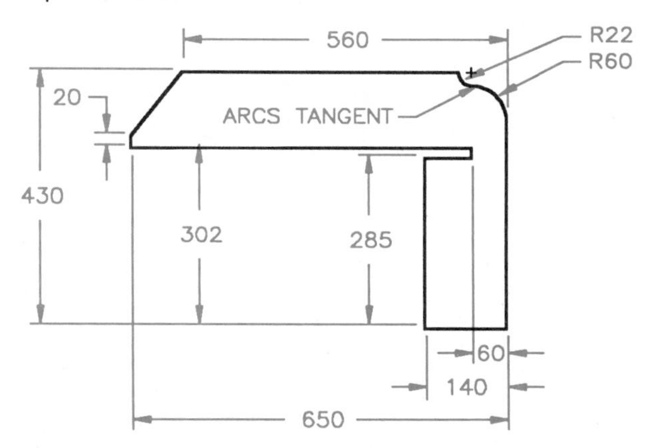

Problem 3–15 Circle and arc object lines and centerlines (in.)

Part Name: Bogie Lock
Material: .25-in.-Thick Mild Steel
Courtesy Allied Systems Company.

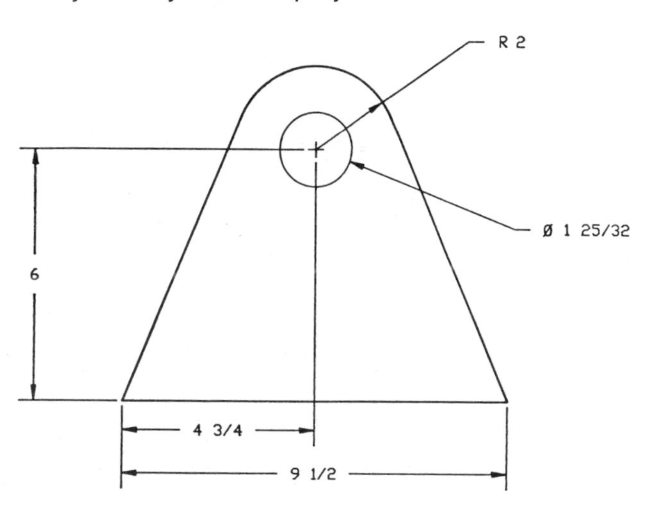

Problem 3–16 Circle, object lines, and centerlines (in.)

Part Name: Stove Back
Material: .25-in.-Thick Mild Steel

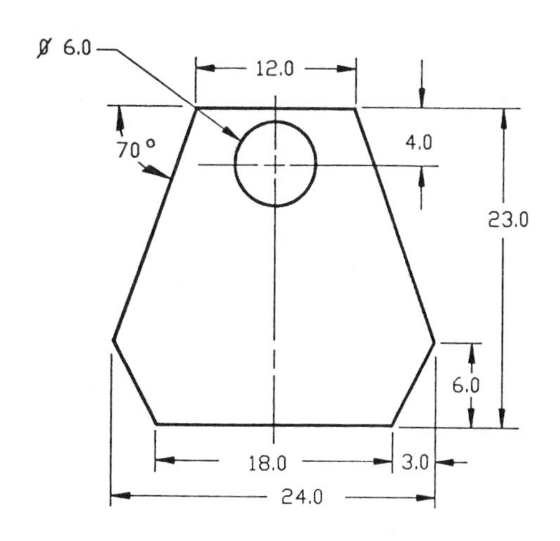

Problem 3–17 Circle and arc object lines and centerlines (in.)

Part Name: Bogie Lock
Material: .25-in.-Thick Mild Steel
Courtesy Allied Systems Company.

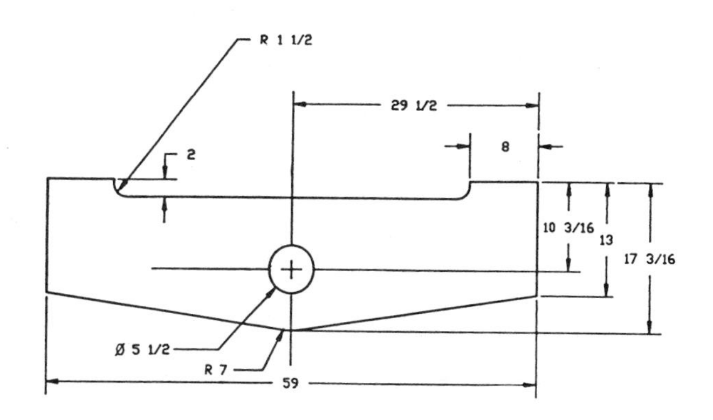

Problem 3–18 Circle and arc object lines and centerlines (metric)

Part Name: T-Slot Cleaner
Material: 6-mm-Thick Cold Rolled Steel

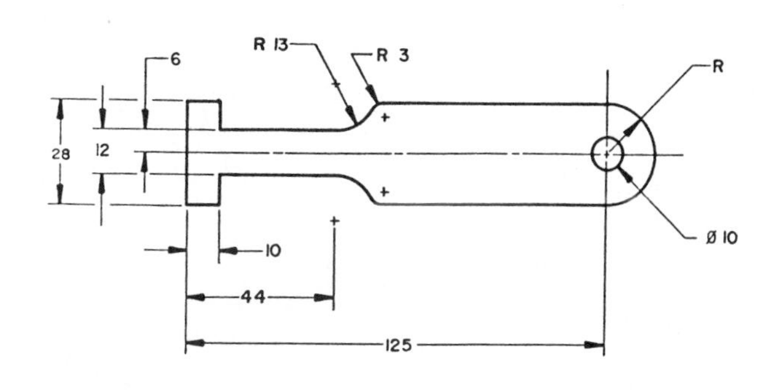

NOTES:
1. BREAK ALL SHARP EDGES.
2. STAMP NAME ON HANDLE.

Problem 3–19 Circle and arc object lines and centerlines (in.)

Part Name: T-Slot Cleaner (inch)
Material: .25-in.-Thick Cold Rolled Steel

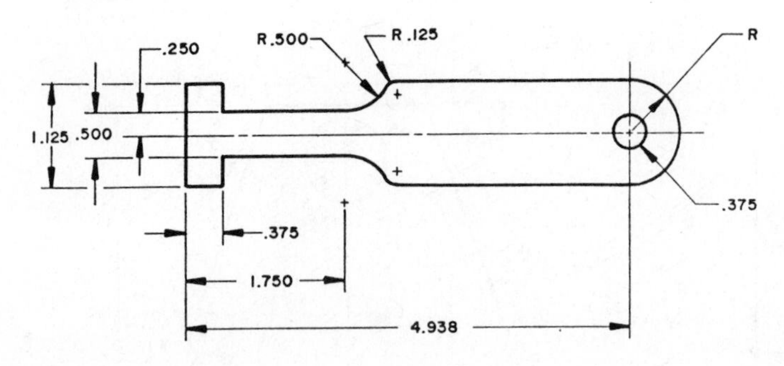

NOTES:
1. BREAK ALL SHARP EDGES.
2. STAMP NAME ON HANDLE.

Problem 3–20 Circle and arc object lines and centerlines (in.)

Part Name: Plate
Material: .125-in.-Thick Aluminum

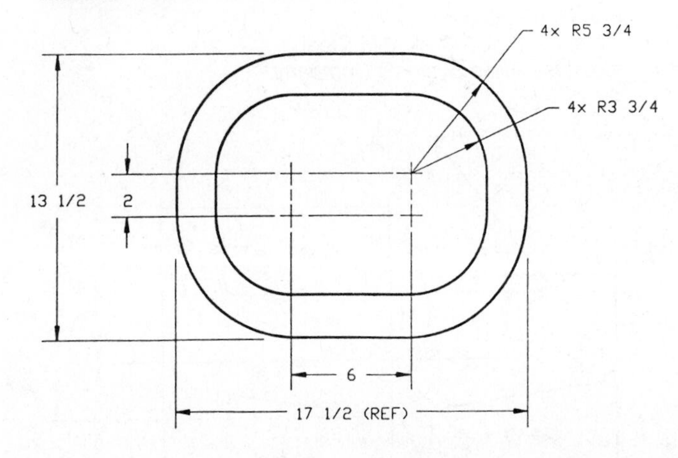

Problem 3–21 Circle and arc object lines and centerlines (in.)

Part Name: Gasket
Material: .062-in.-Thick Neoprene

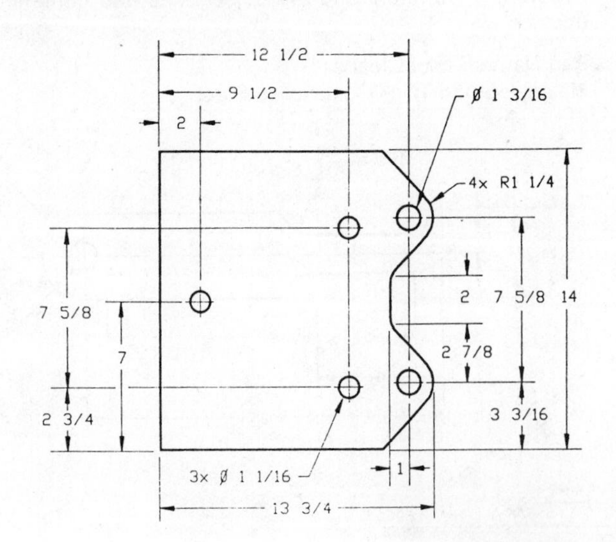

Problem 3–22 Circle and arc object lines and centerlines (in.)

Part Name: Gasket
Material: .062-in.-Thick Cork
Gasket for hydraulic pump.

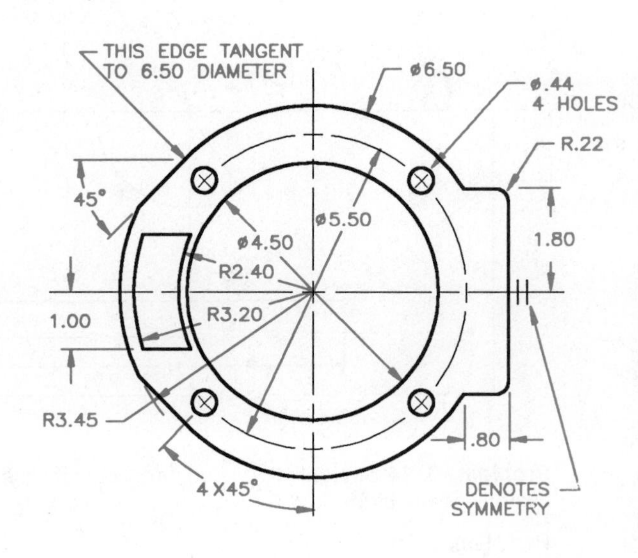

Problem 3–23 Arc object lines, centerlines, and phantom lines (in.)

Part Name: Bogie Lock
Material: .25-in.-Thick Mild Steel
Connect the leader lines and place the notes on the drawing.
Courtesy Allied Systems Company.

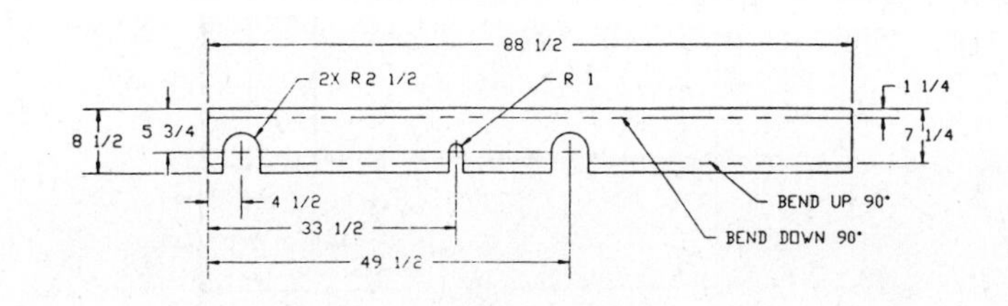

Problem 3–24 Object lines and hidden lines (in.)

Part Name: V-block
Material: 4.00-in.-Thick Mild Steel

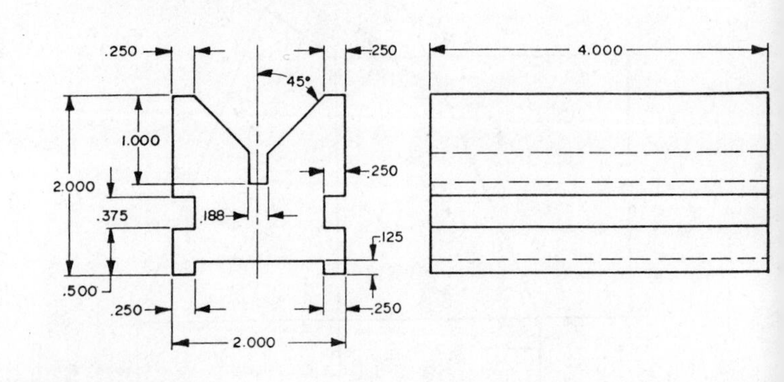

PART 4. DRAWING WITH THE COMPUTER PROBLEMS

DIRECTIONS

The problems in this section are representative of the skills required when using a computer graphics system. The use of the Cartesian coordinate system is stressed because of its universal use in computer-aided drafting. Problems 3–25, 3–26, and 3–27 test this skill of locating points.

Problem 3–25 You will need graph paper with 10 squares per inch for this problem. The 11 points given below each have an X and Y coordinate. Plot each of the points on the grid paper using Cartesian coordinates and connect them with a straight line. The origin of your coordinate system does not have to be the extreme lower left corner of the paper.

POINT	X	Y
1	0	1.5
2	3	1.5
3	3	0
4	7.12	0
5	7.12	2.25
6	6.18	2.25
7	6.18	.75
8	3.93	.75
9	3.93	2.25
10	0	2.25
11	0	1.5

Problem 3–26 The two objects given below (objects A and B) are similar to the previous problem, but each has two views, front and right side. Use 10-squares-to-the-inch graph paper and follow the same procedure you used in problem 3–25. The lower left corner of each view should be the origin point for that view. Each view should have its own separate Cartesian coordinate system.

		FRONT		RIGHT SIDE	
Object A	**Point**	**X**	**Y**	**X**	**Y**
	1	0	0	0	0
	2	3	0	2.75	0
	3	3	3.6	2.75	3.6
	4	0	3.6	0	3.6
Object B	1	0	0	0	0
	2	4	0	3.8	0
	3	4	2.1	3.8	4.2
	4	3	2.1	3.8	4.2
	5	3	4.2	0	2.1
	6	1	4.2	0	0
				Move to:	
	7	1	2.1	0	2.1
	8	0	2.1	3.8	2.1
	9	0	0		

Problem 3–27 The following problem contains drawing data given in both incremental and polar coordinates. The points should be plotted and drawn on 10-squares-to-the-inch grid paper.

A. INCREMENTAL COORDINATES			B. POLAR COORDINATES		
Point	X	Y	Point	Angle	Radius
1	0	0	1	0	0
2	3	0	2	0	1.25
3	0	2	3	90	1
4	–1	0	4	0	1.25
5	0	1	5	270	1
6	–1	0	6	0	1.50
7	0	–2	7	90	2
8	–1	0	8	180	2.75
9	0	–1	9	90	1
			10	180	1.75
			11	270	3

Problem 3–28 Arcs, circles, and centerlines (inch)

Part Name: Bracket
Material: Mild Steel (MS)
Courtesy Allied Systems Company.

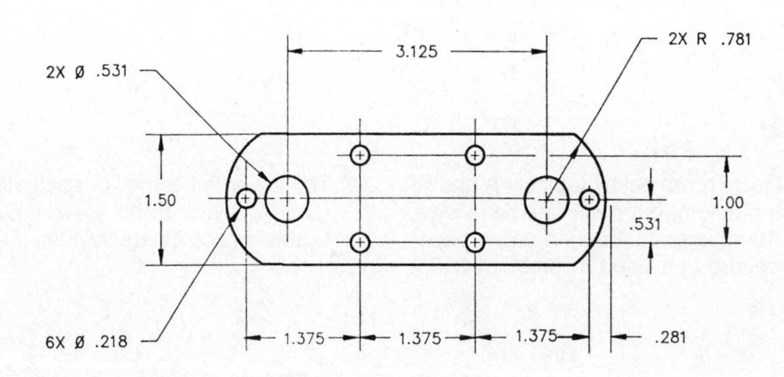

CHAPTER 4

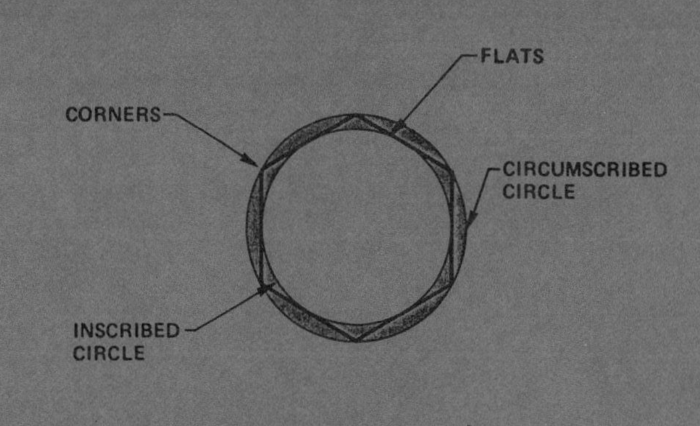

Geometric Construction

LEARNING OBJECTIVES

After completing this chapter you will use manual or computer-aided drafting to:

- Draw parallel and perpendicular lines.
- Construct bisectors and divide lines and spaces into equal parts.
- Draw polygons.
- Draw tangencies.
- Draw ellipses.
- Solve an engineering problem by making a formal drawing with geometric constructions from an engineer's sketch or layout.

Machine parts and other manufactured products are made up of different geometric shapes ranging from squares and cylinders to complex irregular curves. Geometric constructions are methods that may be used to draw various geometric shapes or to perform drafting tasks related to the geometry of product representation and design. The proper use of geometric constructions requires a fundamental understanding of plane geometry. When using geometric construction techniques extreme accuracy and the proper use of drafting instruments are important. Pencil or compass leads should be sharp and instruments should be in good shape. Always use very lightly drawn construction lines for all preliminary work.

When computers are used in drafting, the task of creating most geometric construction related drawings becomes much easier, although the theory behind the layout of geometric shapes and related constructions remains the same.

LINE CHARACTERISTICS

A straight line segment is a line of any given length, such as lines A–B shown in Figure 4–1.

A curved line may be in the form of an arc with a given center and radius or an irregular curve without a defined radius as shown in Figure 4–2.

Two or more lines may intersect at a point as in Figure 4–3. The opposite angles of two intersecting lines are equal; a = a, b = b.

Parallel lines are lines equidistant throughout their length, and if they were to extend indefinitely they would never cross. (See Figure 4–4.)

Perpendicular lines intersect at a 90° angle as shown in Figure 4–5.

GEOMETRIC SHAPES

Angles

Angles are formed by the intersection of two lines. Angles are sized in degrees (°). Components of a degree are minutes and seconds. There are 60 minutes (') in one degree and there are 60 seconds (") in one minute. 1° = 60', 1' = 60". The four basic types of angles are shown in Figure 4–6. A straight angle equals 180° while a right angle equals 90°. An acute angle contains less than 90° and an obtuse angle contains more than 90° but less than 180°.

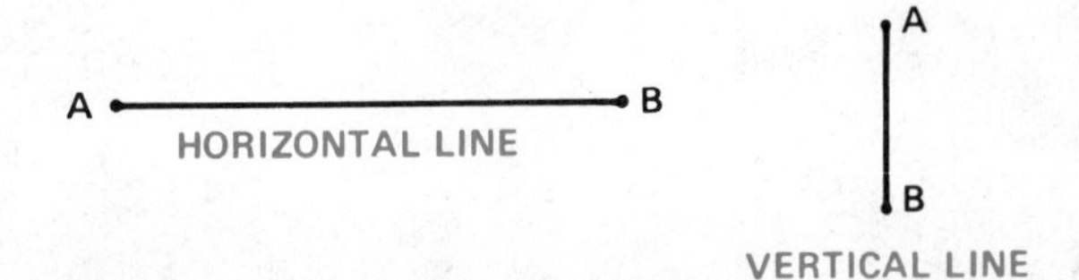

Figure 4–1 Horizontal and vertical lines.

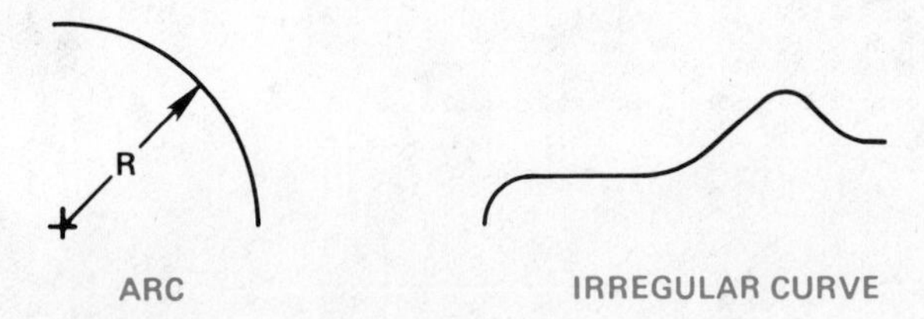

Figure 4–2 Arc and irregular curve.

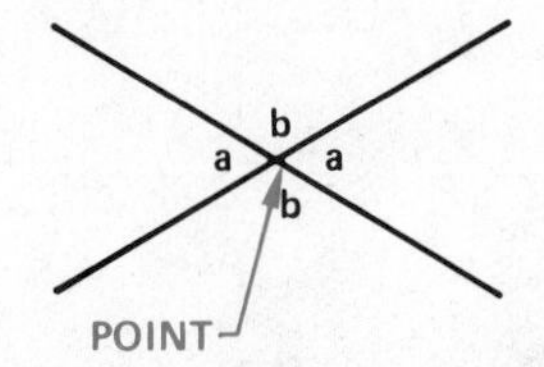

Figure 4–3 Intersecting lines.

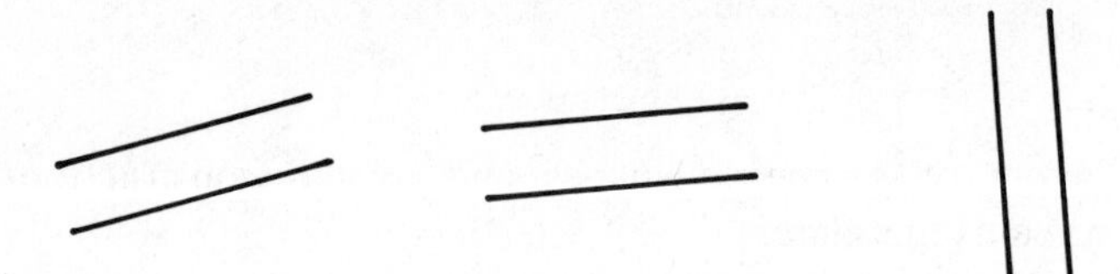

Figure 4–4 Parallel lines.

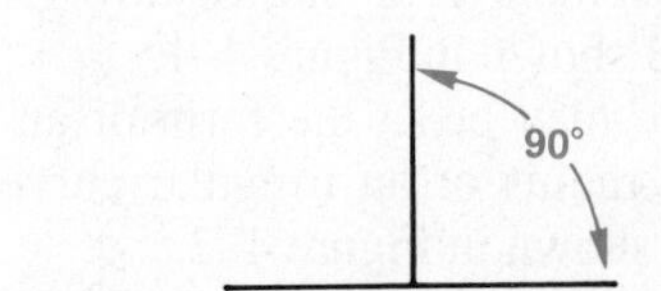

Figure 4–5 Perpendicular lines.

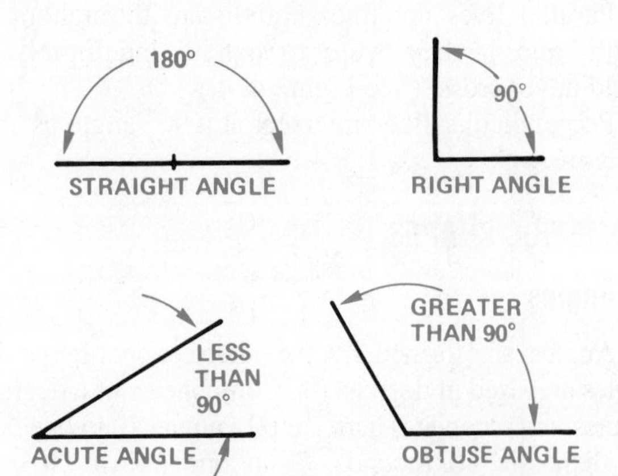

Figure 4–6 Types of angles.

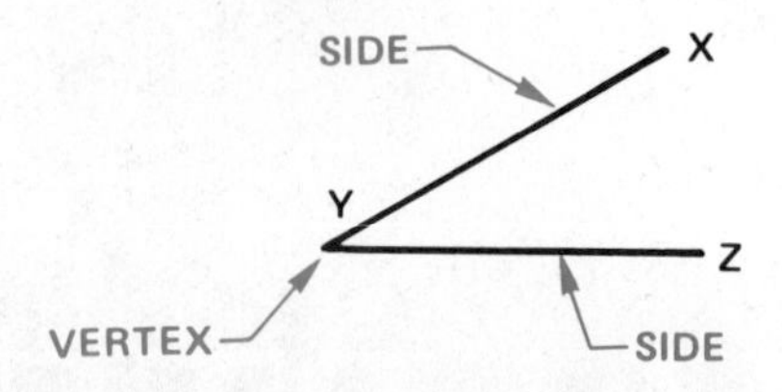

Figure 4–7 Parts of an angle.

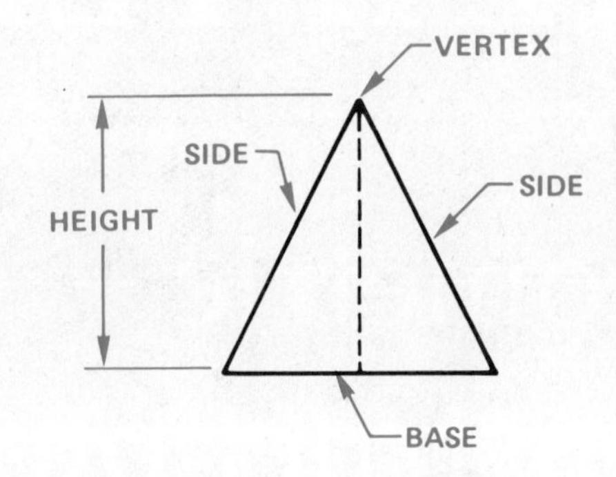

Figure 4–8 Parts of a triangle.

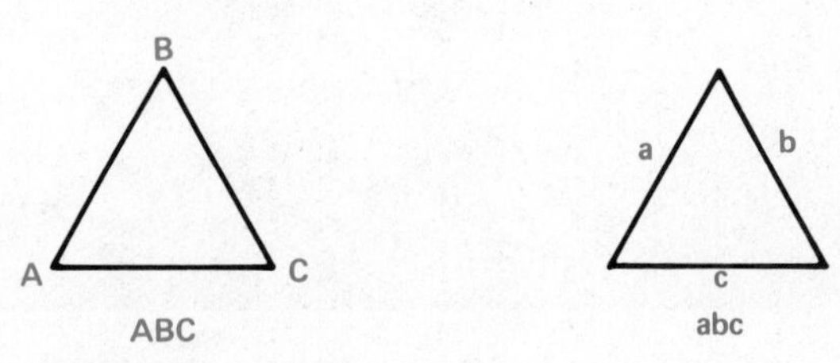

Figure 4–9 Labeling triangles.

The parts of an angle are shown in Figure 4–7. An angle is labeled by giving the letters defining the line ends, and vertex, with the vertex always between the ends, such as XYZ where Y is the vertex.

Triangles

A triangle is a geometric figure formed by three intersecting lines creating three angles. The sum of the interior angles always equals 180°. The parts of a triangle are shown in Figure 4–8. Triangles are labeled by lettering the vertex of each angle such as ABC, or by labeling the sides abc as shown in Figure 4–9.

There are three kinds of triangles: acute, obtuse, and right as shown in Figure 4–10. Two special acute triangles are the equilateral, which has equal sides and angles, and the isosceles, which has two equal sides and angles. An isosceles triangle may also be obtuse. In a scalene triangle, no sides or angles are equal.

Right triangles have certain unique geometric characteristics. Two internal angles equal 90° when added. The side opposite the 90° angle is called the hypotenuse, as seen in Figure 4–11.

Quadrilaterals

Quadrilaterals are four-sided polygons that may have equal or unequal sides or interior angles. The sum of the

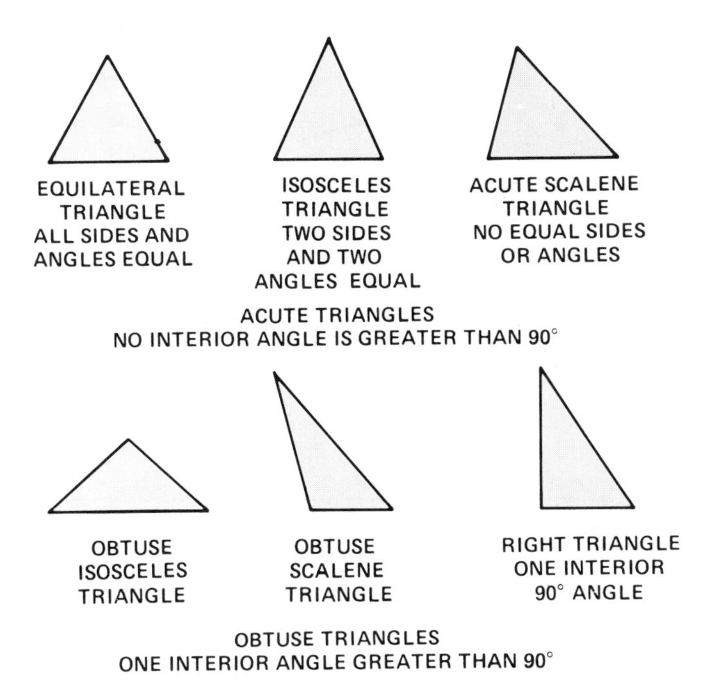

Figure 4–10 Types of triangles.

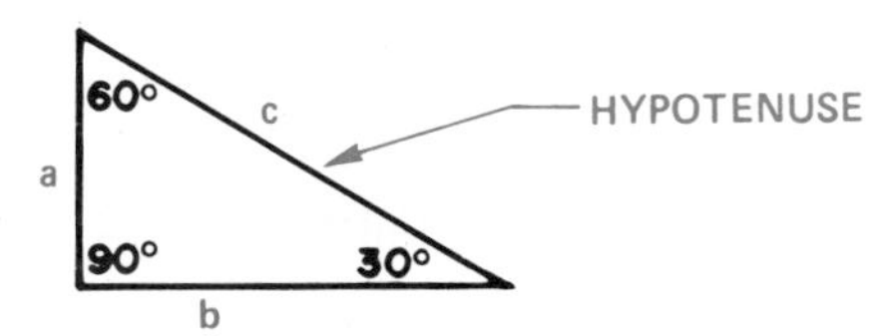

Figure 4–11 Right triangles.

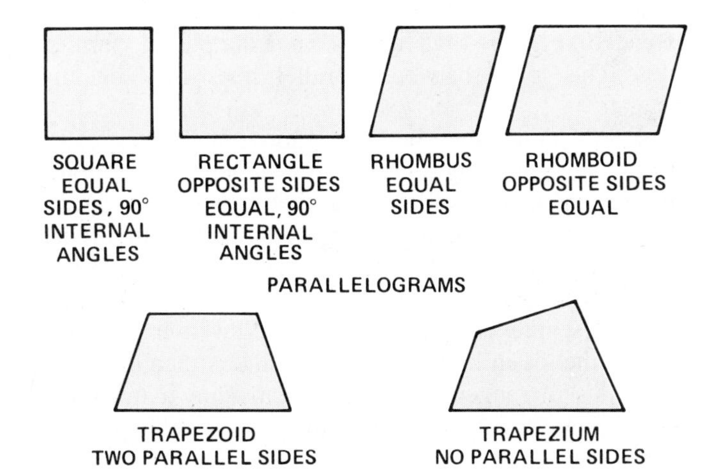

Figure 4–12 Quadrilaterals.

interior angles is 360°. Quadrilaterals with parallel sides are called parallelograms. (See Figure 4–12.)

Regular Polygons

Some of the most commonly drawn geometric shapes are regular polygons. Regular polygons have equal sides and equal internal angles. Polygons are closed figures with any number of sides, but no less than three. The relationship of a circle to a regular polygon is that a circle may be drawn to touch the corners (circumscribed) or the sides, known as flats, (inscribed) of a regular polygon. (See Figure 4–13.) This relationship is an advantage in constructing regular polygons. Some common regular polygons are shown in Figure 4–14.

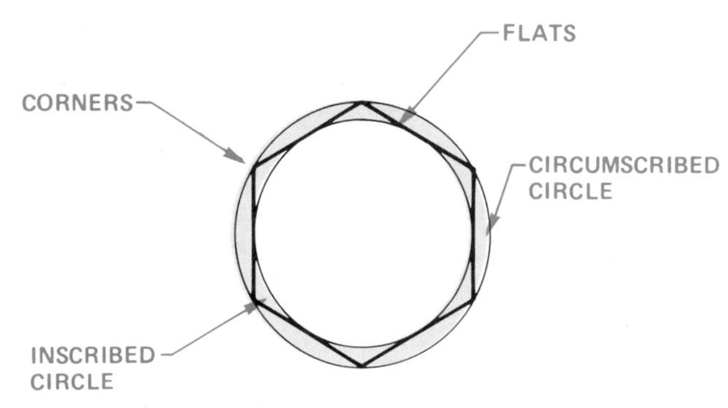

Figure 4–13 Regular polygon.

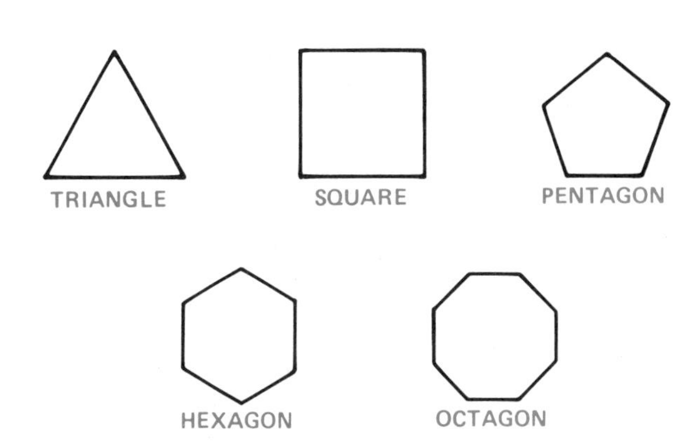

Figure 4–14 Common regular polygons.

Circles

A circle is a closed curve with all points along the curve at an equal distance from a point called the center. The circle has a total of 360°. The circumference is the distance around the circle. The radius is the distance from the center to the circumference. The diameter is the distance across the circle through the center. Figure 4–15 shows the parts of a circle and circle relationships.

An arc is part of the circumference of a circle. The arc may be identified by a radius, an angle, or a length (Figure 4–16).

Tangents

Straight or curved lines are tangent to a circle or arc when the line touches the circle or arc at only one point. If a line were to connect the center of the circle or arc to the point of tangency, the tangent line and the line from the center would form a 90° angle. (See Figure 4–17.)

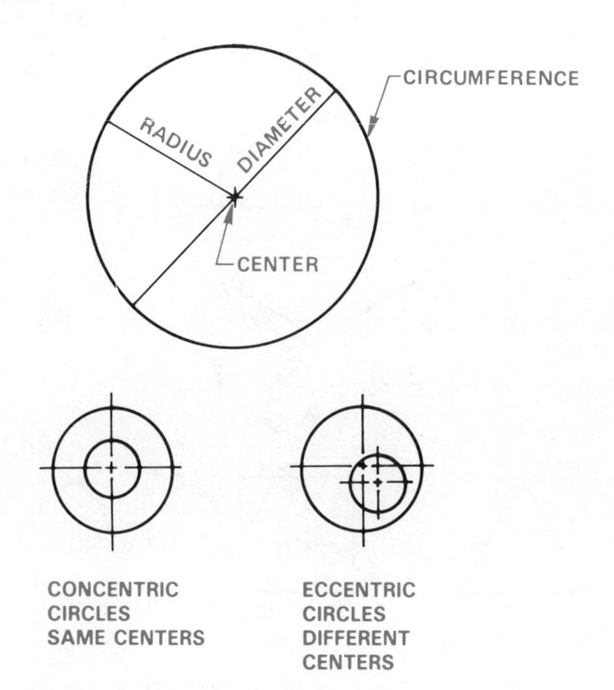

Figure 4–15 Parts of a circle.

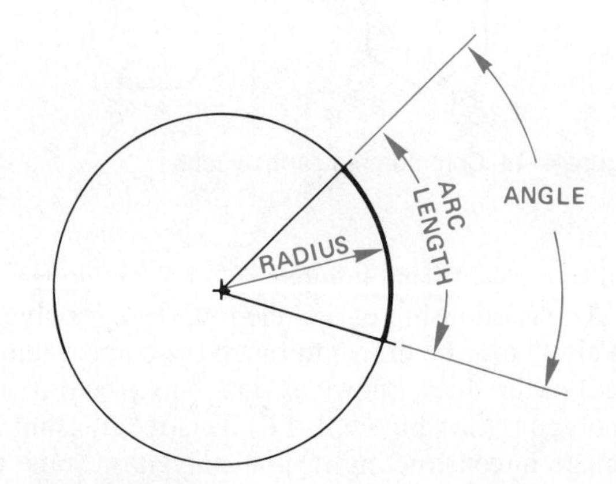

Figure 4–16 Arc.

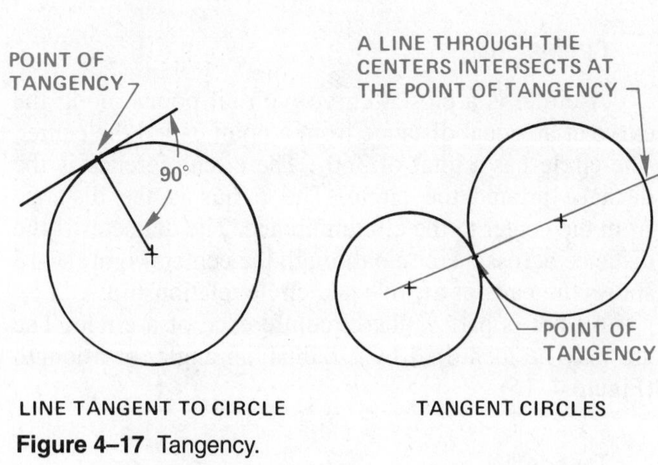

Figure 4–17 Tangency.

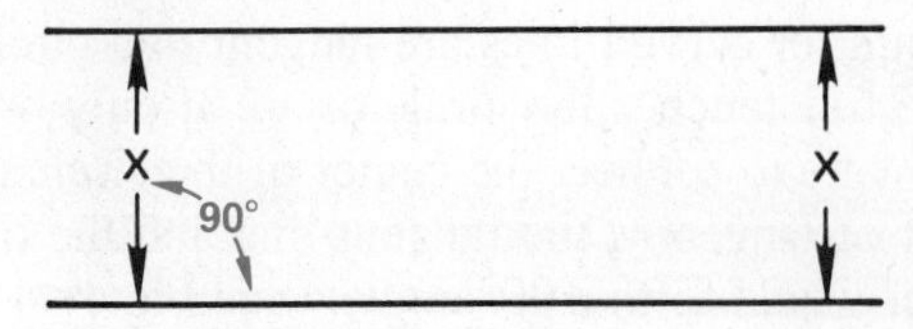

Figure 4–18 Parallel lines.

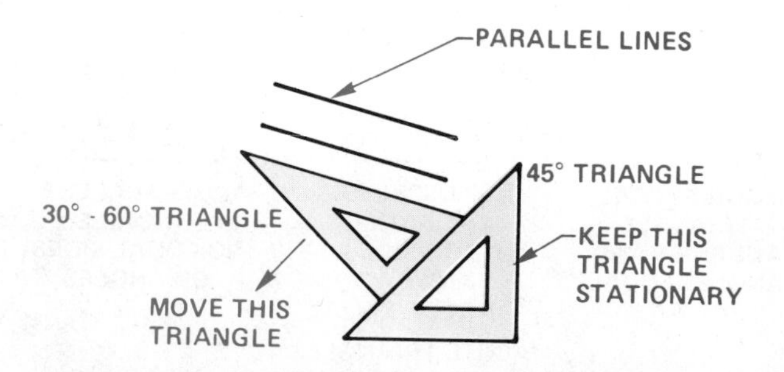

Figure 4–19 Constructing parallel lines with angles.

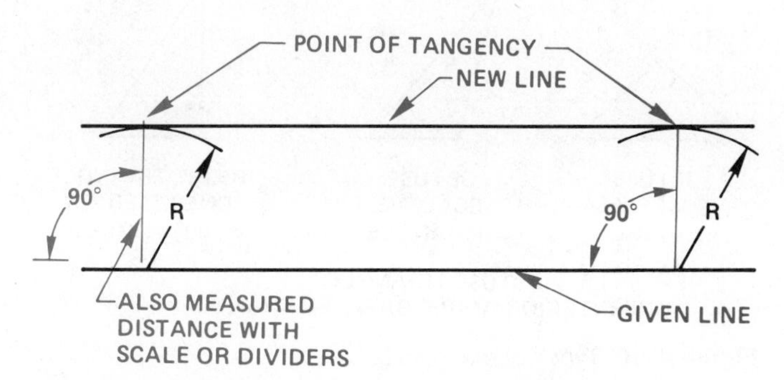

Figure 4–20 Constructing parallel lines with straightedge and compass.

COMMON GEOMETRIC CONSTRUCTIONS

Parallel Lines

Parallel lines are lines evenly spaced at all points along their length and will not intersect even when extended. Figure 4–18 shows an example of parallel lines. The space between parallel lines may be any distance.

Parallel lines may be drawn horizontally, vertically, or at any angle using a drafting machine. When a drafting machine is not available, parallel lines may be drawn with the aid of standard triangles as shown in Figure 4–19. Parallel lines may be drawn with a straightedge and compass when the distance between lines is established. Use the compass radius to draw arcs near the ends of the given line. The parallel line is then drawn at the points of tangency of the arcs as shown in Figure 4–20. Draw very light construction lines for all construction layout.

Parallel or concentric arcs may be drawn where the distance between arcs is equal. Establish parallel arcs by adding the radius of Arc 1 (R_1) to the distance between the arcs (X), as shown in Figure 4–21.

Concentric circles are parallel circles drawn from the same center. The distance between circles is equal, as shown in Figure 4–22.

Parallel irregular curves are drawn any given distance (R) apart by using a compass set at the given distance. A series of arcs is lightly drawn all along the given curve at the required distance. The line to be drawn parallel is then constructed with an irregular curve

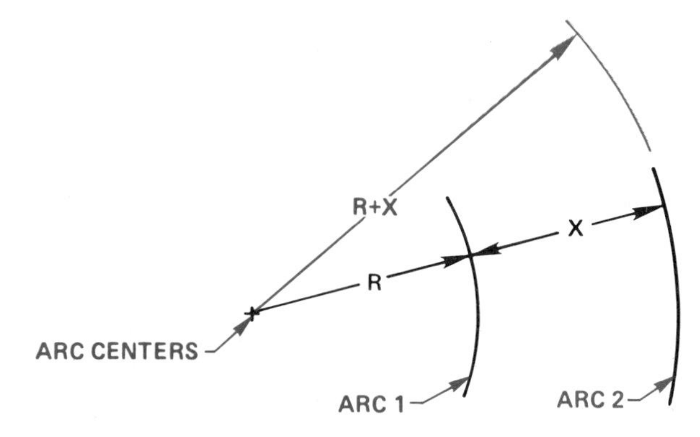

Figure 4–21 Parallel arcs.

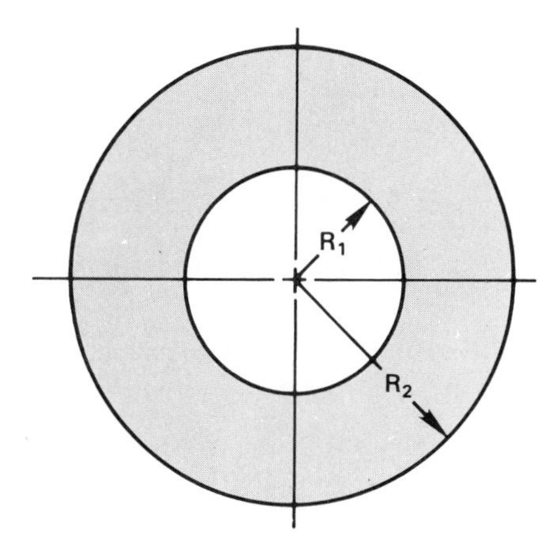

Figure 4–22 Concentric circles.

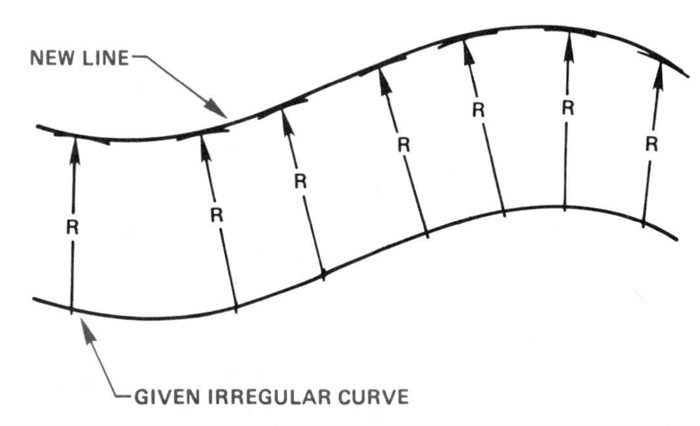

Figure 4–23 Constructing parallel irregular curves.

at the points of tangency of the construction arcs. (See Figure 4–23.)

Perpendicular Lines

Perpendicular lines intersect at 90°, or a right angle, as seen in Figure 4–24. Perpendicular lines are drawn with the horizontal and vertical scales of a drafting machine, even if the scales are set at an angle away from horizontal, as seen in Figure 4–25. Perpendicular lines can also be drawn with a straightedge and triangle or two triangles as shown in Figure 4–26.

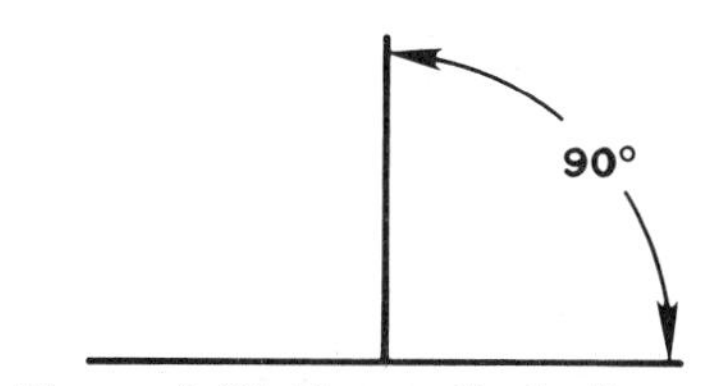

Figure 4–24 Perpendicular lines.

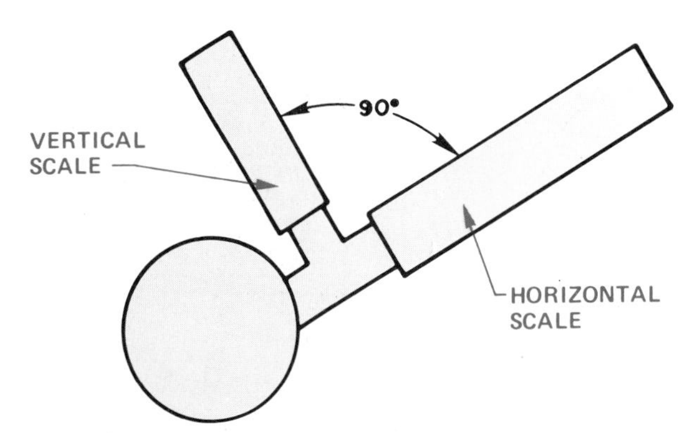

Figure 4–25 Perpendicular drafting machine scales.

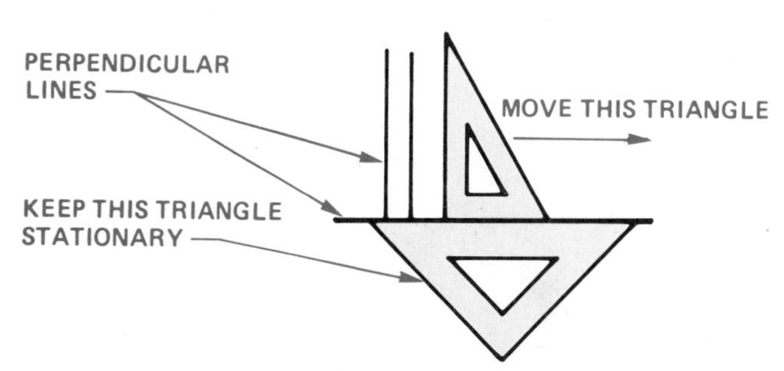

Figure 4–26 Perpendicular lines with triangles.

Perpendicular Bisector

The perpendicular bisector of a line may be obtained using a straightedge and a compass.

Step 1 Set the compass radius more than halfway across the given line segment and draw two intersecting arcs from the line ends.

Step 2 Connect the points where the arcs intersect, as in Figure 4–27.

Bisecting an Angle

Any given angle may be divided into two equal angles by using a compass and straightedge.

Step 1 Adjust the compass to any radius and draw an arc as shown in Figure 4–28.

Step 2 At the points where the arc intersects the sides of the angle, draw two arcs of equal radius.

Step 3 Connect a straight line from the vertex of the angle to the point of intersection of the two arcs.

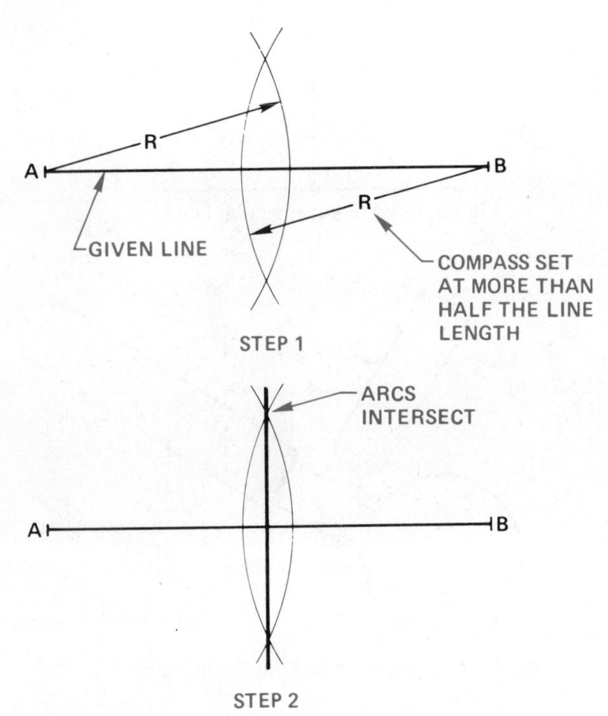

Figure 4–27 Constructing a perpendicular bisector.

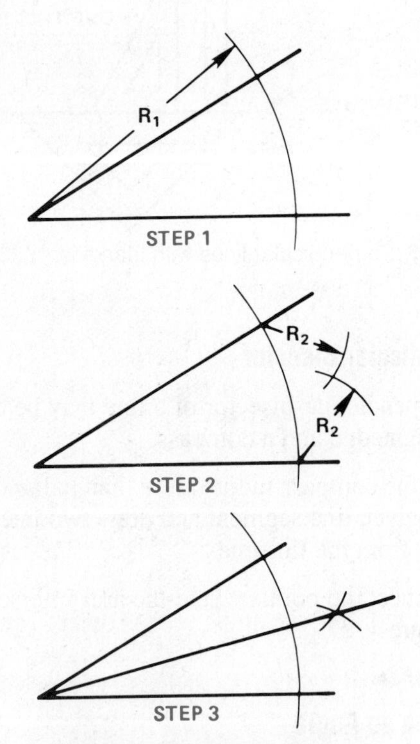

Figure 4–28 Bisecting an angle.

Transferring an Angle

A given angle may be transferred to a new location as shown in Figure 4–29. This method also works well for transferring a triangle.

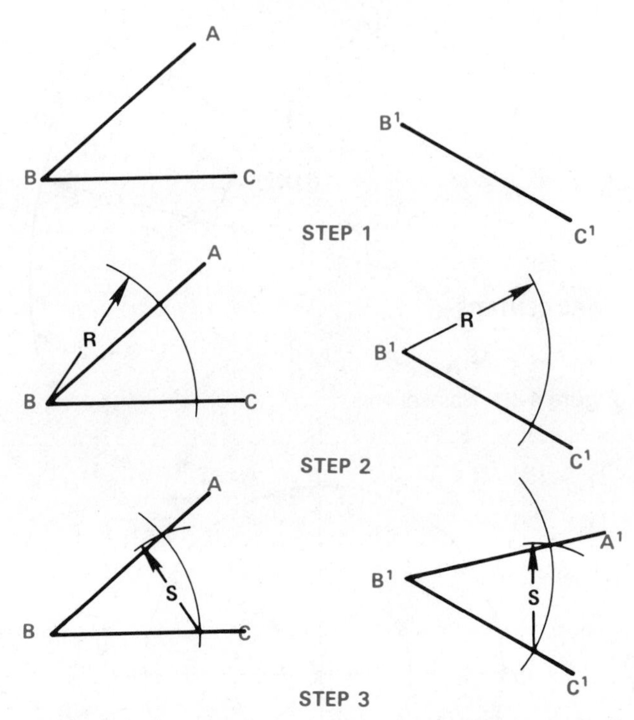

Figure 4–29 Transferring a given angle to a new location.

Step 1 Given angle ABC, draw line B^1C^1 in the new location and position.

Step 2 Draw any arc, R, with its center at B on the given angle. Transfer and draw arc R with its center at point B^1.

Step 3 Using the intersection of one side of the given angle and arc R as center, set a compass distance to the opposite intersection of arc R. Then transfer the new radius, S, to the new position using the intersection of B^1C^1 and arc R as center. Connect a line from B^1 to the intersection of the two arcs, R and S, as shown in Figure 4–29.

Dividing a Line Into Equal Parts

Any given line can be divided into any number of equal parts. Divide given line AB, shown in Figure 4–30, into eight equal parts.

Step 1 Draw a construction line at any angle from either end of line AB as shown in Figure 4–31.

Step 2 Use a scale or dividers to divide the angled construction line into eight equal parts. Use any size division that will extend down the angled line to approximately the length of the given line. (See Figure 4–32.)

Step 3 Connect the last point (8) of the construction line to point B of the given line. Then draw lines parallel to line 8B from each of the numbered points on the construction line to the

Figure 4–30 Given line AB to divide into equal parts.

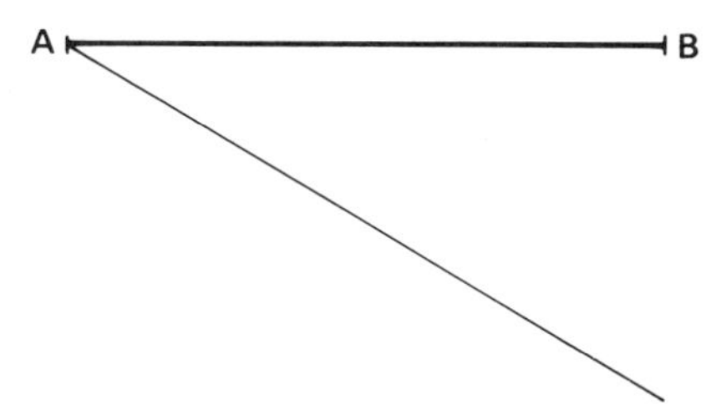

Figure 4–31 Angled construction line — step 1.

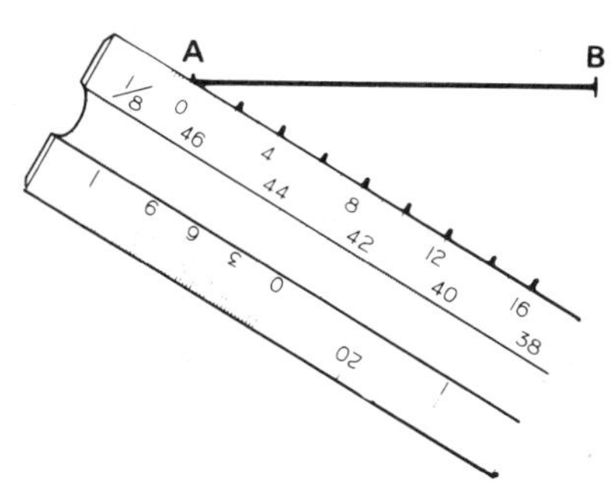

Figure 4–32 Divide angled line into required number of equal parts — step 2.

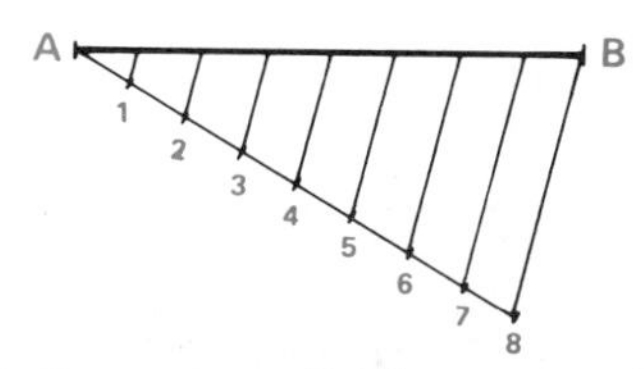

Figure 4–33 Connect parallel line segments — step 3.

given line AB. (See Figure 4–33.) You now have the given line AB, divided into eight equal parts. This same process may be used for any number of equal parts.

Dividing a Space Into Equal Parts

You can divide any given space into any number of equal parts. Given the space shown in Figure 4–34, divide it into 12 equal parts.

Step 1 Between the lines that establish the given space, place a scale with the required number of increments (12) so that the zero on the scale is on either line and the 12 is on the other line. Mark each increment. (See Figure 4–35.)

Step 2 Remove the scale. Each increment you have marked has divided the given space into 12 equal spaces. Example 4–36 shows parallel lines drawn at each mark to complete the space division.

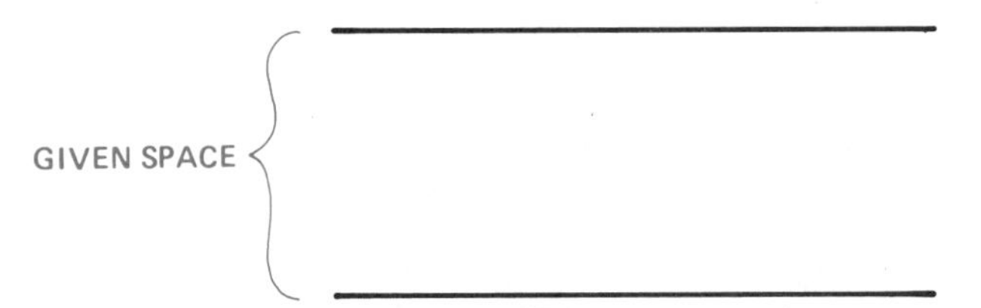

Figure 4–34 Dividing a given space into equal parts.

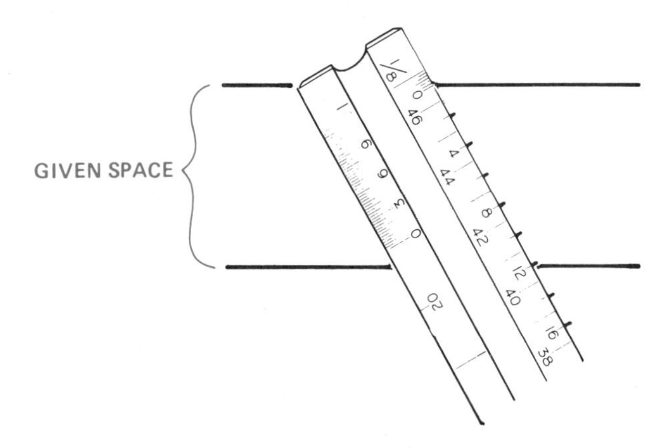

Figure 4–35 Place a scale with required number of increments between the lines — step 1.

Figure 4–36 Given space divided into equal parts with parallel lines — step 2.

CONSTRUCTING POLYGONS

Triangles

The following technique is known as triangulation. Using given triangle sides x, y, and z, in Figure 4–37, draw a triangle as shown in the following steps:

Step 1 Lay out one of the given sides. Select the side that will be the base; z has been chosen as shown in Figure 4–38. From one end of line z strike an arc equal in length to one of the other lines, x, for example, as in Figure 4–38.

Step 2 From the other end of line z strike an arc with a radius equal to the remaining line, y. Allow this arc to intersect the previous arc. Where the two arcs cross, draw a line to the end of the base line, as in Figure 4–39, to complete the triangle.

Any given polygon may be constructed or transferred to a new location using this triangulation method.

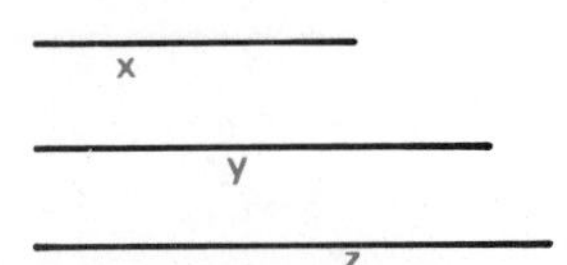

Figure 4–37 Construct a triangle given three sides.

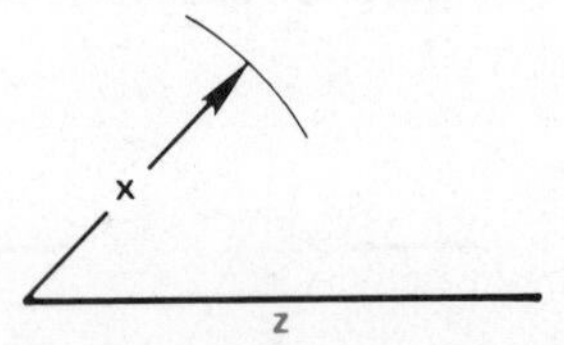

Figure 4–38 Layout line z and swing arc x — step 1.

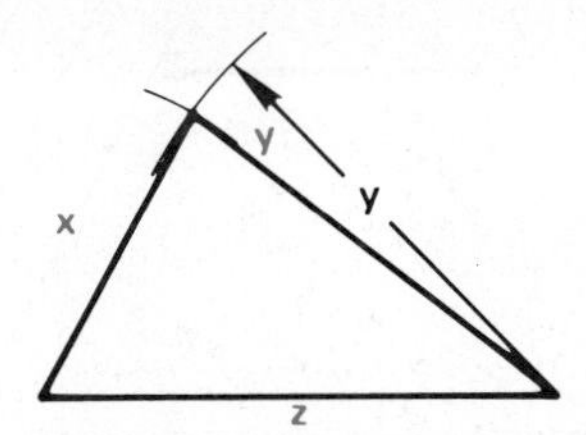

Figure 4–39 Swing arc y — step 2.

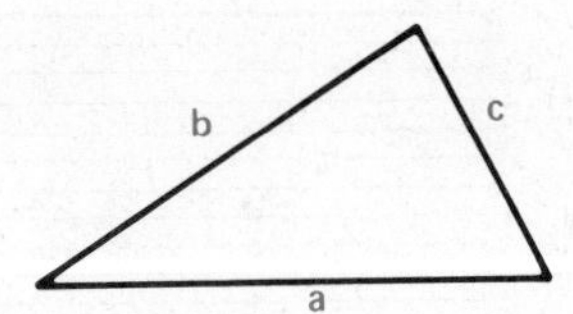

Figure 4–40 Transfer a given triangle to a new location.

Given the triangle, abc, in Figure 4–40, transfer it to a new location as shown in the following steps:

Step 1 Transfer one of the sides of the triangle to the new location by measuring its length or using dividers. (See Figure 4–41.)

Step 2 From one end of line a in the new location, draw an arc equal in length to line b. From the other end of line a, draw an intersecting arc equal in length to line c. (See Figure 4–42(a).)

Step 3 Connect both ends of line a to the intersection of the arcs to form the triangle in its new position as shown in Figure 4–42(b).

A right angle may be drawn when the length of the two sides adjacent to the 90° angle are given as shown in Figure 4–43.

Step 1 Draw line a perpendicular to line b. (See Figure 4–44.)

Step 2 Connect a line from the end of line a to the end of line b to establish the right triangle. (See Figure 4–45.)

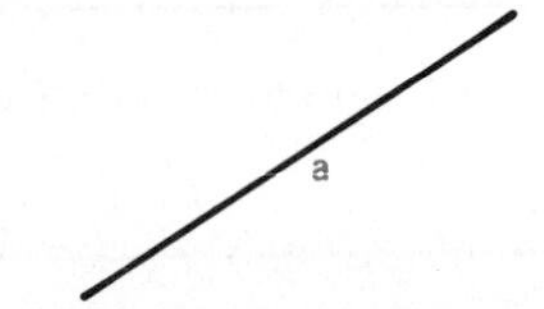

Figure 4–41 Establish a new location — step 1.

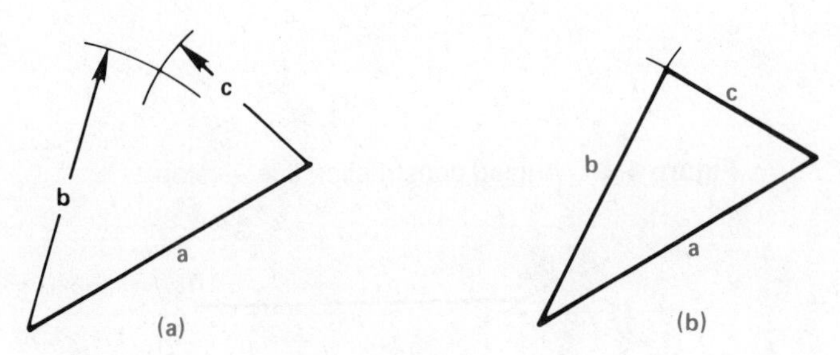

Figure 4–42 (a) Swing arcs b and c — step 2; (b) connect the ends of line a to intersection of arcs — step 3.

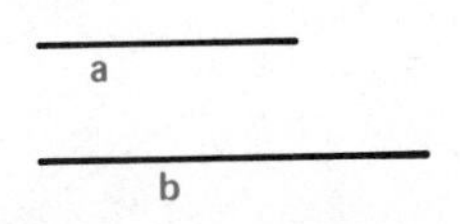

Figure 4–43 Construct a right angle given two sides.

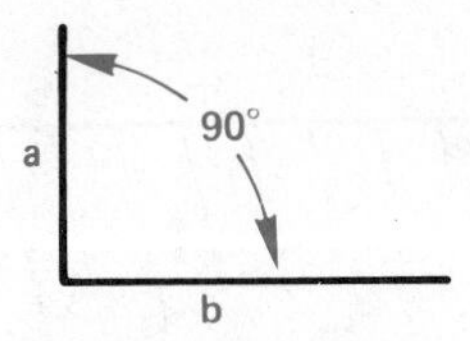

Figure 4–44 Draw line a perpendicular to line b — step 1.

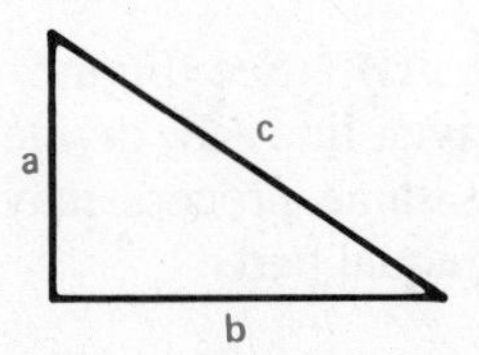

Figure 4–45 Connect hypotenuse — step 2.

Draw a right triangle given the length of the hypotenuse and one side as in Figure 4–46.

Step 1 Draw line c and establish its center. You can use the perpendicular bisector method to find the center. (See Figure 4–47.)

Step 2 From the center of line c, draw a 180° arc with the compass point at the line center and the compass lead at the line end. (See Figure 4–48.)

Step 3 Set the compass with a radius equal in length to the other given line, a. From one end of line c draw an arc intersecting the previous arc as shown in Figure 4–49.

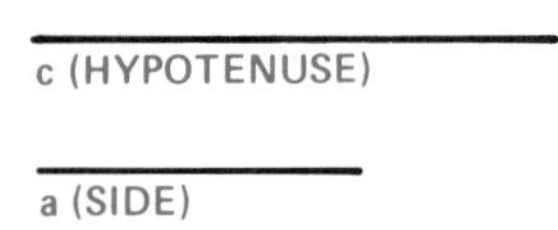

Figure 4–46 Construct a right triangle given a side and hypotenuse.

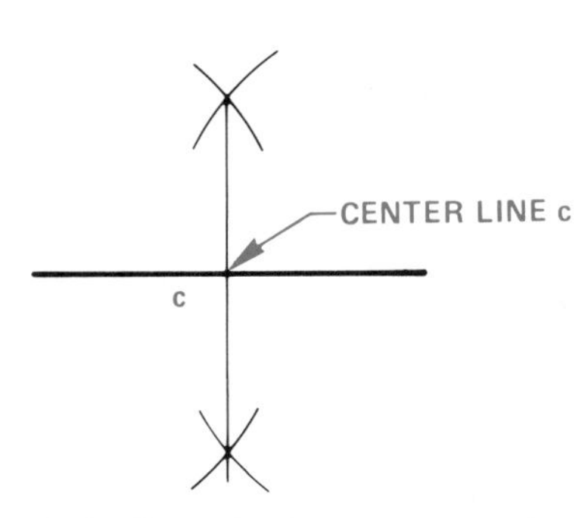

Figure 4–47 Draw the hypotenuse and establish its center — step 1.

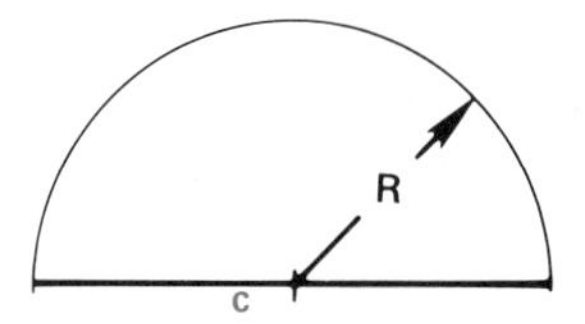

Figure 4–48 Draw a 180° arc from the center — step 2.

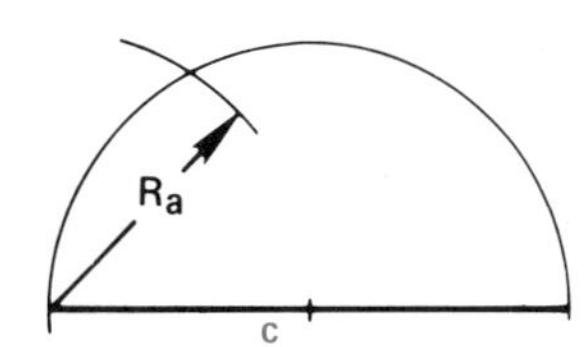

Figure 4–49 Use radius R_a to establish the end of line a — step 3.

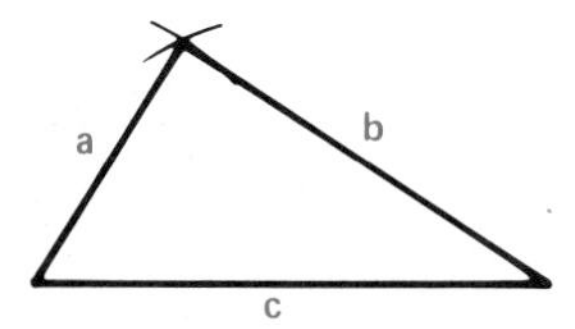

Figure 4–50 Complete the right triangle — step 4.

Step 4 From the intersection of the two arcs in step 3, draw two lines connecting the ends of line c to complete the required right triangle. (See Figure 4–50.)

An equilateral triangle may be drawn, given the length of one side, as demonstrated in the following steps.

Step 1 Draw the given length. Then set a compass with a radius equal to the length of the given line. Use this compass setting to draw intersecting lines from the ends of the given line as in Figure 4–51.

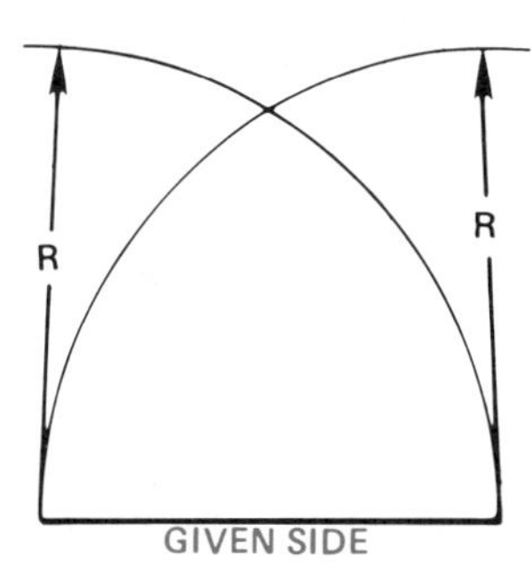

Figure 4–51 To construct an equilateral triangle given one side, swing equal arcs R — step 1.

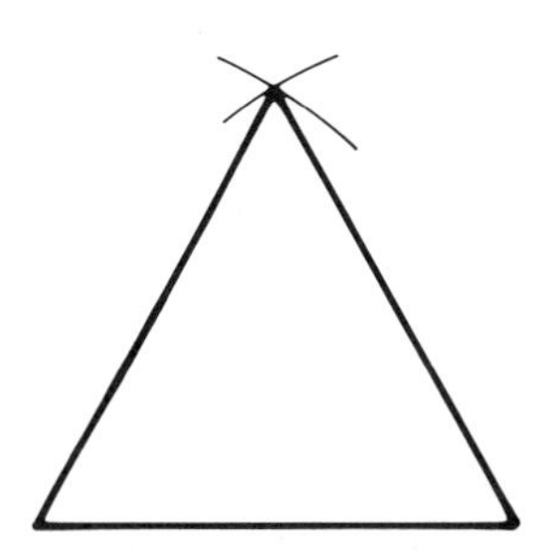

Figure 4–52 Complete the equilateral triangle — step 2.

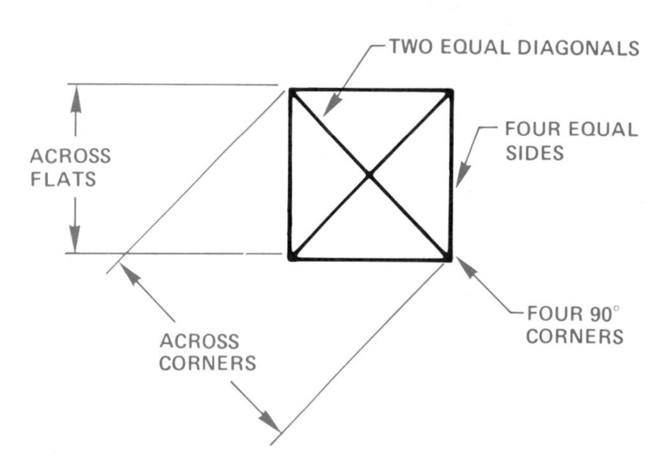

Figure 4–53 Elements of a square.

Step 2 Connect the point of intersection of the two arcs to the ends of the given side to complete the equilateral triangle as shown in Figure 4–52.

Squares

Square-head bolts or square nuts are often drawn in conjunction with manufactured parts. Use geometric constructions when necessary; however, when practical use square templates or special square-head bolt and nut templates. There are several methods that may be used to draw a square, each of which relates to the characteristics of the square shown in Figure 4–53.

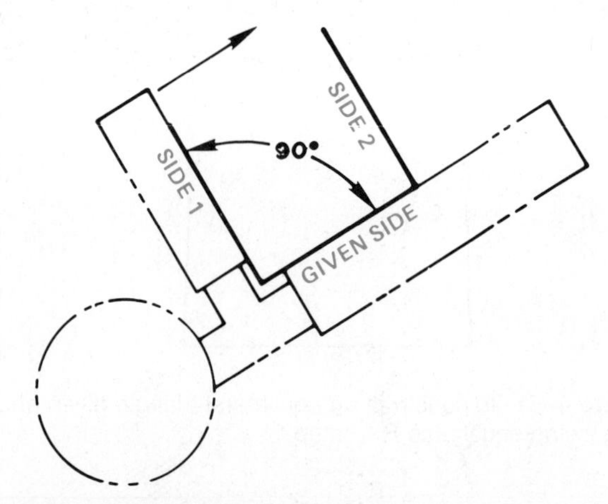

Figure 4–54 Use the drafting machine to establish 90° angles and sides — step 1.

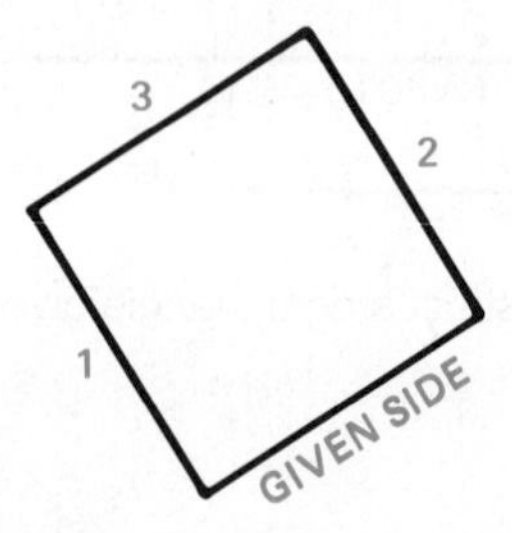

Figure 4–55 Complete the square by drawing side 3 — step 2.

Draw a square with the length of one side given as follows:

Step 1 Draw the length of the given side. Then with the drafting machine, or other appropriate method of obtaining a 90° angle, draw lines from each end equal in length to the given side. (See Figure 4–54.)

Step 2 Connect the end of sides 1 and 2 drawn in step 1 to complete the square with side 3, shown in Figure 4–55.

Draw a square with the distance across the flats or across the corners given.

Step 1 Draw a circle equal in diameter to the distance across the flats, Figure 4–56(a), and a circle with a diameter equal to the distance across the corners, Figure 4–56(b). Notice that the position of the centers dictates the position of the squares.

Step 2 Draw 45° lines tangent to the circle and extending to the centerlines as seen in Figure 4–57(a). Draw lines inside the circle connecting the intersections of the centerlines and the circle as seen in Figure 4–57(b).

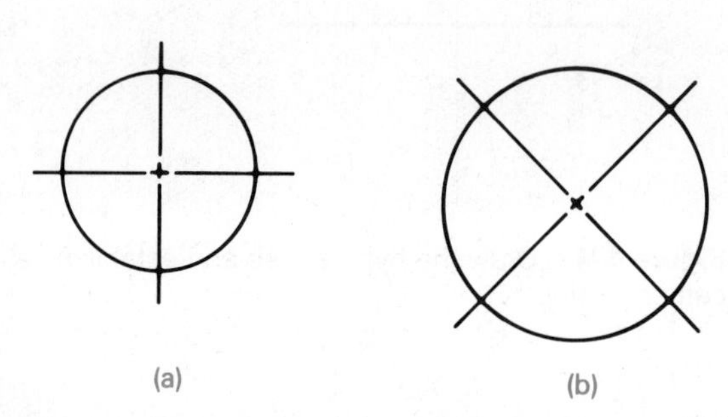

Figure 4–56 Draw a circle to construct a square with flats or corners given — step 1.

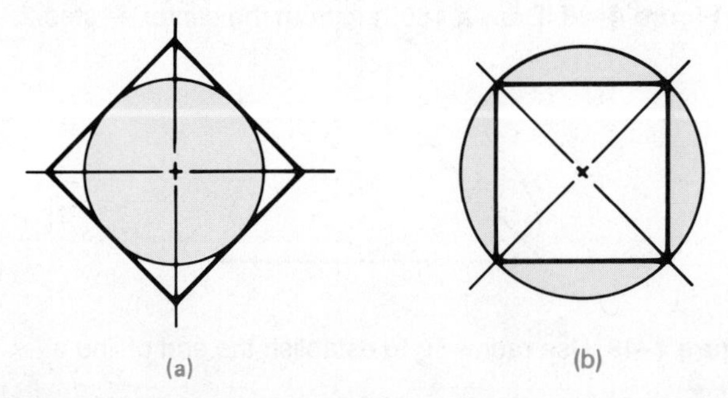

Figure 4–57 A circumscribed square at (a), and an inscribed square at (b) — step 2.

Regular Hexagon

Hexagons are six-sided polygons. Each side of a regular hexagon is equal and its interior angles are 120° as shown in Figure 4–58. Hexagons are commonly used as the shape for the heads of bolts and for nuts. Generally the dimension given for the size of a hexagon is the distance across the flats. The distances across the flats and across the corners are shown in Figure 4–58.

There are two easy geometric construction methods to draw a hexagon. One gives the distance across the flats; the other gives the distance across the corners. When practical use a hexagon template or a special hex-head bolt, screw, or hex nut template.

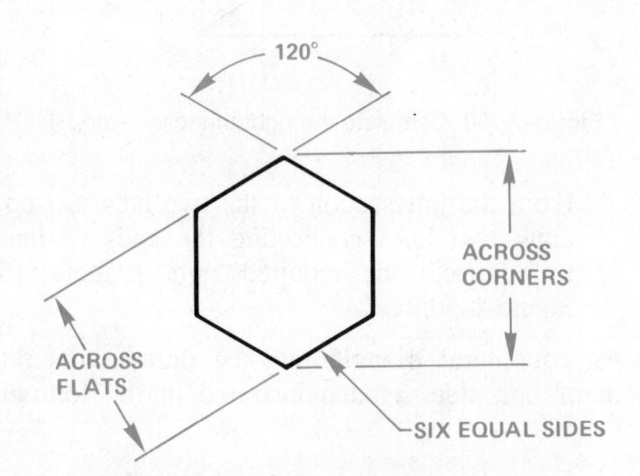

Figure 4–58 Elements of a regular hexagon.

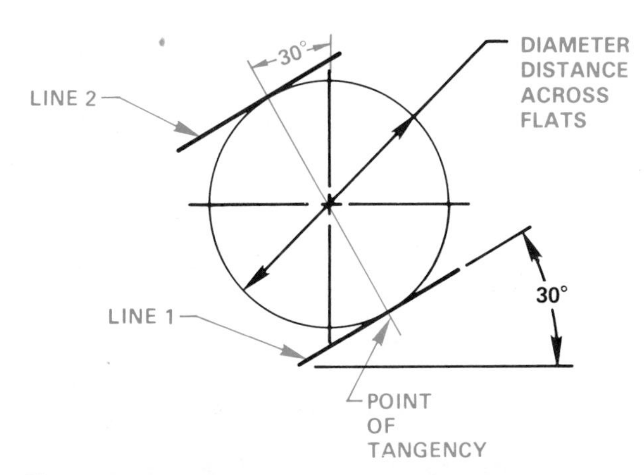

Figure 4–59 Constructing a hexagon — steps 1 and 2.

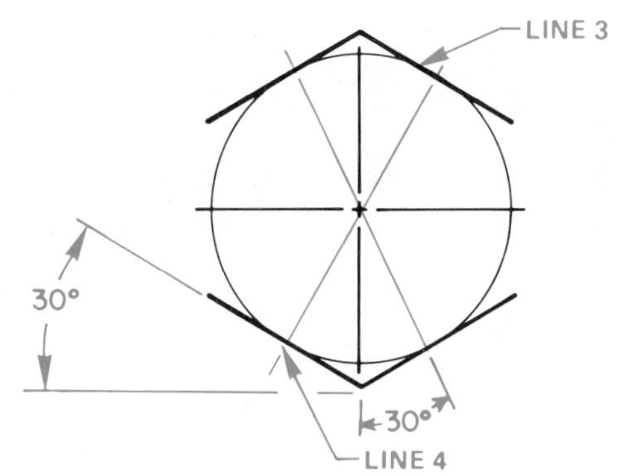

Figure 4–60 Step 3.

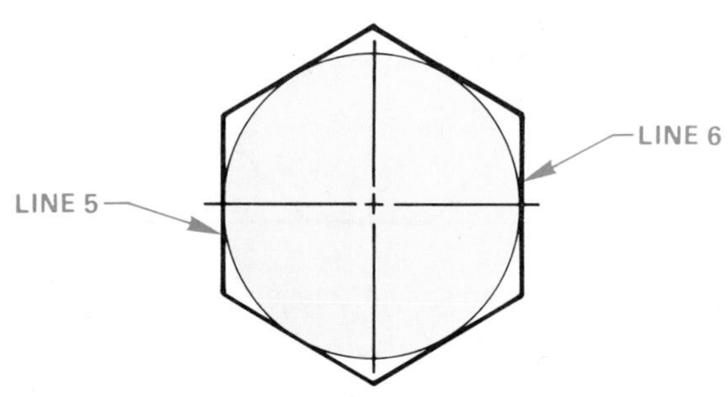

Figure 4–61 Step 4.

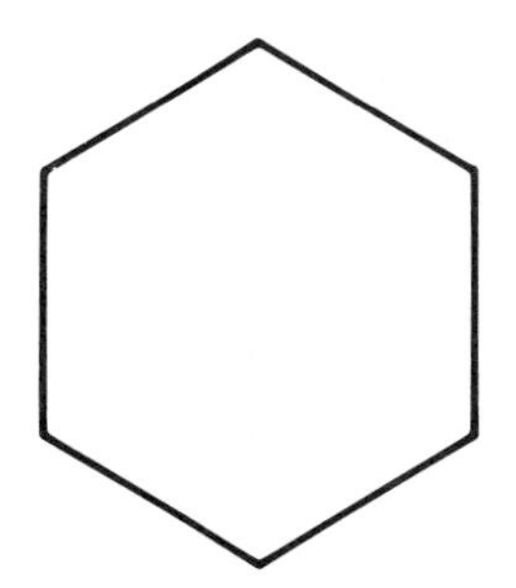

Figure 4–62 Complete the hexagon — step 5.

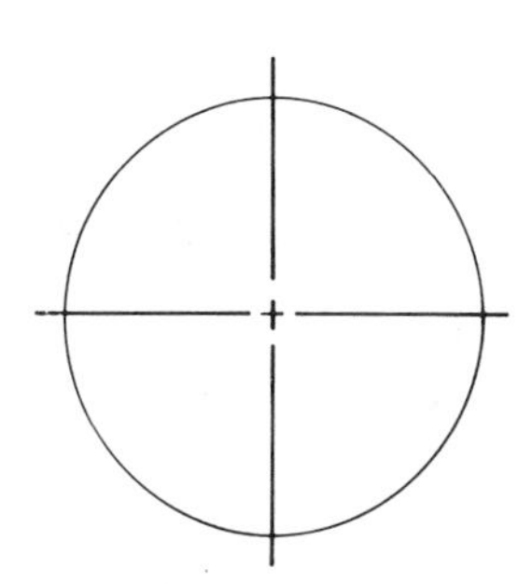

Figure 4–63 Construct a hexagon given the distance across the corners — step 1.

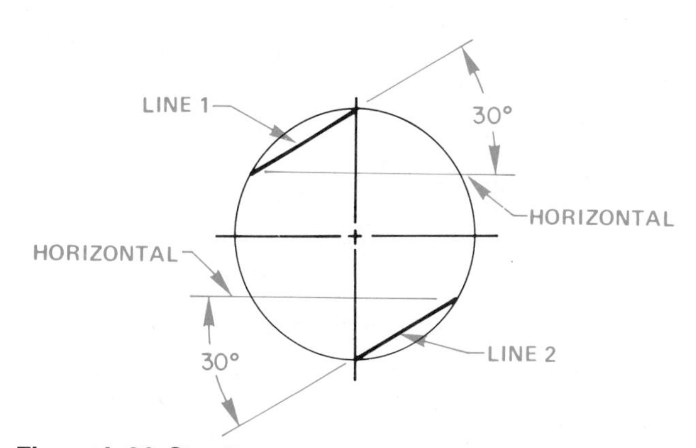

Figure 4–64 Step 2.

Across the Flats. Given the distance across the flats, do the following to construct a hexagon.

Step 1 Lightly draw a circle with the given distance as the diameter. (See Figure 4–59.)

Step 2 Set your drafting machine at 30° from horizontal or use a triangle. Then lightly draw lines 1 and 2 tangent to the circle as seen in Figure 4–59.

Step 3 Now set the angle at 30° the other way from horizontal and draw lines 3 and 4 tangent to the circle as shown in Figure 4–60.

Step 4 Draw lines 5 and 6 vertical and tangent to the circle as in Figure 4–61.

Step 5 Finally, darken the six lines to create the required hexagon. (See Figure 4–62.)

Across the Corners. Given the distance across the corners of a hexagon, use the following construction steps:

Step 1 Lightly draw centerlines in the desired location of the hexagon center. Then lightly draw a circle with the diameter equal to the given distance across the corners. Use construction lines as shown in Figure 4–63.

Step 2 From the location where the vertical centerline touches the circle, draw two lines 30° from horizontal, inside the circle as shown in Figure 4–64.

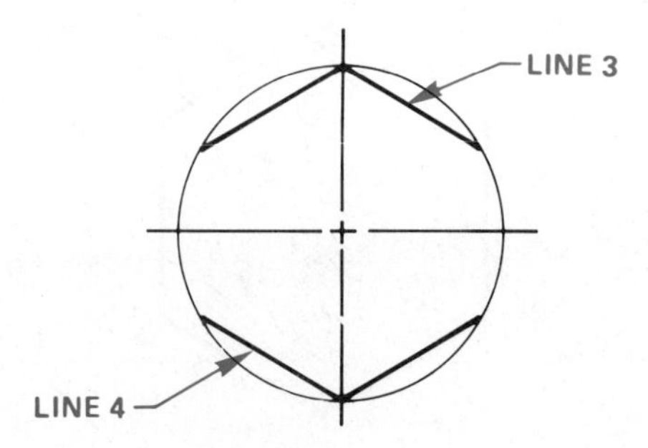

Figure 4–65 Step 3.

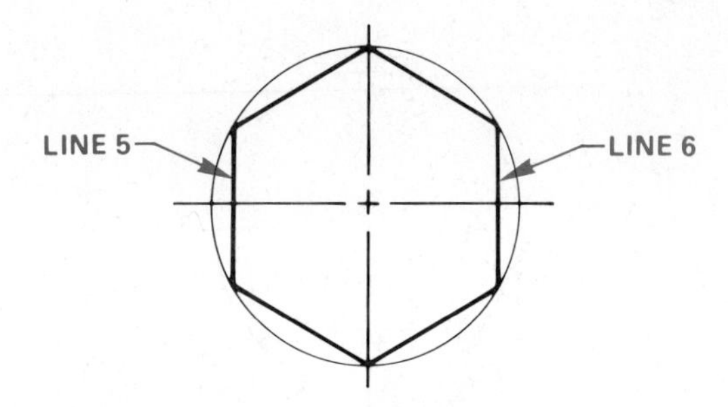

Figure 4–66 Step 4.

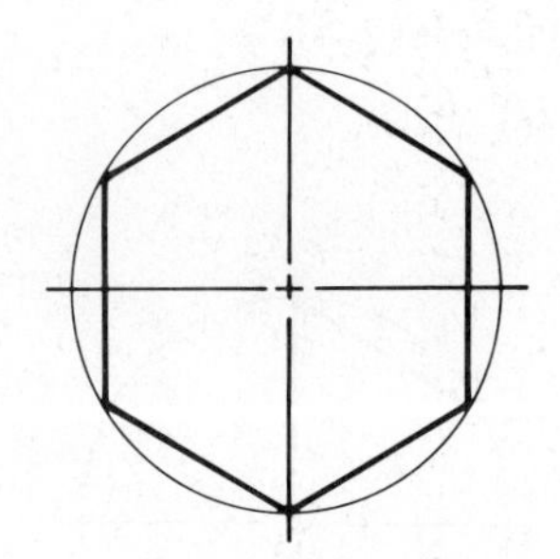

Figure 4–67 Step 5.

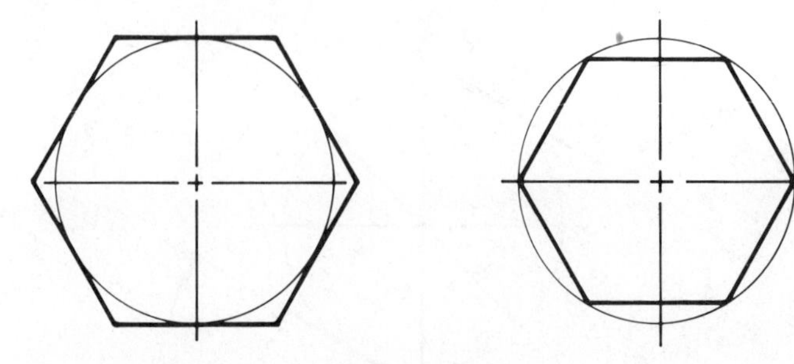

Figure 4–68 Alternate hexagon positions.

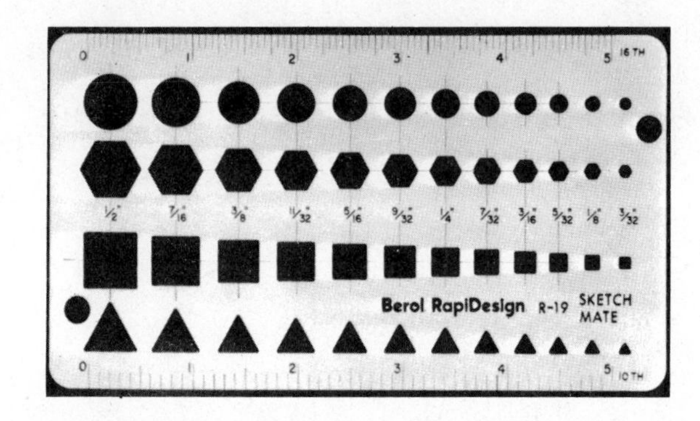

Figure 4–69 General purpose template. *Courtesy Berol USA RapiDesign.*

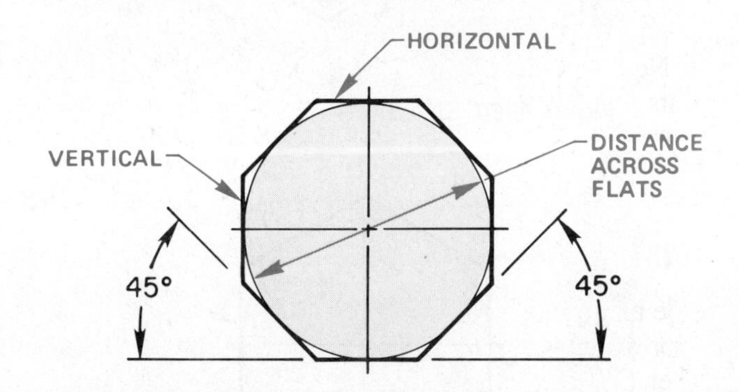

Figure 4–70 Elements of an octagon.

Step 3 Draw lines 3 and 4 at 30° from horizontal in the other direction, as seen in Figure 4–65.

Step 4 Draw two vertical lines, shown as lines 5 and 6 in Figure 4–66.

Step 5 Finally, darken object lines to create the required hexagon as shown in Figure 4–67.

The position of a hexagon may be rotated by establishing the six sides in an alternate relationship to the circle centerlines, as shown in Figure 4–68. Be sure that the included angles are 120° and each side is equal in length. The template shown in Figure 4–69 is a handy aid in drawing a variety of shapes, including the hexagon.

Regular Octagon

The octagon may be drawn using the same principle applied to the construction of a hexagon. A regular octagon has eight equal sides; therefore, if you use a 45° line as shown in Figure 4–70 you can easily draw an octagon about the given circle with a diameter the distance across the flats.

Polygons

A polygon is any closed plane geometric figure with three or more sides or angles. Hexagons and octagons are polygons. You can draw any polygon (if you know the distance across the flats and the number of sides) by the same method used to draw the hexagon and octagon across the flats.

There are 360° in a circle; therefore, if you need to draw a 12-sided polygon divide 360° by 12 (360° ÷ 12 = 30°) to determine the central angle of each side. Since there are 30° between each side of a 12-sided polygon, divide a circle with a diameter equal to the distance across the flats into equal 30° parts as shown in Figure 4–71.

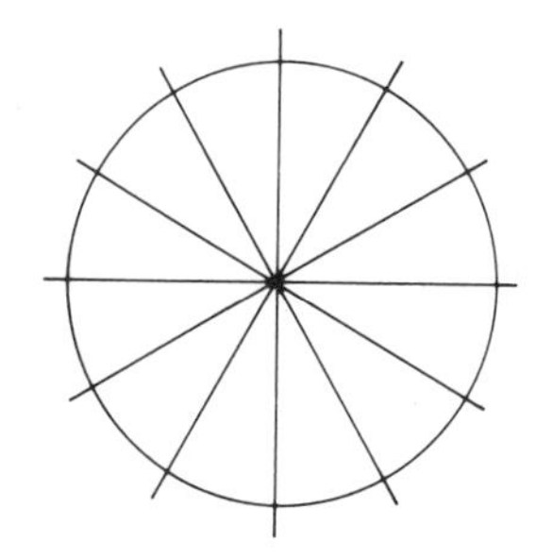

Figure 4–71 Constructing a regular polygon with 12 sides.

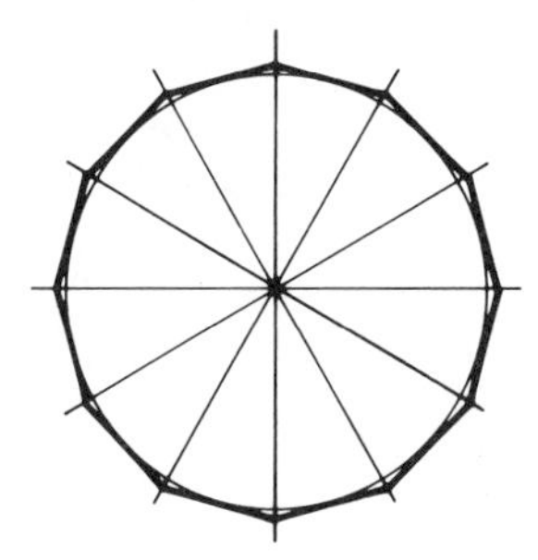

Figure 4–72 Connecting the 12 radial lines.

Now, connect the 12 radial lines with line segments that are tangent to each arc segment, as shown in Figure 4–72.

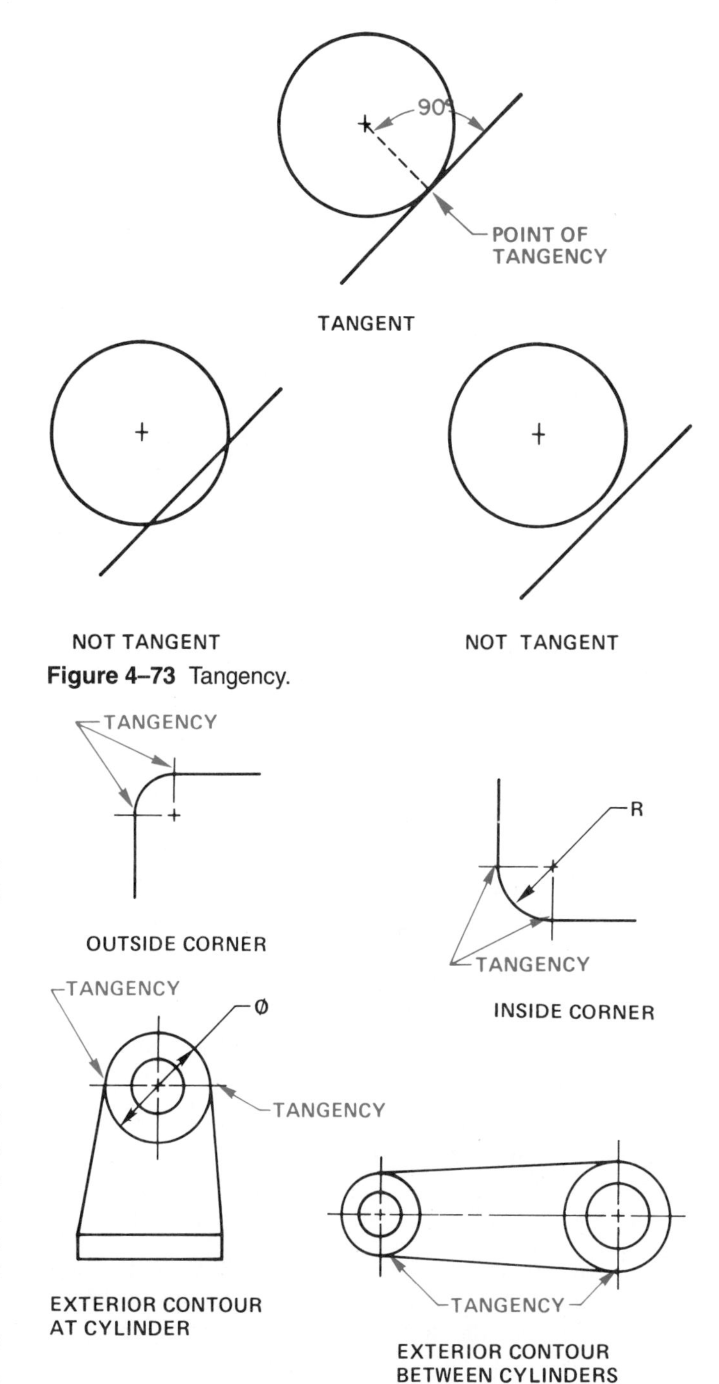

Figure 4–73 Tangency.

Figure 4–74 Common tangencies.

CONSTRUCTING CIRCLES AND TANGENCIES

To be tangent to a circle or arc, a line must touch the circle or arc at only one point, and a line drawn from the center of the circle or arc must be perpendicular to the tangent line at the point of tangency. (See Figure 4–73.) Lines drawn tangent to circles or arcs are common in mechanical drafting. It is important that tangent lines be drawn with a smooth transition at the point of tangency. Figure 4–74 shows some common tangency examples.

When drafting tangent features it is usually best to lightly draw the centerlines, then draw the arcs or circles, and finally draw the tangent lines. It is easier to draw a line tangent to a circle or arc than it is to draw a circle or arc tangent to a line. (See Figure 4–75.) When possible, use a template to draw the arcs and circles. Keep in mind that template circles are calibrated in diameters while arcs have radius dimensions. So, for example, a .50-in. radius would require using a 1.00-in. diameter circle. The combination of templates and a drafting machine will provide the easiest and fastest results; use a compass only if necessary. Figure 4–76 shows the relationship between a well-drawn and a poorly drawn tangency.

Draw an arc tangent to a given acute or obtuse angle in much the same way as previously shown for lines and circles. Given the acute and obtuse angles in Figure 4–77, draw an arc with a 12-mm radius tangent to the sides of each angle.

Step 1 Select a circle template with a 24-mm diameter circle. Move the template into position at each angle, leaving a small space for the lead or pen point thickness, and draw the arc as shown in Figure 4–78.

Step 2 Remove the template and draw the sides of each angle to the points of tangency of the arcs. (See Figure 4–79.)

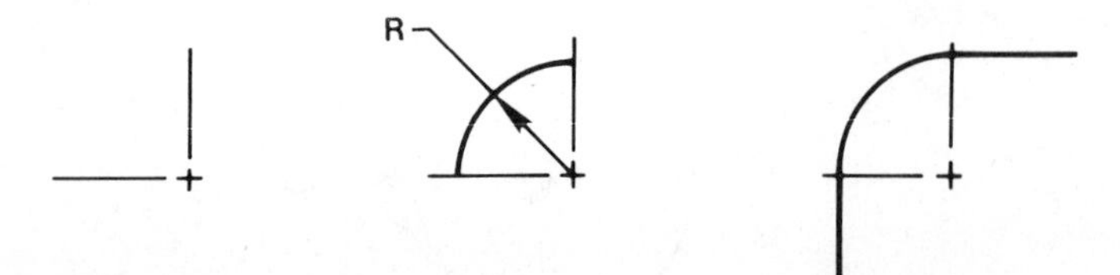

Figure 4–75 Drawing lines tangent to an arc and an arc at a 90° corner.

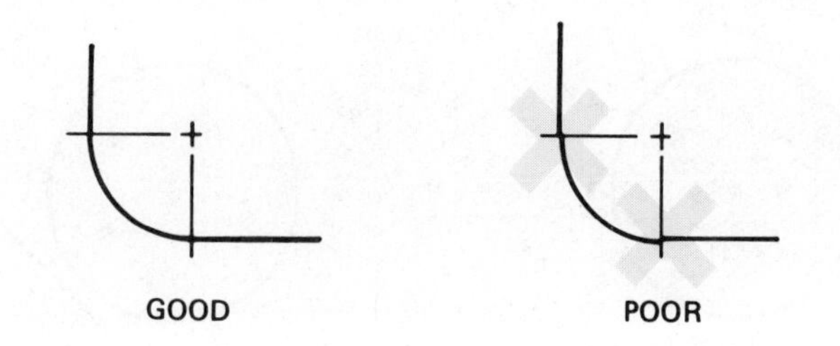

Figure 4–76 Good and poor tangency examples.

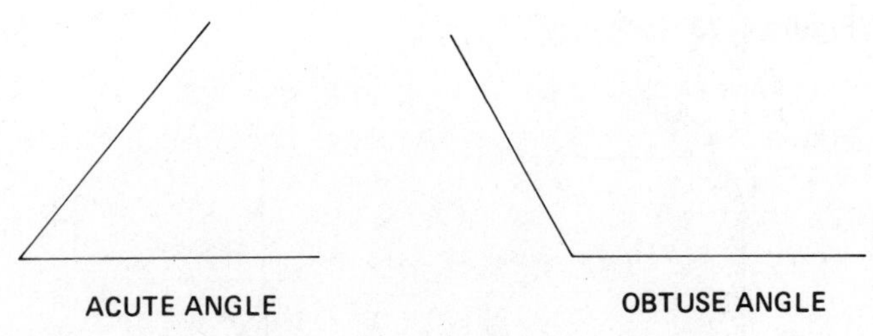

Figure 4–77 Construct an arc tangent to given angles.

If the arc center is required for dimensioning purposes use the following procedure.

Step 1 Establish the arc centers by lightly drawing lines parallel to the given sides at a distance from the sides equal to the radius of the arc. In this example the radius is 12 mm. (See Figure 4–80.)

Step 2 Set the compass at the given radius, 12 mm, and draw two arcs tangent to the given sides. Connect the sides to the points of tangency of the arcs as shown in Figure 4–81.

Arc Tangent to a Given Line and a Circle

The important point to remember is that the distance from the point of tangency to the center of the tangent arc is equal to the radius of the tangent arc. Given the machine part in Figure 4–82, drawn an arc tangent to the cylinder and base with a 24-mm radius.

Step 1 Draw a line parallel to the base at a distance equal to the radius of the arc, 24 mm. Then draw an arc that intersects the line drawn with a radius equal to the given arc plus the radius of the cylinder (24 + 11 = 35 mm). (See Figure 4–83.)

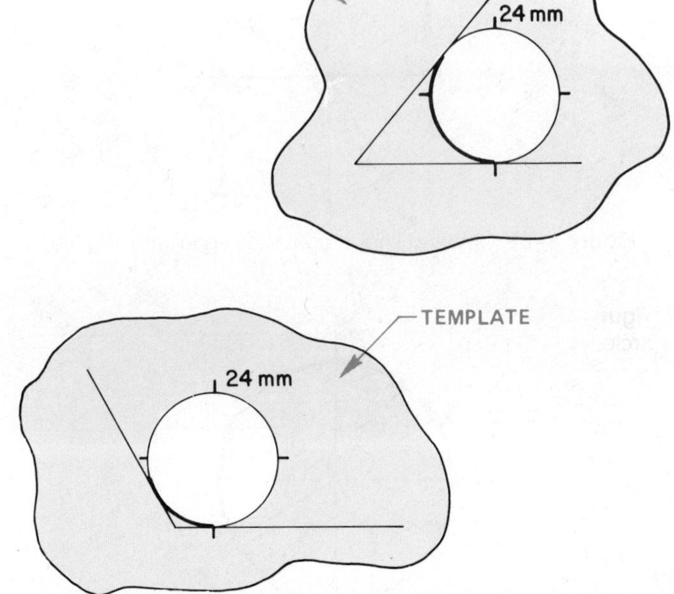

Figure 4–78 Draw a tangent arc with a template — step 1.

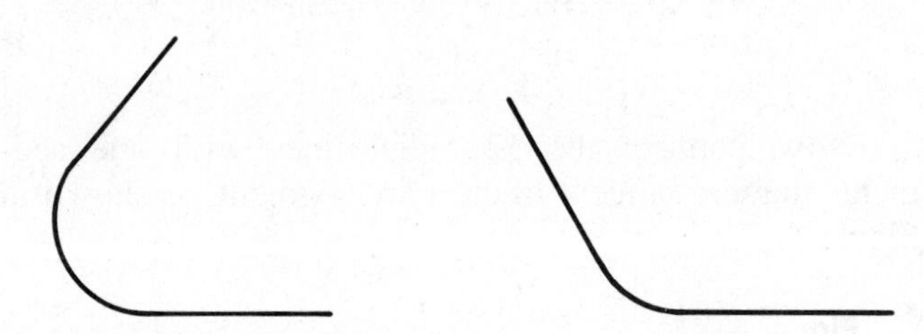

Figure 4–79 Completed tangent arcs — step 2.

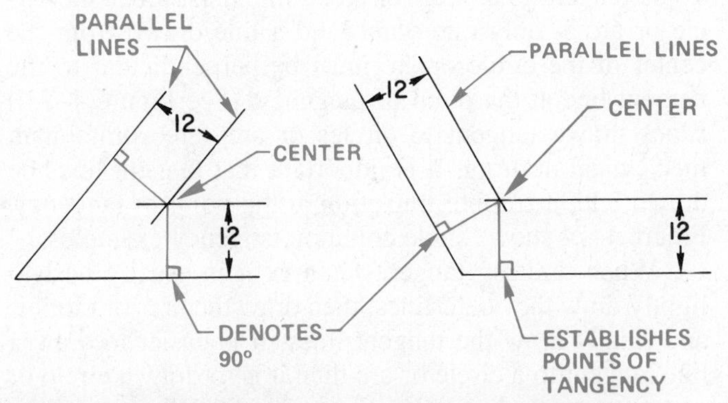

Figure 4–80 Drawing tangent arcs with a compass; this procedure also works for arcs at 90° corners — step 1.

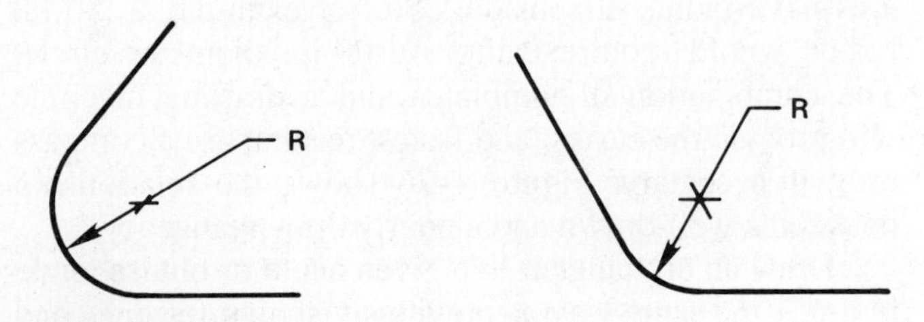

Figure 4–81 Completed tangent arcs — step 2.

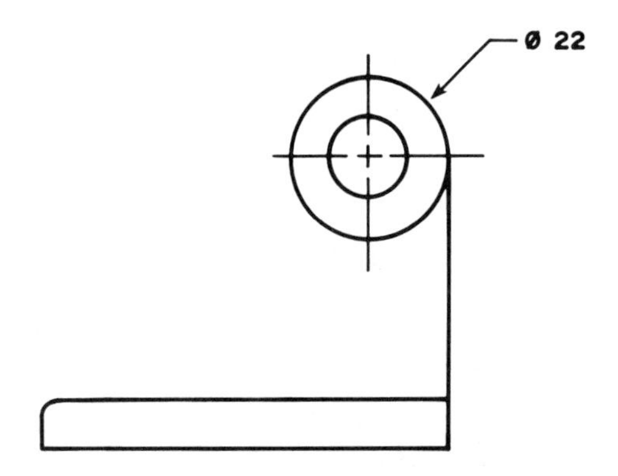

Figure 4–82 Construct an arc tangent to a given line and a circle.

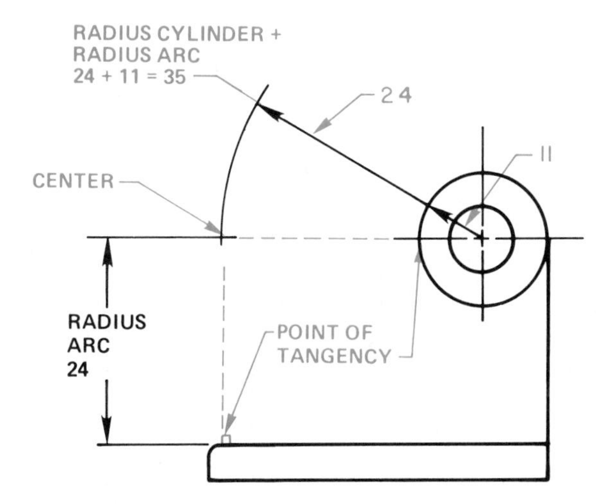

Figure 4–83 Step 1.

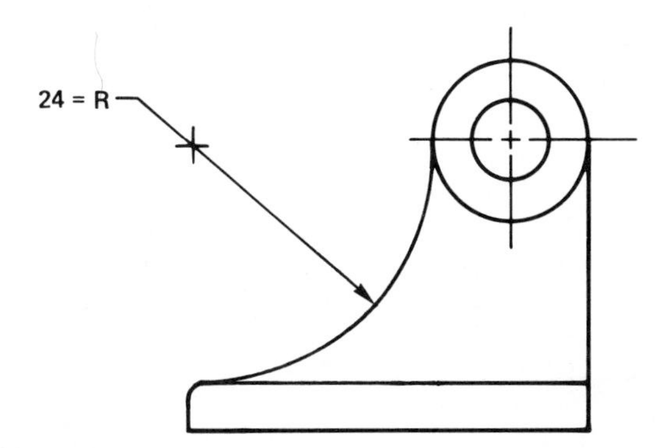

Figure 4–84 Step 2.

Step 2 Use a compass or a template to draw the required radius from the center point established in Step 1. (See Figure 4–84.)

The previous example demonstrated an internal arc tangent to a circle or arc and a line. A similar procedure is used for an external arc tangent to a given circle and line as shown in Figure 4–85. In this case the radius of the cylinder (circle) is subtracted from the arc radius to obtain the center point.

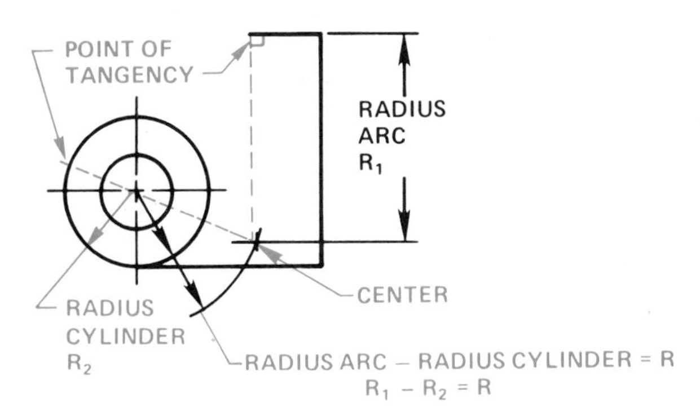

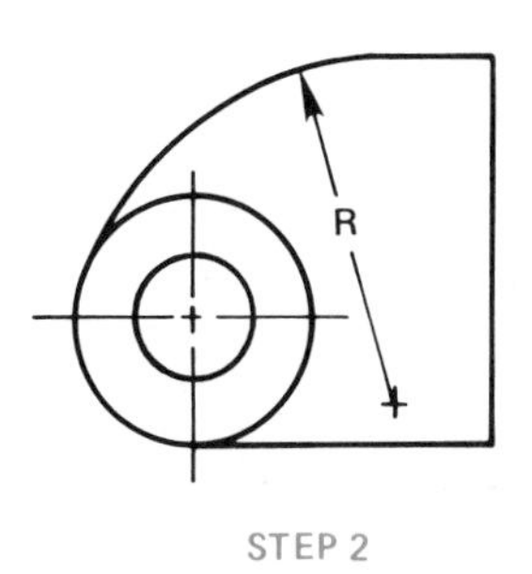

Figure 4–85 Constructing an external arc tangent to a line and cylinder.

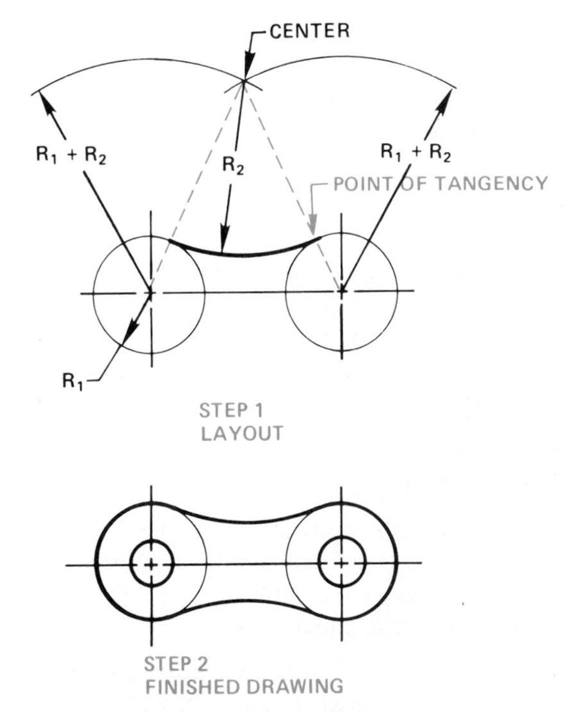

Figure 4–86 Chain link.

The following examples, Figures 4–86 through 4–88, are situations that commonly occur on machine parts where an arc may be tangent to given cylindrical or arc shapes. Keep in mind that the methods used to draw the tangents are the same as just described. The key is that the center of the required arc is always placed at a

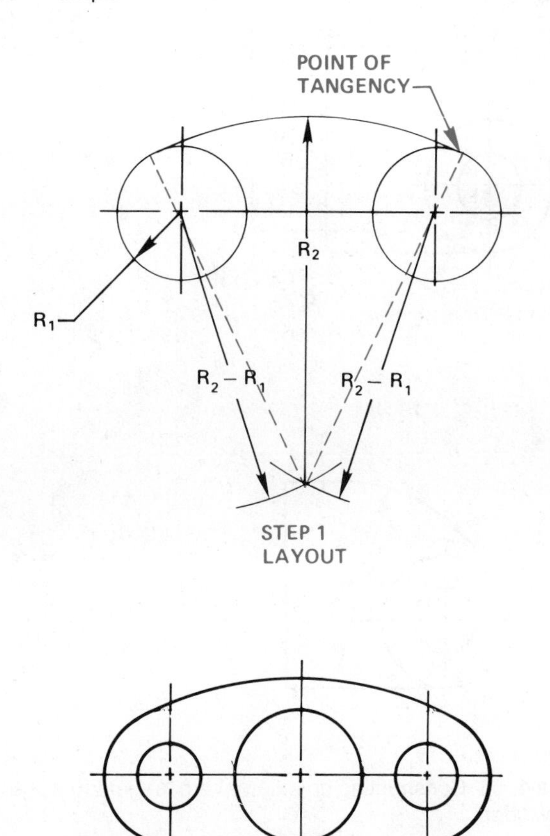

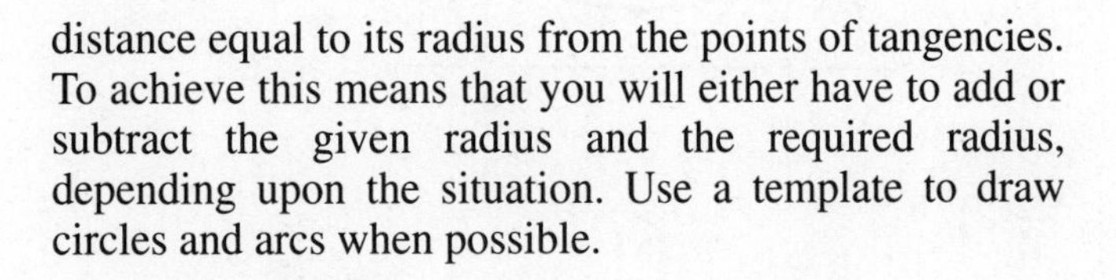

Figure 4–87 Gasket.

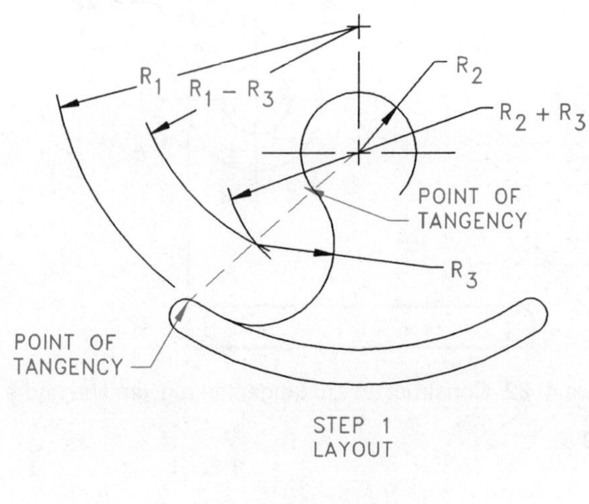

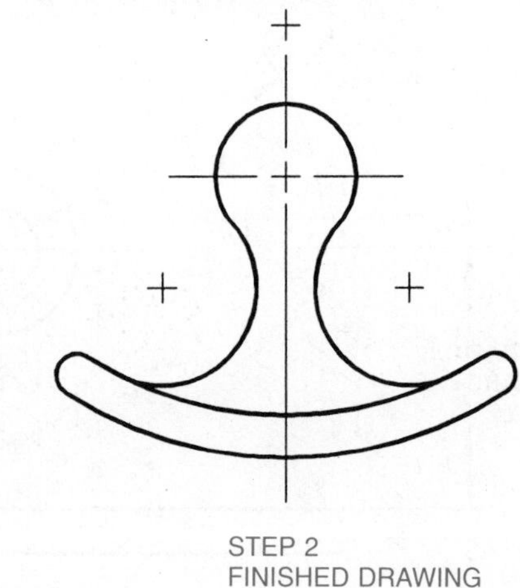

Figure 4–88 Hammer head.

distance equal to its radius from the points of tangencies. To achieve this means that you will either have to add or subtract the given radius and the required radius, depending upon the situation. Use a template to draw circles and arcs when possible.

Ogee Curve

An *S* curve, commonly called an ogee curve, occurs in situations where a smooth contour is needed between two offset features. The following steps show how to obtain an S curve with equal radii.

Step 1 Given the offset points A and B in Figure 4–89, draw a line between points A and B and bisect that line to find point C.

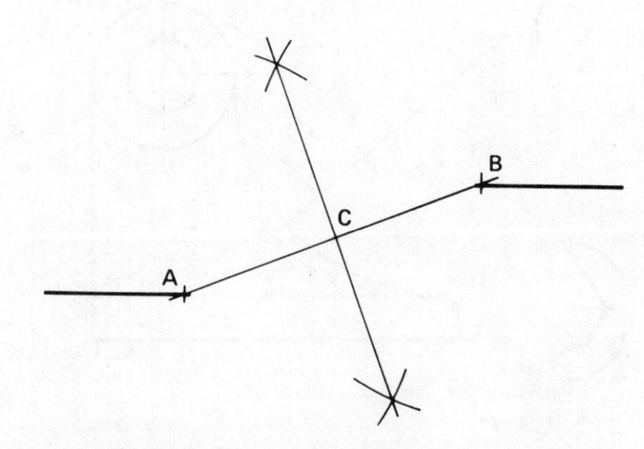

Figure 4–89 Constructing an S curve — step 1.

Step 2 Draw the perpendicular bisector of lines AC and CB. From point A draw a line perpendicular to the lower line that will intersect the AC bisector at X. From point B draw a line perpendicular to the upper line that will intersect the CB bisector at Y. (See Figure 4–90.)

Step 3 With X and Y as the centers, draw a radius from A to C and a radius from B to C as shown in Figure 4–91.

Step 4 If the S curve line, AB, represents the centerline of the part, then develop the width of the part

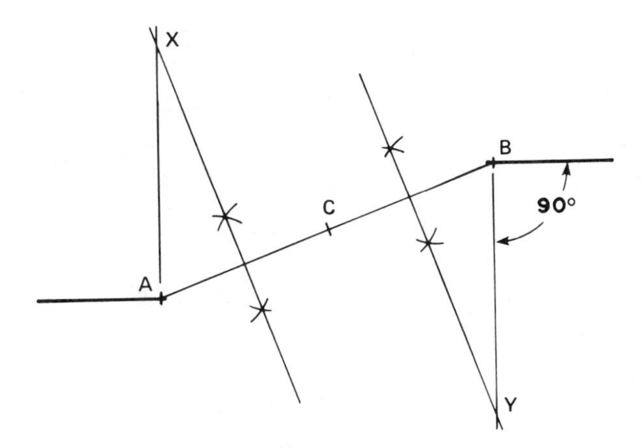

Figure 4–90 Constructing an S curve — step 2.

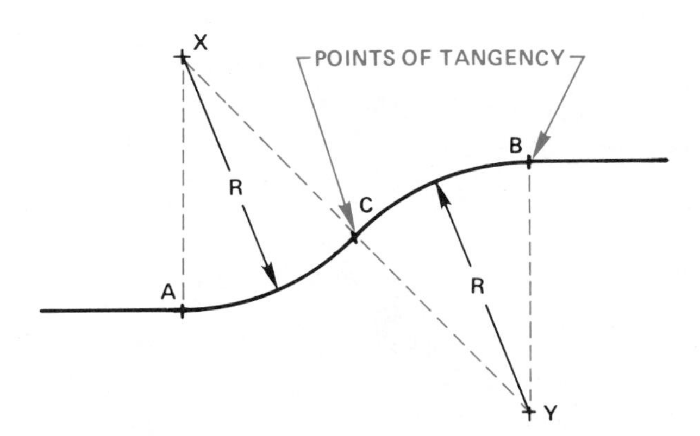

Figure 4–91 Constructing an S curve — step 3.

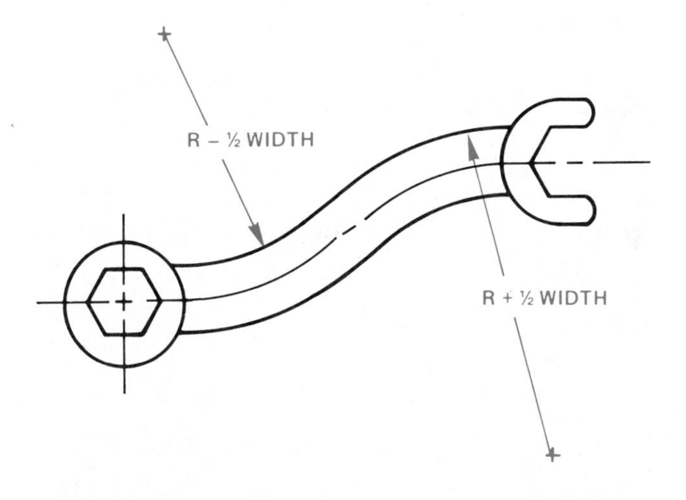

Figure 4–92 A wrench with a complete S curve — step 4.

parallel to the centerline using concentric arcs to complete the drawing. (See Figure 4–92.)

When an S curve has unequal radii the procedure is similar to the previous example, as shown in Figure 4–93.

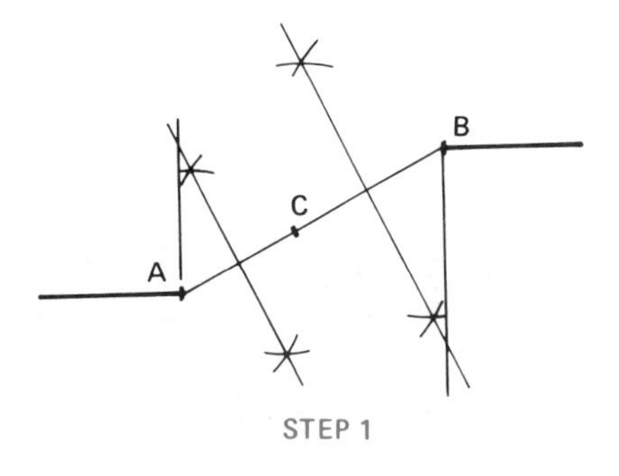

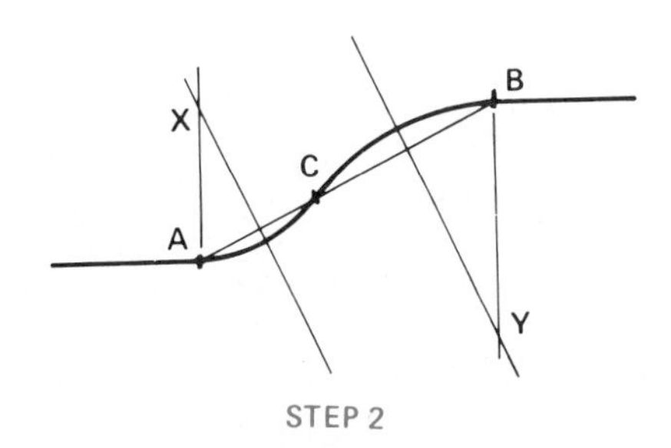

Figure 4–93 Constructing an S curve with unequal radii.

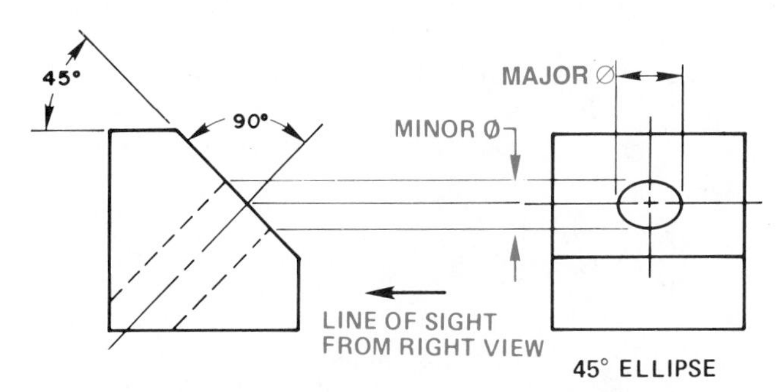

Figure 4–94 Elliptical view.

CONSTRUCTING AN ELLIPSE

When a circle is viewed at an angle, an ellipse is observed. For example, if a surface with a through hole is inclined 45°, the representation would be a 45° ellipse, as shown in Figure 4–94. An ellipse has a major diameter and a minor diameter. In Figure 4–94 the minor diameter is established by projecting the top and bottom edge from the slanted surface. The major diameter, in this case, is equal to the diameter of the circle.

If possible, always use a template to draw elliptical shapes. Ellipse templates are available from 10° to 85°. A zero-degree ellipse is a line and a 90° ellipse is a circle. The ellipse shown in Figure 4–94 could have been drawn by projecting the centerlines and then lining up the center marks of a 45° ellipse template. See Figure 4–99.)

DRAWING GEOMETRIC SHAPES

The process used for drawing geometric shapes involves three steps:

1. select the command
2. locate the feature
3. input the parameters, that is, draw the feature

Circle. A circle, for example, may be drawn several different ways. After selecting the CIRCLE command, the information that you have about the circle will determine the method by which you draw it. The methods are shown in Figure 4–95, (a) through (d). The first method (a) requires that two points, the center (1) and a point on the circle (2), be digitized before the circle can be drawn. The second, example (b), requires that two opposite points (1 and 2) on the circumference (the diameter) be digitized. The third, example (c), requires that three points be digitized on the circumference. The final method (d) requires that the center point be digitized and the radius be entered as a number selected from the keypad on the menu tablet, or typed from the keyboard.

Ellipse. Ellipses are quick and easy to draw on a CADD system. Most programs have a command such as ELLIPSE. In this command you are asked for the major and minor diameters and the computer automatically draws an ellipse with the information.

Arc. Arcs are constructed in the same manner as circles. But arcs are only portions of circles and therefore the two end points of the arc must be digitized. The methods of constructing an arc are shown in Figure 4–96, (a) through (d). The fourth method (d) has two variations, each requiring numerical input. The first variation requires point number 1 (beginning point) to be digitized, as well as point number 2, a location on the arc. Then a numerical value for the radius and length of the arc must be entered from the menu keypad or from the keyboard. The second variation requires the same two points, but instead of entering the radius, the center point location of the arc is entered at the keypad.

Tangencies. Tangencies are easy to draw with the computer. Most CADD programs have a command or option such as TANGENT, which allows you to pick a circle or arc and automatically draw a line or another circle or arc tangent to the selected object.

Polygon. A polygon can be drawn using the same methods as the CIRCLE command, but some additional information is needed. Since polygons have corners, or vertices, the computer needs to know where the first vertex should be located. A prompt asking for the location will appear on the screen and the answer should be in degrees. Figure 4–97 shows two examples of polygons that have been given different starting angles. The final bit of information the computer needs is the number of sides. After this number is entered from the menu or keyboard, the polygon is drawn.

Rectangle. A rectangle is one of the easiest shapes to draw. Only two points are needed. The first is one corner of the rectangle, which may be any of the four corners. Once you digitize this corner, the direction location of the second point has already been determined. The second point is the corner of the rectangle opposite the first corner that was digitized. The computer then connects these two points with the straight lines needed to form the rectangle. Figure 4–98 illustrates this process. Of course rectangles and boxes can be drawn using the LINE command, but it takes longer.

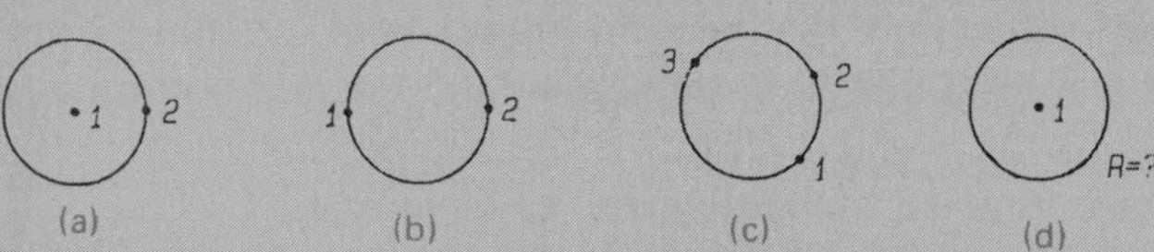

Figure 4–95 Four ways to digitize a circle.

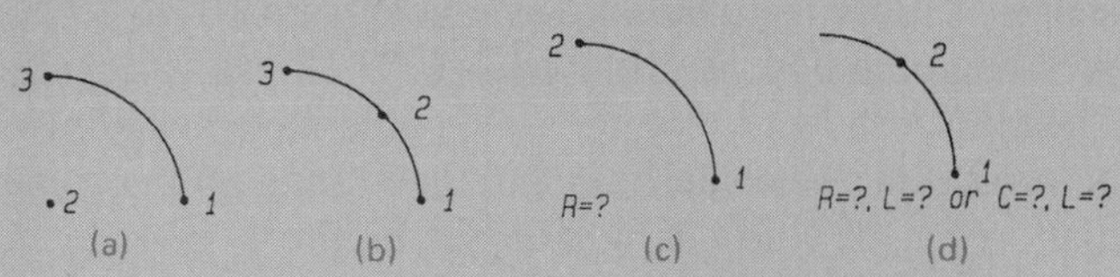

Figure 4–96 Four ways to digitize an arc.

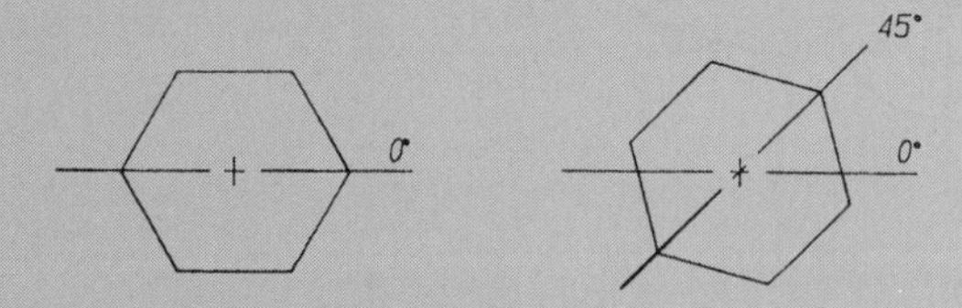

Figure 4–97 Polygon starting angles.

Figure 4–98 Rectangle construction using opposite corners.

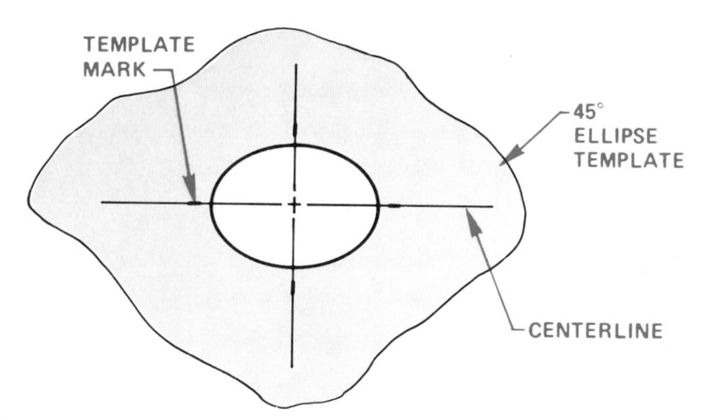

Figure 4–99 Drawing an ellipse with a template.

If the elliptical shape does not exactly fit the available template calibrations, then use one that is close. How close depends on your company standards. Most companies would rather have the drafters draw a close representation than take the time to lay out an ellipse geometrically. When templates are not available or are not close enough for the particular ellipse that must be drawn, several methods can be used to draw elliptical shapes. One of the most practical is the concentric circle method described in the following steps.

Step 1 Use construction lines to draw two concentric circles, one with a diameter equal to the minor ellipse diameter and the other with a diameter equal to the major ellipse diameter. (See Figure 4–100.)

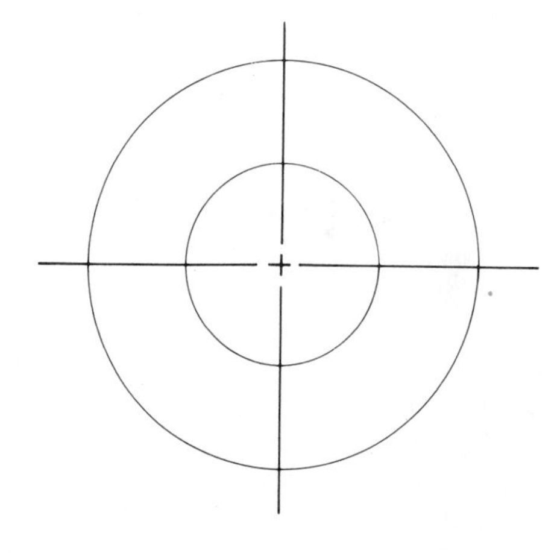

Figure 4–100 Constructing an ellipse, concentric circle method — step 1.

Step 2 Use the drafting machine or triangles to divide the circles into at least 12 equal parts; 30° and 60° each way from horizontal. More parts will give better accuracy, but will also consume more time. (See Figure 4–101.)

Step 3 From the points of intersection of each line at the circumference of the circles, draw horizontal lines from the minor diameter and vertical intersecting lines from the major diameter, essentially creating a series of right triangles, although the intersection points only are needed. (See Figure 4–102.)

Step 4 Using an irregular (French) curve carefully connect the points created in Step 3 to complete the ellipse as shown in Figure 4–103.

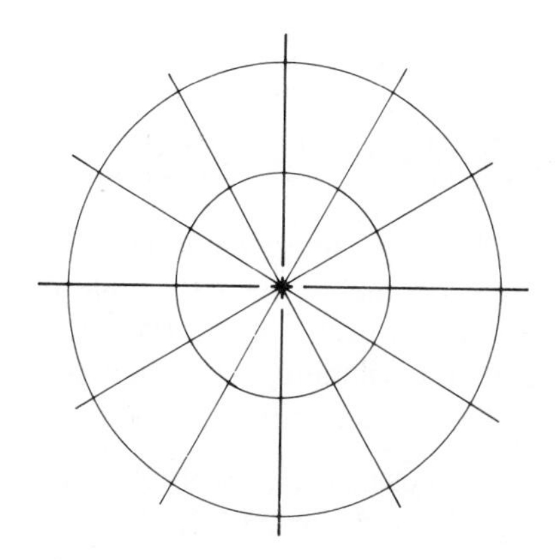

Figure 4–101 Concentric circle method — step 2.

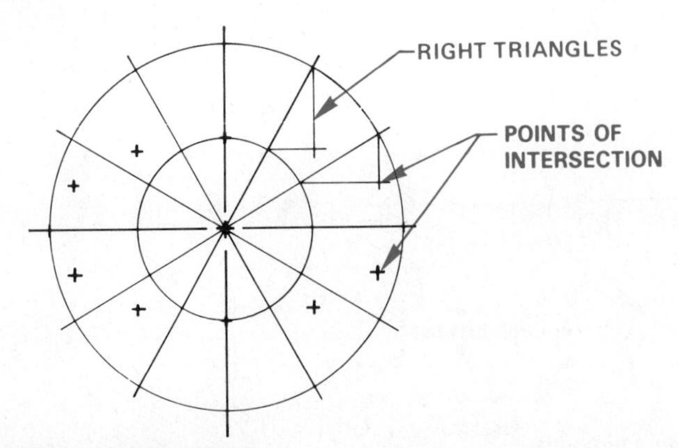

Figure 4–102 Step 3.

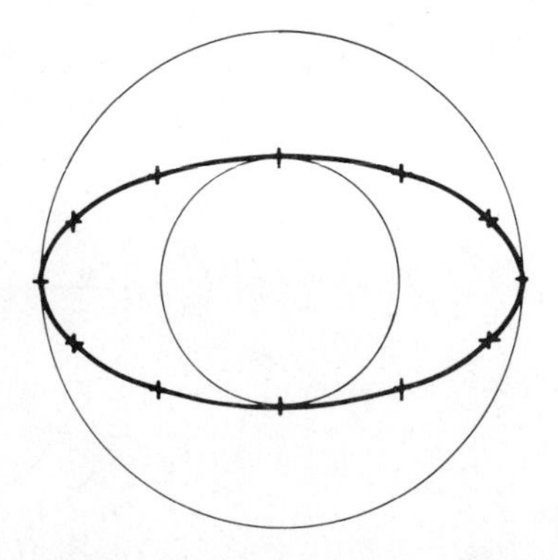

Figure 4–103 Complete ellipse — step 4.

PROFESSIONAL PERSPECTIVE

Geometric constructions are found in nearly every engineering drafting assignment. Some geometric constructions are as simple as a radius corner while others are very complex. As a professional drafter, you should be able to quickly identify the type of geometric construction involved in the problem and solve it with one of the techniques you have learned. When you solve these problems, accuracy and carefully drawn construction lines are critical. Use the best tools available. For example, a professional drafter hardly ever draws an ellipse using construction methods; he or she uses the large variety of ellipse templates available. Always use a template if you can, although, if all else fails, you do know how to construct the ellipse. Even though the geometric construction methods are presented for the manual drafter, the principles are the same for the CADD drafter. You should always refer to these principles and techniques to insure that your CADD application has performed the task correctly. You will find with CADD that some of the very time-consuming constructions, such as dividing a space into equal parts or drawing a pentagon, are not only extremely fast but they are almost perfectly accurate.

ENGINEERING PROBLEM

You should always approach an engineering drafting problem in a systematic manner. Plan out how you propose to solve the problem. As an entry-level drafter you need to do this planning with sketches and written notes. These sketches and notes help you decide:

- The scale to use to effectively display the drawing. If the scale is too small, complex detail may be lost, but if the scale is too large, the drawing may take up too much space and cause you to spend too much time.
- What paper size or CADD limits should be established. This depends on the drawing scale, the amount of detail, and the amount of specifications that go with the drawing.

After these decisions are made, you need to proceed with the problem in a logical manner. Refer to the engineer's sketch shown in Figure 4–104 as you follow these layout steps for either manual drafting or CADD:

Step 1 Do all preliminary manual work using construction lines so that if you make an error, it is easy to erase. Begin by establishing the centers of

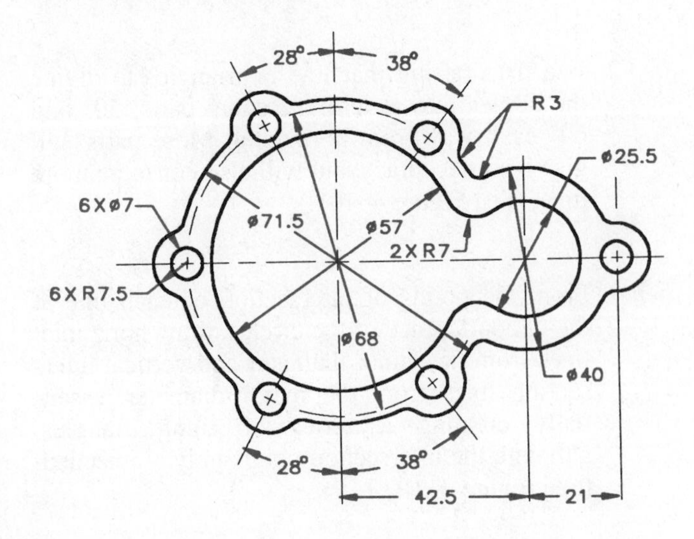

Figure 4–104 Engineer's sketch.

the ∅ 57 and ∅ 25.5 circles and then draw the circles.

Step 2 Draw the parallel outside circles of ∅ 71.5 and ∅ 40.

Step 3 Locate and draw the 6 × R 7.5 arcs and then using tangencies blend the R 3 radii with the outside arcs.

Step 4 Draw the 2 × R 7 arcs tangent to the large inside arcs.

Step 5 Draw the centerlines for the 6 × ∅ 7 circles and then draw the circles in place.

PROFESSIONAL HINT

Always draw the construction centerlines for circles and arcs at 90° angles before drawing the circle or arc using a template. If you draw circles and arcs with a template using arc centerlines as shown in the proposed drawing, your circles and arcs will be off center. The construction centerlines are very lightly drawn or drawn with light blue lead for construction purposes only.

Step 6 Darken all thin lines and then darken all object lines. If there are arcs and straight lines, draw the arcs first and then blend the straight lines into the tangencies. The finished drawing is shown in Figure 4–105.

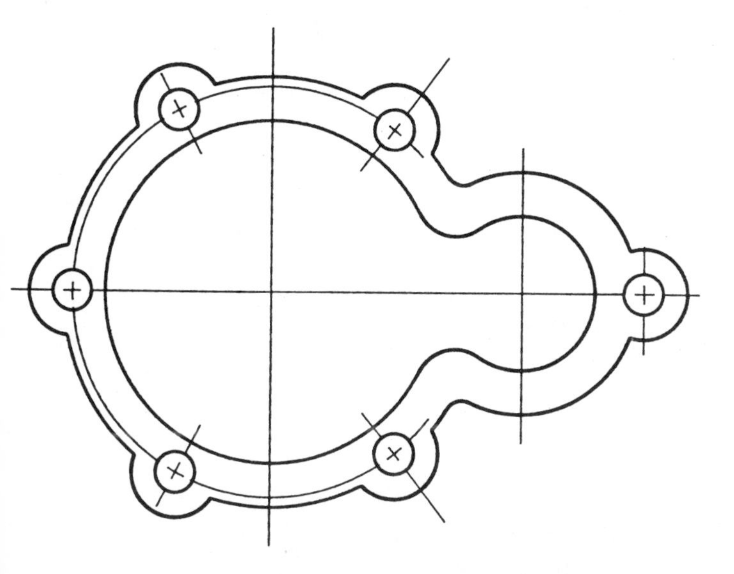

Figure 4–105 The complete drawing for the engineer's sketch.

A Layout Approach for CADD Applications

Make a form like the example given in Figure 4–106 to list all of your drawing steps. Write the command in the left column. Choose from the following list of commands and provide the information indicated.

POINT — Cartesian coordinate location
SCALE — numerical size
ORIGIN — coordinates
CIRCLE — center, diameter
ARC — center, beginning and ending angles (degrees)
POLYGON — center, radius flat to flat (RF), or radius point to point (RP)
DASHED LINE — end points
TEXT (NOTES) — 1) height, width, slant angle; 2) angle direction; 3) lower left or lower right coordinates

Indicate the type of line in column 2, the coordinates in column 3, and additional parameters (diameter, radius, angles, etc.) in column 4.

SAMPLE FORM

Student Name________ File Name________
Scale________ Date________

① COMMAND	② LINE TYPE	③ X	Y	④ PARAMETERS
Point	*Solid*	*0*	*0*	
"	*"*	*2*	*0*	
"	*"*	*2*	*1*	
"	*"*	*0*	*1*	
Circle	*Solid*	*1*	*.5*	*Dia. = .25*
Arc	*"*	*0*	*.5*	*R=.5, 90°- 180°*

Figure 4–106 CADD planning form.

GEOMETRIC CONSTRUCTION PROBLEMS

DIRECTIONS

1. Problems may be completed using manual or computer-aided drafting depending on your course guidelines.
2. Use very lightly drawn lines (construction lines) for all preliminary work and darken in only the completed product. *Do not* erase the construction lines when you complete the problems. This will help your instructor observe your drawing technique.
3. Geometric construction problems are presented as written instructions, drawings, or a combination of both. If the problems are presented as drawings without dimensions, transfer the drawing from the text, using dividers and scales, to your drawing sheet. Draw dimensioned problems full scale (1:1) unless otherwise specified. Draw all object, hidden, and centerlines. Do not draw dimensions.

Problem 4–1 Draw two tangent circles with their centers on a horizontal line. Circle 1 has a 64-mm diameter and circle 2 has a 50-mm diameter.

Problem 4–2 Make a perpendicular bisector of a horizontal line that is 79 mm long.

Problem 4–3 Make an angle 48° with one side vertical. Bisect the angle.

Problem 4–4 Divide a 96-mm line into 7 equal spaces.

Problem 4–5 Draw two parallel horizontal lines each 50 mm long with one 44 mm above the other. Divide the space between the lines into 8 equal spaces.

Problem 4–6 Transfer triangle a,b,c to a new position with side c 45° from horizontal.

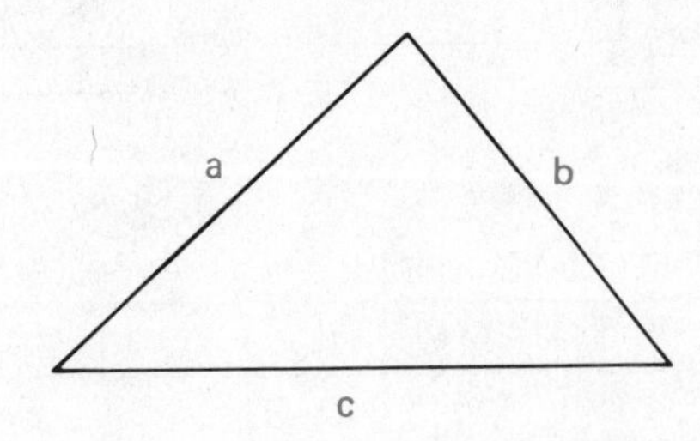

Problem 4–7 Make a right triangle with one side 35 mm long and the other side 50 mm long.

Problem 4–8 Make a right triangle with the hypotenuse 75 mm long and one side 50 mm long.

Problem 4–9 Make an equilateral triangle with 65-mm sides.

Problem 4–10 Make a square with a distance of 50 mm across the flats.

Problem 4–11 Draw a hexagon with a distance of 2.5 in. across the flats.

Problem 4–12 Draw a hexagon with a distance of 2.5 in. across the corners.

Problem 4–13 Draw an octagon with a distance of 2.5 in. across the flats.

Problem 4–14 Draw a rectangle 50 mm x 75 mm with 12-mm radius tangent corners.

Problem 4–15 Draw two separate angles, one a 30° acute angle and the other a 120° obtuse angle. Then draw a .5-in. radius arc tangent to the sides of each angle.

Problem 4–16 Given the following incomplete part, draw an inside arc with a 2.5-in. radius tangent to ∅ A and line B.

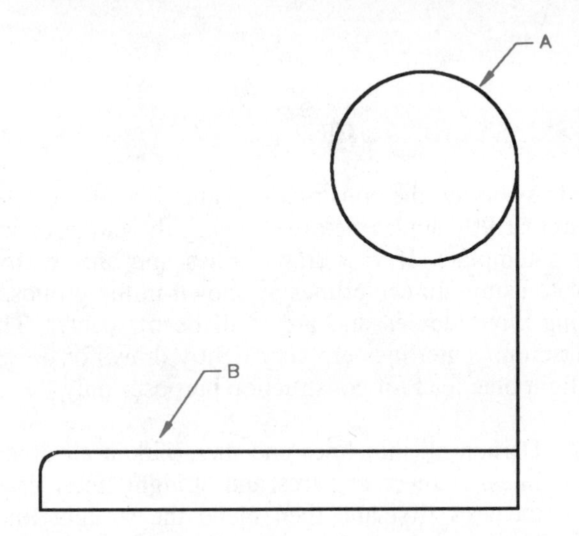

Problem 4–17 Draw two circles with 25-mm diameters and 38 mm between centers on horizontal. On the upper side between the two circles draw an inside radius 44 mm. On the lower side draw an outside radius 76 mm.

Problem 4–18 Draw an S curve with equal radii between points A and B given below:

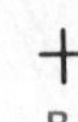
B

A
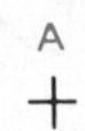

Problem 4–19 Draw two lines, one on each side of the ogee curve drawn in problem 4–18 and .5 in. away.

Problem 4–20 Make an ellipse with a 38-mm minor diameter and a 50-mm major diameter.

Problem 4–21 Corner arcs (in.)

Part Name: Plate Spacer
Material: .250-Thick Aluminum

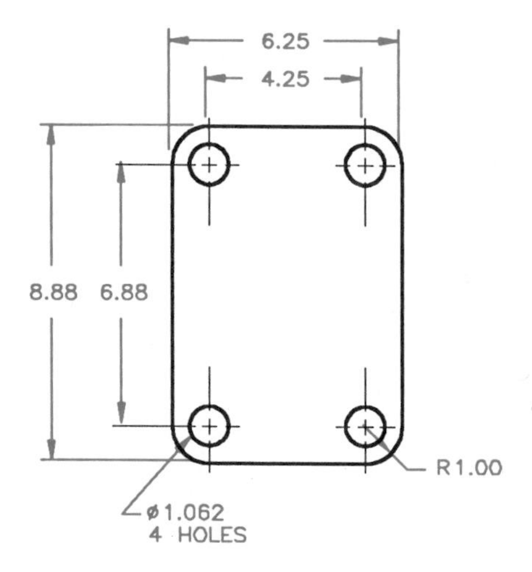

Problem 4–22 Circles (in.)

Part Name: Flange
Material: .50-Thick Cast Iron (CI)

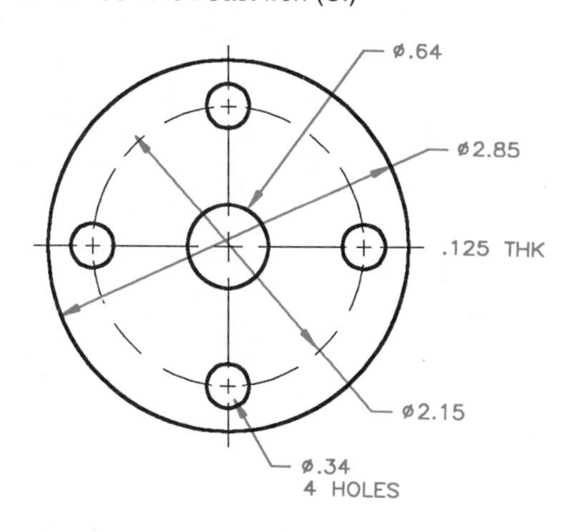

Problem 4–23 Hexagon (in.)

Part Name: Sleeve
Material: Bronze
Draw the hexagon with circles view only.

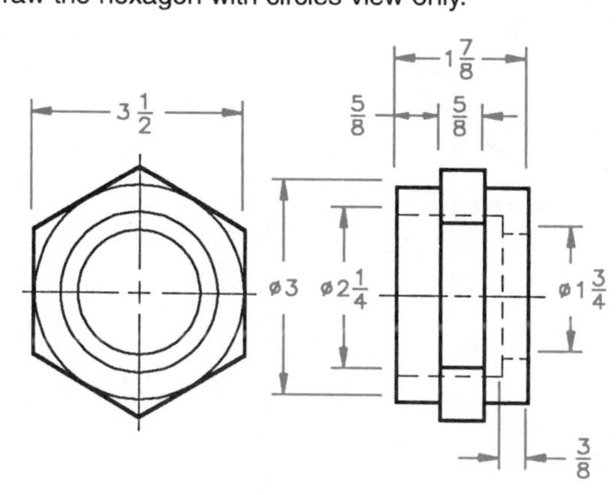

Note: this engineer's sketch shows dimensions to hidden features, which is not a standard practice.

Problem 4–24 Tangencies (metric)

Part Name: Coupler
Material: SAE 1040

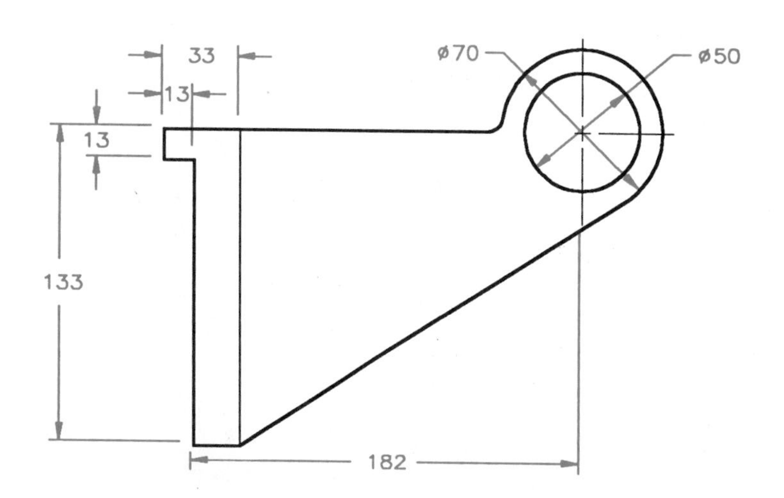

Problem 4–25 Tangencies (in.)

Part Name: Hanger
Material: Mild Steel (MS)

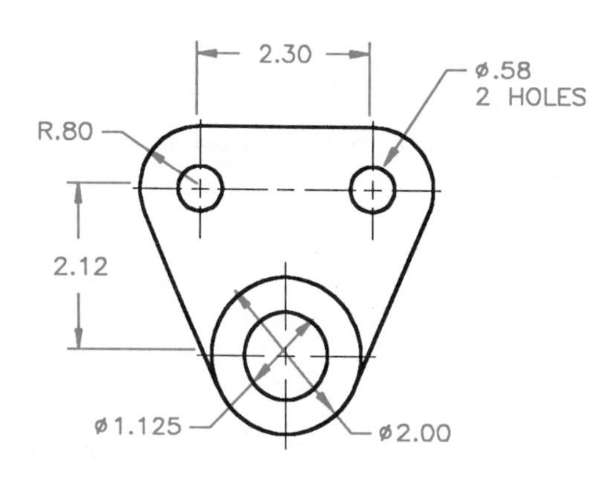

Problem 4–26 Tangencies (in.)

Part Name: Bracket
Material: Aluminum

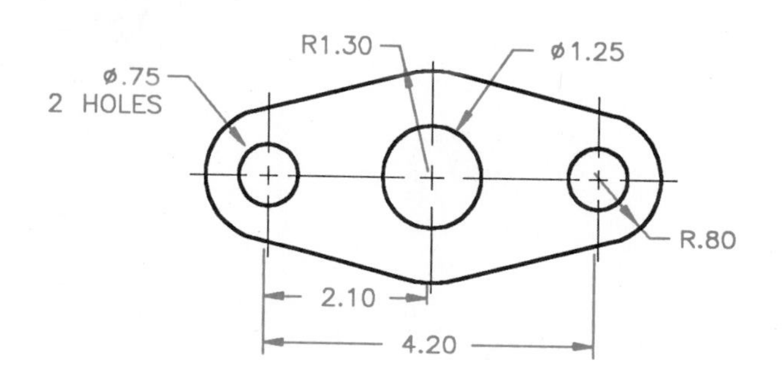

Problem 4–27 Arcs and tangencies (in.)

Part Name: Gusset
Material: .250-Thick Aluminum
Specific instructions: sizes are given as limit dimensions. Limit dimensioning is discussed in Chapter 8. For now, when a dimension is shown as .531/.468, for example, you produce the drawing using an even numeral between the two. In this case you draw using a dimension of .500. *Courtesy TEMCO*

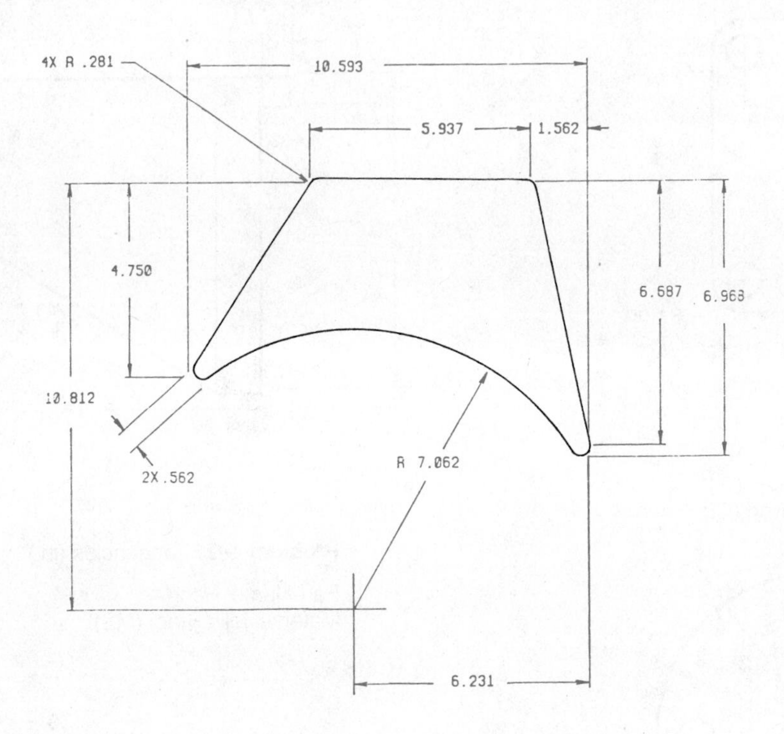

Problem 4–28 Circle and arc object lines and centerlines (metric)

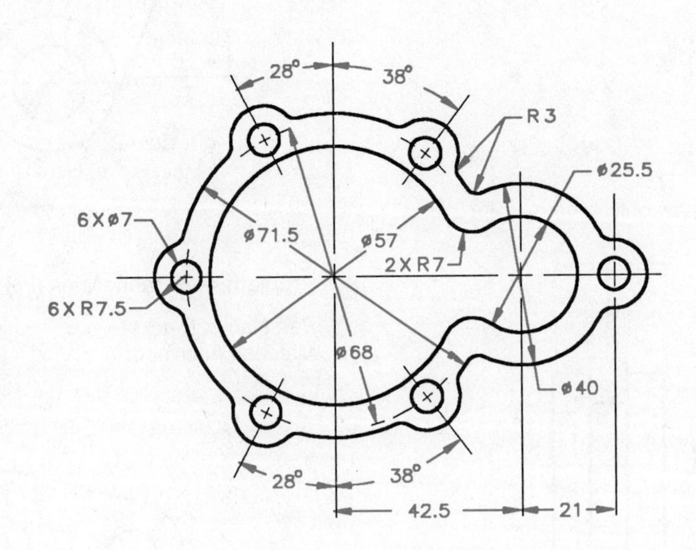

CHAPTER 5

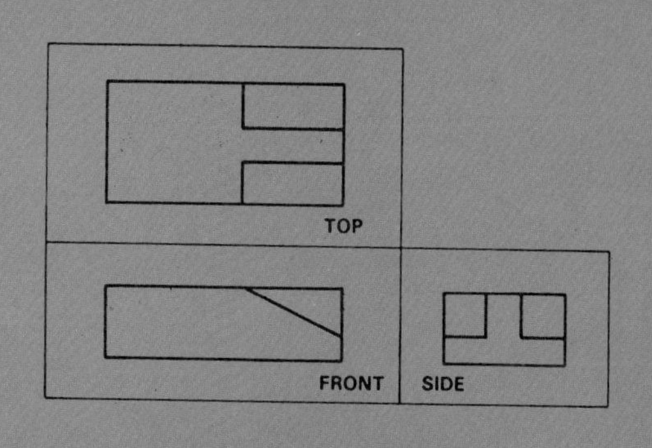

Multiviews and Auxiliary Views

LEARNING OBJECTIVES

After completing this chapter you will:

- Prepare single- and multiview drawings.
- Select appropriate views for presentation.
- Draw view enlargements.
- Establish runouts.
- Explain the difference between first- and third-angle projection.
- Draw first and secondary auxiliary views.
- Prepare formal drawings from an engineer's sketch and actual industrial layouts.

ANSI This chapter is developed in accordance with the ANSI standard for multiview presentation, titled *Multi and Sectional View Drawings*, ANSI Y14.3–1975. This standard is available from the American National Standards Institute, 1430 Broadway, New York, NY 10018. The content of this discussion provides an in-depth analysis of the techniques and methods of multiview presentation.

Orthographic projection is any projection of the features of an object onto an imaginary plane called a plane of projection. The projection of the features of the object is made by lines of sight that are perpendicular to the plane of projection. When a surface of the object is parallel to the plane of projection, the surface appears in its true size and shape on the plane of projection. In Figure 5–1, the plane of projection is parallel to the surface of the object. The line of sight (projection from the object) is perpendicular to the plane of projection. Notice also that the object appears three-dimensional (width, height, and depth) while the view on the plane of projection has only two dimensions (width and height). In situations where the plane of projection is not parallel to the surface of the object, the resulting orthographic view is foreshortened, or shorter than true length. (See Figure 5–2.)

MULTIVIEWS

Multiview projection establishes views of an object projected upon two or more planes of projection by using orthographic projection techniques. The result of multiview projection is a multiview drawing. A multiview drawing represents the shape of an object using two or more views. Consideration should be given to the choice and number of views used so, when possible, the surfaces of the object are shown in their true size and shape.

It is often easier for an individual to visualize a three-dimensional picture of an object than it is to visualize a two-dimensional drawing. In mechanical drafting, however, the common practice is to prepare completely dimensioned and detailed drawings using two-dimensional views, known as *multiviews*. Figure 5–3 shows an object represented by a three-dimensional drawing, also called a *pictorial*, and three two-dimensional views, or multiviews, also known as orthographic projection. The multiview method of drafting represents the *shape* description of the object.

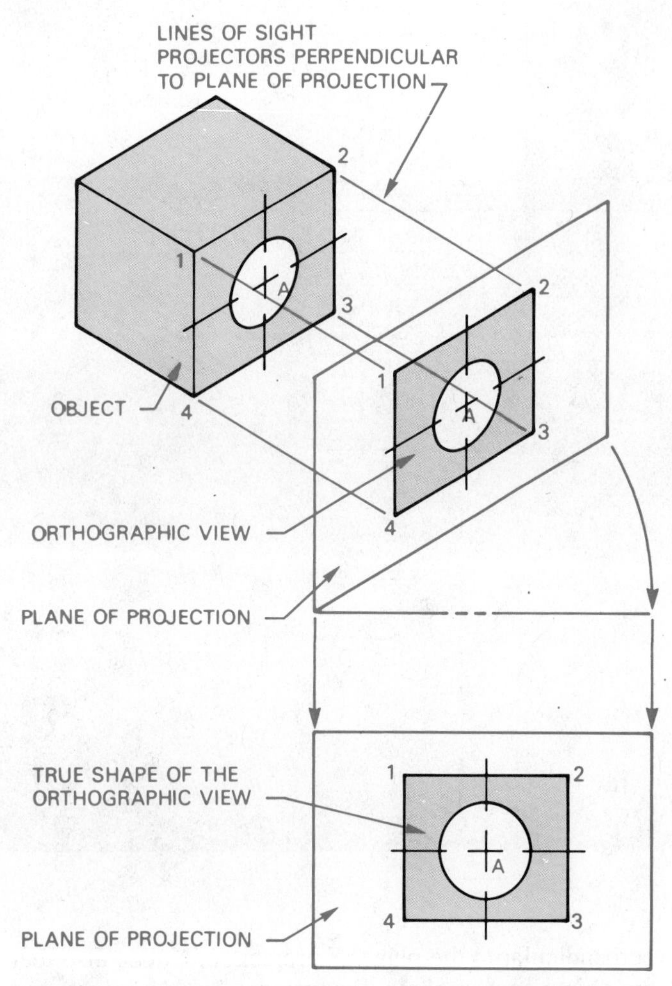

Figure 5–1 Orthographic projection to form orthographic view.

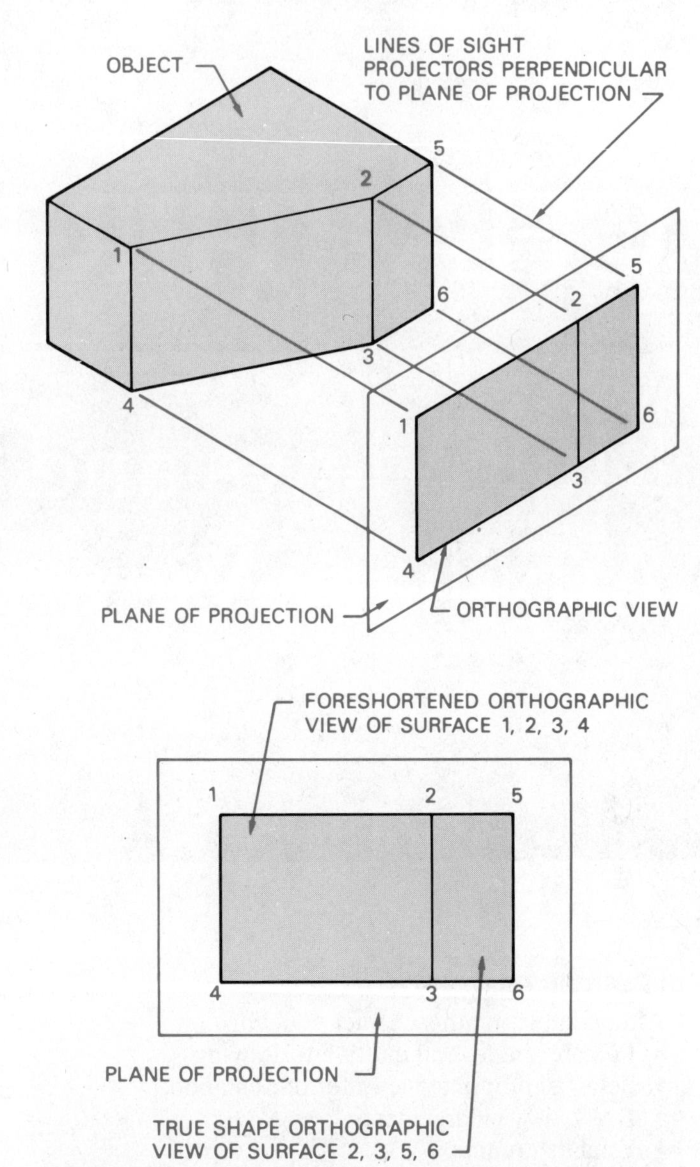

Figure 5–2 Projection of a foreshortened orthographic surface.

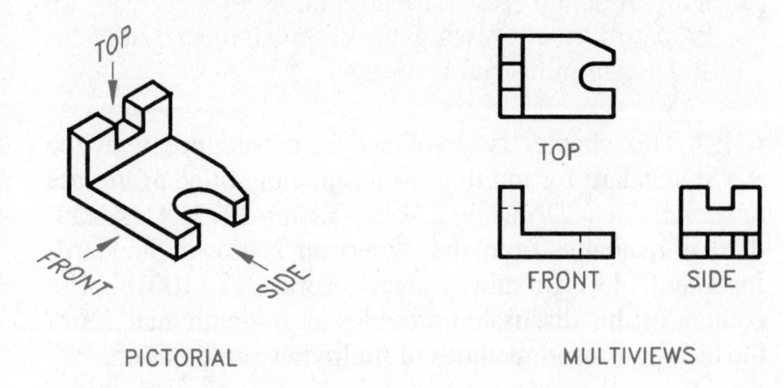

Figure 5–3 Pictorial vs. multiview.

Glass Box

If we place the object in Figure 5–3 in a glass box so the sides of the glass box are parallel to the major surfaces of the object, we can project those surfaces onto the sides of the glass box and create multiviews. Imagine the sides of the glass box are the planes of projection that were previously discussed. (See Figure 5–4.) If we look at all sides of the glass box, we have six total views: front, top, right-side, left-side, bottom, and rear. Now unfold the glass box as if the corners were hinged about the front (except the rear view) as demonstrated in Figure 5–5. These hinge lines are commonly called *fold lines*.

Completely unfold the glass box onto a flat surface and you have the six views of an object represented in multiview. Figure 5–6 shows the glass box unfolded. Notice that the views are labeled front, top, right, left, rear, and bottom. It is in this arrangement that the views are always found when using multiviews in third-angle projection. *Third-angle projection* is the principal multiview projection used by United States industries. First-angle projection is commonly used in European countries.

Analyze Figure 5–6 in more detail so you can see the items that are common between views. Knowing how to identify features of the object that are common between views will aid you in the visualization of multiviews. Notice in Figure 5–7 that the views are aligned. The top

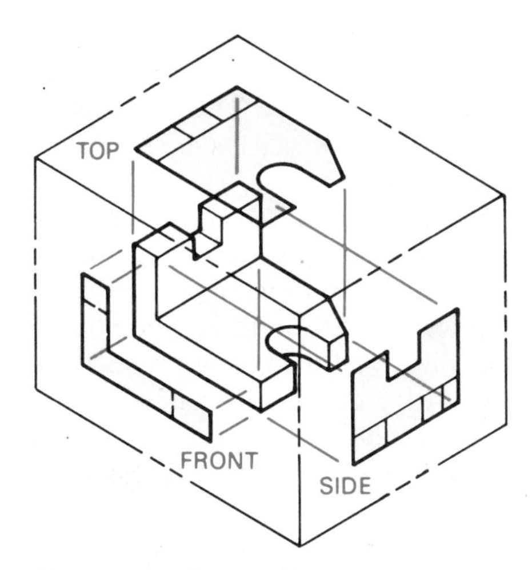

Figure 5–4 Glass box.

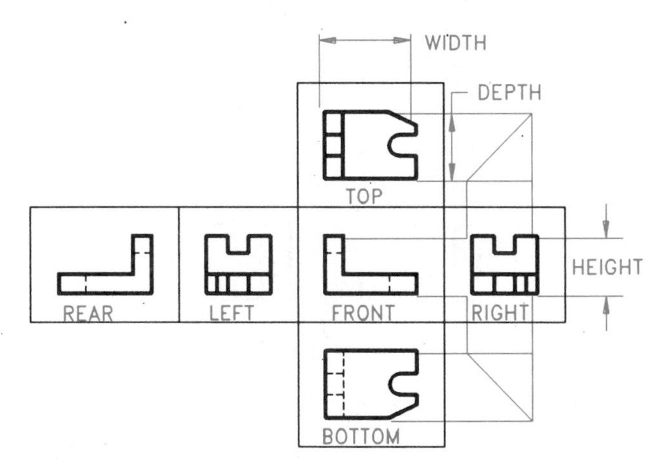

Figure 5–7 View alignment.

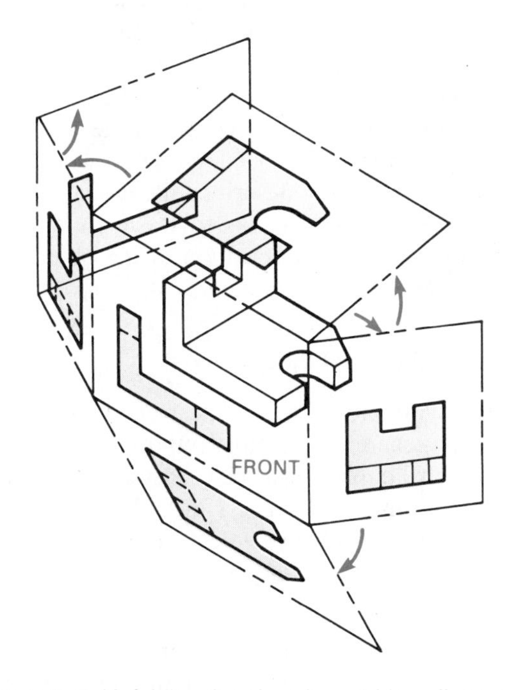

Figure 5–5 Unfolding the glass box at hinge lines, also called fold lines.

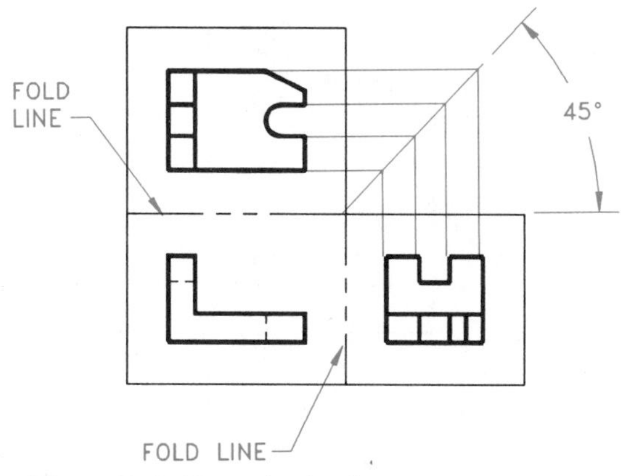

Figure 5–8 45° projection line.

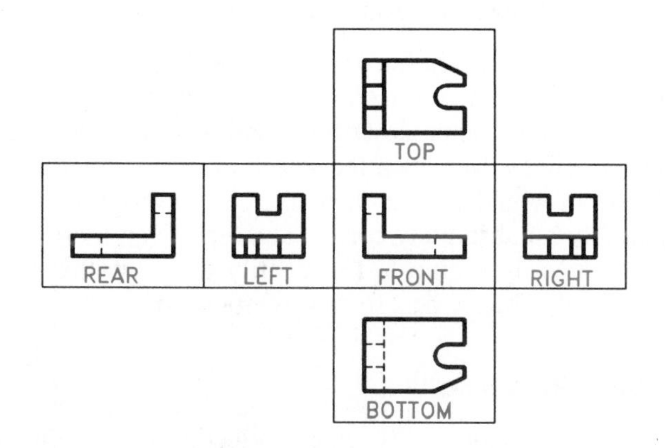

Figure 5–6 Glass box unfolded.

view is directly above and the bottom view is directly below the front view. The left-side view is directly to the left, while the right-side view is directly to the right of the front view. This format allows the drafter to project points directly from one view to the next to help establish related features on each view.

Now, look closely at the relationship among the front, top, and right-side views. (A similar relationship exists with the left-side view.) Figure 5–8 shows a 45° line projected from the corner of the fold, or reference line (hinge), between the front, top, and side views. This 45° line is used as an aid in projecting views. All of the features established on the top view can be projected to the 45° line and then down onto the side view. This projection works because the depth dimension is the same between the top and side views. The reverse is also true. Features from the side view may be projected to the 45° line and then over to the top view.

The same concept of projection achieved in Figure 5–8 using the 45° line also works by using a compass at the intersection of the horizontal and vertical fold lines. The compass establishes the common relationship between the top and side views as shown in Figure 5–9. Another method commonly used to transfer the size of

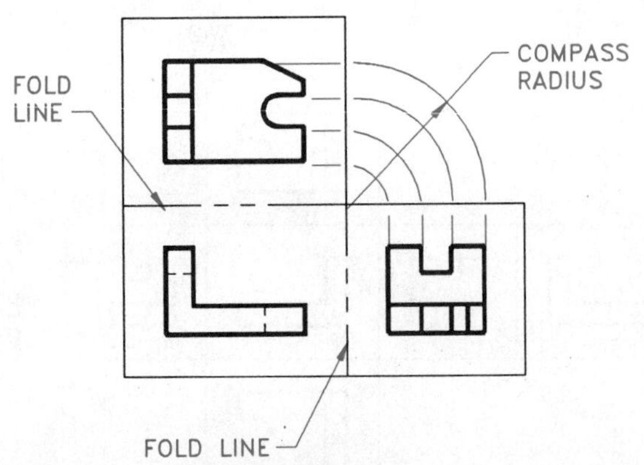

Figure 5–9 Projection with a compass.

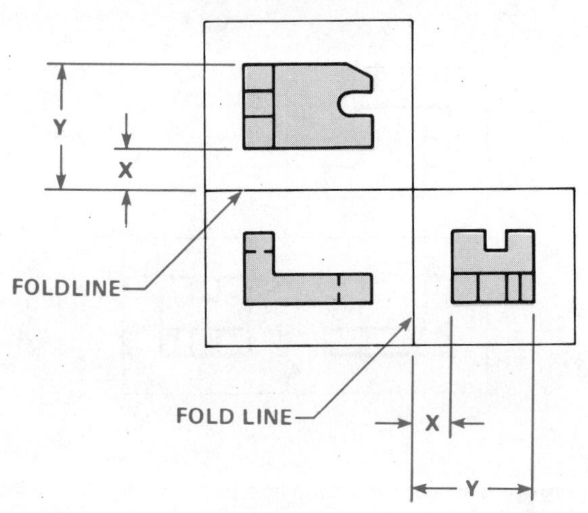

Figure 5–10 Using dividers to transfer view projections.

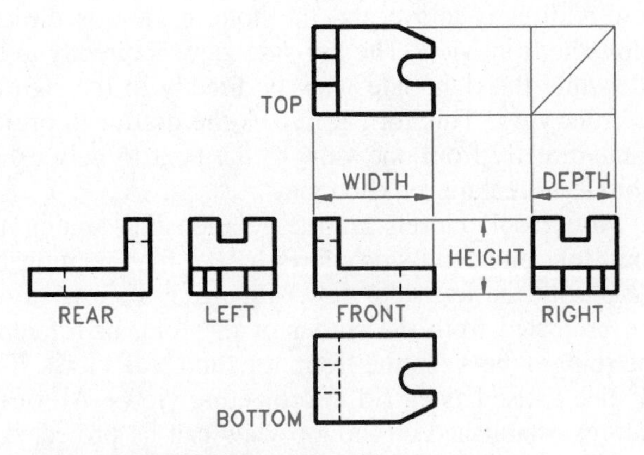

Figure 5–11 Multiview orientation.

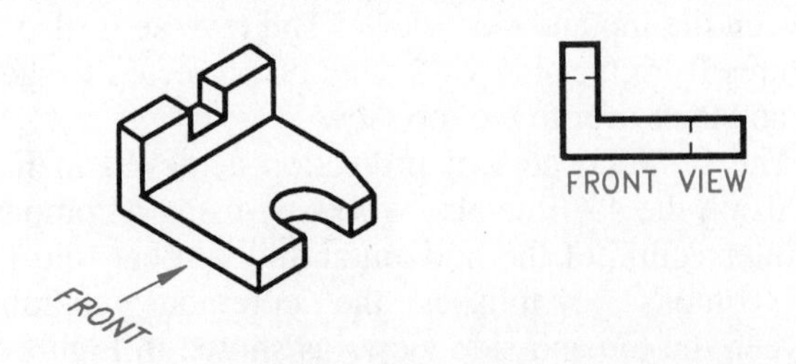

Figure 5–12 Front view selection.

features from one view to the next is to use dividers to transfer distances from the fold line of the top view to the fold line of the side view. The relationships between the fold line and these two views are the same, as shown in Figure 5–10.

The front view is usually the most important view and the one from which the other views are established. There is always one dimension common between adjacent views. For example, the width is common between the front and top views and the height between the front and side views. Knowing this allows you to relate information from one view to another. Take one more look at the relationship among the six views shown in Figure 5–11.

VIEW SELECTION

Although there are six primary views that you can select to completely describe an object, it is seldom necessary to use all six. As a drafter, you must decide how many views are needed to properly represent the object. If you draw too many views you are wasting time, which costs your employer money. If you draw too few views, you may not have completely described the object. The manufacturing department then has to waste time trying to determine the complete description of the object, which again costs your employer money.

Selecting the Front View

Usually, you should select the front view first. The front view is generally the most important view and, as you learned from the glass box description, the front view is the origin of all other views. There is no exact way for everyone to always select the same front view, but there are some guidelines to follow. The front view should:

- represent the most natural position of use.
- provide the best shape description or most characteristic contours.
- have the longest dimension.
- have the fewest hidden features.
- be the most stable and natural position.

Take a look at the pictorial drawing in Figure 5–12. Notice the front view selection. This front view selection violates the best shape description and fewest hidden features guidelines. However, the selection of any other view as the front would violate other rules, so in this case there is possibly no absolutely correct answer. Given the pictorial drawings in Figure 5–13, identify the view that you believe would be the best front view for each.

Other View Selection

Use the same rules when selecting other views needed as you do when selecting the front view:

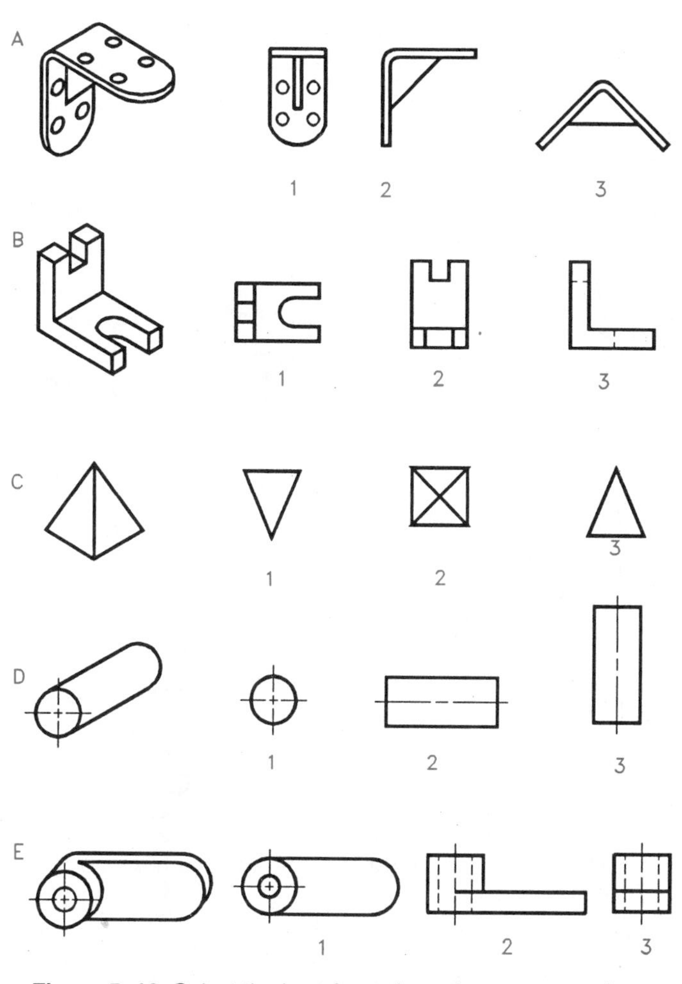

Figure 5–13 Select the best front views that correspond to the pictorial drawings at the left. You may make a first and second choice, if you wish.

- most contours
- longest side
- least hidden features
- best balance or position

Given the six views of the object in Figure 5–14, which views would you select to completely describe the object? If your selection is the front, top, and right side, then you are correct. Let's take a closer look. Figure 5–15 shows the selected three views. The front view shows the best position and the longest side, the top view clearly represents the angle and the arc, and the right-side view shows the notch. Any of the other views have many more hidden features. You should always avoid hidden features if you can.

It is not always necessary to select three views. Some objects can be completely described with two views or even one view. When selecting fewer than three views to describe an object, you must be careful which view other than the front view you select. (See Figure 5–16.)

A. 2,1. B. 3,2. C. 3. D. 2,1. E. 2,1.

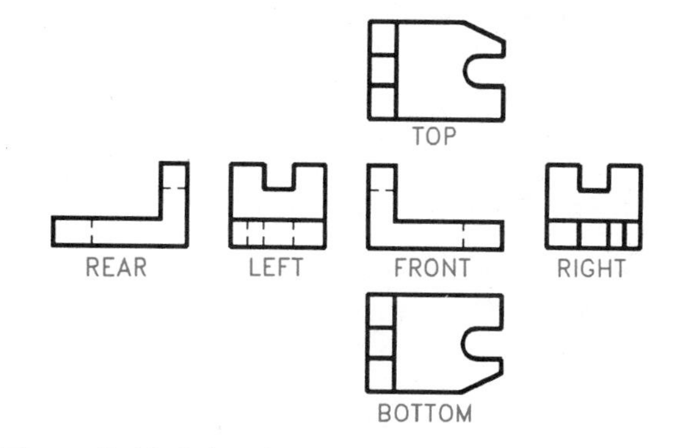

Figure 5–14 Select the necessary views to describe the object.

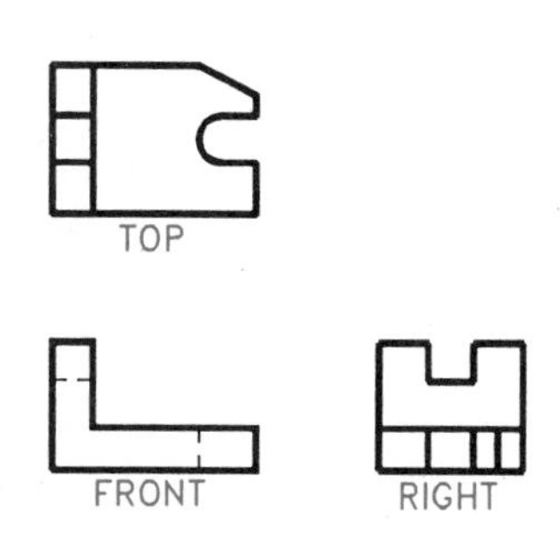

Figure 5–15 The selected views.

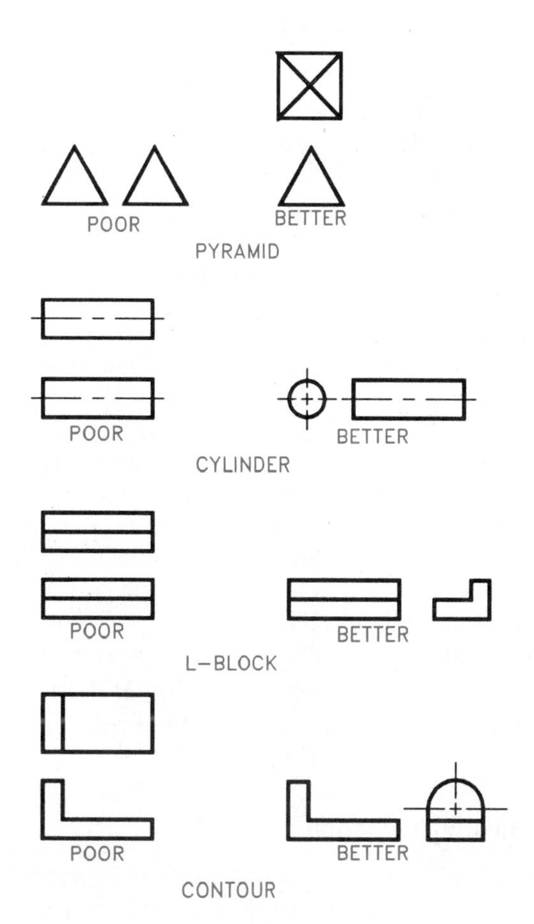

Figure 5–16 Selecting two views.

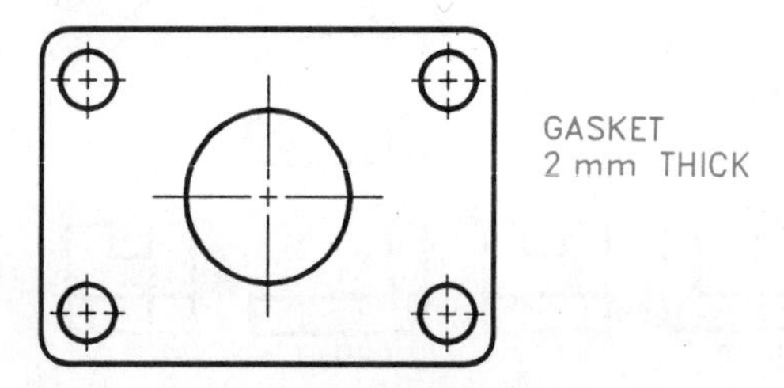

Figure 5–17 One-view drawing.

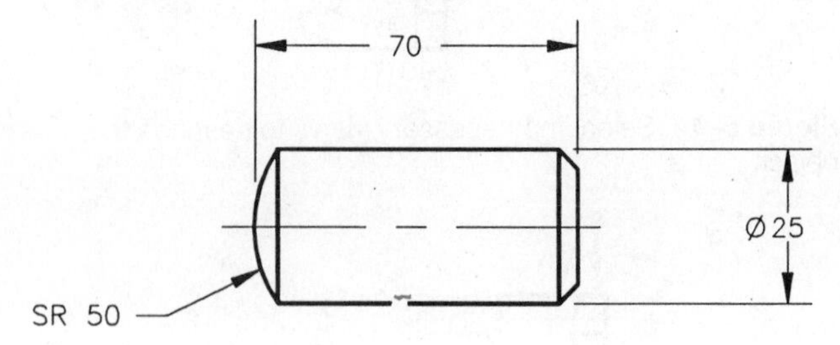

Figure 5–18 One-view drawing.

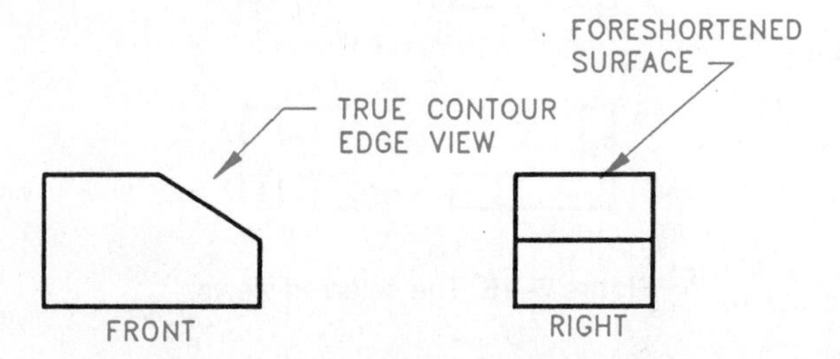

Figure 5–19 Contour representation.

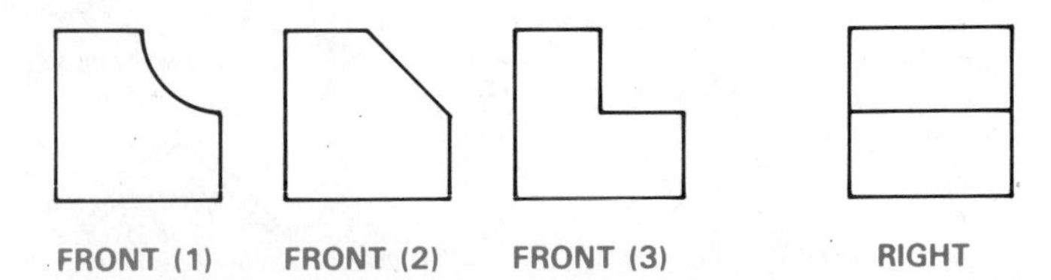

Figure 5–20 Select the front view that properly goes with the given right-side view.

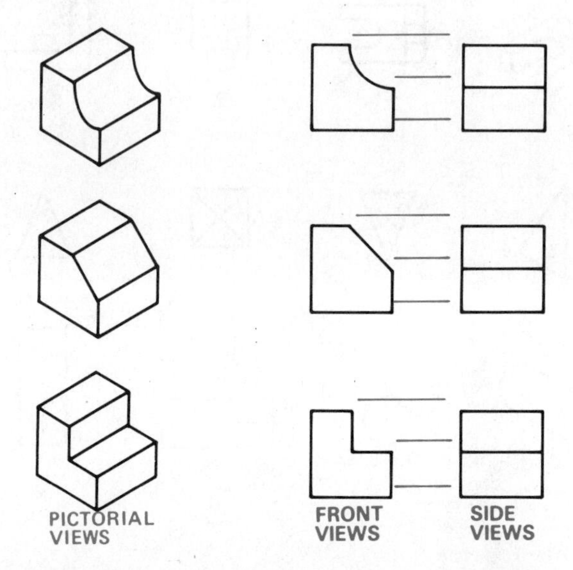

Figure 5–21 The importance of a view that clearly shows the contour of a surface.

One-view drawings are also often practical. When an object has shape and a uniform thickness, more than one view is unnecessary. Figure 5–17 shows a gasket drawing where the thickness of the part is identified in the materials specifications of the title block. The types of parts that may fit into this category include gaskets, washers, spacers, or similar features.

Although two-view drawings are generally considered the industry's minimum recommended views for a part, other objects that are clearly identified by dimensional shape may be drawn with one view, as in Figure 5–18.

In Figure 5–18, the shape of the pin is clearly identified by the 25-mm diameter. In this case, the second view would be a circle and would not necessarily add any more valuable information to the drawing. The primary question to keep in mind is, can the part be easily manufactured, without confusion, from the drawing? If there is any doubt, then the adjacent view should probably be drawn. As an entry-level drafter, you should ask your drafting supervisor to clarify the company policy regarding the number of views to be drawn.

Contour Visualization

Some views do not clearly identify the shape of certain contours. You must then draw the adjacent views to visualize the contour. (See Figure 5–19.)

The edge view of a surface shows that surface as a line. The true contour of the slanted surface is seen as an edge in the front view of Figure 5–19 and describes the surface as an angle, while the surface is foreshortened, slanting away from your line of sight, in the right-side view. In Figure 5–20, select the front view that properly describes the given right-side view. All three front views in Figure 5–20 could be correct. The side view does not help the shape description of the front view contour. (See Figure 5–21.)

Cylindrical shapes appear round in one view and rectangular in another view, as seen in Figure 5–22. Both views in Figure 5–22 may be necessary as one shows the diameter shape and the other shows the length. The ability to visualize from one view to the next is a critical skill for a drafter. You may have to train yourself to look at two-dimensional objects and picture three-dimensional shapes. You can also use some of the techniques discussed here to visualize features from one view to another.

Partial Views

When symmetrical objects are drawn in limited space or when there is a desire to save valuable drafting time, partial views may be used. The top view in Figure

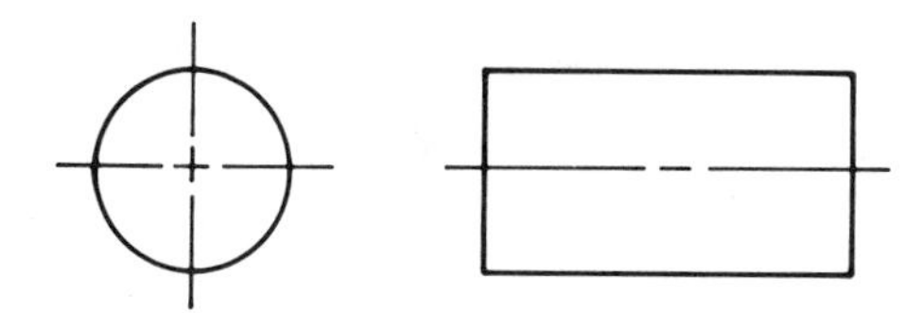

Figure 5–22 Cylindrical shape representation.

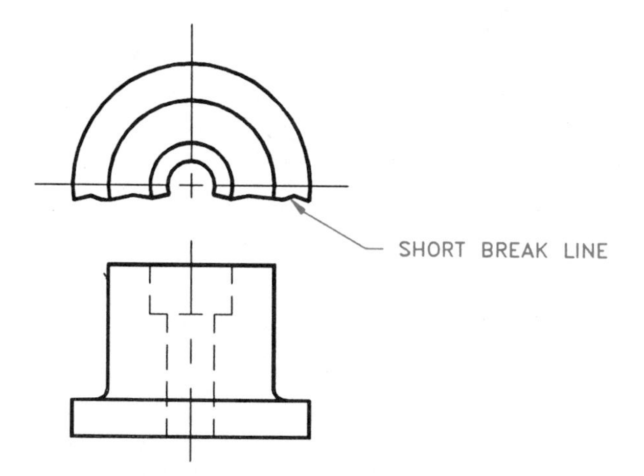

Figure 5–23 Partial view.

5–23 is a partial view. Notice how the short break line is used to show that a portion of the view is omitted. Some drafters may elect to stop the view at the centerline and therefore not use the short break line. Caution should be exercised when using partial views, as confusion could result in some situations. If the partial view reduces clarity, then draw the entire view.

View Enlargement

In some situations, when part of a view has detail that cannot be clearly dimensioned due to the drawing scale or complexity, a view enlargement may be used. To establish a view enlargement, a thick phantom line circle is placed around the area to be enlarged. This circle is broken at a convenient location and an identification letter (.18 to .25 in. in height) is centered in the break. Arrowheads are then placed on the line adjacent to the identification letter, as shown in Figure 5–24A. An enlarged view of the detailed area is then shown in any convenient location in the field of the drawing. The view identification and scale is then placed below the enlarged view. (See VIEW A, Figure 5–24.)

PROJECTION OF CIRCLES AND ARCS

Circles on Inclined Planes

When the line of sight in multiviews is perpendicular to a circular feature, such as a hole, the feature appears round, as shown in Figure 5–25. When a circle is projected onto an inclined surface, its view becomes elliptical in shape, as shown in Figure 5–26. The ellipse shown in the top and right-side views of Figure 5–26 was established by projecting the major diameter from the top to the side view and the minor diameter to both views from the front view, as shown in Figure 5–27. The major diameter in this example is the hole diameter.

The rectangular areas obtained from the two diameters in the top and right-side views of Figure 5–27 provide the parameters of the ellipse to be drawn in these views. The easiest method of drawing the ellipse is to obtain an ellipse template that has a shape that fits within the area.

Ellipse templates, as discussed in Chapter 1, designate elliptical shapes by the angle of the circle in relationship to the line of sight. So if your slanted surface is 30°, then a 30° ellipse template would be used to draw the

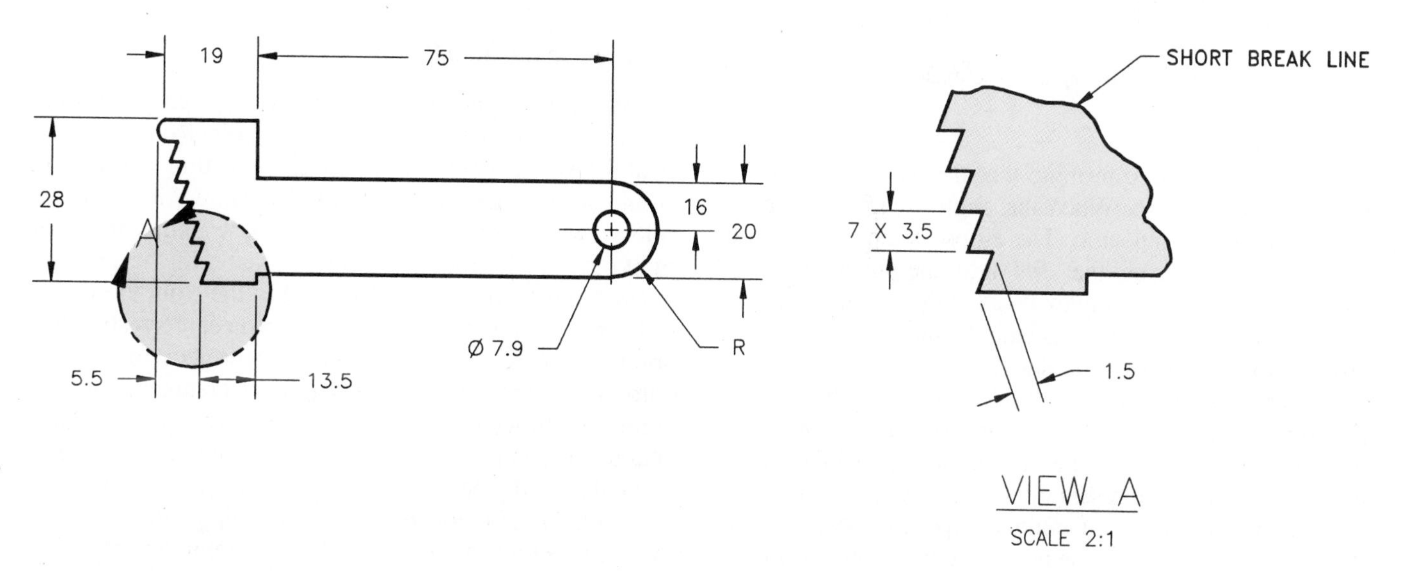

Figure 5–24 View enlargement example.

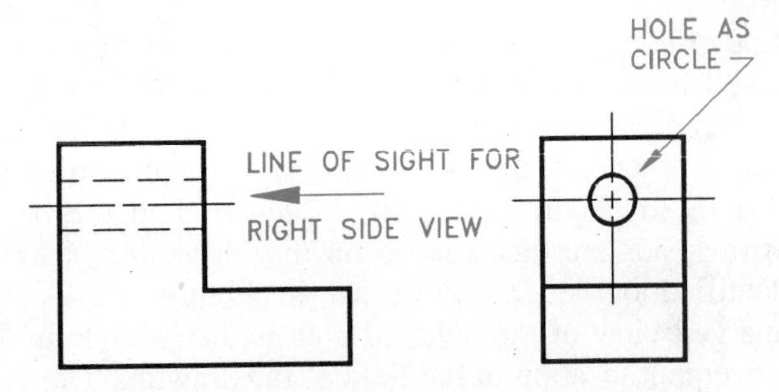

Figure 5–25 View of a hole projected as a circle and its hidden view through the part.

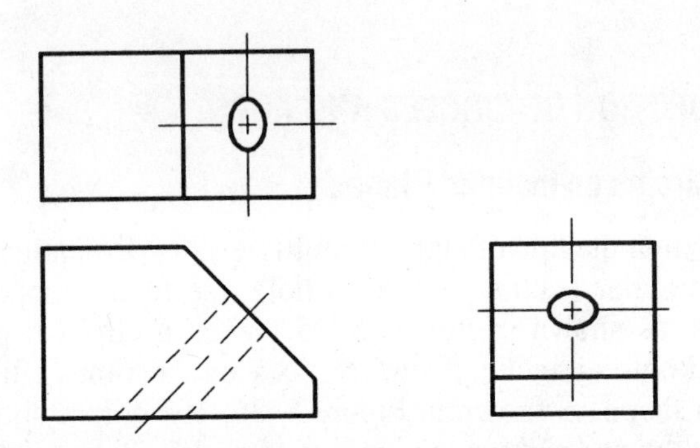

Figure 5–26 Hole represented on an inclined surface as an ellipse.

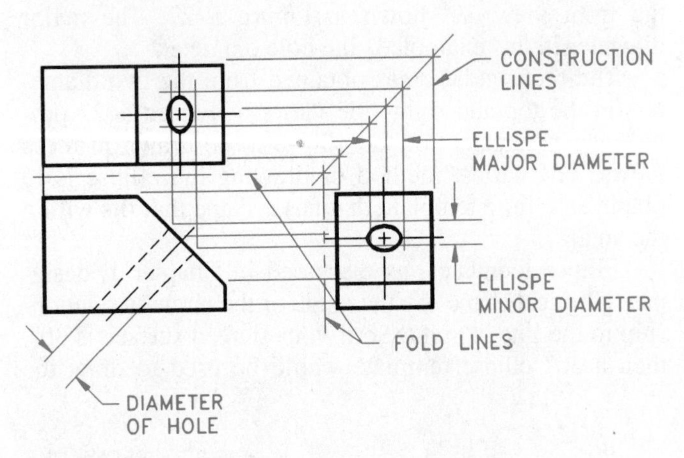

Figure 5–27 Establish an ellipse in the inclined surface.

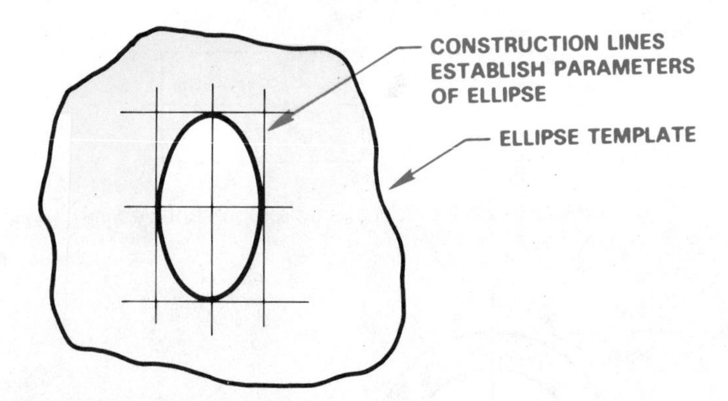

Figure 5–28 Ellipse template.

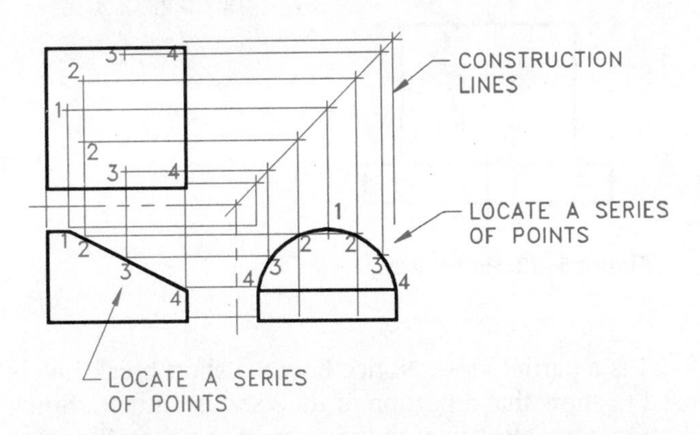

Figure 5–29 Locating an inclined surface in multiview.

elliptical shape. Keep in mind that the other view would require a 60° ellipse. When the angle is 45°, then both views would be the same. The ellipse template selected should have a shape that fits into the rectangle constructed, as shown in Figure 5–28. If you cannot find a representative ellipse that fits exactly, then use the closest size possible, or verify the drawing procedure with your supervisor. Establishing an elliptical shape with geometric construction is much more time-consuming but may be necessary when templates are not available. The elliptical shapes are a representation and may not have to be perfect, depending upon company policy. Also, notice that the hidden lines for the hole have been omitted in the top and side views of Figure 5–27. These lines would required a great amount of extra work without adding significantly to the function of the drawing. Such a practice of omitting lines for an accurate representation may be used with caution. Remember that the drawing must easily communicate the manufacture of the part.

Arcs on Inclined Planes

When a curved surface from an inclined plane must be drawn in multiview, a series of points on the curve establishes the contour. Begin by selecting a series of points around the curved contour as shown in the right-side view of Figure 5–29. Project these points from the right side to the inclined front view. From the point of intersection on the inclined surface in the front view, project lines to the top view. Then project corresponding points from the right side to the 45° projection line and onto the top view. Corresponding lines create a pattern of points, as shown in the top view of Figure 5–29. You may also use a compass or dividers to transfer measurements from the fold lines, as previously discussed. After the series of points is located in relationship to the front and top views, connect the points with an irregular curve. See the completed curve in Figure 5–30.

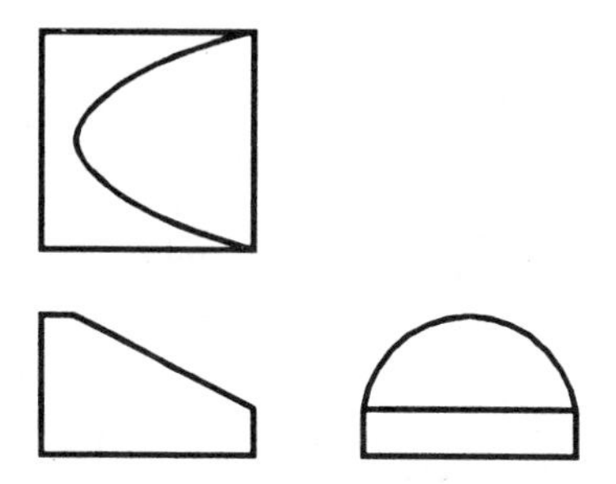

Figure 5–30 Completed curve.

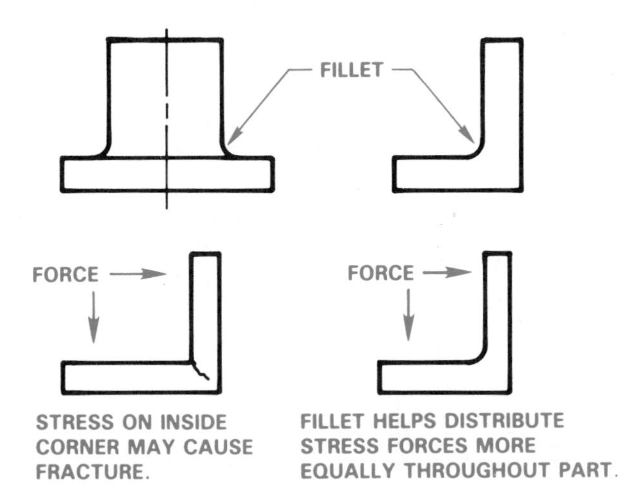

Figure 5–31 Fillets.

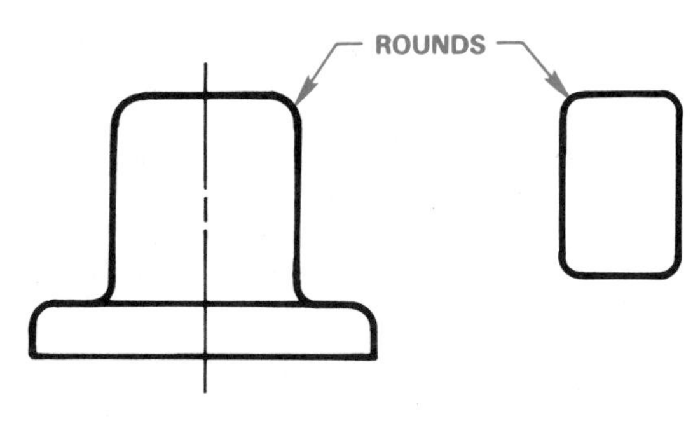

Figure 5–32 Rounds.

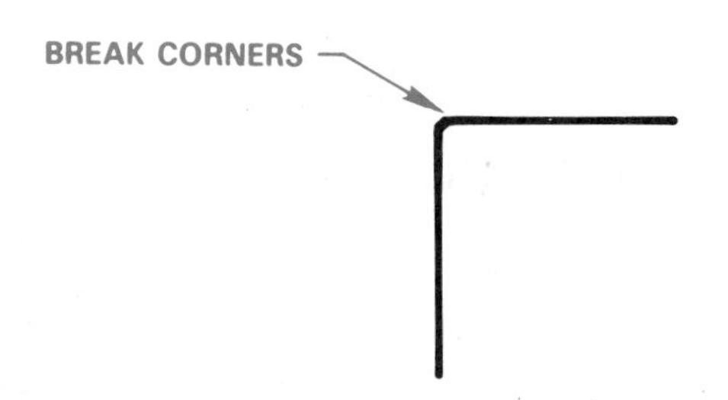

Figure 5–33 Break corner.

Fillets and Rounds

Fillets are slightly rounded inside curves at corners, generally used to ease the machining of inside corners or to allow patterns to release more easily from casting and forgings. Fillets may also be designed into a part to allow additional material on inside corners for stress relief. (See Figure 5–31.)

Other than the concern of stress factors on parts, certain casting methods require that inside corners have fillets. The size of the fillet often depends on the precision of the casting method. For example, very precise casting methods may have smaller fillets than green-sand casting, where the exactness of the pattern requires large inside corners.

Rounds are rounded outside corners that are used to relieve sharp exterior edges. Rounds are also necessary in the casting and forging process for the same reasons as fillets. Figure 5–32 shows rounds represented in views.

A machined edge will cause sharp corners, which may be desired in some situations. However, if these sharp corners are to be rounded, the extent of roundness depends upon the function of the part. When a sharp corner has only a slight relief, it may be referred to as a break corner, as shown in Figure 5–33.

FILLETS

Fillets and rounds can be drawn with the FILLET or FILLET/MODIFY commands. Intersecting lines should already be on the screen such as those in Figure 5–34(a). The FILLET command asks for the two intersecting lines to be digitized. They may be highlighted on the screen. The next prompt may be to indicate the general position of the radius center. (See Figure 5–34(b).) Finally, numerical input is required for the radius of the fillet and the fillet is then drawn as shown in Figure 5–34(c). If the intersecting lines are not automatically removed, they may have to be deleted with a command such as CLIP or TRIM.

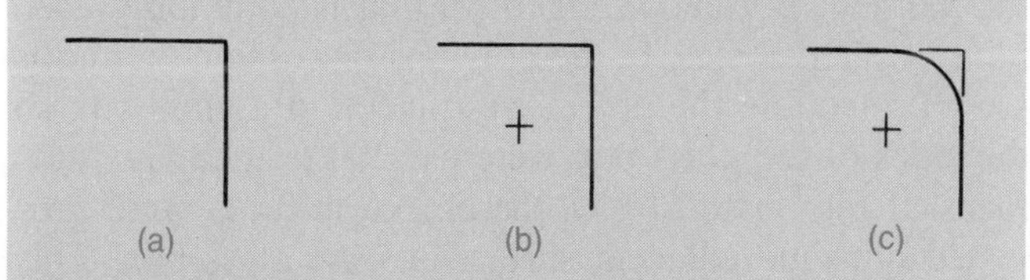

Figure 5–34 Constructing fillets and rounds using intersecting lines.

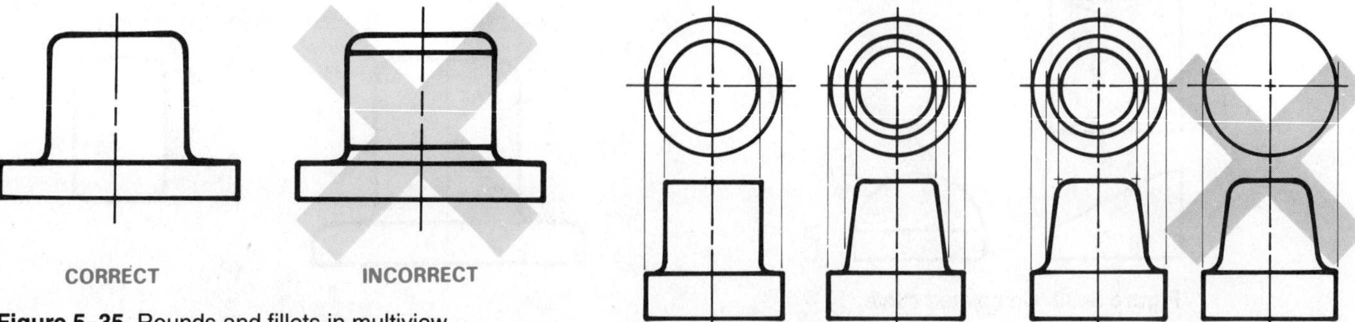

Figure 5–35 Rounds and fillets in multiview.

Figure 5–36 Rounded curves and cylindrical shapes in multiview.

Rounded Corners in Multiview

An outside or inside rounded corner of an object is represented in multiview as a contour only. The extent of the round or fillet is not projected into the view as shown in Figure 5–35. Cylindrical shapes may be represented with a front and top view, where the front identifies the height and the top shows the diameter. Figure 5–36 shows how these cylindrical shapes should be represented in multiview. Figure 5–37 shows the representation of the contour of an object as typically represented in multiview.

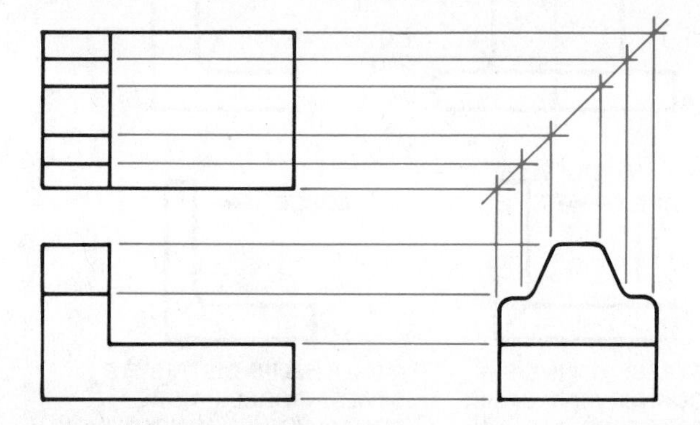

Figure 5–37 Contour in multiview.

Runouts

The intersections of features with circular objects are projected in multiview to the extent where one shape runs into the other. The characteristics of the intersecting features are known as runouts. The runout of features intersecting cylindrical shapes is projected from the point of tangency of the intersecting feature, as shown in Figure 5–38. Notice also that the shape of the runout varies when drawn at the cylinder depending on the shape of the intersecting feature. Rectangular-shaped features have a fillet at the runout, while curved (elliptical or round) features contour toward the centerline at the runout. Runouts may also exist when a feature such as a web intersects another feature, as shown in Figure 5–39.

THIRD-ANGLE PROJECTION

The method of multiview projection described in this chapter is also known as third-angle projection. This is the method of view arrangement that is commonly used in the United States. In the previous discussion on multiview projection, the object was placed in a glass box so the sides of the glass box were parallel to the major surfaces of the object. Next, the object surfaces were projected onto the adjacent surfaces of the glass box. This achieved the same effect as if the viewer's line of sight were perpendicular to the surface of the box and looking directly at the object, as shown in Figure 5–40. With the multiview concept in mind, assume an area of space is divided into four quadrants, as shown in Figure 5–41.

If the object were placed in any of these quadrants, the surfaces of the object would be projected onto the adjacent planes. When placed in the first quadrant, the method of projection is known as first-angle projection. Projections in the other quadrants are termed second-, third-, and fourth-angle projections. Second- and fourth-angle projections are not used, though first- and third-angle projections are very common.

Third-angle projection, as commonly used in the United States, is achieved when we take the glass box from Figure 5–40 and place it in quadrant three from Figure 5–41. Figure 5–42 shows the relationship of the glass box to the projection planes in the third-angle projection. In this quadrant, the projection plane is between the viewer's line of sight and the object. When the glass box in the third-angle projection quadrant is unfolded, the result is the multiview arrangement previously discussed and shown in Figure 5–43.

A third-angle projection drawing may be accompanied by a symbol on or adjacent to the drawing title block. The standard third-angle projection symbol is shown in Figure 5–44.

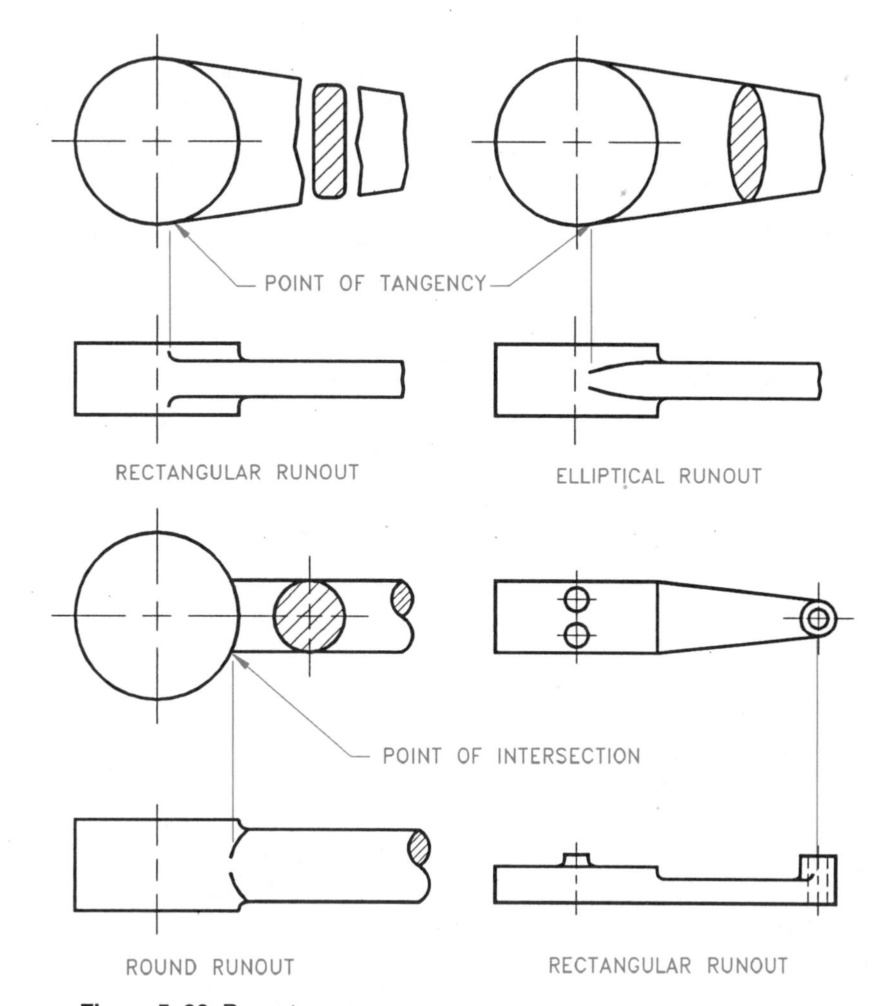

Figure 5–38 Runouts.

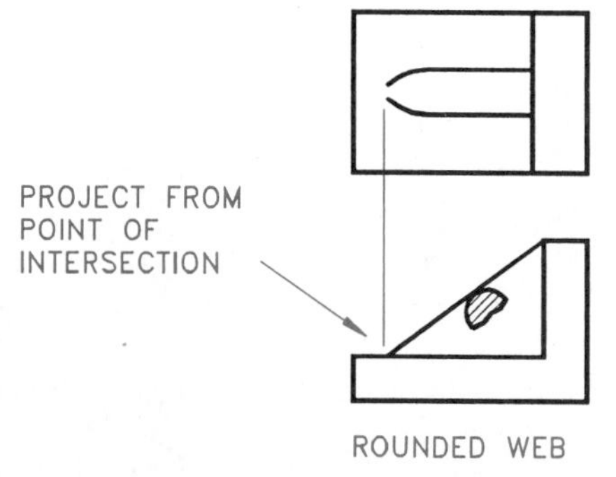

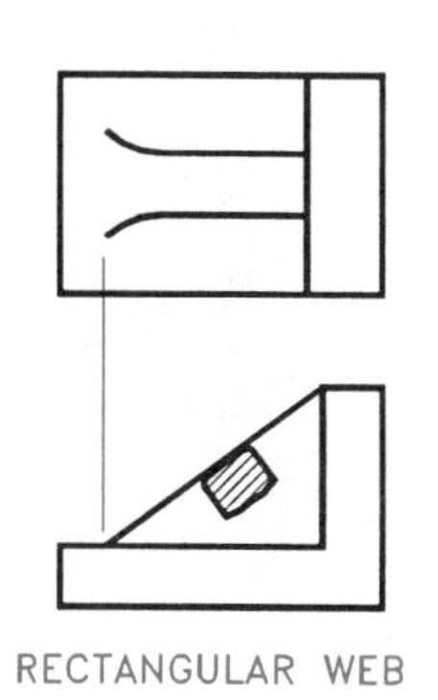

Figure 5–39 Other types of runouts.

FIRST-ANGLE PROJECTION

First-angle projection is commonly used in Europe and other countries of the world. This method of projection places the glass box in the first quadrant. Views are established by projecting surfaces of the object onto the surface of the glass box. In this projection arrangement, however, the object is between the viewer's line of sight and the projection plane, as you can see in Figure 5–45. When the glass box in the first-angle projection quadrant is unfolded, the result is the multiview arrangement shown in Figure 5–46.

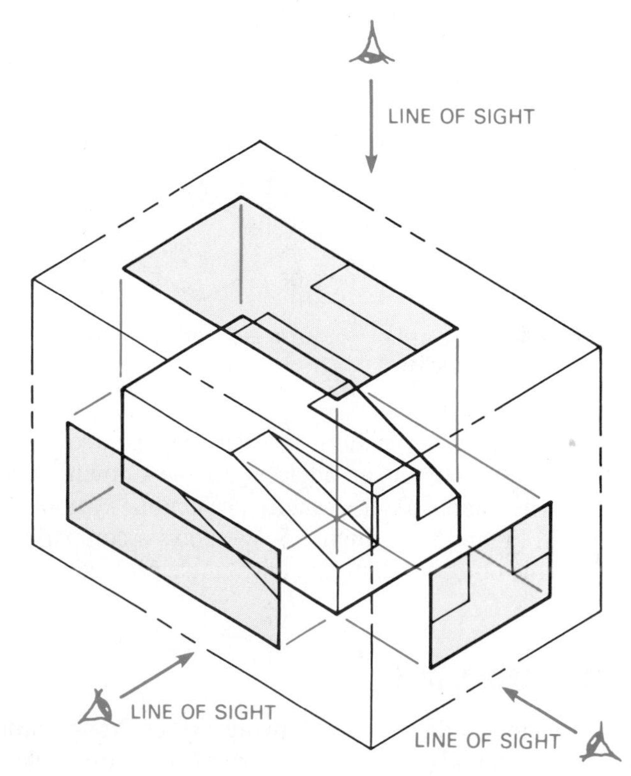

Figure 5–40 Glass box in third-angle projections.

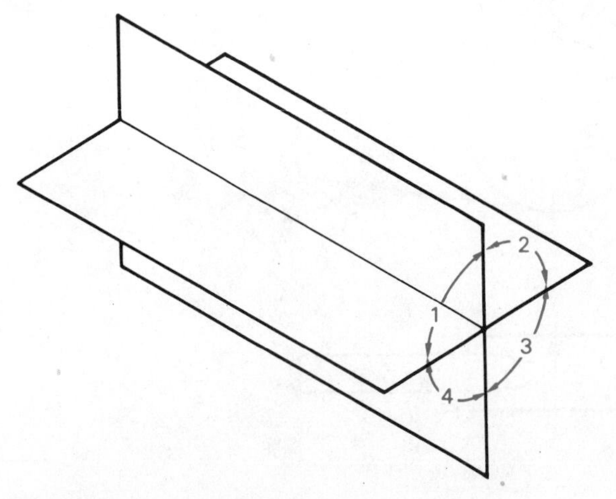

Figure 5–41 Quadrants of spatial visualization.

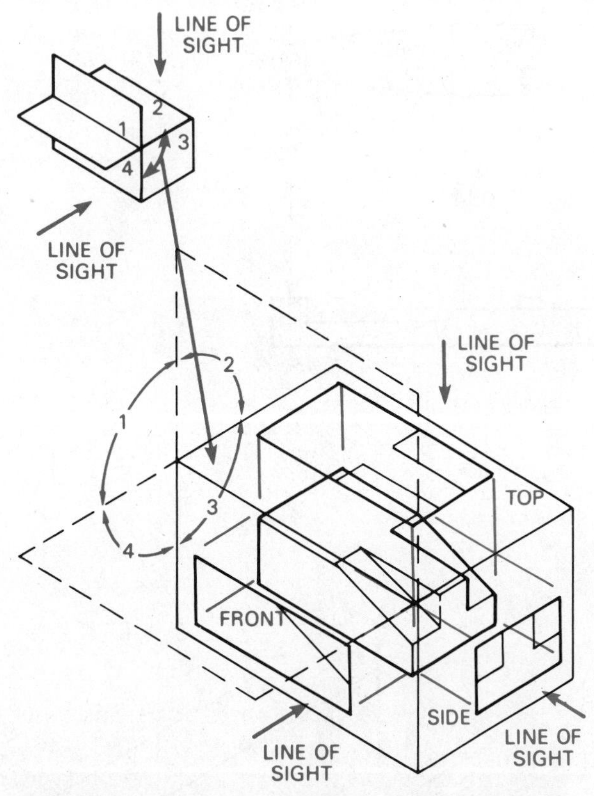

Figure 5–42 Glass box placed in the third quadrant for third-angle projection.

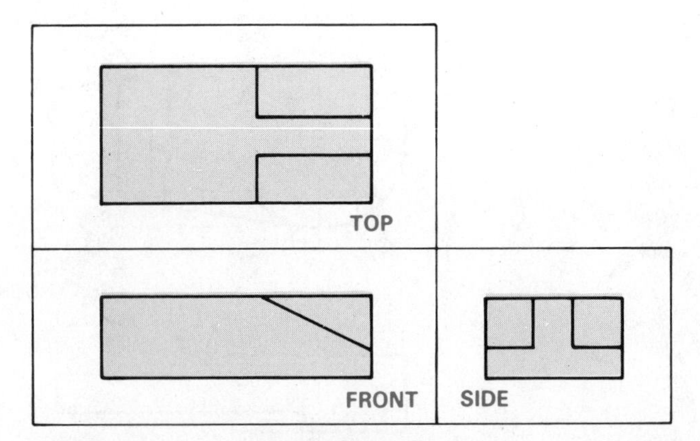

Figure 5–43 Third-angle projection.

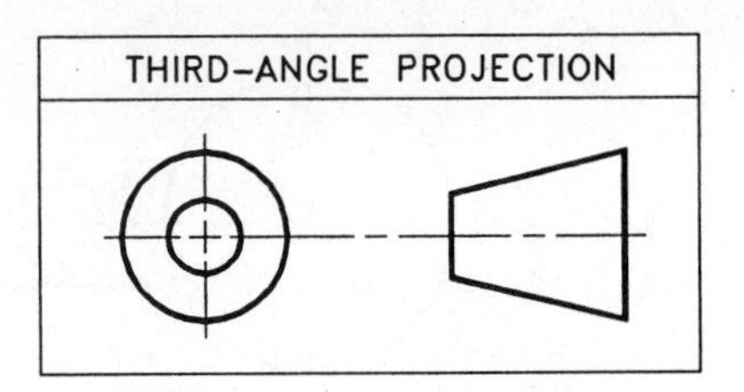

Figure 5–44 Third-angle projection symbol.

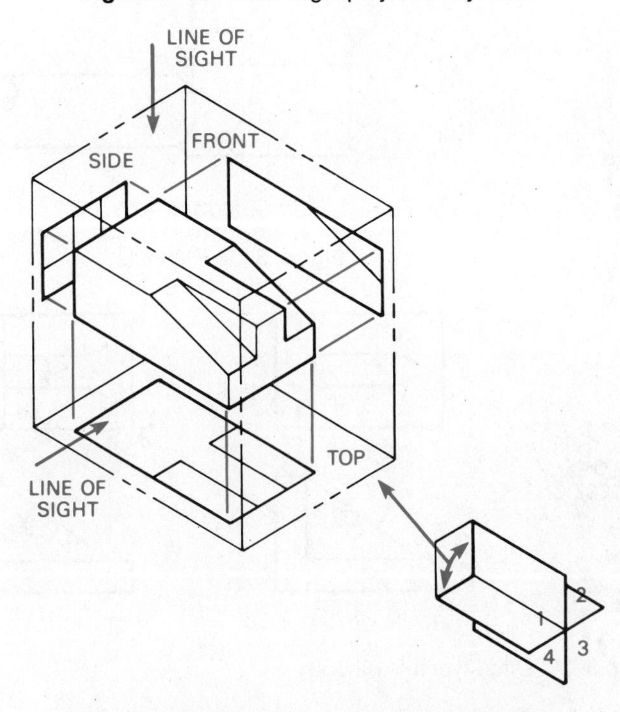

Figure 5–45 Glass box in first-angle projection.

A first-angle projection drawing may be accompanied by a symbol on or adjacent to the drawing title block. The standard first-angle projection symbol is shown in Figure 5–47. Figure 5–48 shows a comparison of the same object in first- and third-angle projections.

LAYOUT TECHNIQUES

Many factors influence the drawing layout. Your prime objective should be a clear and easy-to-read interpretation of selected views with related information. Although this chapter deals with multiview presentation, it is not totally realistic to consider view layout without thinking about the effects of dimension placement on the total drawing. Chapter 8 correlates the multiview drawings of shape description with dimensioning, known as size description.

The initial steps in view layout should be performed using rough sketches. An experienced drafter may be able to lay out a drawing while thinking about what the completed drawing should look like. However, this technique sometimes causes more work in the long run. By

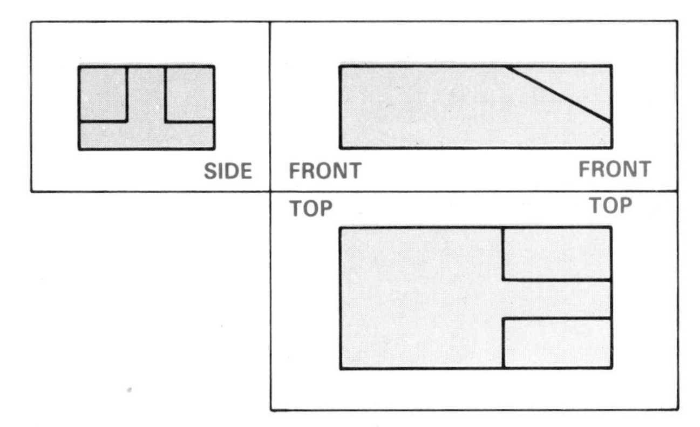

Figure 5–46 First-angle projection.

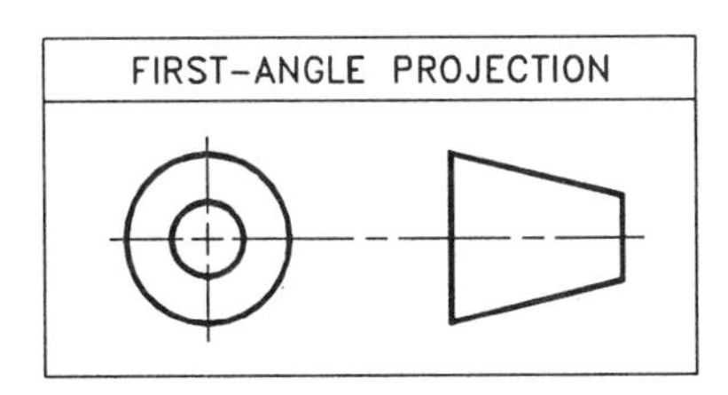

Figure 5–47 First-angle projection symbol.

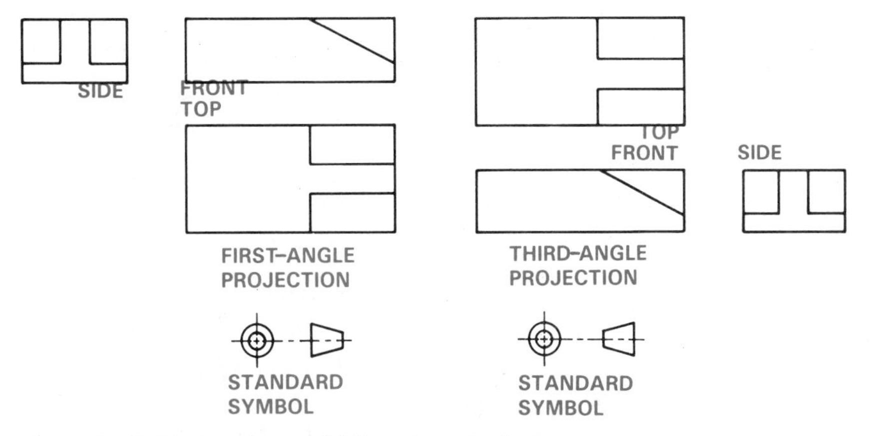

Figure 5–48 First-angle and third-angle projections.

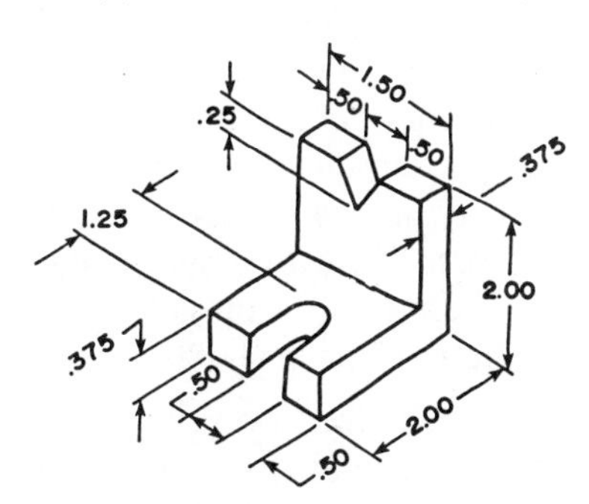

Figure 5–49 Engineering sketch.

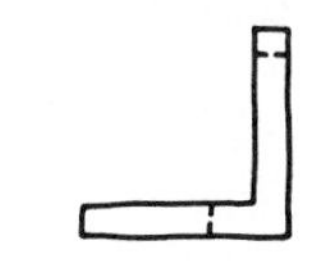

Figure 5–50 Sketch the front view.

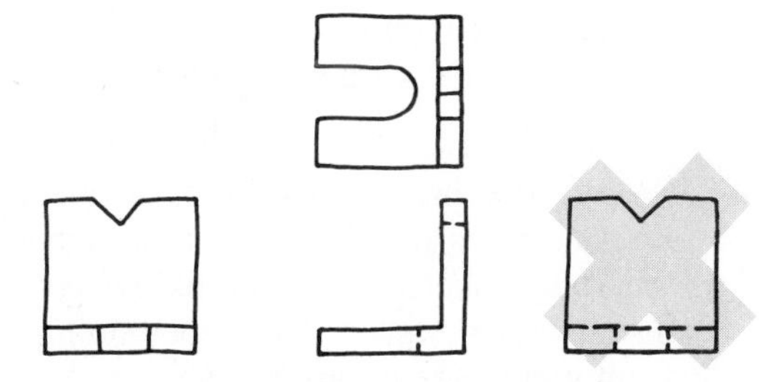

Figure 5–51 Sketch the required views.

using rough sketches, you can analyze which views you need before you begin working on vellum or Mylar®.

Sketching the Layout

Consider the engineering sketch in Figure 5–49 as you evaluate the proper view layout.

Step 1 Select the front view using the rules discussed in this chapter. Sketch the front view that you have picked. Try to keep your sketch proportional to the actual object, as in Figure 5–50.

Step 2 Select the other views needed to completely describe the shape of the V-block mount, as shown in Figure 5–51.

The front, top, and left-side views clearly define the shape of the V-block mount. Now lay out the formal drawing using the sketch as a guide. Several factors must be considered before you begin:

1. Size of vellum or Mylar® sheet available
2. Scale of the drawing
3. Number and size of views
4. Amount of blank space required for future revisions
5. Dimensions and notes (not drawn at this time)

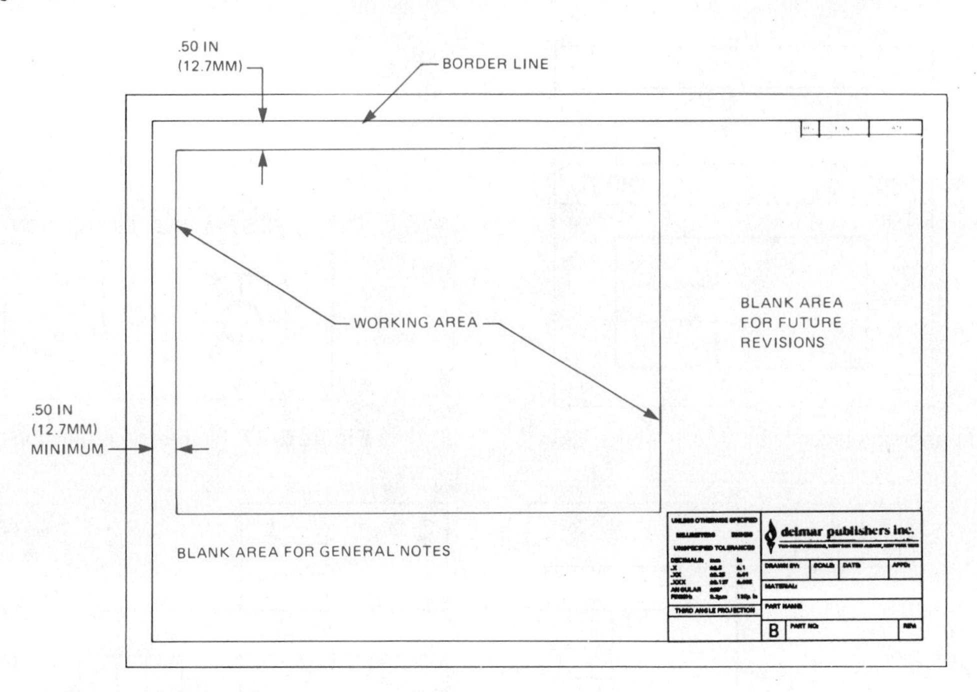

Figure 5–52 Recommended working area.

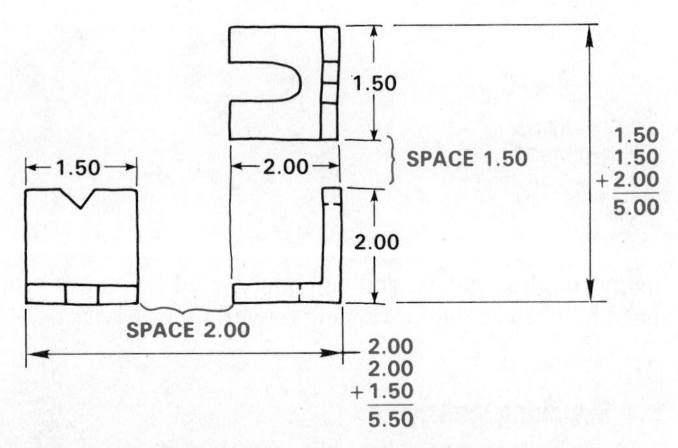

Figure 5–53 Rough sketch and overall dimensions and selected space between views.

Drawing the Layout

Step 1 Use a B-size (11 × 17 in.) drawing sheet. The recommended working area is shown in Figure 5–52. The amount of blank area on a drawing depends upon company standards. Some companies want the drawing to be easy to read with no crowding; others may want as much information as possible on a sheet. Generally, .50 in. should be the minimum space between the drawing and border line. An area for future revision generally (but not always) should be left between the title block and upper right corner. An area for general notes should be available adjacent to the title block for ANSI standard layout or in the upper left corner for military standard layout. The remaining area is the space available for drawing. On Figure 5–52, this area is about 10 × 6 in.

Step 2 After determining the approximate working area, use your rough sketch as a guide to establish the actual size of the drawing by adding the overall dimensions and the space between views. (See Figure 5–53.) The amount of space selected will not crowd the views. The amount to select is an arbitrary decision and one where the drafting technician must use good judgment. Keep in mind that for now we will not consider dimensions when evaluating space requirements. The effect of dimensions on drawing layouts will be discussed in Chapter 8.

Step 3 Now, in each direction subtract the total drawing size from the total space available and divide by two. This gives you the boundaries of the drawing area. Use construction lines to block out the total drawing areas that you have selected. (See Figure 5–54.)

Calculations:

Total space height	6.00
Total drawing height	–5.00
	1.00 ÷ 2 = .50
Total space width	10.00
Total drawing width	–5.50
	4.50 ÷ 2 = 2.25

Step 4 Within the area established, use construction lines to block out the views that you selected in the rough sketch. Use multiview projection as discussed in this chapter, beginning with the front view. (See Figure 5–55.)

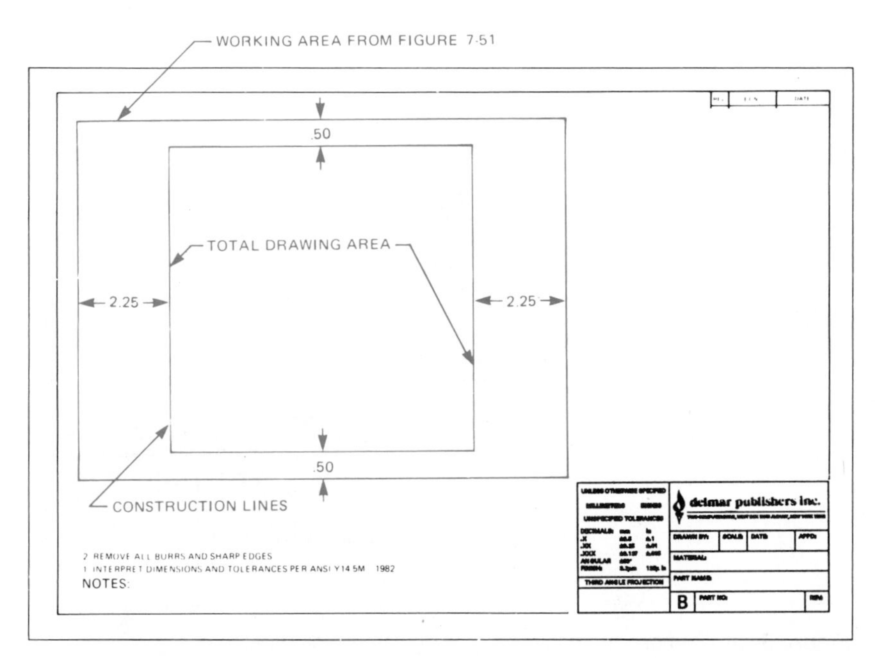

Figure 5–54 Lay out the total drawing area.

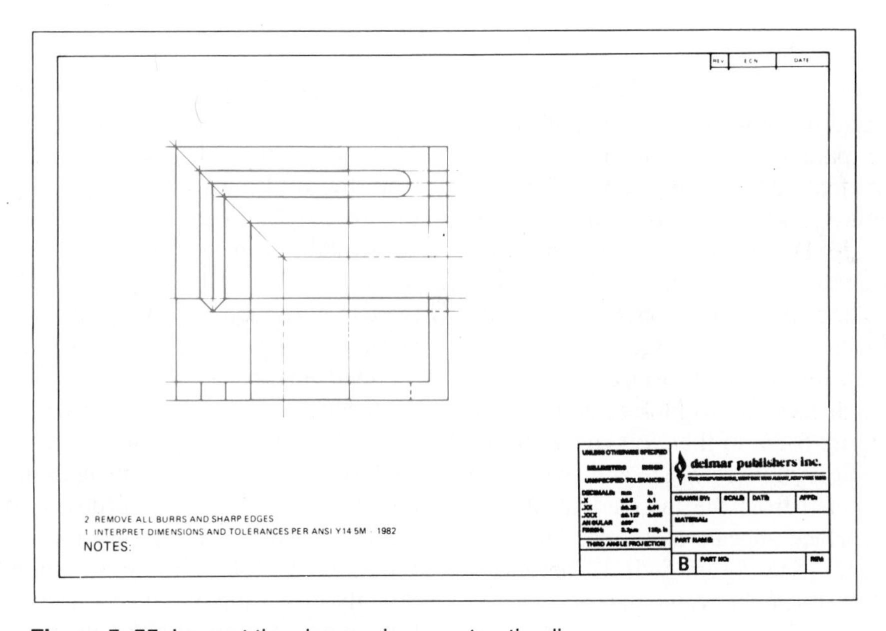

Figure 5–55 Lay out the views using construction lines.

The construction lines, if properly drawn, will not have to be erased. Remember that construction lines are drawn very lightly with a 6H, 4H, or light-blue lead. In any case, the construction lines should not reproduce in the diazo process, if properly drawn.

Step 5 Complete your drawing by using proper techniques to draw the finish lines. To help keep the final drawing neat, remember to:

a. Work from top to bottom.
b. Work from left to right if right handed or from right to left if left handed.
c. Draw thin lines first.
d. Draw circles and arcs next.
e. Draw object lines.
f. Do all lettering and place a clean paper under your hand while lettering.
g. Avoid moving equipment or your hands over completed areas.
h. Keep your equipment and hands clean.
i. Cover large completed areas with blank paper to help avoid smudging.

Figure 5–56 shows the completed drawing.

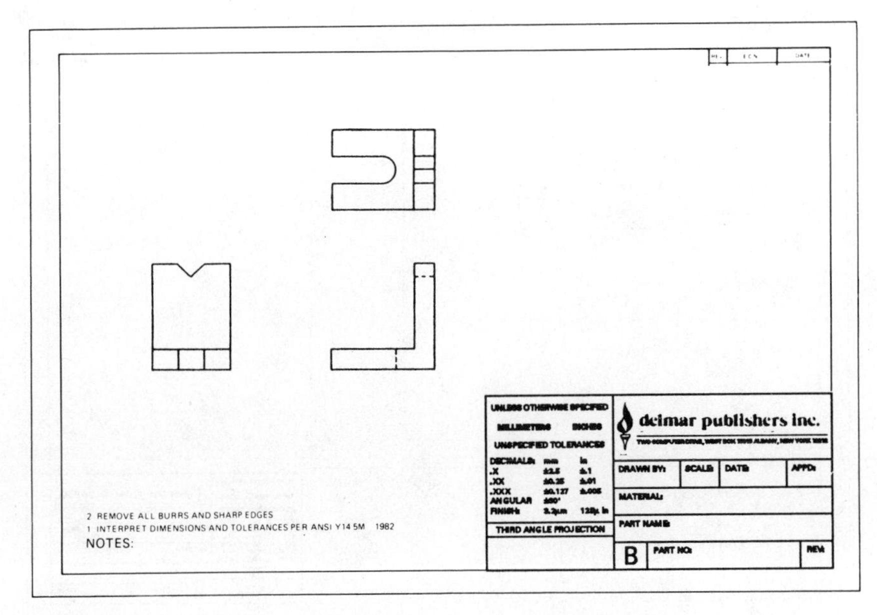

Figure 5–56 Completed multiview drawing without dimensions.

AUXILIARY VIEWS

Auxiliary views are used to show the true size and shape of a surface that is not parallel to any of the six principal views. When a surface feature is not perpendicular to the line of sight, the feature is said to be foreshortened, or shorter than true length. These foreshortened views do not give a clear or accurate representation of the feature. It is not proper to place dimensions on foreshortened views of objects. Figure 5–57 shows three views of an object with an inclined foreshortened surface.

An auxiliary view allows you to look directly at the inclined surface in Figure 5–57 so that you can view the surface and locate the hole in its true size and shape. An auxiliary view is projected from the inclined surface in the view where that surface appears as a line. The projection is at a 90° angle. (See Figure 5–58.) The height dimension, *H*, is taken from the view that shows the height in its true length.

Notice in Figure 5–58 that the auxiliary view shows only the true size and shape of the inclined surface. This is known as a partial auxiliary view. A full auxiliary view, Figure 5–59, also shows all the other features of the object projected onto the auxiliary plane. Normally the information needed from the auxiliary view is the inclined surface only, and the other areas do not usually add clarity to the view. However, each object must be considered separately. Usually, you do not need to draw those areas that do not add clarity to the drawing.

Look at the glass box principle that was discussed earlier as it applies to auxiliary views, as shown in Figure 5–60. Figure 5–61 shows the glass box unfolded. Notice that the fold line between the front view and the auxiliary view is parallel to the edge view of the slanted surface.

The auxiliary view is projected from the edge view of the inclined surface, which establishes true length. The true width of the inclined surface may be transferred from the fold lines of the top or side views as shown in Figure 5–62. Projection lines are drawn as very light construction lines and should not reproduce. Hidden lines are generally not shown on auxiliary views unless the use of hidden lines helps clarify certain features. The auxiliary view may be projected directly from the inclined surface, as in Figure 5–58. When this is done, a centerline or projection line may continue between the views to indicate alignment and view relationship. This centerline or projection line is often used as an extension line for dimensioning purposes. If view alignment is clearly obvious, the projection line may be omitted. When it is not possible to align the auxiliary view directly from the inclined surface, then viewing-plane lines may be used and the view placed in a convenient location on the drawing (Figure 5–63(a)).

The viewing-plane lines are labeled with letters so the view can be clearly identified. This is especially necessary when several viewing-plane lines are used to label different auxiliary views. Place views in the same relationship as the viewing-plane lines indicate. Do not rotate the auxiliary but leave it in the same relationship as if it were projected from the slanted surface. Where multiple views are used, orient views from left to right and from top to bottom.

Figure 5–63(b) shows some other examples of partial auxiliary views in use. It is important to visualize the relationship of the slanted surfaces, edge view, and auxiliary view. If you have trouble with visualization, it is possible to establish the auxiliary view through the

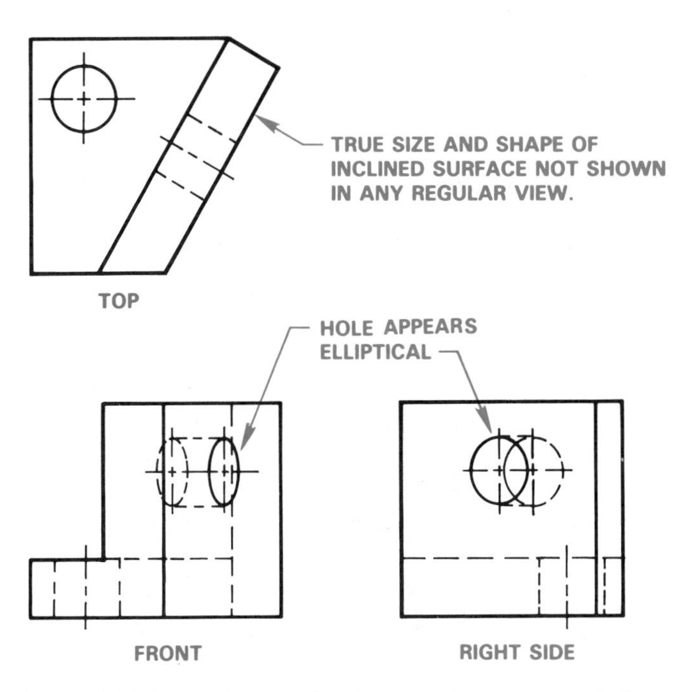

Figure 5–57 Foreshortened surface auxiliary view needed.

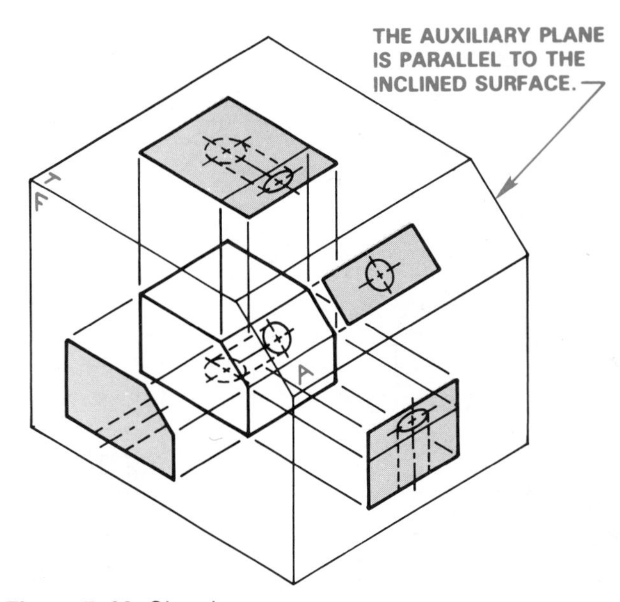

Figure 5–60 Glass box.

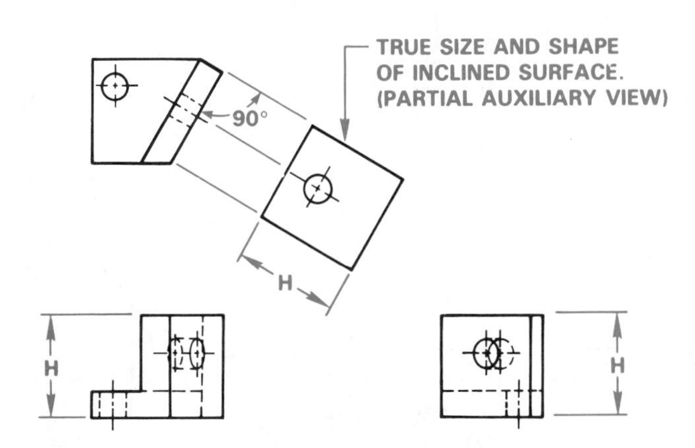

Figure 5–58 Partial auxiliary view.

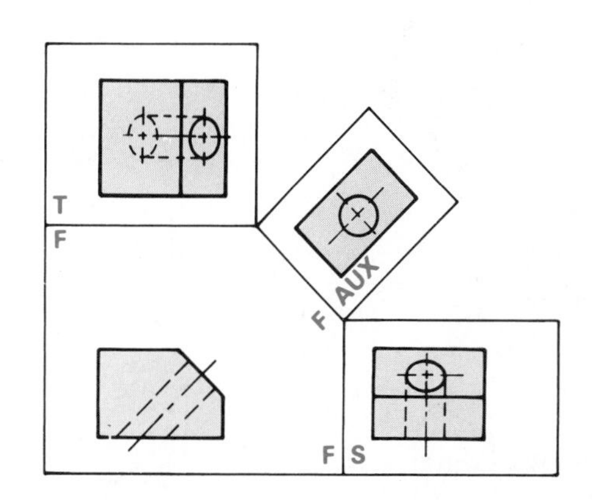

Figure 5–61 Glass box unfolded.

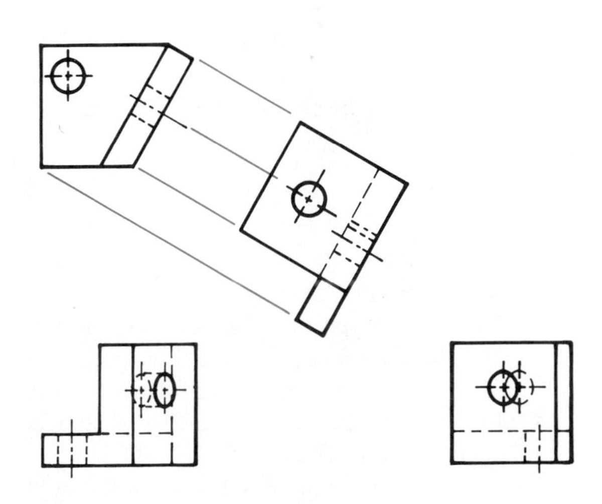

Figure 5–59 Complete auxiliary view.

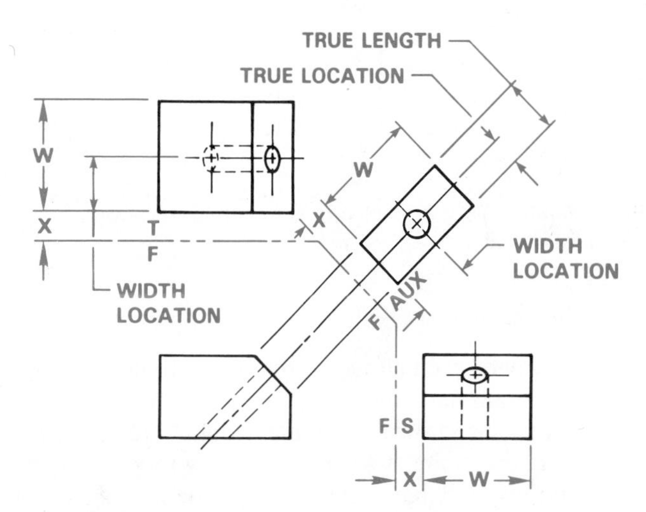

Figure 5–62 Establishing auxiliary view with fold lines.

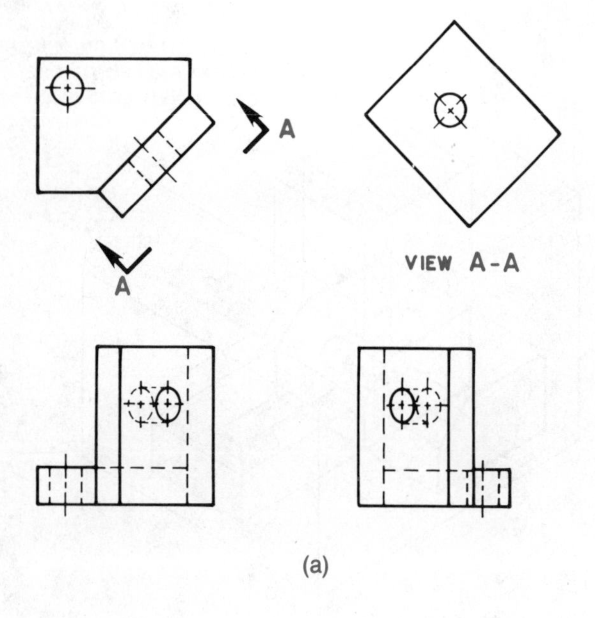

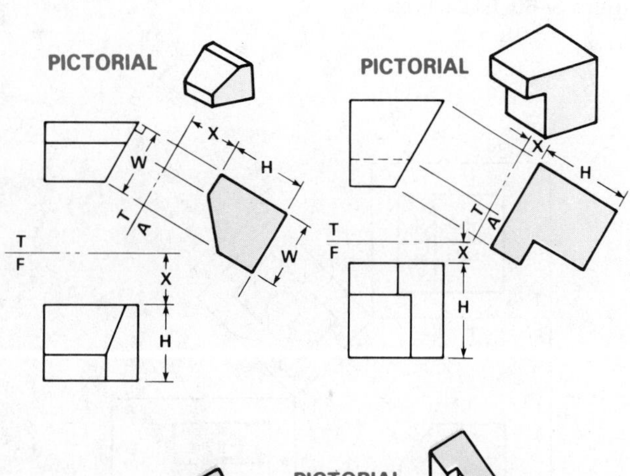

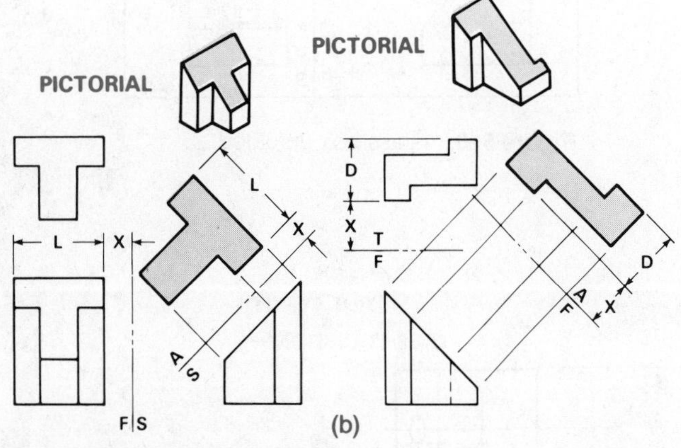

Figure 5–63 (a) Establishing auxiliary view with viewing plane; (b) partial auxiliary views.

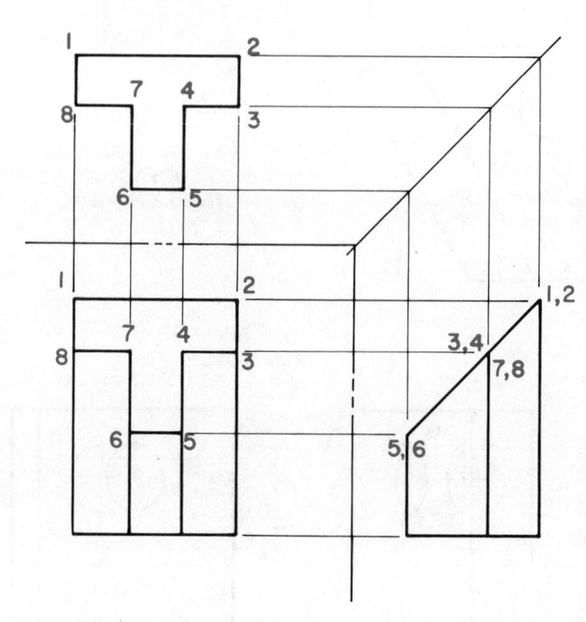

Figure 5–64 Multiview layout — step 1.

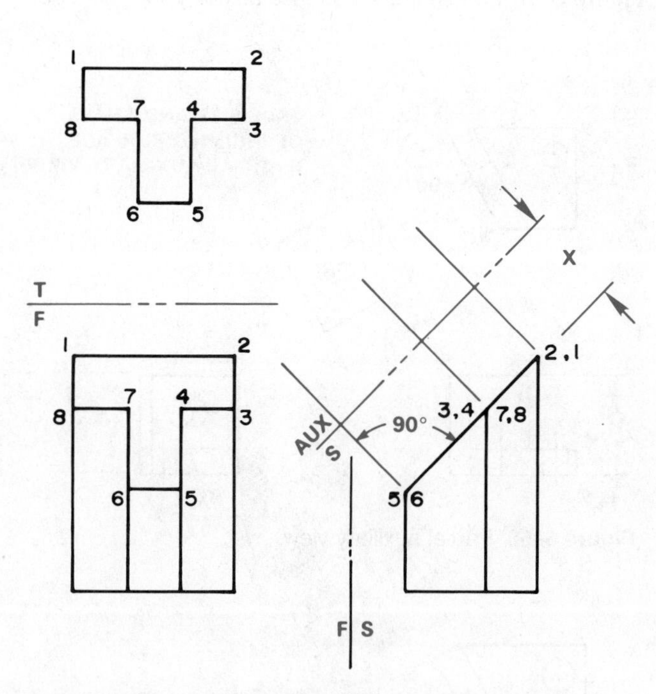

Figure 5–65 Establish the auxiliary fold line — step 2.

mechanics of view projection. Use the following steps:

Step 1 Number each corner of the inclined view so that the numbers coincide from one view to the other, as shown in Figure 5–64. Carefully project one point at a time from view to view. Some points will have two numbers depending upon the view.

Step 2 With the corresponding points numbered in each view, draw an auxiliary fold line parallel to the edge view of the slanted surface. The auxiliary fold line may be any convenient distance from the edge view. Project the points on the edge view perpendicular (90°) across the auxiliary fold line. (See Figure 5–65.)

MULTIVIEW AND AUXILIARY VIEW DRAWING

So far in your study you have learned that multiview and auxiliary view drawings are the orthographic projection of two or more planes at right angles to each other. One important aspect of this technology is accuracy. As a manual drafter, the accuracy depends on your sharp pencil and a keen eye for technique. Even solutions in the best manual have a degree of error due to line widths or the slight tilt of the pencil in relation to the paper. Multiview and auxiliary view drawings produced with CADD are, on the other hand, nearly perfect. The reason for this perfection has to do with the capabilities of the hardware and software, and the fact that the computer is actually displaying the mathematical counterpart of the graphic analysis. In other words, the views are a representation of the mathematical coordinates of the problem being presented. Therefore, the drawing accuracy is only controlled by the accuracy potential of the computer. Figure 5–66 shows a multiview drawing on a CADD monitor.

Most CADD systems provide the drafter with "drawing aids" such as a screen grid similar to the light-blue lines preprinted on vellum to assist the manual drafter. The screen grid is displayed as dots. These grid dot patterns may be displayed in any convenient increment to assist in accurately placing the views. The drafter may also establish designated increments for the screen cursor crosshairs to snap. This allows the drafter to accurately determine distances on the drawing without guesswork. Another asset of the CADD system is the display of drawing distance and coordinates. For example, when you draw a line on the screen, the computer displays the exact line length. Drawing multiviews with these kinds of drawing aids makes the job easy and accurate. When drawing auxiliary views, the CADD drawing aids are flexible and may be rotated at the angle of the auxiliary view to assist in accurately laying out the view as shown in Figure 5–67.

A CADD system also allows you to work in layers and colors for a clearer picture of the problem and solution. For example, you can set a layer for projection lines shown in any color, such as red. This helps you visualize the entire drawing and allows you to accurately project a feature from one view to the next. The layers may be turned on or off at your convenience. When you complete the drawing and you are ready to make a plot, all you have to do is turn off the layer with projection lines and the drawing is ready to plot.

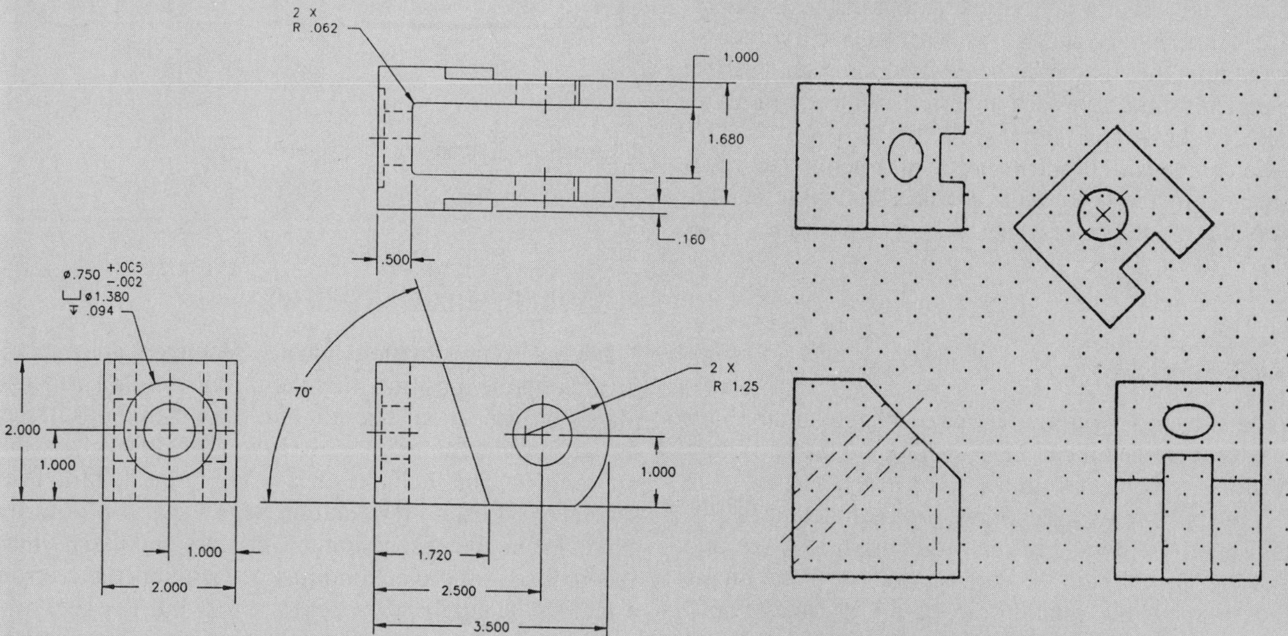

Figure 5–66 Multiview drawing.

Figure 5–67 Using the rotated grid points to help draw the auxiliary view.

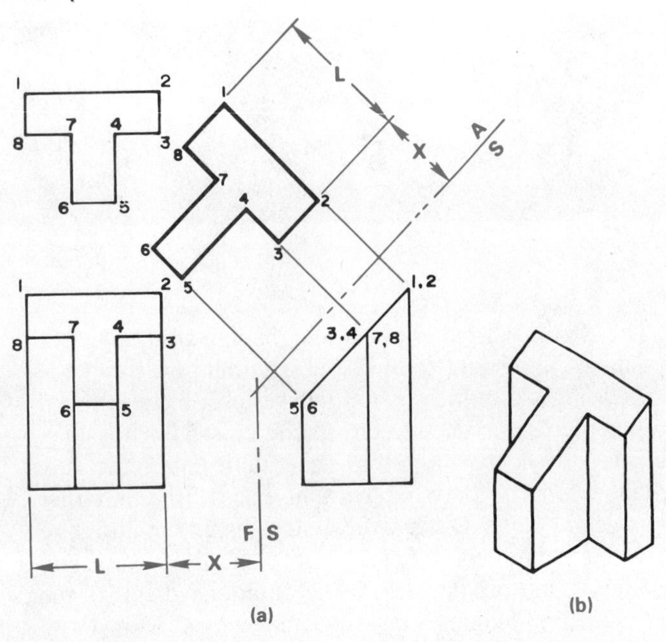

Figure 5–68 (a) Layout auxiliary view — step 3; (b) freehand sketch to help visualize object.

Step 3 Establish the distance to each point from the adjacent view and transfer the distance to the auxiliary view as shown in Figure 5–68(a). Sometimes it is helpful to sketch a small pictorial to assist in visualization. (See Figure 5–68(b).)

PLOTTING CURVES IN AUXILIARY VIEWS

Curves are plotted in auxiliary views in the same manner as those shapes described previously. When corners exist, they are used to lay out the extent of the auxiliary surface. An auxiliary surface with irregular curved contours requires that the curve be divided into elements so the element points can be transferred from one view to the next, as shown in Figure 5–69.

The contour of elliptical shapes may be plotted, or if the shape coincides with an available elliptical template, a template should always be used to save time. (See Figure 5–70.)

ENLARGEMENTS

In some situations you may need to enlarge an auxiliary view so a small detail can be shown more clearly. Figure 5–71 shows an object with a foreshortened surface. The two principal views are clearly dimensioned at a 1:1 scale. The 1:1 scale is too small to clarify the shape and size of the slot through the part. A viewing-plane line is placed to show the relationship of the auxiliary view. The auxiliary view can then be drawn in any convenient location at any desired scale; in this case 2:1.

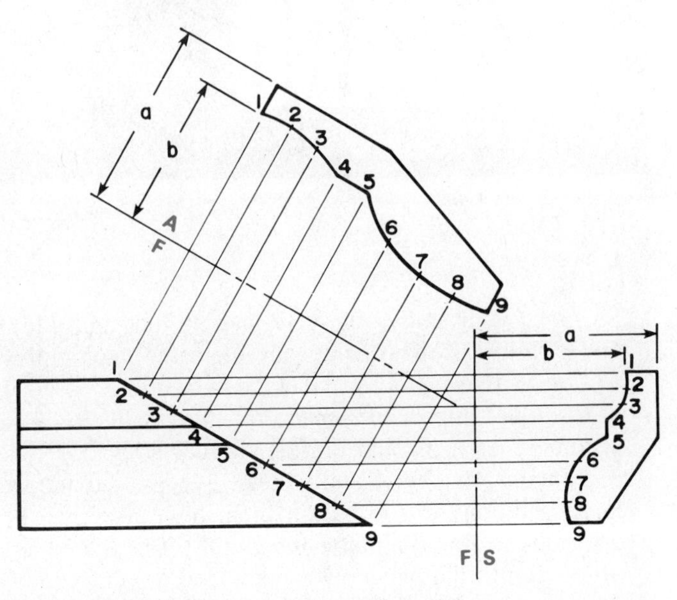

Figure 5–69 Plotting curves in auxiliary view.

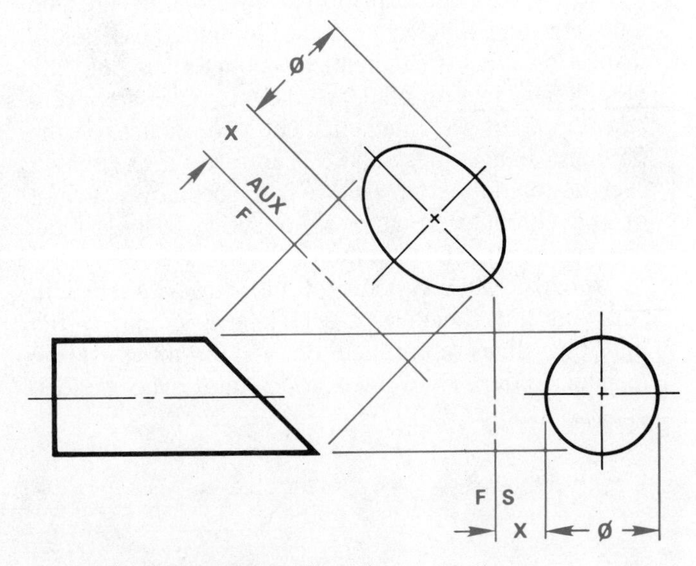

Figure 5–70 Elliptical auxiliary view.

SECONDARY AUXILIARY VIEWS

There are some situations when a feature of an object is in an oblique position in relationship to the principal planes of projection. These inclined, or slanting, surfaces do not provide an edge view in any of the six possible multiviews. The inclined surface in Figure 5–72 is foreshortened in each view, and an edge view also does not exist. From the discussion on primary auxiliary views you realize that projection must be from an edge view to establish the auxiliary.

In order to obtain the true size and shape of the inclined surface in Figure 5–72, a secondary auxiliary

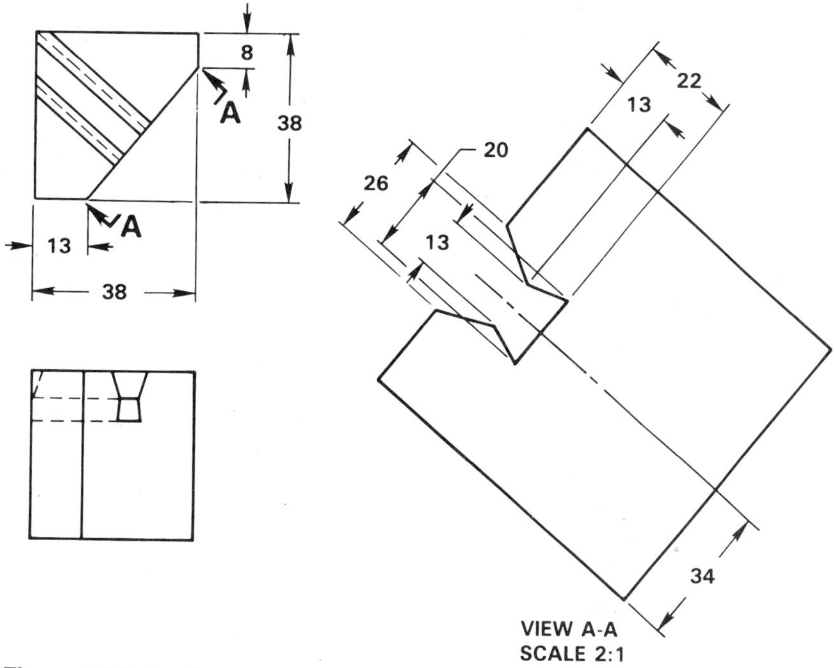

Figure 5–71 Auxiliary view enlargement.

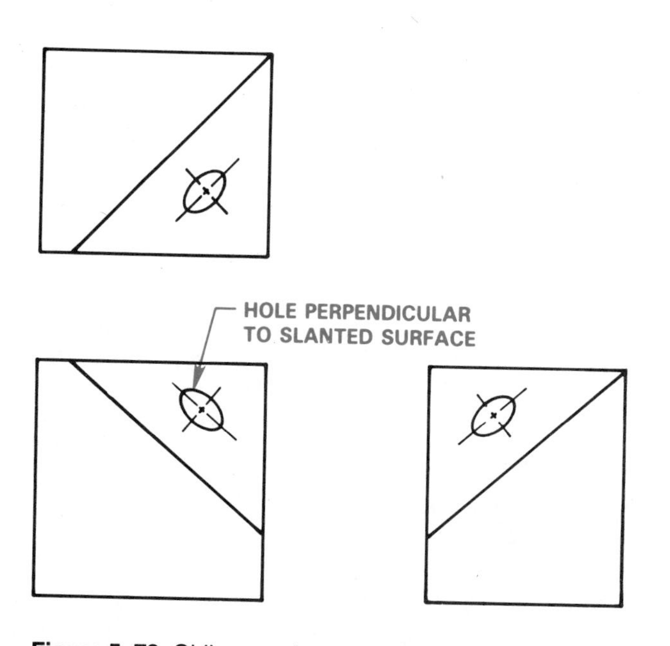

Figure 5–72 Oblique surface; there is no edge view in the principal views.

view is needed. The following steps may be used to prepare a secondary auxiliary.

Step 1 Only two principal views are necessary to work from, as the third view does not add additional information. Establish an element in one view that is true in length, as shown in Figure 5–73. Label the corners of the inclined surface and draw a fold line perpendicular to one of the true-length elements.

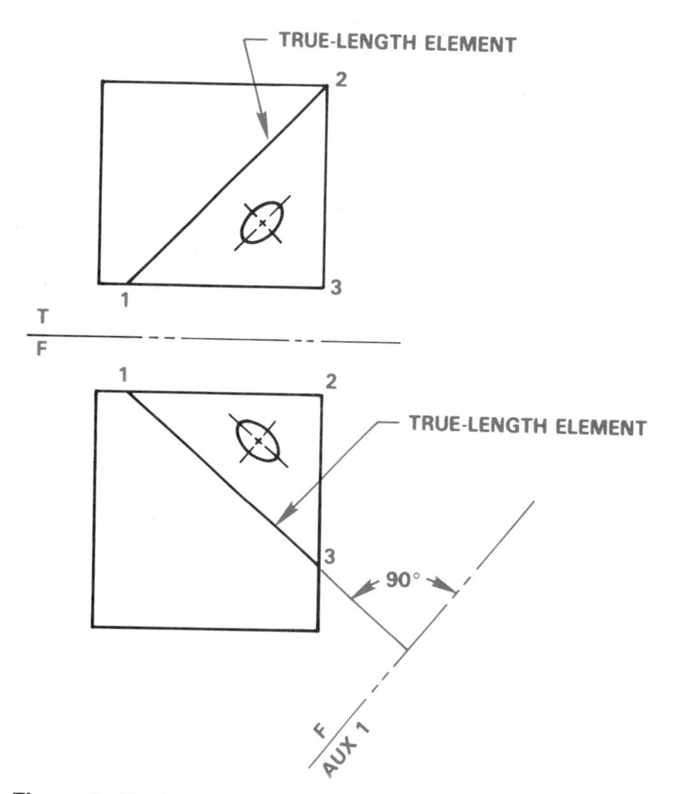

Figure 5–73 Draw a fold line perpendicular to a true length element — step 1.

Step 2 The purpose of this step is to establish a primary auxiliary view that displays the slanted surface as an edge. Project the slanted surface onto the primary auxiliary plane, as shown in Figure 5–74. Doing so results in the inclined surface appearing as an edge view or line.

Step 3 Now, with an edge view established, the next step is the same as the normal auxiliary view procedure. Draw a fold line parallel to the edge view. Project points from the edge view perpendicular to the secondary fold line to establish points for the secondary auxiliary, as shown in Figure 5–75.

In Figure 5–75, the primary auxiliary view established all corners of the inclined surface in a line, or edge view. This edge view is necessary so the perpendicular line of projection (sight) for the secondary auxiliary assists in establishing the true size and shape of the surface. In many situations both the primary and secondary auxiliary views are used to establish the relationship between features of the object. (See Figure 5–76.) The primary auxiliary view shows the relationship of the

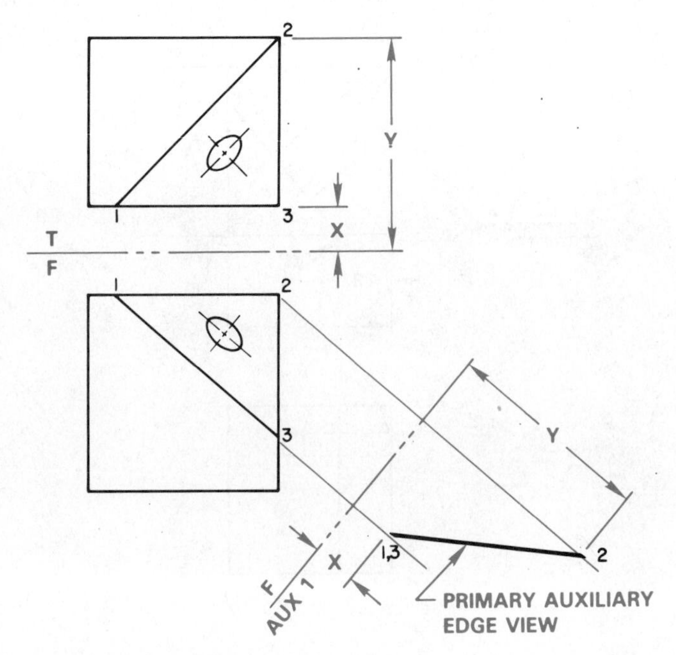

Figure 5–74 Primary auxiliary edge view of oblique surface — step 2.

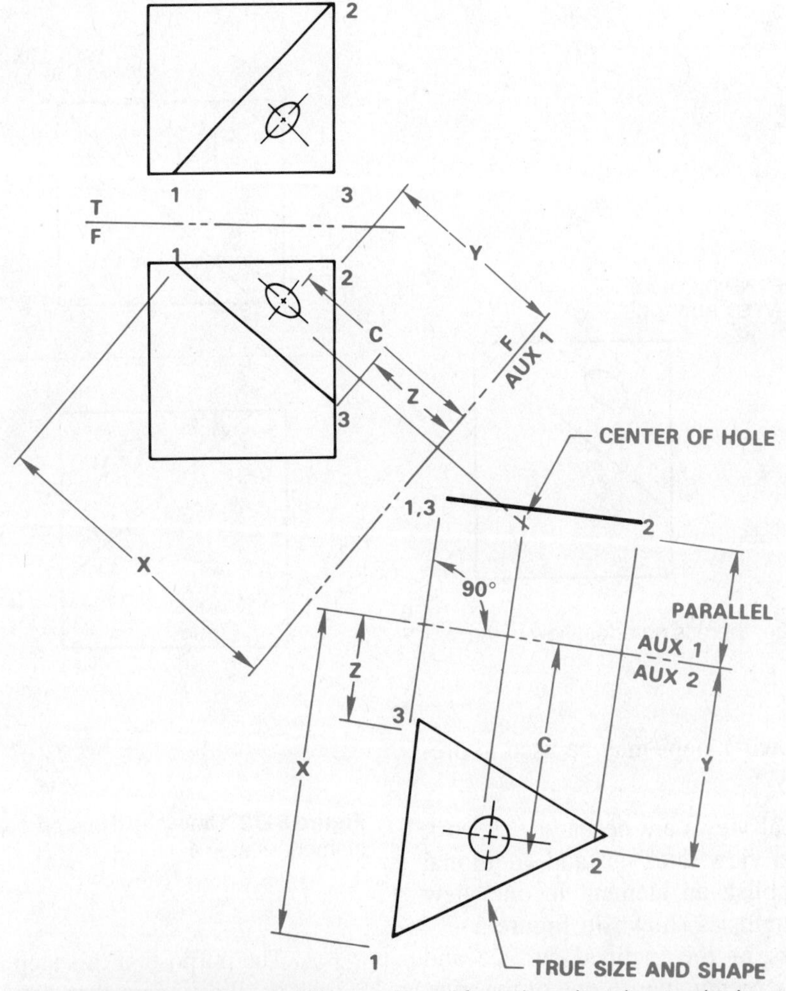

Figure 5–75 Secondary auxiliary projected from the edge view to obtain the true size and shape of the oblique surface — step 3.

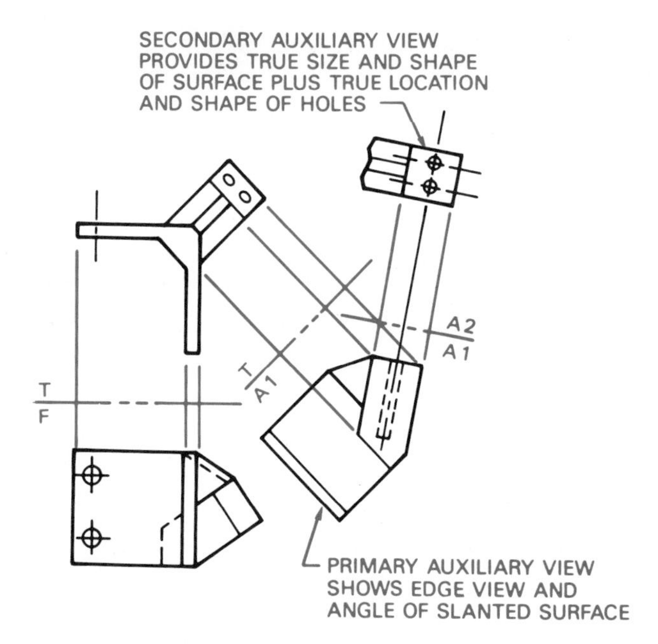

Figure 5–76 Primary and secondary auxiliary views.

inclined feature to the balance of the part, and the secondary auxiliary view shows the true size and shape of the inclined features plus the true location of the holes.

PROFESSIONAL PERSPECTIVE

For many people, one of the most difficult aspects of engineering drafting is the need to visualize two-dimensional views. This "spacial" ability is natural for some, while others must carefully analyze every part of a multiview drawing in order to fully visualize the product. Ideally, you should be able to look at the multiviews of an object and readily formulate a pictorial representation in your mind. The principal reason that multiviews are aligned is to assist in the visualization and interpretation process. The front view is the most important view, and that is why it contains the most significant features. As you look at the front view, you should be able to gain a lot of information about the object, then visually project from the front view to other views to gain an understanding of the entire product.

It is up to the drafter to select the best views to completely describe the object. Too few views make the object difficult or impossible to interpret, while too many views make the drawing too complex and waste valuable drafting time. If you follow the guidelines set up in this chapter, you should be able to successfully complete the task. Remember, always begin with a sketch. With the sketch you can quickly determine view arrangement and spacing. Without the preliminary sketch, you might use a lot of drafting time and end up discovering that the layout does not work as you expected.

ENGINEERING PROBLEM

Working with auxiliary views is a little more complex than dealing with the normal multiviews. The auxiliary view is often difficult to place in relationship to the other views. Space requirements many times cause the auxiliary view to interfere with other views. Whenever possible, it is best to project the auxiliary view directly from the inclined surface with one projection line connecting the inclined surface to the auxiliary view. This arrangement makes it easier for the reader to correctly interpret the relationship of the views. This method is shown in the layout steps provided in Figures 5–64 through 5–65 and Figure 5–68.

When drawing space is limited, it is possible to use the viewing-plane method to display the auxiliary view. This technique is shown in Figure 5–63. The advantage of the viewing-plane method is that it allows you to place the auxiliary view in any convenient place on the drawing. Multiple auxiliary views should be placed in a group arranged from right to left and from top to bottom on the drawing. One unorthodox method that is often used by entry-level drafters is to place viewing planes all over the drawing and shift views from their normal position. This practice should be avoided, because the normal multiview projection is always best. You should follow the same layout procedure outlined for multiviews in Figures 5–49 through 5–56.

The steps summarized here work the same for manual drafting or CADD:

- Select and sketch the front view.
- Select and sketch the other principal views in proper relationship to the front view, eliminating any unnecessary views.
- An additional step is to sketch the auxiliary view or views in their proper positions.
- Establish the working area and the sheet size to be used.
- Determine the total drawing area.
- Lay out the views using construction lines.
- Complete the formal drawing.

MULTIVIEWS AND AUXILIARY VIEW PROBLEMS

PART 1

DIRECTIONS

Missing lines. Determine the missing lines in Problems 5–1 to 5–8 and redraw the following objects in multiview. Do not draw the pictorial; it is only given to aid visualization. Use a scale or dividers to transfer dimensions to your drawing from the given sketches.

Problem 5–1 Pocket Block

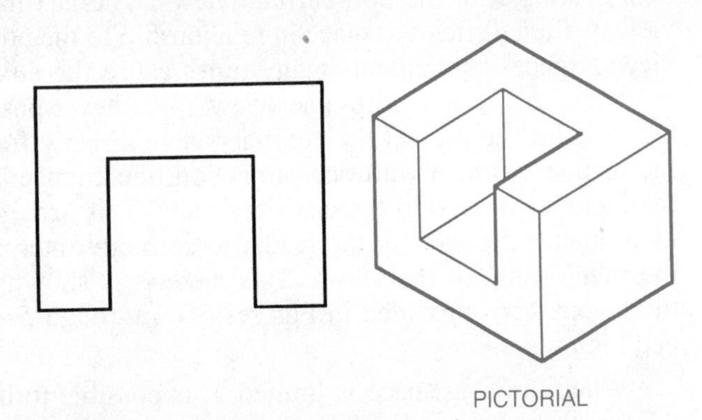

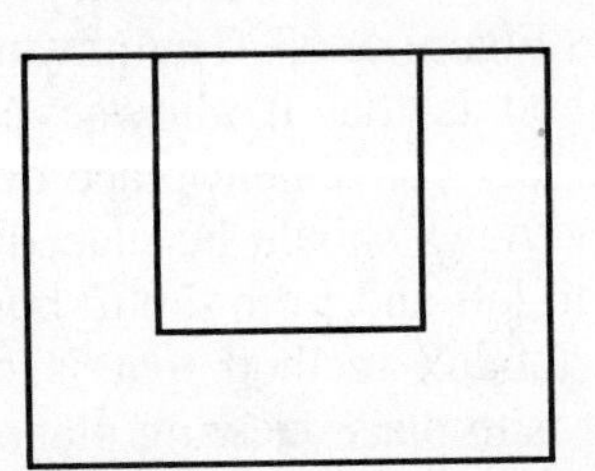

Problem 5–2 Angle Gage

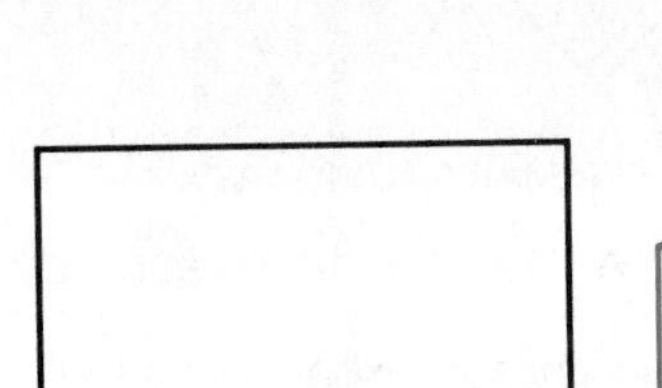

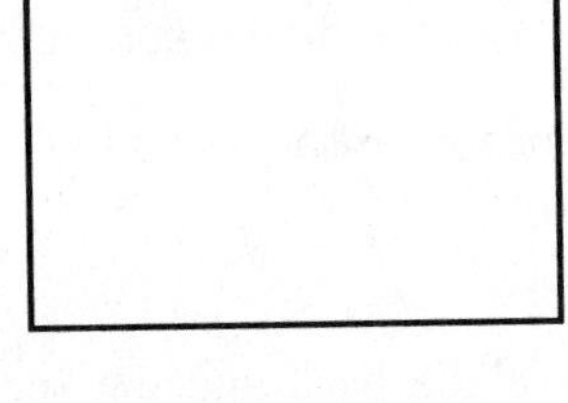

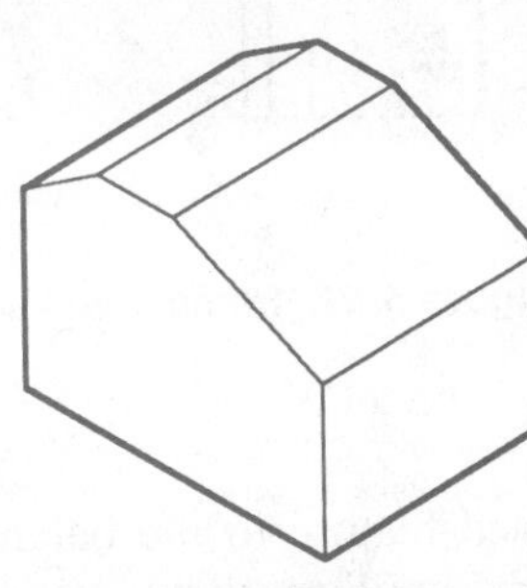

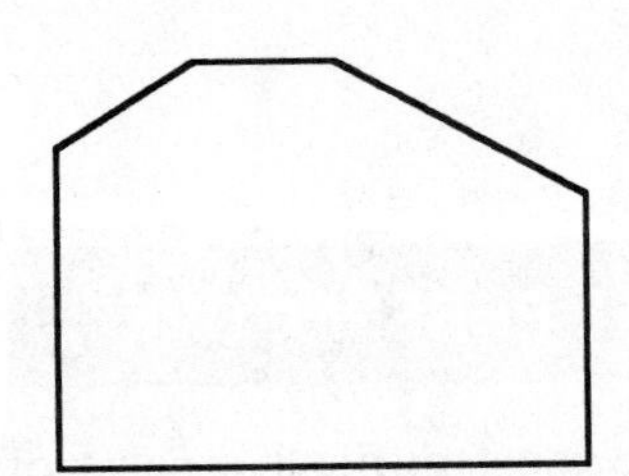

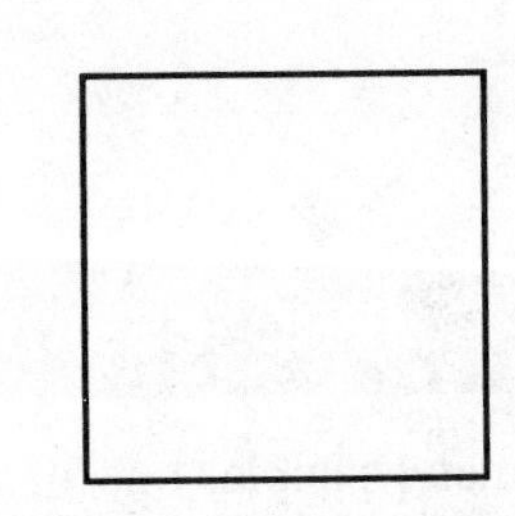

Problem 5–3 Base

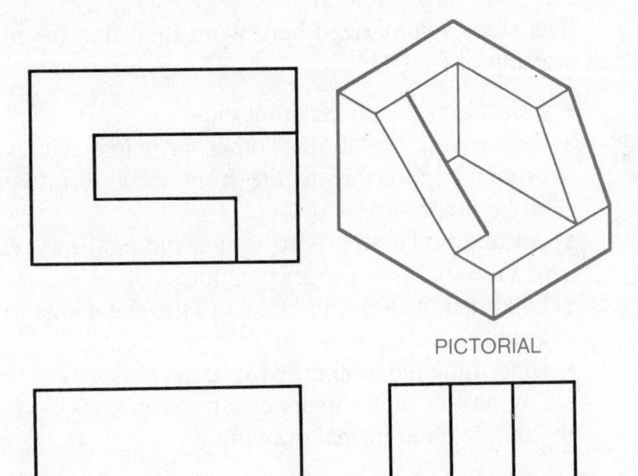

Problem 5–4 Corner Block

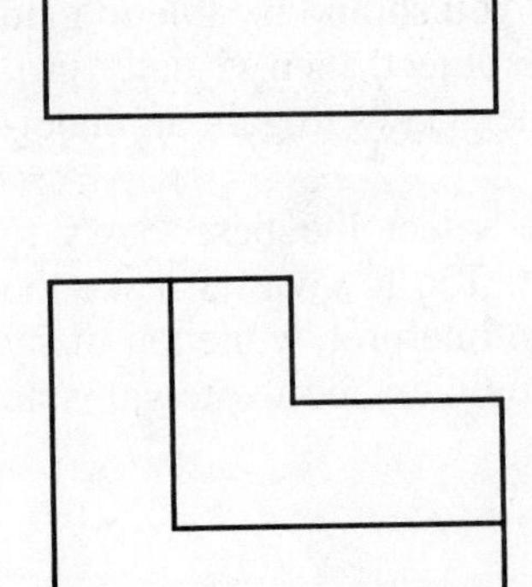

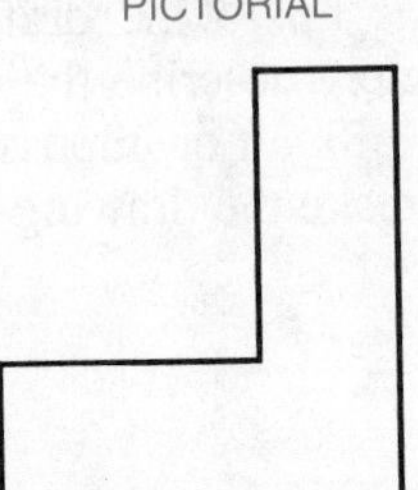

Problem 5–5 Cylinder Block

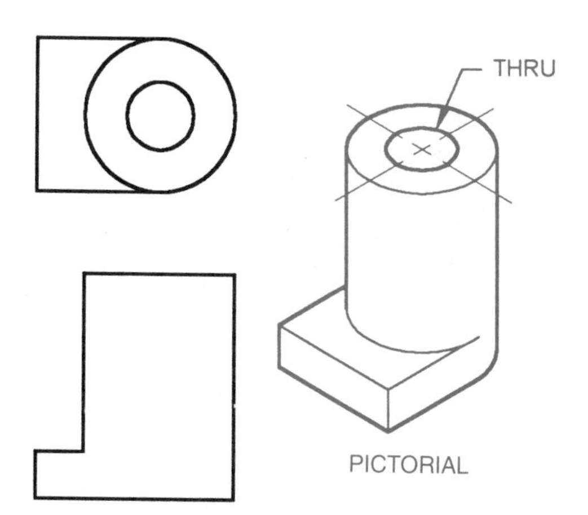

Problem 5–6 Shaft Block

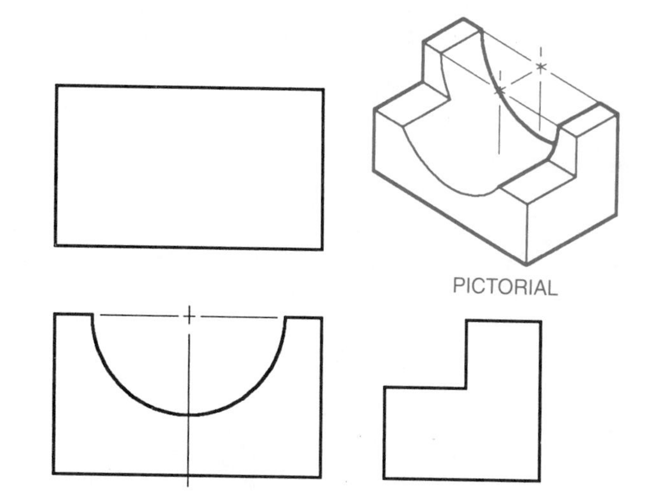

Problem 5–7 Gib

PICTORIAL

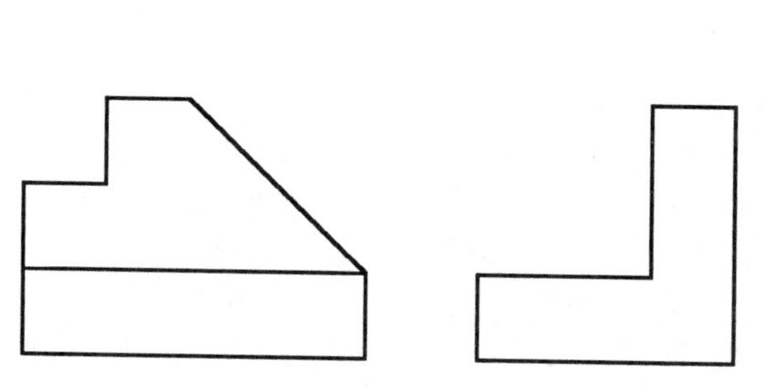

Problem 5–8 Eccentric

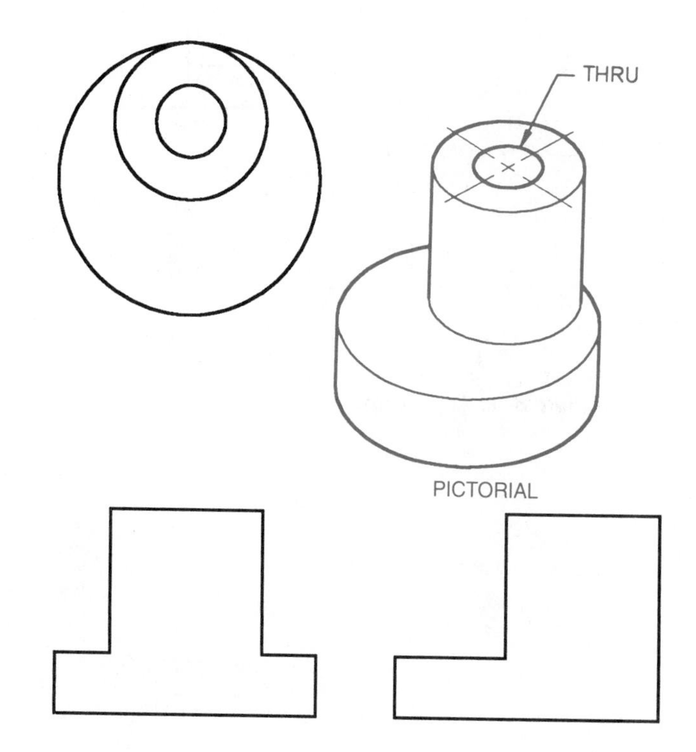

DIRECTIONS

Missing views. Determine the missing views for Problems 5–9 to 5–12 and redraw the objects in multiview. Do not draw the pictorial; it is only given to aid visualization. Use a scale or dividers to transfer dimensions to your drawing from the given sketches.

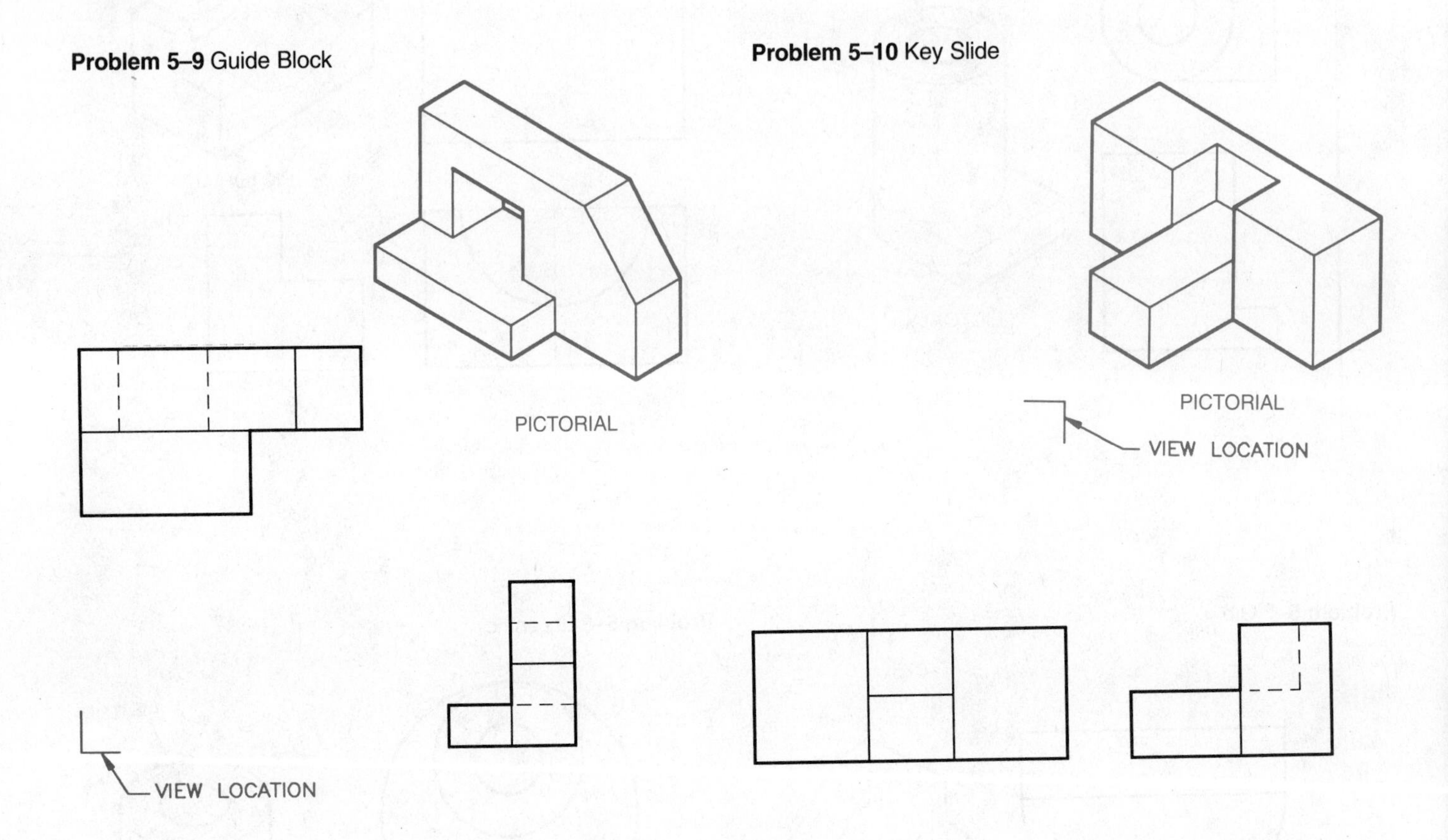

Problem 5–11 Angle Bracket

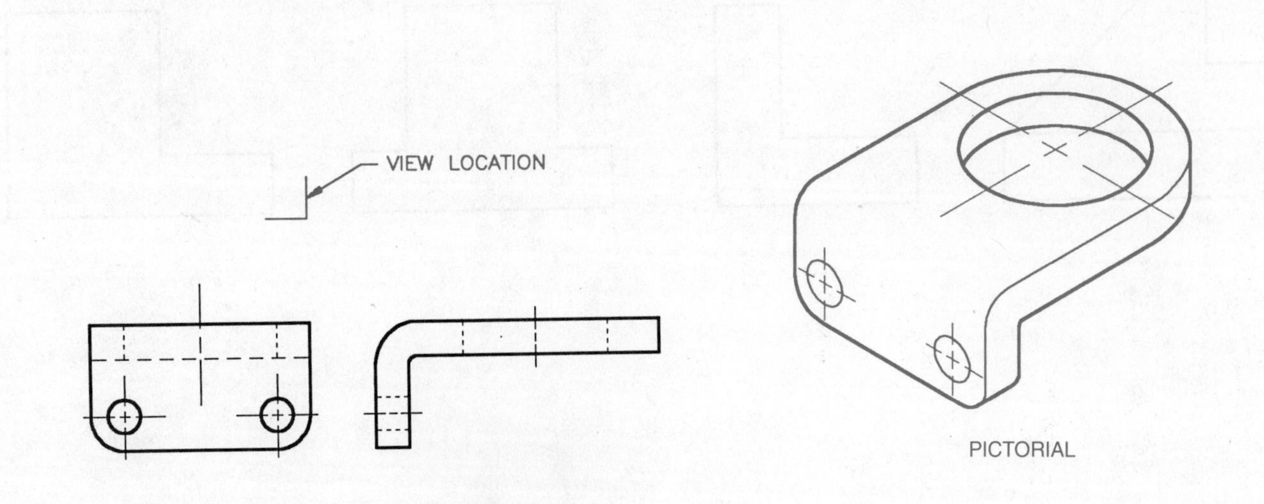

Problem 5–12 Cleves

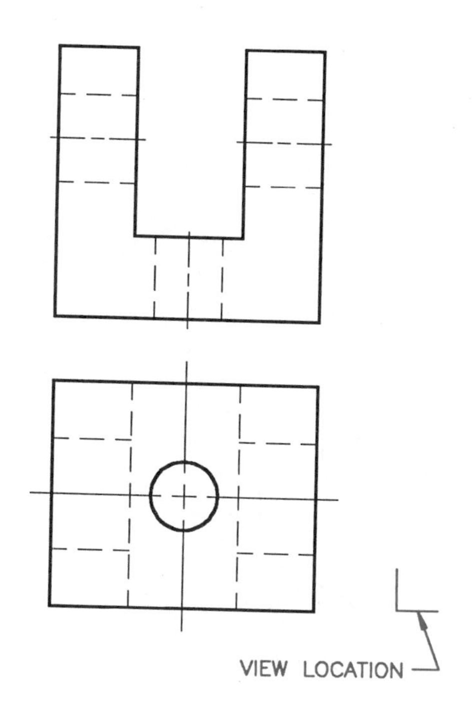

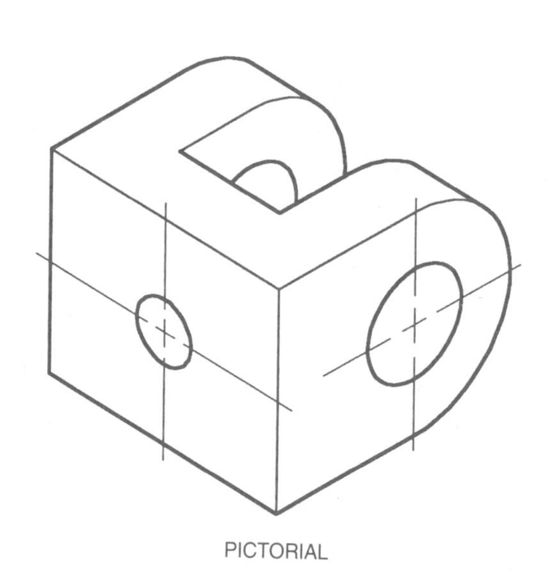

PICTORIAL

DIRECTIONS

Multiviews. Redraw the objects in Problems 5–13 and 5–14 in multiview. The proposed view locations are given. Use dimensions given on pictorial sketches. Do not dimension views. Do not draw the pictorial view.

Problem 5–13 V-block (metric)

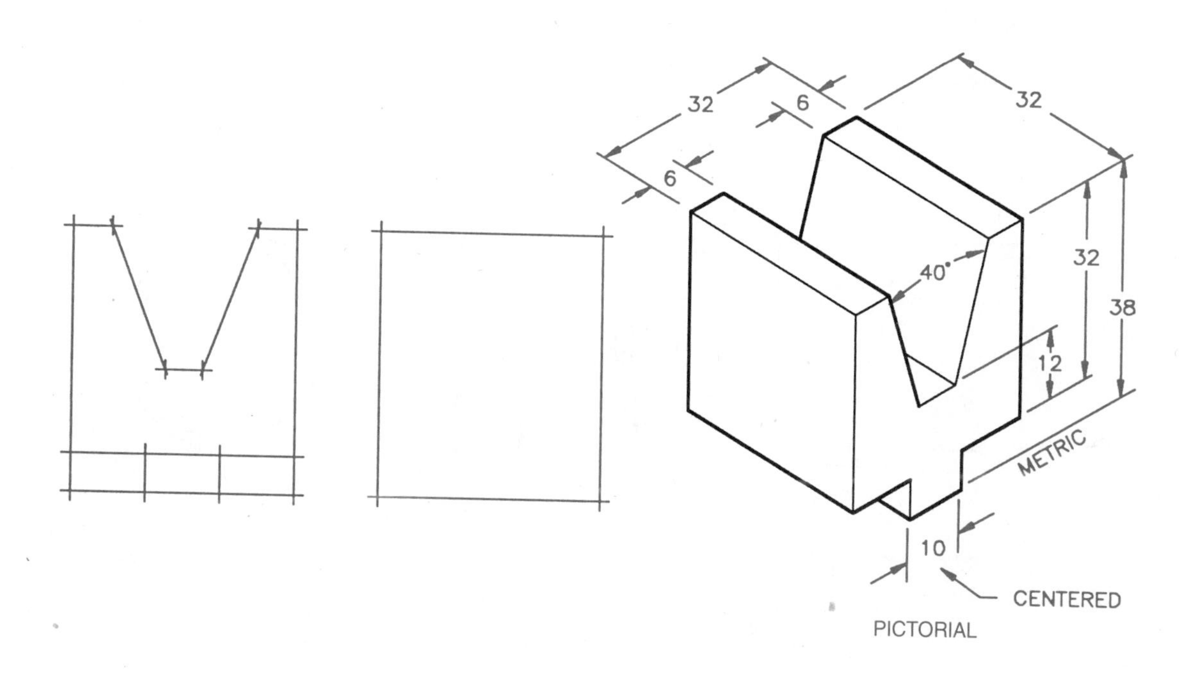

PICTORIAL

Problem 5–14 Guide Base (in.)

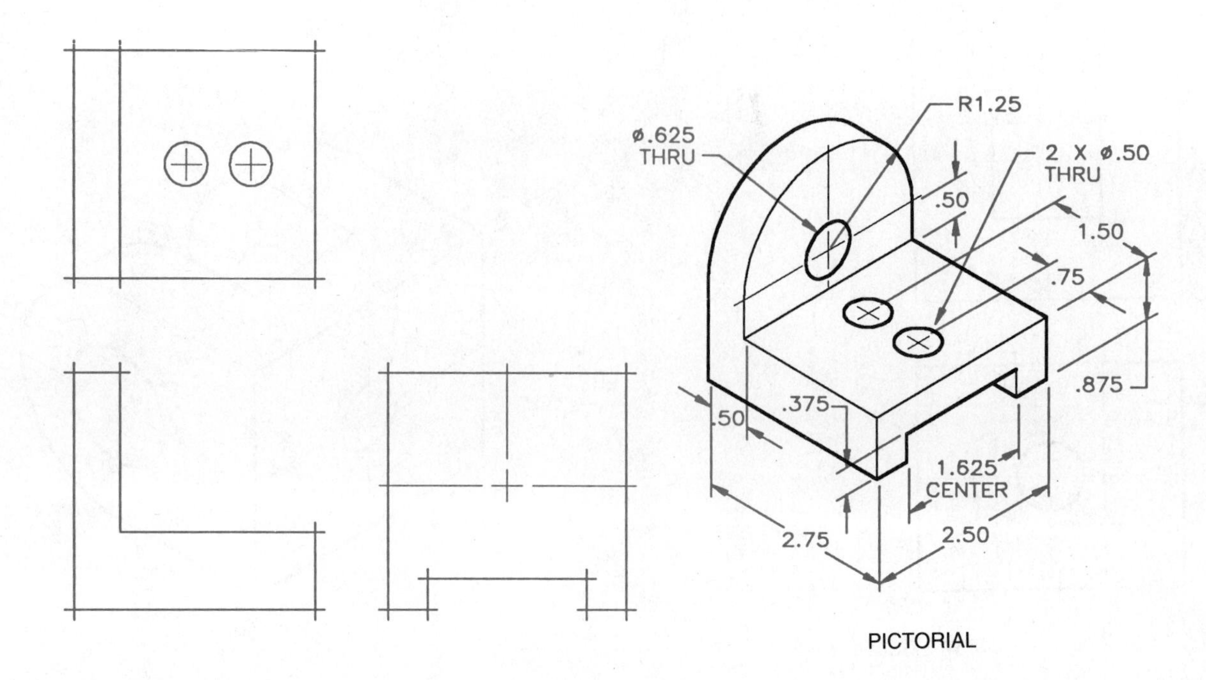

PART 2

DIRECTIONS

Problems 5–15 to 5–25 contain one or two views. Draw the views shown without dimensions. Place notes, if given, in lower left corner of your drawing.

Problem 5–15 Cylindrical object (in.)

Part Name: Sleeve Bearing
Material: Phosphor Bronze
Courtesy Production Plastics.

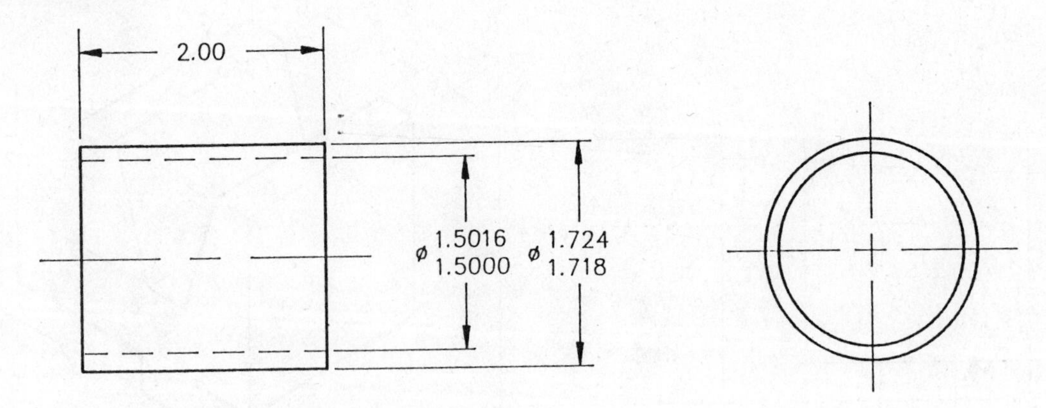

2. DEBURR & BREAK ALL SHARP EDGES.
1. INTERPRET DWG PER ANSI Y14.5M.

NOTES:

Problem 5–16 Arc and hole (in.)

Part Name: Door Lock
Material: Mild Steel

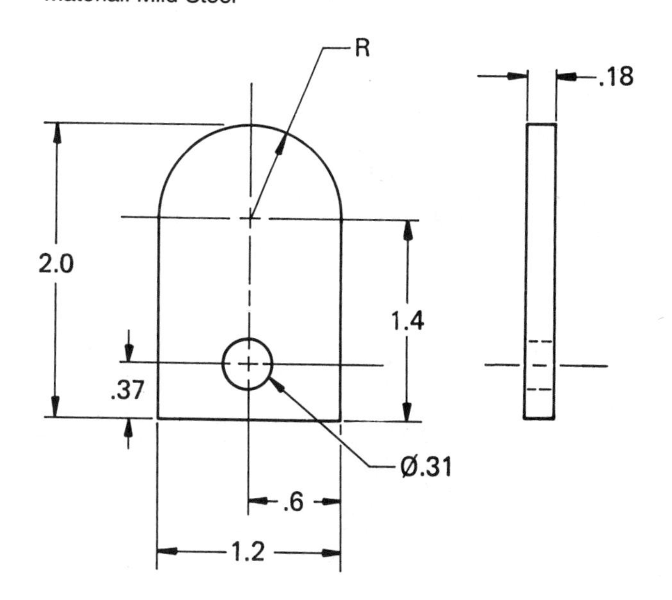

Problem 5–17 Circles and spherical radius (in.)

Part Name: Spherical Insert
Material: Bronze
Courtesy Production Plastics.

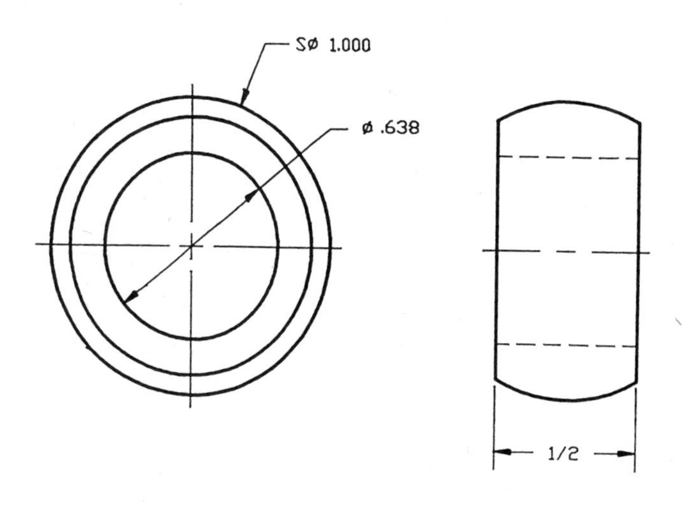

UNLESS OTHERWISE SPECIFIED:

ALL FRACTIONAL DIMENSIONS ±1/32
ALL TWO-PLACE DECIMALS ±.010
ALL THREE-PLACE DECIMALS ±.005

Problem 5–18 Planes and slot (in.)

Part Name: Step Block
Material: Mild Steel

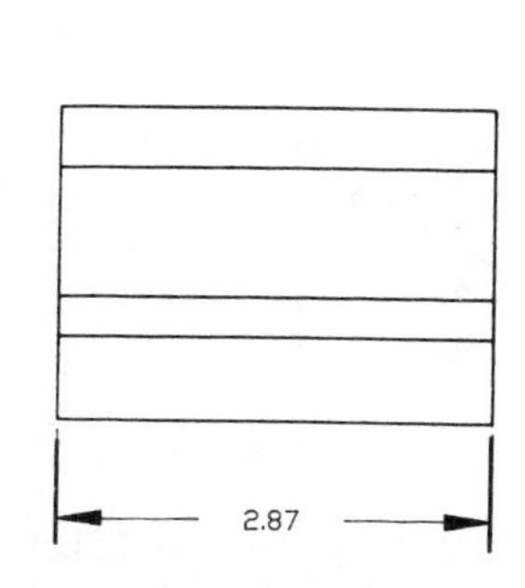

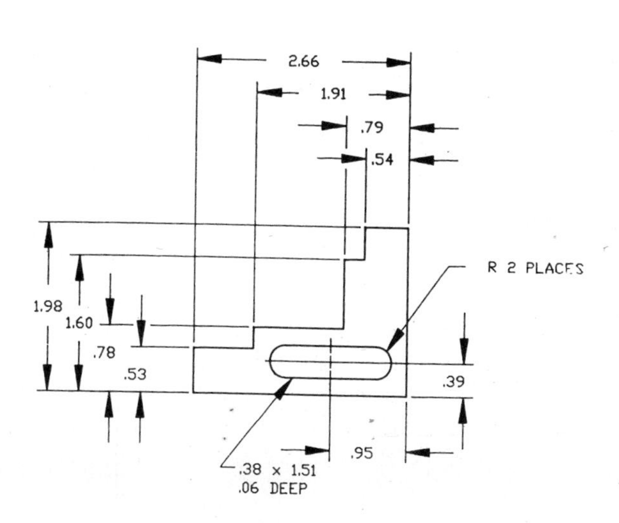

Problem 5–19 Cylinders and circles (in.)

Part Name: Roll End Bearing
Material: Phosphor Bronze
Courtesy Production Plastics.

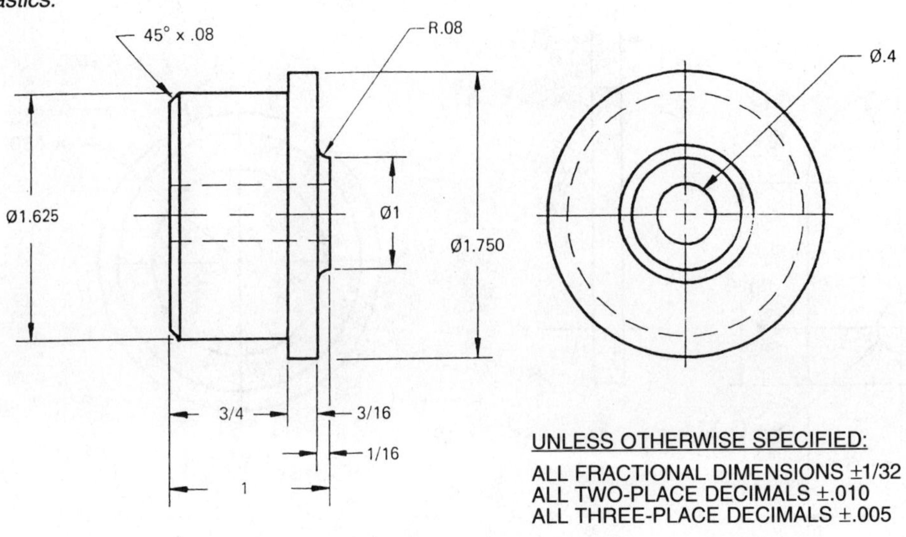

UNLESS OTHERWISE SPECIFIED:
ALL FRACTIONAL DIMENSIONS ±1/32
ALL TWO-PLACE DECIMALS ±.010
ALL THREE-PLACE DECIMALS ±.005

Problem 5–20 Circle arcs and planes (in.)

Part Name: V-block Clamp
Material: SAE 1080

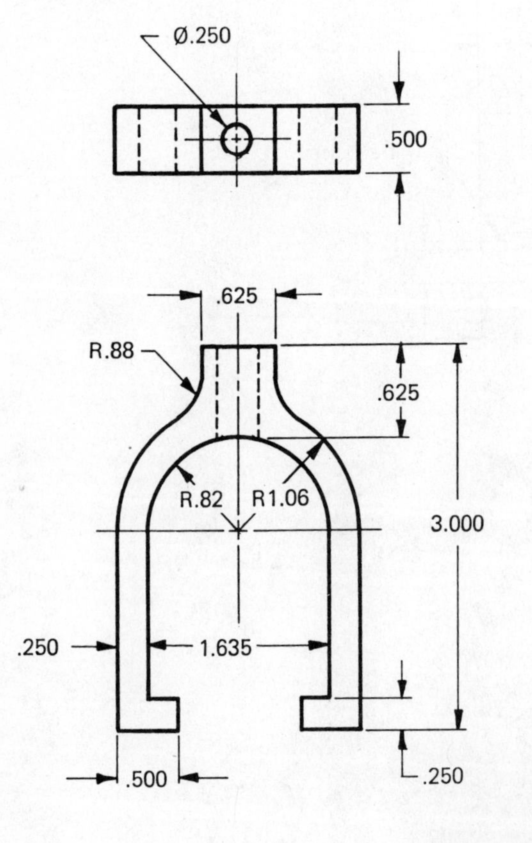

Problem 5–21 View enlargement (in.)

Part Name: Drill Gauge
Material: 16-GA Mild Steel

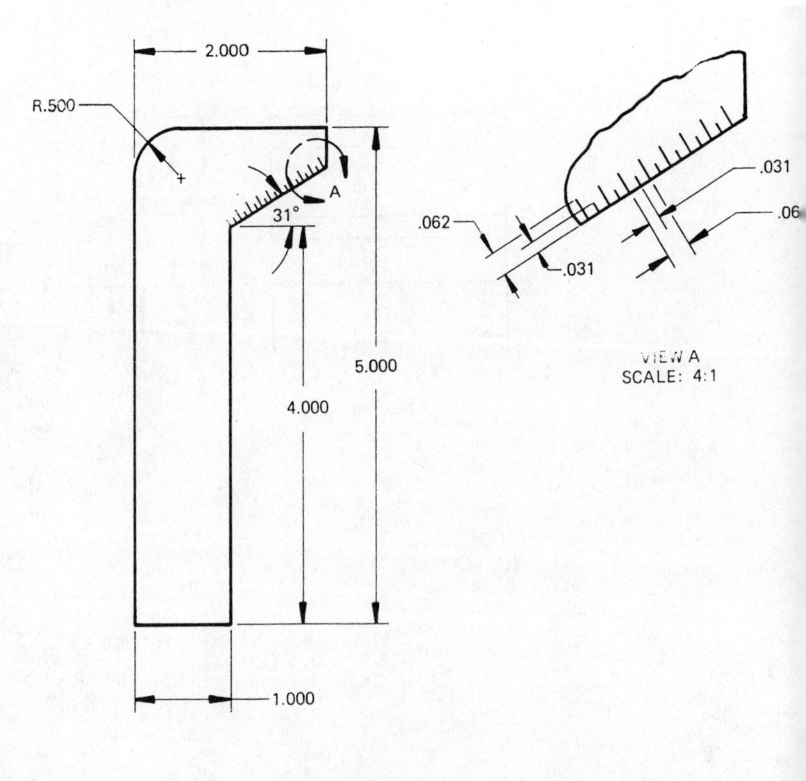

Problem 5–22 Multiple features (in.)

Part Name: Lock Ring
Material: SAE 1020

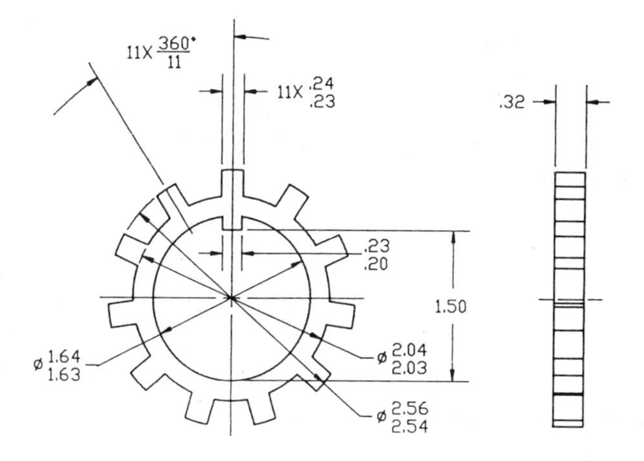

Problem 5–23 Angles and holes (in.)

Part Name: Bracket
Material: SAE 1020

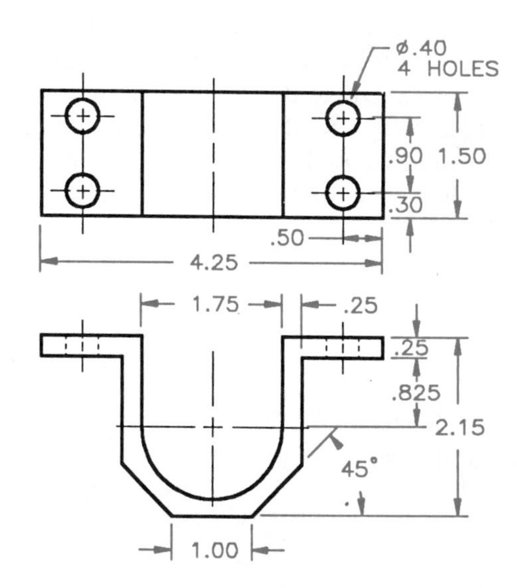

Problem 5–24 Single view (metric)

Part Name: Symmetrical Chart Plate
Material: 10-mm-thick Stainless Steel

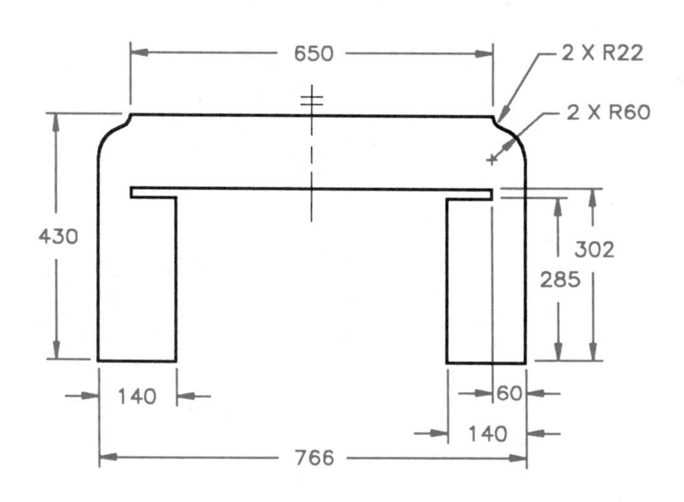

Problem 5–25 Single view (metric)

Part Name: Body Flange Gasket
Material: 0.4-mm Nitrile Rubber
Unspecified Radii: 3 mm
Courtesy Vellumoid, Inc.

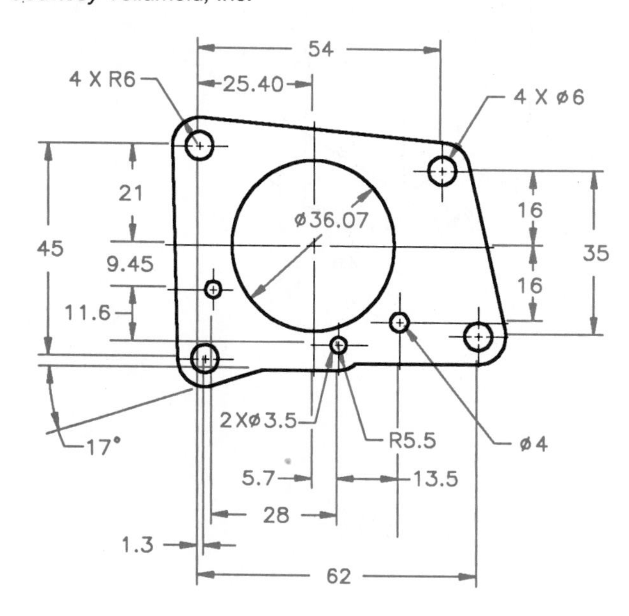

PART 3

DIRECTIONS

1. Determine the front view, then establish the other views, if any, of Problems 5–26 to 5–37. Use manual drafting or CADD to draw complete multiviews for each object as specified by your course objectives.
2. Make a multiview sketch of the selected problems as close to correct proportions as possible. This will help you determine sheet size and view spacing for your final layout.
3. Use your sketch as a guide to draw an original multiview drawing on an adequately sized sheet at an appropriate scale.

Problem 5–26 Multiviews (metric)

Part Name: Guide Rail
Material: SAE 4310

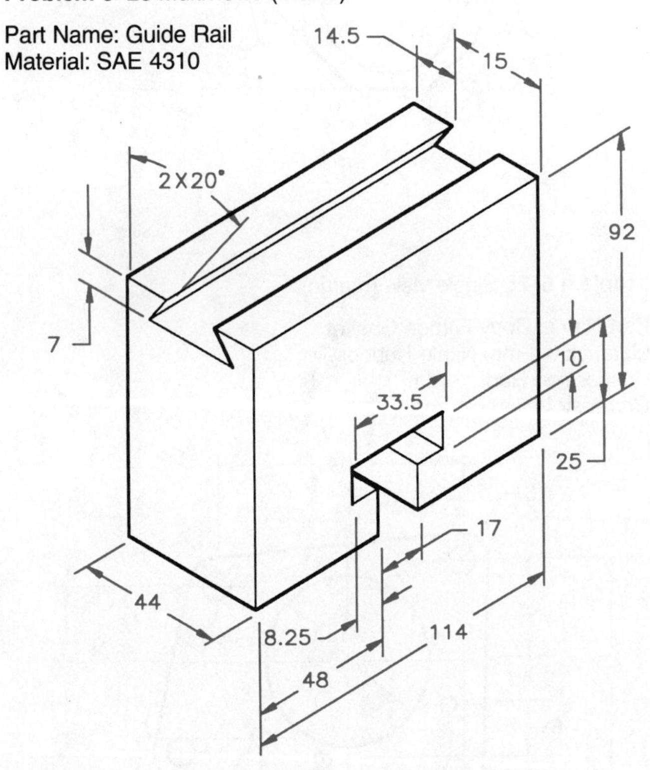

Problem 5–27 Multiviews (in.)

Part Name: Support Bracket
Material: Plastic

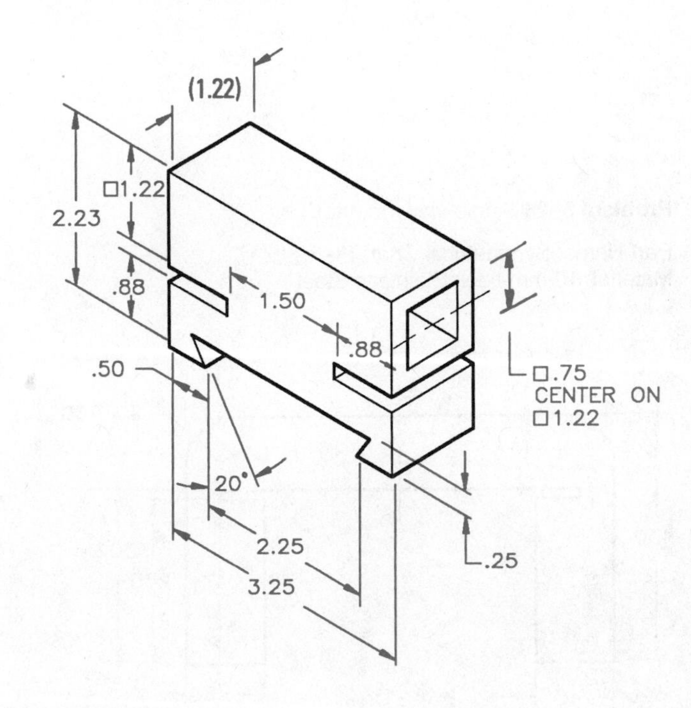

Problem 5–28 Multiviews (in.)

Part Name: V-block
Material: A-Steel

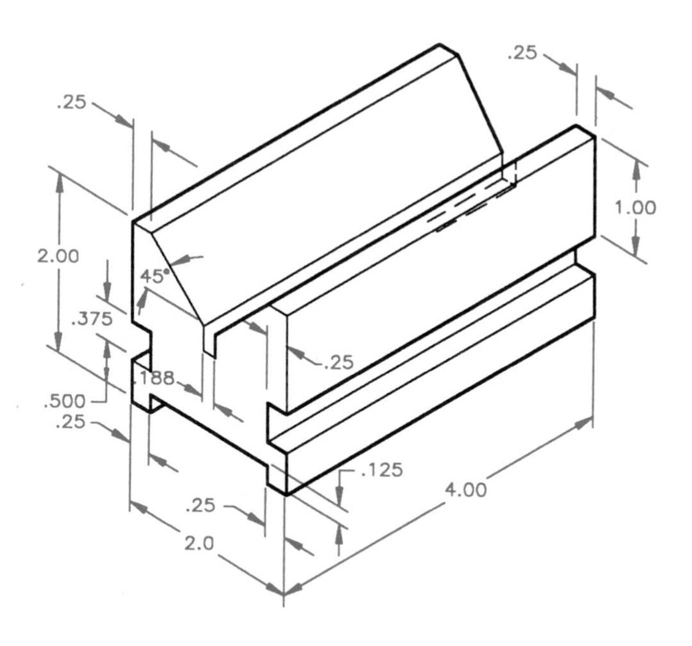

Problem 5–29 Multiviews, angles, and holes (metric)

Part Name: Angle Bracket
Material: Mild Steel

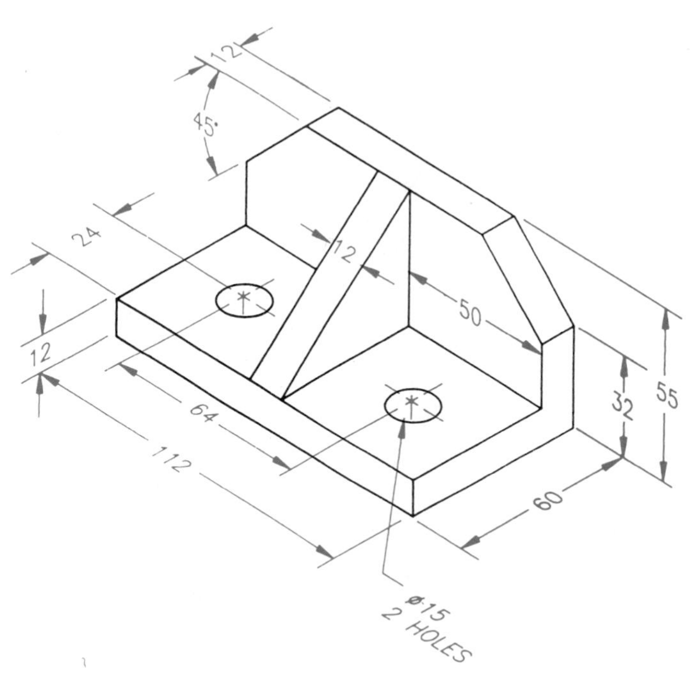

Problem 5–30 Multiviews (in.)

Part Name: Support
Material: Mild Steel

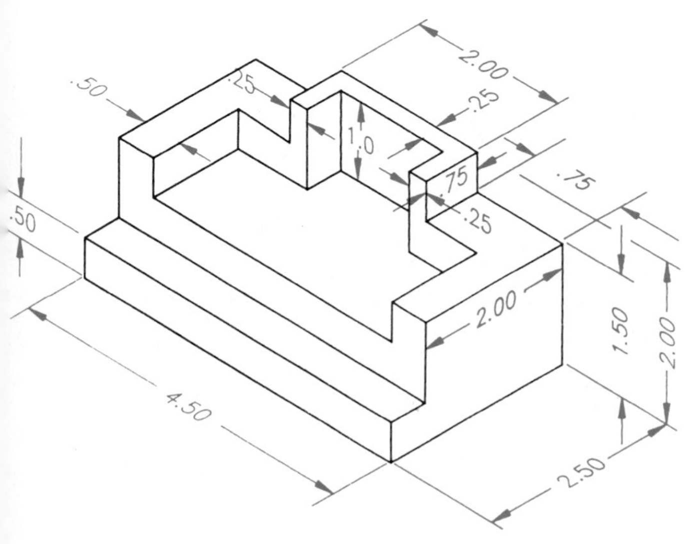

Problem 5–31 Multiviews, arcs, holes, and angles (in.)

Part Name: Journal Bracket
Material: Mild Steel

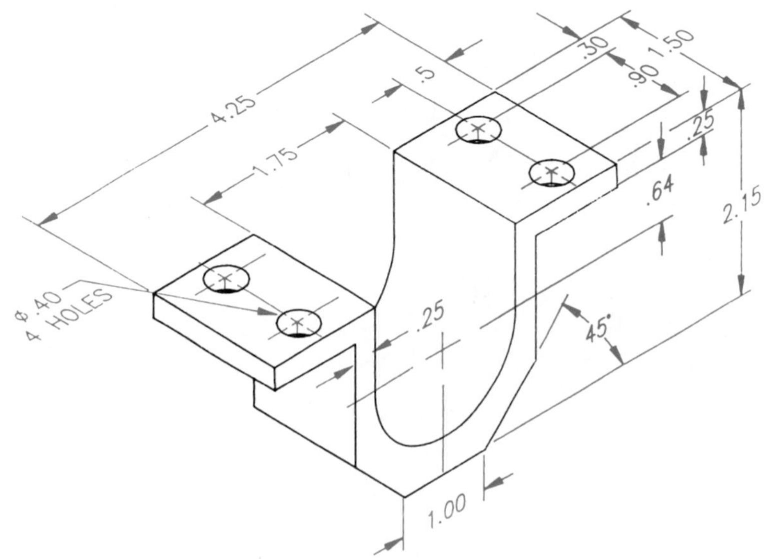

Problem 5–32 Multiviews, circles, and arcs (in.)

Part Name: Chain Link
Material: SAE 4320

Problem 5–33 Multiviews, circles, and arcs (in.)

Part Name: Pivot Bracket
Material: Cold-Rolled Steel

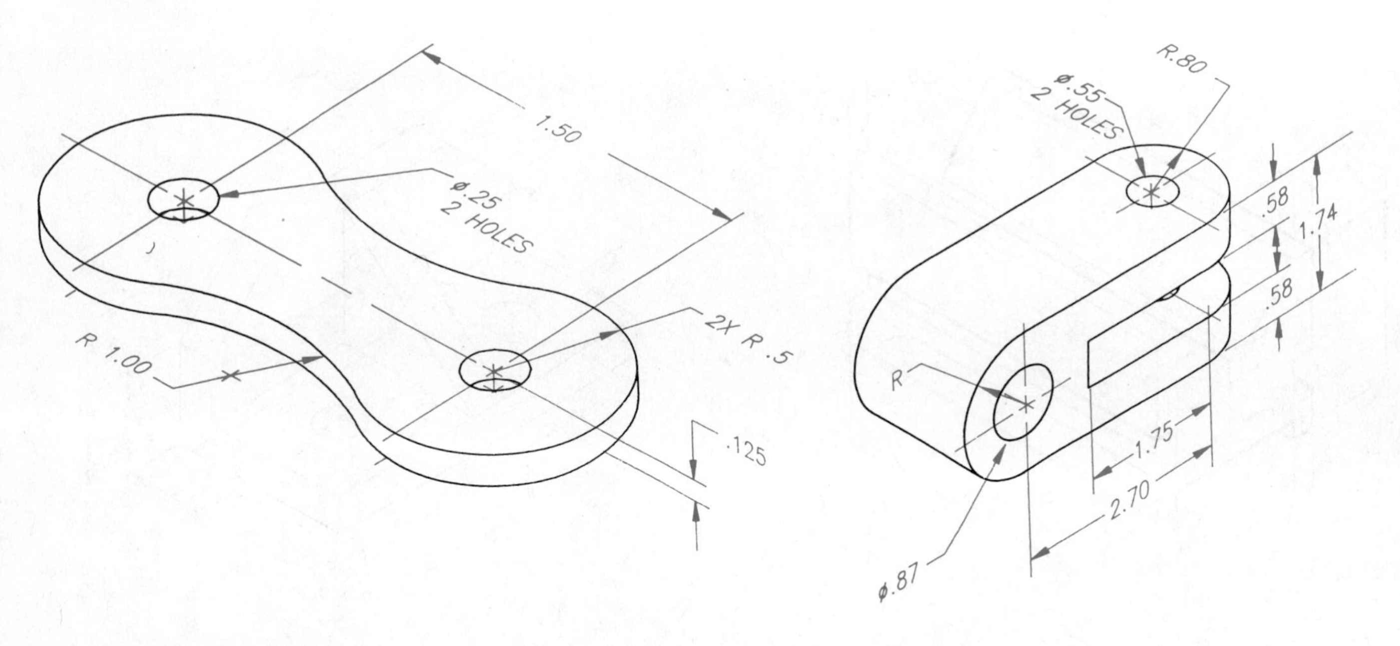

Problem 5–34 Multiview arcs and circles (metric)

Part Name: Hinge Bracket
Material: Cast Aluminum

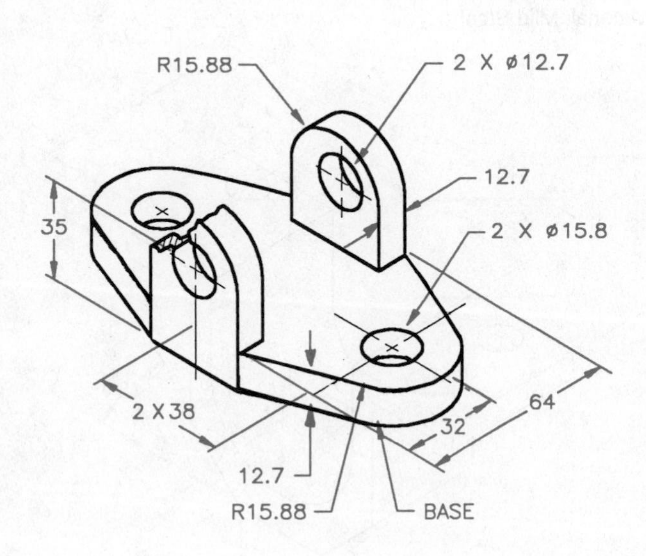

Problem 5–35 Multiview circles and arcs (in.)

Part Name: Bearing Support
Material: SAE 1040

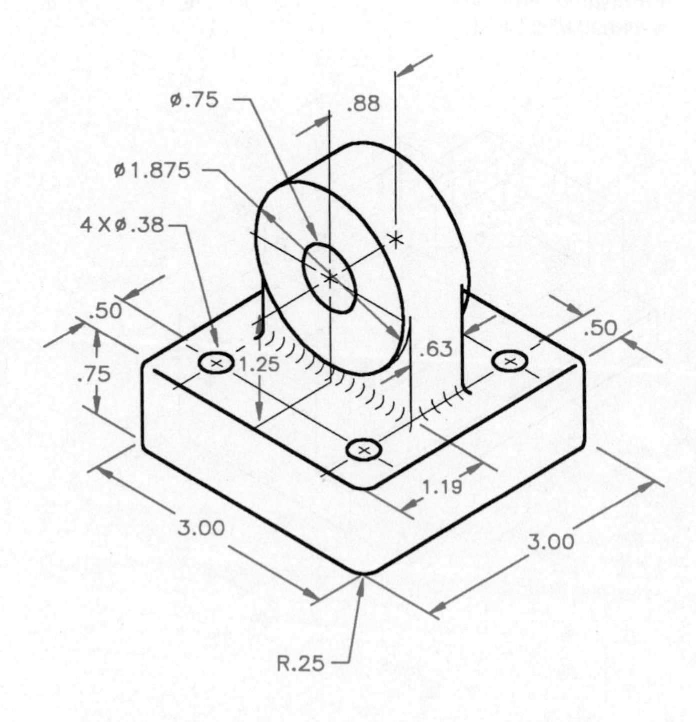

Problem 5–36 Multiviews (in.)

Part Name: Mounting Bracket
Material: SAE 1020
All Fillets R.13

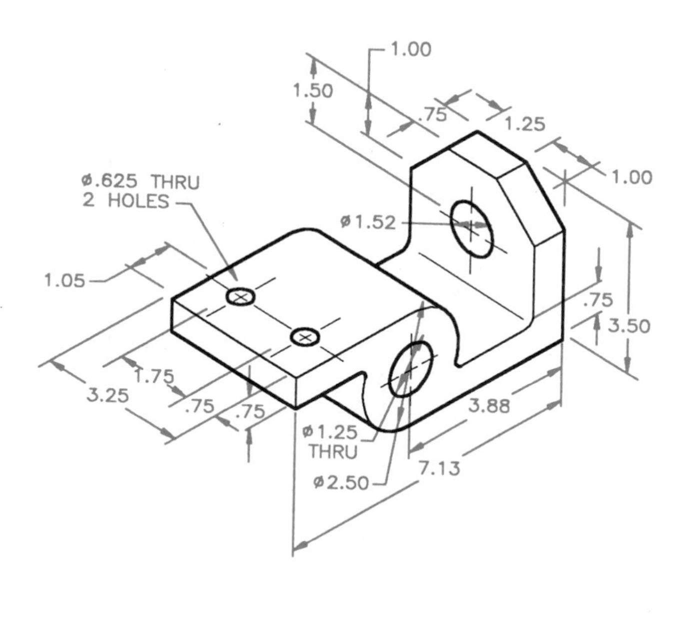

Problem 5–37 Multiviews (in.)

Part Name: Support Base
Material: SAE 1040

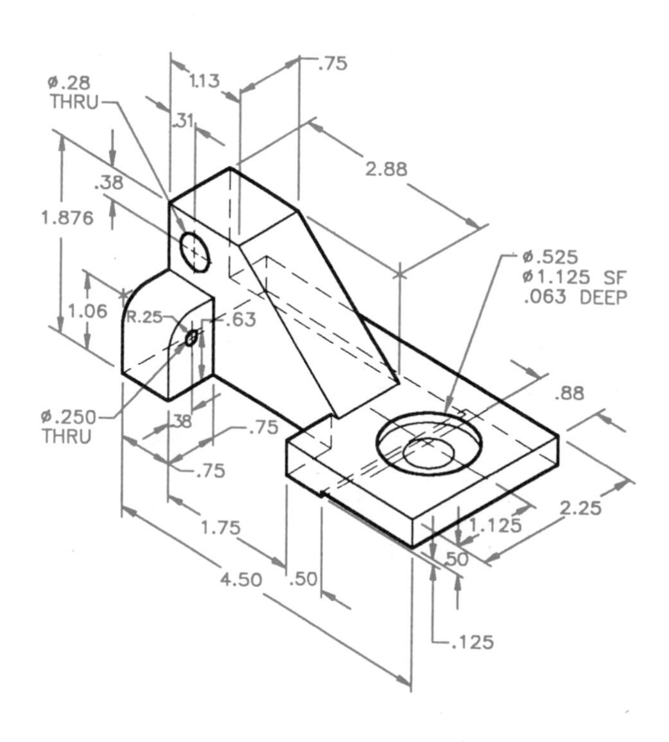

PART 4

DIRECTIONS

Missing lines in auxiliary views. Given the sketches for Problems 5–38 to 5–46, determine the missing lines in the partial auxiliary views. Redraw the objects in multiview, including the correct auxiliary views. Use a scale or dividers to transfer dimensions from the given sketches to your drawing.

Problem 5–38

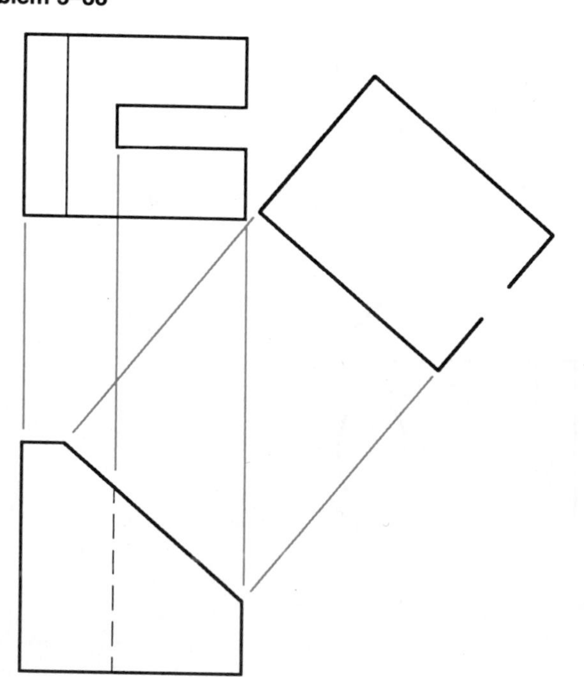

Problem 5–39

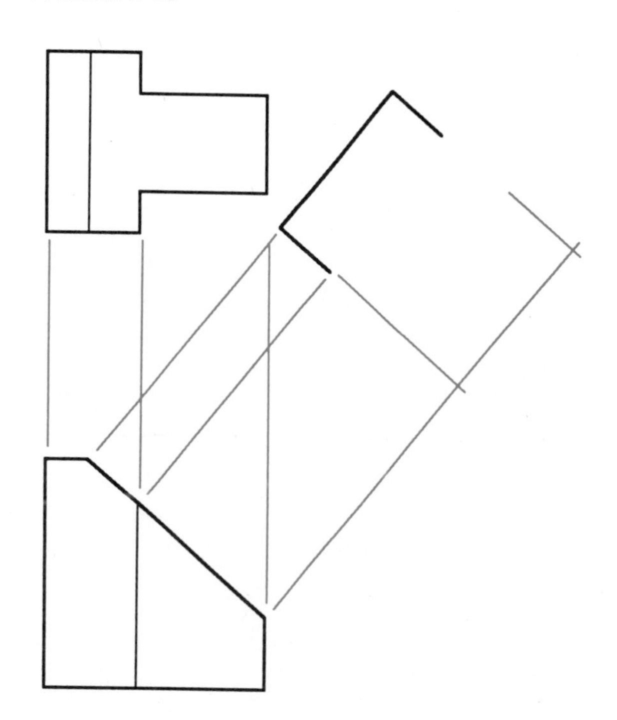

Problem 5–40

Problem 5–41

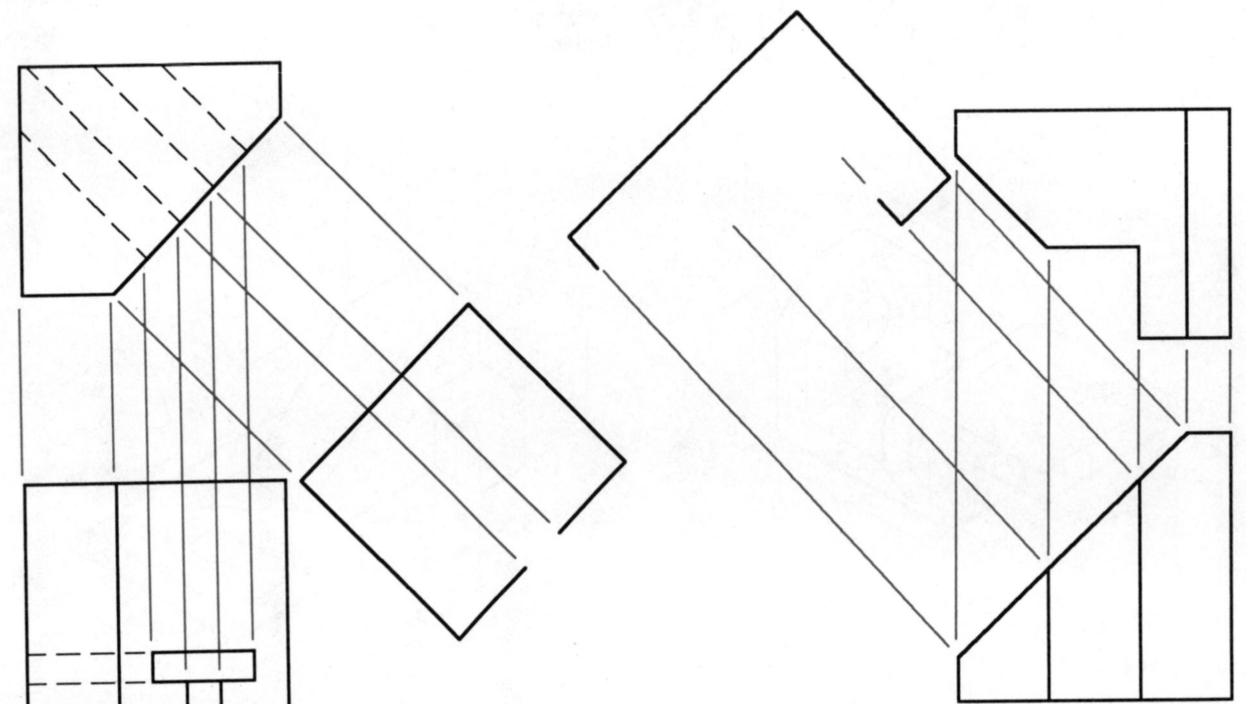

Problem 5–42

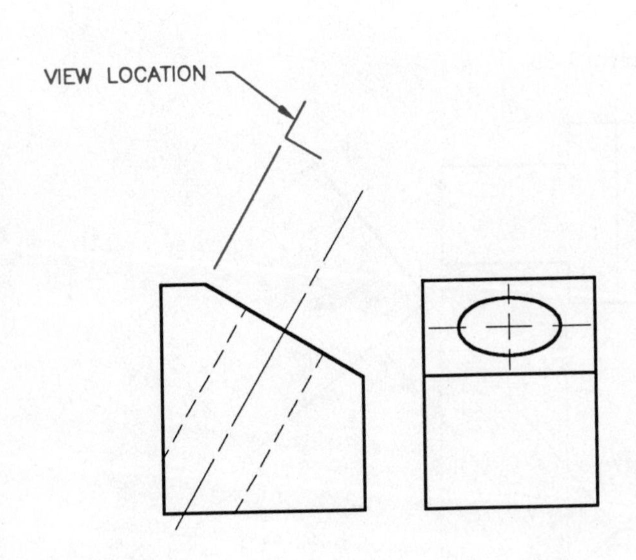

Problem 5–43

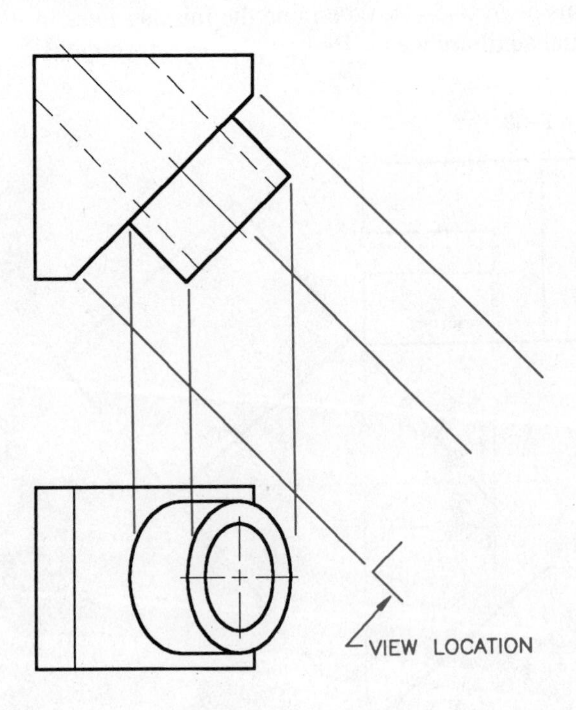

Problem 5–44

Problem 5–45

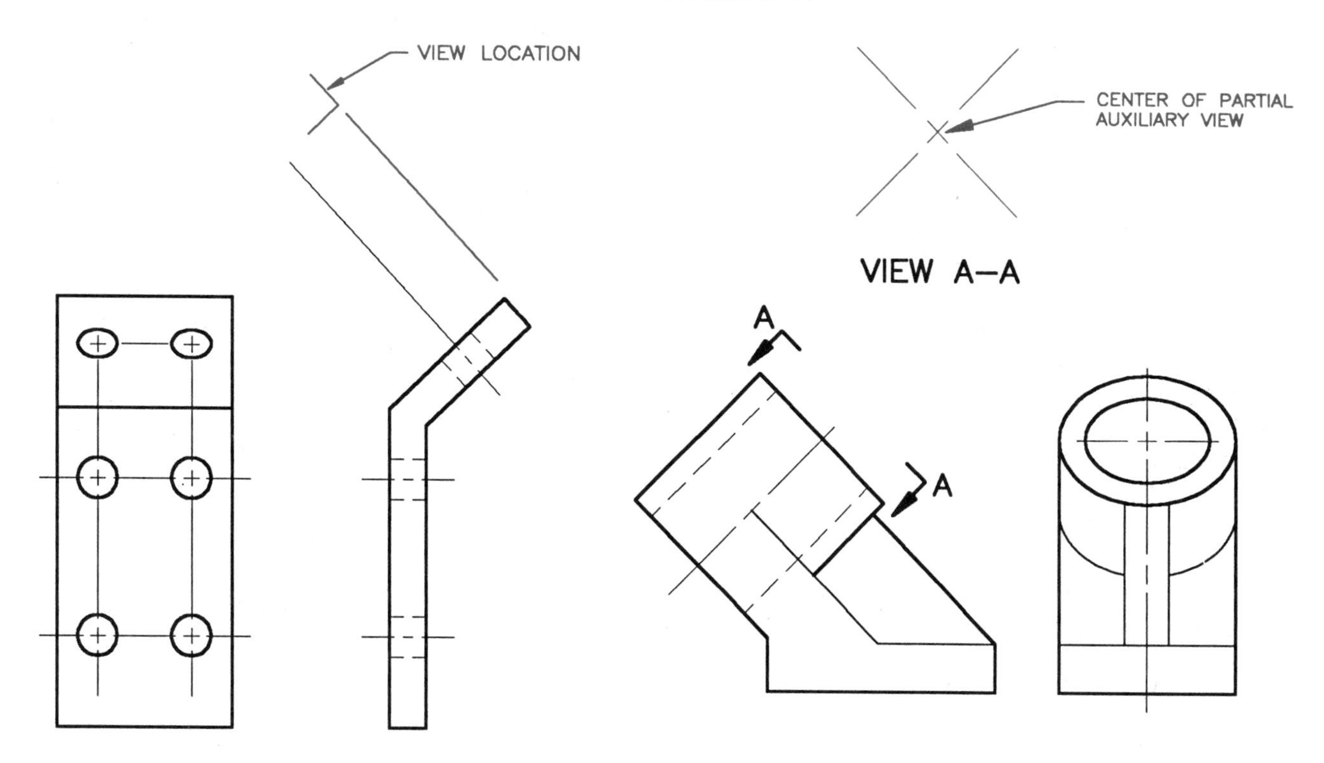

Problem 5–46

Title: Angle V-block (in.)
Material: SAE 4320

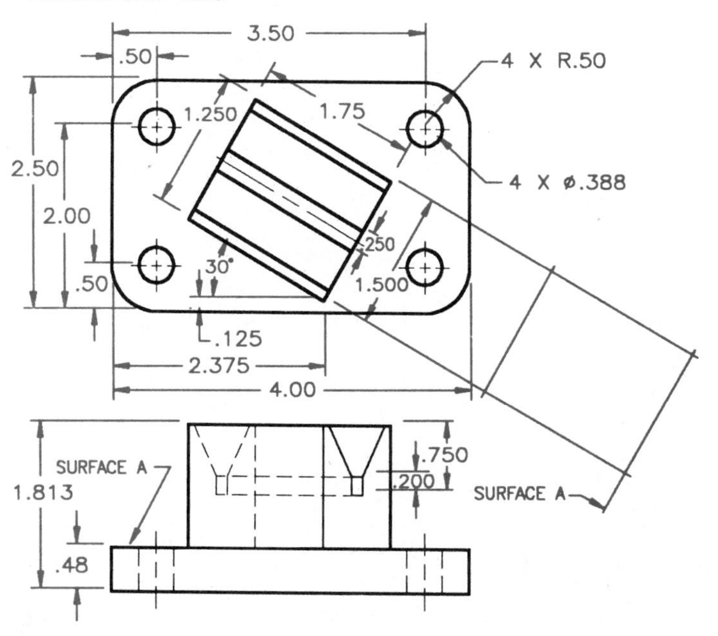

Problem 5–47 Primary auxiliary view (metric)

Part Name: Belt Guide
Material: Aluminum

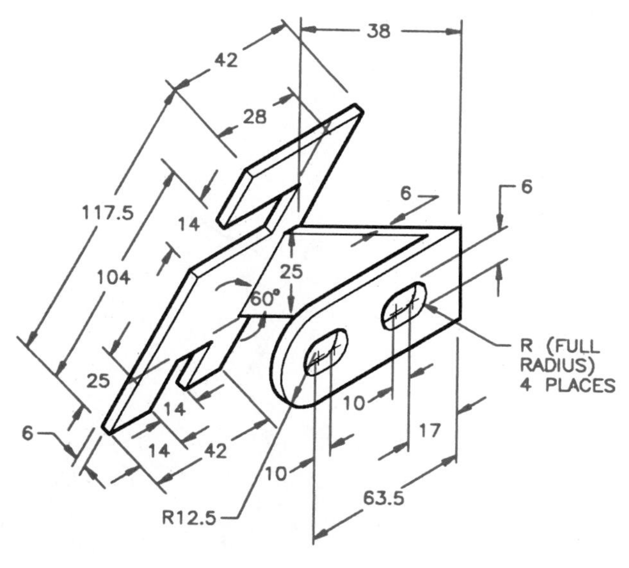

PART 5

DIRECTIONS

1. Make a preliminary sketch of the objects in Problems 5–47 to 5–50. Try to determine which view would be the best front view. Sketch at least one view showing the true shape of the slanted surface(s). Be sure to relate that view correctly to the standard views.
2. Using your sketch as a guide, make a multiview drawing that includes the necessary standard views and the full or partial auxiliary view(s) needed. Use a proper scale on the appropriate size drawing sheet. Use computer-aided drafting if required by your course.

Problem 5–48 Primary auxiliary view (metric)

Part Name: Pivot Link
Material: Aluminum Number 195

Problem 5–49 Secondary auxiliary view (metric)

Part Name: Skewed Face Plate
Material: SAE 1040

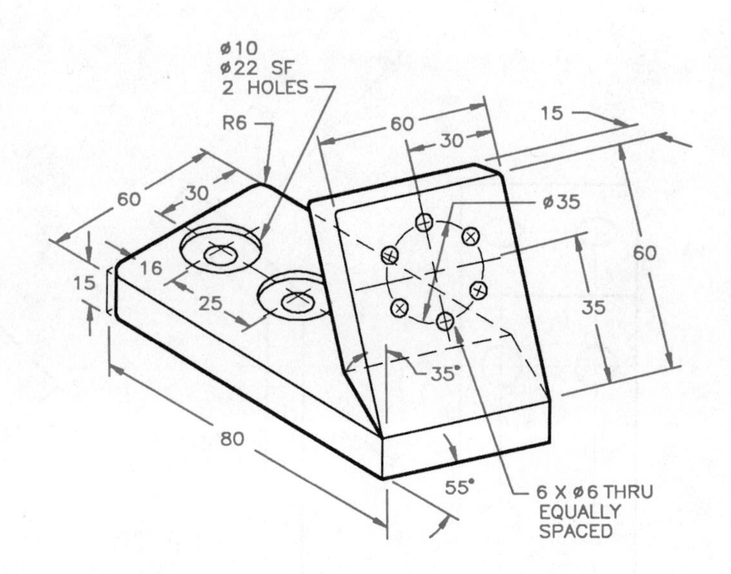

Problem 5–50 Auxiliary views (metric)

Part Name: Chassis Switch Plate
Material: 4-mm-thick .416 Stainless Steel

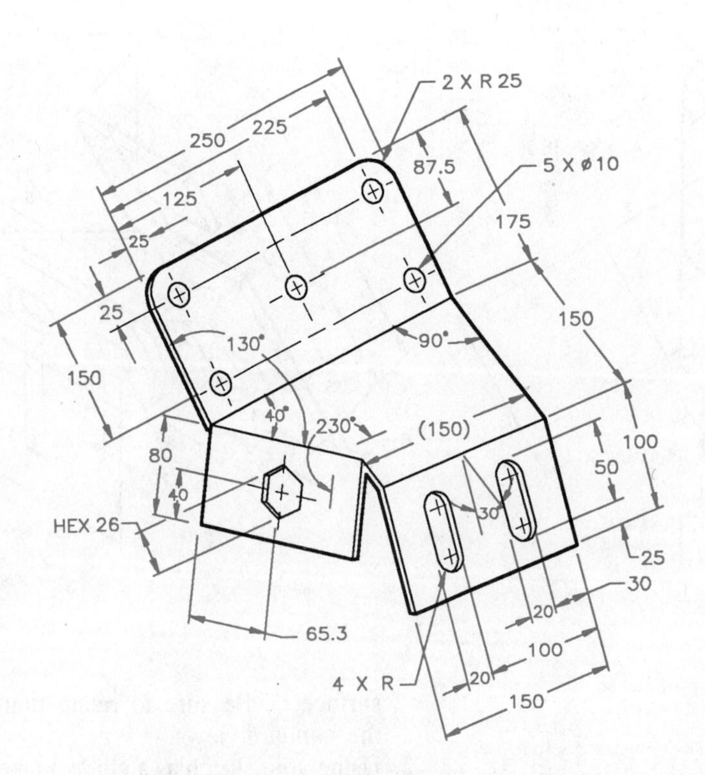

CHAPTER 6

Descriptive Geometry

LEARNING OBJECTIVES

After completing this chapter you will:

- Solve point projection problems.
- Draw required views of lines and find true lengths.
- Find the missing view of a plane and establish the true size and shape.
- Determine the slope of a plane.
- Draw the true size and shape of irregular shapes.
- Evaluate solving engineering problems with problem-solving techniques.

In descriptive geometry, problems are graphically solved by using the relationships among points, lines, planes, and curved surfaces. These relationships can be found by making projections from views of these points, lines, and surfaces. This chapter will use problem-solving principles similar to those explained in Chapter 5, but will also discuss some additional concepts. Descriptive geometry principles are most valuable for determining the angle between two lines, two planes, or a line and a plane, or for locating the intersection between two planes, a cone and a plane, or two cylinders.

The solution of descriptive geometry problems is entirely graphical, therefore the accuracy of the answer depends on the accuracy used in making the solution, and, sometimes, the method of solution. For example, the degree of accuracy in locating the intersection of two lines depends on the angle between the lines. The degree of accuracy increases as the angle approaches 90°, as shown in Figure 6–1. Often when the intersecting lines are nearly parallel, another method of avoiding this difficulty can be found.

Points on a line should be as far apart as possible when the points are used to determine the direction of the line, as shown in Figure 6–2. Always use the method that produces the most accuracy.

One good way to learn the techniques of descriptive geometry is to study how to manipulate a point within

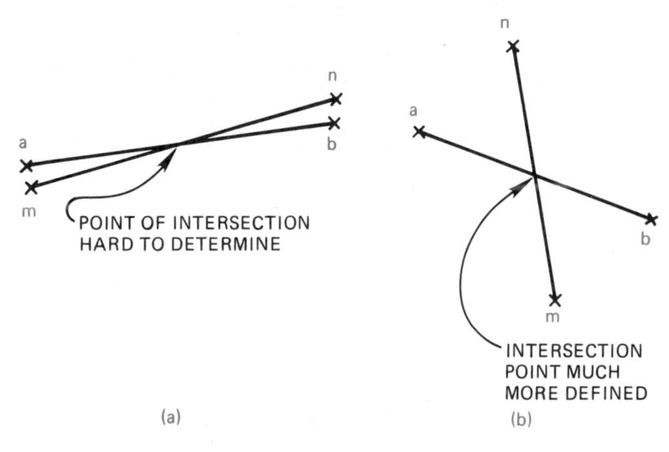

Figure 6–1 Accuracy of intersecting lines.

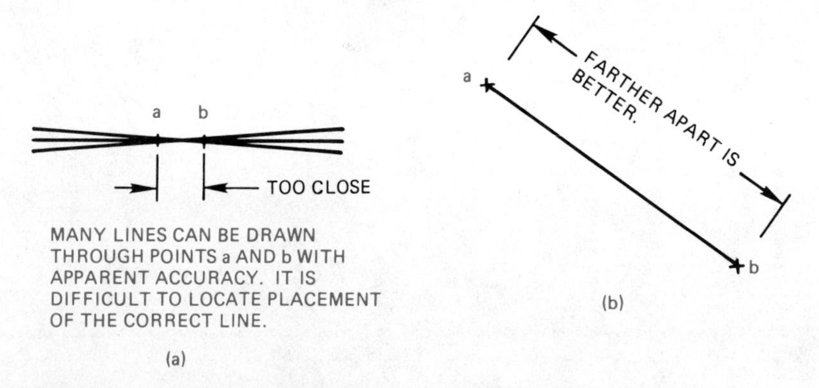

Figure 6–2 Determining the direction of a line.

the multiview box theory. Add to that the projection of straight lines (two points can define a line), the projection of planes (lines can define planes), and developing planes (surfaces combine to form an object), and you will have mastered the basics of descriptive geometry.

In addition to the practical applications of descriptive geometry, these concepts provide valuable training in the skills of visualization of points, lines, and surfaces in space. If visualization is difficult, then memorizing the basic steps in solving descriptive geometry may be necessary.

There are several ways to solve descriptive geometry problems. An attempt has been made to include at least one good method and, in some cases, the two most understandable solutions for each situation.

STANDARDS OF PROJECTION

The following elements are illustrated in Figure 6–3, which is a pictorial drawing of some basic descriptive geometry terminology. Figure 6–4 illustrates the same multiview box of planes unfolded.

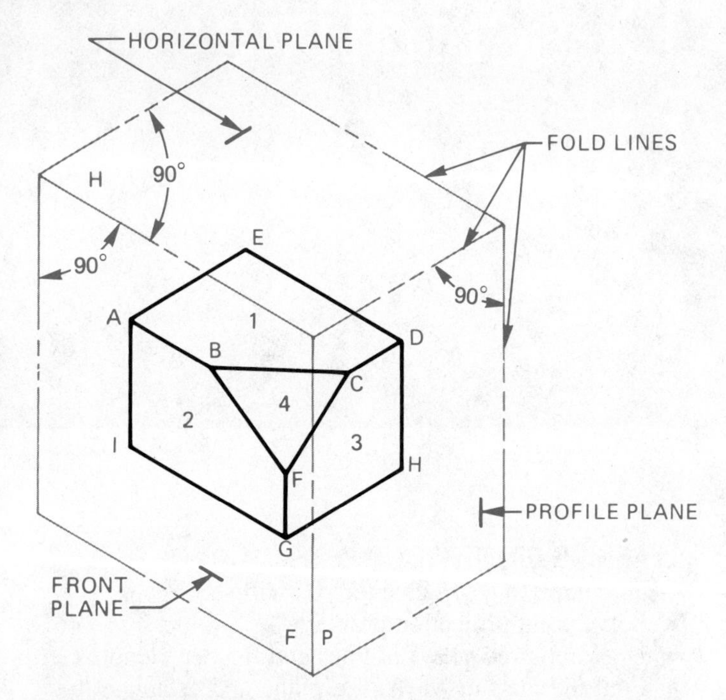

Figure 6–3 Basic descriptive geometry terminology illustrated on the multiview box.

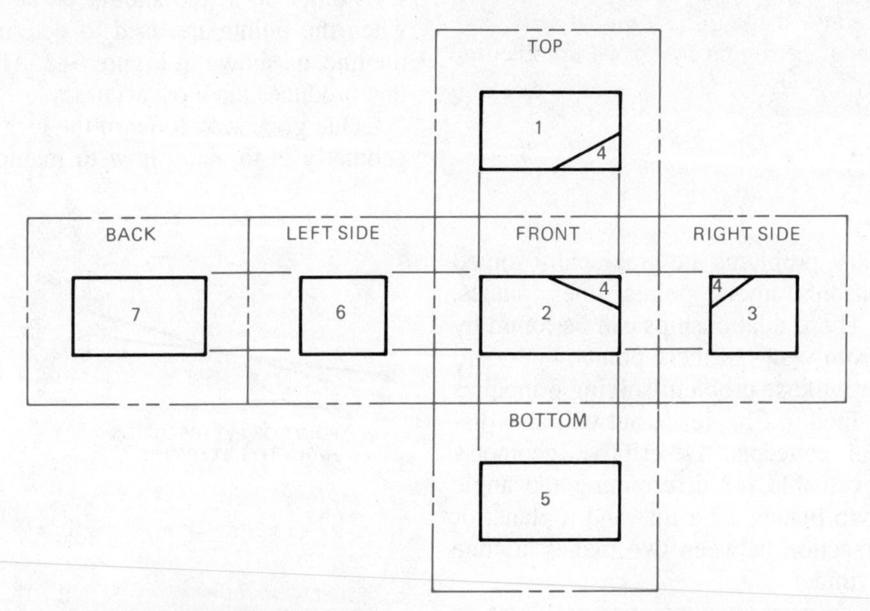

Figure 6–4 Unfolded multiview box.

Definitions

1. *Line of sight* is an imaginary straight line from the eye of the observer to a point on the object. All lines of sight for a particular view are assumed to be parallel, which means that the observer is either an infinite distance away, or the observer occupies a slightly different position when looking at each point on the object.
2. A *projection plane* is an imaginary surface on which the view of the object is projected and drawn. This surface is imagined to exist between the object and the observer. The observer's line of sight is *always* perpendicular to the projection plane.
3. A *fold line* is the line of intersection between two projection planes. A fold line appears as an edge view of a projection plane when that projection plane is folded back 90° from the plane being viewed. It is the line on which one plane is folded to bring it into the plane of the adjacent (touching) projection plane.
4. *Projection lines* (projectors) are straight lines, drawn perpendicular (at 90°) to the fold lines, that connect the projection of a point in a view to the projection of the same point in an adjacent view. These lines are required for obtaining the point locations in the adjacent view(s).
5. A *front projection plane* shows the projected size of the front of an object. The lines of sight are horizontal and the projection plane is vertical in an elevation view.
6. A *profile projection plane* shows the projected size of the vertical side of an object. The lines of sight are horizontal and the projection plane is vertical in an elevation view.
7. A *horizontal,* or a *top projection, plane* shows the projected size of the top of an object. The lines of sight are vertical and the projection plane is level with the horizon in an elevation view.
8. *Adjacent views* are views that are next to each other. Their projection planes are perpendicular to each other in space. They have a common fold line between them. Two views of any point lie on the same projection line, which is 90° to the fold line between the two adjacent views. Given any two adjacent views of any object, you can find all the dimensions of that object.

PROJECTION OF A POINT

Figure 6–5 shows a point that has been projected onto three mutually perpendicular projection planes. The actual point itself is shown by a small six-pointed cross. Any method can be used to identify a point; the idea is to

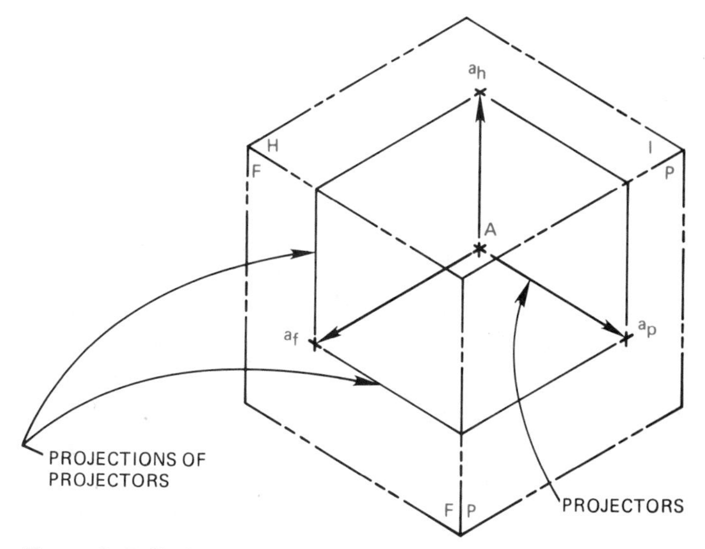

Figure 6–5 Projection of a point.

be consistent. In this chapter, the real point is identified by a capital letter. The letter A was chosen at random in Figure 6–5. Everywhere point A is projected, a cross and a lower-case letter with an attached subscript lower-case letter (a_f) is used. The subscript lower-case letter corresponds to the plane onto which the point was projected. For example, a_f is the projection of point A on the front projection plane. In Figure 6–6(b), the fold lines are identified by two upper-case letters corresponding to the two adjacent projection planes. Right angle symbols have been used in this example to remind you that the projectors of point A project to the projection planes at 90° angles to the plane — always!

Given the horizontal and front views of point A, Figure 6–6(b) shows the sequence used in locating the correct position in the profile view.

Step 1 Draw a line from a_h to a_f, then draw a projector from a_f perpendicular through fold line F–P. Measure the distance, D, from point a_h to the fold line H–F.

Step 2 Dividers can be used to transfer this distance, D, from fold line F–P along the projector of a_f into the profile view. Identify the point with a cross and the letters a_p.

In Figure 6–7, the same method is used to locate a point in an auxiliary view. In this example, the location of the fold line H–A was chosen arbitrarily, but later on, the exact placement of the auxiliary fold line H–A will be important when working with lines and planes. To project a point through successive auxiliary views, use the fold lines for reference measurements as shown in Figure 6–8. The distance D is determined first, then distance X is established. Notice that in each situation, a view has been "skipped" to determine the correct distances.

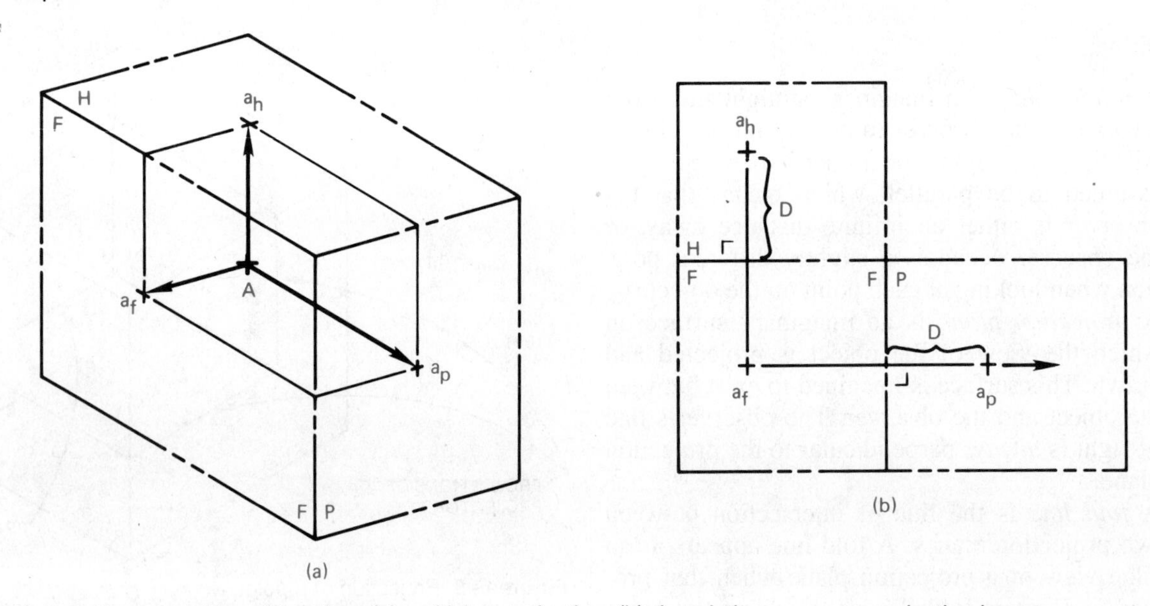

Figure 6–6 Locating a point in a third view: (a) multiview projection; (b) descriptive geometry projection layout.

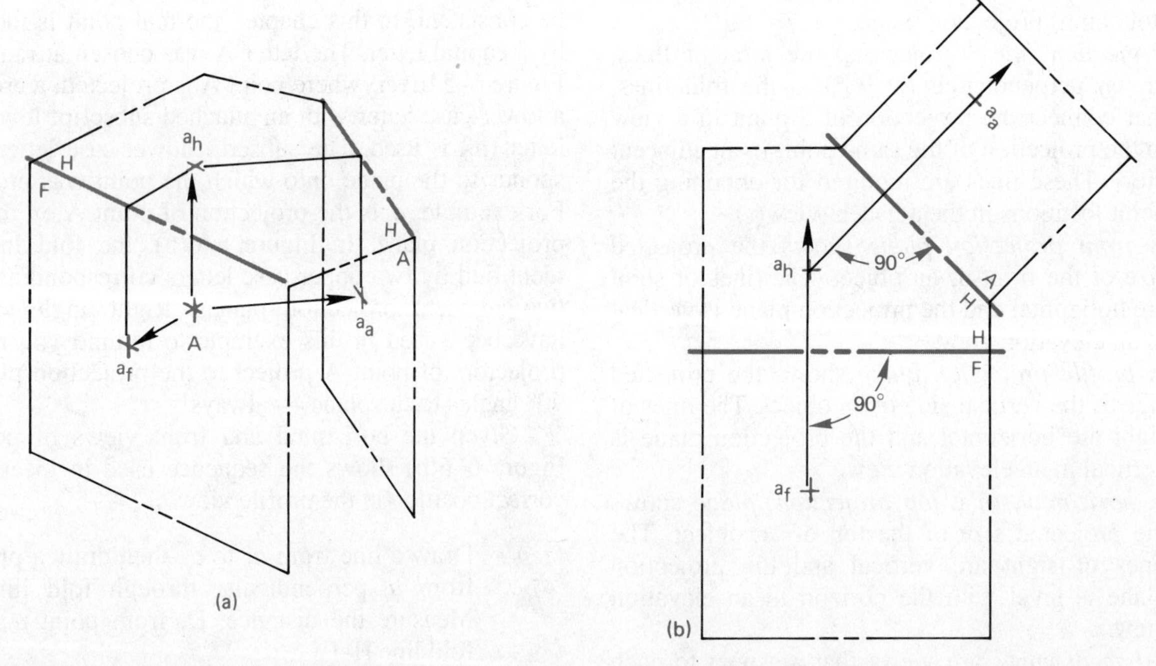

Figure 6–7 Locating a point in an auxiliary view: (a) multiview projection; (b) descriptive geometry projection layout.

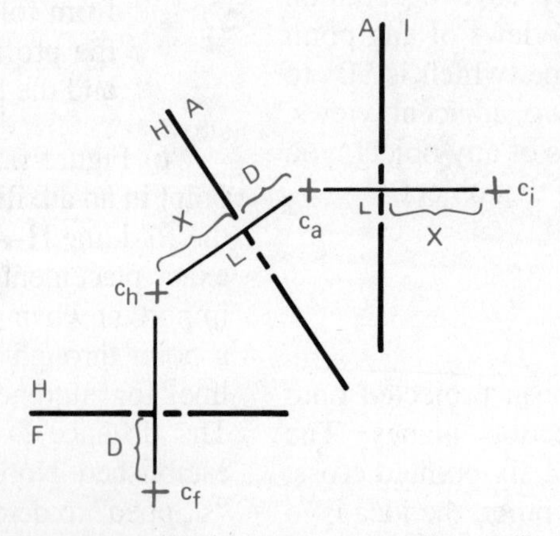

Figure 6–8 Projection of a point into successive auxiliary views.

DESCRIPTIVE GEOMETRY

So far in your study you have learned that descriptive geometry is the graphical analysis of points, lines, or planes in space represented by the orthographic projection on two planes at right angles to each other. This is a technical way of saying that you are drawing the orthographic or multiviews of the points, lines, or planes in the engineering problem. One aspect of this technology that has been stressed in this chapter is accuracy. For manual drafters, accuracy depends on a sharp pencil and a keen eye for detail and exactness. Even the best manual solutions have a degree of error due to line widths or the slight tilt of the pencil to paper. Descriptive geometry solutions with CADD are, on the other hand, nearly perfect. The reason for this perfection has to do with the capabilities of the hardware and software, and the fact that you are actually displaying the mathematical counterpart of the graphic analysis. In other words, the points, lines, or planes are a representation of the mathematical coordinates of the problem being presented. Therefore, the solution's accuracy is only controlled by the accuracy potential of the computer. A CADD system also allows you to work in layers and colors for a clearer picture of the problem and solution. For example, you can set a layer for projection lines where these lines show in red. This may help you visualize the entire problem and keep your solution clearer.

LINES IN SPACE

Since a line can be represented by two points that are connected, the two end points of a line can be projected on their respective projectors into each projection plane to form the view of that line. Figures 6–9 and 6–10 show two different lines projected in the multiview box concept and their respective descriptive geometry layout.

CLASSIFICATION OF LINES

In general, lines are classified or named according to the plane or planes of projection to which they are parallel. These classifications include *horizontal*, *front*, and *profile lines*, as well as *vertical* lines, which are parallel to both the front and profile planes. The lines will appear in true length on the planes of projection to which they are parallel. *Oblique* lines are parallel to none of the projection planes. Figures 6–11 through 6–14 show how these lines are represented in multiview boxes and their respective descriptive geometry layouts.

Parallel Lines

When lines are parallel, they are an equal distance from each other throughout their length. Figure 6–15

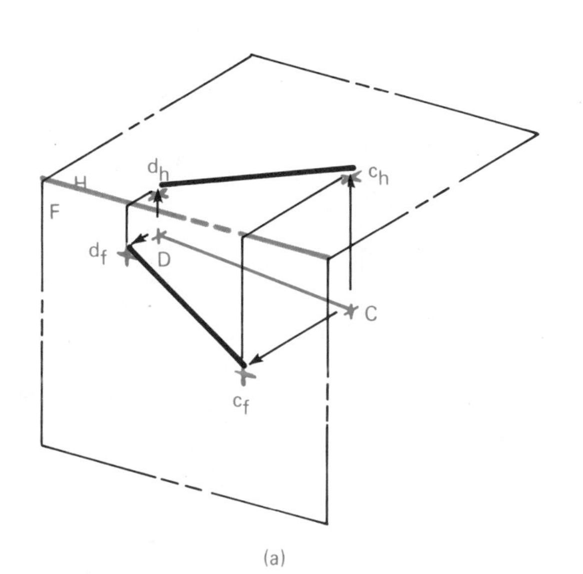

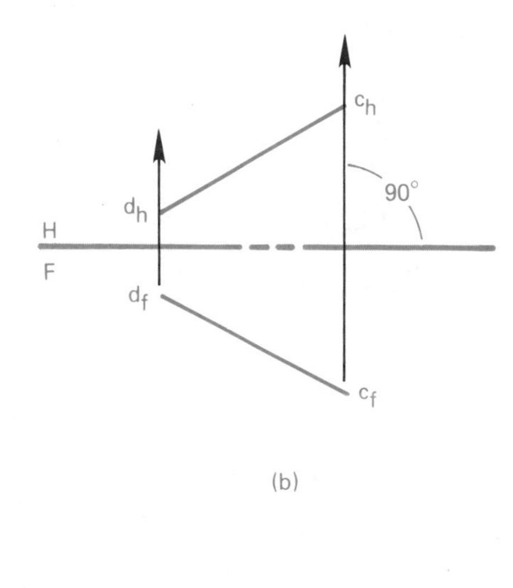

Figure 6–9 Projection of line in two views: (a) multiview projection; (b) descriptive geometry layout.

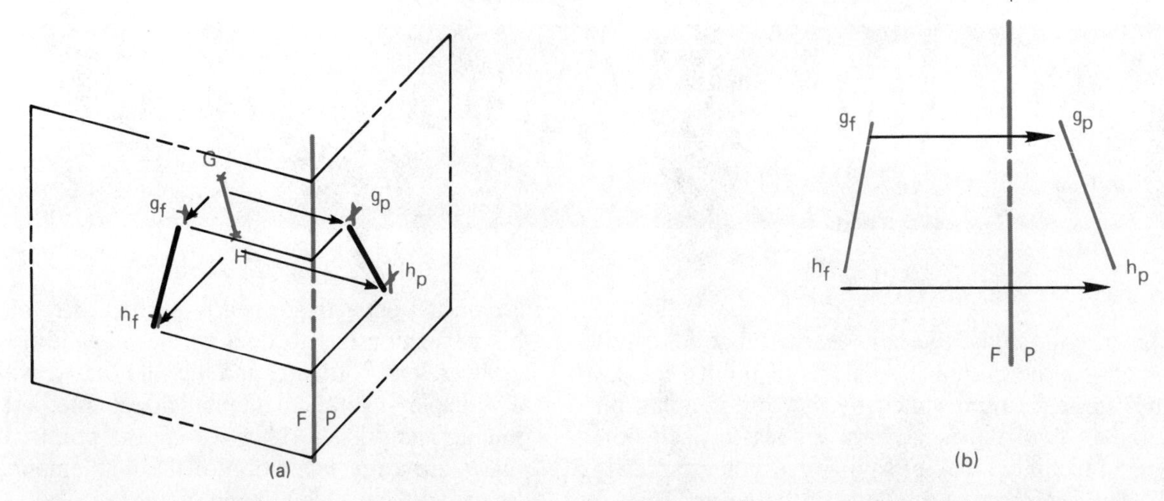

Figure 6–10 Projection of a line in two views: (a) multiview projection; (b) descriptive geometry layout.

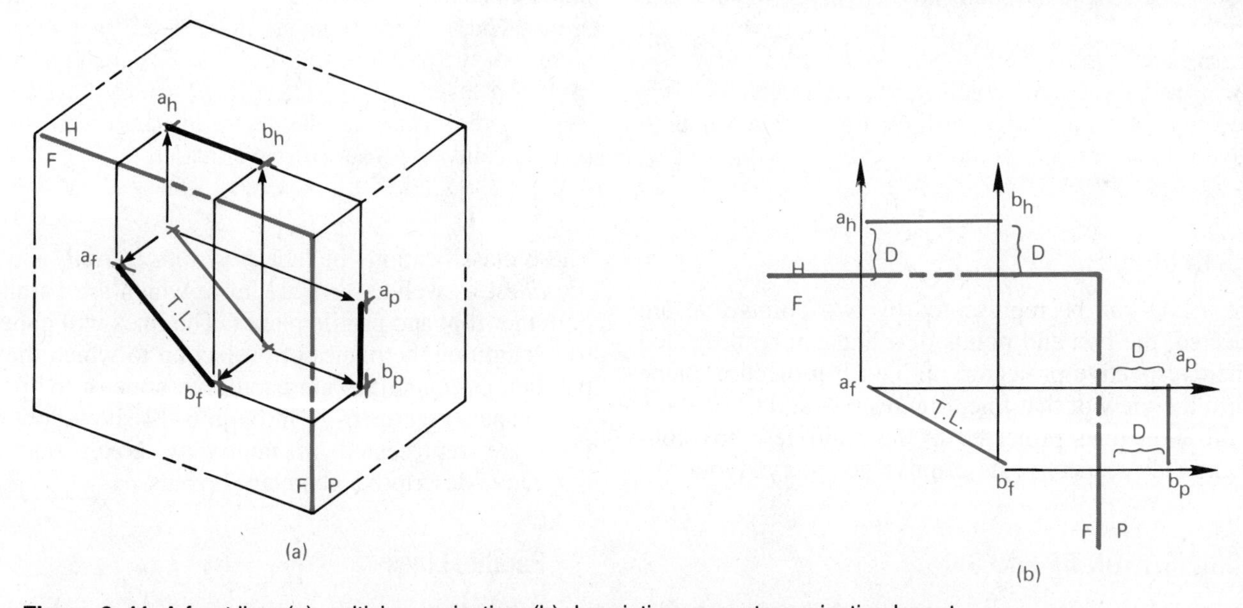

Figure 6–11 A front line: (a) multiview projection; (b) descriptive geometry projection layout.

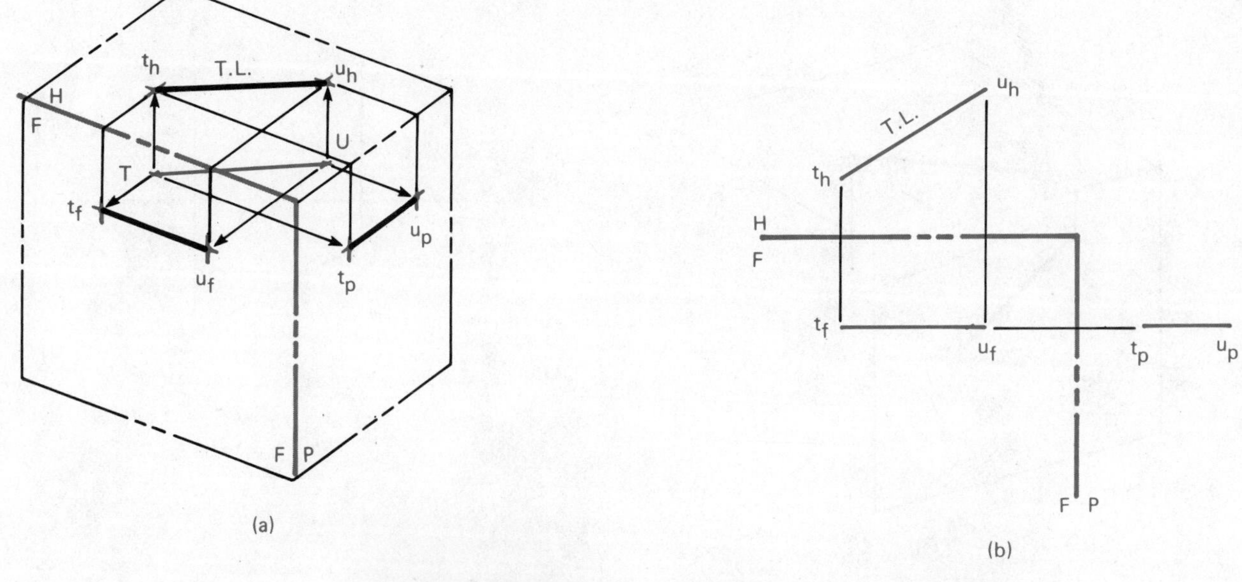

Figure 6–12 A horizontal line: (a) multiview projection; (b) descriptive geometry projection layout.

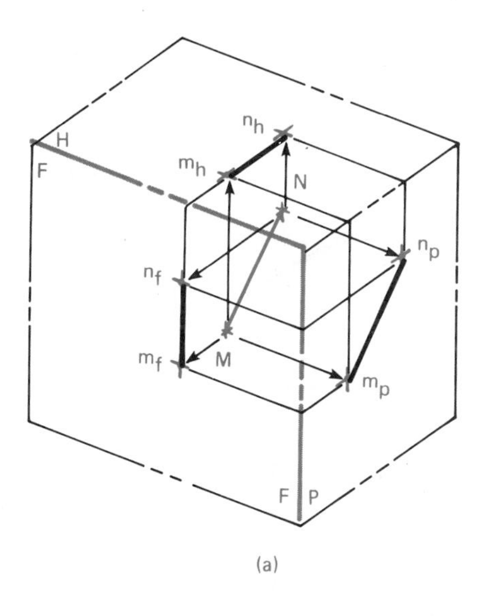

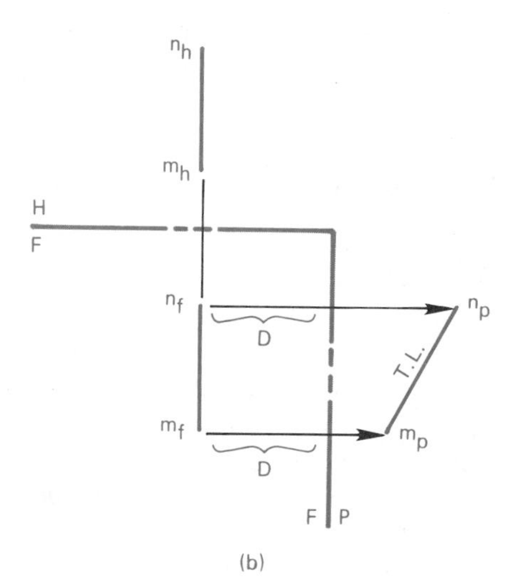

Figure 6–13 A profile view: (a) multiview projection; (b) descriptive geometry projection layout.

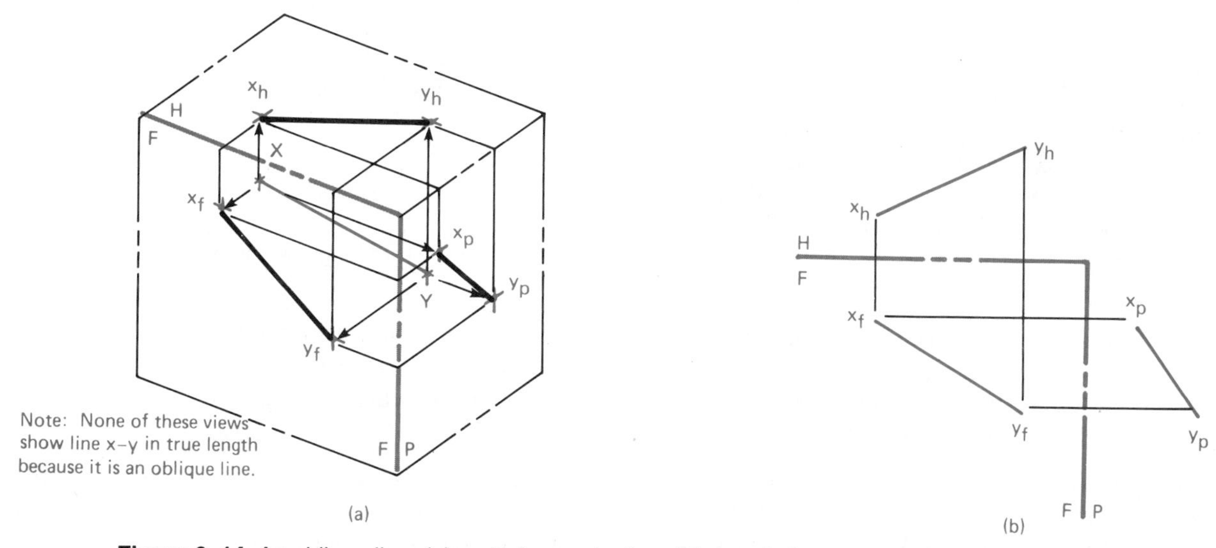

Figure 6–14 An oblique line: (a) multiview projection; (b) descriptive geometry projection layout.

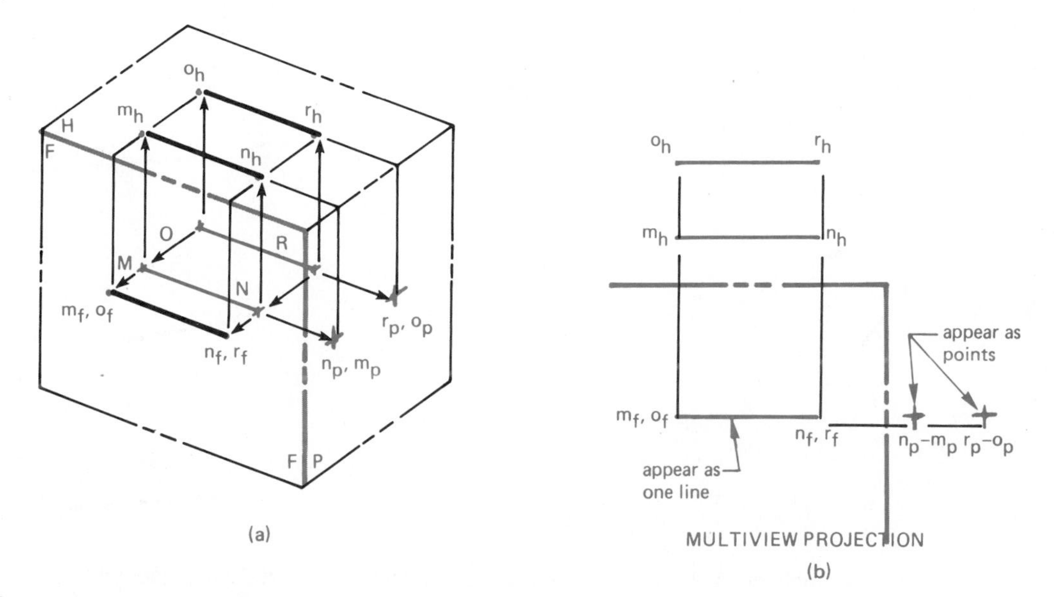

Figure 6–15 Parallel lines: (a) multiview projection; (b) descriptive geometry projection layout.

shows that the parallel lines M–N and O–R will appear parallel in all projected views with only two exceptions. They will not appear parallel when one line is behind the other as shown in the front view, where they appear as one line. And they will not appear parallel when they are seen in views showing the ends of the lines. In this example, they appear as points in the profile view.

True-Length Lines

The *true length* of a line is the actual straight-line distance between its two ends. If a line appears shorter than it really is, it is called foreshortened. The true length of any line is found only on a plane of projection parallel to that line. In Figure 6–16, line A–B is parallel to the front projection plane and shows true length in the front view. Likewise in Figure 6–17, line C–D is parallel to the profile projection plane and shows in true length in the profile view. Note the label, T.L., on the projected true-length line.

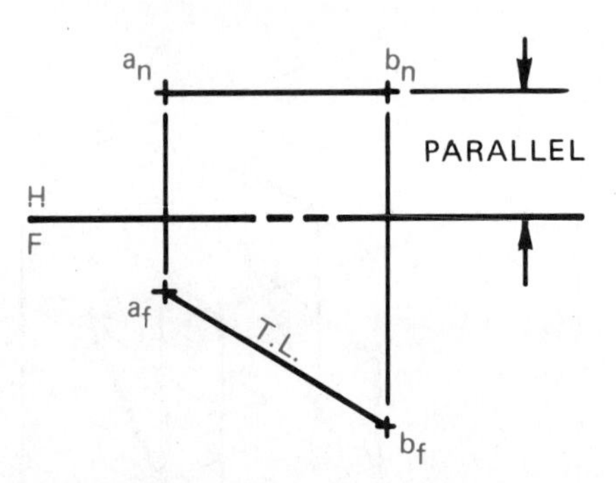

Figure 6–16 True-length line in front view.

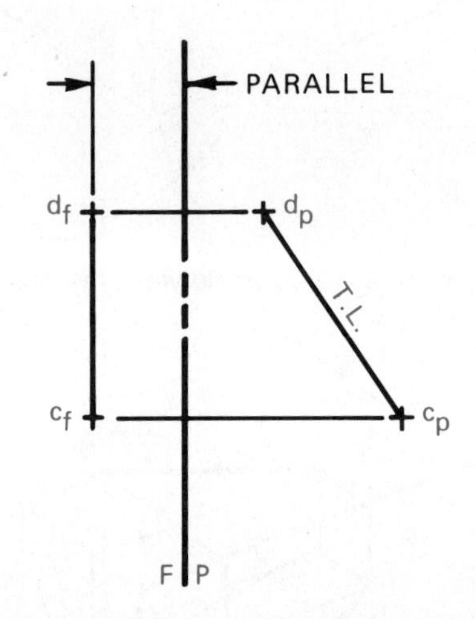

Figure 6–17 True-length line in profile view.

Finding the True Length of an Oblique Line

When a line appears foreshortened in all of the given projected views, there are two good methods used to find its true length. One is for the viewer to move around the line until the line appears parallel to the viewer's line of sight, as shown in Figure 6–18. To do this graphically, follow these steps:

Step 1 Create an auxiliary view of the line by placing the fold line H–A or F–A of the auxiliary view parallel to either of the original views of the line S–T, as shown in Figure 6–19(a) or 6–19(b).

Step 2 Project the end points S and T of the line S–T into the auxiliary view H–A or F–A with the correct distance measurements X and Y taken from the fold line H–F to points s_f and t_f, as shown in Figure 6–20(a). Figure 6–20(b) shows the correct distance measurements E and D for the new projection of the line s_a–t_a from the fold line H–F to points s_h and t_h.

Step 3 Transfer the distances found for each point from the fold line H–A or F–A along their respective projectors into auxiliary view A. If drawing accuracy is good, the line s_a–t_a will be in true length in the auxiliary view, as shown in Figure 6–20(a) or 6–20(b).

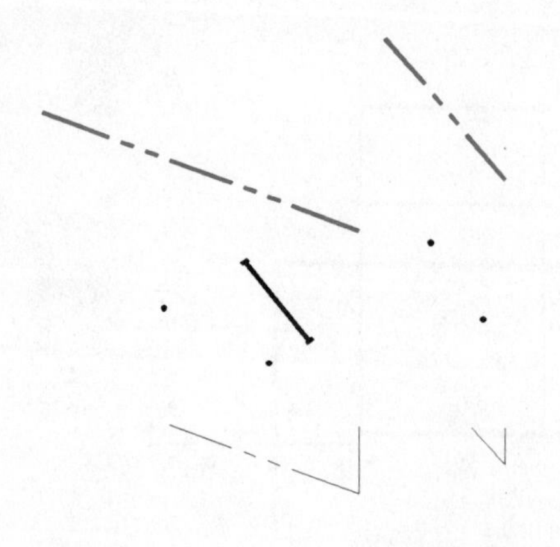

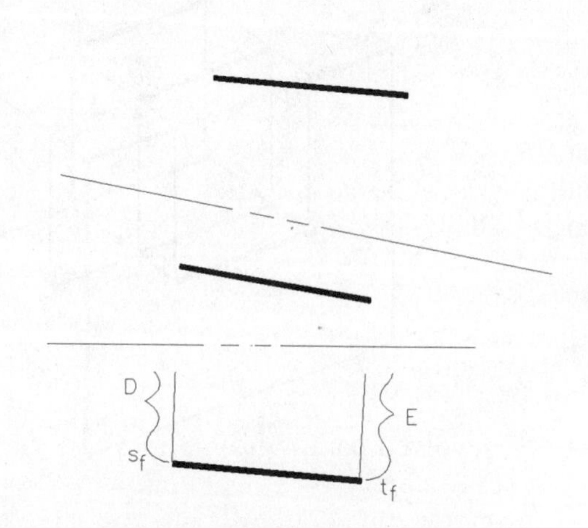

Figure 6–18 Finding the true length of an oblique line: (a) multiview projection; (b) descriptive geometry projection layout.

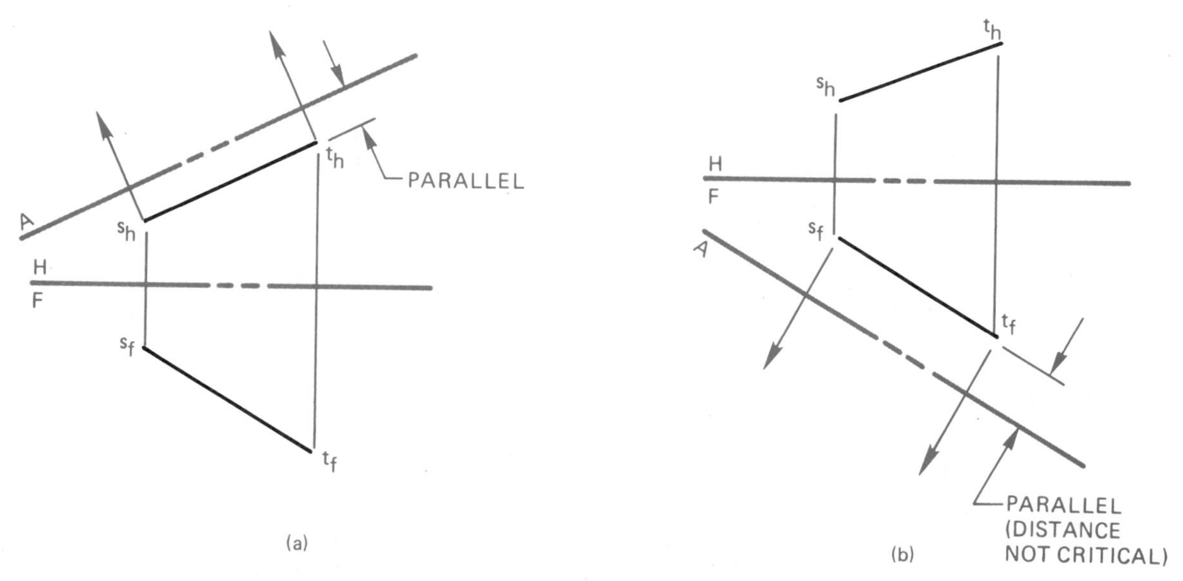

Figure 6–19 First step in finding the true length on an oblique line: (a) projecting from the top view; (b) projecting from the front view.

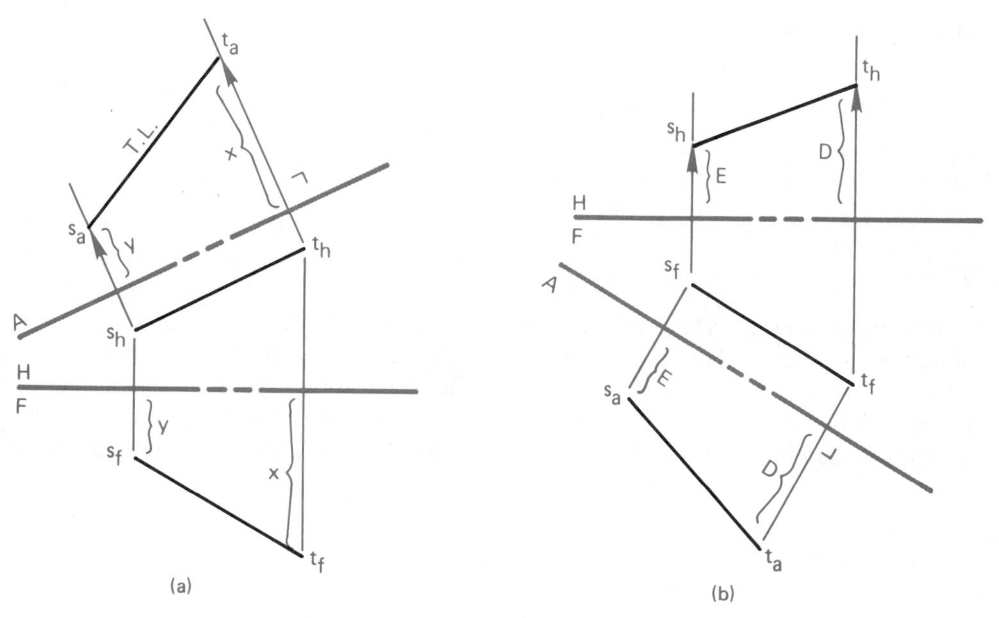

Figure 6–20 Final steps in finding the true length of an oblique line: (a) projecting from the top view; (b) projecting from the front view.

A second method, which is sometimes more accurate, would be to revolve the line until it is parallel to one of the existing projection planes. In this case, the line is moved, not the viewer. The true length of the line is found without the need for an auxiliary view. Figure 6–21(a) shows how to graphically find the true length of a line by this method.

Revolve one end of the line C–D. Point D is revolved parallel to the fold line H–F using the other end of the line, point C, as the axis. This can be done in either the top or front view. The top view was chosen in Figure 6–21(a), revolving point d_h around point c_h. Two new points d_{h1} and d_{h2} are possible and either one can be used.

Step 2 Project the new point d_{h1} perpendicular to the fold line H-F and beyond (down) at least the distance of the original point d_f into the front view F as shown in Figure 6–21(b).

Step 3 In the front view F, project the point d_f *parallel* to the fold line H–F until it crosses or touches the d_{h1} point projected from the top view. This intersection will create the new point d_{f1}. The distance between point c_f and the new point d_{f1} is the true length of line C–D.

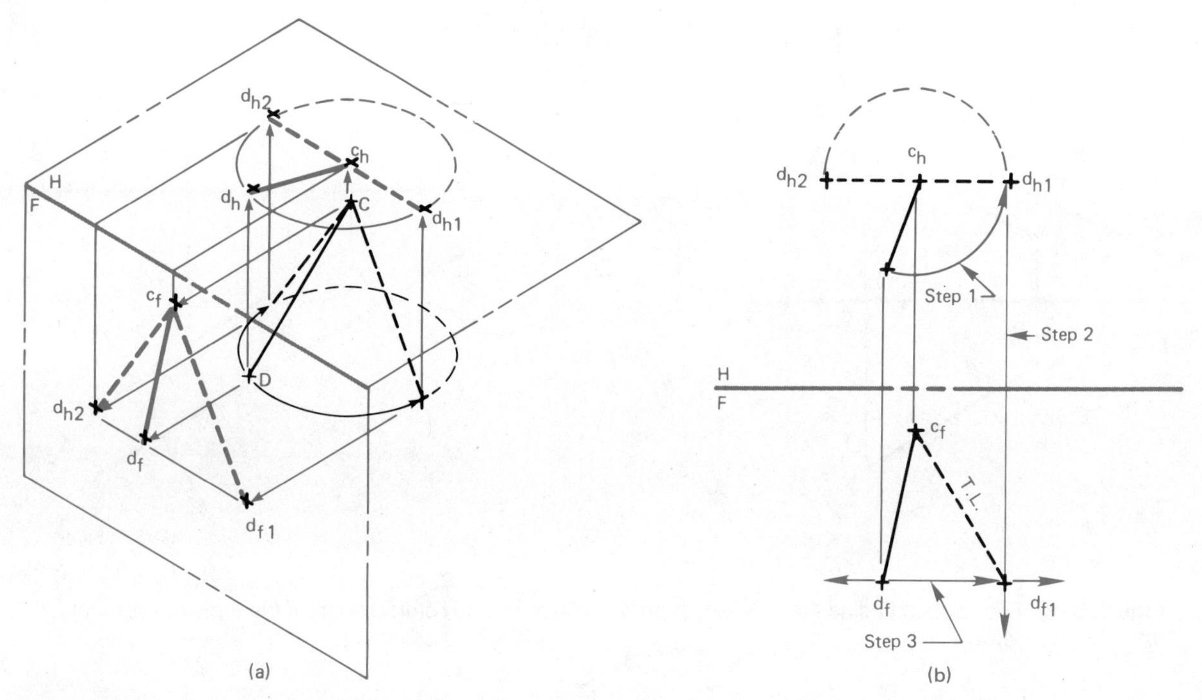

Figure 6–21 Finding the true length of a line by revolution: (a) multiview projection; (b) descriptive geometry projection layout.

Figure 6–22 shows how to revolve a line M–N about either end point.

End View of a Line

The end view of a line is a point. There are several problem-solving situations, such as finding the true shape of planes, where this concept will be necessary to know. The next adjacent view of a line shown in true length will show that line as a point, as seen in Figure 6–23.

PLANES

Without question, problems involving plane surfaces are some of the most common found in industry. The basic principles involving planes are applicable in most industrial fields. A plane is a flat surface that is not curved or warped, and in which a straight line will lay. Figure 6–24 shows the formation of a general plane surface that actually represents the surface of the object by using two sets of parallel lines. Two intersecting lines can form a plane,

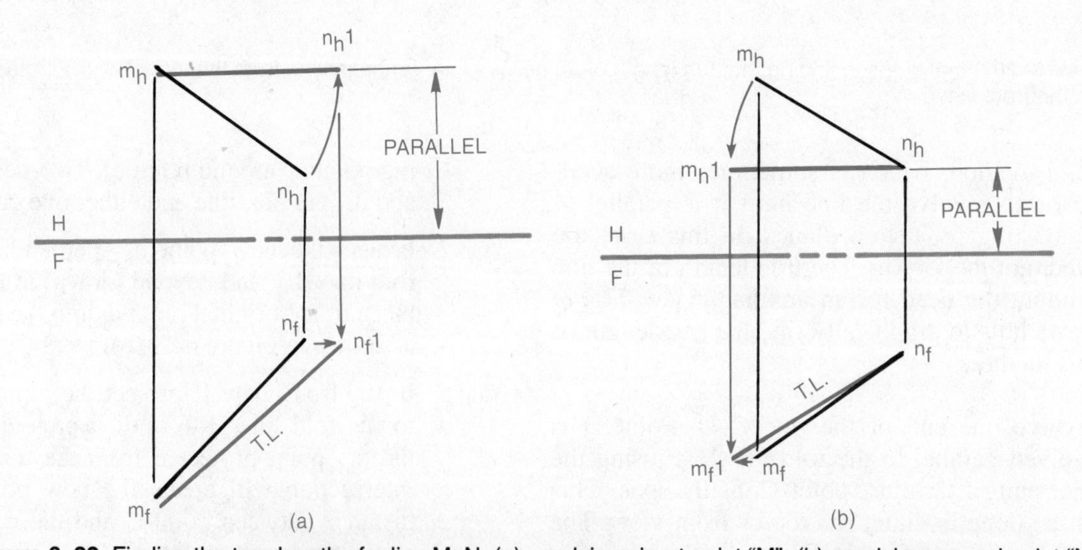

Figure 6–22 Finding the true length of a line M–N: (a) revolving about point "M"; (b) revolving around point "N."

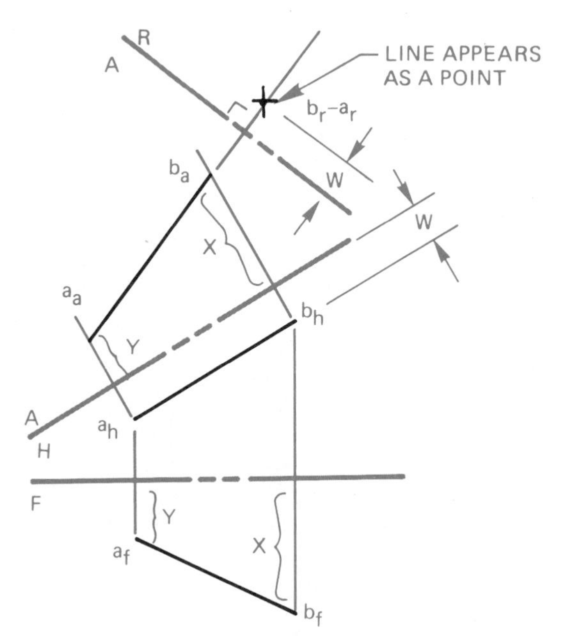

Figure 6–23 Descriptive geometry projection layout of the end of a line.

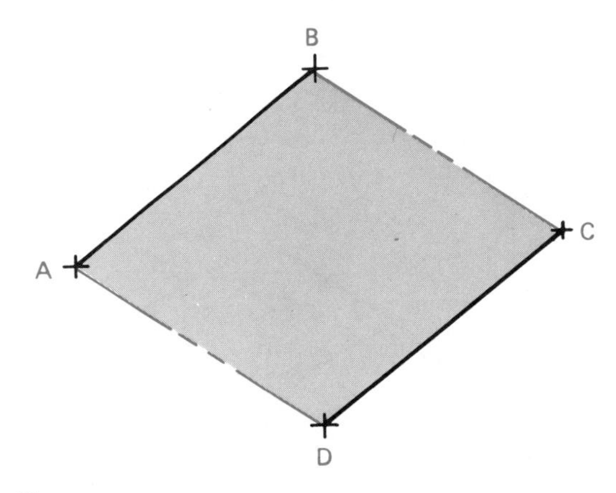

Figure 6–24 A plane represented by two sets of parallel lines.

as seen in Figure 6–25. In Figure 6–26, three points *not* in a straight line can determine a plane, while Figure 6–27 shows a fourth method using one straight line and one point not on the line to create a plane.

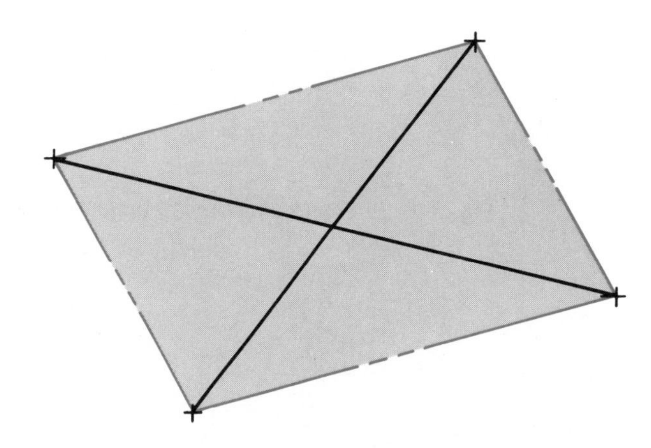

Figure 6–25 A plane represented by two intersecting lines.

Classification of Plane Surfaces

Normal Planes. Plane surfaces are classified similar to lines in that they are called *normal planes* if the plane is parallel to any one of the primary projection planes. A plane parallel to the front projection plane, such as plane 2 in Figure 6–28, is called a front plane and is shown in true shape in the front view. Plane 1 is parallel to the top projection plane and is called a horizontal plane. It appears in true shape in the top view. Plane 3 is parallel to the profile projection plane so it is called a profile plane. It is shown in true shape in the profile view.

Oblique Planes. Oblique planes are not parallel to any of the primary projection planes, such as plane 4 in Figure 6–28. Some or none of the dimensions will be true in oblique planes.

True Shape. True shape is the actual or real size of the plane. True dimensions can be established on a true shape. In Figure 6–29, plane ABC is parallel to the horizontal projection plane and is called a horizontal plane.

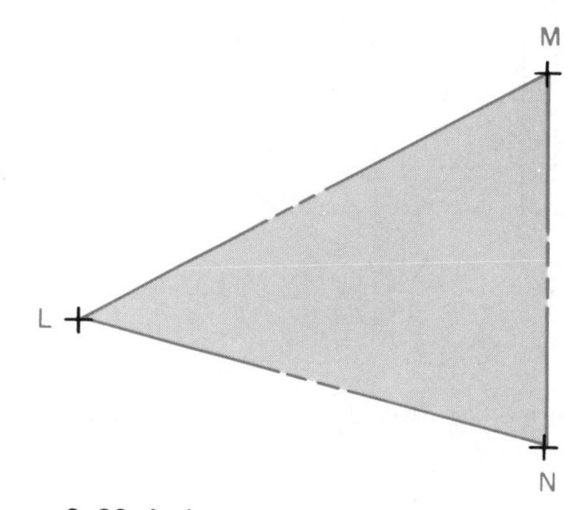

Figure 6–26 A plane represented by three points not in a straight line.

Locating a Plane in Space

A plane is a combination of points and/or lines. Therefore, locating a plane is merely a matter of projecting to and connecting a few more points in the third view, as shown in Figure 6–30.

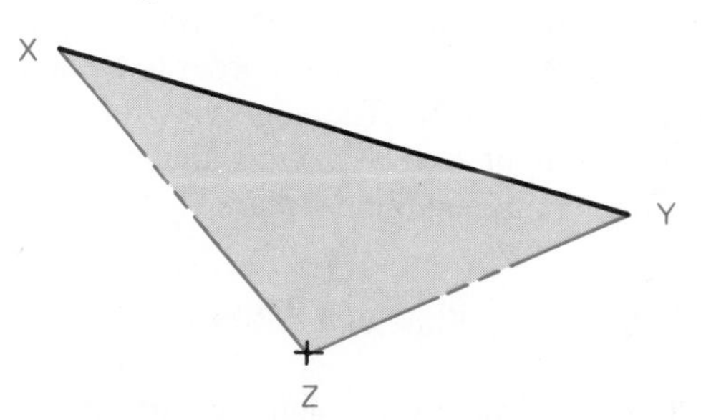

Figure 6–27 A plane represented by one straight line and one point not on the line.

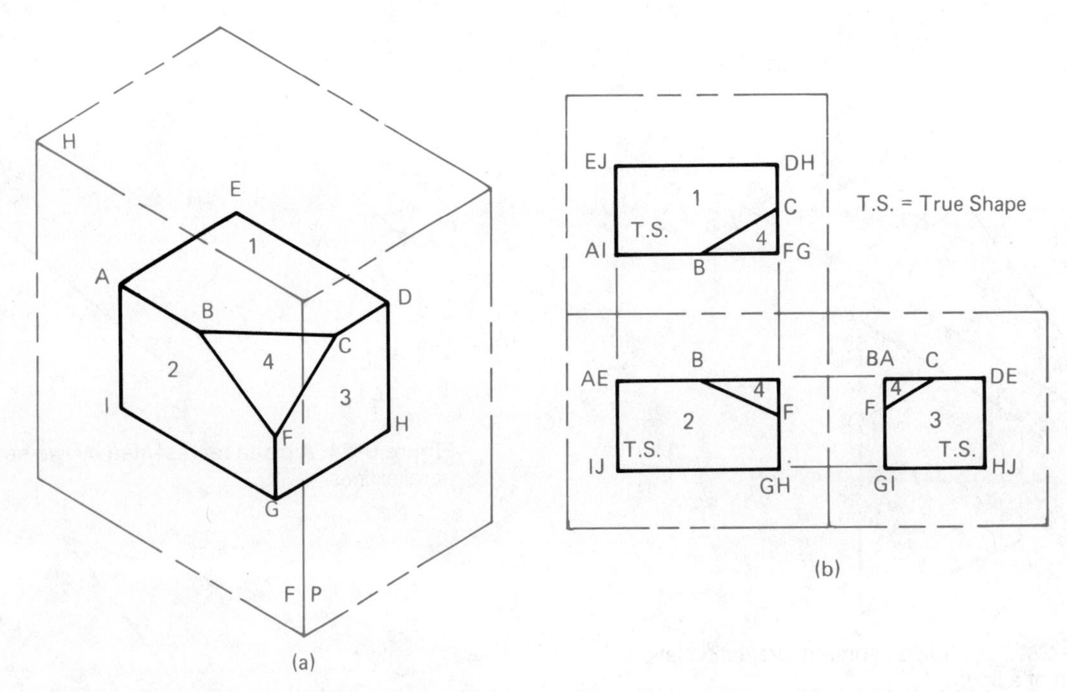

Figure 6–28 Planes in multiview projection: (a) multiview projection box; (b) unfolded multiview projection box.

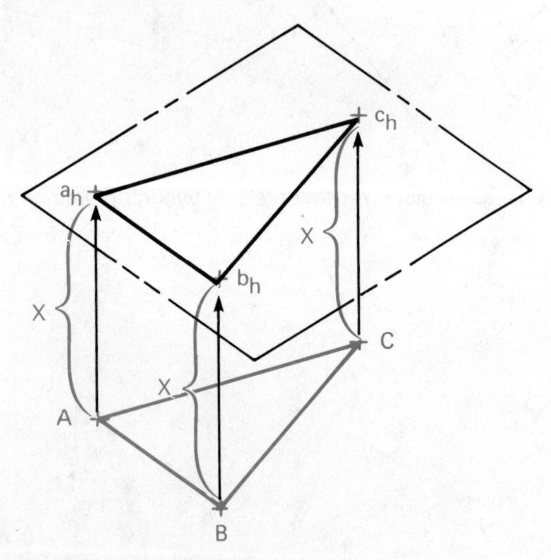

Figure 6–29 True shape of a horizontal plane.

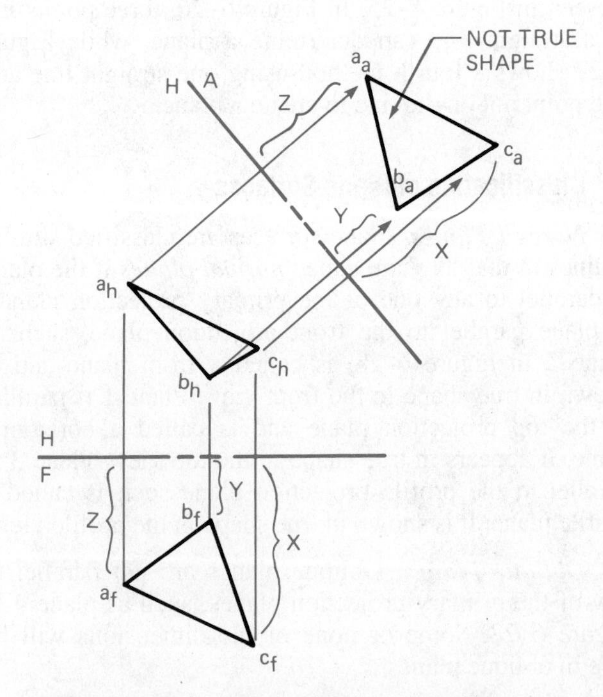

Figure 6–30 Locating a plane in an auxiliary view.

Finding the Edge View of a Plane

Knowing how to find the edge view of a plane is an essential concept to understand in order to be able to find the true shape of a plane. The edge view of a plane will appear as a straight line in a view that shows the end view of a line contained on that plane.

Finding an Edge View of a Plane

Step 1 Locate a line on the plane in true length. Figure 6–31 shows one method in locating a true-length line on an oblique plane WST. Given two views of the oblique plane, a horizontal line x_f–t_f is drawn parallel to fold line H–F. This line could be anywhere on the plane as long as it is drawn parallel to the fold line H–F. However, better accuracy will result when the line drawn is located where the end points of the line have the most distance between them. In this example, using the existing point t_f makes use of the greatest distance. Notice the second horizontal

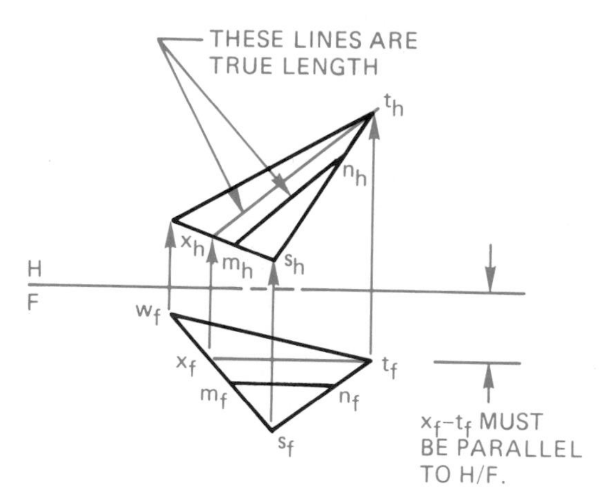

Figure 6–31 Finding a true-length line on an oblique plane.

line drawn, m_f–n_f, was not used because the result could be less accurate in the final solution. When line x_f–t_f is projected correctly into the top view, the line x_h–t_h is in true length in that view.

Step 2 The true length line x_h–t_h has already been established for plane WST in Figure 6–32 as was done in the previous explanation. It already has been established that a true-length line viewed from one end will appear as a point. The fold line H–A is drawn perpendicular to the true-length line x_h–t_h. When the true-length line is projected into the new view, it will appear as point t_a,x_a. When the rest of the points, w_a and s_a, are projected into the new view, they will form line w_a–t_a–x_a–s_a. This line represents the edge view of the plane WST. Figure 6–33 shows that in a normal plane, such as plane ABC, the second view of the plane will already appear as an edge view, and in this case, line a_h–c_h will be in true length in the top view and the plane ABC will be in true shape in the front view.

Finding the True Shape of a Plane

In order to find the true shape of any plane surface — normal, inclined, or oblique — an *edge view* of the plane surface must first exist or be found. Using the same plane as shown in Figure 6–32 and working from the edge view found, Figure 6–34 shows how to find the true shape of plane WST. The approach here is to make sure the new fold line A–I for the additional auxiliary view I is PARALLEL to the edge view of the plane. Once the fold line is drawn correctly, project all of the points of the plane into view I. The true shape of the plane will appear here.

If there are irregular or circular shapes such as holes or other various design features on the true shape surface, these can be projected back into the other views. To do this, establish various points onto the edges of those

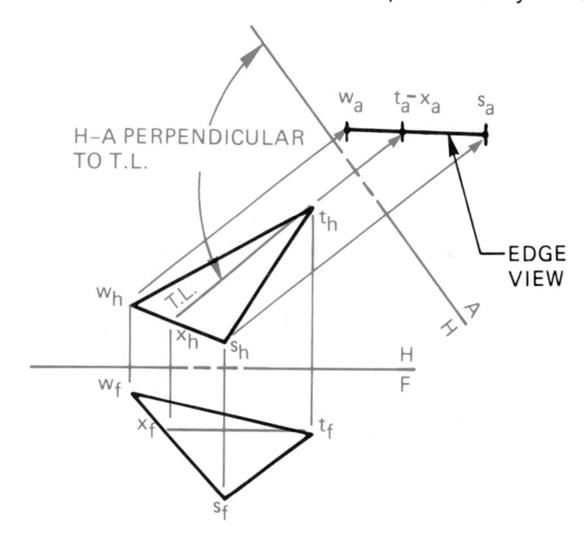

Figure 6–32 Finding the edge view of an oblique plane.

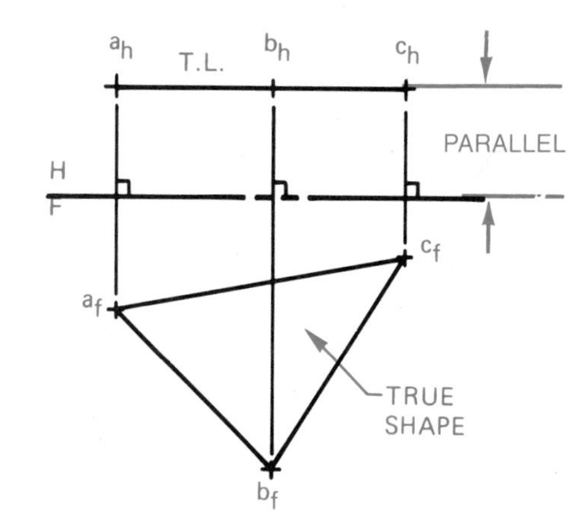

Figure 6–33 Edge view of a normal plane.

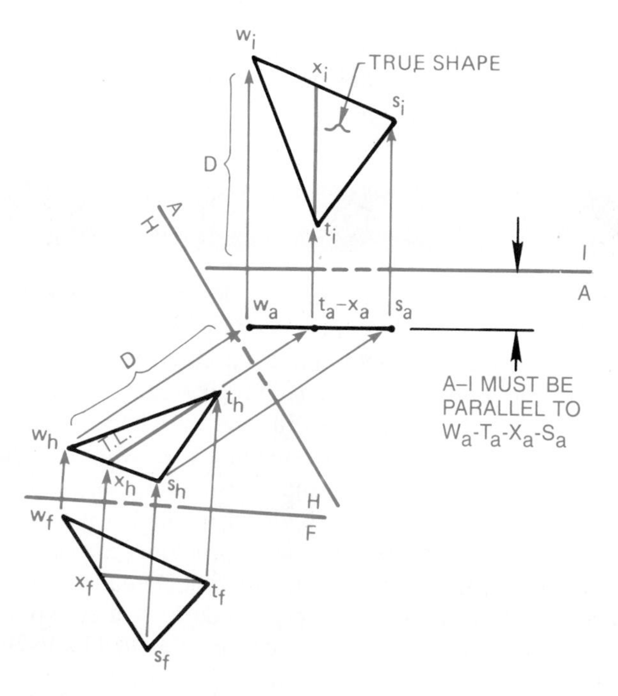

Figure 6–34 Finding the true shape of a plane.

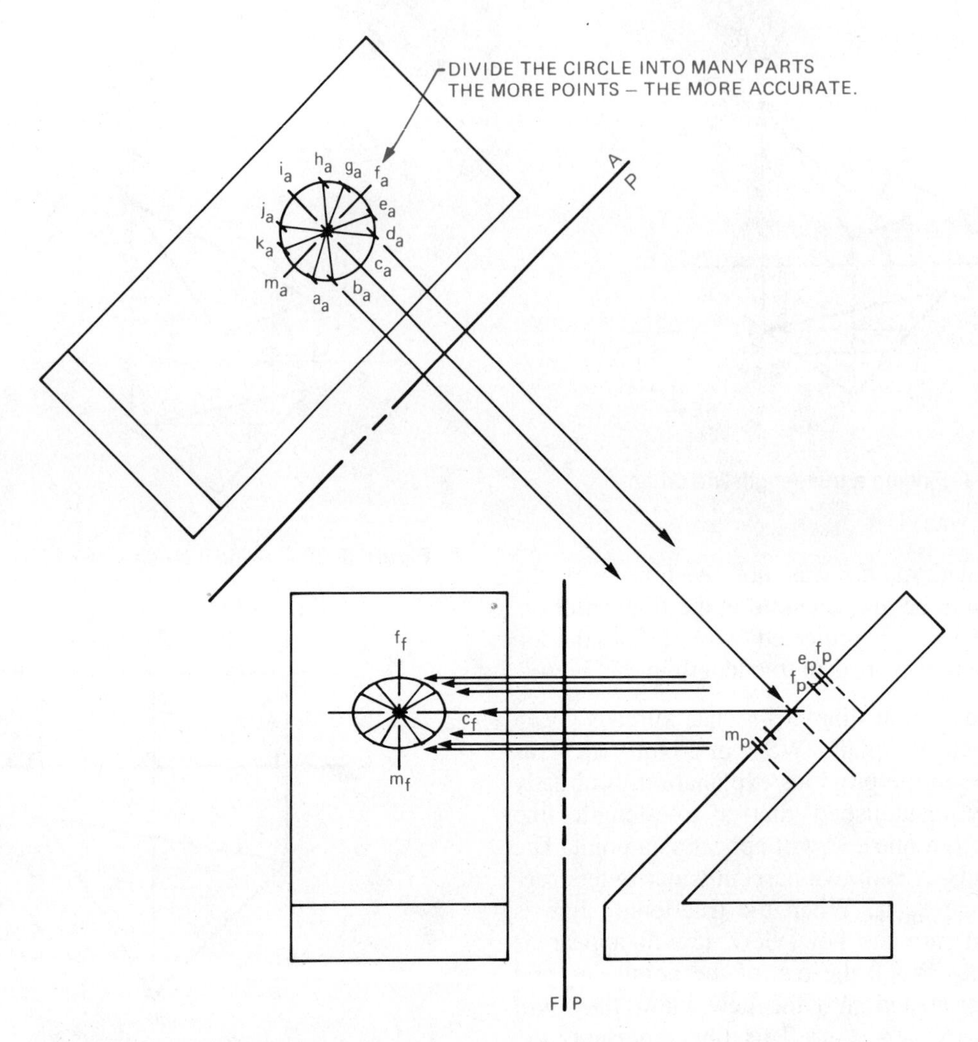

Figure 6–35 Finding the true shapes of irregular features.

features as shown in Figure 6–35. The more points used, the more accurate the projection of the features obtained back into the original views.

PROFESSIONAL PERSPECTIVE

Finding the correct answers to descriptive geometry problems is mainly dependent on two things: the method used to solve the problems and the accuracy in constructing the drawing. Different techniques can be used to graphically solve each problem, and the challenge is to select the best method. There also must be extreme accuracy taken when drawing and constructing the various lines. For people who tend to visualize objects and situations easily, descriptive geometry problems are not too difficult to solve. Others who do not visualize easily need to follow a step-by-step procedure — a cookbook approach — and memorize specific rules to obtain the correct answers. Either way, solving descriptive geometry problems can and should be fun and intriguing.

ENGINEERING PROBLEM

A SIX-STEP PROBLEM-SOLVING SYSTEM

When trying to solve descriptive geometry problems, or any other types of problems, following the rules may not always work. To help you arrive at correct solutions, it is important to use some method for problem solving. The basic six-step system that follows is an example of just one that will work for you in many problem-solving situations.

Step 1 Identify and define the problem.

Step 2 Research and generate solutions.

Step 3 Evaluate possible solutions.

Step 4 Use the best solution(s).

Step 5 Evaluate the best solution(s).

Step 6 Finalize the solution.

Identify and Define the Problem

The clearer and more exact this statement is, the closer a solution will be. This is the most important step of this six-step system. You have to clearly know what you are trying to solve before you can solve it.

Example 1. The solution for "Design a clock!" would be vastly different from "Design a device that pinpoints the time of day by the second, minute, and hour, the day of the month, month of the year; operates under water; has a cadmium cell electric power source that runs one year per charge; will not corrode; is lightweight (½ oz); and can be strapped to your wrist comfortably."

Example 2. Consider the statement "I'm hungry!" as opposed to "I want a double cheeseburger and fries." Both problems are similar. However, the more exact the problem statement, the easier it is to solve. In addition, the solution is more easily judged as to its effectiveness.

Example 3. Sometimes a problem can be overstated. "Mark off a football field over there and make it exactly 100 yd. long." What are the tolerances: ±1 ft, ±1 yd, ±.001 in.? If the distance had to be paced off, it would just be luck to get within a yard. Using a transit would possibly narrow down the accuracy to a couple of inches. Even at a couple of inches, the actual marking lines would alter the limits. Emphasize what you are sure of and leave open to question the factor(s) that is (are) variable.

Research and Generate Solutions

Example 1. "What is the volume in cubic inches of a particular irregular iron casting?" Some possible solutions:

1. Break the piece down into geometric solids, calculate the volume of each piece, and add the volumes to find the total.
2. Fill a bucket of water to the top, place the piece in the water, catch and measure the runoff.
3. Weigh the piece. Use a weight-to-volume conversion formula to find the volume.

Example 2. "An alarm clock is needed to wake you up in the morning in order for you to get to work on time." Possible solutions include:

1. Buy an alarm clock.
2. Borrow one!
3. Build a timer that knocks a bucket of water onto your bed at the appropriate time.
4. Stop getting up on time.
5. Get fired or quit the job.

Evaluate Possible Solutions

By looking at all the pros and cons of each solution, it is possible to select the most appropriate one for the stated problem. If no solution seems to work, reexamine the statement or come up with more solutions. In example 2, some of the solutions listed as possible may sound absurd. But quitting the job could lead to another type of work not dependent on an alarm clock, and more satisfying. If the solutions are technical solutions, as in example 1 of "Research and Generate Solutions," the solution(s) chosen may be determined by the accuracy required or the equipment available.

Use the Best Solutions

Try the solution chosen, either in actuality or with models and prototypes.

Evaluate the Best Solution(s)

Did the solution(s) do what was needed? If not, reevaluate the possible solutions and select another. If the solution worked, congratulations!

Finalize the Solution

When the solution is obtained, draw it if necessary and get it ready for production.

SAMPLE DESCRIPTIVE GEOMETRY PROBLEM 1 (SOLVED BY THE SIX-STEP METHOD)

1. *Identify and define the problem.* Find the total length of the ridge and hips of the roof on the house in Figure 6–36 in order to cover them with the proper length of sheet metal flashing. Restated, find the true length of the ridge and hips shown.
2. *Research and generate solutions.* To find the true lengths of the lines, many methods can be used.
   a. Fold-line technique: Find or create a view in which the line you want to be shown in true length is parallel to a fold line. The adjacent view will show that line in true length.
   b. Revolving method: The line is revolved within one of the two given views parallel to the fold line. The other view will show that line in true length.
   c. Using trigonometry: Apply the knowledge of adjoining sides and angles.
   d. Approximation: Compare the lengths to a known distance and guess.
   e. Build a prototype or model and measure it.
   f. Climb up on the roof and measure the actual distance.
3. *Evaluate the possible solutions.* Assume there is an accurately scaled two-view drawing to work from, as in Figure 6–37. The quickest solutions would be 2(a) and (b), listed above. The choice would depend on which one is easier.

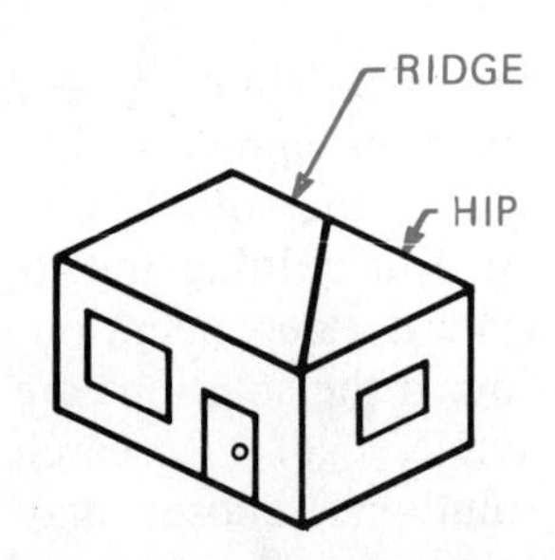

Figure 6–36 Problem setup.

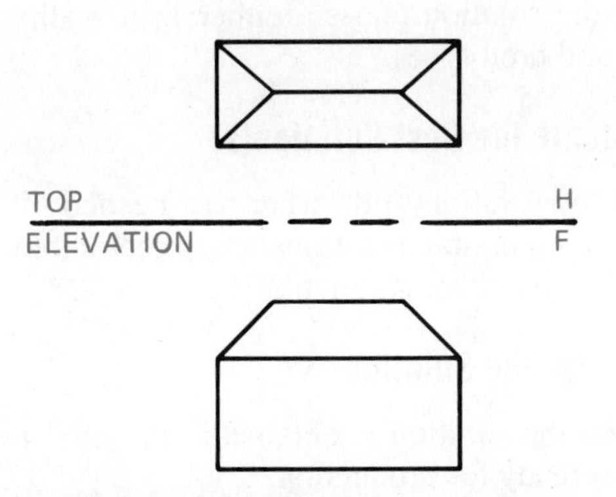

Figure 6–37 Scaled drawing of a house.

4. *Use the best solution.* The ridge is already shown in true length in either view because the ridge line is parallel to the fold line H–F. The hips will need to be revolved parallel to the fold line in either view to find their true lengths. See Figure 6–38.
5. *Evaluate the best solution.* Did it work? The answer is yes, plus it was quick and simple.
6. *Finalize the solution.* In this case, the solution worked. Therefore, the problem is solved.

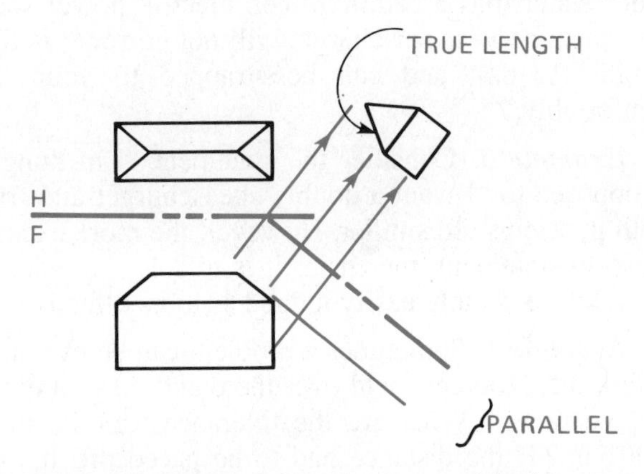

Figure 6–38 Finding the true length of the hip.

DESCRIPTIVE GEOMETRY PROBLEMS

DIRECTIONS

Please read the problems carefully before you begin working. Your instructor may assign one or more of the following problems. Use a separate sheet of drawing paper with graph lines on it so that the proper setup can be duplicated where necessary for each problem. Use one sheet per problem.

Problem 6–1 Point projection. Correctly project point d into the profile view P.

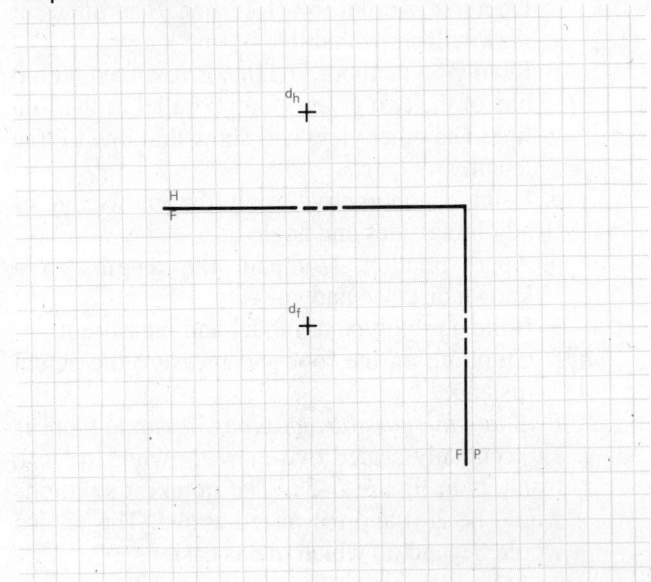

Problem 6–2 Point projection. Correctly project point r into the auxiliary view A.

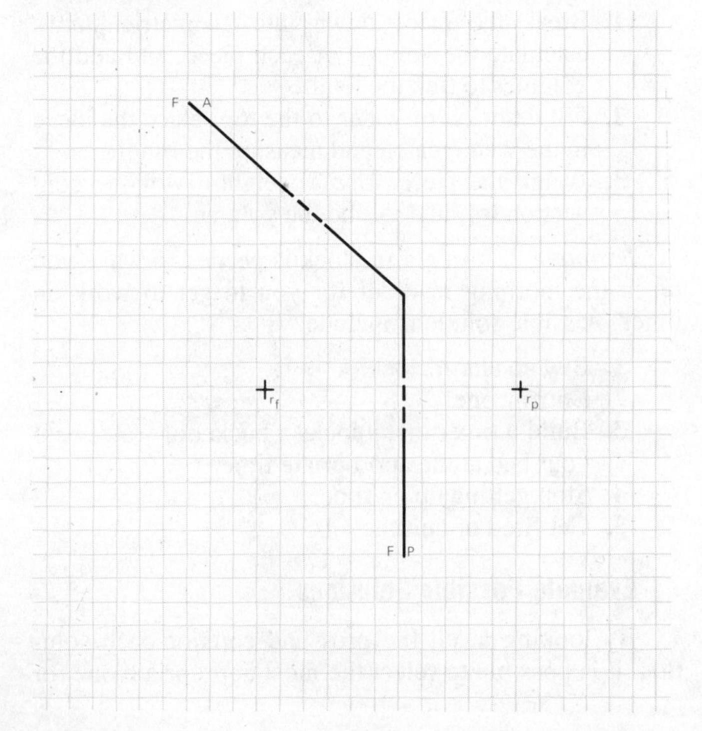

Problem 6–3 Point projection. Correctly project point t into the auxiliary view A.

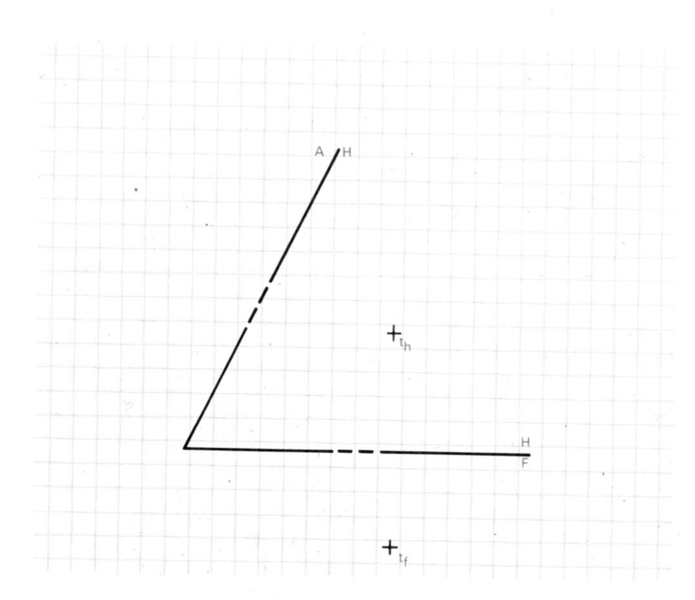

Problem 6–4 Point projection. Correctly project point f into the auxiliary view G.

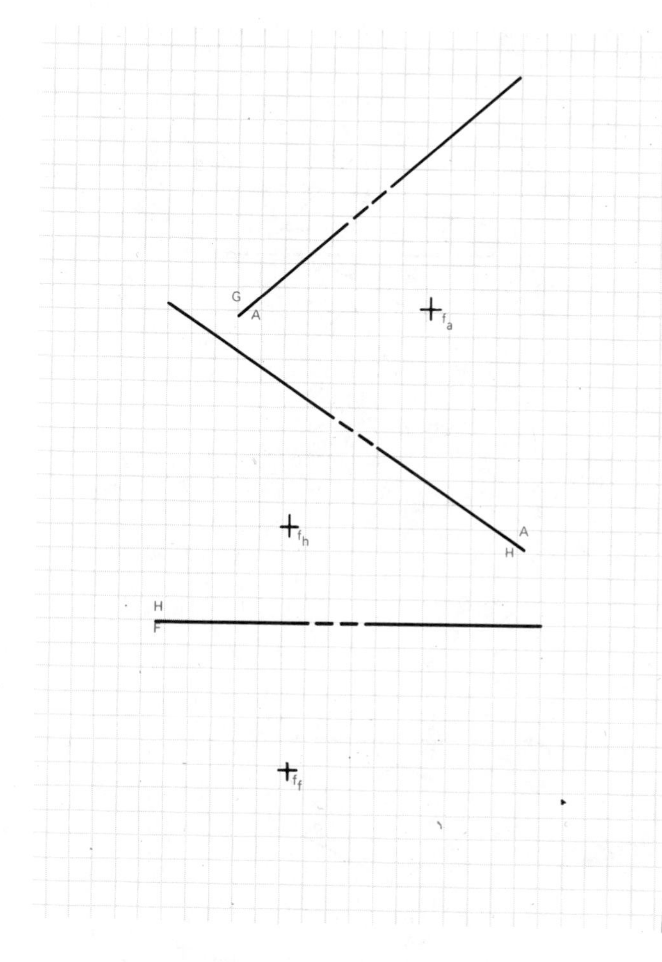

Problem 6–5 Point projection. Correctly project point w into all the required auxiliary views.

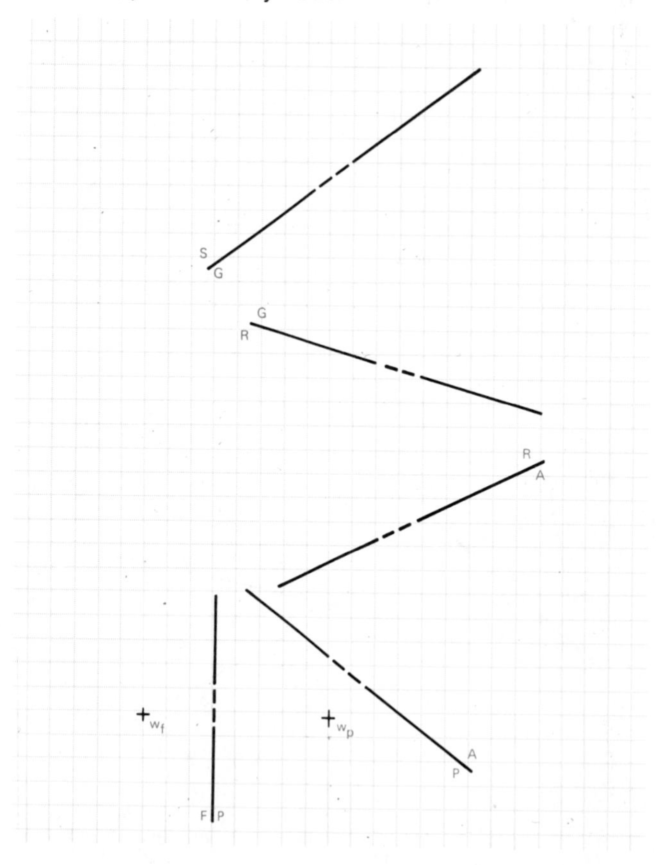

Problem 6–6 Line projection. Complete the missing views of (a), (b), (c), and (d). If the true length of the lines appears in any of the views, notate it.

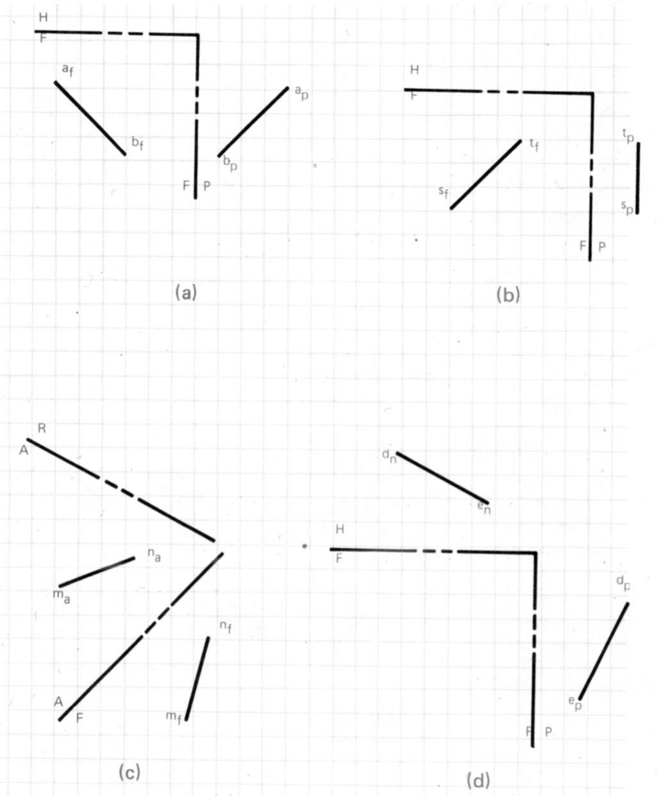

Problem 6–7 True length. Find the true length of each of the four lines in parts (a), (b), (c), and (d). Identify the true lengths with T.L.

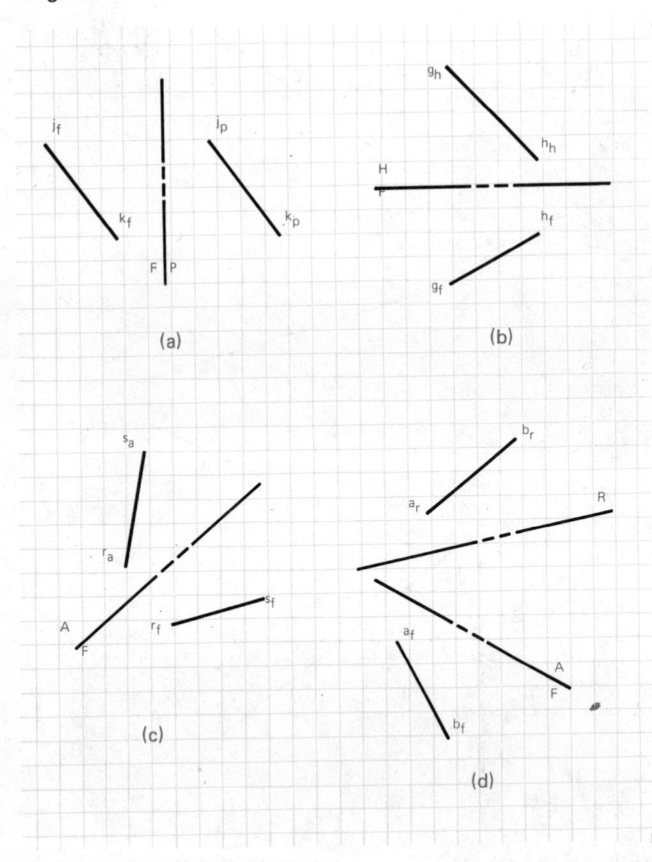

Problem 6–8 End view. Find the end view of each of the four lines in parts (a), (b), (c), and (d).

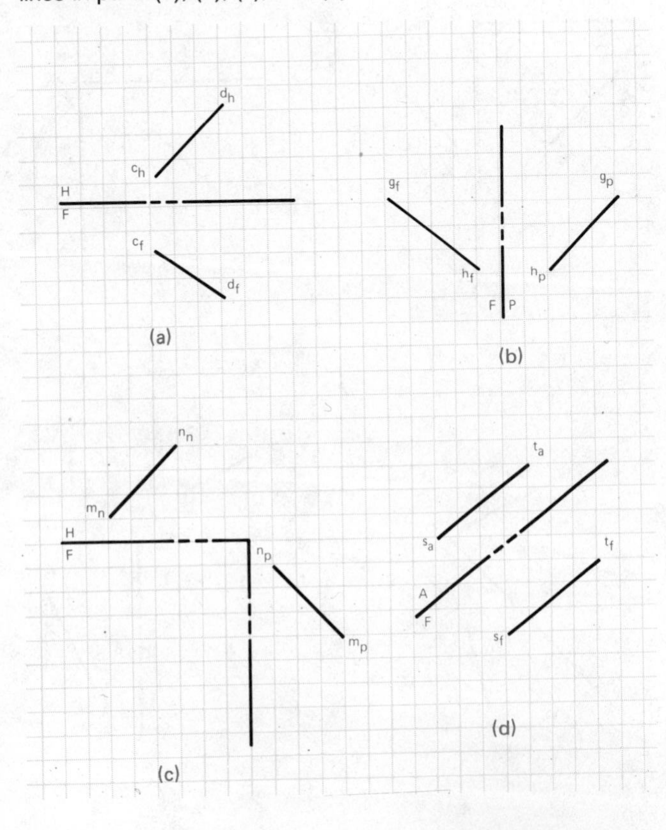

Problem 6–9 Plane projection. Find the missing view of each of the four planes in parts (a), (b), (c), and (d).

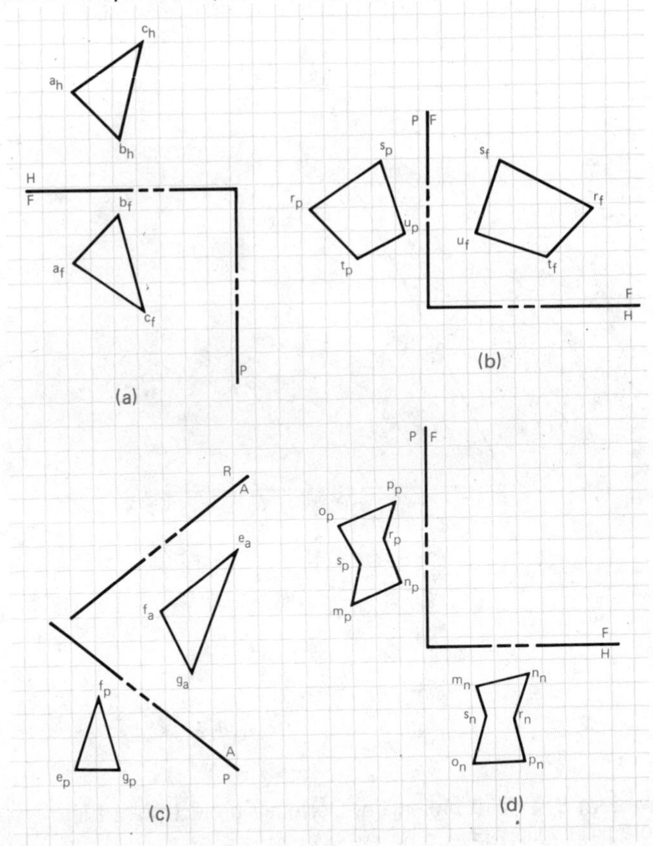

Problem 6–10 Plane edge view. Find the edge view of the planes in parts (a) and (b).

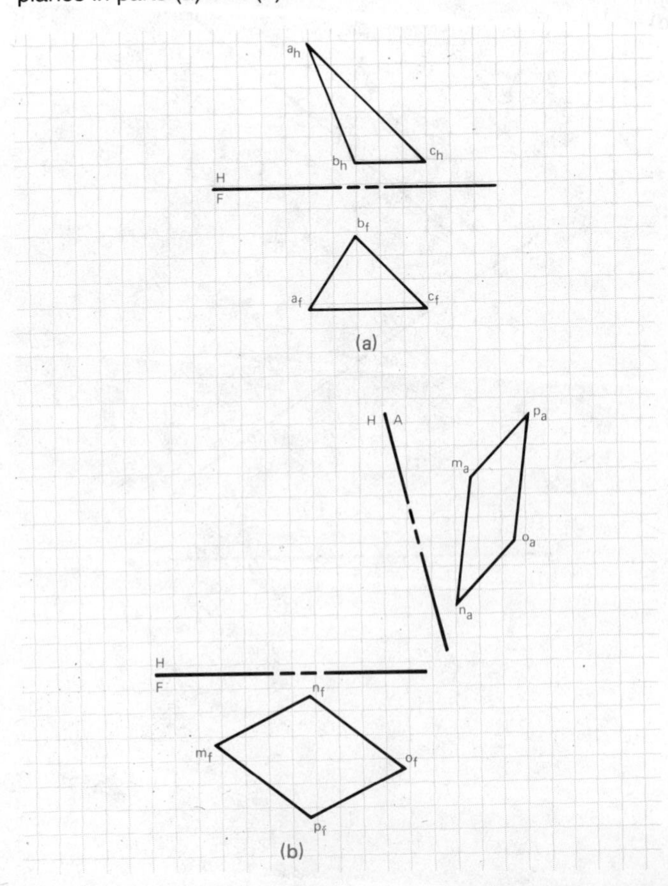

Problem 6–11 Plane true shape. Find the true shape of each of the planes in parts (a) and (b).

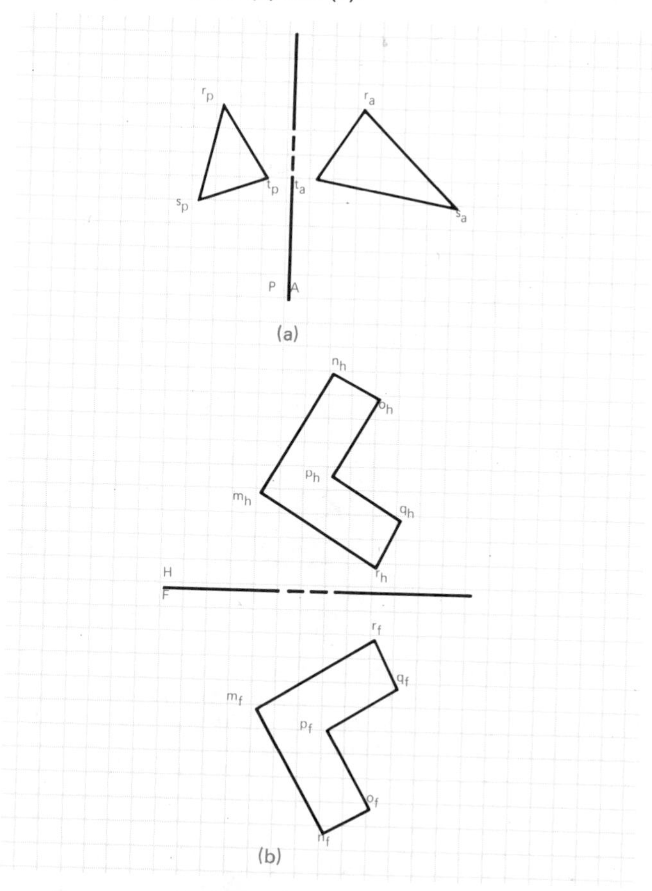

Problem 6–12 Multiview drawings. Draw the front, top, and right-side views of each of the four blocks (a), (b), (c), and (d) using the given dimensions.

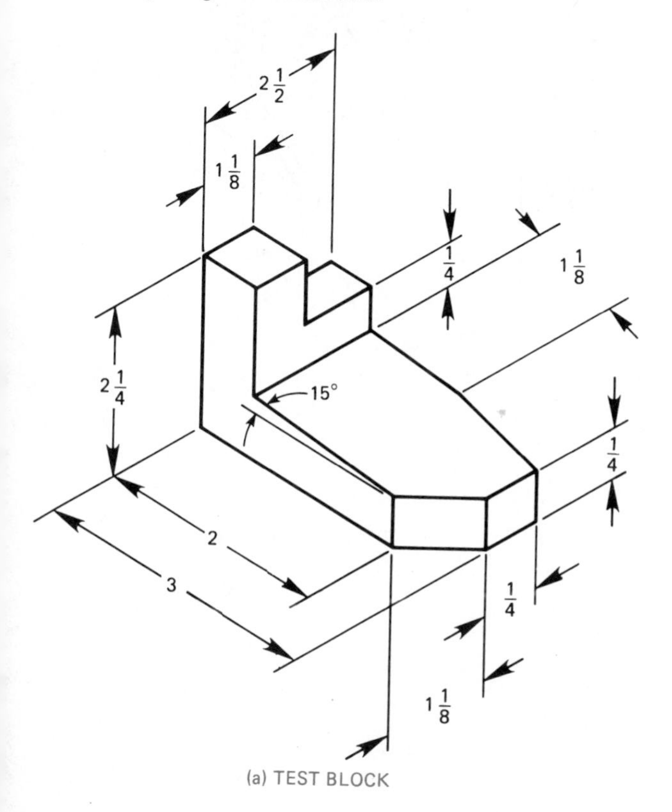

(a) TEST BLOCK

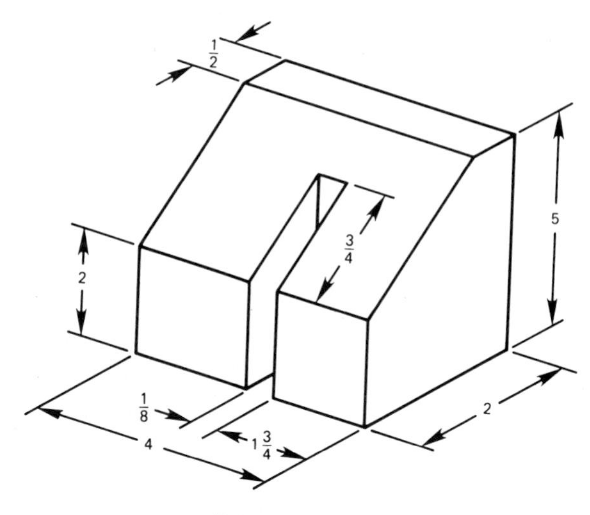

(b) WIRE GUIDE BLOCK

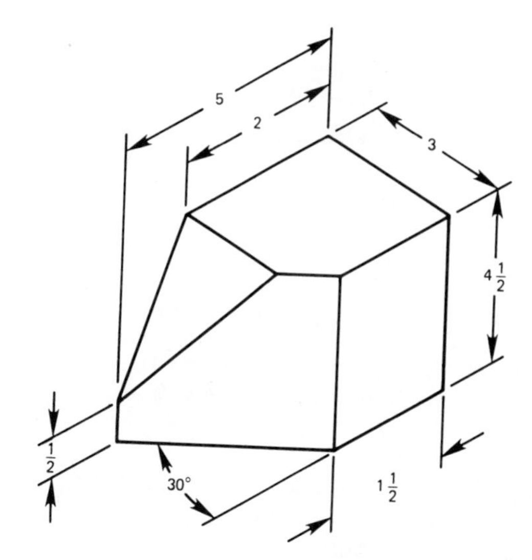

(c) JIG PROTOTYPE BLOCK

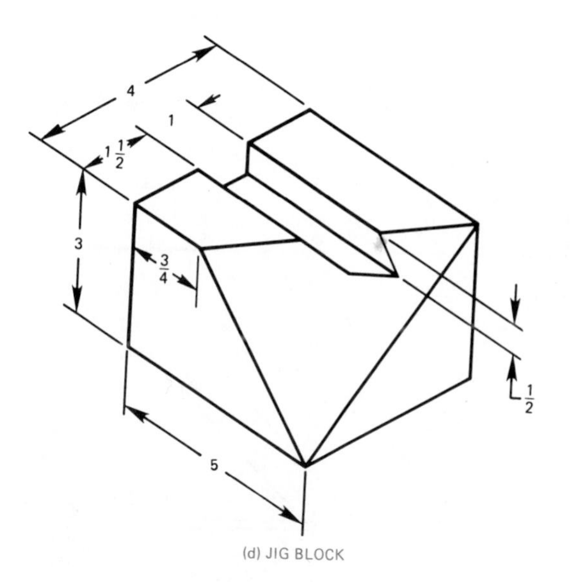

(d) JIG BLOCK

Problem 6–13 True shape. The sloping sides of the concrete bridge pier all have the same slope as shown. Draw views showing the true shape and size of planes A and C.

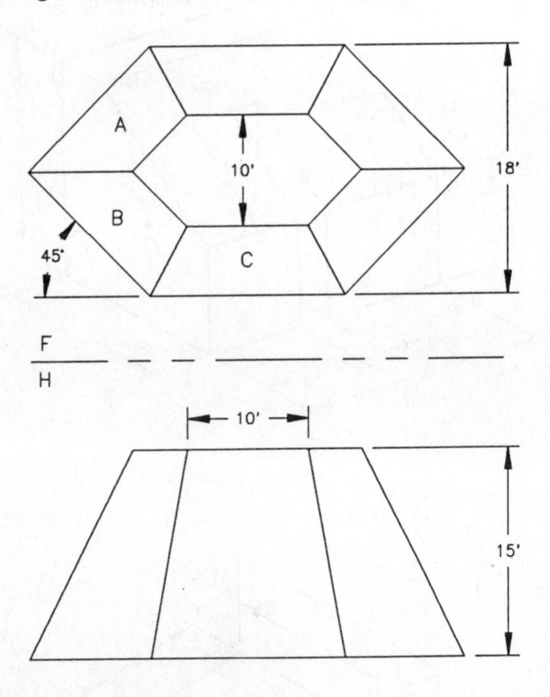

Problem 6–14 True shape. The ends of a wooden timber, 12 in. square, must be cut to fit against a wall and floor as shown. The completed top view shows that the two faces, or sides, of the timber are in vertical planes. Draw a new view of the timber showing one of its vertical faces in true size. Draw another view showing one of the sloping faces in true size.

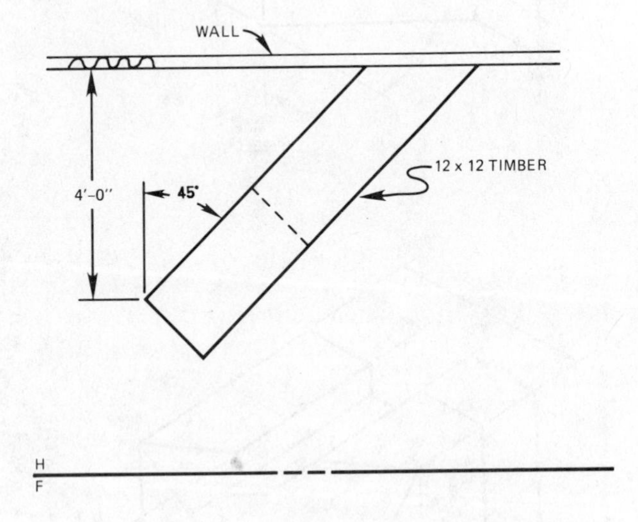

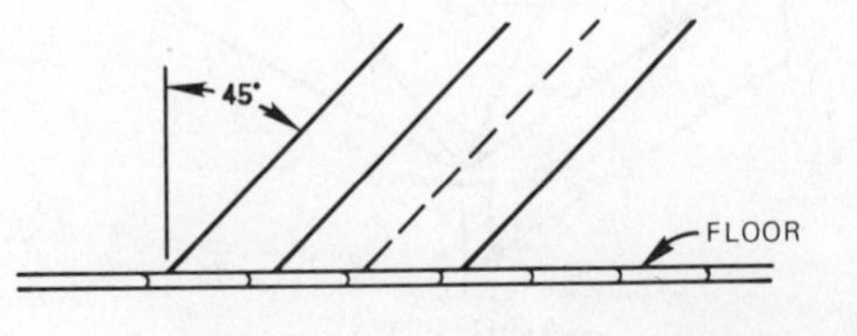

SCALE: 1 in. = 1 ft.

Problem 6–15 True shape — irregular shapes. From the given engineer's sketch, draw the top and front views of the cable anchor bracket. In the necessary auxiliary view, only the inclined part of the bracket need be drawn.

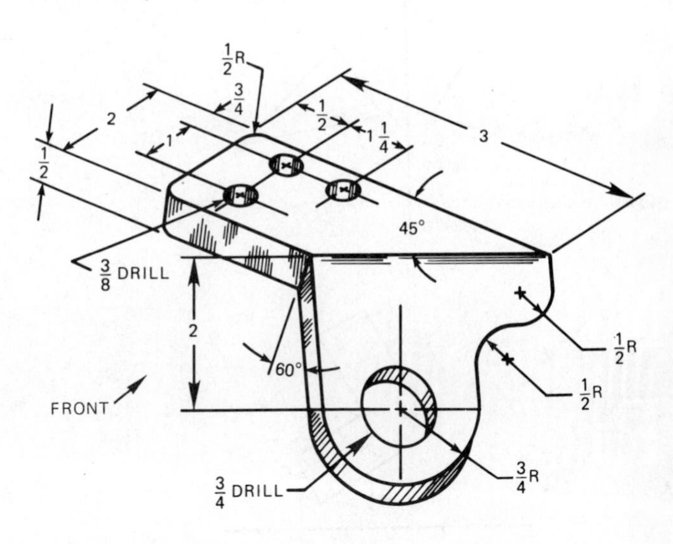

Problem 6–16 True shape — irregular shapes. From the given engineer's sketch, draw the top and front views of the angle bracket. In the necessary auxiliary view, only the inclined part of the bracket need be drawn.

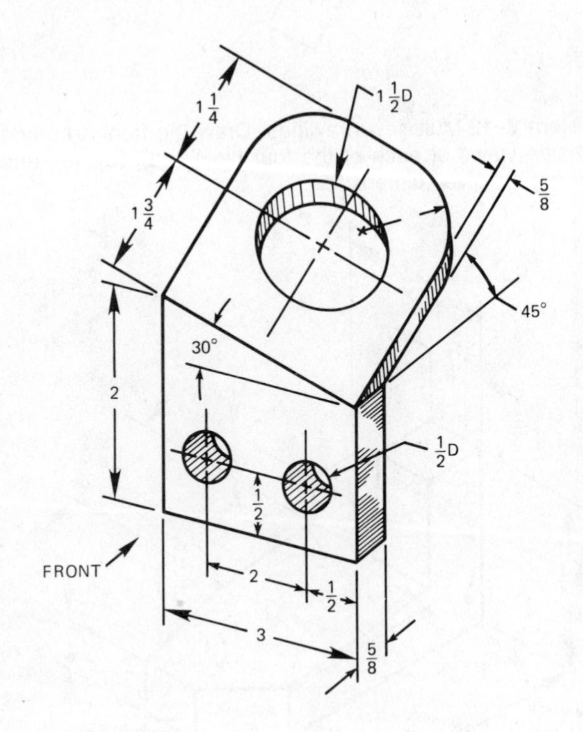

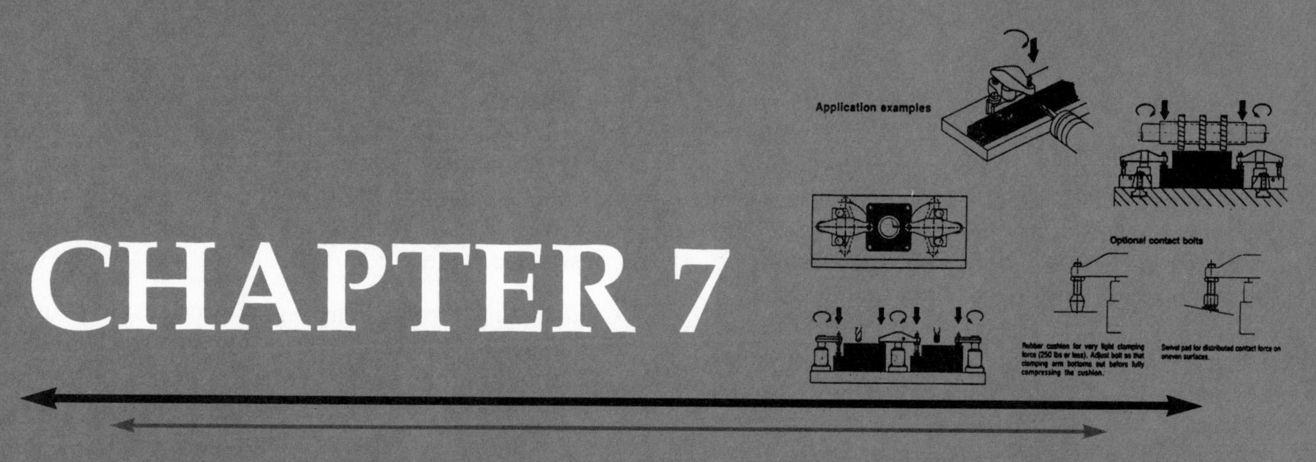

CHAPTER 7

Manufacturing Processes

LEARNING OBJECTIVES

After completing this chapter, you will:

- Define and describe various manufacturing materials, material terminology, numbering systems, and material treatment.
- Discuss casting processes and terminology.
- Explain the forging process and terminology.
- Describe manufacturing processes.
- Define and draw the representation of various machined features.
- Explain tool design and drafting practices.
- Discuss the statistical process quality control assurance system.
- Evaluate the results of an engineering and manufacturing problem.

A number of factors influence product manufacturing. Beginning with research and development, a product should be designed to meet a market demand, have good quality, and be economically produced. The sequence of product development begins with an idea and results in a marketable commodity, as shown in Figure 7–1.

MANUFACTURING MATERIALS

There is a wide variety of materials available for product manufacturing that fall into three general categories: metal, plastic, and inorganic materials. Metals are classified as *ferrous*, *nonferrous*, and *alloys*. Ferrous metals contain iron, such as cast iron and steel. Nonferrous metals do not have iron content; for example, copper and aluminum. Alloys are a mixture of two or more metals.

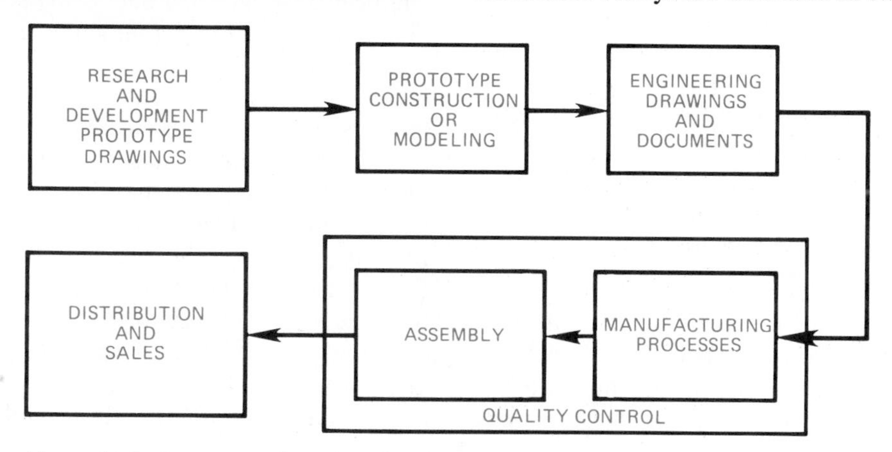

Figure 7–1 Sequence of product development.

Plastics or *polymers* have two types of structure: thermoplastic and thermoset. *Thermoplastic* material may be heated and formed by pressure and upon reheating, the shape can be changed. *Thermoset* plastics are formed into a permanent shape by heat and pressure and may not be altered by heating after curing. Plastics are molded into shape and only require machining for tight tolerance situations or when holes or other features are required that would be impractical to produce in a mold. It is common practice to machine some plastics for parts, such as gears and pinions.

Inorganic materials include carbon, ceramics, and composites. Carbon and graphite are classified together and have properties that allow molding by pressure. These materials have low *tensile strength* (ability to be stretched) and high compressive strength with increasing strength at increased temperatures. Ceramics are clay, glass, refractory, and inorganic cements. Ceramics are very hard, brittle materials that are resistant to heat, chemicals, and corrosion. Clay and glass materials have a crystalline structure, while *refractories* must be bonded together by applying temperatures. Due to their great heat resistance, refractories are used for high-temperature applications, such as furnace liners. *Composites* are two or more materials that are bonded together by adhesion. *Adhesion* is a force that holds together the molecules of unlike substances when the surfaces come in contact. These materials generally require carbide cutting tools or special methods for machining.

Cast Iron

There are several classes of cast iron, including gray, white, chilled, alloy, malleable, and nodular cast iron. Cast iron is primarily an alloy of iron and 1.7%–4.5% of carbon, with varying amounts of silicon, manganese, phosphorus, and sulfur.

Gray Cast Iron. Gray iron is a popular casting material for automotive cylinder blocks, machine tools, agricultural implements, and cast iron pipe. Gray cast iron is easily cast and machined. It contains 1.7%–4.5% carbon and 1%–3% silicon.

ANSI/ASTM The American National Standards Institute (ANSI) and American Society for Testing Materials (ASTM) specifications A48–76 group gray cast iron into two classes: easy to manufacture (20A, 20B, 20C, 25A, 25B, 25C, 30A, 30B, 30C, 35A, 35B, 35C); and more difficult to manufacture (40B, 40C, 45B, 45C, 50B, 50C, 60B, 60C). The prefix denotes the minimum tensile strength in thousand pounds per square inch.

White Cast Iron. White cast iron is extremely hard, brittle, and has almost no *ductility* (the ability to be stretched, drawn, or hammered thin without breaking). Caution should be exercised when using this material, because thin sections and sharp corners may be weak and the material is less resistant to impact uses. This cast iron is suited for products with more compressive strength requirements than gray cast iron (compare over 200,000 pounds per square inch [psi] with 65,000–160,000 psi). White cast iron is used where high wear resistance is required.

Chilled Cast Iron. When gray iron castings are chilled rapidly, an outer surface of white cast iron results. This material has the internal characteristics of gray cast iron and the surface advantage of white cast iron.

Alloy Cast Iron. Elements such as nickel, chromium, molybdenum, copper, or manganese may be alloyed with cast iron to increase the properties of strength, wear resistance, corrosion resistance, or heat resistance. Alloy iron castings are commonly used for such items as pistons, crankcases, brake drums, and crushing machinery.

Malleable Cast Iron. The term *malleable* means the ability to be hammered or pressed into shape without breaking. Malleable cast iron is produced by heat-treating white cast iron. The result is a stronger, more ductile, and shock-resistant cast iron that is easily machinable.

ANSI/ASTM The specifications for malleable cast iron are found in ANSI/ASTM A47–77.

Nodular Cast Iron. Special processing procedures, along with the addition of magnesium or cerium bearing alloys, result in a cast iron with spherical-shaped graphite rather than flakes, as in gray cast iron. The results are iron castings with greater strength and ductility. Nodular cast iron may be chilled to form a wear-resistant surface ideal for use in crankshafts, anvils, wrenches, or heavy-use levers.

Steel

Steel is an alloy of iron containing 0.8%–1.5% carbon. Steel is a readily available material that may be worked in either a heated or cooled state. The properties of steel can be changed by altering the carbon content and heat treating. *Mild steel* (MS) is low in carbon (less than 0.3%), and is commonly used for forged and machined parts, but may not be hardened. *Medium carbon steel* (0.3%–0.6% carbon) is harder than mild steel yet remains easy to forge and machine. *High carbon steel* (0.6%–1.50% carbon) may be hardened by heat treating, but is difficult to forge, machine, or weld.

Hot-rolled steel (HRS) characterizes steel that is formed into shape by pressure between rollers or by forging when in a red-hot state. When in this hot condition, the steel is easier to form than when it is cold. An added advantage of hot forming is a consistency in the grain structure of the steel, which results in a stronger, more ductile metal. The surface of hot-rolled steel is rough, with a blue-black oxide buildup. The term *cold-rolled steel* (CRS) implies the additional forming of steel after initial hot rolling. The cold-rolling process is used to clean up hot-formed steel, provide a smooth, clean surface, insure dimensional accuracy, and increase the tensile strength of the finished product.

Steel alloys are used to increase such properties as hardness, strength, corrosion resistance, heat resistance, and wear resistance. Chromium steel is the basis for stainless steel and is used where corrosion and wear resistance is required. Manganese alloyed with steel is a purifying element that adds strength for parts that must be shock and wear resistant. Molybdenum is added to steel when the product must retain strength and wear resistance at high temperatures. When tungsten is added to steel, the result is a material that is very hard and ideal for use in cutting tools. Tool steels are high in carbon and/or alloy content so that the steel will hold an edge when cutting other materials. When the cutting tool requires deep cutting at high speed, then the alloy and hardness characteristics are improved for a classification known as *high-speed steel.* Vanadium alloy is used when a tough, strong, nonbrittle material is required.

Steel castings are used for machine parts where the use requires heavy loads and the ability to withstand shock. These castings are generally stronger and tougher than cast iron. Steel castings have uses as turbine wheels, forging presses, gears, machinery parts, and railroad car frames.

Steel Numbering Systems. **AISI/SAE** The American Iron and Steel Institute (AISI) and Society of Automotive Engineers (SAE) provide similar steel numbering systems. Steels are identified by four numbers, except for some chromium steels, which have five numbers. For a steel with the identification SAE 1020 the first two numbers (10) identify the type of steel and the last two numbers (20) denote the approximate amount of carbon in hundredths of a percent, (0.20% carbon). The letter L or B may be placed between the first and second pair of numbers. When this is done, the L means that lead is added to improve machinability, and the B identifies a boron steel. The prefix E denotes that the steel is made using the electric furnace method. The prefix H indicates that the steel is produced to hardenability limits. Steel that is degassed and deoxidized before solidification is referred to as *killed steel* and is used for forging, heat treating, and difficult stampings. Steel that is cast with little or no degasification is known as *rimmed steel*, and has applications where sheets, strips, rods, and wires with excellent surface finish or drawing requirements are needed. General applications of SAE steels are shown in Figure 7–2. For a more in-depth analysis of steel and other metals, refer to the *Machinery's Handbook.**

*Erick Oberg, Franklin D. Jones, and Holbrook L. Horton, *Machinery's Handbook*, 23rd ed. (New York: Industrial Press Inc., 1988).

Hardening of Steel. The properties of steel may be altered by *heat treating*. Heat treating is a process of heating and cooling steel using specific, controlled conditions and techniques. When steel is initially formed and allowed to cool naturally, it is fairly soft. *Normalizing* is a process of heating the steel to a specified temperature and then allowing the material to cool slowly by air, which brings the steel to a normal state. In order to harden the steel, the metal is first heated to a specified temperature which varies with different steels. Next the steel is *quenched*, which means to cool suddenly by plunging into water, oil, or other liquid. Steel may also be *case hardened* using a process known as *carburization.* Case hardening refers to the hardening of the surface layer of the metal. Carburization is a process where carbon is introduced into the metal by heating to a specified temperature range while in contact with a solid, liquid, or gas material consisting of carbon. This process is often fol-

Application	SAE No.	Application	SAE No.
Adapters	1145	Chain pins, transmission	4320
Agricultural steel	1070	" " "	4815
" "	1080	" " "	4820
Aircraft forgings	4140	Chains, transmission	3135
Axles, front or rear	1040	" "	3140
" " "	4140	Clutch disks	1060
Axle shafts	1045	" "	1070
" "	2340	" "	1085
" "	2345	Clutch springs	1060
" "	3135	Coil springs	4063
" "	3140	Cold-headed bolts	4042
" "	3141	Cold-heading steel	30905
" "	4063	Cold-heading wire or rod	rimmed*
" "	4340	" " "	1035
Ball-bearing races	52100	Cold-rolled steel	1070
Balls for ball bearings	52100	Connecting-rods	1040
Body stock for cars	rimmed*	" "	3141
Bolts, anchor	1040	Connecting-rod bolts	3130
Bolts and screws	1035	Corrosion resisting	51710
Bolts, cold-headed	4042	" "	30805
Bolts, connecting-rod	3130	Covers, transmission	rimmed*
Bolts, heat-treated	2330	Crankshafts	1045
Bolts, heavy-duty	4815	"	1145
" " "	4820	"	3135
Bolts, steering-arm	3130	"	3140
Brake levers	1030	"	3141
" "	1040	Crankshafts, Diesel engine	4340
Bumper bars	1085	Cushion springs	1060
Cams, free-wheeling	4615	Cutlery, stainless	51335
" "	4620	Cylinder studs	3130
Camshafts	1020	Deep-drawing steel	rimmed*
"	1040	" " "	30905
Carburized parts	1020	Differential gears	4023
" "	1022	Disks, clutch	1070
" "	1024	" "	1060
" "	1320	Ductile steel	30905
" "	2317	Fan blades	1020
" "	2515	Fatigue resisting	4340
" "	3310	" "	4640
" "	3115	Fender stock for cars	rimmed*
" "	3120	Forgings, aircraft	4140
" "	4023	Forgings, carbon steel	1040
" "	4032	" " "	1045
" "	1117	Forgings, heat-treated	3240
" "	1118	" " "	5140

* The "rimmed" and "killed" steels listed are in the SAE 1008, 1010 and 1015 group. See general description of these steels.

Figure 7–2 General applications for SAE steels. *Reprinted, by permission, from Oberg, Jones, and Horton,* Machinery's Handbook, *23rd ed. (New York: Industrial Press Inc., 1988), pp. 446–48.*

Application	SAE No.	Application	SAE No.
Forgings, heat-treated	6150	Key stock	1030
Forgings, high-duty	6150	“ “	2330
Forgings, small or medium	1035	“ “	3130
Forgings, large	1036	Leaf springs	1085
Free-cutting carbon steel	1111	“ “	9260
“ “ “ “	1113	Levers, brake	1030
Free-cutting chro.-ni. steel	30615	“ “	1040
Free-cutting mang. steel	1132	Levers, gear shift	1030
“ “ “ “	1137	Levers, heat-treated	2330
Gears, carburized	1320	Lock-washers	1060
“ “	2317	Mower knives	1085
“ “	3115	Mower sections	1070
“ “	3120	Music wire	1085
“ “	3310	Nuts	3130
“ “	4119	Nuts, heat-treated	2330
“ “	4125	Oil-pans, automobile	rimmed*
“ “	4320	Pinions, carburized	3115
“ “	4615	“ “	3120
“ “	4620	“ “	4320
“ “	4815	Piston-pins	3115
“ “	4820	“ “	3120
Gears, heat-treated	2345	Plow beams	1070
Gears, car and truck	4027	Plow disks	1080
“ “ “ “	4032	Plow shares	1080
Gears, cyanide-hardening	5140	Propeller shafts	2340
Gears, differential	4023	“ “	2345
Gears, high duty	4640	“ “	4140
“ “ “	6150	Races, ball-bearing	52100
Gears, oil-hardening	3145	Ring gears	3115
“ “ “	3150	“ “	3120
“ “ “	4340	“ “	4119
“ “ “	5150	Rings, snap	1060
Gears, ring	1045	Rivets	rimmed*
“ “	3115	Rod and wire	killed*
“ “	3120	Rod, cold-heading	1035
“ “	4119	Roller bearings	4815
Gears, transmission	3115	Rollers for bearings	52100
“ “	3120	Screws and bolts	1035
“ “	4119	Screw stock, Bessemer	1111
Gears, truck and bus	3310	“ “ “	1112
“ “ “ “	4320	“ “ “	1113
Gear shift levers	1030	Screw stock, open hearth	1115
Harrow disks	1080	Screws, heat-treated	2330
“ “	1095	Seat springs	1095
Hay-rake teeth	1095	Shafts, axle	1045

Application	SAE No.	Application	SAE No.
Shafts, cyanide-hardening	5140	Steel, cold-heading	30905
Shafts, heavy-duty	4340	Steel, free-cutting carbon	11111
“ “ “	6150	“ “ “ “	1113
“ “ “	4615	Steel, free-cutting chro.-ni.	30615
“ “ “	4620	Steel, free-cutting mang.	1132
Shafts, oil-hardening	5150	“ “ “ “	0000
Shafts, propeller	2340	Steel, minimum distortion	4615
“ “	2345	“ “ “	4620
“ “	4140	“ “ “	4640
Shafts, transmission	4140	Steel, soft ductile	30905
Sheets and strips	rimmed*	Steering arms	4042
Snap rings	1060	Steering-arm bolts	3130
Spline shafts	1045	Steering knuckles	3141
“ “	1320	Steering-knuckle pins	4815
“ “	2340	“ “ “	4820
“ “	2345	Studs	1040
“ “	3115	“	1111
“ “	3120	Studs, cold-headed	4042
“ “	3135	Studs, cylinder	3130
“ “	3140	Studs, heat-treated	2330
“ “	4023	Studs, heavy-duty	4815
Spring clips	1060	“ “ “	4820
Springs, coil	1095	Tacks	rimmed*
“ “	4063	Thrust washers	1060
“ “	6150	Thrust washers, oil-harden.	5150
Springs, clutch	1060	Transmission shafts	4140
Springs, cushion	1060	Tubing	1040
Springs, leaf	1085	Tubing, front axle	4140
“ “	1095	Tubing, seamless	1030
“ “	4063	Tubing, welded	1020
“ “	4068	Universal joints	1145
“ “	9260	Valve springs	1060
“ “	6150	Washers, lock	1060
Springs, hard-drawn coiled	1066	Welded structures	30705
Springs, oil-hardening	5150	Wire and rod	killed*
Springs, oil-tempered wire	1066	Wire, cold-heading	rimmed*
Springs, seat	1095	“ “ “	1035
Springs, valve	1060	Wire, hard-drawn spring	1045
Spring wire	1045	“ “ “ “	1055
Spring wire, hard-drawn	1055	Wire, music	1085
Spring wire, oil-tempered	1055	Wire, oil-tempered spring	1055
Stainless irons	51210	Wrist-pins, automobile	1020
“ “	51710	Yokes	1145
Steel, cold-rolled	1070		

*** The "rimmed" and "killed" steels listed are in the SAE 1008, 1010 and 1015 group. See general description of these steels.**

Figure 7–2 Continued

lowed by quenching to enhance the hardening process. *Tempering* is a process of reheating a normalized or hardened steel through a controlled process of heating the metal to a specified temperature, followed by cooling at a predetermined rate to achieve certain hardening characteristics. For example, the tip of a tool may be hardened while the balance of the tool remains unchanged.

Under certain heating and cooling conditions and techniques, steel may also be softened using a process known as *annealing*.

Hardness Testing. There are several methods of checking material hardness. The techniques have common characteristics based on the depth of penetration of a measuring device or other mechanical systems that evaluate hardness. The Brinell and Rockwell hardness tests are popular. The Brinell test is performed by placing a known load using a ball of a specified diameter, in contact with the material surface. The diameter of the resulting impression in the material is measured and the Brinell Hardness Number (BHN) is then calculated. The Rockwell hardness test is performed using a machine that measures hardness by determining the depth of penetration of a spherical-shaped device under controlled conditions. There are several Rockwell hardness scales depending on the type of material, the type of penetrator, and the load applied to the device. A general or specific note on a drawing that requires a hardness specification may read: CASE HARDEN 58 PER ROCKWELL “C” SCALE. For additional information, refer to the *Machinery's Handbook*.

Nonferrous Metals

Metals that do not contain iron have properties that are better suited for certain applications where steel may not be appropriate.

Aluminum. Aluminum is corrosion resistant, lightweight, easily cast, conductive of heat and electricity, may be easily *extruded* (shaped by forcing through a die), and is very malleable. Pure aluminum is seldom used, but alloying it with other elements provides materials that have an extensive variety of applications. Some aluminum alloys lose strength at temperatures generally above 121° C (250° F), while they gain strength at cold temperatures. There are a variety of aluminum alloy numerical designations used. A two- or three-digit num-

COMPUTER NUMERICAL CONTROL MACHINE TOOLS (CNC)

Computer-aided design and drafting (CADD) has a direct link to computer-aided manufacturing (CAM) in the form of *computer numerical control* (CNC) machine tools. A flow chart for the computer numerical control process is shown in Figure 7–3. Figure 7–4 shows a CNC machine. In most cases, the drawing is generated in a computer and this information is sent directly to the machine tool for production. In some applications, information in the computer is transferred to a punched numerical control tape. The numerical control tape may be used as a backup system to the CNC, or the tape may be used as a copy for the numerical control program. Nearly all types of machine tools have been designed to operate from numerical control tape. However, this method is becoming obsolete.

Many CNC systems offer a microcomputer base that incorporates a CRT (monitor) display and full alphanumeric keyboard, making programming and on-line editing easy and fast. For example, data input for certain types of machining may result in the programming of one of several identical features while the other features are oriented and programmed automatically, such as the five equally spaced blades of the centrifugal fan shown in Figure 7–5. Among the advantages of CNC machining are increased productivity, reduction of production costs, and manufacturing versatility.

The drafter has a special challenge when preparing drawings for CNC machining. The drawing method must coordinate with a system of controlling a machine tool by instructions in the form of numbers. The types of dimensioning systems that relate directly to CNC applications are tabular, arrowless, datum, and related coordinate systems. The main emphasis is to coordinate the dimensioning system with the movement of the machine tools. This can be best accomplished with datum dimensioning systems in which each dimension originates from a common beginning point. Programming the CNC system requires that cutter compensation be provided for contouring. This task is automatically computer-calculated in some CNC machines, as shown in Figure 7–6. Definitions and examples of the dimensioning systems are discussed in Chapter 8, Dimensioning.

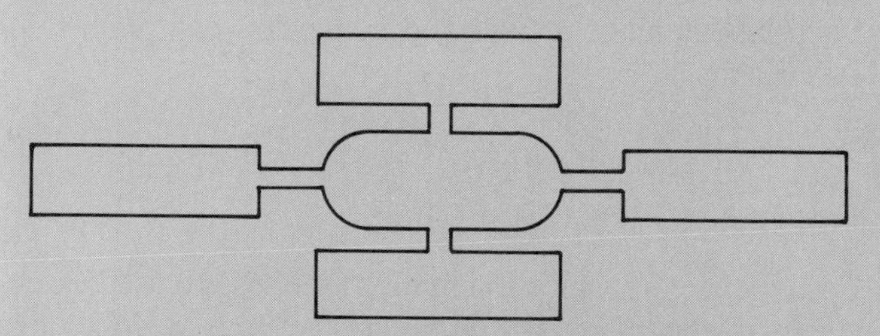

Figure 7–3 The computer numerical control (CNC) process.

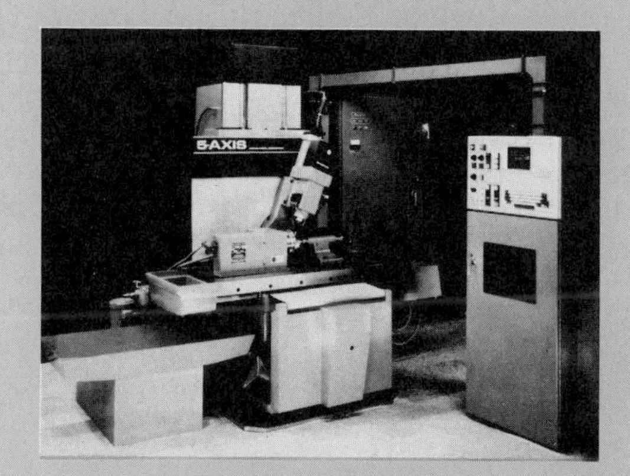

Figure 7–4 CNC machine. *Courtesy Boston Digital Corporation.*

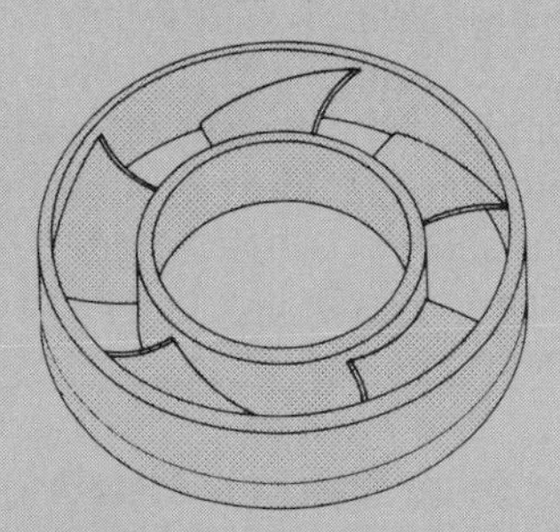

Figure 7–5 CNC programming of a part with five equally spaced blades. This figure demonstrates the CNC programming of one of the five equally spaced blades and the automatic orientation and programming of the other four blades. *Courtesy Boston Digital Corporation.*

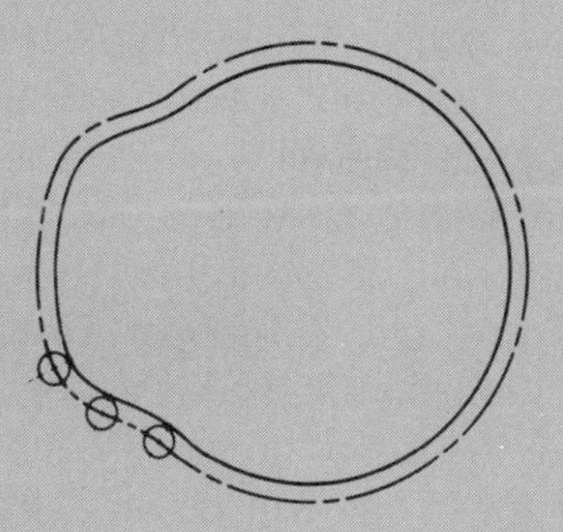

Figure 7–6 Automatic cutter compensation for profile machining. *Courtesy Boston Digital Corporation.*

ber is used, and the first digit indicates the alloy type, as follows: 1 = 99% pure, 2 = copper, 3 = manganese, 4 = silicon, 5 = magnesium, 6 = magnesium and silicon, 7 = zinc, 8 = other. For designations above 99%, the last two digits of the code are the amount over 99%. For example, 1030 denotes 99.30% aluminum. The second digit is any number between 0 and 9 where 0 means no control of specific impurities and numbers 1–9 identify control of individual impurities.

Copper Alloys. Copper is easily rolled and drawn into wire, has excellent corrosion resistance, is a great electrical conductor, and has better ductility than any metal except for silver and gold. Copper is alloyed with many different metals for specific advantages, which include improved hardness, casting ability, machinability, corrosion resistance, elastic properties, and lower cost.

Brass. Brass is a widely used alloy of copper and zinc. Its properties include corrosion resistance, strength, and ductility. For most commercial applications, brass has about a 90% copper and 10% zinc content. Brass may be manufactured by any number of processes including casting, forging, stamping, or drawing. Its uses include valves, plumbing pipe and fittings, and radiator cores. Brass with greater zinc content may be used for applications requiring greater ductility, such as cartridge cases, sheet metal, or tubing.

Bronze. Bronze is an alloy of copper and tin. Tin in small quantities adds hardness and increases wear resistance. Tin content in coins and medallions, for example, ranges from 4% to 8%. Increasing amounts of tin also improve the hardness and wear resistance of the material, but cause brittleness. Phosphorus added to bronze (phosphor bronze) increases its casting ability and aids in the production of more solid castings, which is important for thin shapes. Other materials, such as lead, aluminum, iron, and nickel, may be added to copper for specific applications.

Precious and Other Specialty Metals

Precious metals include gold, silver, and platinum. These metals are valuable because they are rare, costly to produce, and have specific properties that influence use in certain applications.

Gold. Gold for coins and jewelry is commonly hardened by adding copper. Gold coins, for example, are 90% gold and 10% copper. The term *carat* is used to refer to the purity of gold, where 1/24 gold is one carat. Therefore, 24 × 1/24 or 24 carats is pure gold. Fourteen-carat gold, for example, is 14/24 gold and 10/24 copper. Gold is extremely malleable, corrosion resistant, and is the best conductor of electricity. In addition to use in jewelry and coins, gold is used as a conductor in some electronic circuitry applications. Gold is also sometimes used in applications where resistance to chemical corrosion is required.

Silver. Silver is alloyed with 8%–10% copper for use in jewelry and coins. Sterling silver for use in such items as eating utensils and other household items is 925/1000 silver. Silver is easy to shape, cast, or form, and the finished product may be polished to a high-luster finish. Silver has uses similar to gold because of corrosion resistance and ability to conduct electricity.

Platinum. Platinum is rarer and more expensive than gold. Industrial uses include applications where corrosion resistance and a high melting point are required. Platinum is used in catalytic converters because it has the unique ability to react with and reduce carbon monoxide and other harmful exhaust emissions in automobiles. The high melting point of platinum makes it desirable in certain aerospace applications.

Columbium. Columbium is used in nuclear reactors because it has a very high melting point, 2403° C (4380° F), and is resistant to radiation.

Titanium. Titanium has many uses in the aerospace and jet aircraft industries because it has the strength of steel, the approximate weight of aluminum, and is resistant to corrosion and temperatures up to 427° C (800° F).

Tungsten. Tungsten has been used extensively as the filament in light bulbs because of its ability to be drawn into very fine wire and its high melting point. Tungsten, carbon, and cobalt are formed together under heat and pressure to create tungsten carbide, the hardest man-made material. Tungsten carbide is used to make cutting tools for any type of manufacturing application. Tungsten carbide saw blade inserts are used in saws for carpentry so the cutting edge will last longer. Such blades make a finer and faster cut than plain steel saw blades.

MANUFACTURING PROCESSES

Casting, forging, and machining processes are used extensively in the manufacturing industry. It is a good idea for the entry-level drafter or pre-engineer to be generally familiar with types of casting, forging, and machining processes, and to know how to prepare related drawings. A number of methods may be used by industry to prepare casting, forging, and machining drawings. For this reason, it is best for the beginning drafter to remain flexible and adapt to the standards and techniques used by the specific company. As the drafter gains knowledge of company products, processes, and design goals, he or she may begin to produce designs. It is not uncommon for a drafter to become a designer after three years of practical experience.

Castings

Castings are the end result of a process called founding. *Founding*, or *casting*, as the process is commonly called, is the pouring of molten metal into a hollow or wax-filled mold. The mold is made in the shape of the desired casting. There are several casting methods used in industry. The results of some of the processes are castings that are made to very close tolerances and with smooth finished surfaces. In the simplest terms, castings are made in three separate steps: (1) a pattern that is the same shape as the desired finished product is constructed; (2) using the pattern as a guide, a mold is made by packing sand or other material around the pattern; (3) when the pattern is removed from the mold, molten metal is poured into the hollow cavity. After the molten metal solidifies, the surrounding material is removed and the casting is ready for cleanup or machining operations.

Sand Casting. Sand casting is the most commonly used method of making castings. There are two general types of sand castings: green sand and dry sand molding. *Green sand* is a specially refined sand that is mixed with specific moisture, clay, and resin, which work as binding agents during the molding and pouring procedures. New sand is light brown in color; the term green sand refers to the moisture content. In the *dry sand* molding process, the sand does not have any moisture content. The sand is bonded together with specially formulated resins. The end result of the green sand or the dry sand molds is the same.

Sand castings are made by pounding or pressing the sand around a split pattern. The first or lower half of the pattern is placed upside down on a molding board, then sand is pounded or compressed around the pattern in a box called a *drag*. The drag is then turned over, and the second or upper half of the pattern is formed when another box, called a *cope*, is packed with sand and joined to the drag. A fine powder is used as a parting agent between the cope and drag at the parting line. The parting line is the separating joint between the two parts of the pattern or mold. The entire box, made up of the cope and drag, is referred to as a *flask*. (See Figure 7–7.) Before the molten metal can be poured into the cavity, a passageway for the metal must be made. The passageway is called a *runner* and *sprue*. The location and design of the sprue and runner are important to allow for a rapid and continuous flow of metal. Additionally, vent holes are established to allow for gases, impurities, and metal to escape from the cavity. Finally, a riser (or group of risers) is used, depending on the size of the casting, to allow for the excess metal to evacuate from the mold and, more importantly, to help reduce shrinking and incomplete filling of the casting. (See Figure 7–8.) After the casting has solidified and cooled, the filled risers, vent holes, and runners are removed.

Cores. In many situations a hole or cavity is desired in the casting to help reduce the amount of material removal later or to establish a wall thickness. When this is necessary, a core is used. Cores are made from either clean sand mixed with binders, such as resin, and baked

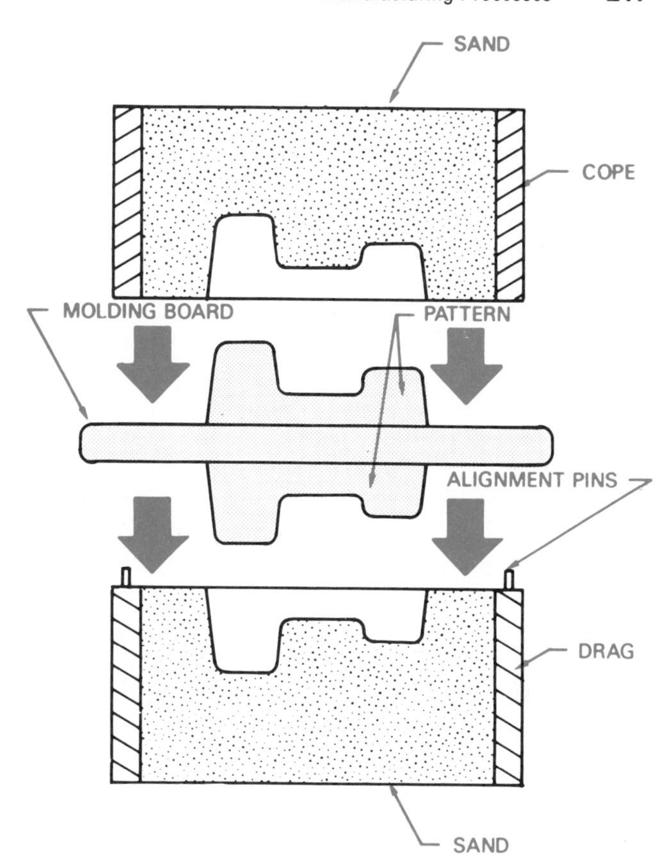

Figure 7–7 Components of the sand casting process.

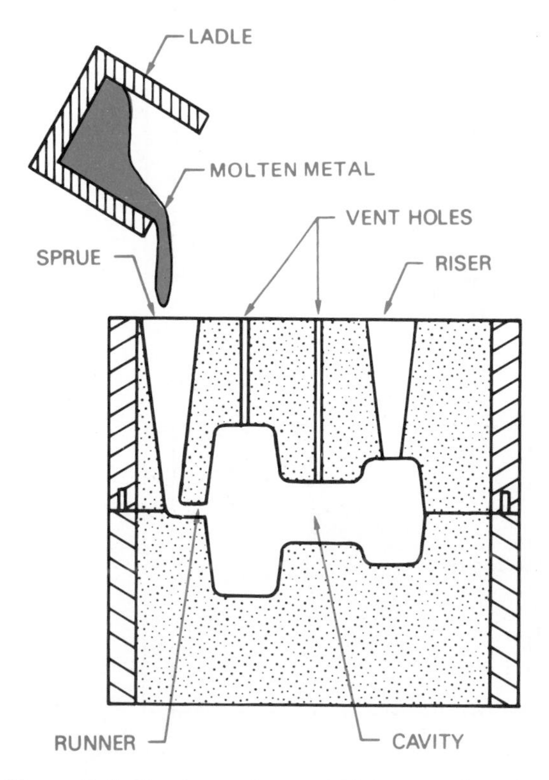

Figure 7–8 Pouring molten metal into a sand casting mold.

in an oven for hardening or ceramic products (when a more refined surface finish is required). When the pattern is made, a place for positioning the core in the mold is established; this is referred to as the core print. After the mold is made, the core is then placed in position in the mold at the core print. The molten metal, when poured into the mold, flows around the core. After the metal has cooled, the casting is removed from the flask and the core is cleaned out, usually by shaking or tumbling. Figure 7–9 shows cores in place. Cored features help reduce casting weight and save on machining costs. Cores used in sand casting should generally be more than one inch (25.4 mm) in cross section. Cores used in precision casting methods may have much closer tolerances and fine detail. Certain considerations must be taken for supporting very large or long cores when placed in the mold. Usually in sand casting, cores require extra support when they are three times longer than the cross-sectional dimension. Depending on the casting method, the material, the machining required, and the quality, the core holes should be a specified dimension smaller than the desired end product if the hole is to be machined to its final dimension. Cores in sand castings should be between .125 to .5 in. (3.2 to 12.7 mm) smaller than the finished size.

Centrifugal Casting. Objects with circular or cylindrical shapes lend themselves to *centrifugal casting*. In this casting process, a mold is revolved very rapidly while molten metal is poured into the cavity. The molten metal is forced outward into the mold cavity by centrifugal forces. No cores are needed because the fast revolution holds the metal against the surface of the mold. This casting method is especially useful for casting cylindrical shapes such as tubing, pipes, or wheels. (See Figure 7–10.)

Die Casting. Some nonferrous metal castings are made using the *die casting* process. Zinc alloy metals are the most common, although brass, bronze, aluminum, and other nonferrous products are also made using this process. Die casting is the injection of molten metal into a steel or cast iron die under high pressure. The advantage of die casting over other methods, such as sand casting, is that castings can be produced quickly and economically on automated production equipment. When multiple dies are used, a number of parts can be cast in one operation. Another advantage of die casting is that high-quality precision parts can be cast with fine detail and a very smooth finish.

Permanent Casting. Permanent casting refers to a process in which the mold can be used many times. This type of casting is similar to sand casting in that molten metal is poured into a mold. It is also similar to die casting because the mold is made of cast iron or steel. The

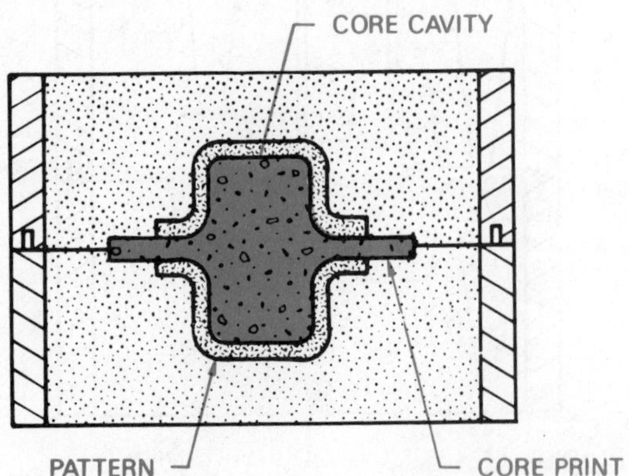

Figure 7–9 Cores in place.

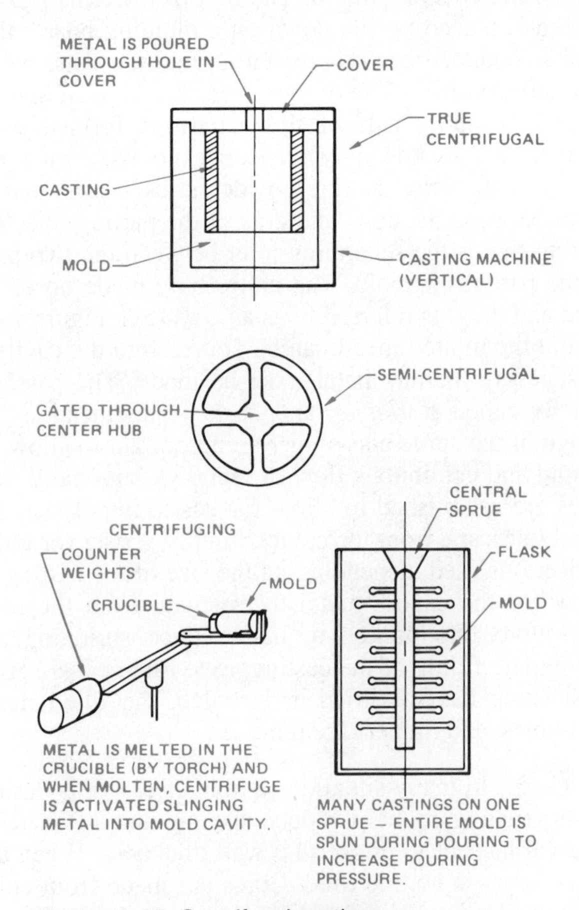

Figure 7–10 Centrifugal casting.

The result of permanent casting is a product that has better finished qualities than can be gained by sand casting.

Investment Casting. *Investment casting* is one of the oldest casting methods. It was originally used in France for the production of ornamental figures. The process used today is a result of the *ciré perdue*, or *lost-wax,* casting technique that was originally used. The reason that investment casting is also called lost-wax casting is that the pattern is made of wax. This wax pattern allows for the development of very close tolerances, fine detail, and precision castings. The wax pattern is coated with a ceramic paste. The shell is allowed to dry and then is baked in an oven to allow the wax to melt and flow out; thus "lost" as the name implies. The empty ceramic mold has a cavity that is the same shape as the precision wax pattern. This cavity is then filled with molten metal. When the metal solidifies, the shell is removed and the casting is complete. Generally very little cleanup or finishing is required on investment castings. (See Figure 7–11.)

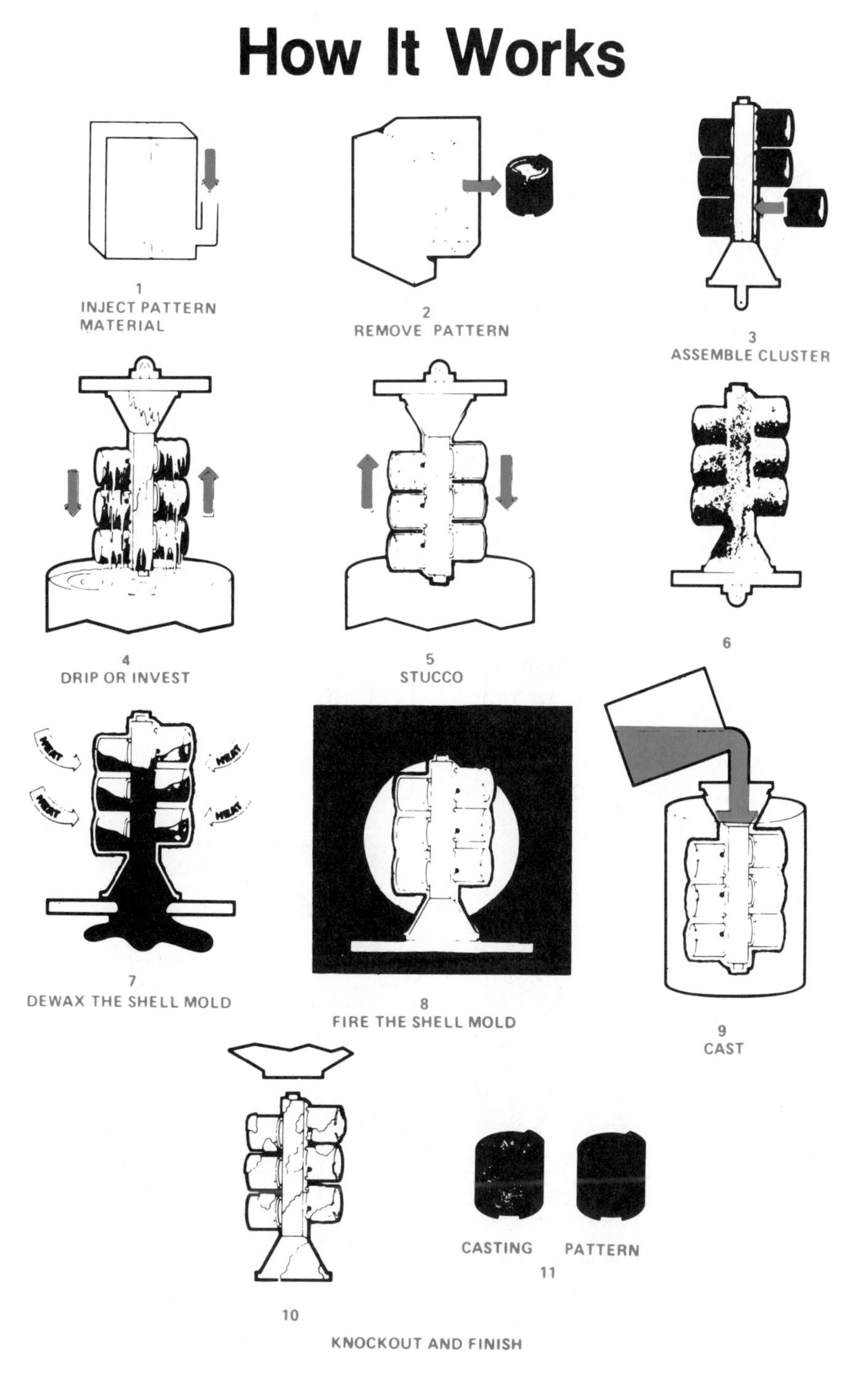

Figure 7–11 The investment casting process. *Courtesy Precision Castparts Corporation.*

Forgings

Forging is a process of shaping malleable metals by hammering or pressing between dies that duplicate the desired shape. The forging process is shown in Figure 7–12. Forging may be accomplished on hot or cold materials. Cold forging is possible on certain materials or material thicknesses where hole punching or bending is the required result. Some soft, nonferrous materials may be forged into shape while cold. Ferrous materials such as iron and steel must be heated to a temperature that results in an orange-red or yellow color. This color is usually achieved between 982° to 1066° C. Forging is used for a large variety of products and purposes. The advantage of forging over casting or machining operations is that the material is not only shaped into the desired form, but in the process it retains its original grain structure. Forged metal is generally stronger and more ductile than cast metal, and exhibits a greater resistance to fatigue and shock than machined parts. Notice in Figure 7–13 that the grain structure of the forged material remains parallel to the contour of the part, while the machined part cuts through the cross section of the material grain.

Hand Forging. Hand forging is an ancient method of forming metals into desired shapes. The method of heating metal to a red color and then beating it into shape is called *smithing* or, more commonly, *blacksmithing*. Blacksmithing is used only in industry for finish work, but is still used for horseshoeing and the manufacture of specialty ornamental products.

Machine Forging. Types of machine forging include upset, swaging, bending, punching, cutting, and welding. *Upset* forging is a process of forming metal by pressing along the longitudinal dimension to decrease the length while increasing the width. For example, bar stock is

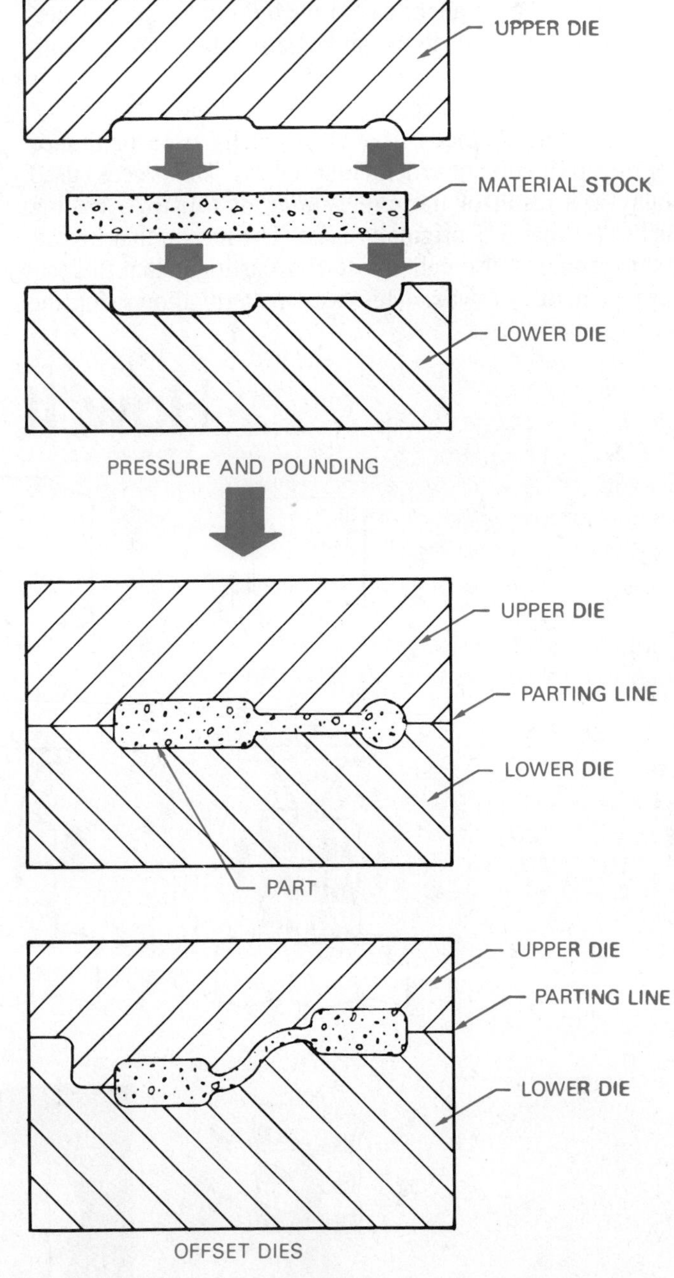

Figure 7–12 The forging process.

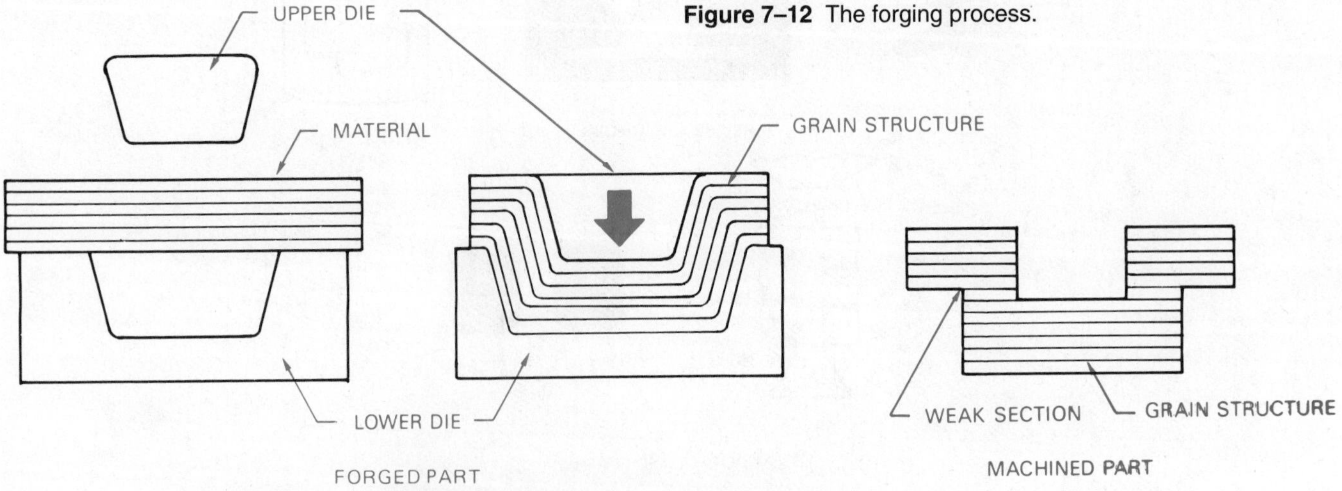

UPPER DIE
MATERIAL
GRAIN STRUCTURE
LOWER DIE
FORGED PART
WEAK SECTION
GRAIN STRUCTURE
MACHINED PART

Figure 7–13 Forging compared to machining.

upset forged by pressing dies together from the ends of the stock to establish the desired shape. *Swaging* is the forming of metal by using concave tools or dies that result in a reduction in material thickness. *Bending* is accomplished by forming metal between dies, changing it from flat stock to a desired contour. Bending sheet metal is a cold forging process in which the metal is bent in a machine called a break. *Punching* and *cutting* are performed when the die penetrates the material to create a hole of any desired shape and depth or to remove material by cutting away. In forge *welding*, metals are joined together under extreme pressure. Material that is welded in this manner is very strong. The resulting weld takes on the same characteristics as the metal before joining.

Mass production forging methods, as shown in Figure 7–14, allow for the rapid production of the high-quality products shown in Figure 7–15. In machine forging, the dies are arranged in sequence so that the finished forging is done in a series of steps. Complete shaping may take place after the material has been moved through several stages. Additional advantages of machine forging include:

1. The part is formed uniformly throughout the length and width.
2. The greater the pressure exerted on the material, the greater the improvement of the metallic properties.
3. Fine grain structure is maintained to help increase the part's resistance to shock.
4. A group of dies may be placed in the same press.

Figure 7–14 Machine forging. *Courtesy Muller Weingarten Corp.*

MACHINE PROCESSES

The concepts covered in this chapter serve as a basis for many of the dimensioning practices presented in Chapter 8. A general understanding of machining processes and the drawing representations of these processes is a necessary prerequisite to dimensioning practices. Problem applications are provided in the following chapters because a knowledge of dimensioning practice is important and necessary to be able to complete manufacturing drawings.

Mechanical drafting and manufacturing are very closely allied. The mechanical drafter should have a general knowledge of manufacturing methods, including machining processes. It is common for the drafter to consult with engineers and machinists regarding the best methods to implement a drawing from the design to the manufacturing processes. Part of design problem solving is to create a design that is not only functional, but one that can also be manufactured using available technology and at a cost that justifies the end product. The solutions to these types of concerns depend on how familiar the drafter and designer are with the manufacturing capabilities of the company. For example, if the designer over-tolerances a part or feature, the result could be a rejection of the part by manufacturing or an expensive machining operation. The drafter should also know something about how machining processes operate so that drawing specifications do not call out something that is not feasible to manufacture. The drafter should be familiar with the

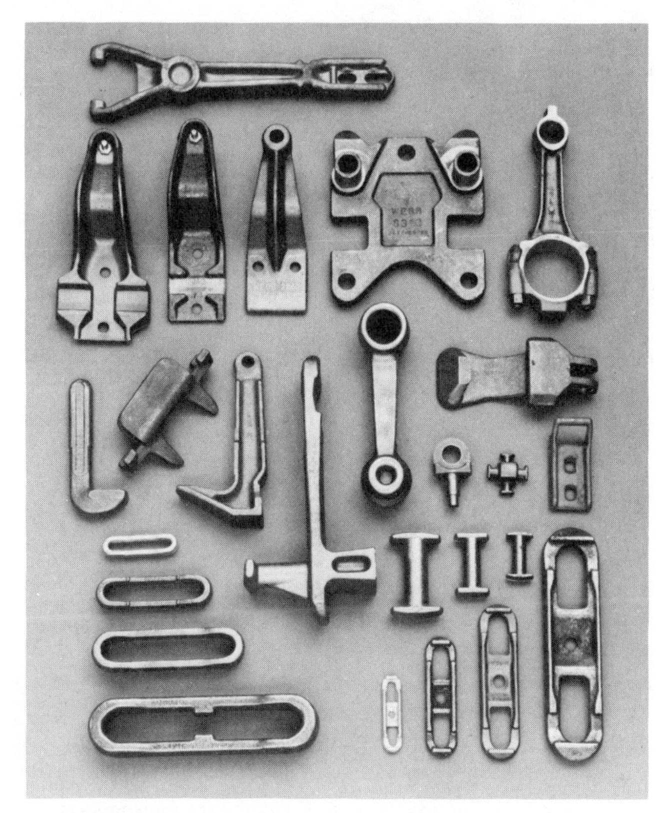

Figure 7–15 Forged products. *Courtesy Jarvis B. Webb Co.*

processes that are used in machine operations so the notes that are placed on a drawing conform to the proper machining techniques. Notes for machining processes are given on a drawing in the same manner that they are performed in the shop. For example, the note for a counterbore is given as the diameter and depth of the hole (first process) and the diameter and depth of the counterbore (second process). Also, the drafter should know that the specification given will, in fact, yield the desired result. One interesting aspect of the design drafter's job is the opportunity to communicate with people in manufacturing and come up with a design that will enable the product to be easily manufactured.

Machine Tools

Drilling Machine. The *drilling machine*, often referred to as a *drill press* (Figure 7–16), is commonly used to machine drilled holes. Drilling machines are also used to perform other operations, such as reaming, boring, countersinking, counterboring, and tapping. During the drilling procedure, the material is held on a table while the drill or other tool is held in a revolving spindle over the material. When drilling begins, a power or hand-feed mechanism is used to bring the rotating drill in contact with the material. Mass production drilling machines are designed with multiple spindles. Automatic drilling procedures are available on turret drills. These turret drills allow for automatic tool selection and spindle speed. Several operations can be performed in one setup — for example, drilling a hole to a given depth and tapping the hole with a specified thread.

Figure 7–16 Drilling machine. *Courtesy Delta International Machining Corp.*

Grinding Machine. A *grinding machine* uses a rotating abrasive wheel, rather than a cutting tool, for the purpose of removing material. (See Figure 7–17.) The grinding process is generally used when a smooth, accurate surface finish is required. Extremely smooth surface finishes can be achieved by honing or lapping. *Honing* is a fine abrasive process often used to establish a smooth finish inside cylinders. *Lapping* is the process of creating a very smooth surface finish using a soft metal impregnated with fine abrasives, or fine abrasives mixed in a coolant that floods over the part during the lapping process.

Lathe. One of the earliest machine tools, the *lathe* (Figure 7–18) is used to cut material by turning cylindrically shaped objects. The material to be turned is held between two rigid supports, called *centers*, or in a holding device called a *chuck* or *collet*, as shown in Figure 7–19. The material is rotated on a spindle while a cutting tool is brought into contact with the material. The cutting tool is supported by a tool holder on a carriage that slides along a bed as the lathe operation continues. Figure 7–20 shows a drilling operation performed on a lathe. The turret is used in mass production manufacturing where one machine setup must perform several operations. A turret lathe is designed to carry several cutting tools in place of the lathe tailstock or on the lathe carriage. The operation of the turret provides the operator with an automatic selection of cutting tools at preestablished fabrication stages. Figure 7–21 shows an example of eight turret stations and the tooling used.

Figure 7–17 Grinding machine. *Courtesy Litton Industrial Automation.*

Figure 7–18 Lathe. *Courtesy Hardinge Brothers, Inc.*

Figure 7–20 Drilling operation performed on a lathe. *Courtesy The Cleveland Twist Drill Company.*

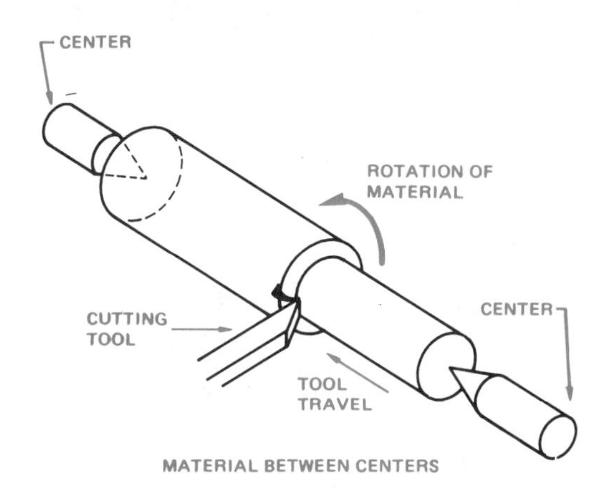

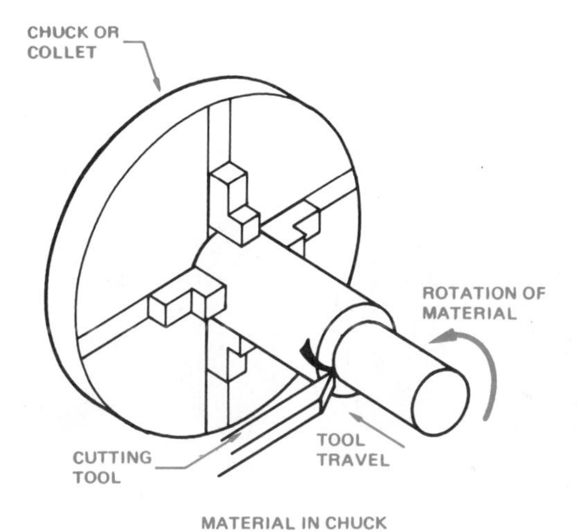

Figure 7–19 How material is held in a lathe.

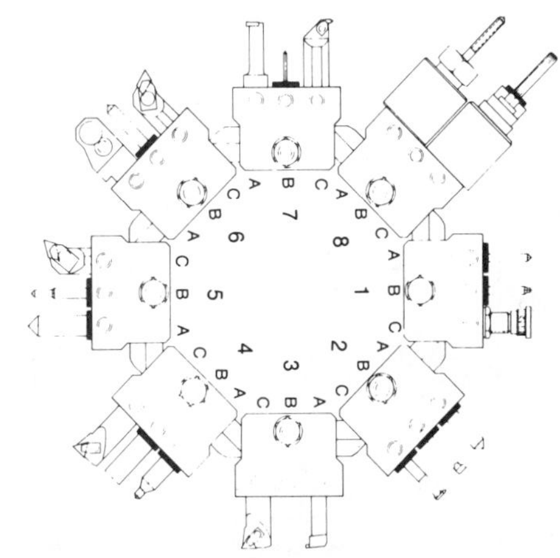

STATION	TOOLING Model	TOOLING Description
1A		Drill with Bushing
1B		Center Drill with Bushing
1C	T20-⅝	Adjustable Revolving Stock Stop
2A		Drill with Bushing
2B		Boring Bar with Bushing
2C		Threading Tool with Bushing
3A		Grooving Tool
3C		Insert Turning Tool
4A		Center Drill with Bushing
4B		Flat Bottom Drill with Bushing
4C		Insert Turning Tool
5A		Drill with Bushing
5B		Step Drill with Bushing
5C		Insert Turning Tool
6A	T8-⅝	Knurling Tool
6B		Drill with Bushing
6C		Insert Turning Tool
7A		Grooving Tool
7B		Drill with Bushing
7C		Insert Threading Tool
8A	TT-⅝	"Collet Type" Releasing Tap Holder
		Tap Collet
		Tap
8C	TE-⅝	Tool Holder Extension
	T19-⅝	Floating Reamer Holder
		Reamer With Bushing

Figure 7–21 A turret with eight tooling stations and the tools used at each station. *Courtesy Toyoda Machinery USA Inc.*

Milling Machine. The *milling machine* (see Figure 7–22) is one of the most versatile machine shop tools. The milling machine uses a rotary cutting tool to remove material from the work. The two general types of milling machines are *horizontal* and *vertical* mills. The difference is in the position of the cutting tool, which may be mounted on either a horizontal or vertical spindle. In the operation, the work is fastened to a table that is mechanically fed into the cutting tool, as shown in Figure 7–23. There are a large variety of milling cutters available that influence the flexibility of operations and shapes that can be performed using the milling machine. Figure 7–24 shows a few of the milling cutters available. Figure 7–25 shows a series of milling cutters grouped together to perform a milling operation. End milling cutters, as shown in Figure 7–26, are designed to cut on the end and the sides of the cutting tool. Figure 7–27 shows an end mill in operation. Milling machines that are commonly used in high-production manufacturing often have two or more cutting heads that are available to perform multiple operations. The machine tables of standard horizontal or vertical milling machines move from left to right (x-axis), forward and backward (y-axis), and up and down (z-axis), as shown in Figure 7–28.

The Universal Milling Machine. Another type of milling machine, known as the *universal milling machine*, has table action that includes x-, y-, and z-axis movement plus angular rotation. The universal milling machine looks much the same as other milling machines,

Figure 7–23 Material removal with a vertical milling machine cutter. *Courtesy The Cleveland Twist Drill Company.*

Figure 7–22 Milling machine: (a) horizontal mill; (b) close-up of milling cutter.

Figure 7–24 Milling cutters. *Courtesy The Cleveland Twist Drill Company.*

Figure 7–25 A grouping of horizontal milling cutters for a specific machining operation. *Courtesy The Cleveland Twist Drill Company.*

Figure 7–26 End cutter. *Courtesy The Cleveland Twist Drill Company.*

Figure 7–27 End mill operation. *Courtesy The Cleveland Twist Drill Company.*

but has the advantage of additional angular table movement as shown in Figure 7–29. This additional table movement allows the universal milling machine to produce machined features, such as spirals, not possible on conventional machines.

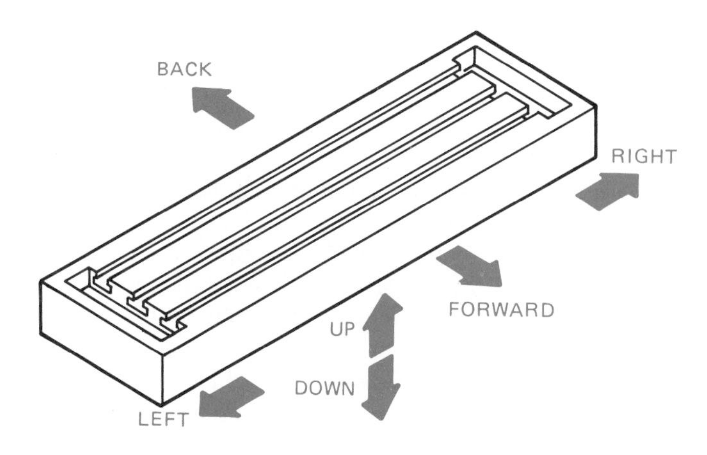

Figure 7–28 Table movements on a standard milling machine.

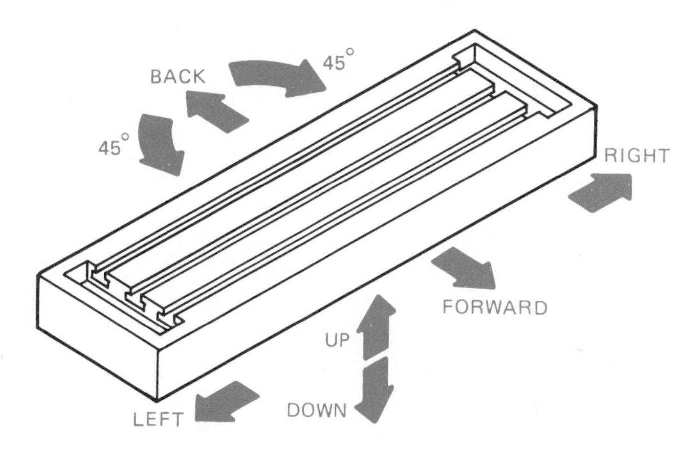

Figure 7–29 Table movements on a universal milling machine.

Saw Machines. Saw machines may be used as cutoff tools to establish the length of material for further machining, or saw cutters can be used to perform certain machining operations such as cutting a narrow slot *(kerf).*

Two types of machines that function as cutoff units only are the *power hacksaw* and the *band saw*. These saws are used to cut a wide variety of materials, including aluminum, brass, or steel, and are used rather than abrasive cutoff saws because abrasive discs are expensive and wear out quickly. The hacksaw, as shown in Figure 7–30, operates using a back-and-forth motion. The fixed blade in the power hacksaw cuts material on the forward motion. The metal cutting band saw is available in a vertical or horizontal design, as shown in Figure 7–31. This type of cutoff saw has a continuous band that runs either vertically or horizontally around turning wheels. Vertical band saws may also be used to cut out irregular shapes.

Saw machines are also made with circular abrasive or metal cutting wheels. The *abrasive saw* may be used for high-speed cutting where a narrow saw kerf is desirable or when very hard materials must be cut. One advantage of the abrasive saw is its ability to cut a variety of materials — from soft aluminum to case-hardened steels. (Cutting a variety of metals on the band or the power hacksaw requires blade and speed changes.) A disadvantage of the abrasive saw is the expense of abrasive discs. Many companies use this saw only when versatility is needed. The abrasive saw is usually found in the grinding room where abrasive particles can be contained, but may also be used in the shop for general purpose cutting. Metal cutting saws with teeth, also known as *cold* saws, are used for precision cutoff operations, cutting saw kerfs, slitting metal, and other manufacturing uses. Figure 7–32 shows a circular saw blade.

Water jet cutting is used on composite materials and thin metal with a computer-controlled 55,000-psi water jet. Cuts are made with holding tolerances of .0008 in. (0.020 mm) and without generating heat.

Shaper. The *shaper* is used primarily for production of horizontal, vertical, or angular flat surfaces. Shapers are generally becoming out of date and are rapidly being replaced by milling machines. A big problem with the shaper in mass production industry is that it is very slow. And this machine tool cuts only in one direction. One of the main advantages of the shaper is its ability to cut irregular shapes that cannot be conveniently reproduced on a milling machine or other machine tools. However, other more advanced multiaxis machine tools are now available that quickly and accurately cut irregular contours.

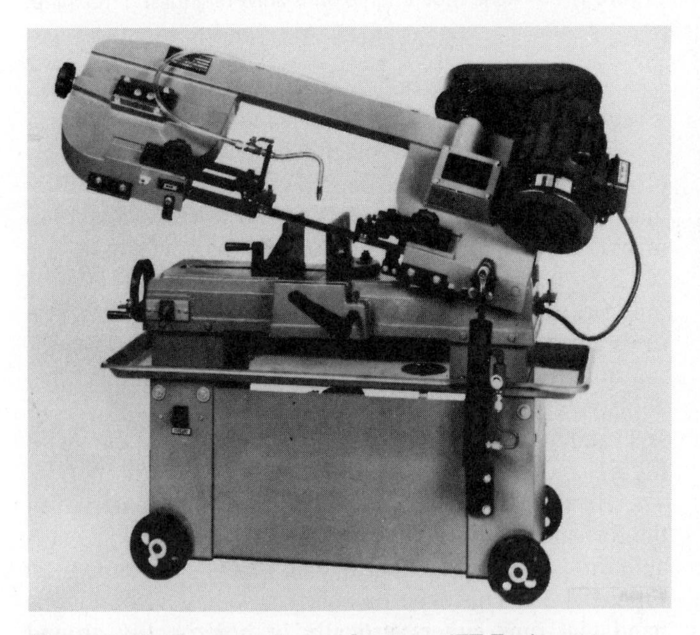

Figure 7–30 Power hacksaw. *Courtesy JET Equipment and Tool.*

Chemical Machining

Chemical machining uses chemicals to remove material accurately. The chemicals are placed on the material to be removed while other areas are protected. The amount of time the chemical remains on the surface

(a)

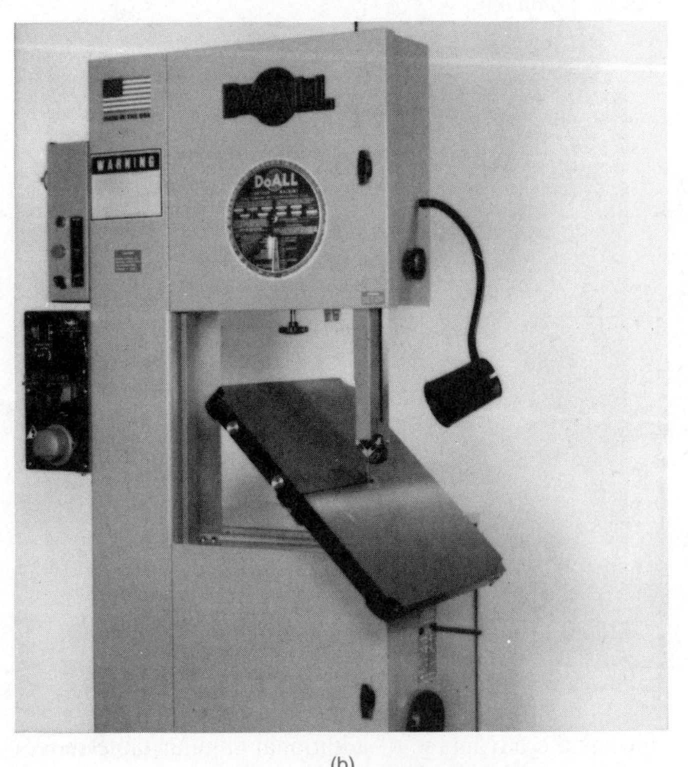

(b)

Figure 7–31 (a) Horizontal band saw, (b) vertical band saw. *Courtesy DoALL Company.*

Figure 7–32 Circular blade saw. *Courtesy The Cleveland Twist Drill Company.*

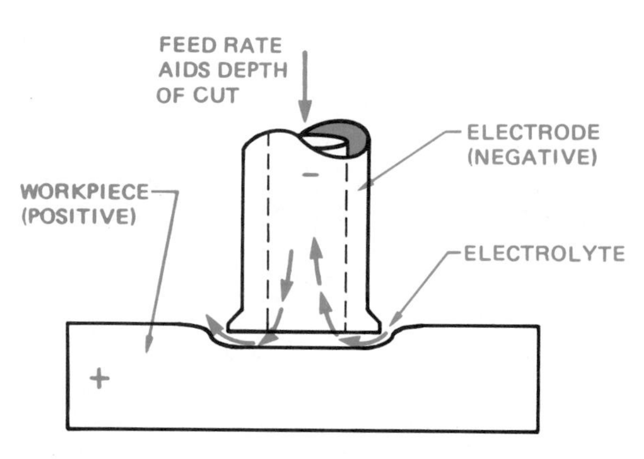

Figure 7–33 Electrochemical machining (ECM).

determines the extent of material removal. This process, also known as chemical milling, is generally used in situations where conventional machining operations are difficult. A similar method, referred to as chemical blanking, is used on thin material to remove unwanted thickness in certain areas while maintaining "foil" thin material at the machined area. Material may be machined to within .00008 in. (0.002 mm) using this technique.

Electrochemical Machining. Electrochemical machining (ECM) is a process in which a direct current is passed through an electrolyte solution between an electrode and the workpiece. Chemical reaction, caused by the current in the electrolyte, dissolves the metal, as shown in Figure 7–33.

Electrodischarge Machining (EDM). In *electrodischarge machining*, the material to be machined and an electrode are submerged in a dielectric fluid which is a nonconductor, forming a barrier between the part and the electrode. A very small gap of about .001 inch is maintained between the electrode and the material. An arc occurs when the voltage across the gap causes the dielectric to break down. These arcs occur about 25,000 times per second, removing material with each arc. The compatibility of the material and the electrode is important for proper material removal. The advantages of EDM over conventional machining methods include its success in machining intricate parts and shapes that otherwise cannot be economically machined, and its use on materials that are difficult or impossible to work with, such as stainless steel, hardened steels, carbides, and titanium.

Electron Beam (EB) Cutting and Machining. In this type of chemical machining, an electron beam generated by a heated tungsten filament is used to cut or "machine" very accurate features into a part. This process may be used to machine holes as small as .0002 in. (0.005 mm) or contour irregular shapes with tolerances of .0005 in. (0.013 mm). *Electron beam cutting* techniques are versatile and may be used to cut or machine any metal or nonmetal.

Ultrasonic Machining. Ultrasonic machining, also known as impact grinding, is a process in which a high-frequency mechanical vibration is maintained in a tool designed to a specific shape. The tool and material to be machined are suspended in an abrasive fluid. The combination of the vibration and the abrasive causes the material removal.

Laser Machining. The laser is a device that amplifies focused light waves and concentrates them in a narrow, very intense beam. The term LASER comes from the first letters of the words "*L*ight *A*mplification by *S*timulated *E*mission of *R*adiation." Using this process, materials are cut or machined by instant temperatures up to 75,000° F (41,649° C). *Laser machining* may be used on any type of material and produces smooth surfaces without burrs or rough edges.

Machined Features and Drawing Representations

The following discussion provides a brief definition of the common manufacturing-related terms. The figures that accompany each definition show an example of the tool, a pictorial of the feature, and the drawing representation. The terms are organized in categories of related features, rather than alphabetical order, as follows: drill, ream, bore, counterbore, countersink, counterdrill, spotface, boss, lug, pad, chamfer, fillet, round, dovetail, kerf, key, keyseat, keyway, neck, spline, threads, T-slot, and knurl.

Drill. A *drill* is used to machine new holes or enlarge existing holes in material. The drilled hole may go through the part, in which case the note THRU can be added to the diameter dimension. When the views of the hole clearly denote that the hole goes through the part,

COMPUTER-INTEGRATED MANUFACTURING

Completely automated manufacturing combines computer-aided design and drafting (CADD), computer-aided engineering (CAE), and computer-aided manufacturing (CAM) into a controlled system known as *computer-integrated manufacturing* (CIM). As shown in Figure 7–34, CIM is the bringing together of all the technologies in a management system that coordinates CADD, CAM, robotics, CNC, and material handling from the beginning of the design process to the shipment of the end product.

Parametrics play an important role in the successful implementation of a CIM system. Computerized parametric design refers to the ability to change a variable and automatically have all of the other values change to fit the revised conditions. In the complete, computer-integrated manufacturing process, all phases of the system must work together or the system does not work as intended. Parametrics can work to keep the entire system in balance. The parametric program keeps all phases of the organization in synchronization. For example, if one element of the system, such as the production of engineering drawings, has scheduling problems, then the entire program from design conception to product shipping is automatically revised to accommodate the situation.

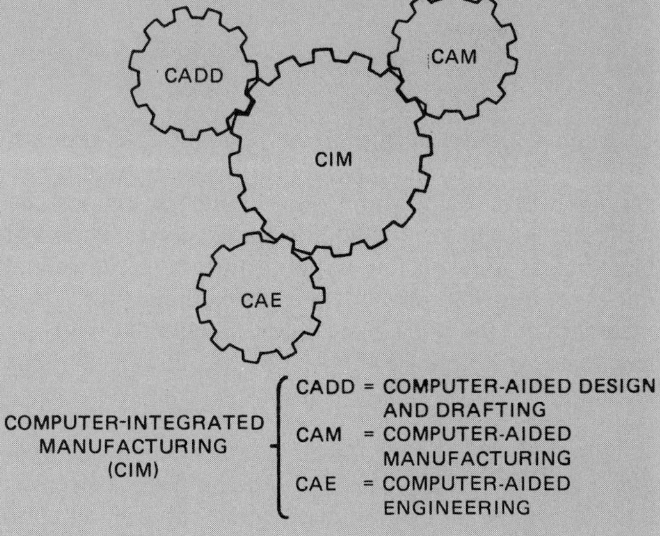

Figure 7–34 Computer-integrated manufacturing (CIM).

then the note THRU may be omitted. When the hole does not go through, the depth must be specified. This is referred to as a *blind hole*. The drill depth is the total usable depth to where the drill point begins to taper. A drill is a conical-shaped tool with cutting edges, normally used in a drill press. The drawing representation of a drill point is a 120° total angle. (See Figure 7–35.)

Ream. The tool is called a reamer. The *reamer* is used to enlarge or finish a hole that has been drilled, bored, or cored. A cored hole is cast in place, as previously discussed. A reamer removes only a small amount of material; for example, .005–.016 in. depending on the size of a hole. The intent of a reamed hole is to provide a smooth surface finish and a closer tolerance than available with the existing hole. A reamer is a conical-shaped tool with cutting edges similar to a drill; however, a reamer will not create a hole as with a drill. Reamers may be used on a drill press, lathe, or mill. (See Figure 7–36.)

Bore. Boring is the process of enlarging an existing hole. The purpose may be to make a drilled or cored hole in a cylinder or part concentric with or perpendicular to other features of the part. A boring tool is used on a lathe for removing internal material. (See Figure 7–37.)

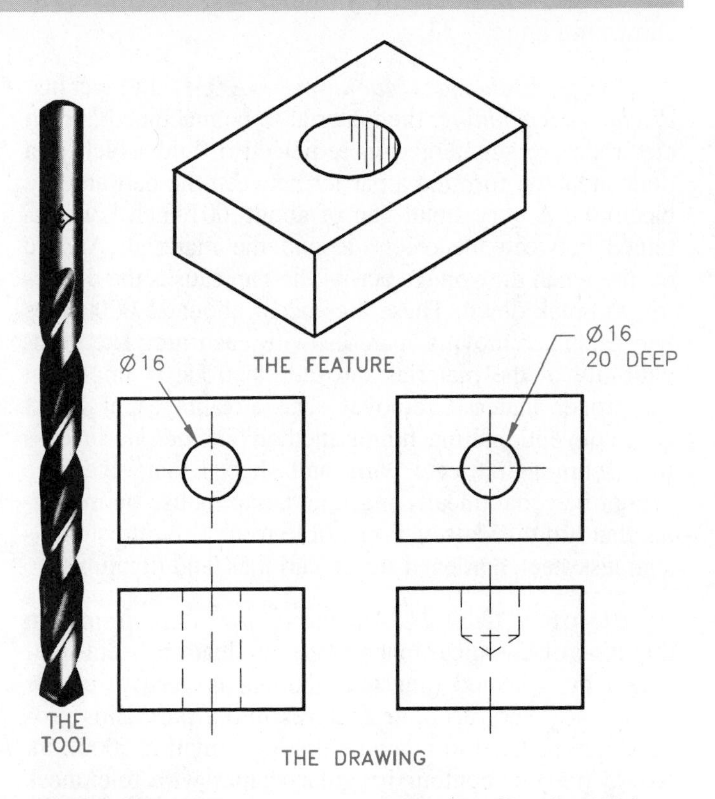

Figure 7–35 Drill. *Tool photo courtesy The Cleveland Twist Drill Company.*

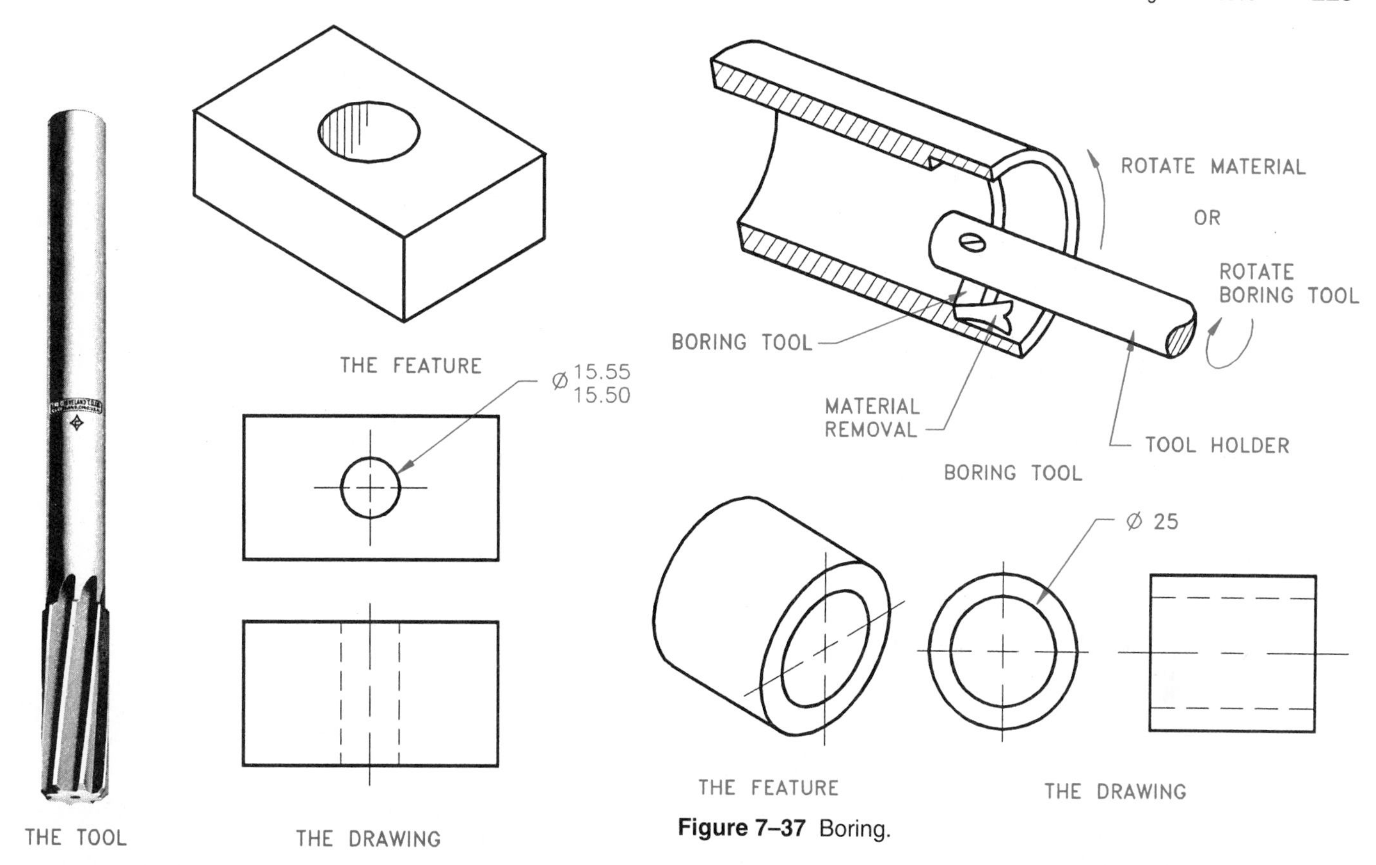

Figure 7–36 Ream. *Tool photo courtesy The Cleveland Twist Drill Company.*

Figure 7–37 Boring.

Counterbore. The *counterbore* is used to enlarge the end(s) of a machined hole to a specified diameter and depth. The machined hole is made first, and then the counterbore is aligned during the machining process by means of a pilot shaft at the end of the tool. Counterbores are usually made to recess the head of a fastener below the surface of the object. The drafter should be sure that the diameter and depth of the counterbore are adequate to accommodate the fastener head and fastening tools. (See Figure 7–38.)

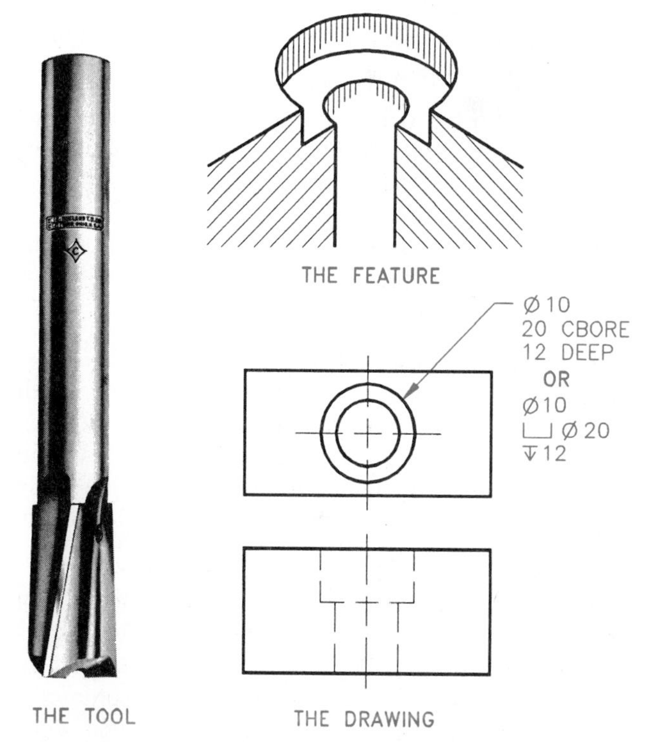

Figure 7–38 Counterbore. *Tool photo courtesy The Cleveland Twist Drill Company.*

Countersink. A *countersink* is a conical feature in the end of a machined hole. Countersinks are used to recess the conically shaped head of a fastener, such as a flat-head machine screw. The drafter should specify the countersink note so that the fastener head will be recessed slightly below the surface. (See Figure 7–39.)

Counterdrill. A *counterdrill* is a combination of two drilled features. The first machined feature may go through the part, while the second feature is drilled, to a given depth, into one end of the first. The result is a machined hole that looks similar to a countersink-counterbore combination. The angle at the bottom of the counterdrill is a total of 120°, as shown in Figure 7–40.

Spotface. A *spotface* is a machined, round surface on a casting, forging, or machined part on which a bolt head or washer can be seated. Spotfaces are similar in characteristics to counterbores, except that a spotface is generally only about 2 mm or less in depth. Rather than a depth specification, the dimension from the spotface surface to the opposite side of the part may be given. This is also

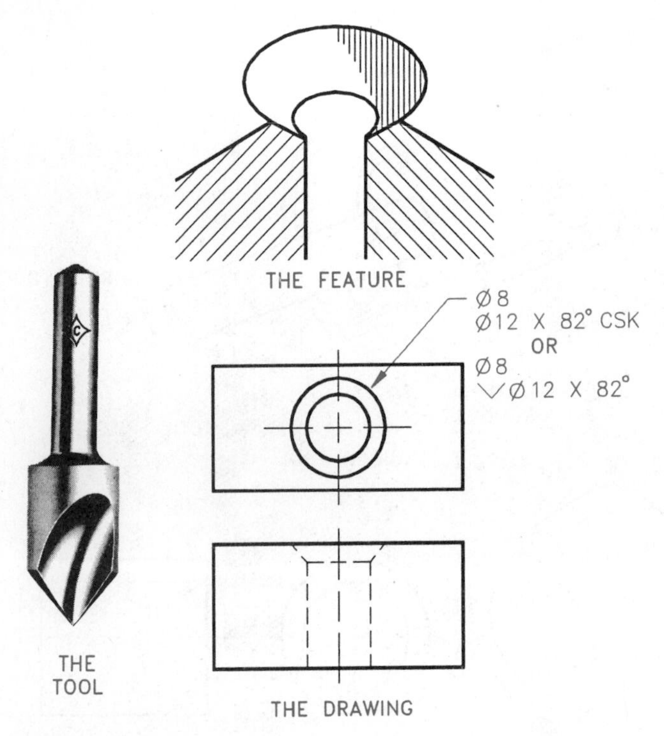

Figure 7–39 Countersink. *Tool photo courtesy The Cleveland Twist Drill Company.*

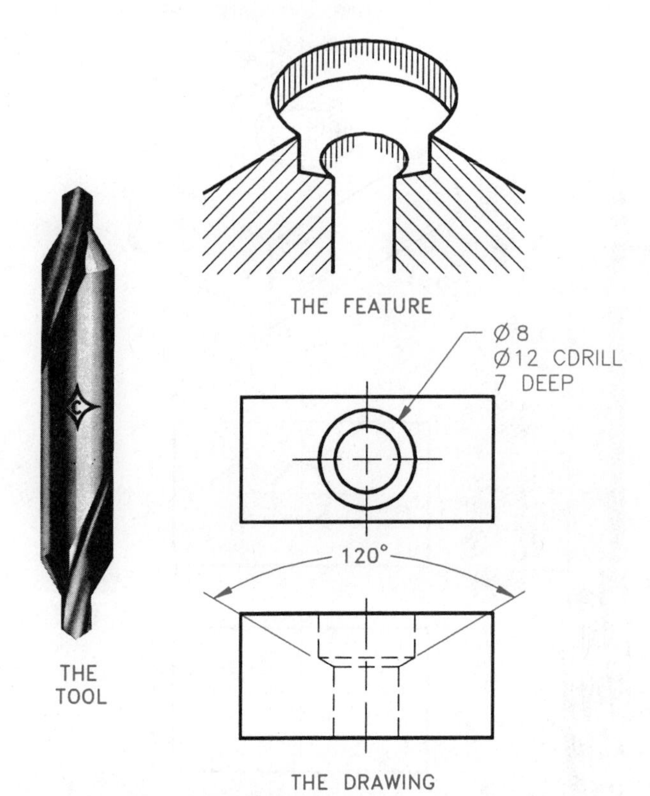

Figure 7–40 Counterdrill. *Tool photo courtesy The Cleveland Twist Drill Company.*

true for counterbores; however, the depth dimension is commonly provided in the note. When no spotface depth is given, the machinist will spotface to a depth that establishes a smooth cylindrical surface. (See Figure 7–41.)

Boss. A *boss* is a circular pad on forgings or castings that projects out from the body of the part. While more closely related to castings and forgings, the surface of the boss is often machined smooth for a bolt head or washer surface to seat on. Also, the boss commonly has a hole machined through it to accommodate the fastener's shank. (See Figure 7–42.)

Lug. Generally cast or forged into place, a *lug* is a feature projecting out from the body of a part, usually rectangular in cross section. Lugs are used as mounting brackets or function as holding devices for machining operations. Lugs are commonly machined with a drilled hole and a spotface to accommodate a bolt or other fastener. (See Figure 7–43.)

Pad. A *pad* is a slightly raised surface projecting out from the body of a part. The pad surface can be any size or shape. The pad may be cast, forged, or machined into place. The surface is often machined to accommodate the mounting of an adjacent part. A boss is a type of pad, although the boss is always cylindrical in shape. (See Figure 7–44.)

Chamfer. A *chamfer* is the cutting away of the sharp external or internal corner of an edge. Chamfers may be used as a slight angle to relieve a sharp edge, or to assist

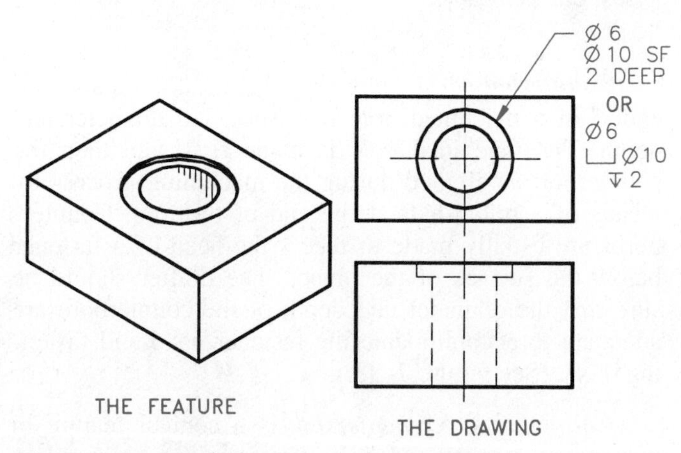

Figure 7–41 Spotface.

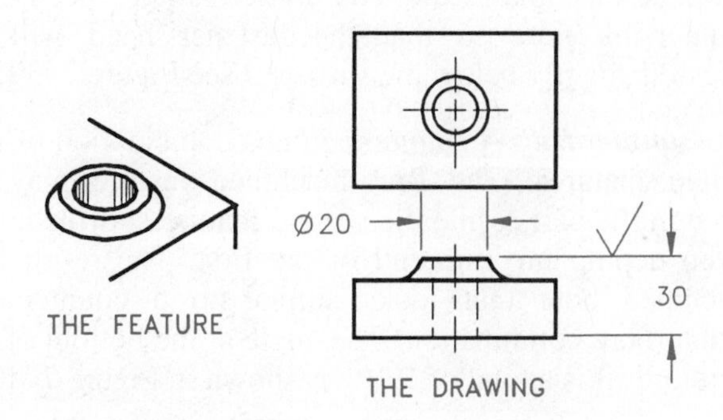

Figure 7–42 Boss.

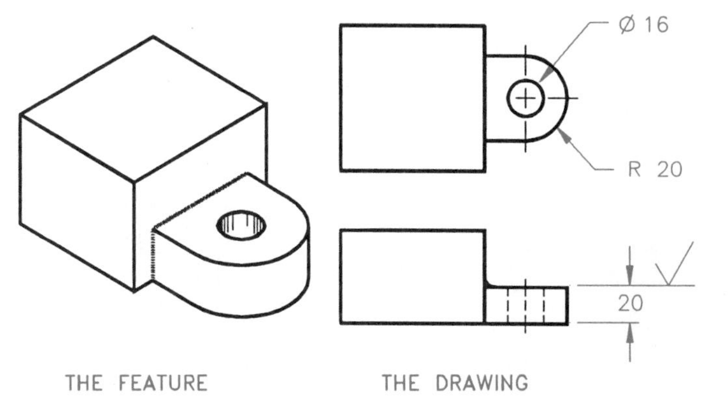

Figure 7–43 Lug.

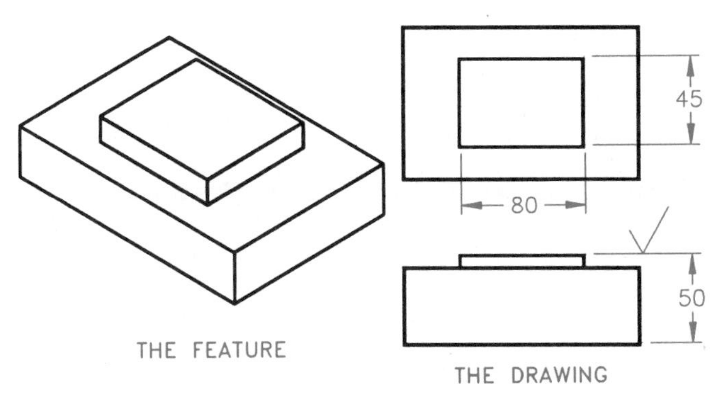

Figure 7–44 Pad.

the entry of a pin or thread into the mating feature. (See Figure 7–45.) Verify alternate methods of dimensioning chamfers in Chapter 8.

Fillet. A *fillet* is a small radius formed between the inside angle of two surfaces. Fillets are often used to help reduce stress and strengthen an inside corner. Fillets are common on the inside corners of castings and forgings to strengthen corners. Fillets are also used to help a casting or forging release a mold or die. Fillets are arcs given as radius dimensions. The fillet size depends on the function of the part and the manufacturing process used to make the fillet. (See Figure 7–46.)

Round. A *round* is a small-radius outside corner formed between two surfaces. Rounds are used to refine sharp corners, as shown in Figure 7–47. In some situations when a sharp corner must be relieved and a round is not required, a slight corner relief may be used, referred to as a *break corner*. The note BREAK CORNER may be used on the drawing. Another option is to provide a note that specifies REMOVE ALL BURRS AND SHARP EDGES. Burrs are machining fragments that are often left on a part after machining.

Dovetail. A *dovetail* is a slot with angled sides that may be machined at any depth and width. Dovetails are commonly used as a sliding mechanism between two mating parts. (See Figure 7–48.)

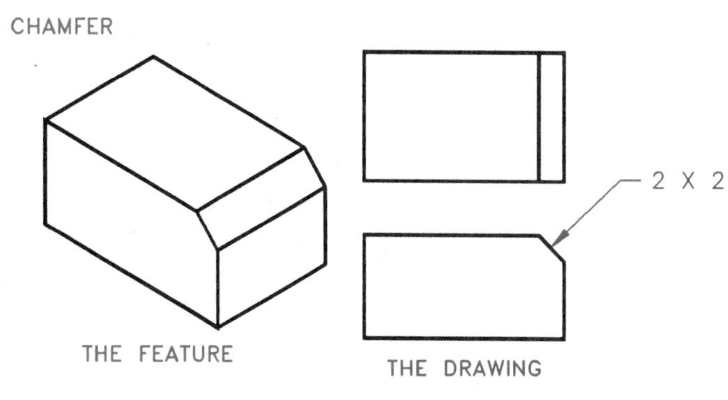

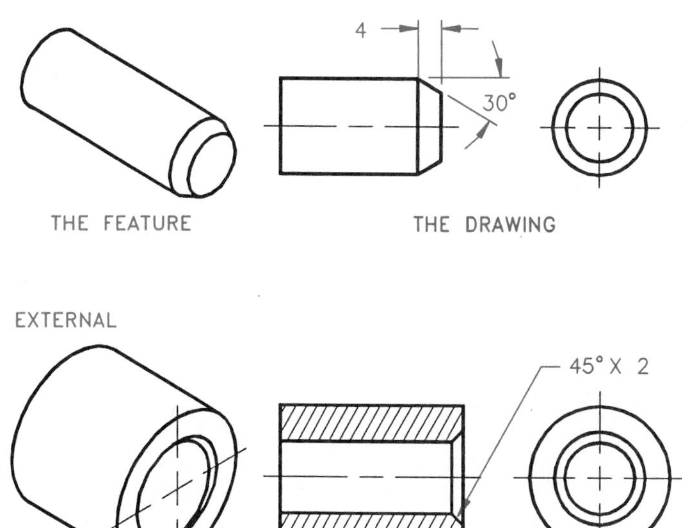

Figure 7–45 Chamfers.

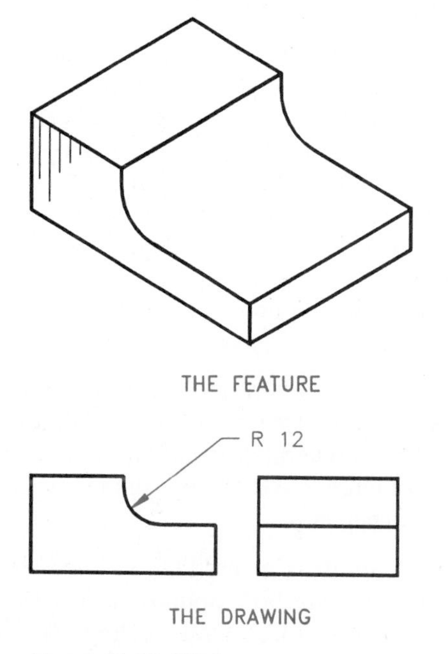

Figure 7–46 Fillet.

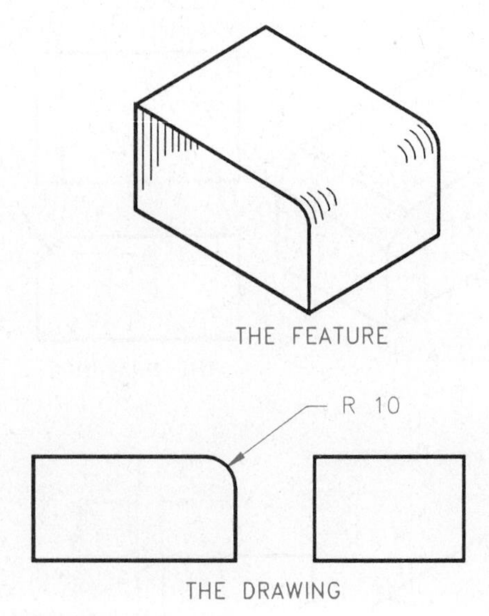

Figure 7–47 Round.

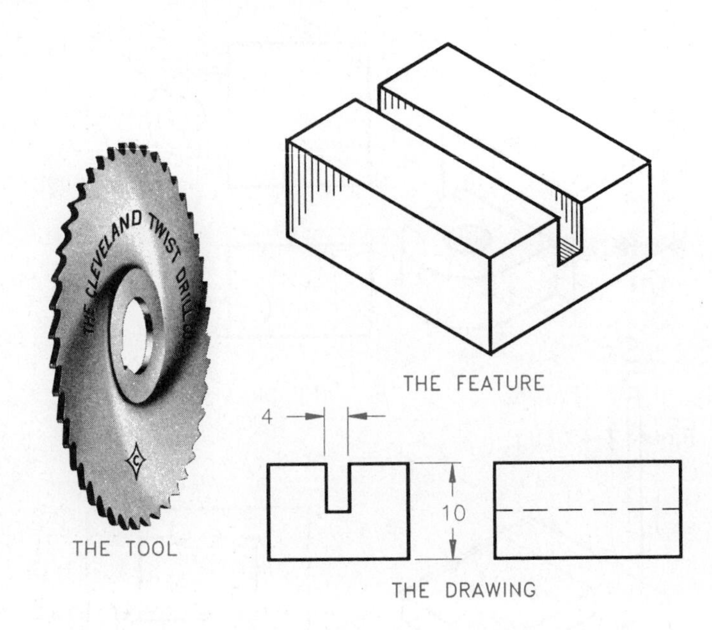

Figure 7–49 Kerf. *Tool photo courtesy The Cleveland Twist*

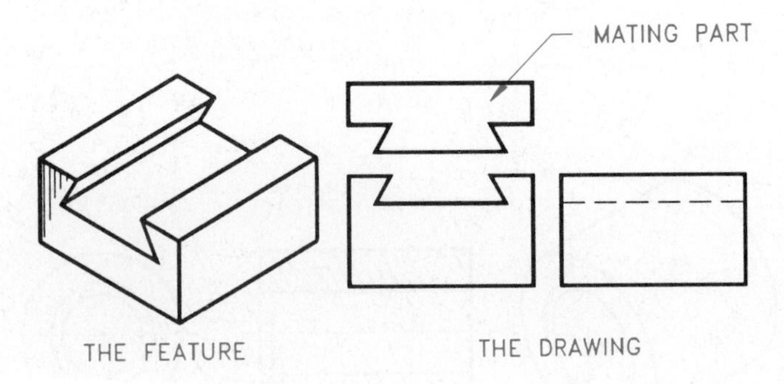

Figure 7–48 Dovetail.

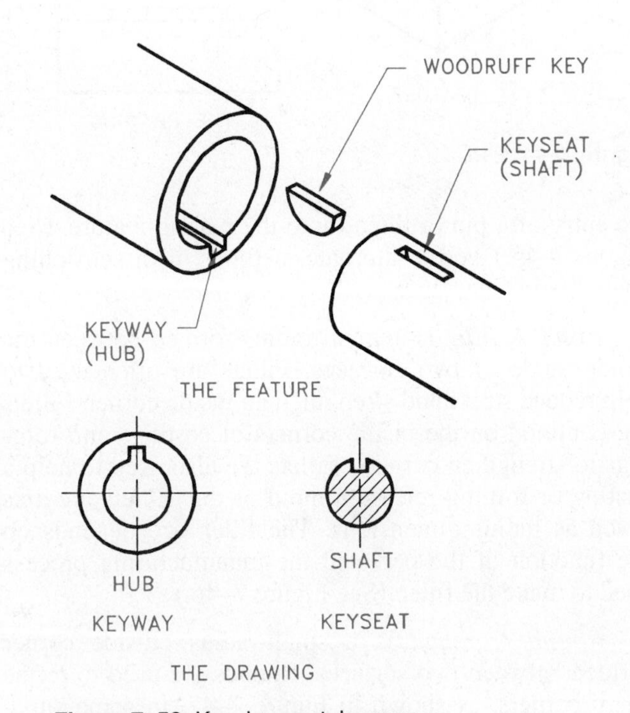

Figure 7–50 Key, keyseat, keyway.

Kerf. A *kerf* is a narrow slot formed by removing material while sawing or using some other machining operation. (See Figure 7–49.)

Key, Keyseat, Keyway. A *key* is a machine part that is used as a positive connection for transmitting torque between a shaft and a hub, pulley, or wheel. The key is placed in position in a *keyseat*, which is a groove or channel cut in a shaft. The shaft and key are then inserted into a hub, wheel, or pulley where the key mates with a groove, called a *keyway*. There are several different types of keys. The key size is often determined by the shaft size. (See Figure 7–50.) Types of keys and key sizes are discussed in Chapter 9, Fasteners and Springs.

Neck. A *neck* is the result of a machining operation that establishes a narrow groove on a cylindrical part or object. There are several different types of neck grooves, as shown in Figure 7–51. Dimensioning necks is clearly explained in Chapter 8.

Spline. A *spline* is a gear-like, serrated surface on a shaft and in a mating hub. Splines are used to transmit torque and allow for lateral sliding or movement between two shafts or mating parts. A spline can be used to take the place of a key when more torque strength is required or when the parts must have lateral movement. (See Figure 7–52.)

Threads. There are many different forms of threads available that are used as fasteners to hold parts together, to adjust parts in alignment with each other, or to transmit power. Threads that are used as fasteners are commonly referred to as *screw threads*. *External threads* are thread forms on an external feature, such as a bolt or shaft. The machine tool used to make external threads is commonly called a *die*. Threads may be machined on a

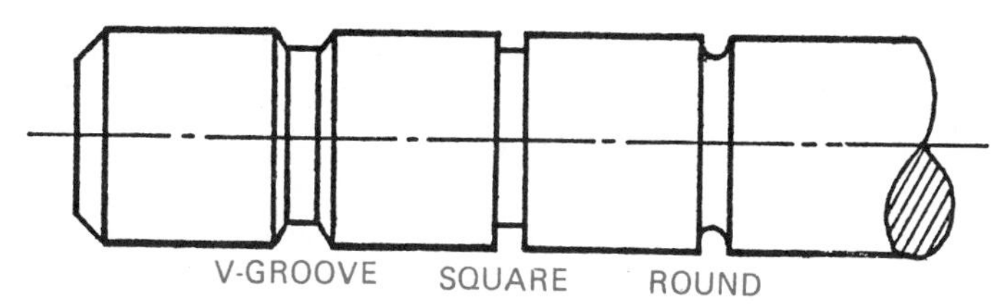

Figure 7–51 Neck.

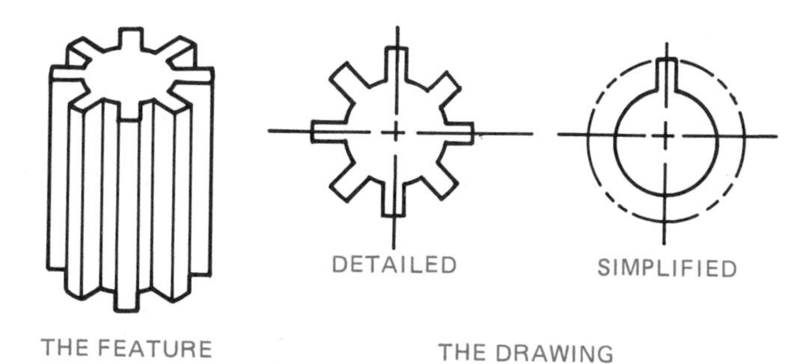

Figure 7–52 Spline.

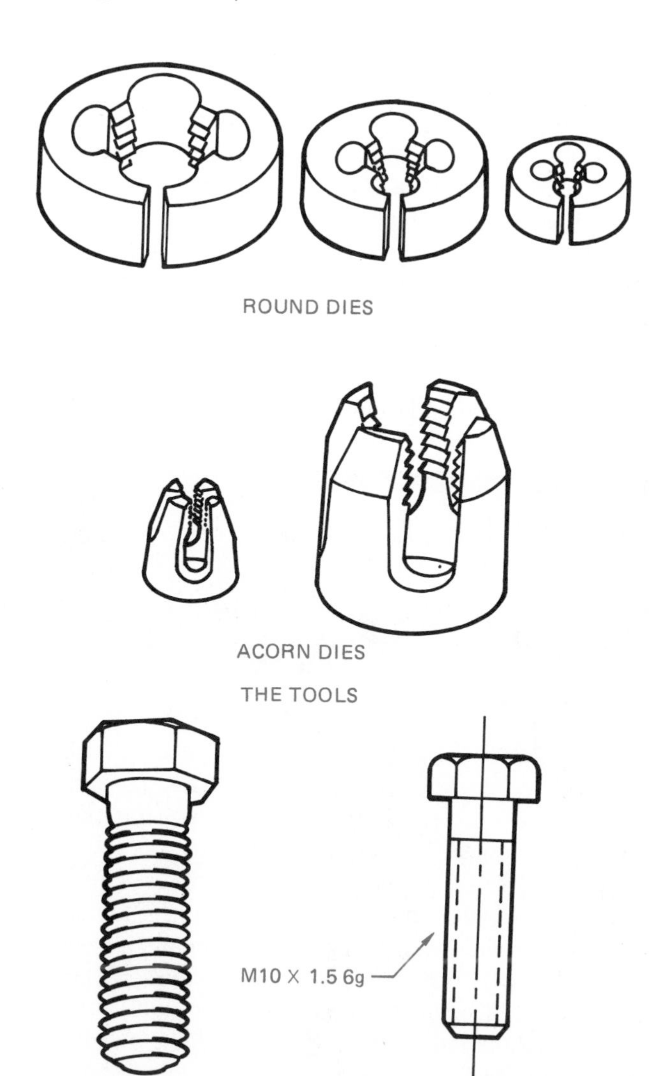

Figure 7–53 External thread, produced by a die.

lathe using a thread-cutting tool. (See Figure 7–53.) *Internal threads* are threaded features on the inside of a hole. The machine tool that is commonly used to cut internal threads is called a *tap*. (See Figure 7–54.)

T-slot. A *T-slot* is a slot of any dimension that is cut to resemble a "T." The T-slot may be used as a sliding mechanism between two mating parts. (See Figure 7–55.)

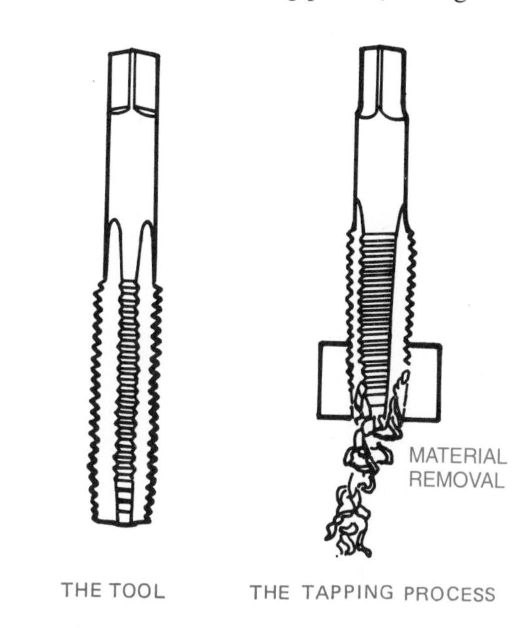

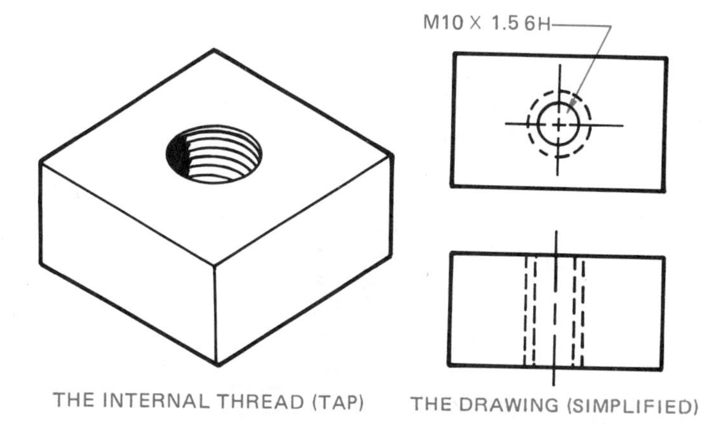

Figure 7–54 Internal thread, produced by a tap.

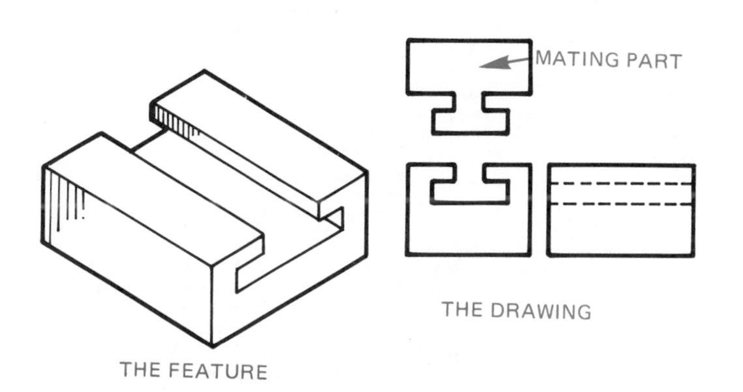

Figure 7–55 T-slot.

Knurl. Knurling is a cold forming process used to uniformly roughen a cylindrical or flat surface with a diamond or straight pattern. Knurls are often used on handles or other gripping surfaces. Knurls may also be used to establish an interference (press) fit between two mating parts. (See Figure 7–56.)

Surface Texture. Surface texture, or *surface finish*, is the intended condition of the material surface after manufacturing processes have been implemented. Surface texture includes such characteristics as roughness, waviness, lay, and flaws. Surface roughness is one of the most common characteristics of surface finish, which consists of the finer irregularities of the surface texture due, in part, to the manufacturing process. The surface roughness is measured in micrometers (μ m) or microinches (μ in.). Micro (μ) means millionth. The roughness averages for different manufacturing processes are shown in Figure 7–57. Surface finish is considered a drafting specification and therefore is discussed in detail in Chapter 8.

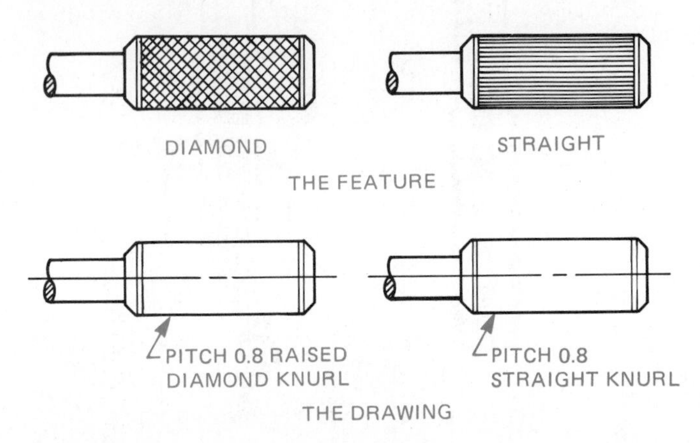

Figure 7–56 Knurl.

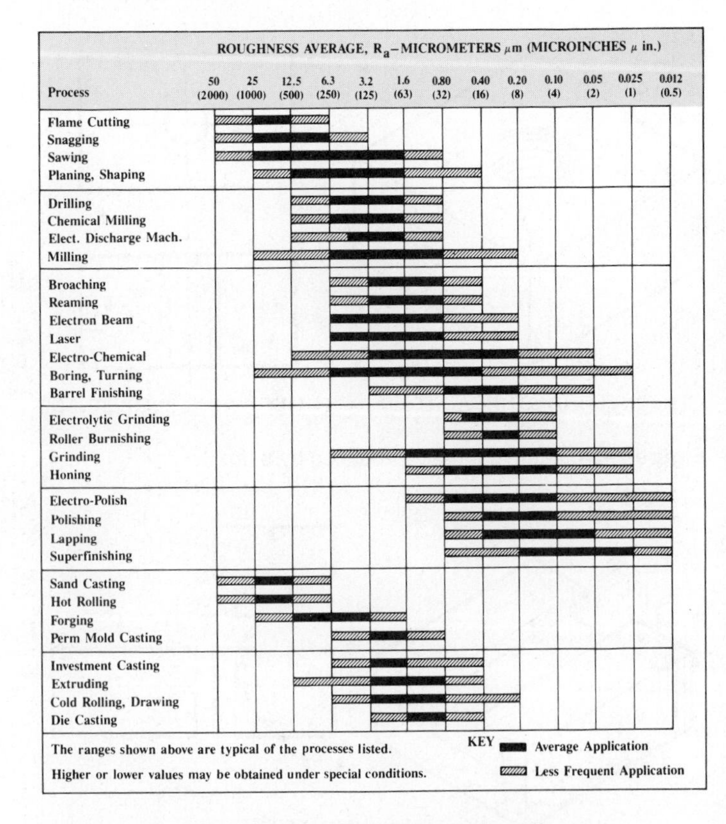

Figure 7–57 Surface roughness produced by different manufacturing processes. *Reprinted, by permission, from Oberg, Jones, and Horton,* Machinery's Handbook, *23rd ed. (New York: Industrial Press Inc., 1988), figure 5, p. 705.*

Design and Drafting of Machined Features

Drafters should gain a working knowledge of the machining processes and capabilities of their companies. Drawings should be prepared that will allow machining within the capabilities of the machinery available. If the local machinery will not produce parts that have certain functional requirements, then the machining may need to be performed elsewhere, or the company may have to purchase the necessary equipment. However, the first consideration should be looking for the least-expensive method to obtain the desired result. Avoid overmachining. Machining processes are expensive, so drawing requirements that are not necessary for the function of the part should be avoided. For example, surface finishes become more expensive to machine as the roughness height decreases. So if a surface roughness of 125 microinches is adequate, then do not use a 32-microinch specification just because you like smooth surfaces. For example, note the difference between 63- and 32-microinch finishes. A 63-microinch finish is a good machine finish that may be performed using sharp tools at high speeds with extra fine feeds and cuts. The 32-microinch callout, on the other hand, requires extremely fine feeds

JIG AND FIXTURE DESIGN

Computer-aided design and drafting (CADD) jig and fixture design programs make it possible for the designer or engineer to electronically construct the tooling for a specific application directly on the computer terminal. As with other CADD programs, tooling component libraries, consisting of fully detailed and accurate jig and fixture components, are available. The speed of tooling design sharply increases as compared to manual methods. Items such as clamps, locator pins, rests, or fixture bases can be retrieved from the library and inserted into your design quickly by using the specific functions provided by the software.

and cuts on a lathe or milling machine and in many cases requires grinding. The 32-microinch finish is more expensive to perform.

In a manufacturing environment where cost and competition are critical considerations, a great deal of thought must be given to meeting the functional and appearance requirements of a product for the least possible cost. It generally does not take very long for an entry-level drafter to pick up these design considerations by communicating with the engineering and manufacturing departments. Many drafters become designers, checkers, or engineers within a company by learning the product and being able to implement designs based on the company's manufacturing capabilities.

TOOL DESIGN

In most production machining operations, special tools are required to either hold the workpiece or guide the machine tool. Tool design involves knowledge of kinematics (study of mechanisms), machining operations, machine tool function, material handling, and material characteristics. *Tool design* is also known as *jig and fixture design*. In mass production industries, jigs and fixtures are essential to insure that each part is produced quickly and accurately within the dimensional specifications. These tools are used to hold the workpiece so that machining operations are performed in the required positions. Application examples are shown in Figure 7–58. Jigs are either fixed or moving devices that are used to hold the workpiece in position and guide the cutting tool. Fixtures do not guide the cutting tool, but are used in a fixed position to hold the workpiece. Fixtures are often used in the inspection of parts to insure that the part is held in the same position each time a dimensional or other type of inspection is made.

Jig and fixture drawings are prepared as an assembly drawing where all of the components of the tool are shown as if they were assembled and ready for use, as shown in Figure 7–59. Components of a jig or fixture often include such items as fast-acting clamps, spring-loaded positioners, clamp straps, quick-release locating pins, handles, and knobs, or screw clamps, as shown in Figure 7–60. Normally the part or workpiece is drawn in position using phantom lines, in a color such as red, or a combination of phantom lines and color.

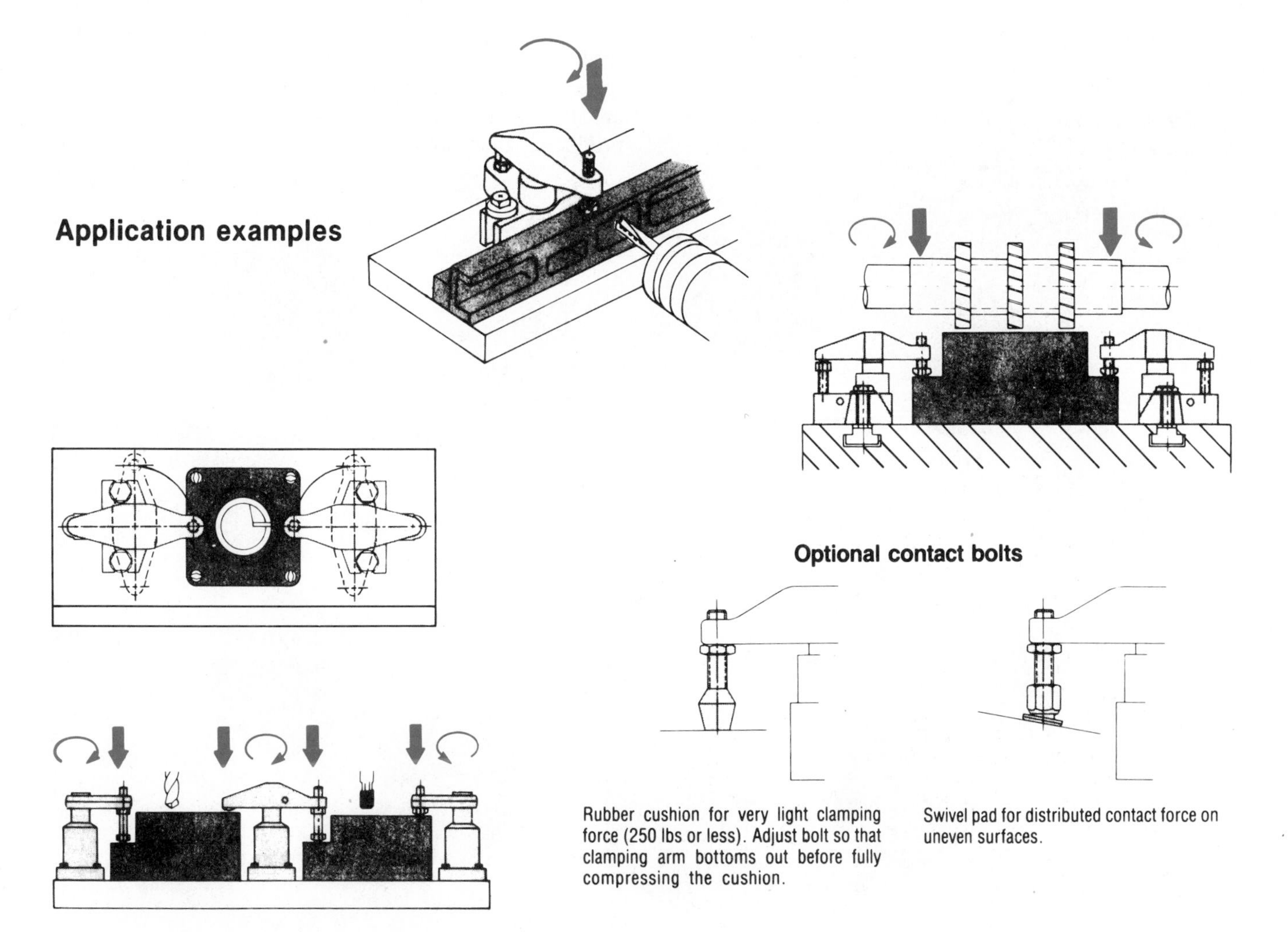

Figure 7–58 Fixture application examples. *Courtesy Carr Lane Manufacturing Co.*

STATISTICAL PROCESS CONTROL (SPC)

A system of quality improvement is helpful to anyone who turns out a product, or is engaged in a service and wishes to improve the quality of work, and at the same time increase the output; all with less labor, and at reduced cost. Competition is here to stay, regardless of the nature of the business. Improved quality means less waste and less rework, resulting in increased profits and an improved market position. Customarily, many managers see quality as a drag on profits. Quality is often placed after cost and delivery, because some managers believe that high quality can only be achieved through costly and slow inspection processes. Many managers are now seeing the influence of quality on sales because high quality has become an important criterion in their customers' purchase decisions. In addition, poor quality is expensive. It is estimated that between 15% and 40% of the American manufacturer's product cost is a result of unacceptable output. Regardless of the goods or service produced, it is always less costly to do it right the first time. Improved quality improves productivity, increases sales, reduces cost, and improves profitability. The net result is continued business success.

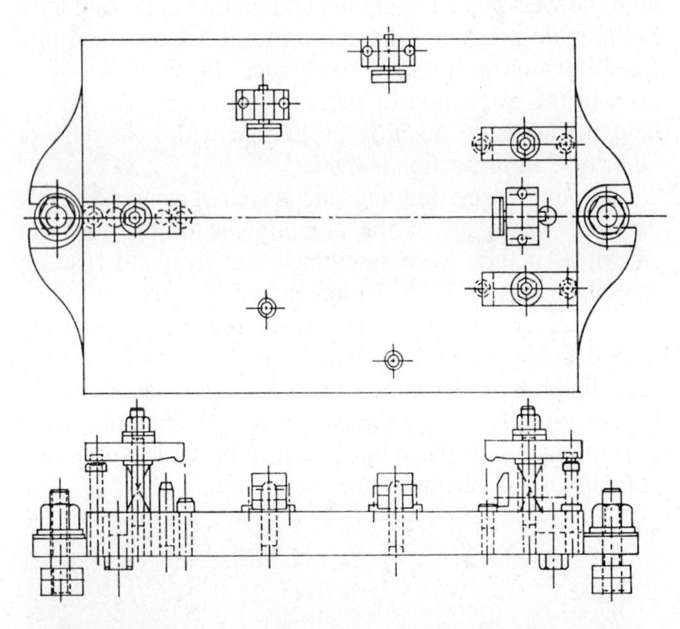

Figure 7–59 Fixture assembly drawing. *Courtesy Carr Lane Manufacturing Co.*

Traditionally, a type of quality control/detection system has been used in most organizations in the United States. This system comprises customer demand for a product, which is then manufactured in a process made up of a series of steps or procedures. Input to the process includes machines, materials, work force, methods, and environment, as shown in Figure 7–61. Once the product or service is produced, it goes to an inspection operation where decisions are made to ship, scrap, rework, or otherwise correct any defects discovered (if they are discovered). In actuality, if nonconforming products are being produced, then some are being shipped. Even the best

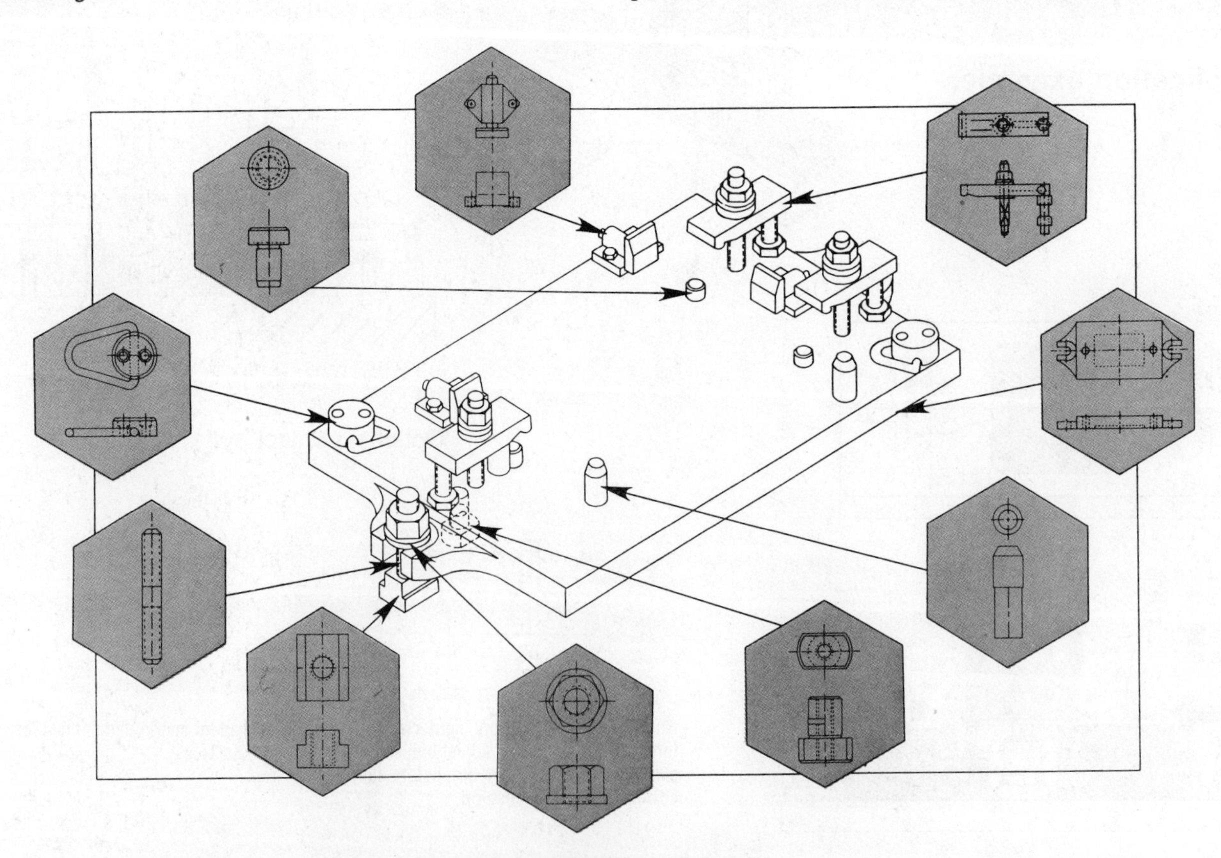

Figure 7–60 Fixture components. *Courtesy Carr Lane Manufacturing Co.*

inspection process screens out only a portion of the defective goods. Problems inherent in this system are that it doesn't work very well and it is costly. And American businesses have become accustomed to accepting these limitations as the "costs of doing business."

The most effective way to improve quality is to alter the production process, the system, rather than the inspection process. This entails a major shift in the entire organization from the detection system to a prevention mode of operation. In this system, the elements (inputs, process, product or service, customer) remain the same, but the inspection method is significantly altered or eliminated. A primary difference between the two systems is that in the prevention system, statistical techniques and problem-solving tools are used to monitor, evaluate, and provide guidance for adjusting the process to improve quality. *Statistical process control* (SPC) is a method of monitoring a process quantitatively and using statistical signals to either leave the process alone, or change it. (See Figure 7–62.) It involves several fundamental elements:

1. The process, product, or service must be measured. It can be measured using either variables (a value that varies), or attributes (a property or characteristic) from the data collected. The data should be collected as close to the process as possible. If you are collecting data on a particular dimension of a manufactured part, it should be collected by the machinist who is responsible for holding that dimension.
2. The data can be analyzed using control charting techniques. Control charting techniques use the natural variation of a process, determining how much the process can be expected to vary if the process is operationally consistent. The control charts are used to evaluate whether the process is operating as designed, or if something has changed.
3. Action is taken based on signals from the control chart. If the chart indicates that the process is *in control* (operating consistently), then the process is left alone at this point. On the other hand, if the process is found to be *out of control* (changing more than its normal variability allows), then action is taken to bring it back into control. It is also important to determine how well the process meets specifications and how well it accomplishes the task. If a process is not in control, then its ability to meet specifications is constantly changing. Process capability cannot be evaluated unless the process is in control. Process improvement generally involves changes to the process system that will improve quality or productivity or both. Unless the process is consistent over time, any actions to improve it may be ineffective.

Manufacturing quality control often uses computerized monitoring of dimensional inspections. When this is done, a chart is developed that shows feature dimensions obtained at inspection intervals. The chart shows the

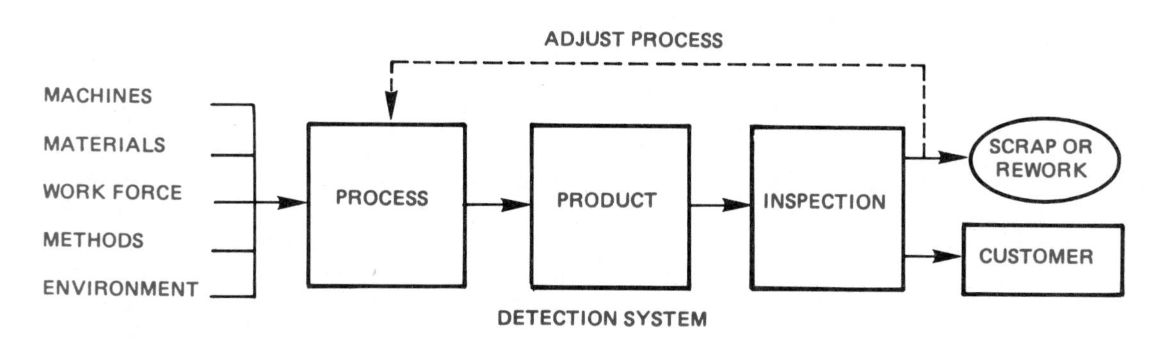

Figure 7–61 Quality control/detection system.

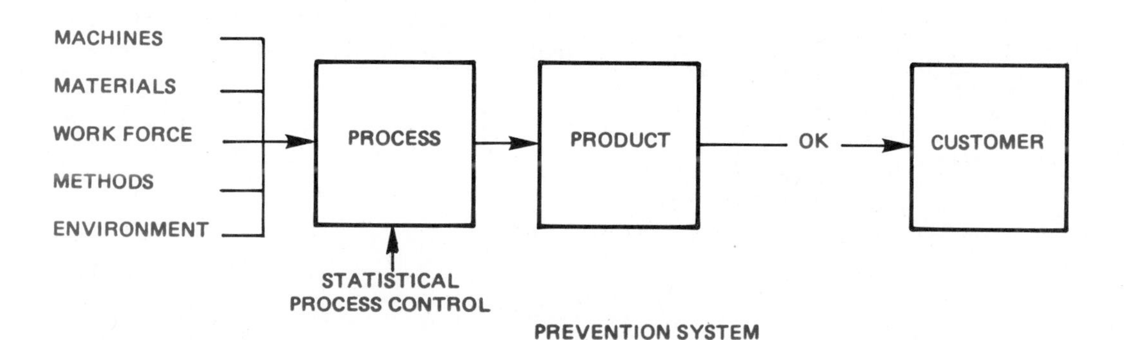

Figure 7–62 Quality control/prevention system.

expected limits of sample averages as two dashed parallel horizontal lines, as shown in Figure 7–63. It is important not to confuse control limits with tolerances — they are not related to each other. The control limits come from the manufacturing process as it is operating. For example, if you set out to make a part 1.000 ± .005 in., and you start the run and periodically take five samples and plot the averages ($\bar{x}$) of the samples, the sample averages will vary less than the individual parts. The control limits represent the expected variation of the sample averages if the process is stable. If the process shifts or a problem occurs, the control limits signal that change. Notice in Figure 7–63 that the $\bar{x}$ values represent the average of each five samples; $\bar{\bar{x}}$ is the average of averages over a period of sample taking. The *upper control limit* (UCL) and the *lower control limit* (LCL) represent the expected variation of the sample averages. A sample average may be "out of control" yet remain within tolerance. During this part of the monitoring process, the out-of-control point represents an individual situation that may not be a problem; however, if samples continue to be measured out of control limits, then the process is out of control (no longer predictable) and therefore may be producing parts outside the specification. Action must then be taken to bring the process back into statistical control or 100% inspection must be resumed. The SPC process only works when a minimum of twenty-five sample means are in control. When this process is used in manufacturing, part dimensions remain within tolerance limits and parts are guaranteed to have the quality designed.

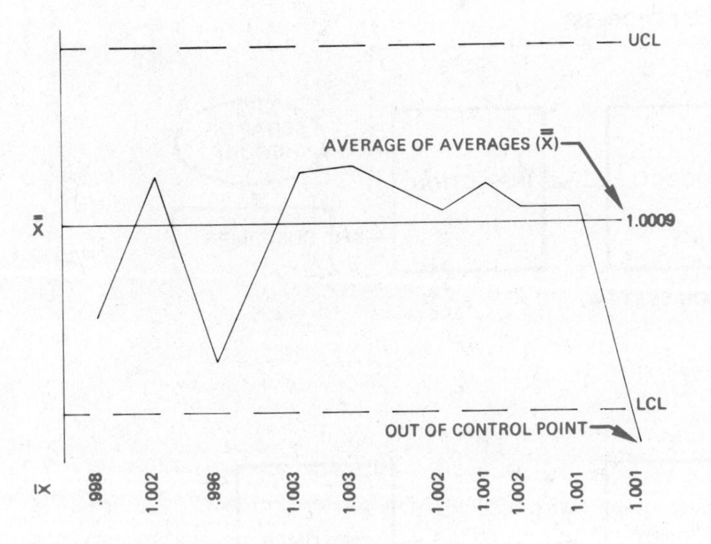

Figure 7–63 Quality control chart.

PROFESSIONAL PERSPECTIVE

This text is designed to introduce you to the many different aspects of engineering drafting. The drafting technology changes slightly from one field to the next. Design and representation methods differ between fields; however, a knowledge of manufacturing processes is an important prerequisite to the understanding of why and how products are designed and engineered.

ENGINEERING PROBLEM

Your company is using the detection system for quality control. This is the way quality control has always been done, but now the management finds that a competitor is beginning to undercut product prices. You go undercover and find that the other company scraps only about 2% of its product due to defects, while your company scraps or reworks around 20% of its product. The other company is getting your business because it has:

- improved quality.
- increased output.
- reduced costs.
- increased sales.
- improved profitability.

Additionally, you find that the other company has gone to a CAD/CAM system for its design and manufacturing processes. It has also implemented statistical process control, which has cut the inspection department from five people to one, a professional for the administration of the system. When you return to management with this information, you propose that they change to a computerized system for design and manufacturing, and a prevention system for quality control. You even prepare an illustration showing how this works. (See Figure 7–62.) The management of your company thanks you for your work, but dismisses the report as being too costly to implement, reminding you that reworking and scrapping parts is just part of the cost of doing business. Unfortunately, a year later the local newspaper reports that your company has gone out of business.

MANUFACTURING PROCESSES PROBLEMS

This chapter is intended as a reference for manufacturing processes. The concepts discussed serve as a basis for further study in the following chapters. A thorough understanding of dimensioning practices is necessary before complete manufactured products can be drawn. Problem assignments ranging from basic to complex manufacturing drawings are assigned in the following chapters.

Problem 7–1 Make a two-view drawing of each of the following machined features using manual or computer-aided drafting as required by your instructor. Prepare the drawings on 8½" × 11" vellum unless otherwise specified by your instructor. Dimensioning may be omitted; however, all line representations should be properly drawn using correct techniques. Machined features to be drawn:

1. Counterbore
2. Chamfer
3. Countersink
4. Counterdrill
5. Drill (not through the material)
6. Fillet
7. Round
8. Spotface
9. Dovetail
10. Kerf
11. Keyseat
12. Keyway
13. T-slot
14. Knurl

Problem 7–2 The following topics require research and/or industrial visitations. It is recommended that you research current professional magazines, visit local industries, or interview professionals in the related field. Your reports should emphasize the following:

- Product
- The process
- Special manufacturing considerations
- The link between manufacturing and engineering
- Current technological advances

Select one or more of the topics listed below or as assigned by your instructor and write a 500-word report for each.

1. Casting
2. Forging
3. Conventional machine shop
4. Computer numerical control (CNC) machining
5. Surface roughness
6. Tool design
7. Chemical machining
8. Electrochemical machining (ECM)
9. Electrodischarge machining (EDM)
10. Electron beam (EB) machining
11. Ultrasonic machining
12. Laser machining
13. Statistical process control (SPC)
14. Plastic
15. Cast iron
16. Steel
17. Aluminum
18. Copper alloys
19. Precious and other specialty metals

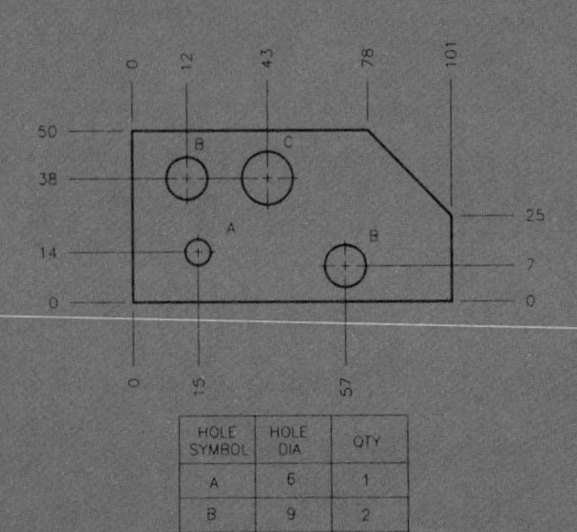

CHAPTER 8

Dimensioning

LEARNING OBJECTIVES

After completing this chapter you will:

- Identify and use common dimensioning systems.
- Explain and apply dimensioning standards as interpreted by ANSI Y14.5M–1982.
- Apply proper specific notes for manufacturing features.
- Place proper general notes on a drawing.
- Interpret and use correct tolerancing techniques.
- Prepare completely dimensioned multiview drawings from engineering sketches and industrial drawings.
- Dimension CAD/CAM machine tool drawings.
- Prepare casting and forging drawings.
- Provide surface finish symbols on drawings.
- Solve tolerance problems including limits and fits.
- Use an engineering problem as the basis for your layout techniques.

A complete detail drawing includes multiviews and dimensions, which provide both shape and size description. There are two classifications of dimensions: size and location. *Size dimensions* are placed directly on a feature to identify a specific size or may be connected to a feature in the form of a note. The relationship of features of an object is defined with *location dimensions*.

Notes are a type of dimension that generally identify the size of a feature or features with more than a numerical specification. For example, a note for a counterbore will give size and identification of the machine process used in manufacturing. There are basically two types of notes: local (or specific) notes and general notes. Local notes are connected to specific features on the views of the drawing. General notes are placed separate from the views and relate to the entire drawing.

It is important for the drafter to effectively combine shape and size descriptions so that the drawing is easy to read and understand. There are many techniques that will help implement this goal. The drafter should carefully evaluate the dimensioning rules while preparing detail drawings and should keep in mind at all times never to crowd information on a drawing. It is better to use larger paper than to crowd the drawing.

DIMENSIONING SYSTEMS

Unidirectional

Unidirectional dimensioning is commonly used in mechanical drafting. It requires that all numerals, figures, and notes be lettered horizontally and be read from the bottom of the drawing sheet. Figure 8–1 shows unidirectional dimensioning in use.

Aligned

Aligned dimensioning requires that all numerals, figures, and notes be aligned with the dimension lines so that they may be read from the bottom (for horizontal dimensions) and from the right side (for vertical dimen-

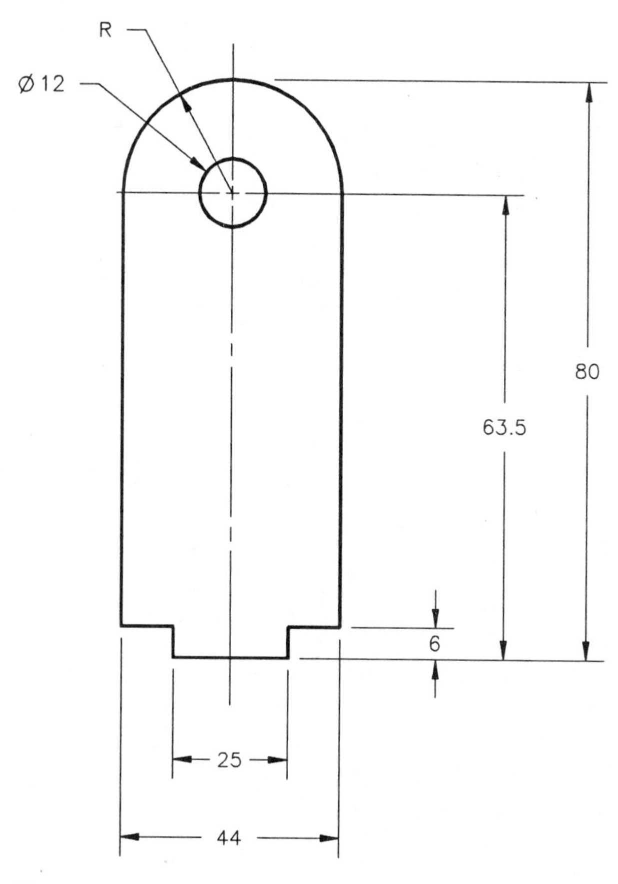

Figure 8–1 Unidirectional dimensioning.

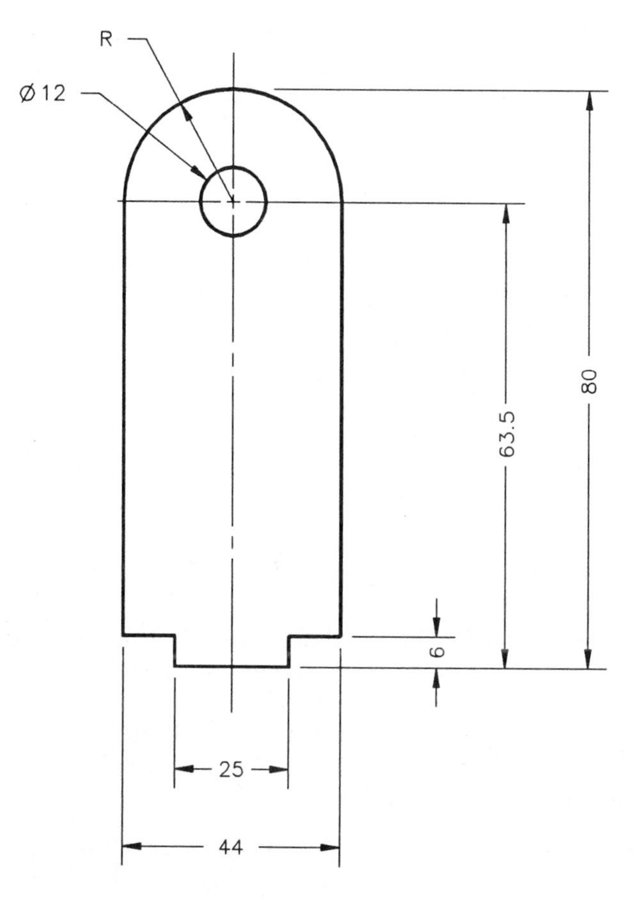

Figure 8–2 Aligned dimensioning.

sions). This method of dimensioning is commonly used in architectural and structural drafting. (See Figure 8–2.)

Tabular

Tabular dimensioning is a system in which size and location dimensions from datums or coordinates (x-, y-, z-axis) are given in a table identifying features on the drawing. Figure 8–3 shows a method of tabular dimensioning.

Arrowless

Also known as dimensioning without dimension lines, *arrowless* dimensioning is similar to tabular dimensioning in that features are identified with letters and keyed to a table. Location dimensions are established with extension lines as coordinates from determined datums. (See Figure 8–4.)

Chart Drawing

Chart drawings are used when a particular part or assembly has one or more dimensions that change depending on the specific application. For example, the diameter of a part may remain constant with several alternate lengths required for different purposes.

The variable dimension is usually labeled on the drawing with a letter in the place of the dimension. The

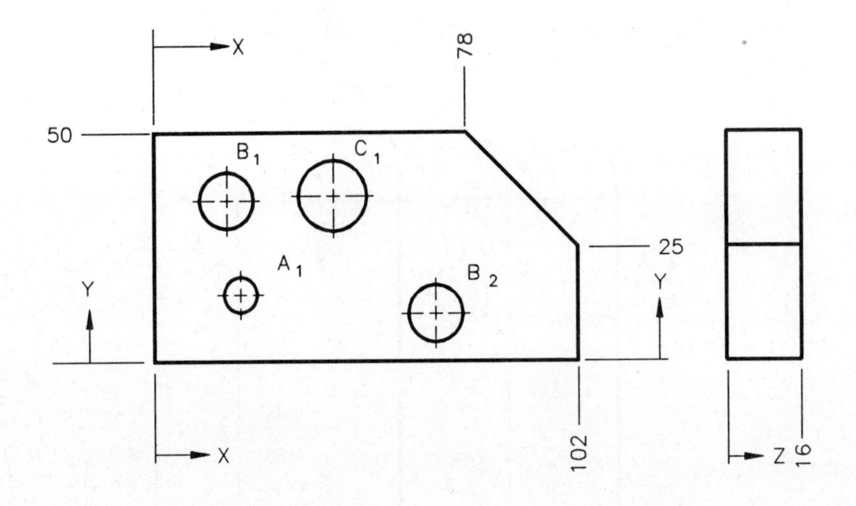

HOLE SYMBOL	HOLE DIA	LOCATION X	X	DEPTH Z
A_1	6	12	14	THRU
B_1	9	12	38	9
B_2	9	57	7	12
C_1	12	43	38	THRU

Figure 8–3 Tabular dimensioning.

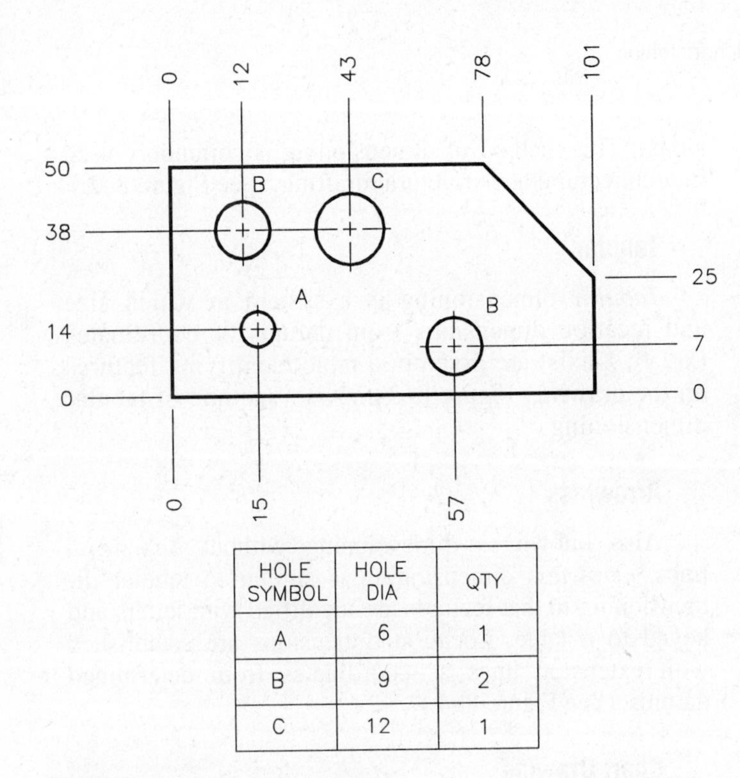

HOLE SYMBOL	HOLE DIA	QTY
A	6	1
B	9	2
C	12	1

Figure 8–4 Arrowless dimensioning.

letter is then placed in a chart where the changing values are identified. Figure 8–5 is a chart drawing that shows two dimensions that have alternate sizes. The view drawn represents a typical part, and the dimensions are

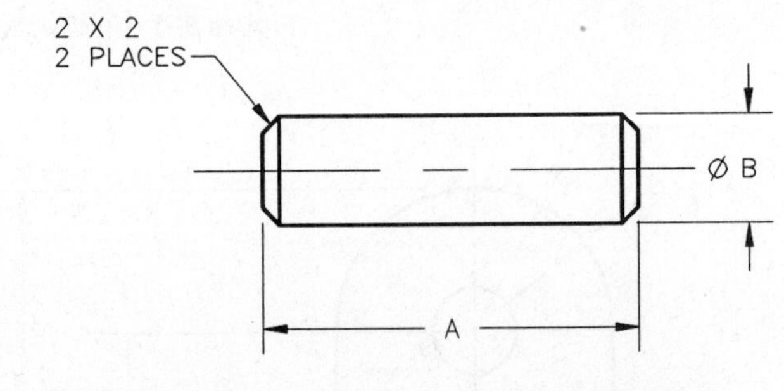

LENGTH A	B=20.3 PART NO	B=38.1 PART NO	B=50.8 PART NO	B=57.2 PART NO
76	DP20.3–76.2	DP38.1–76.2	DP50.8–76.2	DP57.2–76.2
101	DP20.3–101.6	DP38.1–101.6	DP50.8–101.6	DP57.2–101.6
127	DP20.3–127	DP38.1–127	DP50.8–127	DP57.2–127
152	DP20.3–152.4	DP38.1–152.4	DP50.8–152.4	DP57.2–152.4

Figure 8–5 Chart drawing.

labeled A and B. The correlated chart identifies the various lengths (A) available at given diameters (B). The chart in this example also shows purchase part numbers for each specific item. This method of dimensioning is commonly used in vendor or specification catalogues for alternate part identification.

ANSI The ANSI standard document for dimensioning is titled *Dimensioning and Tolerancing* ANSI Y14.5M.

This standard is available from the American National Standards Institute, 1430 Broadway, New York, NY 10018. This standard is all metric as denoted by the M following the number. The following fundamental rules for dimensioning are adapted from ANSI Y14.5M.

1. Each dimension shall have a tolerance, except for those dimensions specifically identified as reference, maximum, minimum, or stock. The tolerance may be applied directly to the dimension or indicated by a general note located in the title block of the drawing.
2. Dimensions for size, form, and location of features shall be complete to the extent that there is full understanding of the characteristics of each feature. Neither scaling nor assumption of a distance or size is permitted.
3. Each necessary dimension of an end product shall be shown. No more dimensions than those necessary for complete definition shall be given. The use of reference dimensions on a drawing should be minimized.
4. Dimensions shall be selected and arranged to suit the function and mating relationship of a part and shall not be subject to more than one interpretation.
5. The drawing should define a part without specifying manufacturing methods. However, in those cases in which manufacturing, processing, quality assurance, or environmental information is essential to the definition of engineering requirements, it shall be specified on the drawing or in document references on the drawing.
6. It is permissible to identify as nonmandatory certain processing dimensions that provide for finish allowance, shrink allowance, and other requirements, provided the final dimensions are given on the drawing. Nonmandatory processing dimensions shall be identified by an appropriate note, such as NONMANDATORY (MFG DATA).
7. Dimensions should be arranged to provide required information for optimum readability. Dimensions should be shown in true profile views and refer to visible outlines.
8. Wires, cables, sheets, rods, and other materials manufactured to gage or code numbers shall be specified by linear dimensions indicating the diameter or thickness. Gage or code numbers may be shown in parentheses following the dimension.
9. A 90° angle is implied where centerlines and lines depicting features are shown on a drawing at right angles and no angle is specified.
10. A 90° basic angle applies where centerlines of features in a pattern or surfaces shown at right angles on the drawing are located or defined by basic dimensions and no angle is specified.
11. Unless otherwise specified, all dimensions are applicable to 20°C (68°F). Compensation may be made for measurements made at other temperatures.

Definitions

Actual Size. The part size as measured after production is known as actual size, also known as produced size.

Allowance. The tightest possible fit between two mating parts. MMC external feature – MMC internal feature = allowance.

Basic Dimension. A basic dimension is considered to be a theoretically exact size, location, profile, or orientation of a feature or point. The basic dimension provides a basis for the application of tolerance from other dimensions or notes. Basic dimensions are drawn with a rectangle around the numerical value. For example: $\boxed{.625}$ or $\boxed{30^\circ}$.

Bilateral Tolerance. A bilateral tolerance is allowed to vary in two directions from the specified dimension. Inch examples are $.250\,^{+.002}_{-.005}$ and $.500 \pm .005$,

metric examples are $12\,^{+0.1}_{-0.2}$ and 12 ± 0.2.

Datum. A datum is considered to be a theoretically exact surface, plane, axis, center plane, or point from which dimensions for related features are established.

Datum Feature. The datum feature is the actual feature of the part that is used to establish a datum.

Dimension. A dimension is a numerical value used on a drawing to describe size, shape, location, or machine process.

Feature. A feature is any physical portion of an object, such as a surface or hole.

Least Material Condition (LMC). Least material condition is the opposite of maximum material condition. The LMC is the lower limit for an external feature and the upper limit for an internal feature.

Limits of Dimension. The limits of a dimension are the largest and smallest possible sizes to which a part may be made as related to the tolerance of the dimension. Consider the following inch dimension and tolerance: .750±.005. The limits of this dimension are calculated as follows: .750 + .005 = .755 upper limit and .750 – .005 = .745 lower limit. For the metric dimension 19.00±0.15, 19.00 + 0.15 = 19.15 is the upper limit, and 19.00 – 0.15 = 18.85 is the lower limit.

Maximum Material Condition (MMC). The maximum material condition, given the limits of the dimension, is the situation where a part contains the most

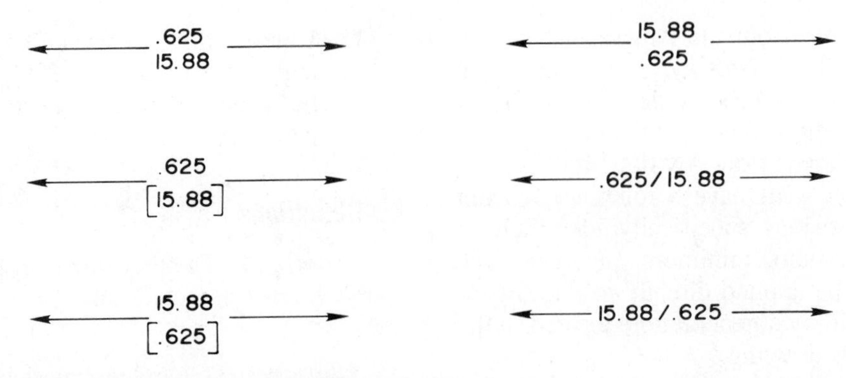

Figure 8–6 Methods of dual dimensioning.

material possible. MMC is the largest limit for an external feature and the smallest limit for an internal feature.

Specified Dimension. The specified dimension is the part of the dimension from which the limits are calculated. For example, the specified dimension of .625±.01 in. is .625. In the following metric example, 15.875±0.254 mm, 15.875 is the specified dimension.

Tolerance. The tolerance of a dimension is the total permissible variation in size. Tolerance is the difference of the lower limit from the upper limit. For example, the limits of .500±.005 in. are .505 and .495 making the tolerance equal to .010 in. The tolerance for the following metric example, 12.7±0.127 mm, is 0.254 mm.

Unilateral Tolerance. A unilateral tolerance is a tolerance that has a variation in only one direction from the specified dimension, as in $.875\begin{smallmatrix}+.000\\-.002\end{smallmatrix}$ in. or $22\begin{smallmatrix}0\\-0.2\end{smallmatrix}$, $22\begin{smallmatrix}+0.2\\0\end{smallmatrix}$.

Units and Dual Dimensioning

Metric units expressed in millimeters or U.S. customary units expressed in decimal inches are considered the standard units of linear measurement on engineering documents and drawings. The selection of millimeters or inches depends on the needs of the individual company. When all dimensions are either in millimeters or inches, the general note, UNLESS OTHERWISE SPECIFIED, ALL DIMENSIONS ARE IN MILLIMETERS (or INCHES), should be lettered on the drawing. Inch dimensions should be followed by *IN.* on predominantly millimeter drawings and *mm* should follow millimeters on predominantly inch-dimensioned drawings.

Another situation where both inches and millimeters are shown for each dimension on a drawing is known as dual dimensioning. Several common methods of dual dimensioning are shown in Figure 8–6. While dual dimensioning is **not** currently an accepted ANSI standard, it is used by some companies where both inch and metric values are needed.

A general note should be placed on the drawing that identifies the method of dual dimensioning used.

For example: $\frac{\text{MILLIMETER}}{\text{INCH}}$, $\frac{\text{INCH}}{\text{MILLIMETER}}$; MILLIMETER/INCH, INCH/MILLIMETER or DIMENSIONS IN [ ] ARE MILLIMETERS (or INCHES).

Decimal Points

Dimension numerals that contain decimal points should be allowed adequate space at the decimal so that there is no crowding with the numerals. A minimum of two-thirds the height of the letters is recommended. The decimal should be clear and bold and in line with the bottom of the numerals. For example, 1.750.

Where millimeter dimensions are less than one, a zero shall precede the decimal, as in 0.8. Decimal inch dimensions do not have a zero before the decimal, as in .75. Proper spacing of numerals with a decimal is shown in Figure 8–7.

Fractions

Fractions are used on engineering drawings, but they are not as common as decimal inches or millimeters. Fraction dimensions generally denote a larger tolerance than decimal numerals. When fractions are used on a drawing the fraction numerals should be the same size as other numerals on the drawing. The fraction bar should be drawn in line with the direction that the dimension reads. For unidirectional dimensioning, the fraction bars will all be horizontal. For aligned dimensioning, the fraction bars will be horizontal for dimensions that read from the bottom of the sheet and vertical for dimensions that read from the right. The fraction numerals should

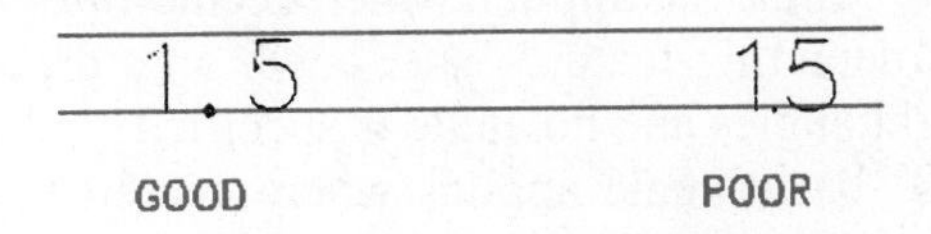

Figure 8–7 Spacing of numerals with decimal.

not be allowed to touch the fraction bar. A space of 1.5 mm or .06 in. is recommended between the bar and the fraction numerals. In a few situations — for example when a fraction is part of a general note, material specification, or title — the fraction bar may be placed diagonally, as shown in Figure 8–8.

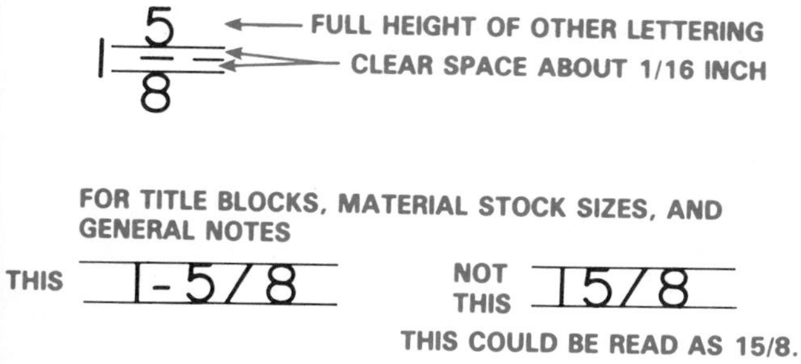

Arrowheads

Arrowheads are used to terminate dimension lines and leaders. Properly drawn arrowheads should be drawn three times as long as they are high. All arrowheads on a drawing should be the same size. Do not use small arrowheads in small spaces. Limited-space dimensioning practice is covered in this chapter.

Some companies require that arrowheads be drawn with an arrow template, while others accept properly drawn freehand arrowheads. (See Figure 8–9.) Individual company preference dictates if arrowheads are filled in solid or left open as shown. Figure 8–10 shows a typical arrowhead template. Notice that only a few of the arrowheads provided are adequate for properly drawn arrowheads. CADD users often prefer the open style, which helps increase regeneration and plotting speed. However, others prefer the appearance of the filled-in arrowhead.

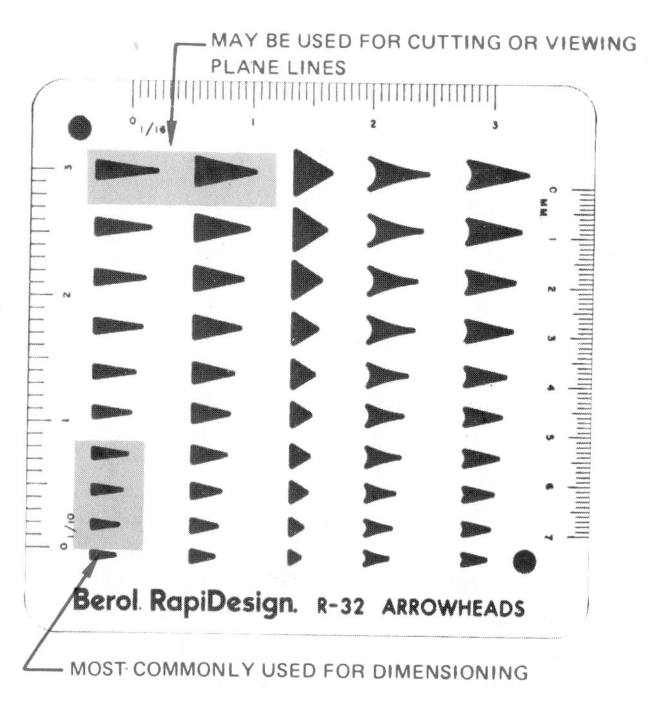

Figure 8–10 Arrowhead template. *Courtesy Berol USA RapiDesign.*

1 5/8 ← FULL HEIGHT OF OTHER LETTERING
← CLEAR SPACE ABOUT 1/16 INCH

FOR TITLE BLOCKS, MATERIAL STOCK SIZES, AND GENERAL NOTES

THIS 1-5/8 NOT THIS 15/8

THIS COULD BE READ AS 15/8.

Figure 8–8 Numerals in fractions.

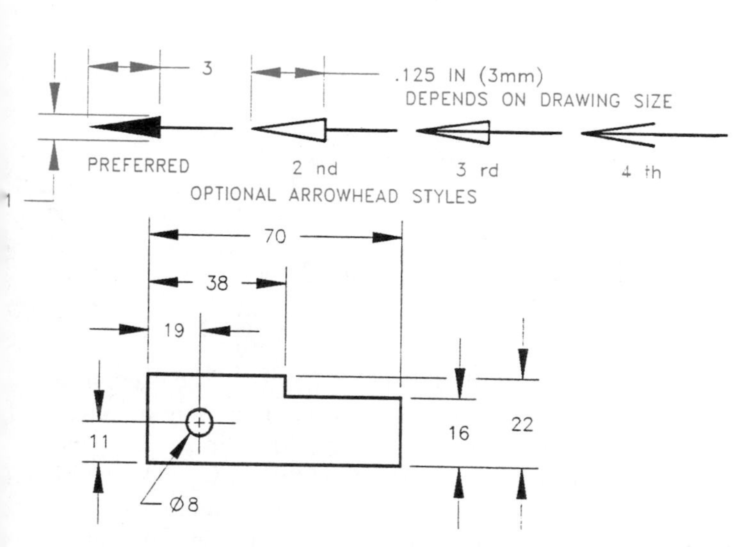

Figure 8–9 Arrowheads.

BASIC DIMENSIONING CONCEPTS

Dimension Line Spacing

Dimension lines should be placed at a uniform distance from the object, and all succeeding dimension lines should be equally spaced. Figure 8–11 shows the minimum acceptable distances for spacing dimension lines. In actual practice the minimum distance is too crowded. Judgment should be used based on space available and information presented. Never crowd dimensions, if at all possible. Always place the smallest dimensions closest to the object and progressively larger dimensions outward from the object. Group dimensions and dimension between views when possible.

Relationship of Dimension Lines to Numerals

Dimension numerals are centered on the dimension line. Numerals are commonly all the same height and are lettered horizontally (unidirectional). A space equal to at least half the height of lettering should be provided between numerals in a tolerance. The numeral, dimension line, and arrowheads should be placed between extension lines when space allows. When space is lim-

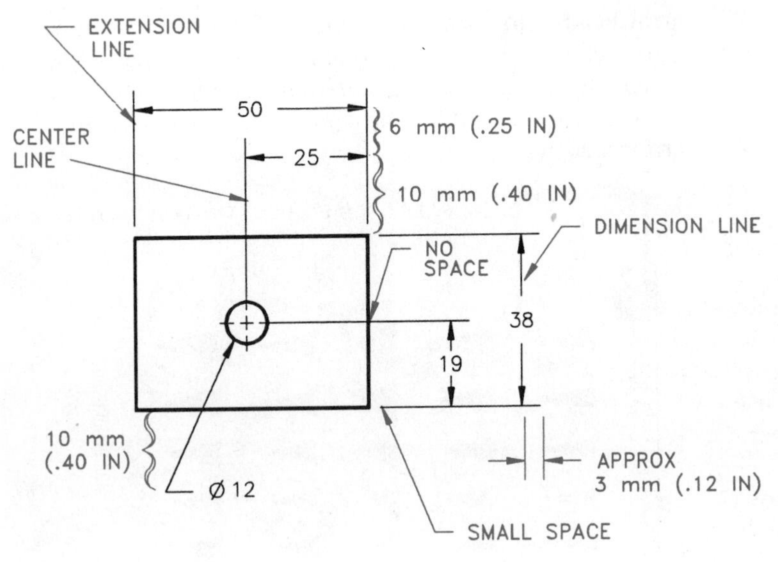

Figure 8–11 Minimum dimension line spacing.

ited, other options should be used. Figure 8–12 shows several dimensioning options. Evaluate each of the examples carefully as you dimension your own drawing assignments. Figure 8–13 shows some correct and incorrect dimensioning practices. Keep in mind that some computer graphics programs do not necessarily acknowledge all of the rules or accepted examples. Some flexibility on your part may be needed to become accustomed to some of the potential differences.

Chain Dimensioning

Chain dimensioning, also known as point-to-point dimensioning, is a method of dimensioning from one feature to the next. Each dimension is dependent on the previous dimension or dimensions. This is a common practice, although caution should be used as the tolerance of each dimension builds on the next, which is known as tolerance buildup, or stacking. See Figure 8–14, which also shows the common mechanical drafting practice of providing an overall dimension while leaving one of the intermediate dimensions blank. The overall dimension is often a critical dimension which should stand independent in relationship to the other dimensions. Also, if all dimensions are given, then the actual size may not equal the given overall dimension due to tolerance buildup. An example of tolerance buildup is when three chain dimensions have individual tolerances of ±0.127 and each feature is manufactured at or toward the +0.127 limit: the potential tolerance buildup is three times +0.127 for a total of 0.381. The overall dimension would have to carry a tolerance of ±0.381 to accommodate this buildup. If the overall dimension is critical, such an amount may not be possible. Thus, either one intermediate dimension should be

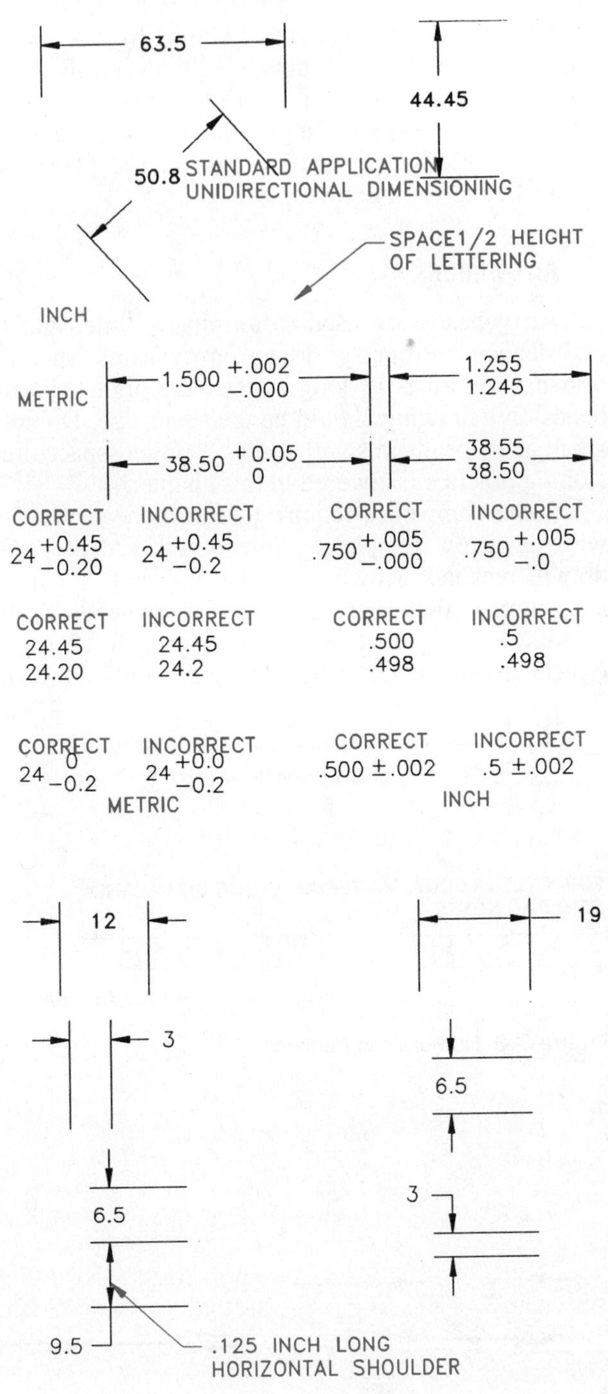

Figure 8–12 Dimensioning applications.

omitted or the overall dimension omitted. The exception to this rule is when a dimension is given only as reference. A reference dimension is enclosed in parentheses, as in Figure 8–15(a). Figure 8–15(b) shows the overall dimension of a symmetrical object as a reference. The symmetrical symbol may also be placed on the centerline as shown.

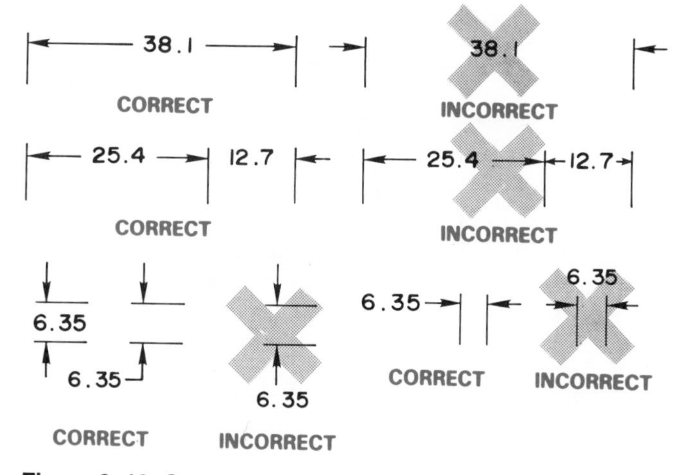

Figure 8–13 Correct and incorrect dimensioning practices.

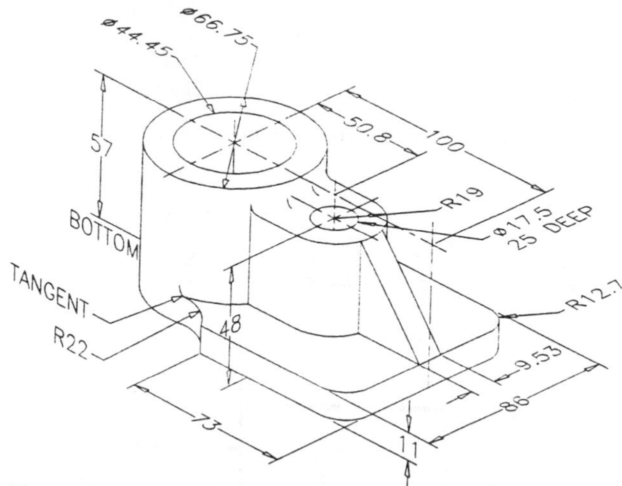

Figure 8–14 Chain dimensioning.

Datum Dimensioning (Stagger Adjacent Dimensions)

Datum dimensioning is a common method of dimensioning machine parts whereby each feature dimension originates from a common surface, axis, or center plane. (See Figure 8–16.) Each dimension in datum dimensioning is independent so there is no tolerance buildup. Figure 8–17 shows how dimensions can be symmetrical about a center plane used as a datum. Also notice in Figure 8–17 how extension lines break when they cross dimension lines, and how dimension numerals are staggered rather than being stacked directly above one another. Always **stagger adjacent dimensions** when possible. Dong so helps clarity and reduces crowding. See CENTER PLANE DATUM illustrated in Figure 8–17.

Dimension cylindrical shapes in the view where the cylinders appear as rectangles. The diameters are identified by the diameter symbol, and the circular view may be omitted. (See Figure 8–18.) Square features may be dimensioned in a similar manner using the square symbol shown in Figure 8–19.

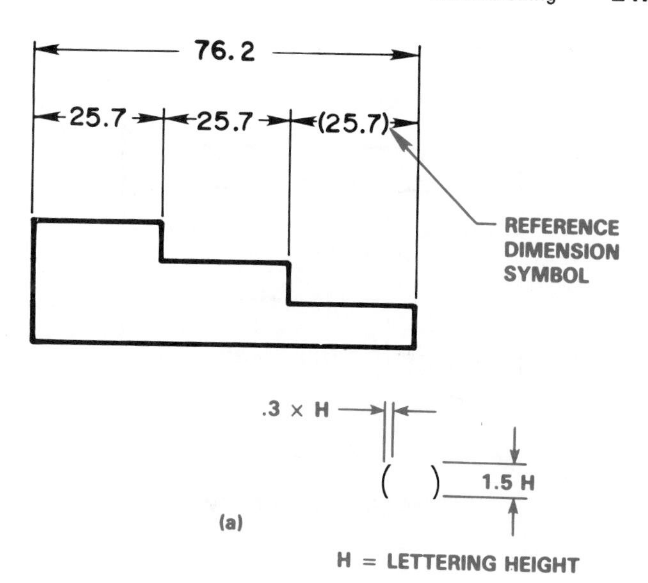

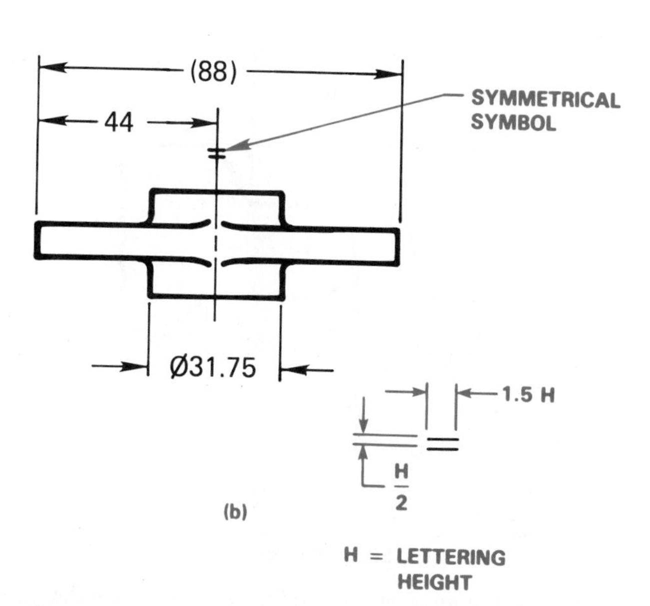

Figure 8–15 (a) Reference dimension; (b) reference dimension on a symmetrical object.

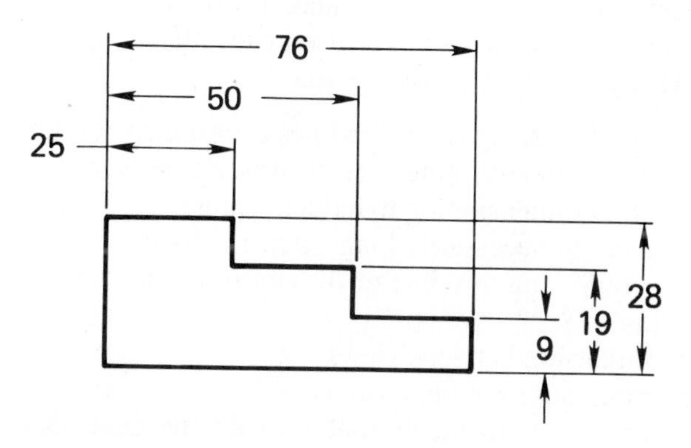

Figure 8–16 Datum dimensioning from a common surface.

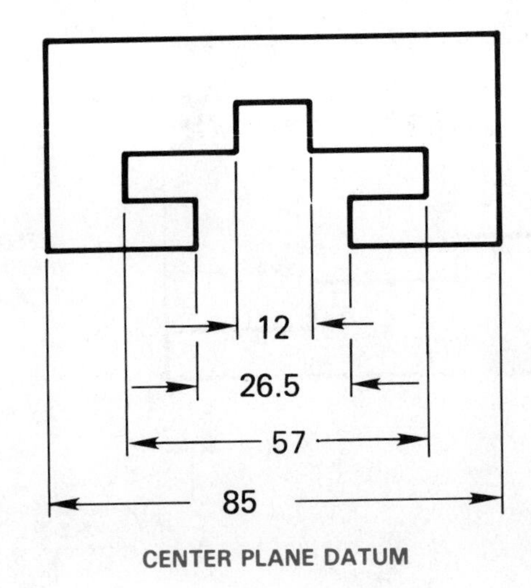

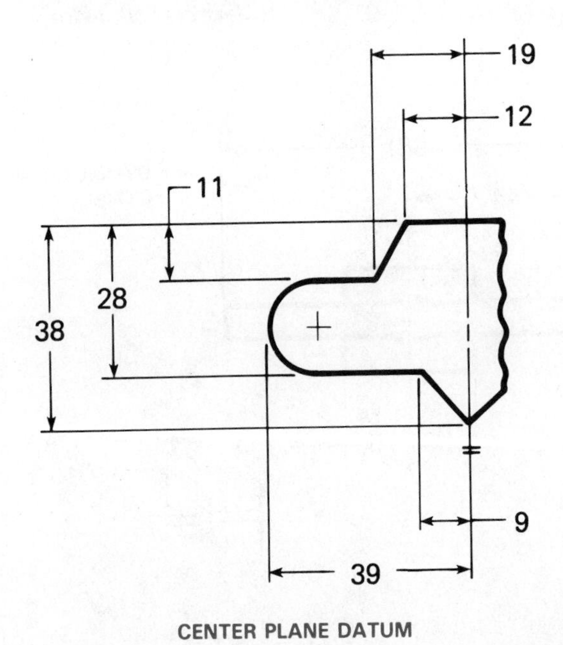

Figure 8–17 Datum dimensioning from a center plane datum.

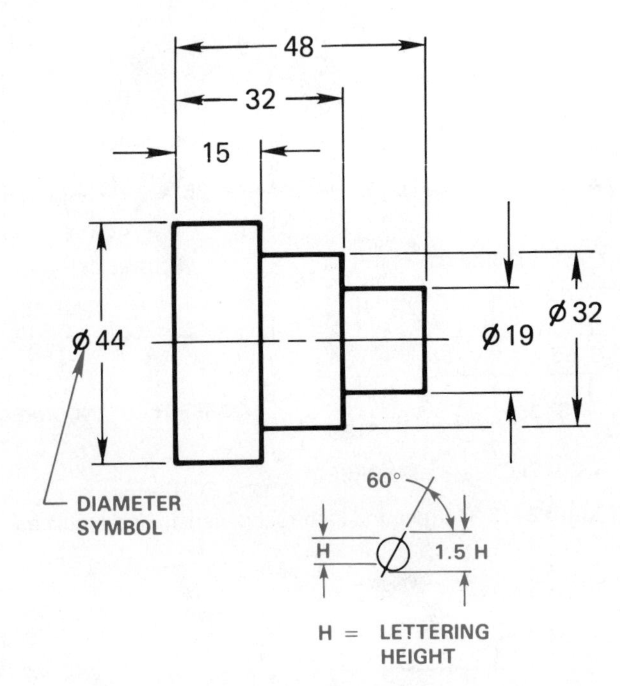

Figure 8–18 Dimensioning cylindrical shapes.

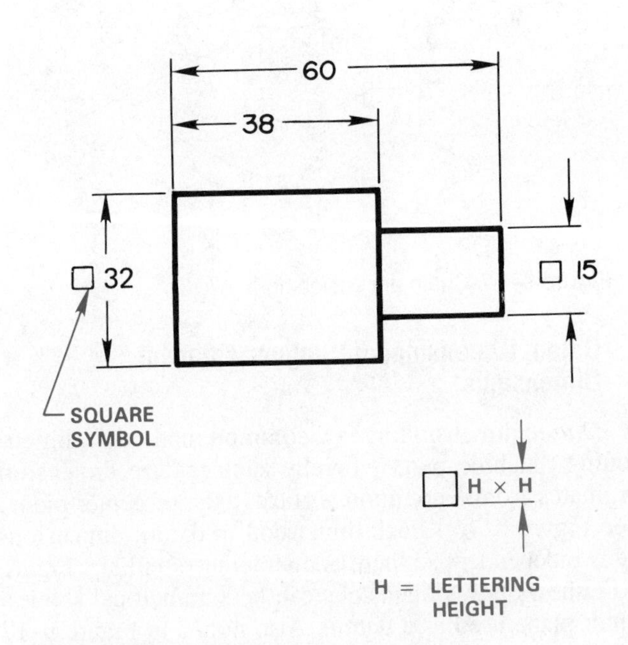

Figure 8–19 Dimensioning square features.

The drawings in Figure 8–20 show both correct and incorrect dimensioning techniques. These are mostly suggested options for correct and incorrect dimensioning. Good judgment should be used with each case.

Always try, when possible, to:

- avoid crossing extension lines over dimension lines
- avoid dimensioning over or through the object
- avoid dimensioning to hidden features
- avoid unnecessarily long extension lines
- avoid using any line of the object as an extension line
- dimension between views
- group adjacent dimensions
- dimension to views that provide the best shape description

Dimensioning Angles

Angular surfaces may be dimensioned as coordinates, as angles in degrees, or as a flat taper. (See Figure 8–21.) Angles are calibrated in degrees, symbol °. (There are 360° in a circle. Each degree contains 60 minutes, symbol '. Each minute has 60 seconds, symbol ". 1° = 60'; 1' = 60".) Notice in the angular method in Figure 8–21, the dimension line for the 45° angle is drawn as an arc. The radius of this arc is centered at the vertex of the angle.

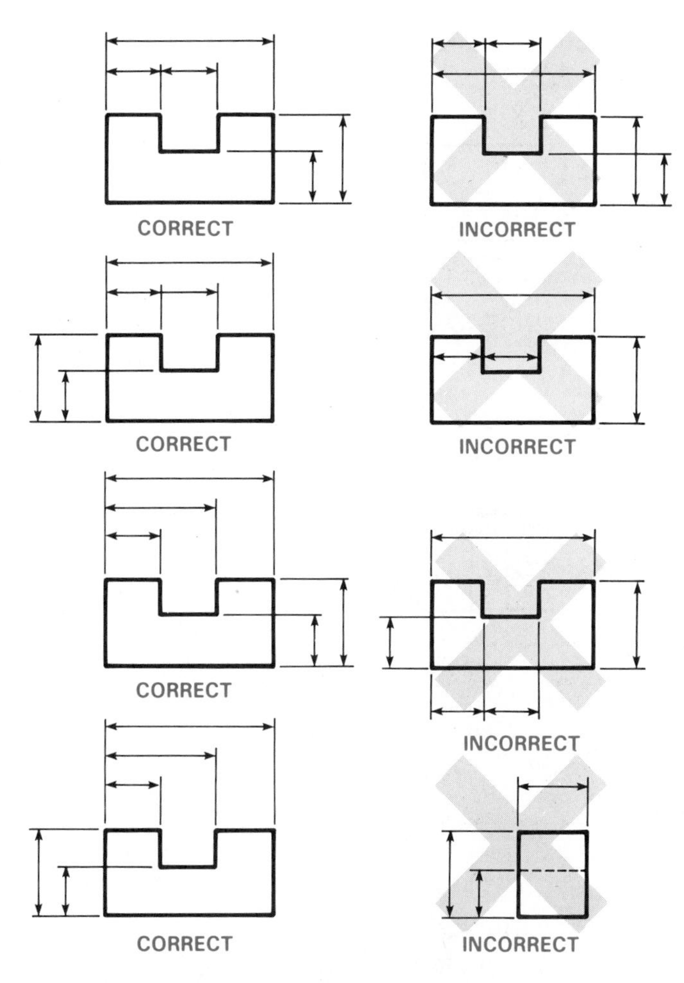

Figure 8–20 Correct and incorrect dimensioning examples.

Dimensioning Chamfers

A chamfer is a slight surface angle used to relieve a sharp corner. Chamfers of 45° may be dimensioned with a note, while other chamfers require an angle and size dimension, as seen in Figure 8–22. A note is used on 45° chamfers because both sides of a 45° angle are equal.

Dimensioning Conical Shapes

Conical shapes should be dimensioned where possible in the view that the cone appears as a triangle, as in Figure 8–23. A conical taper may be treated in one of three possible ways, as shown in Figure 8–24. Other geometric dimensioning methods are possible.

Dimensioning Hexagons and Other Polygons

Dimension hexagons and other polygons across the flats in the views where the true shape is shown. Provide a length dimension in the adjacent view, as shown in Figure 8–25.

Dimensioning Arcs

Arcs are dimensioned with leaders and radius dimensions in the views where they are shown as arcs.

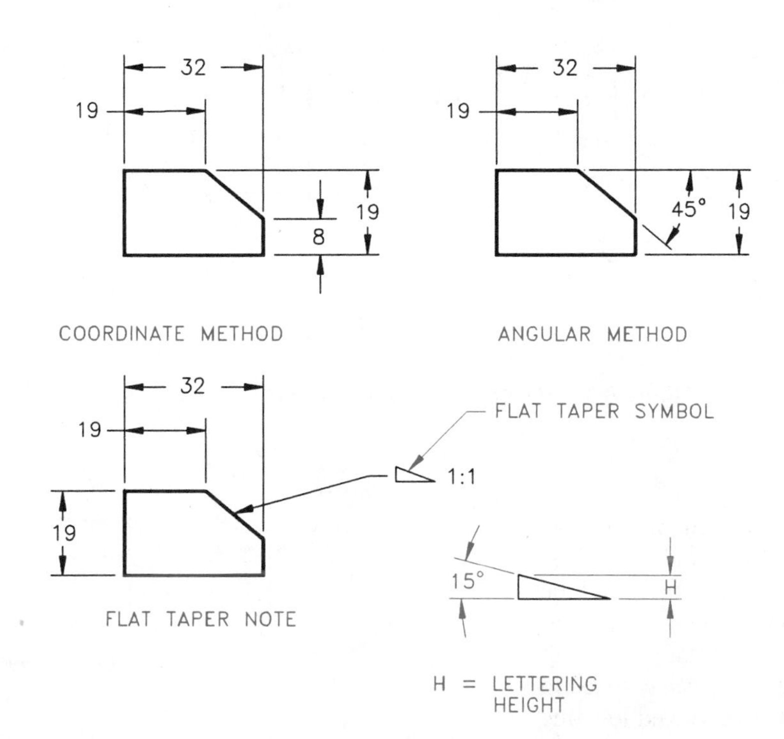

Figure 8–21 Dimensioning angular surfaces.

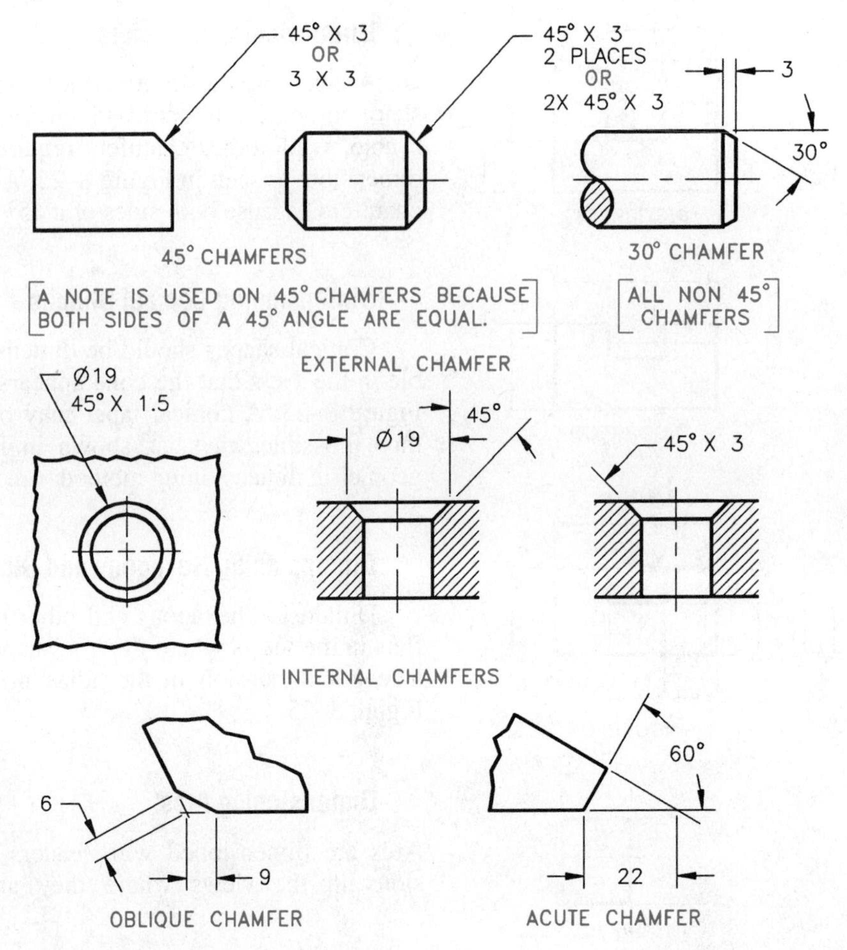

Figure 8–22 Dimensioning chamfers.

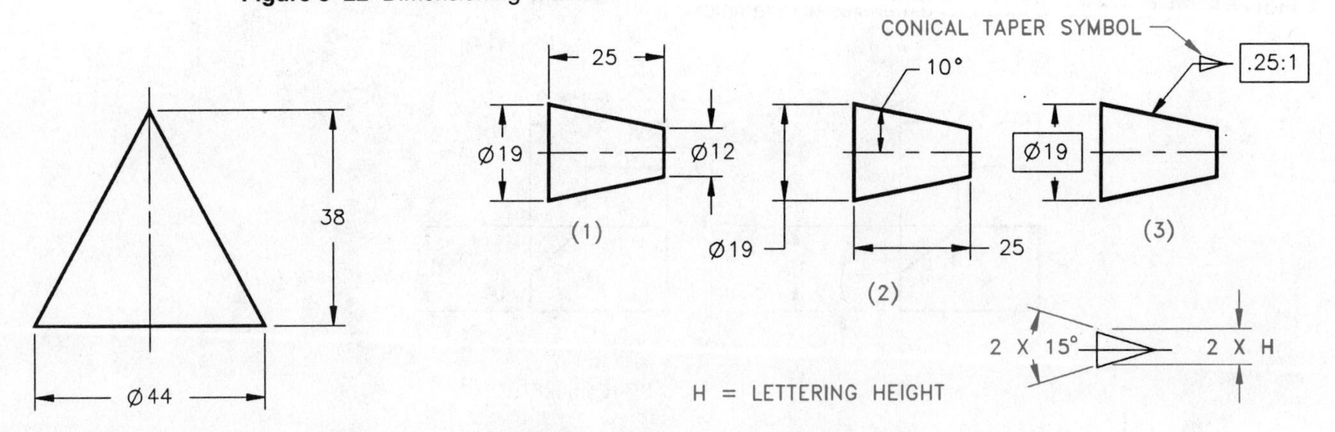

Figure 8–23 Dimensioning conical shapes.

Figure 8–24 Dimensioning conical tapers.

The leader may extend from the center to the arc or may point to the arc, as shown in Figures 8–26 and 8–27. The letter *R* precedes all radius dimensions. Depending on the situation, arcs may be dimensioned with or without their centers located. Figure 8–27 shows a very large arc with the center moved closer to the object. To save space, a break line is used in the leader and the shortened locating dimension. The length of an arc may also be dimensioned one of three ways, as shown in Figure 8–28.

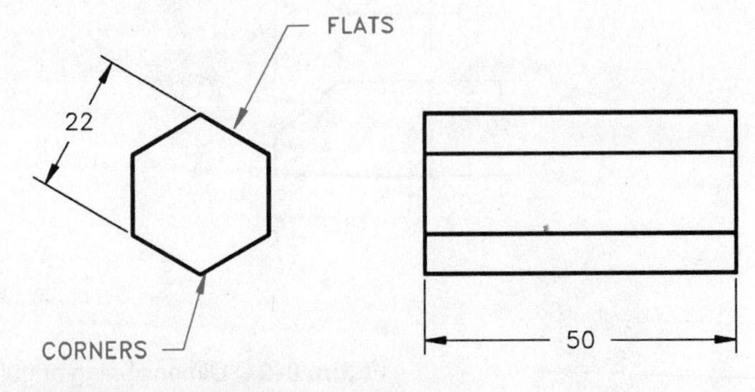

Figure 8–25 Dimensioning hexagons.

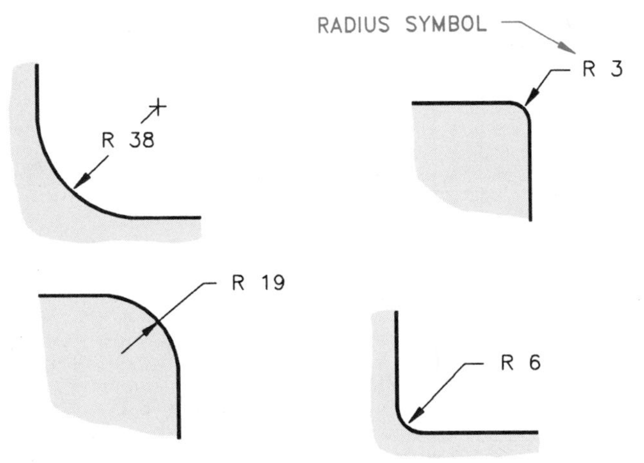

Figure 8–26 Dimensioning arcs — no centers located.

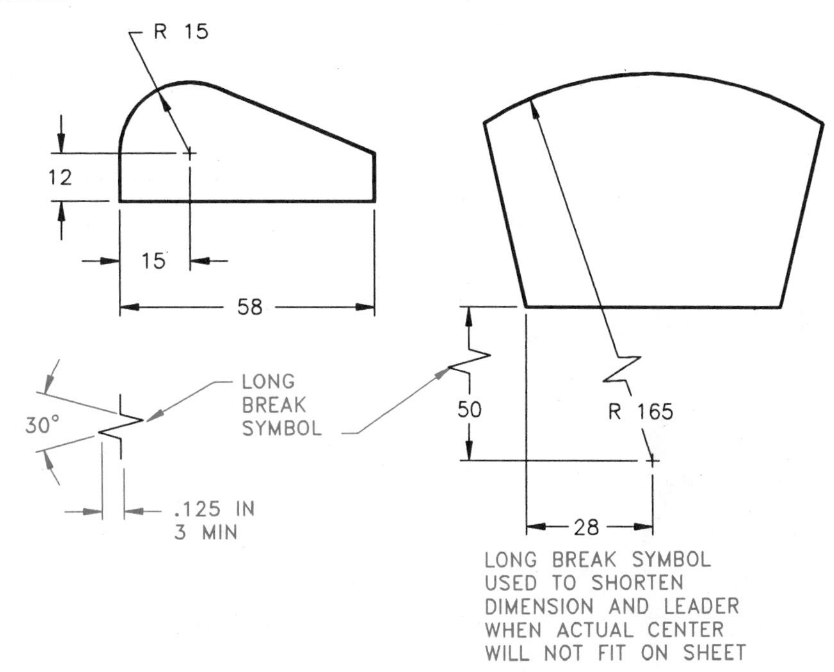

Figure 8–27 Dimensioning arcs — with centers located.

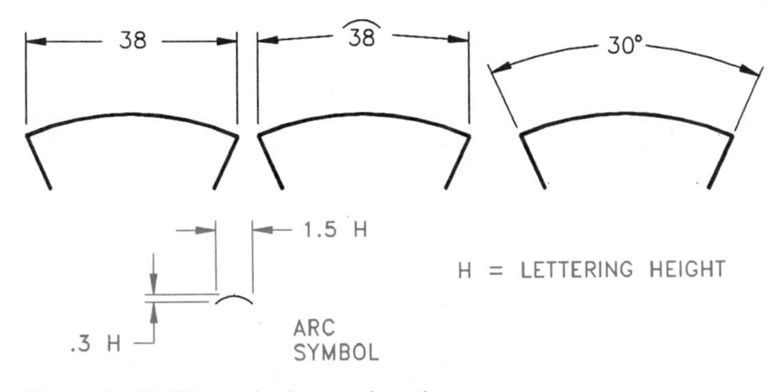

Figure 8–28 Dimensioning arc length.

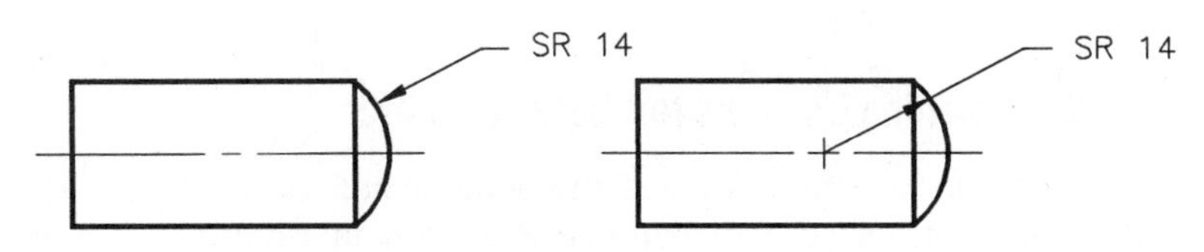

Figure 8–29 Dimensioning a spherical radius.

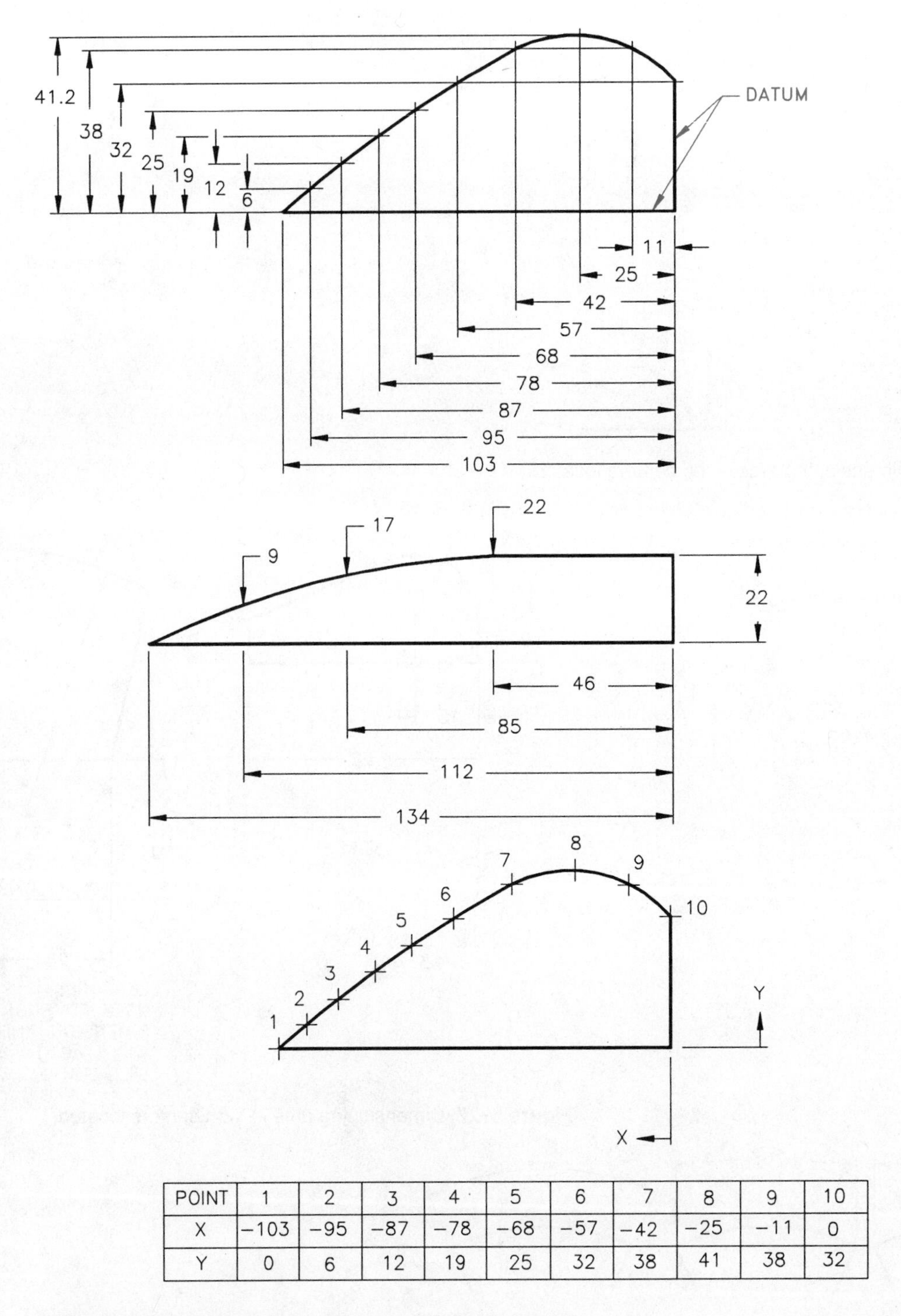

POINT	1	2	3	4	5	6	7	8	9	10
X	−103	−95	−87	−78	−68	−57	−42	−25	−11	0
Y	0	6	12	19	25	32	38	41	38	32

Figure 8–30 Dimensioning contours not defined as arcs.

A spherical radius may be dimensioned with the abbreviation *SR* preceding the numerical value, as shown in Figure 8–29.

Dimensioning Contours Not Defined as Arcs

Coordinates or points along the contour are located from common surfaces or datums as shown in Figure 8–30. Figure 8–31 shows a curved contour dimensioned using oblique extension lines.

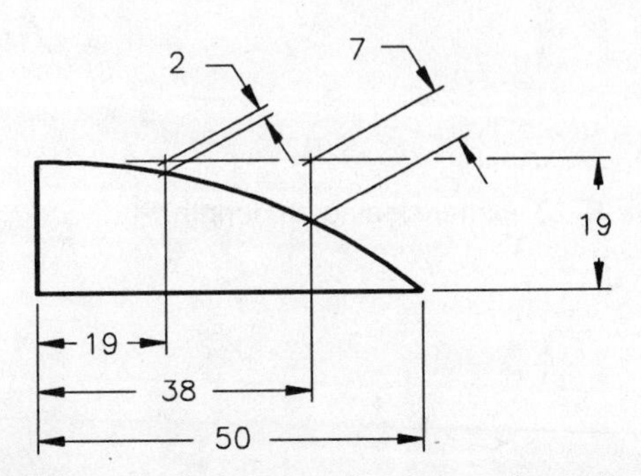

Figure 8–31 Dimensioning a curved contour using oblique extension lines.

DIMENSIONING

The proper placement and use of dimensions is one of the most difficult aspects of drafting. It requires thought and planning. The drafter must first determine the type of dimension that suits the application, then place it on the drawing. The dimensioning options for the CADD drafter are many, and the proper usage of these options should be learned early and practiced whenever possible. A variety of dimensioning possibilities may be located on the menu under the heading dimension parameters, automatic dimensioning, or simply, dimensions. The menu may show each of the possibilities, or they may be listed on the screen after selecting the DIMENSION command.

Two basic types of dimensions are horizontal and vertical. These can be labeled on the menu as HORIZ and VERT, or DMH (dimension horizontal) and DMV (dimension vertical). To place a horizontal dimension, first select the HORIZ command. You may then be prompted to locate one end of the dimension by placing the screen cursor so that the vertical crosshair is aligned with one edge of the feature to be dimensioned, as shown in Figure 8–32(a). Digitize this point and then locate the crosshairs on the opposite end of the feature and digitize that point, as shown in Figure 8–32(b). You have now established the length of the feature. The dimension is calculated by the computer according to the scale and grid spacing that was initially established for the drawing. Next you need to digitize the distance away from the feature the dimension line should be located. After this step is done, the dimension may be drawn on the screen. (See Figure 8–32[c].) The user also may be prompted for additional information such as location of text, number of decimals, location of arrowheads, and length of extension lines. To give you an idea of the amount of information required by the computer to calculate and place a single dimension, here's a list of the possibilities:

1. Beginning point
2. End point
3. Distance from feature
4. Extension lines? (yes or no, on or off) Figure 8–32(d) shows the complete dimension with extension lines drawn.
5. Arrowhead location (inside or outside the extension lines)
6. Text size
7. Text direction
8. Text location (inside, above, or below the dimension line)

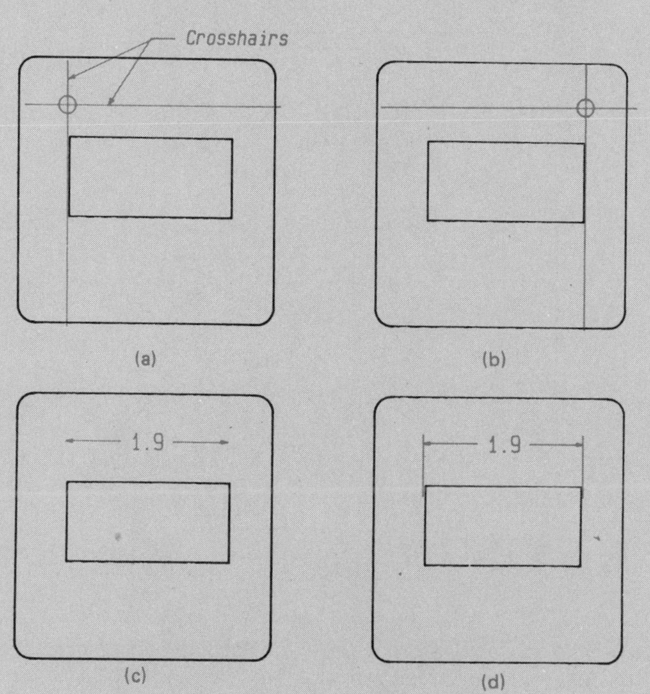

Figure 8–32 Locating a dimension using the screen cursor.

9. Dimension type (linear, datum, radial, or angular dimensioning)
10. Number of decimal places
11. Fractions or decimals
12. Grid spacing
13. Scale
14. Dimension number placement (automatic, semiautomatic, or manual)

Of course, some of these items are established as initializing parameters when you begin your drawing, and some are only required when you select a specific style of dimensioning, but they all affect the dimensions. Therefore, it is important that you learn and practice all the aspects of dimensioning that are a part of your specific CADD system.

Another way to place dimensions on a drawing is to pick the extension line origins on the object. The CADD dimensioning then automatically calculates the extension line offset and places the extension lines, dimension lines, text, and arrowheads. This can be done in either a datum or chain dimensioning format, as shown in Figure 8–33. When you are dimensioning circles, the dimension line and text numeral are automatically placed inside the circle by picking the circle, or you have the option of dimensioning with a leader. Figure 8–34 illustrates some of the options. Dimensioning arcs is also easy. All you have to do is pick the arc to be dimensioned and enter the dimension text; the leader with text numeral is placed on the drawing automatically, as shown in Figure 8–35.

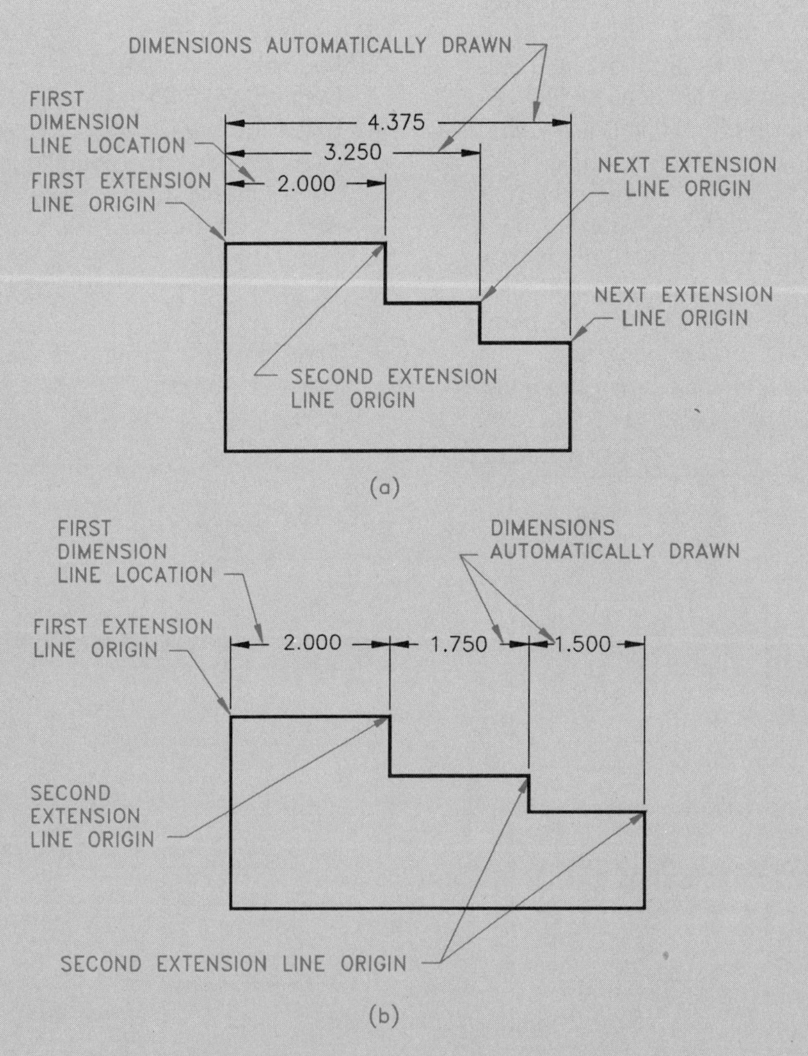

Figure 8–33 Locating dimensions at extension line origins for (a) datum and (b) chain dimensioning.

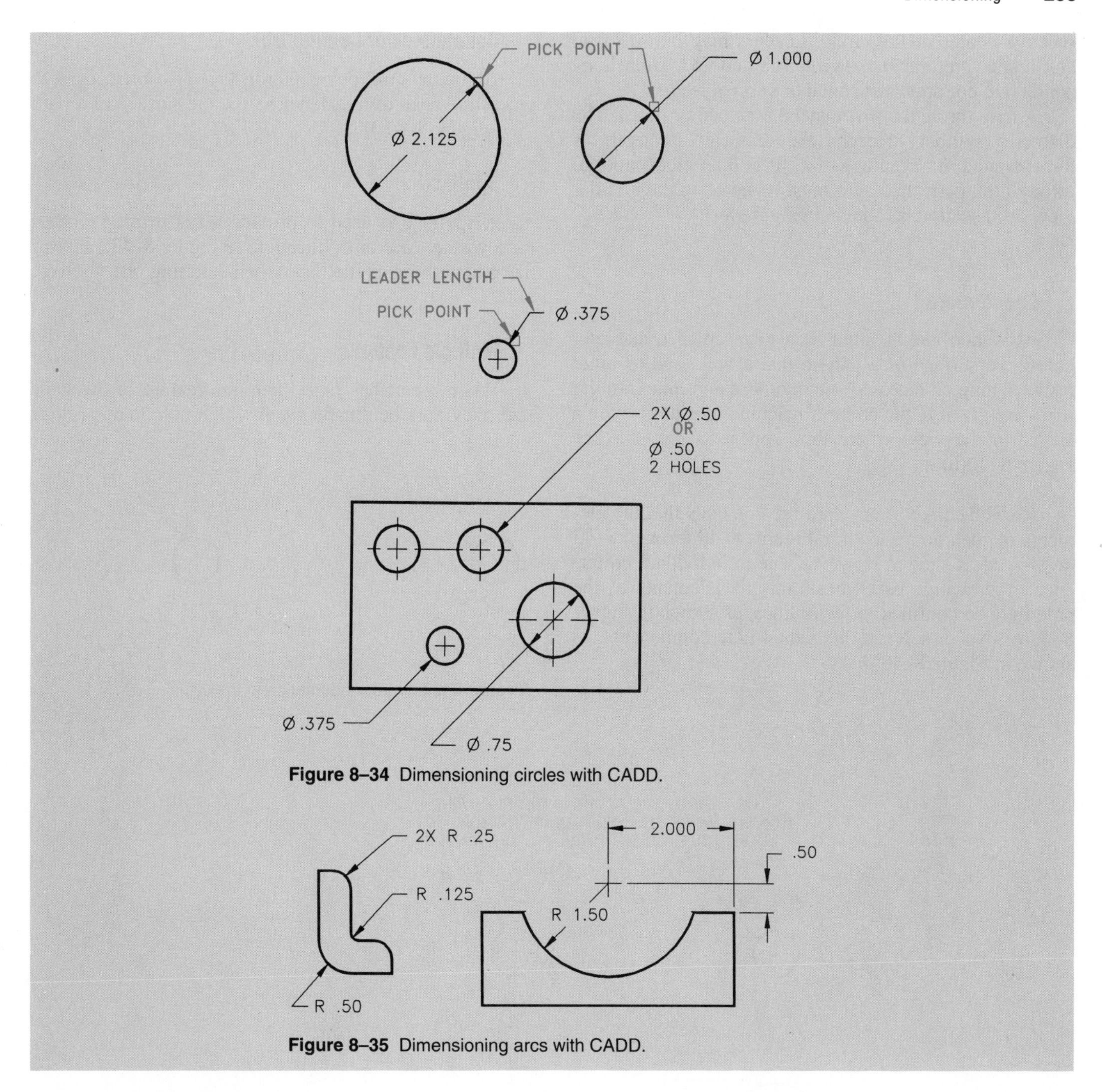

Figure 8–34 Dimensioning circles with CADD.

Figure 8–35 Dimensioning arcs with CADD.

NOTES FOR SIZE FEATURES

Holes

Hole sizes are dimensioned with leaders to the view where they appear as circles, or dimensioned in a sectional view. When leaders are used to establish notes for holes, the shoulder should be centered upon the beginning or the end of the note. When a leader begins at the left side of a note, it should originate at the beginning of the note. When a leader begins at the right side of a note, it should originate at the end of the note. (See Figure 8–36.)

Figure 8–37 shows that the leader should touch the side of the circle and, if it were to continue, would inter-

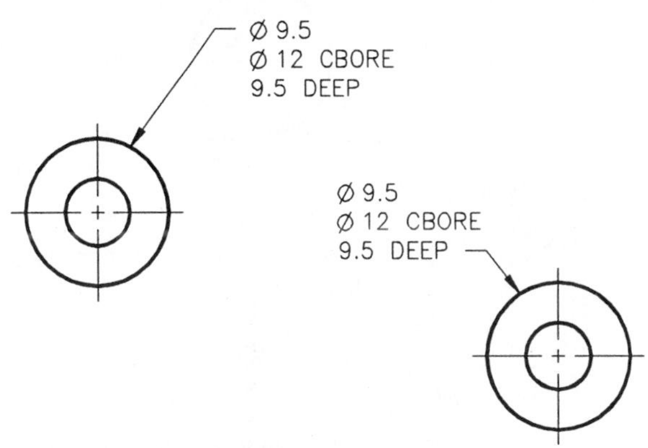

Figure 8–36 Leader orientation to the note.

sect the center of the circle. Leaders may be drawn at any angle, preferably between 15° and 75°, from horizontal. Do not draw horizontal or vertical leaders.

A hole through a part may be assumed or noted. The diameter symbol precedes the diameter numeral, as dimensioned in Figure 8–38. If a hole does not go through the part, the depth must be noted in the circular view or in section, as shown in Figure 8–39.

Counterbore

A counterbore is often used to machine a diameter below the surface of a part so that a bolt head or other fastener may be recessed. Counterbore and other similar notes are given in the order of machine operations with a leader in the view where they appear as circles. (See Figure 8–40(a).)

ANSI The ANSI standard recommends that the elements of each note shown in Figures 8–40 through 8–43 be aligned as shown. However, due to individual preference or drawing space constraints the elements of the note may be confined to fewer lines, as shown in Figure 8–40(b). Never separate individual note components, as shown in Figure 8–40(c).

Countersink or Counterdrill

A countersink, or counterdrill, is also often used to recess the head of a fastener below the surface of a part. (See Figure 8–41.)

Spotface

A spotface is used to provide a flat bearing surface for a washer face or bolthead. (See Figure 8–42.) Follow the counterbore guidelines when lettering the spotface note.

Multiple Features

When a part has more than one feature of the same size, they may be dimensioned with a note that specifies

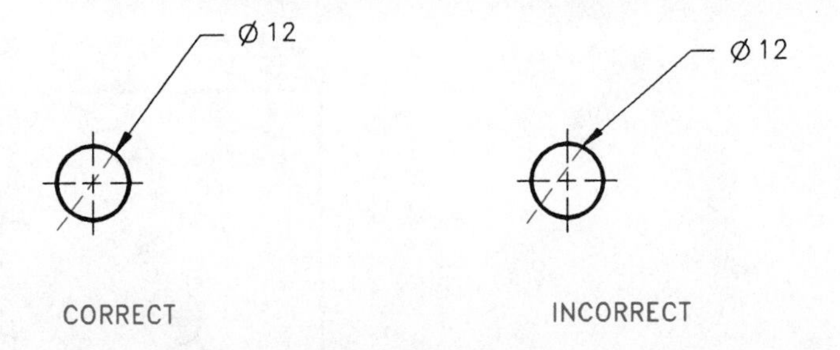

Figure 8–37 Leader orientation to the circle.

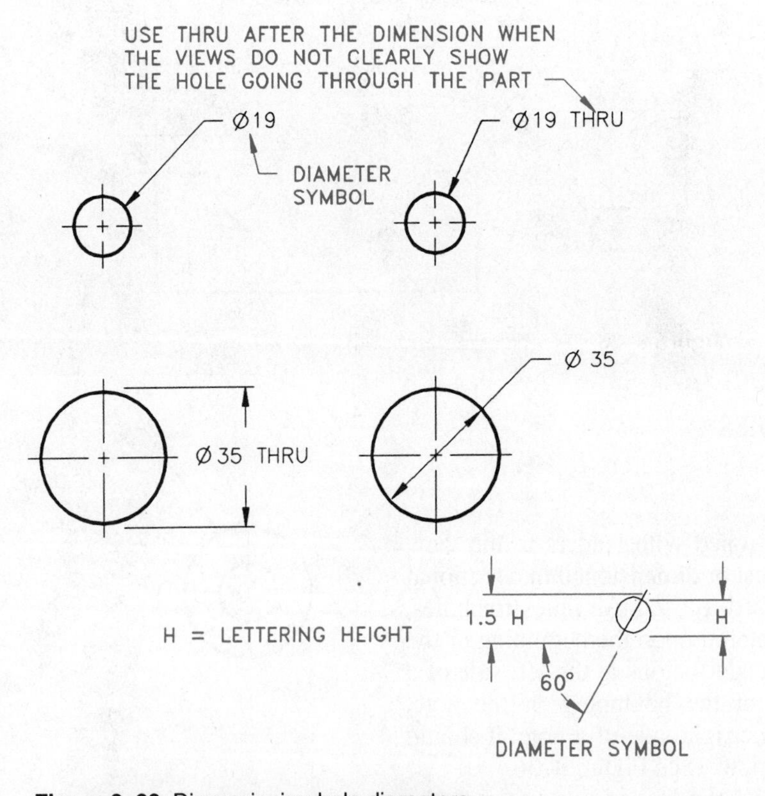

Figure 8–38 Dimensioning hole diameters.

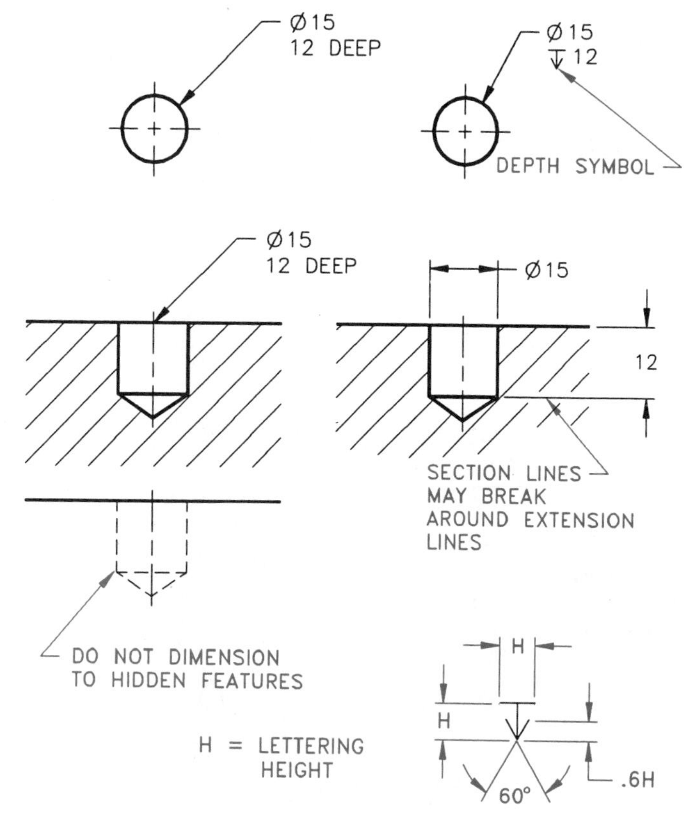

Figure 8–39 Dimensioning hole diameters and depths.

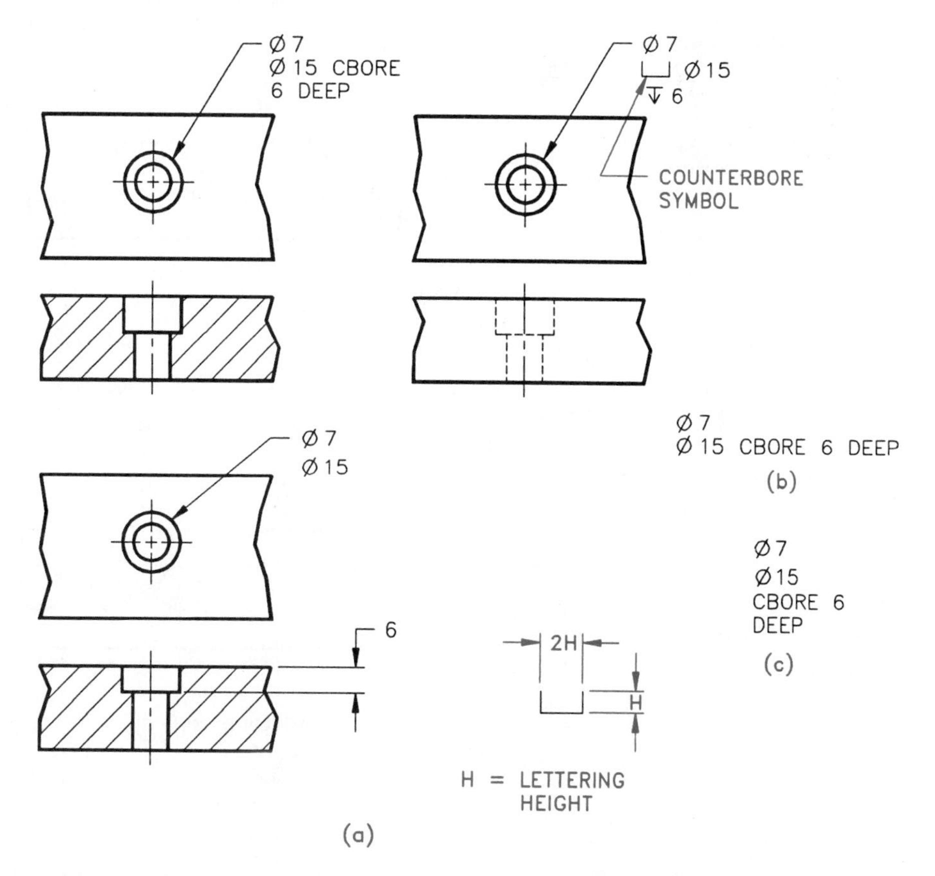

Figure 8–40 (a) Counterbore note; (b) alternate note with elements of note grouped on one line; (c) never split individual note elements.

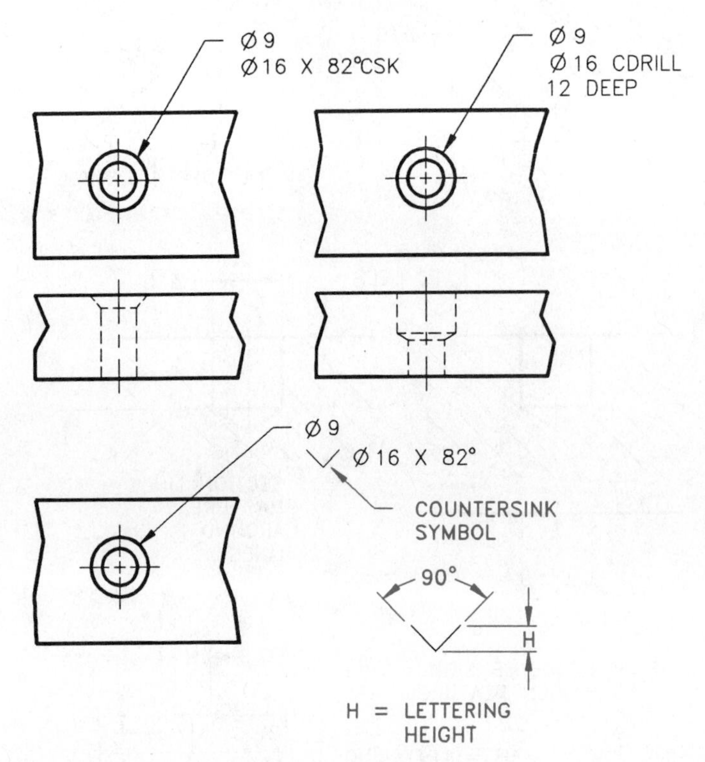

Figure 8–41 Countersink, counterdrill note.

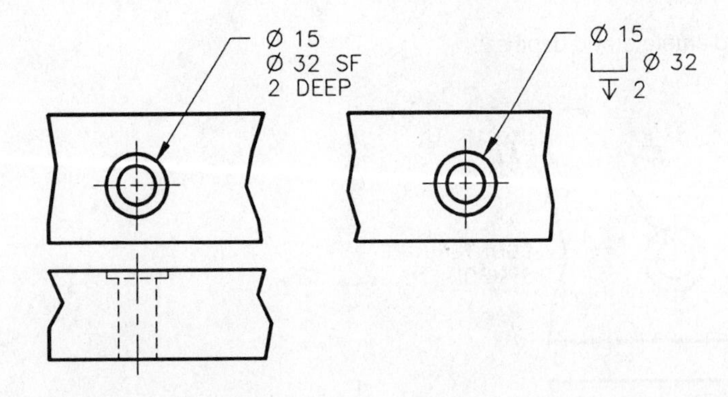

Figure 8–42 Spotface note.

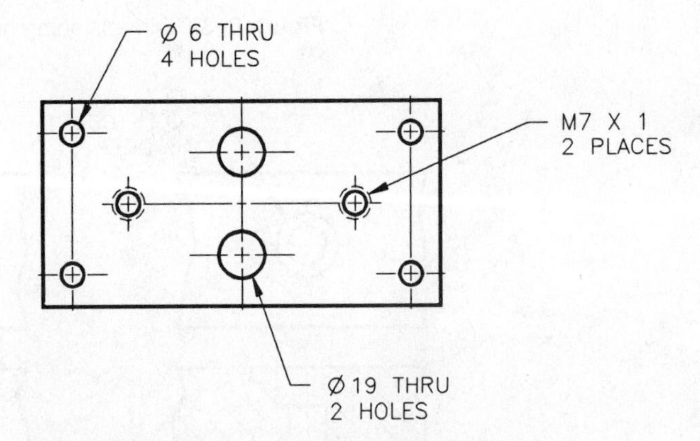

Figure 8–43 Dimension notes for multiple features.

the number of like features, as shown in Figure 8–43. If a part contains several features all the same size, the method shown in Figure 8–44 may be used.

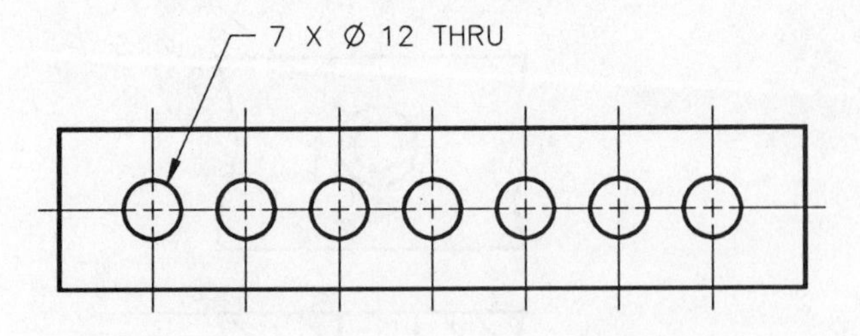

Figure 8–44 Dimension notes for multiple features all of common size.

Slots

Slotted holes may be dimensioned in one of three ways, as shown in Figure 8–45. These methods are used only when the ends are fully rounded and tangent to the sides. When the ends of a slot or external feature have a radius greater than the width of the feature, then the size of the radius must be given, as shown in Figure 8–46.

Keyseats

Keyseats are dimensioned in the view that clearly shows their shape by width, depth, length, and location, as in Figure 8–47.

Knurls

Knurls are dimensioned with notes and leaders that point to the knurl in the rectangular view, as shown in Figure 8–48(a). Although not a recommended ANSI standard, some companies prefer showing a knurl representation on the view as shown in Figure 8–48(b).

Necks and Grooves

Necks and grooves may be dimensioned as shown in Figure 8–49.

LOCATION DIMENSIONS

In general, dimensions either identify a size or a location. The previous dimensioning discussions focused on size dimensions. Location dimensions to cylindrical features, such as holes, are given to the center of the feature in the view where they appear as a circle, or in a sectional view. Rectangular shapes may be located to their sides, and symmetrical features may be located to their centerline or center plane. Some location dimensions also control size.

Locating Holes

Locate a hole center in the view where the hole appears as a circle, as shown in Figure 8–50.

Rectangular Coordinates

Linear dimensions are used to locate features from planes or centerlines, as shown in Figure 8–51.

Polar Coordinates

Angular dimensions locate features from planes or centerlines, as shown in Figure 8–52.

Repetitive Features

Repetitive features may be located by noting the number of times a dimension is repeated, and giving one typical dimension and the total length as reference. This method will be acceptable for chain dimensioning. (See Figure 8–53.)

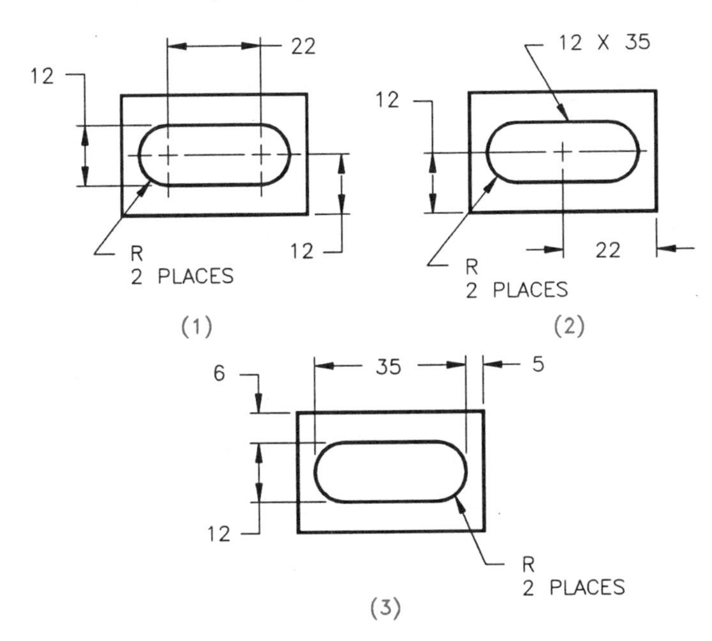

Figure 8–45 Dimensioning slotted holes with full radius ends.

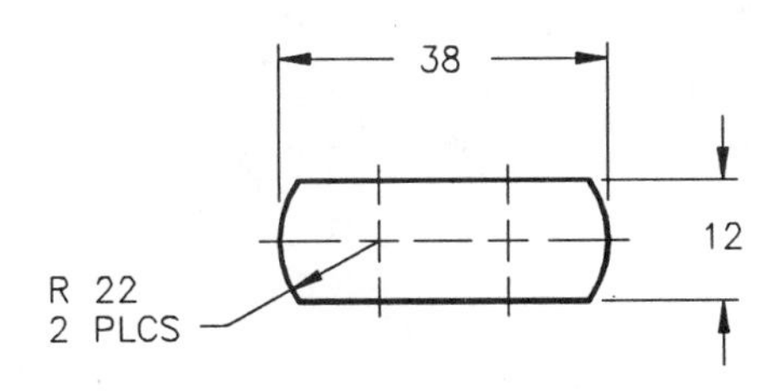

Figure 8–46 Dimensioning slot or external feature with end radius larger than feature width.

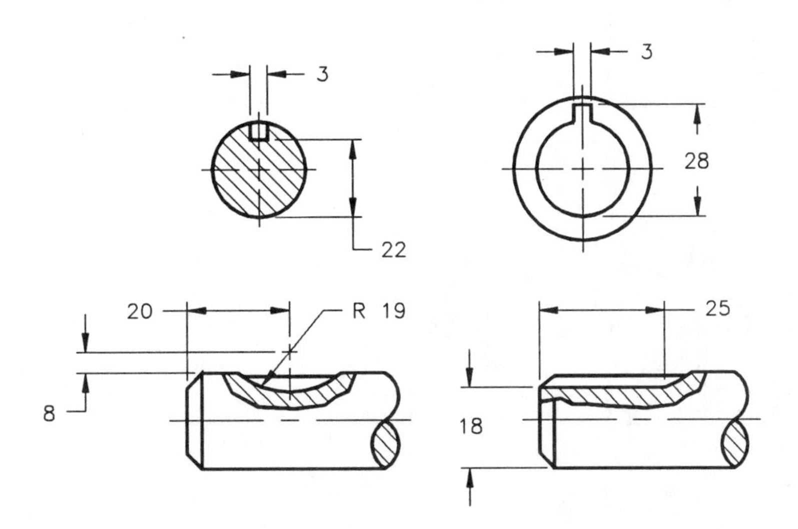

Figure 8–47 Dimensioning keyseats.

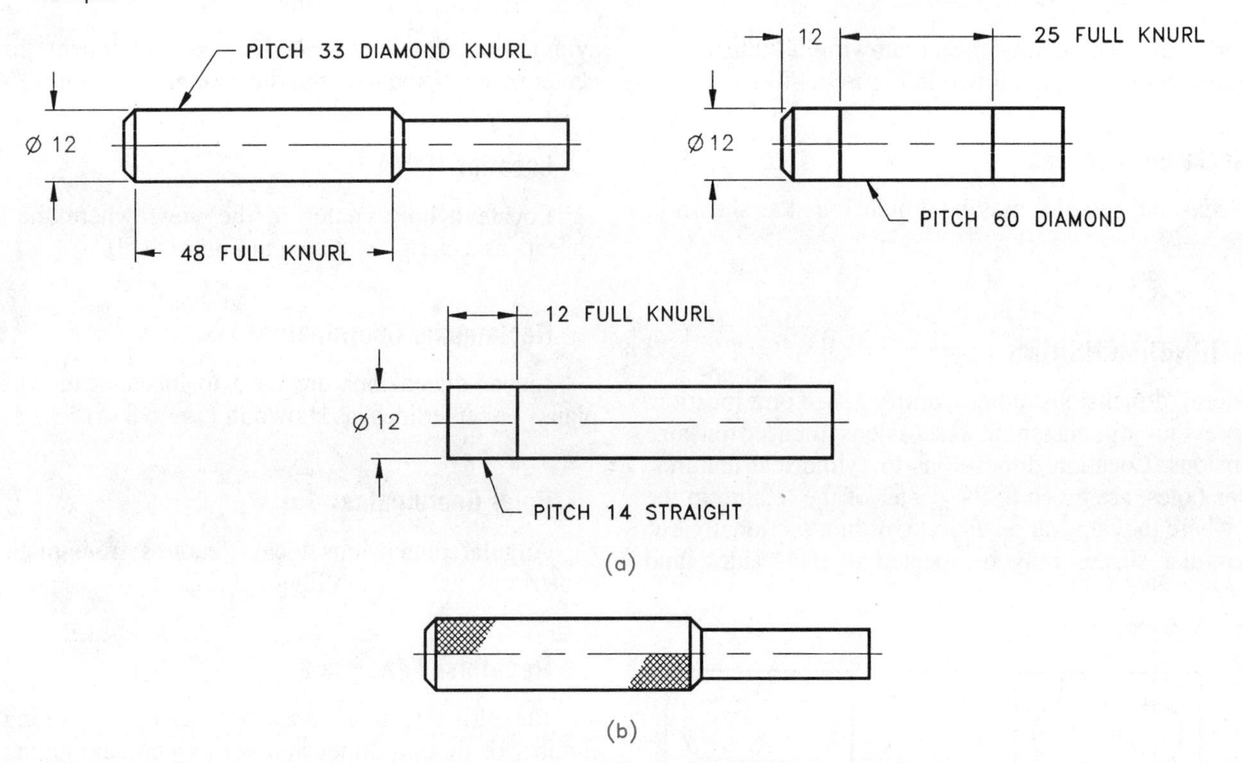

Figure 8–48 (a) Dimensioning knurls; (b) optional knurl representation.

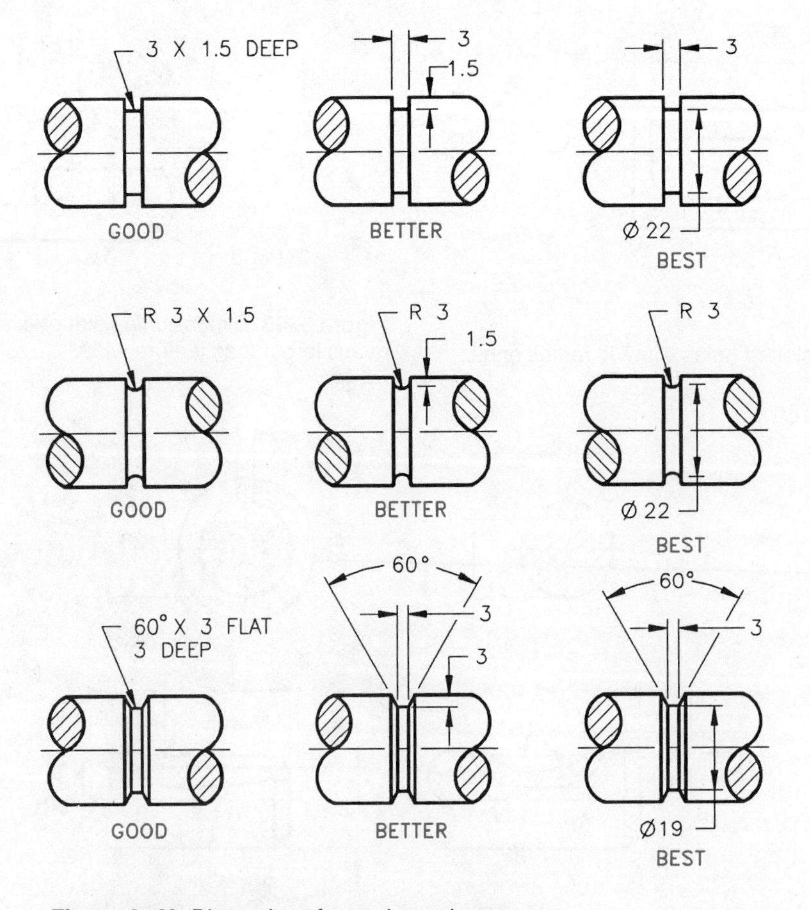

Figure 8–49 Dimensions for necks and grooves.

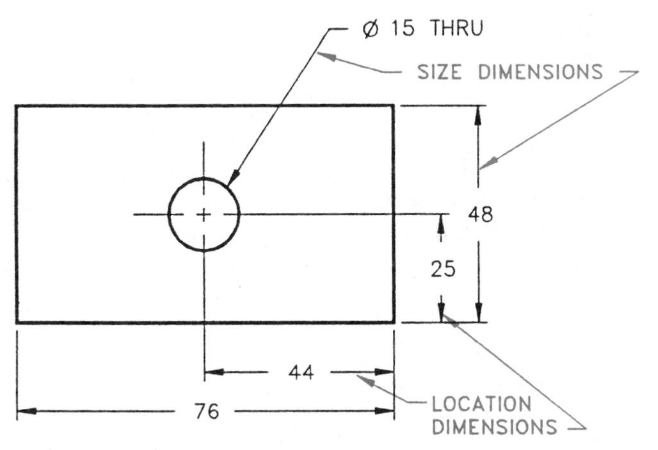

Figure 8–50 Size and location dimensions.

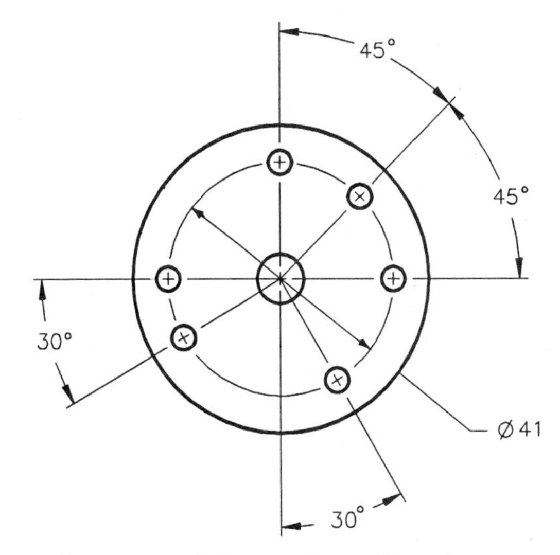

Figure 8–52 Polar coordinate dimensions.

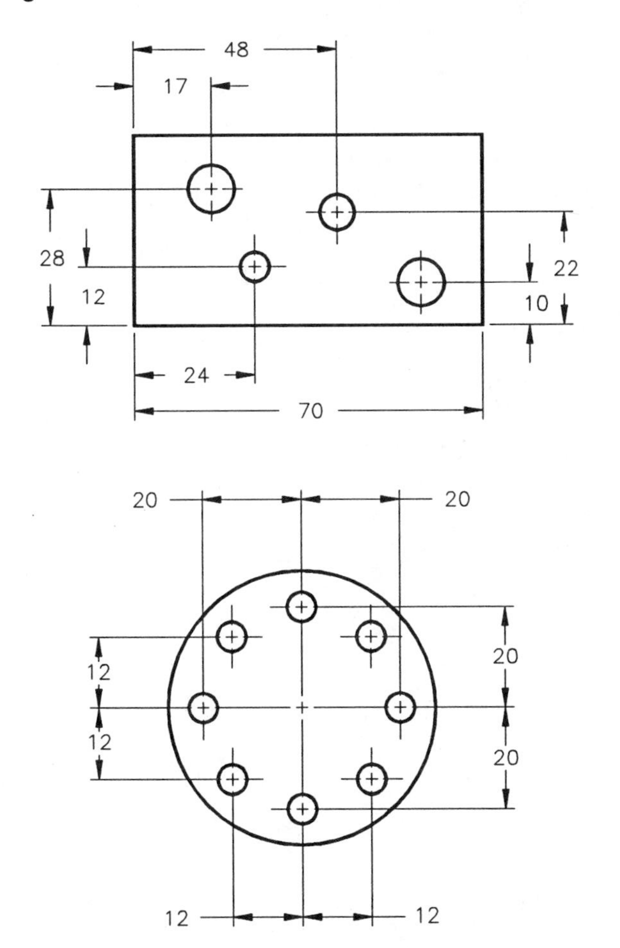

Figure 8–51 Rectangular coordinate location dimensions.

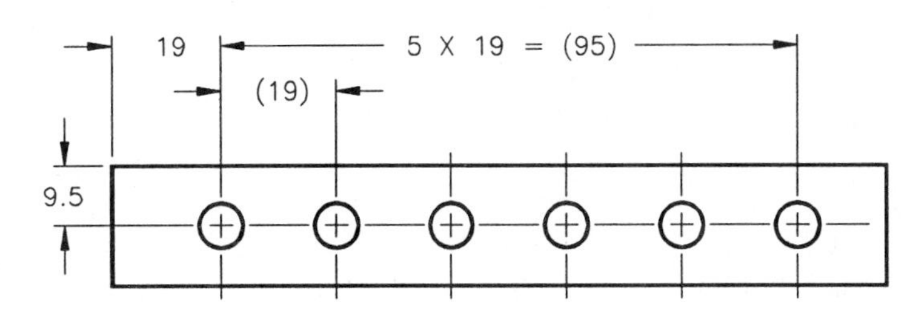

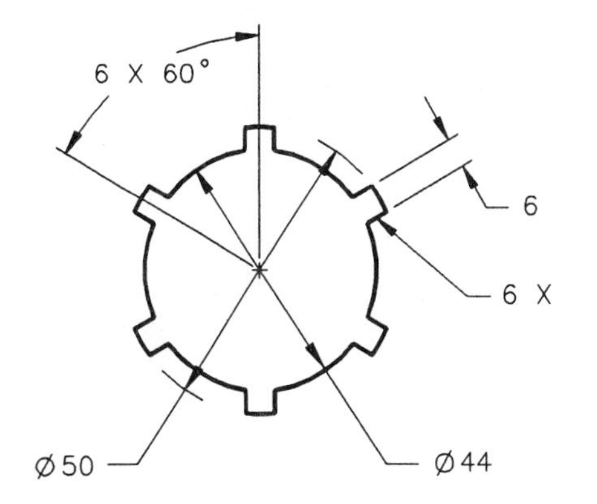

Figure 8–53 Dimensioning repetitive features.

DIMENSION ORIGIN

When the dimension between two features must clearly denote from which feature the dimension originates, the dimension origin symbol may be used. This method of dimensioning means that the origin feature must be established first and the related feature may then be dimensioned from the origin. (See Figure 8–54.)

DIMENSIONING AUXILIARY VIEWS

Dimensions should be placed on views that provide the best size and shape description of an object. In many instances the surfaces of a part are foreshortened and require auxiliary views to completely describe true size, shape, and the location of features. When foreshortened views occur, dimensions should be placed on the auxiliary view for clarity. In unidirectional dimensioning the dimension numerals will be placed horizontally so they read from the bottom of the sheet. When aligned dimensioning is used, the dimension numerals are placed in alignment with the dimension lines. (See Figure 8–55.)

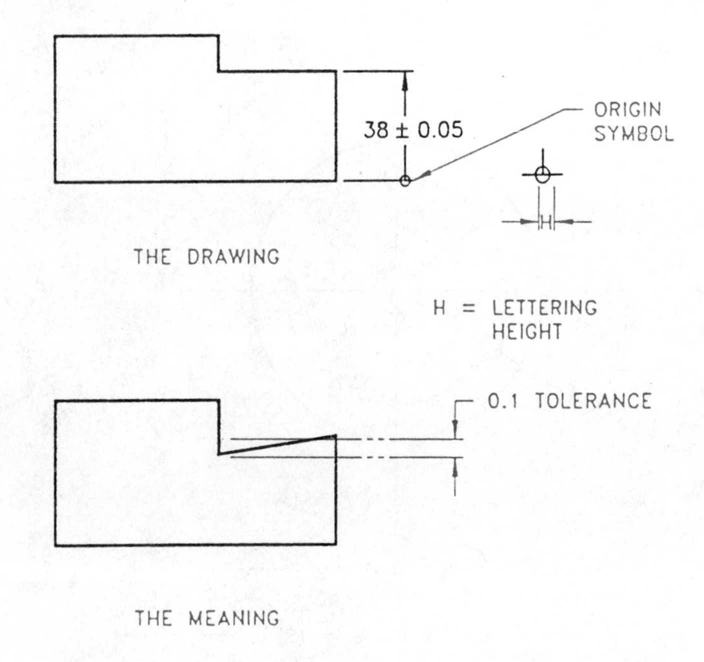

Figure 8–54 Dimension origin.

GENERAL NOTES

The notes that were discussed previously are classified as specific notes because they referred to specific features of an object. General notes, on the other hand, relate to the entire drawing. Each drawing will contain a certain number of general notes either in or near the drawing title block. (See Figure 8–57.) General notes will be concerned with such items as the following:

- Material specifications
- Dimensions: inches or millimeters
- General tolerances
- Confidential note, copyrights, or patents
- Drawn by
- Scale
- Date
- Part name
- Drawing size
- Part number
- Number of revisions
- First-angle or third-angle projection symbol

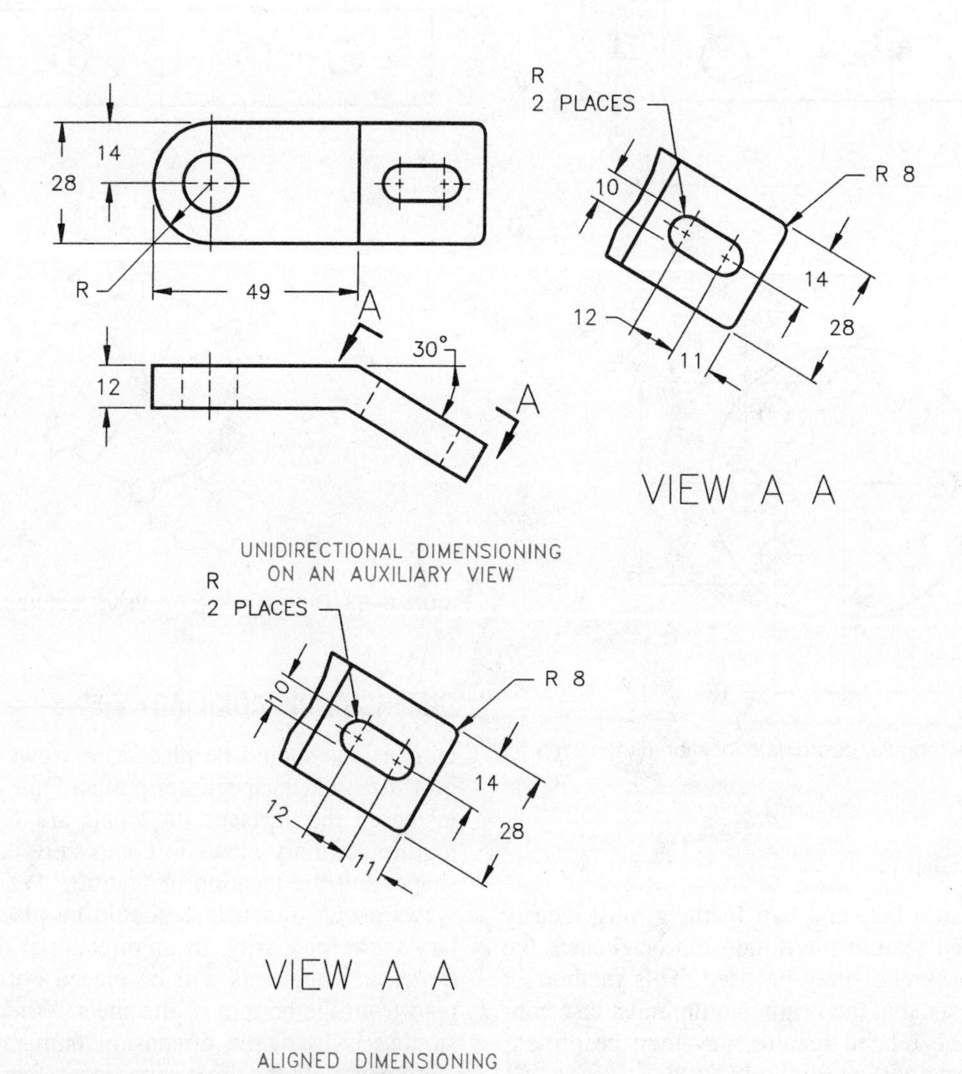

Figure 8–55 Dimensioning auxiliary views.

USING LAYERS FOR DIMENSIONING

A time-tested method of creating several drawings containing different information on the same outline is known as *overlay drafting*. A base drawing is generated, then an overlay is placed over it and a specific type of information is drawn on the overlay. For example, an architectural floor plan is drawn as the base, and then overlays are drawn for the plumbing plan, electrical plan, and so on. The base can be combined with any individual overlay or combinations thereof to create a print of just the information required. Also, several drafters can be given copies of the base and each can create an overlay containing specific information. This enables several drafters to essentially work on the same drawing at the same time; a great time-saver. The final drawing is then composed of several layers or levels.

The concept of layering is a prime component of CADD systems, some having the ability to display over 200 layers. This not only allows different drafters to work on the same drawing at the same time, but also addresses a shortcoming of the computer itself: the speed at which a drawing can be redrawn on the screen. The less information that is on the screen at one time, the less time it takes to be redrawn, and the less clutter there is to get in the way of the drafter. Figure 8–56(a), 8–56(b), and 8–56(c) show a simple example of the layering concept.

Several commands exist to allow the operator to work with the different layers. The LAYER command itself lets the operator assign a layer number to a drawing. RELAYER gives the drafter the chance to change the numbers of the drawing layers, and DISPLAY LAYER, or LAYER CONTROL, provides for the display of any combination of layers.

Color is a valuable tool when working with layers, and more companies are realizing this fact. Many color systems enable the user to assign any color desired to any layer. Some systems have a limited number of colors, while others possess the ability to create thousands of hues. Color gives even more meaning to the layering concept, because several layers displayed on the screen at the same time can be distinguished from each other. The colors on the screen can be transferred to the plotter by use of the PEN command. This command allows specific colors and widths of pens in the pen plotter to be selected.

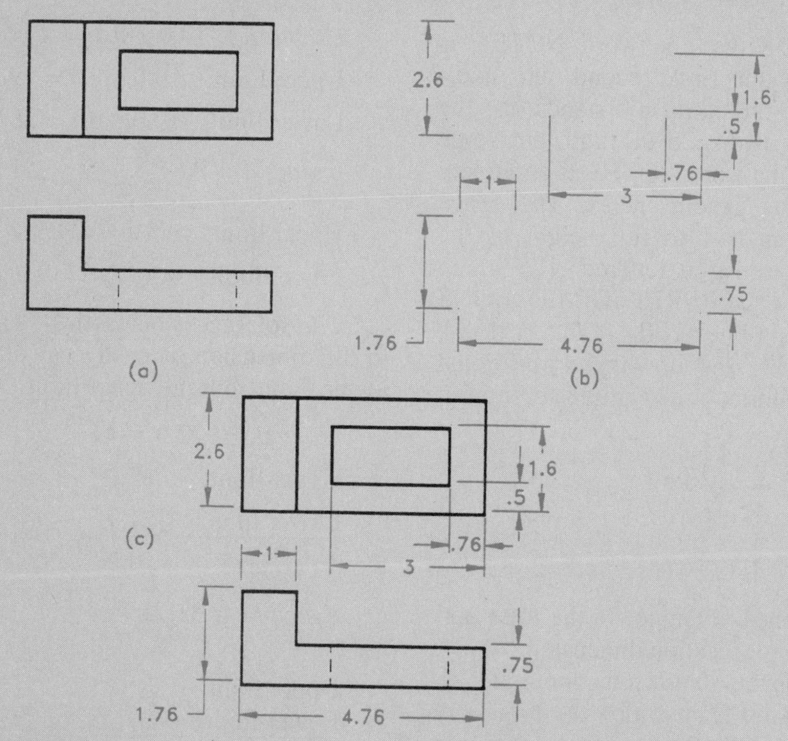

Figure 8–56 Layering enables the drafter to focus on one aspect of the drawing at a time.

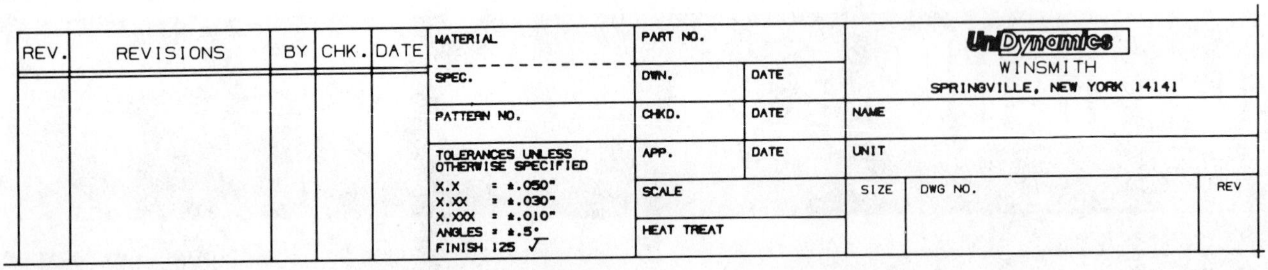

REV.	REVISIONS	BY	CHK.	DATE	MATERIAL	PART NO.		UniDynamics WINSMITH SPRINGVILLE, NEW YORK 14141		
					SPEC.	DWN.	DATE			
					PATTERN NO.	CHKD.	DATE	NAME		
					TOLERANCES UNLESS OTHERWISE SPECIFIED	APP.	DATE	UNIT		
					X.X = ±.050" X.XX = ±.030" X.XXX = ±.010" ANGLES = ±.5° FINISH 125 √	SCALE		SIZE	DWG NO.	REV
						HEAT TREAT				

Figure 8–57 General title block information. *Courtesy UniDynamics Winsmith.*

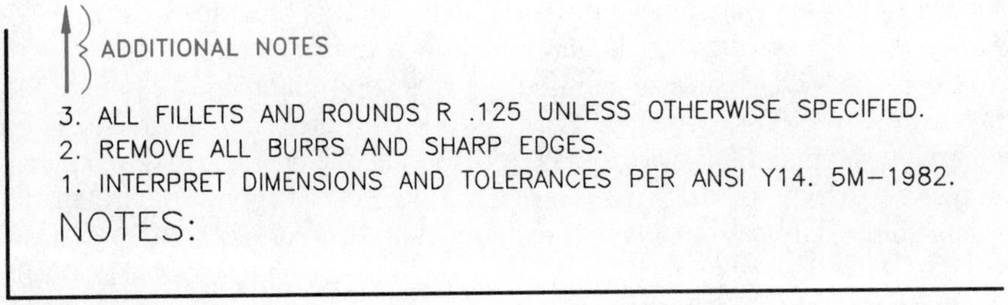

ADDITIONAL NOTES

3. ALL FILLETS AND ROUNDS R .125 UNLESS OTHERWISE SPECIFIED.
2. REMOVE ALL BURRS AND SHARP EDGES.
1. INTERPRET DIMENSIONS AND TOLERANCES PER ANSI Y14. 5M–1982.

NOTES:

Figure 8–58 General notes located in the lower-left corner of the sheet.

ANSI Other general notes are located adjacent to the title block for ANSI drawings. The exact location of the general notes depends on specific company standards. Military standards specify general notes be located near the upper-left corner of the drawing. A common location for general notes is the lower-left corner of the drawing, usually .5 in. each way from the border line. Figure 8–58 shows some general notes that could commonly be included on ANSI standard format drawings. Other common locations for general notes are the lower right corner of the drawing directly above the title block or just to the left of the title block.

Notice in Figure 8–58 that the word NOTES is lettered first followed by the first, second, and additional notes. Depending on company standards, the word NOTES is lettered .18 to .25 in. high, but some companies prefer to omit the word NOTES as an unnecessary preface to obvious general notes. The space between notes is from one-half to full height of the lettering, and the notes are often lettered .125 in. in height. The first note, INTERPRET DIMENSIONS AND TOLERANCES PER ANSI Y14.5M–1982, should be included on all new drawings. Additional notes depend upon the information required to support the drawing.

TOLERANCING

Definitions

As previously mentioned, tolerance is the total permissible variation in a size or location dimension.

A specified dimension, also known as nominal size, is that part of the dimension from which the limits are calculated. For example, 15.8 is the specified dimension of 15.8 ± 0.2.

A bilateral tolerance is allowed to vary in two directions from the specified dimension, as in $6.35^{+0.050}_{-0.127}$ or 12.7 ± 0.127.

A unilateral tolerance varies in only one direction from the specified dimension: $22.225^{\,0}_{-0.127}$ or $19.05^{+0.05}_{\,0}$.

Limits are the largest and smallest possible sizes a feature may be made as related to the tolerance of the dimension.

Example 1: 19.0 ± 0.1

Upper limit: 19.0 + 0.1 = 19.1

Lower limit: 19.0 – 0.1 = 18.9

Example 2: $9.5^{+0}_{-0.5}$

Upper limit: 9.5 + 0 = 9.5

Lower limit: 9.5 – 0.5 = 9.0

The tolerance, being the total permissible variation in the dimension, is easily calculated by subtracting the lower limit from the upper limit.

Example 3: 22.0 ± 0.1

Upper limit:	22.1
Lower limit:	–21.9
Tolerance	0.2

Example 4: $31.75^{+0.10}_{-0}$

Upper limit:	31.85
Lower limit:	–31.75
Tolerance	0.10

All dimensions on a drawing will have a tolerance except reference, maximum, minimum, or stock size dimensions. Dimensions on a drawing may read as in Figure 8–59 with general tolerances specified in the title block of the drawing. General tolerance specifications, as given in a typical industry title block, are shown in Figure 8–60. Using this title block as an example, the tolerances of the dimensions in Figure 8–59 would be as follows (given in inches):

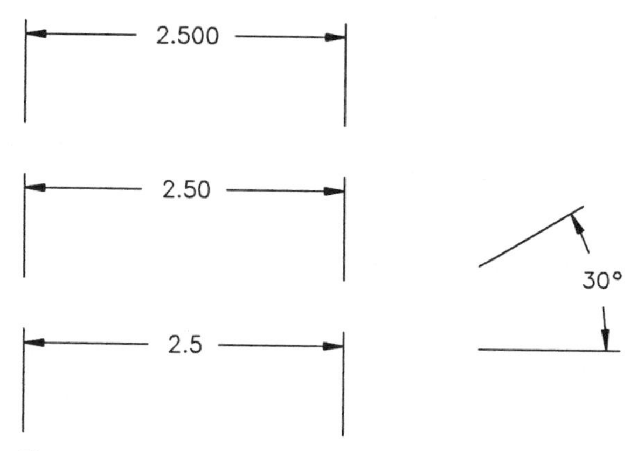

Figure 8–59 Typical drawing dimensions.

2.500 would have .xxx ± .005 applied; 2.500 ± .005, tolerance equals 0.100.

2.50 would have .xx ± .010 applied; 2.50 ± .010, tolerance equals .020

2.5 would have .x ± .020 applied; 2.5 ± .020, tolerance equals .040

30° would have Angles ± 0.5° applied; 30° ± 0.5°, tolerance equals 1°

Dimensions that require tolerances different from the general tolerances given in the title block must be specified in the dimension, as in Figure 8–61.

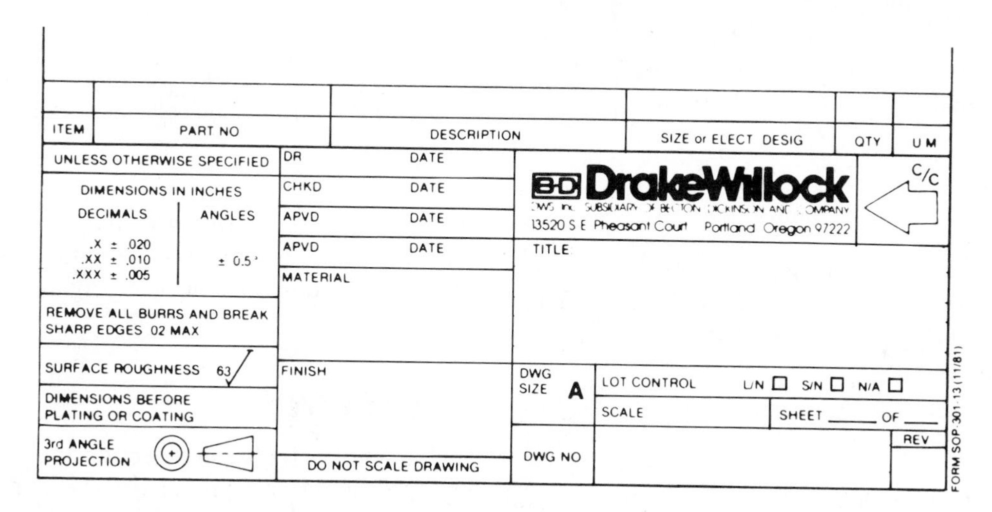

Figure 8–60 General tolerances from a company title block. *Courtesy Drake Willock, DWS, Inc.*

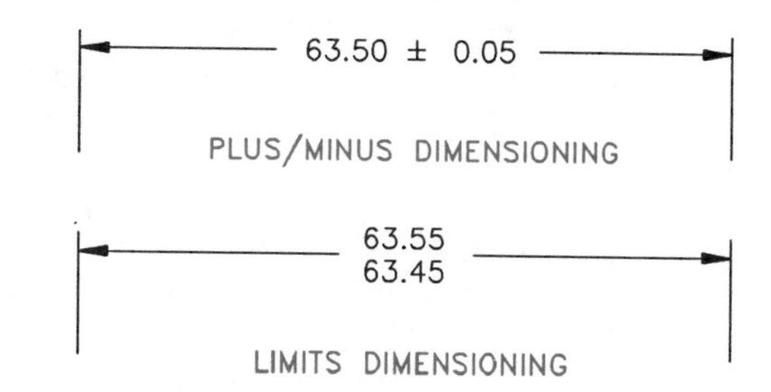

Figure 8–61 Specific tolerance dimensions.

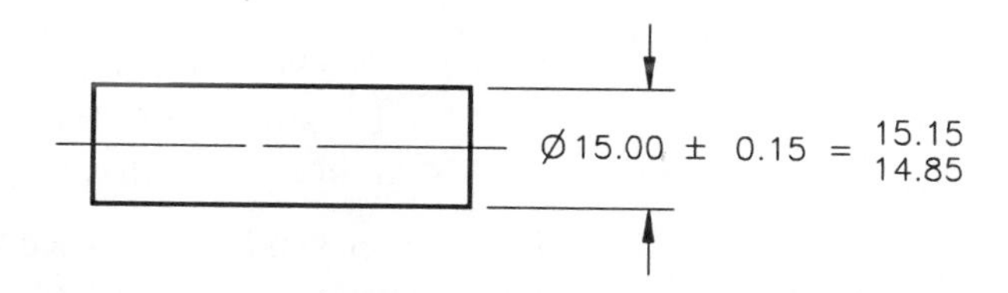

Figure 8–62 Maximum material condition (MMC) of an external feature.

Maximum and Least Material Conditions

Maximum material condition, abbreviated MMC, is the condition of a part or feature when it contains the most amount of material. The key is *most material.* The MMC of an external feature is the upper limit or largest size. (See Figure 8–62.) The MMC of an internal feature is the lower limit or smallest size. (See Figure 8–63.)

The *least material condition*, LMC, is the opposite of MMC. LMC is the least amount of material possible in the size of a part. The LMC of an external feature is its lower limit. The LMC of an internal feature is its upper limit.

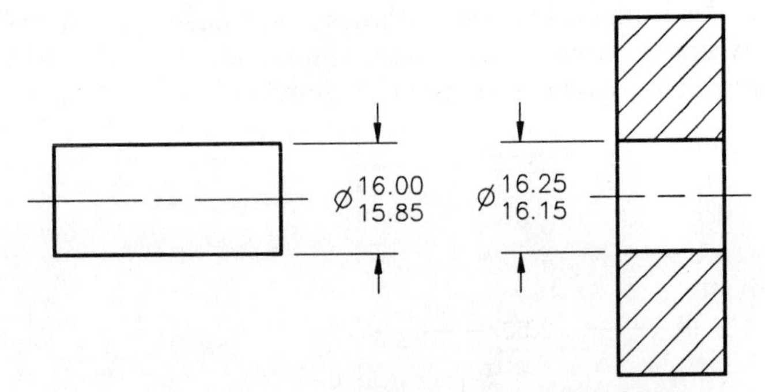

Figure 8–64 A clearance fit between two mating parts.

Clearance Fit

A *clearance fit* is a condition when, due to the limits of dimensions, there will always be a clearance between mating parts. The features in Figure 8–64 have a clearance fit. Notice that the largest size of the shaft is smaller than the smallest hole size.

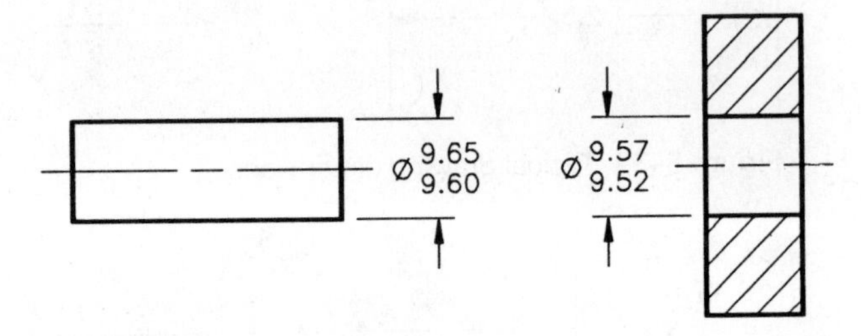

Figure 8–65 An interference fit between two mating parts.

Allowance

The *allowance* of a clearance fit between mating parts is the tightest possible fit between the parts. The allowance is calculated with the formula:

MMC Internal Feature
– MMC External Feature
Allowance

The allowance of the parts in Figure 8–64 is:

MMC Internal Feature	16.15
MMC External Feature	– 16.00
Allowance	0.15

Interference Fit

An interference fit is the condition that exists when, due to the limits of the dimensions, mating parts must be pressed together. Interference fits are used, for example, when a bushing must be pressed onto a housing or when a pin is pressed into a hole. (See Figure 8–65.)

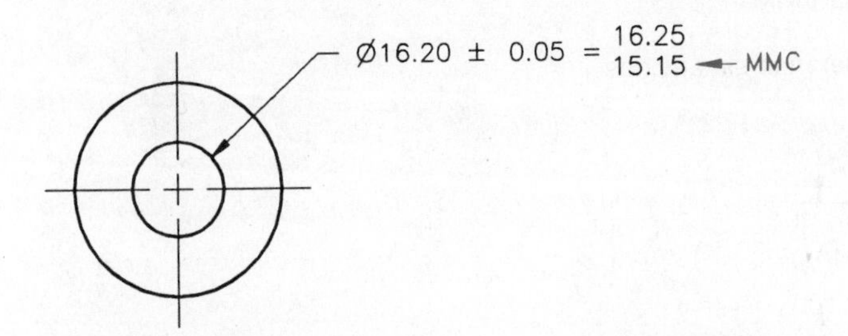

Figure 8–63 Maximum material condition (MMC) of an internal feature.

Types of Fits

Selection of Fits. In selecting the limits of size for any application, the type of fit is determined first based on the use or service required from the equipment being designed; then the limits of size of the mating parts are established to insure that the desired fit will be produced. Theoretically, an infinite number of fits could be chosen, but the number of standard fits described here should cover most applications.

Designation of Standard ANSI Fits. Standard fits are designated by means of the following symbols, which facilitate reference to classes of fit for educational purposes. The symbols are not intended to be shown on manufacturing drawings; instead, sizes should be specified on drawings. The letter symbols used are as follows:

RC Running, or Sliding, Clearance Fit
LC Locational Clearance Fit
LT Transition Clearance, or Interference Fit
LN Locational Interference Fit
FN Force, or Shrink, Fit

These letter symbols are used in conjunction with numbers representing the class of fit: thus FN 4 represents a Class 4 force fit.

Description of Standard ANSI Fits. The classes of fits are arranged in three general groups known as running and sliding fits, locational fits, and force fits.

Running and Sliding Fits (RFC). Running and sliding fits are intended to provide a similar running performance with suitable lubrication allowance, throughout their range of sizes. The clearances for the first two classes, used chiefly as sliding fits, increase more slowly with the diameter than do the clearances for the other classes, so that accurate location is maintained even at the expense of free relative motion. These fits may be described as follows:

- RCI — Close sliding fits are intended for the accurate location of parts that must assemble without perceptible play.
- RC2 — Sliding fits are intended for accurate location, but with greater maximum clearance than class RC1. Parts made to this fit move and turn easily but are not intended to run freely, and in the larger sizes may seize with small temperature changes.
- RC3 — Precision running fits are about the closest fits that can be expected to run freely, and are intended for precision work at slow speeds and light journal pressures, but are not suitable where appreciable temperature differences are likely to be encountered.
- RC4 — Close running fits are intended chiefly for running fits on accurate machinery with moderate surface speeds and journal pressures, where accurate location and minimum play is desired.
- RC5 and RC6 — Medium running fits are intended for high running speeds, or heavy journal pressures, or both.
- RC7 — Free running fits are intended for use where accuracy is not essential, or where large temperature variations are likely to be encountered, or under both these conditions.
- RC8 and RC9 — Loose running fits are intended for use where wide commercial tolerances may be necessary, together with an allowance, on the external member.

Locational Fits (LC, LT, and LN). Locational fits are fits intended to determine only the location of the mating parts; they may provide rigid or accurate location, as with interference fits, or provide some freedom of location, as with clearance fits. Accordingly, they are divided into three groups: clearance fits (LC), transition fits (LT), and interference fits (LN). These fits are described as follows:

- LC — Locational clearance fits are intended for parts which are normally stationary, but which can be freely assembled or disassembled. They range from snug fits for parts requiring accuracy of location, through medium clearance fits for parts such as spigots, to looser fastener fits where freedom of assembly is of prime importance.
- LT — Locational transition fits are a compromise between clearance and interference fits. They are for applications where accuracy of location is important but either a small amount of clearance or interference is permissible.
- LN — Locational interference fits are used where accuracy of location is of prime importance, and for parts requiring rigidity and alignment with no special requirements for bore pressure. Such fits are not intended for parts designed to transmit frictional loads from one part to another by virtue of the tightness of fit. Such conditions are covered by force fits.

Force Fits (FN). Force, or shrink, fits constitute a special type of interference fit normally characterized by maintenance of constant bore pressures throughout its range of sizes. The interference, therefore, varies almost directly with diameter, and the difference between its minimum and maximum values is small so as to maintain the resulting pressures within reasonable limits. These fits are described as follows:

- FN1 — Light drive fits are those requiring light assembly pressures and producing more or less permanent assemblies. They are suitable for thin sections or long fits, or in external cast-iron members.
- FN2 — Medium drive fits are suitable for ordinary steel parts, or for shrink fits on light sections. They are about the tightest fits that can be used with high-grade cast-iron external members.
- FN3 — Heavy drive fits are suitable for heavy steel parts or for shrink fits in medium sections.
- FN4 and FN5 — Force fits are suitable for parts that can be highly stressed, or for shrink fits where the heavy pressing forces required are impractical.

Establishing Dimensions for Standard ANSI Fits. The fit used in a specific situation is determined by the operating requirements of the machine. When the type of fit has been established, the engineering drafter refers to tables that show the standard hole and shaft tolerances for the specified fit. One source of these tables is the *Machinery's Handbook*. Tolerances are based on the type of fit and nominal size ranges, such as 0–.12, .12–.24, .24–.40, .40–.71, .71–1.19, and 1.19–1.97 in. So, if you have a 1-in. nominal shaft diameter and an RC4 fit, refer to Figure 8–66 to determine the shaft and hole limits. The hole and shaft limits for a 1-in. nominal diameter are:

Upper hole limit = 1.000 + .0012 = 1.0012

Lower hole limit = 1.000 + 0 = 1.000

Upper shaft limit = 1.000 – .0008 = .9992

Lower shaft limit = 1.000 – .0016 = .9984

NOMINAL SIZE RANGE IN INCHES	RC4 STANDARD TOLERANCE LIMITS	
	Hole	Shaft
0–.12	+.0006	−.0003
	0	−.0007
.12–.24	+.0007	−.0004
	0	−.0009
.24–.40	+.0009	−.0005
	0	−.0011
.40–.71	+.0010	−.0006
	0	−.0013
.71–1.19	+.0012	−.0008
	0	−.0016
1.19–1.97	+.0016	−.0010
	0	−.0020

Figure 8–66 Standard RC4 fits for nominal sizes ranging from 0 in. to 1.97 in.

You then dimension the hole as ∅ 1.0012–1.000, and the shaft as ∅ .9992–.9984.

Standard ANSI/ISO Metric Limits and Fits. The standard for the control of metric limits and fits is governed by the document ANSI B4.2–1978 (reaffirmed in 1984), *Preferred Metric Limits and Fits.* The system is based on symbols and numbers that relate to the internal or external application and the type of fit. The specifications and terminology for fits are slightly different from the ANSI standard fits previously described. The metric limits and fits are divided into three general categories: clearance fits, transition fits, and interference fits. *Clearance fits* are generally the same as the running and sliding fits explained earlier. With clearance fits, a clearance always occurs between the mating parts under all tolerance conditions. With *transition fits*, a clearance or interference may result due to the range of limits of the mating parts. When interference fits are specified, a press or force situation exists under all tolerance conditions. Refer to Figure 8–67 for the ISO symbol and descriptions of the different types of metric fits.

The metric limits and fits may be designated in a dimension one of three ways. The method used depends on individual company standards and the extent of use of the ISO system. When most companies begin using this system, the tolerance limits are calculated and shown on the drawing followed by the tolerance symbol in parentheses; for example, 25.000–24.979 (25 h7). The symbol in parentheses represents the basic size, 25, and the shaft tolerance code, h7. The term *basic size* denotes the dimension from which the limits are calculated. When companies become accustomed to using the system, they represent dimensions with the code followed by the limits in parentheses, as follows: 25 h7 (25.000–24.979).

TYPE OF FIT	ISO SYMBOL		DESCRIPTION OF FIT
	Hole	Shaft	
Clearance Fit	H11/c11	C11/h11	*Loose running*
	H9/d9	D9/h9	*Free running*
	H8/f7	F8/h7	*Close running*
	H7/g6	G7/h6	*Sliding*
	H7/h6	H7/h6	*Locational clearance*
Transition Fit	H7/k6	K7/h6	*Locational transition*
	H7/n6	N7/h6	*Locational transition*
Interference Fit	H7/p6[1]	P7/h6	*Locational interference*
	H7/s6	S7/h6	*Medium drive*
	H7/u6	U7/h6	*Force*

Figure 8–67 Description of metric fits.

Finally, when a company has used the system long enough for interpreters to understand the designations, the code is placed alone on the drawing, like this: 25 h7. When it is necessary to determine the dimension limits from code dimensions, use the charts in ANSI B4.2–1978 or the *Machinery's Handbook.* For example, if you want to determine the limits of the mating parts with a basic size of 30 and a close running fit, refer to the chart shown in Figure 8–68. The hole limits for the 30-mm basic size in Figure 8–68 are ∅ 30.033–30.000, and the shaft dimension limits are ∅ 29.980–29.959.

DIMENSIONS APPLIED TO PLATINGS AND COATINGS

When platings such as chromium, copper, and brass, or coatings such as galvanizing, polyurethane, and silicone are applied to a part or feature, the specified dimensions should be defined in relationship to the coating or plating process. A general note that indicates that the dimensions

BASIC SIZE	CLOSE RUNNING FIT	
	Hole (H8)	Shaft (f7)
20	20.033	19.993
	20.000	19.959
25	25.033	24.980
	25.000	24.959
30	30.033	29.980
	30.000	29.959
40	40.039	39.975
	40.000	39.950
50	50.039	49.975
	50.000	49.950

Figure 8–68 Tolerances of close running fits for basic sizes ranging from 20 mm to 50 mm.

apply before or after plating or coating is commonly used and will specify the desired variables; for example, DIMENSIONAL LIMITS APPLY BEFORE (AFTER) PLATING (COATING).

MAXIMUM AND MINIMUM DIMENSIONS

In some situations, a dimension with an unspecified tolerance may require that the general tolerance be applied in one direction only from the specified dimension. For example, when a 12-mm radius shall not exceed 12 mm, the dimension may read R 12 MAX. Therefore, when it is desirable to establish a maximum or minimum dimension the abbreviations MAX or MIN may be applied to the dimension. A dimension with a specified tolerance will read as previously discussed; for example, $R12\ {}^{\ 0}_{-0.05}$, or $R12\ {}^{+0.05}_{\ 0}$.

Casting Drawing and Design

The end result of a casting drawing will be the fabrication of a pattern. The preparation of casting drawings depends on the casting process used, the material to be cast, and the design or shape of the part. The drafter will make the drawing of the part the same as the desired end result after the part has been cast. The drafter may need to take certain casting characteristics into consideration, and the pattern maker will need to adjust the size and shape of the pattern to take into account characteristics that the drafter does not intentionally apply on the drawing.

Shrinkage Allowance

When metals are heated and then cooled, the material will shrink until the final temperature is reached. The amount of shrinkage depends on the material used. The shrinkage for most iron is about .125 in. per ft., .250 in. per ft. for steel, .125 to .156 in. per ft. for aluminum, .22 in. per ft. for brass, and .156 in. per ft. for bronze. Values for shrinkage allowance are approximate since the exact allowance depends on the size and shape of the casting and the contraction of the casting during cooling. The drafter normally does not need to take shrinkage into consideration because the pattern maker will use shrink rules that use expanded scales to take into account the shrinkage of various materials.

Draft

Draft is the taper allowance on all vertical surfaces of a pattern, which is necessary to facilitate the removal of the pattern from the mold. Draft is not necessary on horizontal surfaces because the pattern will easily separate from these surfaces without sticking. Draft angles begin at the parting line and taper away from the molding material. (See Figure 8–69.) The draft is added to the minimum design sizes of the product. Draft varies with different materials, size and shape of the part, and casting methods. Little if any draft is necessary in investment casting. The factors that influence the amount of draft are the height of vertical surfaces, the quality of the pattern, and the ease with which the pattern must be drawn from the mold. A typical draft angle for cast iron and steel is .125 in. per ft. Whether the drafter takes draft into consideration on a drawing depends on company standards. Some companies leave draft angles to the pattern maker, while others require drafters to place draft angles on the drawing.

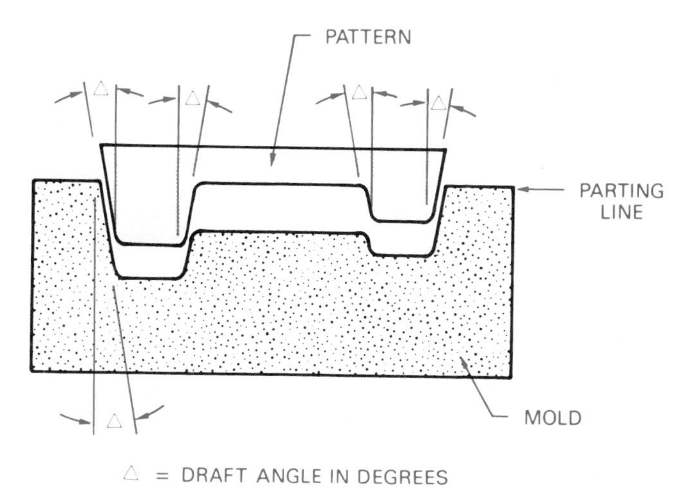

Figure 8–69 Draft angles for casting.

Fillets and Rounds

One of the purposes of fillets and rounds on a pattern is the same as that of draft angles: to allow the pattern to eject freely from the mold. Also, the use of fillets on inside corners helps reduce the tendency of cracks to develop during shrinkage. (See Figure 8–70.) The radius for fillets and rounds depends on the material to be cast, the casting method, and the thickness of the part. The

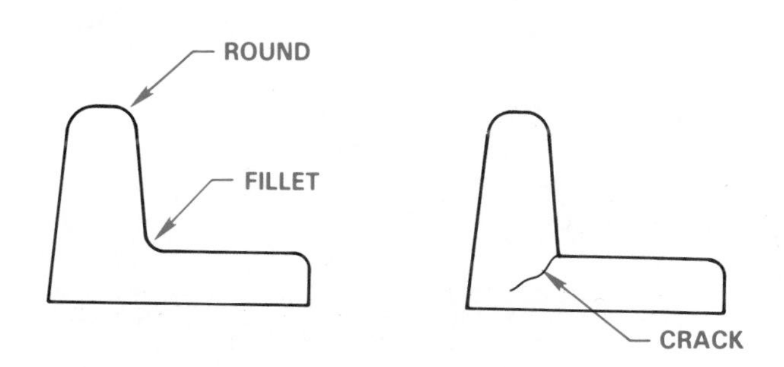

Figure 8–70 Fillets and rounds for castings.

recommended radii for fillets and rounds used in sand casting is determined by part thickness as shown in Figure 8–71.

Machining Allowance

Extra material must be left on the casting for any surface that will be machined. As with other casting design characteristics, the machining allowance depends on the casting process, material, size and shape of the casting, and the machining process to be used for finishing. The standard finish allowance for iron and steel is .125 in., and for nonferrous metals such as brass, bronze, and aluminum .062 in. In some situations the finish allowance may be as much as .5 or .75 in. for castings that are very large or have a tendency to warp.

Other machining allowances may be the addition of lugs, hubs, or bosses on castings that are otherwise hard to hold. The drafter can add these items to the drawing or the pattern maker may add them to the pattern. These features may not be added for product function but they will serve as aids for chucking or clamping the casting in a machine. (See Figure 8–72.)

There are several methods that can be used to prepare drawings for casting and machining operations. The method used depends on company standards. A commonly used technique is to prepare two drawings, one a casting drawing and the other a machining drawing. The casting drawing, as shown in Figure 8–73, shows the part as a casting. Only dimensions necessary to make the casting are shown in this drawing. The other drawing is of the same part, but this time only machining information and dimensions are given. The actual casting goes to the machine shop along with the machining drawing so the features can be machined as specified. The machining drawing is shown in Figure 8–74.

Another method of preparing casting and machining drawings is to show both casting and machining information together on one drawing. This technique requires the pattern maker to add machining allowances. The drawing may have draft angles specified in the form of a note. The pattern maker must add the draft angles to the finished sizes given. With draft angles and finish allowances omitted, the designer or drafter may need to consult with the pattern maker to insure that the casting is properly made. A combination casting/machining drawing is shown in Figure 8–75.

Another technique is to draw the part as a machining drawing and then use phantom lines to show the extra material for machining allowance and draft angles, as shown in Figure 8–76.

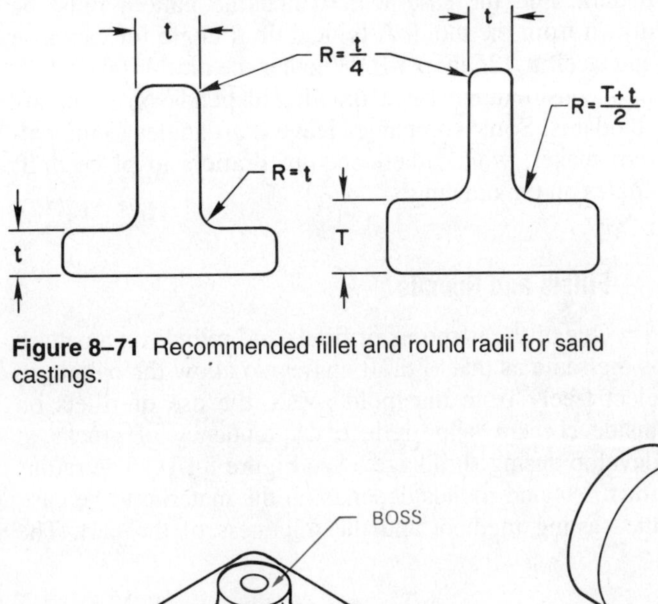

Figure 8–71 Recommended fillet and round radii for sand castings.

Forging, Design, and Drawing

Draft for forgings serves much the same purpose as draft for castings. The draft associated with forging is found in the dies. The sides of the dies must be angled to facilitate the release of the metal during the forging process. If the vertical sides of the dies did not have draft angle, then the metal would become stuck in the die. Internal and external draft angles may be specified differently because the internal drafts in some materials may have to be greater to help reduce the tendency of the part to stick in the die. While draft angles may change

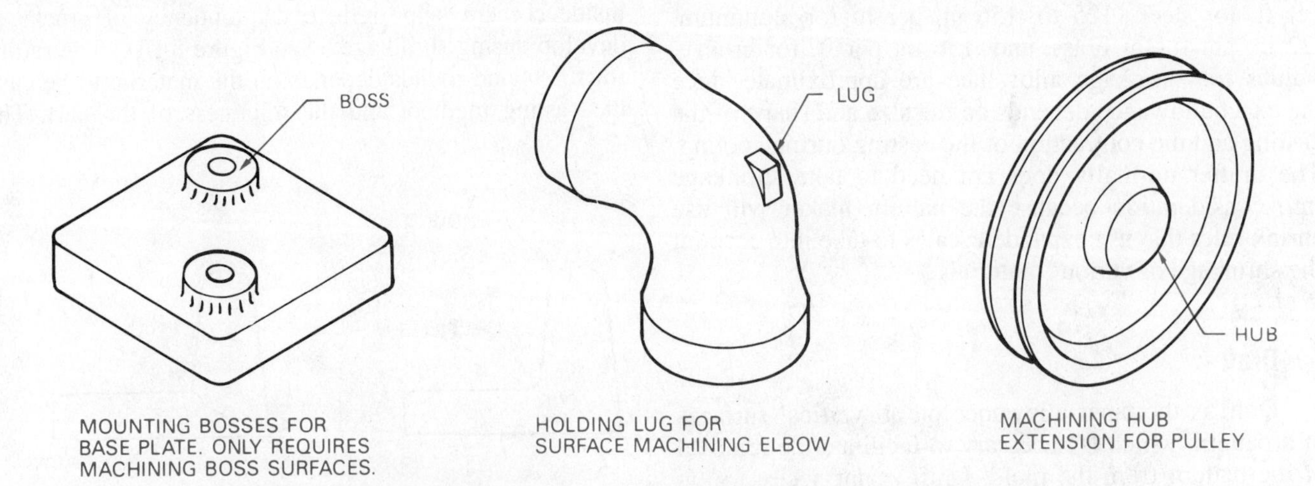

Figure 8–72 Cast features added for machining.

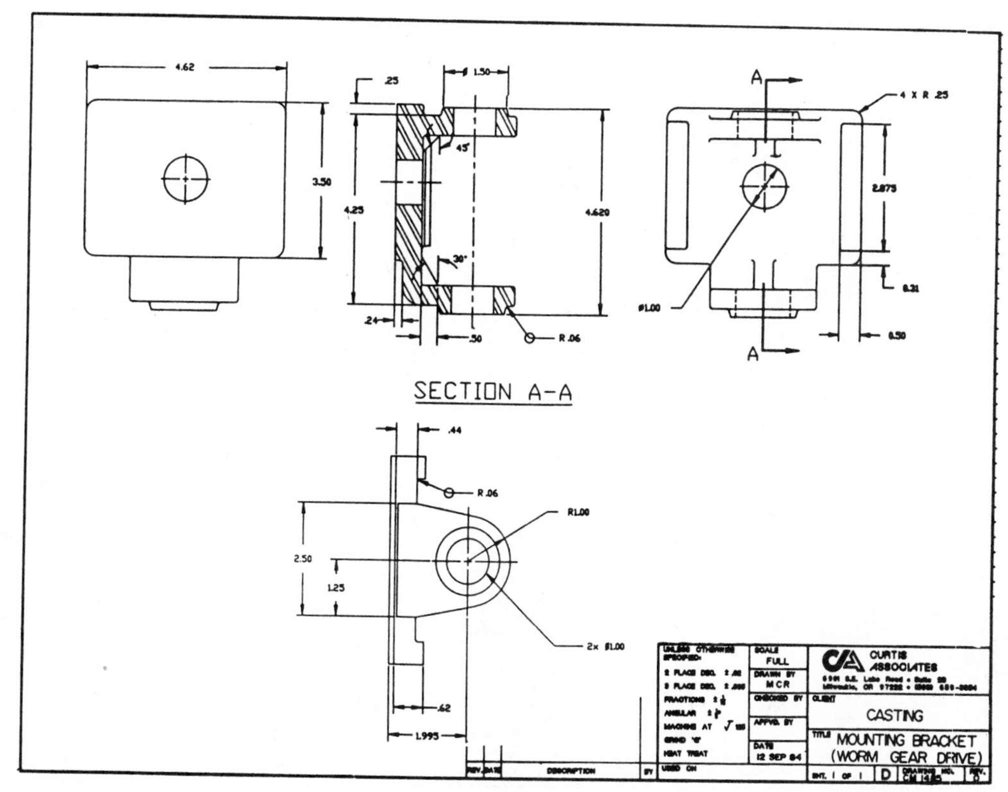

Figure 8–73 Casting drawing. *Courtesy Curtis Associates.*

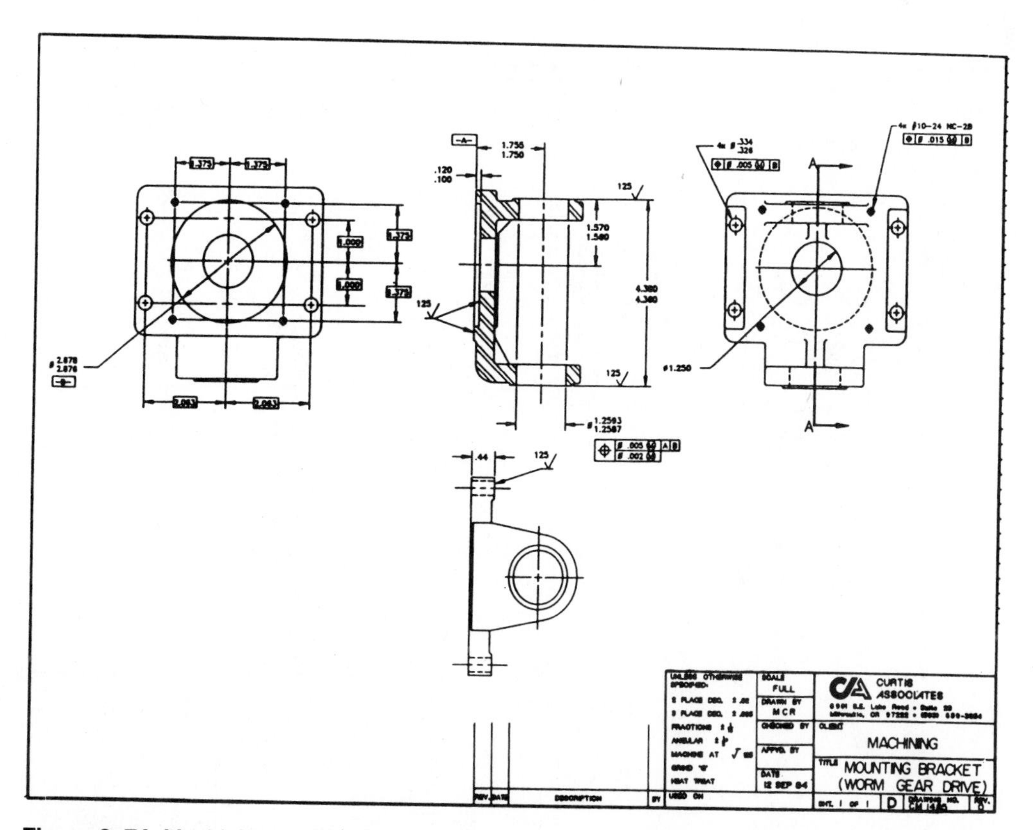

Figure 8–74 Machining drawing for the casting illustrated in Figure 8–73. *Title block courtesy Curtis Associates.*

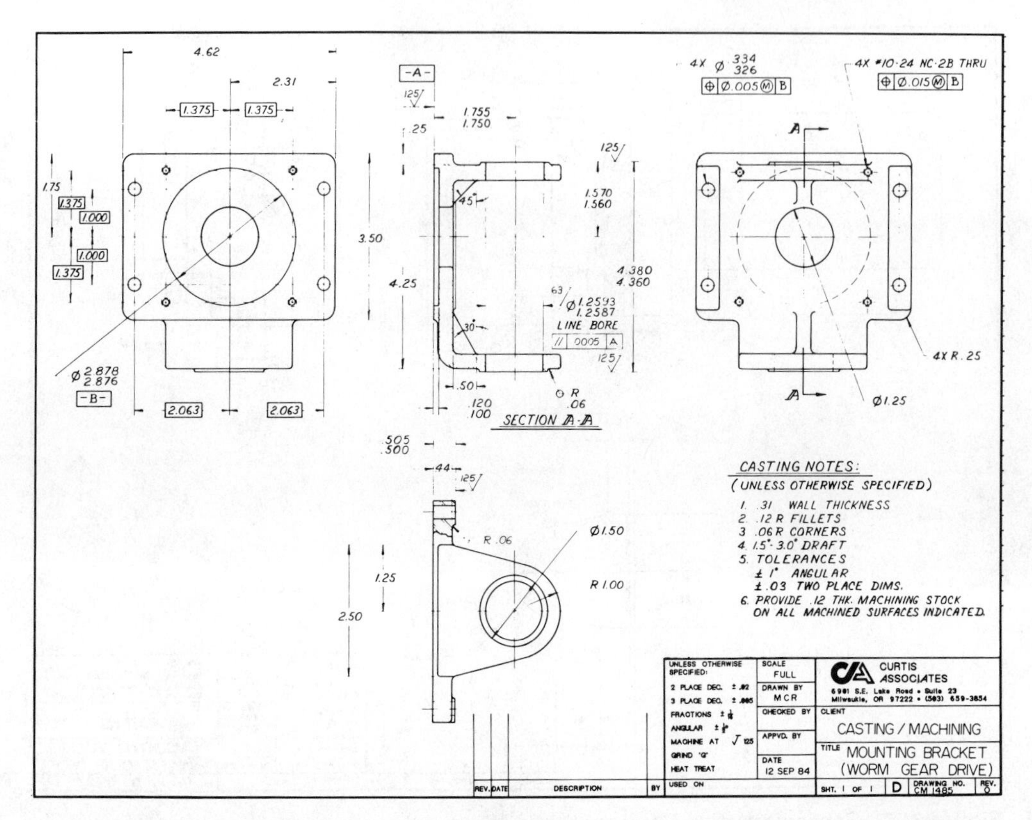

Figure 8–75 Drawing with casting and machining information. *Courtesy Curtis Associates.*

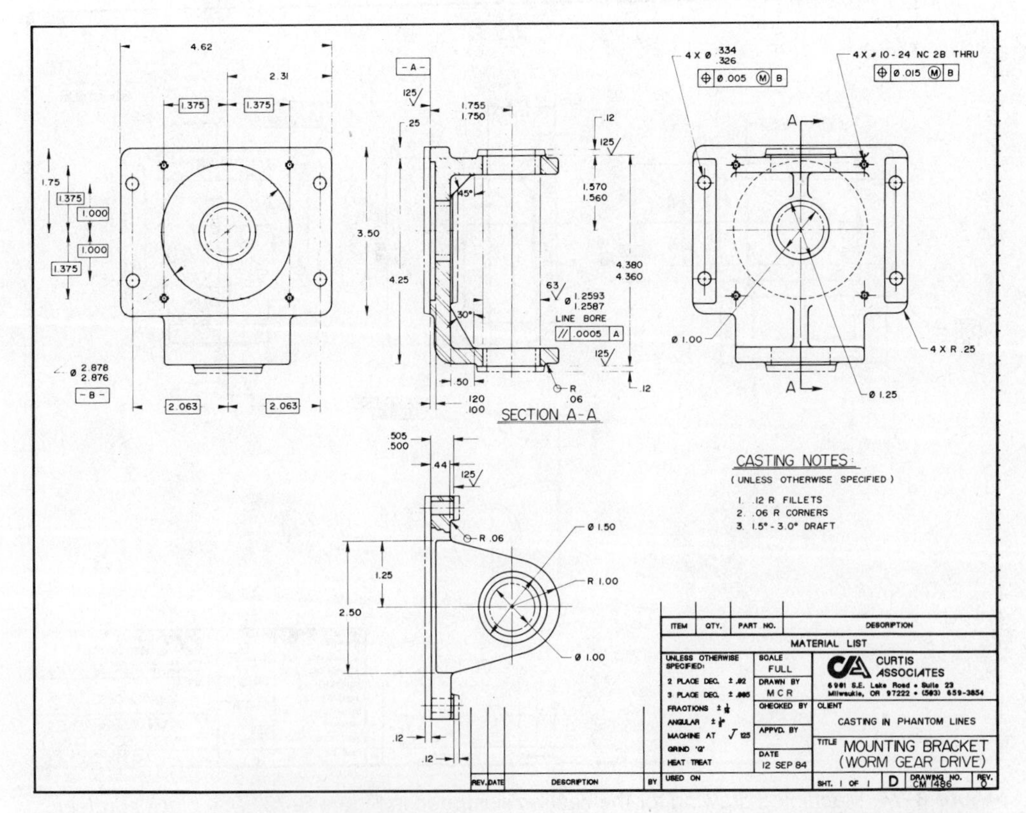

Figure 8–76 Phantom lines used to show machining allowances. *Courtesy Curtis Associates.*

slightly with different materials, the common exterior draft angle recommended is 7°. The internal draft angles for most soft materials is also 7°, but the recommended interior draft for iron and steel is 10°.

The application of fillets and rounds to forging dies is to improve the ejection of the metal from the die. Another reason, similar to that for casting, is increased inside corner strength. One factor that applies to forgings as different from castings is that fillets and rounds that are too small in forging dies may substantially reduce the life of the dies. Recommended fillet and round radius dimensions are shown in Figure 8–77.

Forging Drawings

A number of methods may be used in the preparation of forging drawings. One technique used in forging drawings that is clearly different from the preparation of casting drawings is the addition of draft angles. Casting drawings usually do not show draft angles; forging detail drawings usually do show draft.

Before a forging can be made, the dimensions of the stock material to be used for the forging must be determined. Some companies leave this information to the forging shop to determine; other companies have their engineering department make these calculations. After the stock size is determined, a drawing showing size and shape of the stock material is prepared. The blank material is dimensioned, and the outline of the end product is drawn inside the stock view using phantom lines. (See Figure 8–78.)

Forgings are made with extra material added to surfaces that must be machined. Forging detail drawings may be made to show the desired end product with the outline of the forging shown in phantom lines at areas that require machining. (See Figure 8–79.) Notice the double line around the perimeter denoting draft angle. Another option used by some companies is to make two separate drawings, one a forging drawing and the other a machining drawing. The forging drawing will show all of the views, dimensions, and specifications that relate only to the production of the forging, as shown in Figure 8–80. The machining drawing will not duplicate the information that was provided for the forging. The only information on the machining drawing is views, dimensions, and specifications related to the machining processes, as shown in Figure 8–81.

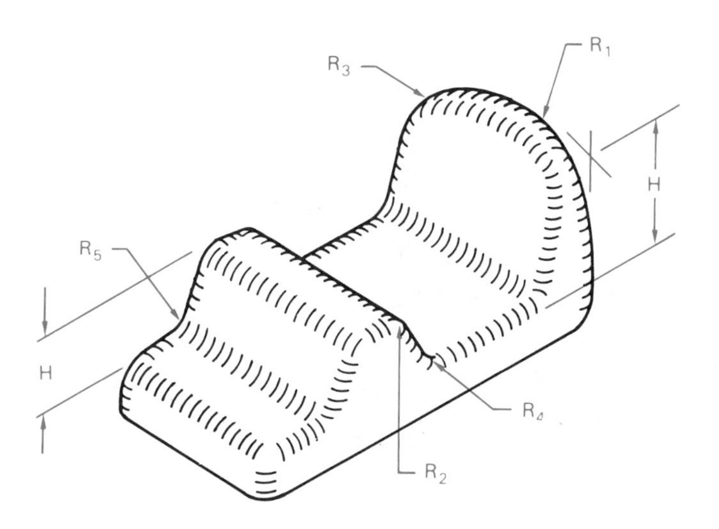

DIMENSIONS IN MILLIMETERS					
H	R_1	R_2	R_3	R_4	R_5
6	1.5	1.5	4.5	3	3
13	1.5	1.5	4.5	3	3
25	3	3	9	6	9
50	4.5	6	13	13	15
75	6	7.5	16	16	25
100	7.5	10.5	25	25	35
125	9	13	23	32	44
150	10.5	15	32	38	50

Figure 8–77 Recommended fillets and rounds for forgings.

MACHINED SURFACES

Surface Finish Definitions

Surface Finish. Surface finish refers to the roughness, waviness, lay, and flaws of a surface. Surface finish is the specified smoothness required on the finished surface of a part that is obtained by machining, grinding, honing, or lapping. The drawing symbol associated with surface finish is shown in Figure 8–82.

Surface Roughness. Surface roughness refers to fine irregularities in the surface finish and is a result of the

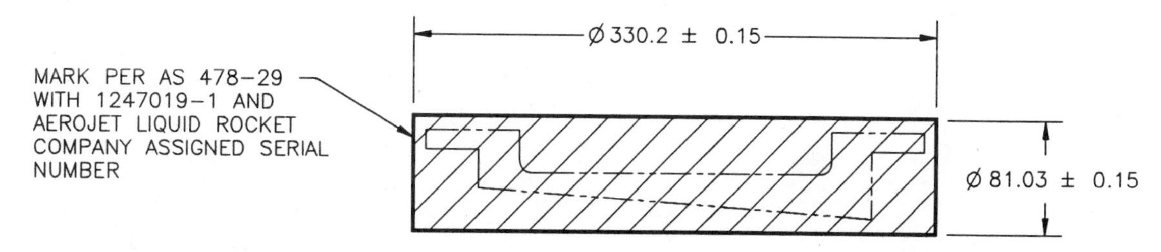

Figure 8–78 Blank material for forging process. *Courtesy Aerojet TechSystems Company.*

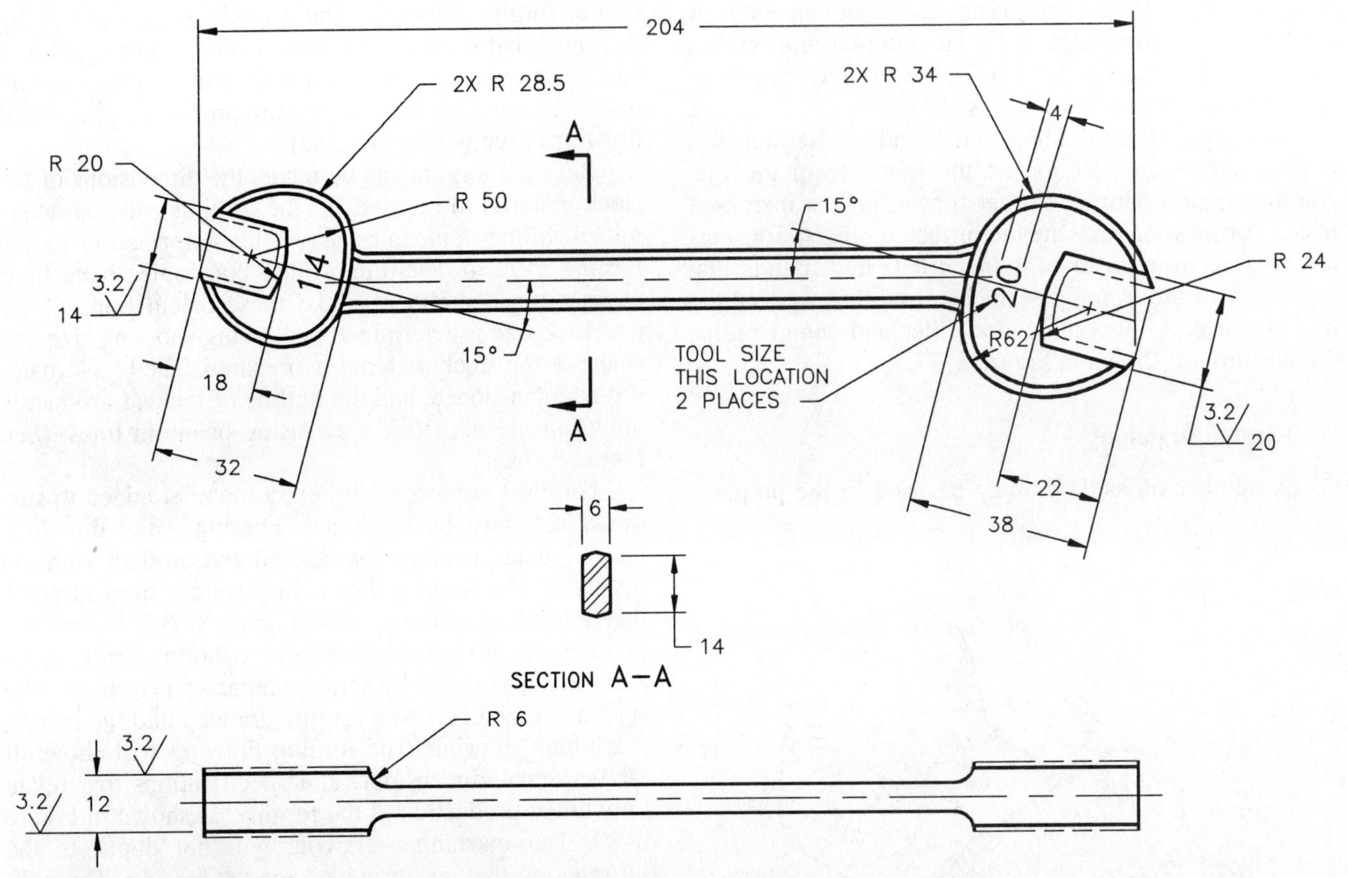

Figure 8–79 Phantom lines used to show machining allowance on a forging drawing.

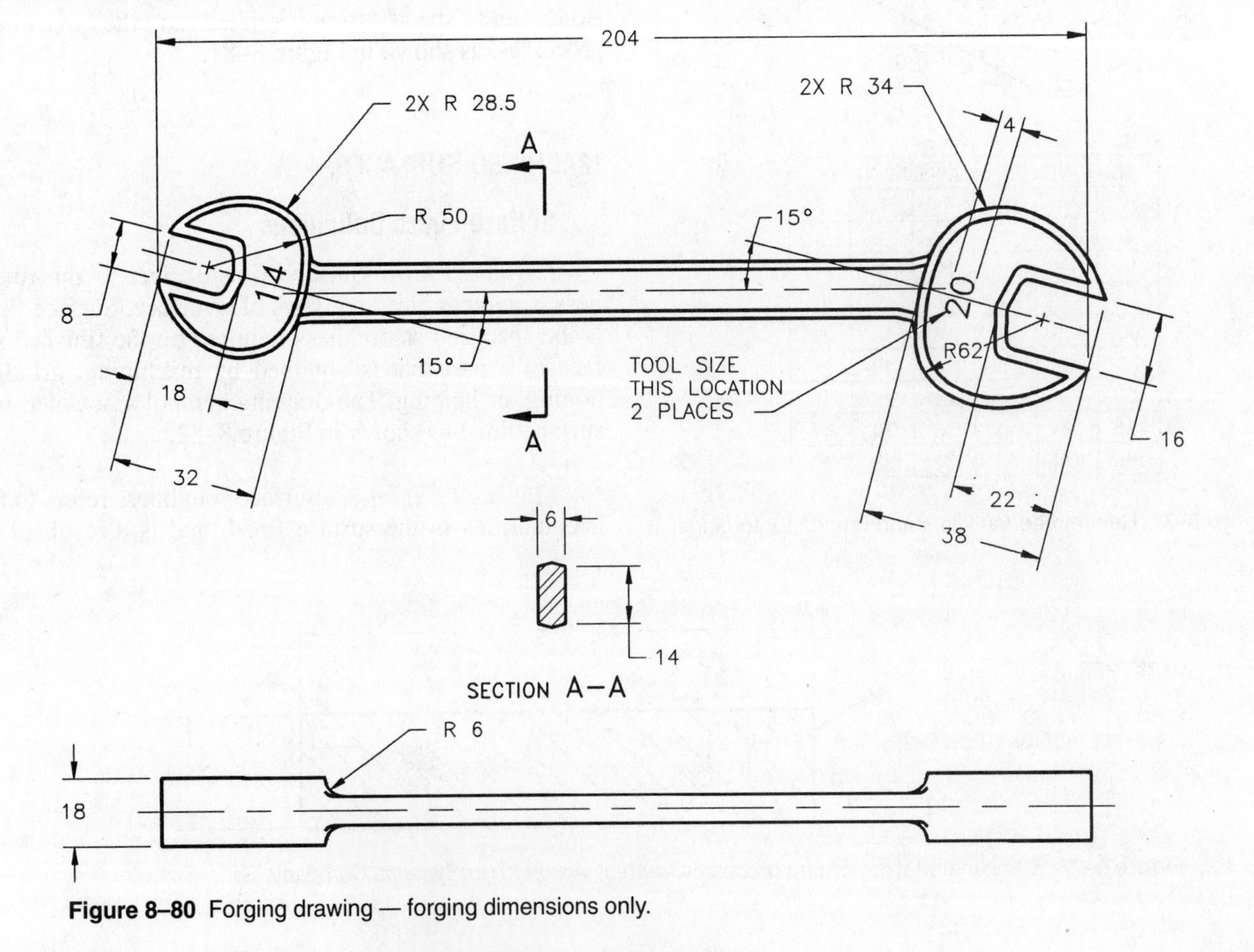

Figure 8–80 Forging drawing — forging dimensions only.

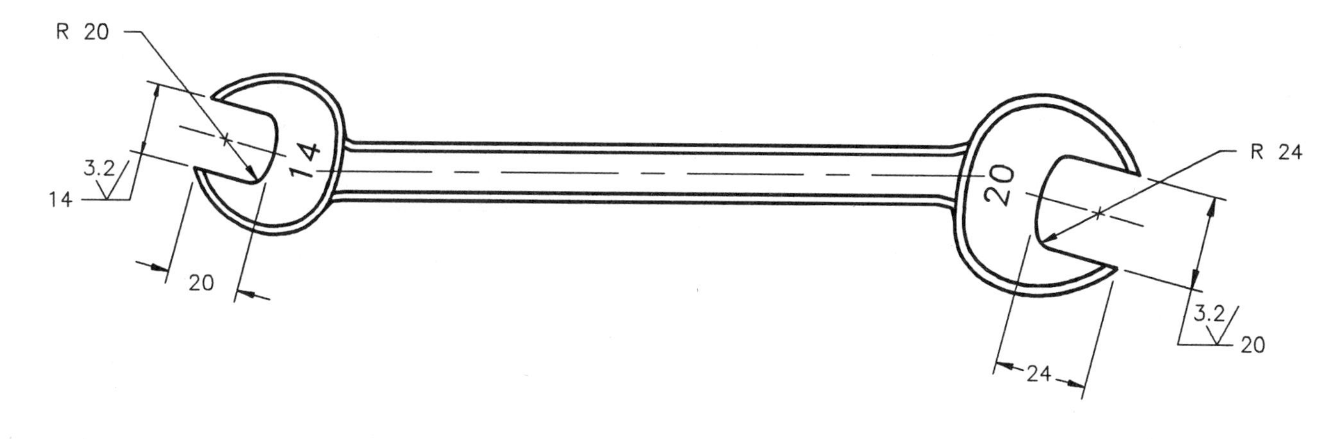

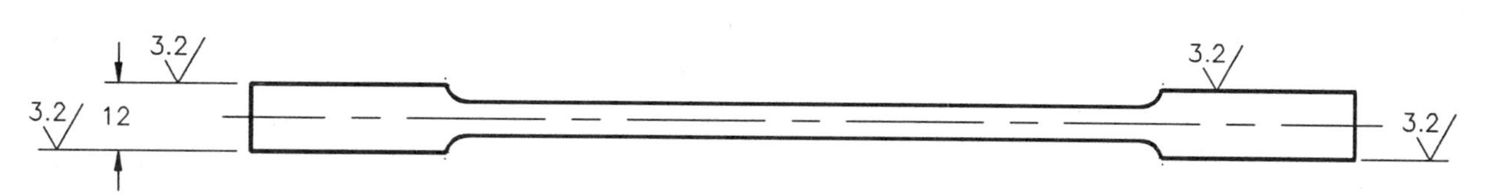

Figure 8–81 Machining drawing — machining dimensions only.

Figure 8–82 Surface finish symbol.

manufacturing process used. Roughness height is measured in micrometers, μm (millionths of a meter), or in microinches, μin (millionths of an inch).

Surface Waviness. Surface waviness is the often widely spaced condition of surface texture usually caused by such factors as machine chatter, vibrations, work deflection, warpage, or heat treatment. Waviness is rated in millimeters or inches.

Lay. Lay is the term used to describe the direction or configuration of the predominant surface pattern. The lay symbol is used if considered essential to a particular surface finish. The characteristic lay symbol may be attached to the surface finish symbol, as shown in Figure 8–83.

Surface Finish Symbol

Some of the surfaces of an object may be machined to certain specifications. When this is done, a surface finish symbol is placed on the view where the surface or surfaces appear as lines (edge view). (See Figure 8–84.) The finish symbol on a machine drawing alerts the machinist that the surface must be machined to the given specification. The finish symbol also tells the pattern or die maker that extra material is required in a casting or forging.

The surface finish symbol is properly drawn using a thin line as detailed in Figure 8–85. The numerals or letters associated with the surface finish symbol should be the same height as the lettering used on the drawing dimensions and notes.

Often only the surface roughness height is used with the surface finish symbol, as for example, the number 3.2 above the finish symbol in Figure 8–86, which denotes 3.2 micrometers. When other characteristics of a surface texture are specified, they are shown in the format represented in Figure 8–87. For example, roughness-width cutoff is a numerical value that establishes the maximum width of surface irregularities to be included in the roughness height measurement. Standard roughness-width cutoff values for inch specifications are .003, .010, .030, .100, .300, and 1.000; .030 is implied when no specification is given.

Figure 8–88 is a magnified pictorial representation of the characteristics of a surface finish symbol. Figure 8–91 shows some common roughness height values in micrometers and microinches, a description of the resulting surface, and the process by which the surface may be produced. When a maximum and minimum limit is specified, the average roughness height must lie within the two limits.

When a standard or general surface finish is specified in the drawing title block or in a general note, then a surface finish symbol without roughness height specified will be used on all surfaces that are the same as the general specification. When a part is completely finished to a given specification, then the general note, FINISH

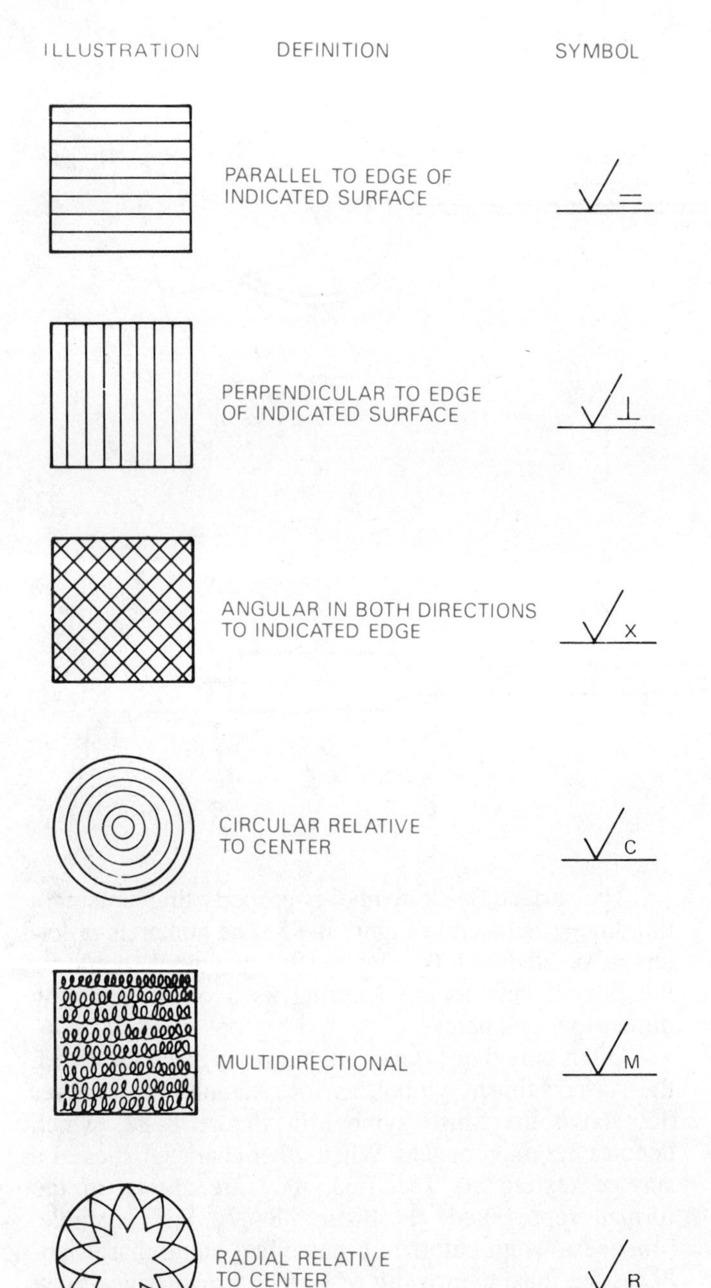

Figure 8–83 Characteristic lay added to the finish symbol.

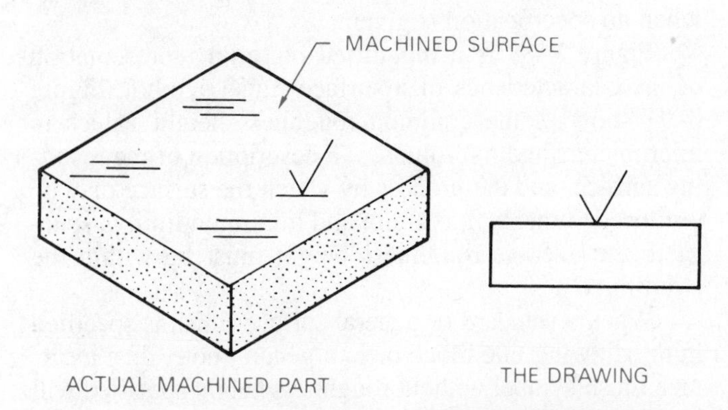

Figure 8–84 Standard surface finish symbol placed on the edge view.

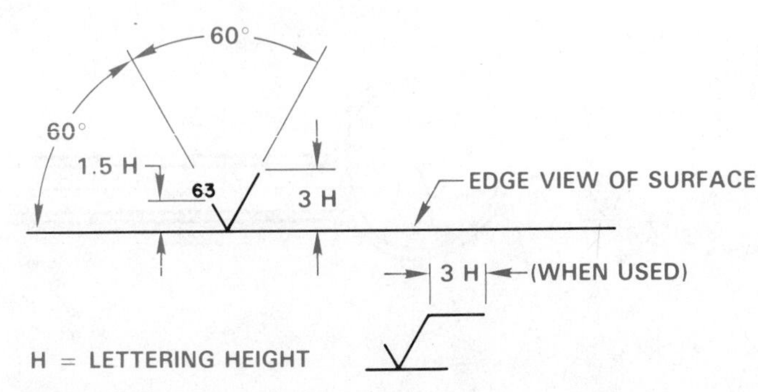

Figure 8–85 Properly drawn surface finish symbol.

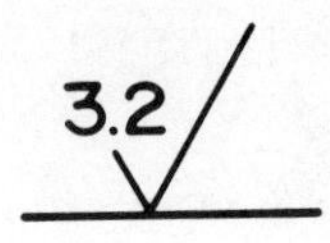

Figure 8–86 Surface finish symbol with surface roughness height given.

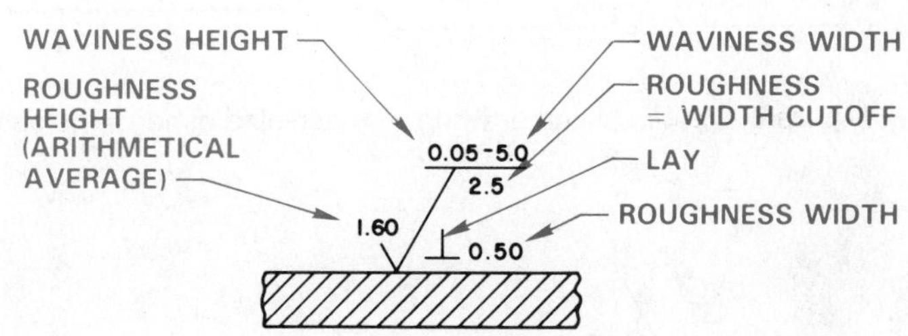

Figure 8–87 Elements of a complete surface finish symbol.

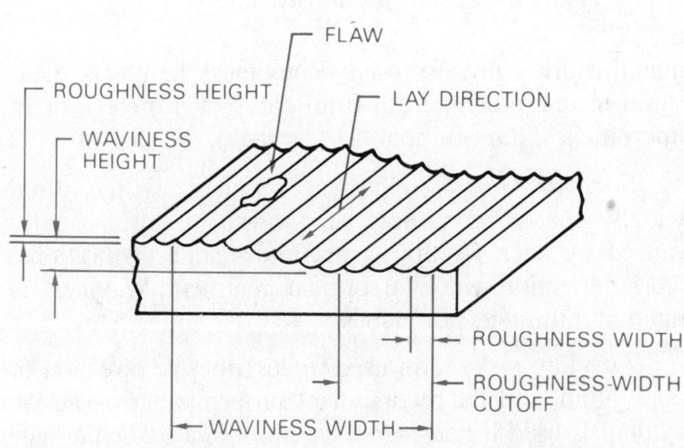

Figure 8–88 Surface finish characteristics magnified.

ALL OVER, or abbreviation FAO, or FAO 125 μIN, may be used. The placement of surface finish symbols on a drawing can be accomplished a number of ways, as shown in Figure 8–92. Additional elements may be applied to the surface finish symbol, as shown in Figure 8–93.

DIMENSIONING SOFTWARE PACKAGES

Most of the major CADD software packages provide for a variety of dimensioning applications depending on the user's need. For example, you may use unidirectional or aligned dimensioning, and datum or chain dimensioning based on your company's standards and procedures. You can also alter the way dimensions are presented by changing any one or more of these variables:

- Space between dimension lines
- Style and height of text
- Size and shape of arrowheads
- Extension line length beyond the last dimension line
- Space between the object and the start of the extension line
- Location of the text in relation to the dimension line

Dimensioning *default* values are normally set to ANSI standards, but the user has the flexibility to change these values for specific applications. A default value refers to a standard preset value. Figure 8–89 shows some of the flexibility available in changing predefined default values.

In addition to the major CADD software packages, there are hundreds of third-party software packages. The term *third-party* refers to programs that support or may be used in conjunction with the main package. In third-party software for dimensioning, the principal emphasis seems to be on datum, tabular, arrowless, and geometric tolerancing packages. These are the types of software that help increase speed and productivity and simplify the dimensioning process. For example, some of the programs have automatic dimensioning, which allows the drafter to pick all of the items for dimensioning and, when finished, press the RETURN key or select AUTOMATIC, depending on the program, and watch everything be dimensioned. Some packages also automatically generate a tabular chart for hole reference. A drawing showing arrowless datum dimensioning is shown in Figure 8–90.

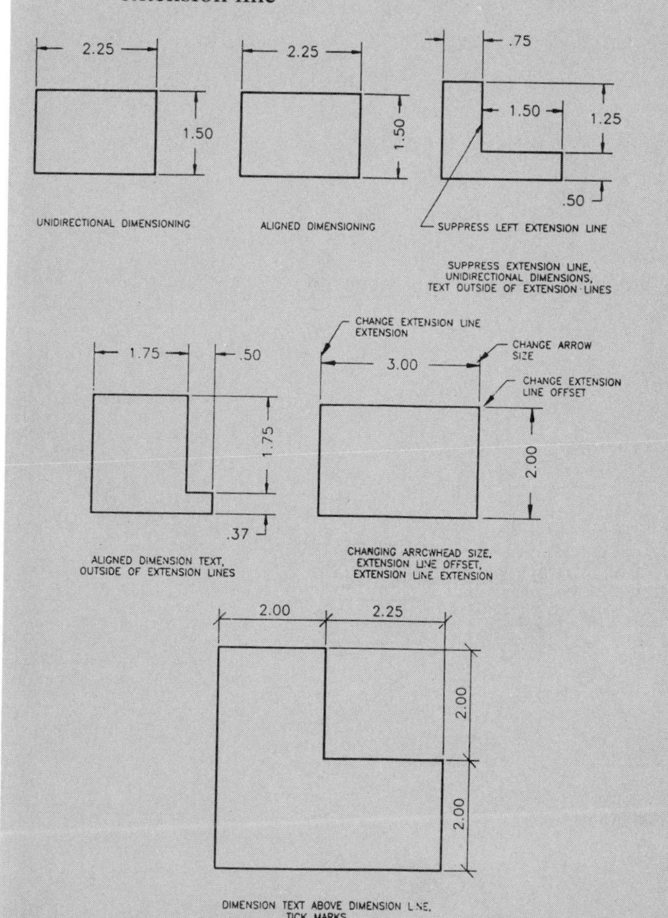

Figure 8–89 Changing dimensioning default values for specific applications.

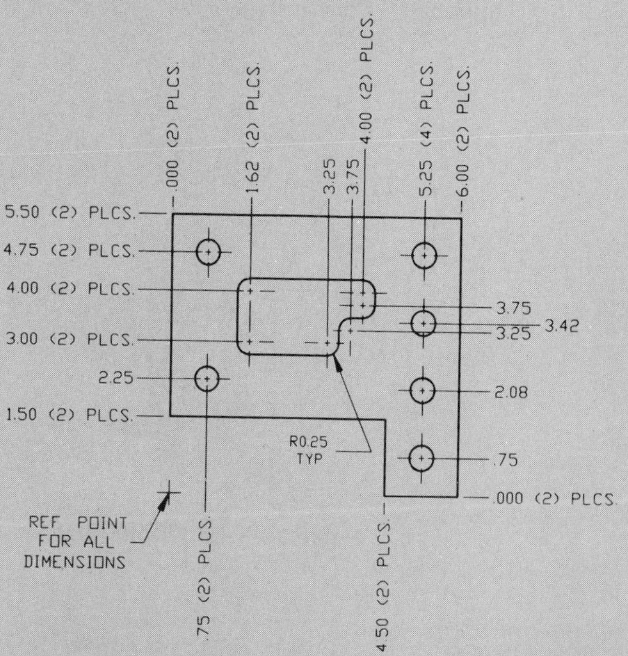

Figure 8–90 Datum dimensioning with a third-party software package. *Courtesy CADMASTER, Inc.*

MICROMETERS	ROUGHNESS HEIGHT RATING MICRO INCHES	SURFACE DESCRIPTION	PROCESS
25	1000	VERY ROUGH	SAW AND TORCH CUTTING, FORGING OR SAND CASTING.
12.5	500	ROUGH MACHINING	HEAVY CUTS AND COARSE FEEDS IN TURNING, MILLING AND BORING.
6.3	250	COARSE	VERY COARSE SURFACE GRIND, RAPID FEEDS IN TURNING, PLANNING, MILLING, BORING AND FILLING.
3.2	125	MEDIUM	MACHINING OPERATIONS WITH SHARP TOOLS, HIGH SPEED, FINE FEEDS AND LIGHT CUTS.
1.6	63	GOOD MACHINE FINISH	SHARP TOOLS, HIGH SPEEDS, EXTRA-FINE FEEDS AND CUTS.
0.80	32	HIGH GRADE MACHINE FINISH	EXTREMELY FINE FEEDS AND CUTS ON LATHE, MILL AND SHAPERS REQUIRED. EASILY PRODUCED BY CENTERLESS, CYLINDRICAL AND SURFACE GRINDING.
0.40	16		
0.20	8	VERY FINE MACHINE FINISH	FINE HONING AND LAPPING OF SURFACE.
0.050 0.100	2-4	EXTREMELY SMOOTH MACHINE FINISH	EXTRA-FINE HONING AND LAPPING OF SURFACE. MIRROR FINISH.
0.025	1	SUPER FINISH	DIAMOND ABRASIVES.

Figure 8–91 Common roughness height values with a surface description and associated machining process.

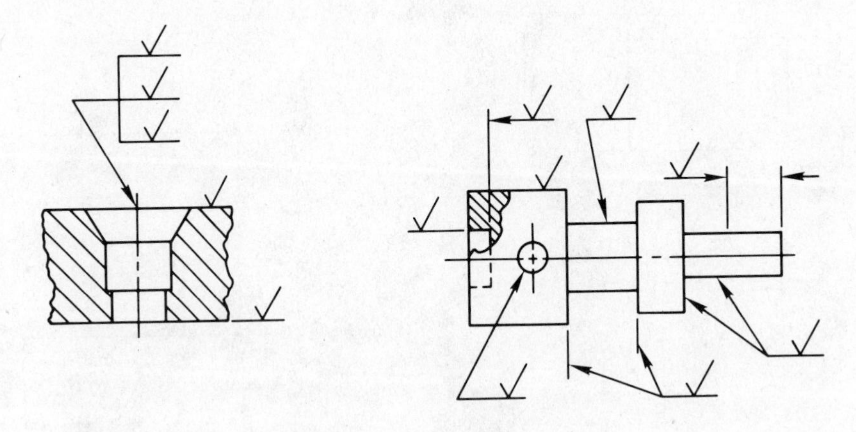

Figure 8–92 Proper placement of surface finish symbols.

DESIGN AND DRAFTING OF MACHINED FEATURES

Drafters should gain a working knowledge of machining processes and the machining capabilities of the company for which they work. Drawings should be prepared that will allow machining within the capabilities of the machinery available. If the local machinery will not produce parts that have certain functional requirements, then

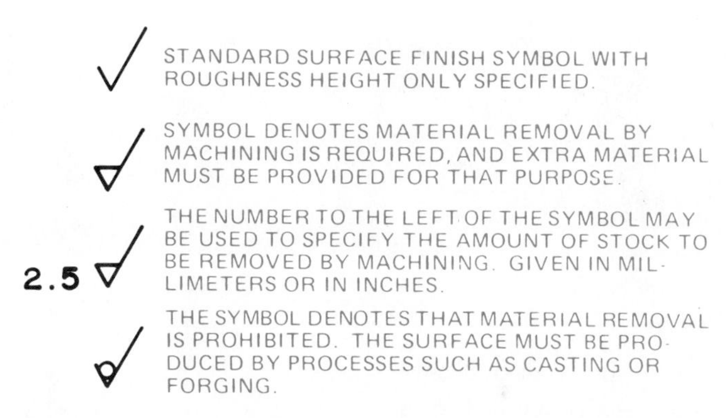

Figure 8–93 Material removal elements added to the surface finish symbol.

the machining may need to be performed elsewhere or the company may have to purchase the proper equipment. However, the first consideration should be the least-expensive method to obtain the desired result. Avoid overmachining. Machining processes are expensive, so do not call for requirements on a drawing that are not necessary for the function of the part. For example, surface finishes become more expensive to machine as the roughness height decreases. So, if a surface roughness of 125 microinches is adequate, then do not use a 32-microinch specification just because you like smooth surfaces. Another example is demonstrated by the difference between the 63- and 32-microinch finish. A 63-microinch finish is a good machine finish that may be performed using sharp tools at high speeds with extra-fine feeds and cuts. The 32-microinch callout, on the other hand, requires extremely fine feeds and cuts on a lathe or milling machine and in many cases requires grinding. The 32 finish is more expensive to perform. In a manufacturing environment where cost and competition are critical considerations, a great deal of thought must be given to meeting the functional and appearance requirements of a product at the least possible cost. It generally does not take very long for an entry-level drafter to pick up these design considerations by communicating with engineering and manufacturing department personnel. Many drafters become designers, checkers, or engineers within a company by learning the product and being able to implement designs based on manufacturing capabilities.

SYMBOLS

Individual industry standards determine whether terms and abbreviations or symbols are used on drawings. With the influence of computer graphics, the use of symbols may become more prevalent. When symbols are drawn manually, a template should be used because many of the symbols are difficult to draw properly freehand, and the template drawing is generally neater and more uniform. Any attempt to draw symbols using drafting tools other than specially designed templates will take too much time. Individual special symbols have been represented and detailed throughout this chapter each time they are introduced.

PROFESSIONAL PERSPECTIVE

The proper placement and use of dimensions is one of the most difficult aspects of drafting. It requires careful thought and planning. First you must determine the type of dimension that fits the application, then place it on the drawing. While the CADD system makes the actual placement of dimensions quick and easy, it does not make the planning process any easier. Making preliminary sketches is very important before beginning a drawing. The preliminary sketch allows you to select and place the views and then place the dimensions. When you were drawing only the views the space requirements were not as critical, but a drawing without dimensions is unrealistic.

The problem with dimensions is that they take up a lot of space. One of the big problems an entry-level drafter faces is determining the space requirements for dimensions. In many cases these requirements are underestimated. As a rule of thumb, if you think the drawing is crowded on a particular size sheet, then play it safe and use the next larger size. Most companies want the drawing information spread out and easy to read, but be careful, because some companies want the drawing crowded with as much information as you can get on a sheet. This text advocates a clean and easy-to-read drawing that is not crowded. Out on the job, however, you must do what is required by your employer. If you are drawing an uncrowded drawing, you should consider leaving about one-quarter of the drawing space clear of view and dimensional information. Usually this space is above and to the left of the title block. This space provides adequate room for general notes and engineering changes.

DIMENSIONING FOR CADD/CAM

The implementation of computer-aided design and drafting (CADD) and computer-aided manufacturing (CAM) in industry is best accomplished when common control of the computer exists between engineering and manufacturing. The success of this automation is, in part, relative to the standardization of operating and documentation procedures. CADD can be accomplished through the same coordinate dimensioning systems previously described in this chapter. Standard dimensioning systems are used to establish a geometric model of the part, which in turn is displayed at a graphics workstation. The data retrieved from this model is the mathematical description of the part to be produced. The drafter must dimension the part completely and accurately so that each contour or geometric shape of the part is continuous. The dimensioning systems that locate features or points on a feature in relation to x-, y-, and z-axes derived from a common origin are most effective, such as datum, tabular, arrowless, and polar coordinate dimensioning. The x-, y-, and z-axes originate from three mutually perpendicular planes which are generally the geometric counterpart of the sides of the part when the surfaces are at right angles as seen in Figure 8–94. If the part is cylindrical, two of the planes intersect at right angles to establish the axis of the cylinder and the third is perpendicular to the intersecting planes as shown in Figure 8–95. The x-, y-, and z-coordinates that are used to establish features on a drawing are converted to x-, y-, and z-axes that correspond to the linear and rotary motions that occur in CAM.

The position of the part in relation to mathematical quadrants determines whether the x, y, and z values are positive or negative. The preferred position is the mathematical quadrant that allows for programming positive commands for the machine tool. Notice in Figure 8–96 that the positive x and y values occur in quadrant 1.

In CAD/CAM and computer-integrated manufacturing (CIM) programs, the drawing is made on the computer screen and sent directly to the computer numerical control machine tool without generating a hard copy of the drawing. In this situation it becomes important for the drafter to understand the machine tool operation. Figure 8–97(a) shows a drawing created for CAD/CAM, and Figure 8–97(b) shows the same drawing as represented on the computer screen

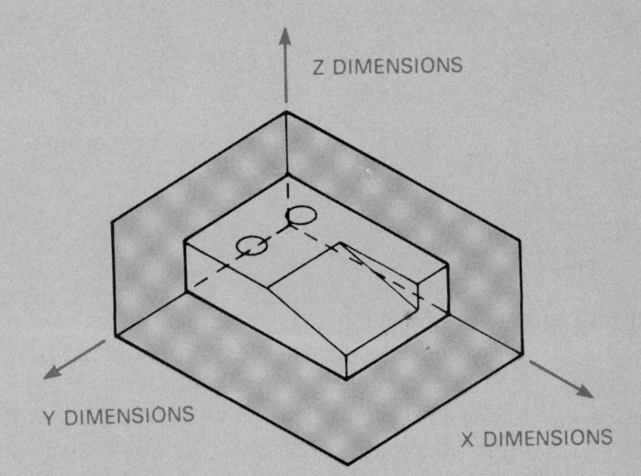

Figure 8–94 Features of a part dimensioned in relation to x-, y-, and z-axes.

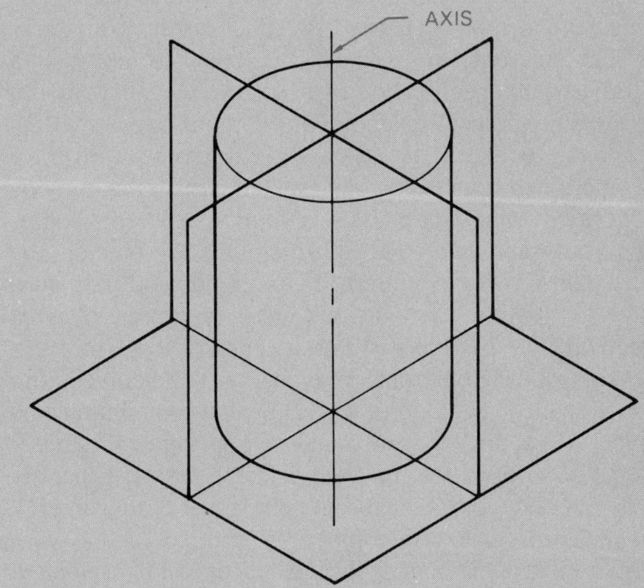

Figure 8–95 The planes related to the axis of a cylindrical feature.

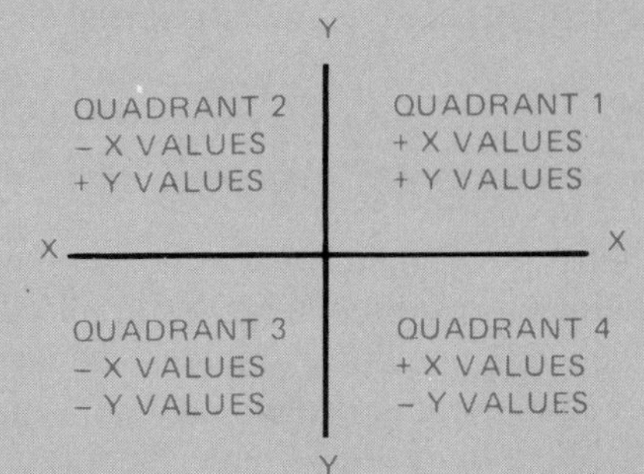

Figure 8–96 Determination of x and y values in related quadrants.

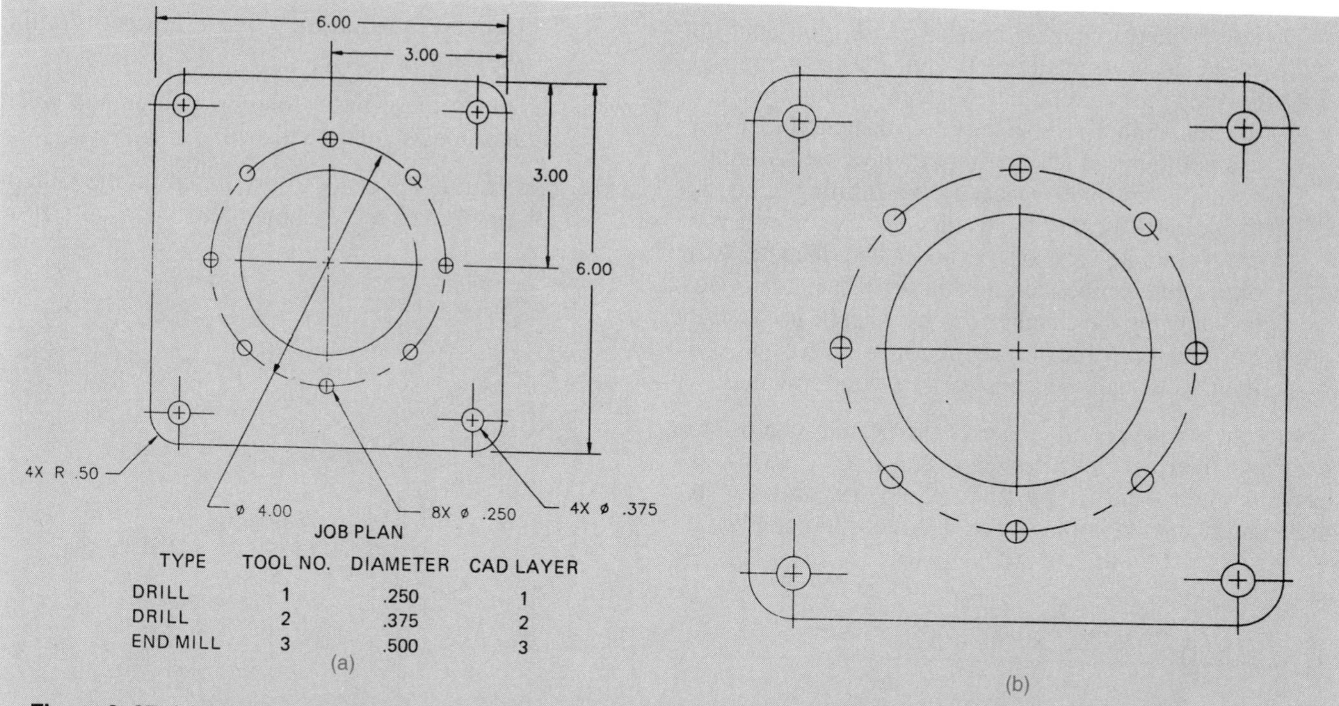

Figure 8–97 (a) A drawing created for CAD/CAM; (b) drawing from (a) represented on the computer screen prior to generating the machine tool program.

prior to generating the machine tool program. Notice, the dimensions are not displayed on the drawing in Figure 8–97(b). The data used to input the information from the original design drawing is used, by the computer, to establish the machine tool paths. The CAD/CAM program also allows the operator to show the tool path, as shown in Figure 8–98. The tool path display allows the operator to determine if the machine tool will perform the assigned machining operation.

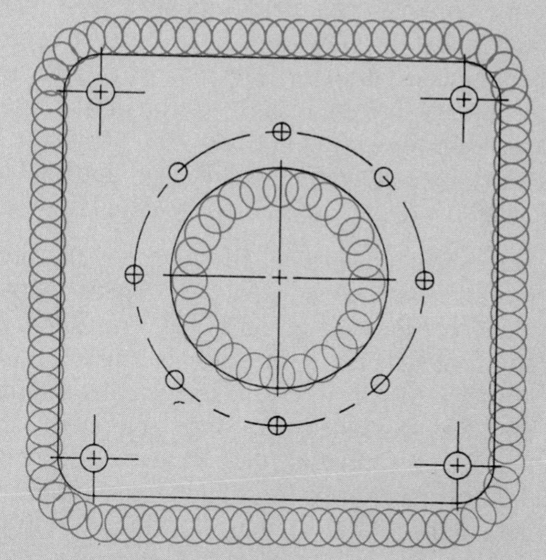

Figure 8–98 The tool path display, shown in color, allows the operator to determine whether the machine tool will perform the assigned machining operation.

Follow the dimensioning rules, guidelines, and examples discussed in this chapter and use proper dimensioning standards. Try as hard as you can to avoid breaking dimensioning standards. Following are some of the pitfalls to watch out for as a beginning drafter:

- Do not crowd dimensions. Keep your dimension line spacing equal and far enough apart to clearly separate the dimension numerals.
- Do not dimension to hidden features. This also means do not dimension to the centerline of a hole in the hidden view. Always dimension to the view where the feature appears as an object line.
- Dimension to the view that shows the most characteristic shape of an object or feature. For example, dimension shapes where you see the contour, dimension holes to the circular view, and dimension cylindrical shapes in the rectangular view.
- Do not stack adjacent dimension numerals; stagger them so they are easier to read.
- Group dimensions as much as you can. It is better to keep dimensions concentrated to one location or one side of a feature rather than spread around the drawing. This makes the drawing easier to read.

- Use dimensioning symbols where you can to speed up drafting time. If you are using manual drafting, use a template to draw symbols or use the nonsymbolic version of preparing notes. If you use computer-aided drafting, always use symbols. Symbols not only speed up the drafting process, they clearly identify the feature. For example, if you draw a single view of a cylinder, the diameter (∅) dimension can be identified in the rectangular view and drawing the circular view may not be necessary.
- Look carefully at the figures in the text and use them as examples as you prepare your drawings.

Try to put yourself in the place of the person who has to read and interpret your drawing. Make the drawing as easy to understand as possible. Keep the drawing as uncluttered and as simple as possible, yet still complete.

ENGINEERING PROBLEM

As mentioned previously in this chapter, a complete detail drawing is made up of multiviews and dimensions. The layout techniques of a detail drawing must include an analysis of how the views and dimensions go together to create the finished drawing. The most effective way to form preliminary layout ideas is through the use of rough sketches as described in Chapter 5.

Consider the engineering sketch in Figure 8–99 as you evaluate how to prepare a complete detail drawing.

Step 1 Select and make a rough sketch of the proper multiviews. Leave plenty of space between views so that dimensions may be added. The engineer's sketch is not always accurate. It is the drafter's responsibility to convert the engineering ideas from the rough stage to a formal drawing. The information may be correct; however, the organization of information and proper layout are the drafter's duties. (See Figure 8–100.)

Step 2 Place dimensions and notes on your multiview sketch. (See Figure 8–101.) Keep the following dimensioning rules in mind as you select and place the dimensions:

- Begin with smallest dimension closest to the object and place dimensions that progressively increase in size further away from the object.
- Do not crowd dimensions.
- Dimension between views where possible.
- Dimension to views that show the best shape of features.
- Group dimensions when possible.
- Do not dimension size or location dimensions to hidden features.
- Stagger adjacent dimension numerals.
- Use leaders to properly label specific notes.
- Convert all information to proper drafting standards.

Evaluate these basic rules as you decide where dimensions should be placed.

Step 3 Determine the scale to use based on the amount of fine detail to be shown and the total size of

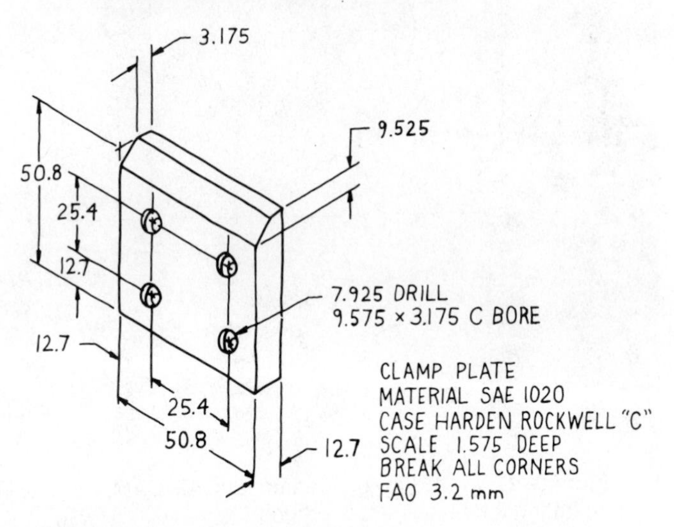

Figure 8–99 Engineering sketch.

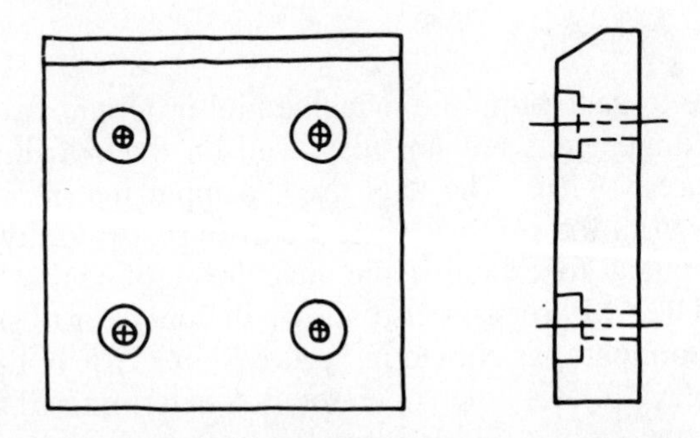

Figure 8–100 Rough sketch of the selected multiviews.

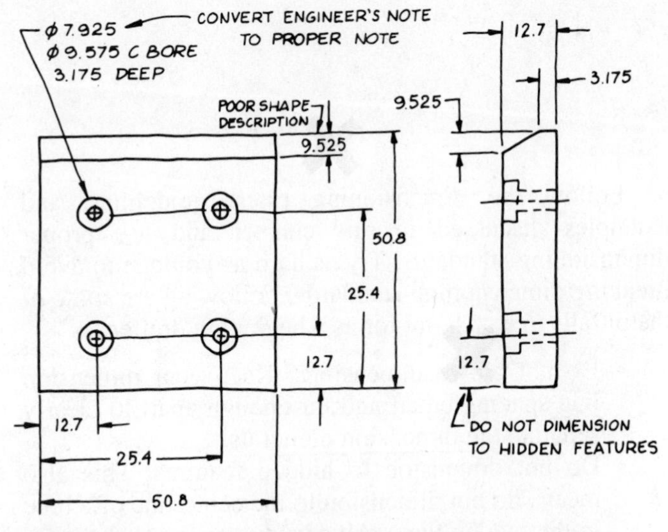

Figure 8–101 Rough sketch with dimensions and notes.

the object.

The clamp plate can easily be drawn full scale.

Step 4 Determine the sheet size to be used based on the drawing scale to be used, the total drawing area (including views and dimensions) and the amount of clear area needed for general notes and future revisions.

See Figures 8–101 and 8–102 as you follow the calculations to determine the length and height of the drawing area needed for the clamp plate.

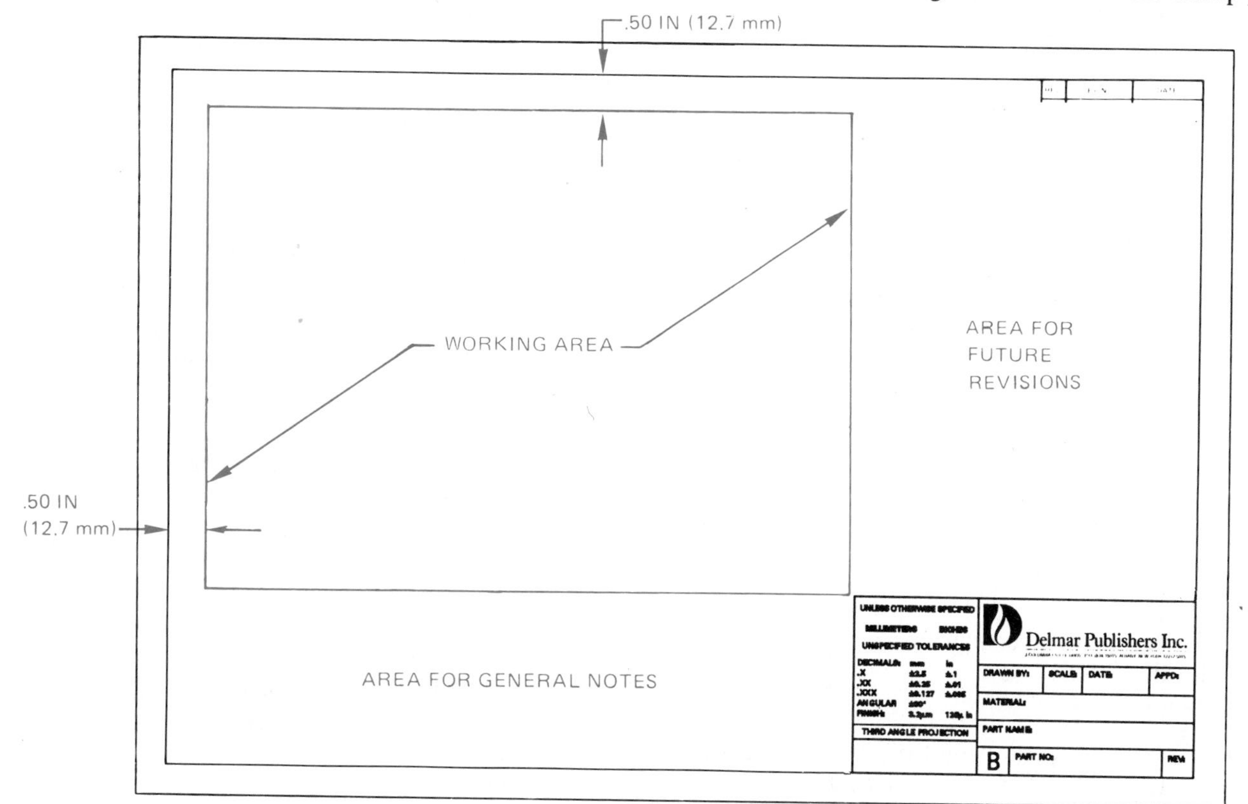

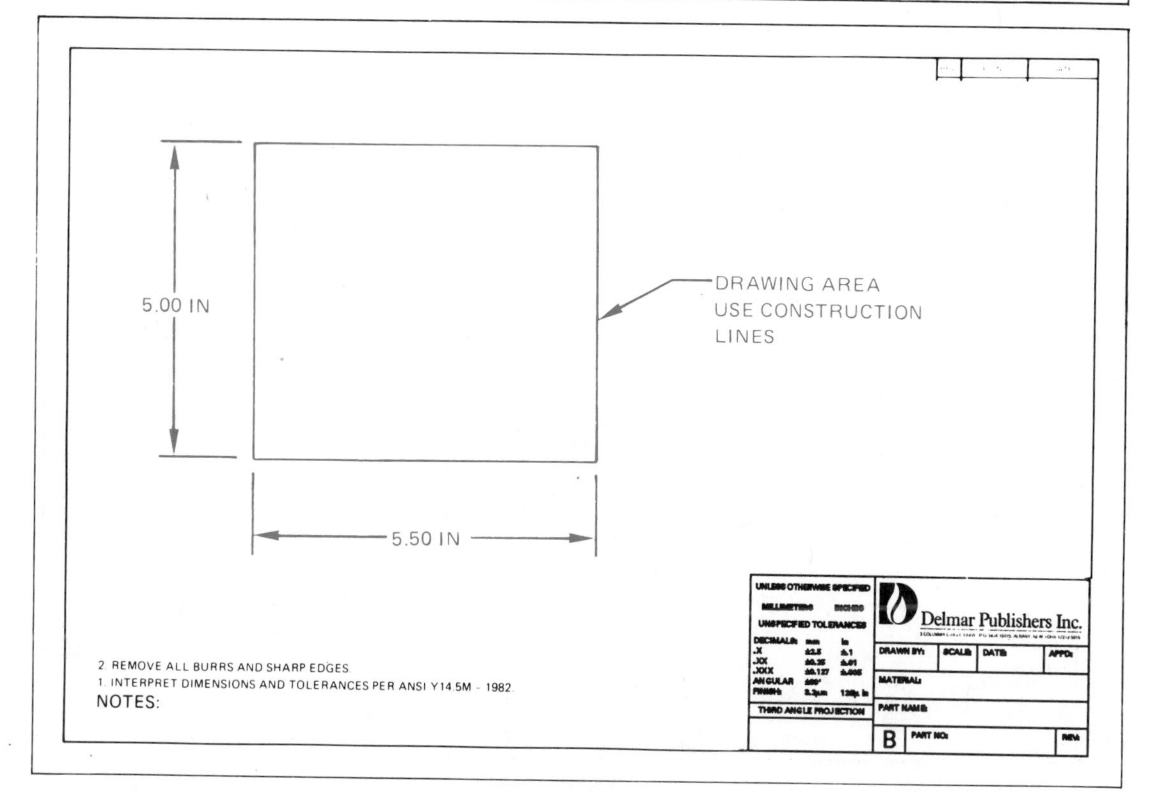

Figure 8–102 Determine the scale, sheet size, and working area.

Length of drawing area (in inches)

Front view width	2.00
Space between views:	
Front view to .50 dimension	.75
.50 to 1.00 dimension	.50
1.00 to 2.00 dimension	.50
2.00 to .375 dimension	.50
.375 dimension to right side view	.75
Right side view depth	+ .50
Drawing area total length	5.50

The space from the view to the first dimension and the spaces between dimensions are drafting decisions. The first space is often, but not always, larger than distances to additional dimension lines. Each dimension line after the first is equally spaced so that dimensions are not crowded.

Height of drawing area (in inches)

Height of views	2.00
Front view to .50 dimension	.75
.50 to 1.00 dimension	.50
1.00 to 2.00 dimension	.50
Right side view to .125 dimension	.75
.125 to .50 dimension	+ .50
There should be enough space for the CBORE note	
Drawing area total height	5.00

The total length and height of the drawing area is 5.50 × 5.00. Keep in mind that these calculations are approximate. There should be adequate space on your drawing for some flexibility. Select a B-size drafting sheet for this drawing. Figure 8–103 shows the 5.50 × 5.00 drawing area blocked out on a B-size sheet. Before you proceed, be sure that your equipment, table, and hands are clean.

Step 5 Use construction lines to draw the selected multiviews, as shown in Figure 8–104.

Step 6 Use construction lines to place hidden, extension, center, dimension, and leader lines, using selected spacing and placement from rough sketch. Also, use guidelines to prepare for the placement of lettering. (See Figure 8–105.)

Step 7 Complete the detail drawing using proper drafting technique in the following sequence:

1. Darken all horizontal thin lines from top to bottom and all vertical thin lines from left to right if right-handed or right to left if left-handed.
2. Draw all circles, arcs, and object lines.
3. Draw all straight object lines in the same order as thin lines.
4. Do all lettering from the top to the bottom of the sheet. Letter all specific and general notes and fill in the title block. To help keep your drawing clean, place a blank sheet of paper under your hand while lettering.
5. Draw all arrowheads.

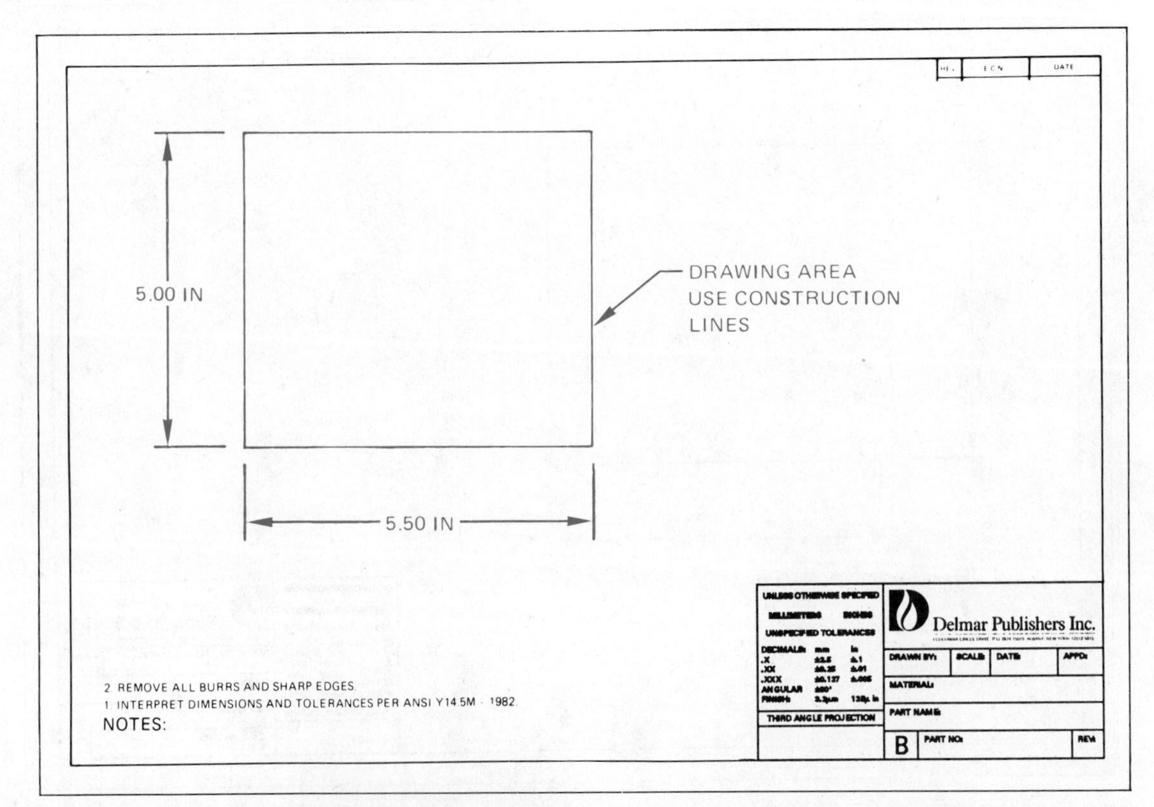

Figure 8–103 Establish the approximate drawing area.

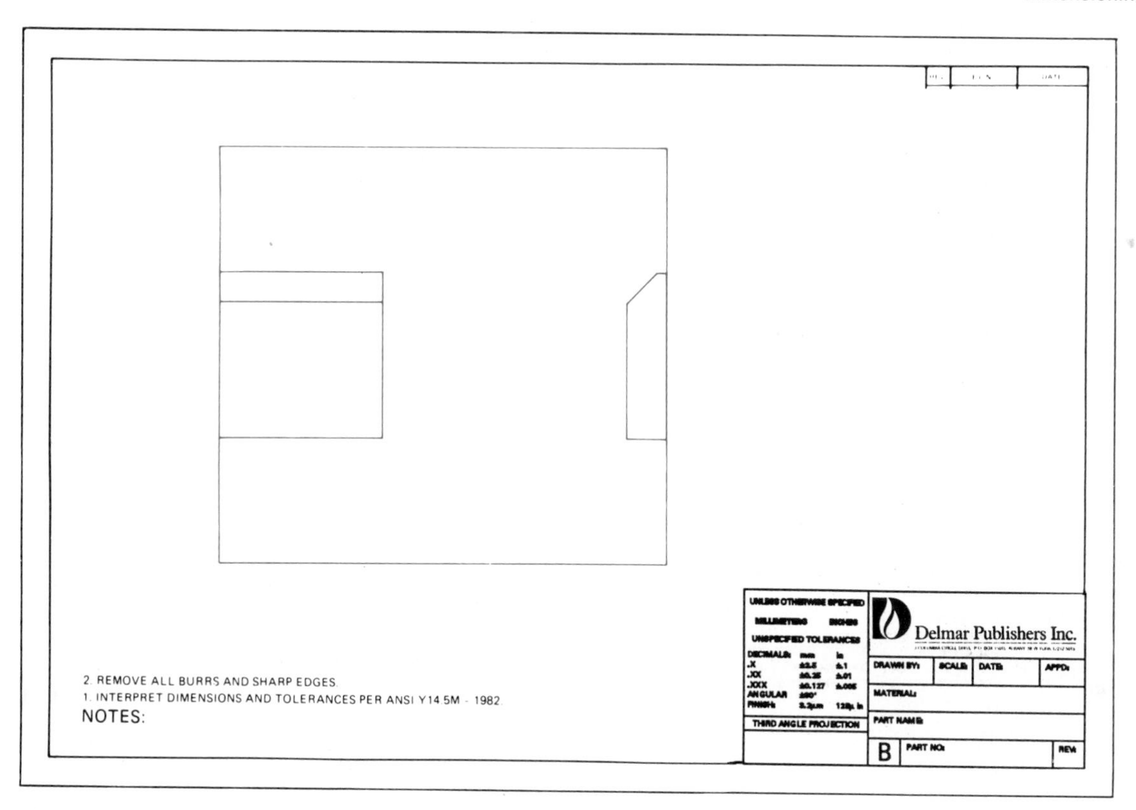

Figure 8–104 Lay out multiviews with construction lines.

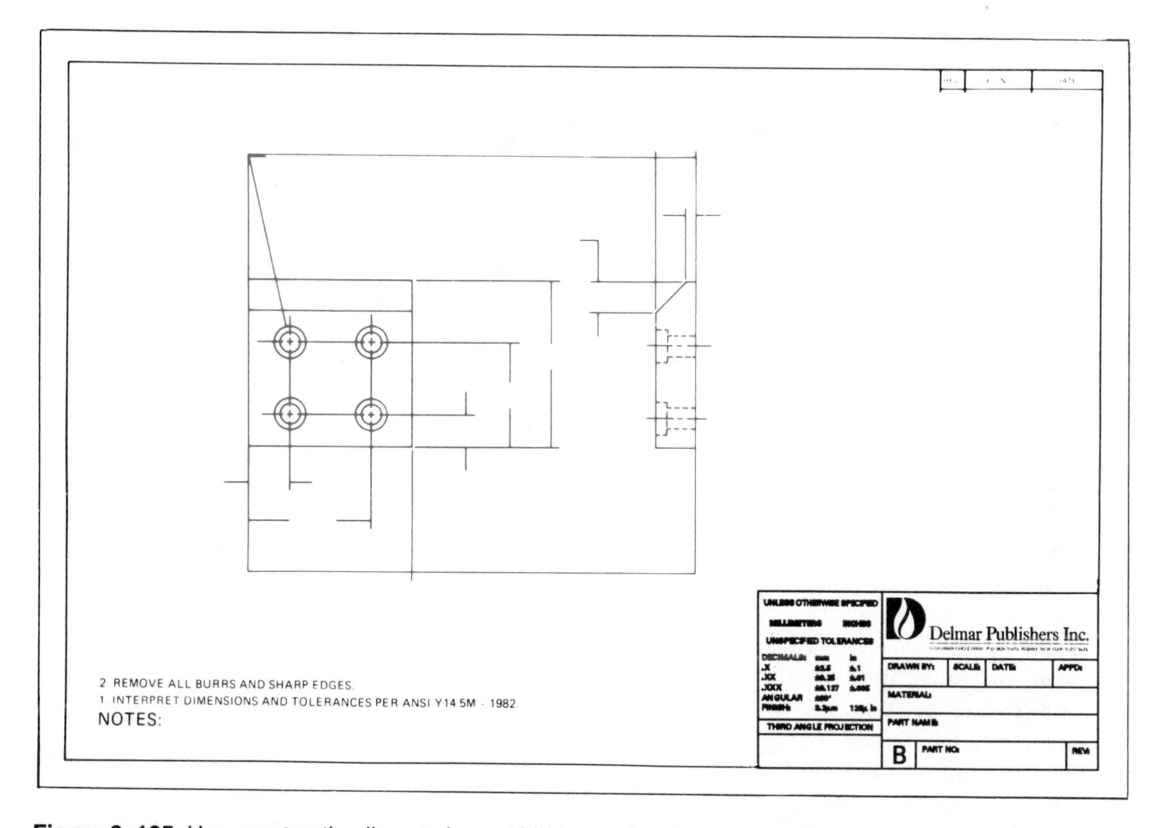

Figure 8–105 Use construction lines to lay out hidden, extension, center, dimension, and leader lines; establish guide lines for lettering.

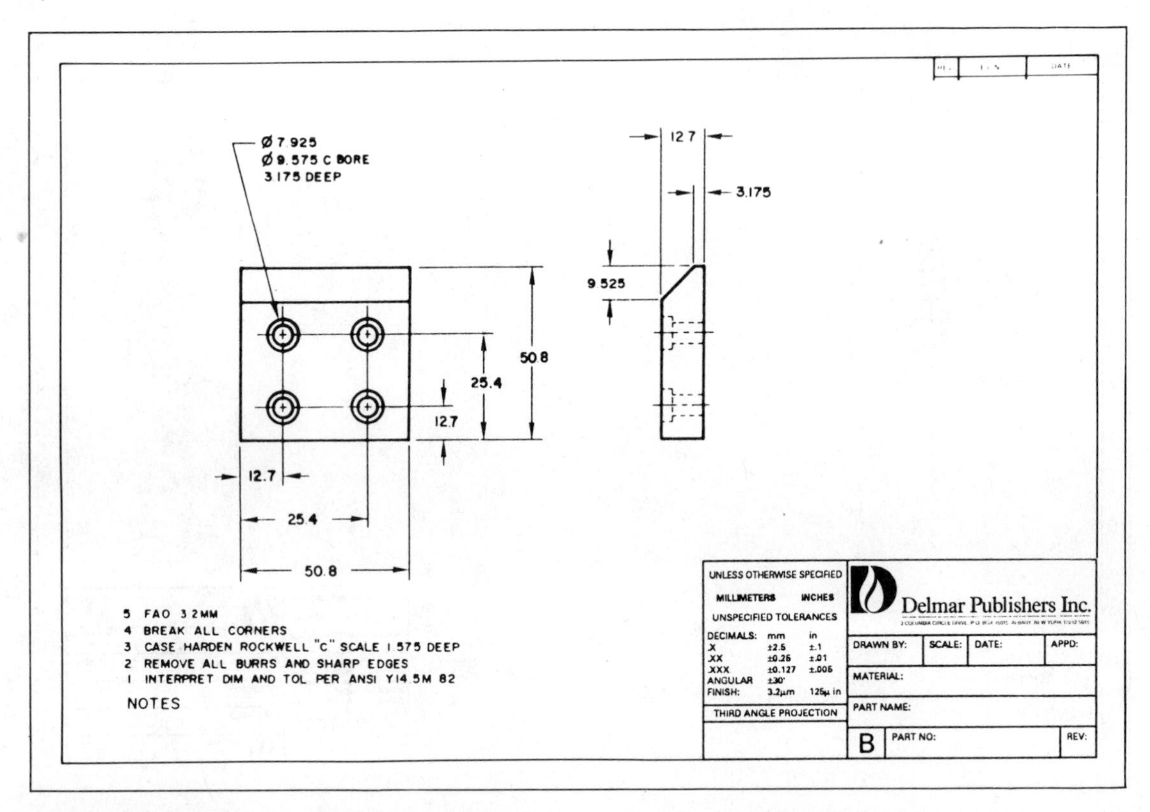

Figure 8–106 Darken all lines and do the lettering to complete the detail drawing.

6. Check your line quality by placing the original over a light table. If light passes through the lines or lettering, the work is not dark enough. Another way to check your work is to run a test-strip diazo copy. If the quality of the print is not adequate, you may need to work more on the drawing. A print will not improve the quality of a drawing.

Figure 8–106 shows the completed detail drawing of the clamp plate.

DIMENSIONING PROBLEMS

DIRECTIONS

1. From the selected sketch, determine which view should be the front view. Then determine which other views, if any, you will need to draw to fully explain the part in a multiview drawing.
2. Make a multiview sketch of the selected problem as close to correct proportions as possible. Be sure to indicate where you intend to place the dimension lines, extension lines, arrowheads, and hidden features, to help you determine the spacing for your final drawing.
3. Using the sketch you have just developed as a guide, make an original multiview drawing on an adequate size drawing sheet and at an appropriate scale. Include all dimensions needed using unidirectional dimensioning. Use computer-aided drafting if specified by your course outline. Include the following general notes at the lower-left corner, .5 in. each direction from the borderline using ⅛-in.-high lettering. Letter the word NOTES with 3/16-in.-high lettering.

2. REMOVE ALL BURRS AND SHARP EDGES.
1. INTERPRET DIMENSIONS AND TOLERANCES PER ANSI Y14.5M–1982 OR 1994 AS APPROPRIATE.

NOTES:

Additional notes may be required depending on the specifications of each individual assignment. Remember that each problem assignment is given as an engineer's rough layout to help simulate actual drafting conditions. Dimensions and views on engineer's sketches may not be placed in accordance with acceptable standards. You will need to carefully review the chapter material when preparing the layout sketch.

UNSPECIFIED TOLERANCES:

DECIMALS	mm	IN.
X	±2.5	±.1
XX	±0.25	±.01
XXX	±0.127	±.005
ANGULAR	±30°	
FINISH	3.2µm	125µIN.

Problem 8–1 Basic practice (metric)

Part Name: Step Block
Material: SAE 1020

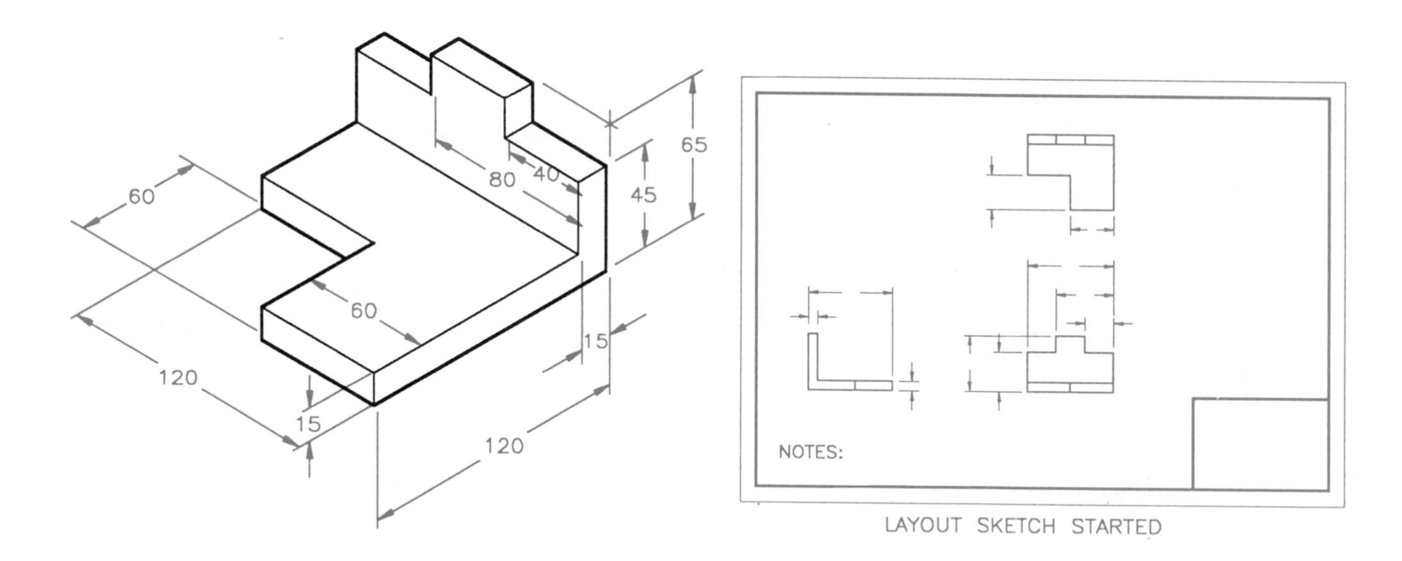

Problem 8–2 Basic practice (metric)

Part Name: Machine Tool Wedge Plate
Material: SAE 4320

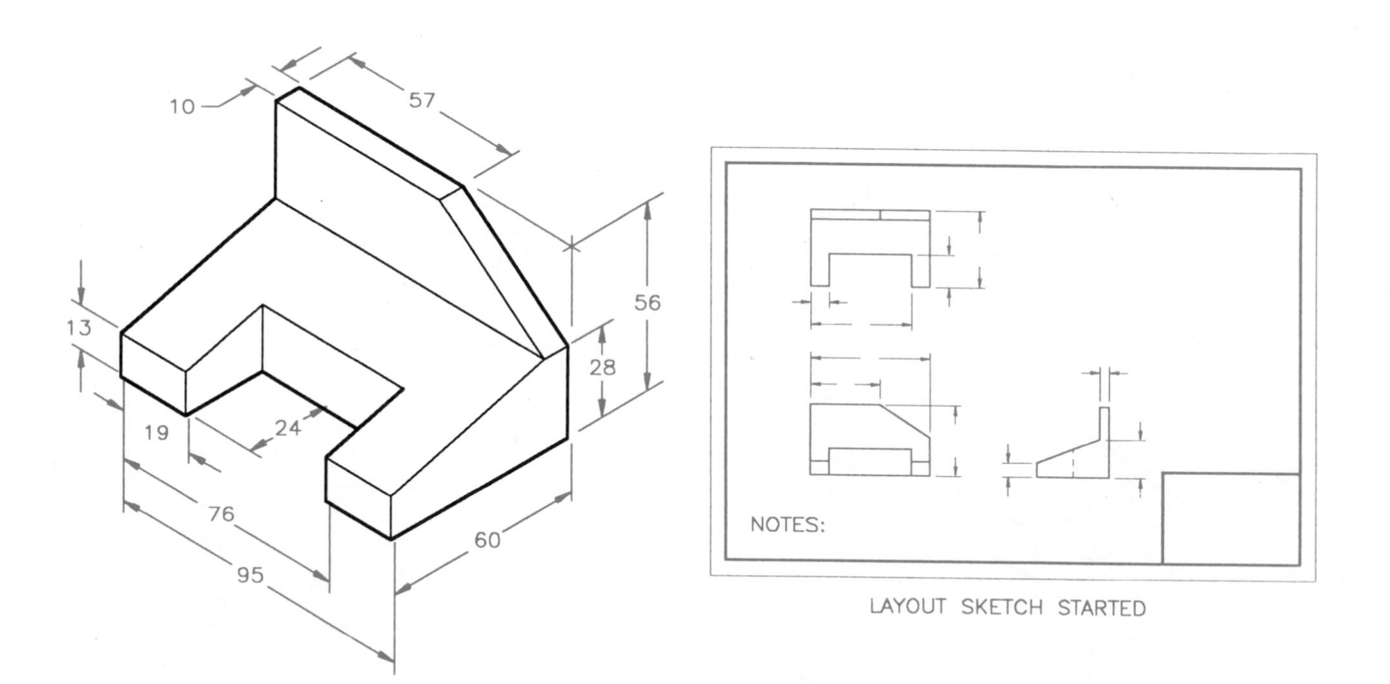

Problem 8–3 Dimensioning basic practice (in.)

Part Name: V-guide
Material: SAE 4320

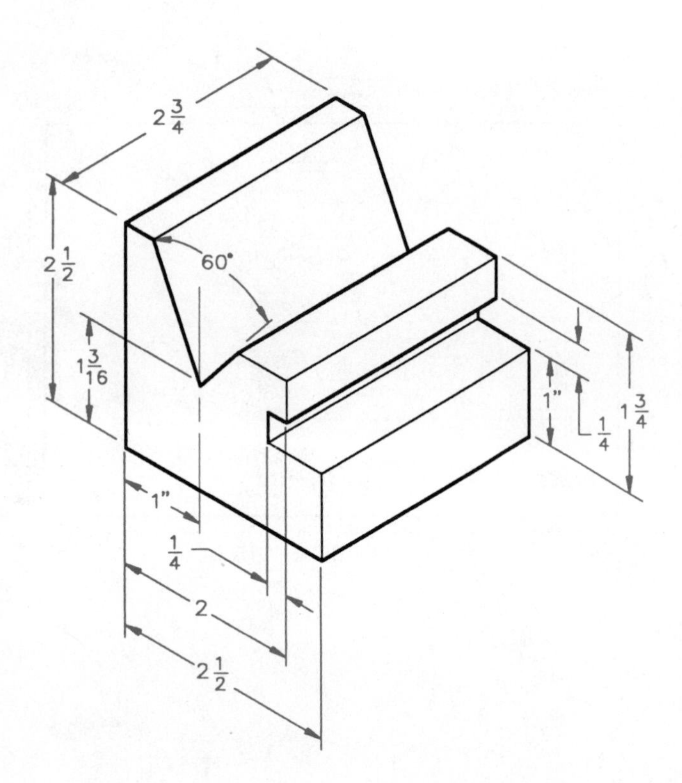

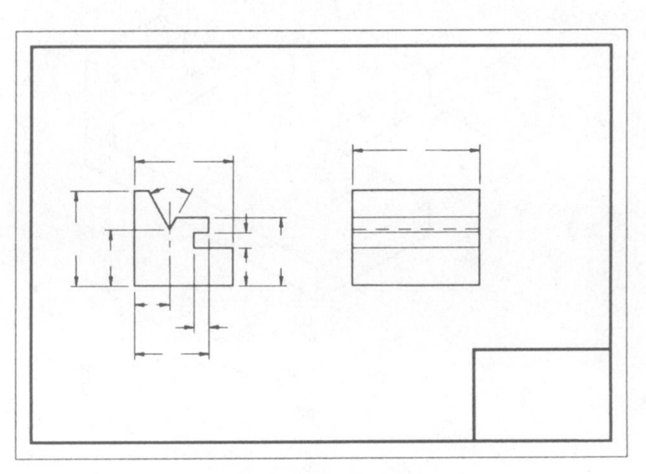

LAYOUT SKETCH
OPTIONAL SINGLE VIEW WITH LENGTH
GIVEN IN GENERAL NOTE OR TITLE BLOCK

Problem 8–4 Circles, arcs, and counterbores (in.)

Part Name: Rest Pad
Material: SAE 1040
Fillets: R .125

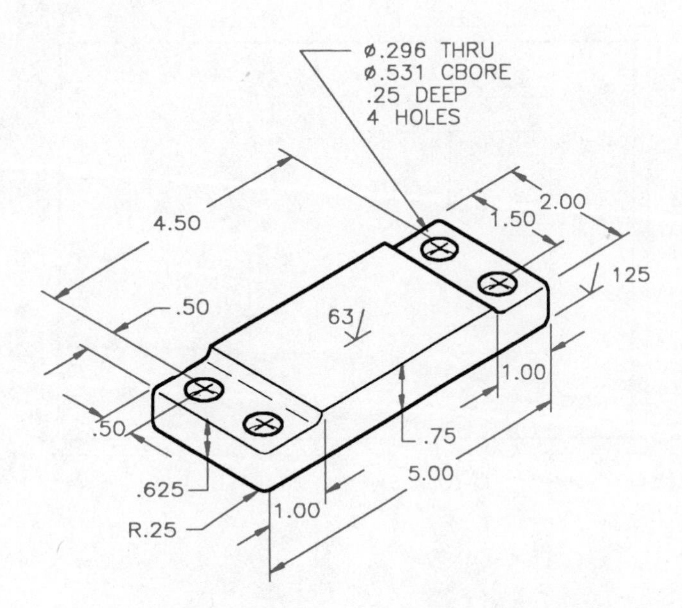

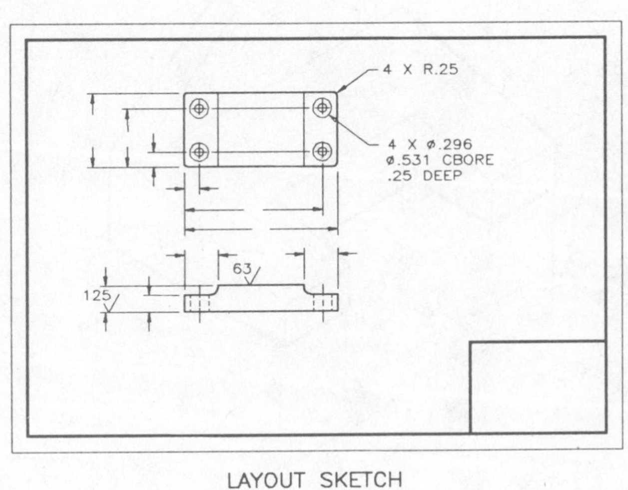

LAYOUT SKETCH

Problem 8–5 Limited space (metric)

Part Name: Angle Bracket
Material: Mild Steel (MS)

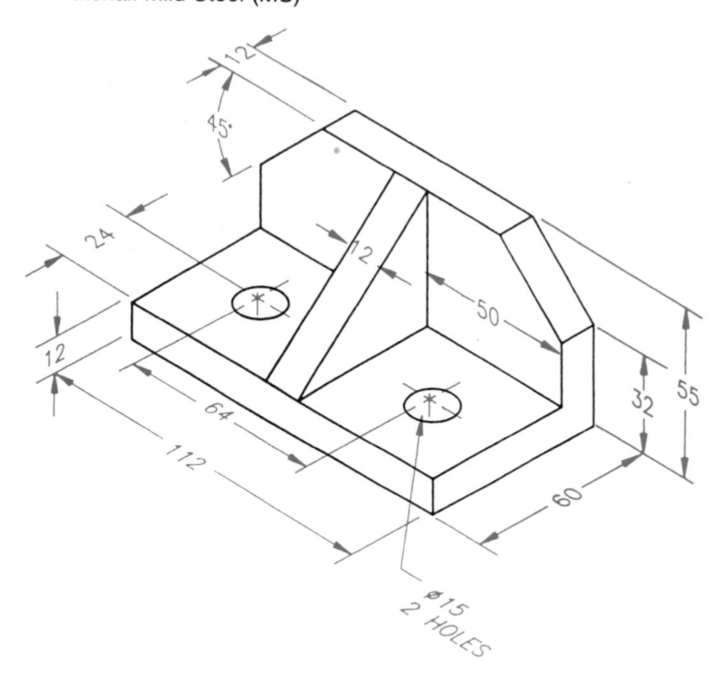

Problem 8–6 Dimensioning contours and limited spaces (in.)

Part Name: Support
Material: Aluminum

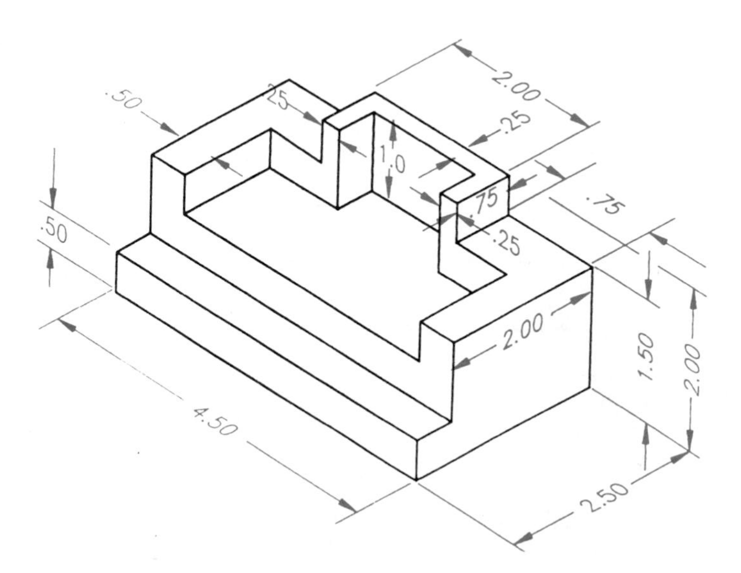

Problem 8–7 Limited spaces (metric)

Part Name: Selector Slide Kicker
Material: Al 1510

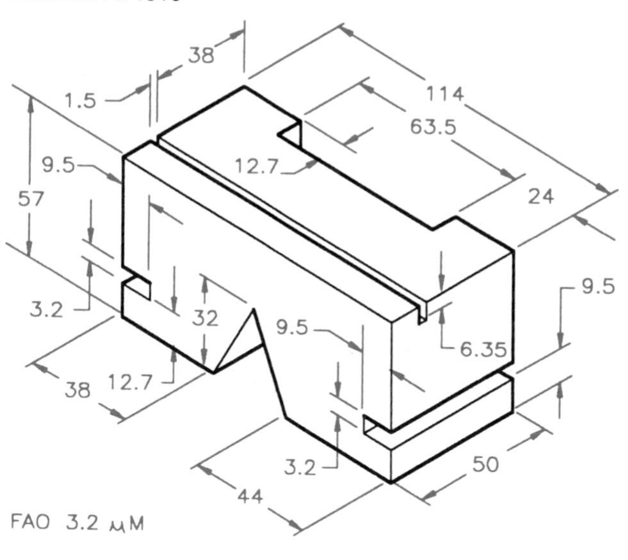

Problem 8–8 Circles and arcs (in.)

Part Name: Chain Link
Material: SAE 4320

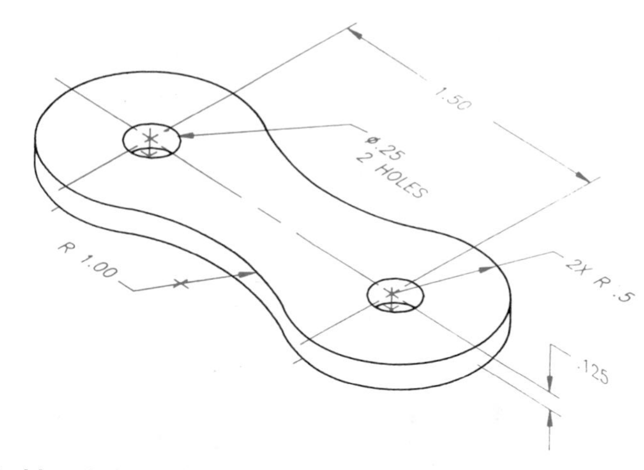

Problem 8–9 Holes and limited space (in.)

Part Name: Pivot Bracket
Material: SAE 1040

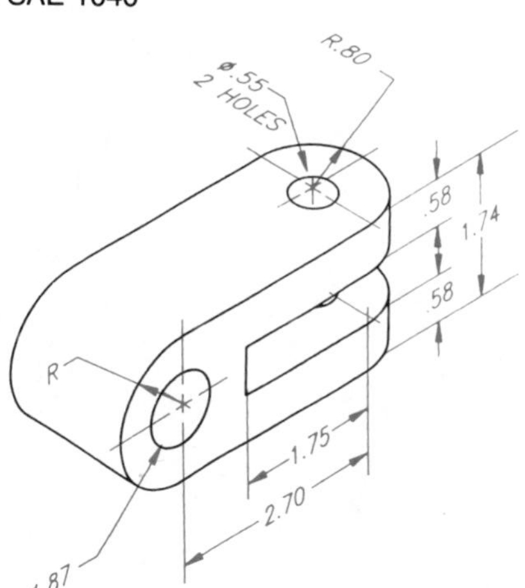

Problem 8–10 Holes, angles, and arcs (in.)

Part Name: Journal Bracket
Material: Cast Iron (CI)

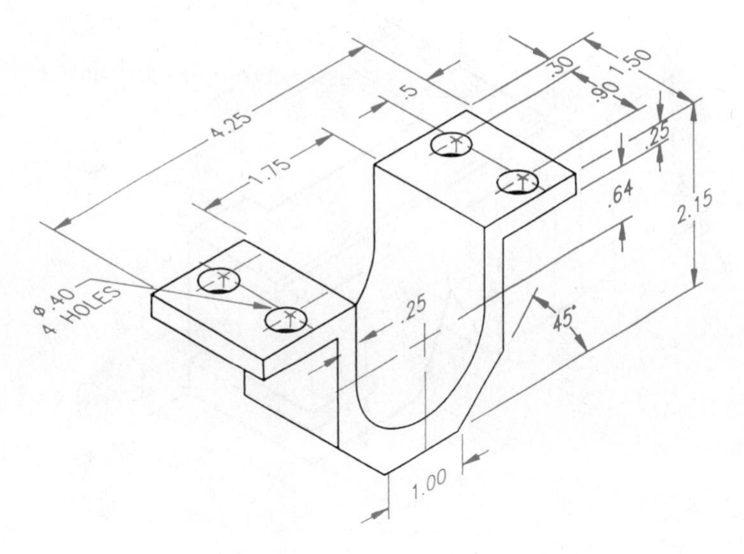

Problem 8–11 Dimensioning angles (in.)

Part Name: Tool Fixture
Material: SAE 4320

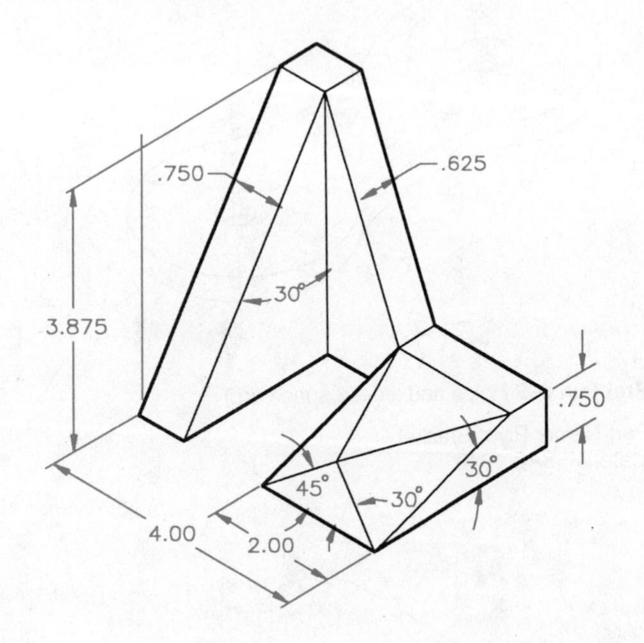

Problem 8–12 Circles and arcs, single view (in.)

Part Name: Thermostat Gauge Standard Motor
Material: Stainless Steel .067/.057 in. Thick
Courtesy Vellumoid Inc.

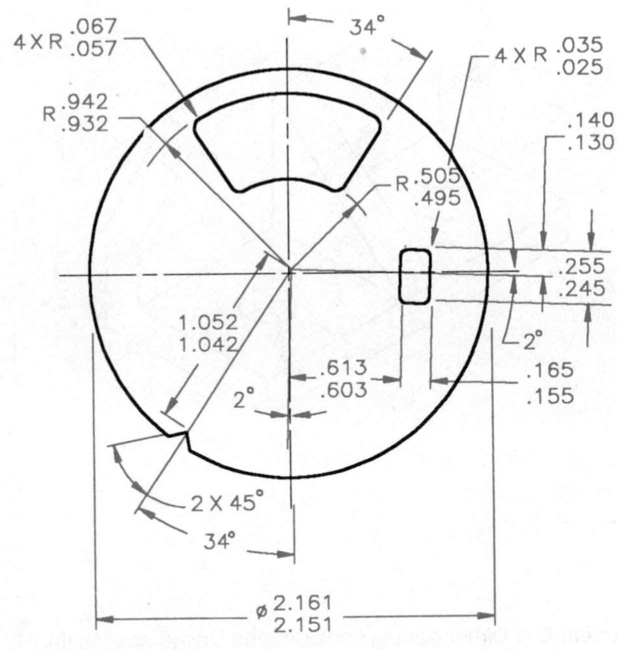

Problem 8–13 Single view (in.)

Part Name: Idler Gear Shaft
Material: MIL-S-7720
Courtesy Aerojet TechSystems Co.

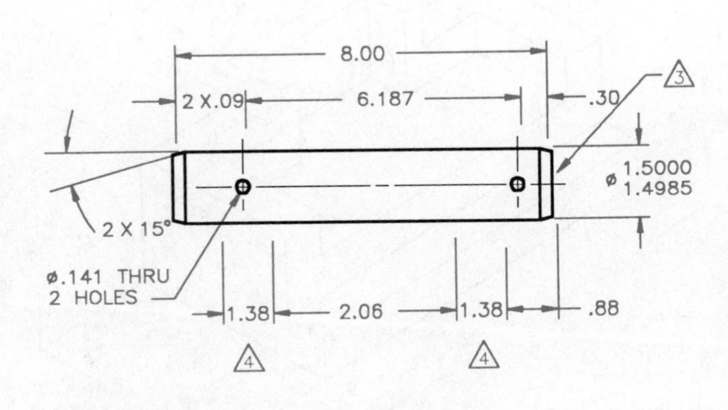

3 MARK WITH 1196975

4 SURFACE FINISH 63 μin. IN AREA INDICATED

Problem 8–14 Circles and arcs (metric)

Part Name: Clamp Jaw
Material: Cast Iron
Fillets and Rounds: R 3 mm

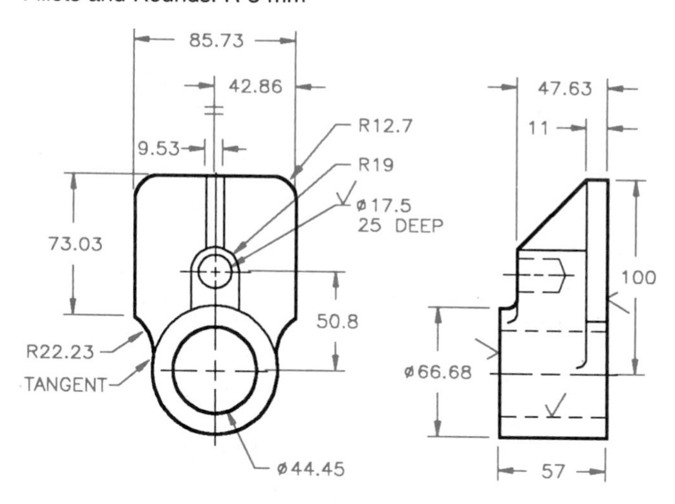

Problem 8–15 Circles and arcs (metric)

Part Name: Hinge Bracket
Material: Cast Aluminum

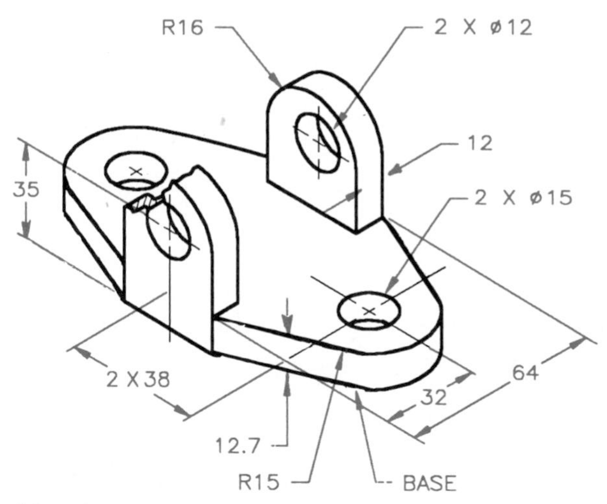

Problem 8–16 Missing view (in.)

Part Name: Bearing Pillow Block
Material: SAE 1040
SPECIFIC INSTRUCTIONS:
Determine the missing view so that holes can be shown and dimensioned as circles. Convert all fractional location dimensions to three-place decimals, and size fractional dimensions to two-place decimals. Verify and add any missing dimensions.
Courtesy Production Plastics, Inc.

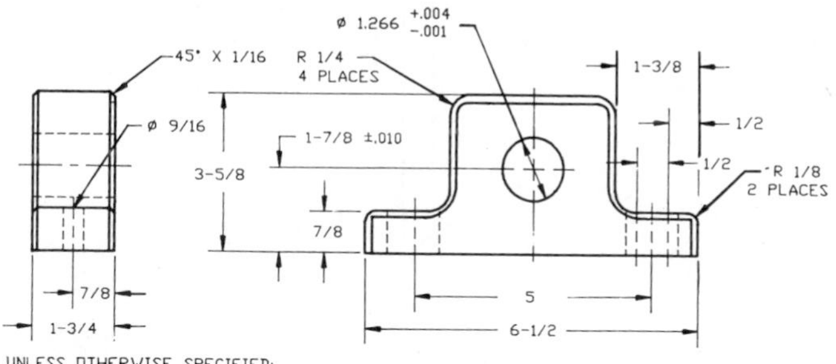

UNLESS OTHERWISE SPECIFIED:
ALL FRACTIONAL DIMENSIONS ±1/32 ALL THREE PLACE DECIMALS ±.005
ALL TWO PLACE DECIMALS ±.010

Problem 8–17 Knurling (in.)

Part Name: Screw Handle
Material: SAE 1020

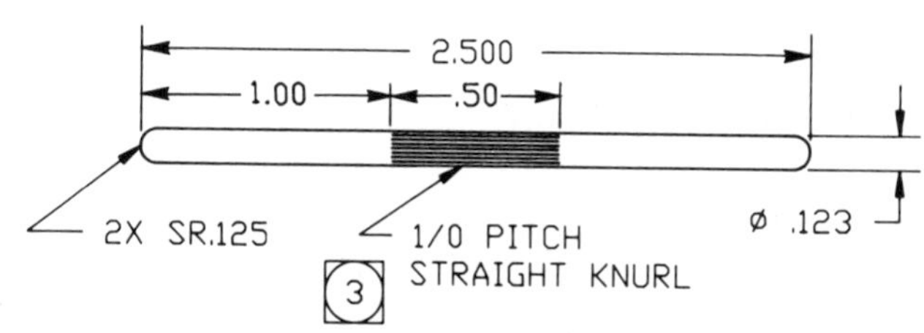

3 KNURL AFTER SIZING,KNURL DIA. TO BE GREATER THAN ⌀.125
2. REMOVE ALL BURRS AND SHARP EDGES.
1. INTERPRET DIMENSIONS AND TOLERANCES PER ANSI Y14.5M-1982.
NOTES

Problem 8–18 Machine features (in.)

Part Name: Spacer
Material: SAE 1030

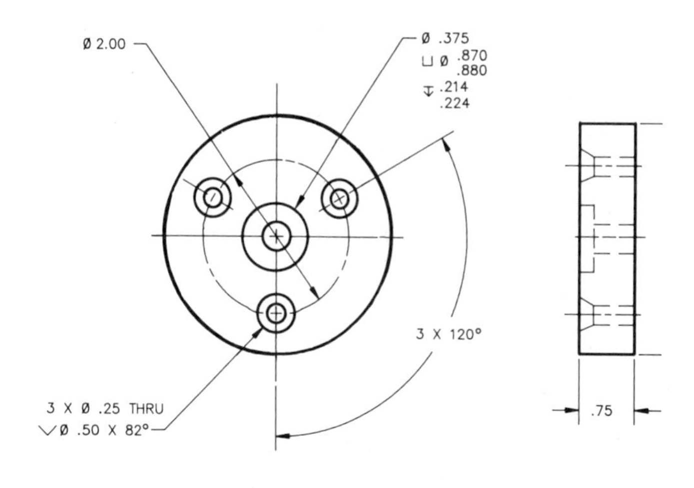

Problem 8–19 Polar dimensioning (in.)

Part Name: Rotating Bracket
Material: .25 THK Stainless Steel

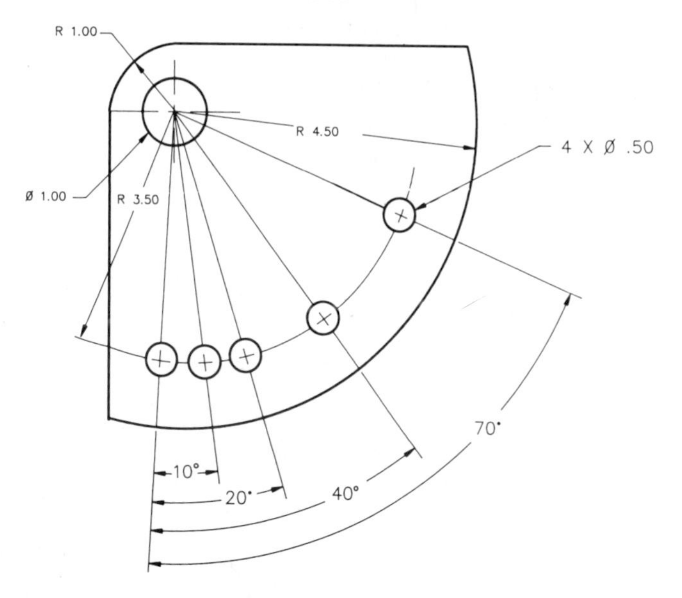

Problem 8–20 Polar coordinate dimensioning (in.)

Part Name: Spacer
Material: Plastic

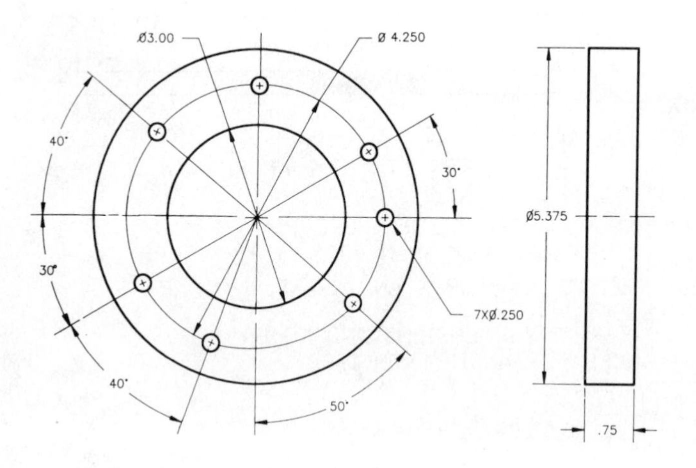

Problem 8–22 Tabular dimensioning (metric)

Part Name: Mounting Base
Material: Stainless Steel

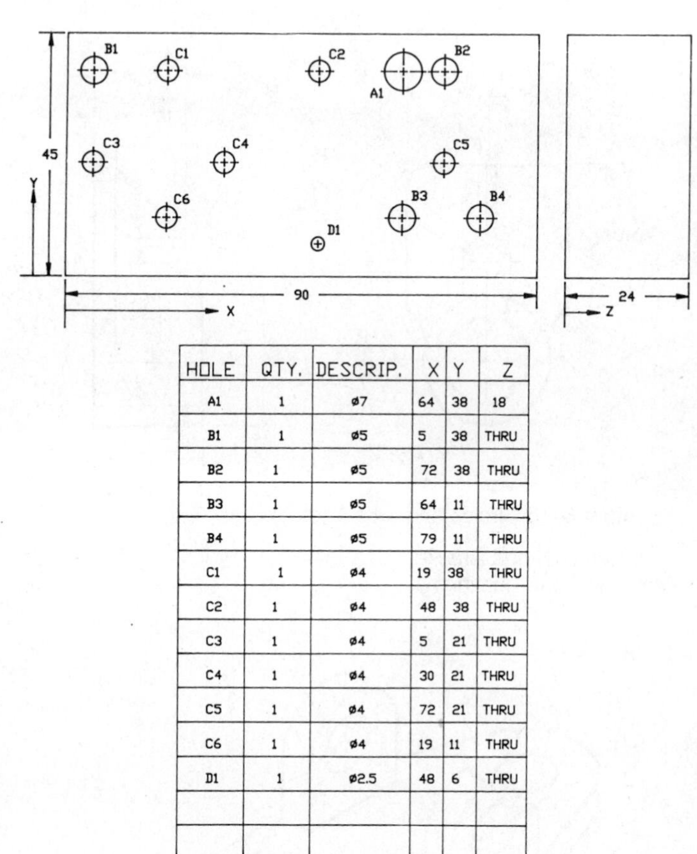

HOLE	QTY.	DESCRIP.	X	Y	Z
A1	1	ø7	64	38	18
B1	1	ø5	5	38	THRU
B2	1	ø5	72	38	THRU
B3	1	ø5	64	11	THRU
B4	1	ø5	79	11	THRU
C1	1	ø4	19	38	THRU
C2	1	ø4	48	38	THRU
C3	1	ø4	5	21	THRU
C4	1	ø4	30	21	THRU
C5	1	ø4	72	21	THRU
C6	1	ø4	19	11	THRU
D1	1	ø2.5	48	6	THRU

Problem 8–21 Repetitive features (in.)

Part Name: Slot Plate
Material: Aluminum

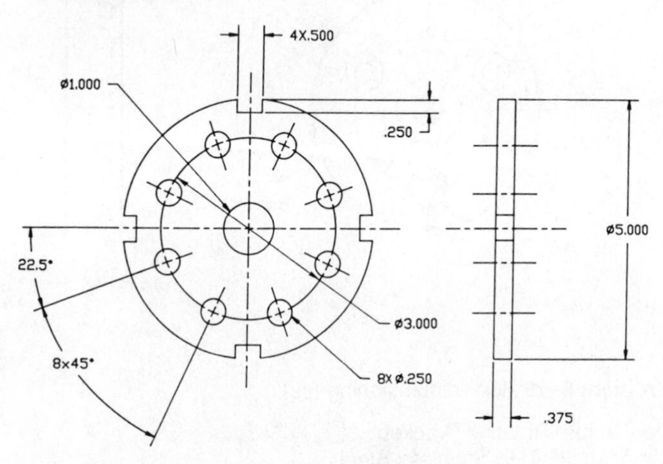

3. HIDDEN LINES REMOVED FOR CLARITY
2. REMOVE ALL BURRS AND SHARP EDGES
1. INTERPRET DIMENSIONS AND TOLERENCE PER ANSI Y14.5–1982

NOTES

Problem 8–23 Dimensioning circles, arcs, and slots (in.)

Part Name: Top Pipe Support Bracket
Material: SAE 1020
Note: All Fillets R .12

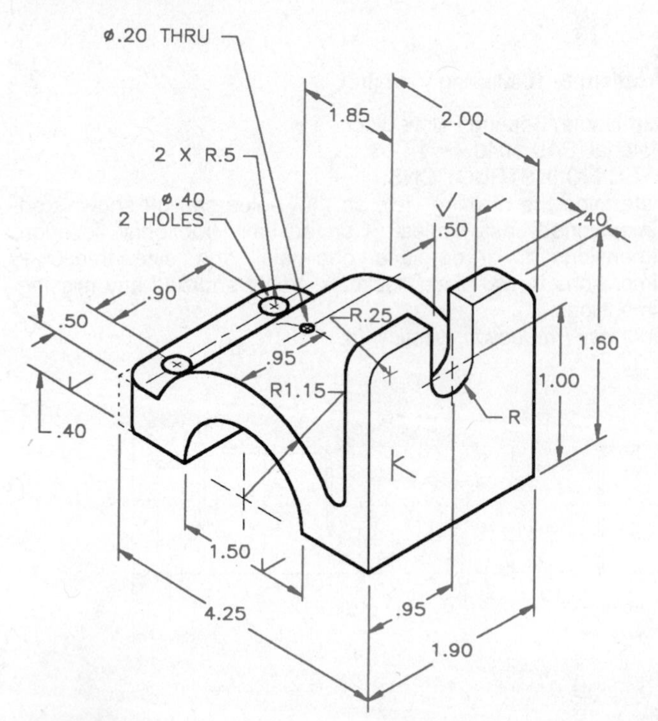

Problem 8–24 Chain dimensioning (metric)

Part Name: Control Housing Cover
Material: Cast Iron

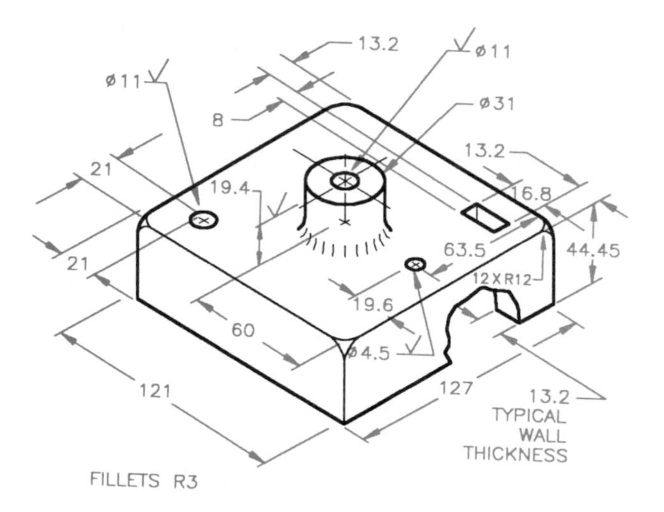

Problem 8–25 Dimensioning compound circles and arcs

Part Name: Multiple Shaft Support
Material: SAE 1030
Note: Fillets and Rounds R .08

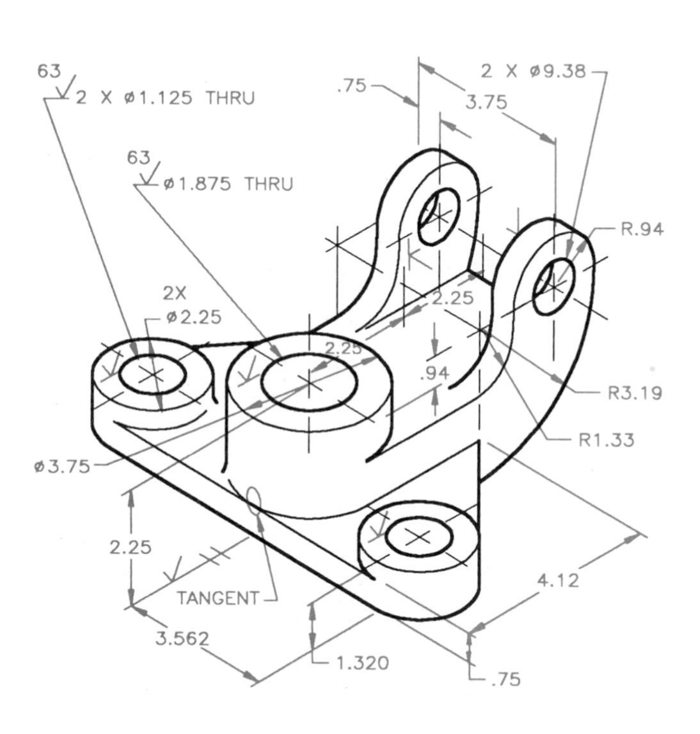

Problem 8–26 Dimensioning auxiliary views

Part Name: Connector
Material: SAE 4320
Finish All Over 63 μin.

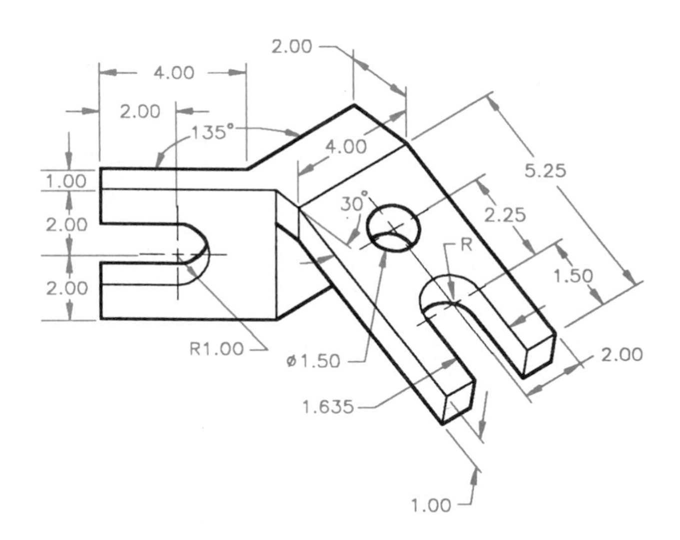

Problem 8–27 Dual dimensioning millimeters [inches] with dimensions given in inches

Part Name: Guide Bracket
Material: SAE 4020
Fillets and Rounds: R .125
SPECIFIC INSTRUCTIONS:
Use dual dimensioning with millimeters followed by inches in brackets [ ].

For example: MILLIMETERS / [INCHES] or MILLIMETERS [INCHES]

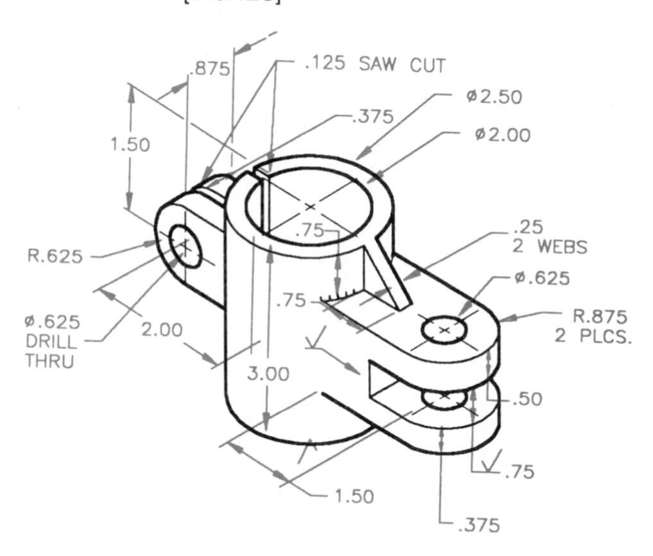

Problem 8–28 Chart drawing (metric)

Part Name: Chart Plate
Material: 10 mm Brass

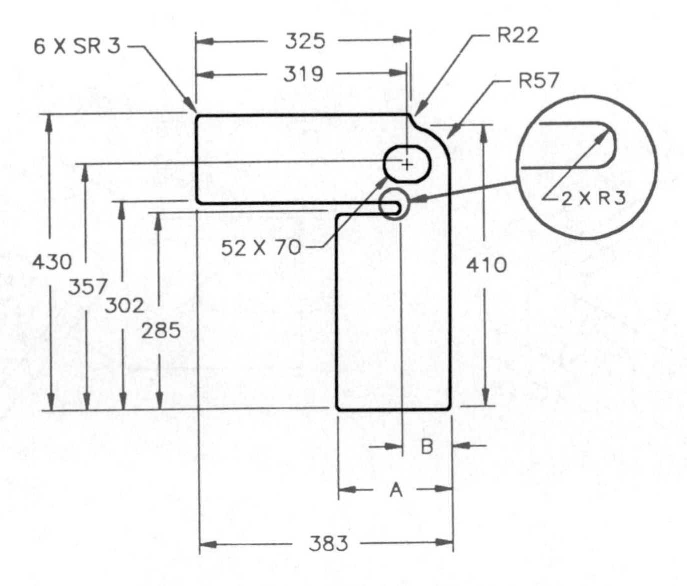

PART NO.	DIM A	DIM B
342142	172	75
342143	140	60
342144	108	45
342145	76	30
342146	44	15

Problem 8–29 Chart drawing

Part Name: Tank Bracket
Material: 11 GA A570-30
Courtesy TEMCO

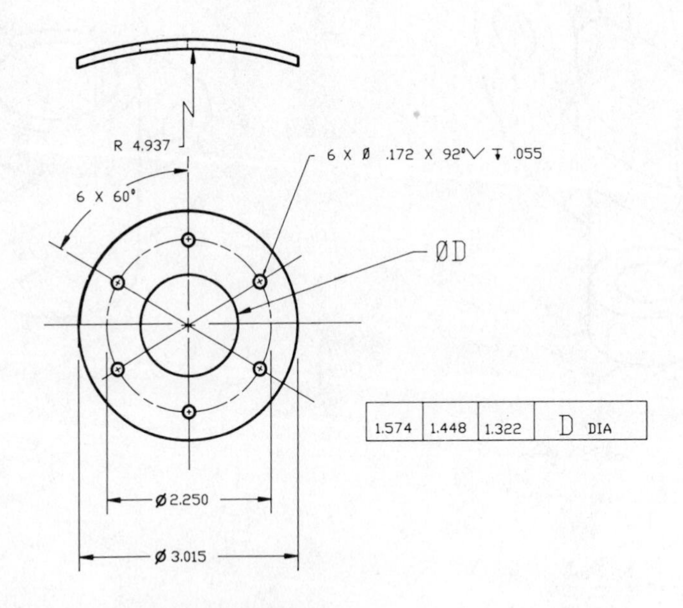

Problem 8–30 Arrowless dimensioning

Part Name: Metering Box Gasket
Material: Neoprene
Notes: Dimensions have been corrected to allow for steel rule blade.
SPECIFIC INSTRUCTIONS:
Convert the engineering sketch to arrowless tabular dimensioning. Set up a table that will identify hole diameters, arc radii, and location dimensions from x- and y-coordinates. Dimensions above Y will be + dimensions and below Y will be – dimensions. Dimensions to the right of X will be + dimensions and to the left of X will be – dimensions.
Courtesy Vellumoid Inc.

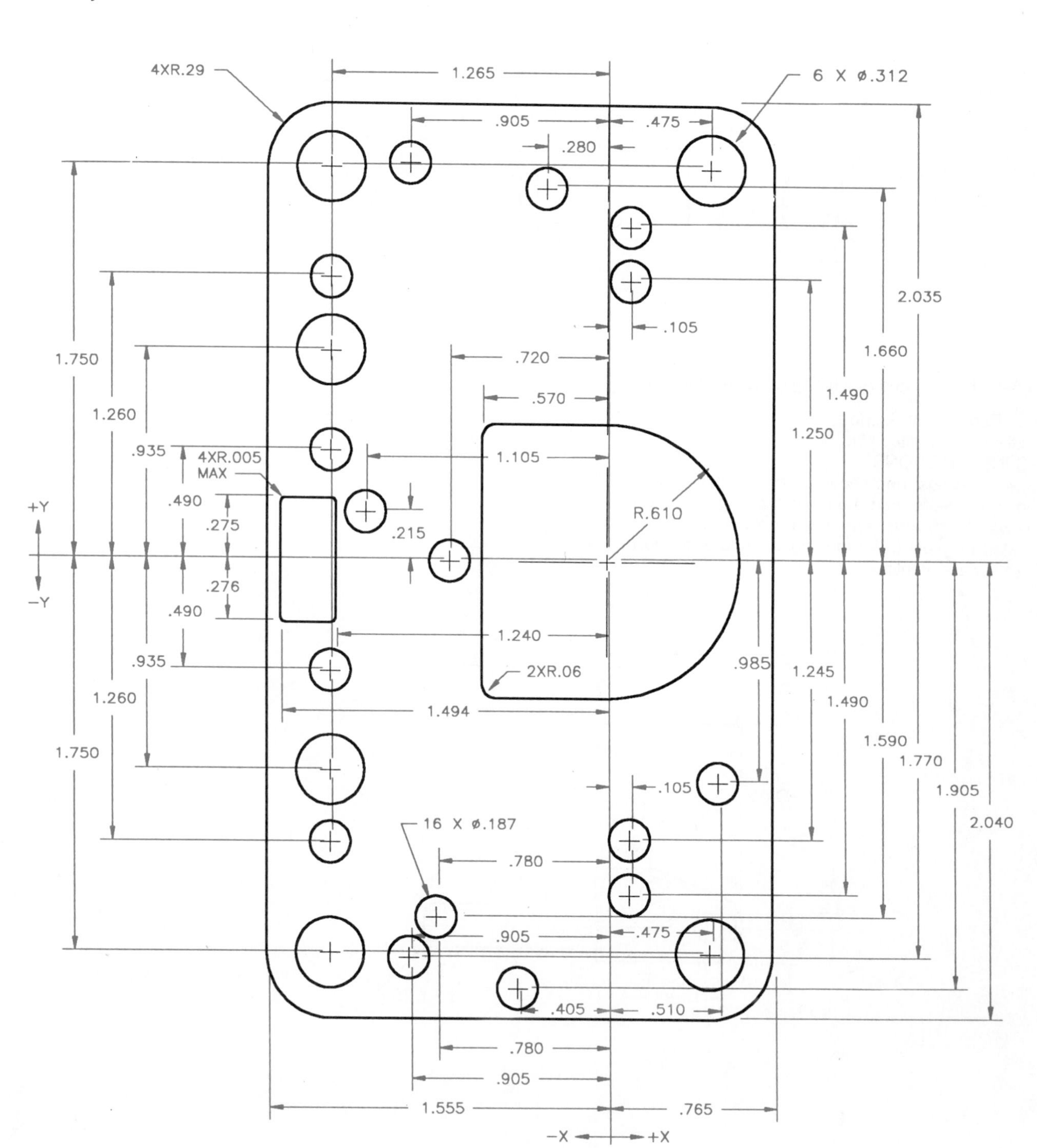

Problem 8–31 Casting drawing (metric)

Part Name: Slider
Material: ASTM 60 CI
Hardness: Brinell 180–220
SPECIFIC INSTRUCTIONS:
Draw part as finished. Include all machining and casting dimensions on one detail drawing. The pattern maker will apply unspecified allowances during production.

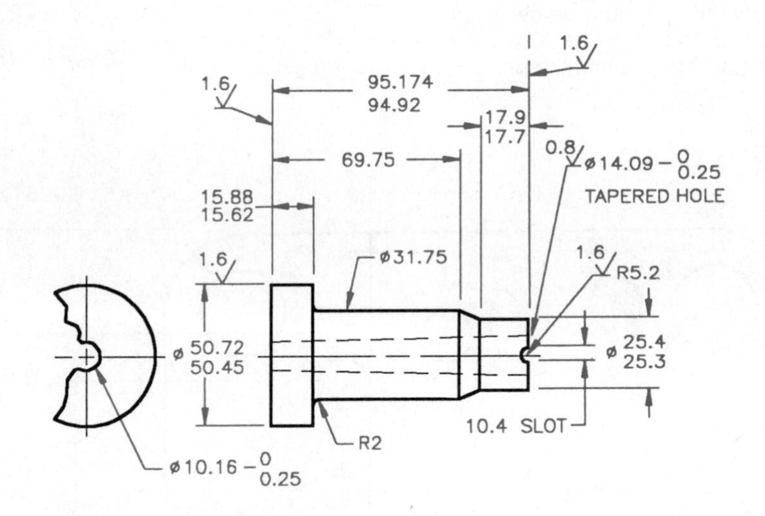

Problem 8–32 Forging and machining drawings (metric)

Part Name: Pump Pivot Support
Material: HRMS (Hot-rolled Mild Steel)
SPECIFIC INSTRUCTIONS:
Prepare two drawings: one showing only forging-related views, dimensions, and notes; and the other showing only machining-related views, dimensions, and notes. Provide the draft angles recommended for steel (refer to chapter). Add 3 mm to forging where finish surface identified.

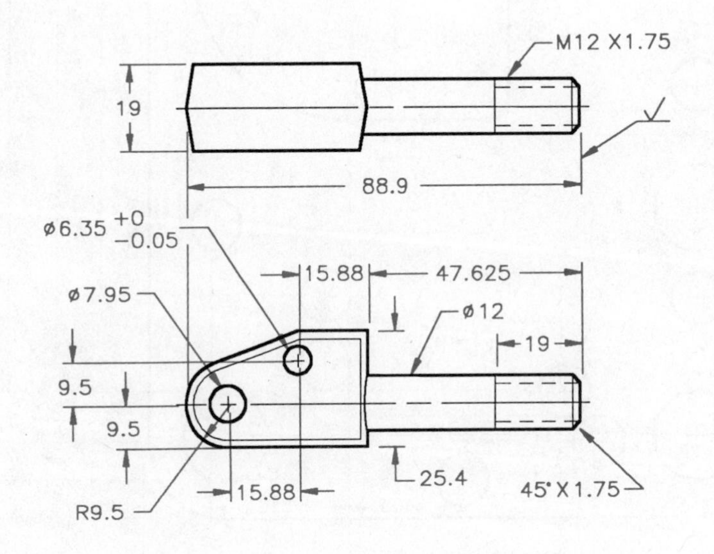

Problem 8–33 Forging drawing

Part Name: Slide
Material: SAE 4150
SPECIFIC INSTRUCTIONS:
Provide forging dimensions only. The machining drawing is located in Problem 11–22, for reference.

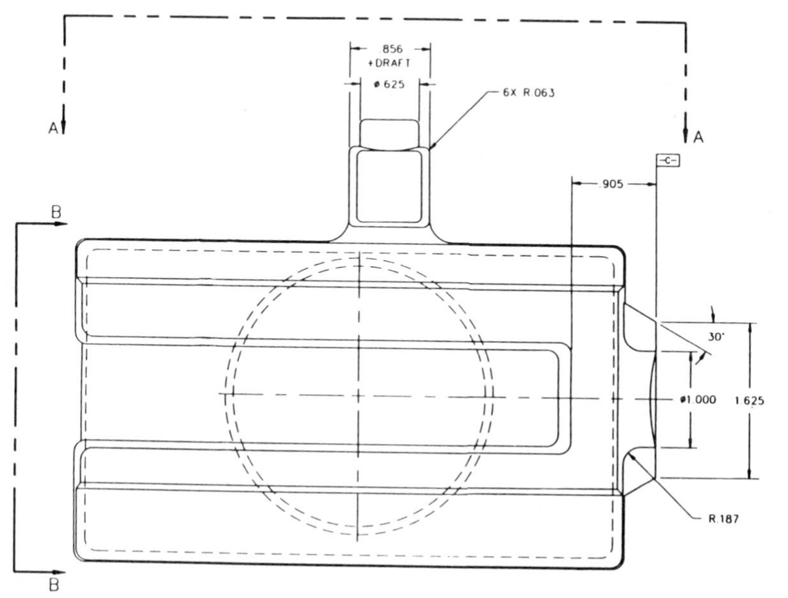

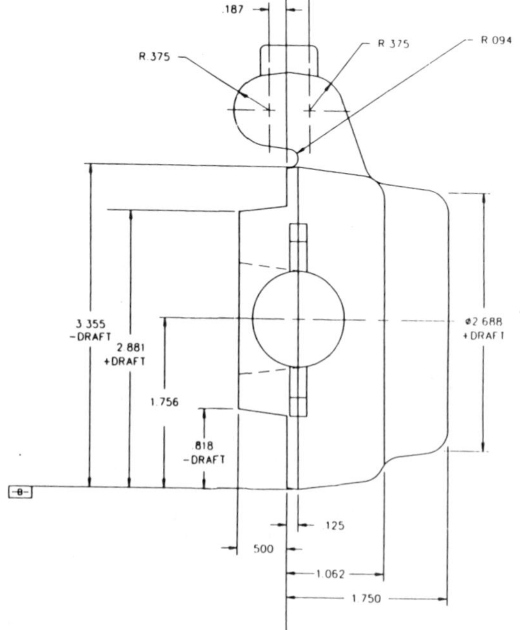

4. ALL DRAFT ANGLES 7°.
3. ALL FILLETS AND ROUNDS R.25.
2. REMOVE ALL BURRS AND SHARP EDGES.
1. INTERPRET ALL DIMENSIONS AND TOLERANCES PER ANSI Y14.5M–1982.
NOTES: UNLESS OTHERWISE SPECIFIED.

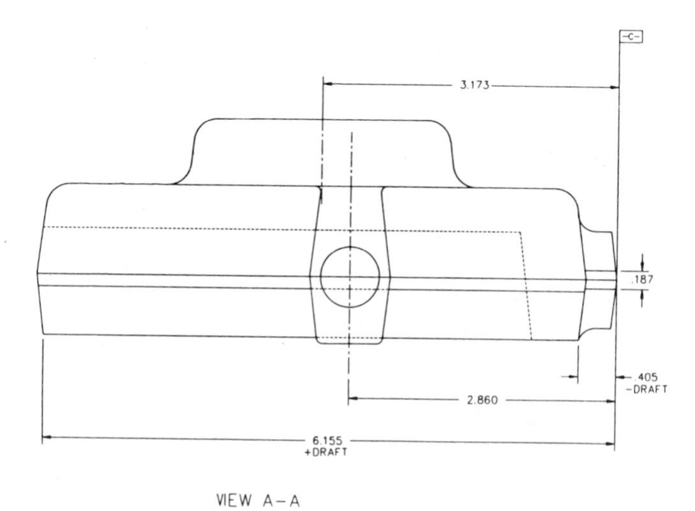

VIEW A–A

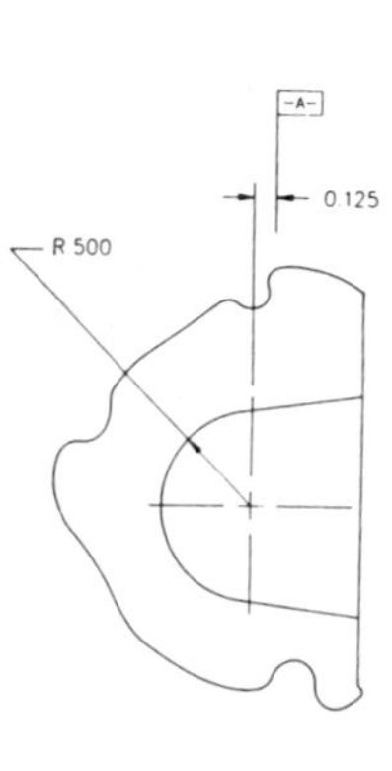

VIEW B–B

Problem 8–34 Forging drawing

Part Name: Slide
Material: SAE 4150
SPECIFIC INSTRUCTIONS:
Provide forging dimensions only. The machining drawing is located in Problem 11–23, for reference.

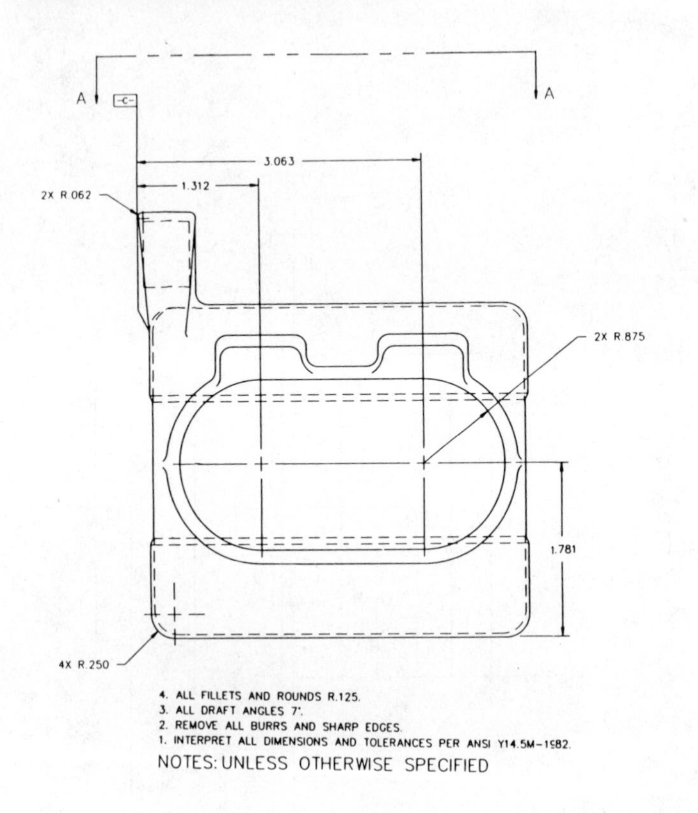

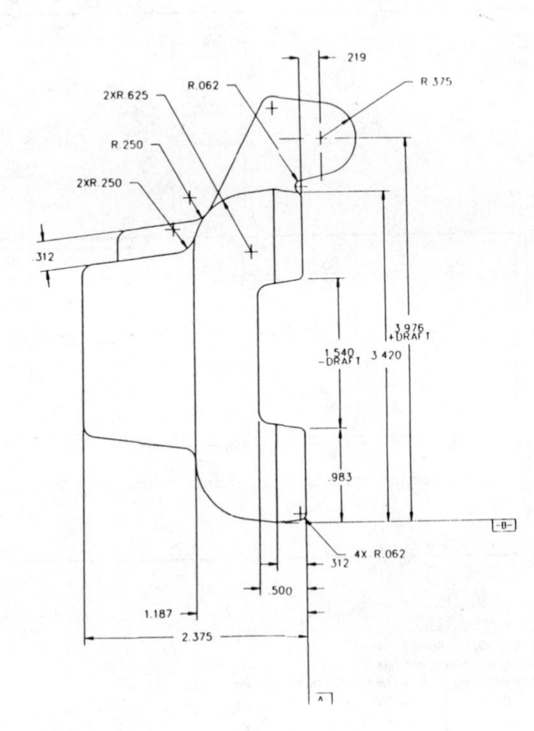

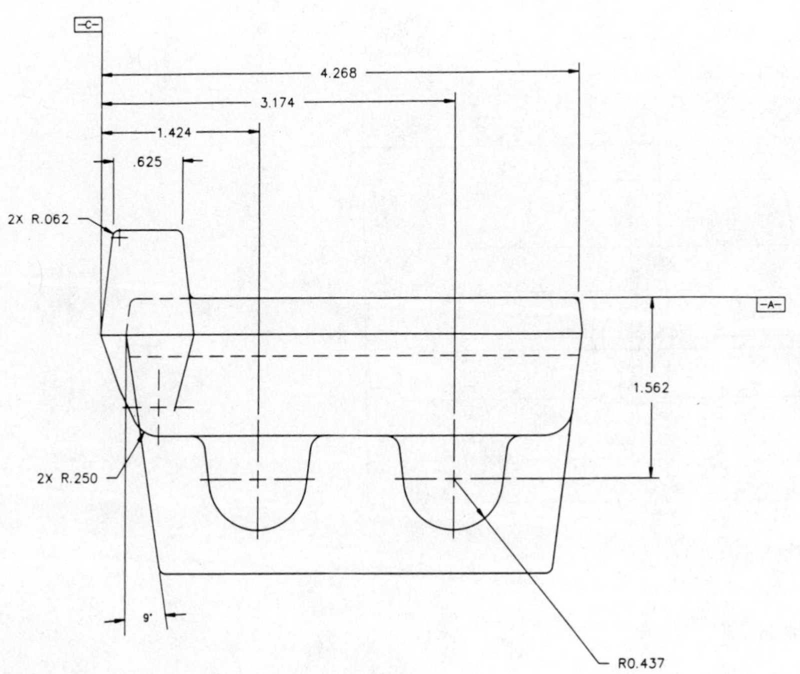

VIEW A–A

Problem 8–35 CAD/CAM

Part Name: Spacer
Material: .75 THK SAE 1030

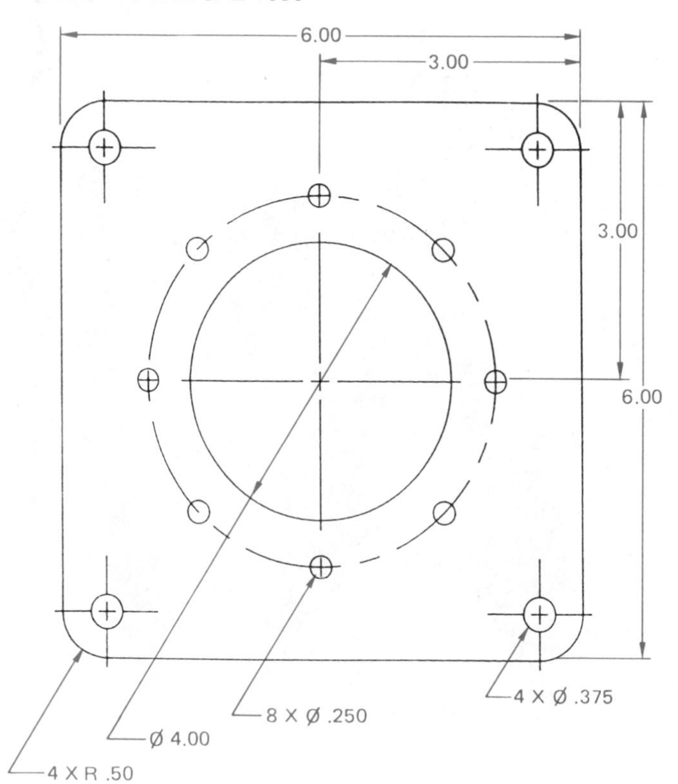

JOB PLAN

TYPE	TOOL NO.	DIAMETER	CAD LAYER
DRILL	1	.250	1
DRILL	2	.375	2
END MILL	3	.500	3

Problem 8–36 CAD/CAM

Part Name: Cover Plate
Material: .500 THK Aluminum

JOB PLAN

TOOL NO.	TYPE	DIAMETER	LENGTH	CAD LAYER
1	END MILL	.75	1.000	1
2	END MILL	.50	1.000	2
3	END MILL	.25	1.000	3
4	SPOT DRILL	.125	1.000	4
5	DRILL	.500	1.000	5
6	DRILL	1.000	1.000	6
	CLAMPS			7

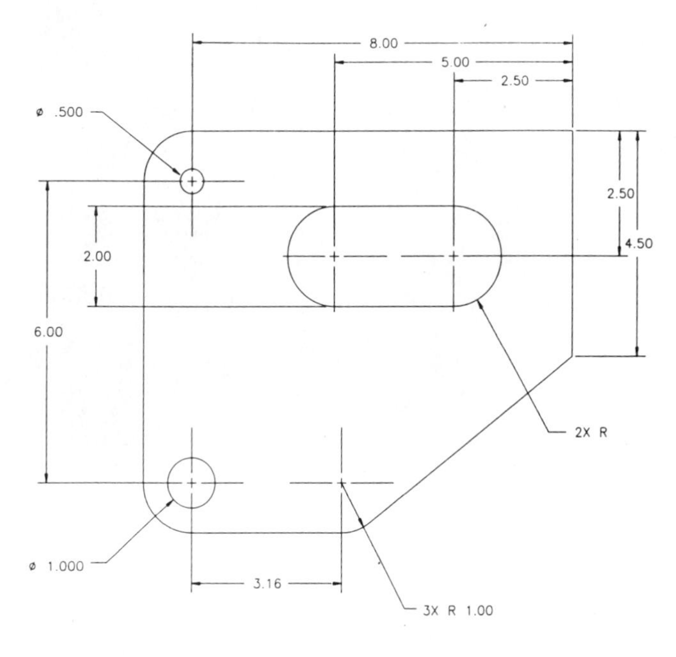

CHAPTER 9

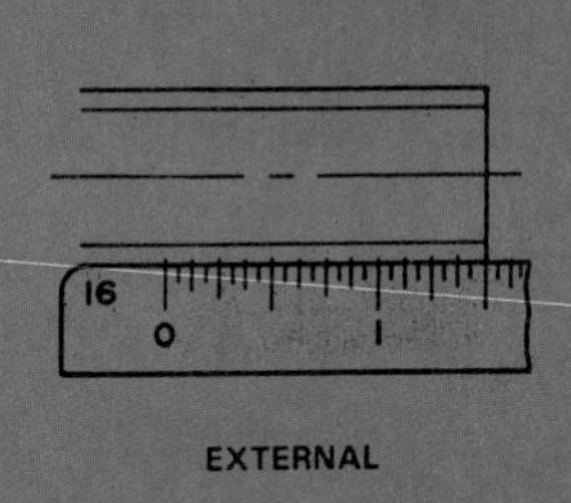

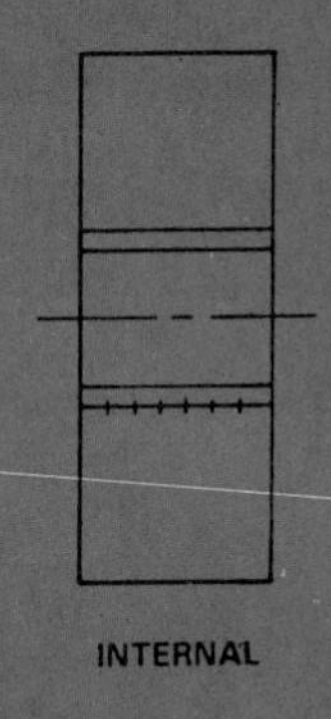

Fasteners and Springs

LEARNING OBJECTIVES

After completing this chapter you will:

- Draw screw thread representations and provide correct thread notes.
- Prepare drawings for fastening devices.
- Draw completely dimensioned spring representations.
- Prepare formal drawings from engineer's sketches and actual industrial layouts.

ANSI This chapter introduces you to the methods of calling for and drafting fasteners and springs. Fasteners include screw threads, keys, pins, rivets, and weldments. There are two types of springs, helical and flat. The American National Standards Institute documents that govern the standards for fasteners and springs are *Screw Thread Representation* ANSI Y14.6–1978, *Screw Thread Representation (Metric Supplement)* ANSI Y14.6aM–1981, *Symbols for Welding and Nondestructive Testing Including Brazing* ANSI/AWS A2.4–1979, and *Mechanical Spring Representation* ANSI Y14.13M–1981.

SCREW-THREAD FASTENERS

The standardization of screw threads was achieved among the United States, United Kingdom, and Canada in 1949. A need for interchangeability of screw-thread fasteners was the purpose of this standardization and resulted in the Unified Thread Series. The Unified Thread Series is now the American standard for screw threads. Prior to 1949 the United States standard was the American National screw threads. The unification standard occurred as a result of combining some of the characteristics of the American National screw threads with the United Kingdom's long accepted Whitworth screw threads. Screw thread systems were revised again in 1974 for metric application. The modifications were minor and based primarily on metric translation. In order to emphasize that the Unified screw threads evolve from inch calibrations, the term Unified Inch Screw Threads is used while the term Unified Screw Threads Metric Translation is used for the metric conversion.

Screw threads are a helix or conical spiral form on the external surface of a shaft or internal surface of a cylindrical hole, as shown in Figures 9–1 and 9–2. Screw threads are used for an unlimited number of services, such as for holding parts together as fasteners, for level-

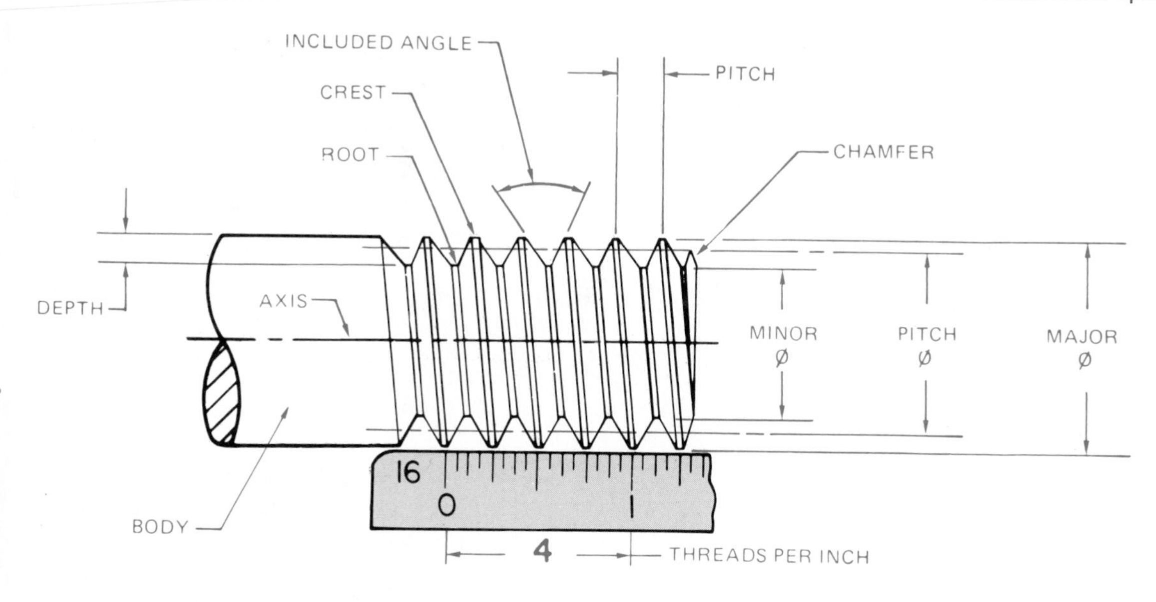

Figure 9–1 External screw thread components.

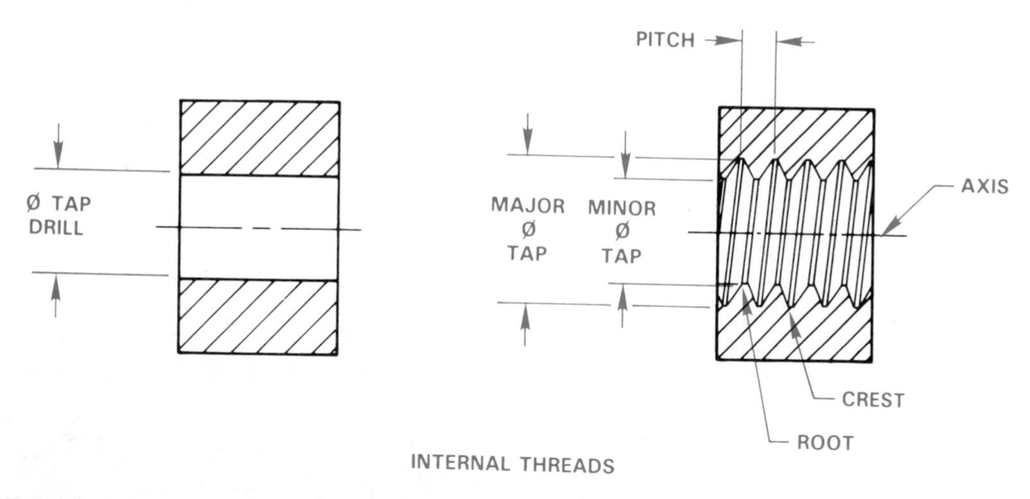

Figure 9–2 Internal screw thread components.

ing and adjusting objects, and for transmitting power from one object or feature to another.

Screw-thread Terminology

Refer to Figure 9–1 and Figure 9–2 as a reference for the following definitions related to external and internal threads.

Axis. The thread axis is the centerline of the cylindrical thread shape.

Body. That portion of a screw shaft that is left unthreaded.

Chamfer. An angular relief at the last thread to help allow the thread to more easily engage with a mating part.

Classes of Threads. A designation of the amount of tolerance and allowance specified for a thread.

Crest. The top of external and the bottom of internal threads.

Depth of Thread. Depth is the distance between the crest and the root of a thread, measured perpendicular to the axis.

Die. A machine tool used for cutting external threads.

Fit. Identifies a range of thread tightness or looseness.

Included Angle. The angle between the flanks (sides) of the thread.

Lead. The lateral distance a thread travels during one complete rotation.

Left-hand Thread. A thread that engages with a mating thread by rotating counterclockwise, or with a turn to the left when viewed toward the mating thread.

Major Diameter. The distance on an external thread from crest to crest through the axis. For an internal thread the major diameter is measured from root to root across the axis.

Minor Diameter. The dimension from root to root through the axis on an external thread and measured across the crests through the center for an internal thread.

Pitch. The distance measured parallel to the axis from a point on one thread to the corresponding point on the adjacent thread.

Pitch Diameter. A diameter measured from a point halfway between the major and minor diameter through the axis to a corresponding point on the opposite side.

Right-hand Thread. A thread that engages with a mating thread by rotating clockwise, or with a turn to the right when viewed toward the mating thread.

Root. The bottom of external and the top of internal threads.

Tap. A tap is the machine tool used to form an interior thread. Tapping is the process of making an internal thread.

Tap Drill. A tap drill is used to make a hole in metal before tapping.

Thread. The part of a screw thread represented by one pitch.

Thread Form. The design of a thread determined by its profile.

Thread Series. Groups of common major diameter and pitch characteristics determined by the number of threads per inch.

Threads per Inch. The number of threads measured in one inch. The reciprocal of the pitch in inches.

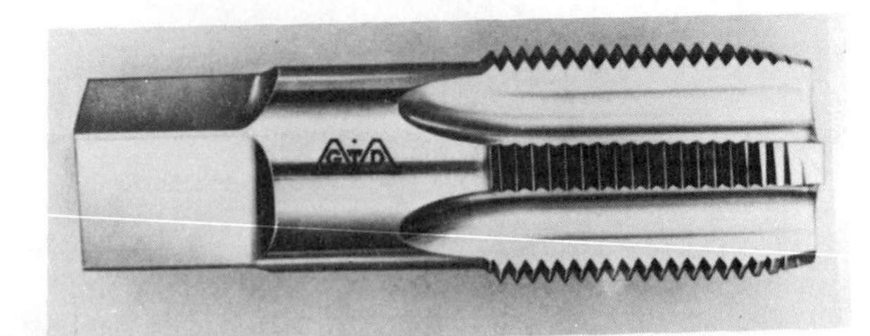

Figure 9–3 Tap. *Courtesy Greenfield Tap & Die, Division of TRW, Inc.*

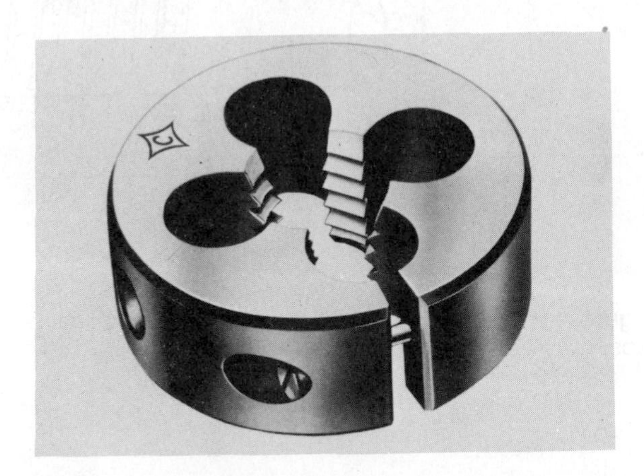

Figure 9–4 Die. *Courtesy Cleveland Twist Drill, an Acme-Cleveland Company.*

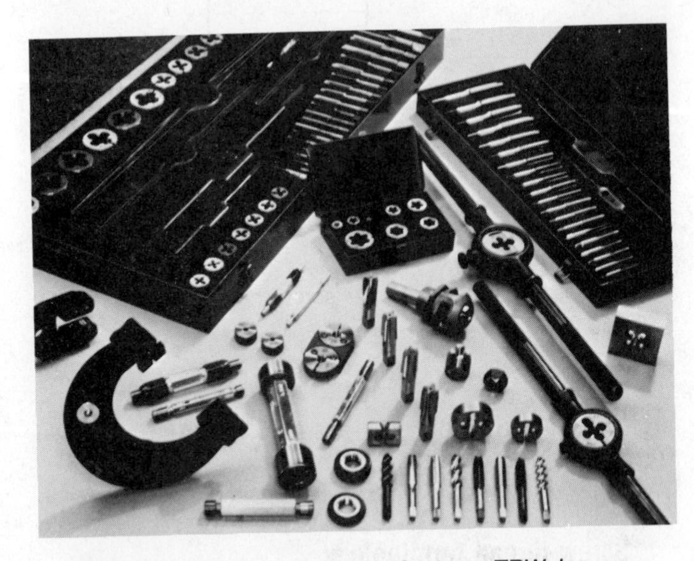

Figure 9–5 Tap and die tools. *Courtesy TRW, Inc.*

THREAD-CUTTING TOOLS

The tap is a machine tool used to form an internal thread as shown in Figure 9–3. A die is a machine tool used to form external threads. (See Figure 9–4.) Figure 9–5 shows a variety of available tap and die tools.

A tap set is made up of a paper tap, a plug tap, and a bottoming tap as shown in Figure 9–6. The taper tap is generally used for starting a thread. The threads are tapered to within ten threads from the end. The tap is tapered so that the tool more evenly distributes the cutting edges through the depth of the hole. The plug tap has the threads tapered to within five threads from the end. The plug tap can be used to completely thread through material or thread a blind hole (a hole that does not go through the material) if full threads are not required all the way to the bottom. The bottoming tap is used when threads are needed to the bottom of a blind hole.

The die is a machine tool used to cut external threads. Thread cutting dies are available for standard thread sizes and designations.

External and internal threads may also be cut on a lathe. A lathe is a machine that holds a piece of material between two centers or in a chucking device. The material is rotated as a cutting tool removes material while traversing along a carriage which slides along a bed. Figure 9–7 shows how a cutting tool can make an external thread.

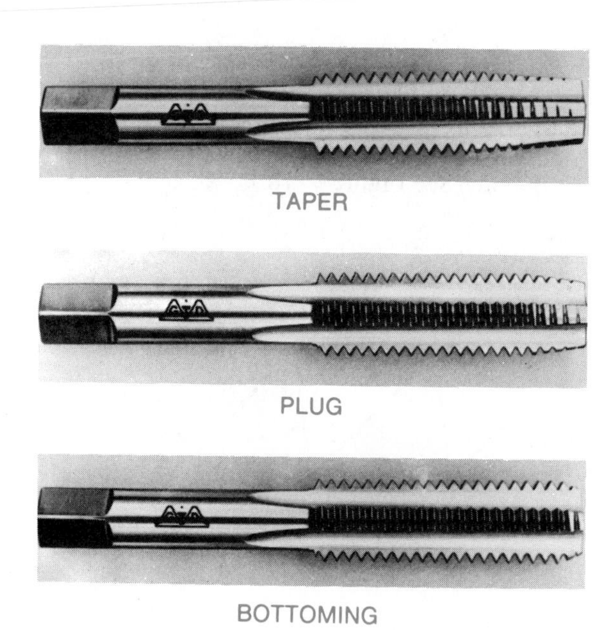

Figure 9–6 Tap set includes taper, plug, and bottoming taps. *Courtesy Greenfield Tap & Die, Division of TRW, Inc.*

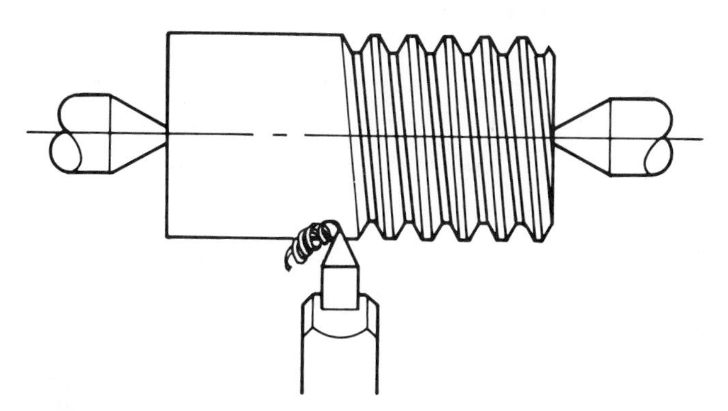

Figure 9–7 Thread cutting on a lathe.

THREAD FORMS

Unified threads are the most common threads used on threaded fasteners. Figure 9–8 shows the profile of a Unified thread.

American National threads, shown in profile in Figure 9–9, are similar to the Unified thread but have a flat root. Still in use today, the American National thread generally replaced the sharp-V thread form.

The *sharp-V thread*, although not commonly used, is a thread that will fit and seal tightly. It is difficult to manufacture since the sharp crests and roots of the threads are easily damaged. (See Figure 9–10.) The sharp-V thread was the original United States standard thread form.

Metric thread forms vary slightly from one European country to the next. The International Organization for Standardization (ISO) was established to standardize metric screw threads. The ISO thread specifications are similar to the Unified thread form. (See Figure 9–11.)

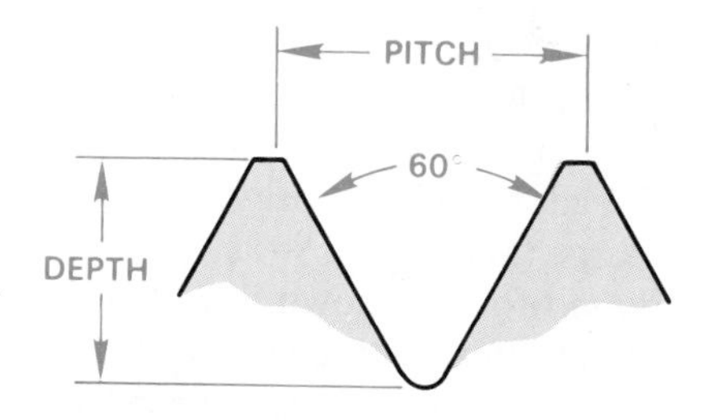

Figure 9–8 Unified thread form.

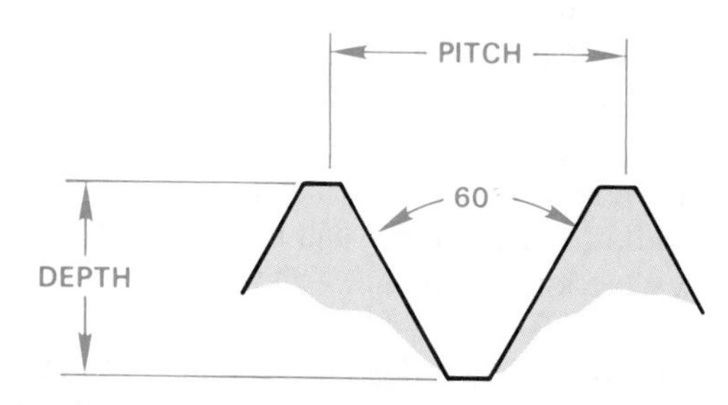

Figure 9–9 American National thread form.

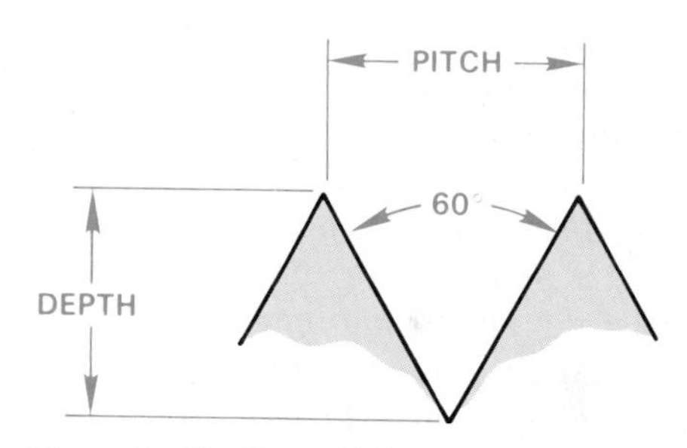

Figure 9–10 Sharp-V thread form.

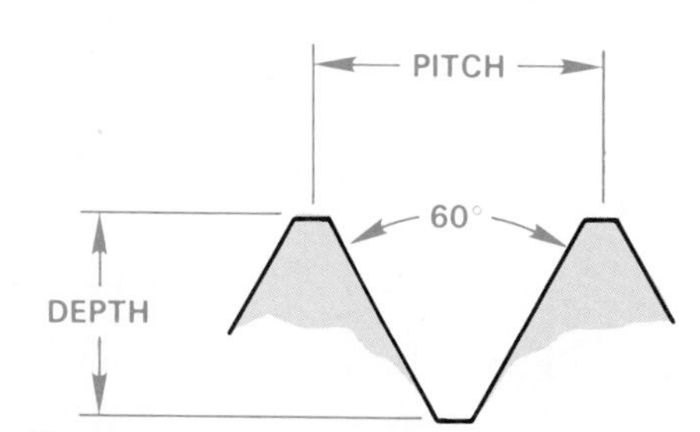

Figure 9–11 Metric thread form.

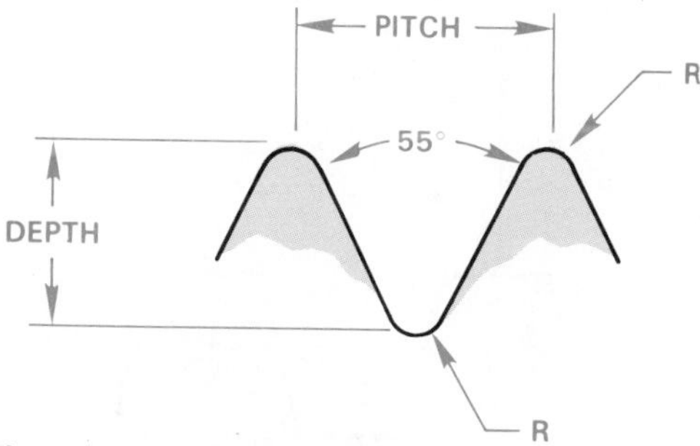

Figure 9–12 Whitworth thread form.

Whitworth threads are the original British standard thread forms developed in 1841. These threads have been referred to as parallel screw threads. The Whitworth thread forms are primarily being used for replacement parts. (See Figure 9–12.)

Square thread forms, shown in Figure 9–13, have a longer pitch than Unified threads. Square threads were developed as threads that would effectively transmit power; however, they are difficult to manufacture because of their perpendicular sides. There are modified square threads with ten-degree sides. The square thread is generally replaced by Acme threads.

Acme thread forms are commonly used when rapid traversing movement is a design requirement. Acme threads are popular on such designs as screw jacks, vice screws, and other equipment and machinery that requires rapid screw action. A profile of the Acme thread form is shown in Figure 9–14.

Buttress threads are designed for applications where high stress occurs in one direction along the thread axis. The thread flank or side that distributes the thrust or force is within 7° of perpendicularity to the axis. This helps reduce the radial component of the thrust. The buttress thread is commonly used in situations where tubular features are screwed together and lateral forces are exerted in one direction. (See Figure 9–15.)

Dardelet thread forms are primarily used in situations where a self-locking thread is required. These threads resist vibrations and remain tight without auxiliary locking devices. (See Figure 9–16.)

Rolled thread forms are used for screw shells of electric sockets and lamp bases. (See Figure 9–17.)

American National Standard taper pipe threads are the standard threads used on pipes and pipe fittings. These threads are designed to provide pressure-tight joints or not, depending on the intended function and materials used. American pipe threads are measured by the nominal pipe size, which is the inside pipe diameter. For example, a ½-in. pipe size has an outside pipe diameter of .840 in. (See Figure 9–18.)

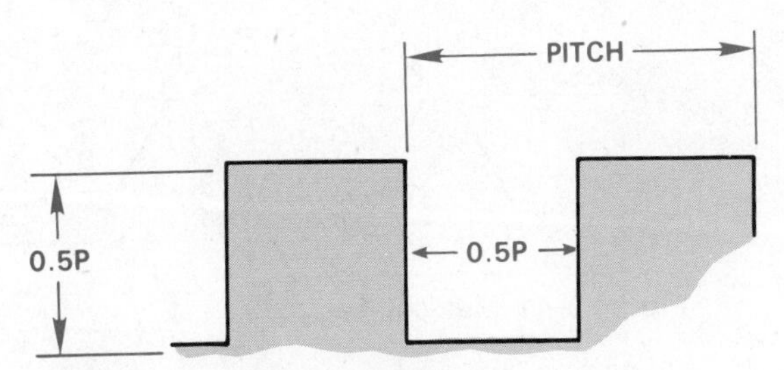

Figure 9–13 Square thread form.

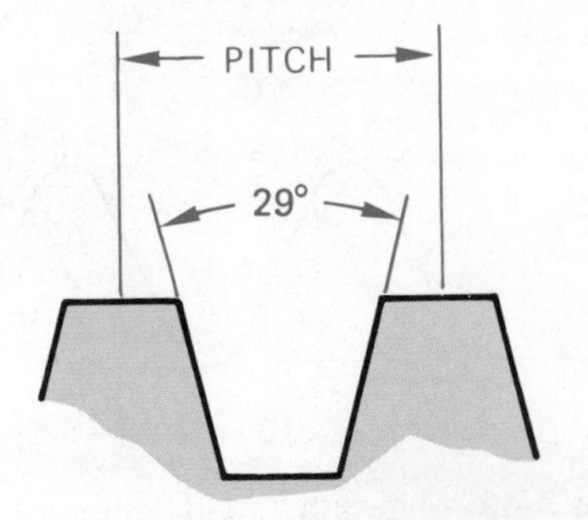

Figure 9–14 Acme thread form.

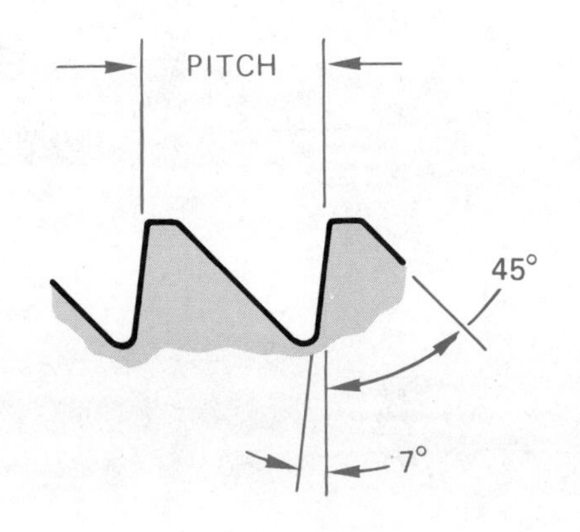

Figure 9–15 Buttress thread form.

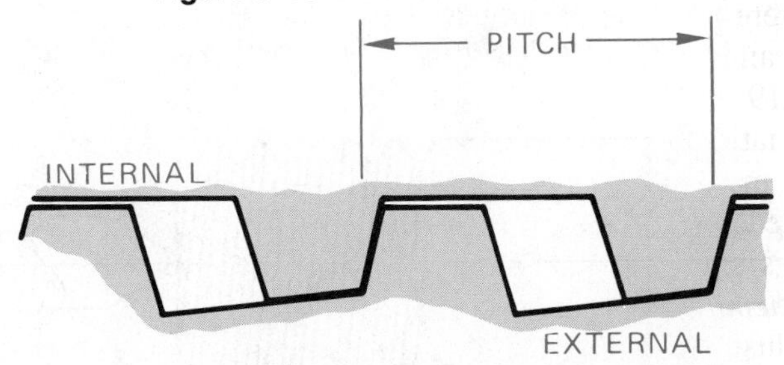

Figure 9–16 Dardelet self-locking thread form.

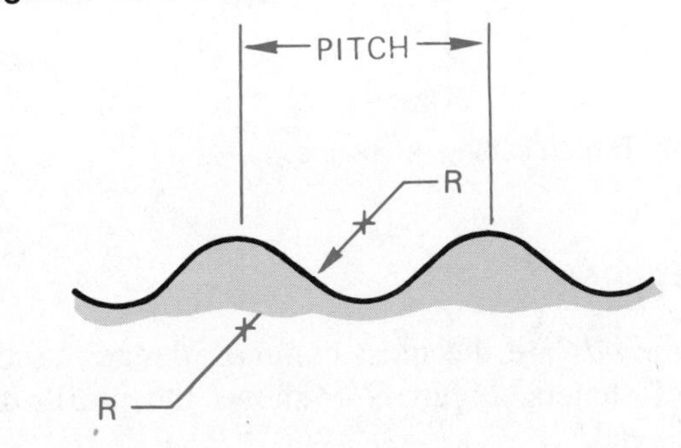

Figure 9–17 Rolled thread form.

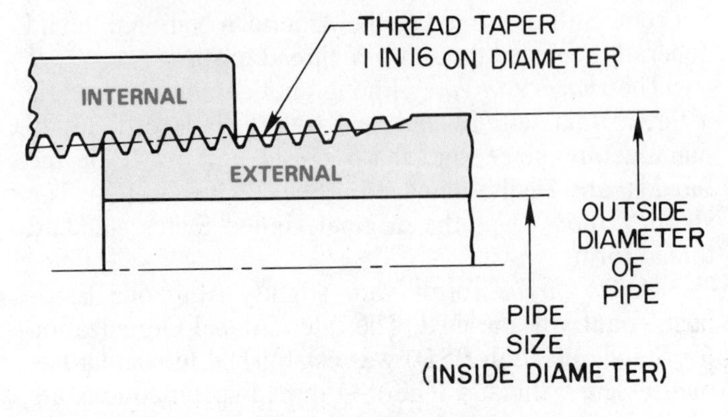

Figure 9–18 American National Standard taper pipe thread.

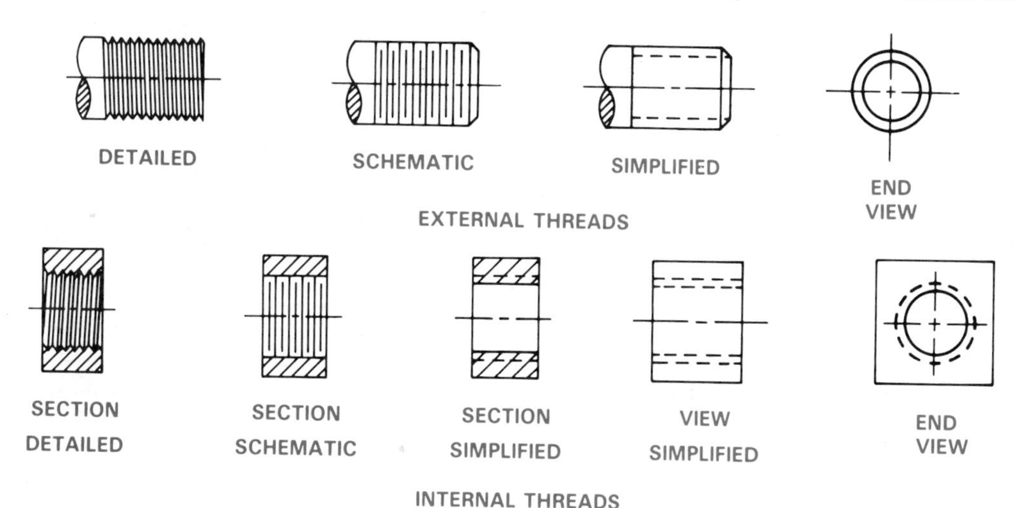

Figure 9–19 Thread representations.

THREAD REPRESENTATIONS

There are three methods of thread representation in use: detailed, schematic, and simplified, as shown in Figure 9–19. The *detailed* representation may be used in special situations that require a pictorial display of threads such as in a sales catalog, or a display drawing. Detailed threads are not common on most manufacturing drawings since they are much too time-consuming to draw. *Schematic* representations are also not commonly used in industry. Although they do not take the time of detailed symbols, they do require extra time to draw. Some companies, however, may continue to use the schematic thread representation.

The actual use and purpose of the drawing will often help determine which thread symbol to use. It may even be possible to mix representations on a particular drawing if clarity is improved. The simplified representation is the most common method of drawing thread symbols. *Simplified* representations clearly describe threads and they are easy and quick to draw. They are also very versatile as they can be used in all situations, while the other representations cannot be used in all situations. Figure 9–20 shows simplified threads in different applications.

Also, notice how the use of a thread chamfer slightly changes the appearance of the thread. Chamfers are commonly applied to the first thread to help accommodate the start of a thread in its mating part.

When an internal screw thread does not go through the part, it is common to drill deeper than the depth of the required thread when possible. This process saves time and reduces the chance of breaking a tap. The thread may go to the bottom of a hole but to produce it requires an extra process using a bottoming tap. Figure 9–21 shows a simplified representation of a thread that does not go through. The bolt should be shorter than the depth of thread so that the bolt does not hit bottom. Notice in Figure 9–21 that the hidden lines representing the major and minor thread diameters are spaced far enough apart to be clearly separate. This spacing is important because

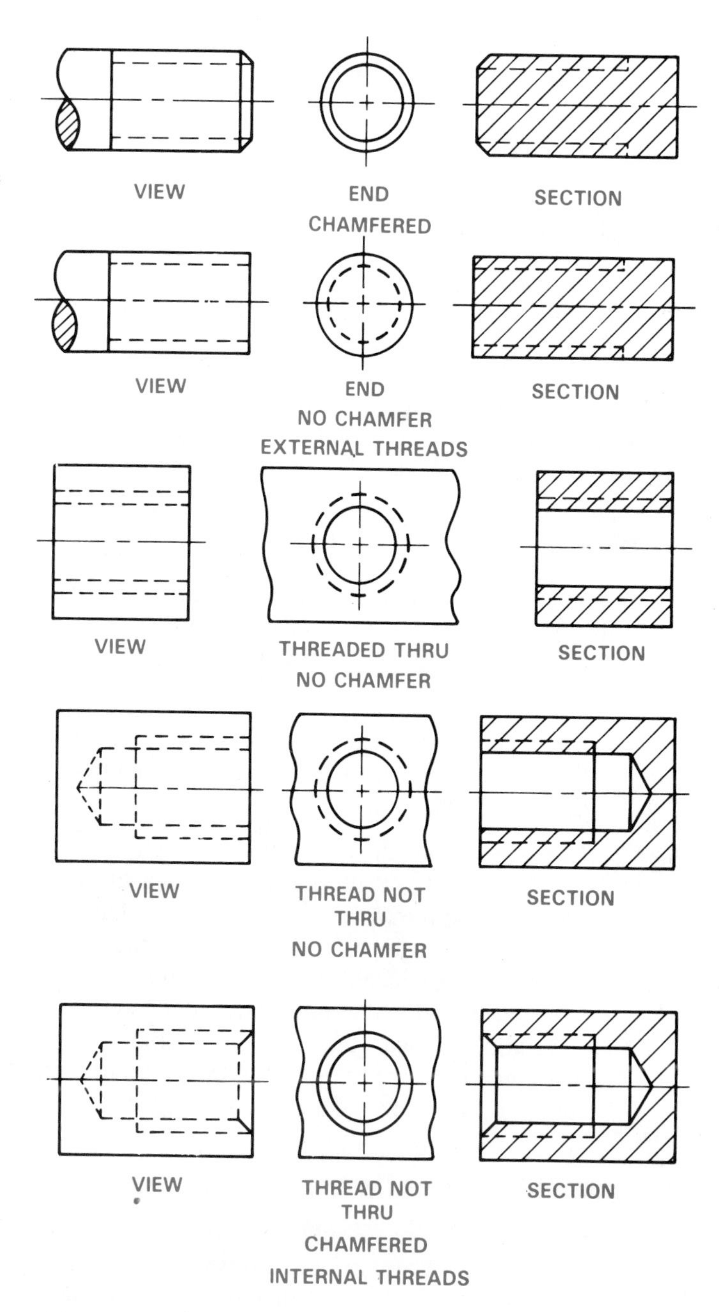

Figure 9–20 Simplified thread representations.

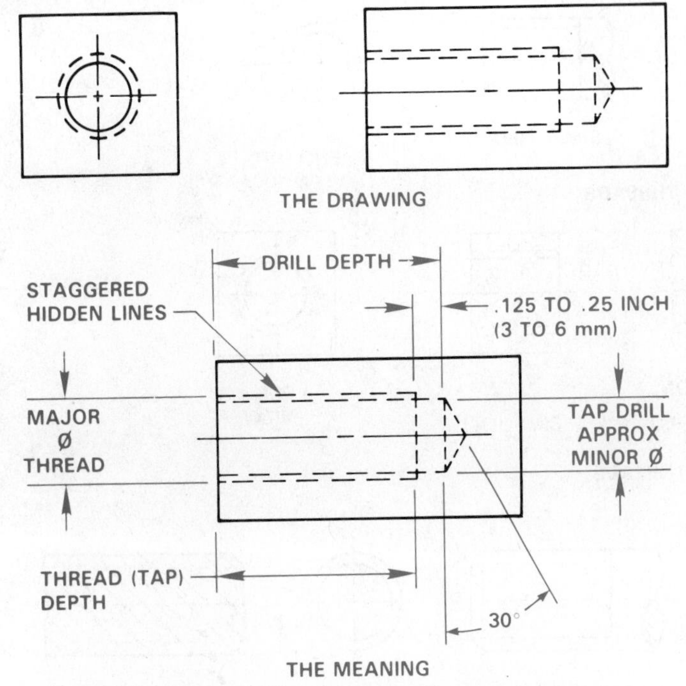

Figure 9–21 The simplified internal thread that does not go through the part.

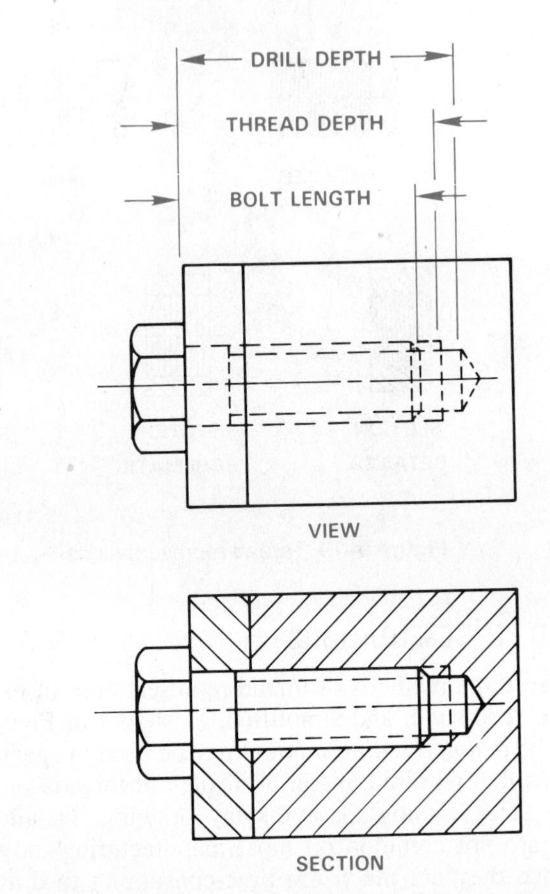

Figure 9–22 The simplified external thread in assembly.

on some threads the difference between the major and minor diameters is very small and, if drawn as they actually appear, the lines would run together. The hidden-line dashes are also drawn staggered for clarity. Figure 9–22 shows a bolt fastener as it would appear drawn in assembly with two parts using simplified thread representation.

Drawing Simplified Threads

Simplified representations are the easiest thread symbols to draw and are the most commonly used in industry. The following steps show how to draw simplified threads.

Step 1 Draw the major thread diameter as an object line for external threads and hidden line for internal threads as in Figure 9–23.

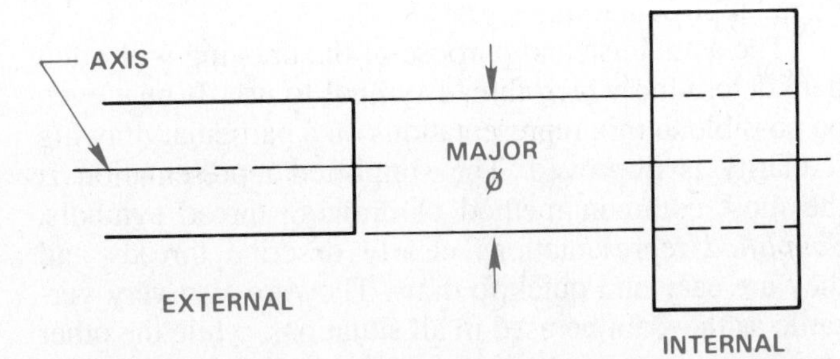

Figure 9–23 The simplified major thread diameter — step 1.

Step 2 Draw the minor thread diameter, which is about equal to the tap drill size (found in a tap drill chart). If the minor diameter and major diameter are too close together, then exaggerate the space. The minor diameter is a hidden line for the external thread and a hidden line staggered with the major diameter lines for the internal thread. The minor diameter is an object line for the internal thread in section. (See Figure 9–24.)

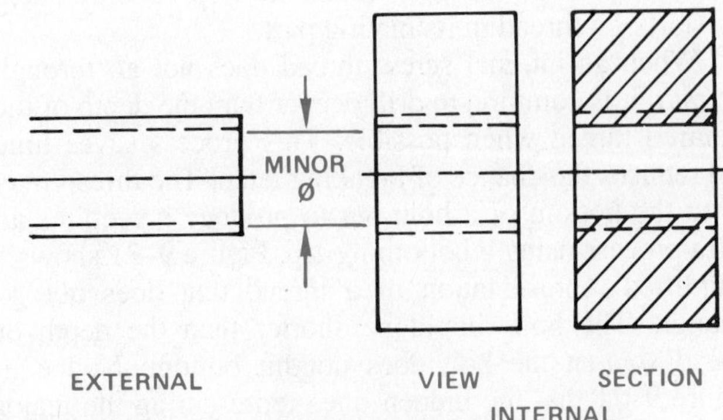

Figure 9–24 The simplified minor diameter — step 2.

Drawing Schematic Threads

Schematic thread representations are drawn to approximate the appearance of threads by spacing lines equal to the pitch of the thread. When there are too many

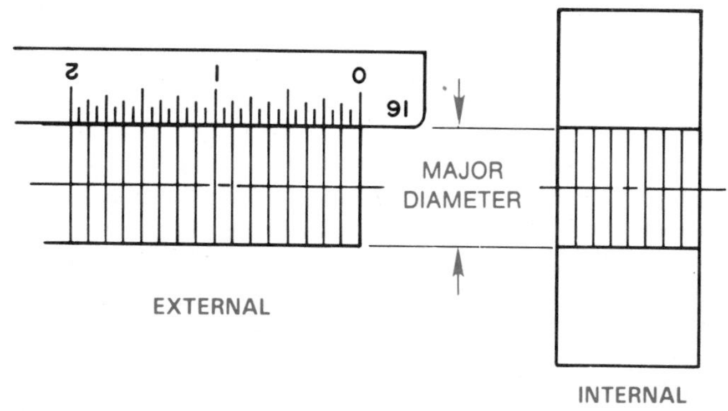

Figure 9–25 The schematic thread — step 1.

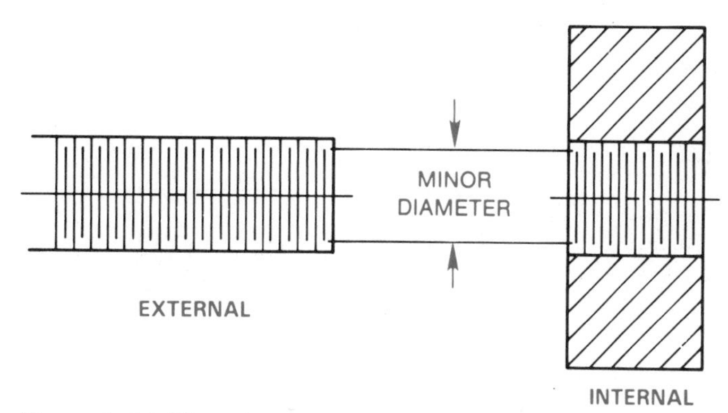

Figure 9–26 The schematic thread — step 2.

threads per inch to draw easily, then the distance can be exaggerated for clarity. The following steps show how to draw schematic thread representations:

Step 1 Draw the major diameter of the thread. Schematic symbols can only be drawn in section for internal threads. Then lay out the number of threads per inch (if convenient and space is available) using a thin line at each space. Figure 9–25 uses eight threads per inch, which lays out easily with the 16 edge of the Architect's scale.

Step 2 Draw a thick line equal in length to the minor diameter between each pair of thin lines drawn in Step 1. (See Figure 9–26.)

Drawing Detailed Threads

Detailed thread representations are the most difficult and time consuming thread symbols to draw. They may be necessary for some applications as they most closely approximate the actual thread. Detailed external and sectioned threads may be drawn but detailed internal threads may not be drawn in multiview. Detailed internal threads may only be drawn in section. The following steps show how to draw detailed thread representations for external and internal threads in section.

Step 1 Use construction lines to lightly draw the major and minor diameters of the thread as shown in Figure 9–27.

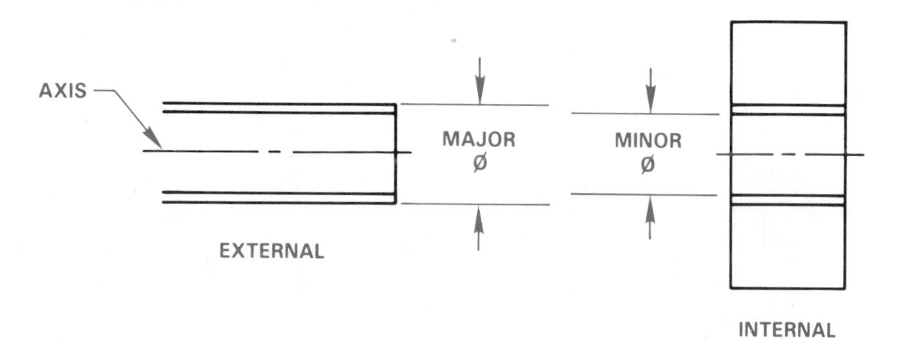

Figure 9–27 The detailed thread — step 1.

Step 2 Divide one edge of the thread into equal parts. In this case we have eight threads per inch so the pitch is .125 in., as shown in Figure 9–29. If the pitch is too small (much less than .125 in.) then exaggerate the distance. Remember, these symbols are representations and while they should have an appearance close to the actual thread, they do not have to be exact if the result would be too difficult and time consuming to draw.

Step 3 Stagger the opposite side one-half pitch and draw parallel thin lines equal to the spaces established in Step 2. (See Figure 9–30.)

Step 4 Draw Vs at 60° to form the root and crest of each thread. Set your vertical machine scale at 30° away from vertical to draw the thread sides. (See Figure 9–31.)

Step 5 Complete the detailed thread representation by connecting the roots of opposite threads by drawing parallel lines as shown in Figure 9–32. Accuracy is very important if detailed threads are to turn out satisfactory.

Detailed thread representations may be used to draw any thread form using the same steps previously shown. The difference occurs in drawing the profile of the

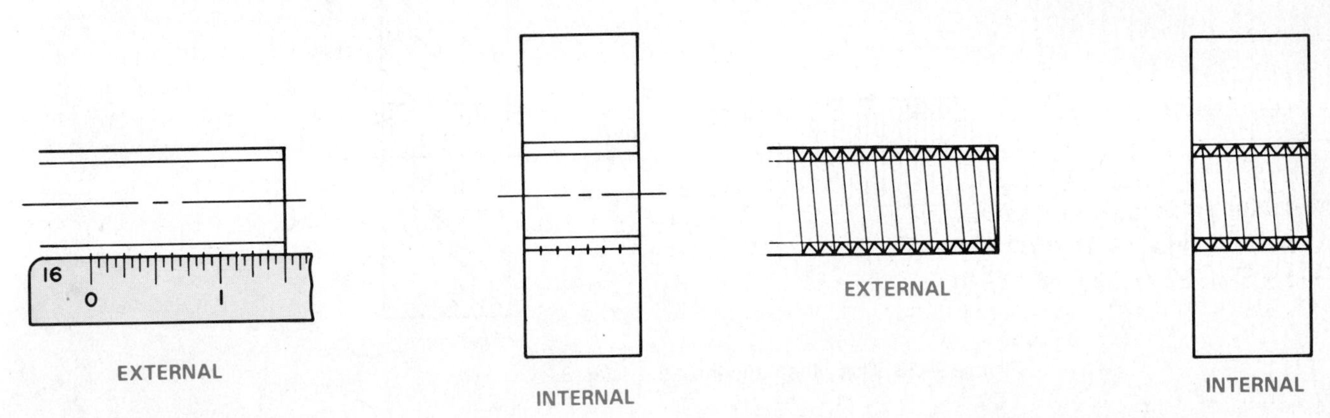

Figure 9–28 The detailed thread — step 2.

Figure 9–30 The detailed thread — step 4.

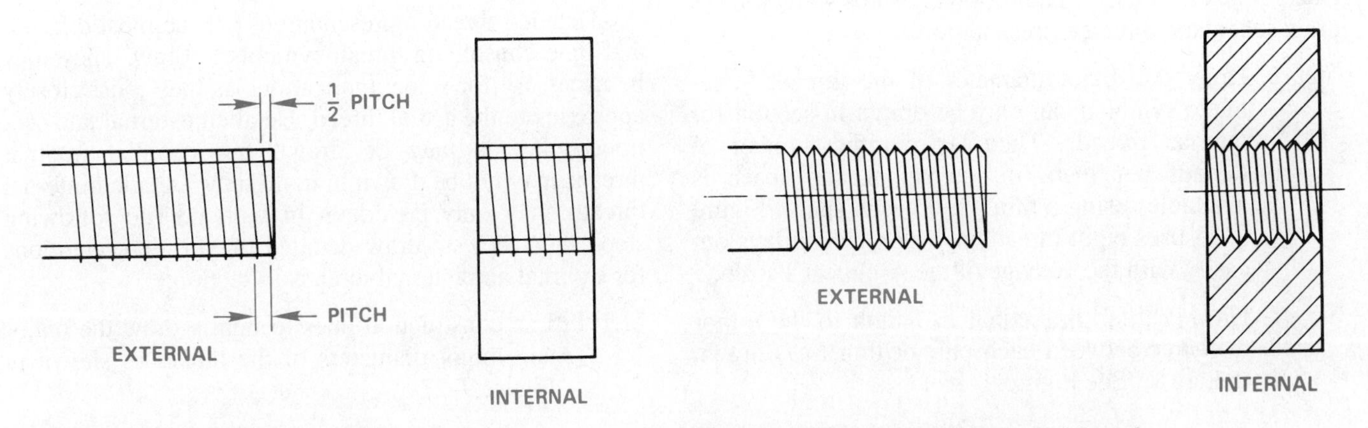

Figure 9–29 The detailed thread — step 3.

Figure 9–31 The complete detailed thread representation — step 5.

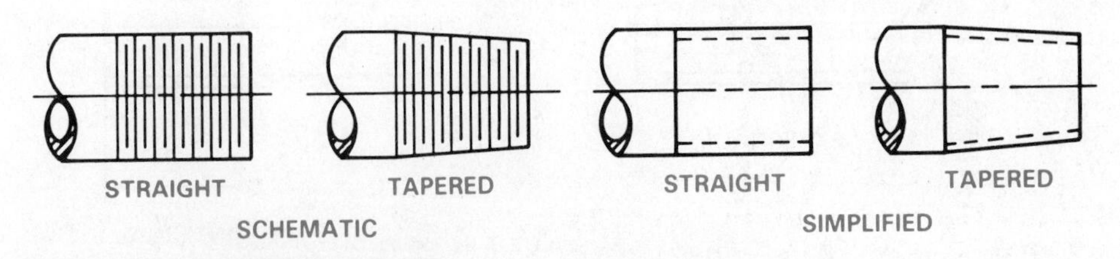

Figure 9–32 Straight and tapered pipe thread representations.

thread. For example, Vs are used to draw Unified, sharp-V, American Standard, or metric threads. The profile will change for other threads as displayed in Figures 9–12 to 9–17. The detailed thread drawings for American National Standard taper pipe threads are the same as Unified threads except that the major and minor diameters taper at a rate of .0625 in. per inch. Figure 9–33 shows that pipe threads may be drawn tapered or straight depending on company preference. The thread note will clearly define the type of thread.

THREAD NOTES

Simplified, schematic, and detailed thread representations clearly show where threads are displayed on a drawing. However, the representations alone do not give the full information about the thread. As the term representation implies, the symbols are not meant to be exact but they are meant to describe the location of a thread when used. The information that clearly and completely identifies the thread being used is the *thread note*. The thread note must be accurate, otherwise the thread will be manufactured incorrectly.

Metric Threads

ISO The metric thread notes shown in Figure 9–33 are the recommended standard as specified by the International Organization for Standardization (ISO). The note components are described below:

M 10 X 1.5—6 H
(A) (B) (C) (D) (E) (F)

(A) M denotes die symbol for ISO metric threads.
(B) The nominal major diameter in millimeters, followed by the symbol X meaning, *by*.
(C) The thread pitch in millimeters, followed by a dash (—).
(D) The number may be a 3, 4, 5, 6, 7, 8, or 9, which identifies the grade of tolerance from fine to coarse. The larger the number, the larger the tolerance. Grades 3 through 5 are fine, and 7 through 9 are coarse. Grade 3 is very fine and grade 9 is very coarse. Grade 6 is the most commonly used and is the medium tolerance metric thread. The grade 6 metric thread is comparable to the class 2 Unified Screw Thread. A letter placed after the number denotes the thread tolerance class of the internal or external thread. Internal threads are designated by upper-case letters, such as G or H, where G denotes a tight allowance and H identifies an internal thread with no allowance. The term allowance refers to the tightness of fit between the mating parts. External threads are defined with lower-case letters such as e, g, or h. For external threads, e denotes a large allowance, g is a tight allowance, and h establishes no allowance. Grades and tolerances below 5 are intended for tight fits with mating parts; those above 7 are a free class of fit intended for quick and easy assembly. When the grade and allowance are the same for both the major diameter and the pitch diameter of the metric thread, then the designation is given as shown, 6H. In some situations where precise tolerances and allowances are critical between the major and pitch diameters, separate specifications could be used, for example 4H 5H, or 4g 5g, where the first group (4g) refers to the grade and allowance of the pitch diameter, and the second group (5g) refers to the grade and tolerance of the major diameter. A fit between a pair of threads is indicated in the same thread note by specifying the internal thread followed by the external thread specification separated by a slash, for example, 6H/6g.
(E) A blank space at (E) denotes a right-hand thread, a thread that engages when turned to the right. A right-hand thread is assumed unless an LH is lettered in this space. LH, which describes a left-hand thread, must be specified for a thread that engages when rotated to the left.
(F) The depth of internal threads or the length of external threads in millimeters is provided at the end of the note. When the thread goes through the part, this space is left blank although some companies prefer to letter the description, THRU.

Remember, the thread note must always be lettered in the order shown in Figure 9–33.

Unified and American National Threads

The thread note shall always be drawn in the order shown in Figure 9–34. The components of the note are described below:

.50—13 UNC—2 A
(A) (B) (C) (D)(E)(F)(G) (H)

(A) The major diameter of the thread in inches followed by a dash (—).

M 10 X 1.5 - 6H
(A) (B) (C) (D) (E) (F)

Figure 9–33 Metric thread note.

.50 - 13 UNC - 2 A
(A) (B) (C) (D)(E) (F) (G) (H)

Figure 9–34 Unified and American National thread note.

(B) Number of threads per inch.

(C) Series of threads are classified by the number of threads per inch as applied to specific diameters and thread forms, such as coarse or fine threads. UNC (in the example) means Unified National Coarse. Others include UNF for Unified National Fine, UNEF for Unified National Extra Fine, or UNS for Unified National Special. The UNEF and UNS thread designations are for special combinations of diameter, pitch, and length of engagement. American National screw threads are identified with UN for external and internal threads, or UNR, a thread designed to improve fatigue strength of external threads only. The series designation is followed by a dash (—).

(D) Class of fit is the amount of tolerance. 1 means a large tolerance, 2 is a general-purpose moderate tolerance, and 3 is for applications requiring a close tolerance.

(E) A means an external thread (shown in the example) while B means an internal thread. (B replaces A in this location.) The A or B may be omitted if the thread is clearly external or internal, as shown on the drawing.

(F) A blank space at (F) denotes a right-hand thread. A right-hand thread is assumed. LH in this space identifies a left-hand thread.

(G) A blank space at (G) denotes a thread with a single lead, that is, a thread that engages one pitch when rotated 360°. If a double or triple lead is required, then the word DOUBLE or TRIPLE must be lettered here.

(H) This location is for internal thread depth or external thread length in inches. When the drawing clearly shows that the thread goes through, this space is left blank. If clarification is needed, then the word THRU may be lettered here.

Other Thread Forms

Other thread forms, such as Acme, are noted on a drawing using the same format. For example, ⅝—8 ACME—2G describes an Acme thread with a ⅝-in. major diameter, 8 threads per inch, and a general purpose (G) class 2 thread fit. For a complete analysis of threads and thread forms refer to the *Machinery's Handbook* published by Industrial Press, Inc.

American National Standard taper pipe threads are noted in the same manner with the letters NPT used to designate the thread form. A typical note may read ¾—14 NPT.

Thread Notes on a Drawing

The thread note is usually applied to a drawing with a leader in the view where the thread appears as a circle for internal threads as shown in Figure 9–35. External threads may be dimensioned with a leader as shown in Figure 9–36, with the thread length given as a dimension or at the end of the note. An internal thread that does not go through the part may be dimensioned as in Figure 9–37. Some companies may require the drafter to indicate the complete process required to machine a thread

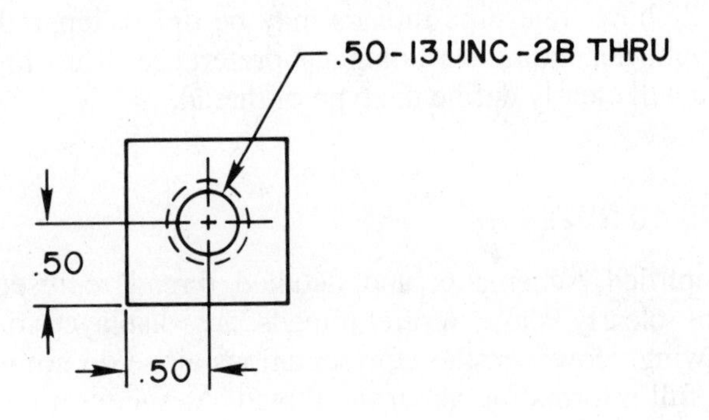

UNIFIED SCREW THREAD

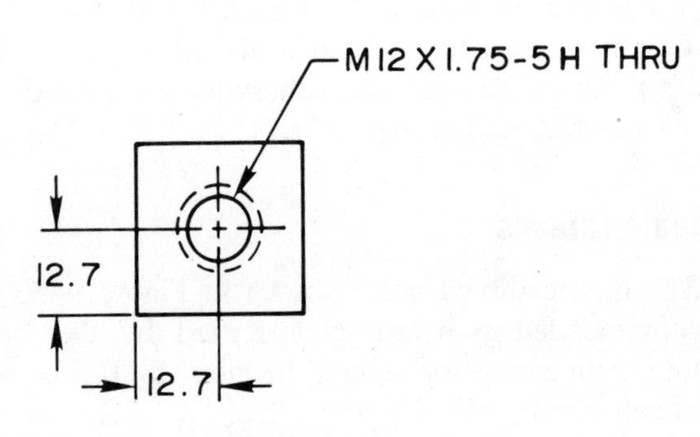

METRIC SCREW THREAD

Figure 9–35 Drawing and noting internal screw threads.

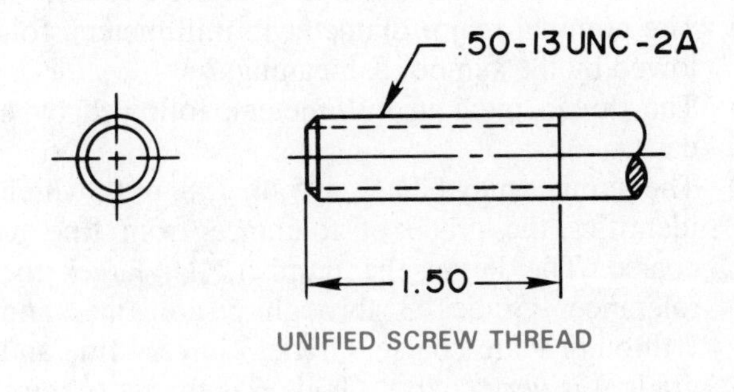

UNIFIED SCREW THREAD

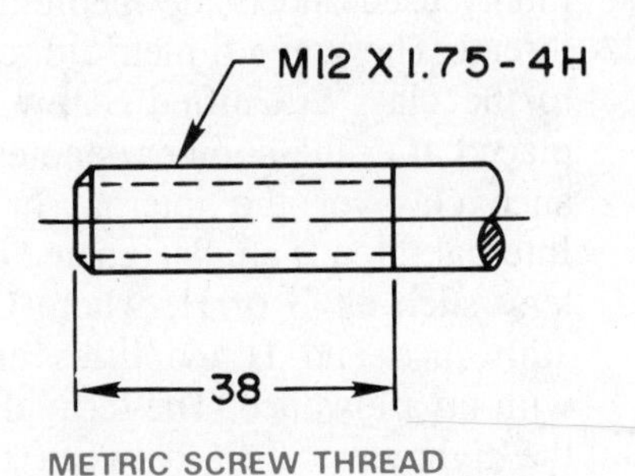

METRIC SCREW THREAD

Figure 9–36 Drawing and noting external screw threads.

including noting the tap drill size, tap drill depth if not through, the thread note, and thread depth if not through. (See Figure 9–38.) A thread chamfer may also be specified in the note as shown in Figure 9–39.

MEASURING SCREW THREADS

When measuring features from prototypes or existing parts, the screw thread size can be determined on a fastener or threaded part by measurement. Measure the major diameter with a vernier caliper or micrometer. Determine the number of threads per inch when a rule or scale is the only available tool by counting the number of threads between inch graduations, as shown in Figure 9–40. The quickest and easiest way to determine the thread specification is with a screw pitch gage, Figure 9–41, which is a set of thin leaves with teeth on the edge of each leaf that correspond to standard thread sections. Each leaf is stamped to show the number of threads per inch. Therefore, if the major diameter measures .625 in. and the number of teeth per inch is 18, then by looking at a thread variation chart you find that you have a .625—18 UNF thread.

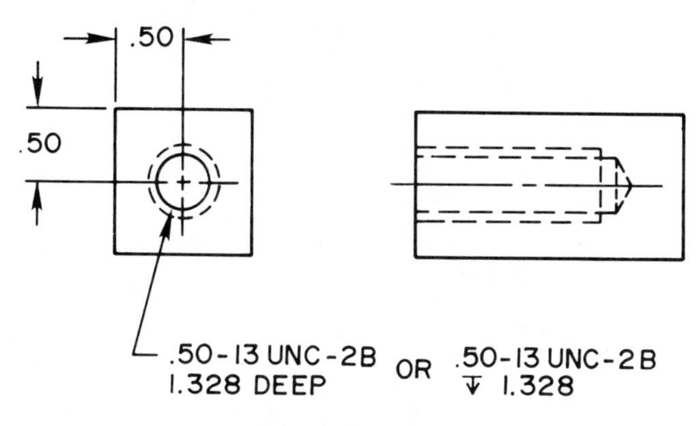

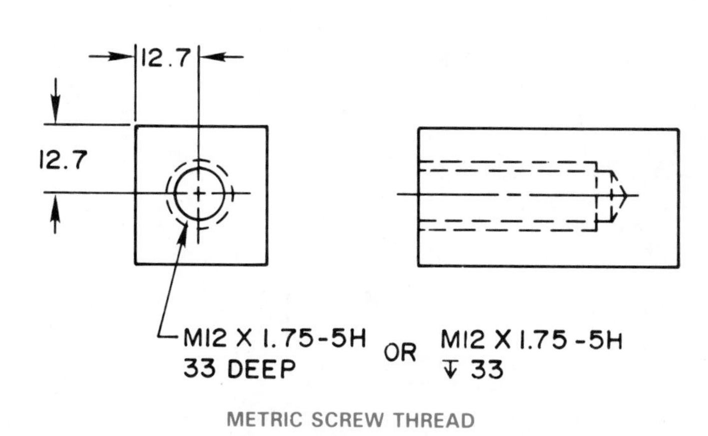

Figure 9–37 Drawing and noting internal screw threads with a given depth.

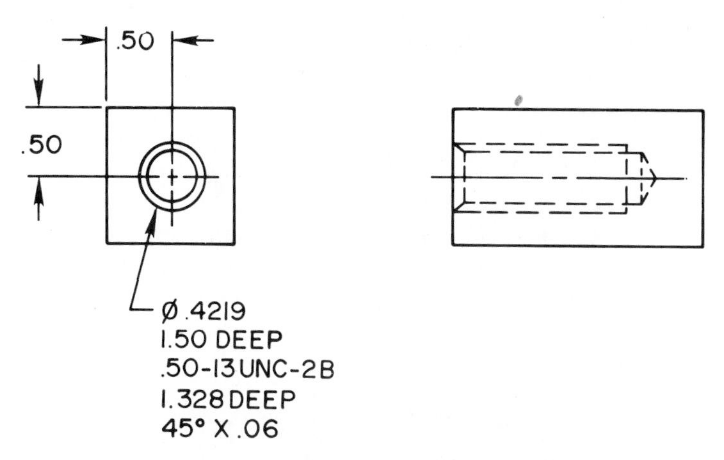

Figure 9–39 Showing drill and thread depth with a chamfer.

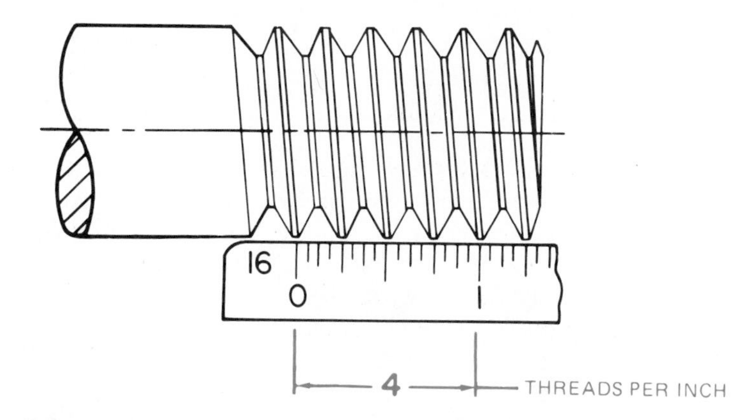

Figure 9–40 Measuring threads with a steel rule.

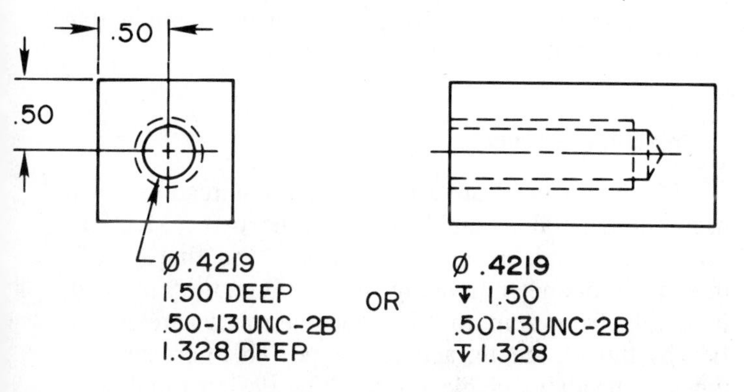

Figure 9–38 Showing drill depth and thread depth.

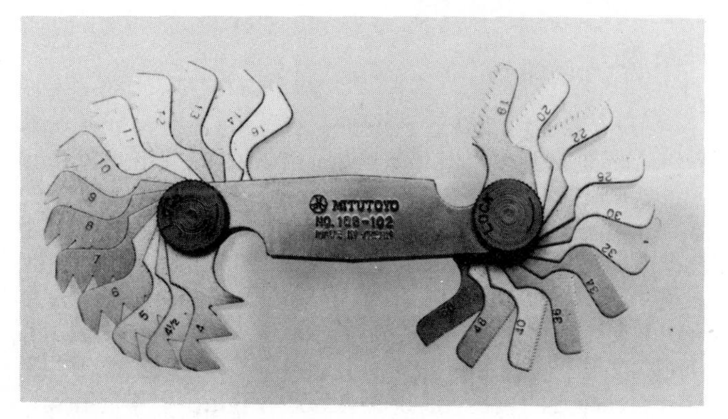

Figure 9–41 Screw pitch gage. *Courtesy MTI Corporation.*

DRAWING FASTENERS

The use of simplified thread representations in manual drafting is mostly due to ease and time savings, while schematic thread representations are used to add a little realism to drawings. Detailed thread representations, on the other hand, are usually avoided unless a specific reason requires this time-consuming drawing method. With CADD the drafter has any type of thread representation available in an instant. And with this flexibility CADD drafters often select the detailed thread because it makes a drawing look more realistic and artistic. However, choosing simplified or schematic thread representations still make sense from an economic standpoint, because screen regeneration and plotting time are less. CADD fastener software allows the user to select the following variables:

- Type of screw, such as machine, cap, or self-tapping.
- Type of nut, if an internal threaded feature.
- Type of head, if any.
- Type of thread representation: simplified, schematic, or detailed.
- Threading of holes or studs is also an option.

After selecting these variables, the drafter gives the thread specifications. The software uses parametrics based on these variables to draw and label the exact thread required in a fraction of the time it would take to manually draw the same fastener. A sample CADD fastener overlay template is shown in Figure 9–28.

Figure 9–42 A CADD fastener template overlay symbols library. *Courtesy Chase Systems.*

THREADED FASTENERS

Bolts and Nuts

A *bolt* is a threaded fastener with a head on one end and is designed to hold two or more parts together with a nut or threaded feature. The *nut* is tightened upon the bolt or the bolt head may be tightened into a threaded feature. Bolts can be tightened or released by torque applied to the head or to the nut. Bolts are identified by a thread note, length, and head type. For example, 5/8—11 UNC—2 X 1½ long hexagon head bolt. Figure 9–43 shows various types of bolt heads. Figure 9–44 shows common types of nuts. Nuts are classified by thread specifications and type. Nuts are available with a flat base or a washer face.

Machine Screws

Machine screws are a popular screw-thread fastener used for general assembly of machine parts. Machine screws are available in coarse (UNC) and fine (UNF) threads, in diameters ranging from .060 in. to .5 in., and in lengths from 1/8 in. to 3 in. Machine screws are specified by thread, length, and head type. There are several types of heads available for machine design flexibility. (See Figure 9–45.)

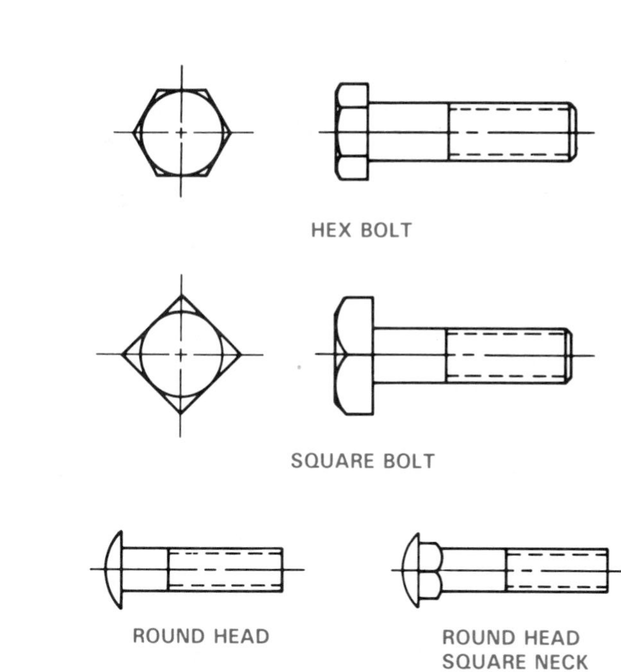

Figure 9–43 Bolt head types.

Cap Screws

Cap screws are fine-finished machine screws that are generally used without a nut. Mating parts are fastened where one feature is threaded. Cap screws have a variety of head types and range in diameter from .060 in. to 4 in., with a large range of lengths. Lengths vary with diameter; for example, lengths increase in 1/16 in. increments for diameters up to 1 in. For diameters larger than 1 in., lengths increase in increments of 1/8 in. or 1/4 in.; the other extreme is a 2-in. increment for lengths over 10 in. Cap screws have a chamfer to the depth of the first thread. Standard cap screw head types are shown in Figure 9–46.

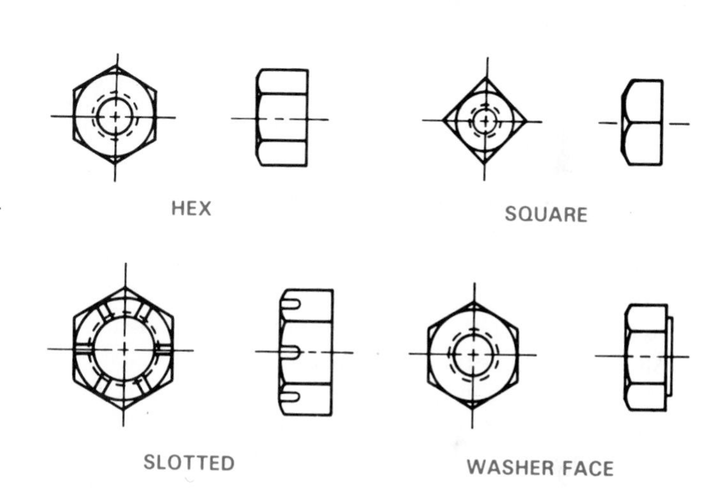

Figure 9–44 Types of nuts.

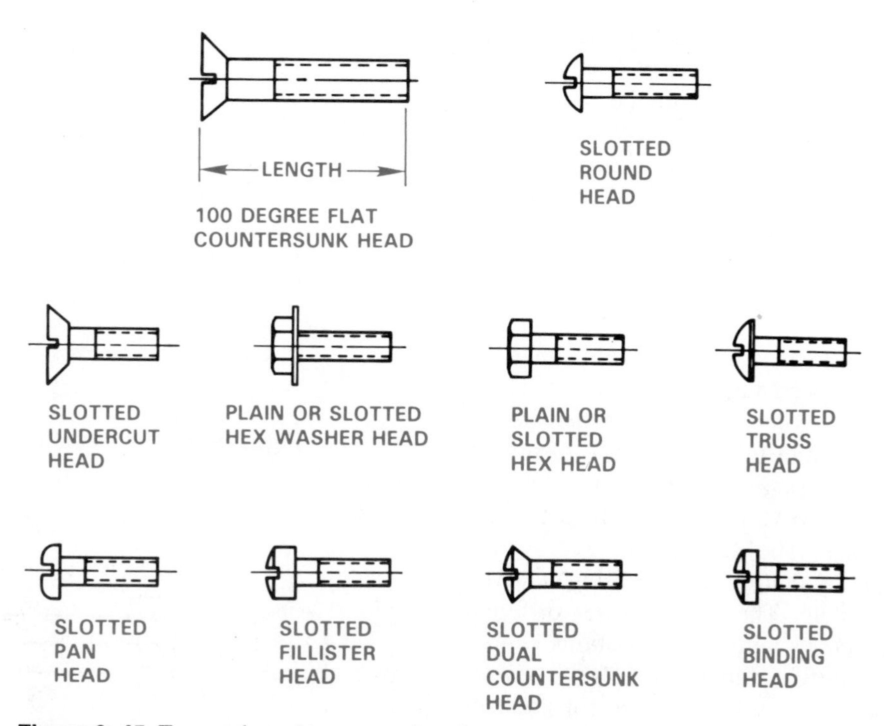

Figure 9–45 Types of machine screw heads.

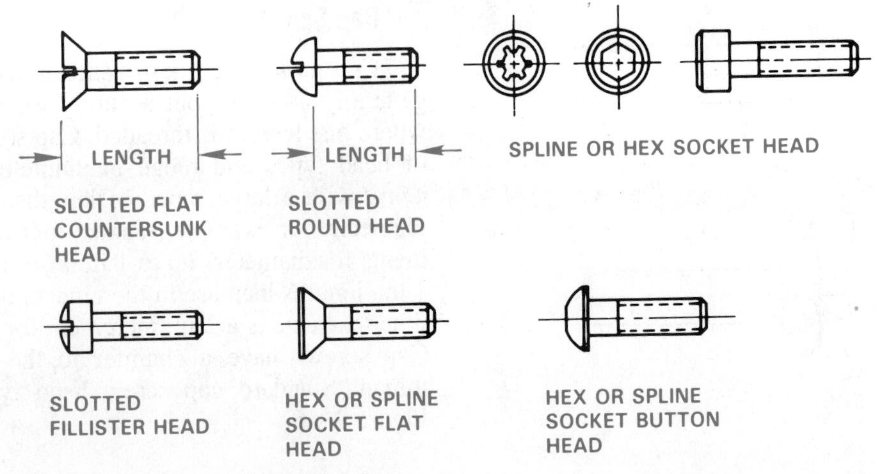

Figure 9–46 Cap screw head styles.

Set Screws

Set screws are used to help prevent rotary motion and to transmit power between two parts such as a pulley and shaft. Purchased with or without a head, set screws are ordered by specifying thread, length, head or headless, and type of point. Headless set screws are available in slotted and with hex or spline sockets. The shape of a set screw head is usually square. Standard square-head set screws have cup points, although other points are available. Figure 9–47 shows optional set types of screws.

Lag Screws and Wood Screws

Lag screws are designed to attach metal to wood or wood to wood. Before assembly with a lag screw, a pilot hole is cut into the wood. The threads of the lag screw then form their own mating thread in the wood. Lag screws are sized by diameter and length. A lag screw is shown in Figure 9–48. Wood screws are similar in function to lag screws and are available in a wide variety of sizes, head styles, and materials.

Self-tapping Screws

Self-tapping screws are designed for use in situations where the mating thread is created by the fastener. These screws are used to hold two or more mating parts when one of the parts becomes a fastening device. A clearance fit is required through the first series of features or parts while the last feature receives a pilot hole similar to a tap drill for unified threads. The self-tapping screw then forms its own threads by cutting or displacing material as it enters the pilot hole. There are several different types of self-tapping screws with head variations similar to cap screws. The specific function of the screw is important as these screws may be designed for applications ranging from sheet metal to hard metal fastening.

Thread Inserts

Screw thread inserts are helically formed coils of diamond-shaped wire made of stainless steel or phosphor bronze. The inserts are used by being screwed into a threaded hole to form a mating internal thread for a threaded fastener. Inserts are used to repair worn or damaged internal threads and to provide a strong thread surface in soft materials. Some screw thread inserts are designed to provide a secure mating of fasteners in situations where vibration or movement could cause parts to

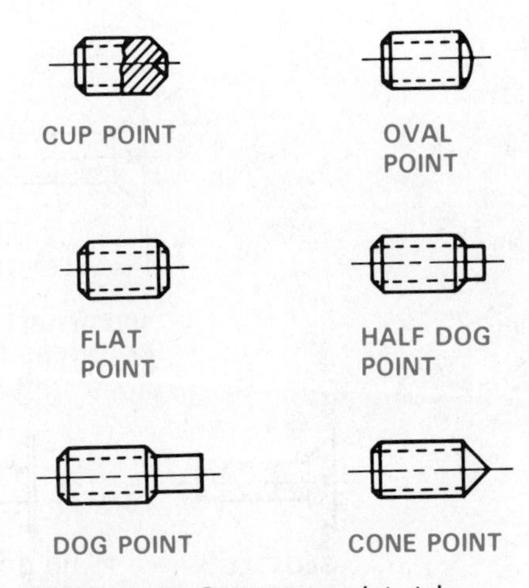

Figure 9–47 Set screw point styles.

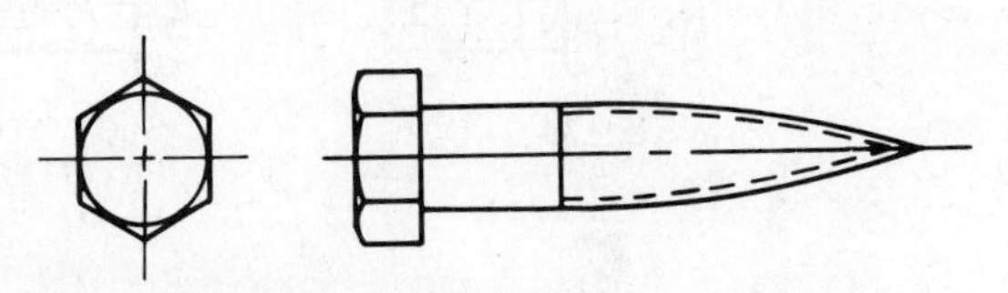

Figure 9–48 Hex head lag screw.

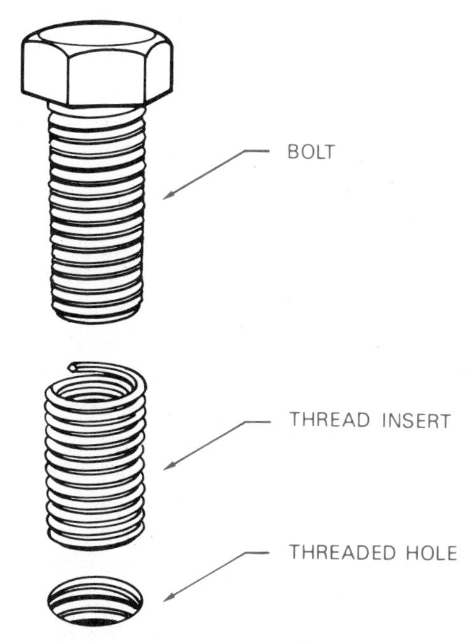

Figure 9–49 Thread insert.

loosen. Figure 9–49 shows the relationship among the fastener, thread insert, and tapped hole.

How to Draw Various Types of Screw Heads

As you have found from the previous discussion, screw fasteners are classified in part by head type, and there is a large variety of head styles. A valuable drafting reference is the *Machinery's Handbook*, which clearly lists specifications for all common head types.

Hexagon head fasteners are generally drawn with the hexagon positioned across the corners vertically in the front view. When projected across the flats the hex head appears like a square head. The following steps show how to draw a hex head bolt. Hex nuts are drawn in the same manner.

Step 1 Use construction lines to draw the end view of the hexagon with the distance across the flats. A ¾-in. nominal thread measures 1⅛ in. across the flats. Position the bolt head so that a front view projection will be across the corners as shown in Figure 9–50.

Step 2 Block out the major diameter and length, and bolt head height. The same ¾-in. hex bolt has a head height, H, of 11/32 in. (See Figure 9–51.)

Step 3 Project the hexagon corners to the front view. Then establish radius, *R*, centers with 60° angles as shown in Figure 9–52. Draw the radii.

Step 4 Draw a 30° chamfer tangent on each side to the small radius arcs. Use object lines to complete both views and draw the thread representation. (See Figure 9–53.)

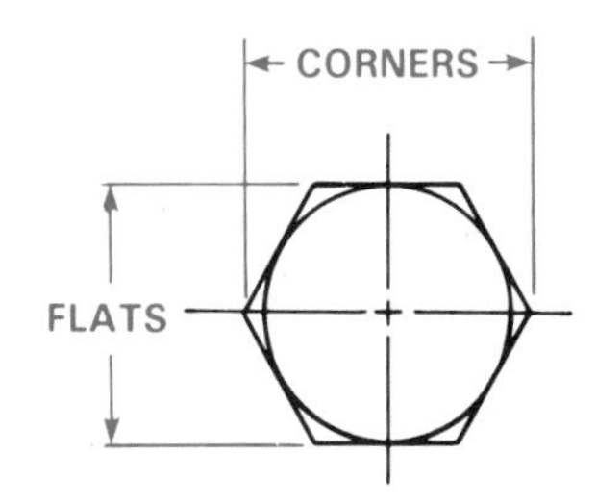

Figure 9–50 The flats and corners of a hexagon — step 1.

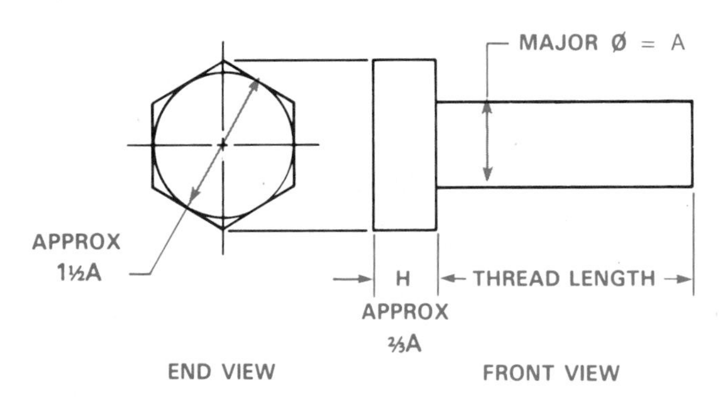

Figure 9–51 Layout with construction lines — step 2.

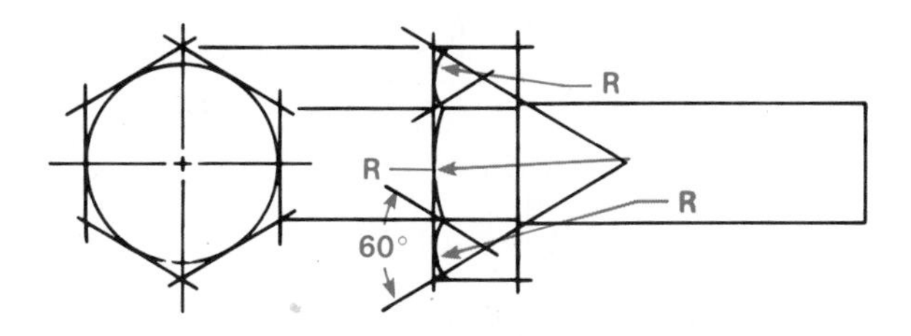

Figure 9–52 Hex head layout — step 3.

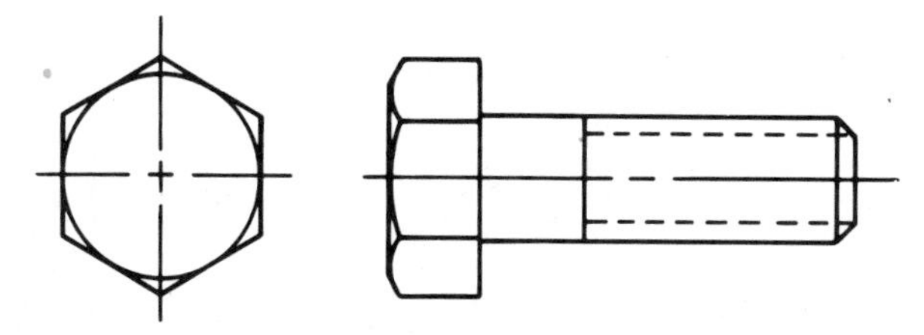

Figure 9–53 Completed drawing — step 4.

Square-head bolts are drawn in the same manner as hex-head bolts.

A few common cap screw heads are shown with approximate layout dimensions in Figure 9–54.

Templates

When bolt or screw heads are drawn only occasionally, then it may be satisfactory to use the layout techniques described in Figures 9–50 through 9–54. However, whether you draw bolt or screw heads a few times or many times each day, it is always better to use

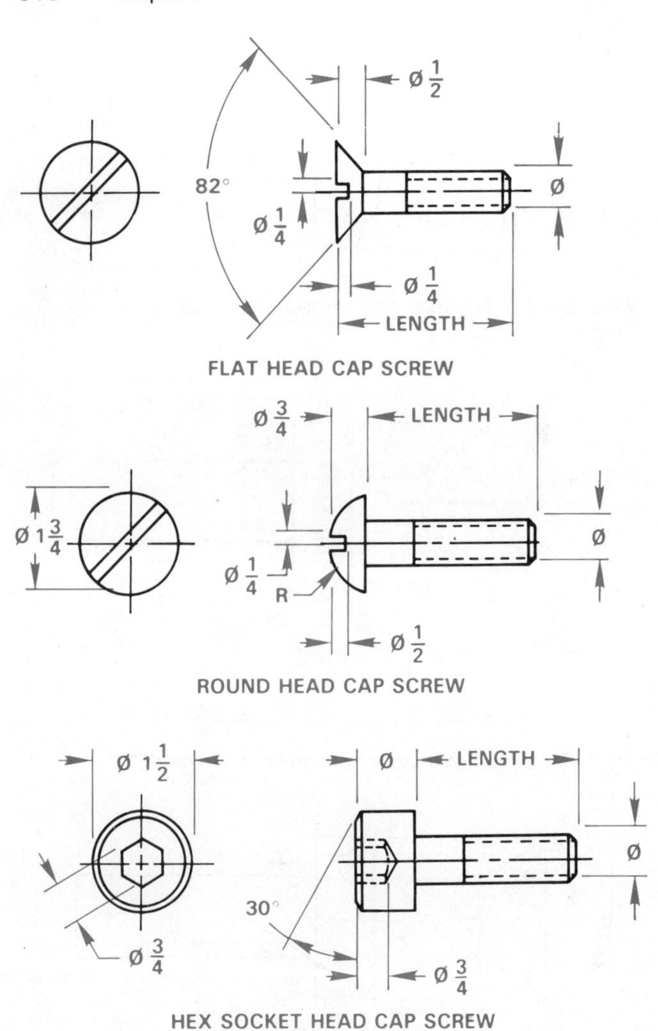

Figure 9–54 Layout specifications for common cap screws.

templates. There are a large variety of templates that are designed to allow you to quickly and easily draw bolt or screw heads. Figure 9–55(a) and Figure 9–55(b) show a variety of available templates that are invaluable for drawing detailed screw threads and head types.

Nuts

A nut is used as a fastening device in combination with a bolt to hold two or more pieces of material together. The nut thread must match the bolt thread for acceptable mating. Figure 9–56 shows the nut and bolt relationship. The hole in the parts must be drilled larger than the bolt for clearance.

There are a variety of nuts in hexagon or square shapes. Nuts are also designed slotted to allow them to be secured with a pin or key. Acorn nuts are capped for appearance. Self-locking nuts are available with neoprene gaskets that help keep the nut tight when movement or vibration is a factor. Figure 9–57 shows some common nuts.

WASHERS

Washers are flat, disk-shaped objects with a center hole to allow a fastener to pass through. Washers are made of metal, plastic, or other materials for use under a nut or bolt head, or at other machinery wear points, to serve as a cushion, or a bearing surface, to prevent leakage or to relieve friction. Several different types of washers serve as a cushion, bearing surface, or locking device. (See Figure 9–58.) Washer thickness varies from .016 in. to .633 in.

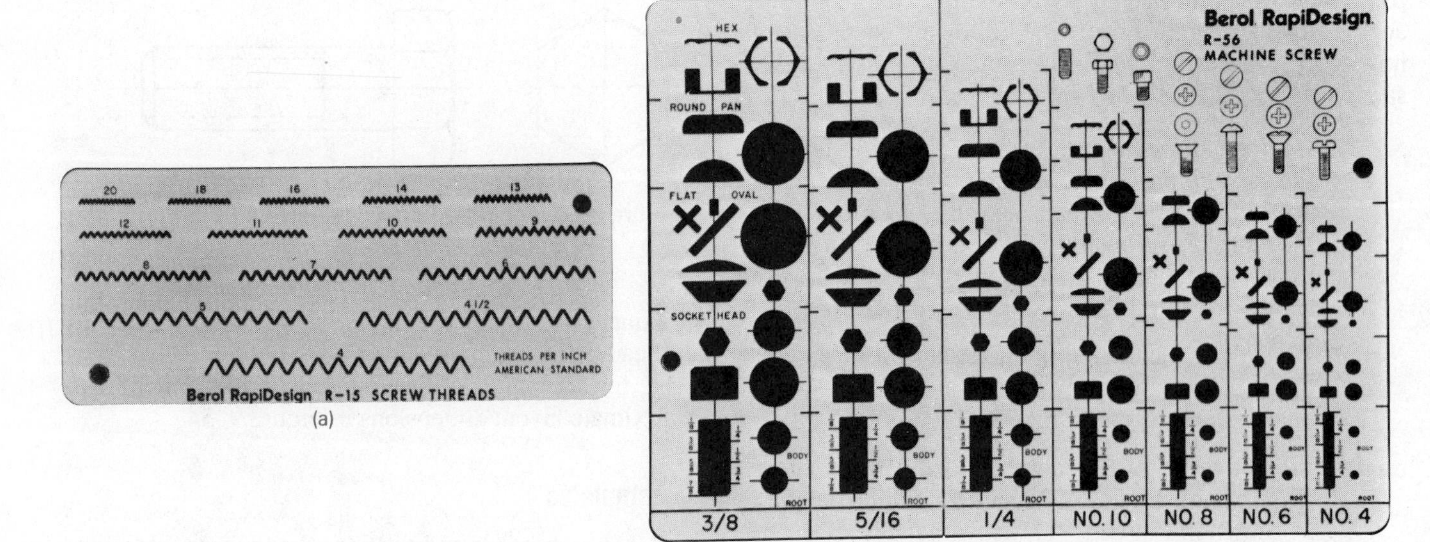

Figure 9–55 (a) Detailed screw-thread template; (b) screw-head templates; many screw and bolt head templates are available, ask your local vendor. *Courtesy Berol USA RapiDesign.*

DOWEL PINS

Dowel pins used in machine fabrication are metal cylindrical fasteners that retain parts in a fixed position or keep parts aligned. Generally, depending upon the function of the parts, one or two dowel pins are sufficient for holding adjacent parts. Dowel pins must generally be pressed into a hole with an interference tolerance of between .0002 in. to .001 in. depending on the material and the function of the parts. Figure 9–59 shows the section of two adjacent parts and a dowel pin. Figure 9–60 is a chart drawing of some standard dowel pins.

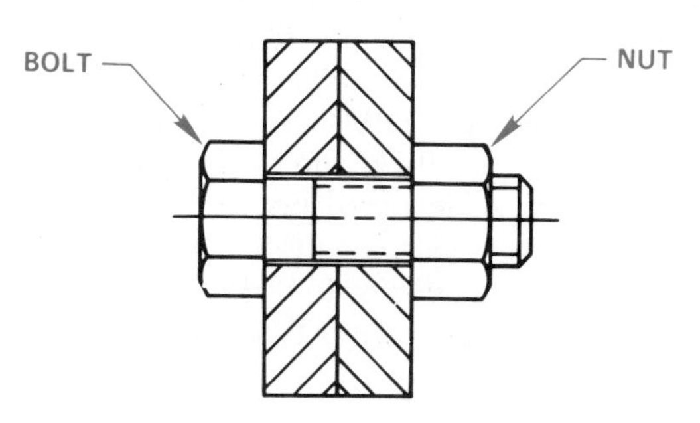

Figure 9–56 Nut and bolt relationship known as a floating fastener.

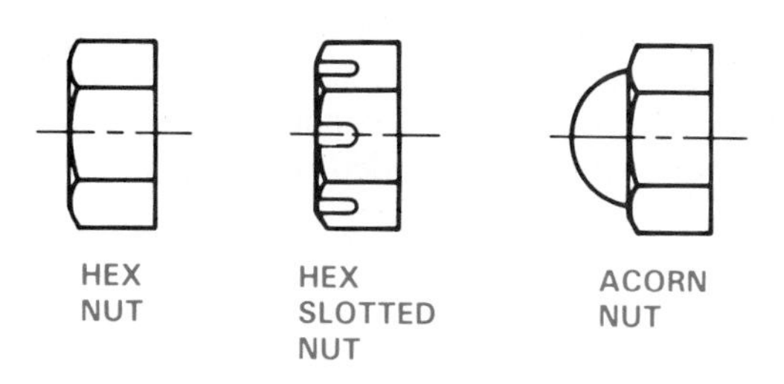

Figure 9–57 Common nuts.

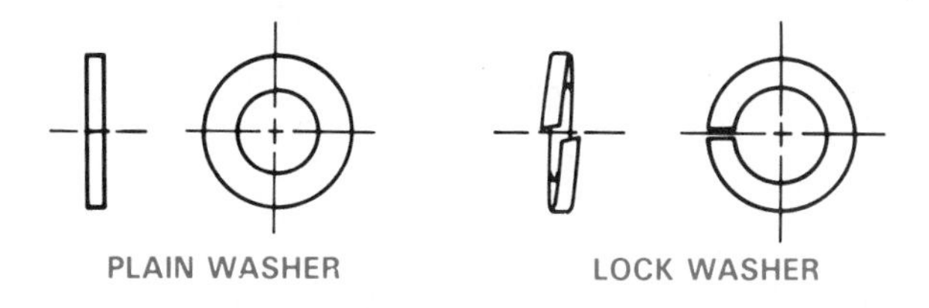

Figure 9–58 Types of washers.

TAPER AND OTHER PINS

For applications that require perfect alignment of accurately constructed parts, *tapered dowel pins* may be better than straight dowel pins. Taper pins are also used for parts that have to be taken apart frequently or where removal of straight dowel pins may cause excess hole wear. Figure 9–61 shows an example of a taper pin assembly. Taper pins, shown in Figure 9–62, range in diameter, *D*, from 7/0, which is .0625 in., to .875 in., and lengths, *L*, vary from .375 in. to 8 in.

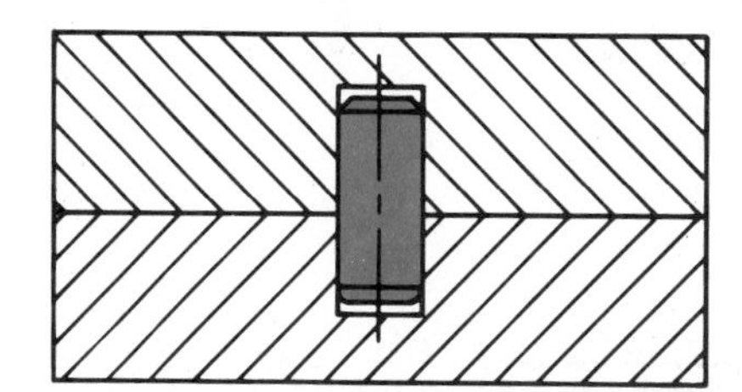

Figure 9–59 Dowel pin in place, sectional view.

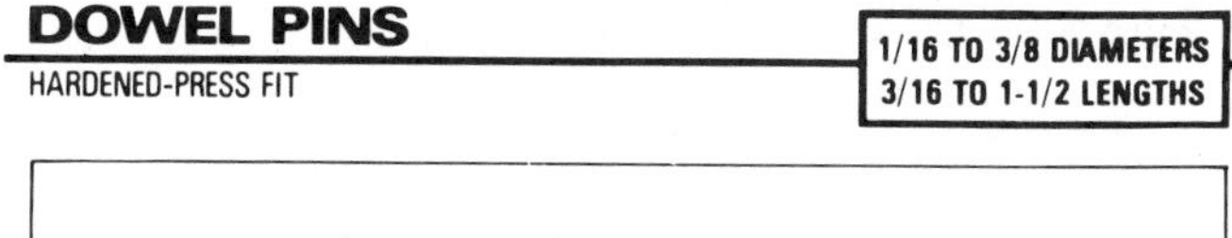

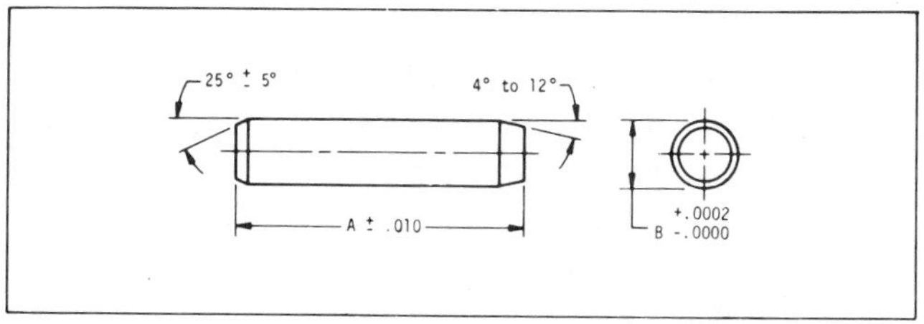

Material: 416 Stainless Steel (Clear Passivate) Hardened to: Rockwell C 36-40
Conforms to specification MS-16555
For 303 stainless pins, please see Cat. No. EPS-A1 and EPS-B1

	1/16 DIA. PIN B = .0626		3/32 DIA. PIN B = .0938		1/8 DIA. PIN B = .1251		5/32 DIA. PIN B = .1563	
A	Cat. No.	Price	Cat. No.	Price	Cat. No.	Price	Cat. No.	Price
3/16	EPS-D1-1	$.09	*EPS-D2-1	$.14	-		-	
1/4	EPS-D1-2	.09	*EPS-D2-2	.15	*EPS-D3-2	$.14	-	
5/16	EPS-D1-3	.10	EPS-D2-3	.11	*EPS-D3-3	.15	-	
3/8	EPS-D1-4	.10	EPS-D2-4	.11	EPS-D3-4	.13	*EPS-D4-4	$.17
7/16	EPS-D1-5	.10	EPS-D2-5	.11	EPS-D3-5	.13	*EPS-D4-5	.17
1/2	EPS-D1-6	.11	EPS-D2-6	.11	EPS-D3-6	.13	EPS-D4-6	.17
5/8	EPS-D1-7	.14	EPS-D2-7	.13	EPS-D3-7	.14	EPS-D4-7	.19
3/4	EPS-D1-8	.14	EPS-D2-8	.13	EPS-D3-8	.15	EPS-D4-8	.19
7/8	*EPS-D1-9	.16	EPS-D2-9	.13	EPS-D3-9	.16	EPS-D4-9	.21
1	-		EPS-D2-10	.14	EPS-D3-10	.18	EPS-D4-10	.23

***Pins not to MS lengths**

Figure 9–60 Dowel pin chart drawing. *Courtesy NORDEX Inc.*

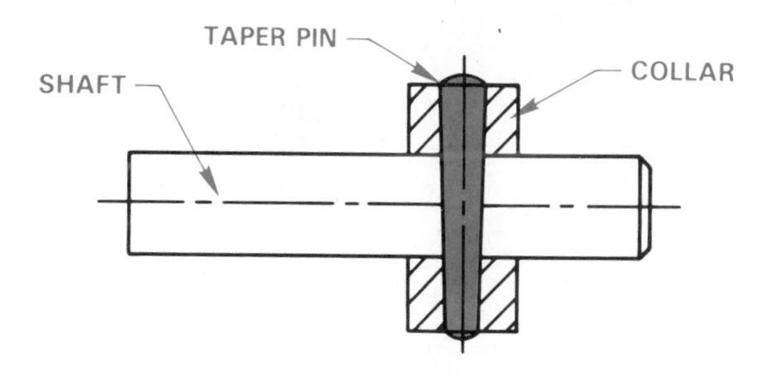

Figure 9–61 Taper pin in assembly, sectional view.

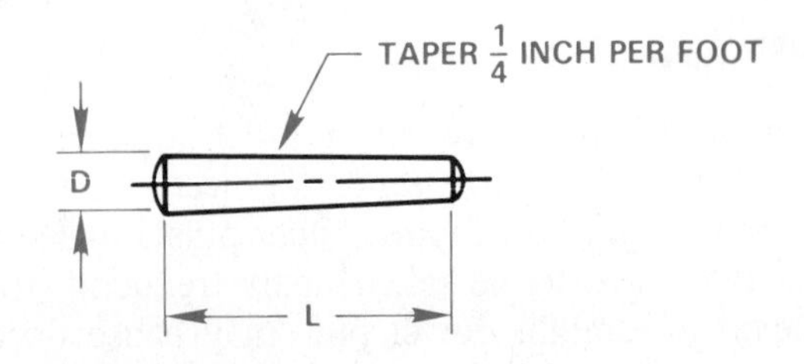

Figure 9–62 Taper pin.

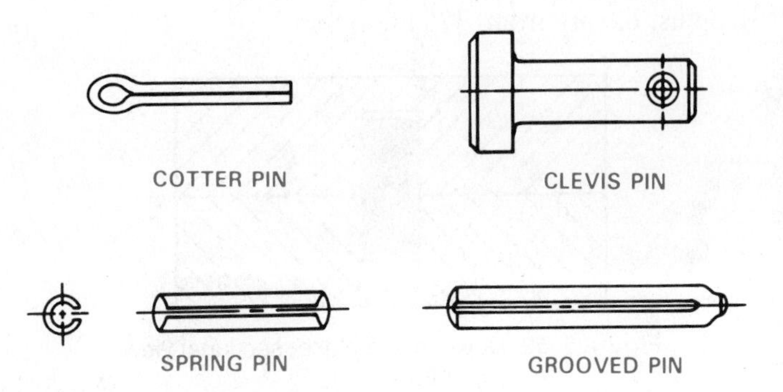

Figure 9–63 Other common pins.

Other types of pins serve functions similar to taper pins, such as holding parts together, aligning parts, locking parts, and transmitting power from one feature to another. Other common pins are shown in Figure 9–63.

RETAINING RINGS

Internal and external retaining rings are available as fasteners to provide a stop or shoulder for holding bearings or other parts on a shaft. They are also used internally to hold a cylindrical feature in a housing. Common retaining rings require a groove in the shaft or housing for mounting with a special plier tool. Also available are self-locking retaining rings for certain applications. (See Figure 9–64.)

KEYS, KEYWAYS, AND KEYSEATS

Standards for keys were established to control the relationship among key sizes, shaft sizes, and tolerances for key applications. A key is an important machine element, which is employed to provide a positive connection for transmitting torque between a shaft and hub, pulley, or wheels. The key is placed in position in a keyseat, which is a groove or channel cut in a shaft. The shaft and key are then inserted into the hub, wheel, or pulley, where the key mates with a groove called a keyway. Figure 9–65 shows the relationship among the key, keyseat, shaft, and hub.

Standard key sizes are determined by shaft diameter. For example, shaft diameters ranging from 7/8 in. to 1 1/4 in. would require a 1/4-in. nominal width key. Keyseat depth dimensions are established in relationship to the shaft diameter. Figure 9–66 shows the standard dimensions for the related features. For a shaft with a 3/4-in. diameter the recommended shaft dimension, *S*, would be .676 in. and the hub dimension, *T*, would be .806 in. when using a rectangular key. Types of keys are shown in Figure 9–67.

RIVETS

A *rivet* is a metal pin with a head used to fasten two or more materials together. The rivet is placed through holes in mating parts and the end without a head extends through the parts to be headed-over (formed into a head)

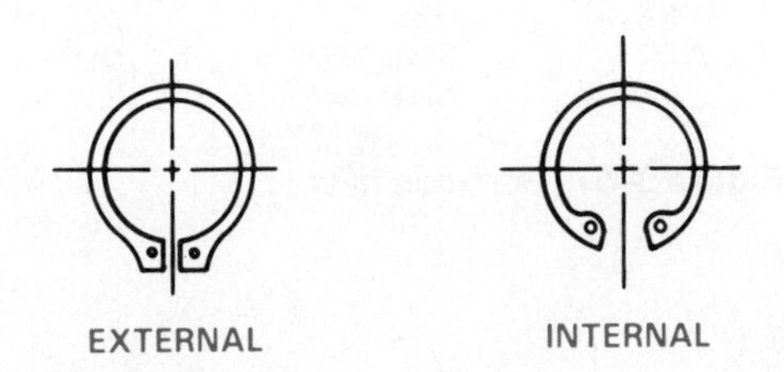

Figure 9–64 Retaining rings.

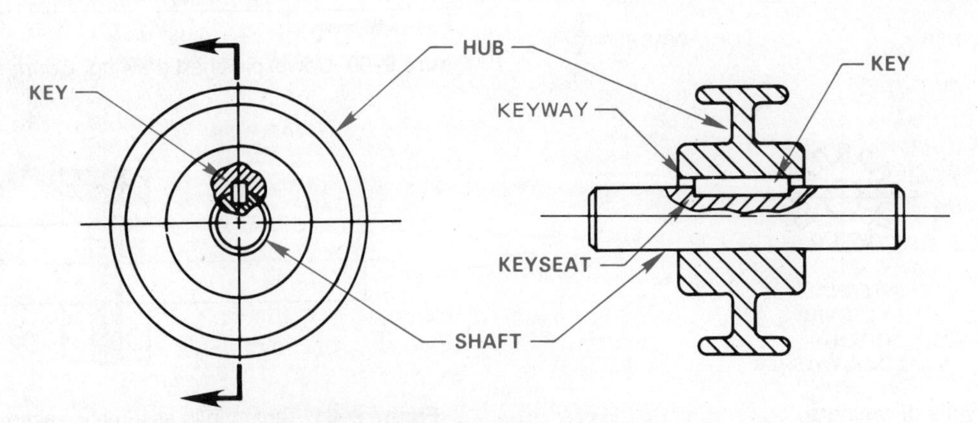

Figure 9–65 Relationship among the key, keyseat, keyway, shaft, and hub.

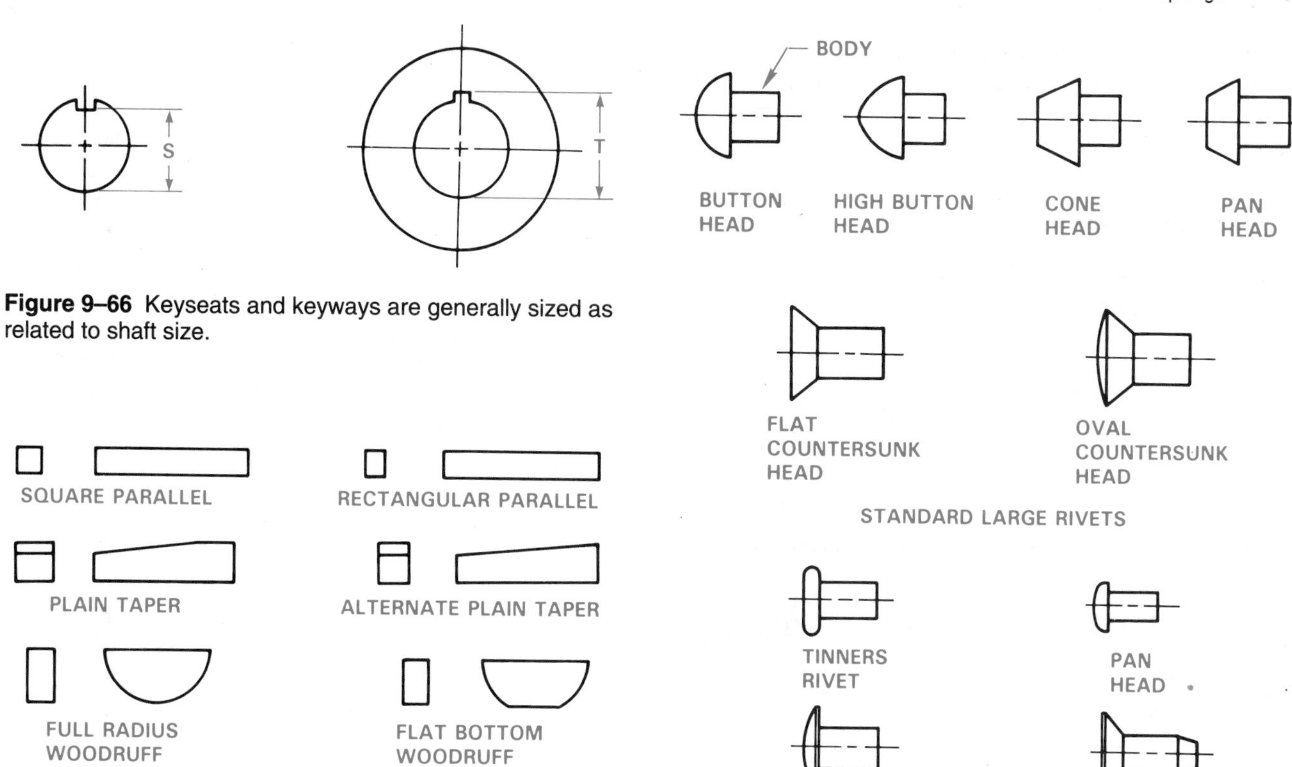

Figure 9–66 Keyseats and keyways are generally sized as related to shaft size.

Figure 9–67 Types of keys.

Figure 9–68 Common rivets.

by hammering, pressing, or forging. The end with the head is held in place with a solid steel bar known as a dolly while a head is formed on the other end. Rivets are classified by body diameter, length, and head type. (See Figure 9–68.)

SPRINGS

A *spring* is a mechanical device, often in the form of a helical coil, that yields by expansion or contraction due to pressure, force, or stress applied. Springs are made to return to their normal form when the force or stress is removed. Springs are designed to store energy for the purpose of pushing or pulling machine parts by reflex action into certain desired positions. Improved spring technology provides springs with the ability to function for a long time under high stresses. The effective use of springs in machine design depends on five basic criteria including material, application, functional stresses, use, and tolerances.

Continued research and development of spring materials have helped to improve spring technology. The spring materials most commonly used include high-carbon spring steels, alloy spring steels, stainless spring steels, music wire, oil-tempered steel, copper-based alloys, and nickel-based alloys. Spring materials, depending on use, may have to withstand high operating temperatures and high stresses under repeated loading.

Spring design criteria are generally based on material gage, kind of material, spring index, direction of the helix, type of ends, and function. Spring wire gages are available from several different sources ranging in diameter from number 7/0 (.490 in.) to number 80 (.013 in.). The most commonly used spring gages range from 4/0 to 40. There are a variety of spring materials available in round or square stock for use depending on spring function and design stresses. The spring index is a ratio of the average coil diameter to the wire diameter. The index is a factor in determining spring stress, deflection, and the evaluation of the number of coils needed and the spring diameter. Recommended index ratios range between 7 and 9 although other ratios commonly used range from 4 to 16. The direction of the helix is a design factor when springs must operate in conjunction with threads or with one spring inside of another. In such situations the helix of one feature should be in the opposite direction of the helix for the other feature. Compression springs are available with ground or unground ends. Unground, or rough, ends are less expensive than ground ends. If the spring is required to rest flat on its end, then ground ends should be used. Spring function depends on one of two basic factors, compression or extension.

Compression springs release their energy and return to their normal form when compressed. Extension springs release their energy and return to the normal form when extended. (See Figure 9–69.)

Spring Terminology

The springs shown in Figure 9–70 show some common characteristics.

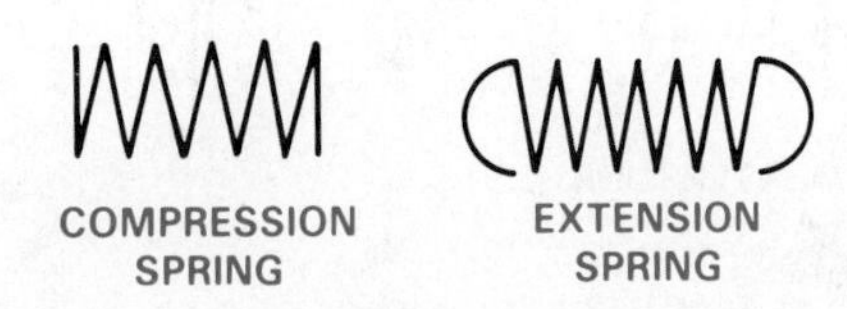

Figure 9–69 Compression and extensions springs.

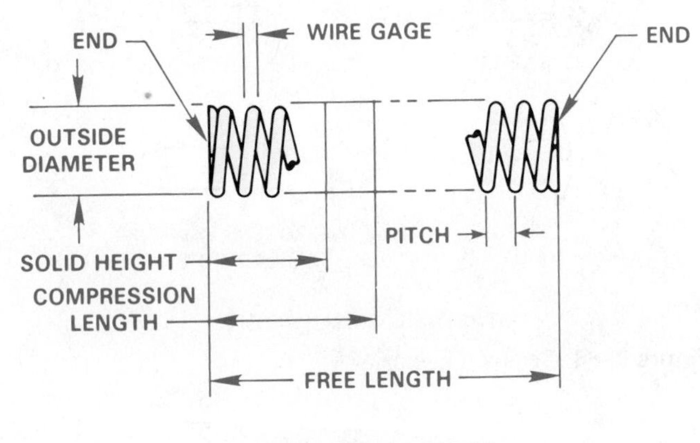

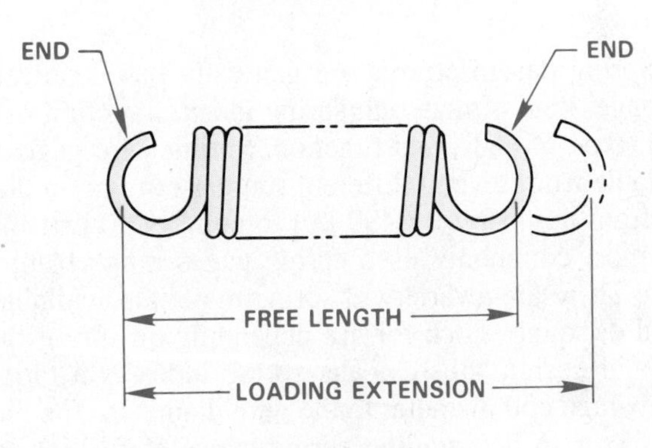

Figure 9–70 Spring characteristics.

Ends. Compression springs have four general types of ends: open or closed ground ends and open or closed unground ends, shown in Figure 9–71. Extension springs have a large variety of optional ends, a few of which are shown in Figure 9–72.

Helix Direction. The helix direction may be specified as right hand or left hand. (See Figure 9–71.)

Free Length. The length of the spring when there is no pressure or stress to affect compression or extension is known as free length.

Compression Length. The compression length is the maximum recommended design length for the spring when compressed.

Solid Height. The solid height is the maximum compression possible. The design function of the spring should not allow the spring to reach solid height when in operation unless this factor is a function of the machinery.

Loading Extension. The extended distance to which an extension spring is designed to operate is the loading extension length.

Pitch. The pitch is one complete helical revolution, or the distance from a point on one coil to the same corresponding point on the next coil.

Torsion Springs

Torsion springs are designed to transmit energy by a turning or twisting action. Torsion is defined as a twisting action that tends to turn one part or end around a longitudinal axis while the other part or end remains fixed. Torsion springs are often designed as antibacklash devices or as self-closing or self-reversing units. (See Figure 9–73.)

Flat Springs

Flat springs are arched or bent flat-metal shapes designed so that when placed in machinery they cause tension on adjacent parts. The tension may be used to level parts, provide a cushion, or position the relative movement of one part to another. One of the most common examples of flat springs is leaf springs on an automobile.

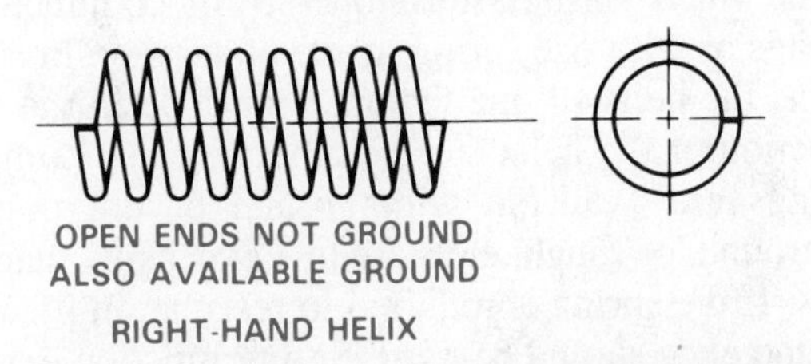

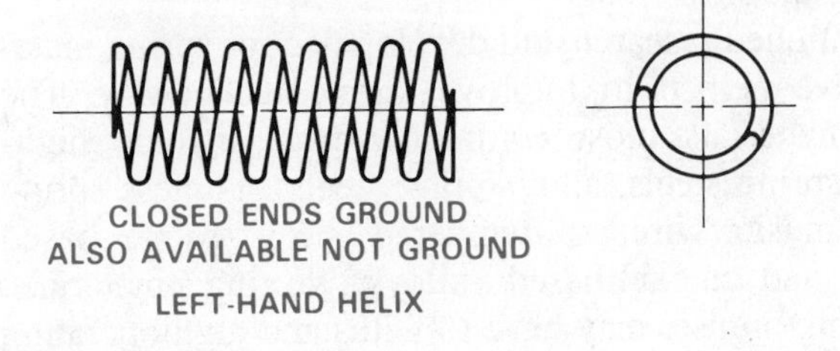

Figure 9–71 Helix direction and compression spring end types.

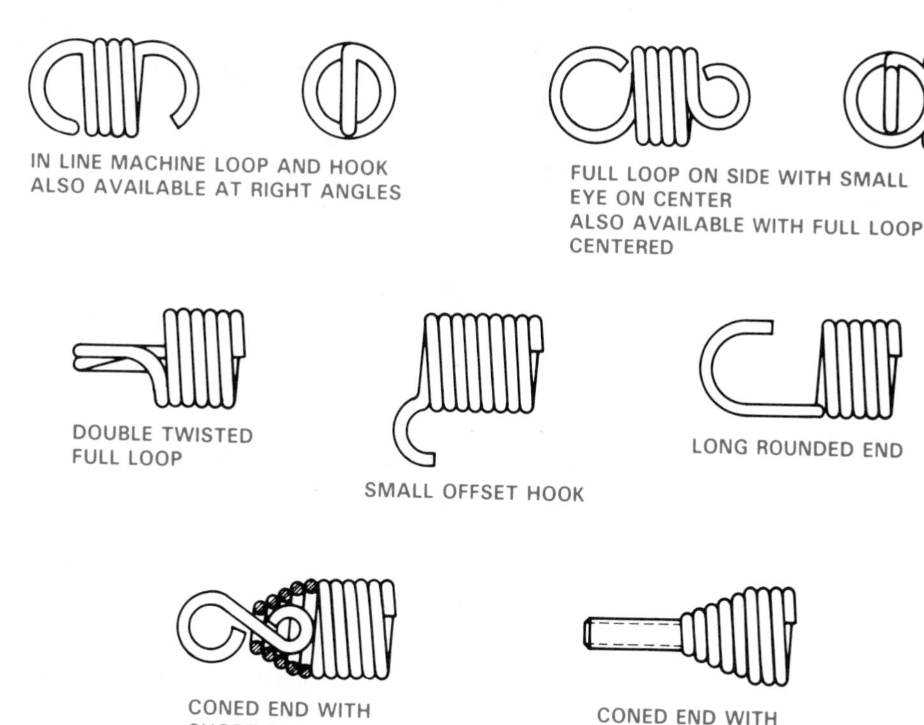

Figure 9–72 Extension spring end types.

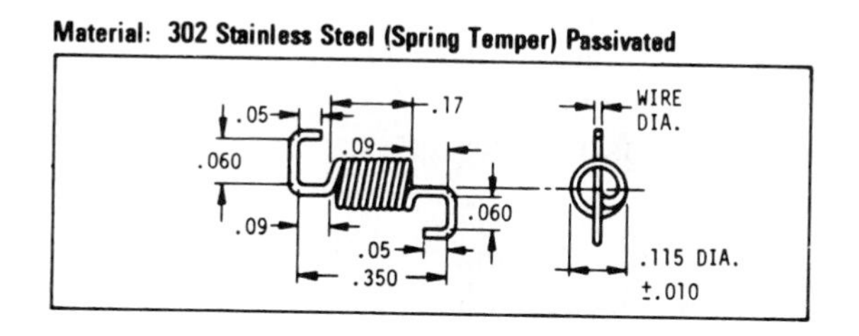

Figure 9–73 Torsion spring, also called antibacklash springs. *Courtesy NORDEX Inc.*

Spring Representations

There are three types of spring representations, detailed, schematic, and simplified as seen in Figure 9–74. Detailed spring drawings are used in situations that require a realistic representation, such as vendors' catalogs, assembly instructions, or detailed assemblies. Schematic spring representations are commonly used on drawings. The single-line schematic symbols are easy to draw and clearly represent springs without taking the additional time required to draw a detailed spring. The use of simplified spring drawings is limited to situations where the clear resemblance of a spring is not necessary. While very easy to draw, the simplified spring symbol must be accompanied by clearly written spring specifications. The simplified spring representation is not very useful in assembly drawings or other situations that require a visual comparison of features.

Figure 9–74 Spring representations.

Spring Specifications

No matter which representation is used several important specifications must accompany the spring symbol. Spring information is generally lettered in the form of a specific or general note.

Spring specifications include outside or inside diameter, wire gage, kind of material, type of ends, surface finish, free and compressed length, and number of coils. Other information, when required, may include spring design criteria, and heat treatment specifications. The information is often provided on a drawing as shown in Figure 9–75. The material note is usually found in the title block.

Drawing Detailed Spring Representations

Detailed spring drawing requires a great degree of accuracy. First determine the outside diameter, number of coils, wire diameter, and free length or compressed length.

The following steps show how to draw a detailed spring representation with the following specifications:

Material: 2.5-mm diameter high-carbon spring steel
Outside Diameter: 16 mm
Free Length: 50 mm
Number of Coils: 6

Step 1 Using construction lines, draw a rectangle equal to the outside diameter (16 mm) wide and the free length (50 mm) long. (See Figure 9–76.)

Step 2 Along one inside edge of the length of the rectangle, lay out seven equally-spaced full circles with a 2.5-mm diameter (wire size). The layout may be done by dividing the length into six equal spaces. (See Figure 9–77.) On the other inside edge, lay out a half circle at each end. Beginning at a distance of ½ P away from one end, draw the first of six full (2.5-mm) circles with equal spaces between them.

Step 3 Connect the circles drawn in Step 2 to make the coils. Draw lines from a point of tangency on one circle to a corresponding point on a circle on the other side. Draw the last element on each side down the edge of the rectangle for ground ends. To draw unground ends, make the last element terminate at the axis of the spring. (See Figure 9–78.)

For detailed coils in a longitudinal sectional view leave the circles from Step 2 and draw that part of the spring that would appear as if the front half were removed as shown in Figure 9–79. Then fill in or section-line the circles.

Drawing Schematic Spring Representations

Schematic spring symbols are much easier to draw than detailed representations while clearly resembling a spring. The following steps show how to draw a schematic spring symbol for the same spring previously drawn.

Step 1 Use construction lines to draw a rectangle equal in size to the outside diameter by the free length as seen in Figure 9–80.

Step 2 Establish seven marks (6 spaces) equally spaced at P distance along one edge of the rectangle. Draw eight marks along the opposite edge

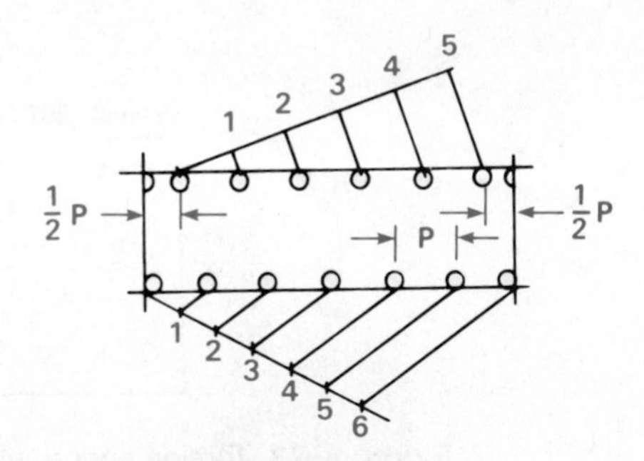

Figure 9–77 Spacing the coils — step 2.

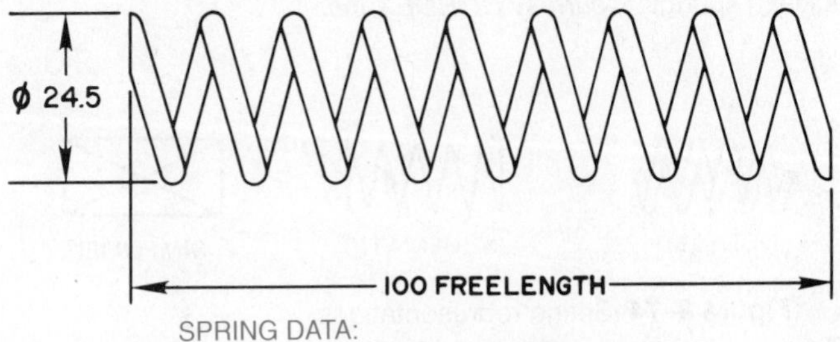

Figure 9–75 Spring drawing with spring data chart.

Figure 9–78 Connect the coils to complete the detailed spring representation — step 3.

Figure 9–79 Detailed spring representation in section.

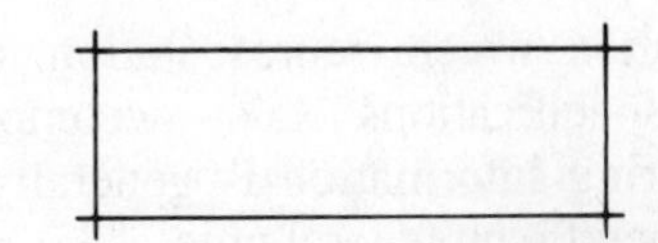

Figure 9–76 Preliminary spring layout — step 1.

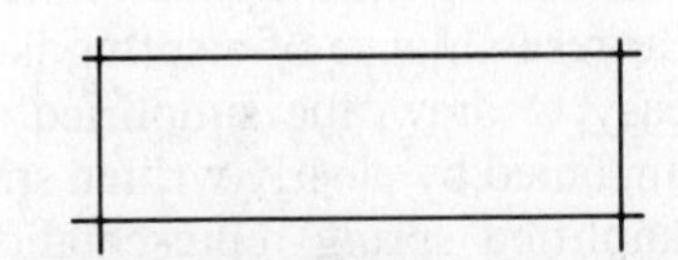

Figure 9–80 Preliminary layout — step 1.

beginning and ending with a space equal to ½ P. (See Figure 9–81.)

Step 3 Beginning on one side, draw the elements of each spring coil as shown in Figure 9–82.

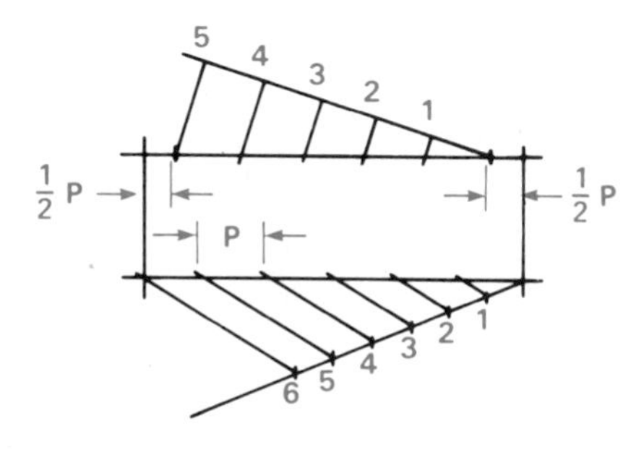

Figure 9–81 Spacing the coils — step 2.

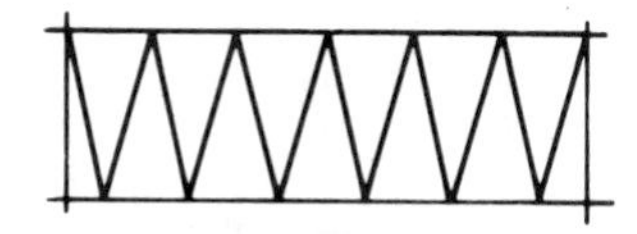

Figure 9–82 Complete the schematic representation — step 3.

DRAWING SPRINGS

Parametric capabilities are the key to drafting and design flexibility with CADD. Parametric means that you have the ability to change a value or values in a situation and automatically have the entire design changed to fit the circumstances. Many CADD packages have this feature available, and CADD programs for spring design are no exception. For instance, if you are designing a compression spring, all you have to do is make the selection from a menu and then enter the following information as prompted by the computer:

- Wire diameter.
- Number of coils.
- Outside diameter.

When you enter this data, the computer calculates the rest of the information, such as free length and compression length, and then asks if you want an external or sectional view. After all the information is complete, the SPRING program automatically draws and dimensions the spring on the screen, followed by prompts for you to rotate or move the spring to the desired location.

Drawing a Simplified Spring Representation

Simplified spring representations are drawn by making a rectangle equal in size to the major diameter and free length and then drawing diagonal lines as shown in Figure 9–83.

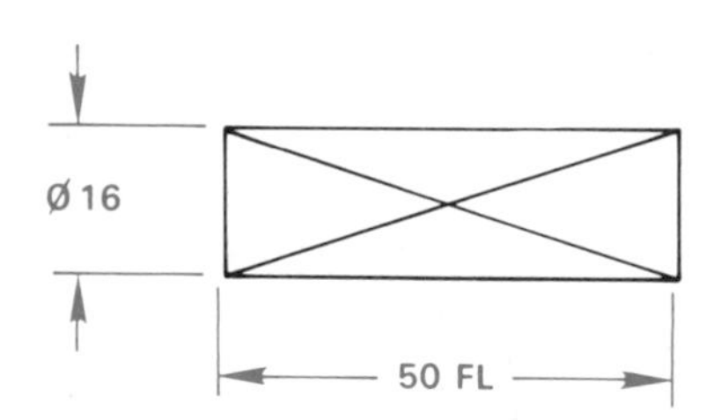

Figure 9–83 Simplified spring representation.

PROFESSIONAL PERSPECTIVE

There are three basic methods for drawing screw thread and spring representations: using construction techniques, using a template, or using CADD. The manual construction techniques should be avoided in most cases because they are very time-consuming. However, in some special applications these methods may be used effectively. The manual drafter should use a template when at all possible to draw screw threads, fastener heads, and spring representations. For the CADD user, there are many software programs available that allow drawing of screw threads in simplified, schematic, or detailed representations. A complete variety of head types can be drawn in addition to springs in simplified, schematic, or detailed representations. This type of CADD software makes an otherwise complex task very simple and saves a lot of drafting time. The entry-level drafter often spends unnecessary time trying to make a screw thread representation look exactly like the real thing. It is important to have an accurate drawing, but with regard to screw threads, the thread note tells the reader the exact thread specifications. This is why threads are shown on a drawing as a representation rather than a duplication of the real thing. The simplified representation is the most commonly used technique, because it is fast and easy to draw.

Your goal as a professional drafter is to make the drawing communicate so that there is no question about what is intended. In simple terms, make the drawing clear and accurate using the easiest method, and make sure the thread note and specifications are complete and correct.

ENGINEERING PROBLEM

What if you are asked to make a drawing from engineer's notes for a fastening device that is needed for one of the products the engineer is designing? The engineer's notes look like Figure 9–84. You may be thinking:

- A 10-gage shaft threaded with 32–UNF–2A. Maybe we can use a 10–32UNF–2A threaded rod and cut to .375 lengths.
- Then we need to chamfer both ends with .015 × 35° and provide a slot in one end that is .030 wide and .047 deep.

No problem, you say. So, you go to work preparing a drawing like the one shown in Figure 9–85. The point here is that the engineer provided a lot of information in the thread note. In fact, standard screws and other fasteners may be completely described without a drawing. A written specification can be used to completely describe an object. For example, 1/2–13UNC–2 × 1.5 LG FILLISTER HEAD MACHINE SCREW.

10 - 32 UNF - 2A THREADED MATERIAL .375 LG.
W/ .015 LATERAL X 35° CHAMFERS ON BOTH ENDS
PROVIDE A .030 WIDE X .047 DEEP SLOT ON ONE END.

Figure 9–84 Engineer's notes.

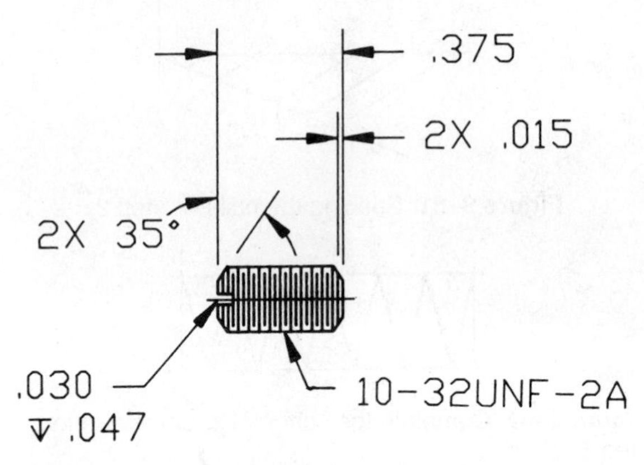

Figure 9–85 CADD drawing from engineer's notes.

FASTENERS AND SPRINGS PROBLEMS

DIRECTIONS

1. From the selected problems determine which views and dimensions should be used to completely detail the part. Use simplified representation unless otherwise specified.
2. Make a multiview sketch, to proper proportions, including dimensions and notes.
3. Using the sketch as a guide, draw an original multiview drawing on an adequately sized drawing sheet. Add all necessary dimensions and notes using unidirectional dimensioning. Use manual or computer-aided drafting as required by your course guidelines.
4. Include the following general notes at the lower left corner of the sheet .5 in. each way from the corner border lines:

   2. REMOVE ALL BURRS AND SHARP EDGES.

   1. INTERPRET DIMENSIONS AND TOLERANCES PER ANSI Y14.5M–1982.

   NOTES:

   Additional general notes may be required depending on the specifications of each individual assignment.

Problem 9–1 (in.)

Part Name: Full Dog Point Gib Screw
Material: 10–32 UNF–2B × .75 Long
Hexagon: .200 across flats × .175 deep

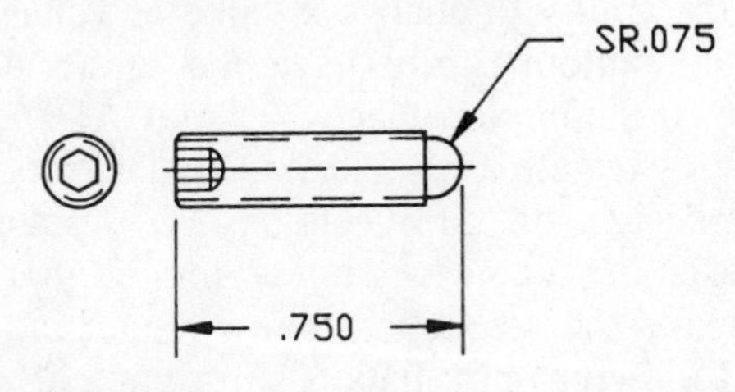

Problem 9–2 (in.)

Part Name: Thumb Screw
Material: SAE 1315 Steel
Note: Provide Medium Diamond Knurl

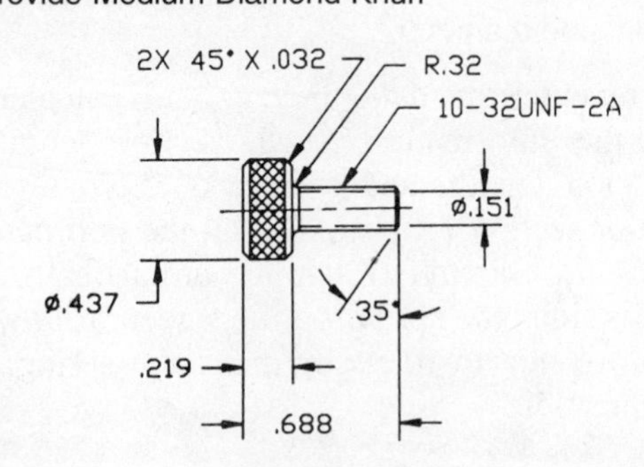

Problem 9–3 (in.)

Part Name: H-Step Threading Screw
Material: SAE 3130

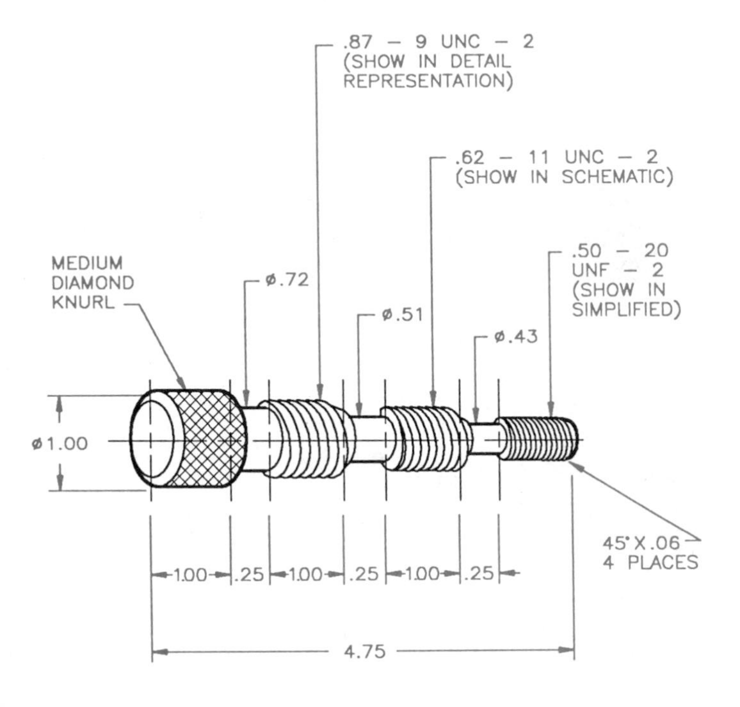

Problem 9–4 (metric)

Part Name: Knurled Hex Soc Hd Step Screw
Material: SAE 1040
Case Harden: 1.6-mm-deep per Rockwell C Scale
Finish: 2 μm; Black Oxide

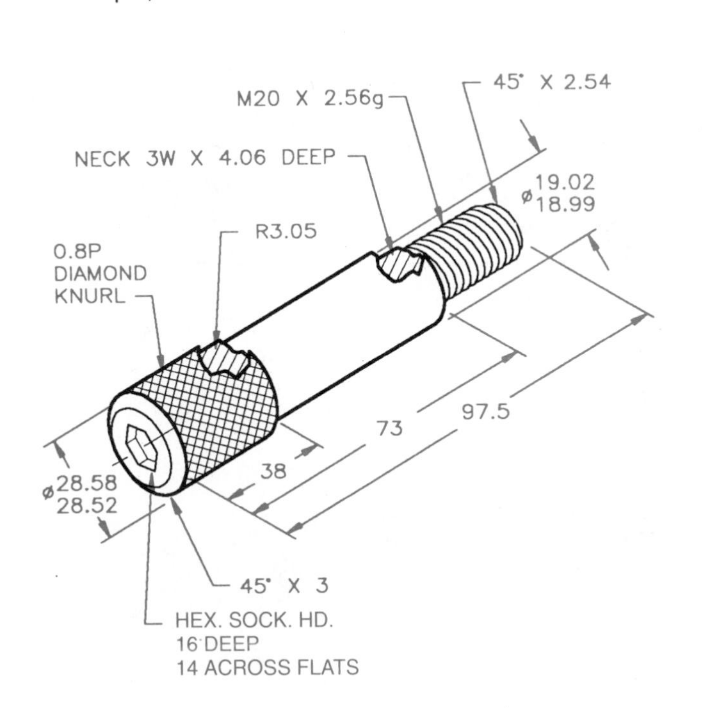

Problem 9–5 (in.)

Part Name: Machine Screw
Material: Stainless Steel
Finish All Over: 2 μm

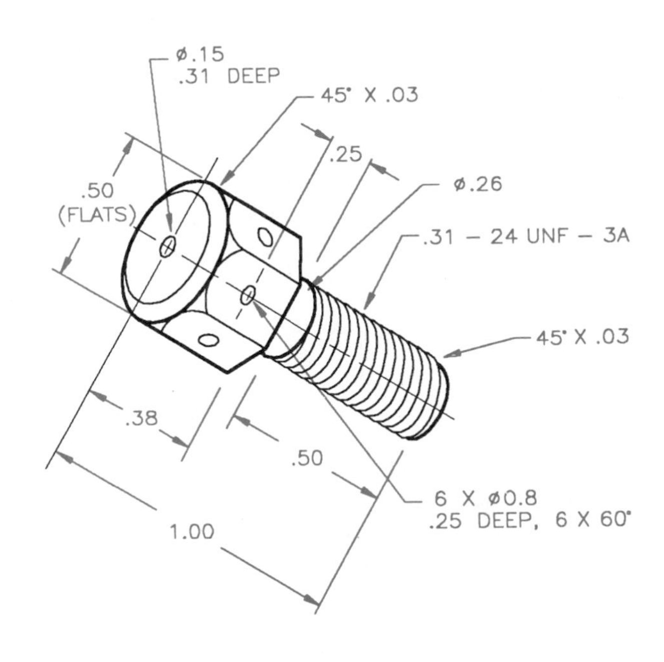

Problem 9–6 (metric)

Part Name: Lathe Dog
Material: Cast Iron

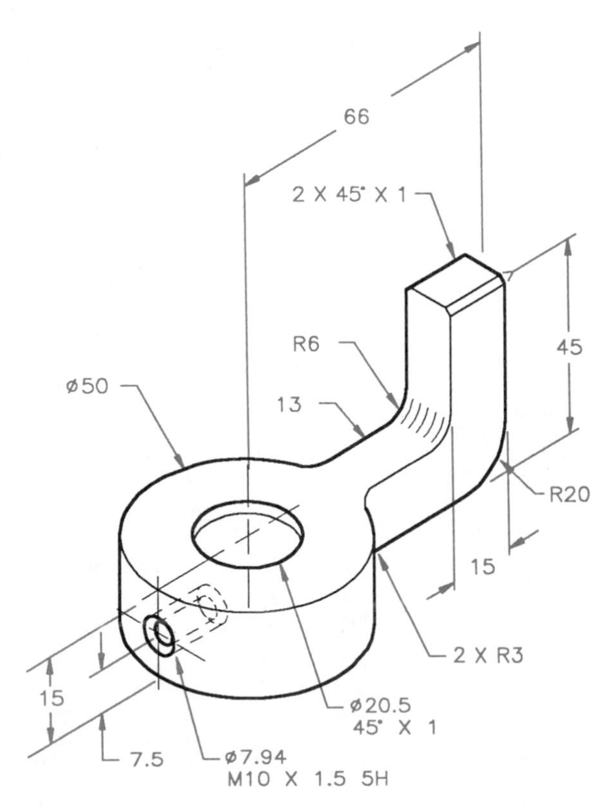

Problem 9–7 (in.)

Part Name: Threaded Step Shaft
Material: SAE 1030

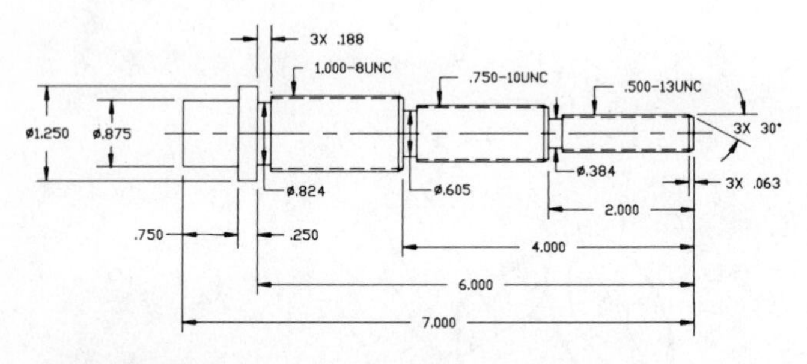

3. TOLERANCES TO BE ±.005.
2. REMOVE ALL BURRS AND SHARP EDGES.
1. INTERPRET ALL DIMENSIONS AND TOLERANCES PER ANSI Y14.5-1982.
NOTES:

Problem 9–8 (in.)

Part Name: Washer Face Nut
Material: SAE 1330 Steel

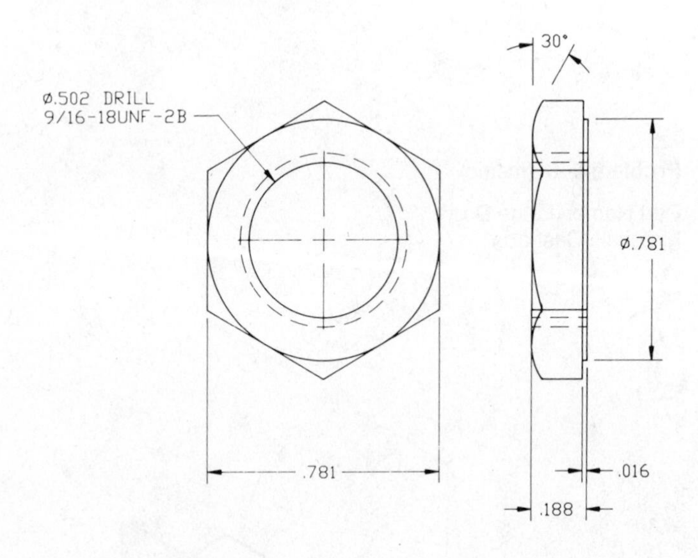

Problem 9–9 (in.)

Part Name: Shoulder Screw
Material: SAE 4320 Steel

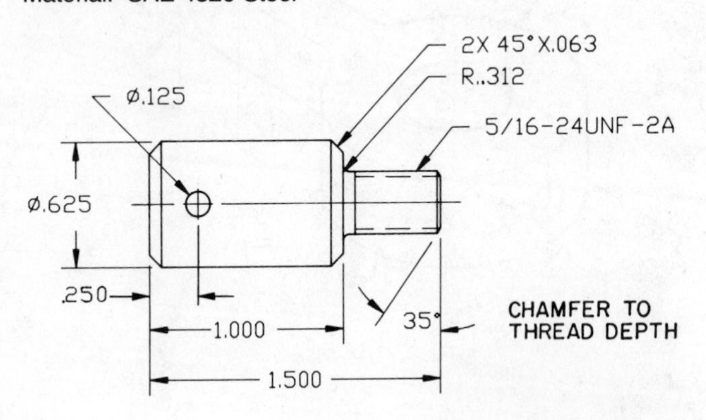

Problem 9–10 (in.)

Part Name: Stop Screw
Material: SAE 4320
Hex Depth: .175
Note: Medium Diamond Knurl

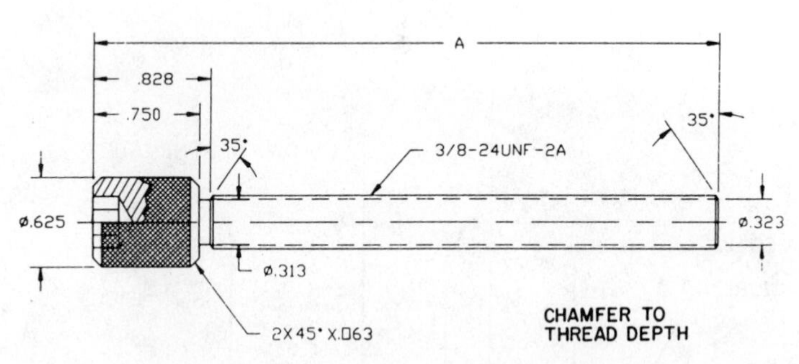

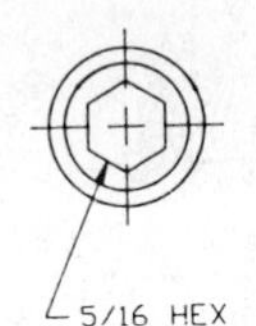

CHART A	
PART#	LENGTH
1DT-1011	4.438
1DT-1012	3.688

Problem 9–11 (in.)

Part Name: Drain Fitting
Material: ∅ 1.25 6061–T6 Aluminum
Courtesy TEMCO.

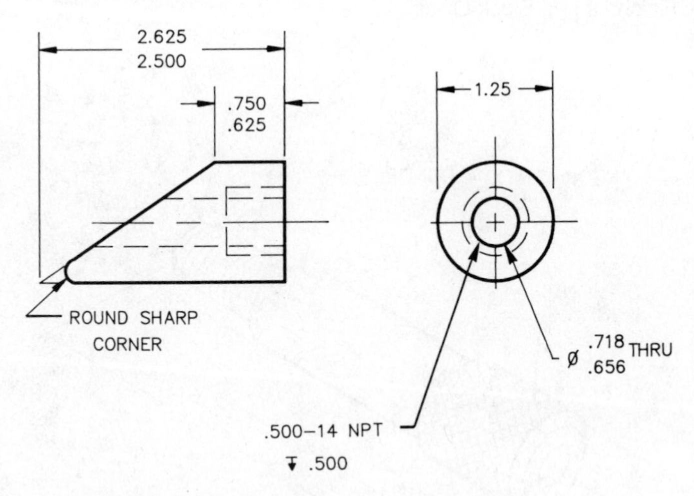

OUTSIDE SURFACE OF COUPLING MUST BE FREE
OF OXIDIZATION AND INK.

Problem 9–12 (metric)

Part Name: Vice Base
Material: Cast Iron

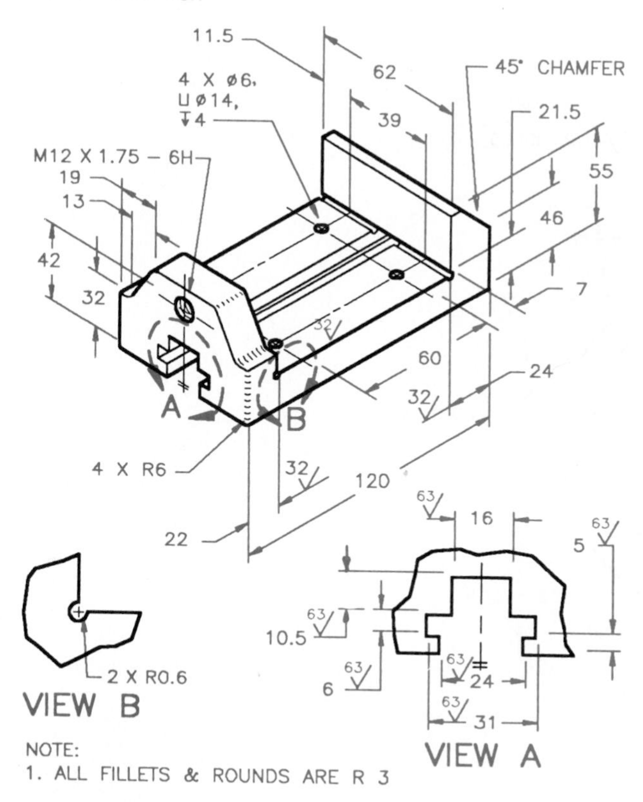

Problem 9–13 (in.)

Part Name: Valve Spring
Material: ∅ .024 Type 302 ASTM A313 Spring Steel
Courtesy TEMCO.

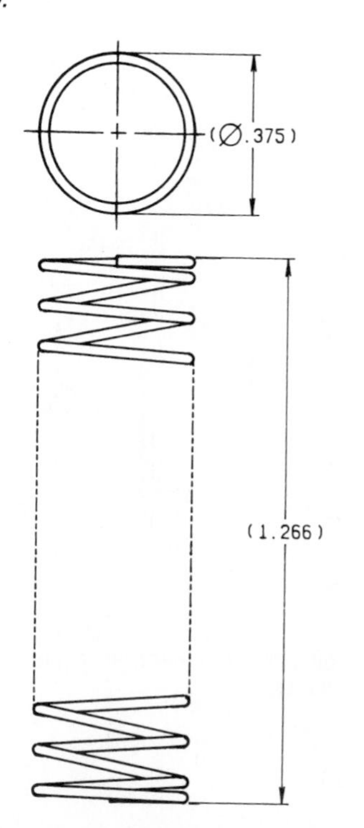

Problem 9–14 (metric)

Part Name: Compression Spring (use schematic representation)
Material: 2.5-mm Steel Spring Wire
Ends: Plain Ground
Outside Diameter: 25
Free Length: 75
Number of Coils: 16
Finish: Chrome Plate

Problem 9–15 (metric)

Part Name: Antibacklash Spring (use detailed representation)
Material: 302 Stainless Steel Spring Tempered 0.38 Diameter
Courtesy Nordex Inc.

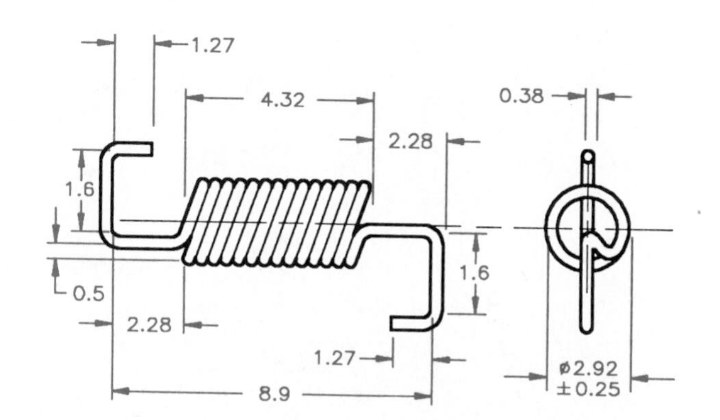

Problem 9–16 (metric)

Part Name: Flat Spring
Material: 3.5-mm Spring Steel
Finish: Black Oxide
Heat Treat: 1-mm-deep Rockwell C Scale

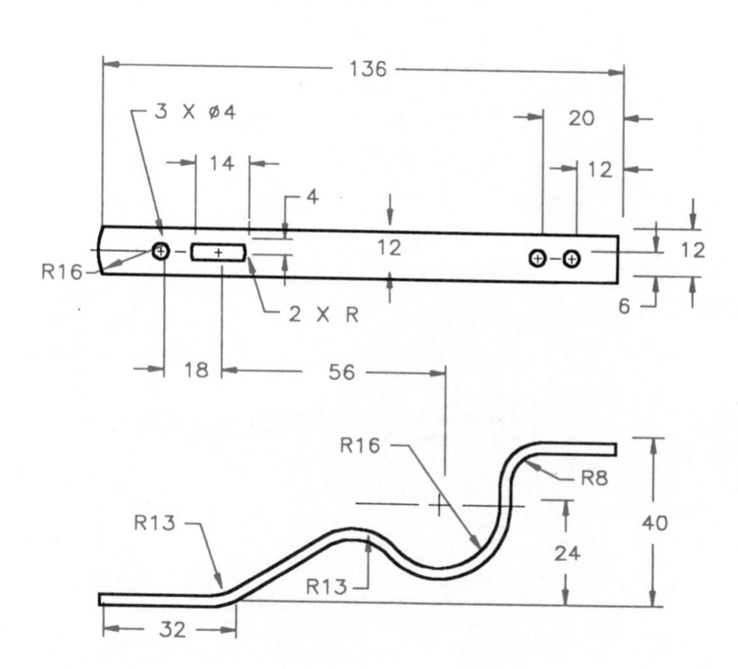

Problem 9–17 (in.)

Part Name: Retaining Ring
Material: Stainless Steel

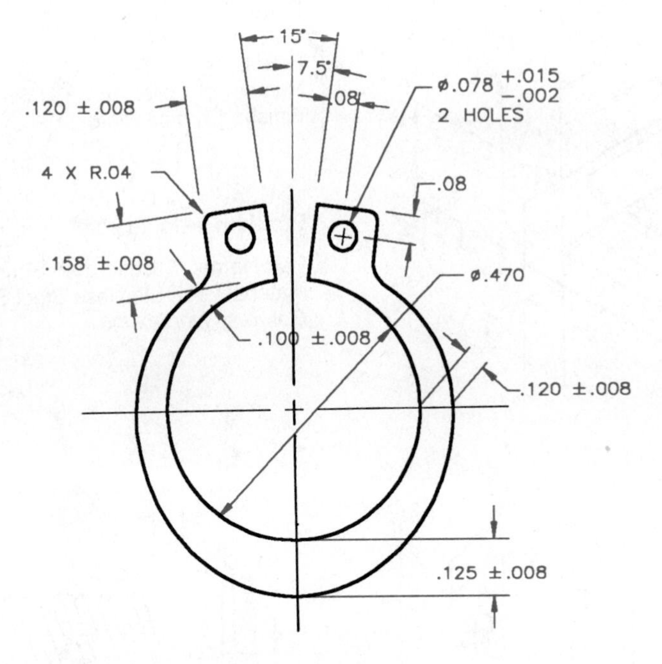

Problem 9–18 (in.)

Part Name: Half Coupling
Material: ∅ 1.250 C1215 Steel
Courtesy TEMCO.

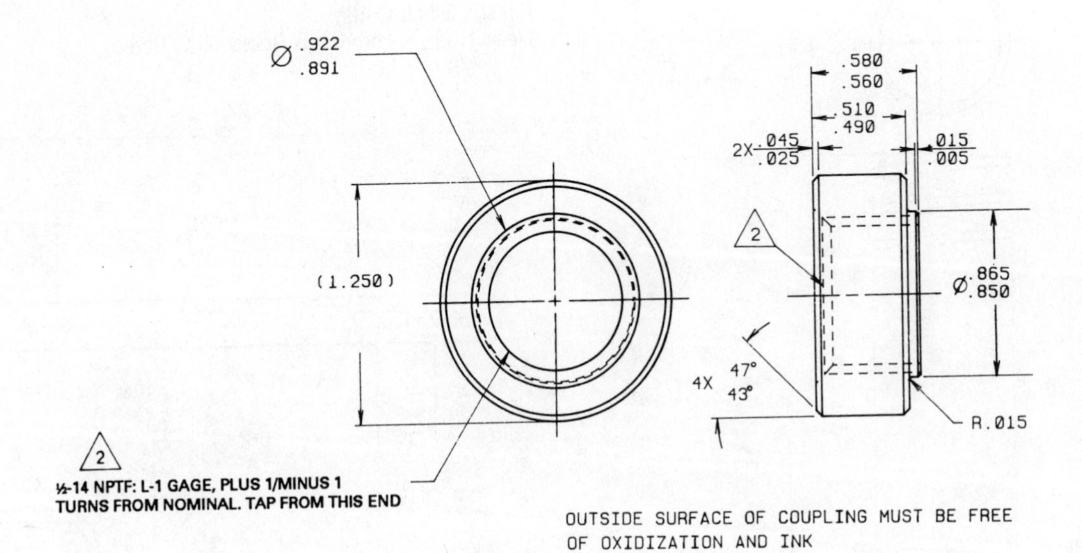

Problem 9–19 (in.)

Part Name: Collar
Material: SAE 1020

.379–.375
2X 45°X.125
2.000
1.051
Ø2.250
Ø1.498 +.003 −.000

Problem 9–20 (in.)

Part Name: Bearing Nut
Material: SAE 1040

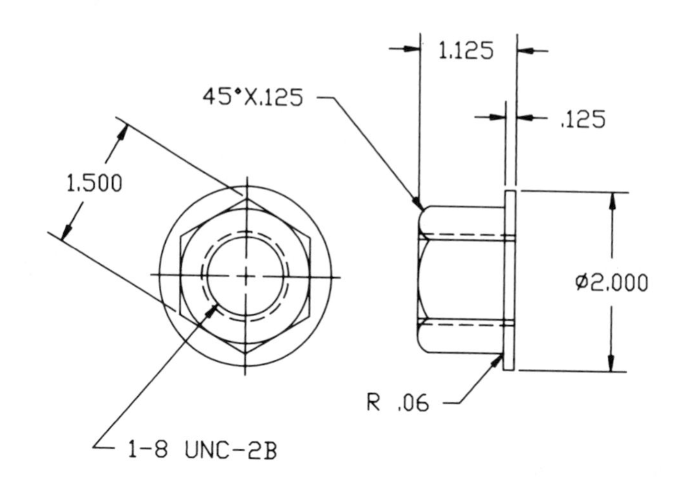

Problem 9–21 (in.)

Part Name: Adjustment Screw
Material: SAE 2010 Steel

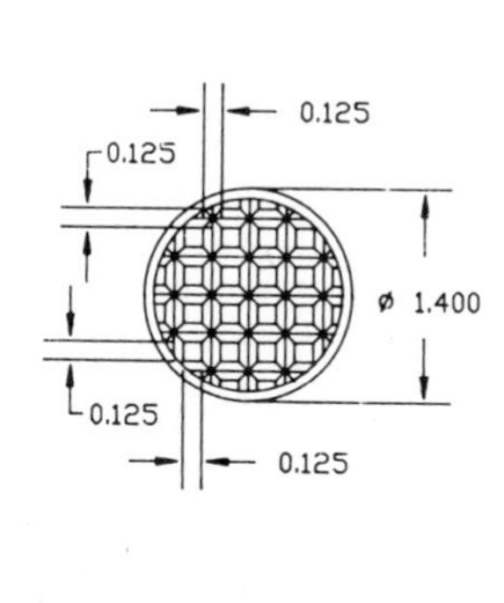

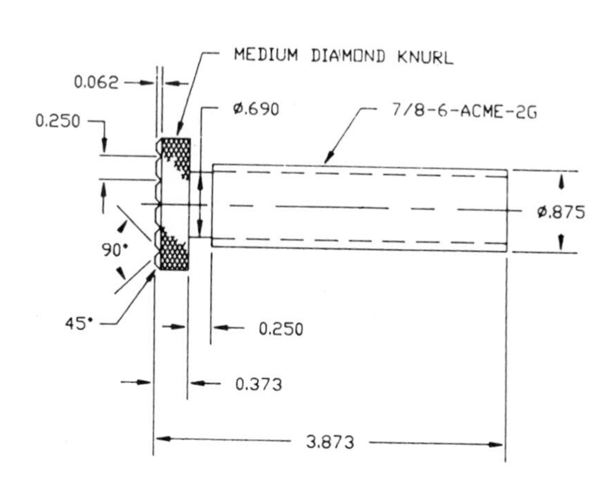

Problem 9–22 (in.)

Part Name: Screw Shaft
Material: 3/4 Hex × 4 1/8 Stock Mild Steel

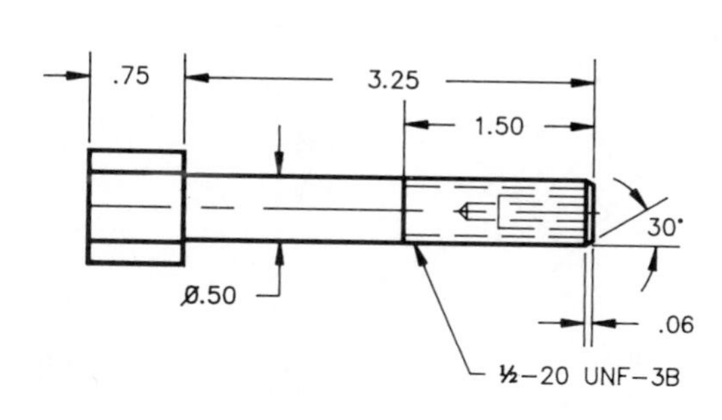

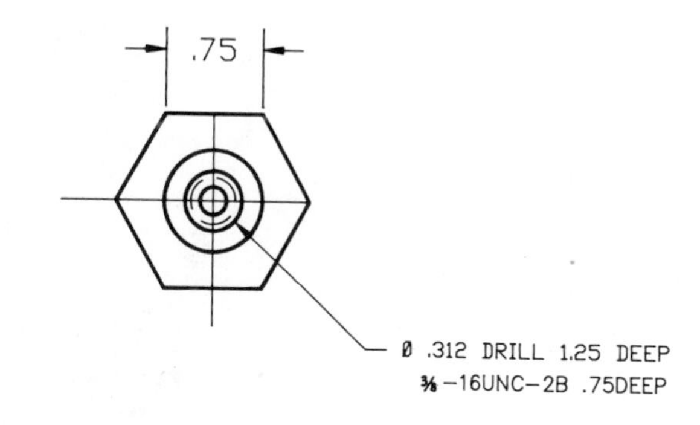

Problem 9–23 (in.)

Part Name: Right Side Member
Material: Aluminum

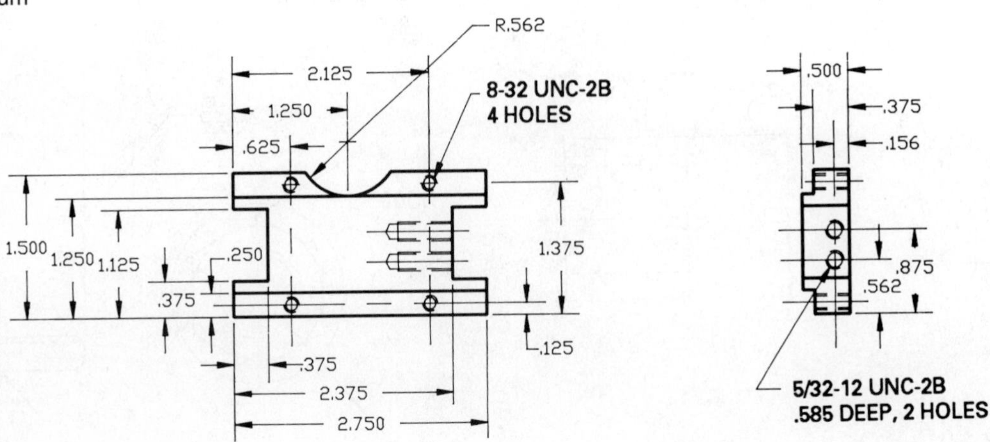

Problem 9–24 (in.)

Part Name: Angled Support Fixture
Material: Mild Steel
Note: Hidden lines omitted for clarity.

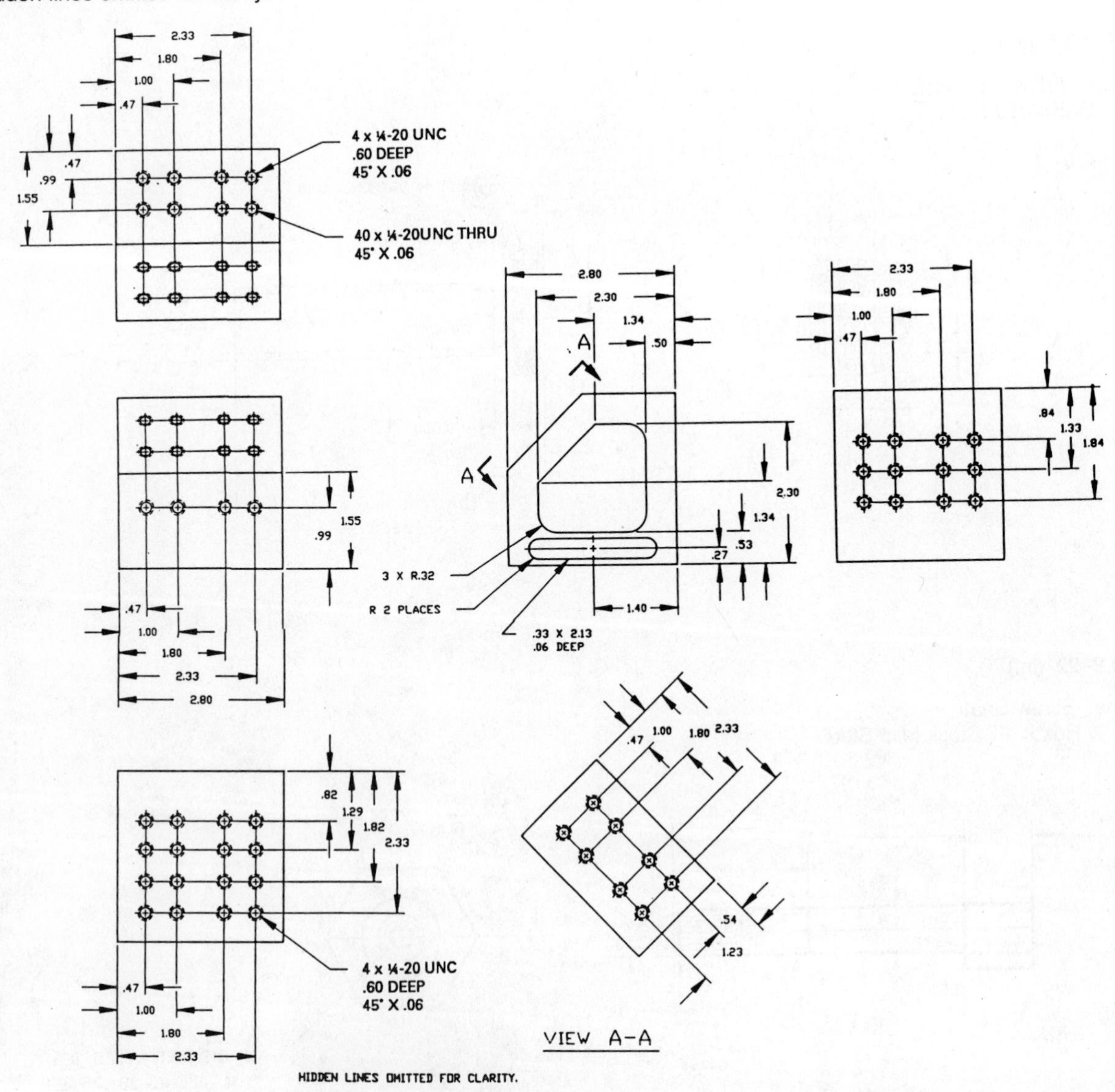

Problem 9–25 (in.)

Part Name: Packing Nut
Material: Bronze

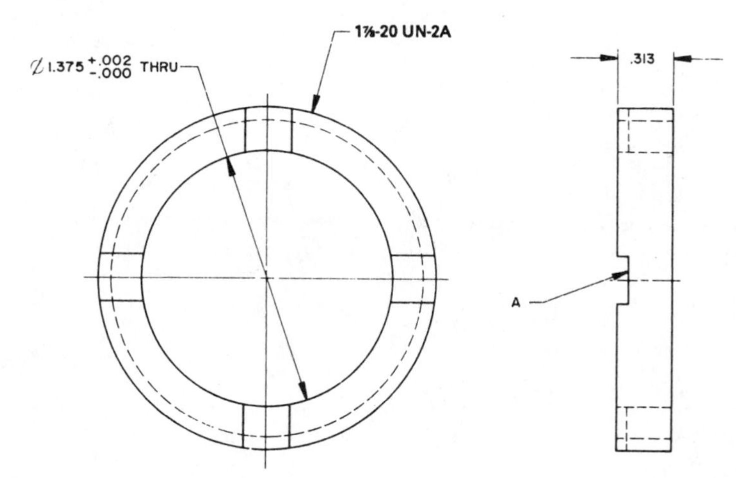

Problem 9–26 (in.)

Part Name: Pump Pivot Support
Material: Cold-rolled Mild Steel

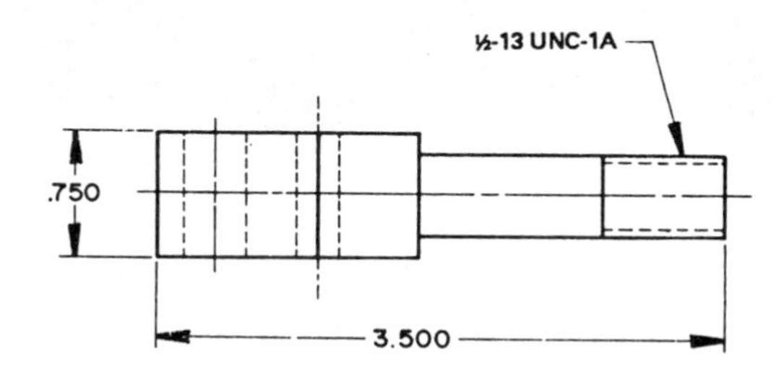

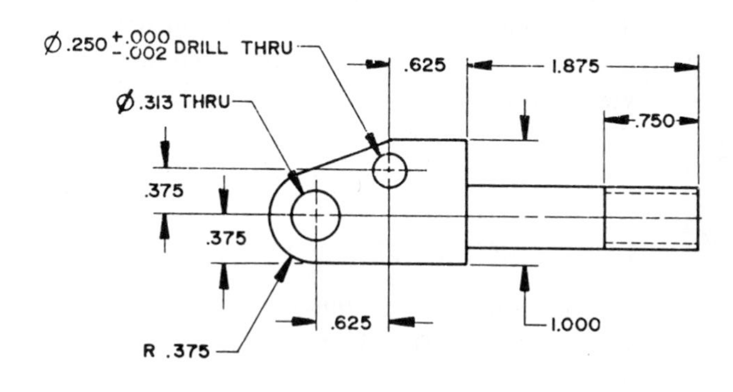

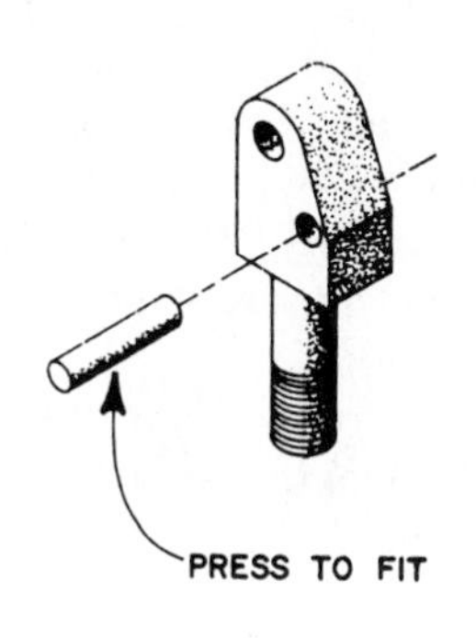

Problem 9–27 (in.)

Part Name: Set Screw
Material: Steel
SPECIFIC INSTRUCTIONS:
Prepare a detailed drawing from the written instructions below.

10-32 UNF-2A THREADED
MATERIAL .375 LG.
W/ .015 LATERAL X 35°
CHAMFERS ON BOTH ENDS
PROVIDE A .030 WIDE X
.047 DEEP SLOT ON
ONE END.

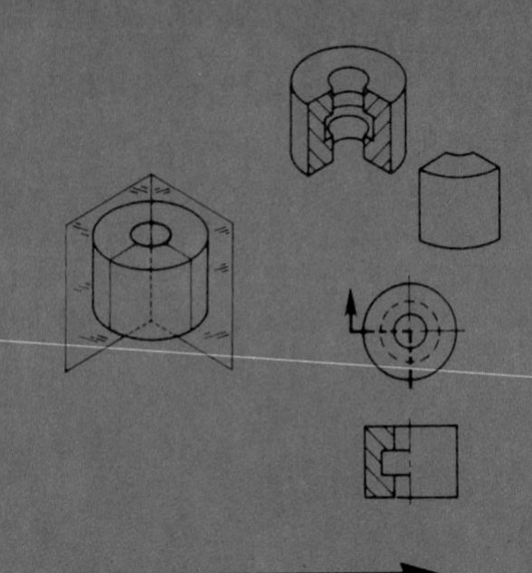

CHAPTER 10

Sections, Revolutions, and Conventional Breaks

LEARNING OBJECTIVES

After completing this chapter you will:

- Draw proper cutting-plane line representations.
- Draw sectional views, including full, half, aligned, broken-out, auxiliary, revolved, and removed sections.
- Prepare drawings with conventional revolutions and conventional breaks.
- Modify the standard sectioning techniques as applied to specific situations.
- Make sectional drawings from given engineer's sketches and actual industrial layouts.

ANSI The American National Standards Institute document that governs sectioning techniques is titled *Multi and Sectional View Drawings* ANSI Y14.3–1975. The content of this chapter is based upon the ANSI standard and provides an in-depth analysis of the techniques and methods of sectional view presentation.

SECTIONING

Sections, or *sectional views*, are used to describe the interior portions of an object that are otherwise difficult to visualize. Interior features that are described using hidden lines may not appear as clear as if they were exposed for viewing as visible features. It is also a poor practice to dimension to hidden features, but the sectional view allows you to expose the hidden features for dimensioning. Figure 10–1 shows an object in conventional multiview representation and using a sectional view. Notice how the hidden features are clarified in the sectional view.

CUTTING-PLANE LINES

The sectional view is obtained by placing an imaginary cutting plane through the object as if you were to cut away the area to be exposed. The adjacent view then becomes the sectional view by removing the portion of the object between the viewer and the cutting plane. (See Figure 10–2).

The sectional view should be projected from the view that has the cutting plane as you would normally project a view in multiview. The *cutting-plane line* is a thick line that represents the cutting plane as shown in Figure 10–2. The cutting-plane line is capped on the ends with arrowheads that show the direction of sight of the sectional view. When the extent of the cutting plane is obvious, only the ends of the cutting-plane line may be used as shown in Figure 10–3. Such treatment of the cutting plane also helps keep the view clear of excess lines.

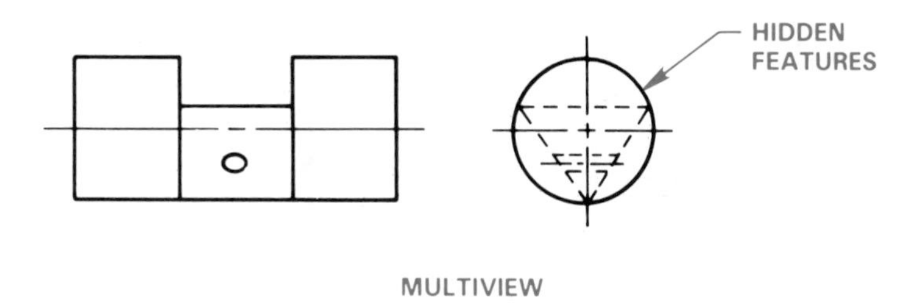

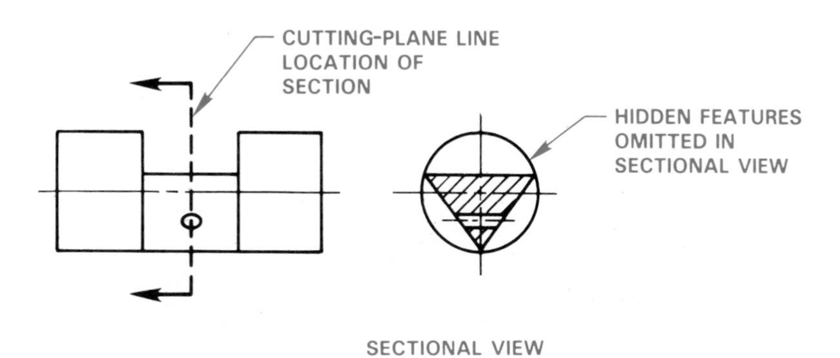

Figure 10–1 Conventional multiview compared to a sectional view.

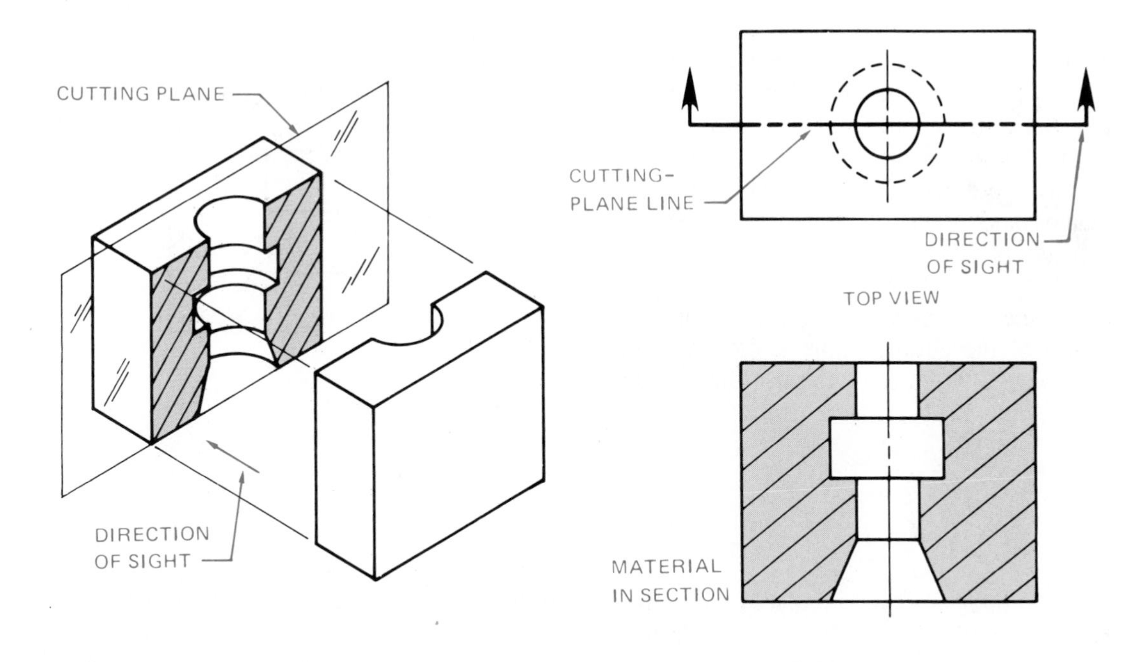

Figure 10–2 Cutting-plane line.

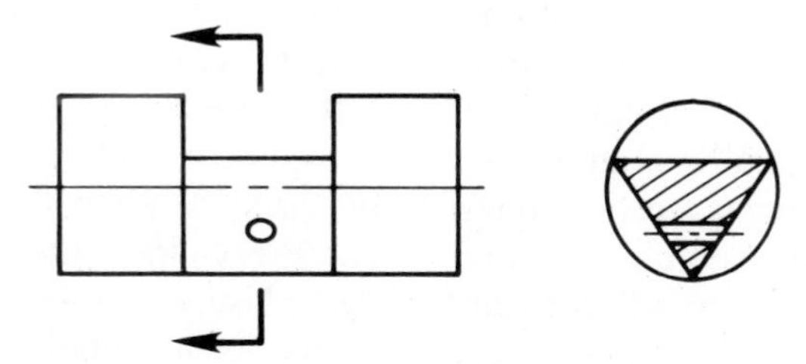

Figure 10–3 Simplified cutting-plane line.

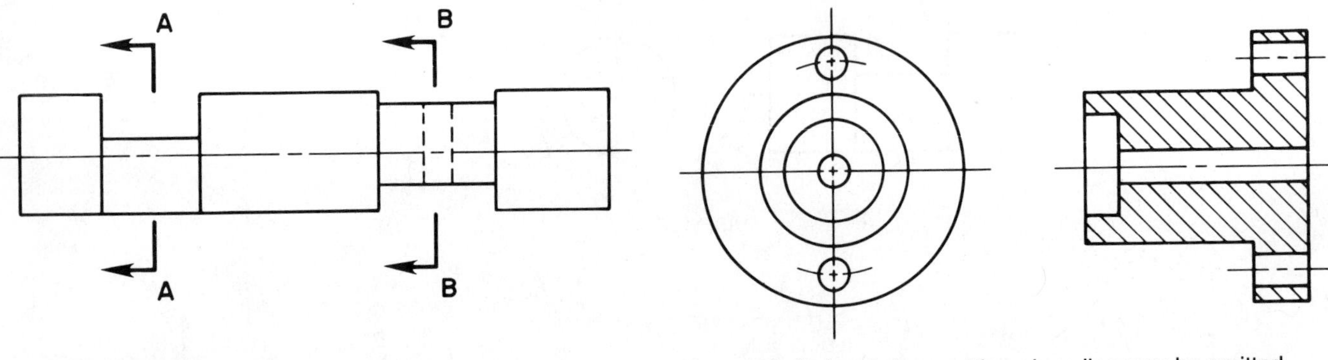

Figure 10–5 An obvious cutting-plane line may be omitted.

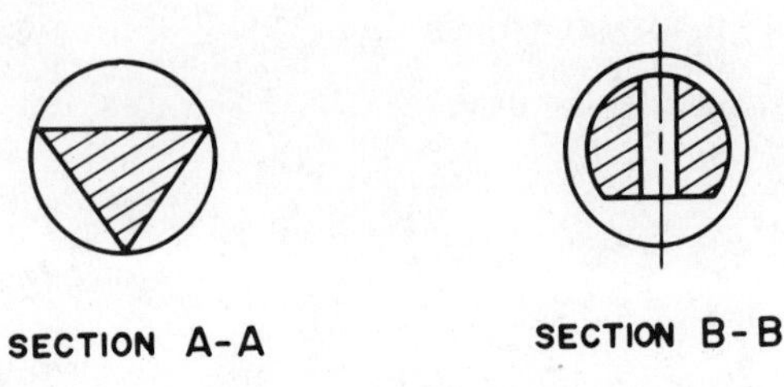

Figure 10–4 Labeled cutting-plane lines and related sectional views.

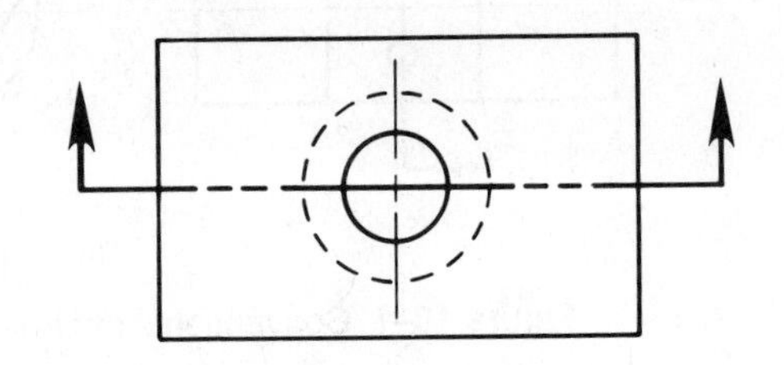

If lack of space restricts the normal placement of a sectional view, the view may be placed in an alternate location. When this is done, the sectional view should not be rotated but should remain in the same orientation as if it were a direct projection from the cutting plane. The cutting planes and related sectional views should be labeled with letters beginning with A, as shown in Figure 10–4.

The cutting-plane line may be completely omitted when the location of the cutting plane is clearly obvious. (See Figure 10–5.) When in doubt, use the cutting-plane line.

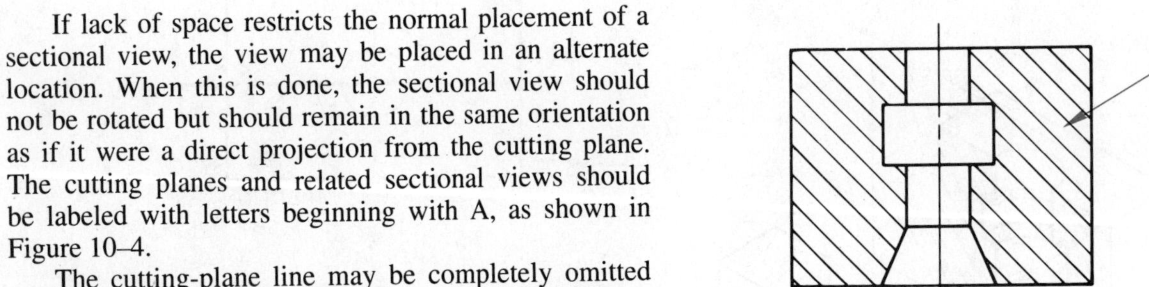

Figure 10–6 Section lines.

SECTION LINES

Section lines are thin lines used in the view of the section to denote where the cutting-plane line has cut through material. (See Figure 10–6.) Also refer to Chapter 3. Section lines are usually drawn equally spaced at 45° but may not be parallel or perpendicular to any line of the object. Any convenient angle may be used to avoid placing section lines parallel or perpendicular to other lines of the object. Angles of 30° and 60° are common. Section lines that are more than 75° or less than 15° from horizontal should be avoided. Section lines must never be drawn either horizontally or vertically. Figure 10–7 shows some common errors in drawing section lines. Section lines should be drawn in opposite directions on adjacent parts, and when several parts are adjacent any suitable angle may be used to make the parts appear clearly separate. When a very large area requires section lining, you may elect to use outline section lining.

Equally spaced section lines denote either a general material designation or cast iron. This method of drawing section lines is quick and easy with the actual material identification located in the drawing title block. The other option is to use coded section lining, which is more time-consuming to draw than general section lining. Coded section lining may be used effectively when a section is taken through an assembly of adjacent parts of different materials as seen in Figure 10–8.

General section lines are evenly spaced. The amount of space between lines is dependent on the size of the part. Very large parts have larger spacing than very small parts. Use your own judgment. The key is to clearly represent section lines without an unnecessary expenditure

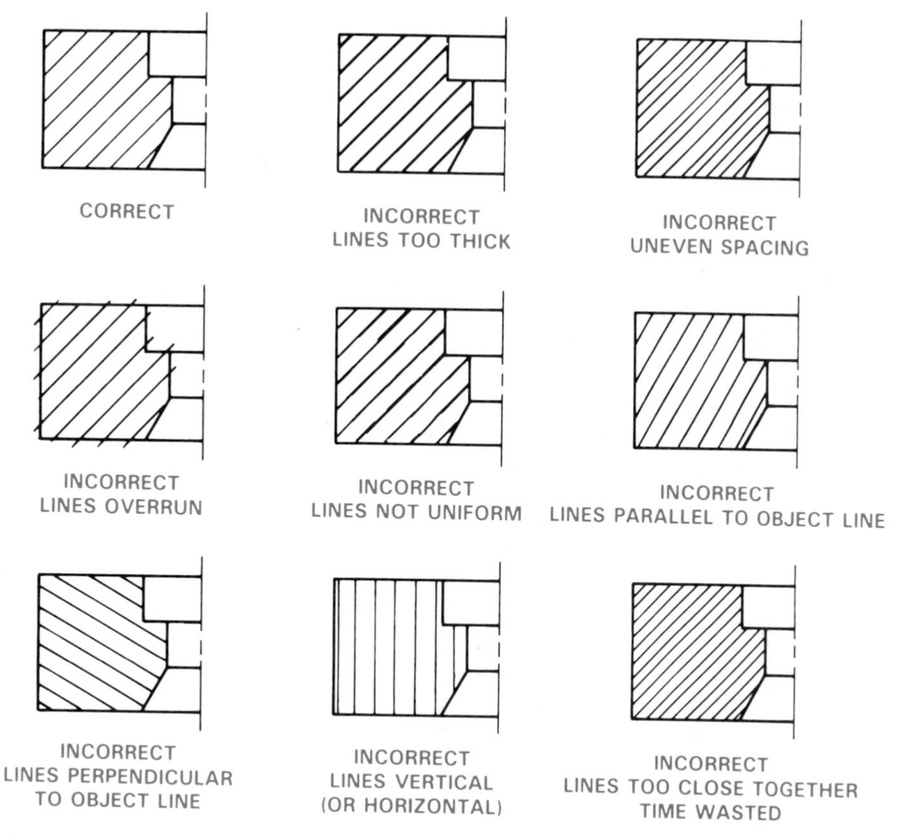

Figure 10–7 Common section line errors.

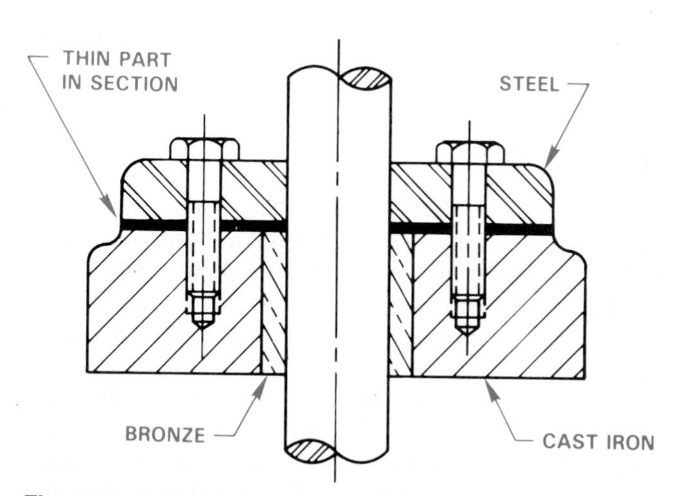

Figure 10–8 Assembly section, coded section lines.

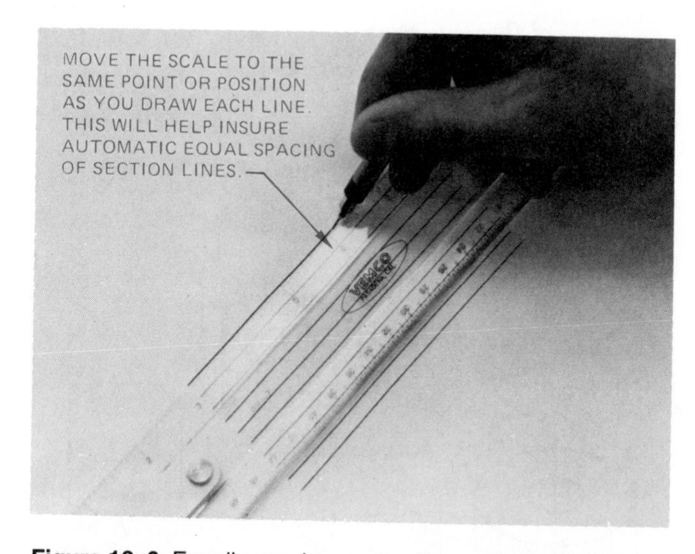

Figure 10–9 Equally spacing section lines with the drafting machine scale.

of time. There are a number of ways to space section lines equally without measuring each space. One good way is to calibrate the space using the same point or mark on the drafting machine scale as you draw each line. This method works with clear plastic scales as shown in Figure 10–9. Another technique is to use the Ames Lettering Guide for drawing equally-spaced section lines. Refer to Chapter 3 to review the use of this tool.

FULL SECTIONS

A *full section* is drawn when the cutting plane extends completely through the object, usually along a center plane as shown in Figure 10–10. The object shown in Figure 10–10 could have used two full sections to further clarify hidden features. In such a case, the cutting planes and related views are labeled. (See Figure 10–11.)

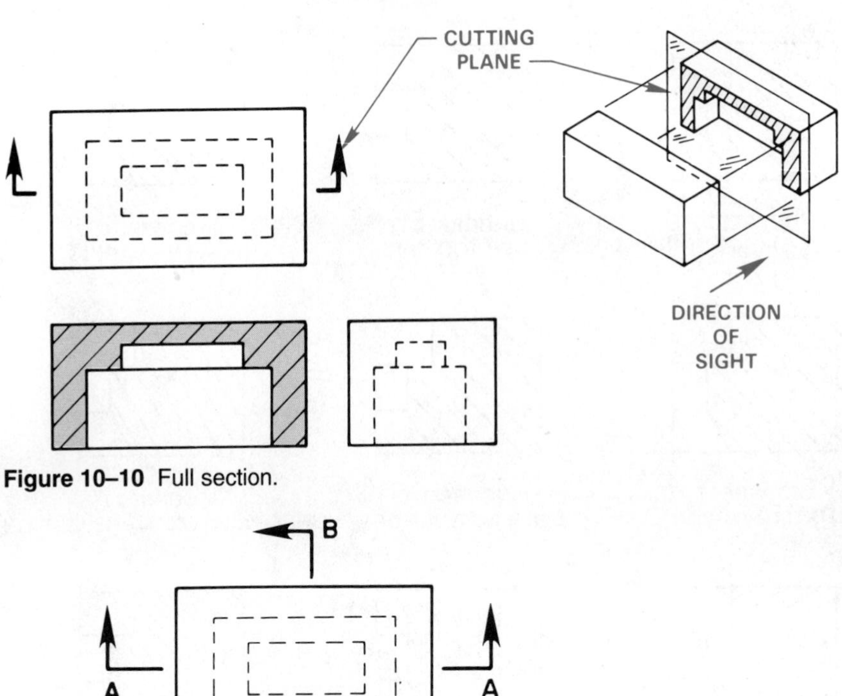

Figure 10–10 Full section.

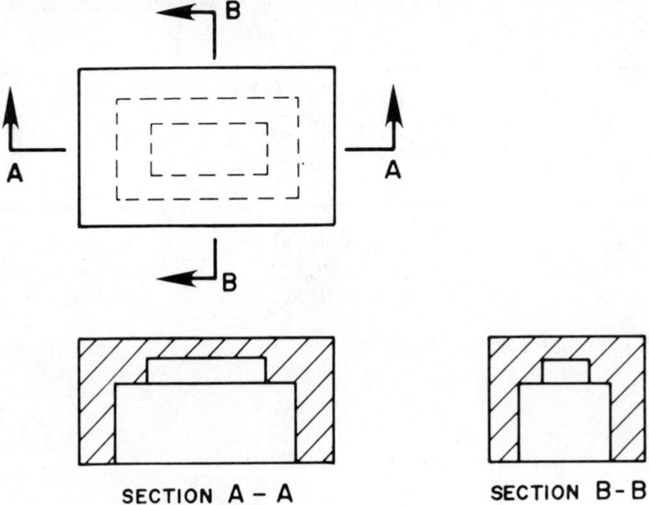

Figure 10–11 Full sections.

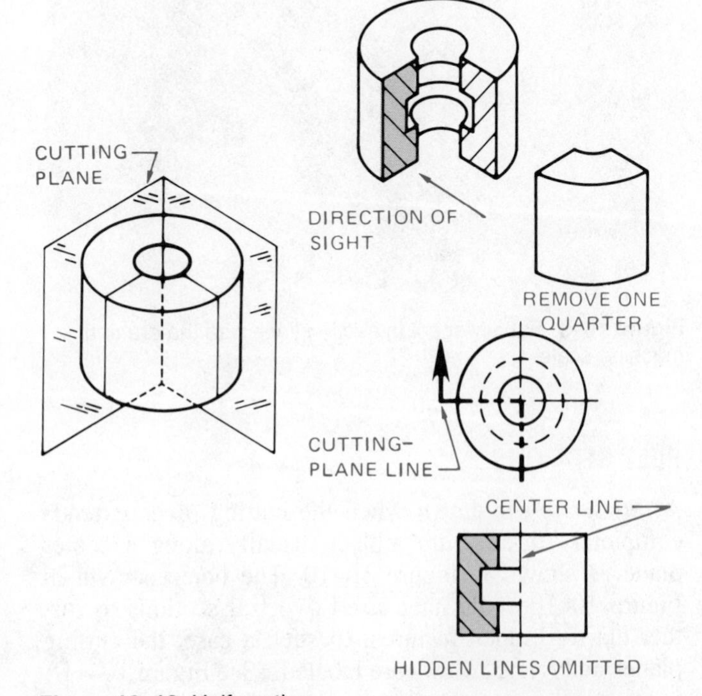

Figure 10–12 Half section.

HALF SECTIONS

A *half section* may be used when a symmetrical object requires sectioning. The cutting-plane line of a half section actually removes one-quarter of the object. The advantage of a half section is that the sectional view shows half of the object in section and the other half of the object as it would normally appear. Thus the name half section. (See Figure 10–12.) Notice that a centerline is used in the sectional view to separate the sectioned portion from the unsectioned portion. Hidden lines are generally omitted from sectional views unless their use improves clarity.

OFFSET SECTIONS

Staggered interior features of an object may be sectioned by allowing the cutting-plane line to *offset* through the features as shown in Figure 10–13. Notice in Figure 10–13 that there is no line in the sectional view indicating a change in direction of the cutting-plane line. Normally the cutting-plane line in an offset section

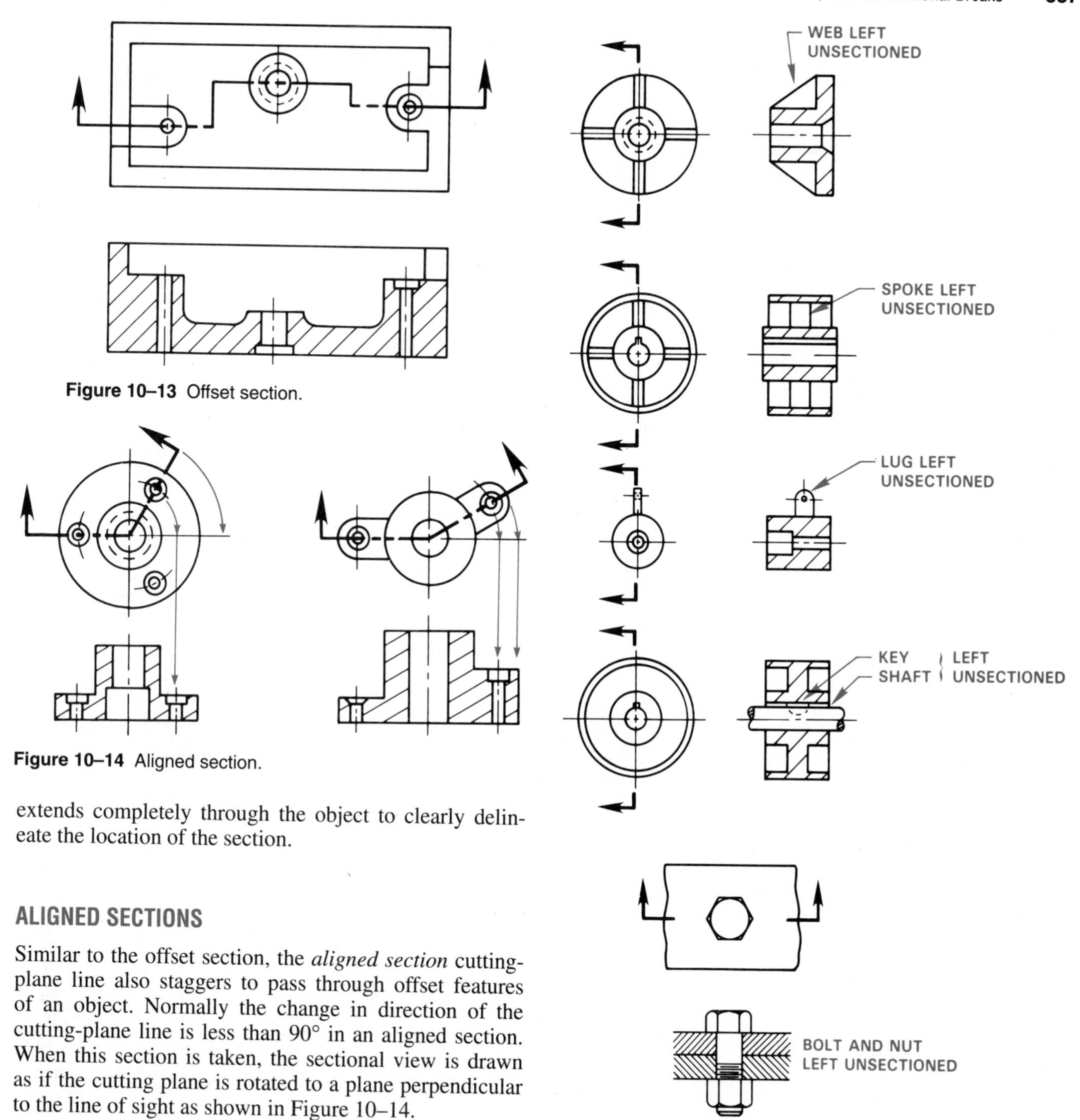

Figure 10–13 Offset section.

Figure 10–14 Aligned section.

Figure 10–15 Certain features are not sectioned when a cutting-plane line passes parallel to their axes.

extends completely through the object to clearly delineate the location of the section.

ALIGNED SECTIONS

Similar to the offset section, the *aligned section* cutting-plane line also staggers to pass through offset features of an object. Normally the change in direction of the cutting-plane line is less than 90° in an aligned section. When this section is taken, the sectional view is drawn as if the cutting plane is rotated to a plane perpendicular to the line of sight as shown in Figure 10–14.

UNSECTIONED FEATURES

Specific features of an object are commonly left unsectioned in a sectional view if the cutting-plane line passes through the feature and parallel to it. The types of features that are left unsectioned for clarity are bolts, nuts, rivets, screws, shafts, ribs, webs, spokes, bearings, gear teeth, pins, and keys. (See Figure 10–15.) When the cutting-plane line passes through the previously described features perpendicular to their axes, then section lines are shown as seen in Figure 10–16.

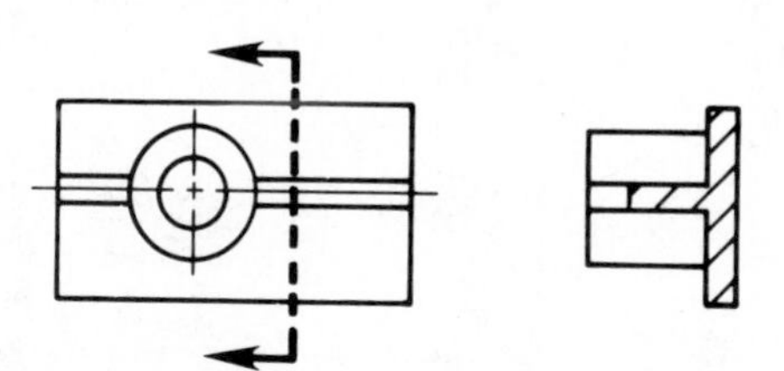

Figure 10–16 Cutting plane perpendicular to normally unsectioned features.

SECTIONING

Everyone should have the chance to spend what often seems like hours drawing section lines in a complex sectional view to really appreciate the speed and accuracy of drawing section lines with a computer. One important aspect of drawing section lines manually is getting the lines uniform and equally spaced. This task is automatic with CADD. Additionally, most CADD drafting packages have more section-line symbols available than you would ever have the need to use. CADD section-line commands such as HATCH are used to select one of a variety of patterns and you have the opportunity to change the section-line scale if you wish. Figure 10–17 shows three different section-line scale factors available.

When using CADD to section-line an area you need to define the boundaries of the area. In other words, tell the computer where the section lines are to be placed so you do not end up with lines in an unwanted area. If you want to section-line inside a circle or square that is an individual unit (entity), then all you need to do is pick the circle or square. However, if each side of the square is an individual entity then you need to place a window around the object to include all lines in the selection. (See Figure 10–18.)

Some CADD programs allow you to draw section lines in an enclosed area simply by picking a point within the area. In AutoCAD, for example, the enclosed area is called a boundary. You get an error message, "BOUNDARY AREA NOT CLOSED," if there is a gap in the boundary. A common boundary gap might be a corner where lines do not meet. You may have to "ZOOM" in on the corner to actually see a very small gap. If this happens, correct the problem and try to section the area again as shown in Figure 10–19.

The computer makes drawing section lines for any drafting field quick and easy. There are standard material section lines for the mechanical drafter, and brick and other patterns for the architect. A large variety of section-line symbols and graphic patterns are available in most CADD packages, or as tablet menu overlays as shown in Figure 10–20.

SCALE = 1 SCALE = 2 SCALE = 3

Figure 10–17 Three different section-line scale factors for CADD applications.

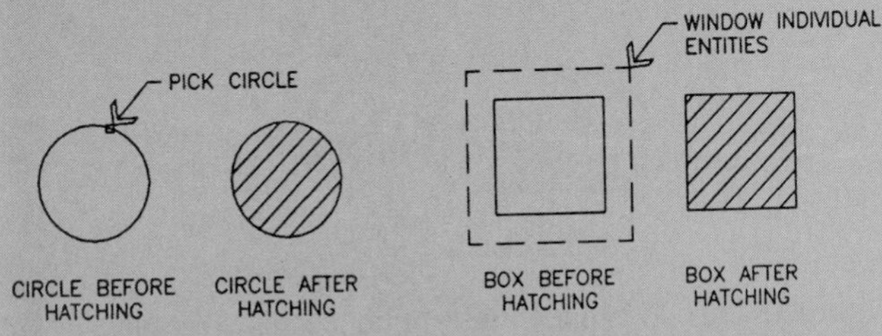

Figure 10–18 Selecting objects for sectioning with CADD.

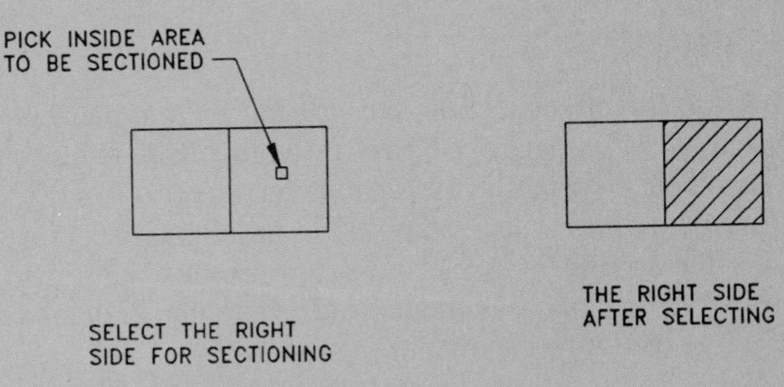

Figure 10–19 Recommended drawing sequence for section-lining adjacent areas with CADD.

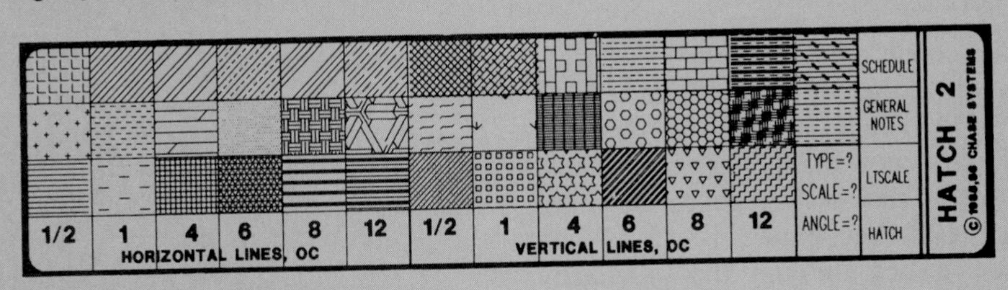

Figure 10–20 CADD tablet overlay for section lines and graphic patterns. *Courtesy Chase Systems.*

CONVENTIONAL REVOLUTIONS

When the true projection of a feature results in foreshortening, the feature should be revolved onto a plane perpendicular to the line of sight as in Figure 10–21. The revolved spoke shown in Figure 10–21 gives a clear representation with a minimum of drafting time. Figure 10–22 shows another illustration of conventional revolution compared to true projection. Notice how the true projection results in a distorted and foreshortened representation of the spoke. The revolved spoke in the preferred view is clear and easy to draw. The practice illustrated in Figure 10–21 and Figure 10–22 also applies to features of unsectioned objects in multiview as shown in Figure 10–23.

BROKEN-OUT SECTIONS

Often a small portion of a part may be broken away to expose and clarify an interior feature. This technique is called a *broken-out section*. There is no cutting-plane line used as you can see in Figure 10–24. A short break line is generally used with a broken-out section.

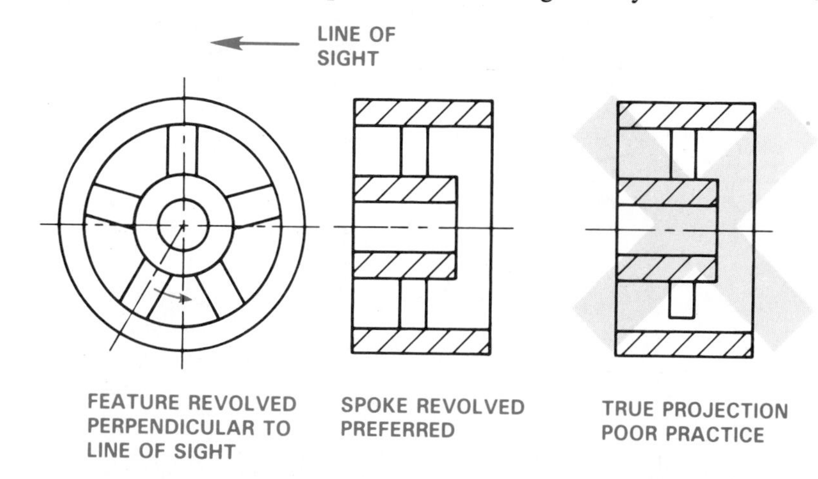

Figure 10–21 Conventional revolution.

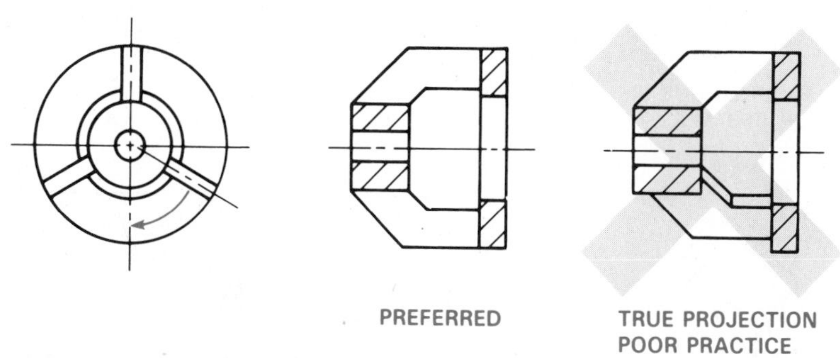

Figure 10–22 Conventional revolution in section.

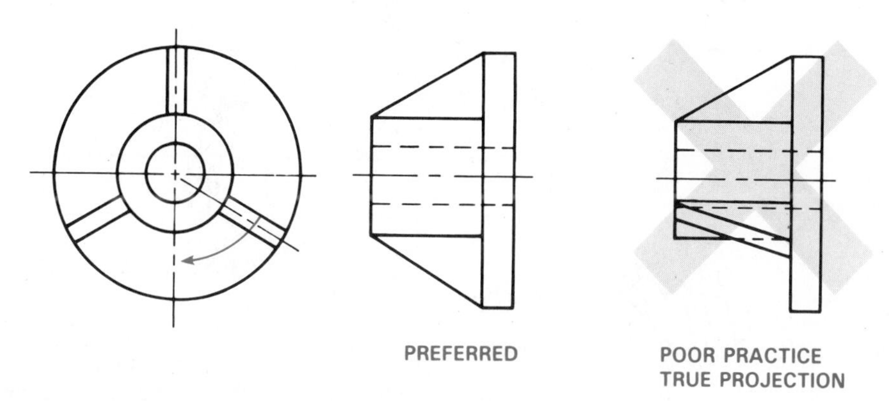

Figure 10–23 Conventional revolution in multiview.

AUXILIARY SECTIONS

A section that appears in an auxiliary view is known as an *auxiliary section*. Auxiliary sections are generally projected directly from the view of the cutting plane. If these sections must be moved to other locations on the drawing sheet, they should remain in the same relationship (not rotated) as if taken directly from the view of the cutting plane. (See Figure 10–25.)

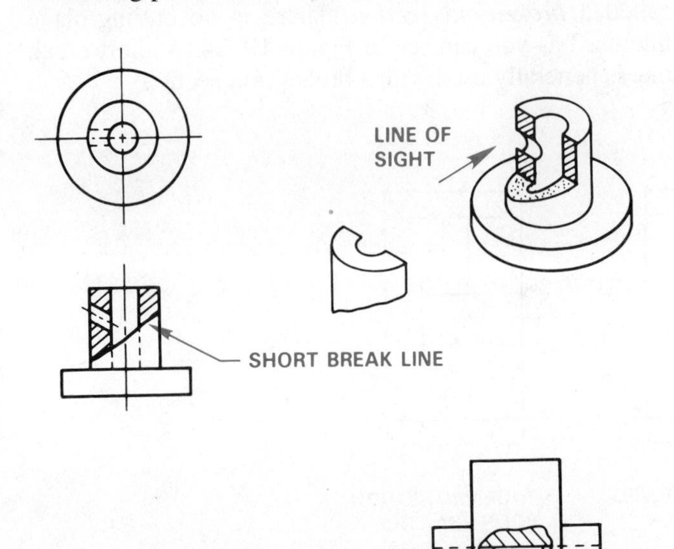

Figure 10–24 Broken-out section.

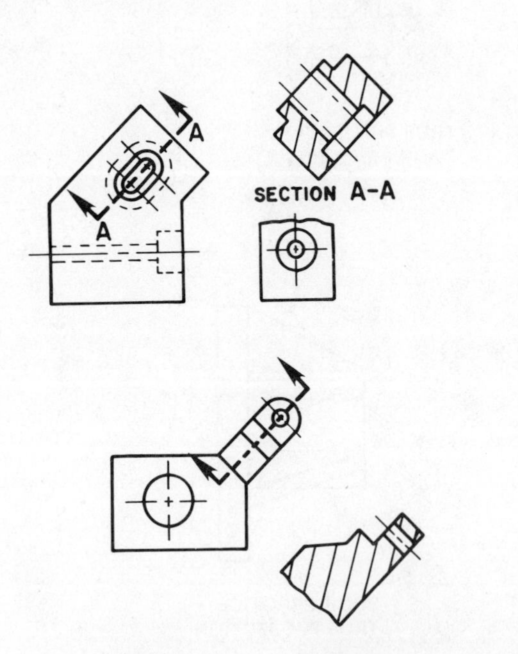

Figure 10–25 Auxiliary sections.

CONVENTIONAL BREAKS

When a long object of constant shape throughout its length requires shortening, *conventional breaks* may be used. These breaks may be used effectively to save time, paper, or space or to increase the scale of an otherwise very long part. Figures 10–26(a), 10–26(b), 10–26(c), and 10–26(d) show typical conventional breaks. Used on metal shapes, the short break line is drawn thick, freehand, and slightly irregular. Notice that the actual length can be given with a long break line used as a dimension line. For wood shapes, the short break line is drawn freehand as a thick, very irregular line.

The break line for solid round shapes may be drawn freehand or with an irregular curve or template. The

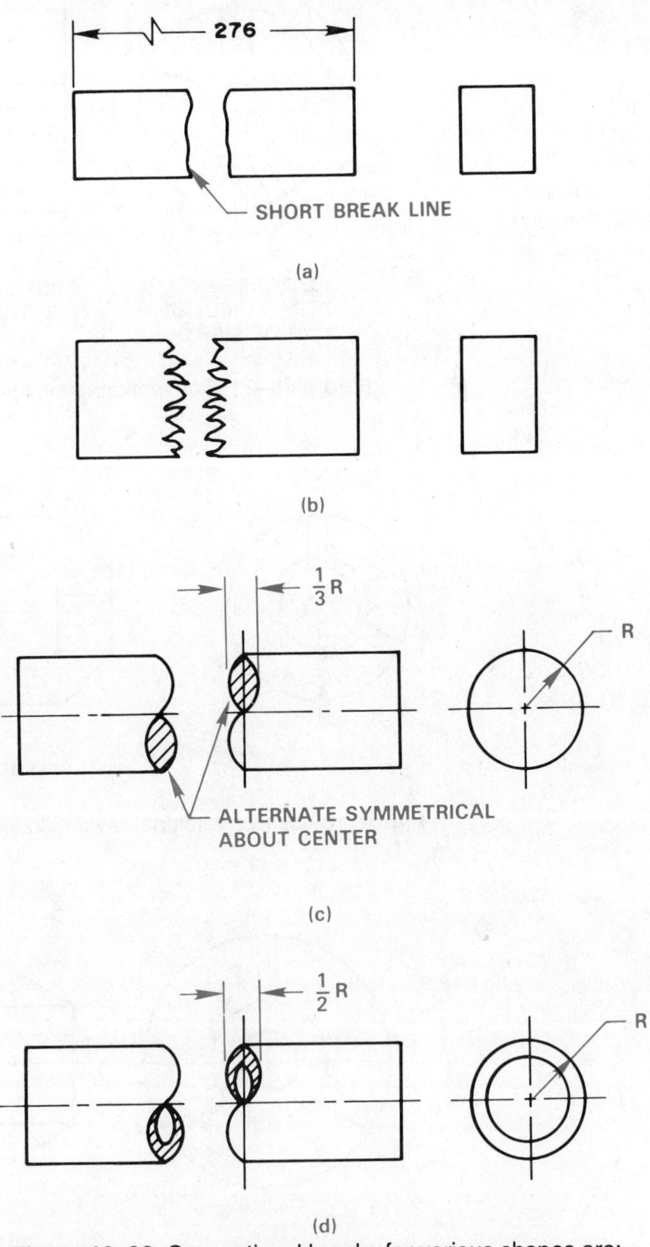

Figure 10–26 Conventional breaks for various shapes are: (a) metal shapes, (b) wood shapes; (c) cylindrical solid shapes; (d) cylindrical tubular shapes.

shape widths should be approximately 1/3 radius and should be symmetrical about the horizontal centerline and the vertical guidelines as shown.

The break lines for tubular round shapes may be drawn freehand or with an irregular curve or template. The total shape widths should be approximately 1/2 radius and should be symmetrical about the horizontal centerline and vertical guidelines as shown.

REVOLVED SECTIONS

When a feature has a constant shape throughout the length that cannot be shown in an external view, a *revolved section* may be used. The desired section is revolved 90° onto a plane perpendicular to the line of sight as shown in Figure 10–27. Revolved sections may be represented on a drawing one of two ways as shown in Figure 10–28(a) and 10–28(b). In Figure 10–28(a) the revolved section is drawn on the part. The revolved section also may be broken away as seen in Figure 10–28(b). The surrounding space may be used for dimensions as shown in Figure 10–29.

Notice in Figure 10–29 that very thin parts in sections may be filled in as opposed to using section lines. This practice is also common when sectioning a gasket or similar feature.

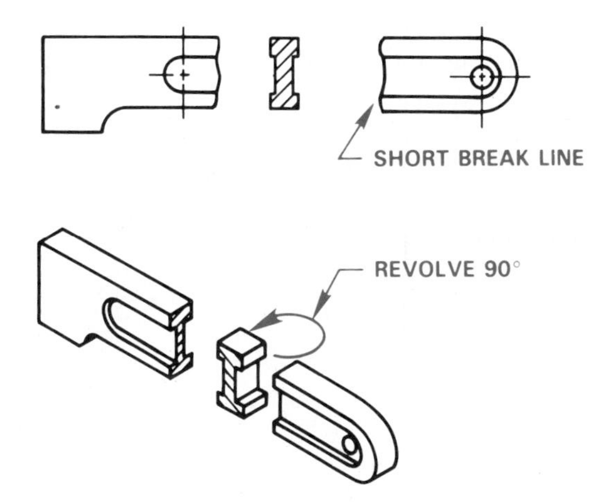

Figure 10–27 Revolved section.

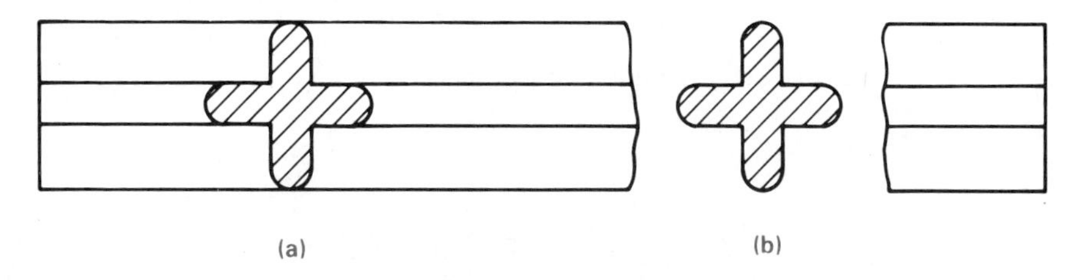

Figure 10–28 (a) Revolved section not broken away; (b) revolved section broken away.

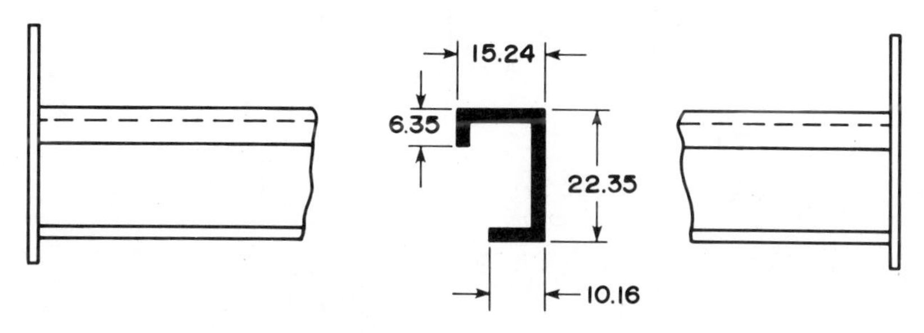

Figure 10–29 Dimensioning a broken-away revolved section.

REMOVED SECTIONS

Removed sections are similar to revolved sections except that they are removed from the view. A cutting-plane line is placed through the object where the section is taken. Removed sections are not generally placed in direct alignment with the cutting-plane line but are placed in a surrounding area as shown in Figure 10–30.

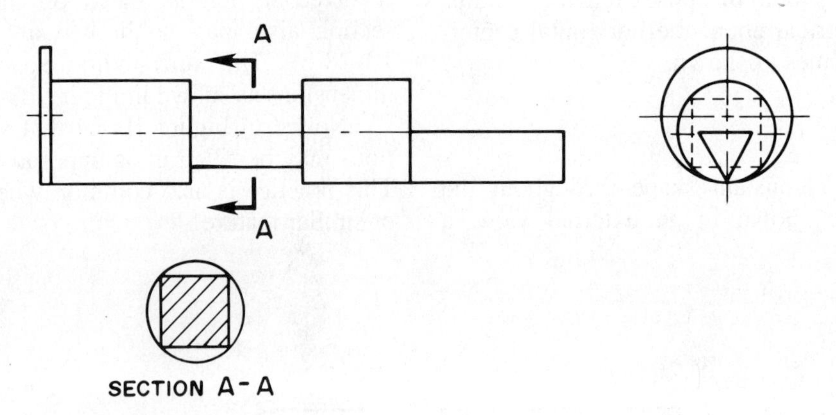

Figure 10–30 Removed section.

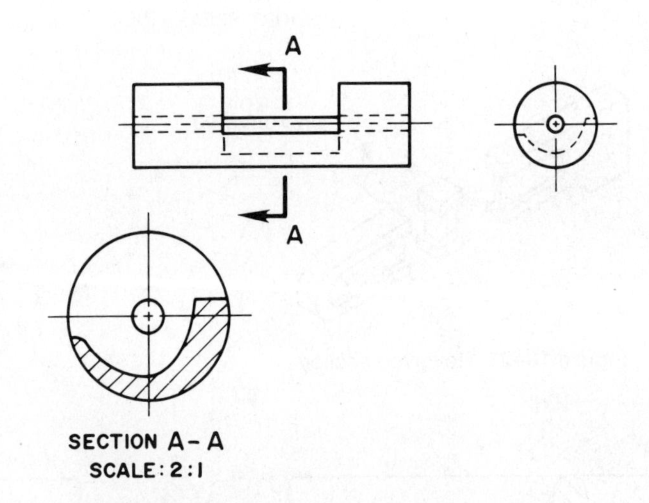

Figure 10–31 Enlarged removed section.

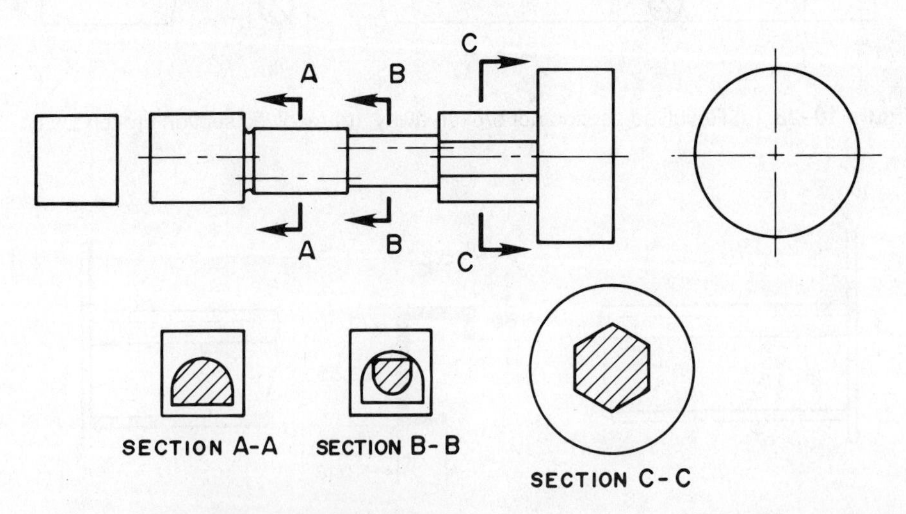

Figure 10–32 Multiple removed sections.

Removed sections may be preferred when a great deal of detail makes it difficult to effectively use a revolved section. An additional advantage of the removed section is that it may be drawn to a larger scale so that close detail may be more clearly identified as shown in Figure 10–31. The sectional view should be labeled, as shown, with Section A–A and the revised scale which in this example is 2:1. The predominant scale of the principal views is shown in the title block.

Multiple removed sections are generally arranged on the sheet in alphabetical order from left to right and top to bottom. (See Figure 10–32.) Notice that hidden lines have been omitted for clarity. The cutting planes and related sections are labeled alphabetically excluding the letters I, O, and Q as they may be mistaken for numbers. When the entire alphabet has been used, label your sections with double letters beginning with AA, BB, and so on.

Another method of drawing a removed section is to extend a centerline adjacent to a symmetrical feature and revolve the section on the centerline as shown in Figure 10–33. The removed section may be drawn at the same scale or enlarged as necessary to clarify detail.

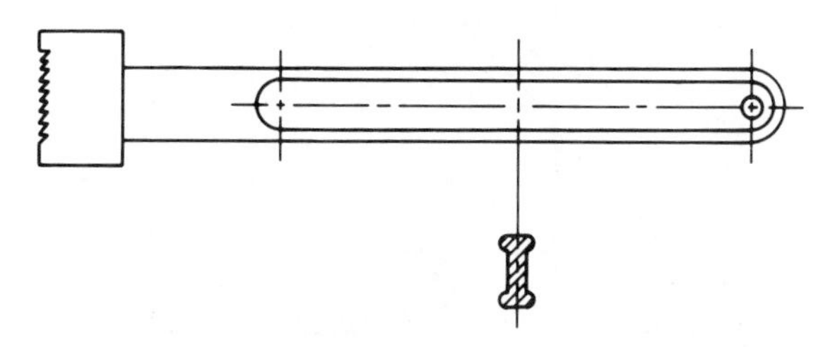

Figure 10–33 Alternate removed section method.

PROFESSIONAL PERSPECTIVE

Your role as an engineering drafter is communication. You are the link between the engineer and manufacturing, fabrication, or construction. Up to this point you have learned a lot about how to perform this communication task, but the hard part comes when you have to apply what you have learned. Your goal is to make every drawing clear, complete, and easy to interpret. There are a lot of factors to consider, including:

- Selection of the front view and related views.
- Deciding if sections are to be used.
- Placement of dimensions in a logical format so that the size description and located features are complete.

This is a difficult job, but you are motivated to do it. And just when you have mastered how to select and dimension multiviews, you are faced with preparing sectional drawings, which means a dozen main options and endless applications. (But don't worry because you have made it this far!) Here are some key points to consider:

- You need to completely show and describe the outside of the part first.
- If there are a lot of hidden lines for internal features, you know immediately that a section is needed, because you cannot dimension to hidden lines.
- Analyze the hidden lines and decide on the type of section that will work best.
- Review each type of sectioning technique until you find one that looks best for the application.

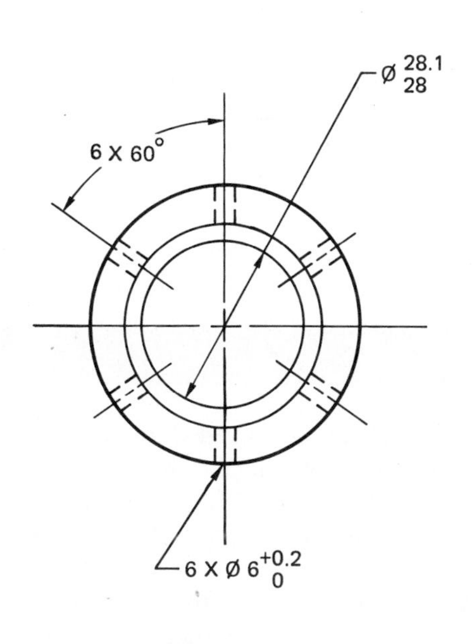

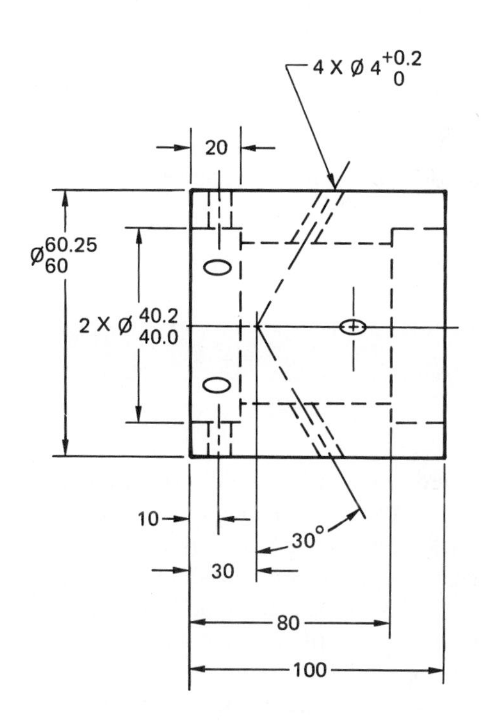

Figure 10–34 Engineer's layout.

- Be sure you clearly label the cutting-plane lines and related multiple sections so they correlate.
- Half sections are good for showing both the outside and inside of the object at the same time, but be cautious about their use, because sometimes they confuse the reader.

You're not alone in this new venture. The problem assignments in this chapter recommend a specific sectioning technique during this learning process.

ENGINEERING PROBLEM

You are working with an engineer on a special project and she gives you a rough sketch of a part from which you are to make a formal drawing. At first glance, you think it is an easy part to draw, but with further study you realize that it is more difficult than it looks. The rough sketch, shown in Figure 10–34, has a lot of hidden features. So your first thought is, "What do I use for the front view?" You decide that a front exterior view would show only the diameter and length. So instead of drawing an outside view, you decide to use a full section to expose all of the interior features and, at the same time, give the overall length and diameter. "This is great!" you say. Then you realize there are 6 holes spaced 60° apart that remain as hidden features in the side (end) view. This is undesirable, because you do not want to section all of the holes. So you decide to use a broken-out section to expose only two holes. This allows you to dimension the 6 × 60° and the 6 × ∅ 6 +0.2 holes all at the same time. Now with all this thinking and planning out of the way, it only takes you three hours to completely draw and dimension the part, and you are ready to give the formal drawing, shown in Figure 10–35, to the checker.

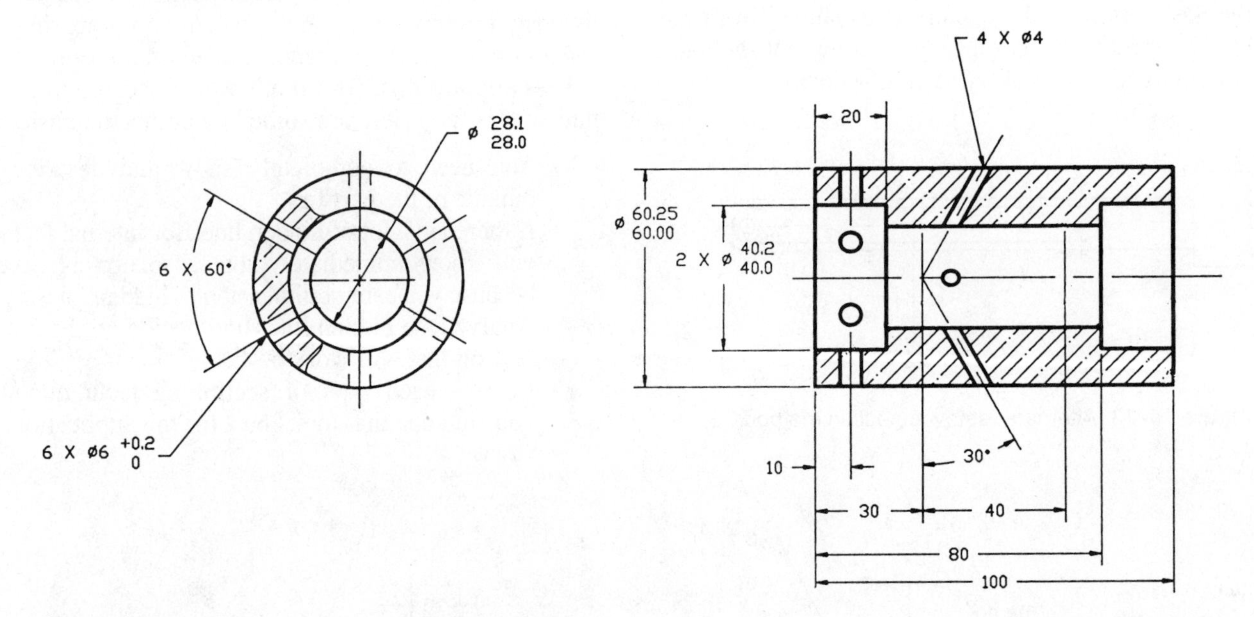

Figure 10–35 CADD solution to engineering problem.

SECTIONING PROBLEMS

DIRECTIONS

1. From the selected engineer's sketch or layout, determine the views and sections that are needed.
2. Make a sketch of the selected views and sections as close to correct proportions as possible. Do not spend a lot of time as the sketch is only a guide. Indicate where the cutting-plane lines and dimensions are to be placed.
3. Using the sketch you have developed as a guide, draw an original sectioned multiview drawing on an adequately sized drawing sheet. Select a scale that properly details the part on the selected sheet size. Use unidirectional dimensioning. Use manual or computer-aided drafting as required by your course guidelines.
4. Include the following general notes at the lower left corner of the sheet .5 in. each way from the corner border lines:

   2. REMOVE ALL BURRS AND SHARP EDGES.

1. INTERPRET DIMENSIONS AND TOLERANCES PER ANSI Y14.5M–1982.

NOTES:
Additional notes may be required depending upon the specifications of each individual assignment. A tolerance block is recommended as shown in problems for Chapter 8 unless otherwise specified.

5. The engineering layouts may not be dimensioned properly. Verify the correct practice before placing dimensions; for example, the diameter symbol should precede the diameter dimension, and leaders should not cross over dimension lines. Check other line and dimensioning techniques for proper standards.

Problem 10–1 Full section (in.)

Part Name: Fitting
Material: Bronze
Finish All Over: 63 µin.

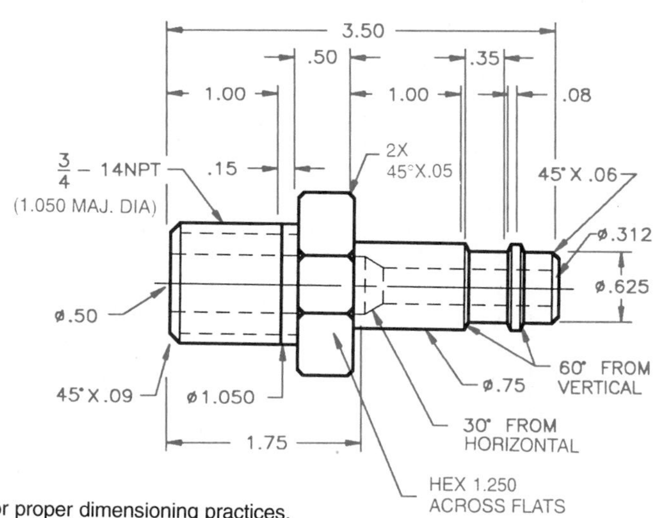

NOTE: Refer to Chapter 8 for proper dimensioning practices. Engineering sketches may not display correct practices. This problem may require a front view, side view showing the hexagon, and a full section to expose the interior features for dimensioning.

Problem 10–2 Full section (in.)

Part Name: Weld Washer Slot
Material: SAE 1040

SPECIFIC INSTRUCTIONS:
Set up dimensions L, T, and W as a chart drawing where the dimensions change for each of the following parts:
Courtesy Production Plastics, Inc.

	L	T	W
PART A	2.125	.500	1.875
PART B	2.250	.625	2.000
PART C	2.500	.750	2.125
PART D	2.625	.875	2.250

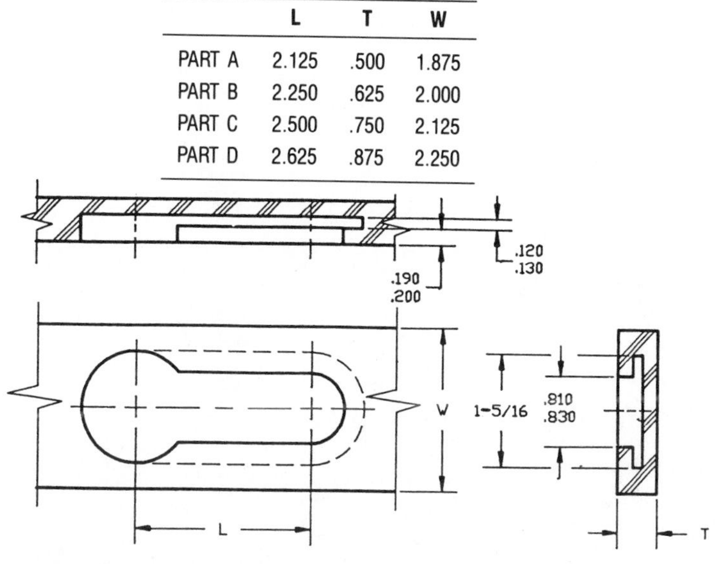

UNLESS OTHERWISE SPECIFIED:

ALL FRACTIONAL DIMENSIONS ±1/32
ALL TWO PLACE DECIMALS ±.010
ALL THREE PLACE DECIMALS ±.005

Problem 10–3 Full section (in.)

Part Name: Weld Coupling
Material: AISI 1010, Killed
SPECIFIC INSTRUCTIONS:
Add diameter symbol to all diameter dimensions.
Courtesy TEMCO.

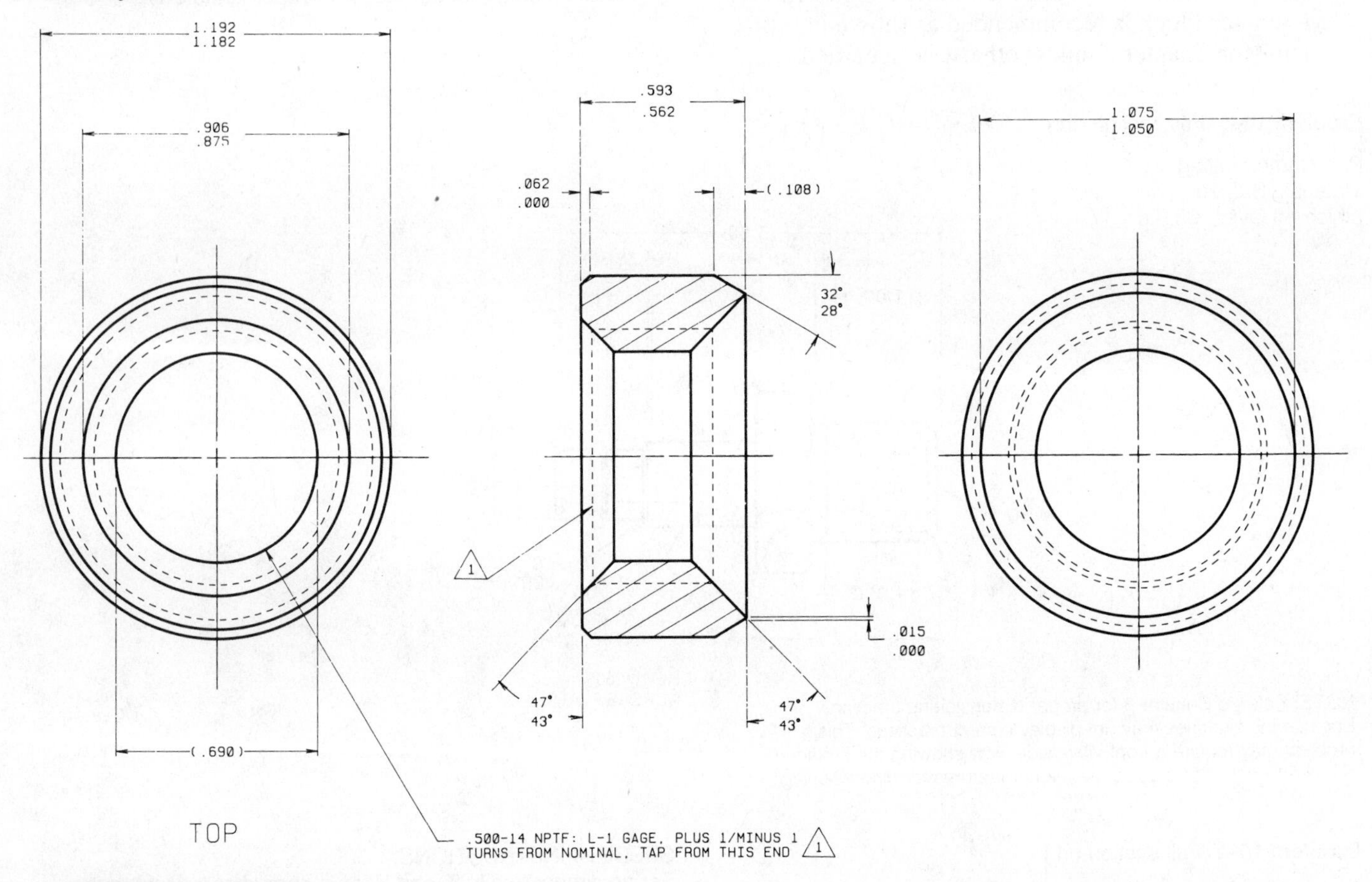

Problem 10–4 Full section (in.)

Part Name: Bolt Guard
Material: Aluminum
Courtesy Stanley Hydraulic Tools, a Division of the Stanley Works.

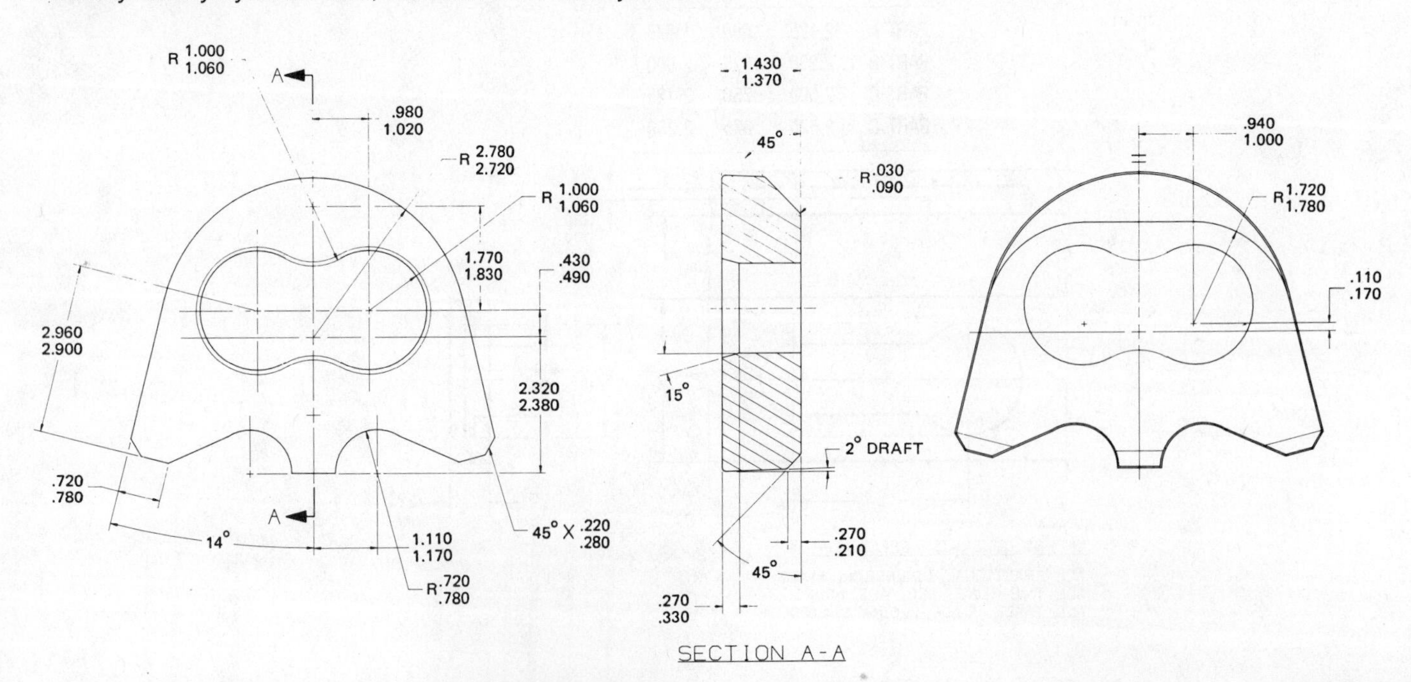

Problem 10–5 Full section and view enlargement (in.)

Part Name: Spring
Material: SAE 1060
Courtesy Stanley Hydraulic Tools, a Division of the Stanley Works.

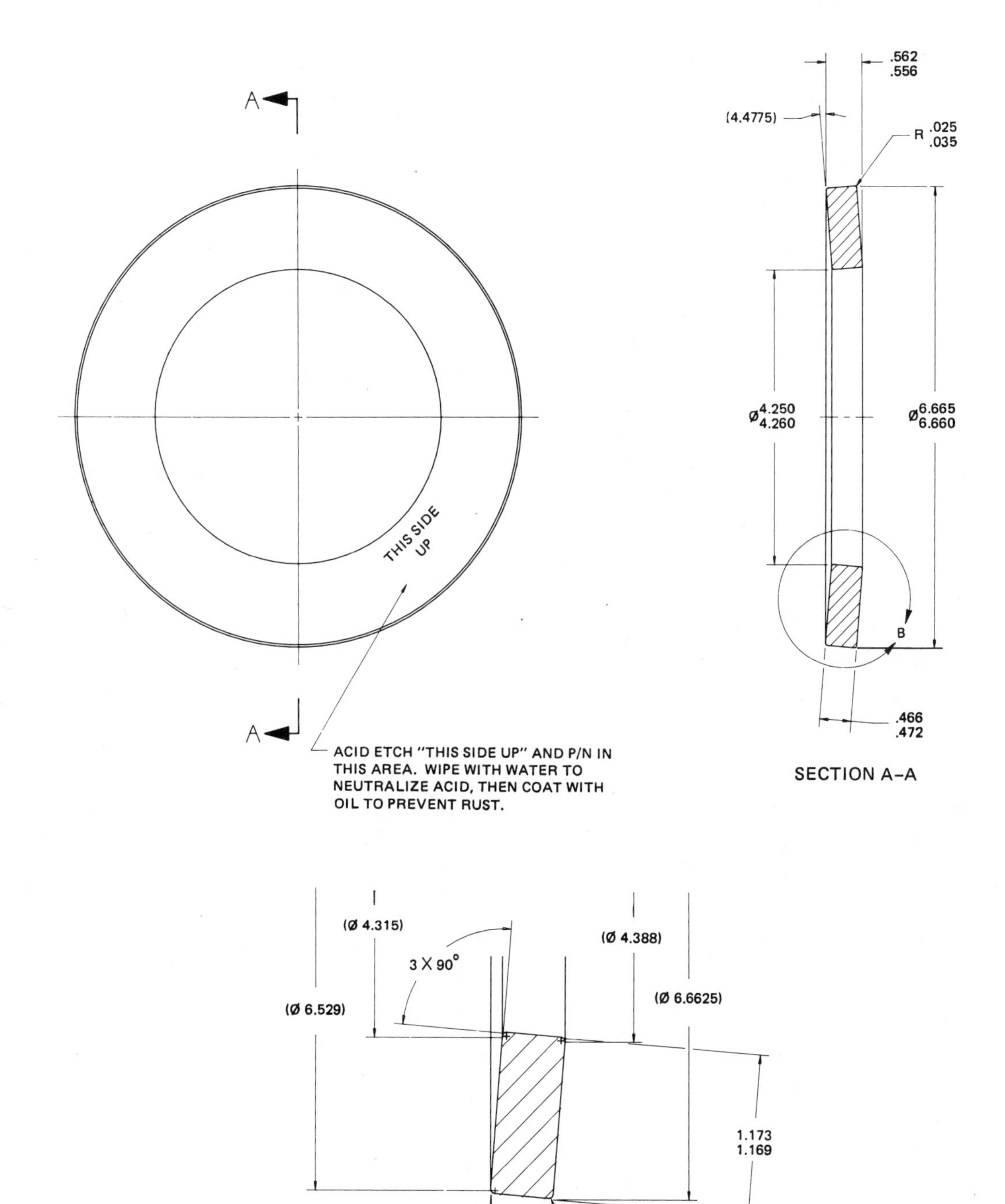

HEAT TREAT:
1. AUSTENITIZE AT 1475°F.
2. DIRECT QUENCH IN AGITATED OIL.
3. TEMPER TO R_cC 44–46.

Problem 10–6 Full section (metric)

Part Name: Hydraulic Valve Cylinder
Material: Phosphor Bronze

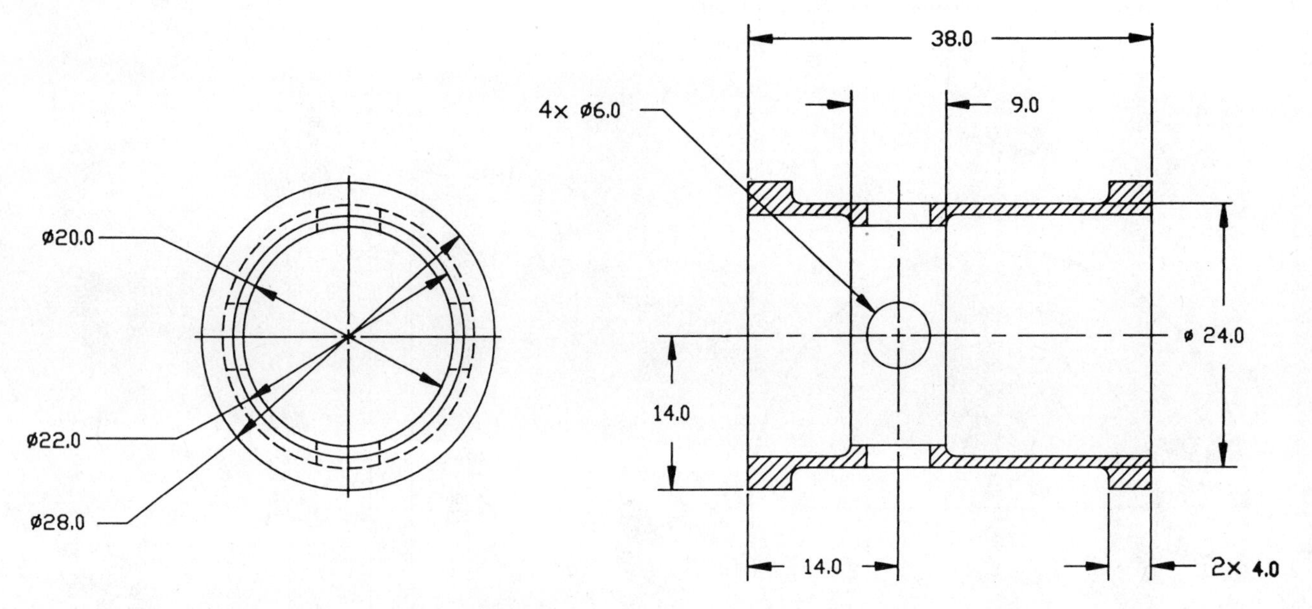

Problem 10–7 Full section (in.)

Part Name: Machine Plate
Material: 6160 T6 Steel

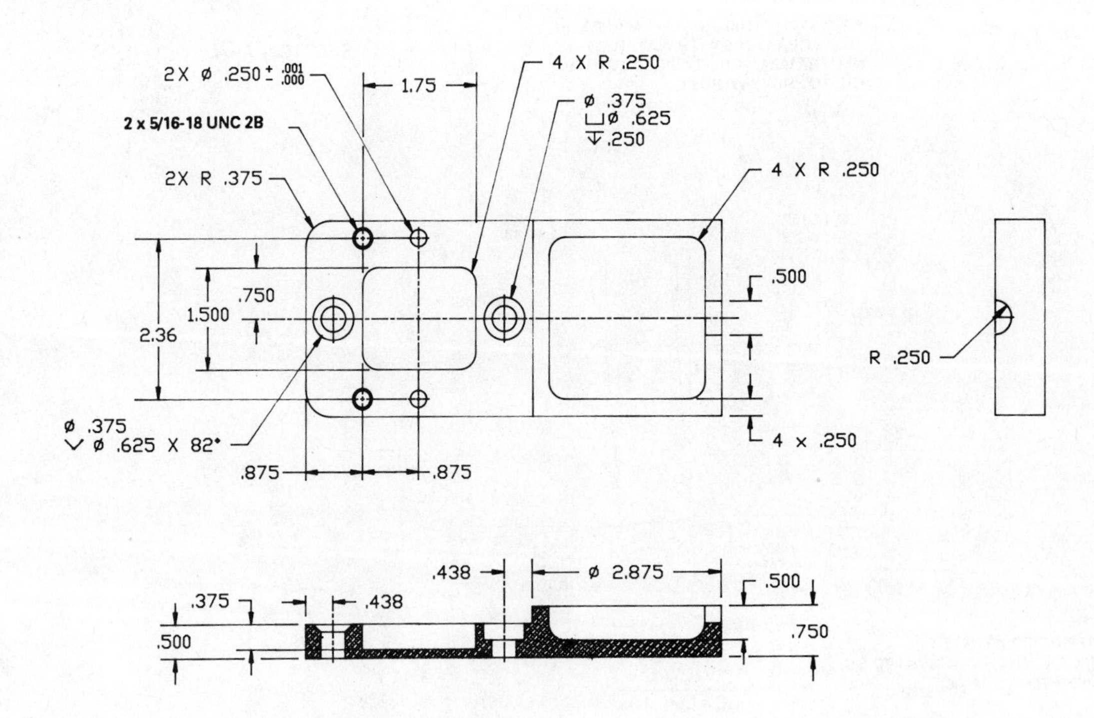

Problem 10–8 Full section and partial auxiliary section (in.)

Part Name: Accumulator Plug
Material: Phosphor Bronze
Courtesy Stanley Hydraulic Tools, a Division of the Stanley Works.

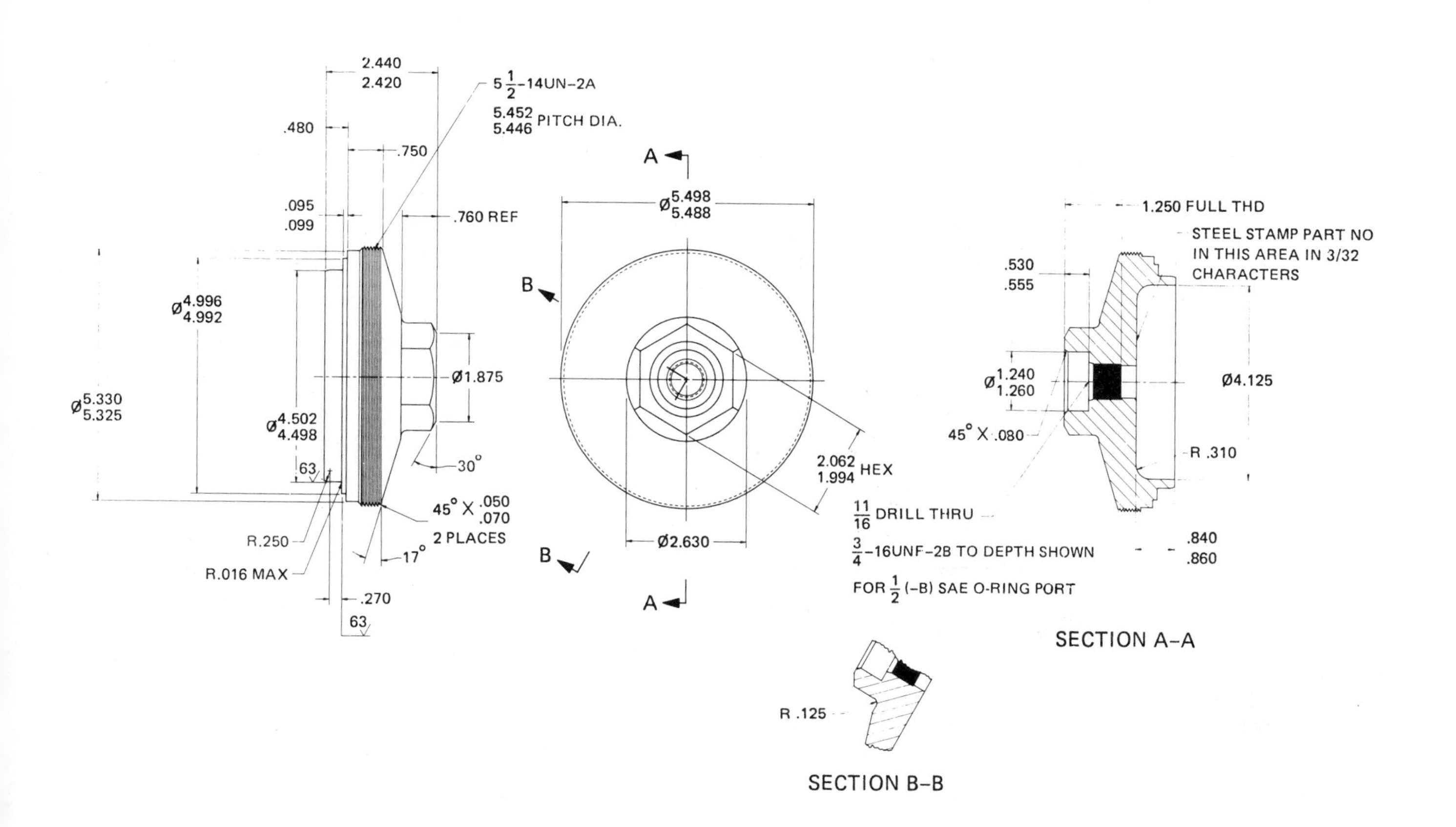

Problem 10–9 Full section (in.)

Part Name: Face Plate
Material: 6160 T6 Steel

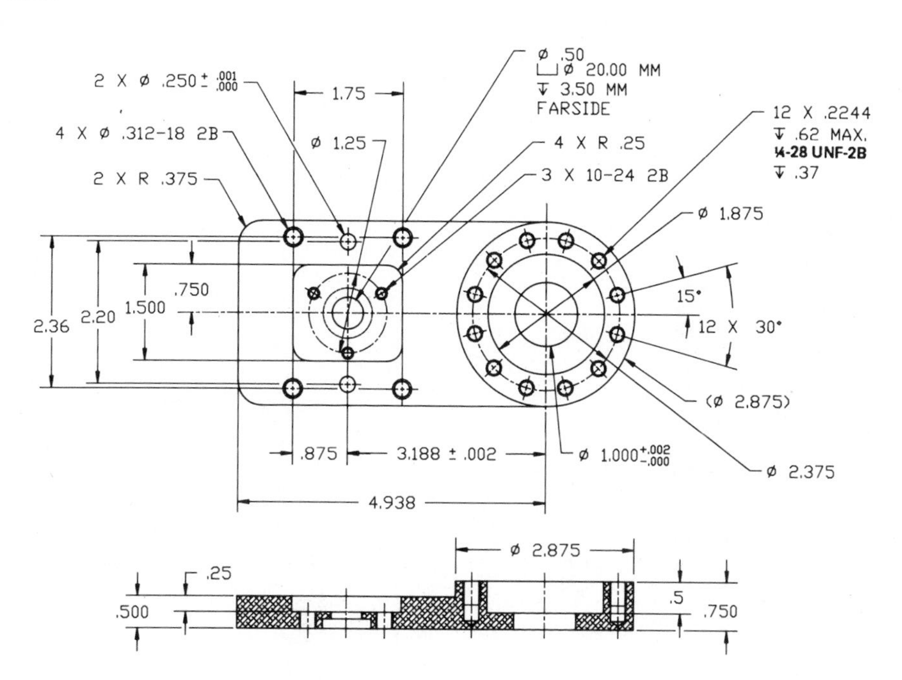

Problem 10–10 Full section, dimensioning cylindrical shapes (in.)

Part Name: Plug
Material: Phosphor Bronze

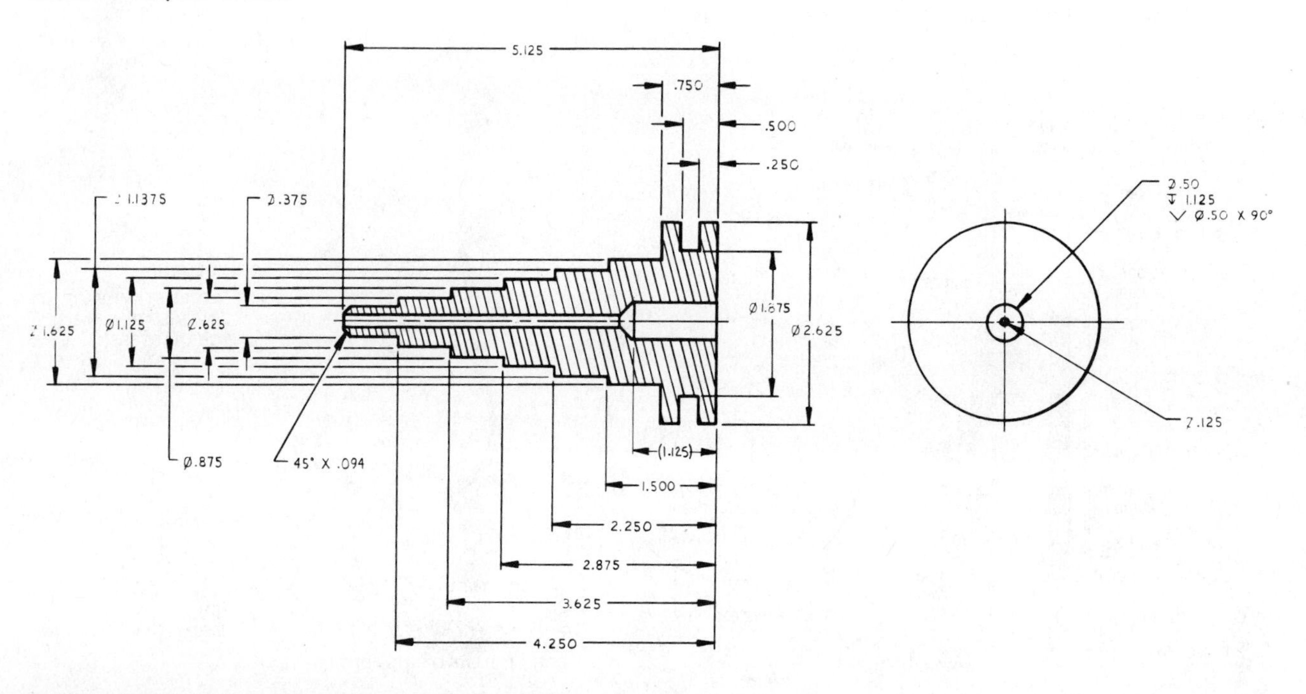

Problem 10–11 Full section and view enlargement (in.)

Part Name: Swivel
Material: SAE 5150
Courtesy Stanley Hydraulic Tools, a Division of the Stanley Works.

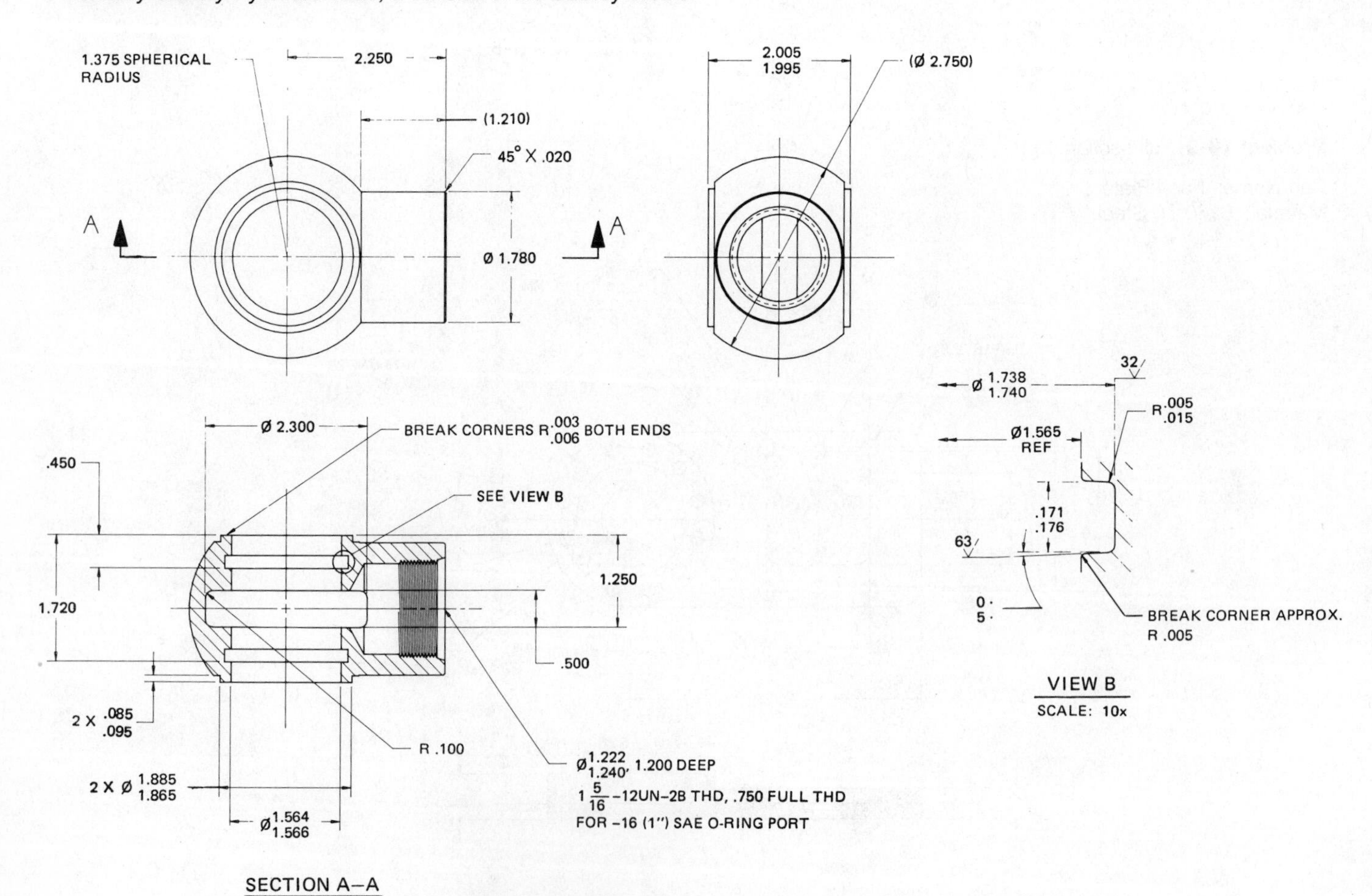

Problem 10–12 Full section (in.)

Part Name: Hub
Material: Cast Iron
SPECIFIC INSTRUCTIONS:
Convert the broken-out section in the given drawing to a full section.

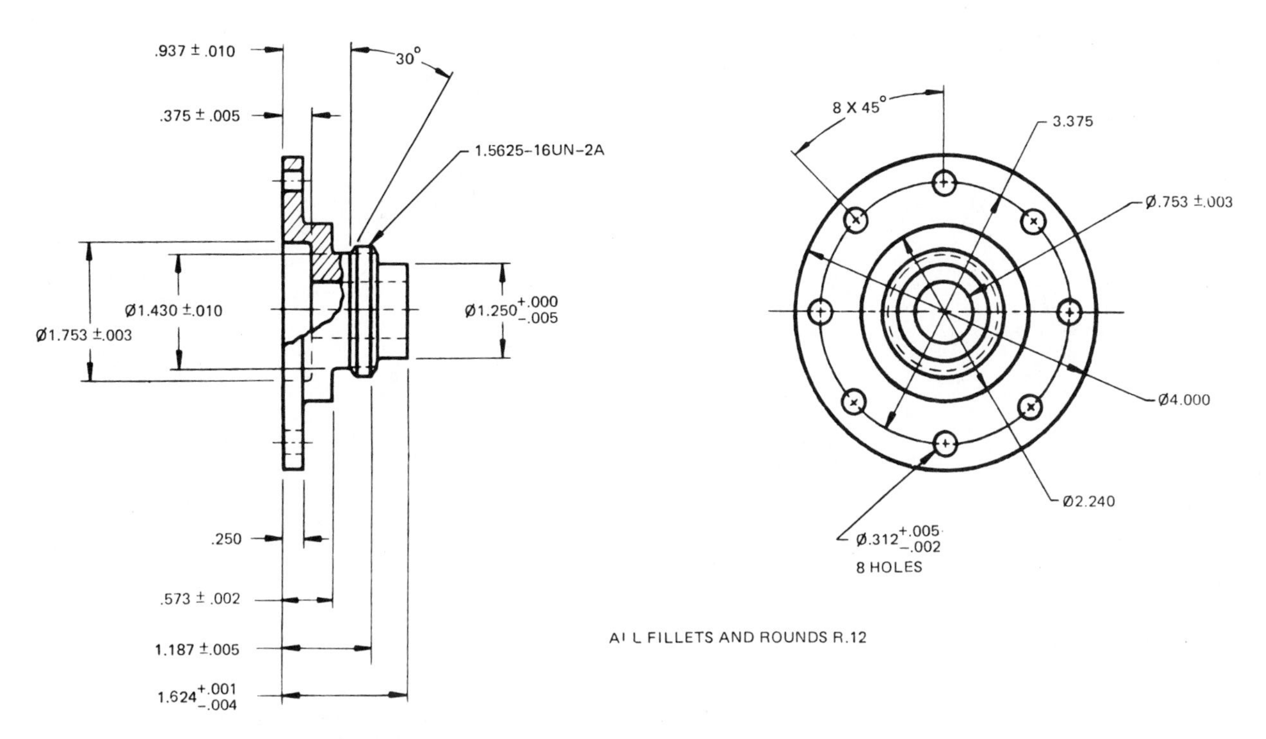

Problem 10–13 Full and broken-out section (metric)

Part Name: Hydraulic Valve Cylinder
Material: Phosphor Bronze

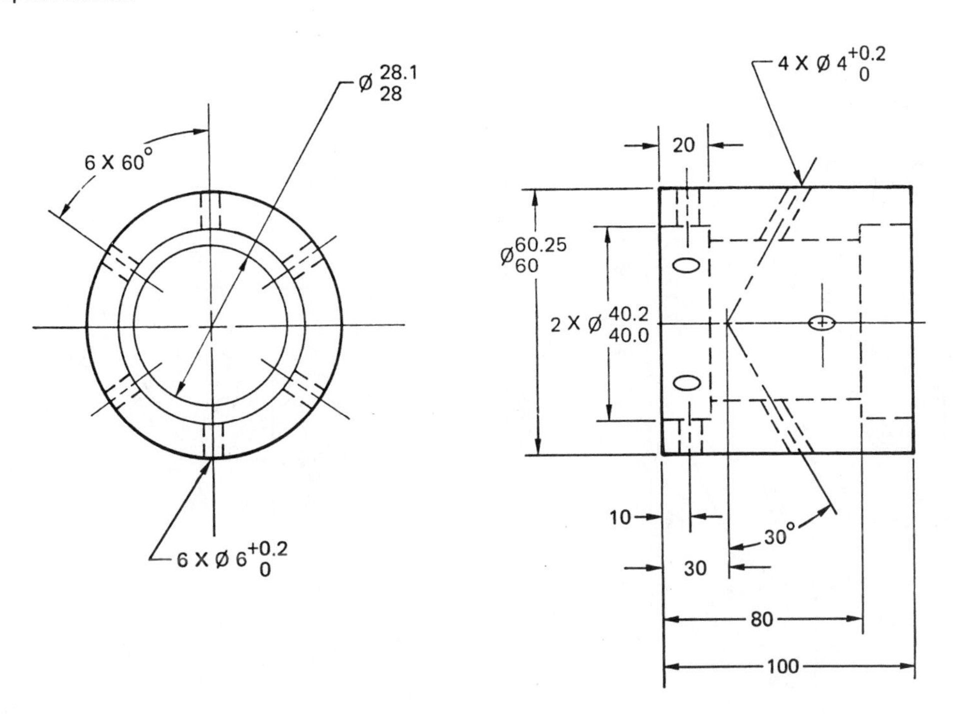

Problem 10–14 Full section, auxiliary section, and view enlargement (in.)

Part Name: Bulk Head
Material: Cast Iron
Courtesy Stanley Hydraulic Tools, a Division of the Stanley Works.

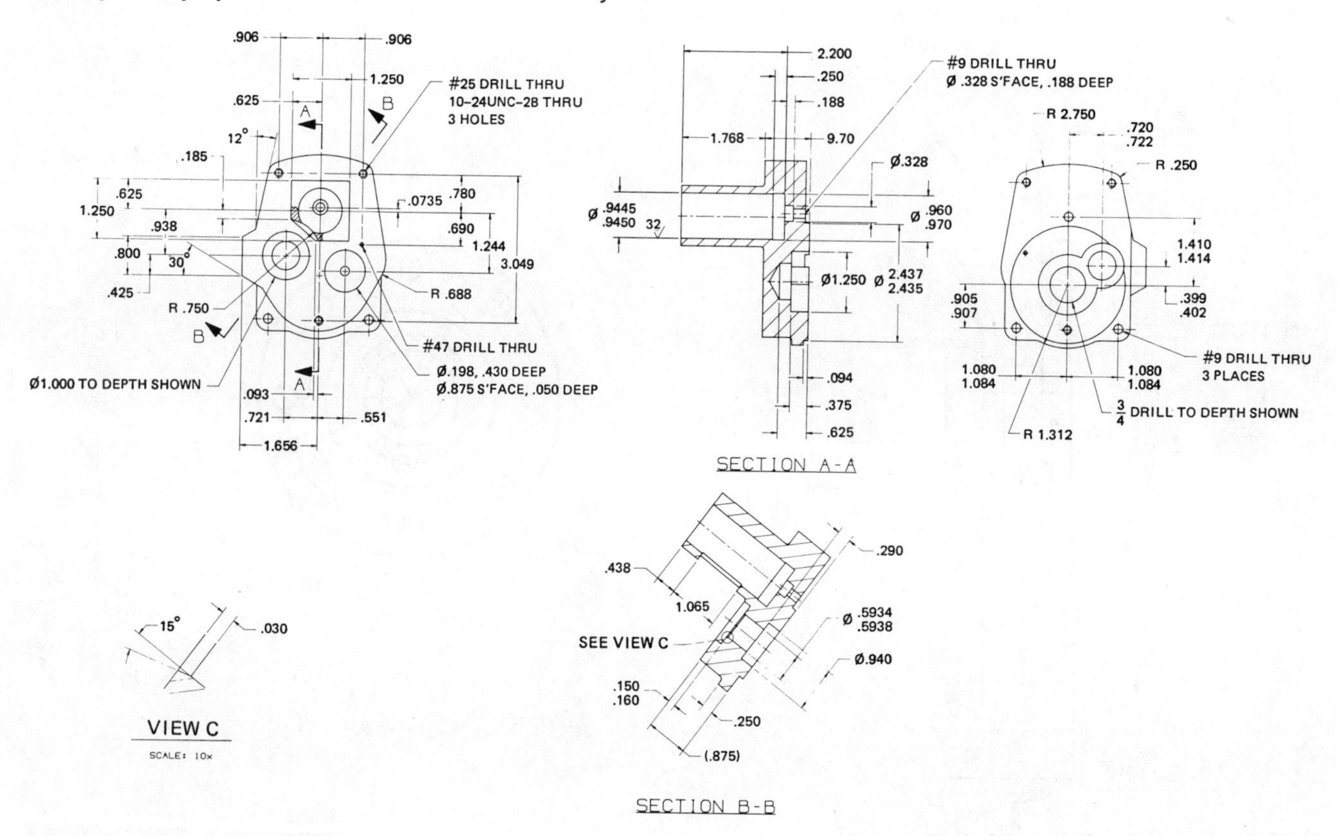

Problem 10–15 Full section (in.)

Part Name: Hanger
Material: SAE 1030
Fillets and Rounds: R.062

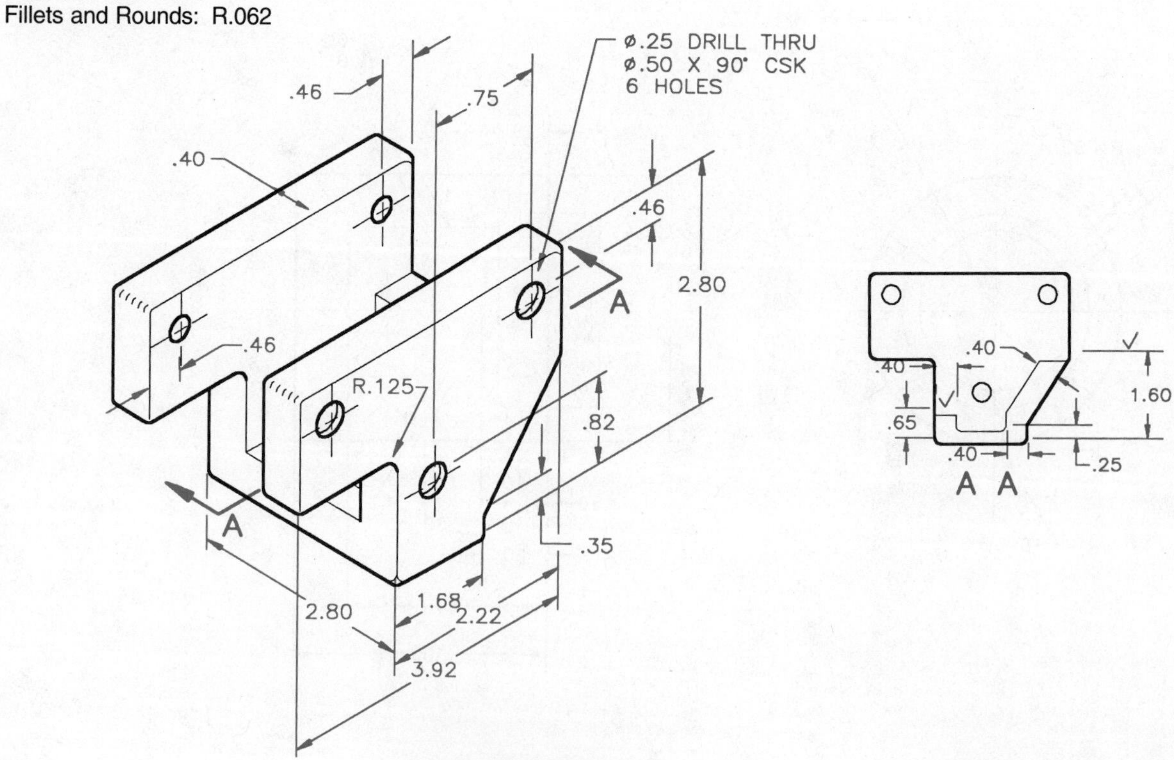

Problem 10–16 Full section (metric)

Part Name: Bearing Housing
Material: SAE 1015
Courtesy Aerojet Techsystems Co.

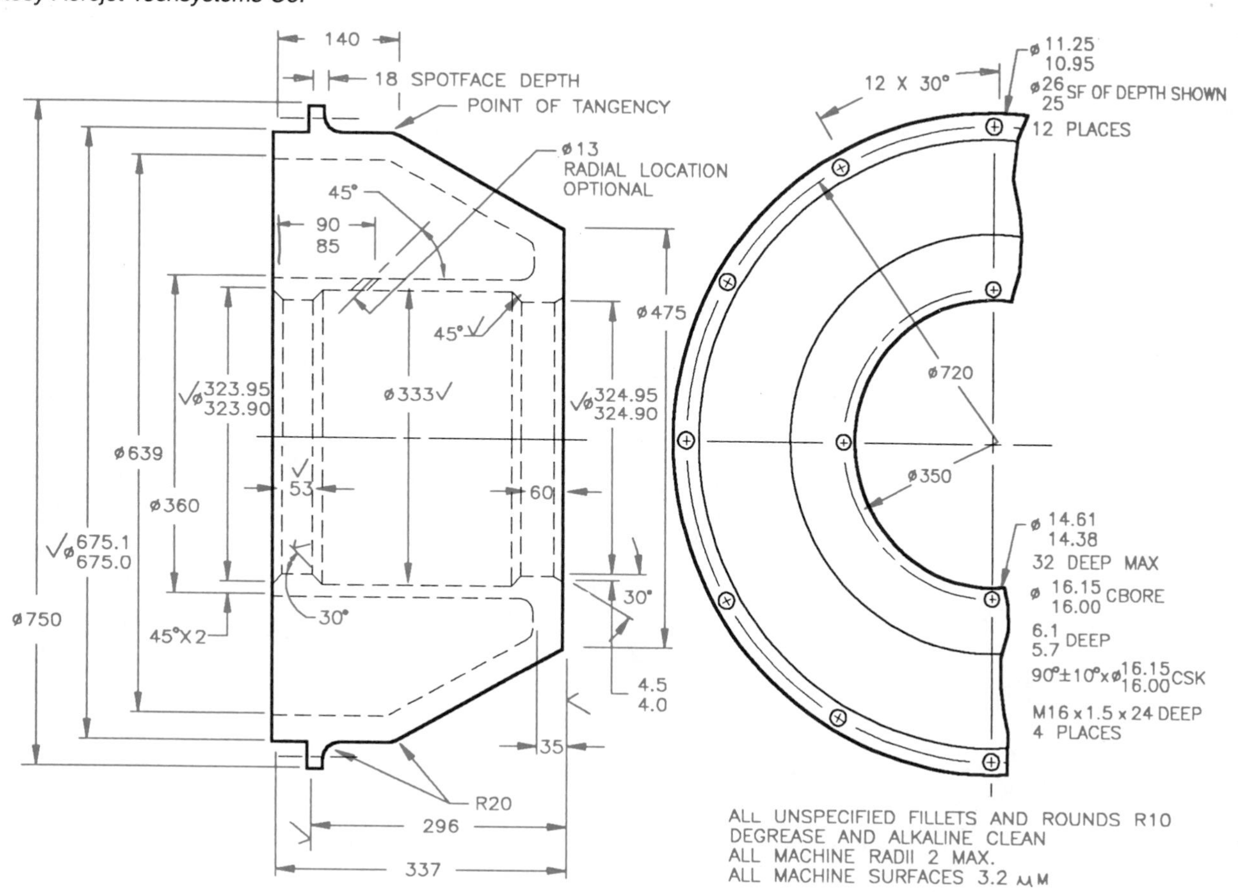

Problem 10–17 Half section (in.)

Part Name: Dial
Material: Bronze

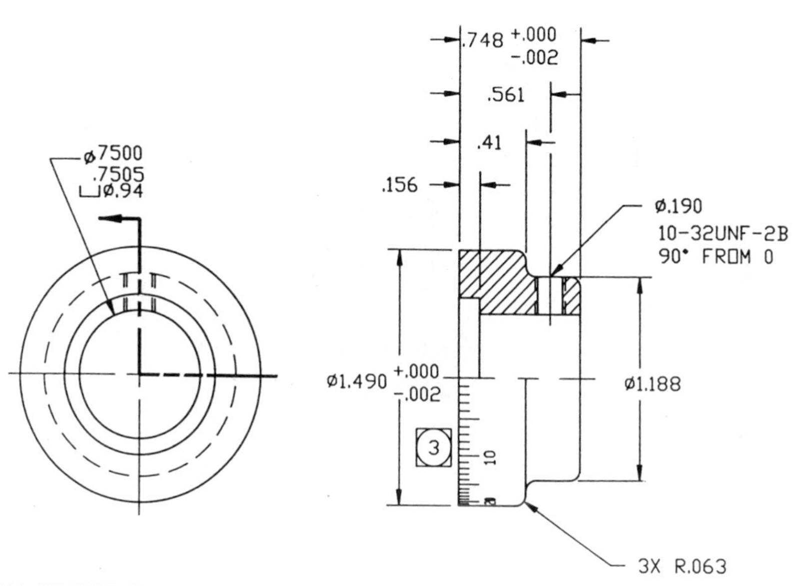

Problem 10–18 Half section (metric)

Part Name: Nozzle Base
Material: Titanium, ASTM-B367 Grade C3
NOTES:
Clean casting by mechanical blasting prior to alpha case removal. Chemically remove alpha case prior to inspection of castings using procedure approved by procurement activity.
Cast surfaces shall be visually inspected and be free from cracks, tears, laps, shrinkage, and porosity.
Radiographic inspect castings per MIL-STD-271 acceptance criteria for porosity and inclusions per ASTM-E-446 and for cavity shrinkage per ASTM-E-192 (plates for .75 wall thickness). Severity levels of casting defects shall be no greater than tabulated below for the indicated casting areas.
Courtesy Aerojet TechSystems Co.

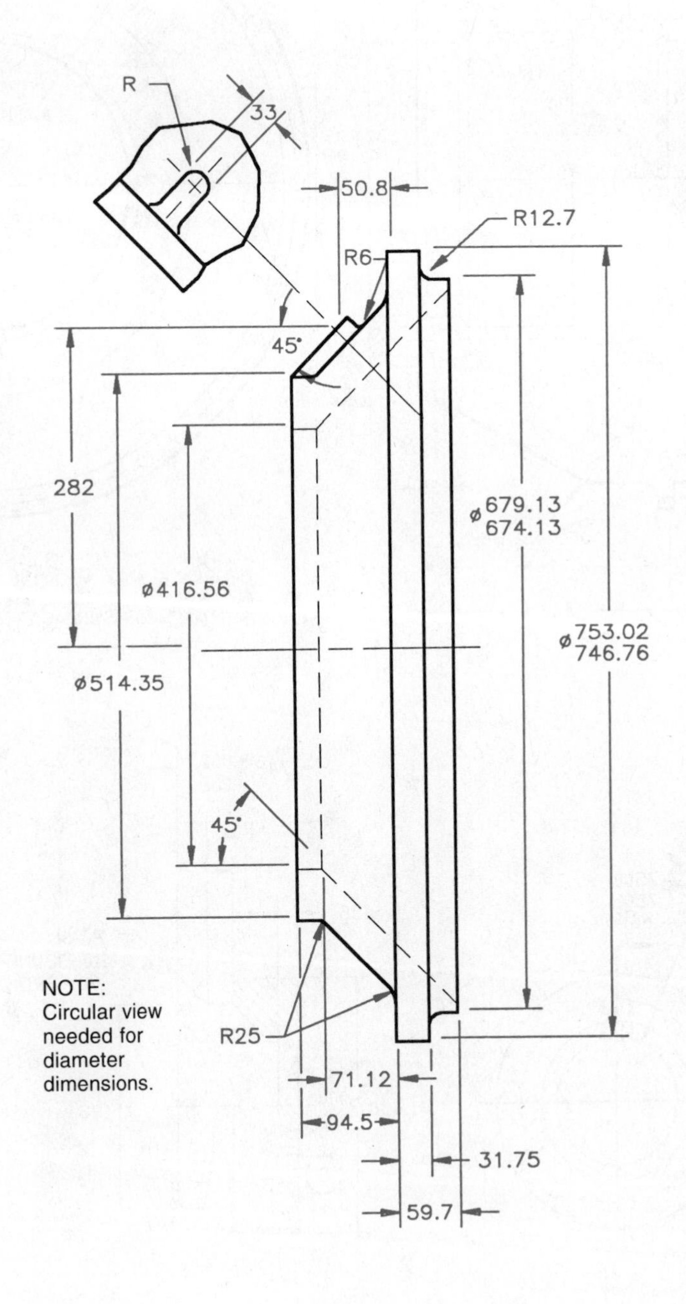

Problem 10–19 Half section and broken-out section (in.)

Part Name: Bench Block
Material: AISI 1018
Case Harden: .020 Deep 59–60 Rockwell C Scale or AISI 4140 Oil Quench 40–45C.

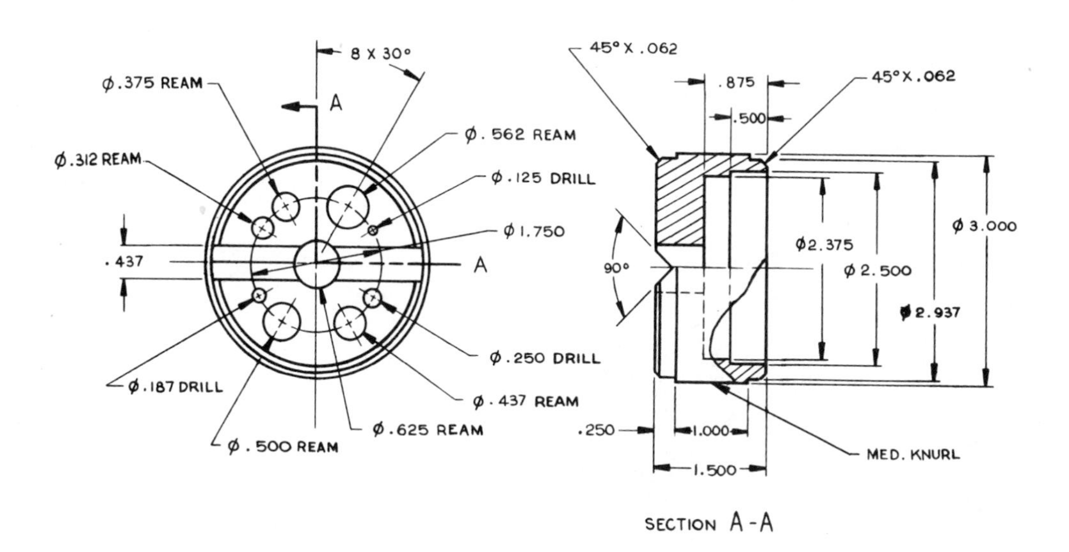

Problem 10–20 Half section (metric)

Part Name: Idler Pully
Material: SAE 4310
Fillets and Rounds: R 3.2 mm

NOTE: Many diameter dimensions should be placed on left and right side views.

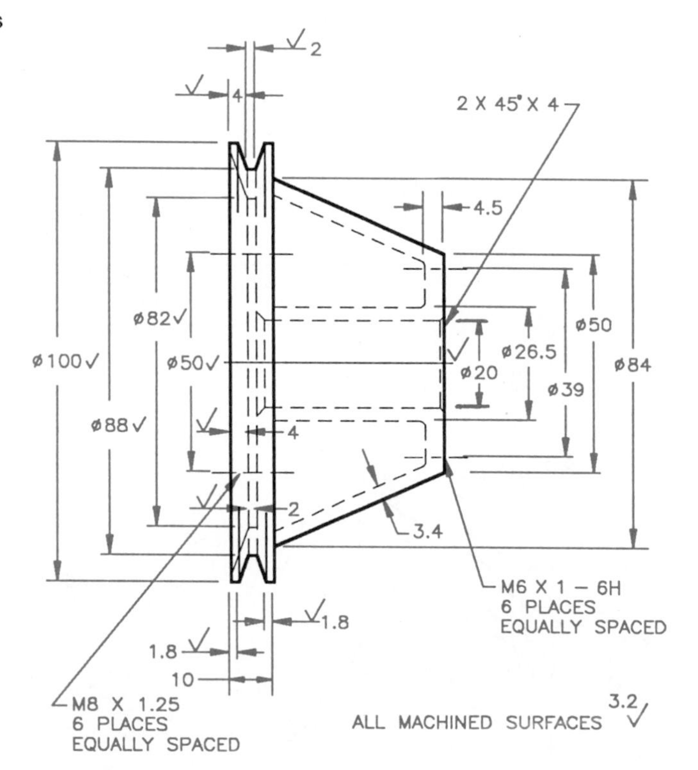

Problem 10–21 Offset section (in.)

Part Name: Die Casting
Material: SAE 6150
SPECIFIC INSTRUCTIONS:
Convert all dimensions to ANSI Y14.5M standards as shown and discussed in this text.
Courtesy Kris Altmiller.

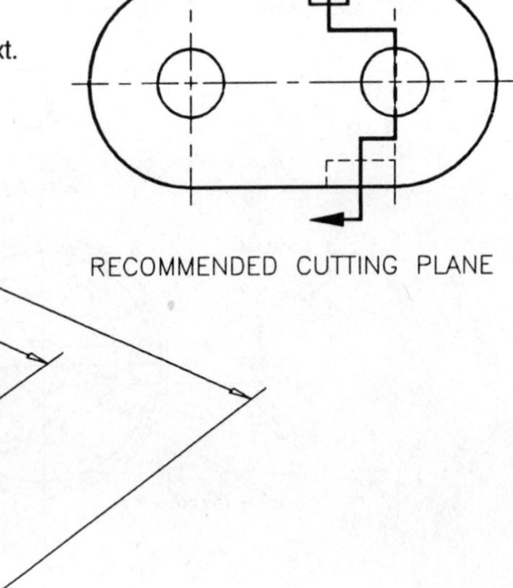

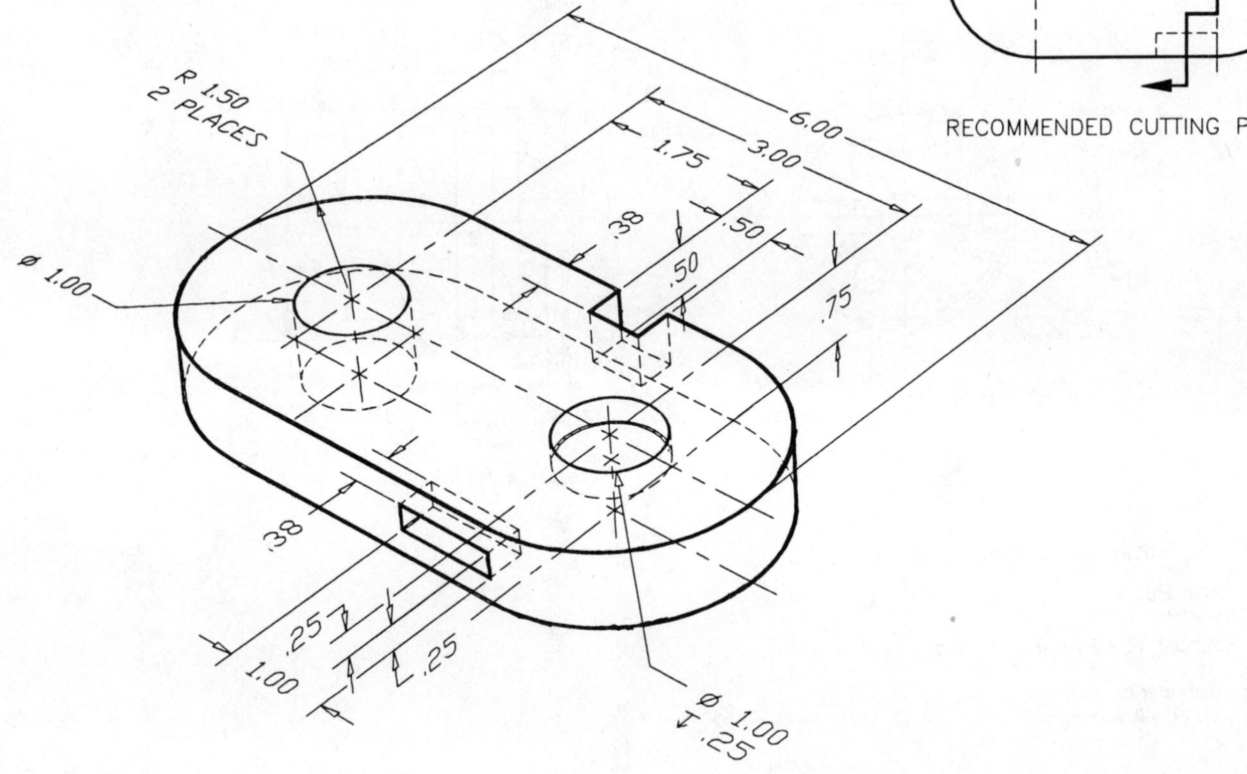

Problem 10–22 Offset section (in.)

Part Name: Drill Plate
Material: SAE 1020
Case Harden: 55 Rockwell C Scale
Fillets and Rounds: R .12
FAO 63 μin.

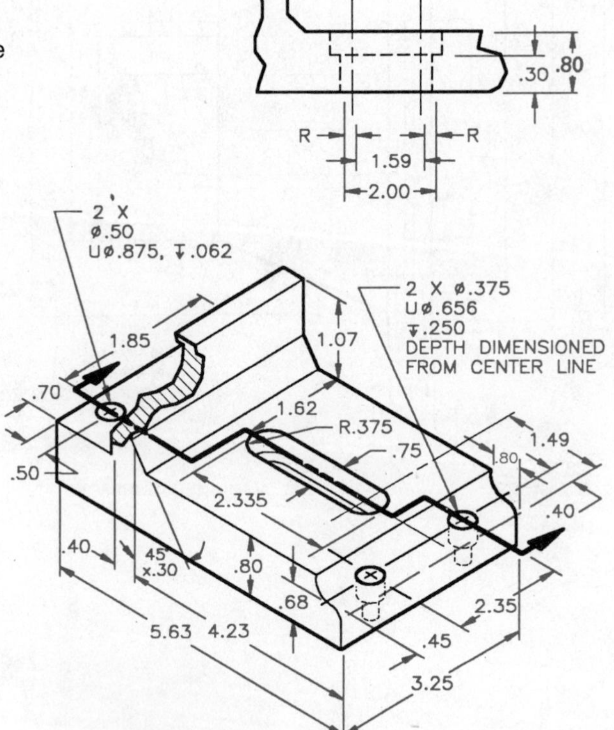

Problem 10–23 Aligned section (in.)

Part Name: Hub
Material: SAE 3145
Fillets and Rounds: R .125

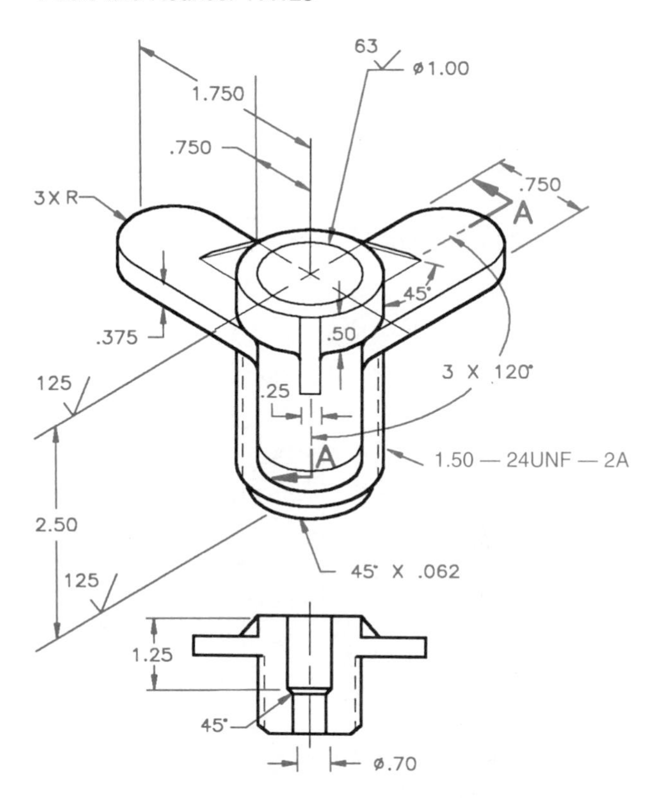

Problem 10–24 Broken-out section (in.)

Part Name: Taper Shaft
Material: SAE 4320
FAO 16 μin.

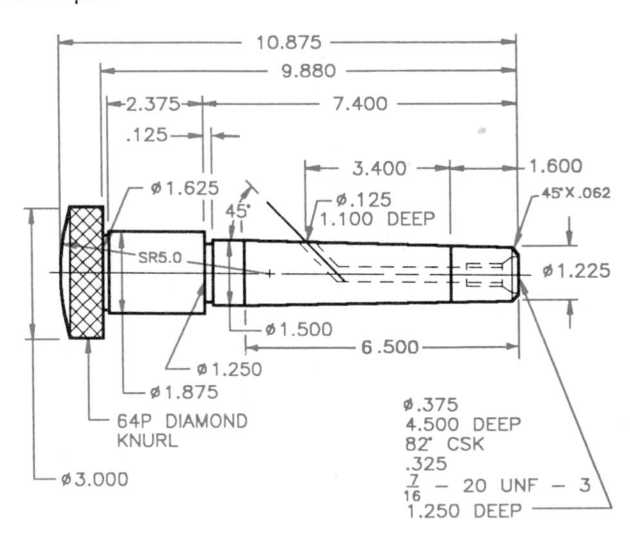

Problem 10–25 Auxiliary section (metric)

Part Name: Gear Base
Material: SAE 2340
Finish All Over: 0.8 μm

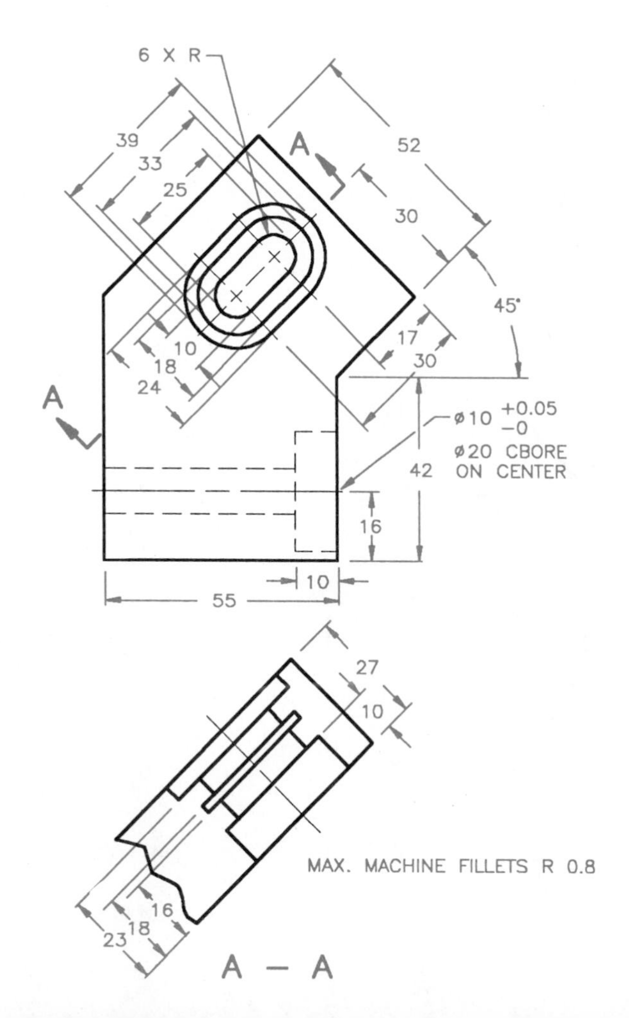

Problem 10–26 Broken-out section, view enlargement (in.)

Part Name: Clamp Cap
Material: Cast Aluminum
Fillets and Rounds: R .06

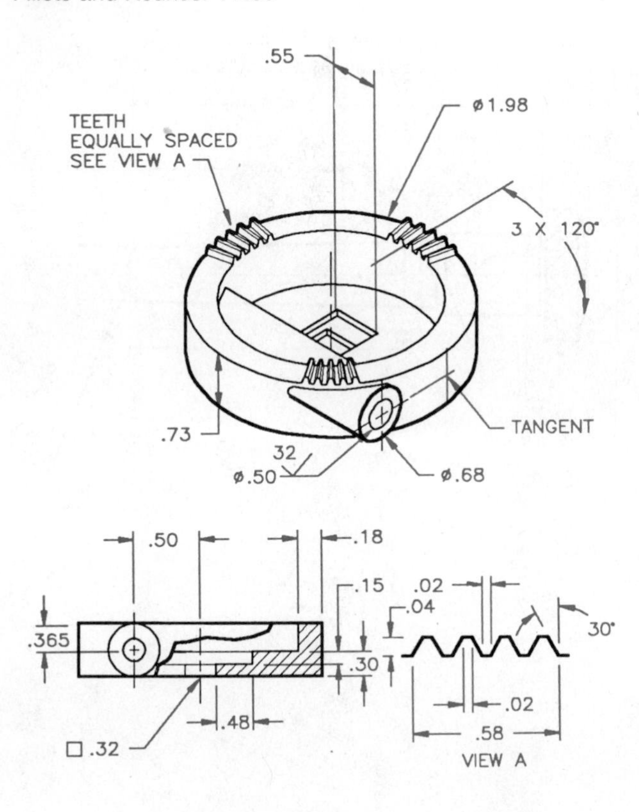

Problem 10–27 Broken-out section (metric)

Part Name: 25-mm 45° Elbow
Material: Cast Iron
Fillets and Rounds: R 4mm
Consider bottom view or auxiliary view for hole pattern dimensions.

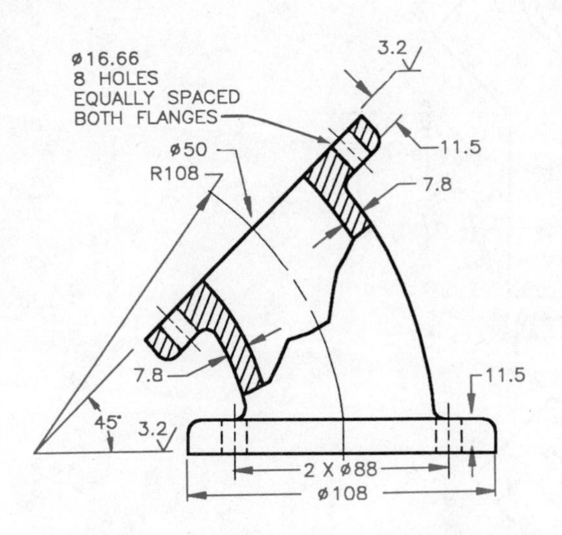

Problem 10–28 Revolved section (in.)

Part Name: End Loading Arm
Material: SAE 2310
Fillets and Rounds: R .25

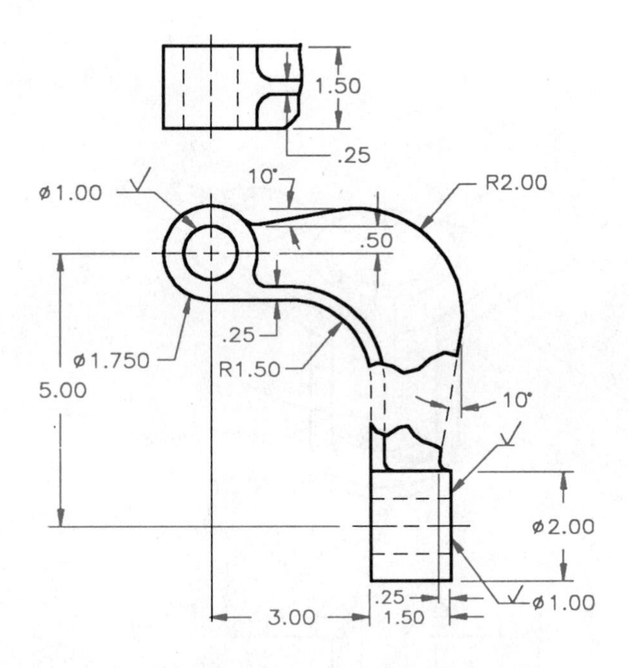

Problem 10–29 Revolved section (in.)

Part Name: Offset Handwheel
Material: Bronze
All Fillets and Rounds: R .12
Finishes: 125 μin.

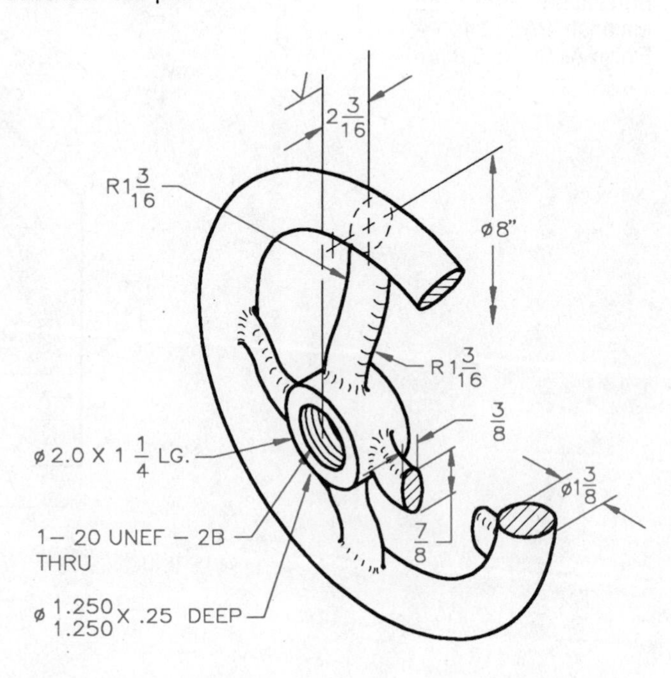

Problem 10–30 Removed sections and view enlargements (in.)

Part Name: Swivel Stem
Material: SAE 4340
Courtesy Stanley Hydraulic Tools, a Division of the Stanley Works.

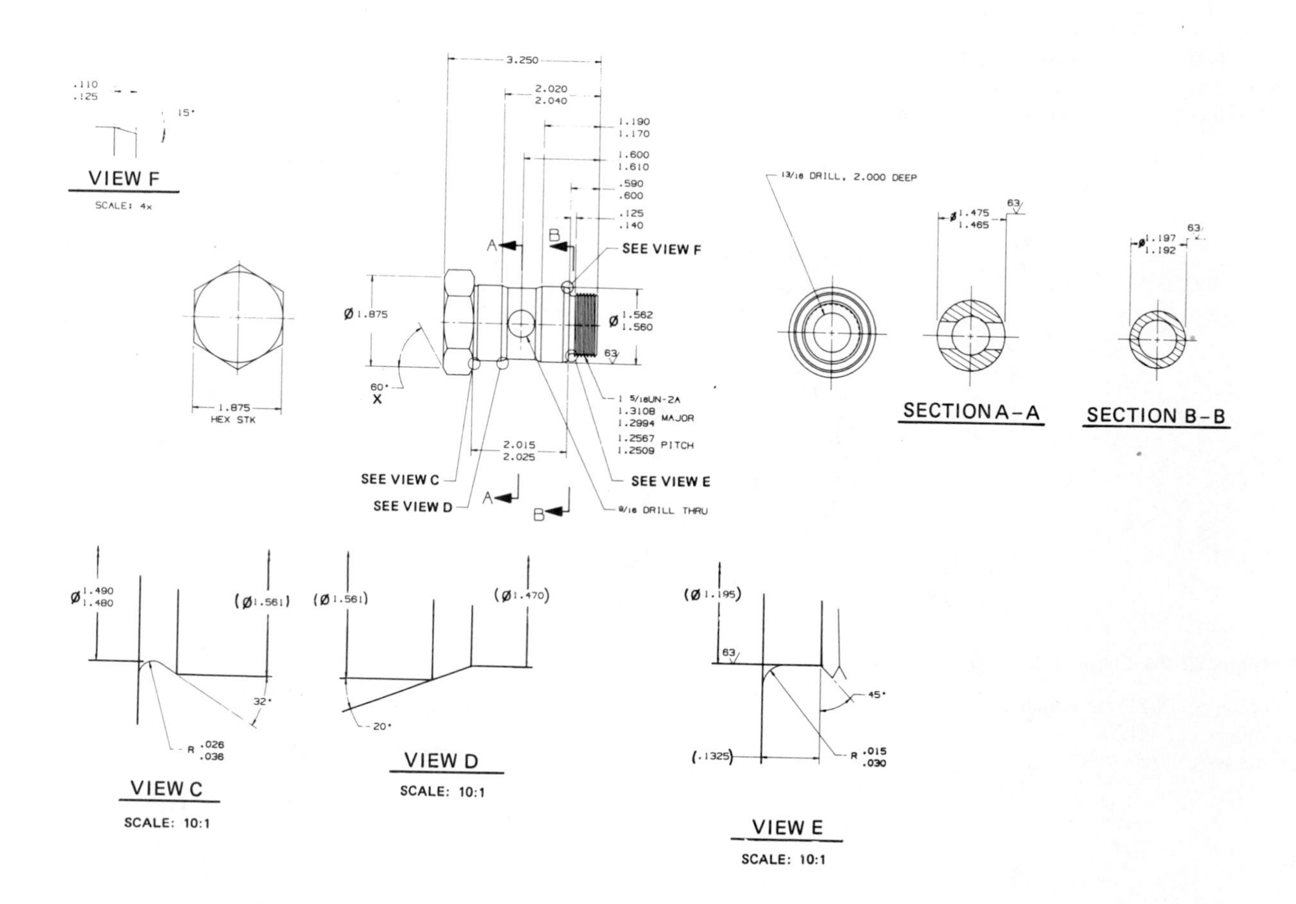

Problem 10–31 Removed section view enlargement (in.)

Part Name: Taper Shaft
Material: SAE 3130

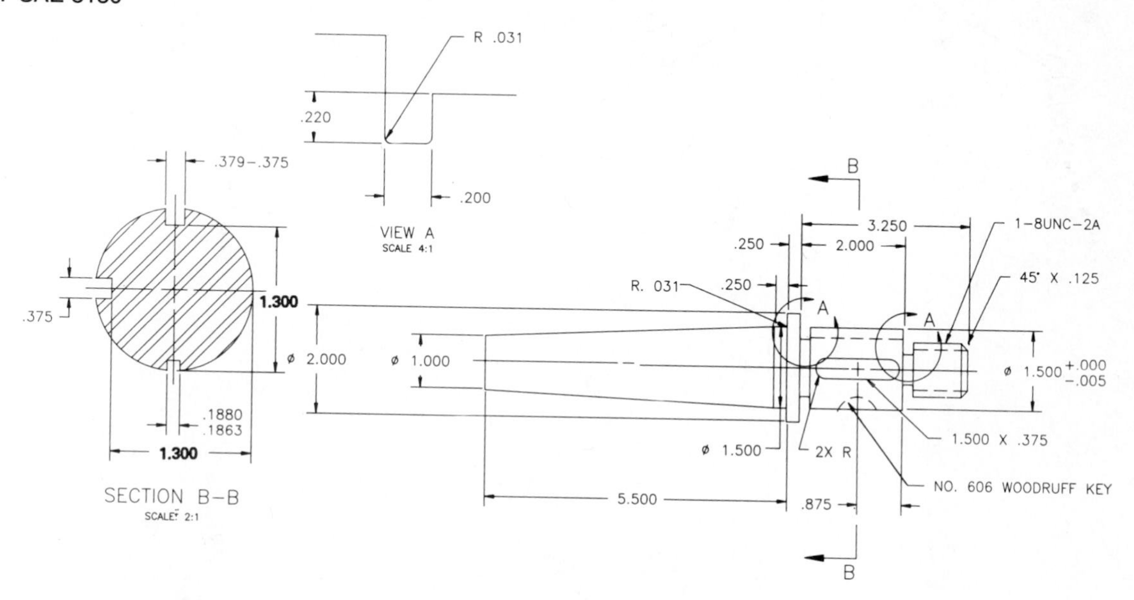

Problem 10–32 Broken-out, revolved section, view enlargement (in.)

Part Name: Pipe Wrench Handle
Material: SAE 5120
Fillets and Rounds: R .06 mm
SPECIFIC INSTRUCTIONS:
Engineering sketch is given in fractional inches, convert all size dimensions to two-place decimals and location dimensions to three-place decimals for final drawing.

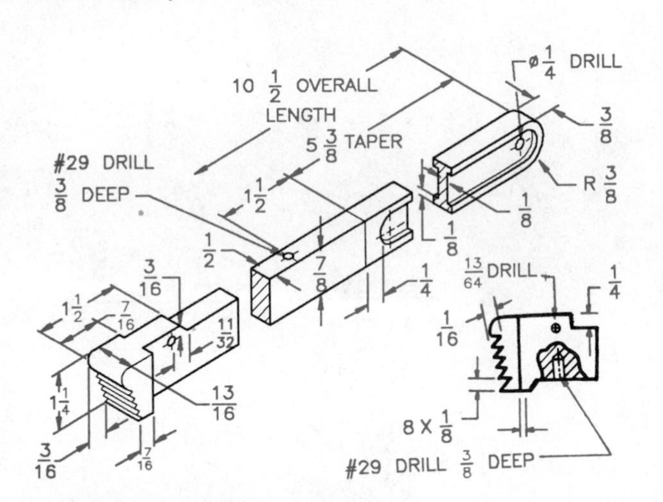

Problem 10–33 Removed sections (metric)

Part Name: Valve Stem
Material: Phosphor Bronze
Finish All Over: 1 μm

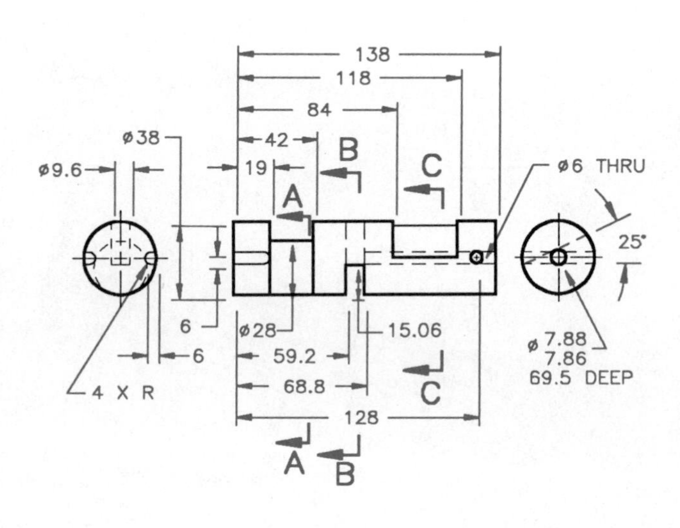

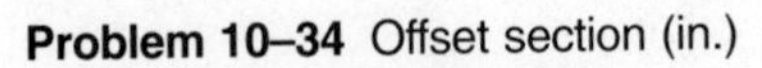

Problem 10–34 Offset section (in.)

Part Name: Die Plate Casting
Material: SAE 3120
Courtesy Kris Altmiller.

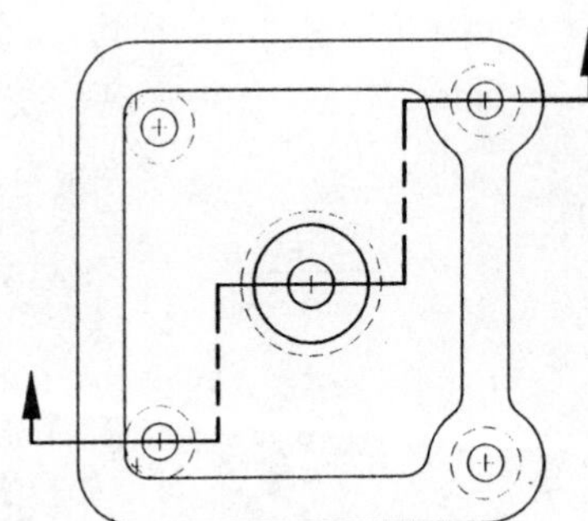

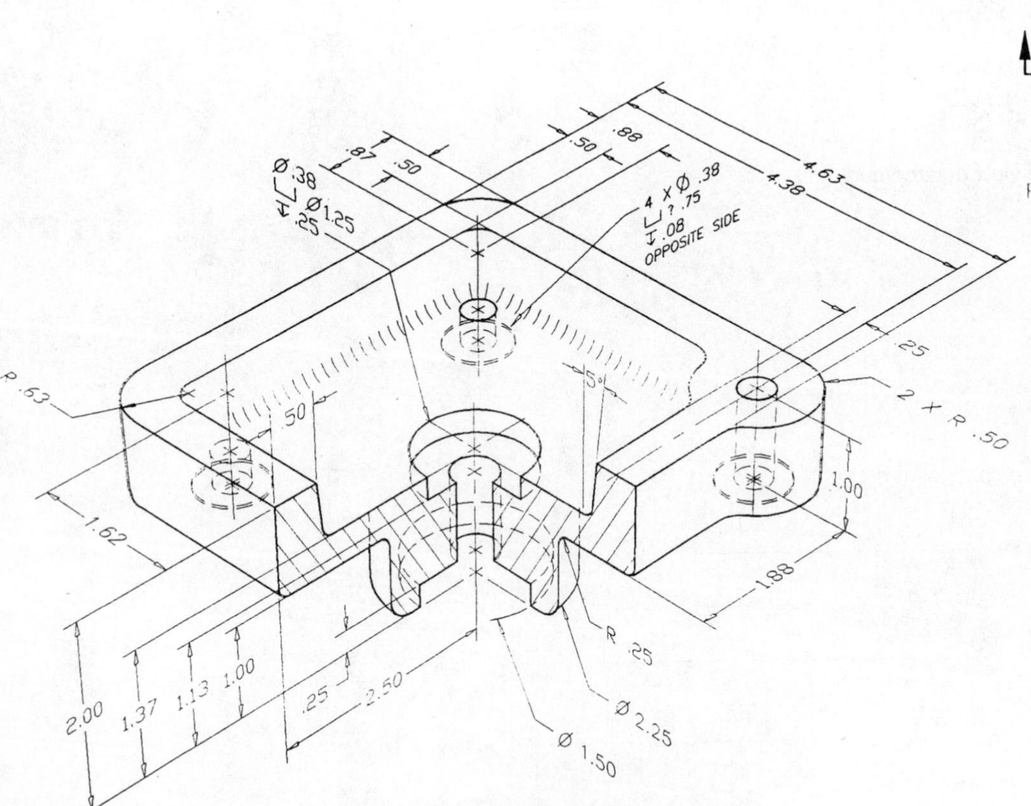

ALL FILLETS AND ROUNDS R .25 UNLESS OTHERWISE SPECIFIED

Problem 10–35 Broken-out section (in.)

Part Name: Slide Bar Connector
Material: SAE 4120
SPECIFIC INSTRUCTIONS:
Convert point-to-point dimensioning to datum dimensioning.

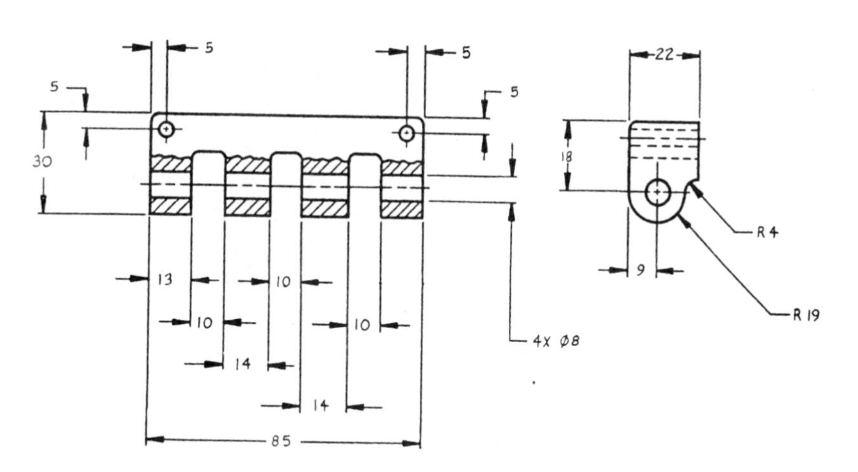

Problem 10–36 Broken-out section (in.)

Part Name: Drain Tube
Material: ∅ 1.00 × .065
Courtesy TEMCO.

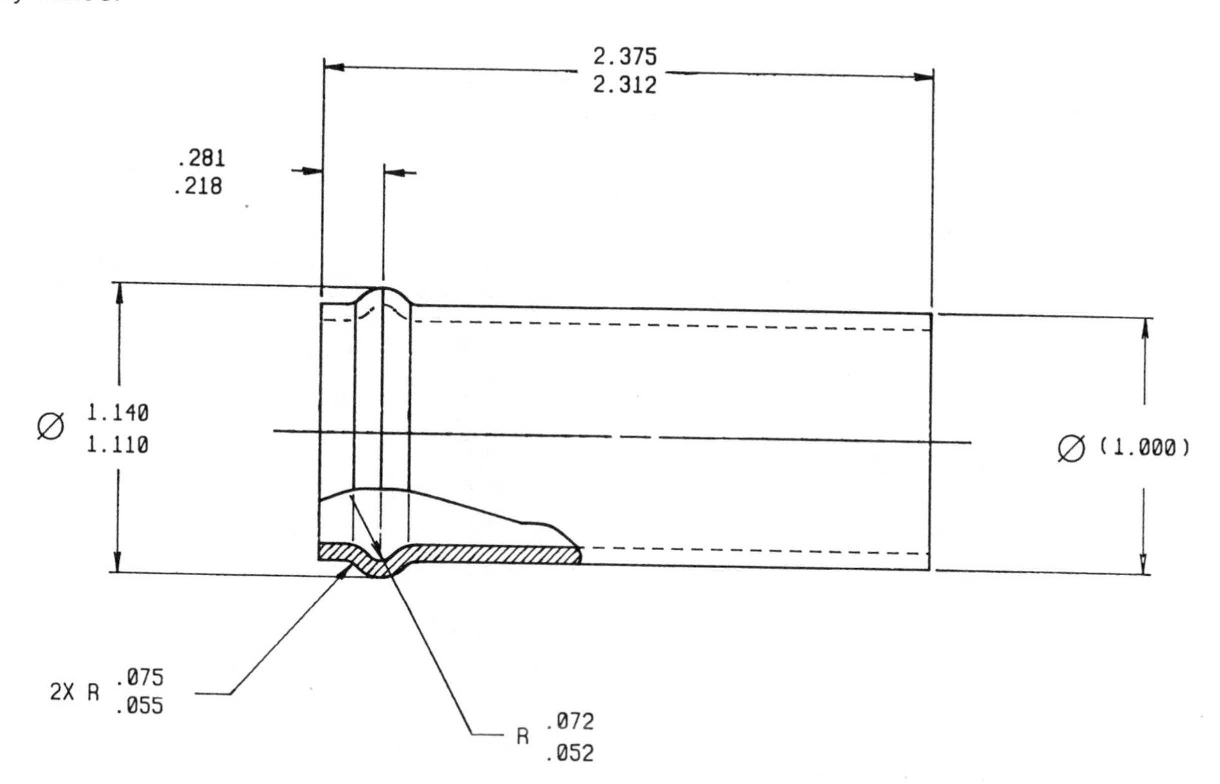

Problem 10–37 Conventional break (in.)

Part Name: Leg
Material: ∅ 2.00 Schedule 40 A120
Courtesy TEMCO.

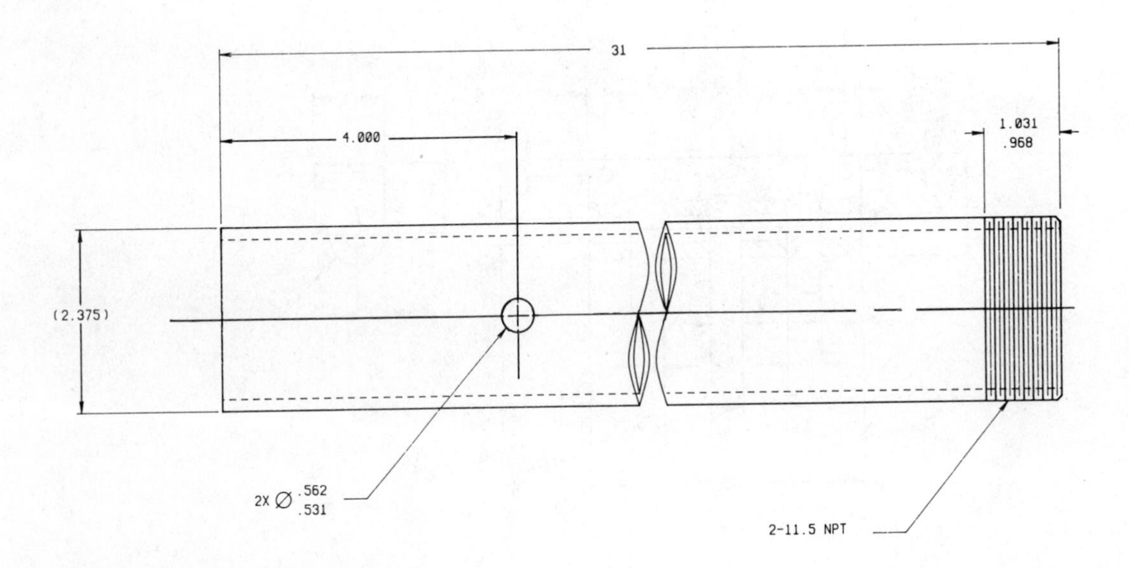

Problem 10–38 Revolved section, conventional revolution (in.)

Part Name: Crank Arm
Material: Cast Steel 80,000
All Fillets and Rounds: R .12

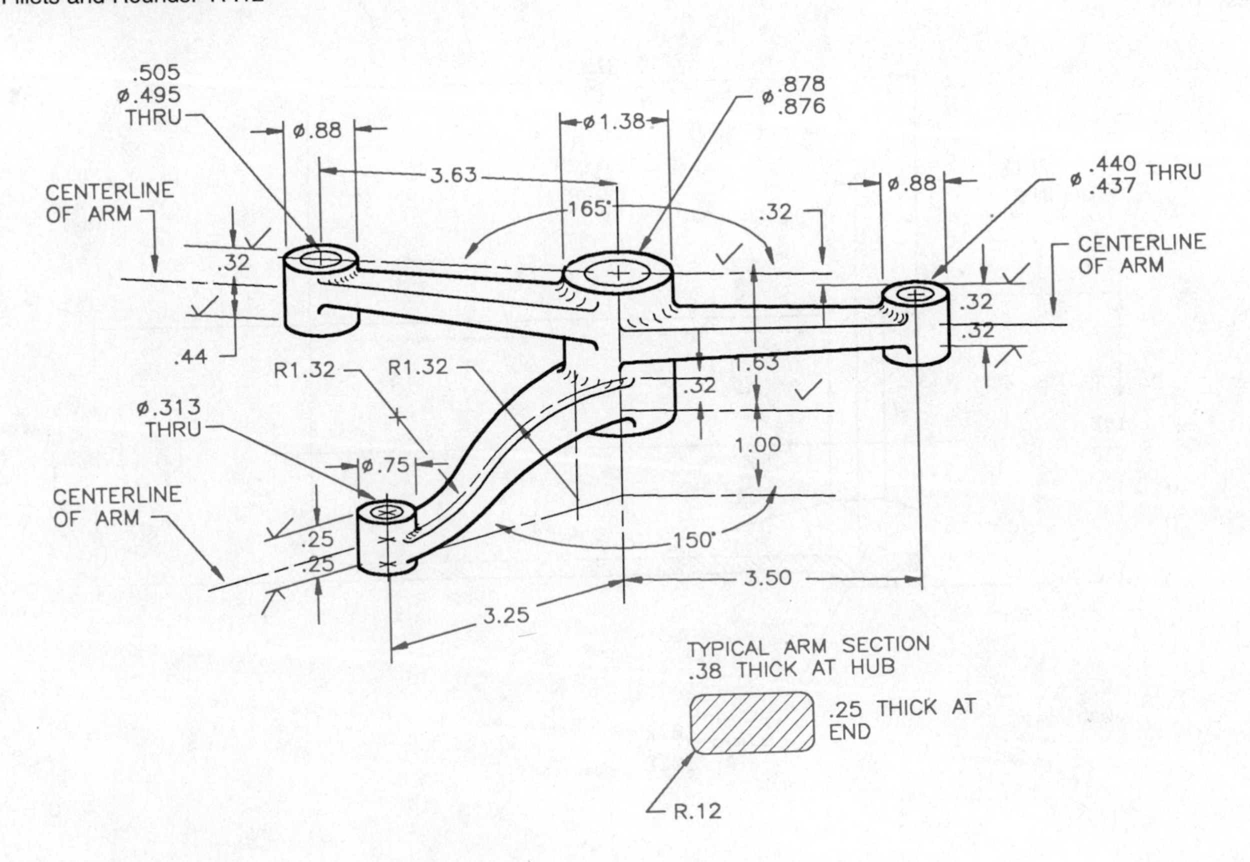

Problem 10–39 Half and full sections (in.)

Part Name: Diffuser Casting
Material: Titanium
SPECIFIC INSTRUCTIONS:
Convert all dimensions to ANSI Y14.5M standards as shown and discussed in this text. Display the casting material using phantom lines as shown in the engineer's layout. Use ANSI standard cutting-plane line.

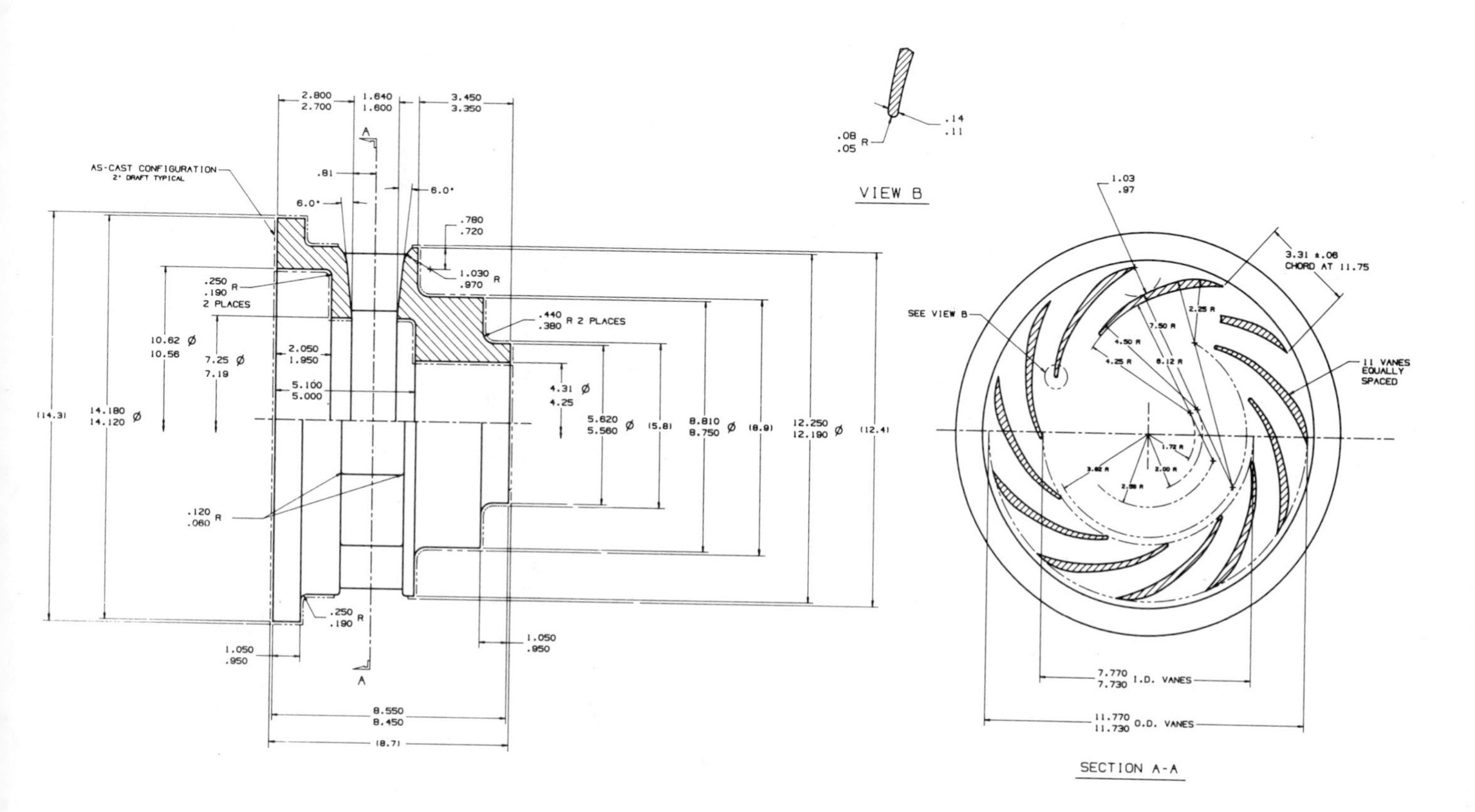

Problem 10–40 Aligned section (in.)

Part Name: Crankshaft Adapter
Material: CI
SPECIFIC INSTRUCTIONS:
1. Draw to ANSI standards.
2. Use ANSI standard cutting-plane line.
Courtesy American Hoist and Derrick Company.

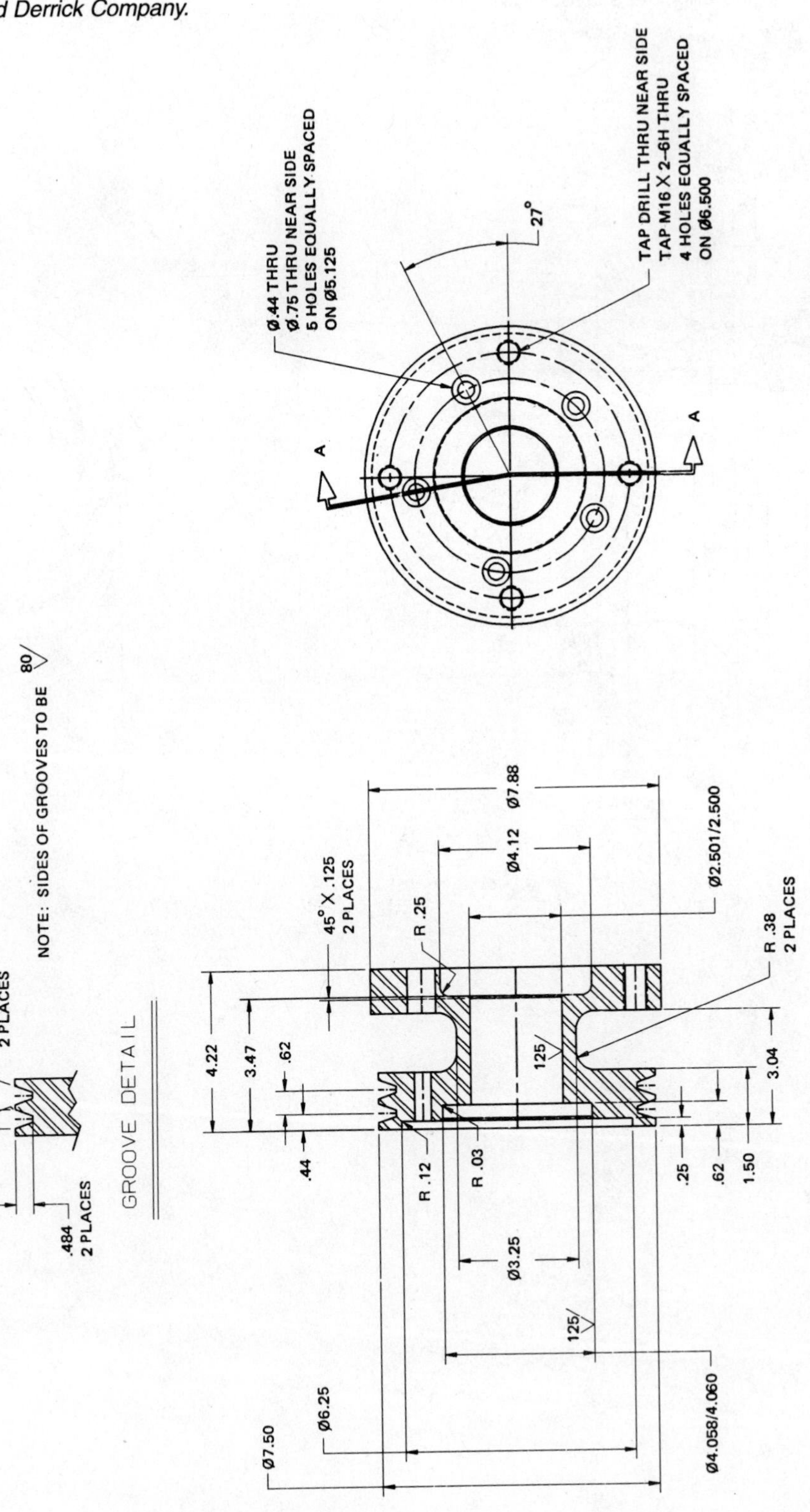

Problem 10–41 Aligned sections (metric)

Part Name: Steering Pump Adapter
Material: HC–80
Courtesy Hyster Company.

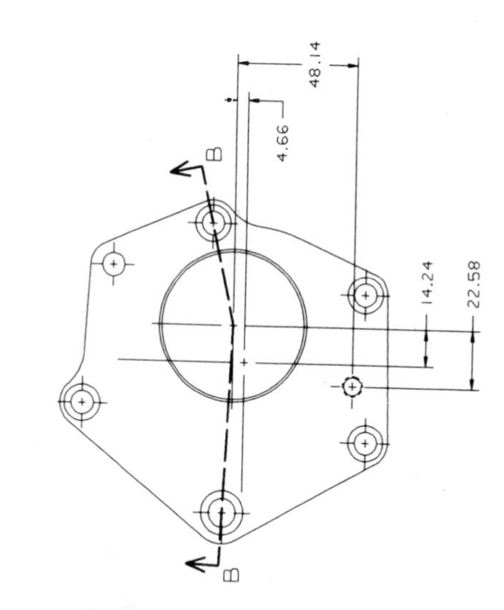

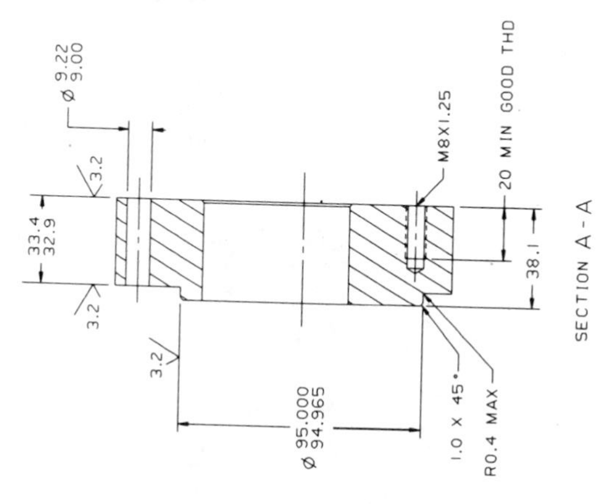

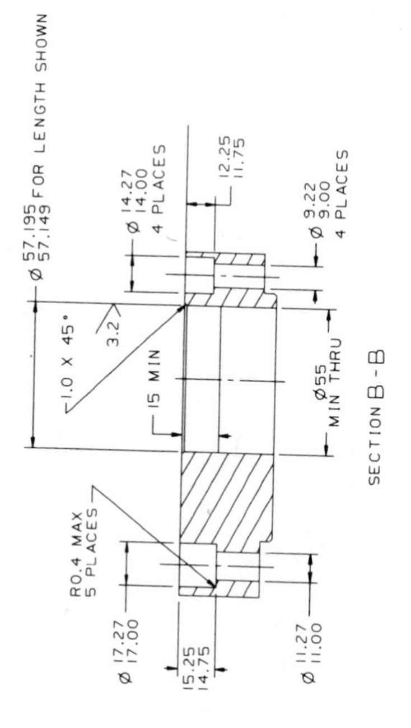

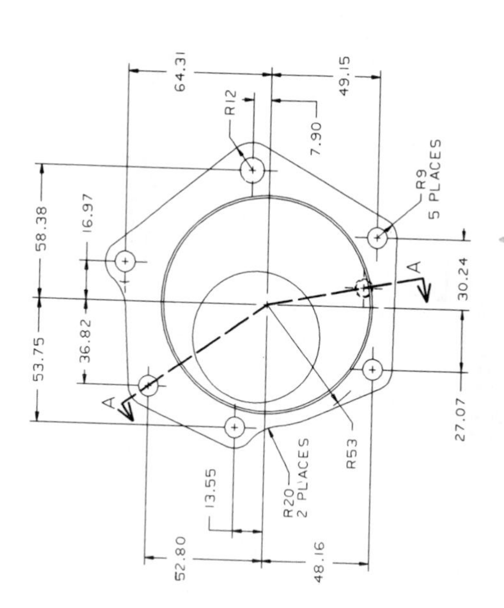

Problem 10–42 Offset section (metric)

Part Name: Pad–Monotrol
Material: Hi-mount Pro-fax 8623
SPECIAL INSTRUCTIONS:
Avoid leaders crossing dimension lines.
Courtesy Hyster Company.

Problem 10–43 Half and full sections (in.)

Refer to the drawing for Problem 10–40

Part Name: Diffuser Casting
Material: Titanium
SPECIFIC INSTRUCTIONS:
Convert all dimensions to ANSI Y14.5M standards as shown and discussed in this text. Casting shown in phantom lines. Omit phantom line technique for showing casting and make two drawings: one a *casting* drawing with only casting information, and the other a *machining* drawing with only machining dimensions and information. Refer to Chapters 7 and 8 for a review of this method.

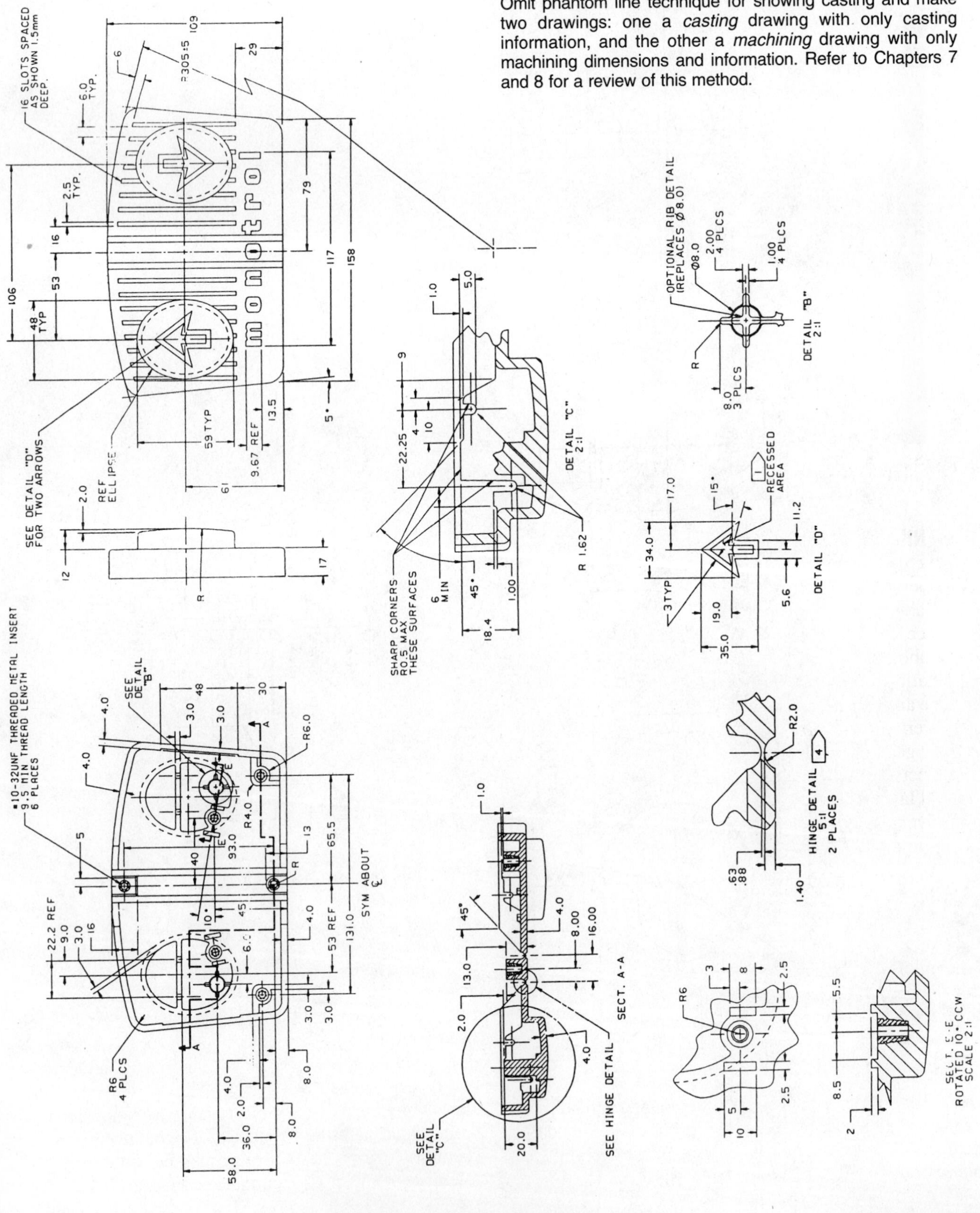

CHAPTER 11

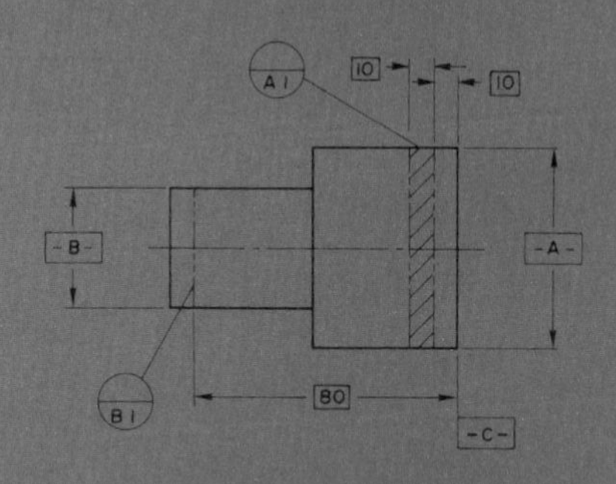

Geometric Tolerancing

LEARNING OBJECTIVES

After completing this chapter, you will:

- Label datum features on a drawing.
- Establish basic dimensions where appropriate.
- Place proper feature control frames on drawings establishing form, profile, orientation, runout, and location of geometric tolerances.
- Use and interpret material condition symbols.
- Determine the virtual condition of features.
- Use geometric tolerancing to completely dimension objects from an engineer's sketch and actual industrial layouts.
- Solve an engineering problem from engineer's notes.

Geometric Tolerancing (GT) is the dimensioning and tolerancing of individual features of a part where the permissible variations relate to characteristics of form, profile, or the relationship between features.

ANSI The standards for geometric tolerancing and dimensioning are governed by the American National Standards document ANSI Y14.5M–1982 titled *Dimensioning and Tolerancing*.

GENERAL TOLERANCING

Tolerancing was introduced in Chapter 8 with definitions and examples of how dimensions are presented on a drawing. This chapter will provide additional analysis of dimensioning characteristics as an introduction into Geometric Tolerancing.

General dimensioning and tolerancing without the addition of GT also applies to permissible variations in form, profile, and location. The degree of form and location control can be increased or decreased by altering the tolerance. General tolerancing does not necessarily take into account the geometric relationship of features of a part or the relationship between mating elements of a product unless expressly stated in note form. Every dimension on a drawing will have a tolerance except for maximum, minimum, and, sometimes, stock dimensions. Stock dimensions need not be toleranced if the mill run tolerances are acceptable. If the stock dimensions are not acceptable, then the stock dimensions must have a tolerance, such as when a precision stamping must have precise stock tolerance. Stock tolerances may be specified on the dimension, in a general note, or in the drawing title block. Dimensions, as you know, are divided into two types: size and location. The limits of a size dimension determine not only the given variation allowed in the size but also the parameter of geometric form. The produced size of the part must also be within the given tolerance at any cross section. (See Figure 11–1.)

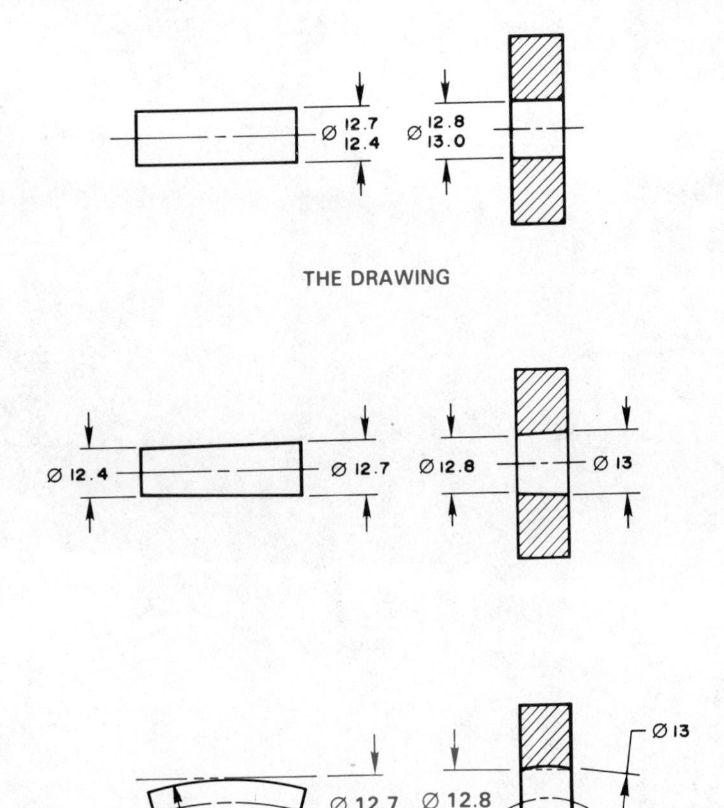

Figure 11–1 Extreme form variations of given tolerances.

SYMBOLOGY

Symbols on drawings represent specific information that would otherwise be difficult and time consuming to duplicate in note form. Symbols must be clearly drawn to required size and shape so that they communicate the desired meaning uniformly. Symbology, as discussed in Chapter 8, is an alternative for longer notes used in general dimensioning. This practice of using symbols to describe feature callouts is an optional practice. Companies that elect to use symbols for notes should not mix lettered notes with symbolic notes. When GT is used, symbols that represent the intended specification must be used. Geometric tolerancing symbols are divided into four basic types:

1. Geometric characteristic symbols, Figure 11–2(a)
2. Material condition (modifying) symbols, Figure 11–2(b)
3. Feature control frame, Figure 11–2(c)
4. Supplementary symbols, Figure 11–2(d)

A variety of professional templates are available to help save time when preparing GT symbols. One is shown in Figure 11–2(e). It is recommended that geometric dimensioning symbols be drawn with a template for good representation, accuracy, and speed. GT symbols may also be easily drawn with computer-aided design equipment.

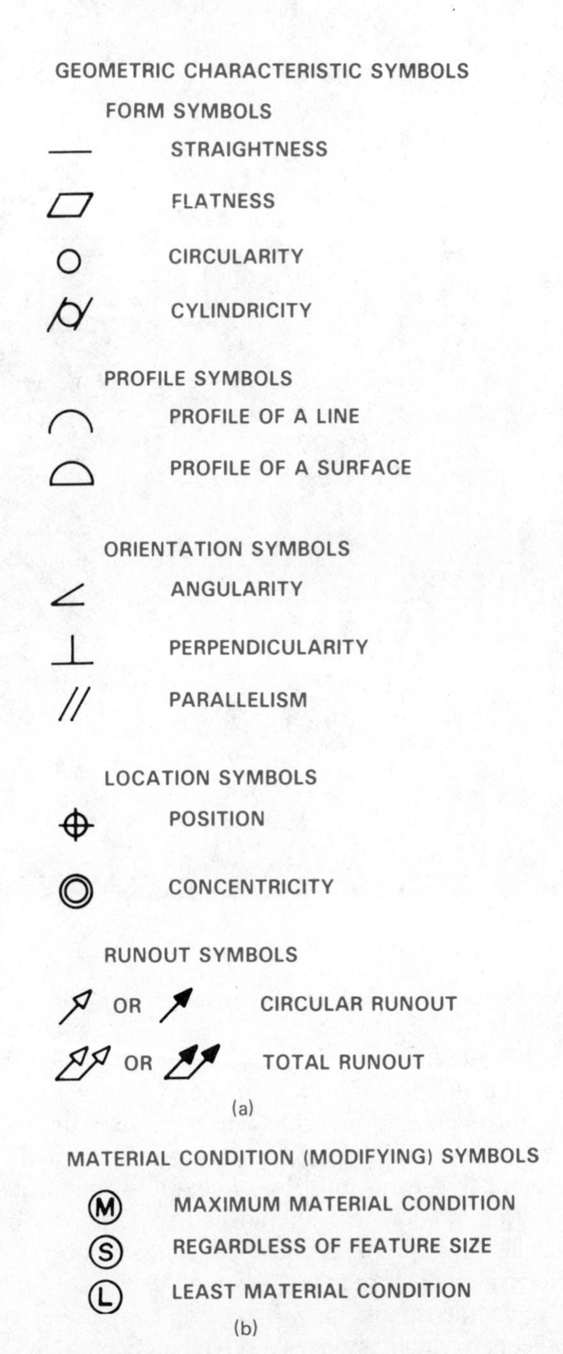

Figure 11–2 (a) Geometric characteristic symbols; (b) material condition (modifying) symbols; (c) feature control frame; (d) supplementary symbols; (e) geometric tolerancing template. *Courtesy Berol USA RapiDesign.*

DATUM FEATURE SYMBOLS

Datums are considered to be theoretically perfect surfaces, planes, points, or axes that are established from the true geometric counterpart of the datum feature. The datum feature, as shown in Figure 11–3, is the actual fea-

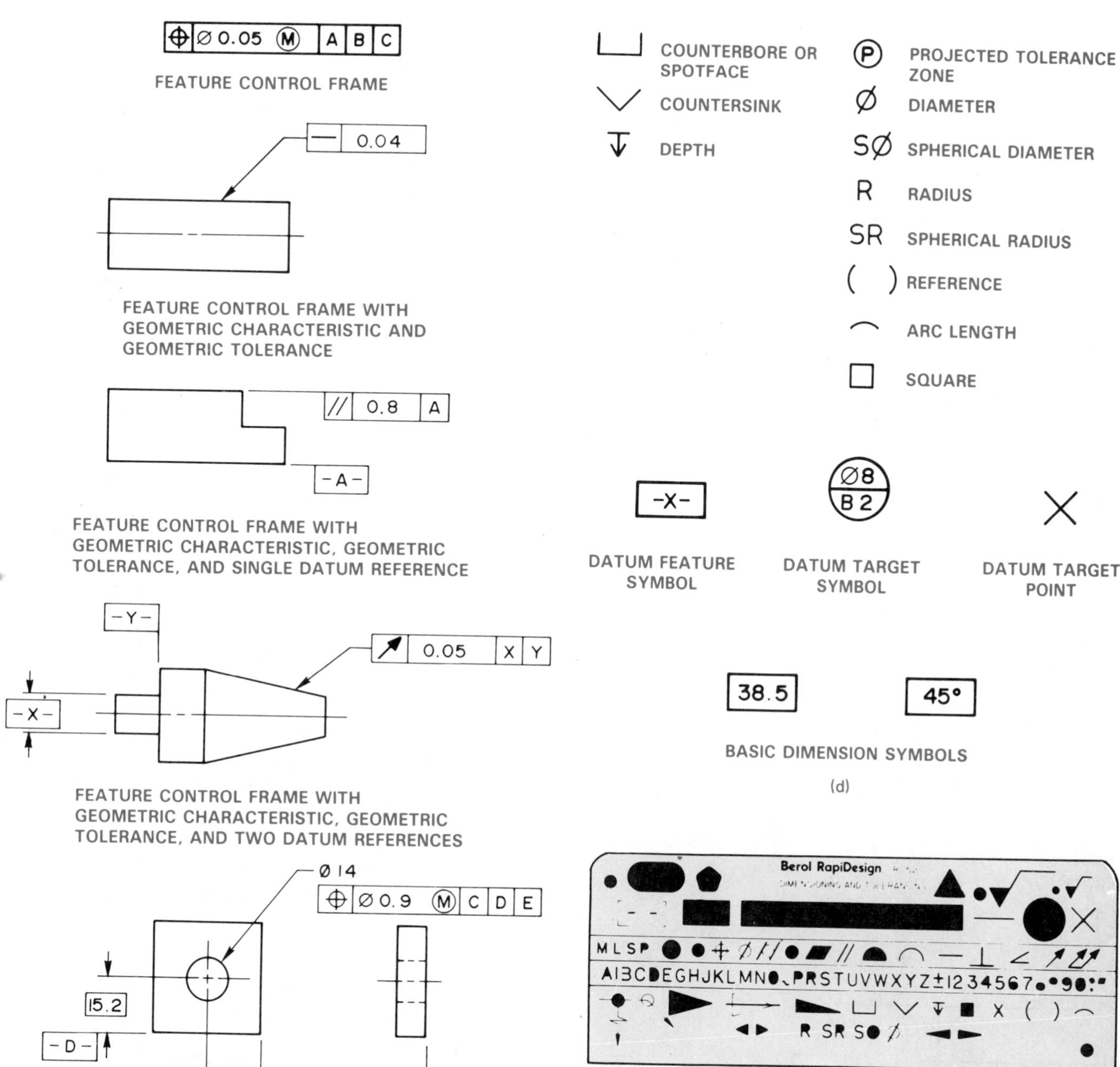

(c)

Figure 11–2 Continued

(d)

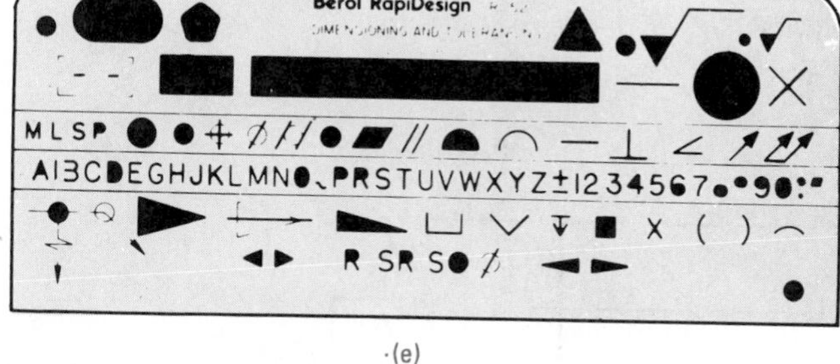

(e)

Figure 11–2 Continued

ture of the part that is used to establish the datum. Datums are used to originate size and location dimensions. Before the use of GT, datums were often implied. When dimensions originated at a common surface, for example, that surface was assumed to be the datum. The only problem with this method was that the manufacturing operation may not have interpreted the implication of the datum in the same way as the engineering department, or the implied datum may have been ignored altogether. With the use of defined datums each part is made with dimensions that begin from the same origin; there is no variation. Datum feature symbols are placed in the view that shows the edge of a surface or are attached to a diameter or symmetrical dimension when associated with a centerline or center plane. (See Figure 11–4.) Datum feature symbols are not placed on a center axis line or center plane line unless the center plane does not have a dimension that clearly differentiates the datum from other features. (See Figure 11–5.) Datum feature symbols are commonly drawn using thin lines with the symbol size related to the size of the lettering on a drawing, as shown in Figure 11–6, unless the template varies from this recommended standard.

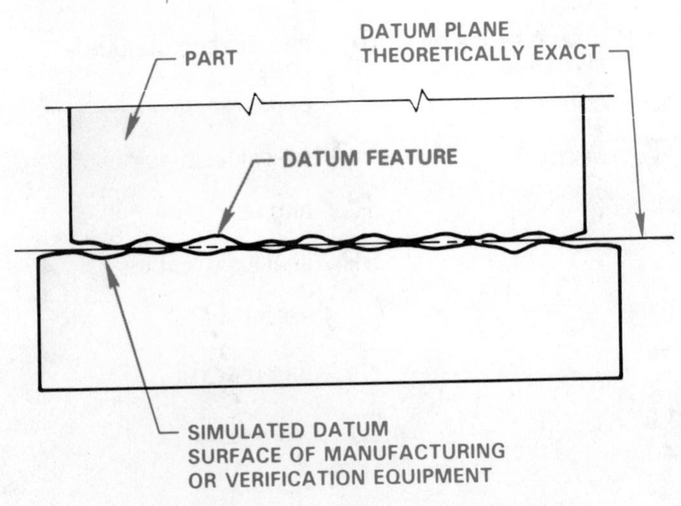

Figure 11–3 Enlarged representation of a datum feature.

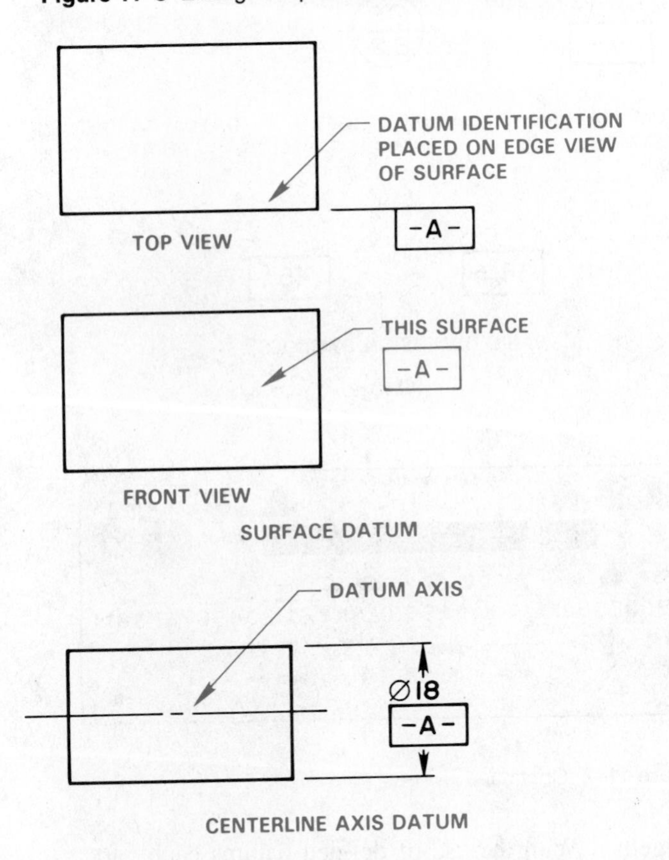

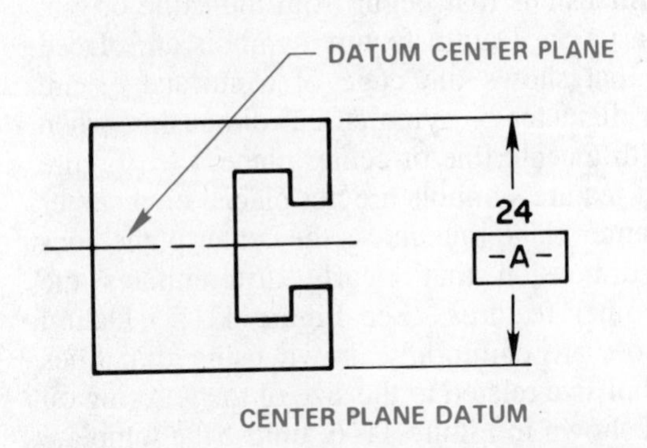

Figure 11–4 Datum identification.

DATUM REFERENCE FRAME

Datum referencing is used to relate features of a part to an appropriate datum or datum reference frame. A datum indicates the origin of a dimensional relationship between a toleranced feature and a designated feature or features on a part. The designated feature serves as a datum feature whereas its true geometric counterpart establishes the datum. As measurements cannot be made from a true geometric counterpart that is theoretical, a datum is assumed to exist in and be simulated by the associated processing equipment. For example, machine tables and surface plates, though not true planes, are of such quality that they are used to simulate the datums from which measurements are taken and dimensions are verified.

Sufficient datum features, those most important to the design of a part, are chosen to position the part in relation to a set of three mutually perpendicular planes, jointly called a *datum reference frame*. This reference frame exists in theory only and not on the part. Therefore, it is necessary to establish a method for simulating the theoretical reference frame from the actual features of the part. This simulation is accomplished by positioning the part on appropriate datum features to adequately relate the part to the reference frame and to restrict the motion of the part in relation to it. (See Figure 11–7.) These planes are simulated in a mutually perpendicular relationship to provide direction as well as the origin for related dimensions and measurements. Thus when a part is positioned on the datum reference frame, dimensions related to the datum reference frame by a feature control frame or note are thereby mutually perpendicular. This theoretical reference frame constitutes the three plane dimensioning system used for datum referencing. Refer to Figure 11–7.

DATUM FEATURES

A datum feature is selected on the basis of its geometric relationship to the toleranced feature and the requirements of the design. To ensure proper part interface and assembly, corresponding features of mating parts are also selected as a datum feature where practical. Datum features must be readily recognizable on the part. Therefore, in the case of symmetrical parts or parts with identical features, physical identification of the datum feature on the part may be necessary. A datum feature should be accessible on the part and be of sufficient size to permit subsequent processing operations.

The three reference planes are mutually perpendicular unless otherwise specified. Planes in the datum reference frame are placed in order of importance. For example, the plane that originates most size and location dimensions, or has the most functional importance, is the primary, or first, datum plane. The next datum is the secondary, or second, datum plane and the third element of

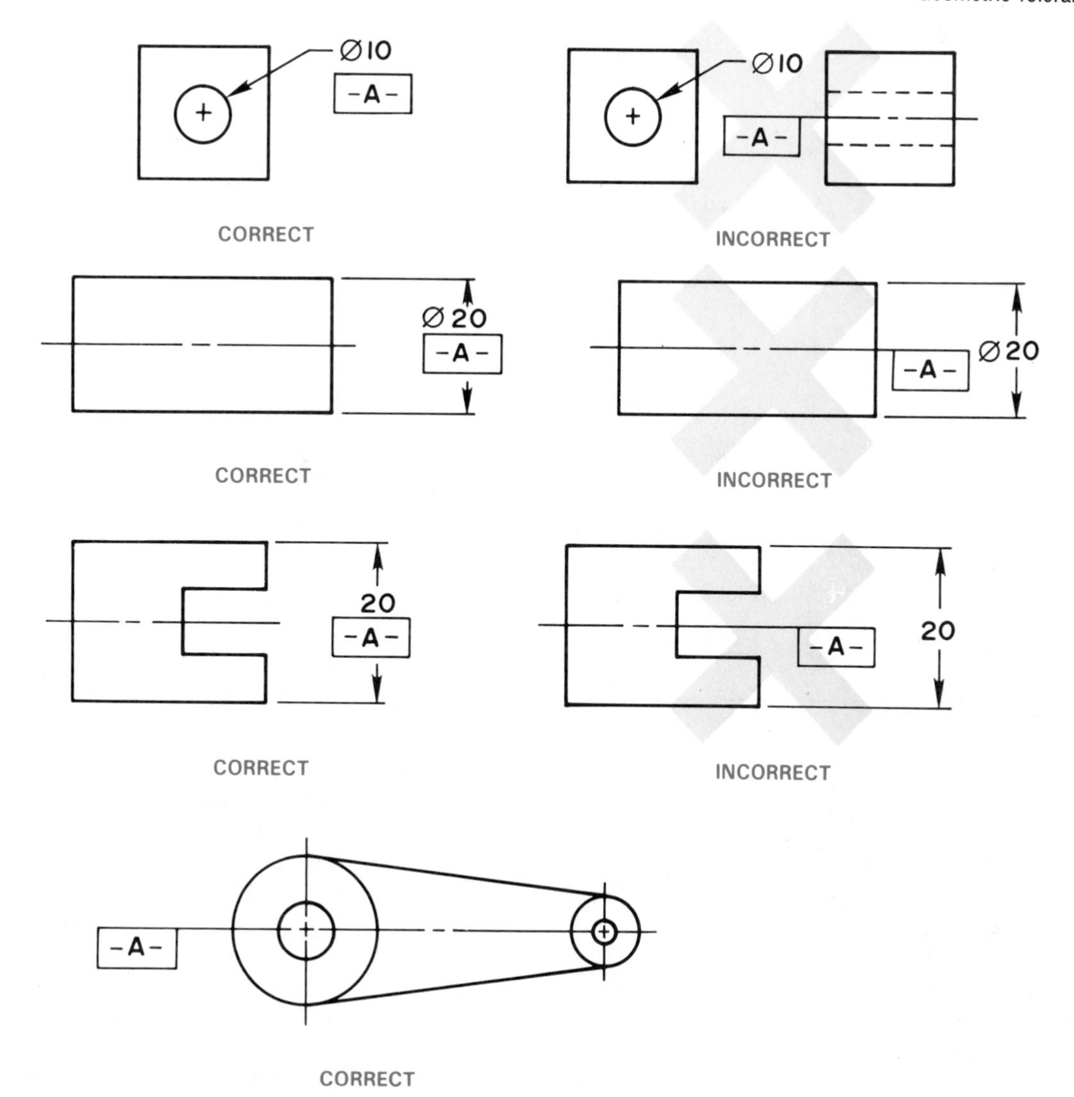

Figure 11–5 Proper placement of centerline and center plane datum feature symbols.

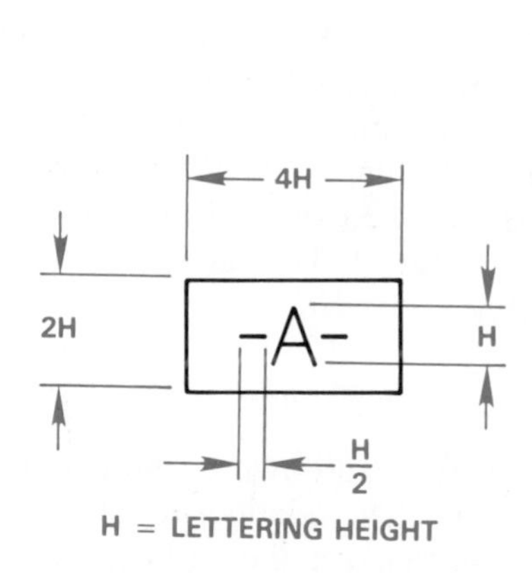

Figure 11–6 Datum feature symbol.

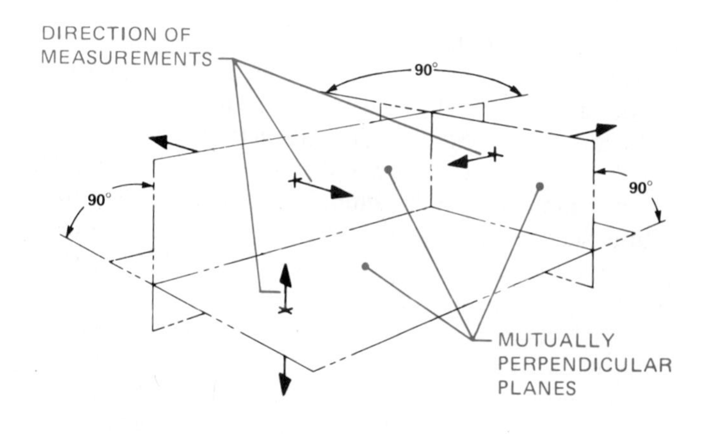

Figure 11–7 Datum reference frame. *Reprinted with permission by the American Society of Mechanical Engineers, ANSI Y14.5M–1982.*

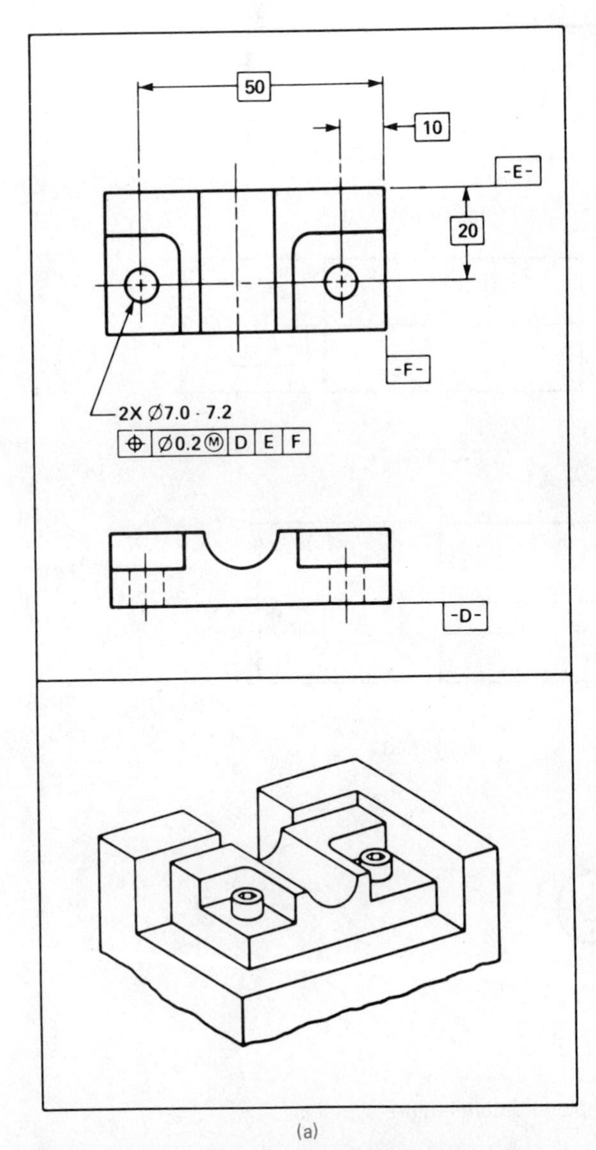

Figure 11–8 (a) Part where datum features are plane surfaces; (b) sequence of datum features relates the part to the datum reference frame. *Reprinted with permission by the American Society of Mechanical Engineers, ANSI Y14.5M–1982.*

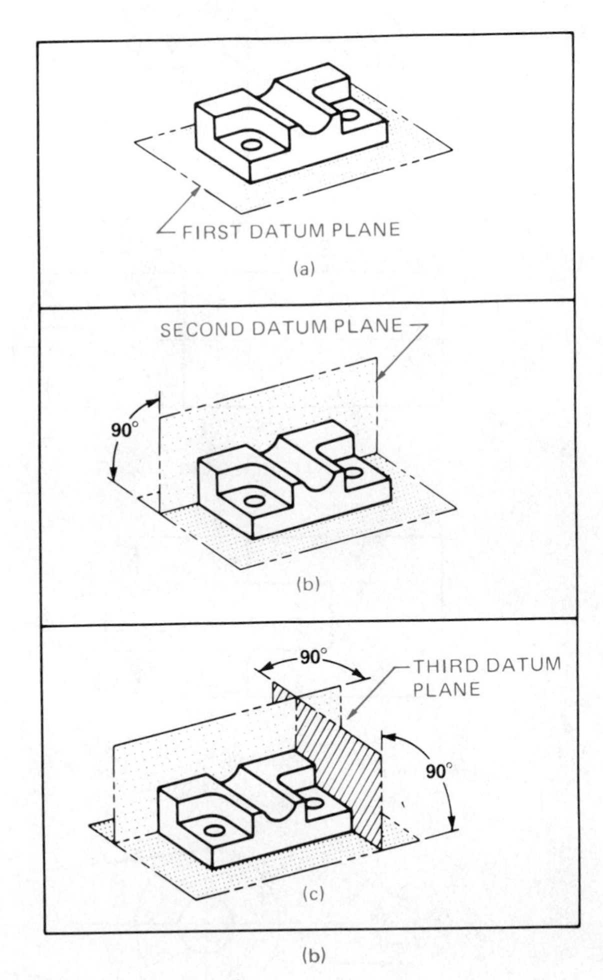

Figure 11–8 Continued

the frame is called the tertiary, or third, datum plane. The desired order of precedence is indicated by entering the appropriate datum reference letters from left to right in the feature control frame. For instructional purposes it may be convenient to label datums as A, B, and C although in industry other letters are also used to identify datums such as D, E, and F, or X, Y, and Z. The letters that should be avoided are O, Q, and I. Figure 11–8(a) shows a part and the planes that are chosen as datum features. Notice in Figure 11–8(b) how the order of precedence of datum features relates the part to the datum reference frame. The datum features are identified as surfaces D, E, and F. These surfaces are most important to the design and function of the part. Surfaces D, E, and F are the primary, secondary, and tertiary datum features respectively since they appear in that order in the feature control frame.

Multiple datum frames may be established for some parts depending on the complexity and function of the part. The relationship between datum frames is often controlled by a representative angle. In this case, datum G is at an angle to the conventional datum reference frame. (See Figure 11–9.)

Partial Surface Datums

In some situations it may be more realistic to apply a datum feature symbol to a portion of a surface rather than the entire surface. For example, when a long part has related features located in one or more concentrated places, then the datum may be located adjacent to the concentration of features. When this is done, a chain line is used to identify the extent of the datum feature. The location and length of the chain line must be dimensioned. (See Figure 11–10.)

Coplanar Surface Datums

Coplanar surfaces are two or more surfaces that are in the same plane. The relationship of coplanar surface datums may characterize the surfaces as one plane or datum in correlated geometric callouts as shown in Figure 11–11.

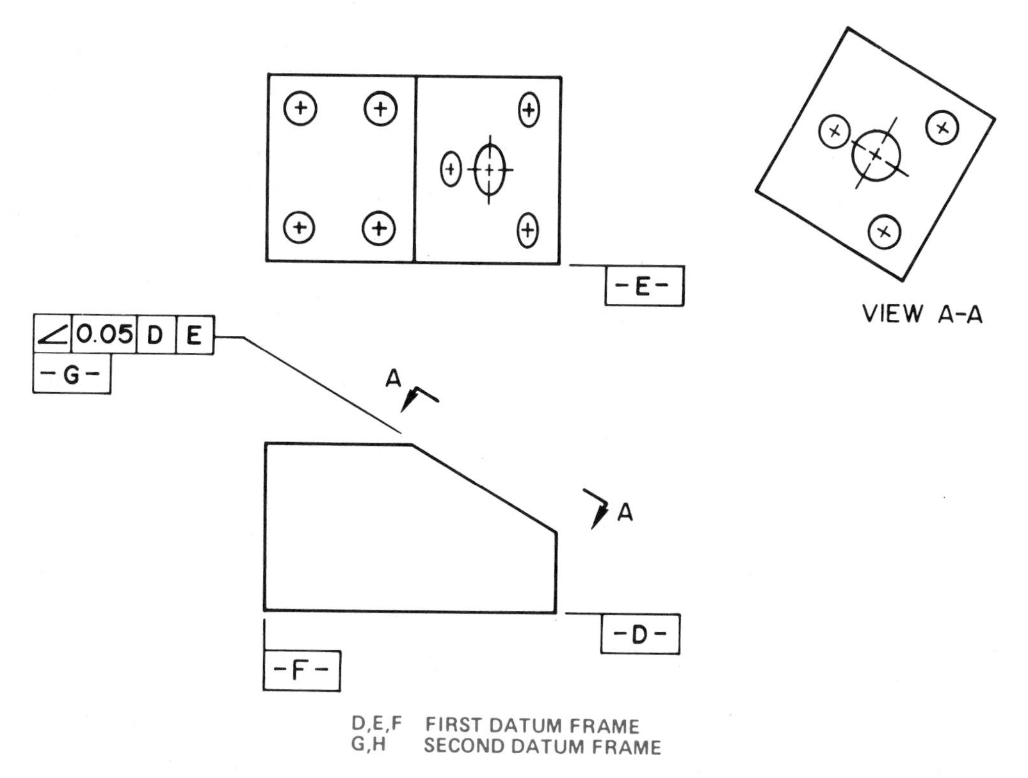

Figure 11–9 Multiple datum frames.

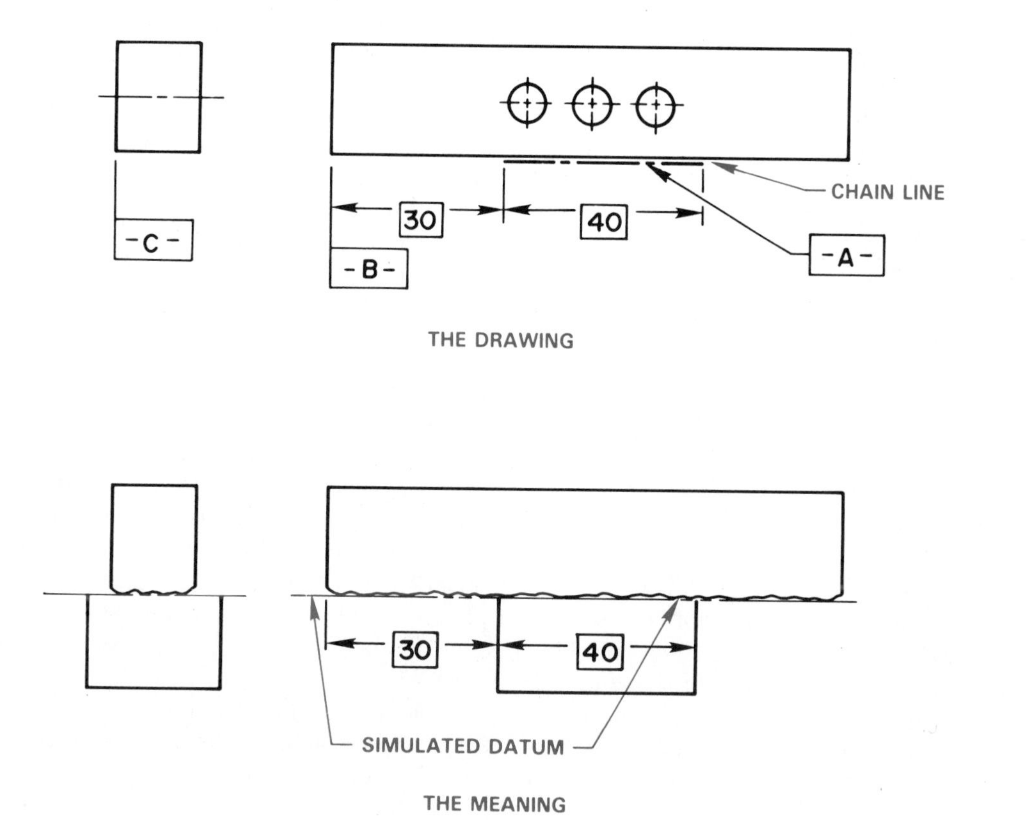

Figure 11–10 Partial surface datums.

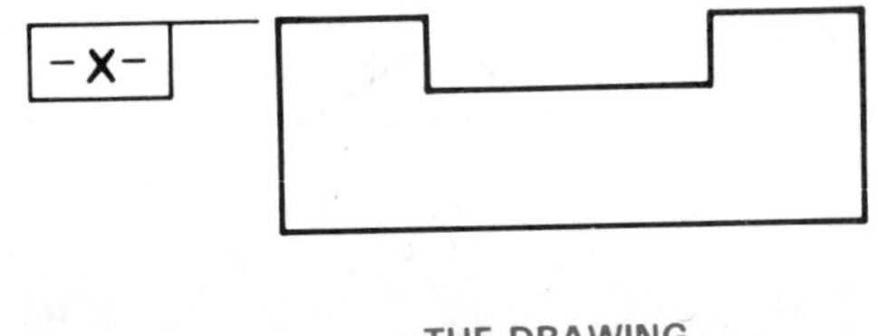

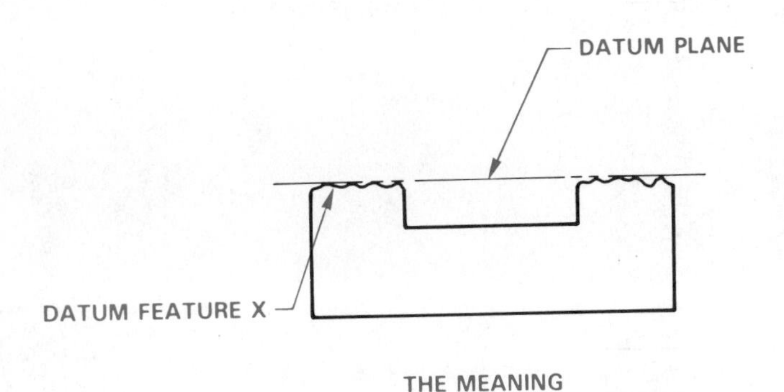

Figure 11–11 Coplanar surface datums.

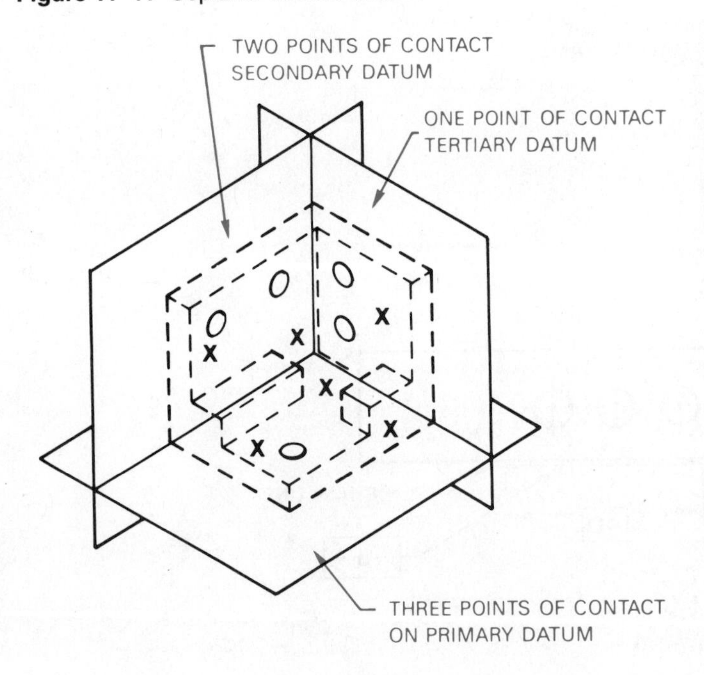

Figure 11–12 Datum frame established by datum target symbols.

DATUM TARGET SYMBOLS

In many situations it is not possible to establish an entire surface or surfaces as datums, or it may not be practical to coordinate a partial datum surface with the part features. When this happens due to the size or shape of the part, then datum targets may be used to establish datum planes. Datum targets are especially useful on parts with surface or contour irregularities such as sand castings or forgings, on some sheet metal parts subject to bowing or warpage, or on weldments where heat can induce warpage. Datum targets are designated points, lines, or surface areas that are used to establish the datum reference frame. The primary datum is established by three points or contact locations. The three locations should be placed on the part so that they form a stable, well-spaced pattern and particularly are not in a line. The secondary datum is created by two points or locations. The tertiary datum is established by one point or location. Remember that the purpose of the datum frame is to prepare a stable position for the part dimensions to be established in relationship to corresponding datums. For this reason, the three points on the primary datum are used to provide stability similar to a three-legged stool. The two points on the secondary datum provide the needed amount of stability when the object is placed against the secondary datum. Finally, the tertiary datum requires only one point of contact to complete the stability between the three datum frame elements. (See Figure 11–12.)

The datum target symbol is drawn as a circle using thin lines and is connected with a leader that points to a target point, line, or surface area. The datum target symbol is divided into two halves by a horizontal line. The top half of the symbol is reserved for identification of the datum target area size when used. The bottom half of the symbol is used for datum target identification. For example, if there are three datum target points on a datum, the first point may be labeled A1, the second A2, and the third A3. The datum target point is located on the surface or edge view from adjacent datums with basic dimensions. (See Figure 11–13.) Chain dimensioning is commonly used to locate datum target points and target areas. The location dimensions must originate from datums. Datum target areas are located to their centers. (See Figure 11–14.) When the surface of a part has an irregular contour or different levels, the contact points, lines, or areas may lie on the same plane and different lengths of locating pins may be used. The pins make point contact as opposed to surface contact; thus the pins are usually rounded.

DATUM AXIS

The datum frame established by a cylindrical object is the base of the part and the two theoretical planes, represented in Figure 11-15 by the X and Y center planes, which cross to establish the datum axis. The actual secondary datum is the cylindrical surface. The center planes located at X and Y are used to indicate the direction of dimensions that originate from the datum axis. (See Figure 11–16.)

Datum target points, lines, or surface areas may also be used to establish a datum axis. A primary datum axis may be established by two sets of three equally spaced

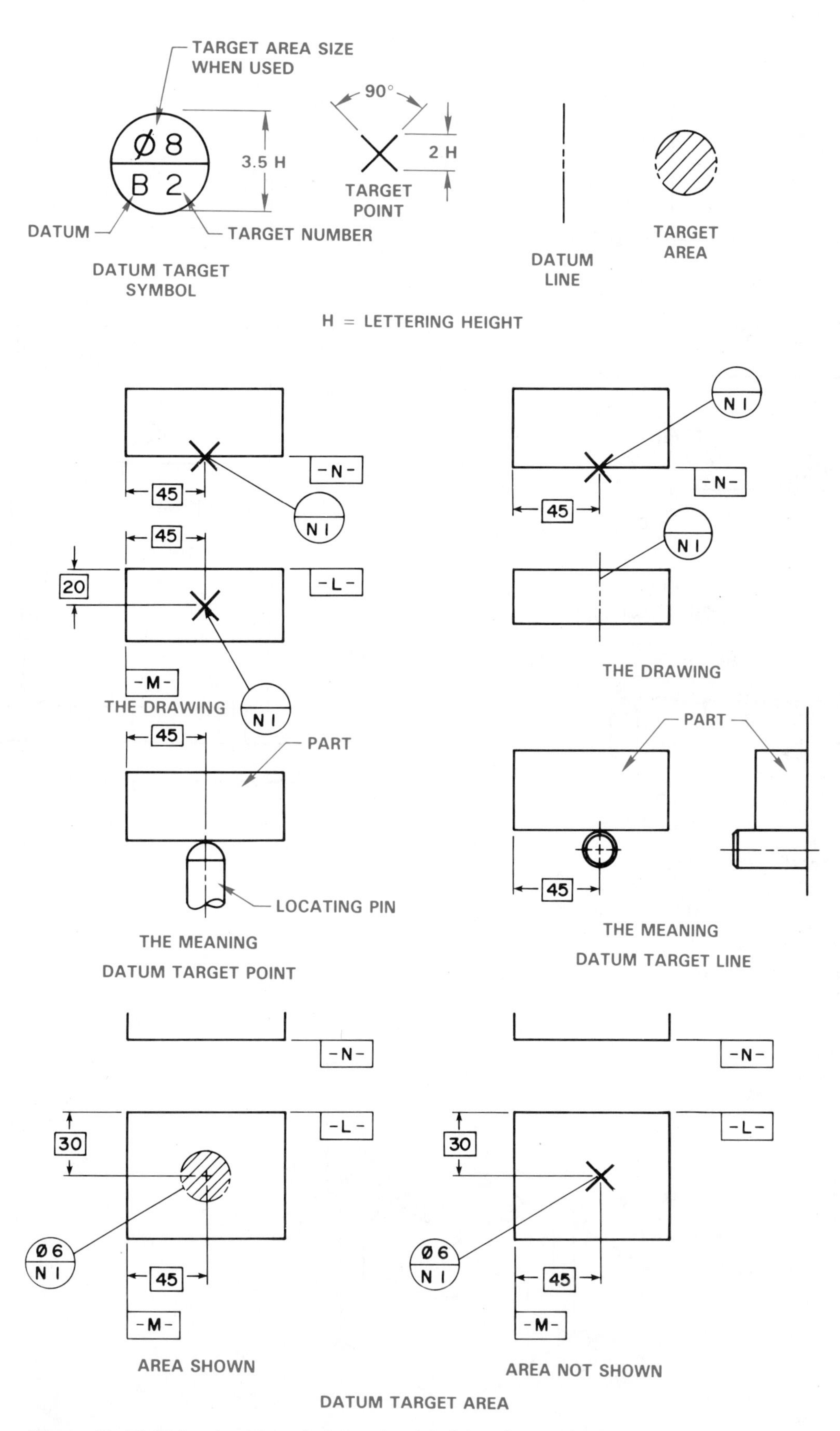

Figure 11–13 Datum target symbol, target point, datum line, and target area.

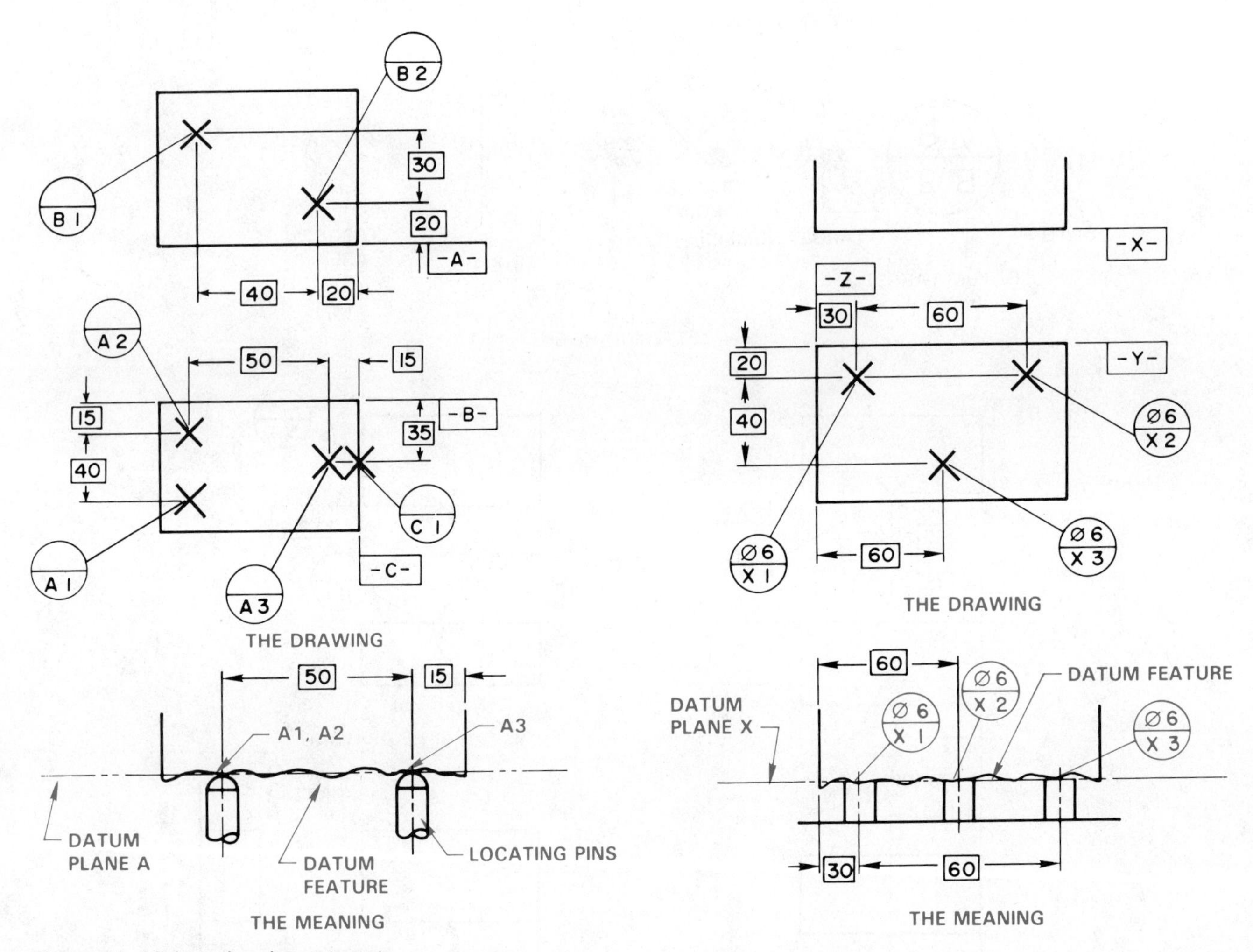

Figure 11–14 Locating datum targets.

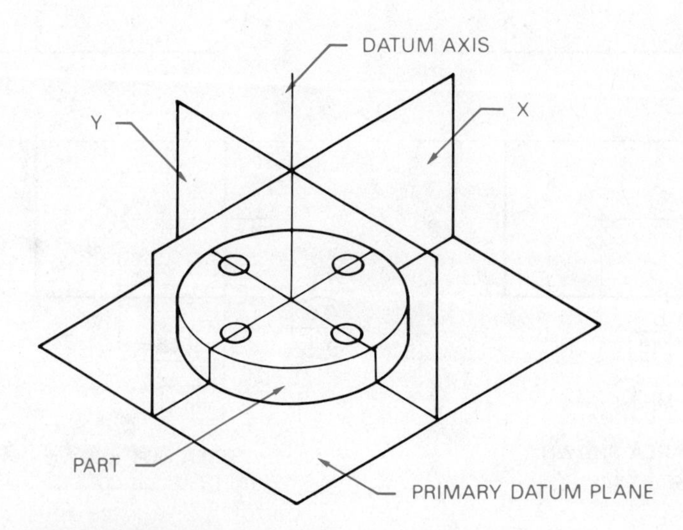

Figure 11–15 Datum frame for a cylindrical object.

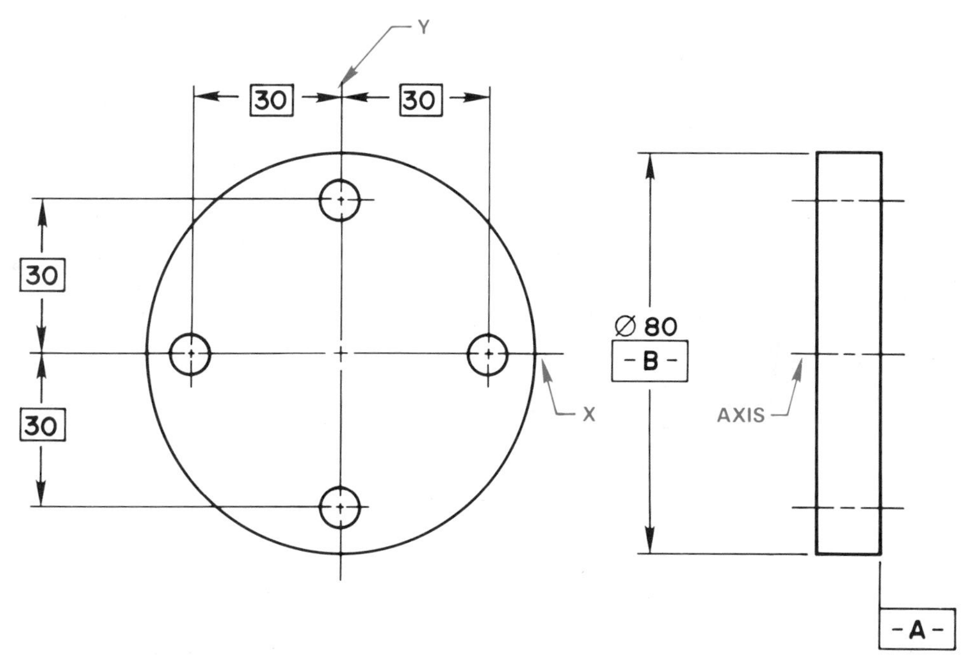

Figure 11–16 Datum axis.

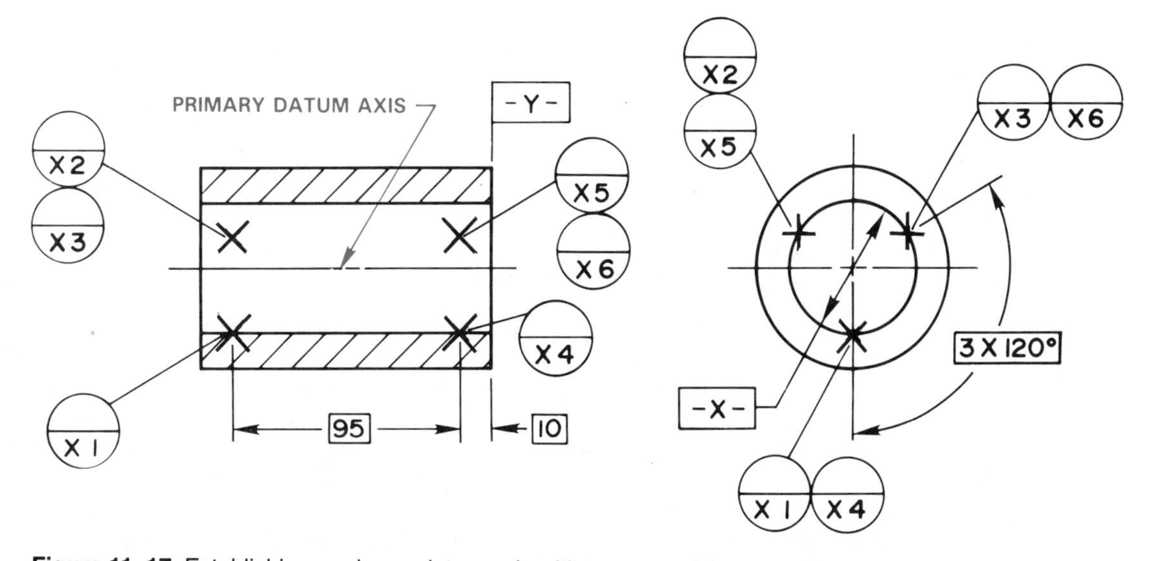

Figure 11–17 Establishing a primary datum axis with two sets of three equally spaced targets.

targets: a set at one end of the cylinder and the other set near the other end as shown in Figure 11–17. When two cylindrical features of different diameters are used to establish a datum axis then the datum target points are identified in correspondence to the adjacent cylindrical datum feature as shown in Figure 11–18. Cylindrical datum target areas and circular datum target lines may also be used to establish the datum axis of cylindrical shaped parts as shown in Figure 11–19. A secondary datum axis may also be established by placing three equally spaced targets on the cylindrical surface. (See Figure 11–20.)

FEATURE CONTROL FRAME

The feature control frame is used to relate a geometric tolerance to a part feature. The elements in a feature control frame must always be in the same order. The most basic format is when a geometric characteristic and related tolerance is applied to an individual feature as shown in Figure 11–21. The next expanded format is when a geometric characteristic, tolerance zone descriptor, and material condition symbol are used in the feature control frame. (See Figure 11–22.) One, two, or three datum references may be included in a feature control

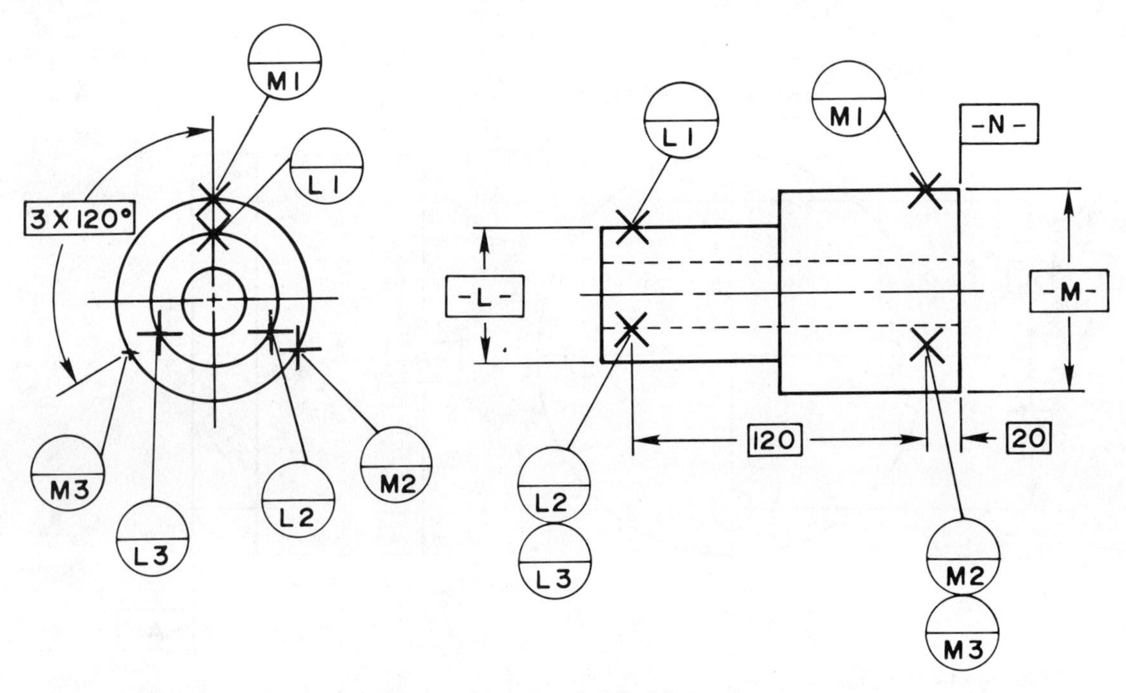

Figure 11–18 Datum targets identified on adjacent cylindrical features.

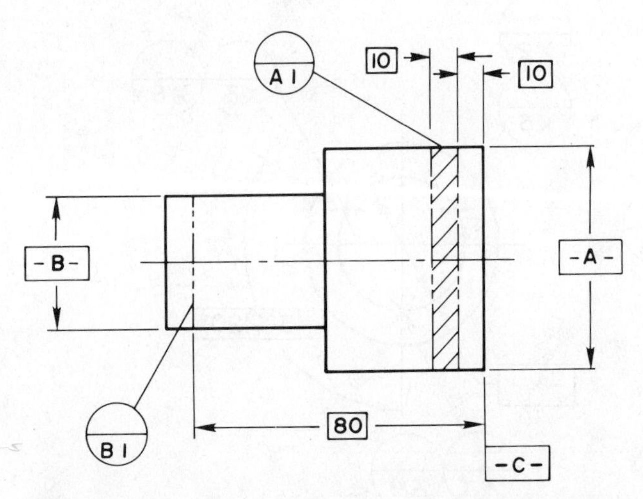

Figure 11–19 Cylindrical datum target areas and circular datum target lines.

frame as shown in Figure 11–23. The material condition symbol that applies to the feature tolerance is always placed after the tolerance. When a material condition symbol is applied to a datum reference, it is placed after the datum reference and in the same compartment as shown in Figure 11–24.

The feature control frame is drawn with thin lines and connected to the feature with a leader or extension line. (See Figure 11–25.) The feature control frame may be combined with a datum feature symbol when the feature controlled by the geometric tolerance also serves as a datum. The datum feature symbol and the datum reference in the feature control frame are considered separately. When a datum feature symbol and a feature control frame are combined, the feature control frame should be shown first. (See Figure 11–26.)

BASIC DIMENSIONS

A basic dimension is defined as any size or location dimension that is used to identify the theoretically exact size, profile, orientation, or location of a feature or datum target. Basic dimensions are the basis from which permissible variations are established by tolerances on other dimensions in notes or in feature control frames. A basic dimension is described on a drawing by placing a thin-line box around the dimension. (See Figure 11–27.)

GEOMETRIC TOLERANCES

Geometric tolerances are divided into five types: form, profile, orientation, runout, and location. These tolerances are subdivided into thirteen characteristics plus modifying terms, all of which will be discussed in this section.

Form Tolerance

A tolerance of form is commonly applied to individual features or elements of single features and is not related to datums. The amount of given form variation must fall within the specified size tolerance zone.

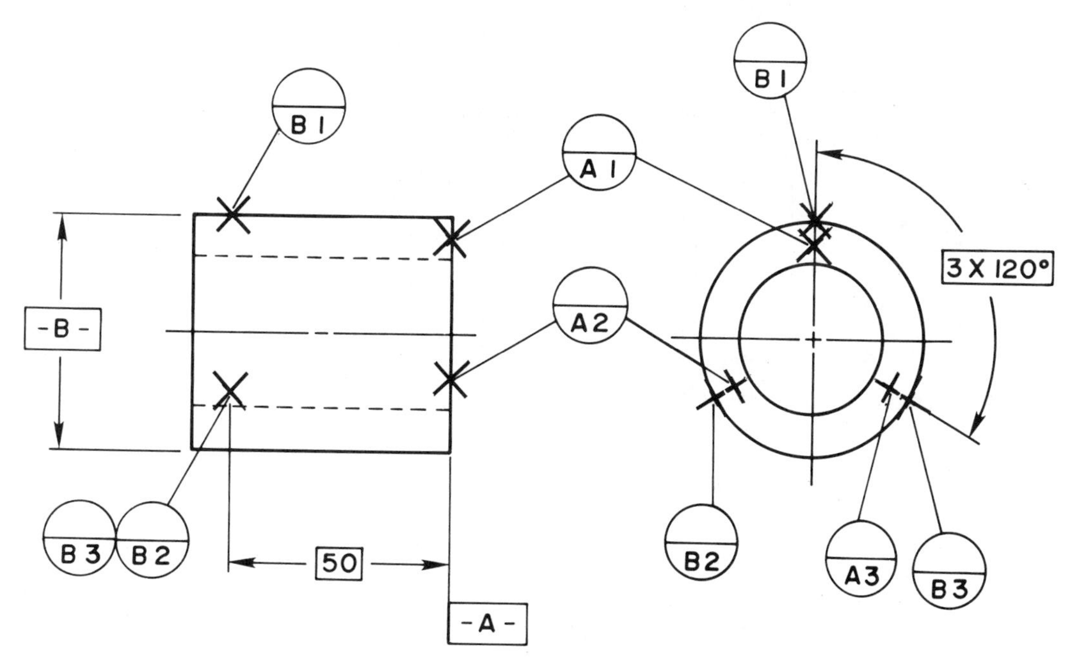

Figure 11–20 Establishing a secondary datum on a cylindrical object with three equally spaced targets.

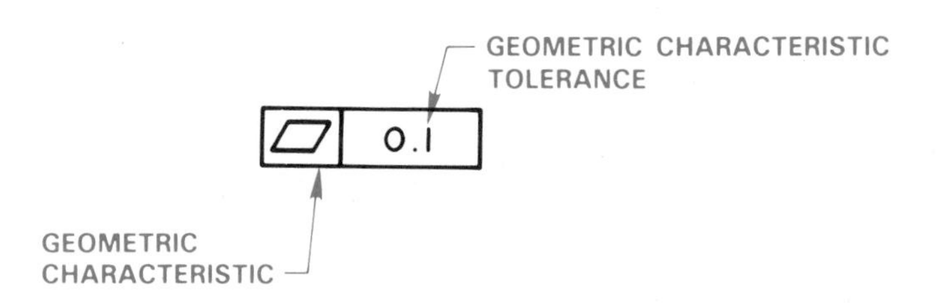

Figure 11–21 Feature control frame with geometric characteristic and related tolerance.

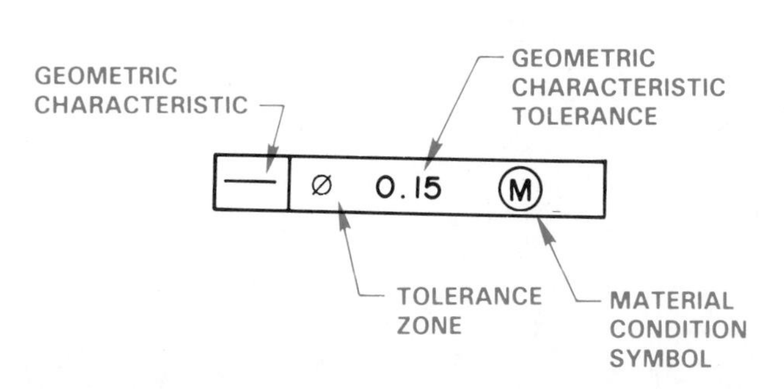

Figure 11–22 Feature control frame with geometric characteristic, tolerance zone descriptor, and material condition symbol.

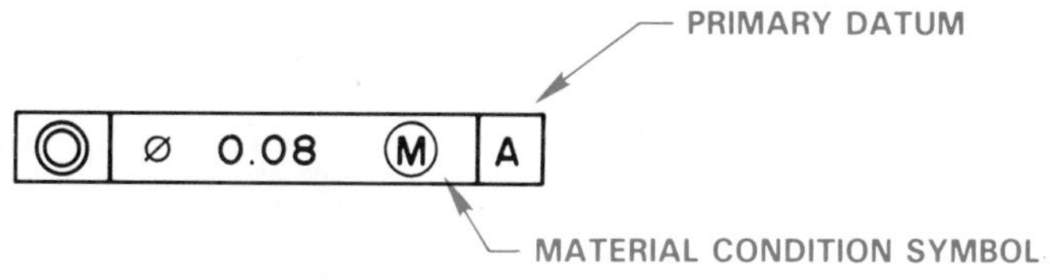

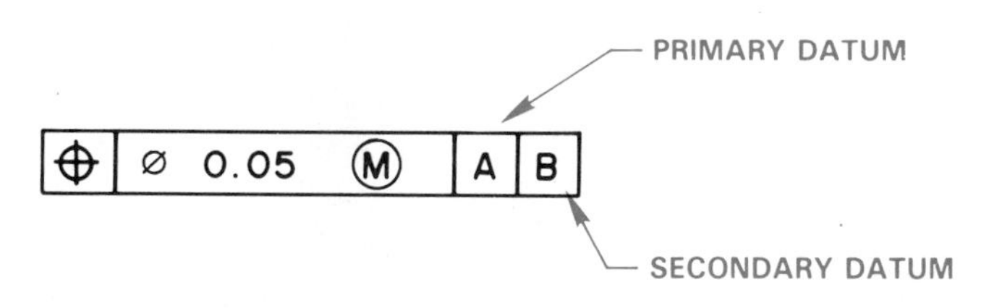

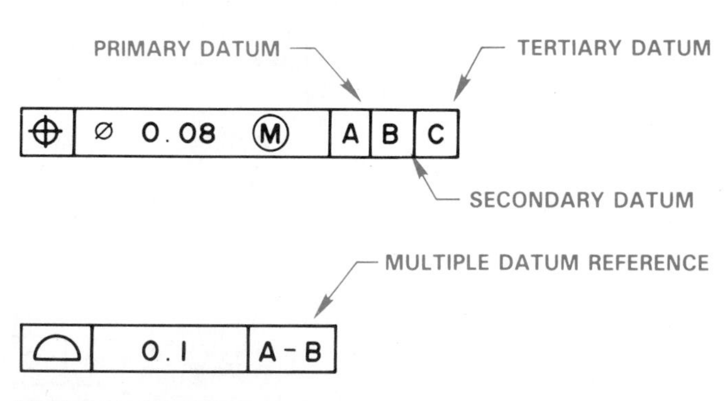

Figure 11–23 Applying one, two, or three datum references to the feature control frame.

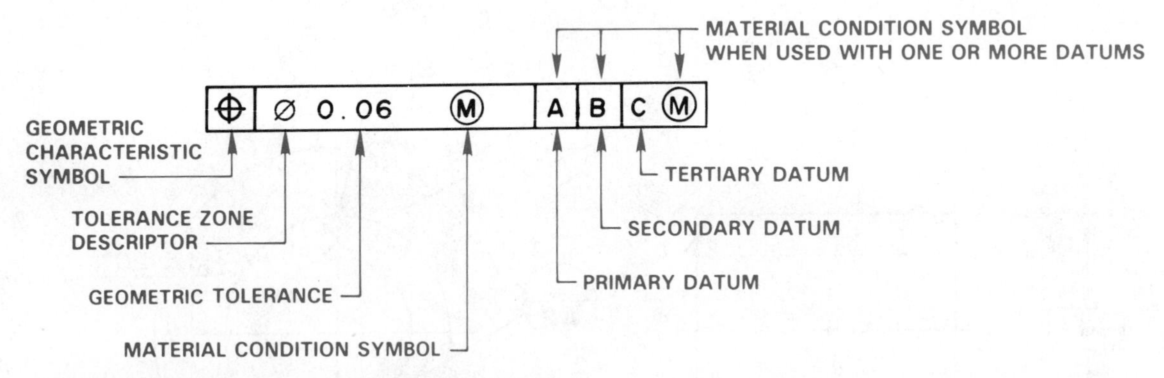

Figure 11–24 Elements of a feature control frame.

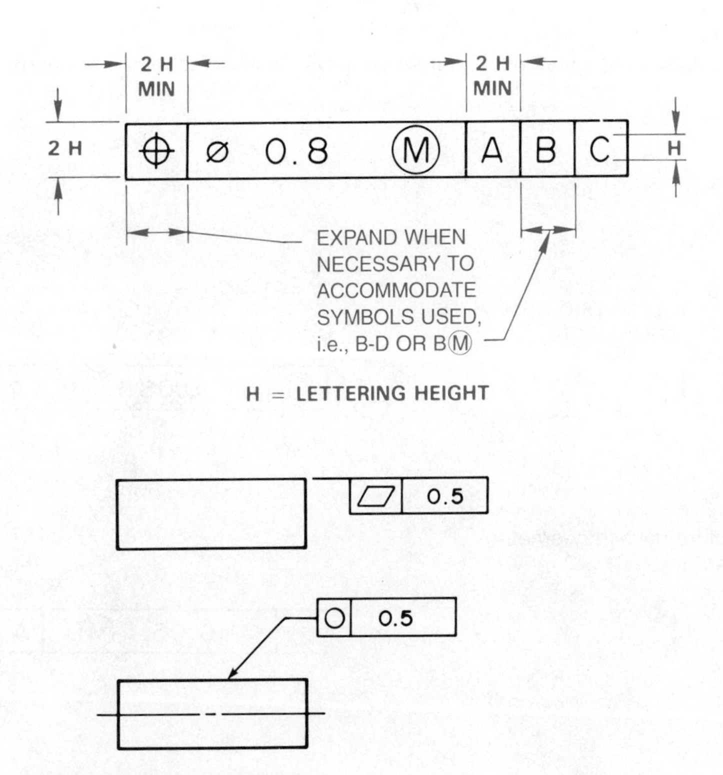

Figure 11–25 Detailed feature control frame and application.

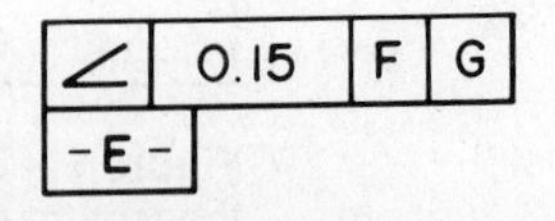

Figure 11–26 Combination of datum feature symbol and feature control frame.

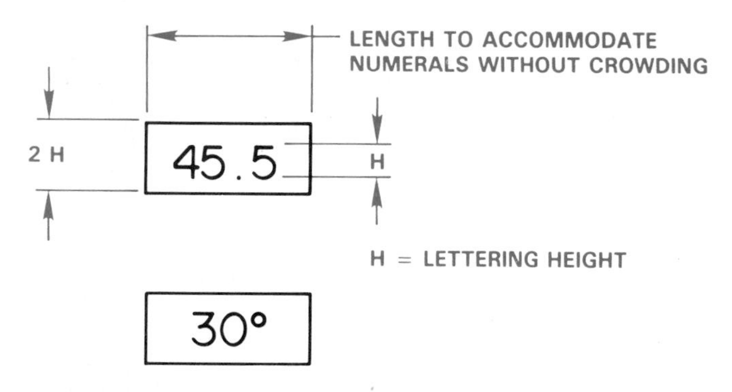

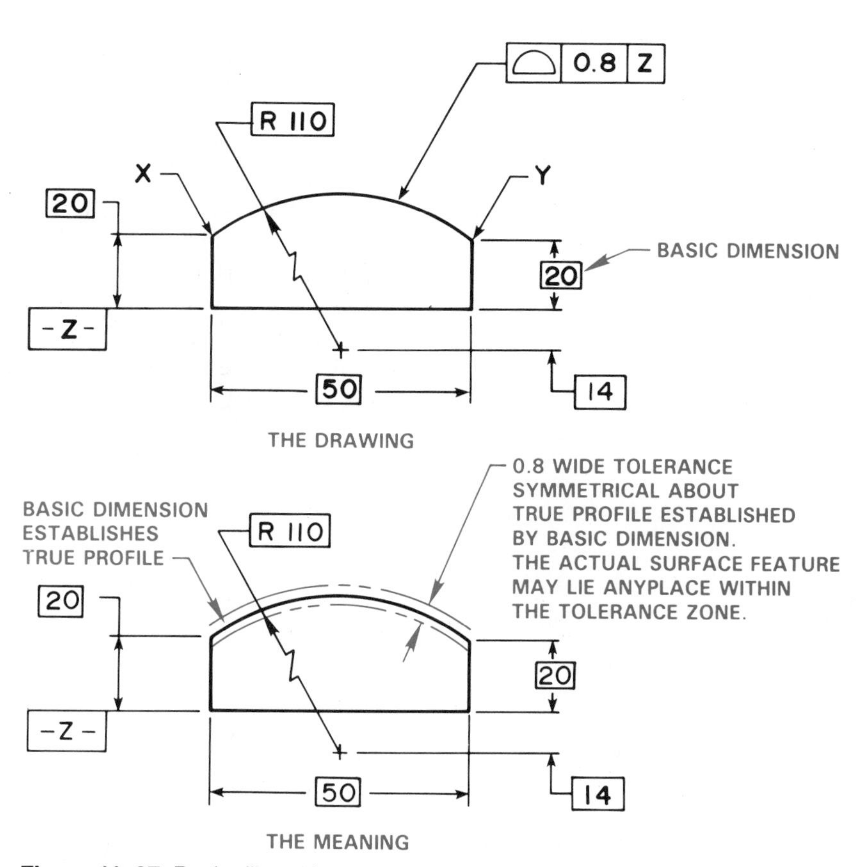

Figure 11–27 Basic dimensions.

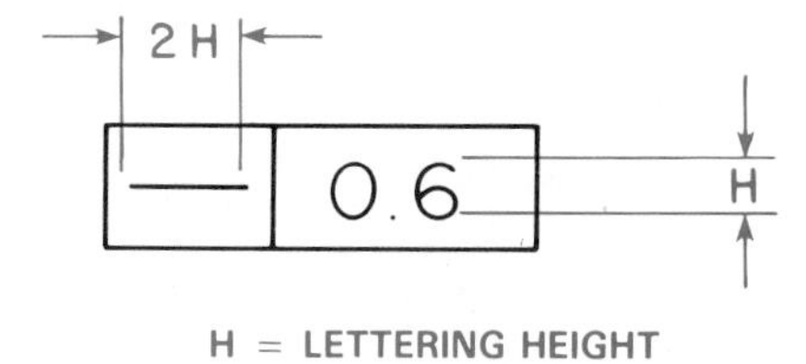

Figure 11–28 Straightness feature control frame.

Straightness. The straightness symbol is detailed in Figure 11–28. Perfect straightness exists when a surface element or the axis of a part is a straight line. A straightness tolerance allows for a specified amount of variation from a straight line. The straightness feature control frame may be attached to the surface of the object with a leader or combined with the diameter dimension of the part. When straightness is connected to the surface of the feature with a leader, surface straightness is implied as shown in Figure 11–29. When the straightness feature control frame is combined with the diameter dimension of the part, then axis straightness is specified. (See Figure 11–30.) While a straightness tolerance is common on cylindrical parts, straightness may also be applied to noncylindrical parts in the same manner. When specifying center feature straightness, a center plane is implied and the diameter tolerance zone descriptor is omitted. The straightness tolerance must always be less than the size tolerance.

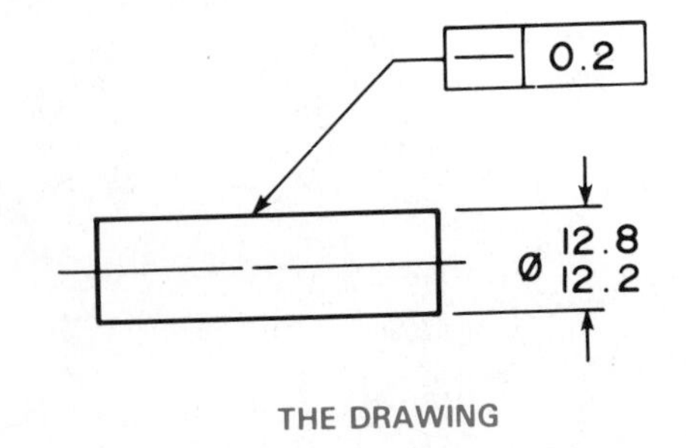

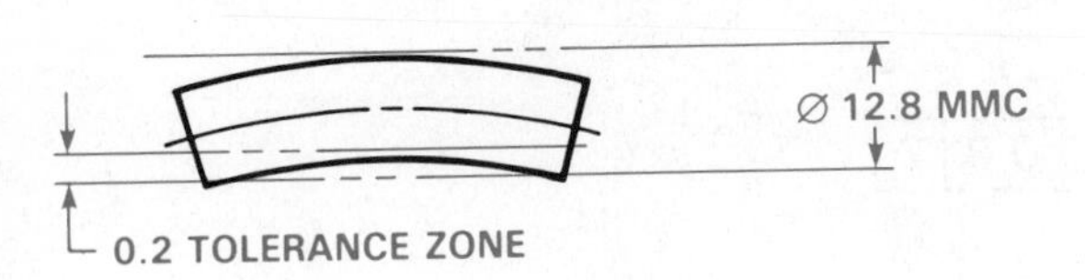

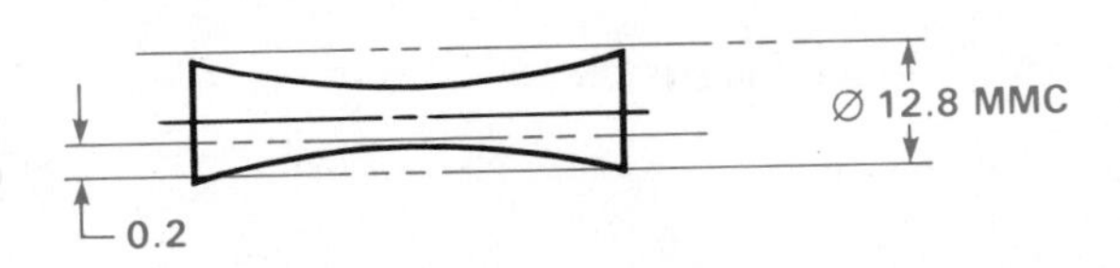

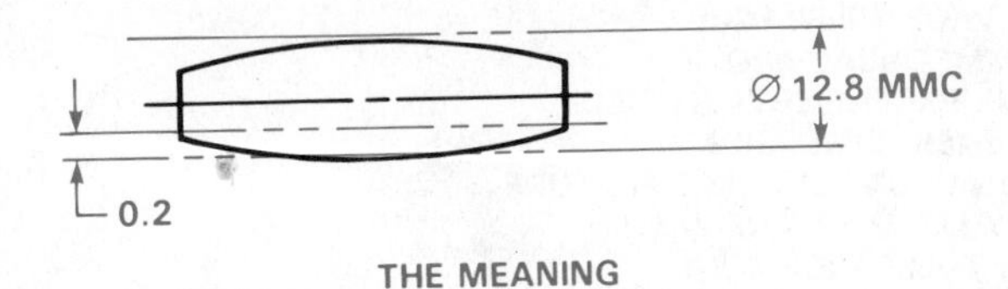

Figure 11–29 Surface straightness.

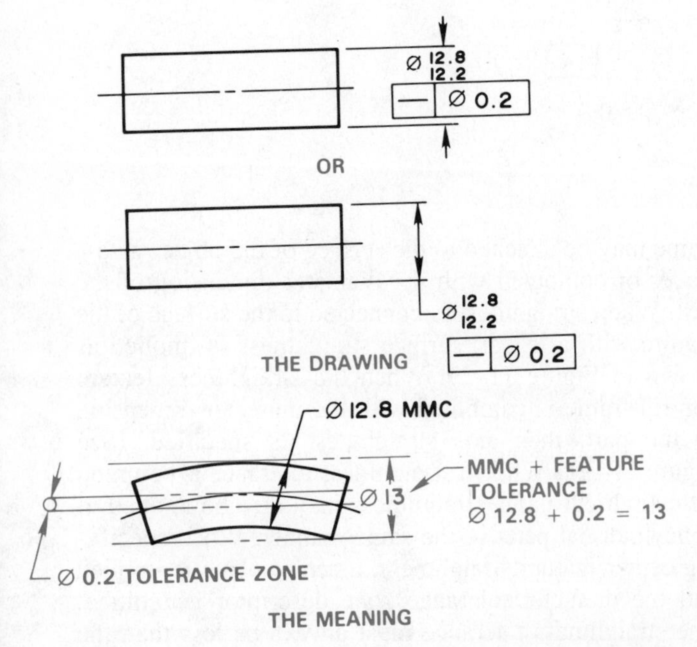

Figure 11–30 Axis straightness.

Unit straightness is a situation where a specified tolerance is given per unit of length, for example 25-mm units, and a greater amount of tolerance is provided over the total length of the part. (See Figure 11–31.) This type of tolerance is not commonly used.

Flatness. The flatness tolerance symbol is detailed in relationship to the feature control frame in Figure 11–32. A surface is considered flat when all of the surface elements lie in one plane. A flatness tolerance callout allows for a specified amount of surface variation from a flat plane. The flatness tolerance zone establishes two parallel planes. The actual surface of the object may not extend beyond the boundary of the tolerance zone and, when associated with the size dimension, the flatness tolerance must be smaller than the size tolerance. The flatness feature control frame may be connected to the edge view of the surface with a leader or with an extension line. (See Figure 11–33.)

Unit flatness may be specified when it is desirable to control the flatness of a given surface area as opposed to the entire surface. A unit flatness callout may be presented with or without a separate tolerance for the total area. Unit flatness, just as unit straightness, should have a total tolerance in order to avoid a situation where the unit tolerance gets out of control. The size of the unit area may be given after the total tolerance specification as shown in Figure 11–34.

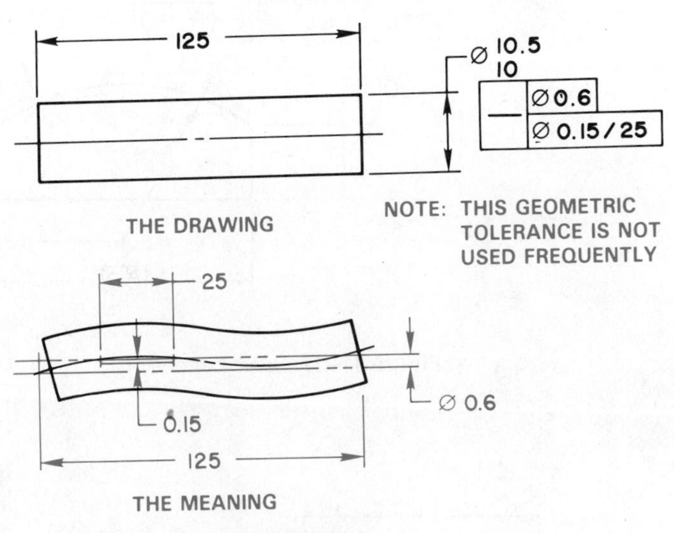

Figure 11–31 Unit straightness.

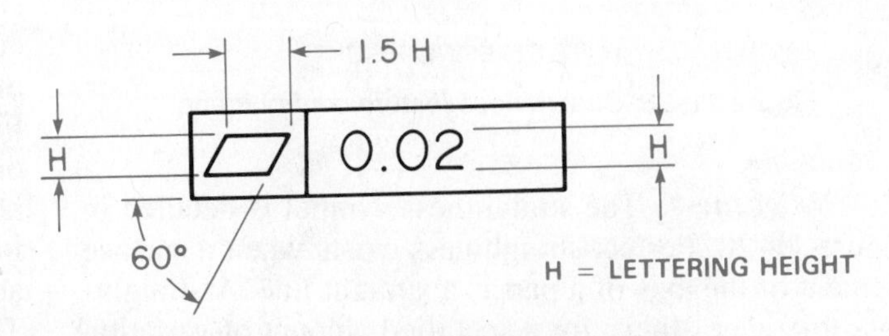

Figure 11–32 Flatness feature control frame.

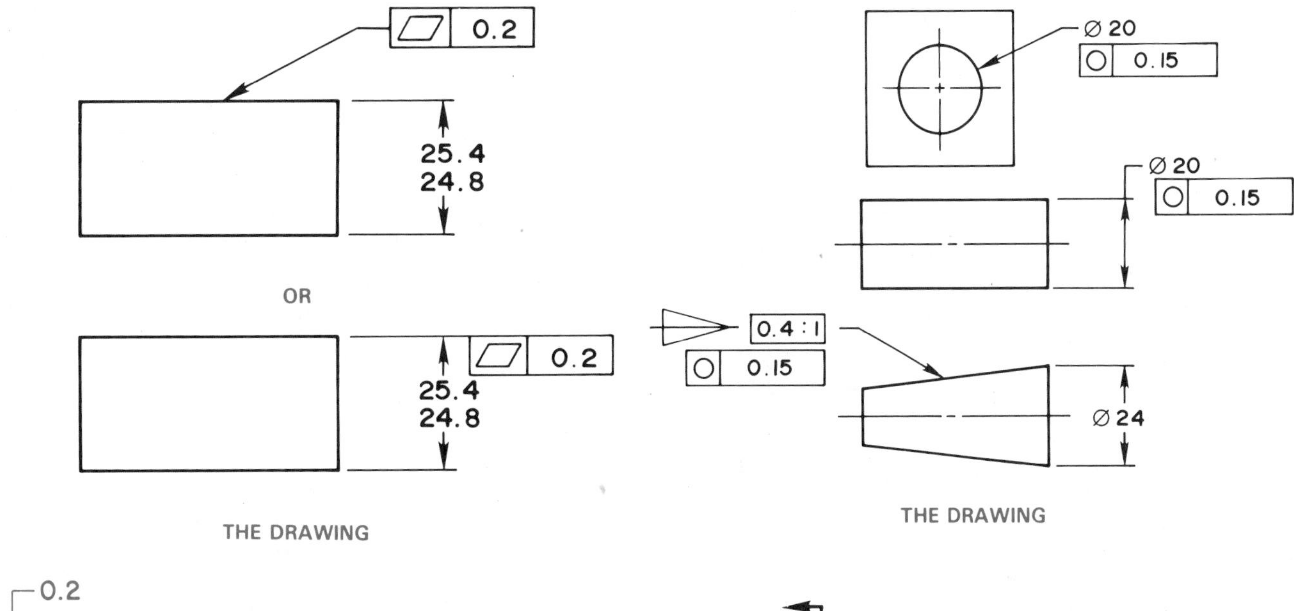

0.2

24.8 25.4

THE MEANING

Figure 11–33 Flatness representation.

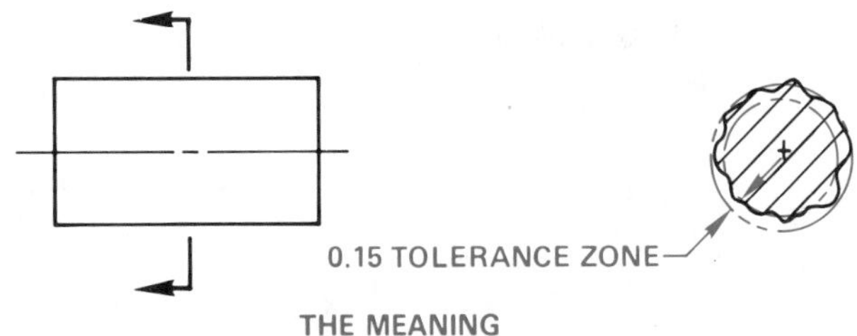

Figure 11–36 Circularity representation.

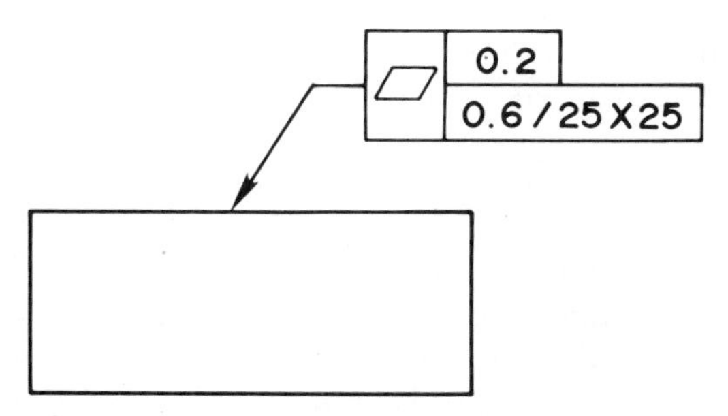

Figure 11–34 Unit flatness.

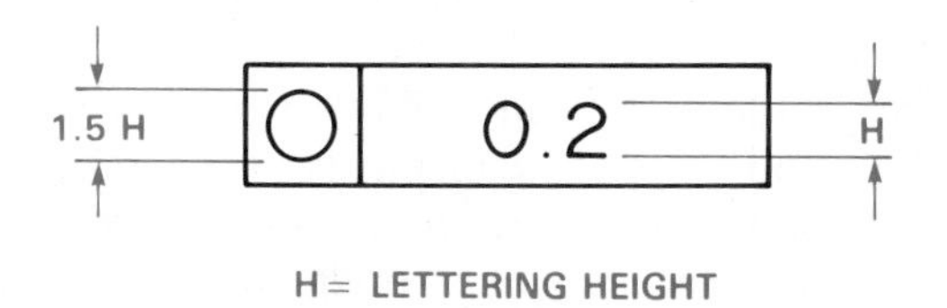

Figure 11–35 Circularity feature control frame.

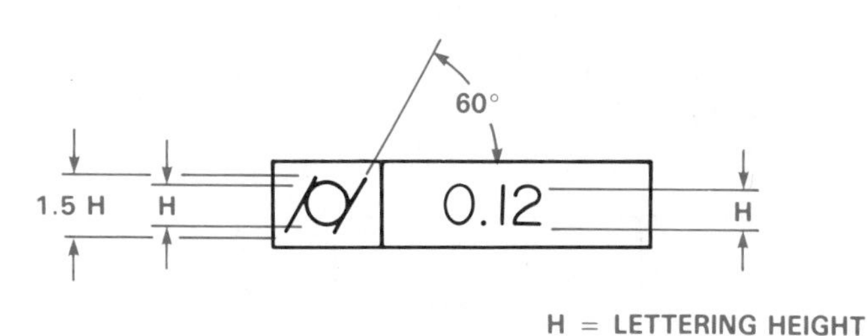

Figure 11–37 Cylindricity feature control frame.

Circularity. The circularity geometric characteristic symbol is detailed in a feature control frame in Figure 11–35. Circularity may be applied to cylindrical, conical, or spherical shapes. Circularity exists when all of the elements of a circle are the same distance from the center. Circularity is a cross-sectional evaluation of the feature to determine if the circular surface lies between a tolerance zone that is made up of two concentric circles. The cross-sectional tolerance zone is established perpendicular to the axis of the part. The term circular or line element will be used throughout this chapter to refer to a cross-sectional or single-line tolerance zone as opposed to a blanket or entire surface tolerance zone. The circularity tolerance zone is a radius dimension. The circularity feature control frame may be connected to the part with a leader in the circular or rectangular view as shown in Figure 11–36. The circularity tolerance must always be less than the size tolerance except for parts subject to free state variations. *Free state variation* is a term used to describe distortion of a part after removal of forces applied during manufacturing.

Cylindricity. The cylindricity geometric characteristic symbol is detailed in Figure 11–37. Cylindricity is similar to circularity in that both have a radius tolerance zone. The difference is that circularity is a cross-sectional tolerance which results in a feature that must lie between two concentric circles, while cylindricity is a

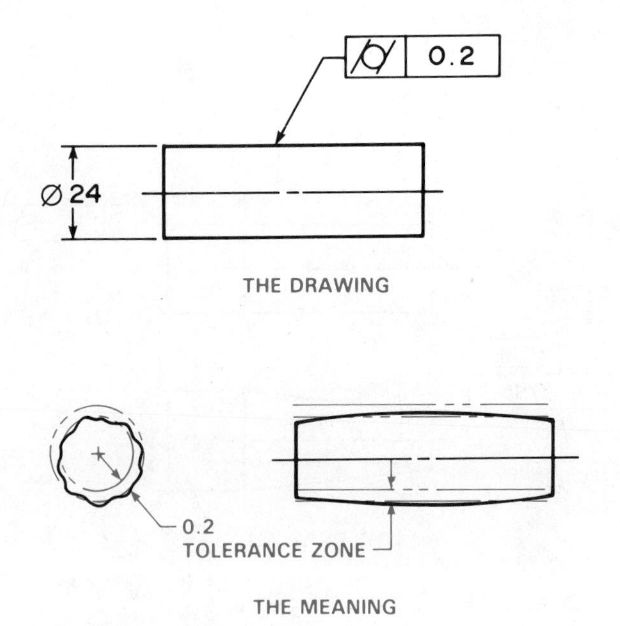

Figure 11–38 Cylindricity representation.

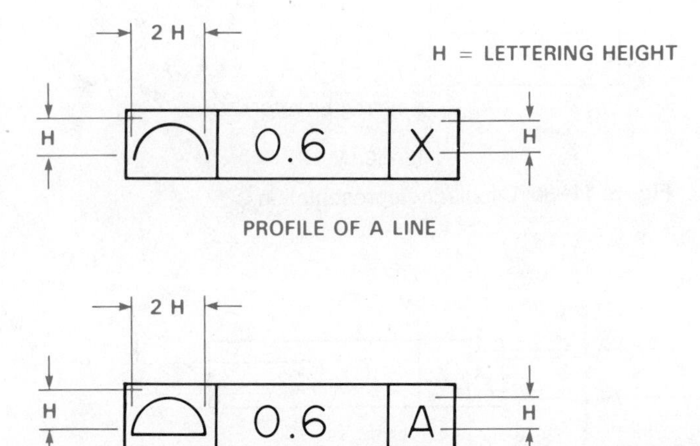

Figure 11–39 Profile of feature control frames.

blanket tolerance that results in a feature lying between two concentric cylinders, which takes into account the straightness of the surface and at the same time is a composite control. The cylindricity feature control frame may be connected with a leader to either the circular or rectangular view. (See Figure 11–38.) The cylindricity tolerance must be less than the size tolerance.

Profile Tolerance

Profile is used to control form or a combination of size, form, and orientation. Profile callouts are commonly used on arcs, curves, or irregular shaped features but can also be applied to plane surfaces. The shape and size of the profile may be defined with basic dimensions or tolerance dimensions. When tolerance dimensions are used to establish profile, the profile tolerance zone must

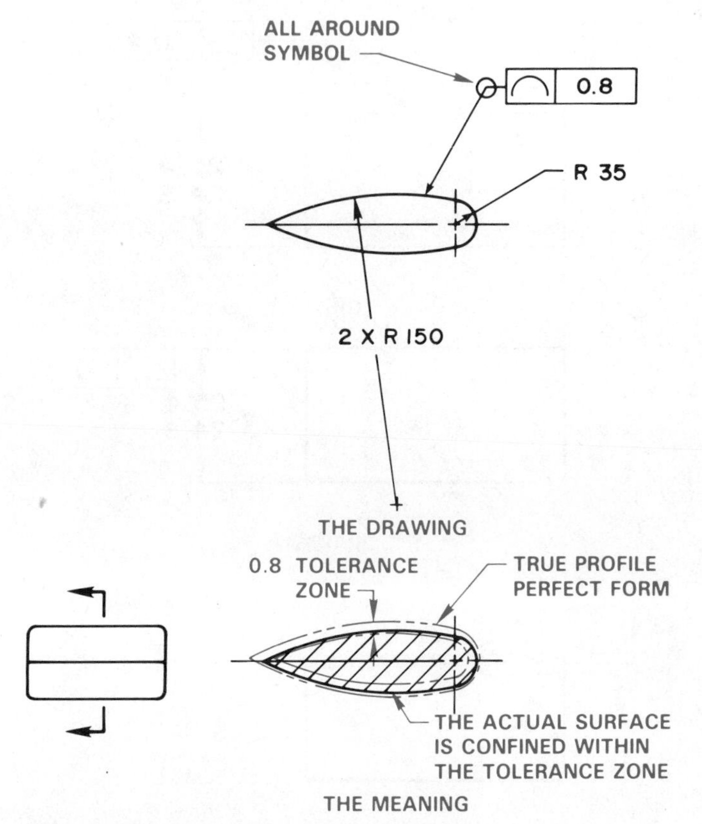

Figure 11–40 Profile of a line all around.

be within the size tolerance. The profile feature control frame is connected to the longitudinal view with a leader line. The two types of profile are profile of a line and profile of a surface. (See Figure 11–39.)

The profile of a line is a cross-sectional or single-line tolerance that extends through the length of the specified feature. The profile of a line may be controlled in relation to datums or without datums. The profile of a line may be all around the part by using the all around symbol on the leader, as seen in Figure 11–40, or between two specified points on the part, as shown in Figure 11–41.

The profile of a surface is a blanket tolerance zone that affects all the surface elements of a feature equally. Profile of a surface is generally referenced to one or more datums so that proper orientation of the profile boundary can be maintained. Profile of a surface may be applied all around a part as in Figure 11–42, or between two specified points as shown in Figure 11–43.

The profile of a line or the profile of a surface is a bilateral tolerance zone when the leader line points to the surface of the part without any additional symbology as was shown in Figures 11–39 through 11–43. A bilateral profile tolerance means that the tolerance zone is split equally on each side of the specified perfect form. Either type of profile callout may also have a unilateral tolerance zone specified. A unilateral tolerance zone will place the entire zone on only one side of the true profile or perfect form. This is accomplished on a drawing by placing a phantom line parallel to the surface where the leader arrowhead touches the part. The phantom line will be placed any clear distance from the part surface and will be either inside or outside depending on the specified direction of the unilateral tolerance from the true

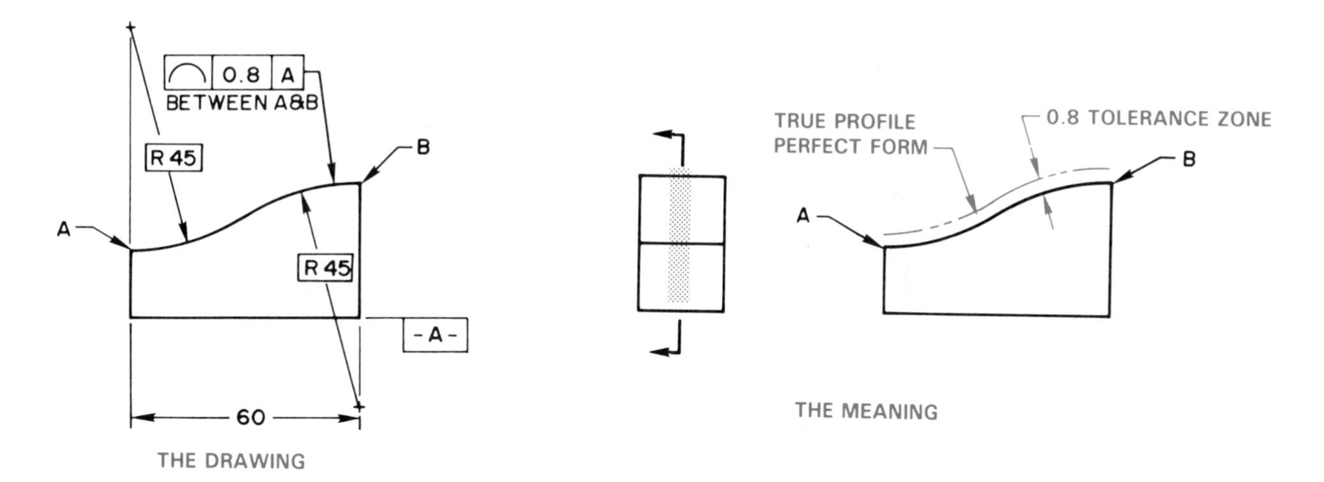

Figure 11–41 Profile of a line between two given points.

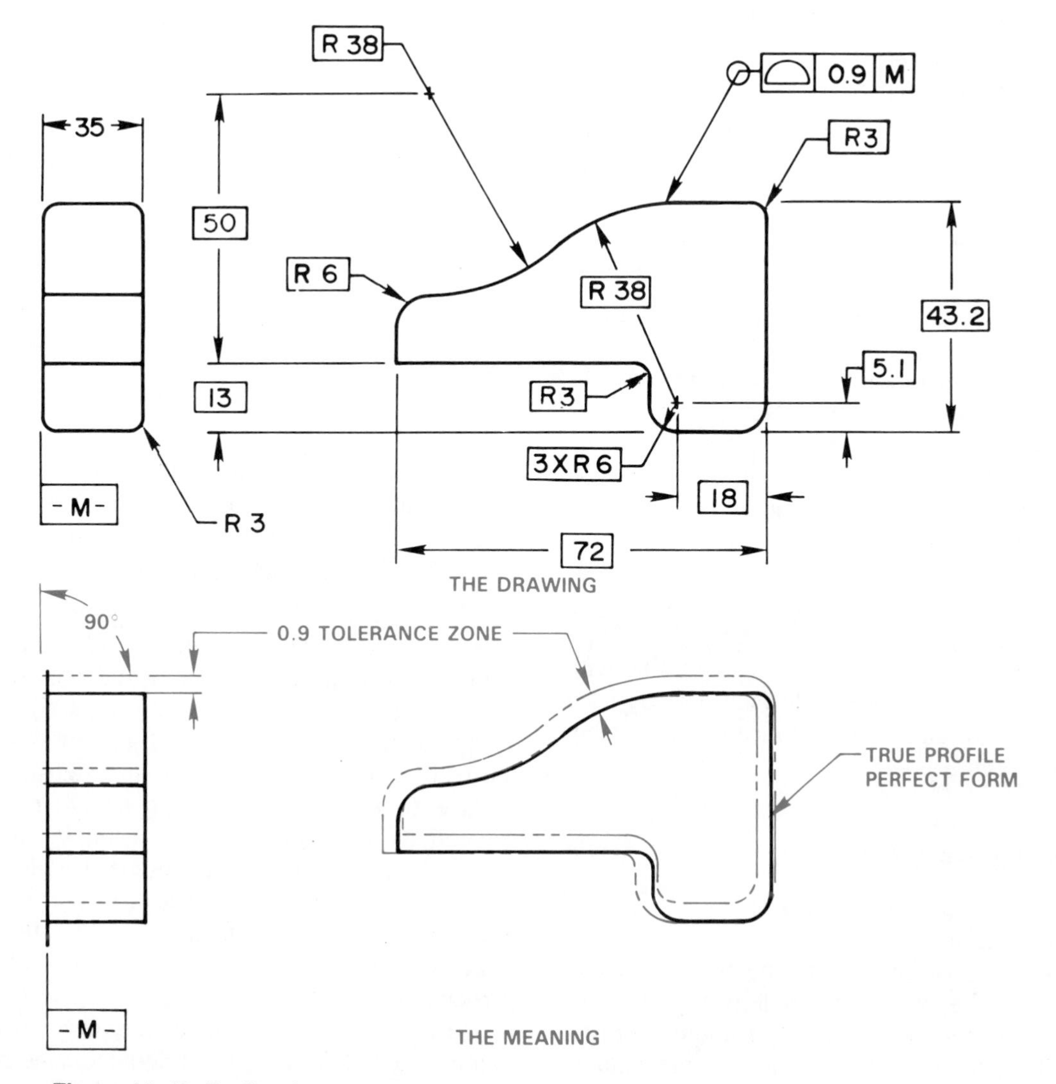

Figure 11–42 Profile of a surface all around.

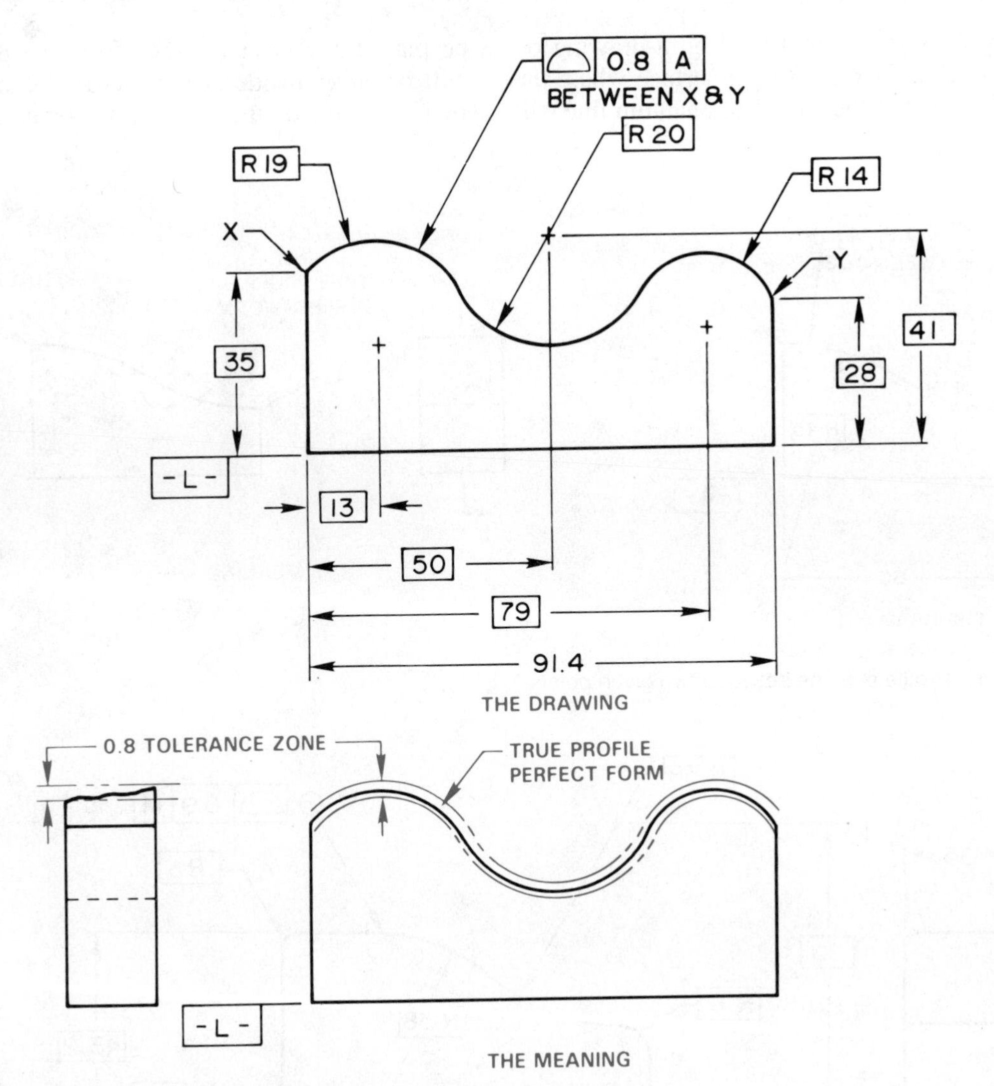

Figure 11–43 Profile of a surface between two given points.

profile. The actual feature will then be confined between the true profile and the given tolerance reference. (See Figure 11–44.)

The profile of coplanar surfaces may also be specified by placing a phantom line between the surfaces in the view where they appear as edges. The profile feature control frame is then connected to the phantom line with a leader. (See Figure 11–45.) Profile may also be used to control the angle of an inclined surface in relationship to a datum. (See Figure 11–46.)

Orientation Tolerance

Orientation tolerances refer to a specific group of geometric characteristics that establish a relationship between the features of an object. The tolerances that orient one feature to another are parallelism, perpendicularity, angularity, and, in some applications, profile. Orientation tolerances require that one or more datums be used to establish the relationship between features. Parallelism, perpendicularity, and angularity also simultaneously control flatness. The geometric tolerance must fall within the size tolerance of the part.

Parallelism. The parallelism geometric characteristic symbol is detailed in Figure 11–47. A parallelism tolerance zone requires that the actual feature is between two parallel planes or lines that are parallel to a datum. The parallelism feature control frame may be attached to the feature with a leader or on an extension line of the surface. (See Figure 11–48.) Unless otherwise specified a parallelism tolerance zone is a blanket tolerance that covers the entire surface. If a single line element is to be specified rather than the surface, then the note, EACH ELEMENT, must be added below the feature control frame. (See Figure 11–49.) This technique also applies to perpendicularity and angularity.

In certain instances parallelism may also be applied to a cylindrical feature. The tolerance zone is a cylindri-

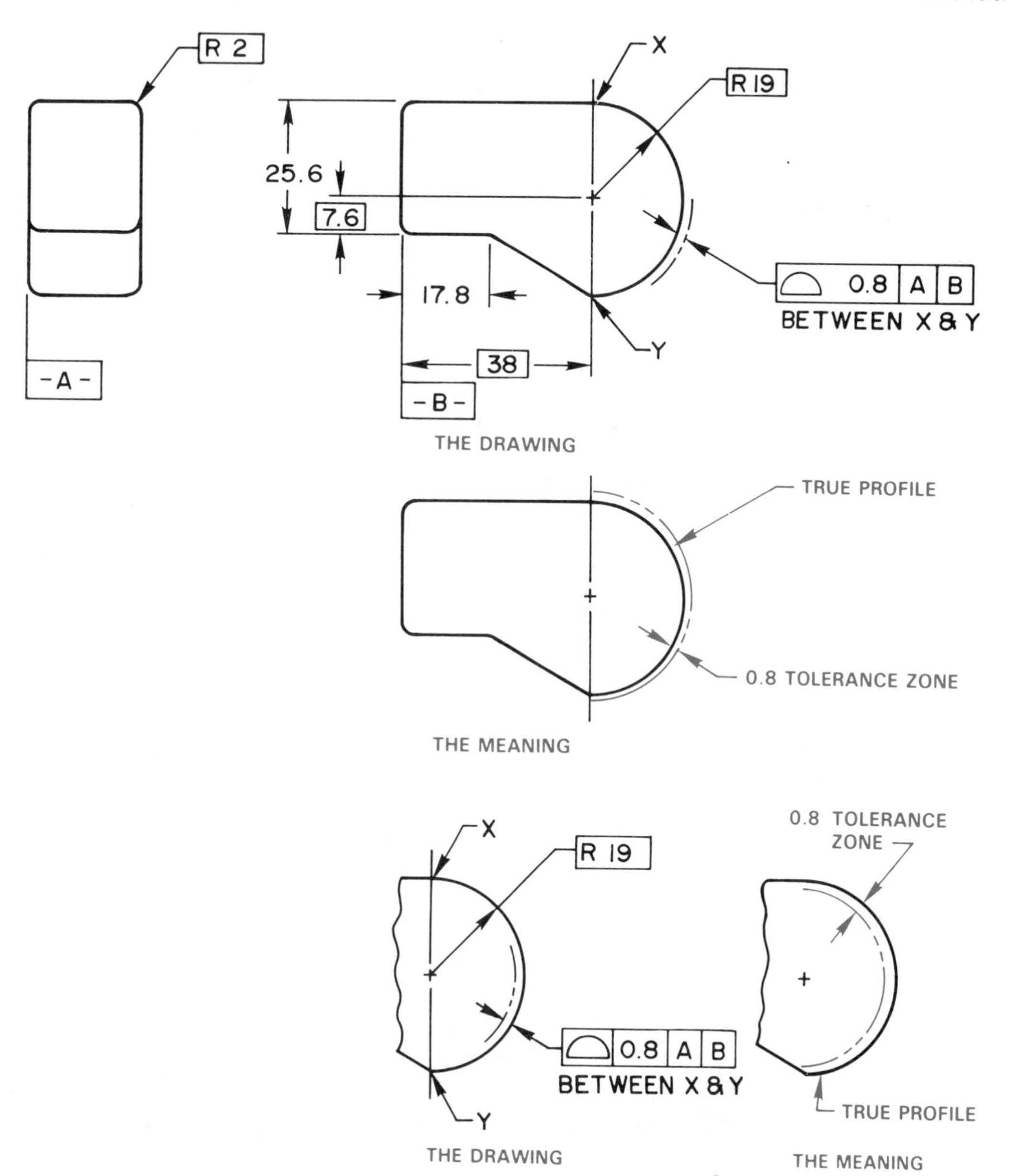

Figure 11–44 A unilateral profile representation.

cal shape that is parallel to a datum axis reference. The feature control frame that relates to an axis specification is attached to the diameter dimension, and a diameter zone descriptor should precede the geometric tolerance. (See Figure 11–50.)

Perpendicularity. The perpendicularity geometric characteristic symbol is detailed in Figure 11–51. A perpendicularity geometric tolerance requires that a given feature be located between two parallel planes or lines, or within a cylindrical tolerance zone that is a basic 90° to a datum. The perpendicularity feature control frame may be connected to the feature surface with a leader or an extension line, or attached to the diameter dimension for axis perpendicularity. (See Figure 11–52.)

Angularity. The angularity geometric characteristic symbol is represented in Figure 11–53. An angularity tolerance zone places a given feature between two parallel planes that are at a specified basic angular dimension from a datum. The basic angle from the datum may be any amount except 90°. The angularity feature control frame may be connected to the surface with a leader or from an extension line, or attached to the diameter dimension for axis angularity. (See Figure 11–54.)

Runout Tolerance

Runout is used to control the relationship of radial features to a datum axis and features that are 90° to a

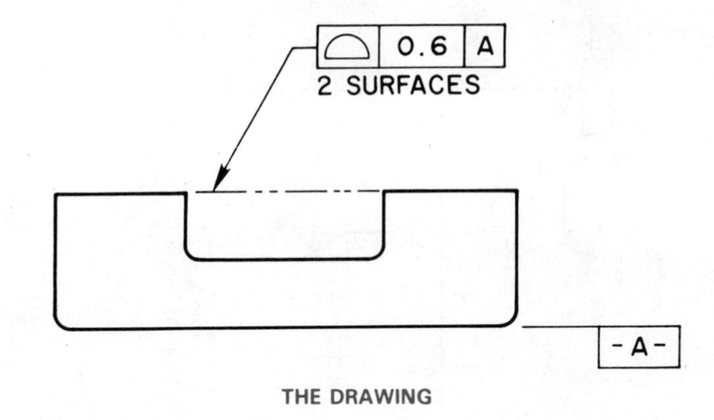

THE DRAWING

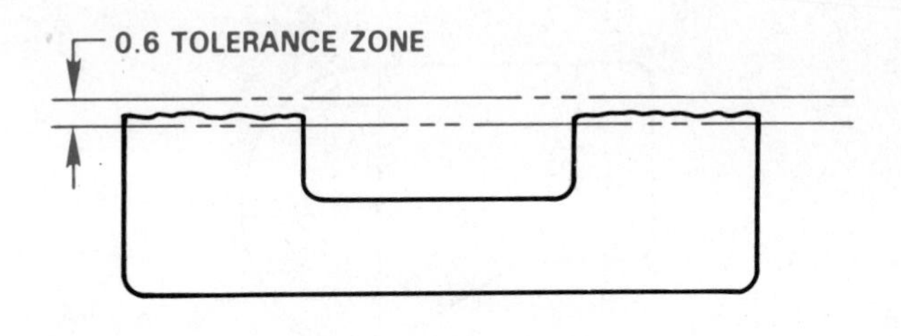

THE MEANING

Figure 11–45 Profile of coplanar surfaces.

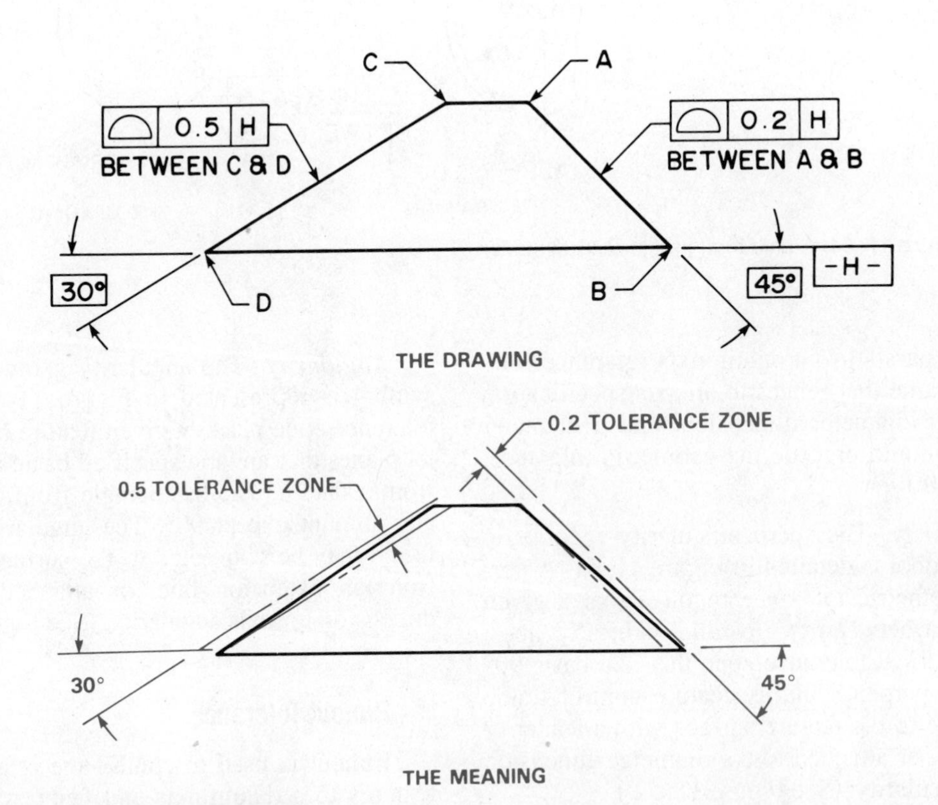

THE MEANING

Figure 11–46 Profile of an inclined surface.

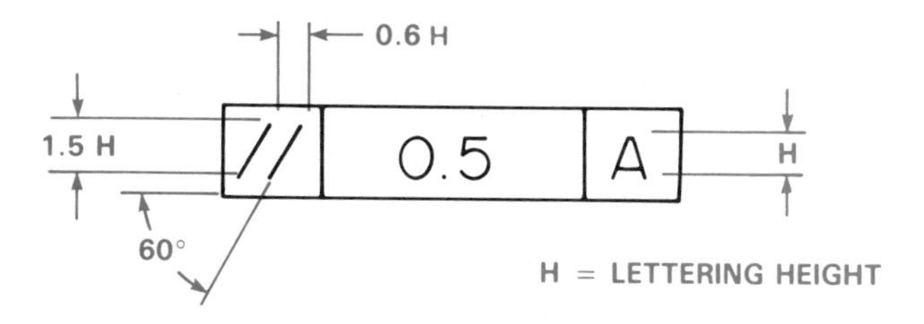

Figure 11–47 Parallelism feature control frame.

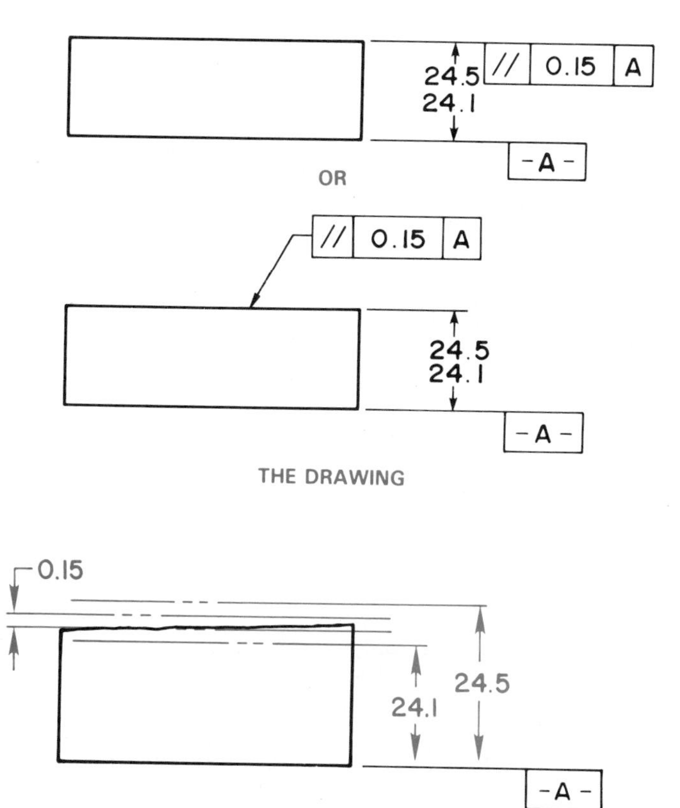

Figure 11–48 Parallelism representation.

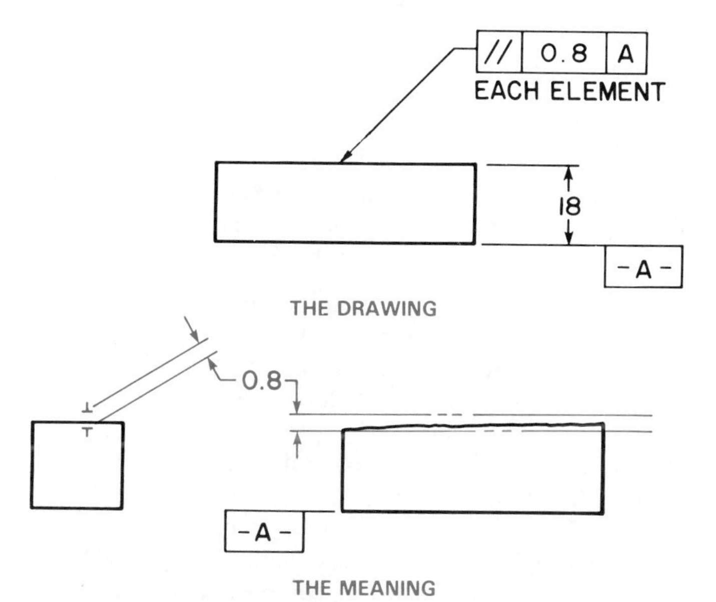

Figure 11–49 Single-line element parallelism.

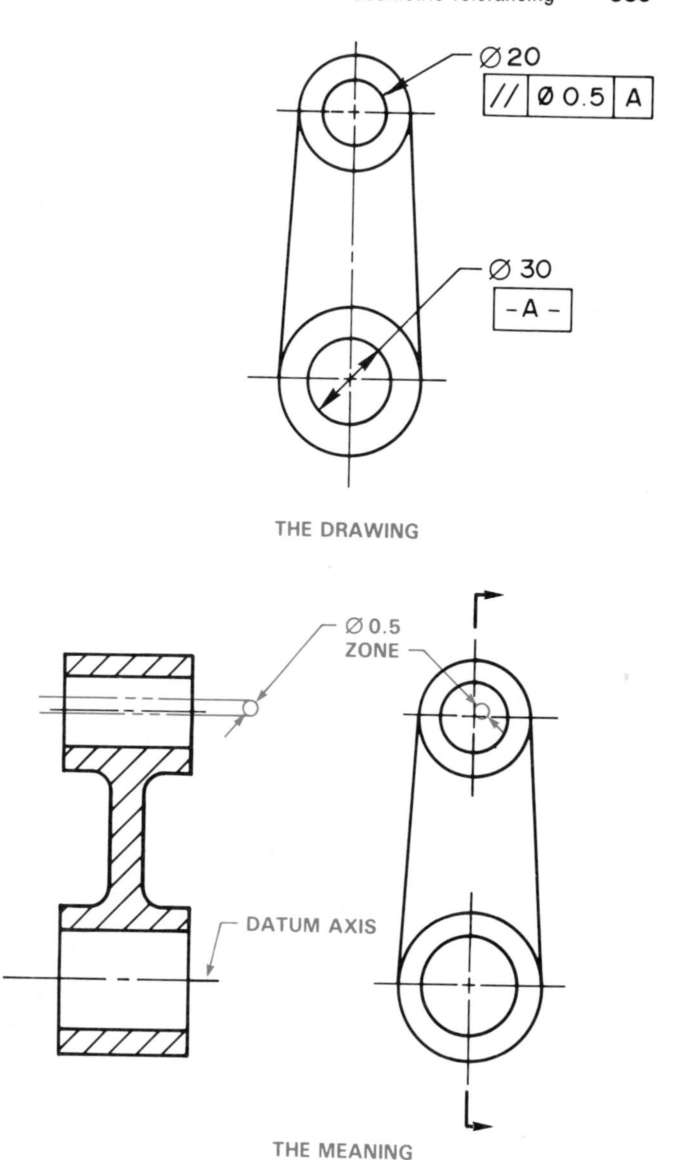

Figure 11–50 Axis parallelism.

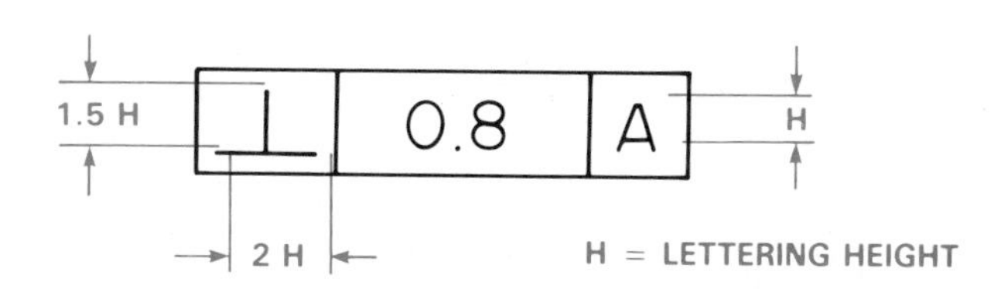

Figure 11–51 Perpendicularity feature control frame.

datum axis. These types of features include cylindrical, tapered, and curved shapes as well as plane surfaces that are at right angles to a datum axis. A runout tolerance is determined when a dial indicator is placed on the surface to be inspected and the part is rotated 360°. The full indicator movement (FIM) on the dial indicator must not exceed the amount of specified tolerance. The runout feature control frame is attached to the required surface with a leader. The runout tolerance applies to the length of the intended surface or until there is a break, or change

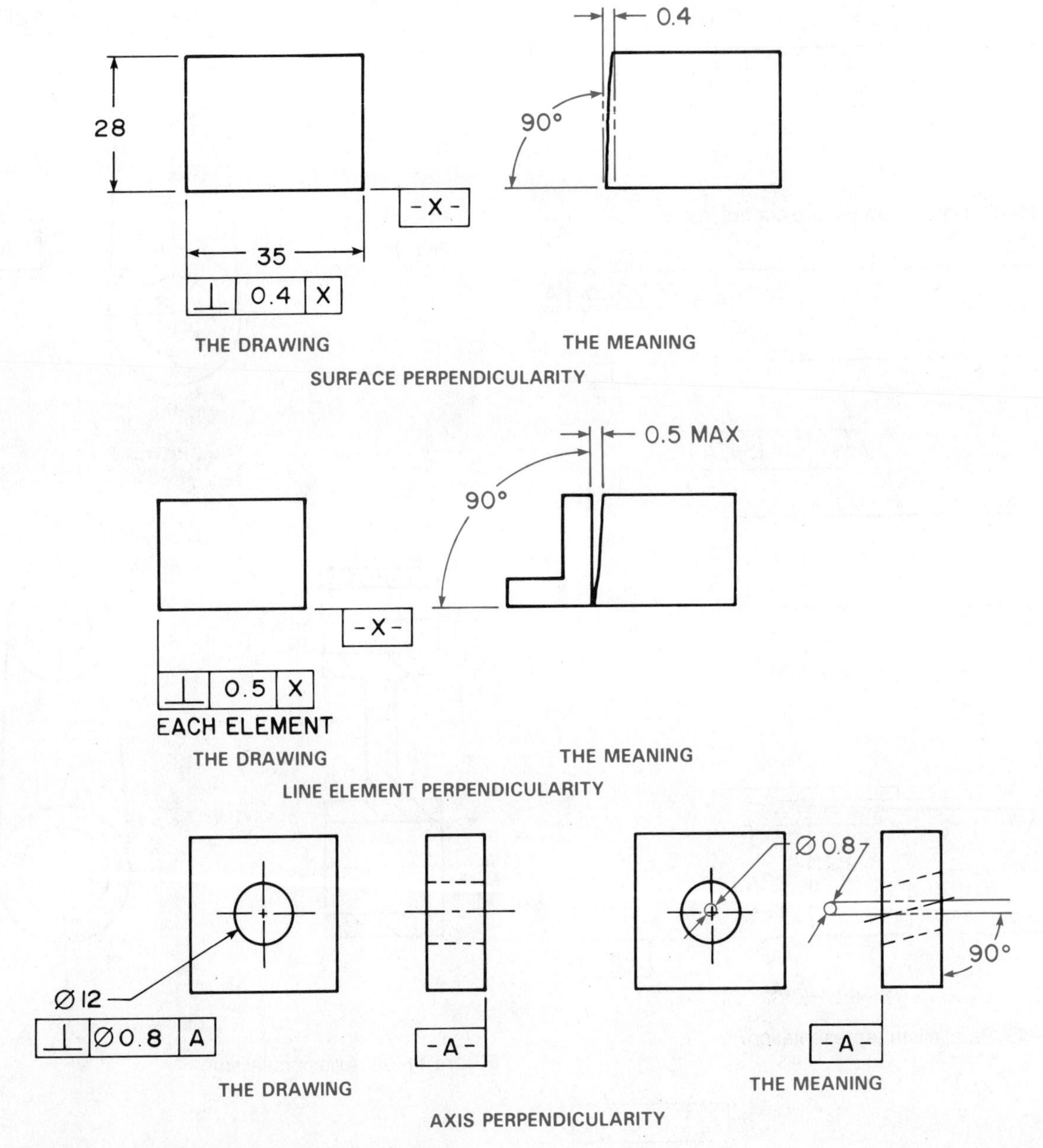

Figure 11–52 Perpendicularity representations.

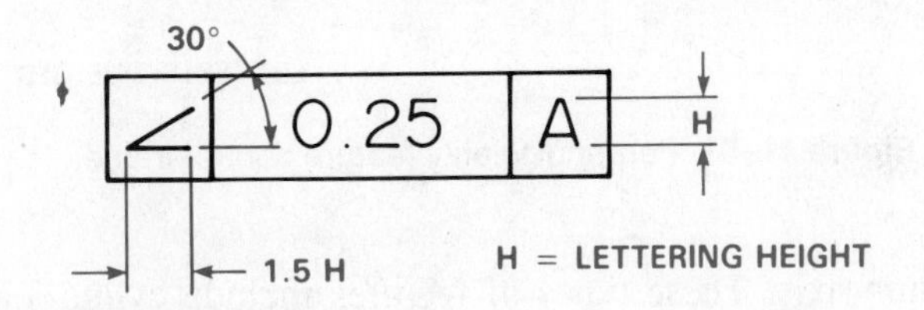

Figure 11–53 Angularity feature control frame.

in shape or diameter of the surface. There are two types of runout, circular and total. (See Figure 11–55.)

Circular Runout. Circular runout provides control of single circular elements of a surface. Circular runout controls circularity and the relationship of the common axes (coaxial) of parts. This tolerance, established by the FIM of a dial indicator, is placed at one location on the feature as the part is rotated 360°. (See Figure 11–56.)

Total Runout. Total runout provides composite control of all surface elements. The tolerance is applied simultaneously to all circular and profile measuring positions as the part is rotated 360°. (See Figure 11–57.) Where applied to surfaces constructed around a datum axis, total runout is used to control cumulative variations of circularity, straightness, coaxiality, angularity, taper, and profile of a surface. Where applied to surfaces constructed at right angles to a datum axis, total runout controls cumulative variations of perpendicularity to detect wobble and flatness, and to detect concavity or convexity.

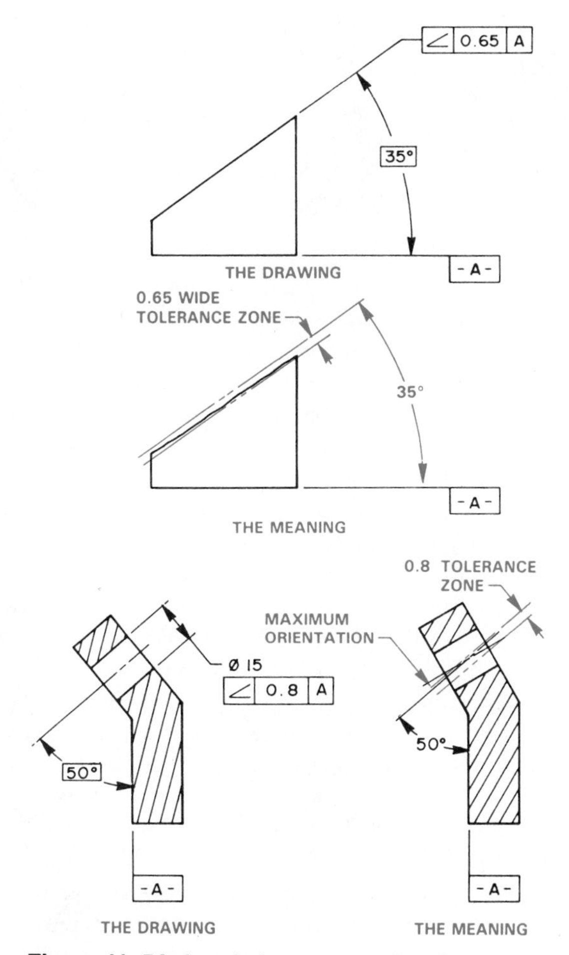

Figure 11–54 Angularity representations.

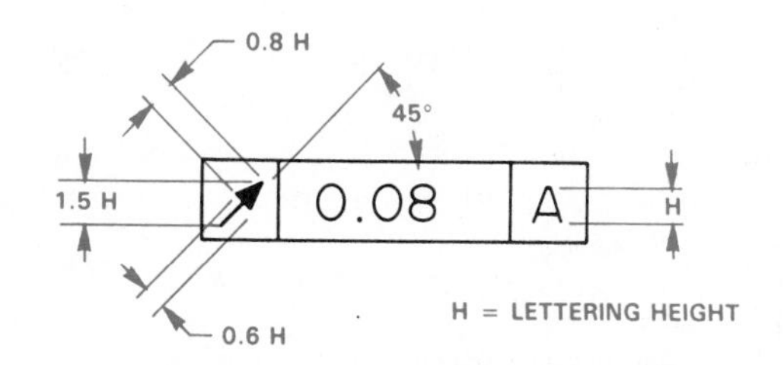

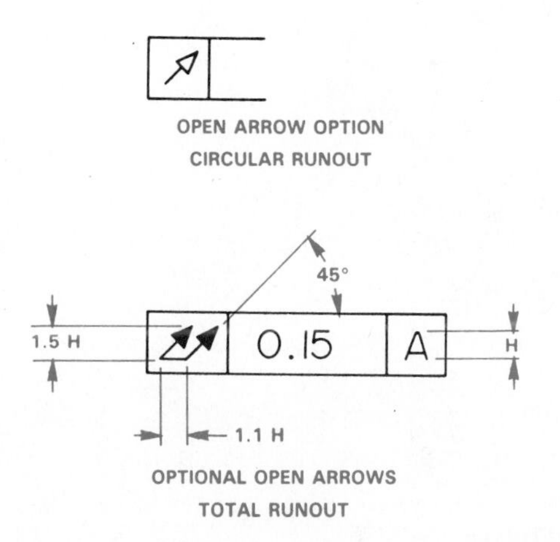

Figure 11–55 Runout feature control frames.

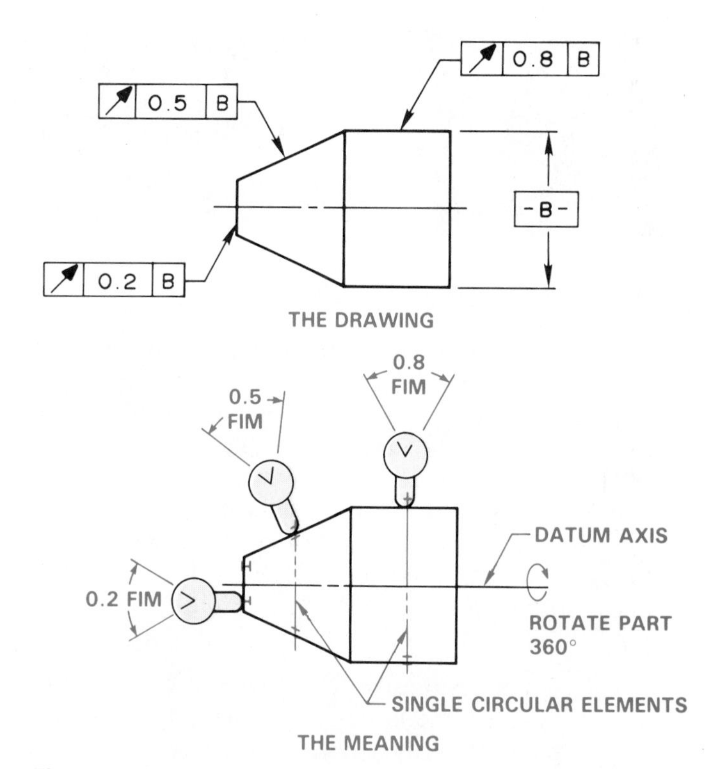

Figure 11–56 Circular runout.

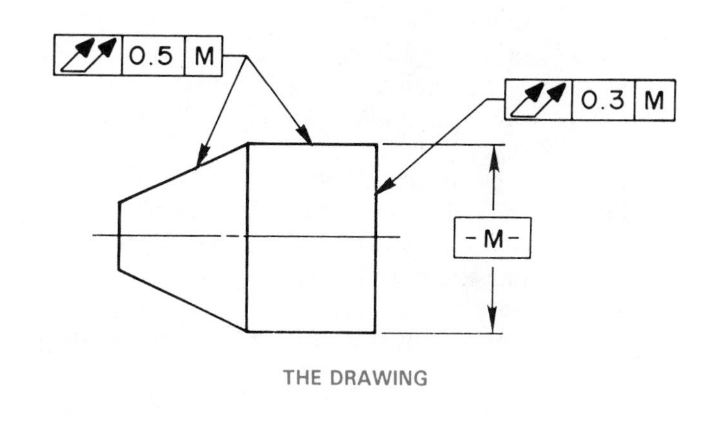

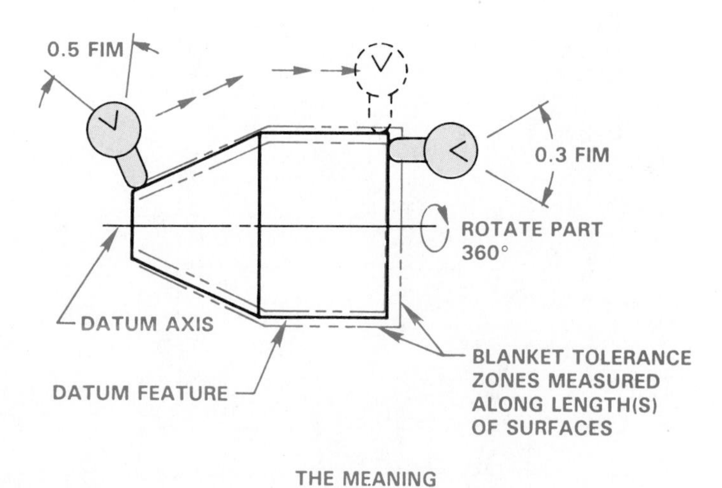

Figure 11–57 Total runout.

GEOMETRIC TOLERANCING

The implementation of geometric tolerancing and dimensioning into a mechanical drafting CADD program is quite practical. The GT symbology makes this application a bonus to the mechanical drafting system. Figure 11–58 shows an example of a menu with a symbol-library template designed to assist the use of GT.

Some dimensioning and tolerancing guidelines for use in conjunction with CADD/CAM are outlined, in part, from ANSI Y14.5M–1982 as follows:

1. Major features of the part should be used to establish the basic coordinate system but are not necessarily defined as datums.
2. Subcoordinated systems that are related to the major coordinates are used to locate and orient features on a part.
3. Define part features in relation to three mutually perpendicular reference planes and along features that are parallel to the motion of CAM equipment.
4. Establish datums related to the function of the part and relate datum features in order of precedence as a basis for CAM usage.
5. Completely and accurately dimension geometric shapes. Regular geometric shapes may be defined by mathematical formulas although a profile feature that is defined with mathematical formulas should not have coordinate dimensions unless required for inspection or reference.
6. Coordinate or tabular dimensions should be used to identify approximate dimensions on an arbitrary profile.
7. Use the same type of coordinate dimensioning system on the entire drawing.
8. Continuity of profile is necessary for CADD. Clearly define contour changes at the change or point of tangency. Define at least four points along an irregular profile.
9. Circular hole patterns may be defined with polar coordinate dimensioning.
10. When possible dimension angles in degrees and decimal parts of degrees; for example, 45°30' = 45.5°.
11. Base dimensions at the mean of a tolerance because the numerical control (NC) programmer will normally split a tolerance and work to the mean. Establish dimensions without limits that conform to the NC machine capabilities and part function where possible. Bilateral profile tolerances are also recommended for the same reason.
12. Geometric tolerancing is necessary to control specific geometric form and location.

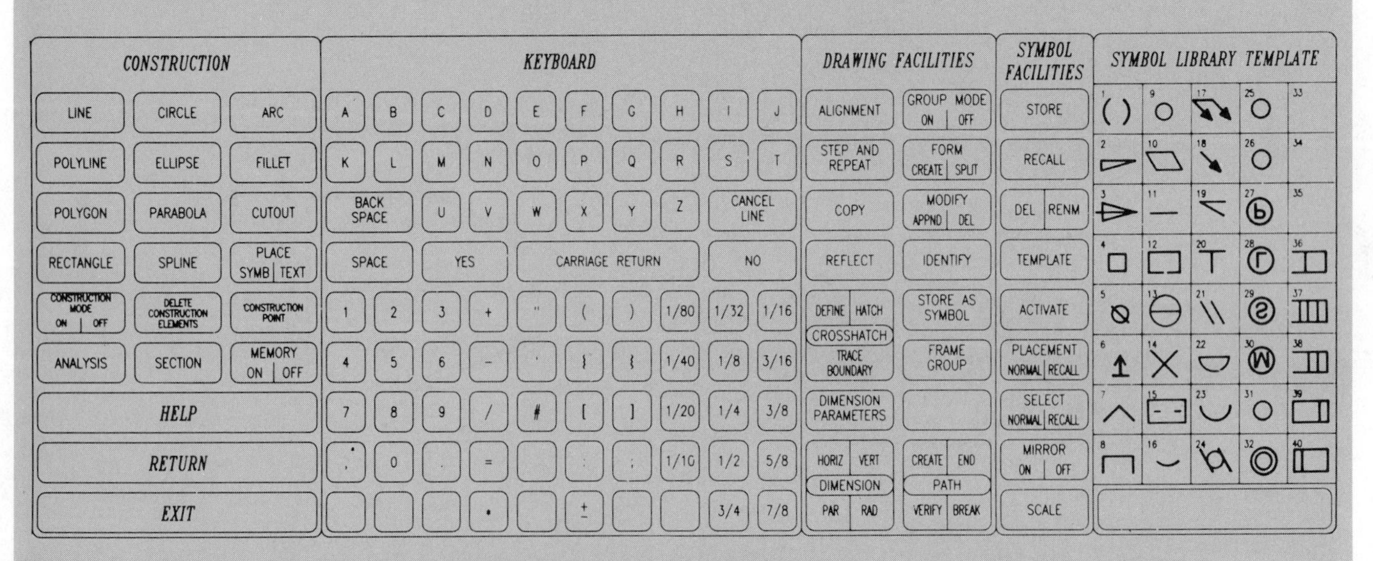

Figure 11–58 CADD menu with GT symbol library. *Courtesy Summagraphics Corporation.*

A portion of a surface may have a specified runout tolerance if it is not desired to control the entire surface. This is done by placing a chain line in the linear view adjacent to the desired location. The chain line is located with basic dimensions as shown in Figure 11–59. When a part has compound datum features, that is, where more than one datum feature is used to establish a common datum, then the combined features are shown separated by a dash in the feature control frame. The datums are of equal importance. (See Figure 11–59.)

MATERIAL CONDITION SYMBOLS

Material condition symbols are used in conjunction with the feature tolerance or datum reference in the feature control frame. The material condition symbols are required to establish the relationship between the size or location of the feature and the geometric tolerance. The use of different material condition symbols will alter the effect of this relationship. The material condition-modifying elements are maximum material condition, MMC; regardless of feature size, RFS; and least material condition, LMC. The standard material condition symbols are shown in Figure 11–60.

Perfect Form Envelope

The form of a feature is controlled by the size tolerance limits. The envelope, or boundary, of these size limits is established at MMC. Remember from the discussion in Chapter 8 that MMC is the largest limit for an external feature and the smallest limit for an internal feature. The key is *most material*. The true geometric form of the feature is at MMC. This is known as the boundary of perfect form. If the part feature is produced at MMC, it is considered to be at perfect form. When it is desired to permit a surface or surfaces of a feature to exceed the boundary of perfect form at MMC, a note such as PERFECT FORM AT MMC NOT REQUIRED is specified, exempting the pertinent size dimension. If a feature is produced at LMC, the opposite of MMC, the form is allowed to vary within the geometric tolerance zone or to the extent of the MMC envelope.

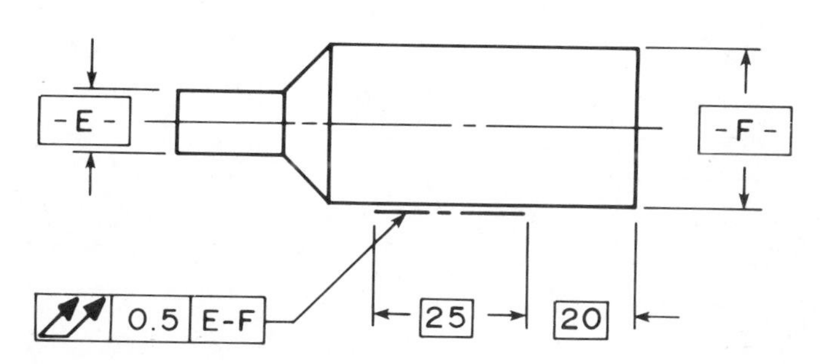

Figure 11–59 Partial surface runout.

Regardless of Feature Size

Regardless of feature size denotes that the geometric tolerance applies at any produced size. The tolerance remains as the specified value regardless of the actual size of the feature. Regardless of feature size is generally implied for all geometric characteristics and related datums. Position tolerances, however, require a specified material condition symbol.

Effect of RFS on Surface Straightness

RFS is implied for the straightness geometric characteristic. The RFS material condition symbol does not have to be placed in the feature control frame unless individual company standards specify its use. When a surface straightness tolerance is used by connecting a leader from the feature control frame to the surface of the part, then the geometric tolerance will remain the same regardless of the feature size, and the actual size may not exceed the perfect form envelope at MMC. An acceptable part may be produced between the given size tolerance, and any straightness irregularity may not be greater than the specified geometric tolerance. (See Figure 11–61.)

Effect of RFS on Axis Straightness

When the straightness tolerance is applied to the axis of the feature by a relationship with the diameter dimension, RFS is implied unless otherwise specified, and the actual feature size plus the geometric tolerance may exceed the MMC perfect form envelope. (See Figure 11–62.)

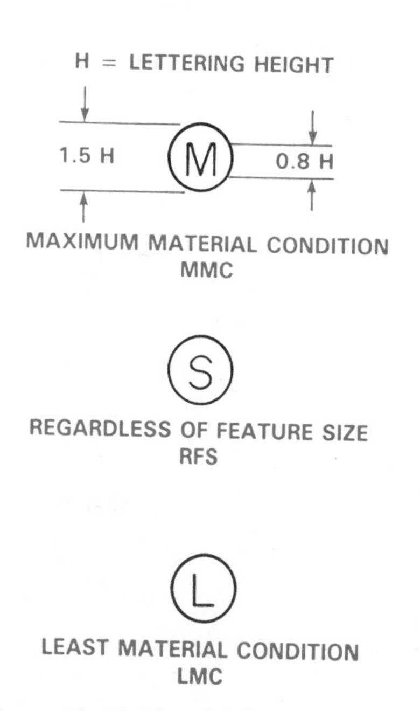

Figure 11–60 Material condition symbols.

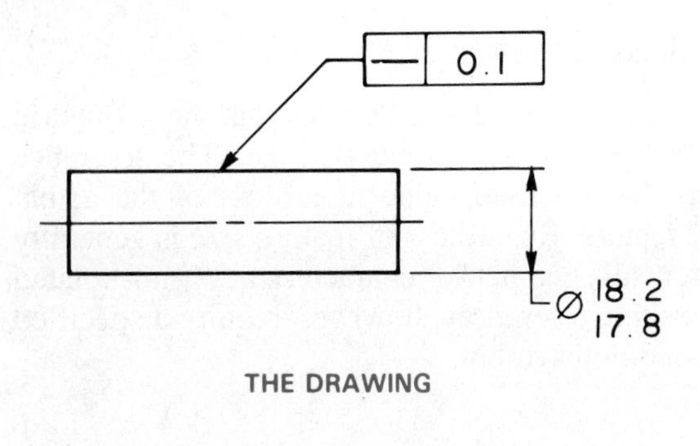

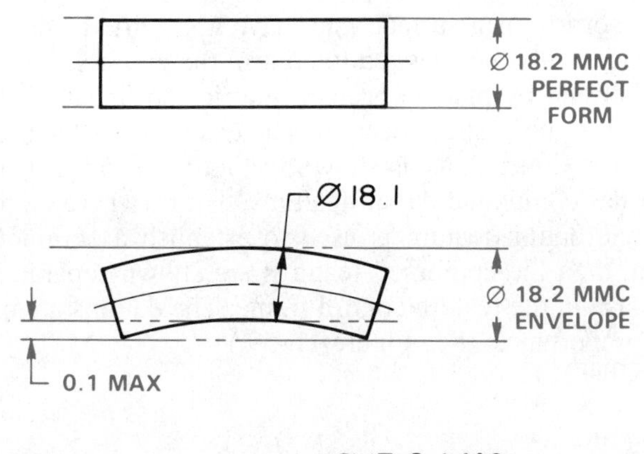

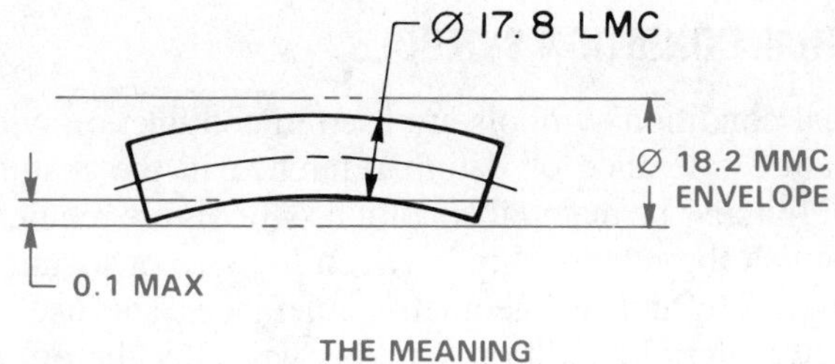

ACTUAL SIZE	GEOMETRIC TOLERANCE
18.2 MMC	0.0
18.1	0.1
18.0	0.1
17.9	0.1
17.8 LMC	0.1

Figure 11–61 Effect of regardless of feature size, RFS, on surface straightness.

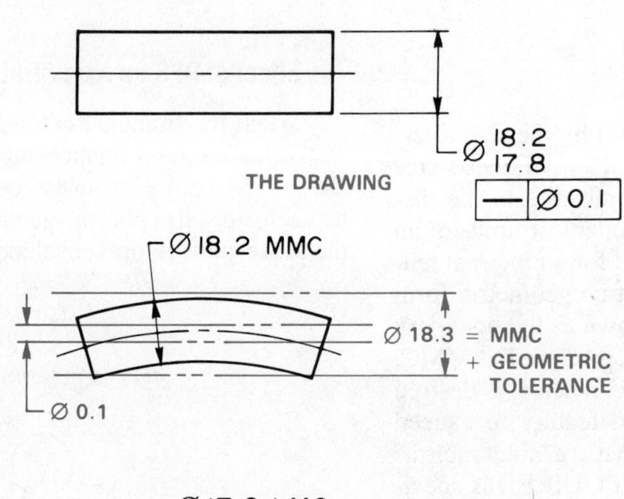

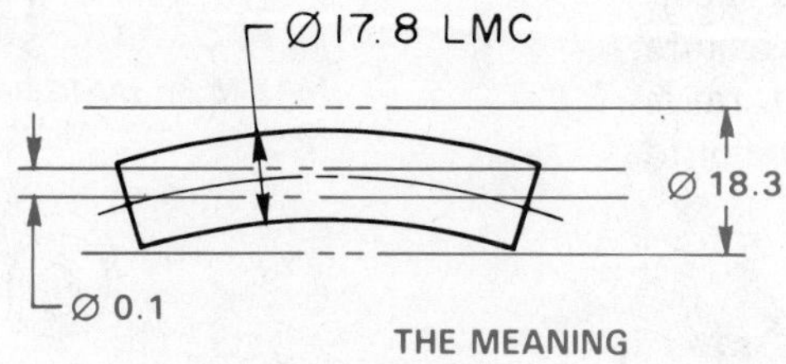

ACTUAL SIZE	GEOMETRIC TOLERANCE
18.2 MMC	0.1
18.1	0.1
18.0	0.1
17.9	0.1
17.8 LMC	0.1

Figure 11–62 Effect of RFS on axis straightness.

Datum Feature RFS

Datum features that are influenced by size variations, such as diameters and widths, are also subject to variations in form. RFS is implied unless otherwise specified. When a datum feature has a size dimension and a geometric form tolerance, the size of the simulated datum is the MMC size limit. This rule applies except for axis straightness where the envelope is allowed to exceed MMC. Figure 11–63 shows the effect of RFS on the primary datum feature with axis and center plane datums. When the datum features are secondary or tertiary then the axis or center plane shall also have an angular relationship to the primary datum. (See Figure 11–64.)

Maximum Material Condition

The use of MMC in conjunction with the geometric tolerance in a feature control frame denotes that the given tolerance is held at the MMC produced size, and as the feature dimension departs from MMC the geometric tolerance is allowed to increase equal to the change from MMC. The maximum amount of change is at the LMC-produced size. MMC must be specified for any geometric characteristic.

Effect of MMC on Form Tolerance

When MMC is specified in conjunction with an axis straightness callout, the MMC feature size envelope is exceeded by the given geometric tolerance. The given geometric tolerance is held at the MMC-produced size, then as the actual produced size departs from MMC the

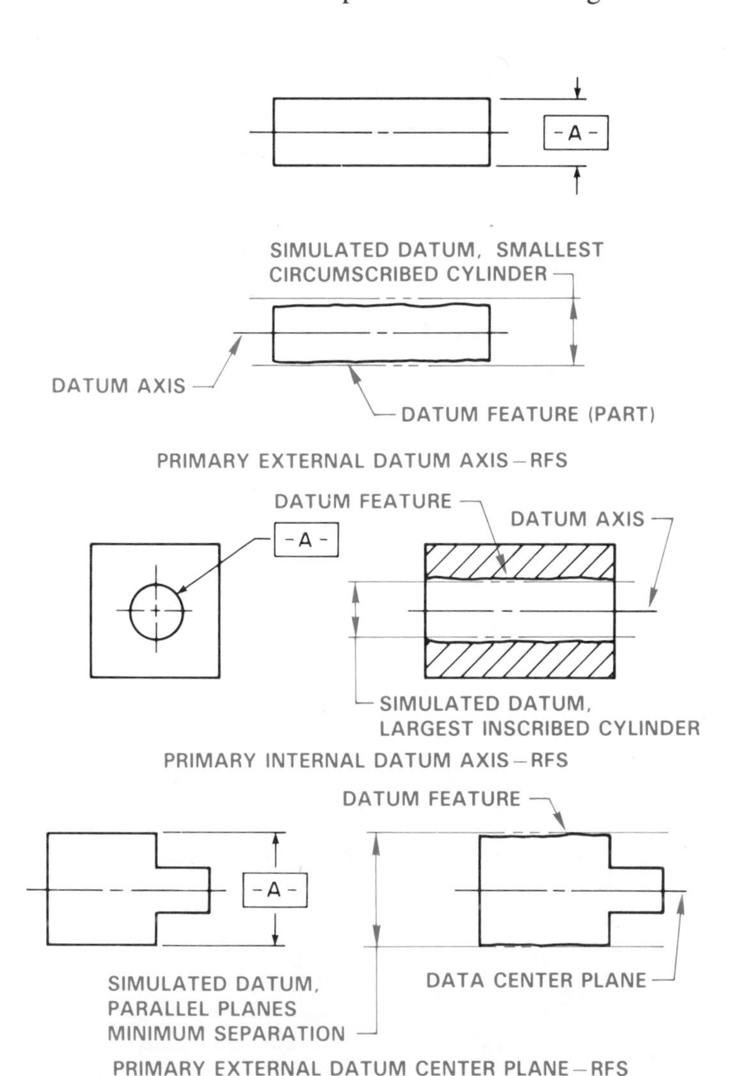

Figure 11–63 Effect of RFS on the primary datum feature with axis and center plane datums.

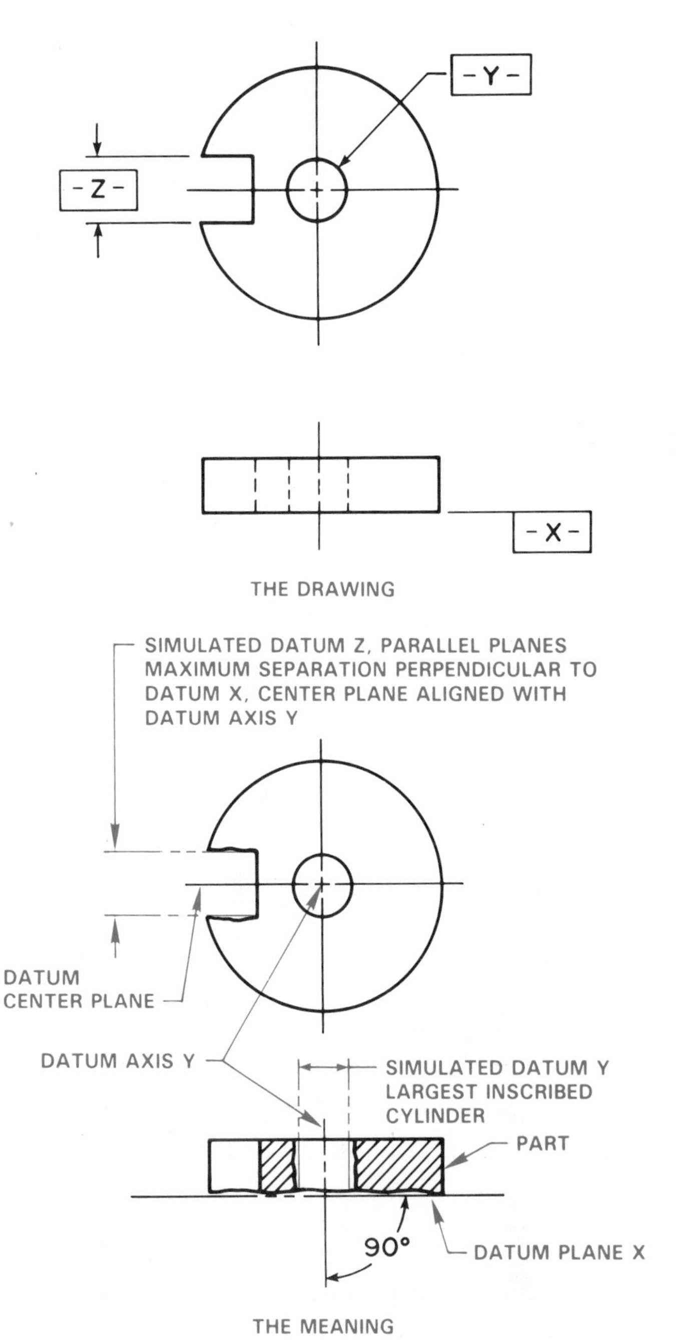

Figure 11–64 The secondary or tertiary datum relationship to the primary datum.

geometric tolerance is allowed to increase equal to the change from MMC to a maximum amount of departure at LMC. (See Figure 11–65.) The same situation may also be demonstrated with a perpendicularity tolerance where a datum axis is perpendicular to a given datum as in Figure 11–66.

Least Material Condition

The application of LMC is not as prevalent as MMC. LMC may be used when it is desirable to control minimum wall thickness. When an LMC material condition symbol is used in conjunction with a geometric tolerance, the specified tolerance is held at the LMC-produced size. Unlike MMC condition, an LMC value is never permitted to be less than LMC. As the actual produced size deviates from LMC toward MMC, the tolerance is allowed an increase equal to the amount of change from LMC. (See Figure 11–67.)

LOCATION TOLERANCE

Location tolerances include concentricity, symmetry, and position. True position is the theoretically exact location of the axis or center plane of a feature. The true position is located using basic dimensions. The locational tolerance specifies the amount that the axis of the feature is allowed to deviate from true position. Concentricity and symmetry imply RFS. Position tolerances must have a correlated material condition symbol: MMC, RFS, or LMC. Reference to true position dimensions must be provided as basic dimensions at each required location on the drawing or a general note may be used to specify

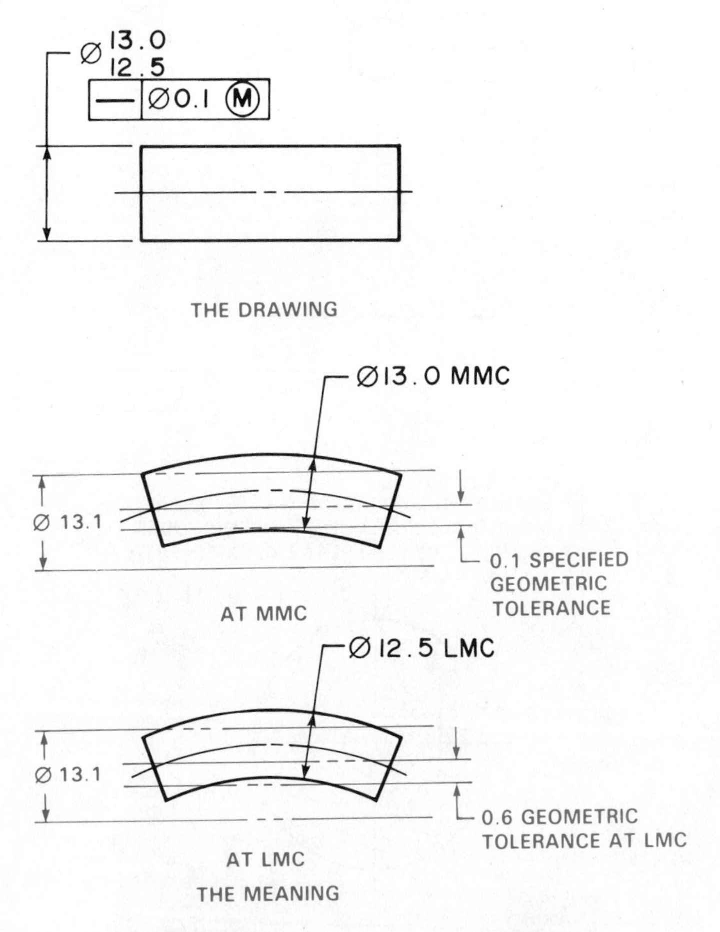

ACTUAL SIZE	GEOMETRIC TOLERANCE
13.0 MMC	0.1
12.9	0.2
12.8	0.3
12.7	0.4
12.6	0.5
12.5 LMC	0.6

Figure 11–65 Effect of maximum material condition, MMC, on form tolerance and axis straightness.

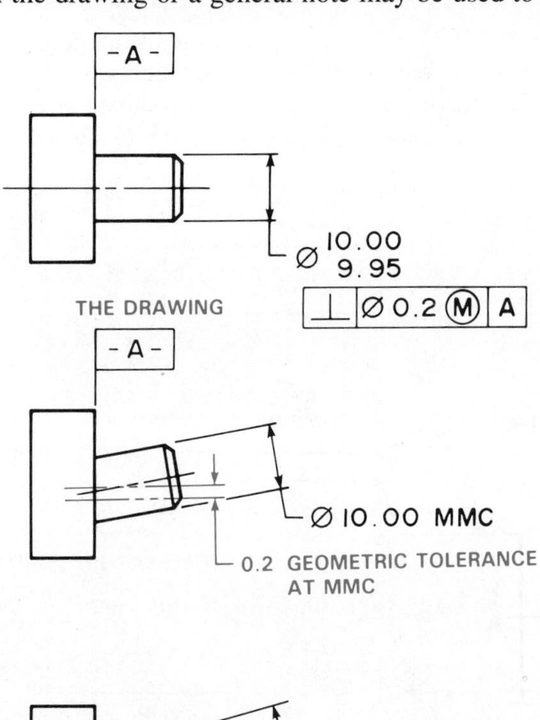

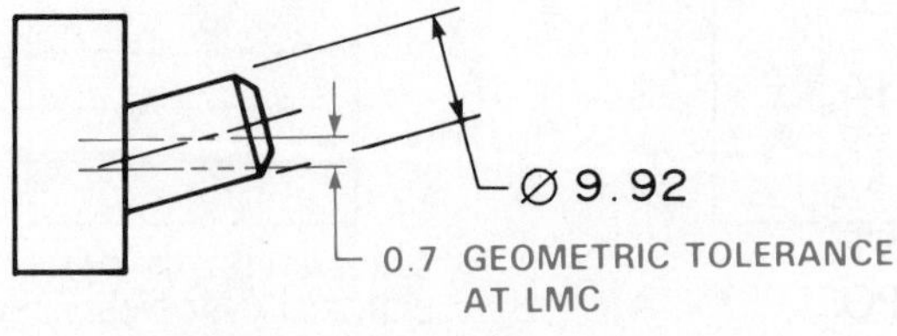

ACTUAL SIZE	GEOMETRIC TOLERANCE
10.00 MMC	0.2
9.99	0.3
9.98	0.4
9.97	0.5
9.96	0.6
9.95 LMC	0.7

Figure 11–66 Effect of MMC on form tolerance and axis perpendicularity.

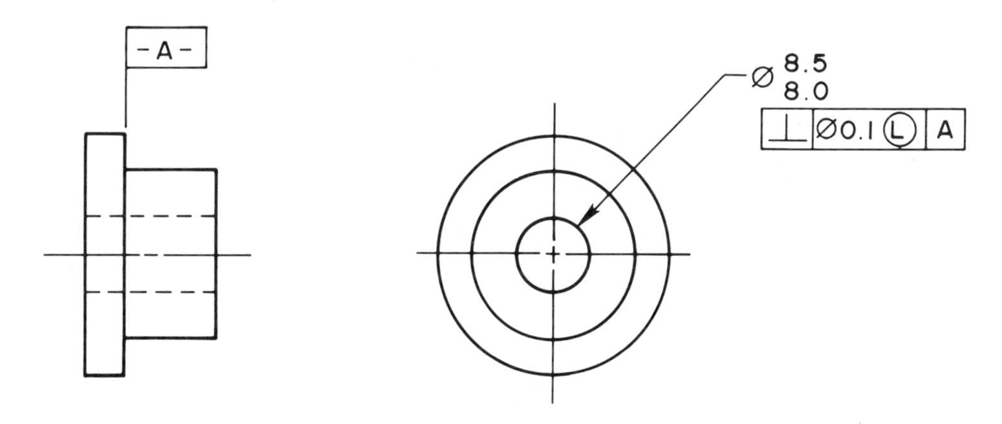

ACTUAL SIZE	GEOMETRIC TOLERANCE
8.0 MMC	0.6
8.1	0.5
8.2	0.4
8.3	0.3
8.4	0.2
8.5 LMC	0.1

Figure 11–67 Application of least material condition, LMC.

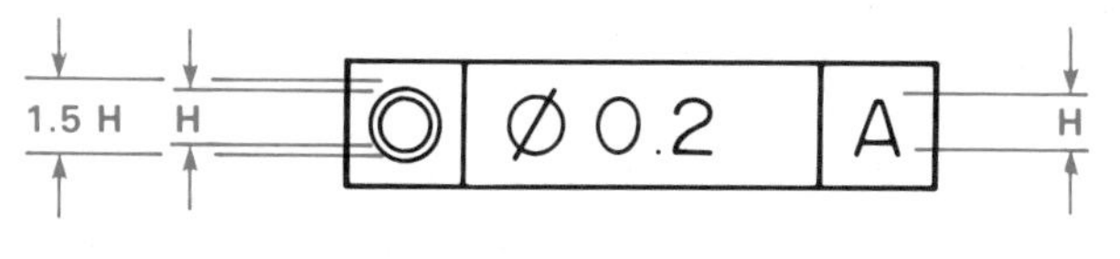

Figure 11–68 Concentricity feature control frame.

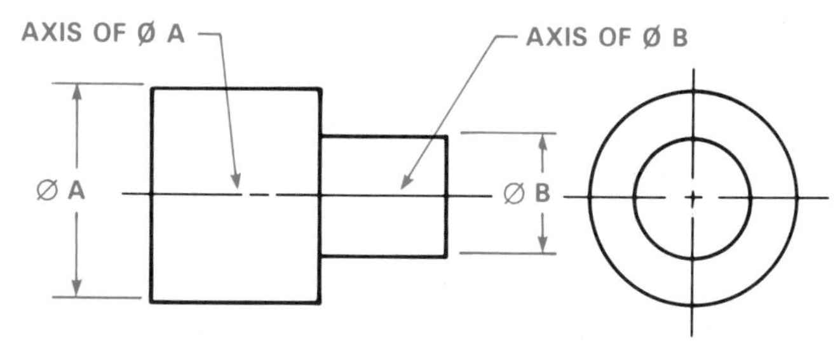

Figure 11–69 Perfect concentricity when both axes coincide.

that UNTOLERANCED DIMENSIONS LOCATING TRUE POSITION ARE BASIC. Basic dimensions are theoretically perfect, so that datum or chain dimensioning may be used equally to locate true position because there is no tolerance buildup. The location feature control frame is generally added to the note or dimension of the related feature.

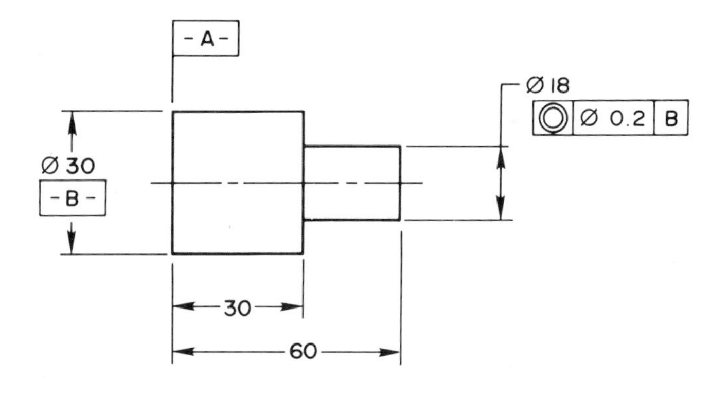

Figure 11–70 Concentricity represented.

Concentricity

The concentricity geometric characteristic symbol is detailed in Figure 11–68. Concentricity is the relationship of the axes of cylindrical shapes. Perfect concentricity exists when the axes of two or more cylindrical features are in perfect alignment. (See Figure 11–69.) A concentricity tolerance allows for a specified amount of deviation of the axes of concentric cylinders as shown in Figure 11–70. The geometric tolerance and related datums imply RFS. It is difficult to control the axis relationship specified by concentricity, so runout will generally be used. Where the balance of a shaft is critical, concentricity may be used.

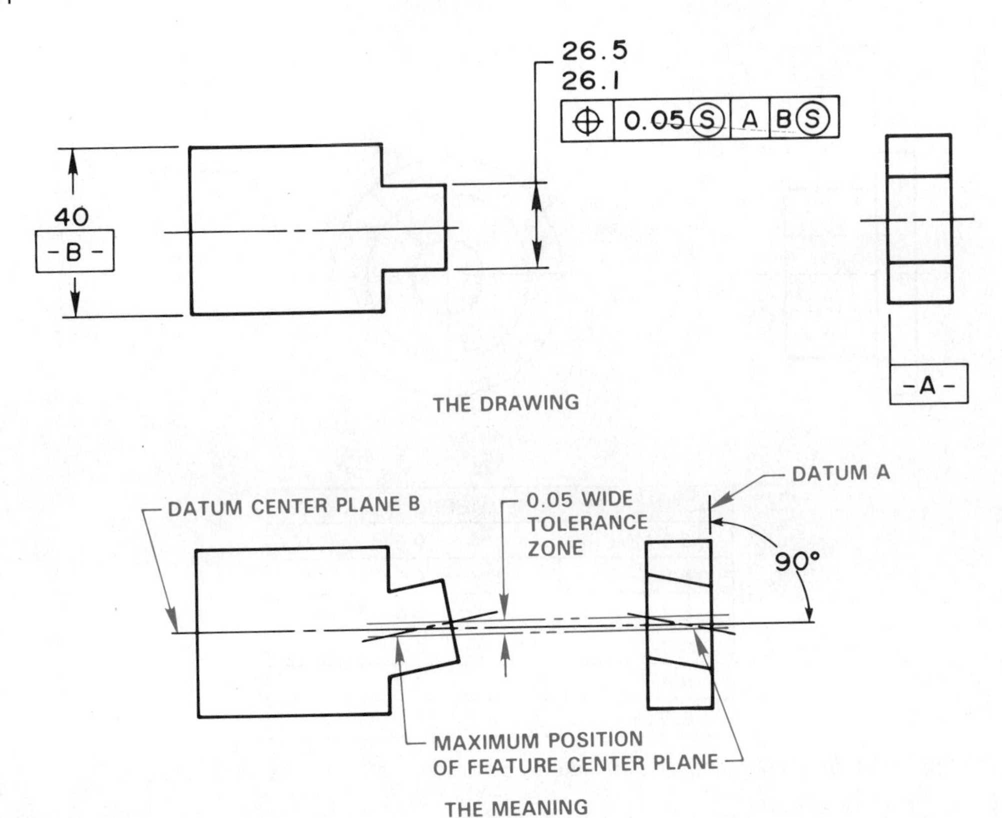

Figure 11–71 Position geometric characteristic specifying symmetry.

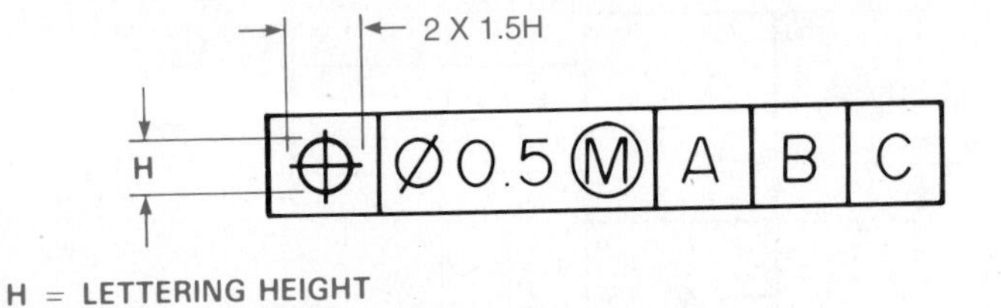

Figure 11–72 Position feature control frame.

Symmetry

Symmetry is the center plane relationship between two or more features. Perfect symmetry exists when the center plane of two or more features is in alignment. The geometric characteristic used to specify symmetry is the position symbol. (See Figure 11–71.) RFS or MMC must be identified in the feature control frame as applied to the feature tolerance or the related datum.

Position

The position geometric characteristic symbol, also used for symmetry, is detailed in Figure 11–72. The material condition symbols Ⓜ, Ⓢ, Ⓛ must accompany the geometric tolerance and the related datum reference. The maximum material condition application is common although some cases require its use regardless of feature size or least material condition. The feature control frame is applied to the note or diameter dimension of the feature. A positional tolerance defines a zone within which the center, axis, or center plane of a feature or size is permitted to vary from the true, theoretically exact, position. Basic dimensions establish the true position from specified datum features and between interrelated features. A positional tolerance is indicated by the position symbol, a tolerance, and appropriate datum references placed in a feature control frame. The location of each feature is given by basic dimensions. Dimensions locating true position must be excluded from the drawing's general tolerances by applying the basic dimension symbol to each basic dimension or by specifying on the drawing or drawing reference the general note, UNTOLERANCED DIMENSIONS LOCATING TRUE POSITION ARE BASIC.

Positional Tolerance at Maximum Material Condition (MMC). Positional tolerance at MMC means that the given tolerance is held at the MMC produced size. Then, as the feature size departs from MMC (gets larger) toward least material condition (LMC), the position tolerance increases equal to the amount of change from MMC to the maximum departure at LMC. Positional tolerance at MMC may be defined by the feature axis or surface. The datum references commonly establish true position perpendicular to the primary datum and the coordinate location dimensions to the secondary and tertiary datums. The position tolerance zone is a cylinder equal in diameter to the given tolerance. The cylindrical tolerance zone extends through the thickness of the part unless otherwise specified. The actual centerline of the feature may be anywhere within the cylindrical tolerance

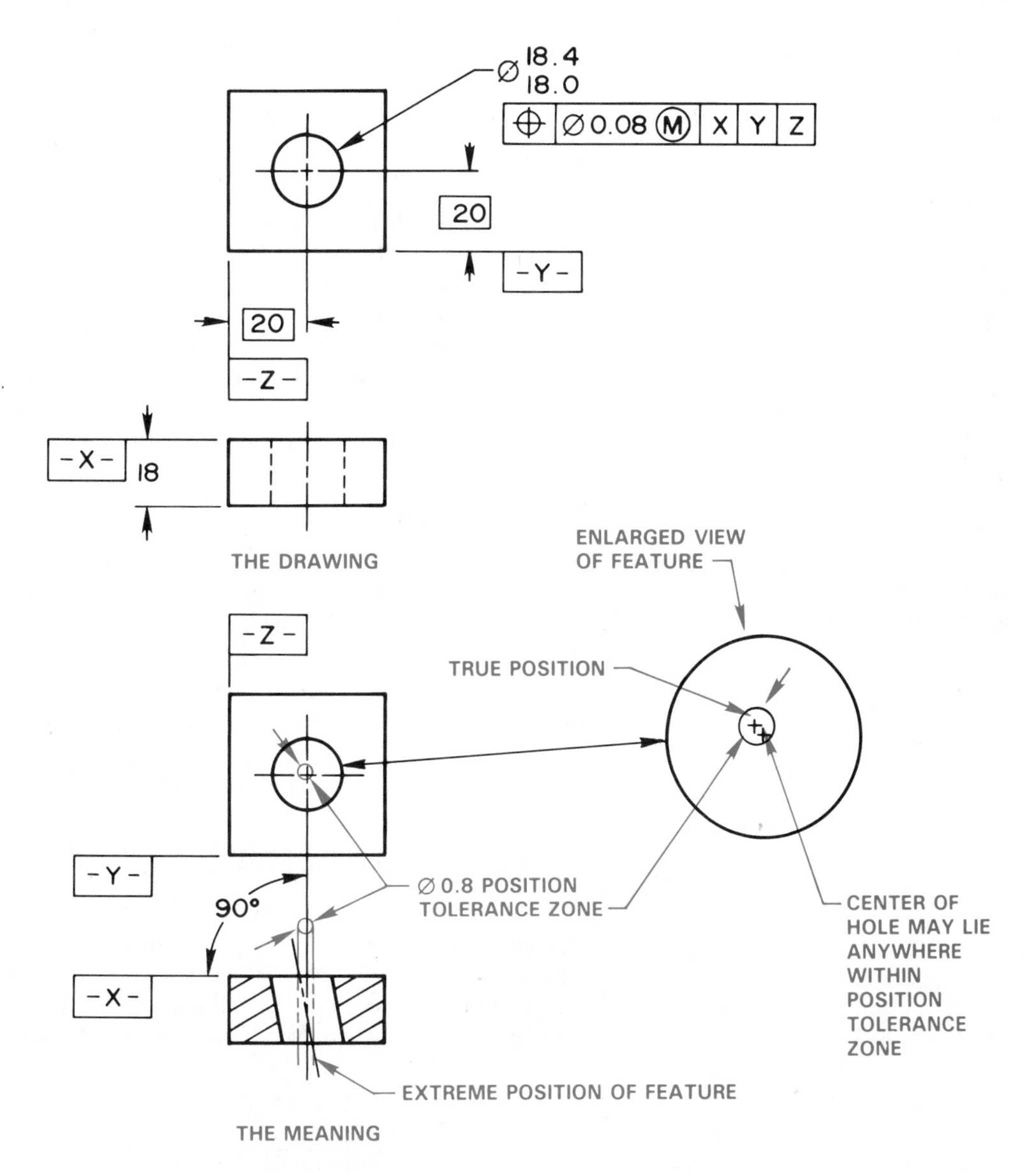

ACTUAL SIZE	POSITIONAL TOLERANCE
18.0 MMC	0.08
18.1	0.18
18.2	0.28
18.3	0.38
18.4 LMC	0.48

Figure 11–73 Positional tolerance at maximum material condition, MMC.

zone. (See Figure 11–73.) Another explanation of position may be related to the surface of a hole. The hole surface may not be inside a cylindrical tolerance zone established by the MMC diameter of the hole less the position tolerance. (See Figure 11–74.)

Zero Positional Tolerance at MMC. When the positional tolerance is associated with the MMC symbol, the tolerance is allowed to exceed the specified amount when the actual feature size departs from MMC. When the actual sizes of features are manufactured very close to MMC, it is critical that the feature axis or surface not exceed the boundaries discussed previously. When parts are rejected due to this situation, it is possible to increase the acceptability of mating parts by reducing the MMC size of the feature to minimum allowance with the mating part and providing a zero positional tolerance at MMC. The positional tolerance is dependent on the feature size. When zero positional tolerance is used, no positional tolerance is allowed when the part is produced at MMC. True position is required at MMC. As the actual size departs from MMC, the positional tolerance increases equal to the amount of change to the maximum tolerance at LMC. (See Figure 11–75.)

Positional Tolerance at RFS. The RFS material condition symbol may be applied to the positional tolerance when it is desirable to maintain the given tolerance at any produced size. This application results in close positional control. (See Figure 11–76.)

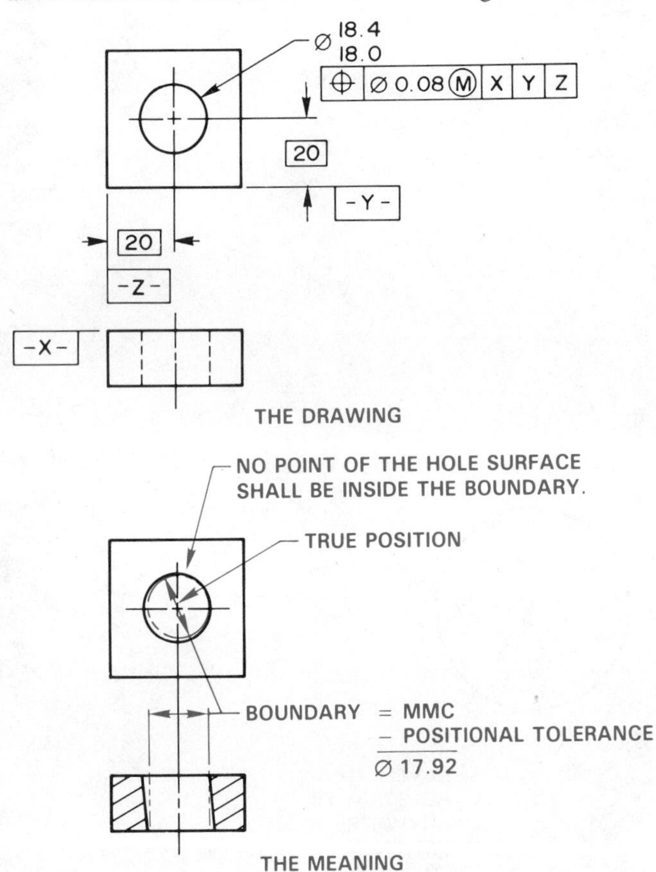

Figure 11–74 Position boundary equals MMC minus position tolerance.

Positional Tolerance at LMC. Positional tolerance at LMC is used to control the relationship of the feature surface and the true position at largest hole size. The function of the LMC specification is generally for the control of minimum-edge distances. When LMC is used the given positional tolerance is held at the LMC-produced size where perfect form is required. As the actual size departs from LMC toward MMC, the positional tolerance zone is allowed to increase equal to the amount of departure. The maximum positional tolerance is at the MMC-produced size. (See Figure 11–77.)

DATUM PRECEDENCE AND MATERIAL CONDITION

The effect of material condition on the datum and related feature may be altered by changing the datum precedence and the applied material condition symbol. The datum precedence is established by the order of placement in the feature control frame. The first datum listed is the primary datum; subsequent datums are secondary and tertiary. Figure 11–78 shows the effect of altering datum precedence and material condition.

POSITION OF MULTIPLE FEATURES

The location of multiple features is handled in a manner similar to the location of a single feature. The true positions of the features are located with basic dimensions using rectangular or polar coordinates. The features are then identified by quantity, size, and position. (See

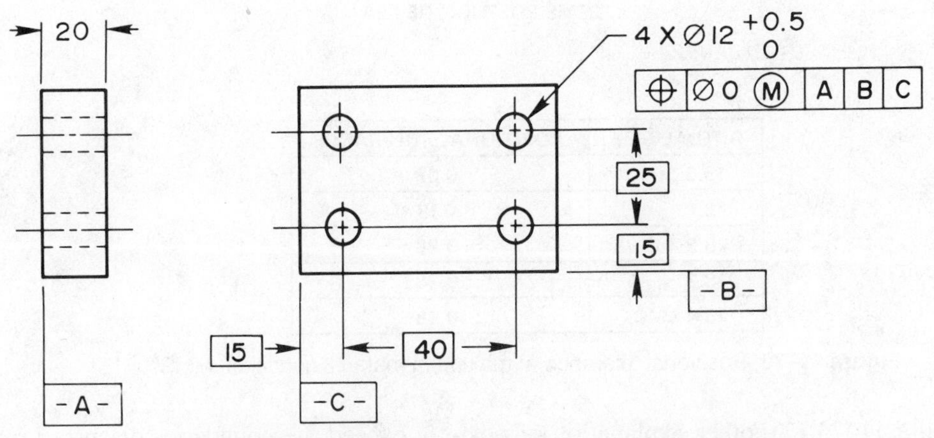

ACTUAL SIZE	POSITIONAL TOLERANCE
12.0 MMC ZERO ALLOWANCE FOR 12mm FASTENER AT MMC	0
12.1	0.1
12.2	0.2
12.3	0.3
12.4	0.4
12.5	0.5

Figure 11–75 Zero positional tolerance at MMC.

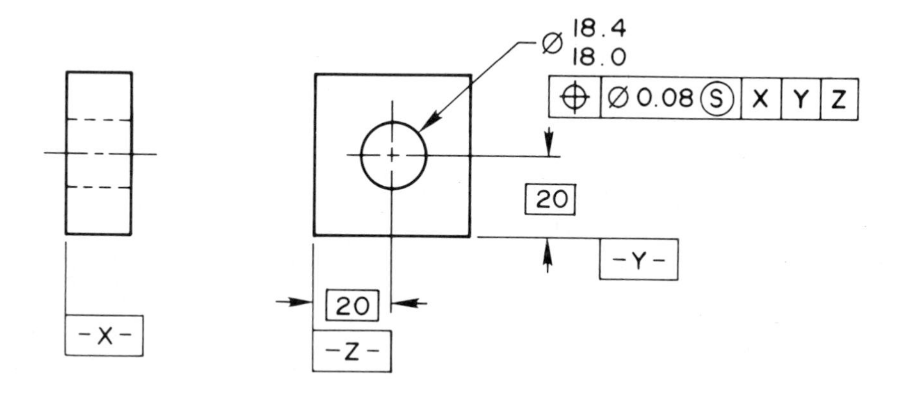

ACTUAL SIZE	POSITIONAL TOLERANCE
18.0 MMC	0.08
18.1	0.08
18.2	0.08
18.3	0.08
18.4 LMC	0.08

Figure 11–76 Positional tolerance at RFS.

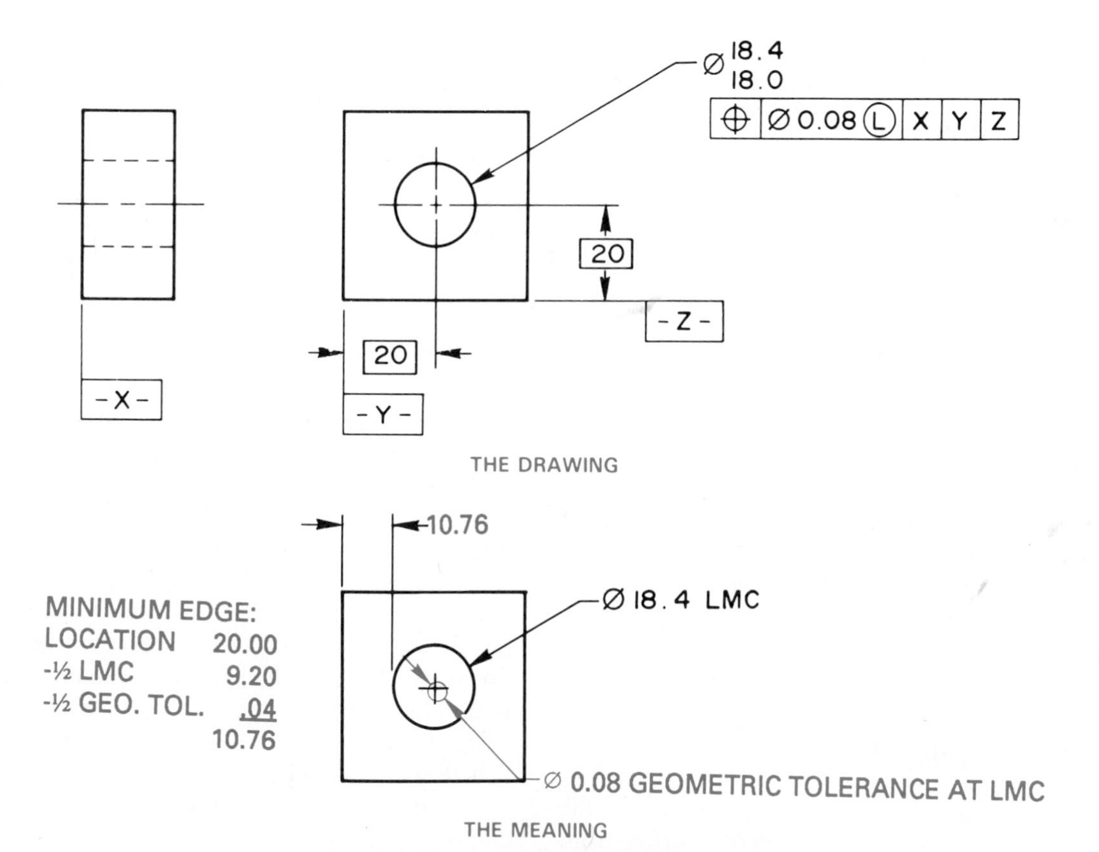

ACTUAL SIZE	POSITIONAL TOLERANCE
18.0 MMC	0.48
18.1	0.38
18.2	0.28
18.3	0.18
18.4 LMC	0.08

Figure 11–77 Positional tolerance at LMC.

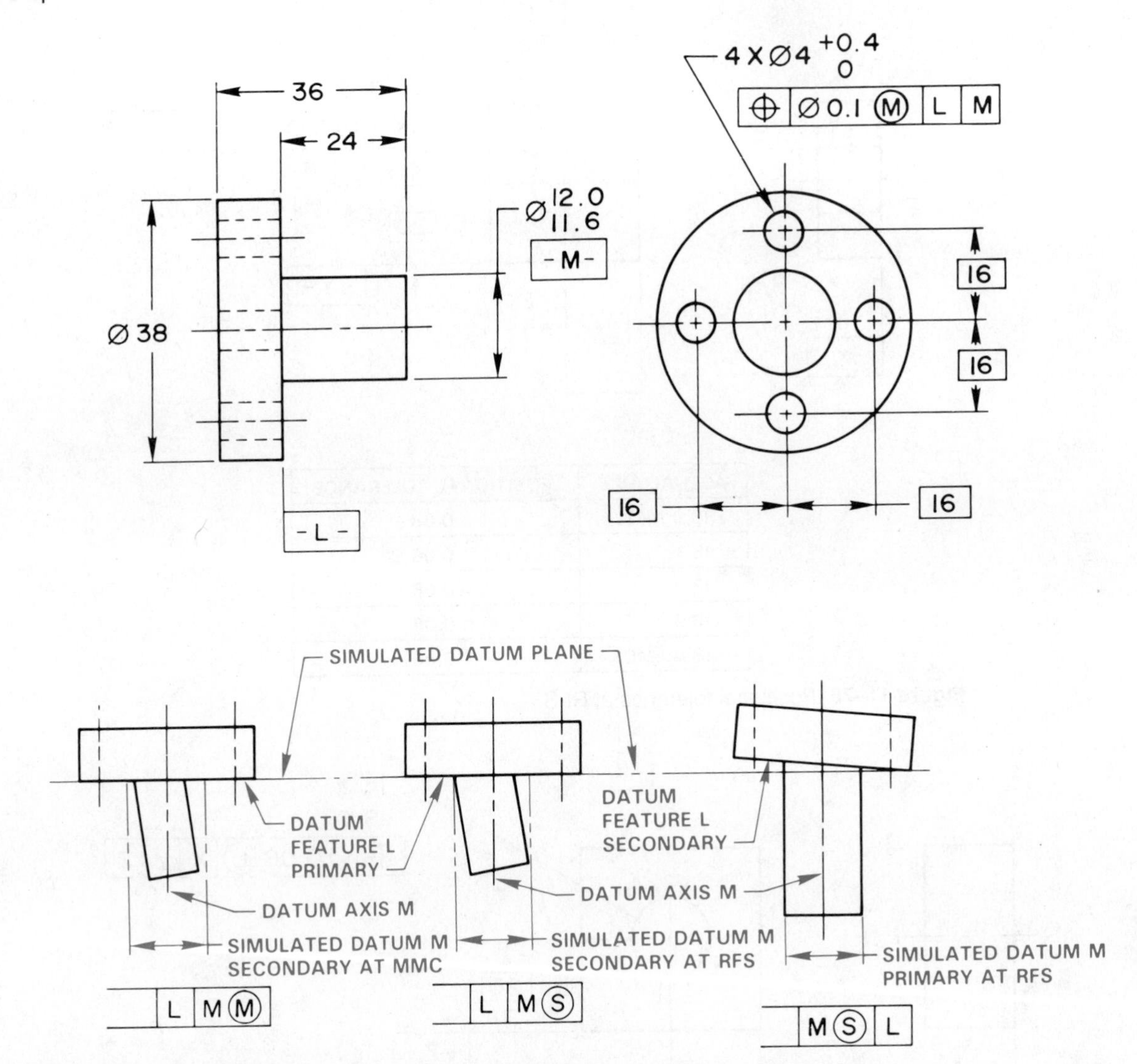

Figure 11–78 The effect of altering datum precedence and material condition.

Figure 11–79.) When two or more separate patterns are referenced to the same datums and with the same datum precedence, then the patterns are functionally the same. Verification of location and size is performed together. If this situation occurs and the interrelationship between patterns is not desired, then the specific note SEP REQT, meaning separate requirement, shall be placed beneath each affected feature control frame.

COMPOSITE POSITIONAL TOLERANCING

In some situations it may be permissible to allow the location of individual features in a pattern to differ from the tolerance related to the items as a group. When this is done, the group of features is given a positional tolerance that is greater in diameter than the zone specified for the individual features. The tolerance zone of the individual features must fall within the group zone. The positional tolerance of the individual elements controls the perpendicularity of the features. An individual tolerance zone may extend partly beyond the group zone only if the feature axis does not fall outside the confines of both zones. When such is the case, the feature control frame is expanded in height and divided into two parts. The upper part, known as the pattern-locating control, specifies the larger positional tolerance for the pattern of features as a group. The lower entry, commonly called the feature-relating control, specifies the smaller positional tolerance for the individual features within the pattern. The pattern-locating control is located first with basic dimensions. The feature-relating control is established at the actual position of the feature center. Only the primary datum is represented in the feature-relating control. (See Figure 11–80.)

Additional Examples of Positional Tolerance

Positional tolerancing of tabs may be accomplished by identifying related datums, dimensioning the relationship between tabs, and providing the number of units followed by the size and feature control frame. (See

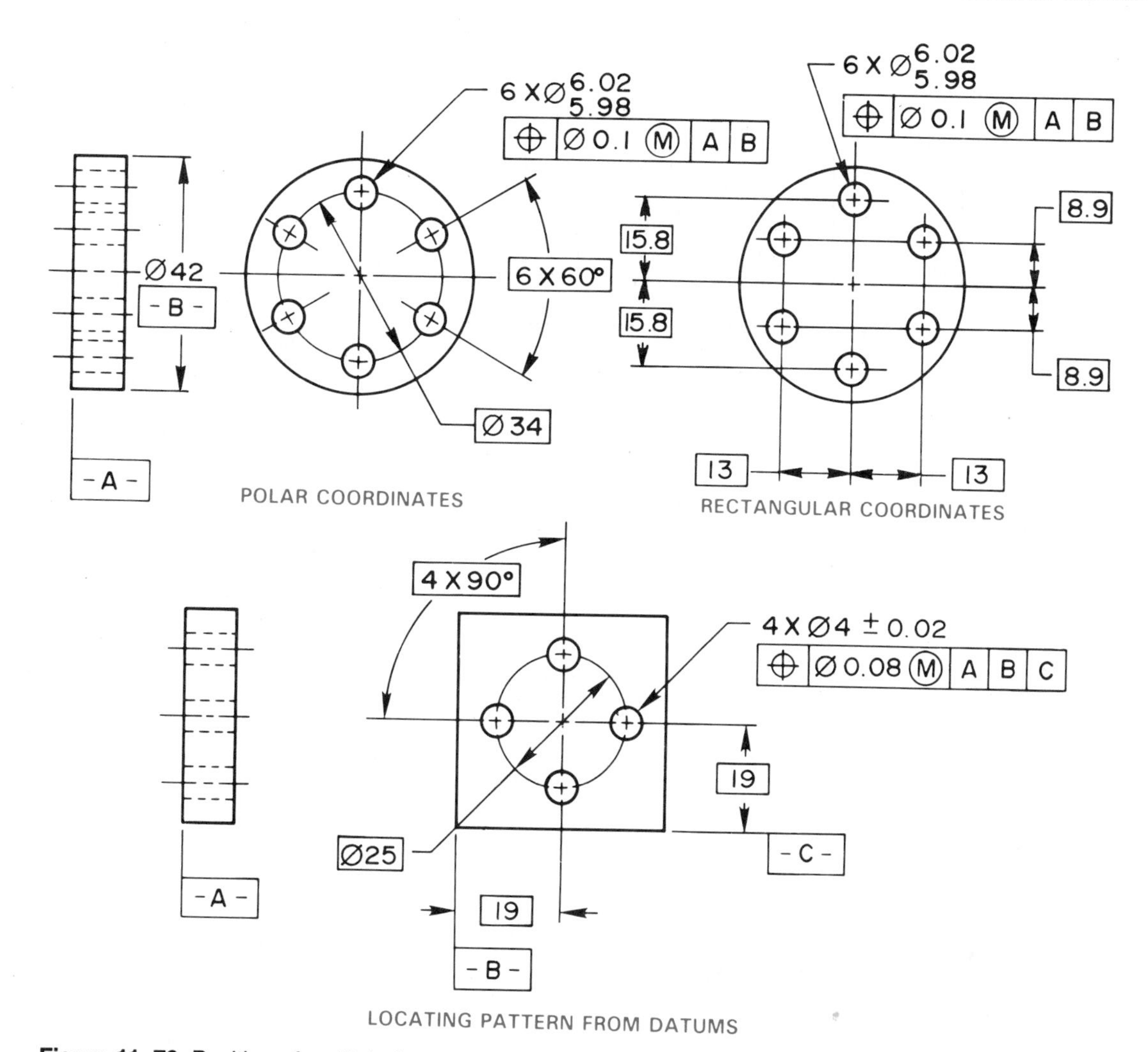

Figure 11–79 Position of multiple features.

Figure 11–81.) This same method may be used for dimensioning the positional tolerance of slots. The positional tolerance of slotted holes may be accomplished by locating the slot centers with basic dimensions and providing a feature control frame to both the length and width of the slotted holes as shown in Figure 11–82.

The orientation or positional tolerance of screw threads applies to the datum axis. The datum feature is established by a pitch diameter cylinder. If a different datum feature is required, the note should be lettered below the feature control frame, for example, MINOR DIA or MAJOR DIA. Similarly, the intended datum feature for gears or splines should be identified below the feature control frame. The options include MAJOR DIA, PITCH DIA, or MINOR DIA.

PROJECTED TOLERANCE ZONE

The standard application of a positional tolerance implies that the cylindrical zone extends through the thickness of the part or feature. In some situations where there is the possibility of interference with mating parts, the tolerance zone could be lengthened to accommodate the axis of the mating part. This type of application is especially useful when the mating features are screws, pins, or studs. These types of conditions are referred to as fixed fasteners because the fastener is fixed in the mating part and there is no clearance allowance. The projected tolerance zone meaning is shown in Figure 11–83. Place the projected tolerance zone symbol below the feature control frame as shown in the top view in Figure 11–84. Place a chain line next to the center line and dimension the "MIN" length of the projection as shown in the front view of Figure 11–84. When calculating the positional tolerance zones that are applied to the parts of a fixed fastener, the following formula may be used: MMC Hole – MMC Fastener (nominal thread size) ÷ 2 = Positional Tolerance Zone of Each Part. In some situations it may be desirable to provide more tolerance to one part than another, for example 60 percent to the threaded part and 40 percent to the unthreaded part.

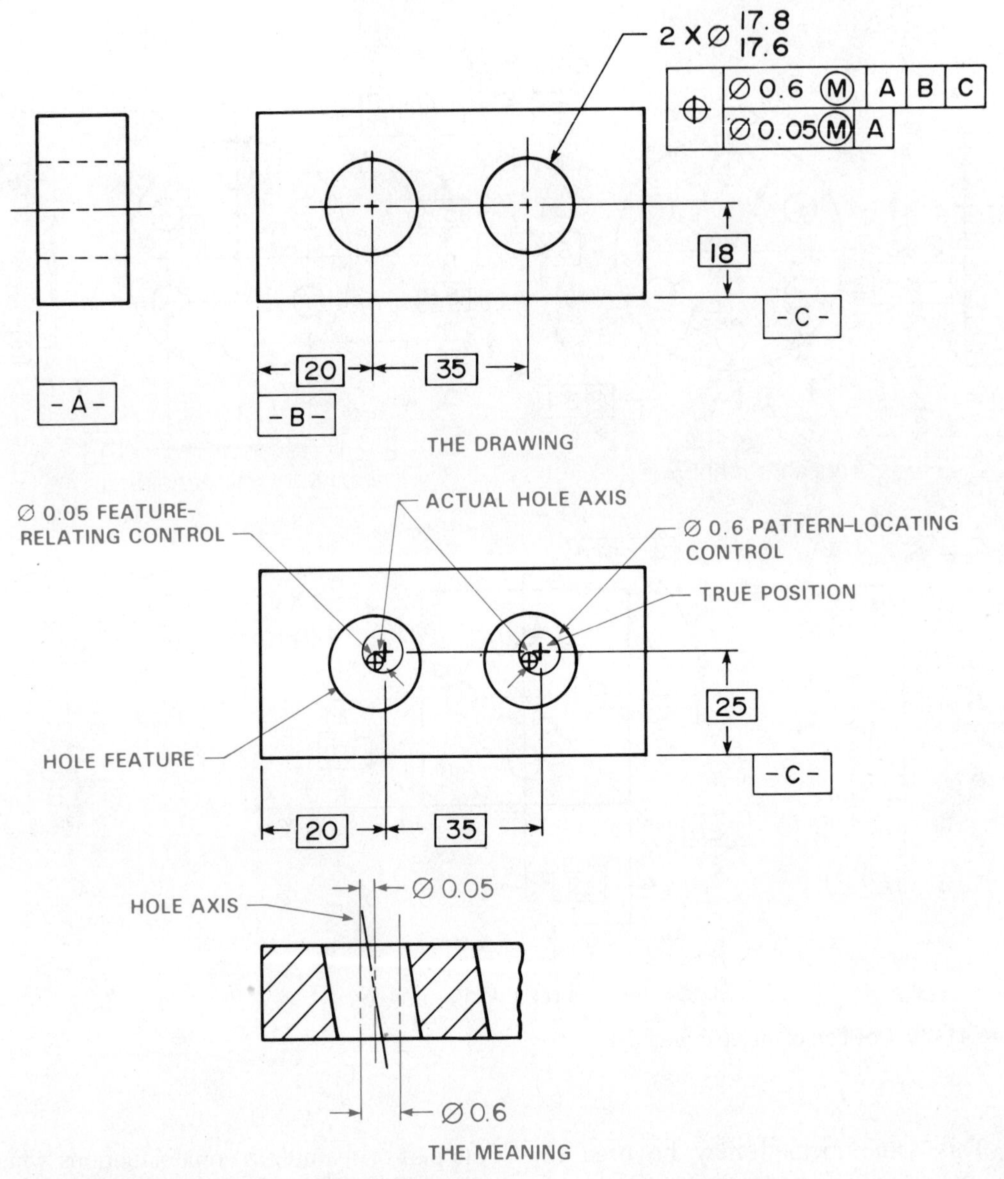

Figure 11–80 Composite positional tolerance.

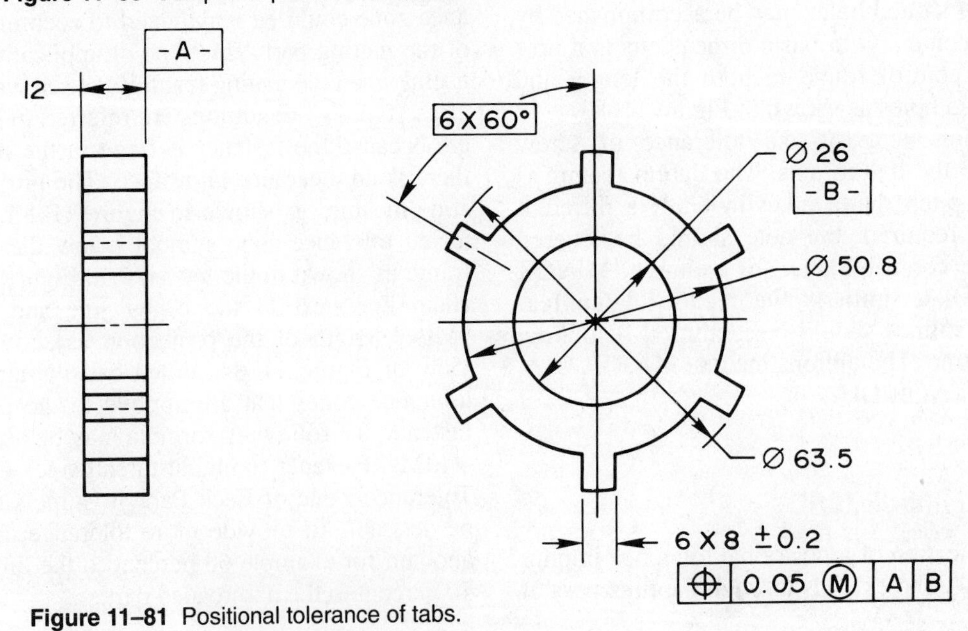

Figure 11–81 Positional tolerance of tabs.

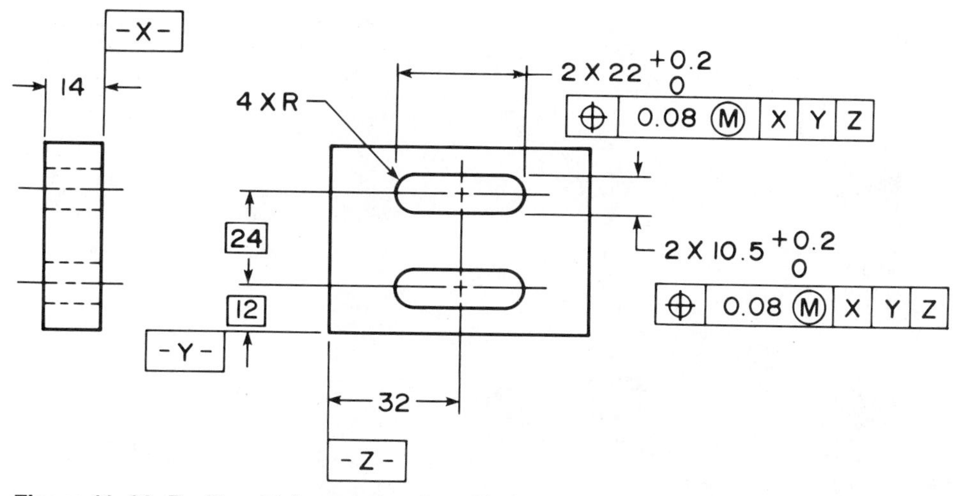

Figure 11–82 Positional tolerance for slotted holes.

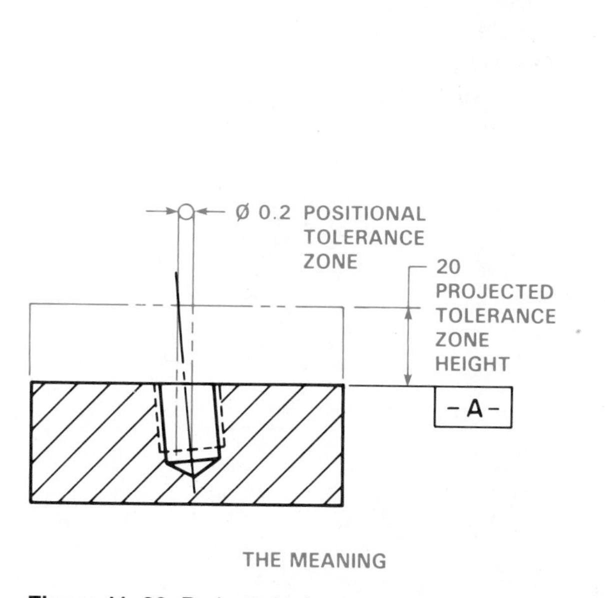

Figure 11–83 Projected tolerance zone.

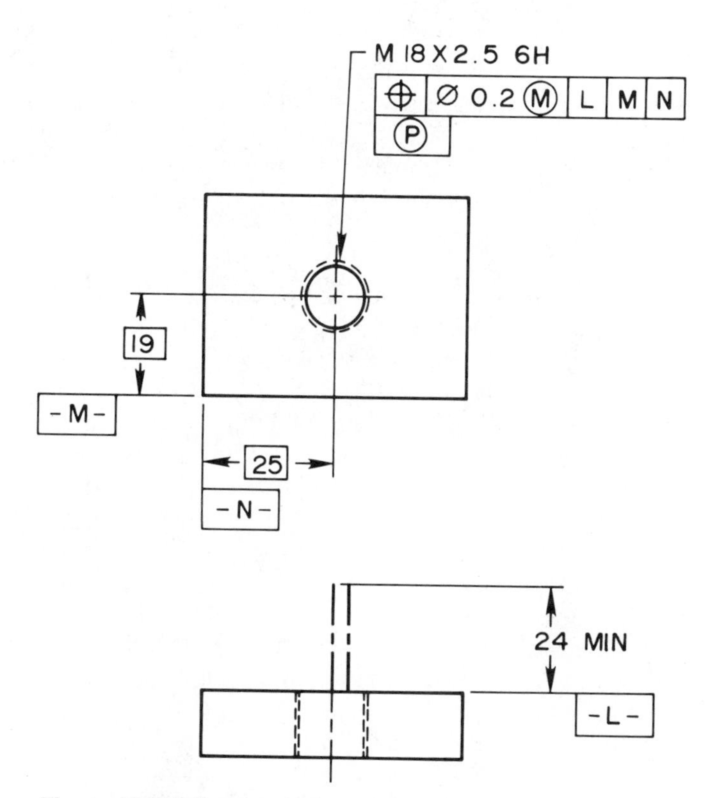

Figure 11–84 Projected tolerance zone, preferred drafting method.

When parts are assembled with fasteners, such as bolts and nuts or rivets, and where all the parts have clearance holes to accommodate the fasteners, the application is referred to as a floating fastener. Floating fasteners require that the fastening device be secured on each side of the part, such as with a bolt and nut. With a fixed fastener, one of the parts is a fastening device. Greater tolerance flexibility with floating fasteners is due to the fastener clearance at each part. When calculating the positional tolerance zone for floating fastener parts, the following formula may be used: MMC Hole – MMC Fastener (nominal thread size) = Positional Tolerance Zone of Each Part.

GEOMETRIC TOLERANCING SYMBOLS

Make a sketch of the desired symbol and decide how it will be placed on a drawing. Name the symbol; this will also be the CADD file name. Select a point on each symbol that will become a convenient point of origin when placing the symbol on a drawing. The point of origin is the location position for placing a symbol on a drawing. Figure 11–85 shows the point of origin determined for a sample of geometric tolerancing symbols.

After you have drawn a specific symbol, store it as a symbol by using such commands as SYMBOL TO DISK, STORE AS SYMBOL, or CREATE SYMBOL. When you have created a group of symbols, it is time to organize them into a symbol library using a command such as CREATE MENU, or TEMPLATE. A common computer prompt will be TYPE THE SYMBOL NAME. Then digitize the box on the template where you want the symbol located. Follow this process until all of the symbols that you created are located on the template. When all of the symbols have been placed on the template, assign a name such as GEOMETRIC TOLERANCING TEMPLATE or LIBRARY. Print out or plot the symbols on a sheet so they correspond to the locations in which they were placed on the template. This new symbol library may then be placed as an overlay on the menu tablet.

Symbol templates may be selected for use by entering a command such as SELECT or ACTIVATE MENU, or by using the pointing device to select the template name from the menu tablet. When the symbol template is ready for use, individual symbols may be entered on a drawing by pressing the pointing device inside the specific box on the symbol library and then pressing the desired location on the drawing. Some systems automatically display the symbol at the cross hairs location on the screen. Then as the point device is moved, the symbol also moves on the screen. When the symbol is positioned in the desired location, a button is pushed that places the symbol. Some symbols may require informational prompts such as ZONE DESCRIPTOR, GEOMETRIC TOLERANCE, MATERIAL CONDITION, and/or DATUM REFERENCE after placement.

If you choose not to design your own custom geometric tolerancing symbols, there are predesigned GT symbol libraries available for most major software packages. Refer to the third-party applications catalog with your software to find any third-party software available. Figure 11–86 shows an example of a tablet menu symbol library overlay designed for GT applications.

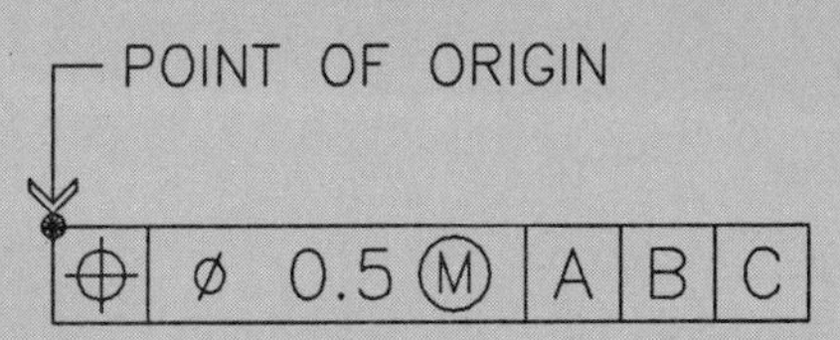

Figure 11–85 The insertion point for geometric tolerancing symbols may be located in any convenient position. This example shows the upper left corner as a convenient insertion point.

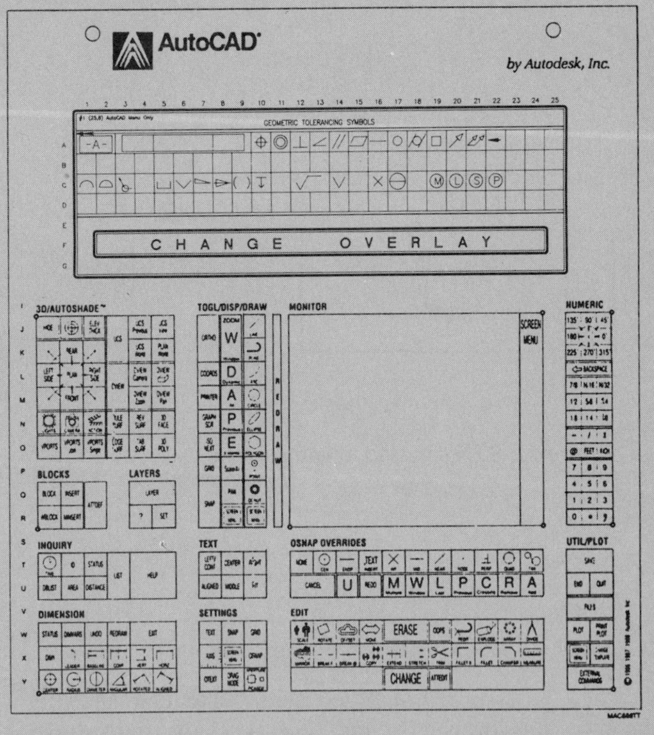

Figure 11–86 A customized tablet menu symbol library overlay for geometric tolerancing symbols. *AutoCAD base template courtesy Autodesk, Inc. Customized tablet menu symbol library overlay for geometric tolerancing symbols courtesy Drafting Technology Services, Inc., Bartlesville, Oklahoma.*

VIRTUAL CONDITION

The tolerances of a feature that relate to size, form, orientation, and location, including the possible application of MMC or RFS, are determined by the function of the part. Consideration must be given to the collective effect of these factors in determining the clearance between mating parts and in establishing gage feature sizes. *Virtual condition* is the resultant boundary of such considerations. Virtual condition is the sole condition where a feature size may be outside of MMC. Controlling the clearance between mating parts is critical to the design process. When features are dimensioned using a combination of size and geometric tolerances, the resulting effects of the specifications should be considered to insure that parts will always fit together. The boundary created by the combined effects of size, MMC, and the geometric tolerance is known as the virtual condition.

When axis straightness is used, the virtual condition determines the amount that the feature will be allowed to exceed the MMC envelope. In Figure 11–87 Virtual Condition = MMC Size + Geometric Tolerance. In the given example the virtual condition of ∅ 18.3 is the smallest diameter that this shaft should be designed to fit within.

When a positional tolerance is applied to an internal feature, the Virtual Condition = MMC Hole – Positional Tolerance. This calculation determines the maximum feature size that should be allowed to fit within the hole. (See Figure 11–88.)

When perpendicularity is applied to an external diameter, such as a pin, the Virtual Condition = MMC Feature + Perpendicularity Geometric Tolerance. The virtual condition determines the smallest acceptable mat-

	GEOMETRIC TOLERANCE	
ACTUAL SIZE	GIVEN RFS IMPLIED	EFFECT IF MMC ADDED
18.2 MMC	0.1	0.1
18.1	0.1	0.2
18.0	0.1	0.3
17.9	0.1	0.4
17.8 LMC	0.1	0.5

Figure 11–87 Virtual condition of shaft equals MMC plus geometric tolerance.

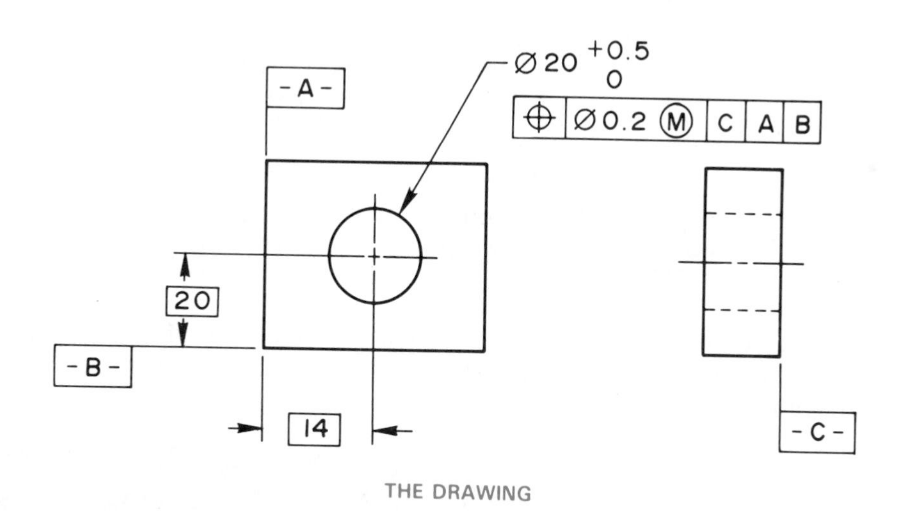

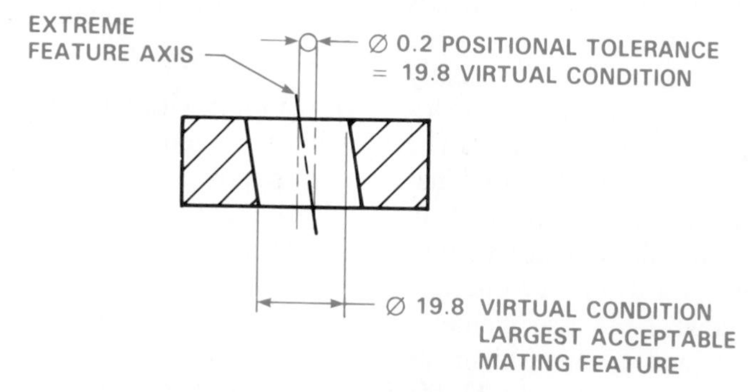

Figure 11–88 Virtual condition of hole equals MMC minus geometric tolerance.

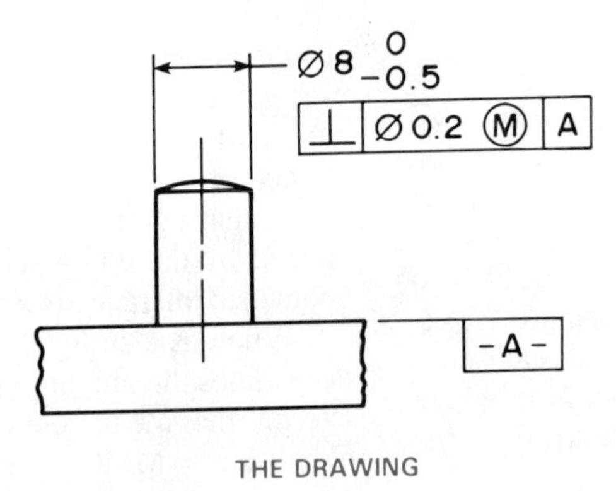

THE DRAWING

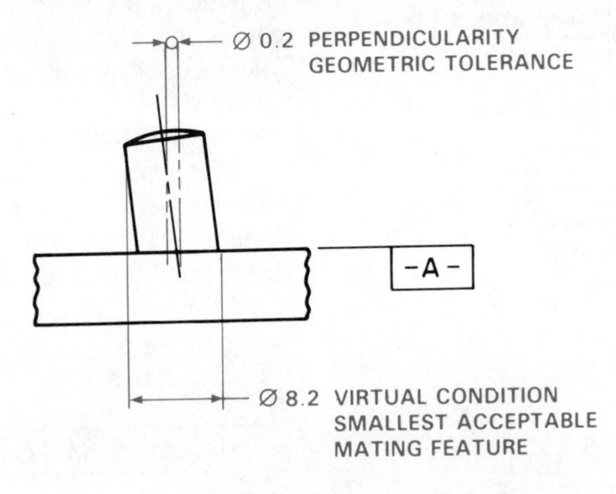

THE MEANING

Figure 11–89 Virtual condition perpendicular pin equals MMC feature plus perpendicularity geometric tolerance.

ing feature that will fit over the given part while maintaining a positive connection between the surfaces at datum A as shown in Figure 11–89.

COMBINATION CONTROLS

In some situations, compatible geometric characteristics may be combined in one feature control frame or separate frames associated with the same surface. This is normally done when the combined effect of two different geometric characteristics and tolerance zones is desired. The profile tolerance may be used to illustrate the combination of geometric characteristics. Profile and parallelism may be combined to control the profile of a surface plus the parallelism of each element to a datum. Profile and runout may be combined to control the line elements within the profile specification and circular elements within the runout tolerance as shown in Figure 11–90. The combined control of parallelism and perpendicularity may be represented on the same feature as shown in Figure 11–91. Other combined geometric characteristics may be used when the callouts are compatible and the design function of the part requires such specific controls.

PROFESSIONAL PERSPECTIVE

Written by Michael A. Courtier, Supervisor/Instructor, CAD/CAM/CAE Employee Training and Development, Boeing Aerospace

In working with engineers and drafters for years on the subject of geometric tolerancing, I find that questions keep popping up in several key areas — datums; limits of size; geometric tolerances as a refinement of size; the tolerance zones (2-D or 3-D, cylindrical or total wide, etc.); and the apparent overlap of controls (circularity, cylindricity, circular runout, total runout, position). It is in these "weak" spots that you should dig a little deeper and fill in your knowledge.

New and old users alike tend to miss the key rules of thumb in selecting *datums*, and most textbooks on the subject don't address this area.

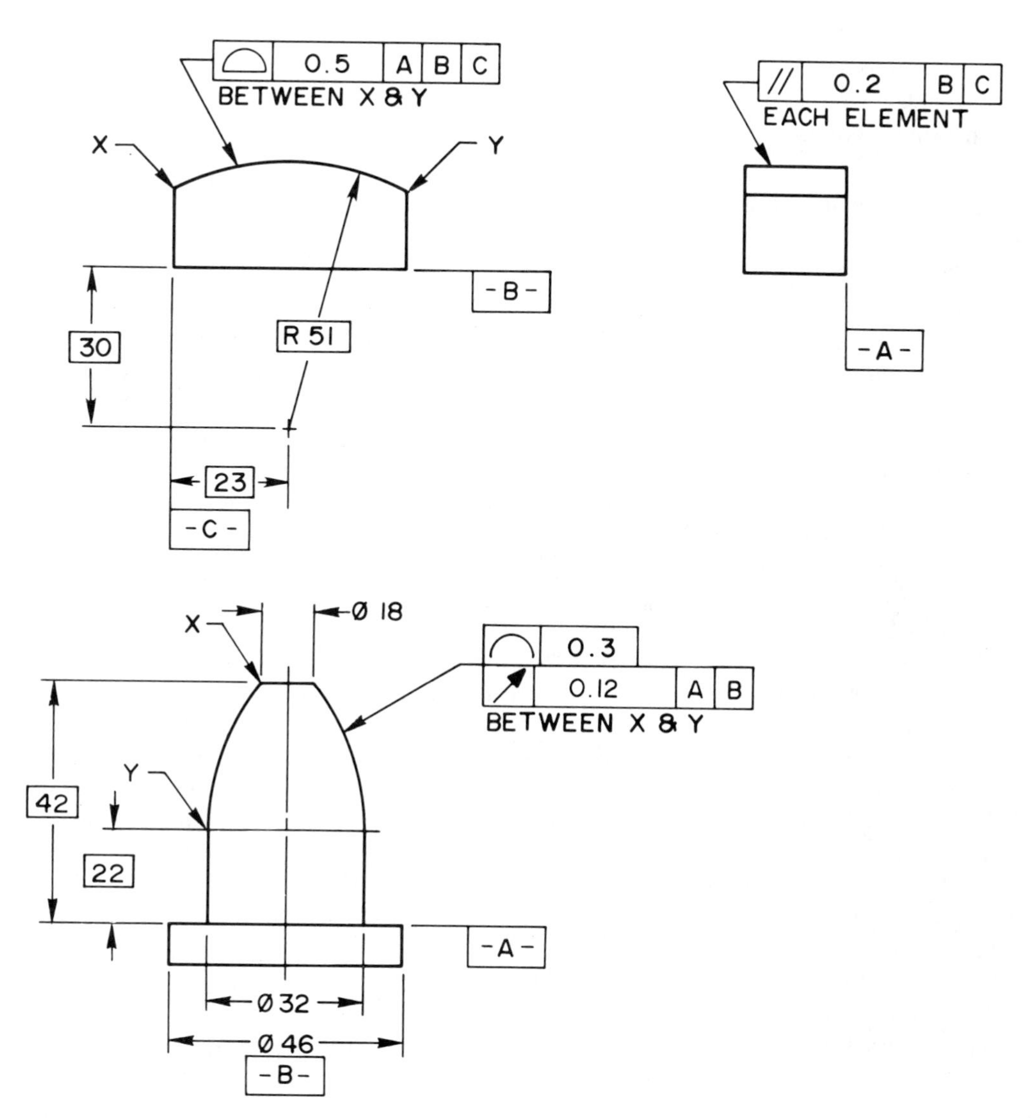

Figure 11–90 Combination controls (profile and runout).

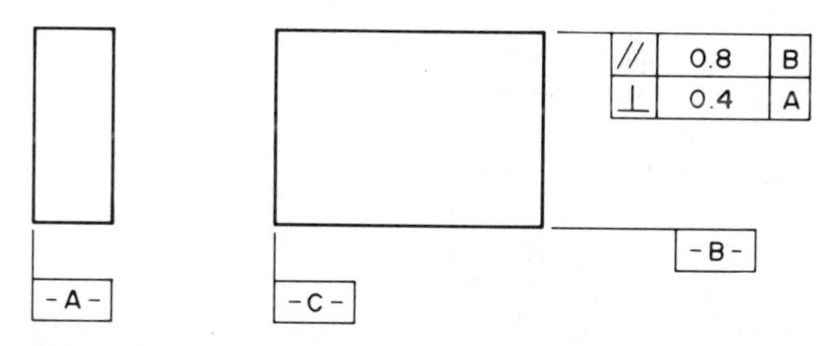

Figure 11–91 Combination controls (parallelism and perpendicularity).

- Make sure you select datums that are real, physical surfaces of the part. (I've seen many poor drawings with datums on centerlines, at the centers of spheres, or even at the center of gravity.)
- Be sure to select large enough datum surfaces. (It's hard to chuck up on a cylinder 1/10" long, or get three noncolinear points of contact on a plane of similar dimensions.)
- Choose datums that are accessible. (If manufacturing or quality assurance has to build a special jig to reach the datum, extra costs are incurred for the ultimate part.)
- Given the above constraints, select functional surfaces and corresponding features of mating parts where possible.

Remember, datums are the *basis* for subsequent dimensions and tolerances, so they should be selected carefully. (You might also be well advised to assign a little form, orientation, or runout control to the datum surface to assure an accurate basis.)

In the area of *limits of size*, there are three possible interpretations of a size dimension. Every class I've taught has been divided on what the proper extent of a size dimension's control should be — the overall size (or geometric form), the size at various cross-sectional checks, or both. The correct answer, per definition in the standard, is **BOTH**! (Study Section 2.7 in ANSI–Y14.5 for further reinforcement.)

The words *refinement of size* are a source of general confusion to new users of geometric tolerancing. Let's say a part is dimensioned on the drawing as .500±.005" thick, with parallelism control to .003" on the upper surface. If a production part has a convex upper surface with low points at .504, then the high point can only be .505, *not* .507. In other words, only .001" of the .003" of parallelism control can be used since parallelism is *a refinement of size within the overall size limits*.

Another one of the often misunderstood aspects of geometric tolerancing is the *tolerance zone*. Many engineers, technicians, drafters, machinists, and inspectors misinterpret the *shape* of the tolerance zone. I recommend that all new users make a chart showing the tolerance zones for each type of tolerance and a sampling of its various applications. This chart should include — at a minimum — the 2-D zones between two lines or concentric circles (e.g., straightness, circularity) versus the 3-D zones between parallel planes (e.g., flatness, perpendicularity); the cylindrical versus the total wide zones (e.g., straightness to an axis versus surface straightness, or position of a hole versus position of a slot); the cross-sectional zones versus the total surface zones (e.g., circularity versus cylindricity); and many, many others. If you can properly visualize the tolerance zone, you're well on your way to the proper interpretation of the geometric tolerance.

One more thing to watch out for is the *apparent overlap of controls*. While it may appear that there is overlap, there are subtle yet distinct differences between all the geometric tolerance controls. Consider cylindrical shapes, for instance. You can use a wide variety of controls for them, including straightness, circularity, cylindricity, concentricity, circular runout, total runout, and position, to name a few. Remember that the first three are form controls, so no datum relationship is controlled. Straightness can control the axis or the surface merely by placing the feature control frame in a different location on the drawing. Straightness is a longitudinal control, whereas circularity is a circular element control (both are 2-D); cylindricity is a 3-D control of the surface relative to itself (no datum). The last four controls are with respect to a datum (or datums), but they have their differences also. Concentricity and the runouts are always RFS, whereas positional tolerancing can be LMC or MMC and get bonus tolerances. Runout is a rotational consideration, while positional tolerancing does not imply rotation.

In summary, ANSI Y14.5M–1982 is a very powerful tool for engineering drawings. Digging deep into the subtleties will separate the amateur from the professional, who knows it will save time and money in every discipline from design to manufacturing to quality assurance testing. Used properly, the engineering design team can convey more information about the overall design to downstream areas or subcontracting concerns that may have no knowledge of the final assembly. The proper assessment of the symbology requires no interpreter for the reader that is well versed in this international sign language called *geometric tolerancing*.

ENGINEERING PROBLEM

There have been some problems in manufacturing one of the parts due to the extreme tolerance variations possible. The engineer requests that you revise the drawing shown in Figure 11–92 and gives you these written instructions:

- Establish datum A with three equally spaced datum target points at each end of the Ø 28.1–28.0 cylinder.
- Establish datum B at the left end surface.
- Make the bottom surfaces of the 2X Ø40.2–40.0 features perpendicular to datum A by 0.06.
- Provide a cylindricity tolerance of 0.3 to the outside of the part.
- Make the 2X Ø40.2–40.0 features concentric to datum A by 0.1.
- Locate the 6X Ø6+0.2 holes with reference to

datum A at MMC and datum B with a position tolerance of 0.05 at MMC.

- Locate the 4X ∅4+0.2 holes with reference to datum A at MMC and datum B with a position tolerance of 0.04 at MMC.

This is an easy job because you did the original drawing on CADD and you have a custom geometric tolerancing package. You go back to the workstation and within one hour you have the check plot shown in Figure 11–93 ready for evaluation.

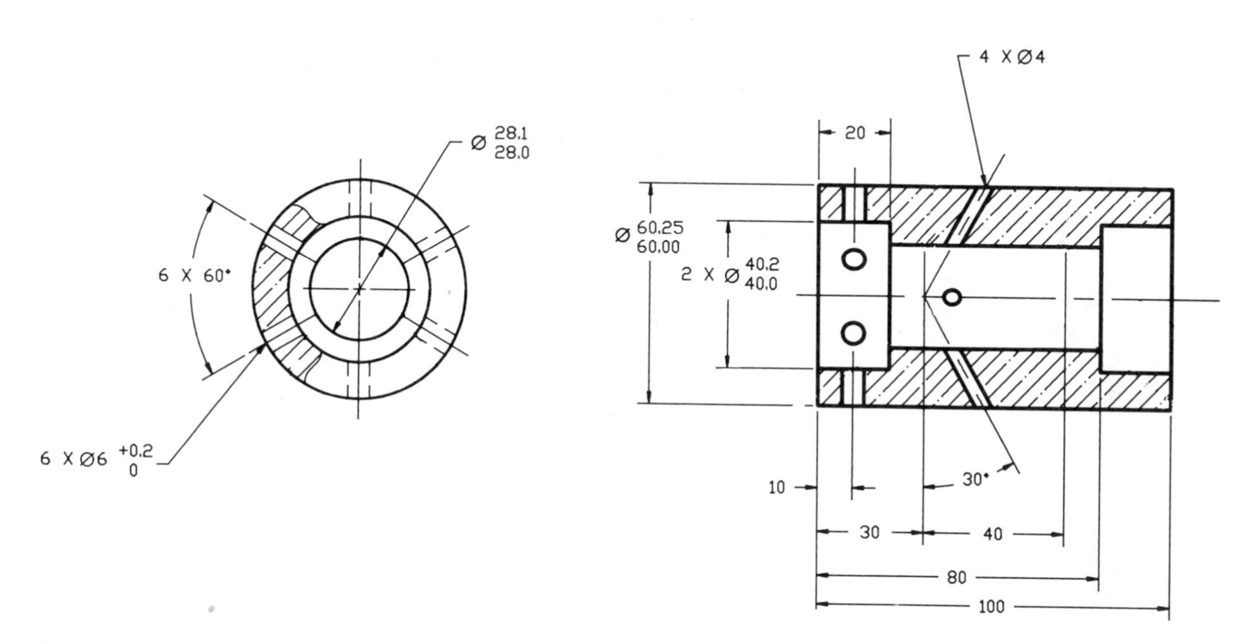

Figure 11–92 The original drawing to be revised with geometric tolerancing added from engineer's notes.

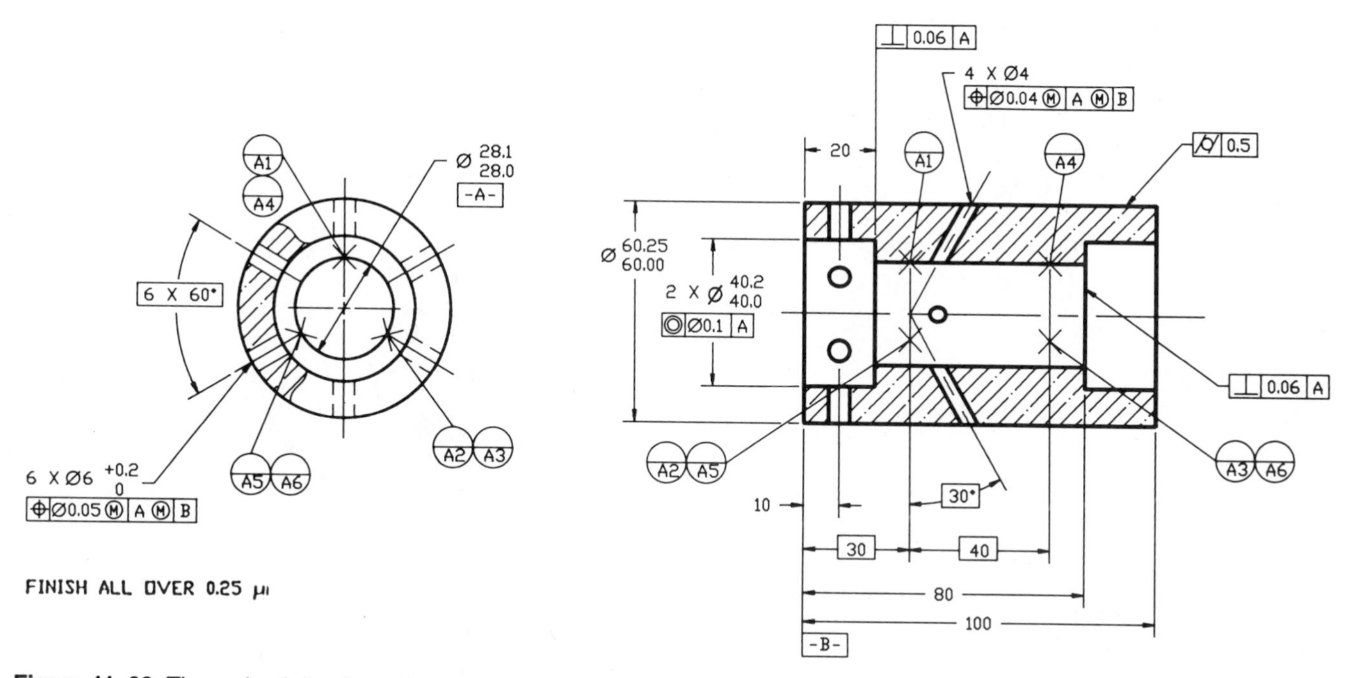

Figure 11–93 The revised drawing with geometric tolerancing added.

GEOMETRIC TOLERANCING PROBLEMS

DIRECTIONS

1. From the selected engineering layouts, determine which views and dimensions should be used to completely detail the part.
2. Make a multiview sketch to proper proportions including dimensions and notes.
3. Using the sketch as a guide, draw an original multiview drawing on an adequately sized drawing sheet. Add all necessary dimensions and notes using unidirectional dimensioning. Use manual or computer-aided drafting as required by your course guidelines.
4. Include the following general notes at the lower left corner of the sheet 0.5 in. each way from the corner border lines:

2. REMOVE ALL BURRS AND SHARP EDGES.
1. INTERPRET DIMENSIONS AND TOLERANCES PER ANSI Y14.5M–1982, OR 1994 IF USED.

NOTES:
Additional general notes may be required depending on the specifications of each individual assignment.

UNSPECIFIED TOLERANCES:

DECIMALS	mm	IN.
X	±2.5	±.1
XX	±0.25	±.01
XXX	±0.127	±.005
ANGULAR	±30°	
FINISH	3.2μm	125μin.

Problem 11–1 Geometric tolerancing (metric)

Part Name: Flow Pin
Material: Bronze
Finish: Finish All Over 0.20 μm.

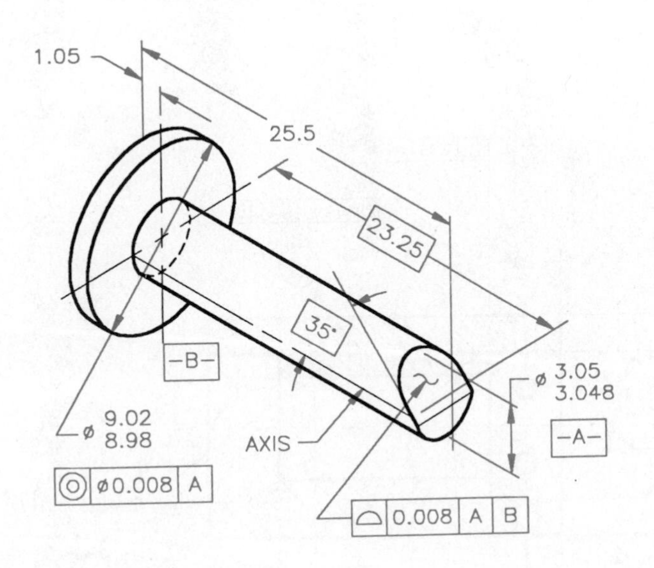

Problem 11–2 Geometric tolerancing (metric)

Part Name: LN2 Test Pump Lock Nut
Material: AMS 5732.
Additional General Notes:
1. 3: Mark per AS478 Class D with 1193125 and applicable dash number.
2. Finish All Over 1.6 μm.

Courtesy Aerojet TechSystems Co.

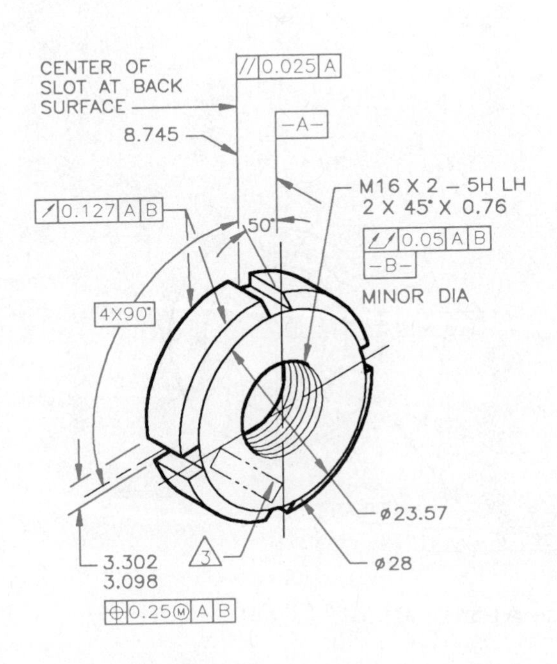

Problem 11–3 (in.)

Part Name: Half Coupling
Material: ∅ 1.250 6061–T6 Aluminum
SPECIFIC INSTRUCTIONS:
Provide MMC material condition after position tolerance except for RFS at threads.
Courtesy TEMCO.

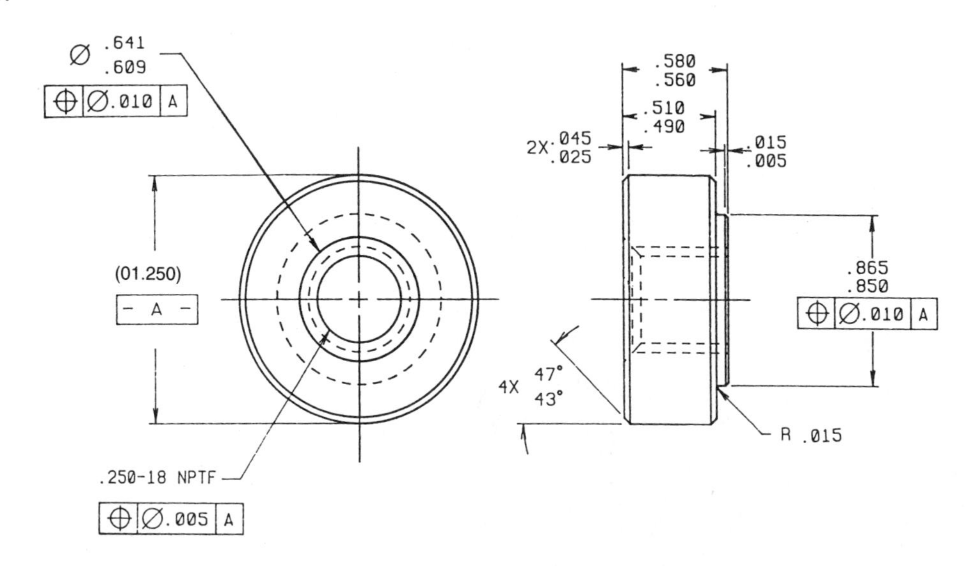

Problem 11–4 (in.)

Part Name: Coupling
Material: AISI 1010, Killed
SPECIFIC INSTRUCTIONS:
Provide MMC material condition after position tolerance except for RFS at threads.
Courtesy TEMCO.

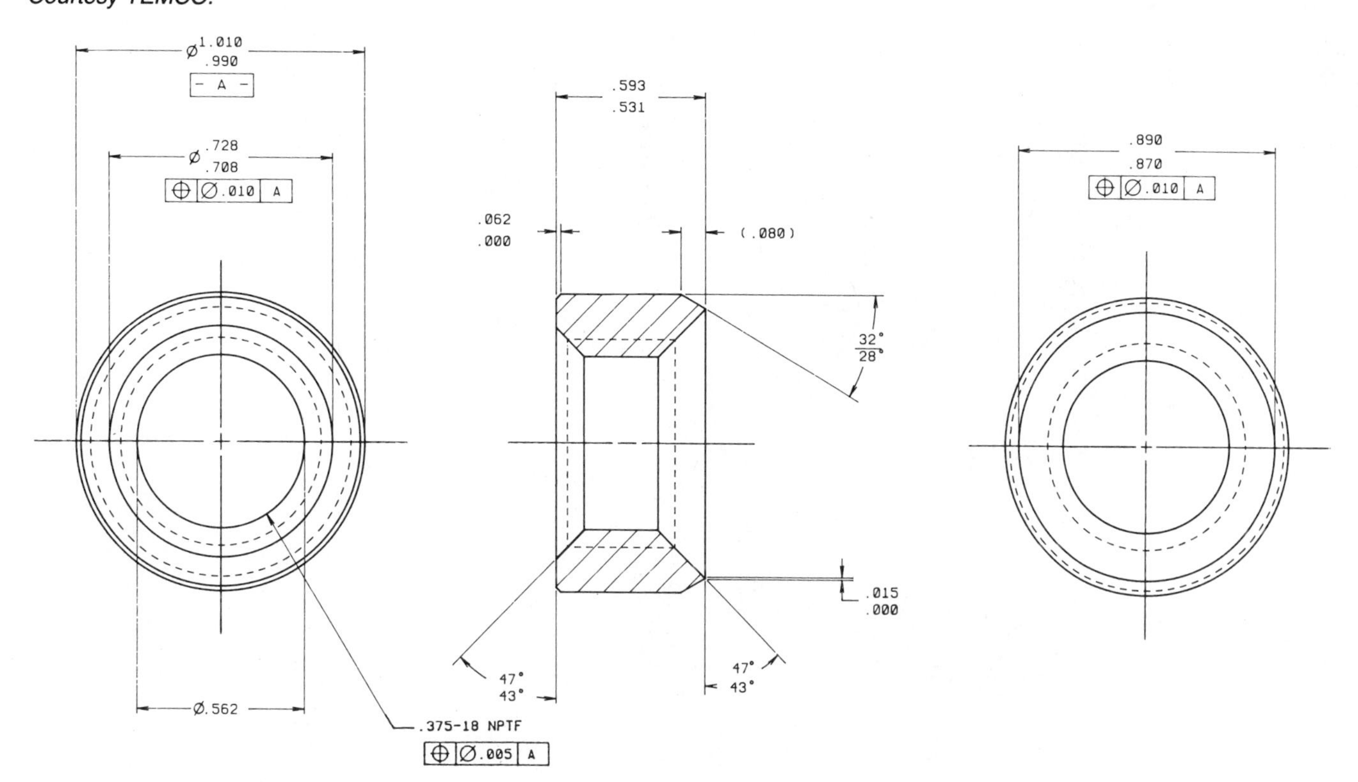

Problem 11–5 (in.)

Part Name: Half Coupling
Material: ∅ 1.625 6061–T6511
SPECIFIC INSTRUCTIONS:
Provide MMC material condition after position tolerance except for RFS at threads.
Courtesy TEMCO.

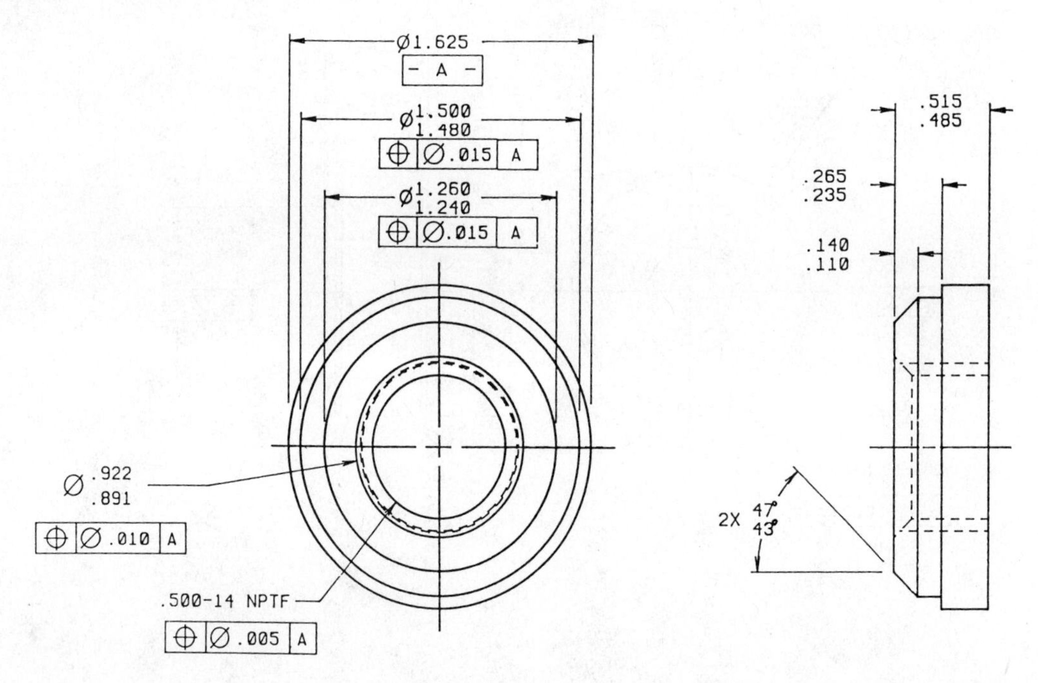

Problem 11–6 (metric)

Part Name: Spline Plate
Material: SAE 3135

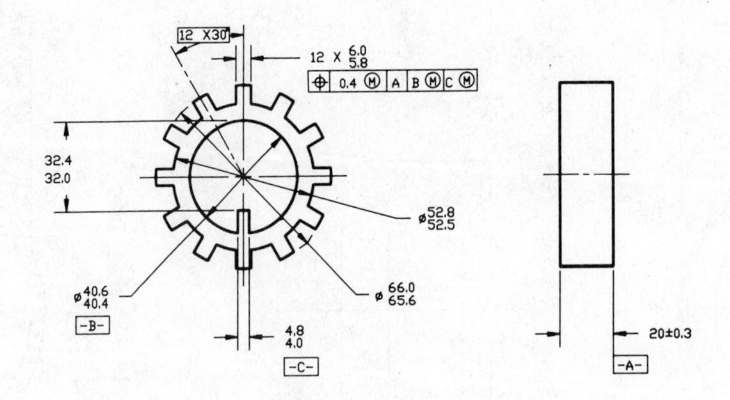

Problem 11–7 (in.)

Part Name: Nut
Material: No. 10 Bronze

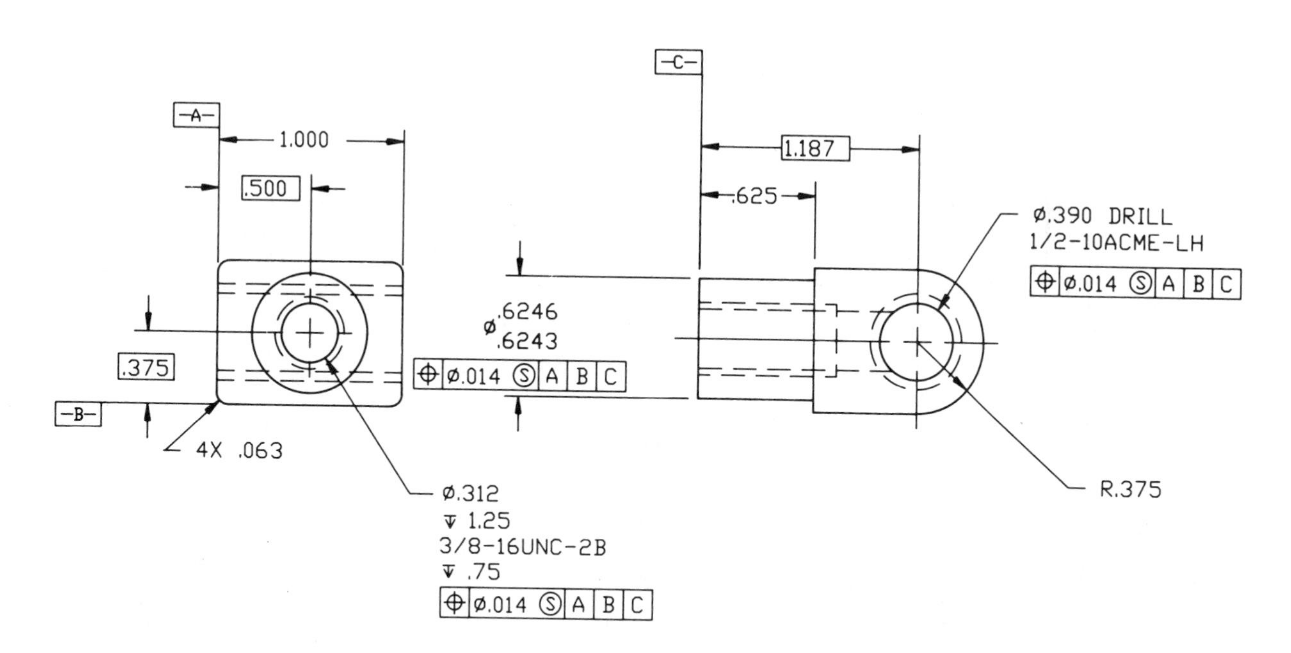

Problem 11–8 (metric)

Part Name: Coupling Bracket
Material: SAE 4310 Steel

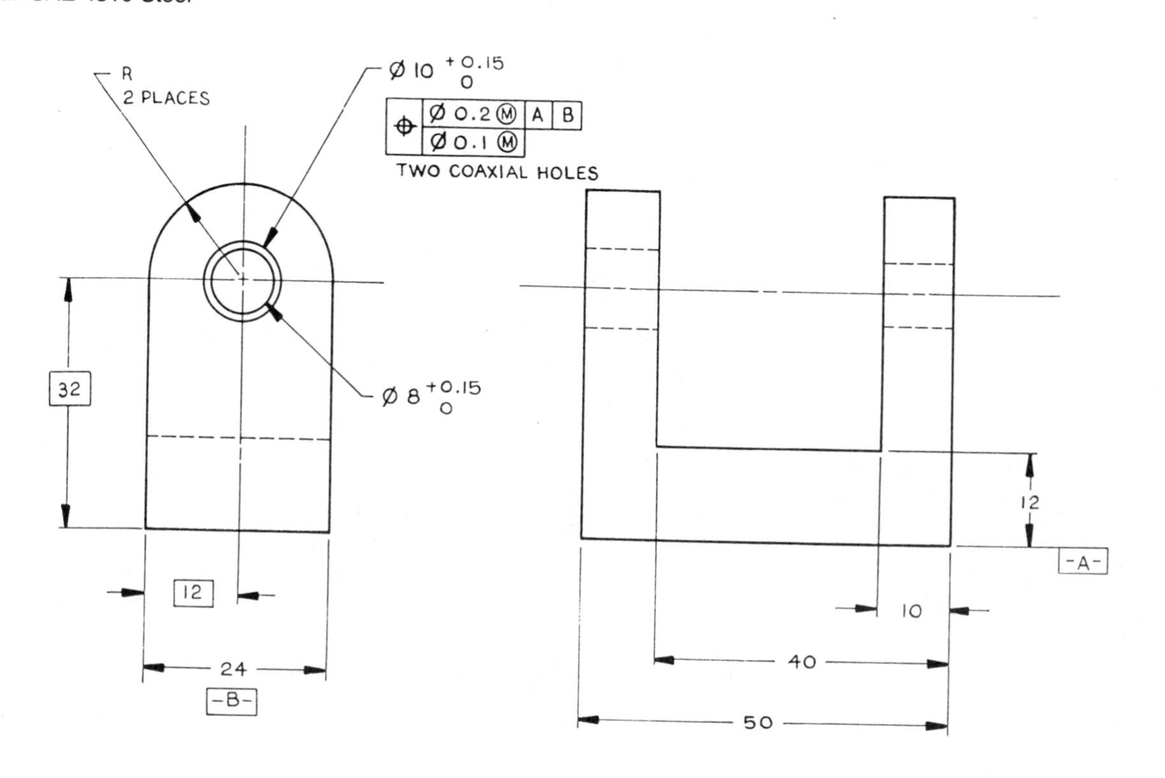

Problem 11–9 (in.)

Part Name: Thrust Washer
Material: SAE 5150

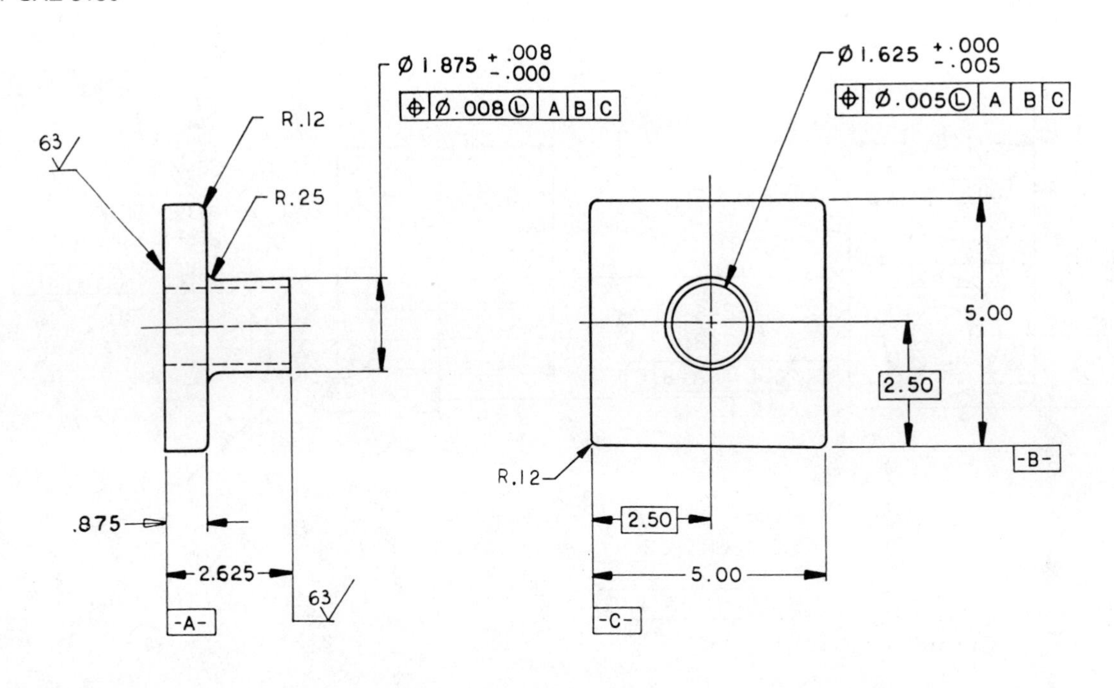

Problem 11–10 (metric)

Part Name: Spacer
Material: SAE 4310

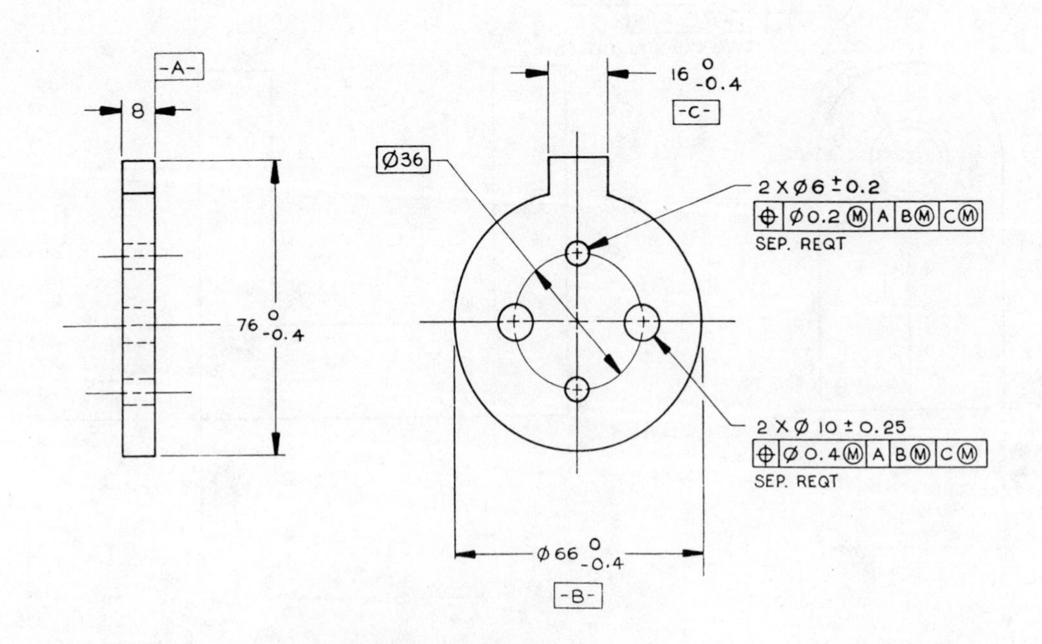

Problem 11–11 (metric)

Part Name: Bearing Support
Material: SAE 1040

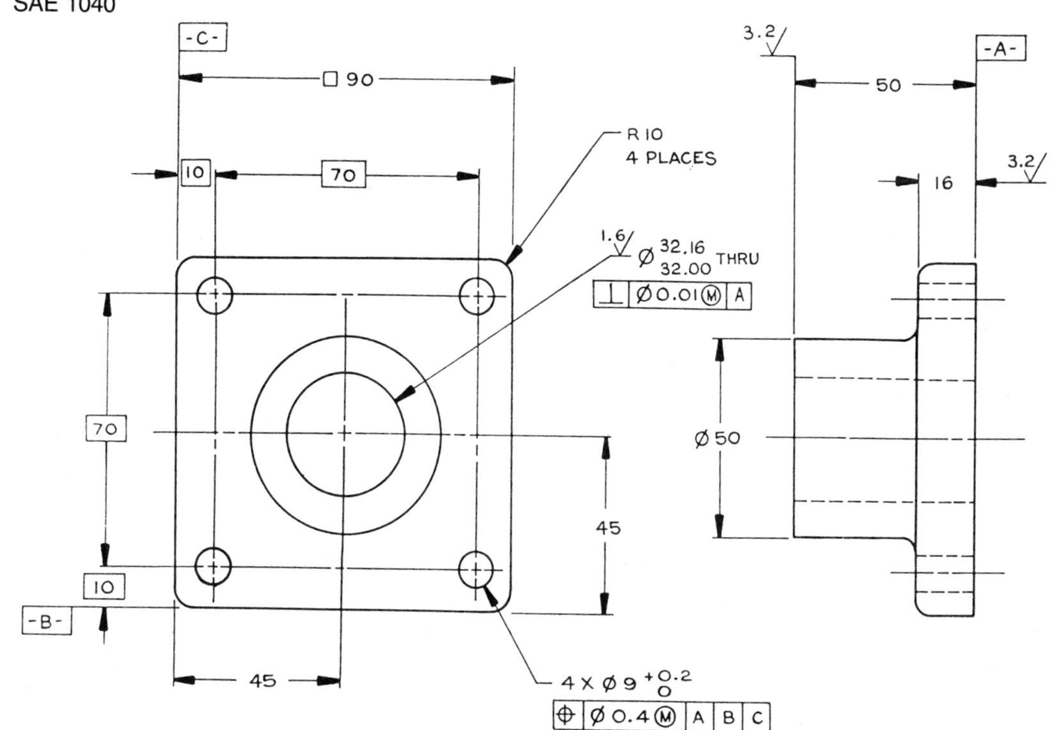

Problem 11–12 (in.)

Part Name: Slide Screw
Material: SAE 4320

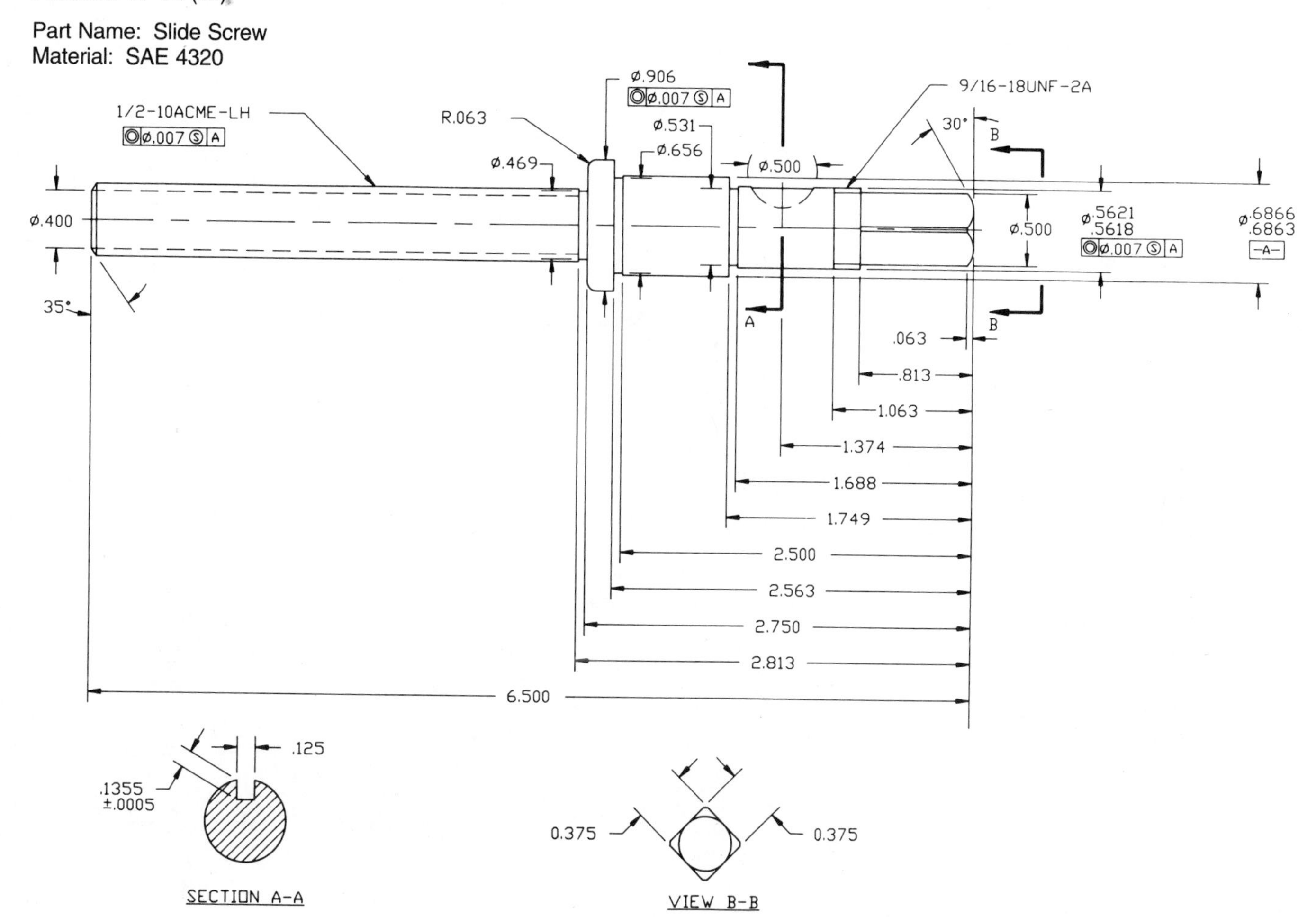

Problem 11–13 (metric)

Part Name: Lock Nut
Material: SAE 3130

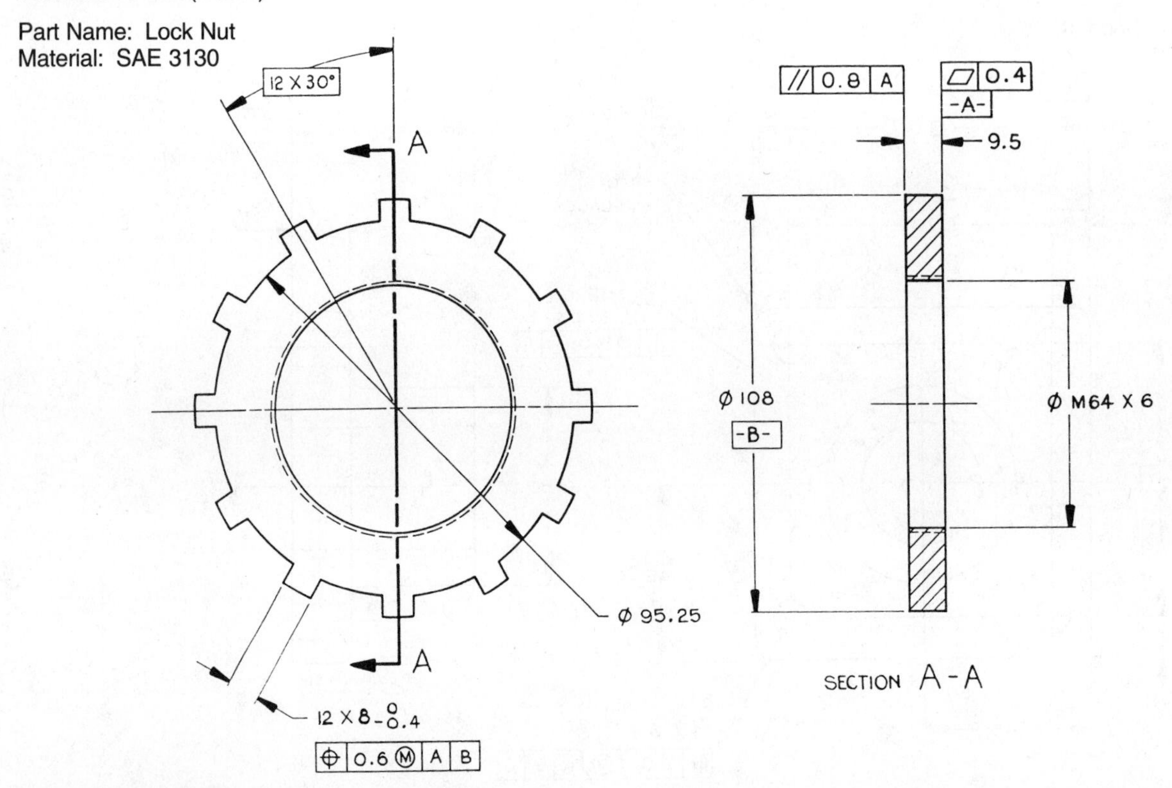

Problem 11–14 (in.)

Part Name: Cover Plate
Material: Phosphor Bronze

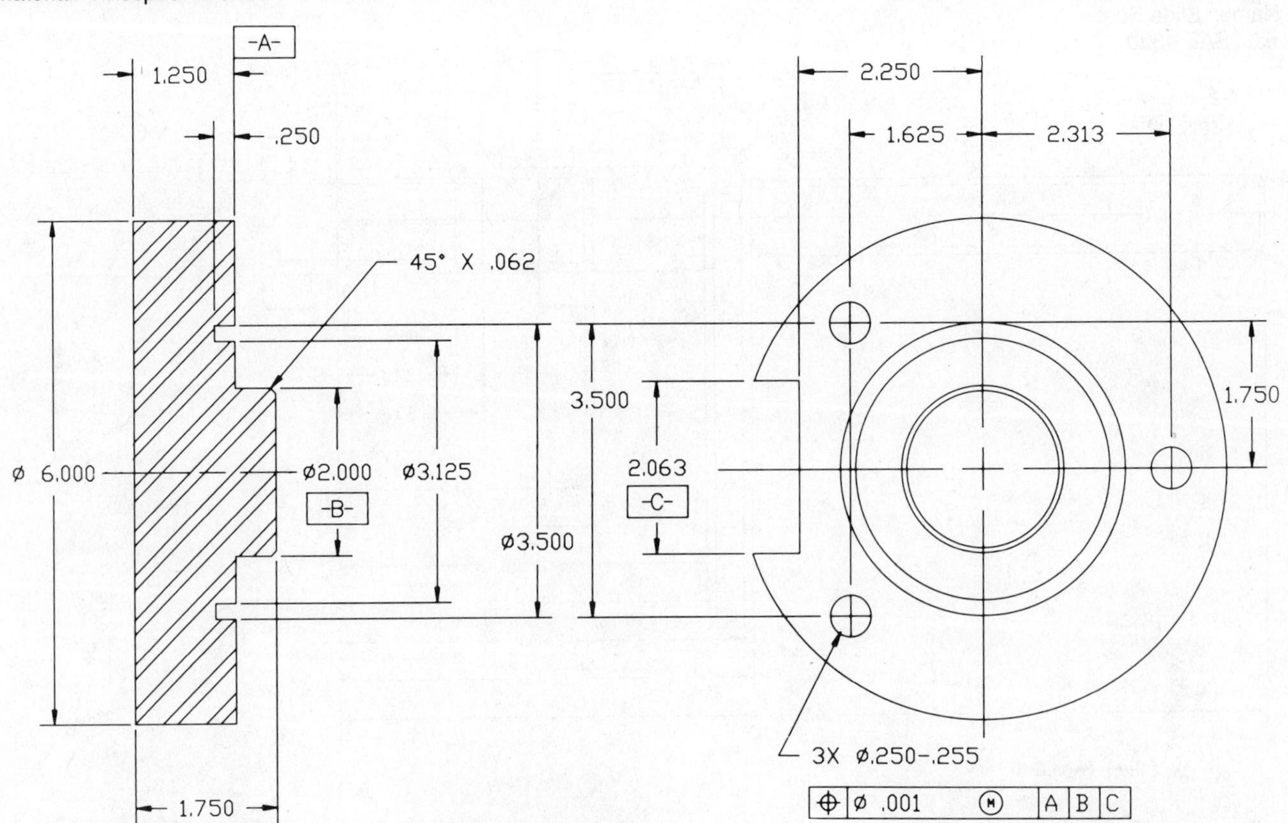

NOTE: ALL UNTOLERANCED DIMENSIONS LOCATING TRUE POSITION ARE BASIC.

Problem 11–15 (in.)

Part Name: Angle Support Mounting
Material: SAE 3110

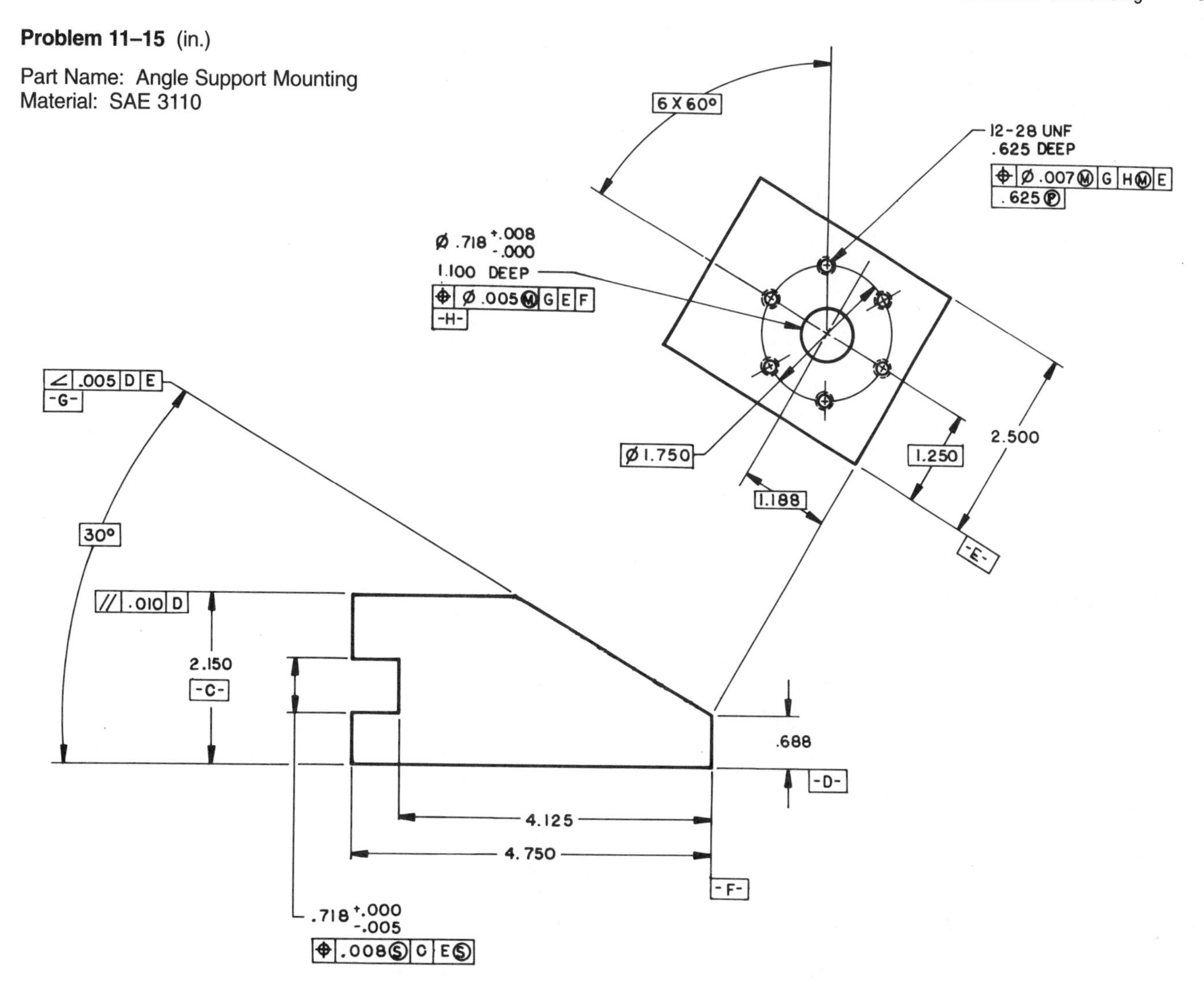

Problem 11–16 (metric)

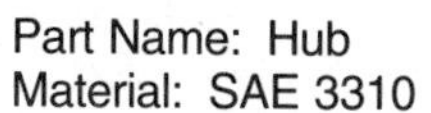

Part Name: Hub
Material: SAE 3310

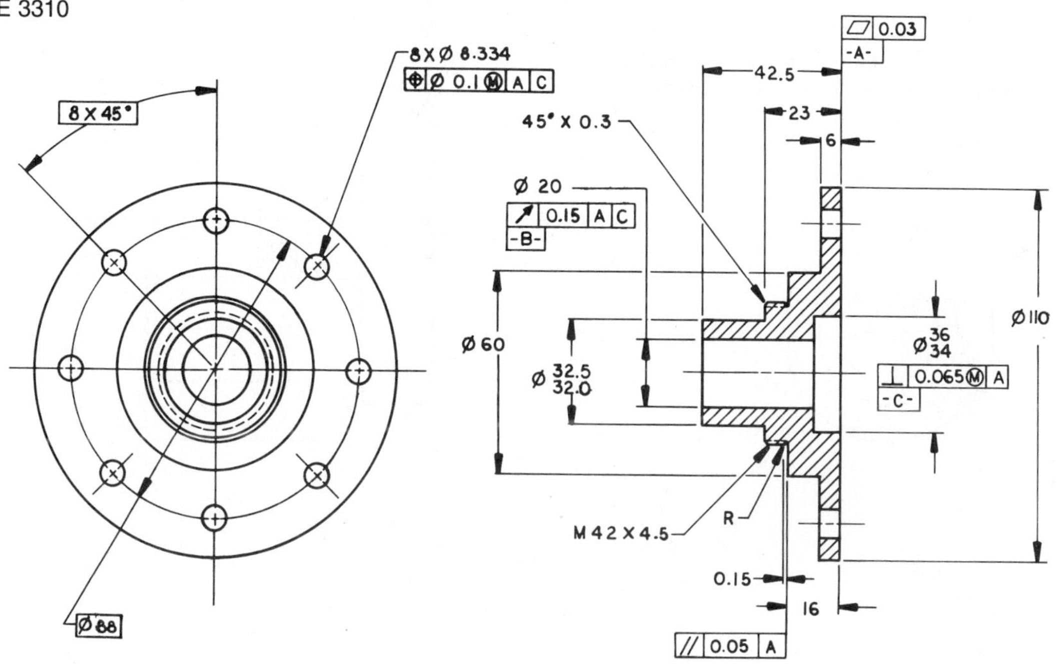

Problem 11–17 Geometric tolerancing (metric)

Part Name: Fixture MIBRDA—1265
Material: SAE 4320
Harden: Brinell 200–240
Additional General Notes:
1. Finish All Over 0.80 μm.

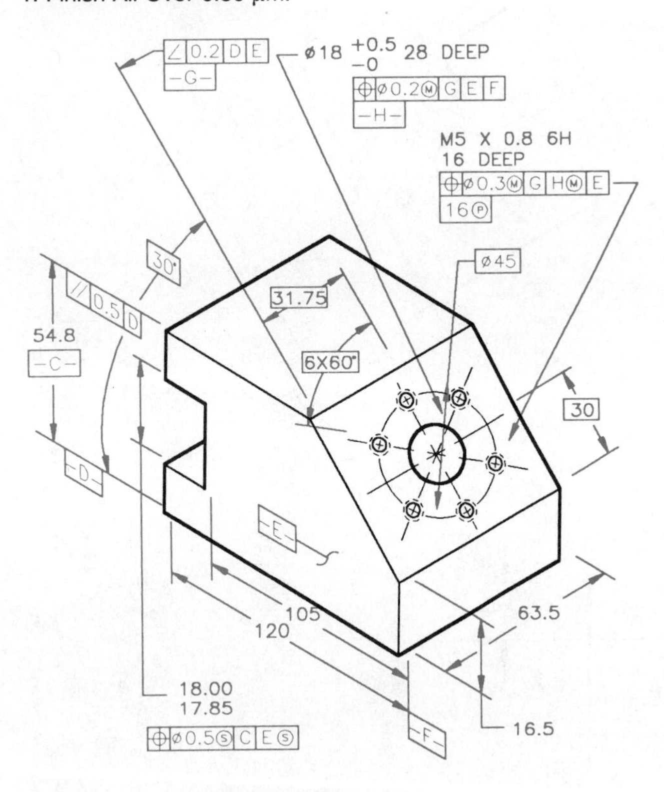

Problem 11–18 Geometric tolerancing (metric)

Part Name: Mounting Bracket
Material: Stainless Steel
Additional General Notes:
1. All Fillets and Rounds R 24.
2. Finish All Over 1.6 μm.

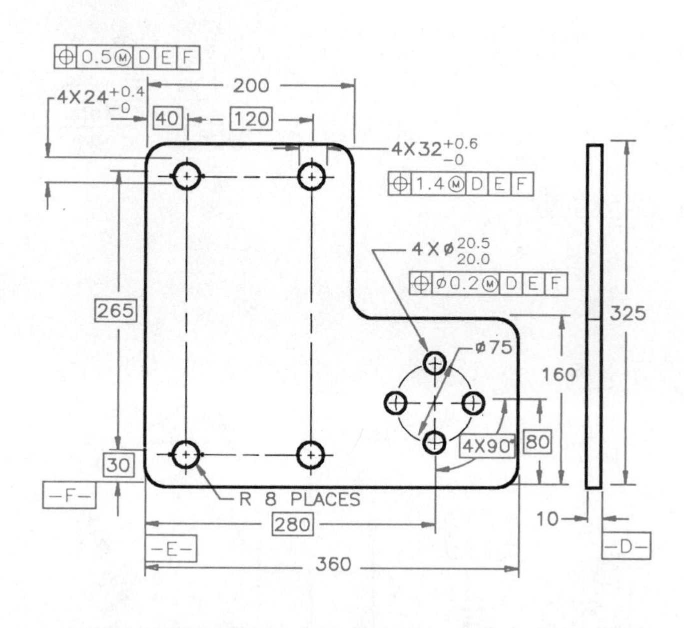

Problem 11–19 Geometric tolerancing (metric)

Part Name: Oscillator Housing
Material: Phosphor Bronze
Additional General Notes:
1. Finish All Over 0.80 μm.
SPECIFIC INSTRUCTIONS:
Use the following engineer's notes to complete the geometric tolerancing of the Oscillator housing:

- Establish datum A with three equally spaced datum target points at each end of the ∅ 28.1–28.0 cylinder.
- Establish datum B at the left end surface.
- Make the bottom surfaces of the 2X ∅40.2–40.0 features perpendicular to datum A by 0.06.
- Provide a cylindricity tolerance of 0.3 to the outside of the part.
- Make the 2X ∅40.2–40.0 features concentric to datum A by 0.1.
- Locate the 6X ∅6+0.2 holes with reference to datum A at MMC and datum B with a position tolerance of 0.05 at MMC.
- Locate the 4X ∅4+0.2 holes with reference to datum A at MMC and datum B with a position tolerance of 0.04 at MMC.

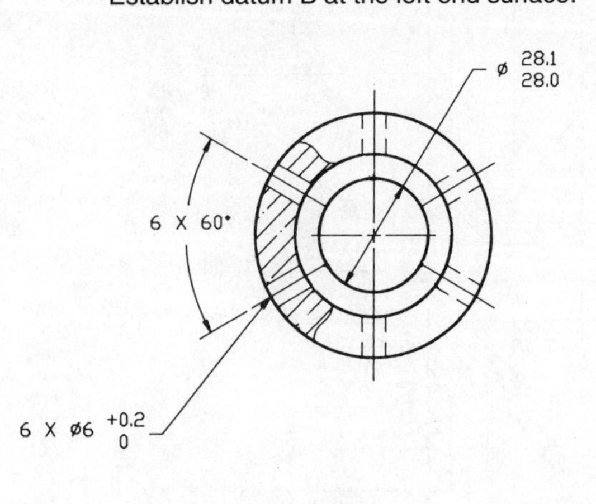

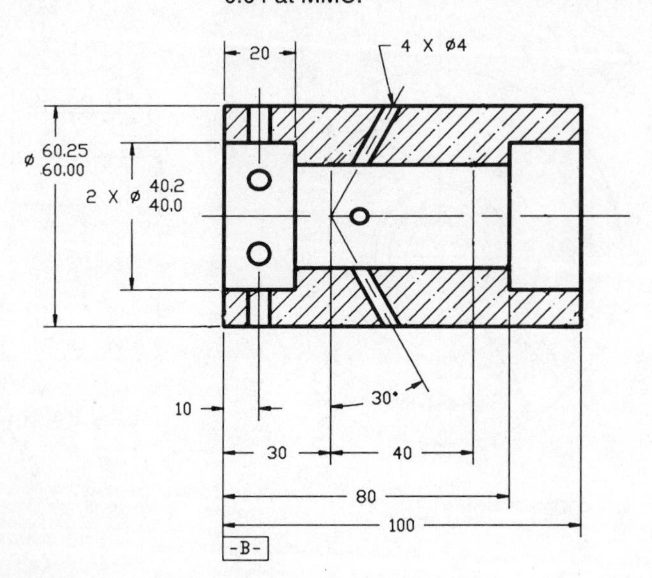

Problem 11–20 Geometric tolerancing (metric)

Part Name: Pinion Gear Shaft
Material: CRES 15–5PH ASTM A564
Additional General Notes:
1. Finish All Over 1.6 μm.
2. Heat Treat Per Mil–H–6875 to H1100 Condition.
3. Penetrant Inspect Finished Part Per Mil–Std–271, Group III. No Evidence of Linear Indications Permitted.
4. Part to be Clean and Free of Foreign Debris.

Courtesy Aerojet Techsystems Co.

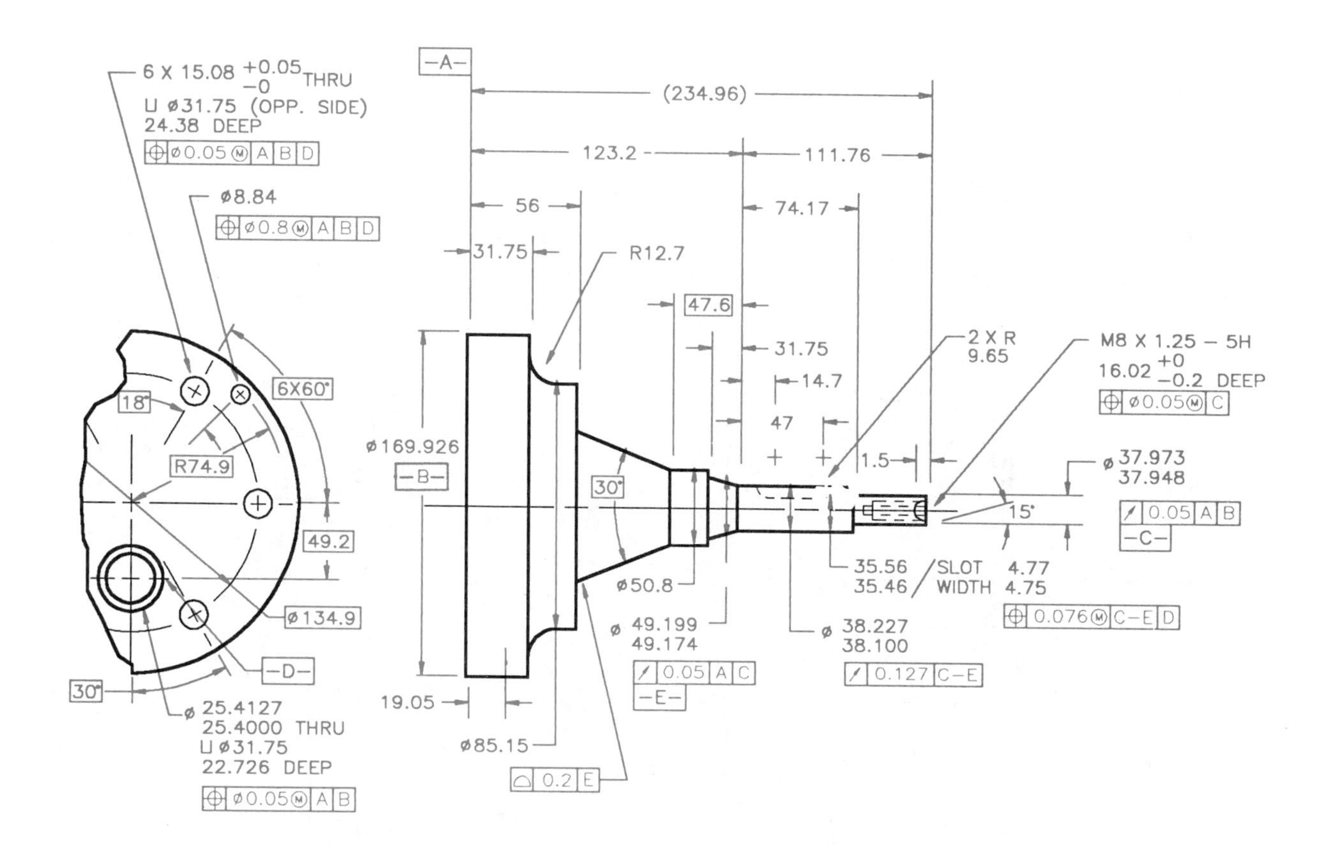

Problem 11–21 (in.)

Part Name: Slide
Material: SAE 5140
Machining Drawing

SPECIFIC INSTRUCTIONS:
Refer to the drawing for Problem 8–33 to determine the forging dimensions. Phantom lines show machining allowance.

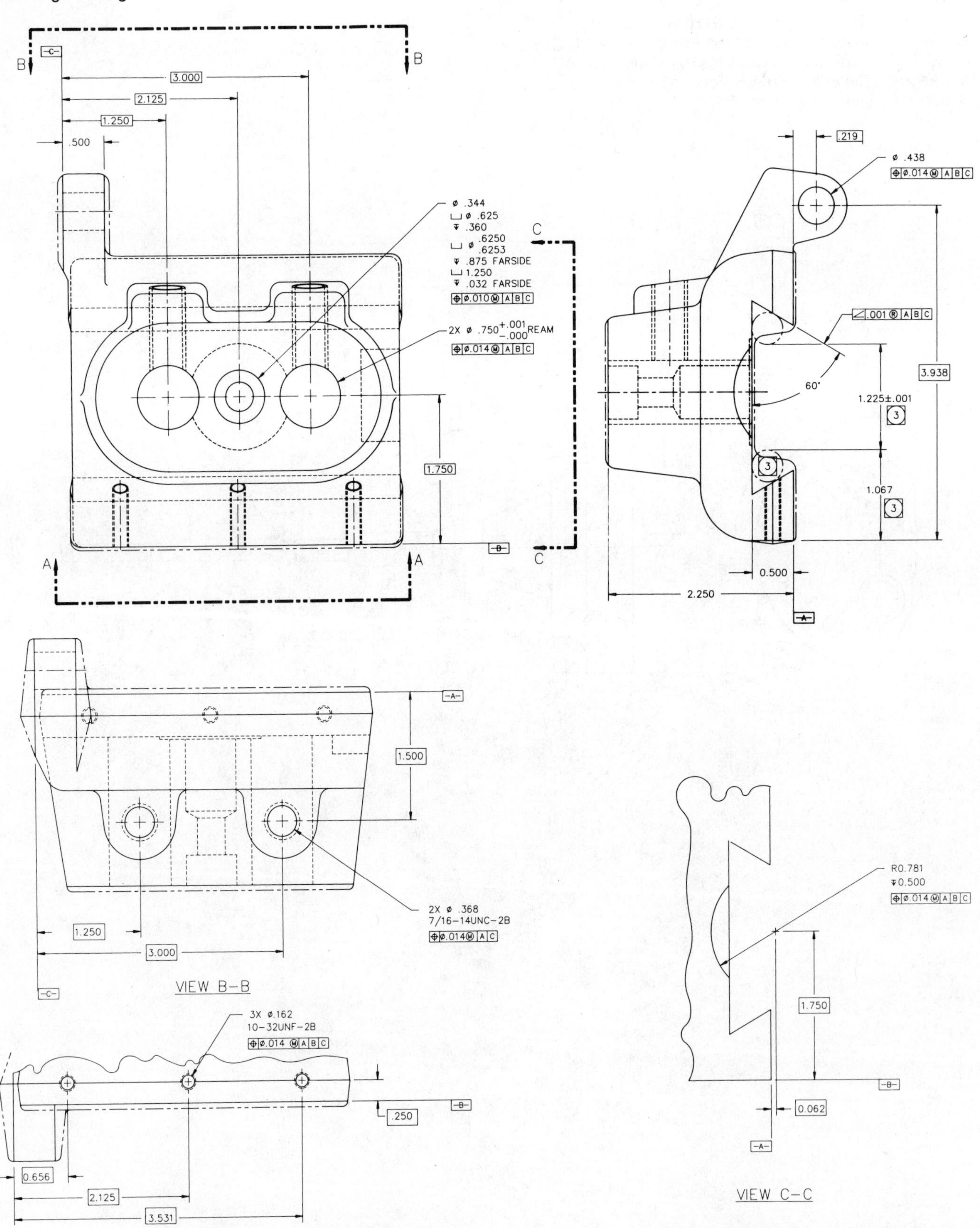

Problem 11–22 (in.)

Part Name: Hub
Material: SAE 4320

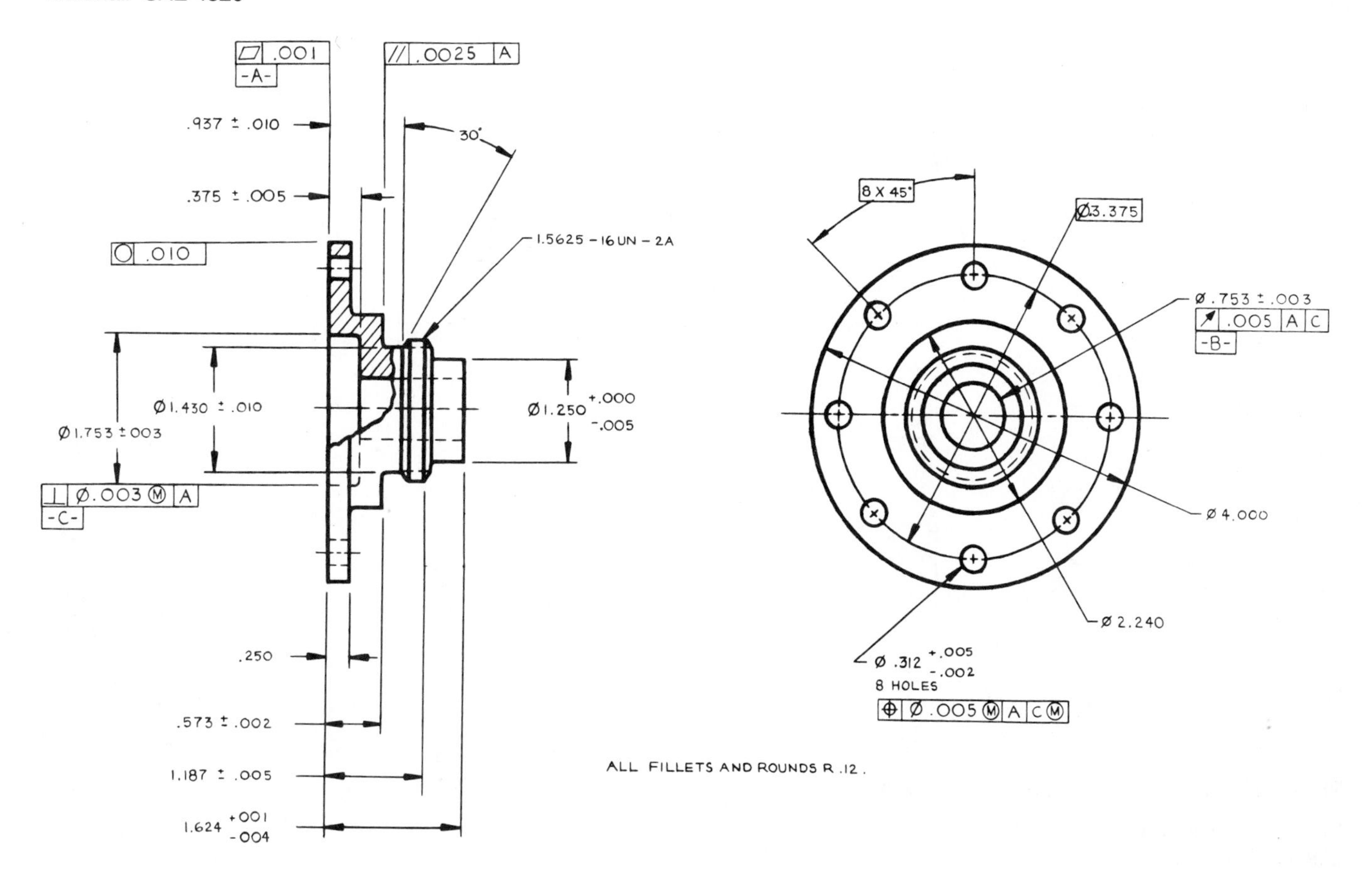

Problem 11–23 (in.)

Part Name: Lock Spacer
Material: SAE 1030

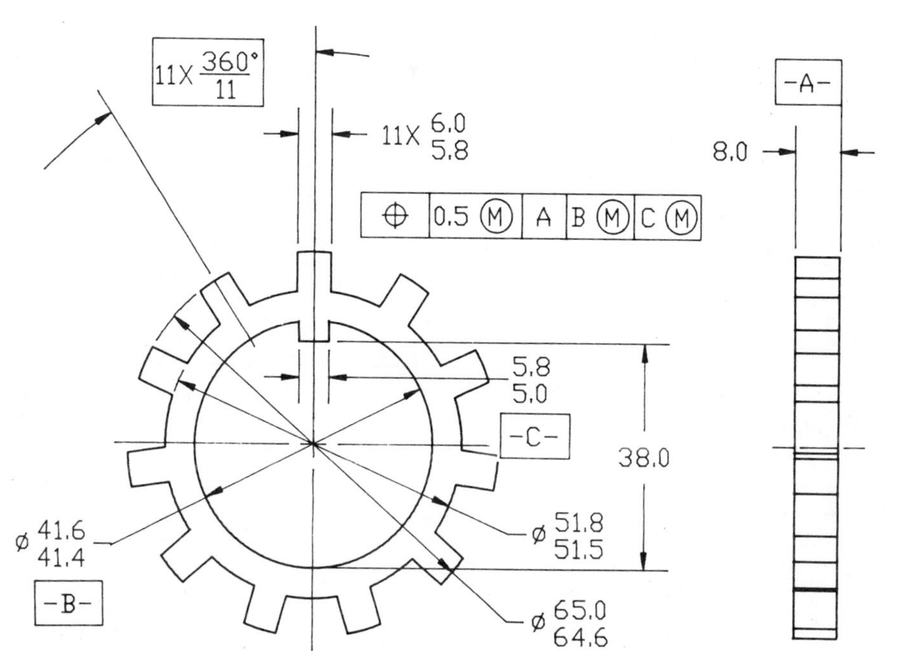

Problem 11–24 (metric)

Part Name: Mounting Plate
Material: SAE 4140

Ø60

Ø10 +0.2 / 0

⊥ Ø0.1 Ⓜ A

8 X Ø6.4 - 6.5

⌖ Ø0.25 Ⓜ A B Ⓜ

8 X ⌴ Ø9.6 - 9.8 ↧ 5.6 -6.0

⌖ Ø0.4 Ⓜ A B Ⓜ

8 × 45°

Ø 74.0 / 73.5

-B-

24 ± 0.5

-A-

FINISH ALL OVER 0.80 μM

Problem 11–25 Geometric tolerancing (metric)

Part Name: Side Panel Mounting Plate
Material: SAE 30308

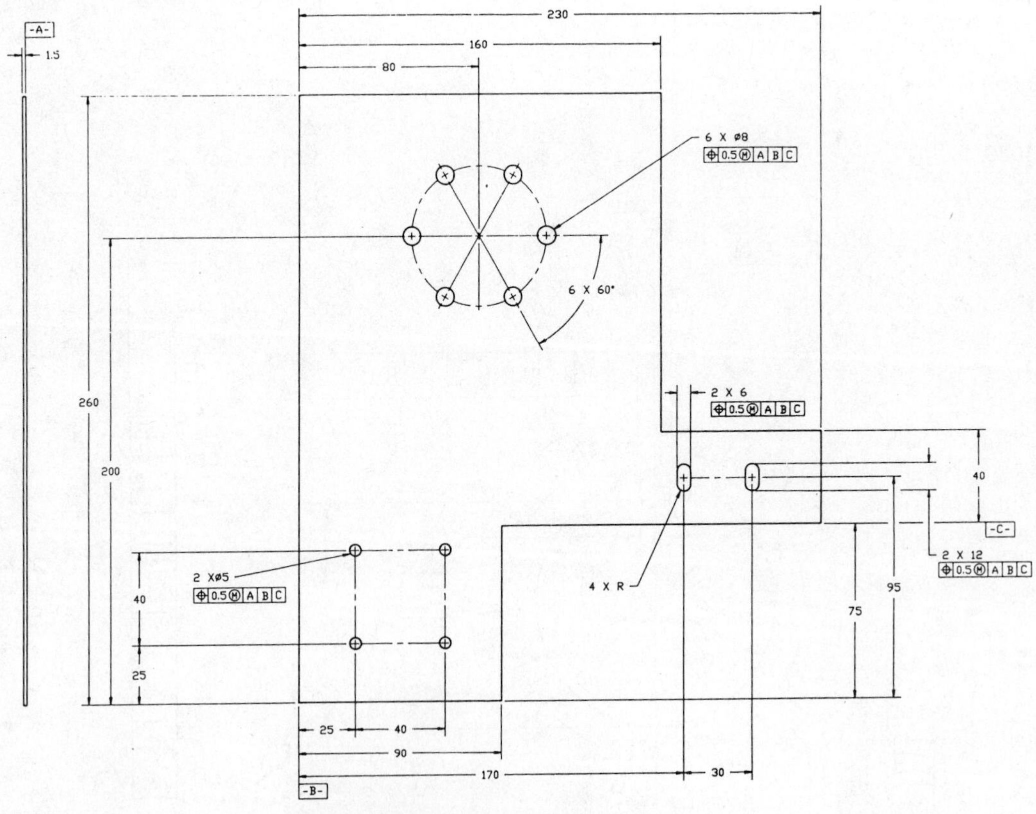

FINISH ALL OVER 0.50 μM

CHAPTER 12

Cams, Gears, and Bearings

LEARNING OBJECTIVES

After completing this chapter, you will:

- Create cam displacement diagrams.
- Design cam profile drawings from previously drawn cam displacement diagrams.
- Make detail gear drawings using simplified representations and gear data charts.
- Establish unknown data for gear trains.
- Calculate bearing information from specifications.
- Design a complete gear reducer from engineering data and sketches.

MECHANISMS

A *mechanism* is an arrangement of parts in a mechanical device or machine. This chapter deals with the design and drafting of elements of a mechanism, including cams, gears, and bearings.

CAMS

A *cam* is a rotating mechanism that is used to convert rotary motion into a corresponding straight motion. The timing involved in the rotary motion is often the main design element of the cam. For example, a cam may be designed to make a follower rise a given amount in a given degree of rotation, then remain constant for an additional period of rotation, and finally fall back to the beginning in the last degree of rotation. The total movement of the cam follower happens in one 360° rotation of the cam. This movement is referred to as the displacement. Cams are generally in the shape of irregular plates, grooved plates, or grooved cylinders. The basic components of the cam mechanism are shown in Figure 12–1.

Cam Types

There are basically three different types of cams: the plate cam, face cam, and drum cam. (See Figure 12–2.) The plate cam is the most commonly used type of cam.

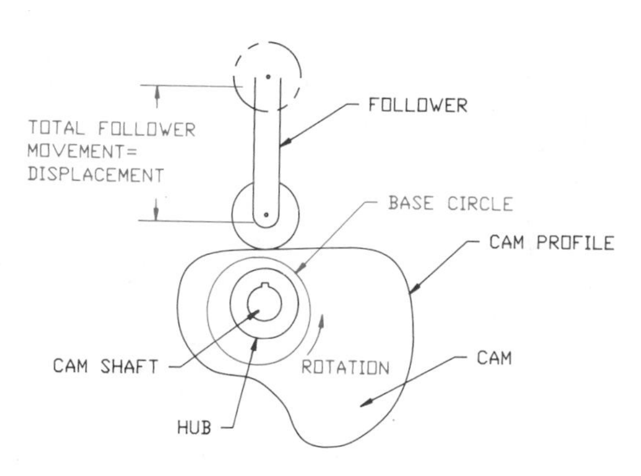

Figure 12–1 Elements of a cam mechanism.

Cam Followers

There are several types of cam followers. The type used depends on the application. The most common type of follower is the roller follower. The roller follower works well at high speeds, reduces friction and heat, and keeps wear to a minimum. The arrangement of the follower in relation to the cam shaft may differ depending on the application. The roller followers shown in Figure 12–3 include the in-line follower where the axis of the follower is in line with the cam shaft; the offset roller follower; and the pivoted follower. The pivoted follower requires spring tension to keep the follower in contact with the cam profile.

Another type of cam follower is the knife-edged follower shown in Figure 12–4(a). This follower is used for only low-speed and low-force applications. The knife-edged follower has a low resistance to wear, but is very responsive and may be effectively used in situations that require abrupt changes in the cam profile.

The flat-faced follower shown in Figure 12–4(b) is used in situations where the cam profile has a steep rise or fall. Designers often offset the axis of the follower. This practice causes the follower to rotate while in operation. This rotating action allows the follower surface to wear evenly and last longer.

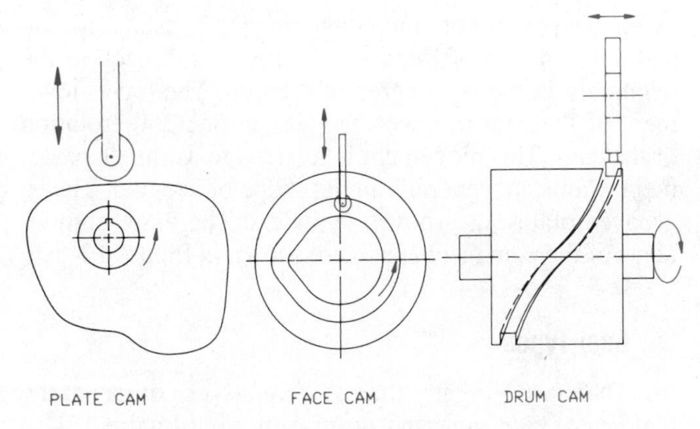

Figure 12–2 Types of cam mechanisms.

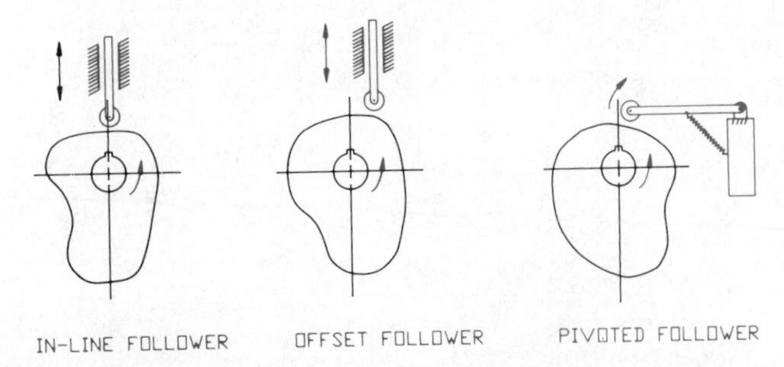

Figure 12–3 Types of cam roller followers.

CAM DISPLACEMENT DIAGRAMS

Cams are generally designed to achieve some type or sequence of a timing cycle in the movement of the follower. There are several predetermined types of motion from which cams are designed. These forms of motion may be used alone, in combination, or custom designed

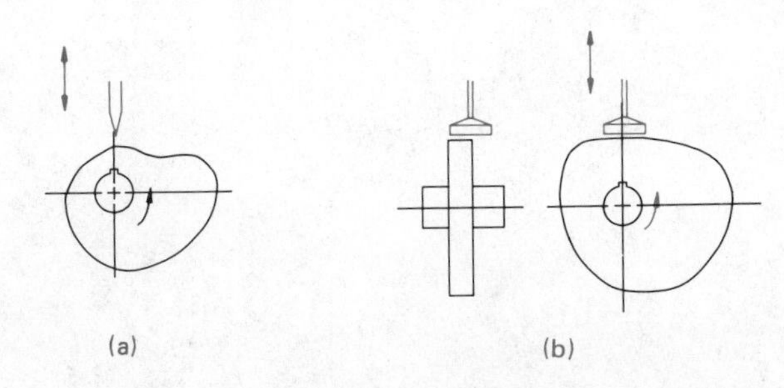

Figure 12–4 (a) Knife-edge cam follower; (b) flat-faced cam follower.

CAMS AND GEARS

Designing cams and gears is easy with CADD if a parametric program is used. In a system of this type all you have to do is change the variables and automatically create a new cam or gear. For cam design the variables are:

- Type of cam motion
- Follower displacement
- The specific rise, dwell, and fall configuration
- Prime circle and base circle diameters
- In-line or offset follower
- Hub diameter, hub projection, face width, shaft and keyway specifications

In gear design the variables are:

- Type of gear
- Pitch diameter, base circle diameter, pressure angle
- Diametral pitch
- Hub diameter, hub projection, face width, shaft and keyway specifications

The program automatically calculates the rest of the data and draws the cam profile or a detailed drawing of the gear in simplified or detailed representation.

to suit specific applications. The following discussion shows you how to set up a cam displacement diagram given a specific type of cam motion. The cam displacement diagram is similar to a graph representing the cam profile in a flat pattern of one complete 360° revolution of the cam. The terms associated with the displacement diagram include *cycle*, *period*, *rise*, *fall*, *dwell*, and *displacement*. A complete cam cycle has taken place when the cam rotates 360°. A period of the cam cycle is a segment of follower operation such as rise, dwell, or fall. Rise exists when the cam is rotating and the follower is moving upward. Fall is when the follower is moving downward. Dwell exists when the follower is constant, not moving either up or down. A dwell is shown in the displacement diagram as a horizontal line for a given increment of degrees. When developing the cam displacement diagram, the height of the diagram is drawn to scale and is equal to the total follower displacement. (See Figure 12–5.) The horizontal scale is equal to one cam revolution or 360°. The horizontal scale may be drawn without scale. Some engineering drafters prefer to make this scale equal to the circumference of the base circle, or any convenient length. The base circle is an imaginary circle with its center at the center of the cam shaft and its radius tangent to the cam follower at zero position. The horizontal scale is then divided into increments of degrees. Each rise and fall is divided into six increments. So, if the follower rises 150°, then each increment is 150°/6 = 25°. If the cam falls between 180° and 360°, this represents a total fall of 180°. The fall increments are 180°/6 = 30° as shown in Figure 12–5.

Simple Harmonic Motion

Simple harmonic motion may be used for high-speed applications if the rise and fall are equal at 180°. Moderate speeds are recommended if the rise and fall are unequal or if there is a dwell in the cycle. This application causes the follower to jump if the speeds are too high.

Draw a cam displacement diagram using simple harmonic motion when the total displacement is 2.00 in. and the cam follower rises the total displacement in 180° and falls back to 0° in 180°. Use the following procedure to set up the displacement diagram:

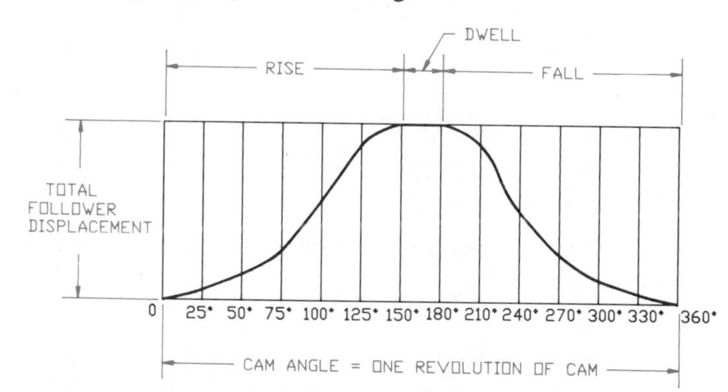

Figure 12–5 Cam displacement diagram.

Step 1 Draw a rectangle equal in height (vertical scale) to the total displacement of 2.00 in. and equal in length (horizontal scale) to 360°. The horizontal scale should have 6°–30° increments for the rise from 0° to 180° and 6°–30° increments for the fall from 180° to 360°. The horizontal scale may be any convenient length. Draw a thin vertical line from each horizontal increment as shown in Figure 12–6.

Step 2 Draw a half circle at one end of the displacement diagram equal in diameter to the rise of the cam. Divide the half circle into six equal parts as shown in Figure 12–7.

Step 3 The cam follower begins its rise at 0°. The rise continues by projecting point 1 on the half circle over to the first (30°) increment on the horizontal scale. Continue this process for points 2, 3, 4, 5, and 6 on the half circle, each intersecting the next increment on the horizontal scale as shown in Figure 12–8.

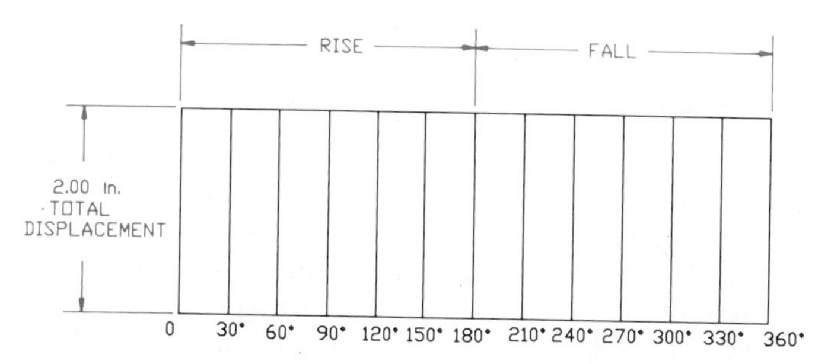

Figure 12–6 Layout for simple harmonic motion cam displacement diagram — step 1.

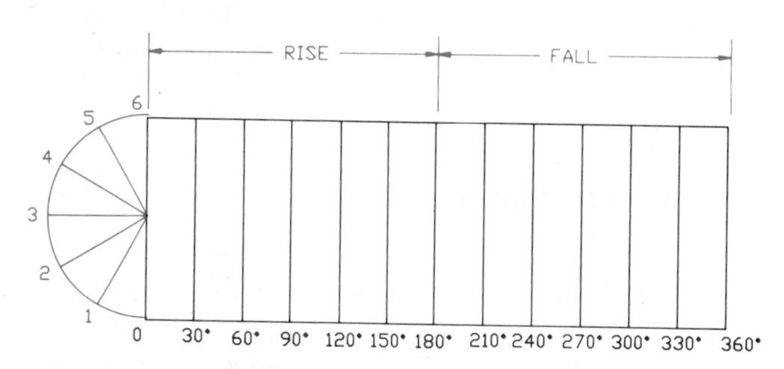

Figure 12–7 Layout for simple harmonic motion cam displacement diagram — step 2.

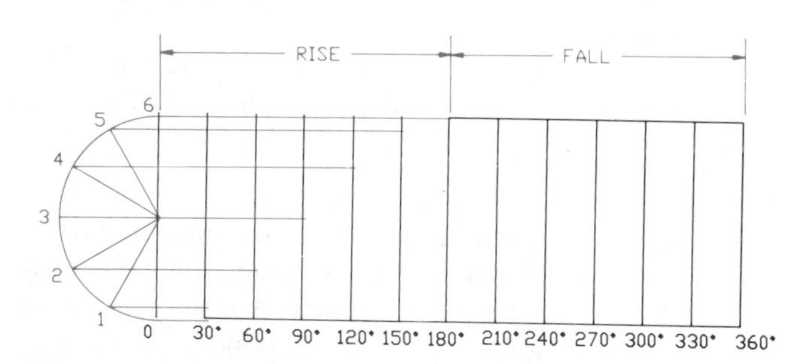

Figure 12–8 Layout for simple harmonic motion cam displacement diagram — step 3.

Step 4 Notice the pattern of points created in the preceding step. Use your irregular curve to carefully connect these points. If you are using CADD, use your curve-fitting command to draw the cam profile as shown in Figure 12–9.

Step 5 Develop the fall profile by projecting the points from the half circle in the reverse order discussed in Step 3. (See Figure 12–10.)

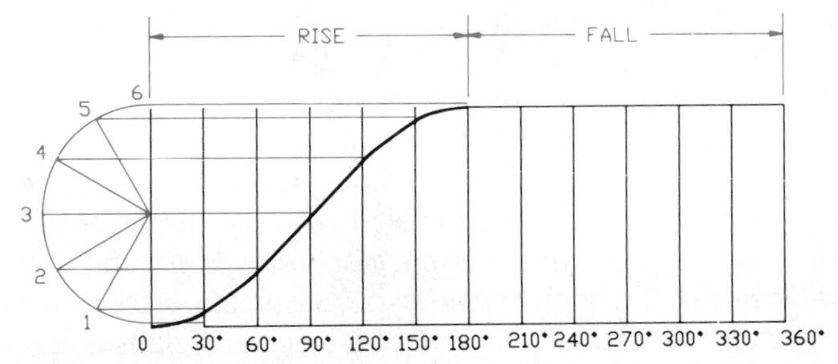

Figure 12–9 Layout for simple harmonic motion cam displacement diagram — step 4.

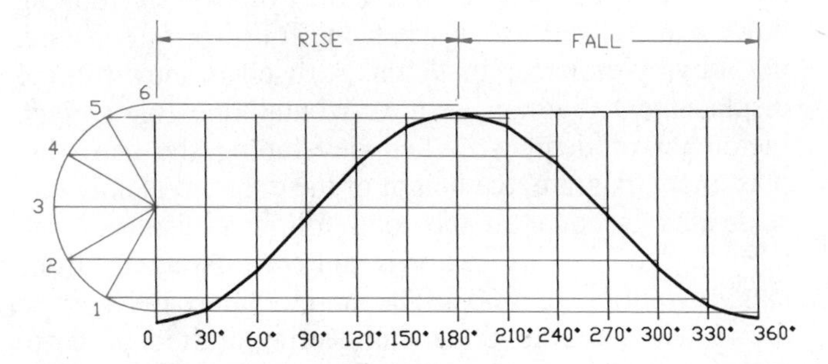

Figure 12–10 Layout for simple harmonic motion cam displacement diagram — step 5.

Constant Velocity Motion

Constant velocity motion is also known as straight-line motion. This is used for the feed control of some machine tools, when it is required for the follower to rise and fall at a uniform rate. Constant velocity motion is only used at slow speeds because of the abrupt change at the beginning and end of the motion period. The displacement diagram is easy to draw. All you have to do is draw a straight line from the beginning of the rise or fall to the end as shown in Figure 12–11.

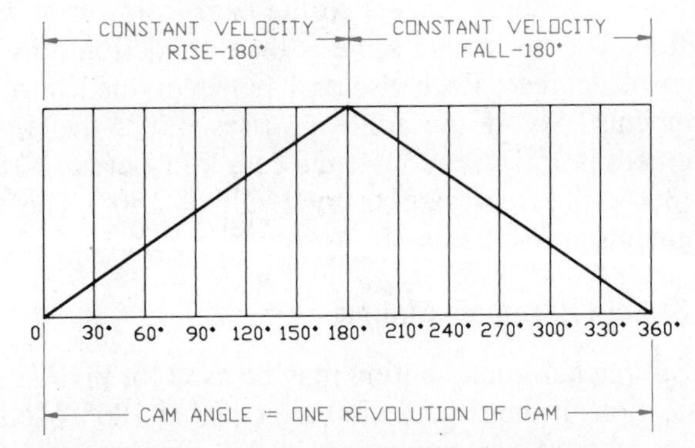

Figure 12–11 Cam displacement diagram for constant velocity motion.

Modified Constant Velocity Motion

Modified constant velocity motion was designed to help reduce the abrupt change at the beginning and end of the motion period. This type of motion may be adjusted to accomplish specific results by altering the degree of modification. This is done by placing a curve at the beginning and end of the rise and fall. The radius of this arc depends on the amount of smoothing required, but the radius normally ranges from one-third to full displacement. If the motion were modified to one-third the displacement, then the cam displacement diagram would be drawn as shown in Figure 12–12.

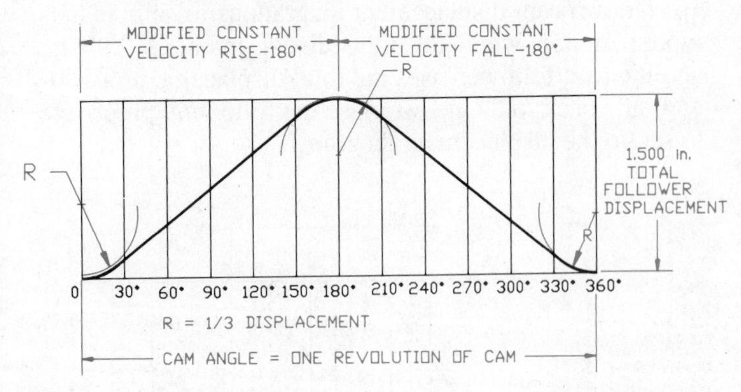

Figure 12–12 Cam displacement program for modified constant velocity motion.

Uniform Accelerated Motion

Uniform accelerated motion is designed to reduce the abrupt change at the beginning and end of a period. It is recommended for moderate speeds, especially when associated with a dwell. The advantage of this motion is its use when constant acceleration for the first half of the rise and constant deceleration for the second half of the rise are required.

Use the following technique to draw a uniform accelerated motion displacement diagram where the follower rises a total of 2.00 in. in 180° and falls back to 0° in 180°:

Step 1 Set up the displacement diagram with the height equal to 2.00 in. total rise and the horizontal scale divided into 30° increments as shown in Figure 12–13. Keep in mind that the horizontal scale is divided into 30° increments if the rise and fall is in 180° each. If the rise, for example, were 120°, then the increments would be 120°/6 = 20° each.

Step 2 Set up a scale with 18 equal divisions at one end of the displacement diagram and mark off the first, fourth, ninth, fourteenth, and seventeenth divisions as shown in Figure 12–14.

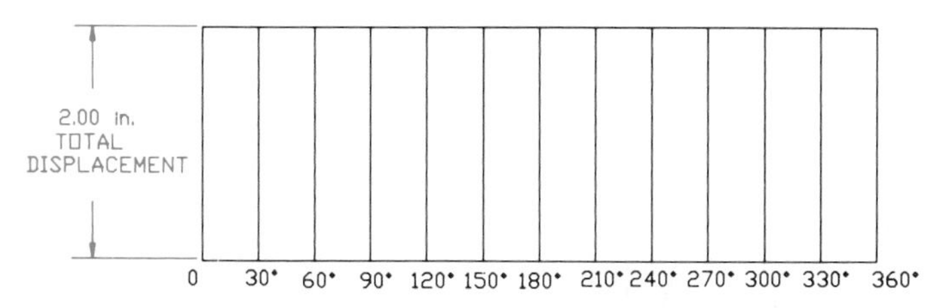

Figure 12–13 Layout for uniform accelerated motion cam displacement diagram — step 1.

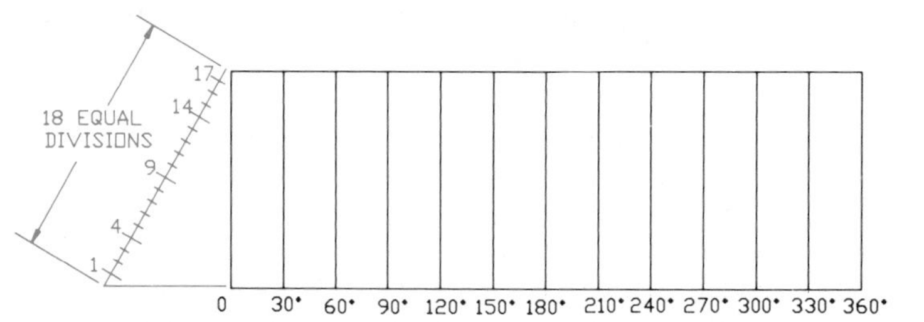

Figure 12–14 Layout for uniform accelerated motion cam displacement diagram — step 2.

Step 3 Establish the rise by projecting from the first division on the scale to the 30° increment on the diagram, then continue with the fourth division to 60°, and so on until each division is used. Continue this same procedure in reverse order to establish the profile of the fall. Connect all of the points to complete the displacement diagram as shown in Figure 12–15.

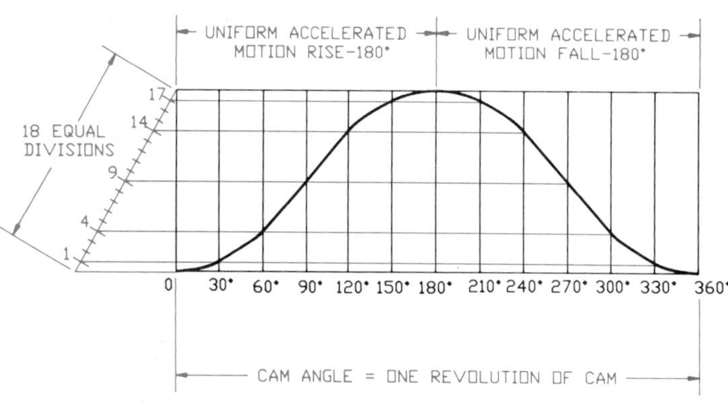

Figure 12–15 Layout for uniform accelerated motion cam displacement diagram — step 3.

Cycloidal Motion

Cycloidal motion is the most popular cam profile development for smooth-running cams at high speeds. The term *cycloidal* comes from the word *cycloid.* A cycloid is a curved line generated by a point on the circumference of a circle as the circle rolls along a straight line. The cycloidal cam motion is developed in this same manner and the result is the smoothest possible cam profile. Cycloidal motion is a little more complex to set up than other types of motion. Use the following procedure to develop a cam displacement diagram for cycloidal motion with a total rise of 2.500 in. in 180°:

Step 1 Begin the displacement diagram with a total rise of 2.500 in. in 180°. Only half of the diagram is shown for this example.

Step 2 Draw a circle tangent to and centered on the total displacement at one end of the diagram. This circle must have a circumference equal to the displacement. Calculate the diameter using the formula $D = C/\pi$. In this case $D = 2.500/3.1414 = 0.7958$. Round off to 0.80 for manual drafting or use as is for a CADD application. Then beginning at the point where the circle is tangent to the diagram, divide the circle into six equal parts and number them as shown in Figure 12–16.

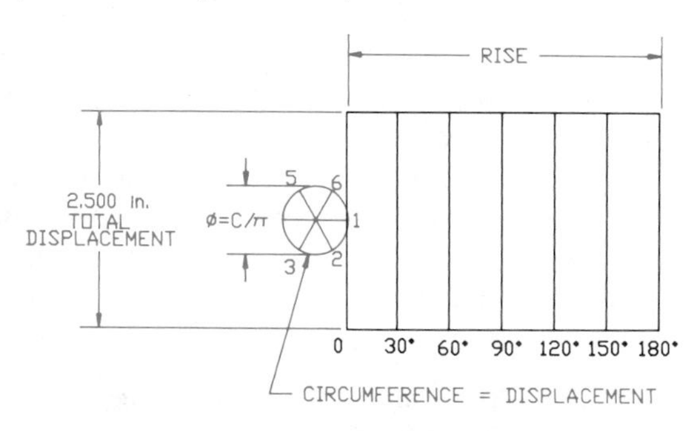

Figure 12–16 Layout for cycloidal motion cam displacement diagram — steps 1 and 2.

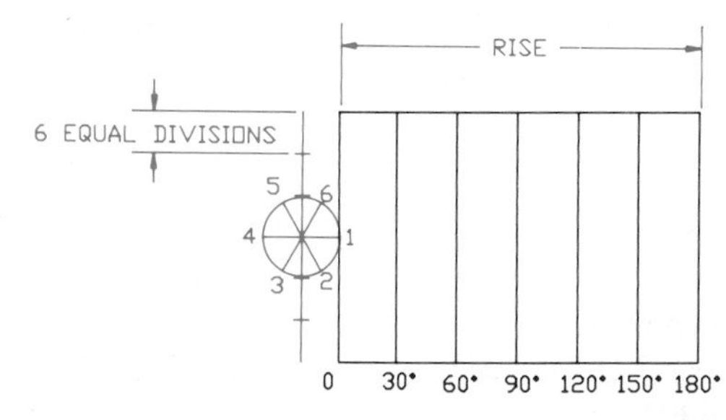

Figure 12–17 Layout for cycloidal motion cam displacement diagram — step 3.

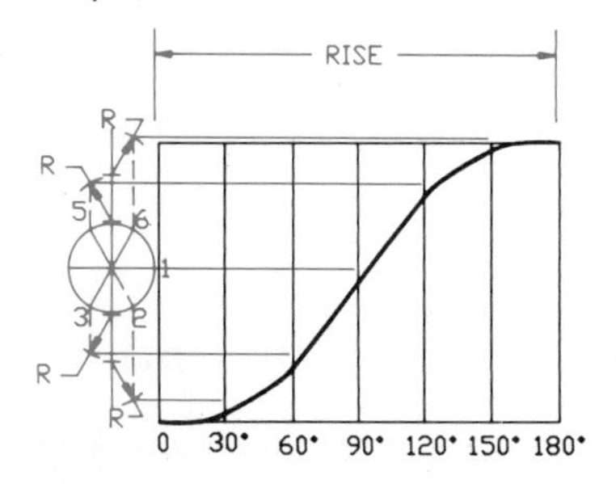

Figure 12–18 Layout for cycloidal motion cam displacement diagram — step 4.

Step 3 Draw a vertical line through the center of the circle equal in length to the displacement. Divide this line into six equal parts as shown in Figure 12–17.

Step 4 Use a radius equal to the radius of the circle to draw arcs from the divisions on the vertical line as shown in Figure 12–18. These arcs should

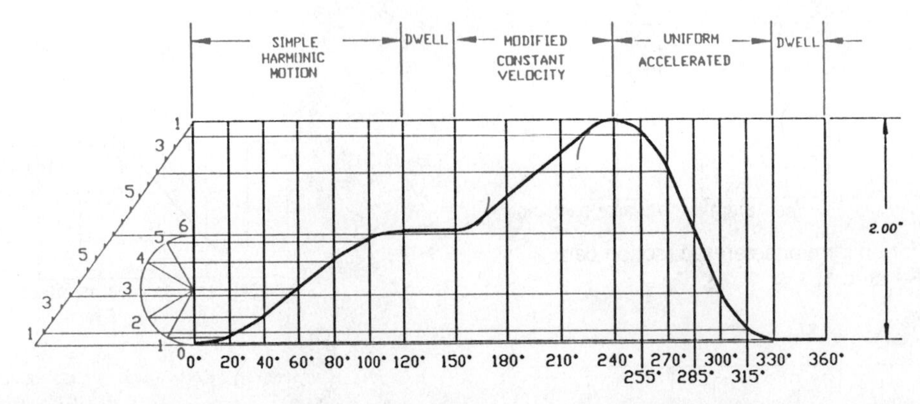

Figure 12–19 The development of a cam displacement diagram with different cam motions.

intersect the dashed lines drawn from points 2, 3, 5, and 6 on the circle. Where the dashed lines and the arcs intersect, draw horizontal lines into the displacement diagram, intersecting the appropriate increment from the horizontal scale. Connect the points of intersection with a smooth curve as shown in Figure 12–18.

Developing a Cam Displacement Diagram with Different Cam Motions

Most cam profiles are not as simple as the preceding examples. Many designs require more than one type of cam motion and may also incorporate dwell. Construct a cam displacement diagram from the following information:

- Total displacement equals 2.00 in.
- Rise 1.00 in. simple harmonic motion in 120°.
- Dwell for 30°.
- Rise 1.00 in. modified constant velocity motion in 90°.
- Fall 2.00 in. uniform accelerated motion in 90°.
- Dwell for 30° through the balance of the cycle.

Look at Figure 12–19 as you review the method of development for each type of cam motion displayed.

CONSTRUCTION OF AN IN-LINE FOLLOWER PLATE CAM PROFILE

Each of the cam displacement diagrams may be used to construct the related cam profiles. The cam profile is the actual contour of the cam. In operation, the cam follower is stationary and the cam rotates on the cam shaft. In cam profile construction, however, the cam is drawn in one position and the cam follower is moved to a series of positions around the cam in relationship to the cam displacement diagram. The following technique is used to draw the cam profile for the cam displacement diagram given in Figure 12–20 and for a cam with a 2.50 in. base circle, a 0.75 in. cam follower, and counterclockwise rotation:

Step 1 Refer to Figure 12–20. Use construction lines for all preliminary work. Draw the cam follower in place near the top of the sheet with phantom lines. Draw the base circle (2.50 in.). The base circle is tangent to the follower at the 0° position. Draw the prime circle 2.50 + .75 = 3.25 in. The prime circle passes through the center of the follower at the 0° position. The cam displacement diagram is placed on Figure 12–20 for easy reference. In actual practice, you may have the displacement diagram next to the cam profile

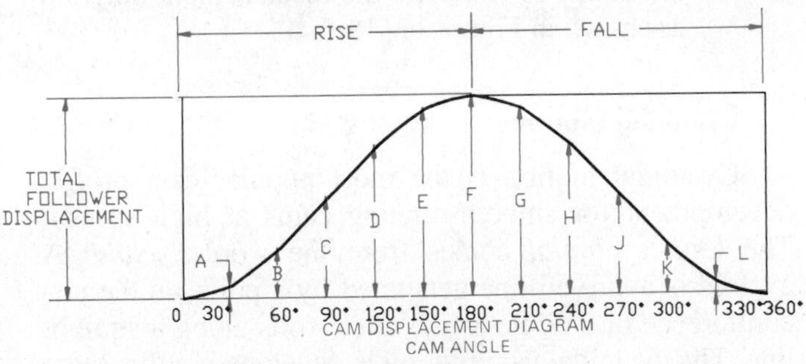

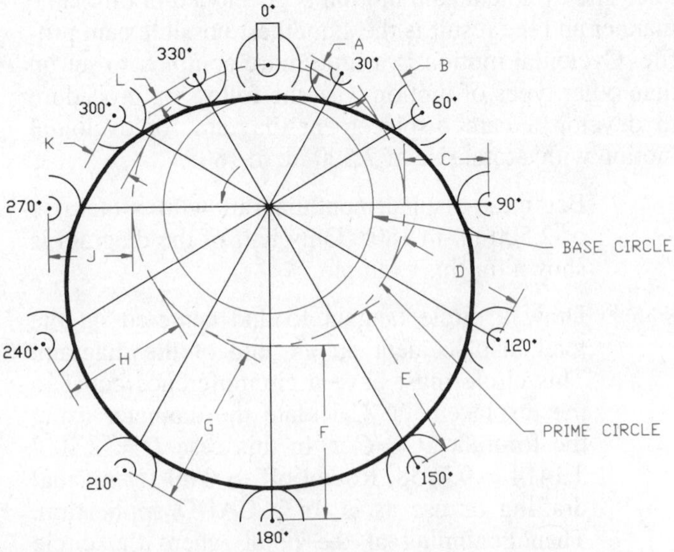

Figure 12–20 Construction of an in-line follower plate cam profile.

drawing or on a CADD layer to be inserted on the screen for reference.

Step 2 Begin working in a direction opposite from the rotation of the cam. Since this cam rotates counterclockwise, work clockwise. Starting at 0° draw an angle equal to each of the horizontal scale angles on the displacement diagram. In this case the angles are 30° increments. This is not true with all cam displacement diagrams. Some have varying increments.

Step 3 Notice the measurements labeled A through L on the displacement diagram in Figure 12–20. Begin by transferring the distance A along the 30° element (in the profile construction drawing) by measuring from the prime circle along this line. This establishes the center of the cam follower at this position. Do the same with each of the measurements B through L located on each corresponding increment.

Step 4 Lightly draw the cam followers in position with centers located at each of the points found in Step 3. (See Figure 12–20.)

Step 5 Draw the cam profile by connecting a smooth curve tangent to the cam followers at each position. This is one place where CADD drafting plays an important role regarding accuracy and speed.

Preparing the Formal Plate Cam Drawing

You are ready to prepare the formal plate cam drawing after you have constructed the plate cam profile using the previously described techniques. The previous drawing may be used because all lines were drawn as construction lines or on a CADD layer that may be turned off when the formal drawing is complete. The information needed on the plate cam drawing includes:

- Cam profile.
- Hub dimensions, including cam shaft, outside diameter, width, keyway dimensions.
- Roller follower placed in one convenient location, such as 60°, using phantom lines.
- The drawing is set up as a chart drawing where A° equals the angle of the follower at each position, and R equals the radius from the center of the cam shaft to the center of the follower at each position.
- A chart giving the values of the angles A and the radii R at each follower position.
- Side view showing the cam plate thickness, and set screw location with thread specification, if used.
- Tolerances, unless otherwise specified.

Establish all measurements for dimensions A and R at each of the follower positions. This may be done graphically by measuring from the profile construction, or mathematically using trigonometry. If a CADD system is

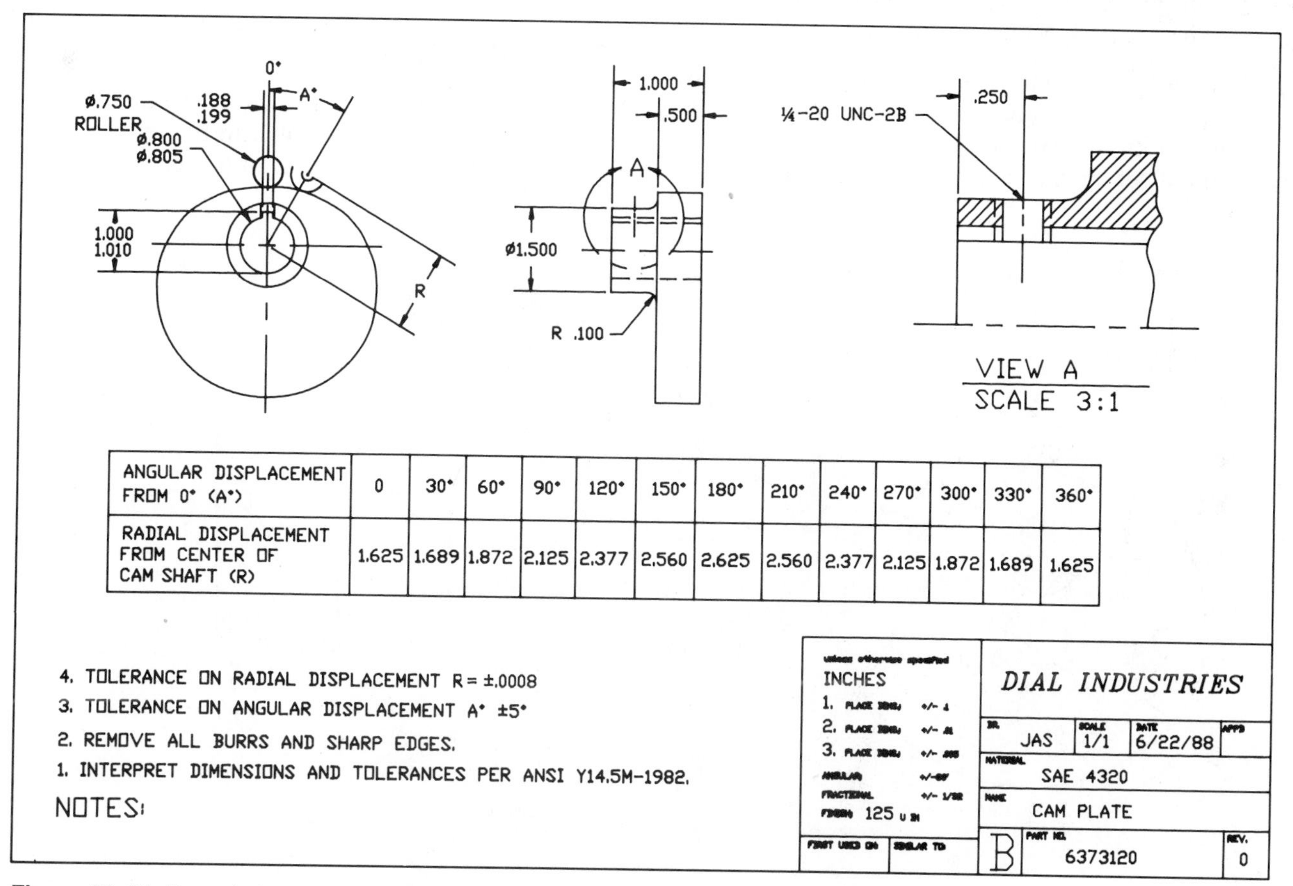

ANGULAR DISPLACEMENT FROM 0° (A°)	0	30°	60°	90°	120°	150°	180°	210°	240°	270°	300°	330°	360°
RADIAL DISPLACEMENT FROM CENTER OF CAM SHAFT (R)	1.625	1.689	1.872	2.125	2.377	2.560	2.625	2.560	2.377	2.125	1.872	1.689	1.625

Figure 12–21 Formal plate cam drawing.

used, the measurements may be taken directly from the layout. The accuracy of the CADD system is excellent. Figure 12–21 shows a formal plate cam drawing.

CONSTRUCTION OF AN OFFSET FOLLOWER PLATE CAM PROFILE

When an offset cam follower is used, the method of construction is a little more complex than the technique used for the in-line follower. For this example, the follower is offset 0.75 in. as shown in Figure 12–22. Prepare the cam profile drawing as follows:

Step 1 Refer to Figure 12–22. Use construction lines or a construction CADD layer for all preliminary work. Draw the cam follower in place near the top of the sheet with phantom lines. Draw the base circle (2.50 in. ∅). The base circle is tangent to the follower at 0° position. Draw the prime circle (2.50 + 0.75 = 3.25 in. ∅). The prime circle passes through the center of the follower at 0° position. Draw the offset circle (1.50 in. ∅). The offset circle is drawn with a radius equal to the follower offset distance. The cam displacement diagram is placed on Figure 12–22 for easy reference. In actual practice, you may have the displacement diagram next to the cam profile drawing or on a CADD layer to be inserted on the screen for reference.

Step 2 Begin working in a direction opposite from the rotation of the cam. Since this cam rotates counterclockwise, work clockwise. Starting at 0° on the offset circle, draw an angle equal to each of the horizontal scale angles on the displacement diagram. In this case the angles are 30° increments. This is not true with all cam displacement diagrams. Some have varying increments.

Step 3 Notice where each of the angle increments drawn in Step 2 intersects the offset circle. At each of these points, draw another line tangent to the offset circle. Make these lines long enough to pass beyond the reference circle.

Step 4 Notice the measurements labeled A through L on the displacement diagram in Figure 12–22. Begin by transferring the distance A along the line tangent to the 30° element by measuring from the prime circle along this line. This establishes the center of the cam follower at this position. Do the same with each of the measurements B through L located on the tangent line from each corresponding increment.

Step 5 Lightly draw the cam followers in position with centers located at each of the points found in Step 4. (Refer to Figure 12–22.)

Step 6 Draw the cam profile by connecting a smooth curve tangent to the cam followers at each position. This is one place where CADD drafting plays an important role regarding accuracy and speed.

DRUM CAM DRAWING

Drum cams are used when it is necessary for the follower to move in a path parallel to the axis of the cam. The drum cam is a cylinder with a groove machined in

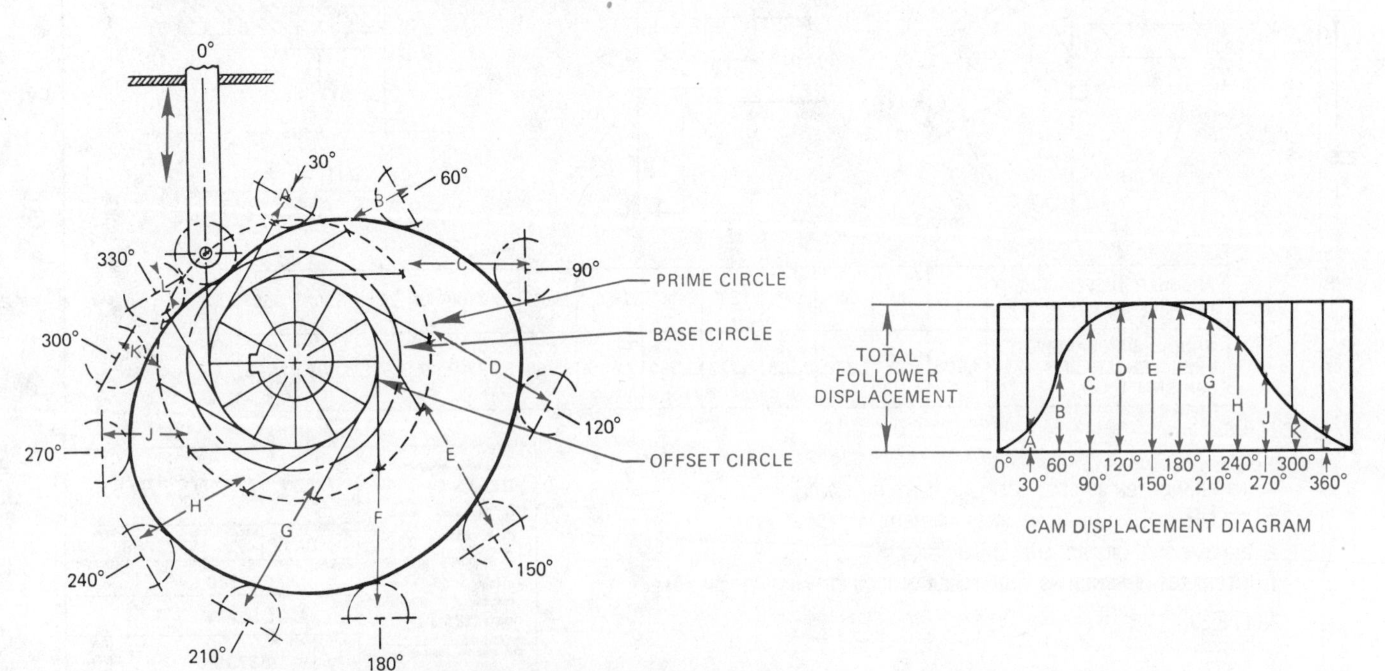

Figure 12–22 Construction of an offset follower plate cam profile.

the surface the shape of the cam profile. The cam follower moves along the path of the groove as the drum is rotated. The displacement diagram for a drum cam is actually the pattern of the drum surface as if it were rolled out flat. The height of the displacement diagram is equal to the height of the drum. The length of the displacement diagram is equal to the circumference of the drum. Refer to the drum cam drawing in Figure 12–23 as you follow the construction steps:

Step 1 Draw the top view showing the diameter of the drum, the cam shaft and keyway, and the roller follower in place at 0°. Draw the front view as shown in Figure 12–23. Draw the outline of the cam displacement diagram equal to the height and circumference of the drum.

Step 2 Draw the roller follower on the displacement diagram at each angular interval.

Step 3 Draw curves tangent to the top and bottom of each roller follower position. These curves represent the development of the groove in the surface of the cam.

Step 4 In the top view, draw radial lines from the cam shaft center equal to the angle increments shown on the displacement diagram. Be sure to lay out the angles in a direction opposite the cam rotation. The points where these lines intersect the depth and the outside circumference of the groove are labeled A, A_1, B, B_1, respectively. Notice that the same corresponding points are labeled on the displacement diagram.

Step 5 From points A, A_1 on the displacement diagram, project horizontally until each point intersects a vertical line from the same corresponding point in the top view. This establishes the points in the front view along the outer and inner edges of the groove, both top and bottom. Continue this process for each pair of points on the drum cam displacement diagram.

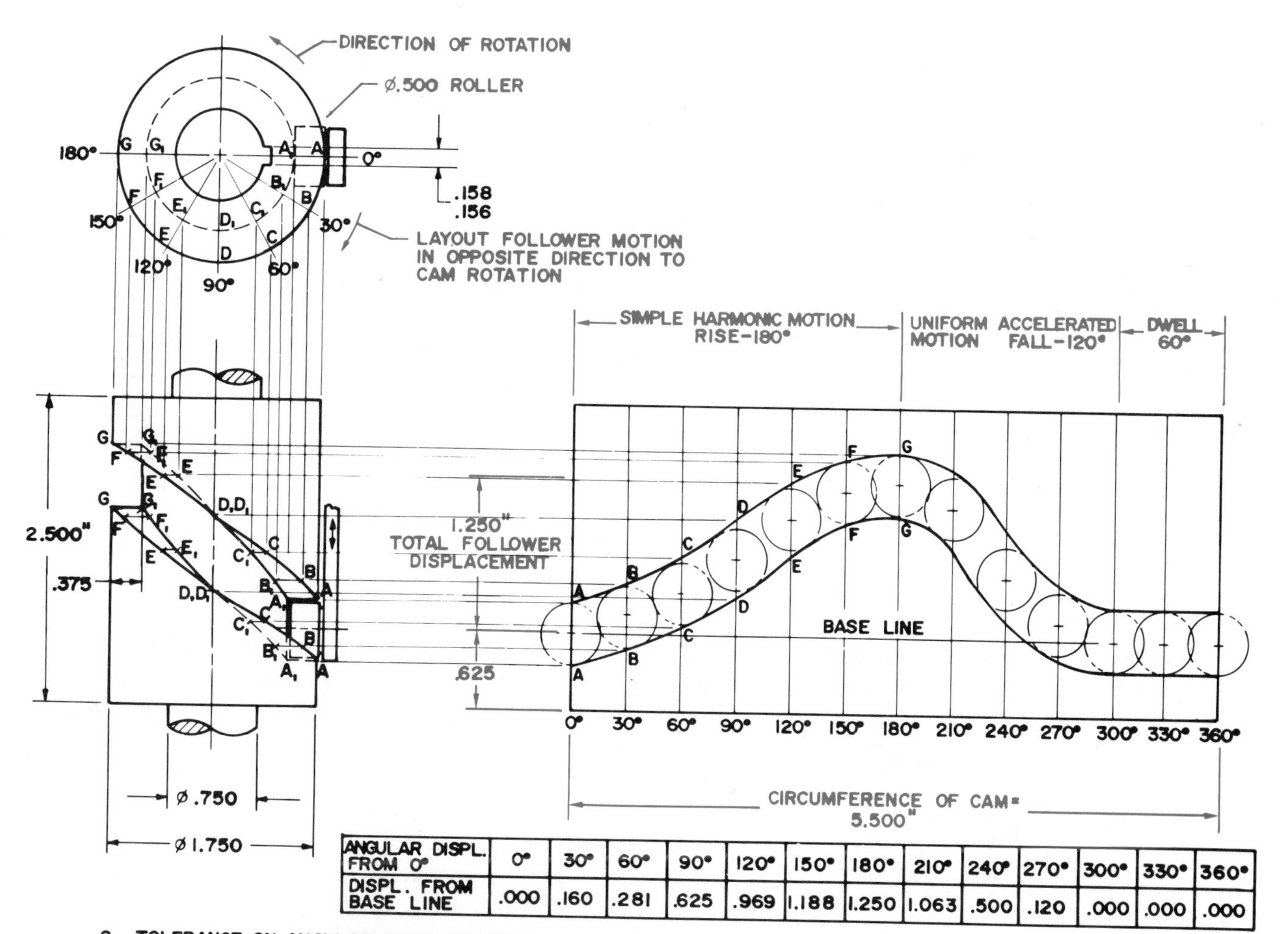

ANGULAR DISPL. FROM 0°	0°	30°	60°	90°	120°	150°	180°	210°	240°	270°	300°	330°	360°
DISPL. FROM BASE LINE	.000	.160	.281	.625	.969	1.188	1.250	1.063	.500	.120	.000	.000	.000

2. TOLERANCE ON ANGULAR DISPLACEMENT ±.5°.
1. TOLERANCE ON DISPLACEMENT FROM BASELINE ±.0008.

NOTES:

Figure 12–23 Construction of a drum cam drawing.

CAM DISPLACEMENT DIAGRAMS AND PROFILES

CADD systems make drawing cam displacement diagrams and cam profiles easy and accurate without the need for mathematical calculations. Constructing the cam displacement diagram and converting the information to the cam profile is the same as with manual drafting except the accuracy is increased substantially. Drawing the cam profile using irregular curves is difficult and time consuming. Most CADD systems have a curve-fitting command that makes it possible to automatically draw the cam profile through the points of tangency at the cam follower positions.

CAM MANUFACTURING

Modern cams are manufactured using computer numerical control (CNC) machining. In many instances the drawing is prepared on a CADD system and transferred to a computer-aided manufacturing (CAM) program for immediate coding for the CNC machine. Some CAM programs allow the designer to create the cam profile and transfer the data to the CNC machine tool without ever generating a drawing. Refer to Chapter 7 on Manufacturing Processes for more information.

GEARS

Gears are toothed wheels used to transmit motion and power from one shaft to another. Gears are rugged and durable and can transmit power with up to 98 percent efficiency with long service life. Gear design involves a combination of material, strength, and wear characteristics. Most gears are made of cast iron or steel, but brass and bronze alloys and plastic are also used for some applications. Gear selection and design is often done through vendors' catalogs or the use of standard formulas. A gear train exists when two or more gears are in combination for the purpose of transmitting power. Generally, two gears in mesh are used to increase or reduce speed, or change the direction of motion from one shaft to another. When two gears are in mesh, the larger is called the gear and the smaller is called the pinion. (See Figure 12–24.)

AGMA/ANSI Gear selection generally follows the guidelines of the American Gear Manufacturers Association (AGMA) or the American National Standards Institute (ANSI).

It is important for engineering drafters to fully understand gear terminology and formulas. However, many drafters will not draw gears, because they are commonly supplied as purchase parts. When this happens, the drafter may be required to make gear selections for specific applications or draw gears on assembly drawings. Details of gears are often drawn using simplified techniques as described in this chapter. In some situations, the gears are drawn as they actually exist for display on assembly drawings or in catalogs. When this is necessary, CADD can make the job easy, increasing productivity as much as 50 times.

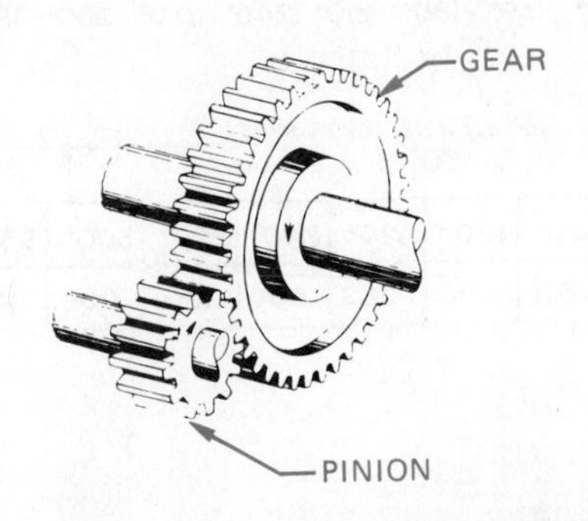

Figure 12–24 The gear and pinion.

GEAR TYPES

The most common and simplest form of gear is the spur gear. This chapter discusses in detail the design, specification, and drafting of spur gears. In addition, explanations are provided regarding bevel gears and worm gears. Gear types are designed based on one or more of the following elements:

- The relationship of the shafts: parallel, intersecting, nonintersecting shafts, or rack and pinion.
- Manufacturing cost.
- Ease of maintenance in service.
- Smooth and quiet operation.
- Load-carrying ability.
- Speed reduction capabilities.
- Space requirements.

Parallel Shafting Gears

Many different types of mating gears are designed with parallel shafts. These include spur and helical gears.

Spur Gears. There are two basic types of spur gears: external and internal spur gears. When two or more spur gears are cut on a single shaft, they are referred to as cluster gears. External spur gears are designed with the teeth of the gear on the outside of a cylinder. (See Figure 12–25.) External spur gears are the most common type of gear used in manufacturing. Internal spur gears have the teeth on the inside of the cylindrical gear. (See Figure 12–26.) The advantages of spur gears over other types is their low manufacturing cost, simple design, and ease of maintenance. The disadvantages include less load capacity and more noise than other types.

Helical Gears. Helical gears have their teeth cut at an angle, which allows more than one tooth to be in contact. (See Figure 12–27.) Helical gears carry more load than equivalent-sized spur gears and operate more quietly and smoothly. The disadvantage of helical gears is that they develop *end thrust.* End thrust is a lateral force exerted on the end of the gear shaft. Thrust bearings are required to reduce the effect of this end thrust. Double helical gears are designed to eliminate the end thrust and provide long life under heavy loads. However, they are more difficult and costly to manufacture. The herringbone gear shown in Figure 12–28 is a double helical gear without space between the two opposing sets of teeth.

Intersecting Shafting Gears

Intersecting shafting gears allow for the change in direction of motion from the gear to the pinion. Different types of intersecting shafting gears include bevel and face gears.

Bevel Gears. Bevel gears are conical in shape, allowing the shafts of the gear and pinion to intersect at 90° or any desired angle. The teeth on the bevel gear have the same shape as the teeth on spur gears except they taper toward the apex of the cone. Bevel gears provide for a speed change between the gear and pinion. (See Figure 12–29.) Miter gears are the same as bevel gears except both the gear and pinion are the same size and are used when shafts must intersect at 90° without speed reduction. Spiral bevel gears have the teeth cut at an angle, which provides the same advantages as helical gears over spur gears.

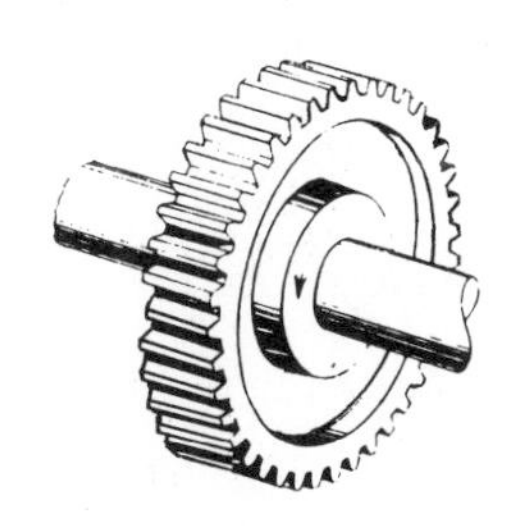

Figure 12–25 External spur gear.

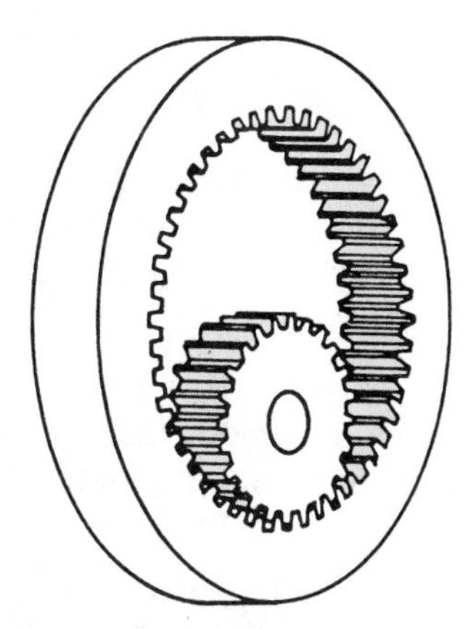

Figure 12–26 Internal spur gear. *Reprinted from Baril, MODERN MACHINING TECHNOLOGY, © 1987 Delmar Publishers Inc.*

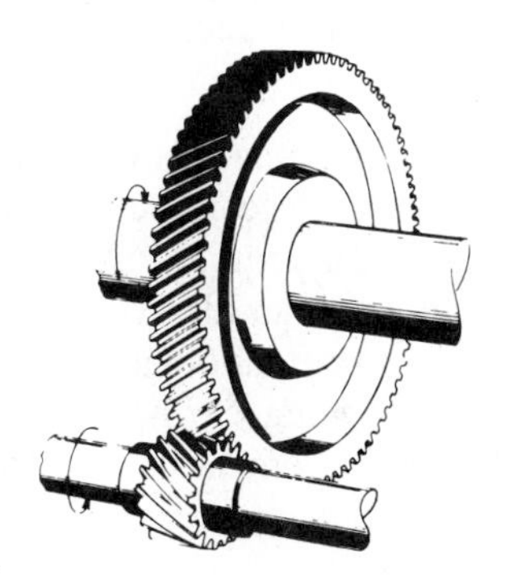

Figure 12–27 Helical gear.

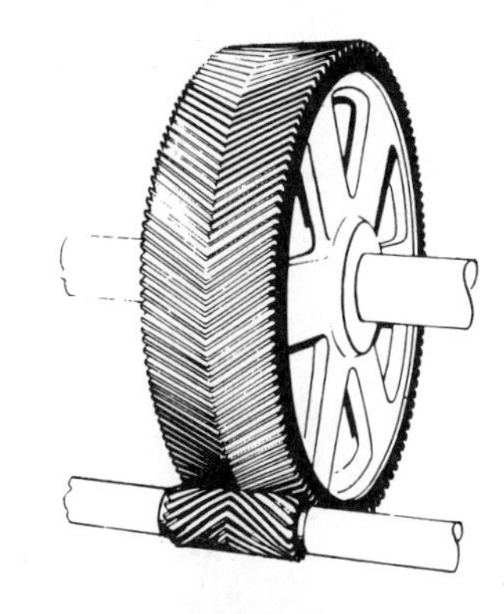

Figure 12–28 Herringbone gear.

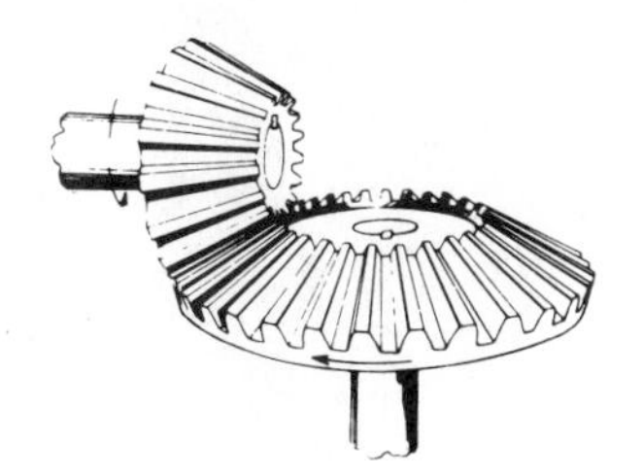

Figure 12–29 Bevel gears.

Face Gears. The face gear is a combination of bevel gear and spur pinion, or bevel gear and helical pinion. This combination is used when the mounting accuracy is not as critical as with bevel gears. The load-carrying capabilities of face gears are not as good as that of bevel gears.

Nonintersecting Shafting Gears

Gears with shafts that are at right angles but not intersecting are referred to as nonintersecting shafts. Gears that fall into this category are crossed helical, hypoid, and worm gears.

Crossed Helical Gears. Also known as right angle helical or spiral gears, crossed helical gears provide for nonintersecting right angle shafts with low load-carrying capabilities. (See Figure 12–30.)

Hypoid Gears. Hypoid gears have the same design as bevel gears except the gear shaft axes are offset and do not intersect. (See Figure 12–31.) The gear and pinion are often designed with bearings mounted on both sides for improved rigidity over standard bevel gears. Hypoid gears are very smooth, strong, and quiet in operation.

Worm Gears. A worm and worm gear are shown in Figure 12–32. This type of gear is commonly used when a large speed reduction is required in a small space. The worm may be driven in either direction. When the gear is not in operation, the worm automatically locks in place. This is a particular advantage when it is important for the gears to have no movement of free travel when the equipment is shut off.

Rack and Pinion

A rack and pinion is a spur pinion operating on a flat straight bar rack. (See Figure 12–33.) The rack and pinion is used to convert rotary motion into straight line motion.

Figure 12–30 Crossed helical gears. *Courtesy Browning Mfg., Division of Emerson Electric Co.*

SPUR GEAR DESIGN

Spur gear teeth are straight and parallel to the gear shaft axis. The tooth profile is designed to transmit power at a constant rate, and with a minimum of vibration and

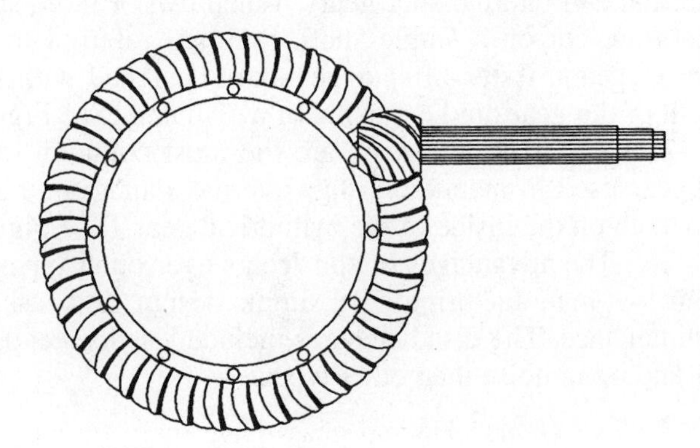

Figure 12–31 CADD drawing of hypoid gears.

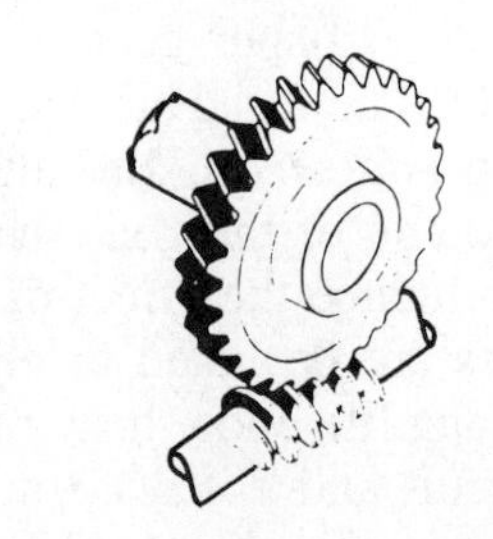

Figure 12–32 Worm and worm gear.

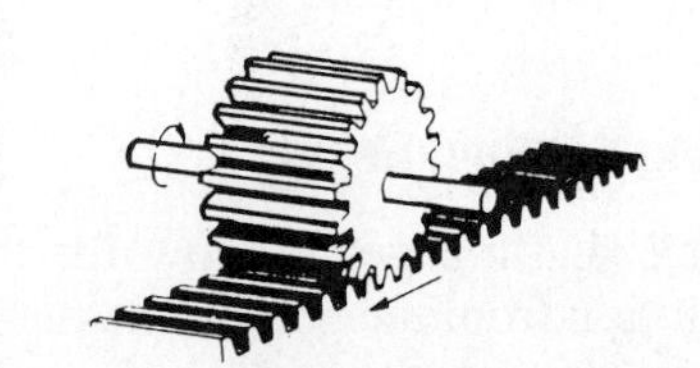

Figure 12–33 Rack and pinion.

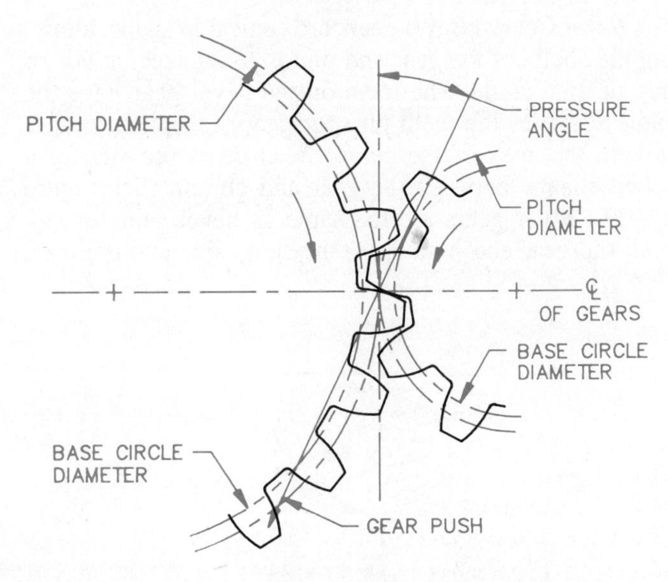

Figure 12–34 The spur gear pressure angle and related terminology.

noise. To achieve these requirements, an *involute curve* is used to establish the gear tooth profile. An involute curve is a spiral curve generated by a point on a chord as it unwinds from the circle. The contour of a gear tooth, based on the involute curve, is determined by a base circle, the diameter of which is controlled by a pressure angle. The pressure angle is the direction of push transmitted from a tooth on one gear to a tooth on the mating gear or pinion. (See Figure 12–34). Two standard pressure angles, 14.5° and 20°, are used in spur gear design. The most commonly used pressure angle is 20° because it provides a stronger tooth for quieter running and heavier load-carrying characteristics. One of the basic rules of spur gear design is to have no fewer than 13 teeth on the running gear and 26 teeth on the mating gear.

Standard terminology and formulas control the drawing requirements for spur gear design and specifications. Figure 12–35 shows a pictorial representation of the spur gear teeth with the components labeled. As an engineering drafter, it is important that you become familiar with the terminology and associated mathematical formulas used to calculate values.

TERM	DESCRIPTION	FORMULA
Pitch Diameter (D)	The diameter of an imaginary pitch circle on which a gear tooth is designed. Pitch circles of two spur gears are tangent.	$D = N/P$
Diametral Pitch (P)	A ratio equal to the number of teeth on a gear per inch of pitch diameter.	$P = N/D$
Number of Teeth (N)	Number of teeth on a gear.	$N = D \times P$
Circular Pitch (p)	The distance from a point on one tooth to the corresponding point on the adjacent tooth, measured on the pitch circle.	$p = 3.1416 \times D/N$ $p = 3.1416/P$
Center Distance (C)	The distance between the axis of two mating gears.	C = sum of pitch DIA/2
Addendum (a)	The radial distance from the pitch circle to the top of the tooth.	$a = 1/P$
Dedendum (b)	The radial distance from the pitch circle to the bottom of the tooth. (This formula is for 20° teeth only.)	$b = 1.250/P$
Whole Depth (h_t)	The full height of the tooth. It is equal to the sum of the addendum and the dedendum.	$h_t = a + b$ $h_t = 2.250/P$
Working Depth (h_k)	The distance that a tooth occupies in the mating space. A distance equal to two times the addendum.	$h_k = 2a$ $h_k = 2.000/P$
Clearance (c)	The radial distance between the top of a tooth and the bottom of the mating tooth space. It is also the difference between the addendum and dedendum.	$c = b - a$ $c = .250/P$
Outside Diameter (D_o)	The overall diameter of the gear. It is equal to the pitch diameter plus two dedendums.	$D_o = D + 2a$
Root Diameter (D_r)	The diameter of a circle coinciding with the bottom of the tooth spaces.	$D_r = D - 2b$
Circular Thickness (t)	The length of an arc between the two sides of a gear tooth measured on the pitch circle.	$t = 1.5708/P$
Chordal Thickness (t_c)	The straight line thickness of a gear tooth measured on the pitch circle.	$t_c = D \sin (90°/N)$
Chordal Addendum (a_c)	The height from the top of the tooth to the line of the chordal thickness.	$a_c = a + t^2/4D$
Pressure Angle (ϕ)	The angle of direction of pressure between contacting teeth. It determines the size of the base circle and the shape of the involute teeth.	
Base Circle Dia. (D_B)	The diameter of a circle from which the involute tooth form is generated.	$D_B = D \cos \phi$

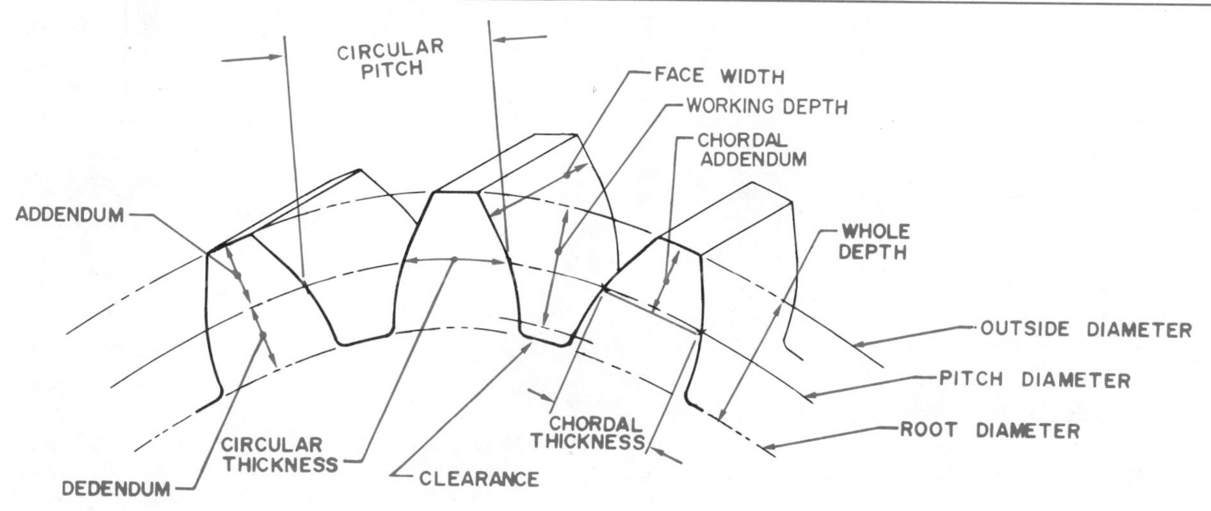

Figure 12–35 Gear terminology and formulas.

Diametral Pitch

The diametral pitch actually refers to the tooth size and has become the standard for tooth size specifications. (See Figure 12–36.) As you look at Figure 12–36, notice how the tooth size increases as the diametral pitch decreases. One of the most elementary rules of gear tooth design is that mating teeth must have the same diametral pitch.

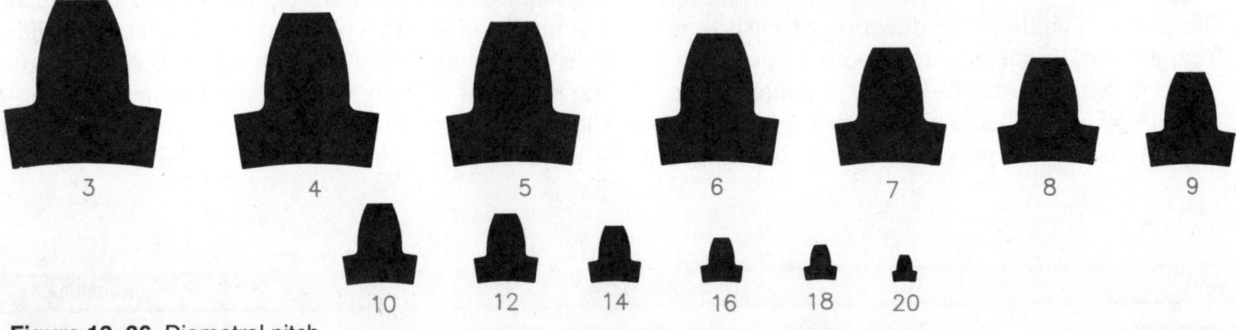

Figure 12–36 Diametral pitch.

CADD *applications*

GEAR TEETH

There are custom CADD programs that allow the gear teeth to be drawn automatically. To do this, the computer prompts the drafter for variables such as pitch diameter, diametral pitch, pressure angle, and number of teeth. The program then automatically notifies the drafter whether the information is accurate, provides all additional data, and creates a detail drawing of the gear. CADD was used to draw the teeth in the gear displayed in Figure 12–37. The CADD drafter may easily draw gears displayed in detailed representation, or use the simplified technique to save regeneration and plotting time.

Some software programs do more than assist in the design process. For example, objects such as gear teeth can be subjected to simulated tests and stress analysis on the computer screen as shown in Figure 12–38.

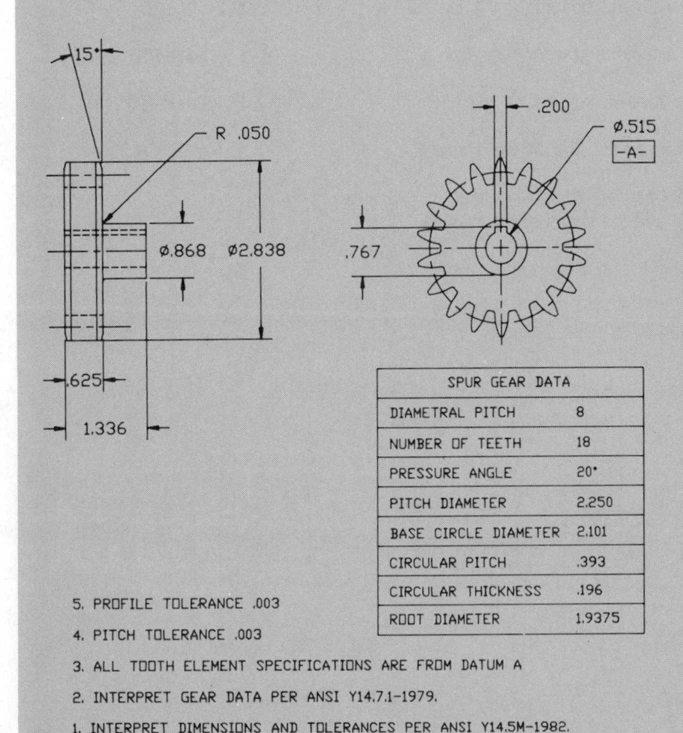

Figure 12–37 Gear detail drawing using CADD to draw the teeth.

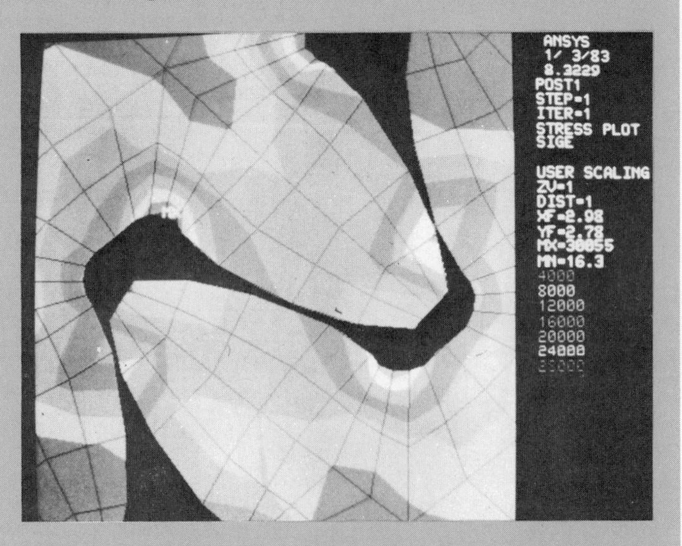

Figure 12–38 Objects such as these gear teeth can be subjected to simulated tests and stress analysis on the computer screen. *Courtesy Swanson Analysis Systems, Inc.*

DESIGNING AND DRAWING SPUR GEARS

ANSI The drafting standard that governs gear drawings is the American National Standards document ANSI Y14.7.1–1971 *Gear Drawing Standards — Part I*. Because gear teeth are complex and time consuming to draw, simplified representations are used to make the practice easier. (See Figure 12–39.) The simplified method shows the outside diameter and the root diameters as phantom lines and the pitch diameter as a centerline in the circular view. In addition, a side view is often required to show width dimensions and related features. (See Figure 12–39(a).) If the gear construction has webs, spokes, or other items that require further clarification, then a full section is normally used. (See Figure 12–39(b).) Notice in the cross section that the gear tooth is left unsectioned and the pitch diameter is shown as a centerline.

When cluster gears are drawn, the circular view may show both sets of gear tooth representations in simplified form, or two circular views may be drawn. (See Figure 12–40.) When cluster gears are more complex than those shown here, multiple views and removed sections may be required.

The gear teeth, as mentioned earlier, are seldom drawn because it takes too much time to draw them. However, one or more teeth may be drawn for specific applications. For example, when a tooth must be in alignment with another feature of the gear, the tooth may be drawn as shown in Figure 12–41.

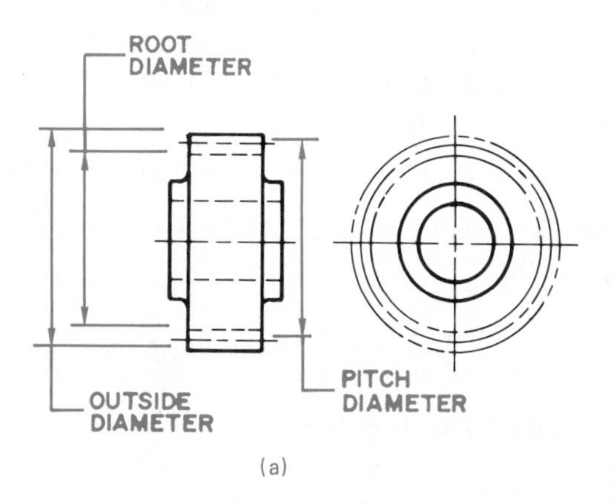

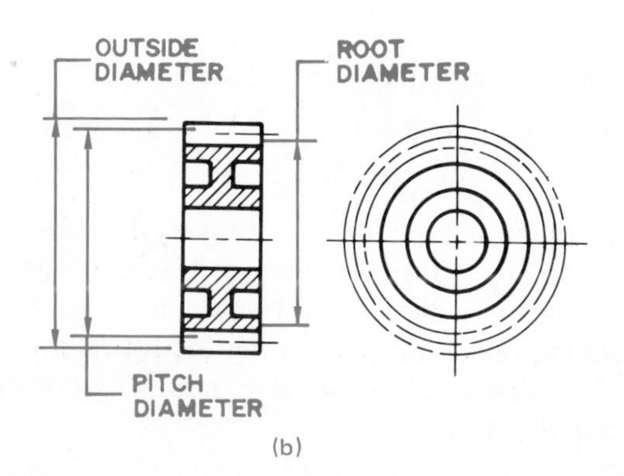

Figure 12–39 Typical spur gear drawings using simplified gear teeth representation.

Gear drawings typically have a chart that shows the manufacturing information associated with the teeth, and with related part detail dimensions placed on the specific views. Figure 12–42 shows all of the information in chart and dimensional form that is traditionally associated with a complete gear drawing.

In some situations, such as pictorial drawings for catalogs, the gear teeth are drawn as they actually exist. To do this the drafter should use a gear tooth template to save time and improve accuracy.

DESIGNING SPUR GEAR TRAINS

A gear train is an arrangement of two or more gears connecting driving and driven parts of a machine. Gear reducers and transmissions are examples of gear trains. The function of a gear train is to:

- transmit motion between shafts.
- decrease or increase the speed between shafts.
- change the direction of motion.

It is important for you to understand the relationship between two mating gears in order to design gear trains.

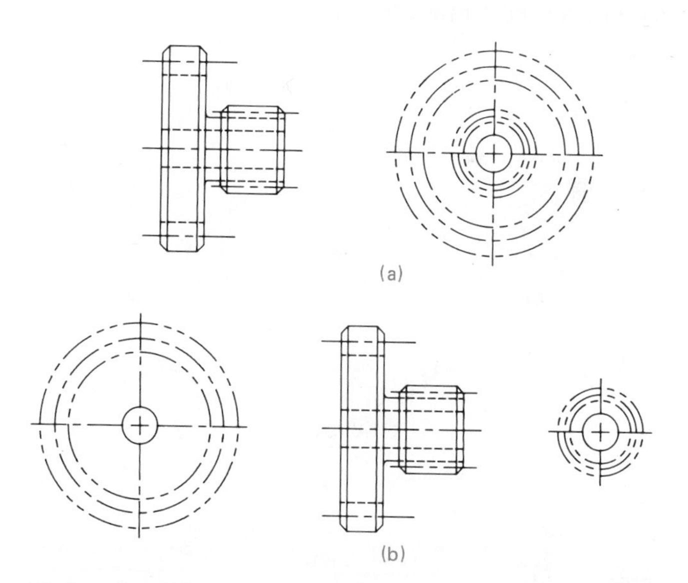

Figure 12–40 Cluster gear drawings, simplified representation.

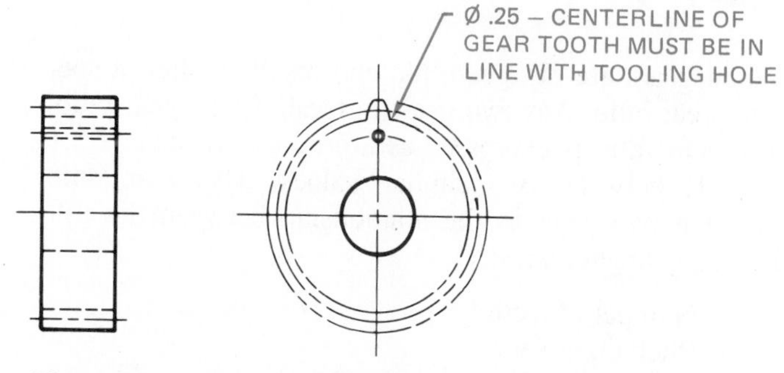

Figure 12–41 Showing the relationship of one gear tooth to another feature on the gear.

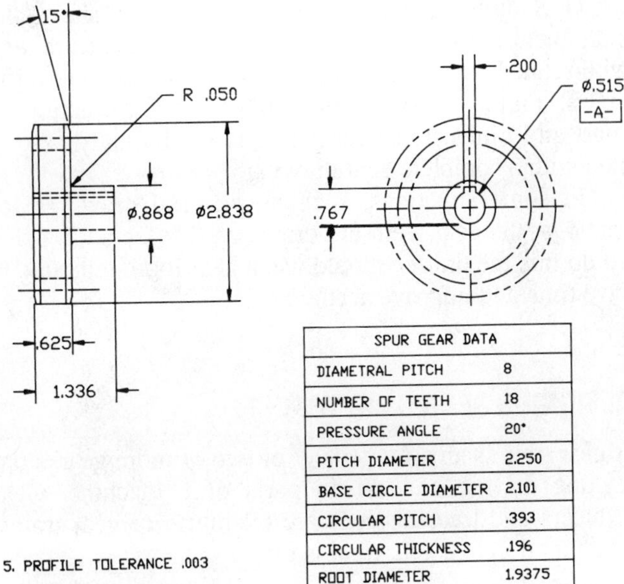

SPUR GEAR DATA	
DIAMETRAL PITCH	8
NUMBER OF TEETH	18
PRESSURE ANGLE	20°
PITCH DIAMETER	2.250
BASE CIRCLE DIAMETER	2.101
CIRCULAR PITCH	.393
CIRCULAR THICKNESS	.196
ROOT DIAMETER	1.9375

5. PROFILE TOLERANCE .003
4. PITCH TOLERANCE .003
3. ALL TOOTH ELEMENT SPECIFICATIONS ARE FROM DATUM A.
2. INTERPRET GEAR DATA PER ANSI Y14.7.1-1979.
1. INTERPRET DIMENSIONS AND TOLERANCES PER ANSI Y14.5M-1982.
NOTES:

Figure 12–42 The complete gear detail drawing with information in chart and dimensional format.

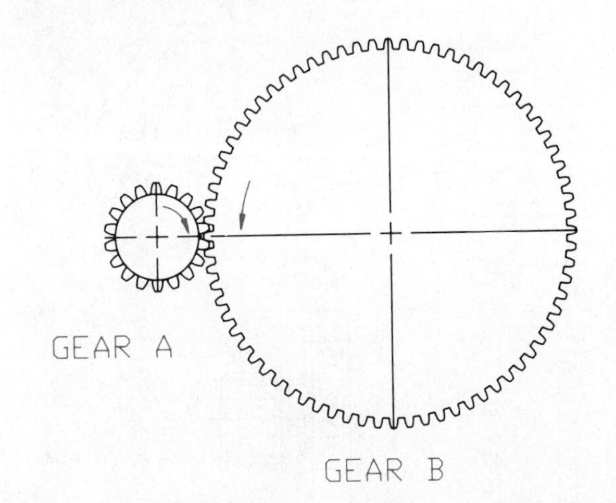

	(D) PITCH DIAMETER	(N) NUMBER OF TEETH	(P) DIAMETRAL PITCH	RPM	DIRECTION
GEAR A	6	18	3	1200	C. WISE
GEAR B	18	54	3	400	C.C.WISE

Figure 12–43 Calculating gear data.

When gears are designed, the end result is often a specific gear ratio. Any two gears in mesh have a gear ratio. The gear ratio is expressed as a proportion, such as 2:1 or 4:1, between two similar values. The gear ratio between two gears is the relationship between the following characteristics:

- Number of teeth
- Pitch diameters
- Revolutions per minute (RPM)

If you have gear A (pinion) mating with gear B, as shown in Figure 12–43, the gear ratio is calculated by dividing like values of the smaller gear into the larger gear as follows:

$$\frac{\text{Number Teeth}_{\text{Gear B}}}{\text{Number Teeth}_{\text{Gear A}}} = \text{Gear Ratio}$$

$$\frac{\text{Pitch Diameter}_{\text{Gear B}}}{\text{Pitch Diameter}_{\text{Gear A}}} = \text{Gear Ratio}$$

$$\frac{\text{RPM}_{\text{Gear A}}}{\text{RPM}_{\text{Gear B}}} = \text{Gear Ratio}$$

Now, calculate the gear ratio for the two mating gears in Figure 12–43 if gear A has 18 teeth, 6-in. pitch diameter, and operates at 1200 RPM, and gear B has 54 teeth, 18-in. pitch diameter, and operates at 400 RPM:

$$\text{Number of Teeth} = 54/18 = 3{:}1$$

$$\text{Pitch Diameter} = 18/6 = 3{:}1$$

$$\text{RPM} = 1200/400 = 3{:}1$$

You can solve for unknown values in the gear train if you know the gear ratio you want to achieve, the number of teeth and pitch diameter of one gear, and the input speed. For example, gear A has 18 teeth, a pitch diameter of 6 in., and in input speed of 1200 RPM, and the ratio between gear A and gear B is 3:1. In order to keep this information well organized, it is recommended that you set up a chart similar to the one shown in Figure 12–43. The unknown values are shown in color for your reference. Determine the number of teeth for gear B as follows:

$$\text{Teeth}_{\text{Gear A}} \times \text{Gear Ratio} = 18 \times 3 = \text{Teeth}_{\text{Gear B}} = 54$$

Or, if you know that gear B has 54 teeth and the gear ratio is 3:1, then:

$$\frac{\text{Teeth}_{\text{Gear B}}}{\text{Gear Ratio}} = \frac{54}{3} = \text{Teeth}_{\text{Gear A}} = 18$$

Determine the RPM of gear B:

$$\frac{\text{RPM}_{\text{Gear A}}}{\text{Gear Ratio}} = \frac{1200}{3} = \text{RPM}_{\text{Gear B}} = 400$$

Determine the pitch diameter of gear B:

$$\text{Pitch Diameter}_{\text{Gear A}} \times \text{Gear Ratio} = 6 \text{ in.} \times 3 = \text{Pitch Diameter Gear B} = 18 \text{ in.}$$

In some situations it will be necessary for you to refer to the formulas given in Figure 12–35 to determine some unknown values. In this case it is necessary to calculate the diametral pitch using the formula P = N/D, where P = diametral pitch, N = number of teeth, and D = pitch diameter:

$$P_{\text{Gear A}} = \frac{N}{D} = \frac{18}{6} = 3$$

Because the diametral pitch denotes the tooth size and the teeth for mating gears must be the same size, the diametral pitch of gear B is also 3.

Keep in mind that the preceding example presents only one set of design criteria. Other situations may be different. Always solve for unknown values based on information that you have, and work with the standard formulas presented in this chapter. Following are some important points to keep in mind as you work with the design of gear trains:

- The RPM of the larger gear is always slower than the RPM of the smaller gear.
- Mating gears always turn in opposite directions. For example, in Figure 12–43, gear B turns counterclockwise if gear A turns clockwise.
- Gears on the same shaft (cluster gears) always turn in the same direction and at the same speed (RPMs).
- Mating gears have the same size teeth (diametral pitch).
- The gear ratio between mating gears is a ratio between the number of teeth, the pitch diameters, and the RPMs.
- The distance between the shafts of mating gears is equal to $\frac{1}{2}D_{Gear\ A} + \frac{1}{2}D_{Gear\ B}$. This distance is 3 + 9 = 12 in. between shafts in Figure 12–43.

Given the gear train shown in Figure 12–44, calculate the unknown values using the formulas and information discussed in this chapter. The unknown values are given in color to help you check your work.

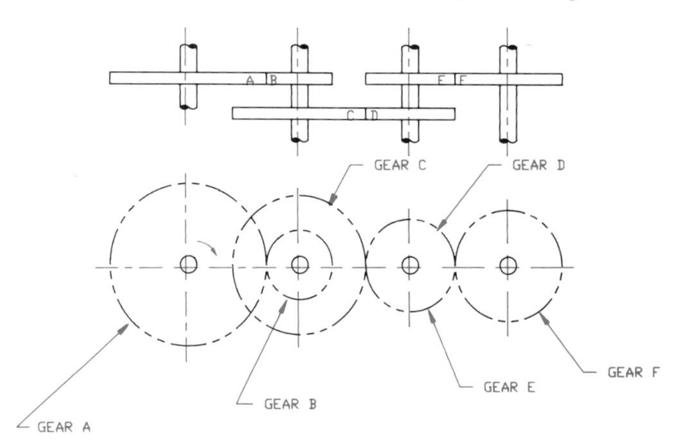

	PITCH DIAMETER (D)	NUMBER OF TEETH (N)	DIAMETRAL PITCH (P)	RPM	DIRECTION	GEAR RATIO	CENTER DISTANCE
GEAR A	7.000	28	4	300	C.WISE	2.33/1	5.000
GEAR B	3.000	12	4	699	C.C.WISE		
GEAR C	6.000	18	3	699	C.C.WISE	1.5/1	5.000
GEAR D	4.000	12	3	1048.5	C.WISE		
GEAR E	4.000	20	5	1048.5	C.WISE	1.2/1	4.400
GEAR F	4.800	24	5	1258.2	C.C.WISE		

Figure 12–44 Calculate the unknown values for a given gear train.

DESIGNING AND DRAWING THE RACK AND PINION

A rack is a straight bar with spur gear teeth used to convert rotary motion to reciprocating motion. Figure 12–45 shows a spur gear pinion mating with a rack. Notice that the circular dimensions of the pinion become linear dimensions on the rack.

When preparing a detailed drawing of the rack, the front view is normally shown with the profile of the first and last tooth drawn. Phantom lines are drawn to represent the top and root, and a centerline is used for the pitch line. (See Figure 12–46.) Related tooth and length dimensions are also placed on the front view and a depth dimension is placed on the side view. A chart is then placed in the field of the drawing to identify specific gear-cutting information.

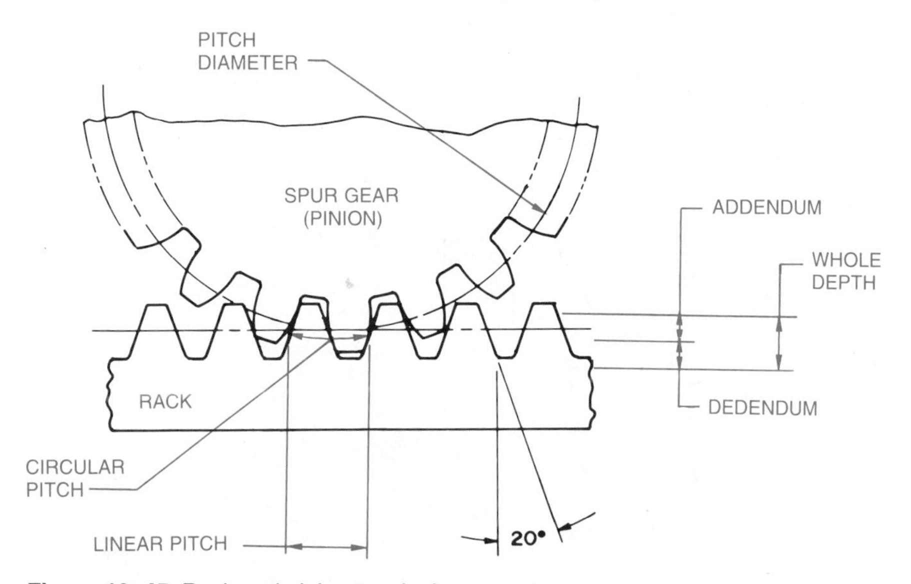

Figure 12–45 Rack and pinion terminology.

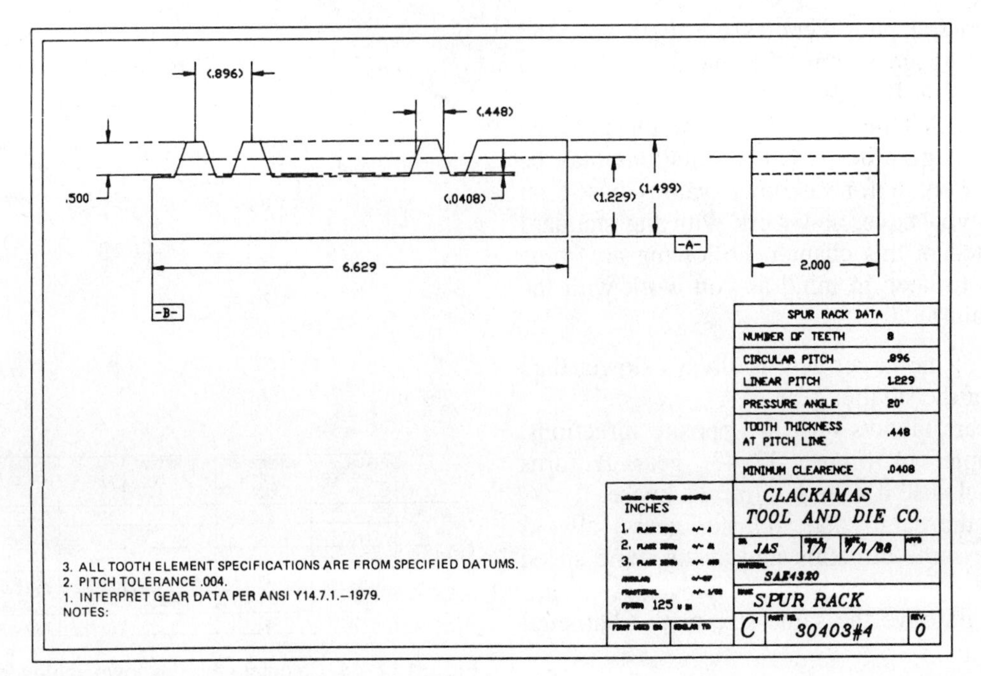

Figure 12–46 A detailed drawing of a rack.

TERM	DESCRIPTION	FORMULA
Pitch Diameter (D)	The diameter of the base of the pitch cone.	$D = N/P$
Pitch Cone	In Figure 14-64, the pitch cone is identified as XYZ.	
Pitch Angle (Ø)	The angle between an element of a pitch cone and its axis. Pitch angles of mating gears depend on their relative diameters (gear a and gear b).	$\tan Ø_a = D_a/D_b$ $\tan Ø_b = D_b/D_a$
Cone Distance (A)	Slant height of pitch cone.	$A = D/2 \sin Ø$
Addendum Angle (δ)	The angle subtended by the addendum. It is the same for mating gears.	$\tan \delta = a/A$
Dedendum Angle (Ω)	The angle subtended by the dedendum. It is the same for mating gears.	$\tan \Omega = b/A$
Face Angle ($Ø_o$)	The angle between the top of the teeth and the gear axis.	$Ø_o = Ø - \delta$
Root Angle ($Ø_r$)	The angle between the bottom of the tooth space and the gear axis.	$Ø_r = Ø - \Omega$
Outside Diameter (D_o)	The diameter of the outside circle of gear.	$D_o = D + 2a(\cos Ø)$
Crown Height (X)	The distance between the cone apex and the outer tip of the gear teeth.	$X = .5\ (D_o)/\tan Ø_o$
Crown Backing (Y)	The distance from the rear of the hub to the outer tip of the gear tooth, measured parallel to the axis of the gear.	
Face Width	A distance that should not exceed one-third of the cone distance (A).	

NOTE: REFER TO THE FORMULAS ON PAGE 437 FOR ADDITIONAL CALCULATIONS.

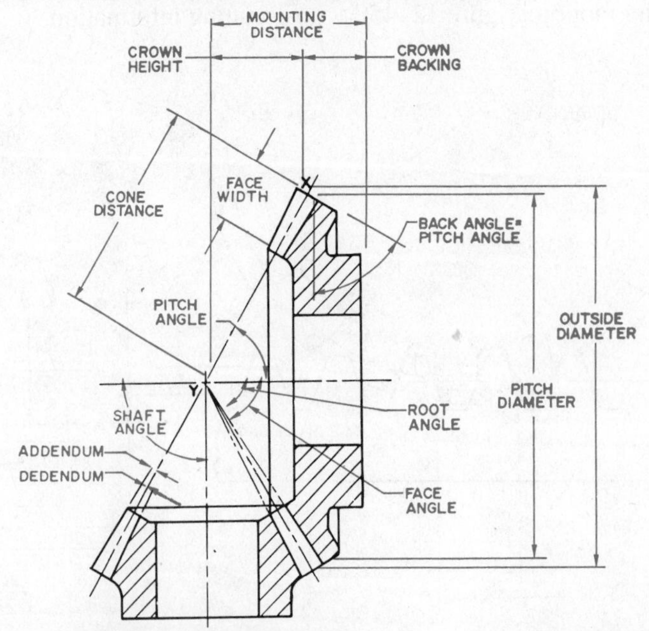

Figure 12–47 Bevel gear terminology and formulas.

DESIGNING AND DRAWING BEVEL GEARS

Bevel gears are usually designed to transmit power between intersecting shafts at 90°, although they may be designed for any angle. Some gear design terminology and formulas relate specifically to the construction of bevel gears. These formulas and a drawing of a bevel gear and pinion are shown in Figure 12–47. Most of the gear terms discussed for spur gears apply to bevel gears. In many cases information must be calculated using the formulas provided in the spur gear discussion shown in Figure 12–35.

The drawing for a bevel gear is similar to the drawing for a spur gear, but in many cases only one view is needed. This view is often a full section. Another view may be necessary to show the dimensions for a keyway. As with other gear drawings a chart is used to specify gear-cutting data. (See Figure 12–48.) Notice in this example that only a partial side view is used to specify the bore and keyway.

DESIGNING AND DRAWING WORM GEARS

Worm gears are used to transmit power between non-intersecting shafts. More important, they are used for large speed reductions in a small space as compared to other types of gears. Worm gears are also strong, move in either direction, and lock in place when the machine is not in operation. A single lead worm advances one pitch with every revolution. A double lead worm advances two pitches with each revolution. As with bevel gears, the worm gear and worm have specific terminology and formulas that apply to their design. The representative worm gear and worm technology and design formulas are shown in Figure 12–49.

When drawing the worm gear, the same techniques are used as discussed earlier. Figure 12–50 shows a worm gear drawing using a half-section method. A side view is provided to dimension the bore and keyway, and a chart is given for gear cutting data.

A detailed drawing of the worm is prepared using two views. The front view shows the first gear tooth on each end with phantom lines between for a simplified representation. The side view is the same as a spur gear drawing with the keyway specifications. The gear-cutting chart is then placed on the field of the drawing or over the title block. Figure 12–51 shows the worm drawing using CADD to provide a detailed representation of the worm teeth in the front view rather than phantom lines as in the simplified representation.

BEARINGS

Bearings are mechanical devices used to reduce friction between two surfaces. They are divided into two large groups known as plain and rolling element bearings. Bearings are designed to accommodate either rotational or linear motion. Rotational bearings are used for radial loads, and linear bearings are designed for thrust loads. Radial loads are loads that are distributed around the shaft. Thrust loads are lateral. Thrust loads apply force to the end of the shaft. Figure 12–52 shows the relationship between rotational and linear motion.

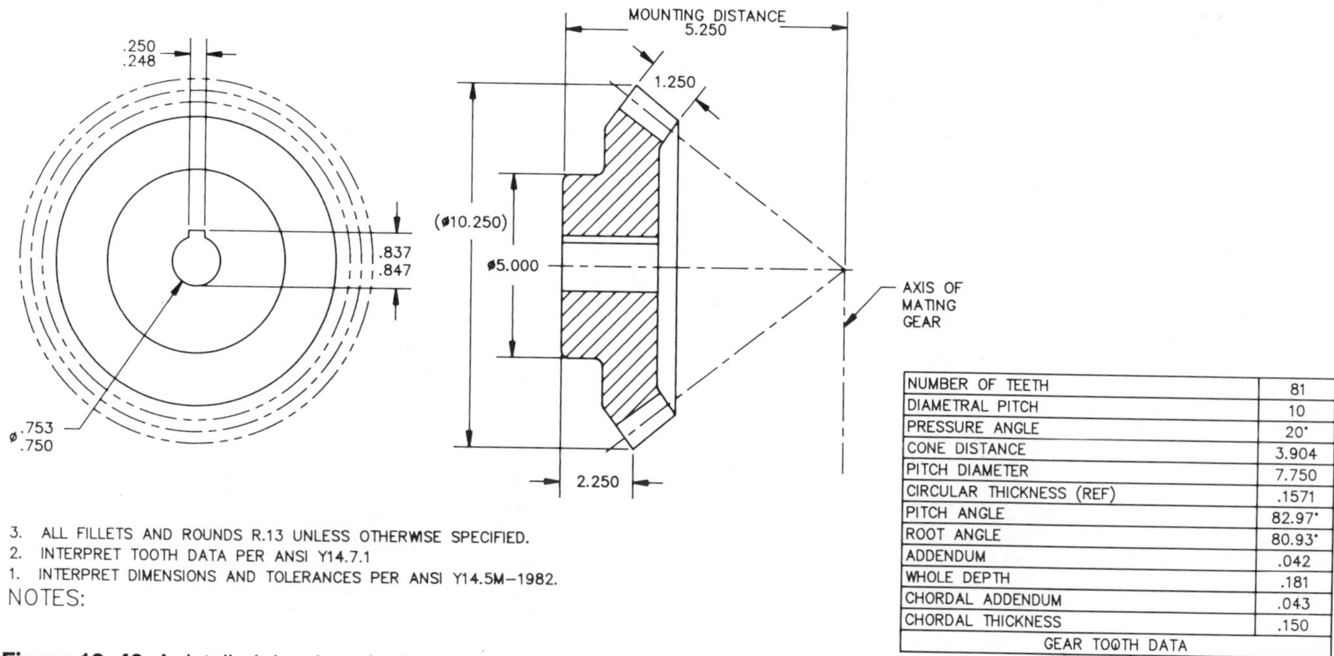

GEAR TOOTH DATA	
NUMBER OF TEETH	81
DIAMETRAL PITCH	10
PRESSURE ANGLE	20°
CONE DISTANCE	3.904
PITCH DIAMETER	7.750
CIRCULAR THICKNESS (REF)	.1571
PITCH ANGLE	82.97°
ROOT ANGLE	80.93°
ADDENDUM	.042
WHOLE DEPTH	.181
CHORDAL ADDENDUM	.043
CHORDAL THICKNESS	.150

Figure 12–48 A detailed drawing of a bevel gear.

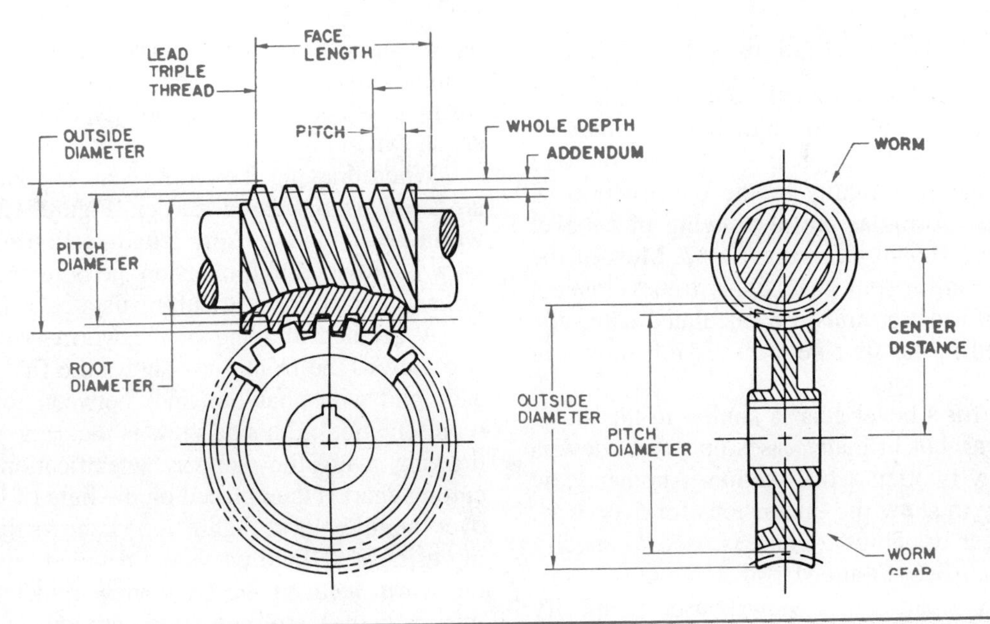

TERM	DESCRIPTION	FORMULA
Pitch Diameter (worm) (D_w)		$D_w = 2C - D_g$
Pitch Diameter (gear) (D_g)		$D_g = 2C - D_w$
Pitch (P)	The distance from one tooth to the corresponding point on the next tooth measured parallel to the worm axis. It is equal to the circular pitch on the worm gear.	$P = L/T$
Lead (L)	The distance the thread advances axially in one revolution of the worm.	$L = D_g/R$ $L = P \times T$
Threads (T)	Number of threads or starts on worm.	$T = L/P$
Gear Teeth (N)	Number of teeth on worm gear.	$N = \pi D_g/P$
Ratio (R)	Divide number of gear teeth by the number of worm threads.	$R = N/T$
Addendum (a)	For single and double threads.	$a = .318P$
Whole Depth (WD)	For single and double threads.	$WD = .686P$

NOTE: REFER TO THE FORMULAS ON PAGE 437 FOR ADDITIONAL CALCULATIONS.

Figure 12–49 Worm and worm gear terminology and formulas.

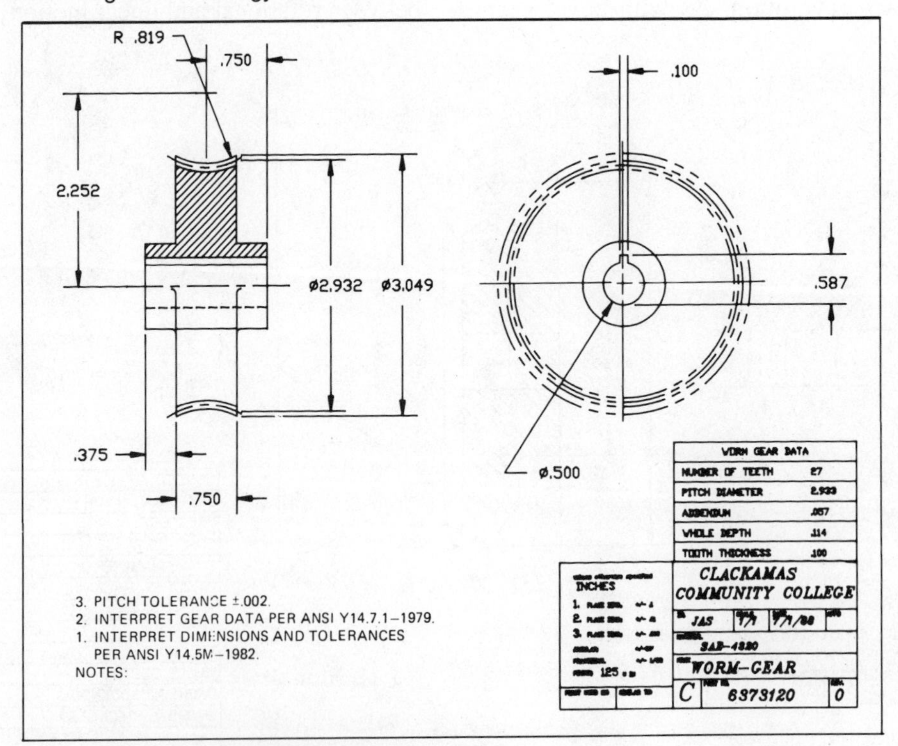

Figure 12–50 A detailed drawing of a worm gear.

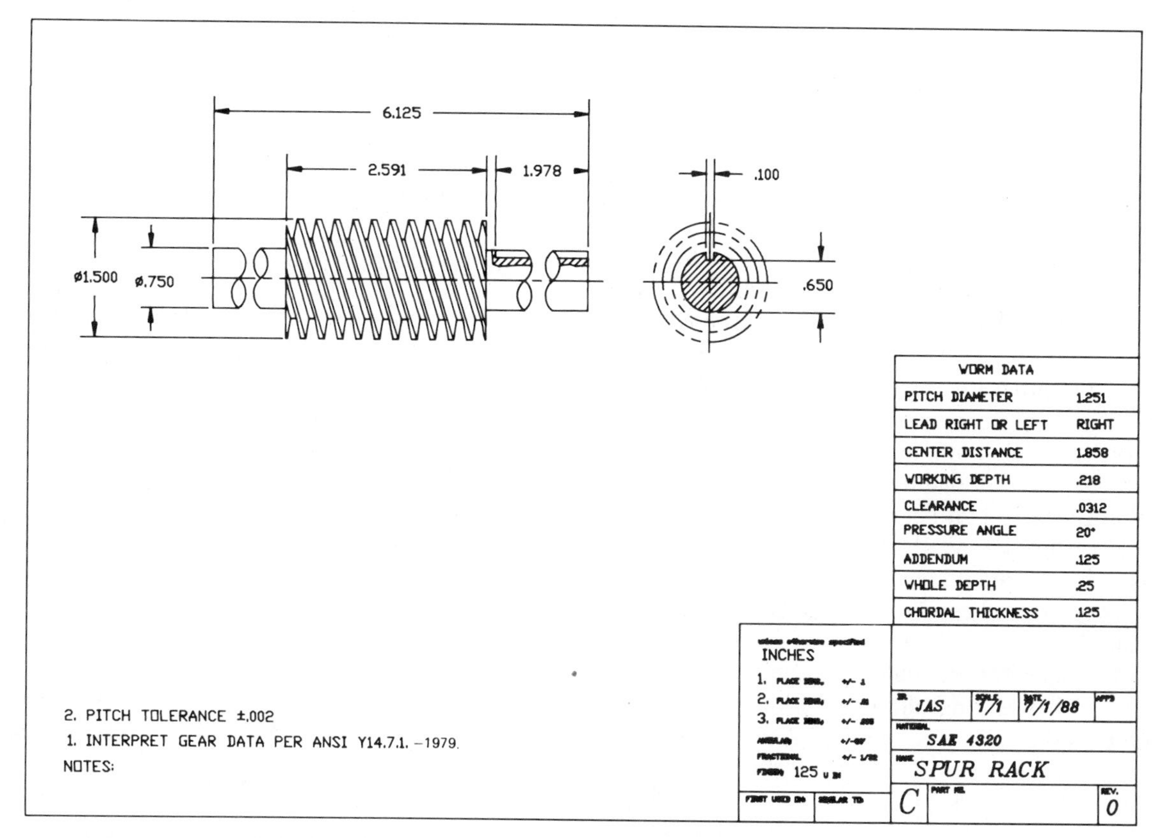

Figure 12–51 A detailed drawing of a worm.

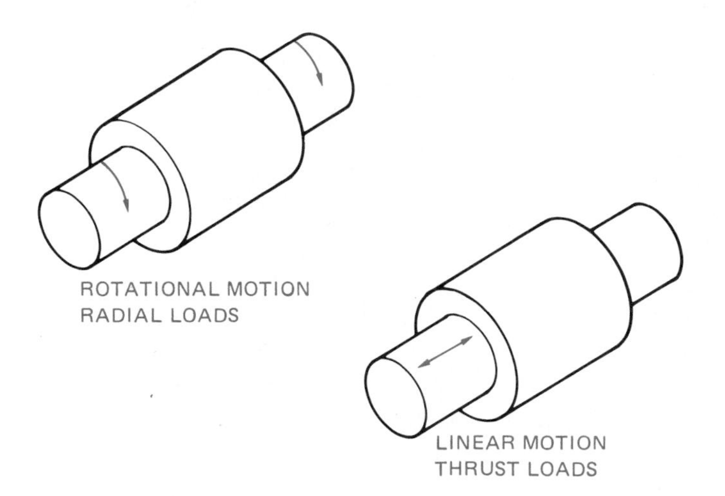

Figure 12–52 Radial and thrust loads.

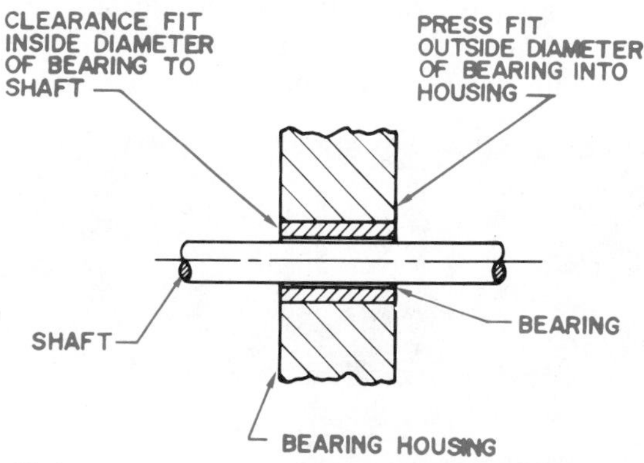

Figure 12–53 Plain bearing terminology and fits.

Plain Bearings

Plain bearings are often referred to as sleeve, journal bearings, or bushings. Their operation is based on a sliding action between mating parts. A clearance fit between the inside diameter of the bearing and the shaft is critical to ensure proper operation. Refer to fits between mating parts in Chapter 8 for more information. The bearing has an interference fit between the outside of the bearing and the housing or mounting device as shown in Figure 12–53.

The material from which plain bearings are made is important. Most plain bearings are made from bronze or phosphor bronze. Bronze bearings are normally lubricated, while phosphor bronze bearings are commonly impregnated with oil and require no additional lubrication. Phosphor bronze is an excellent choice when antifriction qualities are important and where resistance to wear and scuffing are needed.

Rolling Element Bearings

Ball and roller bearings are the two classes of rolling element bearings. Ball bearings are the most commonly used rolling element bearings. In most cases, ball bearings have higher speed and lower load capabilities than roller bearings. Even so, ball bearings are manufactured for most uses. Ball bearings are constructed with two grooved rings, a set of balls placed radially around the rings, and a separator that keeps the balls spaced apart and aligned as shown in Figure 12–54.

Single-row ball bearings are designed primarily for radial loads, but they can accept some thrust loads. Double-row ball bearings may be used where shaft alignment is important. Angular contact ball bearings support a heavy thrust load and a moderate radial load. Thrust bearings are designed for use in thrust load situations only. When both thrust and radial loads are necessary, both radial and thrust ball bearings are used together. Some typical ball bearings are shown in Figure 12–56. Ball bearings are available with shields and seals. A shield is a metal plate on one or both sides of the bearing. The shields act to keep the bearing clean and retain the lubricant. A sealed bearing has seals made of rubber, felt, or plastic placed on the outer and inner rings of the bearing. The sealed bearings are filled with special lubricant by the manufacturer. They require little or no maintenance in service. Figure 12–57 shows the shields and seals used on ball bearings.

Roller bearings are more effective than ball bearings for heavy loads. Cylindrical roller bearings have a high radial capacity and assist in shaft alignment. Needle roller bearings have small rollers and are designed for the highest load-carrying capacity of all rolling element bearings with shaft sizes under 10 in. Tapered roller bearings are used in gear reducers, steering mechanisms, and machine tool spindles. Spherical roller bearings offer the best combination of high load capacity, tolerance to shock, and alignment, and are used on conveyors, transmissions, and heavy machinery. Some common roller bearings are displayed in Figure 12–58.

DRAWING BEARING SYMBOLS

Only engineering drafters who work for a bearing manufacturer normally make detailed drawings of roller or ball bearings. In most other industries, bearings are not drawn because they are purchase parts. When bearings

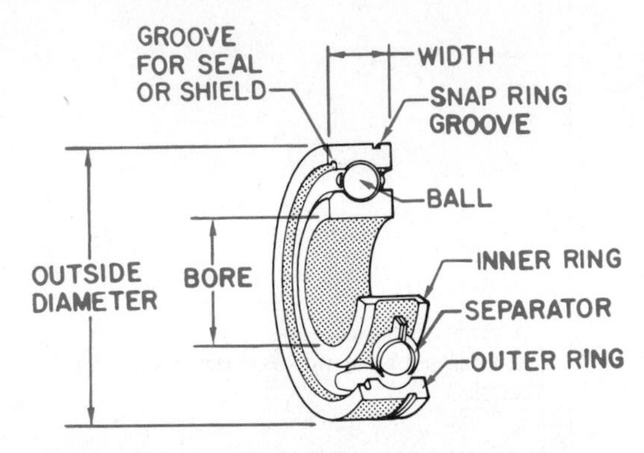

Figure 12–54 Ball bearing components.

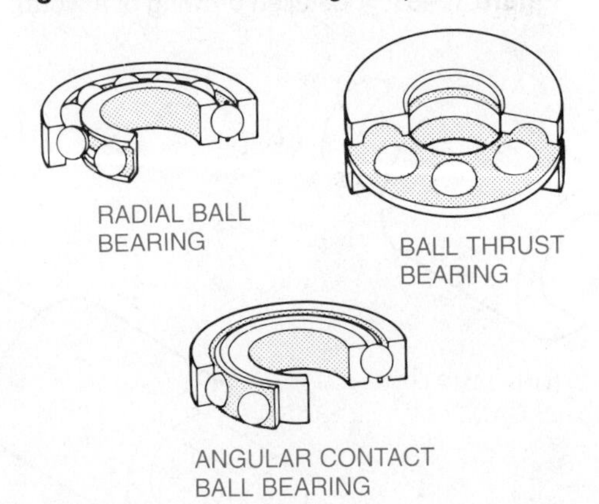

Figure 12–56 Typical ball bearings.

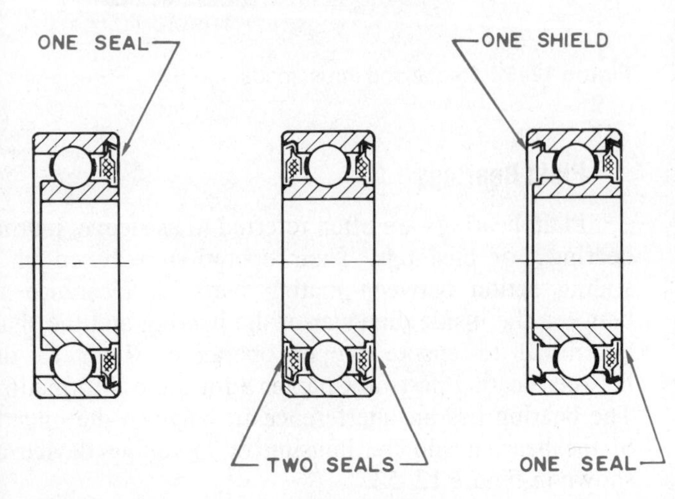

Figure 12–57 Bearing seals and shields.

BEARING SYMBOLS

When CADD is used, the bearing symbols may be drawn and saved in a symbols library for immediate use at any time. This use of CADD helps increase productivity and accuracy over other drafting techniques. (See Figure 12–55.)

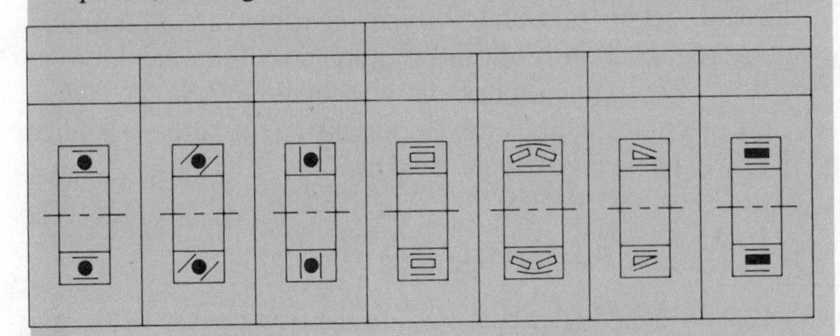

Figure 12–55 Bearing symbols drawn using CADD.

are drawn as a representation on assembly drawings or product catalogs, they are displayed using symbols.

BEARING CODES

Bearing manufacturers use similar coding systems for the identification and ordering of different bearing products. The bearing codes generally contain the following type of information:

- Material
- Bearing type
- Bore size
- Lubricant
- Type of seals or shields

A sample bearing numbering system is shown in Figure 12–59.

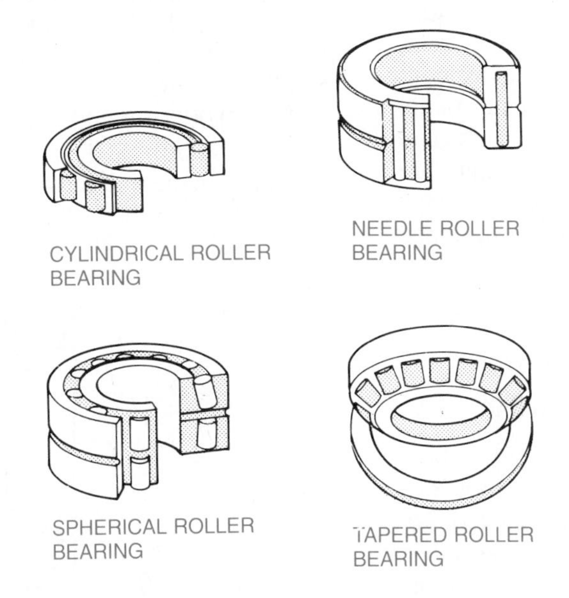

Figure 12–58 Typical roller bearings.

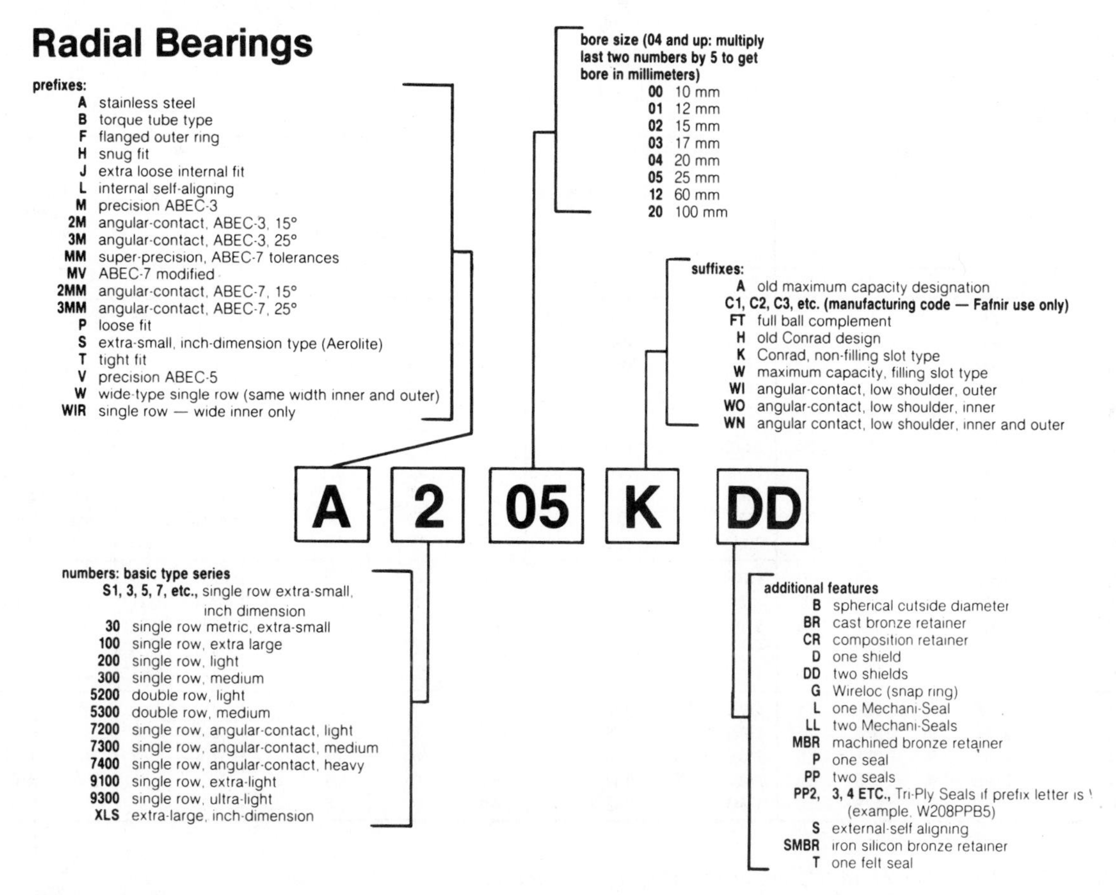

Figure 12–59 A sample bearing numbering system. *Courtesy The Torrington Company.*

BEARING SELECTION

A variety of bearing types are available from manufacturers. Bearing design may differ depending on the use requirements. For example, a supplier may have light, medium, and heavy bearings available. Bearings may have specially designed outer and inner rings. Bearings are available open (without seals or shields), or with one or two shields or seals. Light bearings are generally designed to accommodate a wide range of applications involving light to medium loads combined with relatively high speeds. Medium bearings have heavier construction than light bearings and provide a greater radial and thrust capacity. They are also able to withstand greater shock than light bearings. Heavy bearings are often designed for special service where extra heavy shock loads are required. Bearings may also be designed to accommodate radial loads, thrust loads, or a combination of loading requirements.

Bearing Bore, Outside Diameters, and Width

Bearings are dimensioned in relation to the bore diameter, outside diameter, and width. These dimensions are shown in Figure 12–60. After the loading requirements have been established, the bearing is selected in relationship to the shaft size. For example, if an approximate ∅ 1.5-inch shaft size is required for a medium service bearing, then a vendor's catalog chart similar to the one shown in Figure 12–61 is used to select the bearing.

Referring to the chart shown in Figure 12–61, notice that the first column is the vendor's bearing number, followed by the bore size (B). To select a bearing for the approximate 1.5-inch shaft, go to the chart and pick the bore diameter of 1.5748, which is close to 1.5. This is the 308K bearing. The tolerance for this bore is specified in the chart as 1.5748 + 0.0000 and –0.0005. Therefore, the limits dimension of the bore in this example is 1.5748 – 1.5743. The outside diameter is 3.5433 + 0.0000 – 0.0006 (3.5433 – 3.5427). The width of this bearing is 0.906 + 0.000 – 0.005 (0.906 – 0.901). The fillet radius is the maximum shaft or housing fillet radius

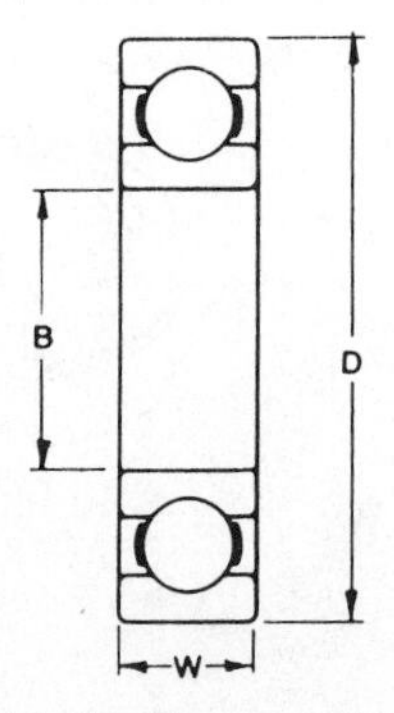

Figure 12–60 Bearing dimensions. *Courtesy The Torrington Company.*

DIMENSIONS — TOLERANCES

Bearing Number	Bore B		tolerance +.0000" +.000 mm to minus		Outside Diameter D		tolerance +.0000" +.000 mm to minus		Width W +.000", −.005" +.00 mm, −.13 mm		Fillet Radius(1)		Wt.		Static Load Rating C_O		Extended Dynamic Load Rating C_E	
	in.	mm	in.	mm	in.	mm	in.	mm	in.	mm	in.	mm	lbs.	kg	lbs.	N	lbs.	N
300K	.3937	10	.0003	.008	1.3780	35	.0005	.013	.433	11	.024	.6	.12	.054	850	3750	2000	9000
301K	.4724	12	.0003	.008	1.4567	37	.0005	.013	.472	12	.039	1.0	.14	.064	850	3750	2080	9150
302K	.5906	15	.0003	.008	1.6535	42	.0005	.013	.512	13	.039	1.0	.18	.082	1270	5600	2900	13200
303K	.6693	17	.0003	.008	1.8504	47	.0005	.013	.551	14	.039	1.0	.24	.109	1460	6550	3350	15000
304K	.7874	20	.0004	.010	2.0472	52	.0005	.013	.591	15	.039	1.0	.31	.141	1760	7800	4000	17600
305K	.9843	25	.0004	.010	2.4409	62	.0005	.013	.669	17	.039	1.0	.52	.236	2750	12200	5850	26000
306K	1.1811	30	.0004	.010	2.8346	72	.0005	.013	.748	19	.039	1.0	.78	.354	3550	15600	7500	33500
307K	1.3780	35	.0005	.013	3.1496	80	.0005	.013	.827	21	.059	1.5	1.04	.472	4500	20000	9150	40500
308K	1.5748	40	.0005	.013	3.5433	90	.0006	.015	.906	23	.059	1.5	1.42	.644	5600	24500	11000	49000
309K	1.7717	45	.0005	.013	3.9370	100	.0006	.015	.984	25	.059	1.5	1.90	.862	6700	30000	13200	58500
310K	1.9685	50	.0005	.013	4.3307	110	.0006	.015	1.063	27	.079	2.0	2.48	1.125	8000	35500	15300	68000
311K	2.1654	55	.0006	.015	4.7244	120	.0006	.015	1.142	29	.079	2.0	3.14	1.424	9500	41500	18000	80000
312K	2.3622	60	.0006	.015	5.1181	130	.0008	.020	1.220	31	.079	2.0	3.89	1.765	10800	48000	20400	90000
313K	2.5591	65	.0006	.015	5.5118	140	.0008	.020	1.299	33	.079	2.0	4.78	2.168	12500	56000	23200	102000
314K	2.7559	70	.0006	.015	5.9055	150	.0008	.020	1.378	35	.079	2.0	5.77	2.617	14300	63000	26000	116000
315K	2.9528	75	.0006	.015	6.2992	160	.0010	.025	1.457	37	.079	2.0	7.00	3.175	16000	71000	28500	125000
316K	3.1496	80	.0006	.015	6.6929	170	.0010	.025	1.535	39	.079	2.0	8.28	3.756	18000	80000	30500	137000
317K	3.3465	85	.0008	.020	7.0866	180	.0010	.025	1.614	41	.098	2.5	11.04	5.008	20000	90000	33500	146000
318K	3.5433	90	.0008	.020	7.4803	190	.0010	.030	1.693	43	.098	2.5	11.29	5.121	22400	98000	35500	156000
320K	3.9370	100	.0008	.020	8.4646	215	.0012	.030	1.8504	47	.098	2.5	15.62	7.085	28500	12700	41500	186000

(1) Maximum shaft or housing fillet radius which bearing corners will clear.

Figure 12–61 Bearing selection chart. *Courtesy The Torrington Company.*

in which the bearing corners will clear. The fillet radius for the 308K bearing is R 0.059. Notice that the dimensions are also given in millimeters.

Shaft and Housing Fits

Shaft and housing fits are important, because tight fits may cause failure of the balls, rollers, or lubricant, or overheating. Loose fits can cause slippage of the bearing in the housing, resulting in overheating, vibration, or excessive wear.

Shaft Fits. In general, for precision bearings, it is recommended that the shaft diameter and tolerance be the same as the bearing bore diameter and tolerance. The shaft diameter used with the 308K bearing is dimensioned ∅ 1.5748 – 1.5743.

Housing Fits. In most applications with rotating shafts, the outer ring is stationary and should be mounted with a push of the hand or light tapping. In general, the minimum housing diameter is 0.0001 larger than the maximum bearing outside diameter and the maximum housing diameter is 0.0003 larger than the minimum housing diameter. With this in mind, the housing diameter for the 308K bearing is 3.5433 + 0.0001 = 3.5434 and 3.5434 + 0.0003 = 3.5437. The housing diameter limits are 3.5437 – 3.5434.

The Shaft Shoulder and Housing Shoulder Dimensions

Next, you should size the shaft shoulder and housing shoulder diameters. The shaft shoulder and housing shoulder diameter dimensions are represented in Figure 12–62 as S and H. The shoulders should be large enough to rest flat on the face of the bearing and small enough to allow bearing removal. Refer to the chart in Figure 12–63 to determine the shaft shoulder and housing shoulder diameters for the 308K bearing selected in the preceding discussion. Find the basic bearing number 308 and determine the limits of the shaft shoulder and

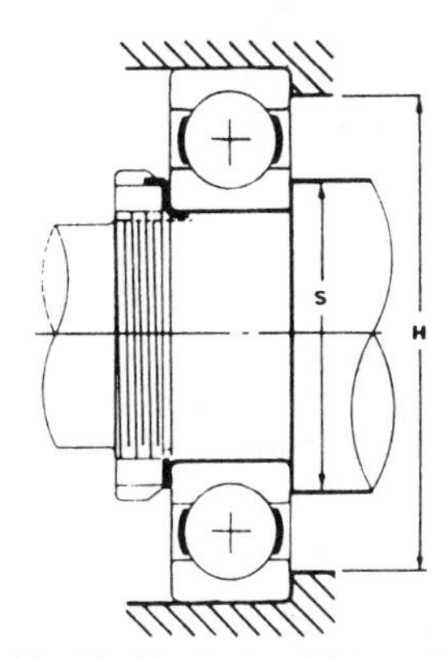

Figure 12–62 Shaft shoulder and housing shoulder dimensions. *Courtesy the Torrington Company.*

Extra-Light • 9100 Series

Basic Bearing Number	Shoulder Diameters shaft, S max.		shaft, S min.		housing, H max.		housing, H min.	
	in.	mm	in.	mm	in.	mm	in.	mm
9100	.52	13.2	.47	11.9	.95	24.1	.91	23.1
9101	.71	18.0	.55	14.0	1.02	25.9	.97	24.6
9102	.75	19.0	.67	17.0	1.18	30.0	1.13	28.7
9103	.81	20.6	.75	19.0	1.30	33.0	1.25	31.8
9104	.98	24.9	.89	22.6	1.46	37.1	1.41	35.8
9105	1.18	30.0	1.08	27.4	1.65	41.9	1.60	40.6
9106	1.38	35.1	1.34	34.0	1.93	49.0	1.88	47.8
9107	1.63	41.4	1.53	38.9	2.21	56.1	2.15	54.6
9108	1.81	46.0	1.73	43.9	2.44	62.0	2.39	60.7
9109	2.03	51.6	1.94	49.3	2.72	69.1	2.67	67.8
9110	2.22	56.4	2.13	54.1	2.91	73.9	2.86	72.6
9111	2.48	63.0	2.33	59.2	3.27	83.1	3.22	81.8
9112	2.67	67.8	2.53	64.3	3.47	88.1	3.42	86.9
9113	2.84	72.1	2.72	69.1	3.66	93.0	3.61	91.7
9114	3.11	79.0	2.91	73.9	4.06	103.1	3.97	100.8
9115	3.31	84.1	3.11	79.0	4.25	108.0	4.16	105.7
9116	3.56	90.4	3.31	84.1	4.65	118.1	4.50	114.3
9117	3.75	95.2	3.50	88.9	4.84	122.9	4.71	119.6
9118	4.03	102.4	3.84	97.5	5.16	131.1	5.13	130.3
9120	4.38	111.3	4.23	107.4	5.55	141.0	5.44	138.2
9121	4.66	118.4	4.53	115.1	5.91	150.1	5.75	146.0
9122	4.91	124.7	4.72	119.9	6.30	160.0	6.18	157.0
9124	5.28	134.1	5.12	130.0	6.69	169.9	6.50	165.1
9126	5.81	147.6	5.51	140.0	7.48	190.0	7.25	184.1
9128	6.06	153.9	5.81	147.6	7.88	200.2	7.68	195.1
9130	6.59	167.4	6.38	162.1	8.39	213.1	8.13	206.5
9132	6.96	176.8	6.56	166.6	9.00	228.6	8.75	222.2
9134	7.56	192.0	7.17	182.1	9.76	247.9	9.44	239.8
9138	8.38	212.9	7.95	201.9	10.95	278.1	10.50	266.7
9140	8.84	224.5	8.35	212.1	11.73	297.9	11.22	285.0

Light • 200, 7200WN Series

Basic Bearing Number	Shoulder Diameters shaft, S max.		shaft, S min.		housing, H max.		housing, H min.	
	in.	mm	in.	mm	in.	mm	in.	mm
200	.56	14.2	.50	12.7	.98	24.9	.97	24.6
201	.64	16.3	.58	14.7	1.06	26.9	1.05	26.7
202	.75	19.0	.69	17.5	1.18	30.0	1.15	29.2
203	.84	21.3	.77	19.6	1.34	34.0	1.31	33.3
204	1.00	25.4	.94	23.9	1.61	40.9	1.58	40.1
205	1.22	31.0	1.14	29.0	1.81	46.0	1.78	45.2
206	1.47	37.3	1.34	34.0	2.21	56.1	2.16	54.9
207	1.72	43.7	1.53	38.9	2.56	65.0	2.47	62.7
208	1.94	49.3	1.73	43.9	2.87	72.9	2.78	70.6
209	2.13	54.1	1.94	49.3	3.07	78.0	2.97	75.4
210	2.34	59.4	2.13	54.1	3.27	83.1	3.17	80.5
211	2.54	64.5	2.41	61.2	3.68	93.5	3.56	90.4
212	2.81	71.4	2.67	67.8	3.98	101.1	3.87	98.3
213	3.03	77.0	2.86	72.6	4.37	111.0	4.19	106.4
214	3.22	81.8	3.06	77.7	4.57	116.1	4.41	112.0
215	3.44	87.4	3.25	82.6	4.76	120.9	4.59	116.6
216	3.69	93.7	3.55	90.2	5.12	130.0	4.93	125.2
217	3.88	98.6	3.75	95.2	5.51	140.0	5.31	134.9
218	4.16	105.7	3.94	100.1	5.91	150.1	5.62	142.7
219	4.38	111.3	4.21	106.9	6.22	158.0	6.06	153.9
220	4.63	117.6	4.41	112.0	6.61	167.9	6.31	160.3
221	4.88	124.0	4.61	117.1	7.01	178.1	6.88	174.8
222	5.13	130.3	4.80	121.9	7.40	188.0	7.06	179.3
224	5.63	143.0	5.20	132.1	7.99	202.9	7.56	192.0
226	6.00	152.4	5.67	144.0	8.50	215.9	8.13	206.5
228	6.50	165.1	6.06	153.9	9.29	236.0	8.81	223.8
230	6.97	177.0	6.46	164.1	10.08	256.0	9.50	241.3
—	—	—	—	—	—	—	—	—
—	—	—	—	—	—	—	—	—
—	—	—	—	—	—	—	—	—

Medium • 300, 7300WN Series

Basic Bearing Number	Shoulder Diameters shaft, S max.		shaft, S min.		housing, H max.		housing, H min.	
	in.	mm	in.	mm	in.	mm	in.	mm
300	.59	15.0	.50	12.7	1.18	30.0	1.15	29.2
301	.69	17.5	.63	16.0	1.22	31.0	1.21	30.7
302	.81	20.6	.75	19.0	1.42	36.1	1.40	35.6
303	.91	23.1	.83	21.1	1.61	40.9	1.60	40.6
304	1.06	26.9	.94	23.9	1.77	45.0	1.75	44.4
305	1.31	33.3	1.14	29.0	2.17	55.1	2.09	53.1
306	1.56	39.6	1.34	34.0	2.56	65.0	2.44	62.0
307	1.78	45.2	1.69	42.9	2.80	71.1	2.72	69.1
308	2.00	50.8	1.93	49.0	3.19	81.0	3.06	77.7
309	2.28	57.9	2.13	54.1	3.58	90.9	3.41	86.6
310	2.50	63.5	2.36	59.9	3.94	100.1	3.75	95.2
311	2.75	69.8	2.56	65.0	4.33	110.0	4.13	104.9
312	2.94	74.7	2.84	72.1	4.65	118.1	4.44	112.8
313	3.19	81.0	3.03	77.0	5.04	128.0	4.81	122.2
314	3.44	87.4	3.23	82.0	5.43	137.9	5.13	130.3
315	3.88	98.6	3.43	87.1	5.83	148.1	5.50	139.7
316	3.94	100.1	3.62	91.9	6.22	158.0	5.88	149.4
317	4.13	104.9	3.90	99.1	6.54	166.1	6.19	157.2
318	4.38	111.3	4.09	103.9	6.93	176.0	6.50	165.1
319	4.63	117.6	4.29	109.0	7.32	185.9	6.88	174.8
320	4.88	124.0	4.49	114.0	7.91	200.9	7.38	187.4
321	5.13	130.3	4.69	119.1	8.31	211.1	7.75	196.8
322	5.50	139.7	4.88	124.0	8.90	226.1	8.25	209.6
324	6.00	152.4	5.28	134.1	9.69	246.1	8.93	226.8
326	6.44	163.6	5.83	148.1	10.32	262.1	9.69	246.1
328	6.93	176.0	6.22	158.0	11.10	281.9	10.38	263.7
330	7.44	189.0	6.61	167.9	11.89	302.0	11.06	280.9
—	—	—	—	—	—	—	—	—
—	—	—	—	—	—	—	—	—
—	—	—	—	—	—	—	—	—

Figure 12–63 Shaft shoulder and housing shoulder dimension selection chart. *Courtesy The Torrington Company.*

the housing shoulder. The shaft shoulder diameter is 2.00 – 1.93 and the housing shoulder diameter is 3.19 – 3.06. Now you are ready to detail the bearing location on the housing drawing and on the shaft drawing. A partial detailed drawing of the shaft and housing for the 308K bearing is shown in Figure 12–64.

Surface Finish of Shaft and Housing

The recommended surface finish for precision bearing applications is 32 microinches (0.80 micrometer) for the shaft finish on shafts under 2 inches in diameter. For shafts over 2 inches in diameter, a 63-microinch (1.6 micrometer) finish is suggested. The housing diameter may have a 125-microinch (3.2 micrometer) finish for all applications.

Bearing Lubrication

It is necessary to maintain a film of lubrication between the bearing surfaces. The factors to consider when selecting lubrication requirements include the:

- type of operation, such as continuous or intermittent.
- service speed in RPMs (revolutions per minute).
- bearing load, such as light, medium, or heavy.

Bearings may also be overlubricated, which may cause increased operating temperatures and early failure. Selection of the proper lubrication for the application should be determined by the manufacturer's recommendations. The ability of the lubricant is due, in part, to *viscosity*. Viscosity is the internal friction of a fluid, which makes it resist a tendency to flow. Fluids with low viscosity flow more freely than those with high viscosity. The chart in Figure 12–65 shows the selection of oil viscosity based on temperature ranges and speed factors.

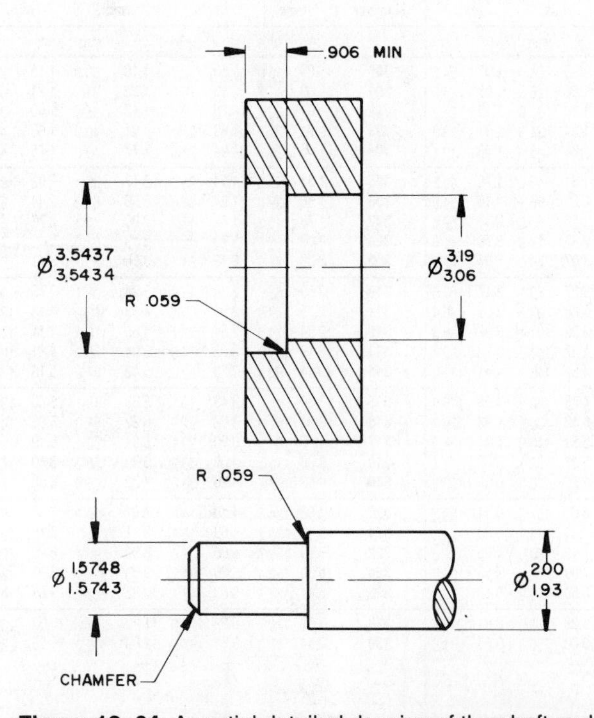

Figure 12–64 A partial detailed drawing of the shaft and housing for the 308K bearing.

Oil Grooving of Bearings

In situations where bearings or bushings do not receive proper lubrication, it may be necessary to provide grooves for the proper flow of lubrication to the bearing surface. The bearing grooves help provide the proper lubricant between the bearing surfaces and maintain adequate cooling. There are several methods of designing paths for the lubrication to the bearing surfaces as shown in Figure 12–66.

Sealing Methods

Machine designs normally include means for stopping leakage and keeping out dirt and other contaminants when lubricants are involved in the machine operation. This is accomplished using static or dynamic sealing devices. Static sealing refers to stationary devices that are held in place and stop leakage by applied pressure. Static seals such as gaskets do not come in contact with the moving parts of the mechanism. Dynamic seals are those that contact the moving parts of the machinery, such as packings.

Gaskets are made from materials that prevent leakage or access of dust contaminants into the machine cavity. Silicone rubber gasket materials are used in applications such as water pumps, engine filter housings, and oil pans. Gasket tapes, ropes, and strips provide good cushioning properties for dampening vibration, and the adhesive sticks well to most materials. Nonstick gasket

Oil Viscosities and Temperature Ranges for Ball Bearing Lubrication

Maximum Temperature Range Degrees F	Optimum Temperature Range Degrees F	Speed Factor, S_i (inner race bore diameter (inches) X RPM) Under 1000	Over 1000
		Viscosity	
−40 to +100	−40 to −10	80 to 90 SSU (at 100 deg. F)	70 to 80 SSU (At 100 deg. F)
−10 to +100	−10 to +30	100 to 115 SSU (at 100 deg. F)	80 to 100 SSU (At 100 deg. F)
+30 to +150	+30 to +150	SAE 20	SAE 10
+30 to +200	+150 to +200	SAE 40	SAE 30
+50 to +300	+200 to +300	SAE 70	SAE 60

Figure 12–65 Selection of oil viscosity based on temperature ranges and speed factors.

materials such as paper, cork, and rubber are available for certain applications. Figure 12–67 shows a typical gasket mounting.

Dynamic seals include packings and seals that fit tightly between the bearing or seal seat and the shaft. The pressure applied by the seal seat or the pressure of the fluid causes the sealing effect. Molded lip packings that provide sealing as a result of the pressure generated by the machine fluid are available. Figure 12–68 shows examples of molded lip packings. Molded ring seals are placed in a groove and provide a positive seal between the shaft and bearing or bushing. Types of molded ring seals include labyrinth, O-ring, lobed ring, and others. Labyrinth, which means maze, refers to a seal that is made of a series of spaced strips that are connected to the seal seat, making it difficult for the lubrication to pass. Labyrinth seals are used in heavy machinery where some leakage is permissible. (See Figure 12–69). The O-ring seal is the most commonly used seal because of its low cost, ease of application, and flexibility. The O-ring may be used for most situations involving rotat-

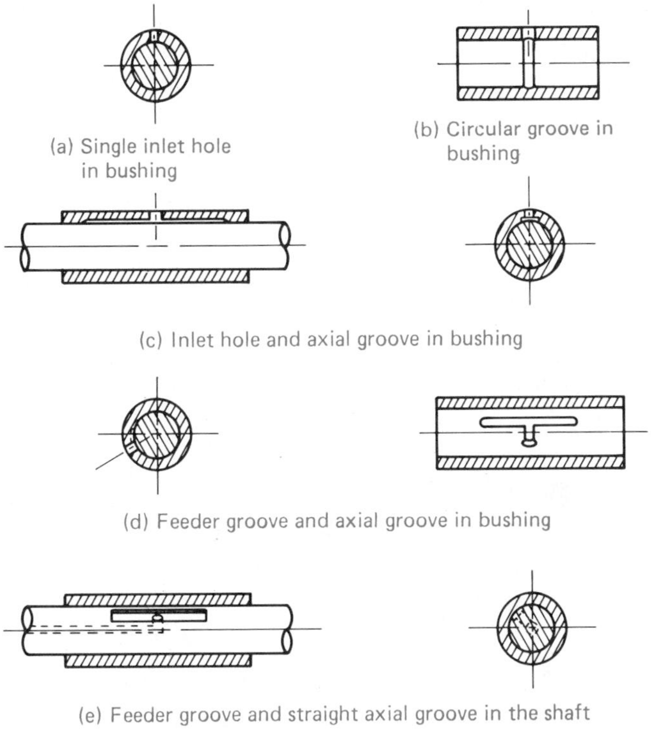

(a) Single inlet hole in bushing

(b) Circular groove in bushing

(c) Inlet hole and axial groove in bushing

(d) Feeder groove and axial groove in bushing

(e) Feeder groove and straight axial groove in the shaft

Figure 12–66 Methods of designing paths for lubrication to bearing surfaces.

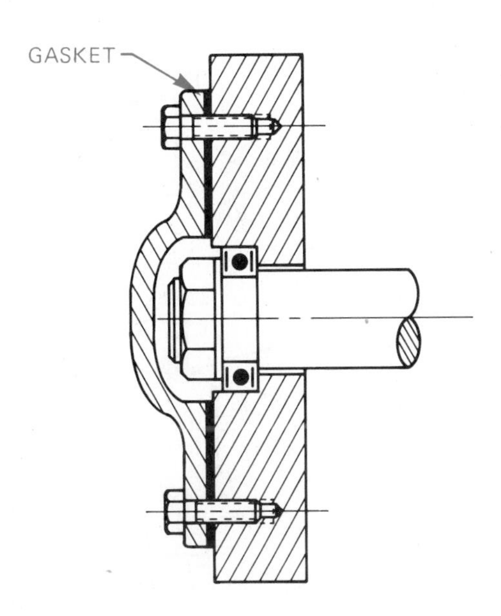

Figure 12–67 Typical gasket mounting.

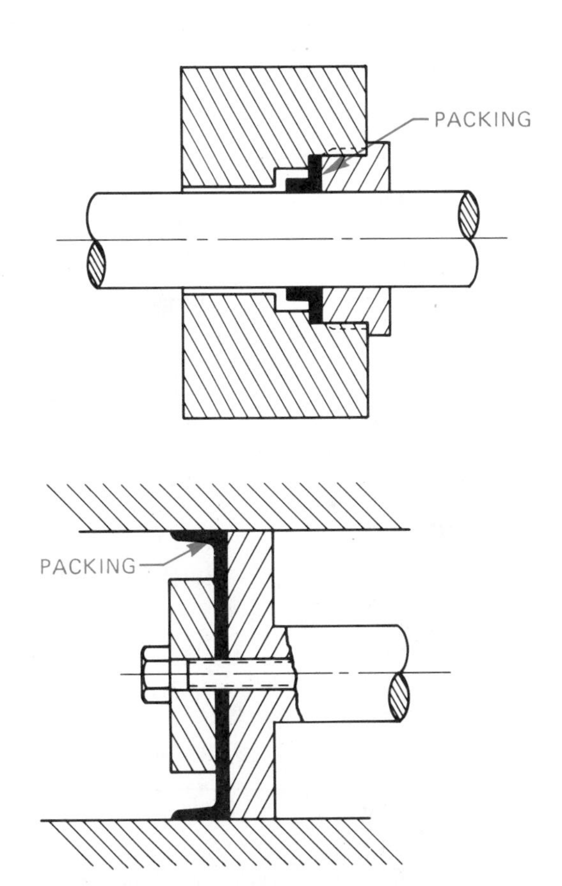

Figure 12–68 Molded lip packings.

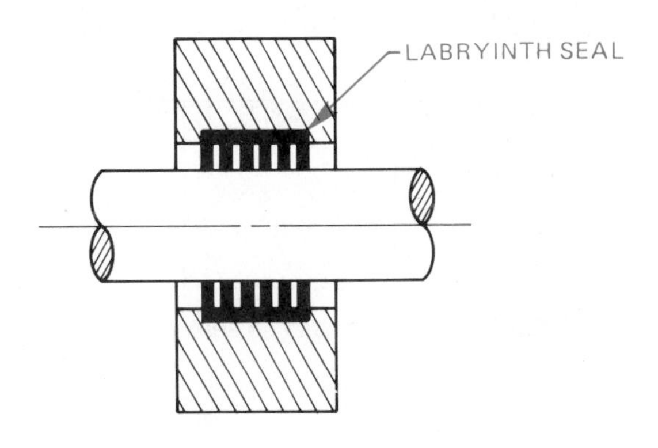

Figure 12–69 Labyrinth steel.

ing or oscillating motion. The O-ring is placed in a groove that is machined in either the shaft or the housing as shown in Figure 12–70. The lobed ring has rounded lobes that provide additional sealing forces over the standard O-ring seal. A typical lobed ring seal is shown in Figure 12–71.

Felt and wool seals are used where economical cost, lubricant absorption, filtration, low friction, and a polishing action are required. However, the ability to completely seal the machinery is not as positive as with the seals described earlier. (See Figure 12–72.)

Bearing Mountings

There are a number of methods used for holding the bearing in place. Common techniques include a nut and lock washer, a nut and lock nut, or a retaining ring. Other methods may be designed to fit the specific application or requirements, such as a shoulder plate. Figure 12–73 shows some examples of mountings.

GEAR AND BEARING ASSEMBLIES

Gear and bearing assemblies show the parts of the complete mechanism as they appear assembled. (See Figure 12–74.) When drawing assemblies, you need to use as few views as possible, but enough to adequately display how all the parts fit together. In some situations all that is needed is a full sectional view that displays all of the internal components. An exterior view such as a front or top view plus a section sometimes works. Dimensions are normally omitted from the assembly unless the

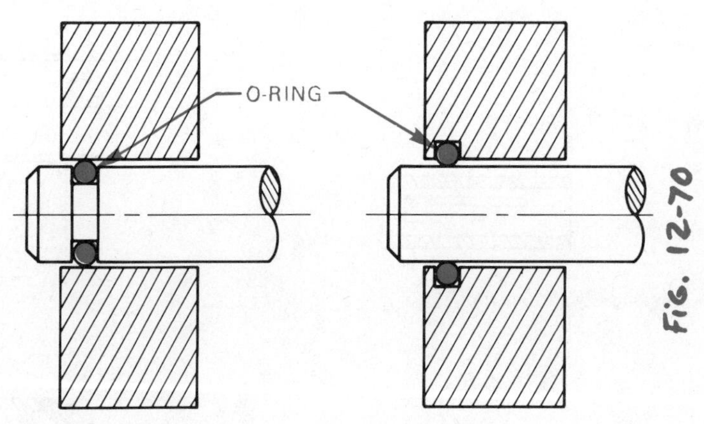

Figure 12–70 O-ring seals.

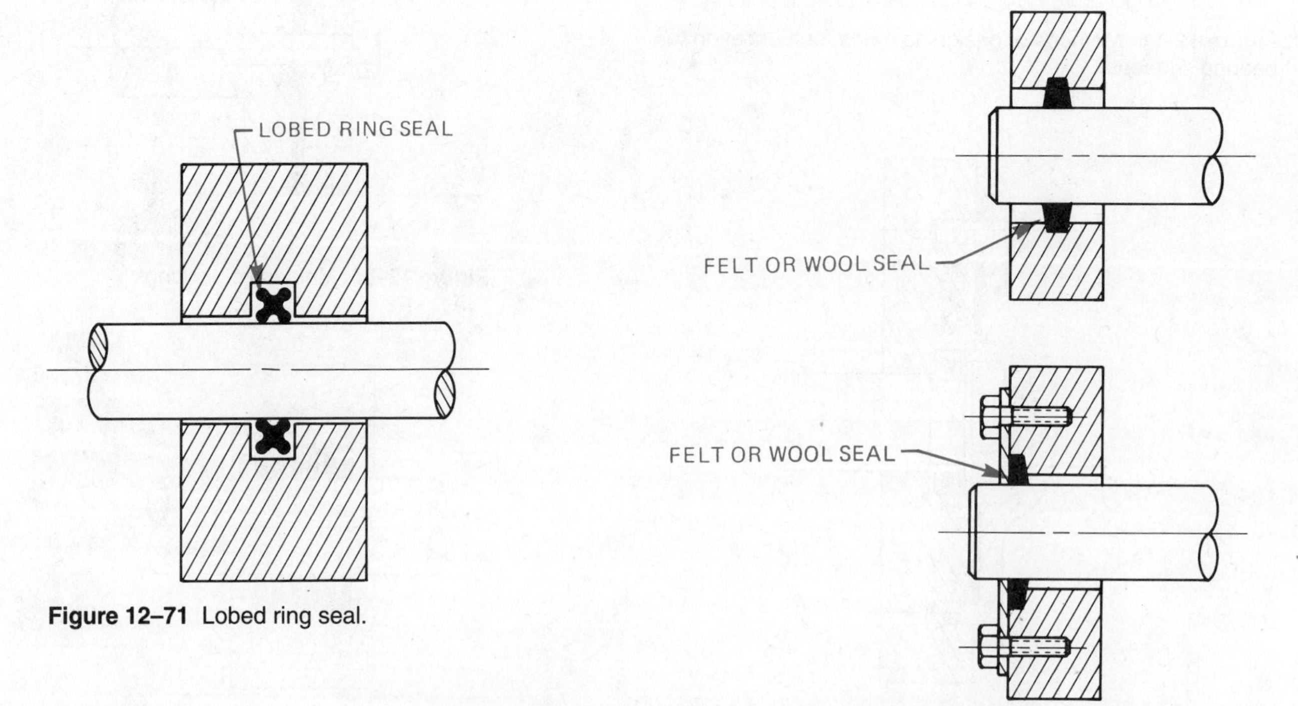

Figure 12–71 Lobed ring seal.

Figure 12–72 Felt and wool seals.

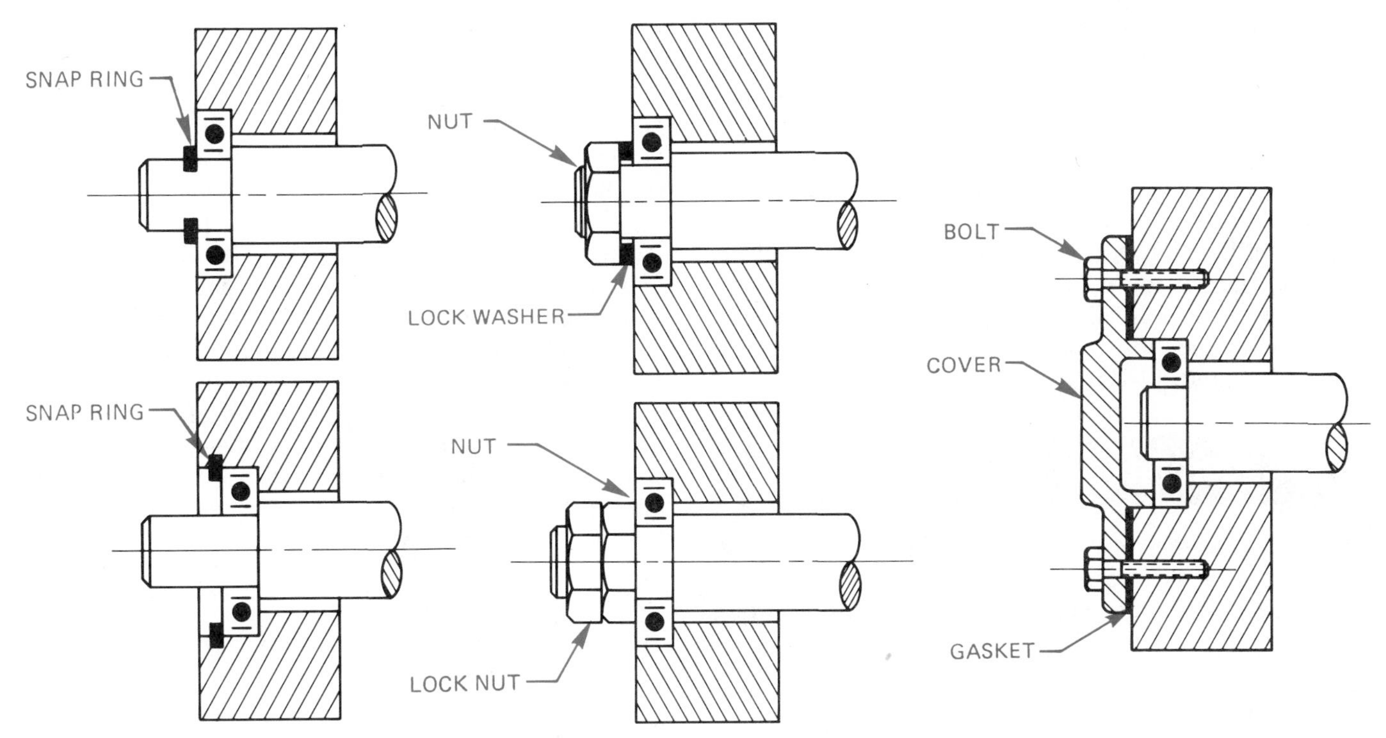

Figure 12–73 Typical bearing mountings.

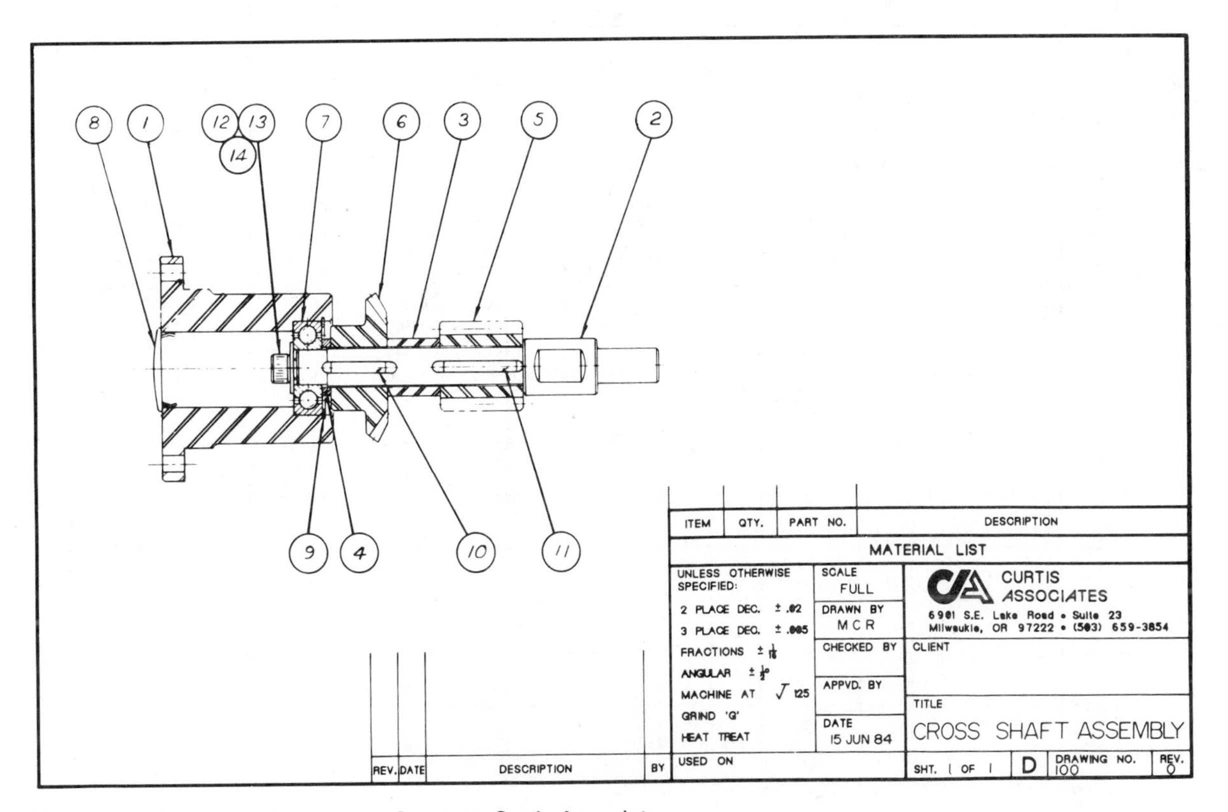

Figure 12–74 Assembly drawing. *Courtesy Curtis Associates.*

dimensions are needed for assembly purposes. For example, when a specific dimension regarding the relationship of one part to another is required to properly assemble the parts, each part is identified with a number in a circle. This circle is referred to as a balloon. The balloons are connected to the part being identified with a leader. The balloons are about one-half inch in diameter. The identification number is quarter-inch-high lettering or text. The balloon numbers correlate with a parts list. The parts list is normally placed on the drawing above or adjacent to the title block, or on a separate sheet as shown in Figure 12–75.

PROFESSIONAL PERSPECTIVE

In every situation regarding gear design there is a need to investigate all of the manufacturing alternatives and provide a solution the customer can afford. Competition is so tough in the manufacturing industry that you should evaluate each design to find the best way to produce the product. The best way to understand this concept as an entry-level engineering drafter is to talk to experienced designers, engineers, and machinists. Go to the shop to see how things are done and determine the drawing requirements directly from the people who know. According to one design engineer, "If you don't know how it is going to be made, you're not a good drafter. Know the manufacturing capabilities of each piece of equipment." In addition to your drafting courses it is a good idea to take some manufacturing technology classes. Math is also an important part of your program. Drafters in this field use a lot of geometry and trigonometry. After you have a strong educational background, the experienced engineer says, "Keep an open mind and look at all the alternatives."

ASSEMBLY Cross Shaft Assembly   USED ON

NUMBER OF UNITS   DATE

HAVE	NEED	P/Ø NO. W/Ø NO.	DET. NO.	PART NO.	DWG.	QTY.	PART NAME	DESCRIPTION	VENDOR
			1		B	1	Bearing Retainer	Ø 3" C.D. Bar	
			2		B	1	Cross Shaft	Ø 3/4" C.D. Bar	
			3		A	1	Spacer	Ø 3/4 O.D. x 11GA Wall Tube x .738 Thick	
			4		A	1	Spacer	Ø 3/4 O.D. x 11GA Wall Tube x .125 Thick	
			5		-	1	Steel Worm 12 D.P. Single Thread	Boston Gear #H 1056 R.H.	
			6		-	1	Bevel Gear 20° P.A. 2" P.D.	Boston Gear #HL 149 Y-G	
			7		-	1	Ball Bearing .4724 Ø Bore	T.R.W. or Equivalent #MRC 201-S22	
			8		-	1	End Plug		
			9		-	1	Snap Ring	Waldes-Truarc #N 5000-125	
			10		-	1	Key Stock	1/8 Sq. x 3/4 Lg.	
			11		-	1	Key Stock	1/8 Sq. x 1 Lg.	
			12		-	1	Socket Head Cap Screw	1/4 UNC x 3/4 Lg.	
			13		-	1	Lockwasher	1/4 Nominal	
			14		-	1	Flat Washer	1/4 Nominal	

Figure 12–75 Parts list. *Courtesy Curtis Associates.*

CAM, GEAR, AND BEARING PROBLEMS

DIRECTIONS

Please read problems carefully before you begin working. Complete each problem on an appropriately sized sheet. Precision work is important for accurate solutions.

CAM DISPLACEMENT DIAGRAMS

Problem 12–1 Construct a cam displacement diagram for a cam follower that rises in simple harmonic motion a total of 2.00 in. in 150°, dwells for 30°, falls 2.00 in. simple harmonic motion in 120°, and dwells for 60°. Draw the horizontal scale 6.00 in.

Problem 12–2 Construct a cam displacement diagram for a cam follower that rises in uniform accelerated motion a total of 2.00 in. in 180°, dwells for 30°, falls 2.00 in. uniform accelerated motion in 120°, and dwells for 30.° Draw the horizontal 6.00 in.

Problem 12–3 Construct a cam displacement diagram for a cam follower that rises in modified constant velocity for 3.00 in. in 180°, falls 3.00 in. modified constant velocity motion in 120°, and dwells for 60°. Use a modified constant velocity motion designed with one-third of the displacement.

Problem 12–4 Construct a cam displacement diagram for a cam follower that rises 2.000 in. cycloidal motion in 120°, dwells for 60°, and falls 2.000 in. in cycloidal motion in 180°.

Problem 12–5 Construct a cam displacement diagram for a cam follower that rises in simple harmonic motion a total of 1.250 in. in 90°, dwells for 60°, rises .750 in. in 45° simple harmonic motion, falls 2.00 in. with cycloidal motion in 120°. Draw the horizontal scale 12 in.

Problem 12–6 Construct a cam displacement diagram for a cam follower that rises in modified constant velocity motion (modified to one-third the displacement) for 3.000 in. in 180°, dwells for 30°, and falls 3.000 in. simple harmonic motion in 30°, and dwells to the end of the cycle. Draw the horizontal scale 12 in.

Problem 12–7 Construct a cam displacement diagram for a cam follower that rises 3.500 in. in 90° cycloidal motion, dwells for 45°, falls 2.500 in. cycloidal motion in 135°, dwells for 30°, falls 1.000 in. simple harmonic motion in 30°, and dwells to the end of the cycle. Draw the horizontal scale 12 in.

Problem 12–8 Construct a cam displacement diagram for a cam follower that rises in cycloidal motion for 3.000 in. in 90°, dwells for 30°, falls 1.000 in. in simple harmonic motion in 90°, dwells for 30°, and falls the remaining 2.000 in. in uniform accelerated motion in 120°. Draw the horizontal scale 12 in.

Problem 12–9 Construct a cam displacement diagram for a cam follower that rises in cycloidal motion a total of 2.1875 in. in 150°, dwells for 30°, falls back to the original level in simple harmonic motion in 150°, then dwells through the remainder of the cycle. Use a 12-in. horizontal scale.

Problem 12–10 Construct a cam displacement diagram for a cam follower that raises a .375 in. diameter in-line roller follower 1.500 in. in uniform accelerated motion in 150°, dwells 45°, falls with modified constant velocity (one-third displacement) in 120°, and dwells the remainder of the cycle.

Problem 12–11 Construct a cam displacement diagram for a cam follower that rises 3.500 in. 90° cycloidal motion, dwells for 30°, falls 2.250 in. cycloidal motion in 150°, falls 1.250 in. simple harmonic motion in 60°, and dwells for 30.° Draw the horizontal scale equal in circumference to a 3.000-in. diameter circle.

CAM PROFILE DRAWINGS

Problem 12–12 Use the displacement diagram constructed in Problem 12–1 and the information given in the illustration below to lay out the plate cam profile drawing. The cam rotates counterclockwise. Make a two-view detailed drawing of the cam and dimension it as shown in this chapter.

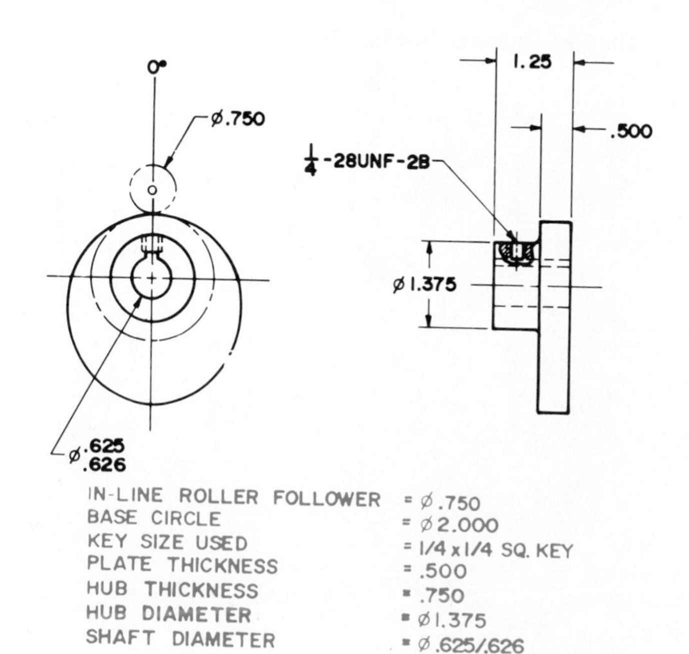

Problem 12–13 Use the displacement diagram constructed in Problem 12–4 and the information given in the illustration for Problem 12–12 to lay out the plate cam profile. The cam is rotating clockwise. Make a two-view detailed drawing of the cam, properly toleranced and dimensioned.

Problem 12–14 Make a two-view detailed drawing of a plate cam using the displacement diagram from Problem 12–10. Completely dimension the drawing using the following information:

The cam rotates counterclockwise.
In-line roller follower = Ø .375
Base circle = Ø 2.750
Shaft = Ø 1.000
Hub diameter = Ø1.750
Hub projection = .500
Plate thickness = .375
All dimensions in inches.

Problem 12–15 Make a two-view detailed drawing of a plate cam using the displacement diagram from Problem 12–6. Completely dimension the drawing using the following information:

The cam rotates counterclockwise.
In-line roller follower = Ø .750
Base circle = Ø 2.000
Shaft = Ø .625
Hub diameter = Ø 1.250
Hub projection = .500
Plate thickness = .500
All dimensions in inches.

Problem 12–16 Make a two-view detailed drawing of a plate cam using the displacement diagram from Problem 12–4. Completely dimension the drawing using the information shown below:

The cam rotates clockwise.

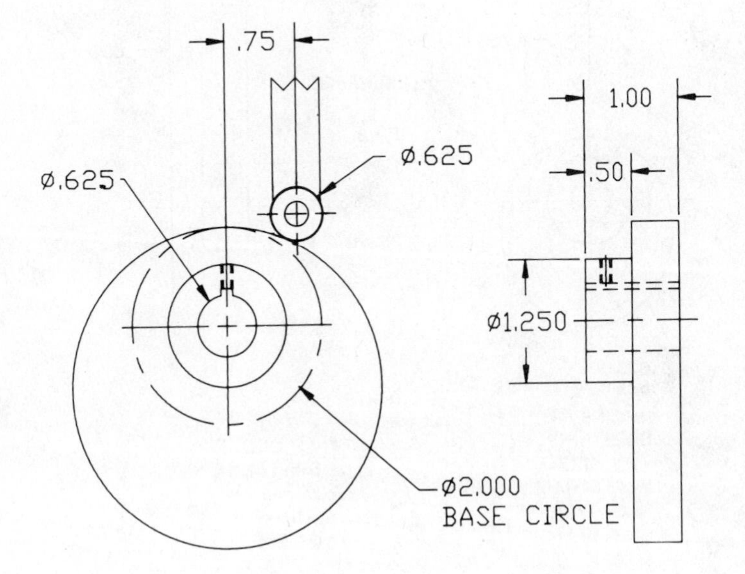

Problem 12–17 Use the displacement diagram constructed in Problem 12–11 and the following illustration to lay out the profile of the groove in the drum cam. The cam rotates clockwise. Make a two-view detailed drawing of the drum cam, with tolerances and dimensions as discussed and shown in this chapter.

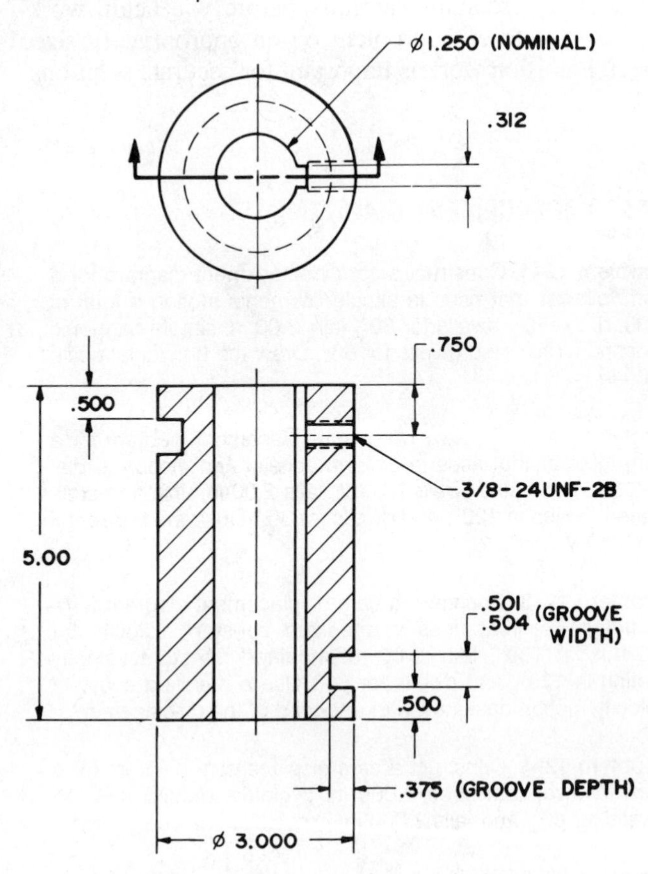

GEAR PROBLEMS

Problem 12–18 Given the gear train and chart shown below, calculate or determine the missing information to complete the chart.

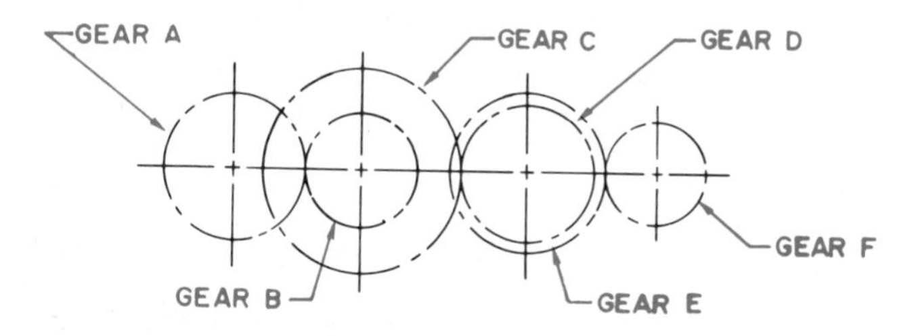

GEAR	DIAMETRAL PITCH (D)	NUMBER OF TEETH (N)	PITCH DIAMETER (P)	RPM	DIRECTION	CENTER DISTANCE	GEAR RATIO
A	4		7.5"	240	CLOCKWISE		
B		18					
C			10.0"	400			
D	5	40					
E	7						
F		14		1500			

Problem 12–19 Given the ten-gear power transmission and chart shown below, calculate or determine the missing information to complete the chart.

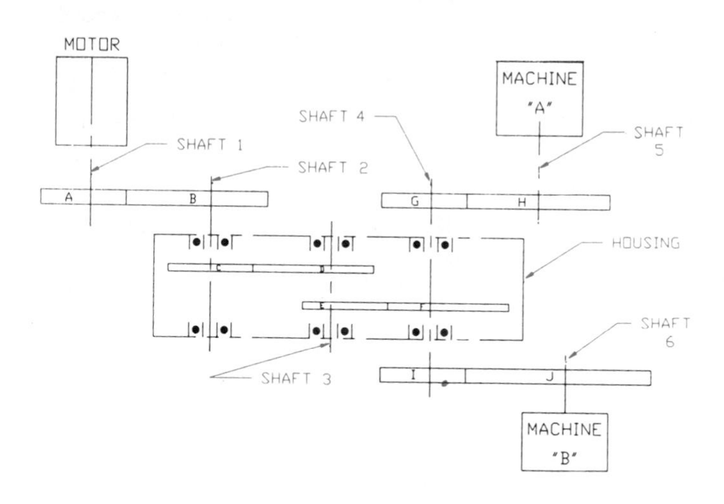

Gear Data Block

Gear	Pitch Diameter	No. of Teeth	Diametral Pitch	RPM	Ctr. Distance Between Mating Gears	Direction
A	3.00			3600	4.00	Counter-clockwise
B			5			
C		48				
D			12	1080		
E	4.00	40				
F		100				
G	5.00				6.00	
H			6			
I			4			
J		40		108		

Problem 12–20 Given the following information, use ANSI standards to make a detailed drawing of the spur gear shown below:

20 teeth	Face width = 2.500 in.
Diametral pitch = 5	Shaft diameter = ∅ 1.125
20° pressure angle	Keyway for a .25 in. square key

Place the centerline of the keyway in line with a radial line through the center of one tooth (one tooth profile needed to show alignment). Include the necessary spur gear data in a chart placed over the title block. Use the formulas given in this chapter to solve for unknown values.

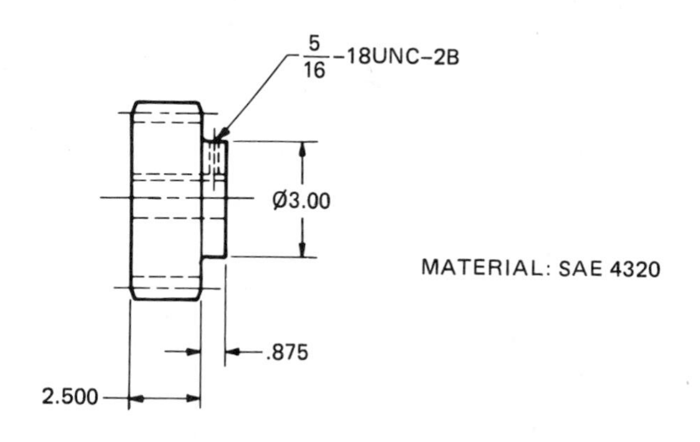

Problem 12–21 Use ANSI standards to make a detailed drawing of a rack that mates with the spur gear in Problem 12–20. The overall length is 24 in.

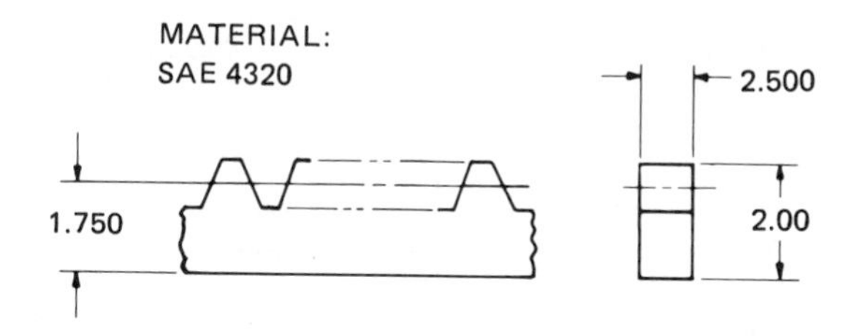

Problem 12–22 Use ANSI standards to make a detailed drawing of a straight bevel gear given the following information and the illustration shown below:

Pitch diameter = ∅ 8.000 in.
Pressure angle = 20°
32 teeth
Diametral pitch = 4
Face width = 1.400
Shaft diameter = ∅ 1.125 in.
Use a .250 square key.

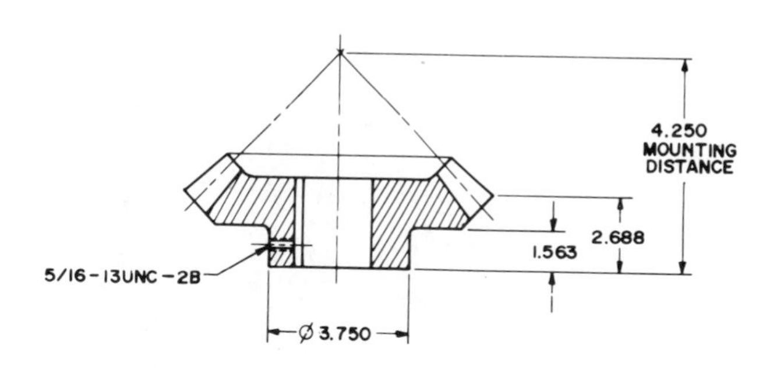

BEARING PROBLEMS

Problem 12–23 Use the charts shown in this chapter to establish the following medium-service bearing dimensions for an approximate ∅ 1.25-in. shaft:

Bearing catalog number ____________
Bore ____________
Outside diameter ____________
Width ____________
Fillet radius ____________
Shaft shoulder diameter ____________
Housing shoulder diameter ____________
Shaft diameter ____________
Housing diameter ____________

Problem 12–24 Use the charts shown in this chapter to establish the following medium-service bearing dimensions for an approximate ∅ 3.5-in. shaft:

Bearing catalog number ____________
Bore ____________
Outside diameter ____________
Width ____________
Fillet radius ____________
Shaft shoulder diameter ____________
Housing shoulder diameter ____________
Shaft diameter ____________
Housing diameter ____________

Problem 12–25 Use the charts shown in this chapter to establish the following medium-service bearing dimensions for an approximate ∅ 20-mm shaft:

Bearing catalog number ____________
Bore ____________
Outside diameter ____________
Width ____________
Fillet radius ____________
Shaft shoulder diameter ____________
Housing shoulder diameter ____________
Shaft diameter ____________
Housing diameter ____________

Problem 12–26 Use the charts shown in this chapter to establish the following medium-service bearing dimensions for an approximate ∅ 60-mm shaft:

Bearing catalog number ____________
Bore ____________
Outside diameter ____________
Width ____________
Fillet radius ____________
Shaft shoulder diameter ____________
Housing shoulder diameter ____________
Shaft diameter ____________
Housing diameter ____________

GEAR DESIGN, BEARING SELECTION, SHAFT DESIGN PROBLEM

Problem 12–27 Design a two-speed gear reducer that will operate eight to ten hours per day and receive moderate shock while in operation. Use the following information:

A 5-hp 1750 electric motor supplies the input power.
There are six gears arranged approximately as shown.
Gear C-D is a cluster gear sliding on the countershaft.
The output speed is 625 RPM when gear C is engaged with gear E.
The output speed is 437.5 RPM when gear D is engaged with gear F.
Gear A has 32 teeth, gear B has 64 teeth, gear C has 25 teeth, and gear D has 24 teeth.

Use the gear and bearing information from this chapter to design the gear reducer and do the following:

1. Determine the diametral pitches for all six gears.
2. Determine the number of teeth for gears E and F.
3. Use tolerances, surface finishes, and fit as discussed in this chapter.
4. Use manufacturer's catalogs shown in this chapter or supplied by your instructor to select standard parts.
5. Make a detailed drawing of the cluster gear, including spur gear data charts for both gears on one sheet.
6. Make detailed drawings of the three shafts — input, output, and countershaft — each on a separate sheet. The shafts are approximately ∅ 1.250 in. Design the shafts based on bearing specifications and fits. Design the keyseats based on the shaft diameter given and as specified in the *Machinery's Handbook* or other source.
7. Make an assembly drawing of the entire product. Prepare a complete parts list placed over the title block. All parts should be ballooned on the drawing to coordinate with the parts list. The dimensions given in the illustration are for reference only. Detailed dimensions are not required on the assembly drawing.

Problem 12–27 Continued.

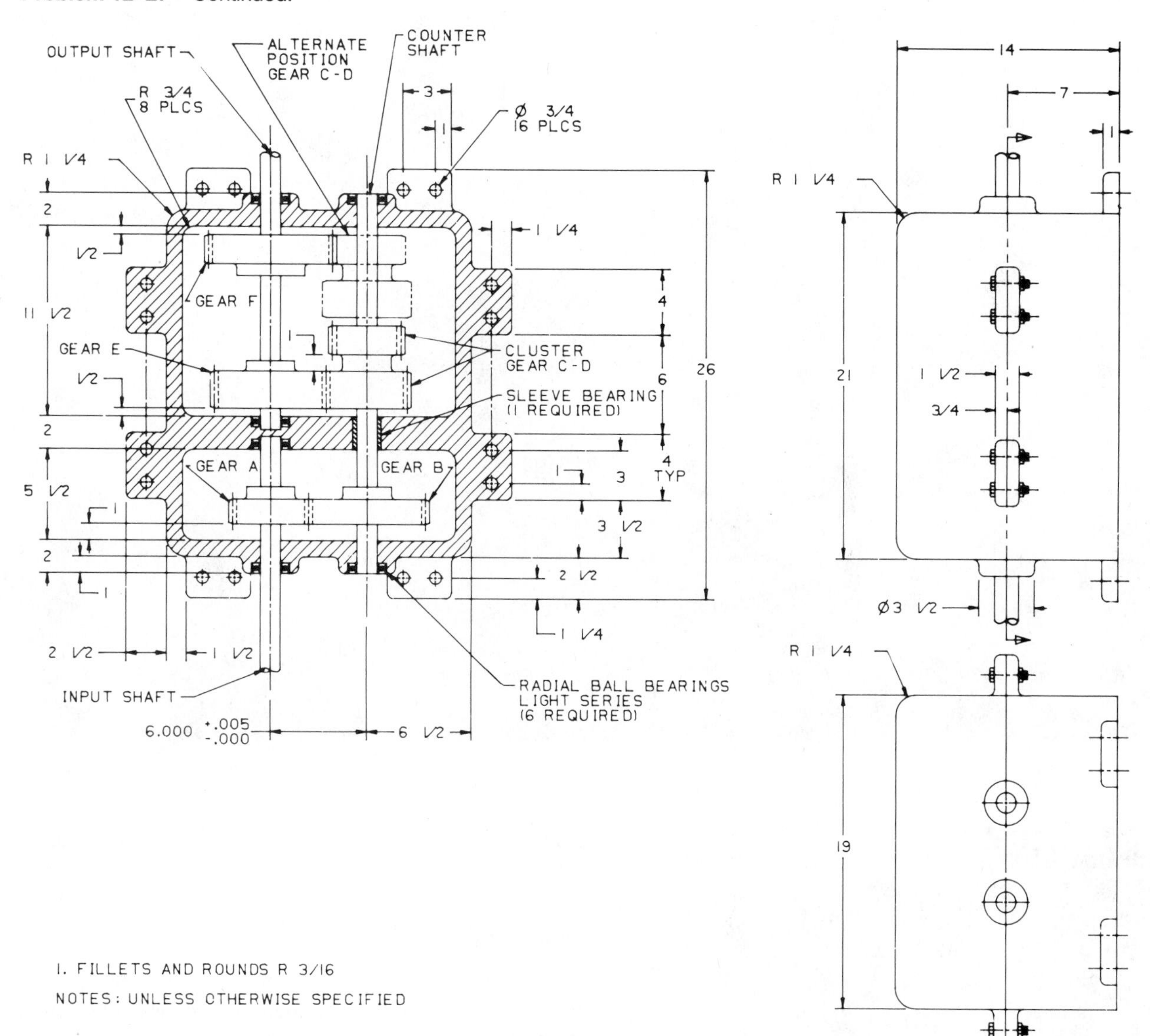

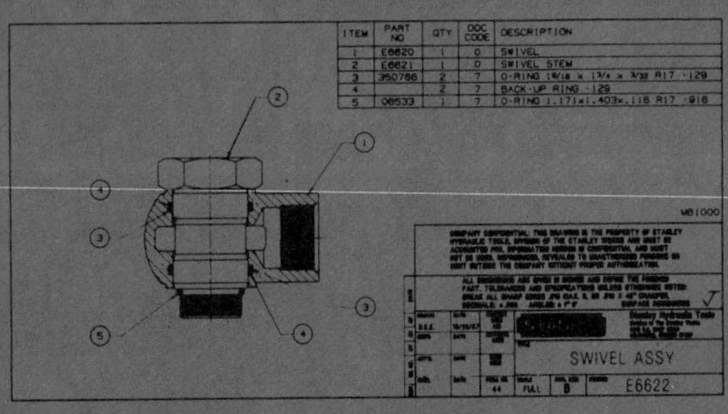

Courtesy Stanley Hydraulic Tools, Division of The Stanley Works.

CHAPTER 13

Working Drawings

LEARNING OBJECTIVES

After completing this chapter, you will:

- Draw complete sets of working drawings, including details, assemblies, and parts lists.
- Prepare written specifications of purchase parts for the parts list.
- Properly group information on the assembly drawing with identification numbering systems.
- Explain the engineering change process.

The drawings that have been shown as examples or assigned as problems in this text are called detail drawings. Detail drawings are the kind of drawings that most entry-level mechanical drafters prepare. When a product is designed and drawings are made for manufacturing, each part of the product must have a drawing. These drawings of individual parts are referred to as detail drawings. Component parts are assembled to create a final product, and the drawing that shows how the parts go together is called the assembly drawing. Associated with the assembly drawing and coordinated to the detail drawings is the parts list. When the detail drawings, assembly drawing, and parts list are combined, they are referred to as a complete set of working drawings. Working drawings, then, are a set of drawings that supply all the information necessary to manufacture any given product. A set of working drawings includes all the information and instructions needed for the purchase or construction of parts and the assembly of those parts into a product.

DETAIL DRAWINGS

Detail drawings, used by workers in manufacturing, are drawings of each part contained in the assembly of a product. The only parts that may not have to be drawn are standard parts. Standard parts are items that can be purchased from an outside supplier more economically than they can be manufactured. Examples of standard, or purchased, parts are common bolts, screws, pins, keys, and any other product that can be purchased from a vendor. Standard parts do not have to be drawn because a written description will clearly identify the part as shown in Figure 13–1. Detail drawings contain some or all of the following items:

1. Necessary multiviews
2. Dimensional information
3. Identity of the part, project name, and part number
4. General notes and specific manufacturing information
5. The material of which the part is made
6. The assembly that the part fits (could be keyed to the part number)
7. Number of parts required per assembly
8. Name of persons who worked on or with the drawing
9. Engineering changes and related information

1/2 - 13UNC-2 X 1.5 LG, SOCKET HEAD CAP SCREW

Figure 13–1 Written description of standard or purchase part.

In general, detail drawings will have information that is classified into three groups:

1. Shape description, which shows or describes the shape of the part
2. Size description, which shows the size and location of features on the part
3. Specifications regarding items such as material, finish, and heat treatment

Figure 13–2 shows an example of a detail drawing.

Sheet Layout

Detail drawings may be prepared with one part per sheet, referred to as a monodetail drawing, as in Figure 13–2, or with several parts grouped on one sheet, which is called a multidetail drawing. The method of presentation depends on the choice of the individual company. The drawing assignments in this text have been done with one detail drawing per sheet, which is common industry practice. Some companies, however, draw many details per sheet. The advantage of one detail per sheet is that each part stands alone so that the drawing of the part can be distributed to manufacturing without several other parts attached. Drawing sheet sizes, then, will vary depending on the part size, scale used, and information presented. This procedure requires that drawings be filed with numbers that allow the parts to be located in relation to the assembly. The advantage of drawing several details per sheet is one of economics. Several details per sheet saves paper and saves drafting time. The drafter may be able to draw several parts on one sheet depending on the size of the parts, the scale used, and the information associated with the parts. If there are six parts on one sheet, then the number of title blocks that must be completed is reduced by five. The company may use one standard sheet size and encourage drafters to place as many parts as possible on one sheet. When this practice is used, there may be a group of sheets with parts detailed for one assembly. The sheet numbers will then correlate the sheets to the assembly, and each part will be keyed to the assembly. The sheets will be given page numbers identifying the page number and the total number of pages in the set. For example, if there are three pages in a set, then the first page is identified as 1 of 3, the second as 2 of 3, and the third as 3 of 3. Figure 13–3 shows an example of a multidetail drawing with several detail drawings on one sheet. Some companies may use both methods at different times depending on the purpose of the drawings and the type of product. For example, it is more common for the parts of a weldment to be drawn grouped on sheets as opposed to one per sheet because the parts may be fabricated at one location in the shop.

Steps in Making a Detail Drawing

This discussion is a brief review of the layout steps covered in Chapters 5 and 8. The best way to begin is to make a rough sketch. The sketch will allow you to put together preliminary thoughts about view and dimension selection. Beginning drafters will probably rely on this aid more than experienced technicians, although it is not uncommon for the most seasoned veteran to use preliminary sketches.

The next consideration may be the selection of paper size. Ideally you should be careful not to crowd the drawing. If in doubt, select the next larger size. The choice of sheet size may also depend on company practice. Some companies may want as much information as possible crowded on each sheet. It may be the goal of every drafter for such a company to place details on every square centimeter of space. Other companies may prefer never to crowd the information on a drawing. Some companies feel that there will be fewer problems in manufacturing if the drawings are not crowded and are easy to read. Another factor involved in the sheet size decision is that extra space should be provided on each drawing for potential future revisions. Most drawings are eventually changed. It is costly to redraw an entire drawing to make changes when a little extra space on the original could have prevented the redrawing. Consider these possibilities when selecting paper size:

1. The number of individual parts to draw
2. The scale of the views
3. The number of views
4. The amount of space for dimensions that the part requires; leave plenty of space for dimensions
5. The number of general notes
6. The need to always leave the area adjacent to the drawing change column clear
7. The room that may be needed for a bill of materials

Detail Drawing Manufacturing Information

Detail drawings may be drawn to suit the needs of the manufacturing processes. A detail drawing may have all of the information necessary to completely manufacture the part — for example, casting and machining information on one drawing. In some situations a completely dimensioned machining drawing may be sent to the pattern or die maker. The pattern or die will then be made to allow for extra material where machined surfaces are specified. When company standards require, two detail drawings may be prepared for each part. One detail will give views and dimensions that are necessary only for

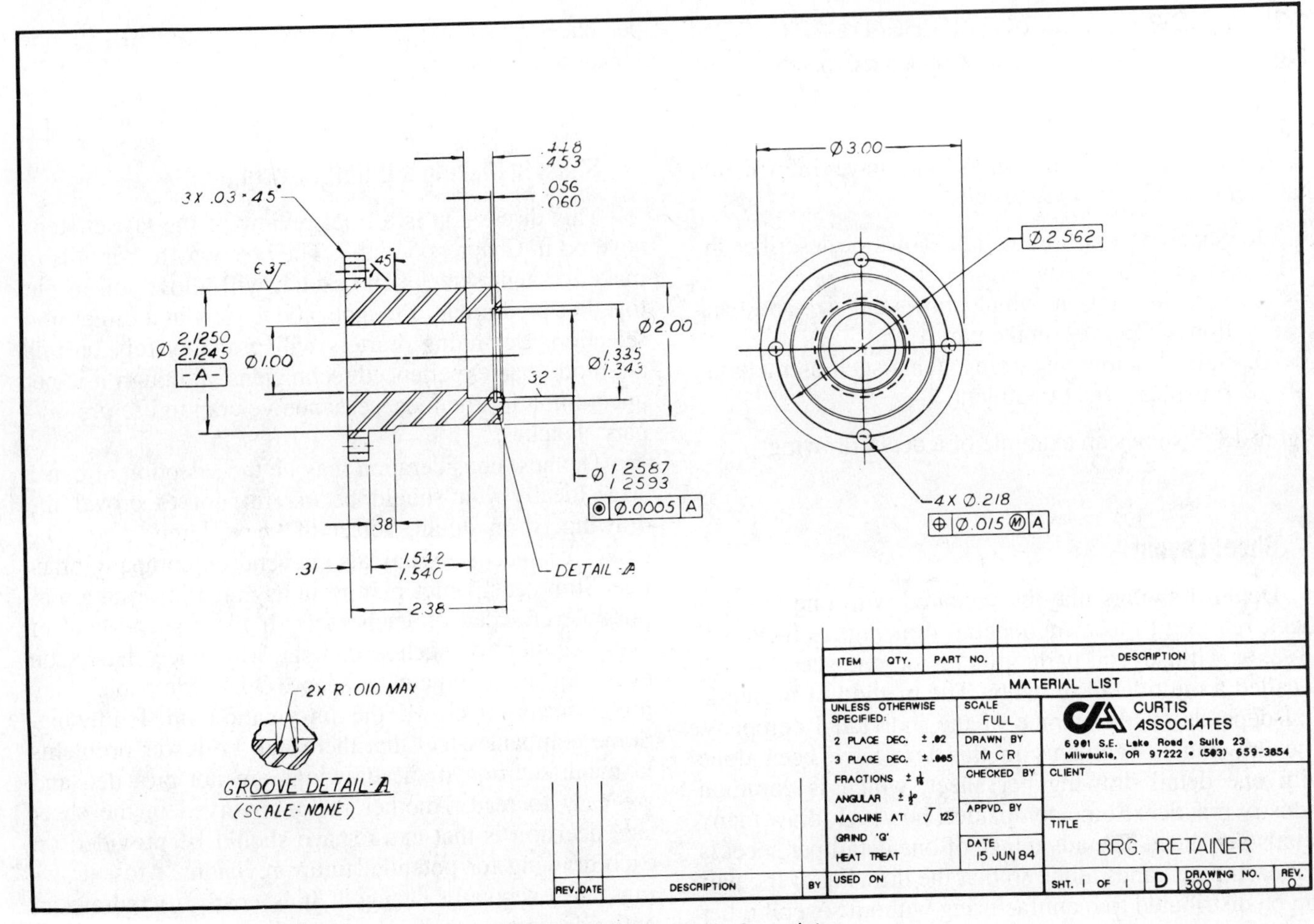

Figure 13–2 Monodetail drawing (one part per sheet). *Courtesy Curtis Associates.*

the casting or forging process. Another detail is drawn that does not give the previous casting or forging dimensions but provides only the dimensions needed to perform the machining operations on the part.

ASSEMBLY DRAWINGS

Most products are composed of several parts. A drawing showing how all of the parts fit together is called an assembly drawing. Assembly drawings may differ in the amount of information provided, and this decision often depends on the nature or complexity of the product. Assembly drawings are generally multiview drawings. The drafter's goal in the preparation of assembly drawings is to use as few views as possible to completely describe how each part goes together. In many cases a single front view will do the job. (See Figure 13–4.) Full sections are commonly associated with assembly drawings because the full section may expose the assembly of most or all of the internal features, as shown in Figure 13–5. If one section or view is not enough to show how the parts fit together, then a number of views or sections may be necessary. In some situations a front view or group of views with broken-out sections is the best method of showing the external features while exposing some of the internal features. (See Figure 13–6.) The drafter must make the assembly drawing clear enough for the assembly department to put the product together. Other elements of assembly drawings that make them different from detail drawings is that they usually contain few or no hidden lines or dimensions. Hidden lines should be avoided on assembly drawings unless absolutely necessary for clarity. The common practice is to draw an exterior view to clarify outside features and a sectional view to expose interior features. Dimensions serve no purpose on an assembly drawing unless the dimensions are used to show the assembly relationship of one part to another. These assembly dimensions are

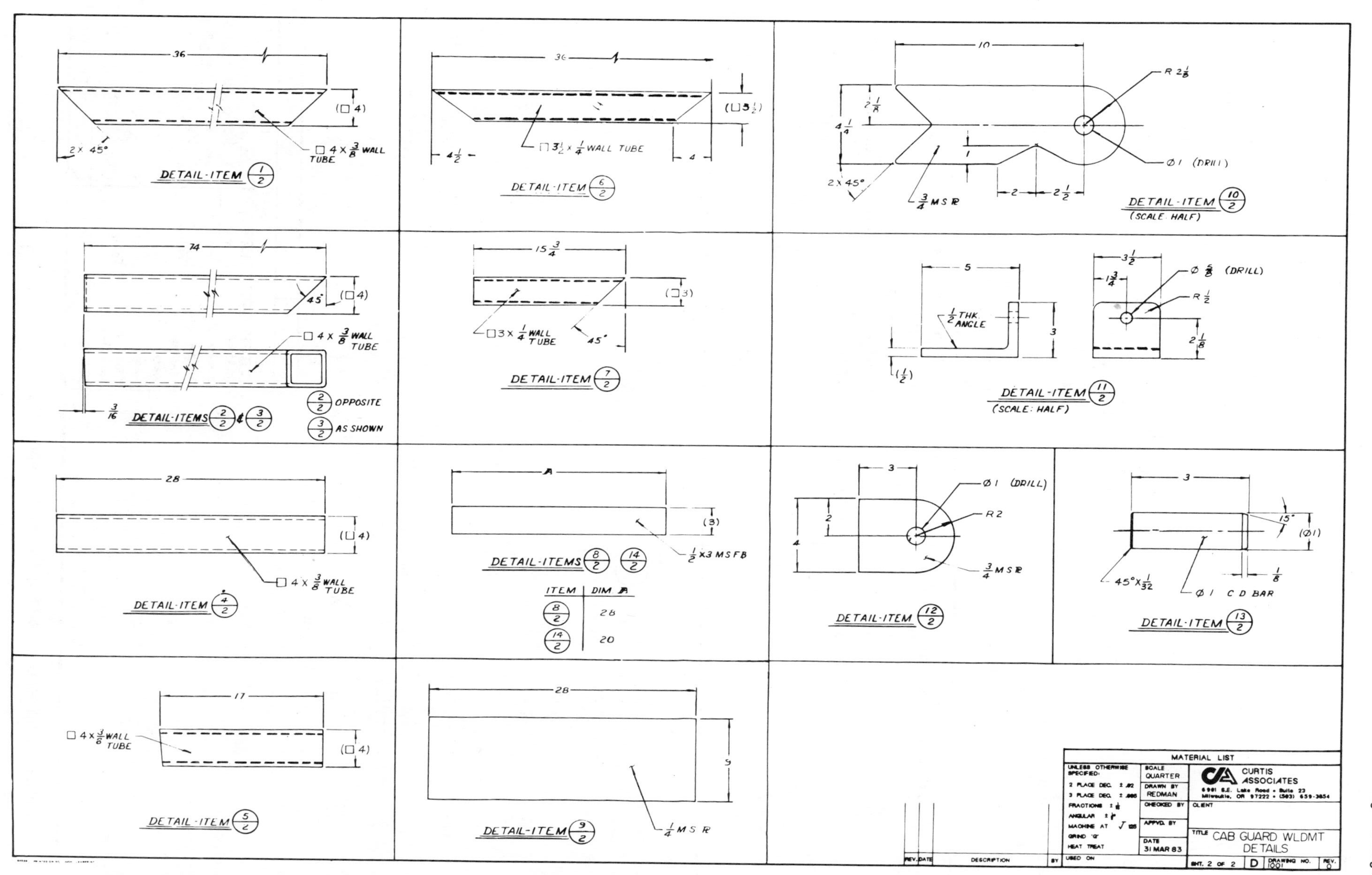

Figure 13–3 Multidetail drawing (several detail drawings on one sheet). *Courtesy Curtis Associates.*

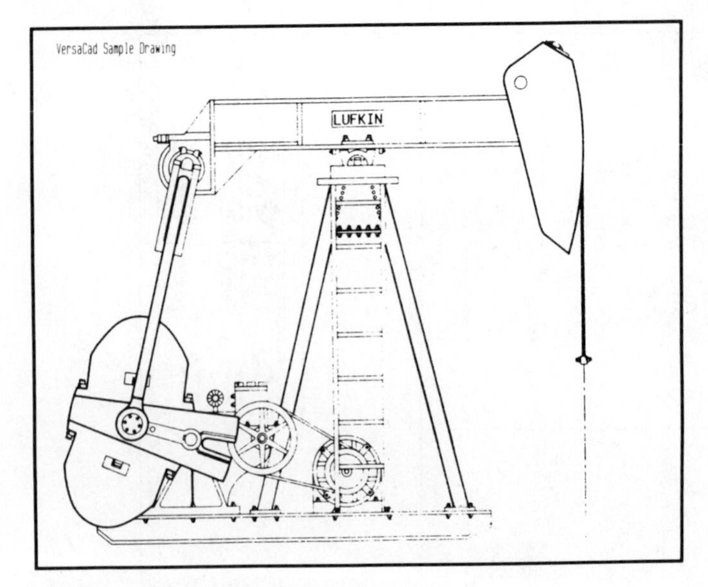

Figure 13–4 CADD drawing of assembly with single front view. *Courtesy T&W Systems, Inc.*

only necessary when a certain distance between parts must exist before proper assembly can take place. Machining processes and other specifications are generally not given on an assembly drawing unless a machining operation must take place after two or more parts are assembled. Other assembly notes may include bolt tightening specifications, assembly welds, or cleaning, painting, or decal placement that must take place after assembly. Figure 13–7 shows a process note applied to an assembly drawing. Some company standards or drawing presentations prefer that assemblies be drawn with sectioned parts shown without section lines. Figure 13–8 shows a full-section assembly with parts left unsectioned.

Assembly drawings may contain some or all of the following information:

1. One or more views
2. Sections necessary to show internal features, function, and assembly
3. Any enlarged views necessary to show adequate detail
4. Arrangement of parts
5. Overall size and specific dimensions necessary for assembly
6. Manufacturing processes necessary for or during assembly
7. Reference or item numbers that key the assembly to a parts list and to the details
8. Parts list or bill of materials

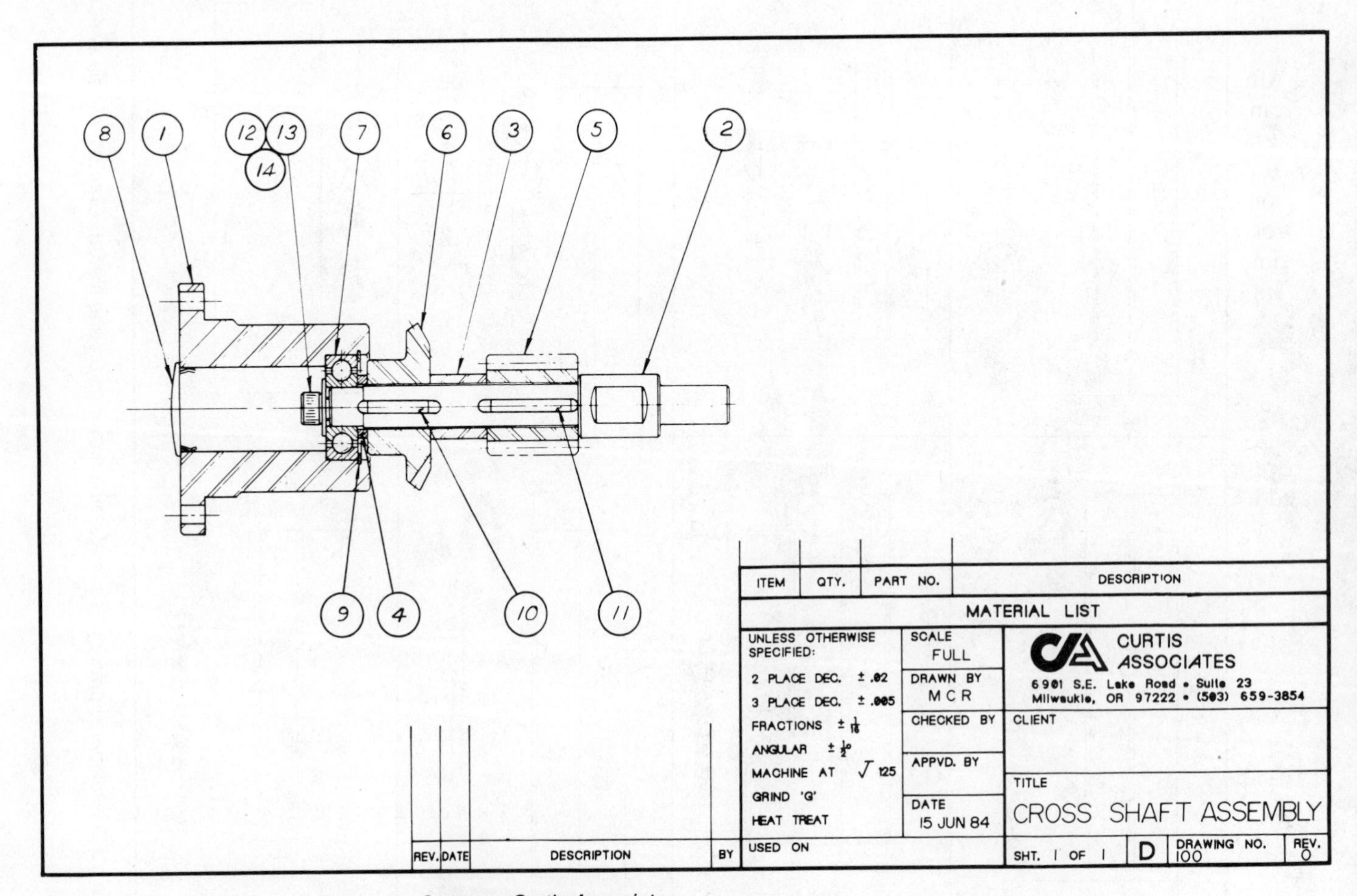

Figure 13–5 Assembly in full section. *Courtesy Curtis Associates.*

CADD *applications*

WORKING DRAWINGS

Using CADD to prepare working drawings may increase your productivity while you are drawing details. Some drafters indicate a 1:1 productivity ratio while others boast up to a 10:1 ratio for CADD versus manual drafting. One engineer explained that it normally takes a drafter 100 hours of training and daily use of the CADD system before he or she is on a 1:1 level of productivity with the manual drafter. After the initial 100 hours, the productivity of the CADD drafter increases. This all depends on some contributing factors such as:

- The complexity of the part
- The experience of the drafter with the CADD system
- The ability of the individual or group to customize the CADD system for specific applications
- The number of common symbols, notes, or details that can be used on a variety of drawings

After all of the detail drawings are made, the drafter can easily and quickly prepare the assembly drawing. It used to take the manual drafter hours or even days to complete the assembly drawing, depending on the complexity of the product. With CADD even the most complex drawings can be done in a short period of time. The assembly should be drawn after the detail drawings of the individual parts are complete because the drafter can use the details to complete the assembly. To do this, follow these helpful techniques.

- Do all detail drawings first.
- Have the views on a layer separate from the dimensions.
- Make the principal view of each detail drawing a BLOCK or SYMBOL that can be sent between drawings.
- Start the assembly drawing by bringing the main detail view into the layout and position it where the other detail views can be conveniently added.
- Bring each detail view into position to begin building the assembly. Use a command such as SCALE to increase or reduce the size of each detail view to fit the scale of the assembly.
- After all of the detail views are assembled, you can use the editing functions to erase, trim, or otherwise clean up the assembly.
- Add balloons, notes, and dimensions (if any).
- Add the parts list. It is best to have a standard parts list format that can be called up and added to any drawing. Then all you have to do is add the text and the drawing is done.

This method saves a lot of time because you do not have to start over again drawing each part in the assembly.

TYPES OF ASSEMBLY DRAWINGS

There are several different types of assembly drawings used in industry:

1. Layout, or design, assembly
2. General assembly
3. Working-drawing or detail assembly
4. Erection assembly
5. Subassembly
6. Pictorial assembly

Layout Assembly

Engineers and designers may prepare a design layout in the form of a sketch or as an informal drawing. These engineering design drawings are used to establish the relationship of parts in a product assembly. From the layout the engineer will prepare sketches or informal detail drawings for prototype construction. This research and development (R and D) is the first step in the process of taking a design from an idea to a manufactured product. Layout, or design, assemblies may take any form depending on the drafting ability of the engineer, the time frame for product implementation, the complexity of the product, or company procedures. In many companies the engineers work with drafting technicians who help prepare formal drawings from engineering sketches or informal drawings. The research and development department is one of the most exciting places for a drafter to be. Figure 13–9 shows a simple layout assembly of a product in the development stage. The limits of operation are shown in phantom lines.

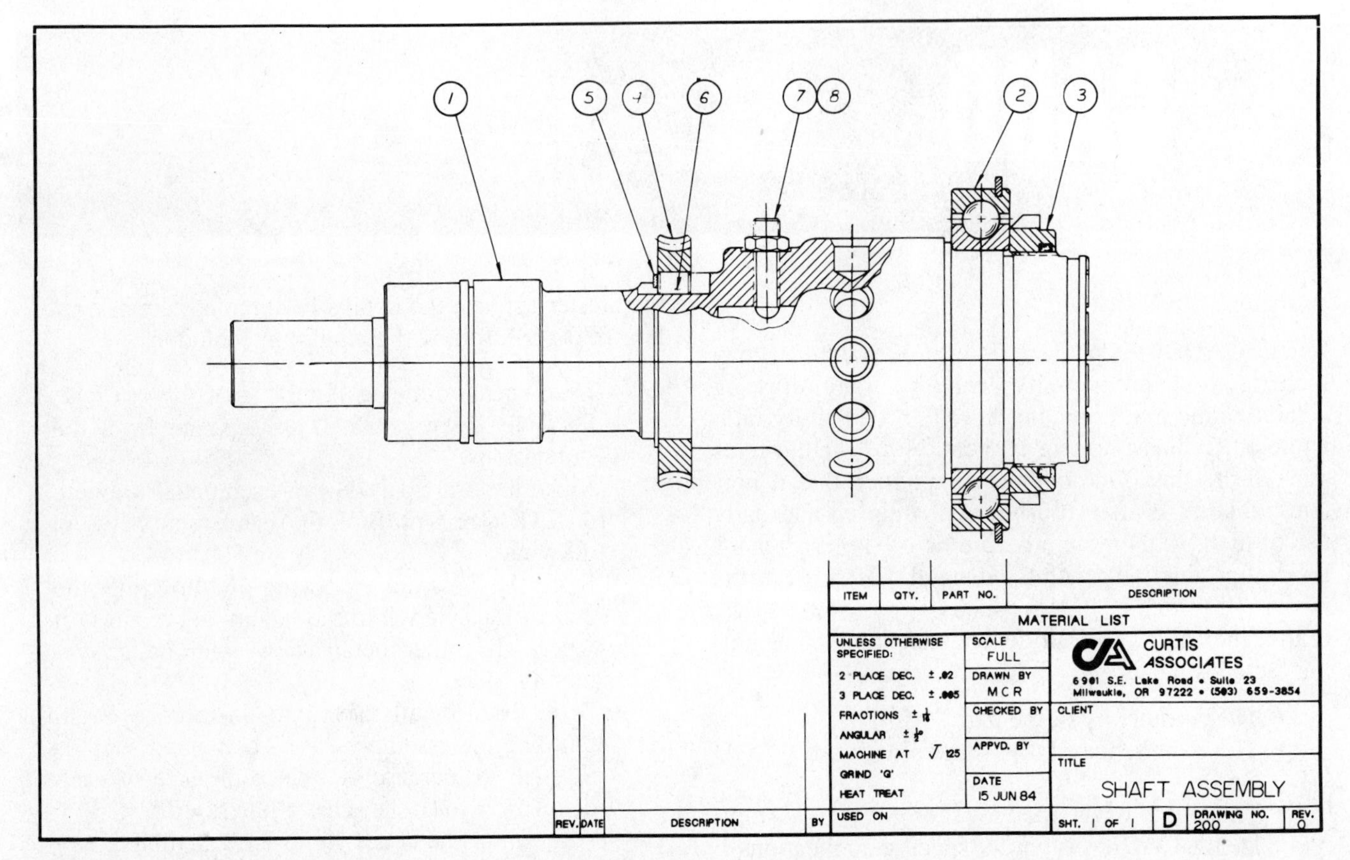

Figure 13–6 Assembly with broken-out sections. *Courtesy Curtis Associates.*

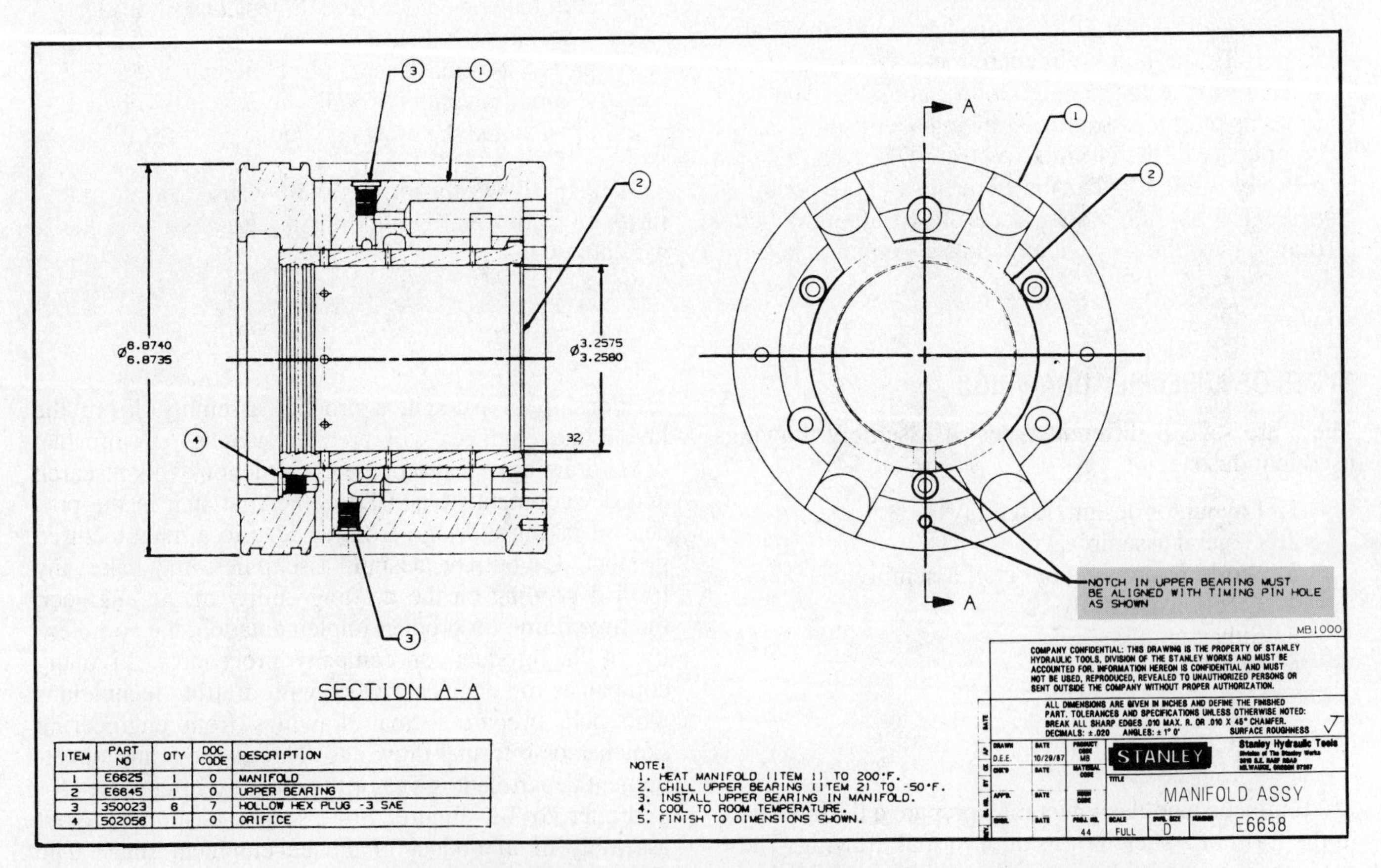

Figure 13–7 Process note on assembly drawing. *Courtesy Stanley Hydraulic Tools, Division of The Stanley Works.*

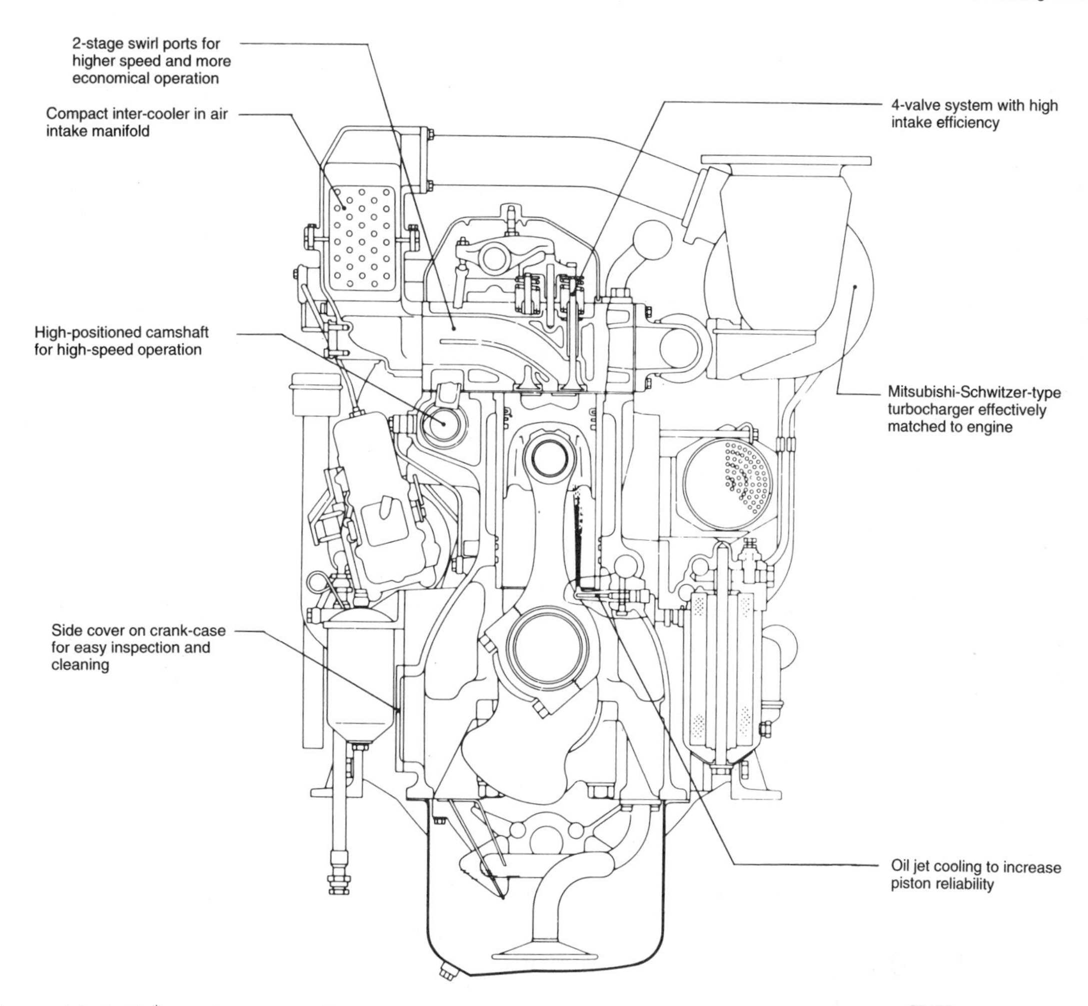

Figure 13–8 Full-section assembly with section lines omitted. *Courtesy Mitsubishi Heavy Industries America, Inc.*

General Assembly

General assemblies are the most common types of assemblies that are used in a complete set of working drawings. A set of drawings contains three parts: detail drawings, an assembly drawing, and a parts list. The assembly drawing, as previously discussed, shows how all of the parts fit together. Figures 13–4 through 13–8 provide examples of general assembly drawings.

Figure 13–9 Layout assembly. The limits of operation are shown in phantom lines.

Working-drawing, or Detail Assembly

When a drawing is created where details of parts are combined on the same sheet with an assembly of those parts, a detail assembly is the result. While this practice is not as common as general assemblies, it is a practice at some companies. The use of working-drawing assemblies may be a company standard, or this technique may be used in a specific situation even when it is not considered a normal procedure at a particular company. The detail assembly may be used when a particular end result dictates that the details and assembly be combined on as few sheets as possible. An example may be a product with few parts that will be produced only once for a specific purpose. (See Figure 13–10.)

Erection Assembly

Erection assemblies generally differ from general assemblies in that dimensions and fabrication specifications are commonly included. Typically associated with products that are made of structural steel, or cabinetry, erection assemblies are used for both fabrication and assembly. Figure 13–11 shows a CADD-produced erec-

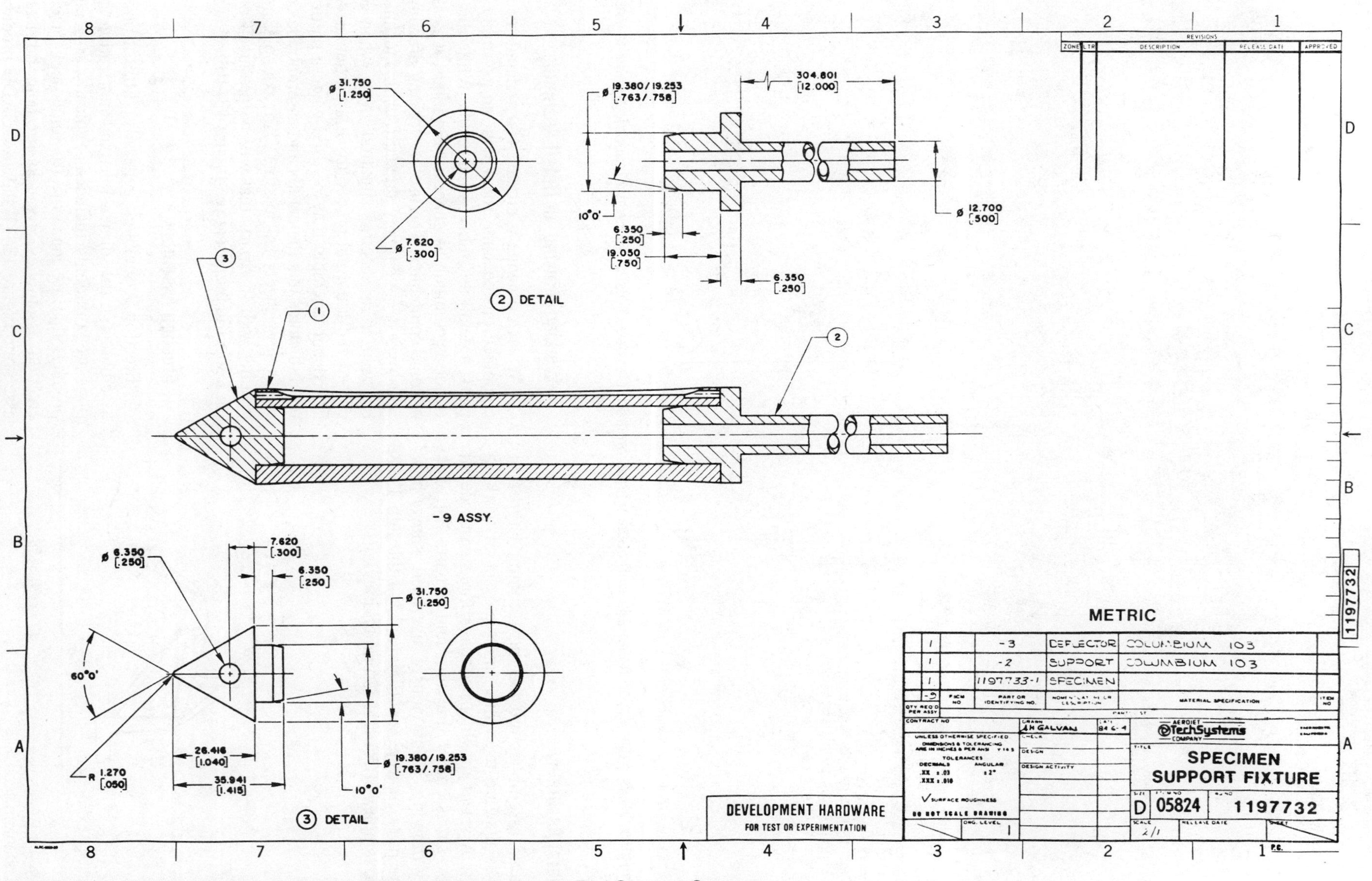

Figure 13–10 Working-drawing, or detail, assembly. *Courtesy Aerojet TechSystems Company.*

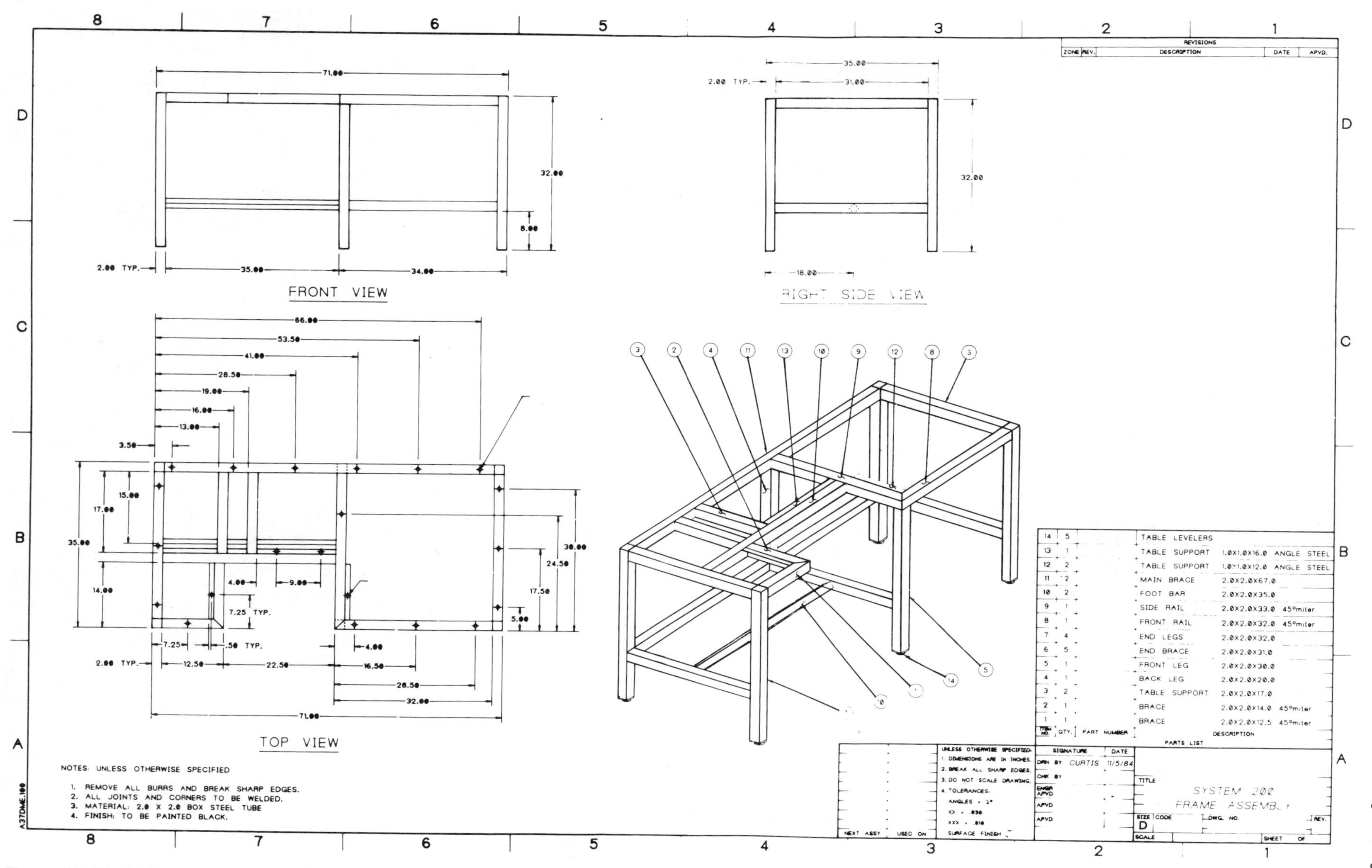

Figure 13–11 CADD drawing of erection assembly. *Courtesy EFT Systems.*

tion assembly with multiviews, fabrication dimensions, and an isometric drawing that also helps display how the parts fit together.

Subassembly

The complete assembly of a product may be made up of several component assemblies. These individual unit assemblies are called subassemblies. A complete set of working drawings may be made up of several subassemblies, each with its own detail drawings, and the general assembly. The general assembly of an automobile, for example, will include the subassemblies of the drive components, the engine components, and the steering column, just to name a few. A subassembly, such as an engine, may be made up of other subassemblies, such as the carburetor or the generator. Figure 13–12 shows a subassembly with a parts list.

Pictorial Assembly

As the name suggests, pictorial assemblies are used to display a pictorial rather than multiview representation of the product, which may be used in other types of assembly drawings. Pictorial assemblies may be made from photographs or artistic renderings. The pictorial assembly may be as simple as the isometric drawing in Figure 13–11, which was used to more clearly assist workers in the assembly of the product. Pictorial assemblies are commonly used in product catalogs or brochures. The purpose of these pictorial representations may be for sales promotion, customer self-assembly, or maintenance procedures. (See Figure 13–13.) Pictorial assemblies may also take the form of exploded technical illustrations, also commonly known as illustrated parts breakdowns. These exploded multiview or isometric pictorials are used in vendors' catalogs and instruction manuals for maintenance or assembly. A pictorial assembly has been used by every individual who has put together a bicycle, model kit, or child's toy. (See Figure 13–14.)

IDENTIFICATION NUMBERS

Identification or item numbers are used to key the parts from the assembly drawing to the parts list. Identification numbers are generally placed in balloons. Balloons are circles that are connected to the related part with a leader line. Several of the assembly drawings in this chapter show examples of identification numbers and balloons. Numbers in balloons are common, although some companies prefer to use identification letters. Balloons are drawn between .375 and 1 in. in diameter depending on the size of the drawing and the amount of

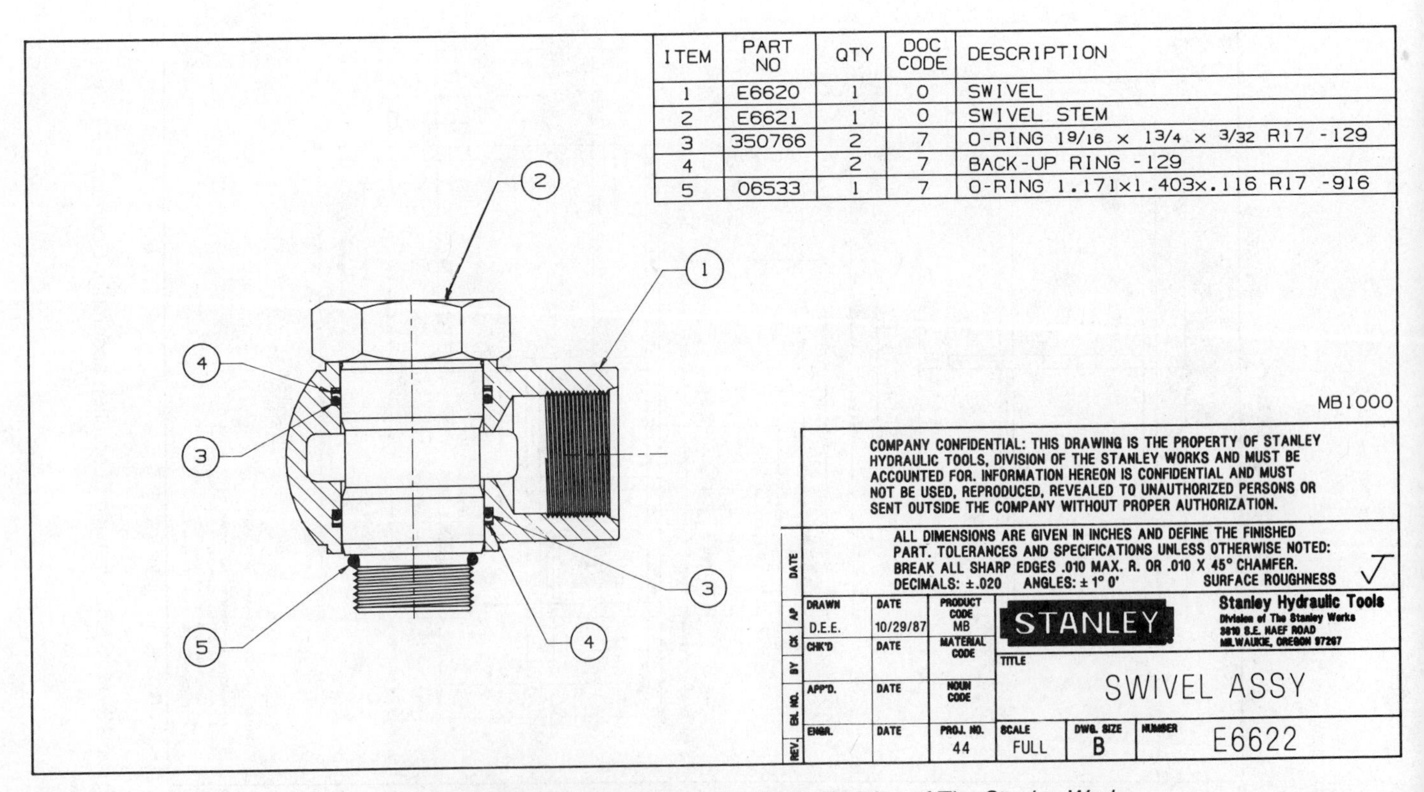

ITEM	PART NO	QTY	DOC CODE	DESCRIPTION
1	E6620	1	0	SWIVEL
2	E6621	1	0	SWIVEL STEM
3	350766	2	7	O-RING 19/16 × 13/4 × 3/32 R17 -129
4		2	7	BACK-UP RING -129
5	06533	1	7	O-RING 1.171×1.403×.116 R17 -916

Figure 13–12 Subassembly with parts list. *Courtesy Stanley Hydraulic Tools, Division of The Stanley Works.*

information that must be placed in the balloon. The leaders that connect the balloons to the parts are thin lines that may be presented in any one of several formats depending on company standards. Figure 13–15 shows the common methods that are used to connect balloon leaders. Notice that the leaders may terminate with arrowheads or dots. Whichever method of connecting balloons is used, the same technique should be used throughout the entire drawing.

The item numbers in balloons should be grouped so that they are in an easy-to-read pattern. This is referred to as information grouping. Good information grouping occurs when balloons are in alignment in a pattern that is easy to follow, as opposed to scattered about the drawing. (See Figure 13–16.) In some situations when a particular group of parts are so closely related that individual identification is difficult, the identification balloons may be grouped adjacent to one another. For example, a cap screw, lock washer, and flat washer may require that the balloons be placed in a cluster or side-by-side, as shown at items 12, 13, and 14 in Figure 13–5.

In some cases the balloons will not only key the detail drawings to the assembly and parts list, they will also key the assembly drawing and parts list to the page on which the detail drawing is found. (See Figure 13–17.) Figure 13–18 is an assembly drawing and parts list that is located on page 1 of a two-page set of working drawings. Notice how the balloons key the parts from the assembly and parts list to the detail drawings found on page 2. The page 2 details are located in Figure 13–3.

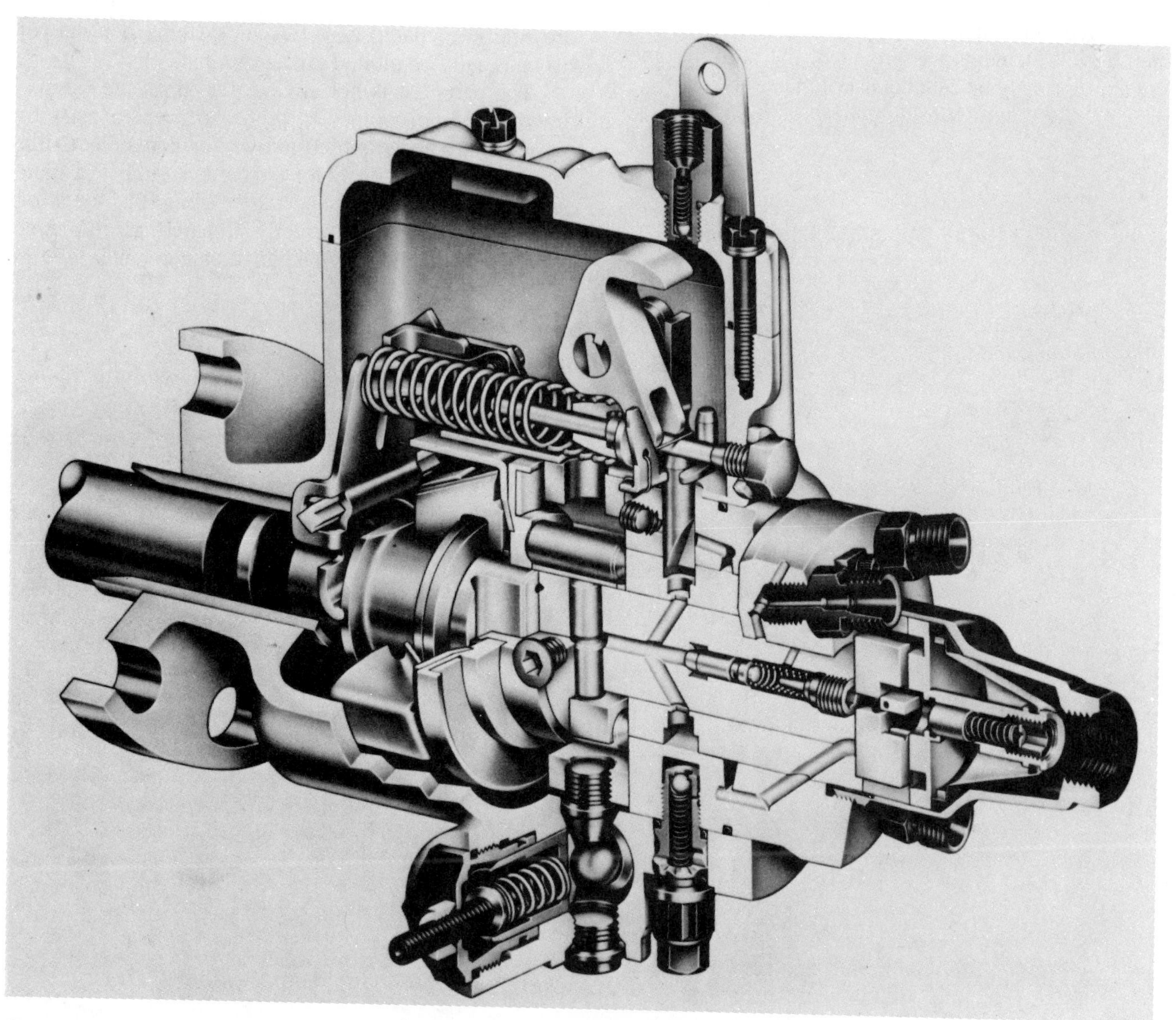

Figure 13–13 Pictorial assembly. *Courtesy Stanadyne Diesel Systems.*

PARTS LISTS

The parts list is usually combined with the assembly drawing yet remains one of the individual components of a complete set of working drawings. The information that is associated with the parts list generally includes:

1. Item number — from balloons
2. Quantity — the amount of that particular part needed for this assembly
3. Part or drawing number, which is a reference back to the detail drawing
4. Description, which is usually a part name or complete description of a purchase part or stock specification including sizes or dimensions
5. Material identification — the material that the part is made of
6. Information about vendors for purchase parts

Parts lists may or may not contain all of the above information, depending on company standards. The elements listed 1 through 4 are the most common items. When all six elements are provided, the parts list may more appropriately be called a list of materials.

The parts list may be drawn on the assembly drawing, as in Figure 13–16 or Figure 13–18. When drawn on the assembly drawing, the parts list may be located above the title block, in the upper right or left corner, or in a convenient location on the drawing field. The location depends on company standards, although the position over the title block is most common. The information on the parts list is usually presented with the first item number followed by consecutive item numbers. When the parts list is so extensive that the columns fill the page, a new group of columns is added adjacent to the first. The reason that parts list data is provided from the bottom of the sheet upward or the top of the sheet downward is so that if additional parts are added to the assembly, space on the parts list will be available.

Some companies prepare parts lists on a computer so that information can be retrieved and edited more easily. When this is done, the parts list may be plotted or typed directly on the drawing original or on an adhesive Mylar® sheet that is added to the drawing. If the entire drawing is prepared on a CADD system, then the parts list can easily be plotted on the original.

The parts list is not always placed on the assembly drawing. Some companies prefer to prepare parts lists on separate sheets, which allows for convenient filing. This method also allows for the parts list to be computer-generated or typed separately from the drawings. Separate parts lists are usually prepared on a computer so that information may be edited conveniently. Another

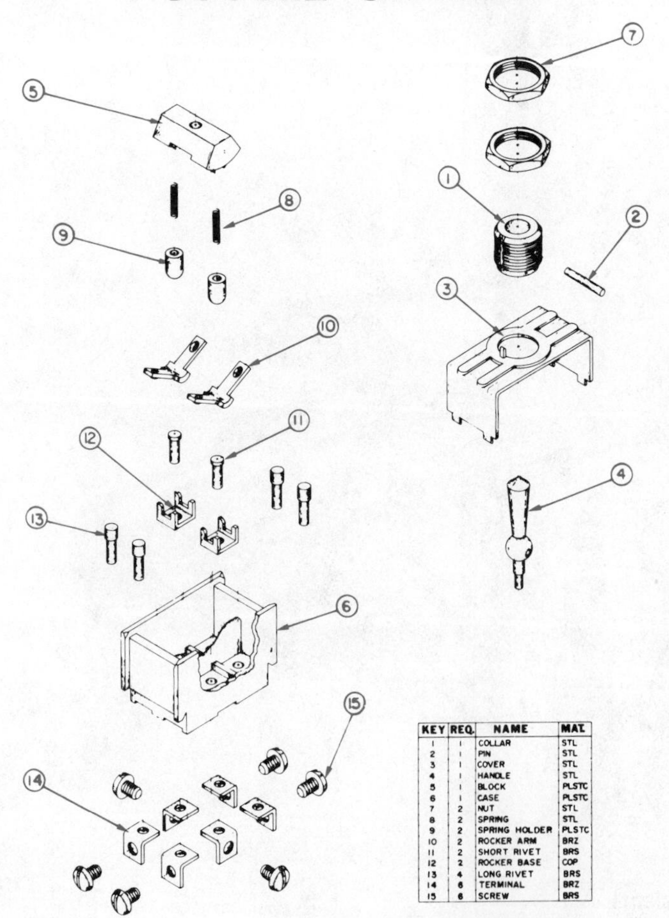

KEY	REQ.	NAME	MAT.
1	1	COLLAR	STL
2	1	PIN	STL
3	1	COVER	STL
4	1	HANDLE	STL
5	1	BLOCK	PLSTC
6	1	CASE	PLSTC
7	2	NUT	STL
8	2	SPRING	STL
9	2	SPRING HOLDER	PLSTC
10	2	ROCKER ARM	BRZ
11	2	SHORT RIVET	BRS
12	2	ROCKER BASE	COP
13	4	LONG RIVET	BRS
14	6	TERMINAL	BRZ
15	6	SCREW	BRS

Figure 13–14 Exploded isometric assembly.

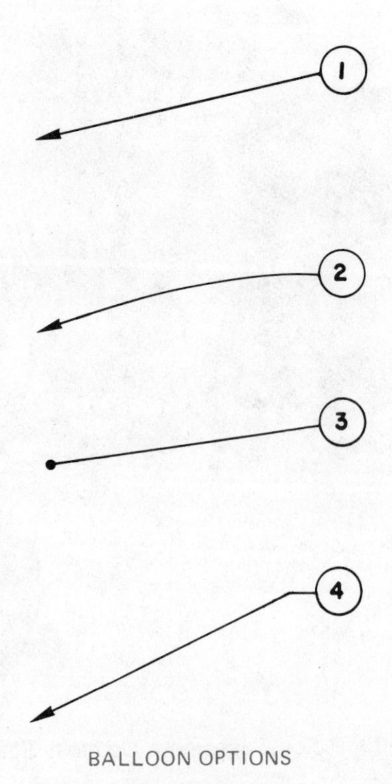

Figure 13–15 Balloons and styles of leaders.

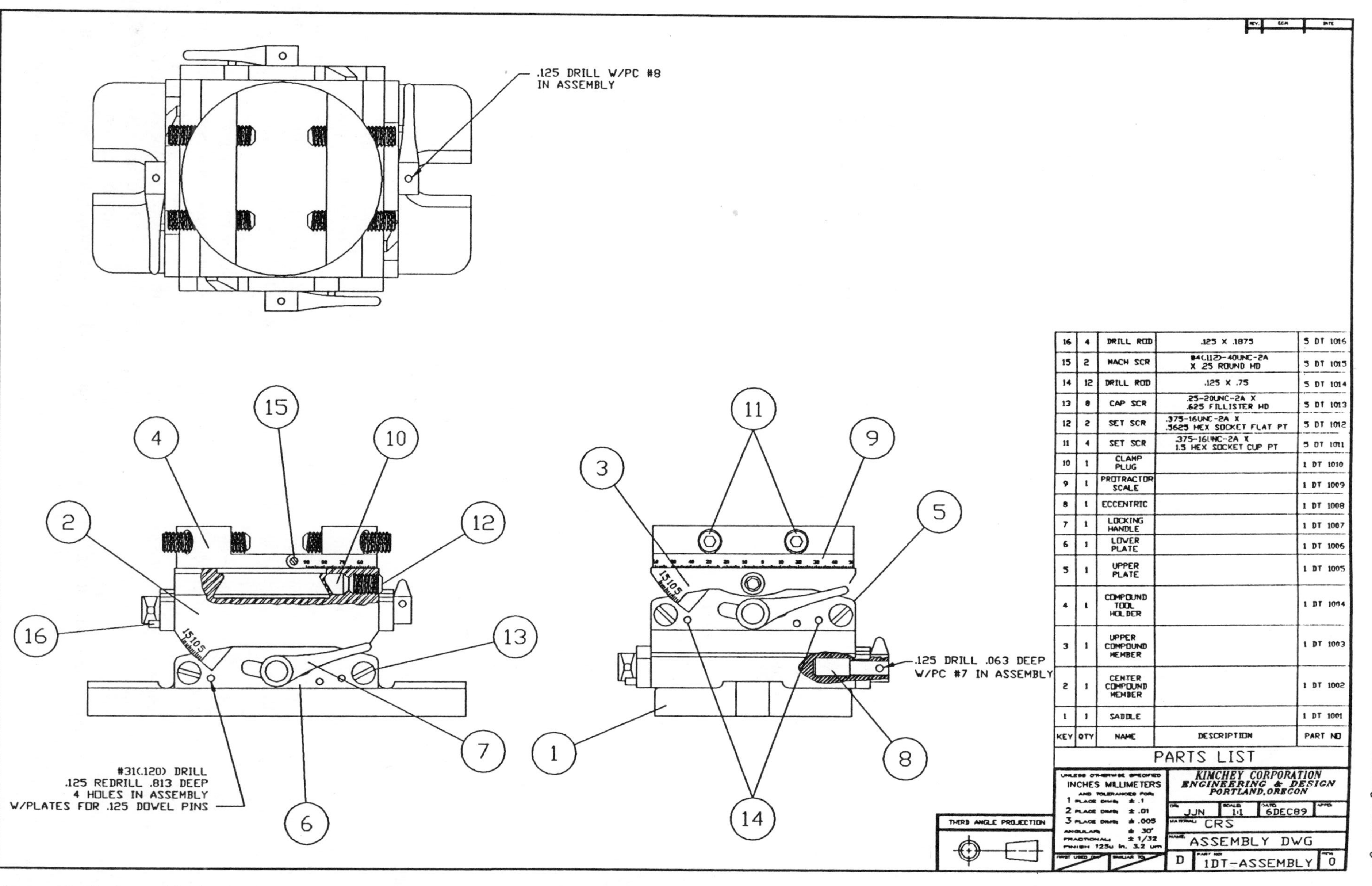

KEY	QTY	NAME	DESCRIPTION	PART NO
16	4	DRILL ROD	.125 X .1875	5 DT 1016
15	2	MACH SCR	#4(.112)-40UNC-2A X .25 ROUND HD	5 DT 1015
14	12	DRILL ROD	.125 X .75	5 DT 1014
13	8	CAP SCR	.25-20UNC-2A X .625 FILLISTER HD	5 DT 1013
12	2	SET SCR	.375-16UNC-2A X .3625 HEX SOCKET FLAT PT	5 DT 1012
11	4	SET SCR	.375-16UNC-2A X 1.5 HEX SOCKET CUP PT	5 DT 1011
10	1	CLAMP PLUG		1 DT 1010
9	1	PROTRACTOR SCALE		1 DT 1009
8	1	ECCENTRIC		1 DT 1008
7	1	LOCKING HANDLE		1 DT 1007
6	1	LOWER PLATE		1 DT 1006
5	1	UPPER PLATE		1 DT 1005
4	1	COMPOUND TOOL HOLDER		1 DT 1004
3	1	UPPER COMPOUND MEMBER		1 DT 1003
2	1	CENTER COMPOUND MEMBER		1 DT 1002
1	1	SADDLE		1 DT 1001

Figure 13–16 Assembly drawing and parts list. *Courtesy Jack Neal, ITT Technical Institute.*

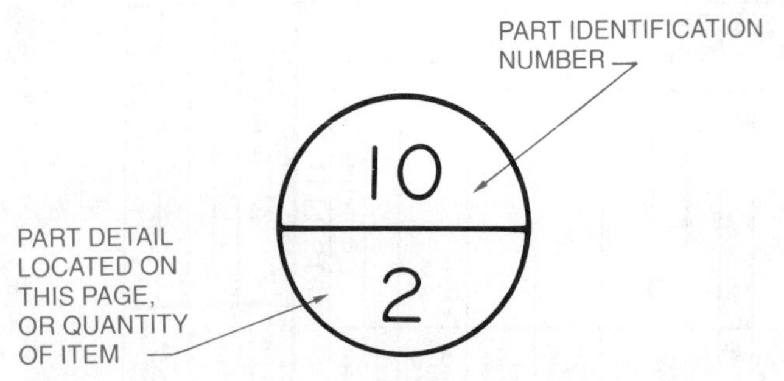

Figure 13–17 Balloon with page identification.

option is to have parts lists typed on a standard parts list form, Figure 13–19.

PURCHASE PARTS

Purchase or standard parts, as previously mentioned, are parts that are manufactured and available for purchase from an outside vendor. It is generally more economical for a company to buy such items than to make them. Parts that are available from suppliers can be found in the *Machinery's Handbook*, *Fastener's Handbook*, or vendors' catalogs. Purchase parts do not require a detail drawing, because a written description will completely describe the part. For this reason, the purchase parts found in a given assembly must be described clearly and completely in the parts list. Some companies have a purchase parts book that is used to record all purchase parts used in their product line. The standard parts book gives a reference number for each part, which is placed on the parts list for convenient identification.

ENGINEERING CHANGES

Engineering change documents are used to initiate and record changes to products in the manufacturing industry. Changes to engineering drawings can be requested from any branch of a company that deals directly with production and distribution of the product. For example, the engineering department may implement product changes as research and development results show a need for upgrading. Manufacturing will request changes as problems arise in product fabrication or assembly. The sales staff may also initiate change proposals that stem from customer complaints. Engineering changes are treated in the same way as original drawings; they are initiated, approved, sent to the drafting department, and filed when completed. It is a good idea to leave at least 4 in. of space below the revision block for future changes.

Engineering Change Request (ECR)

Before a drawing can be changed by the drafting department, an engineering change request (ECR) is needed. This ECR is the document used to initiate a change in a part or assembly. The ECR may come from any one of several sources in the company, such as the engineering, manufacturing, or sales departments, to name just a few. The ECR will usually be attached to a print of the part affected. The print and the ECR will show by sketches and written descriptions what changes will be made. The ECR will also contain a number that will become the reference record of the change to be made. Figure 13–20 shows a sample ECR form. Generally when a drafter receives an ECR, the following procedure is used:

1. Obtain the drawing original, sepia, or computer file.
2. Make the requested change using the same media as that used for the original.
3. Fill out the engineering documents using freehand lettering or a word processor.
4. Distribute a changed copy of the drawing and the completed engineering change forms for release approval.
5. Refile the drawing original.

Engineering Change Notice (ECN)

Records of changes should be kept so that reference can be made between the existing and proposed product. To make sure that records of these changes are kept, special notations are made on the drawing, and engineering change records are kept. These records are commonly known as engineering change notices (ECN) or engineering change orders (ECO).

When a change is to be made, an engineering change request will initiate the change to the drafting department. The drafting technician will then alter the original drawing or computer file to reflect the change request. When the drawing of a part is changed, a revision number is placed next to that change. For example, the first change is numbered R1 or A, the second change is numbered R2 or B, and so on. A circle drawn around the revision number will help the identification stand out clearly from the other drawing numerals. Figure 13–21 shows a part as it exists and also after a change has been made.

When the drafter chooses to make the change by not altering the drawing of the part but only changing the dimension, that dimension is labeled NOT TO SCALE. The method of making the change will be the decision of the drafting department based on the extent of the change and the time required to make it. The not-to-scale symbol may be used to save time. Figure 13–22 shows a change made to a part with the new dimension identified

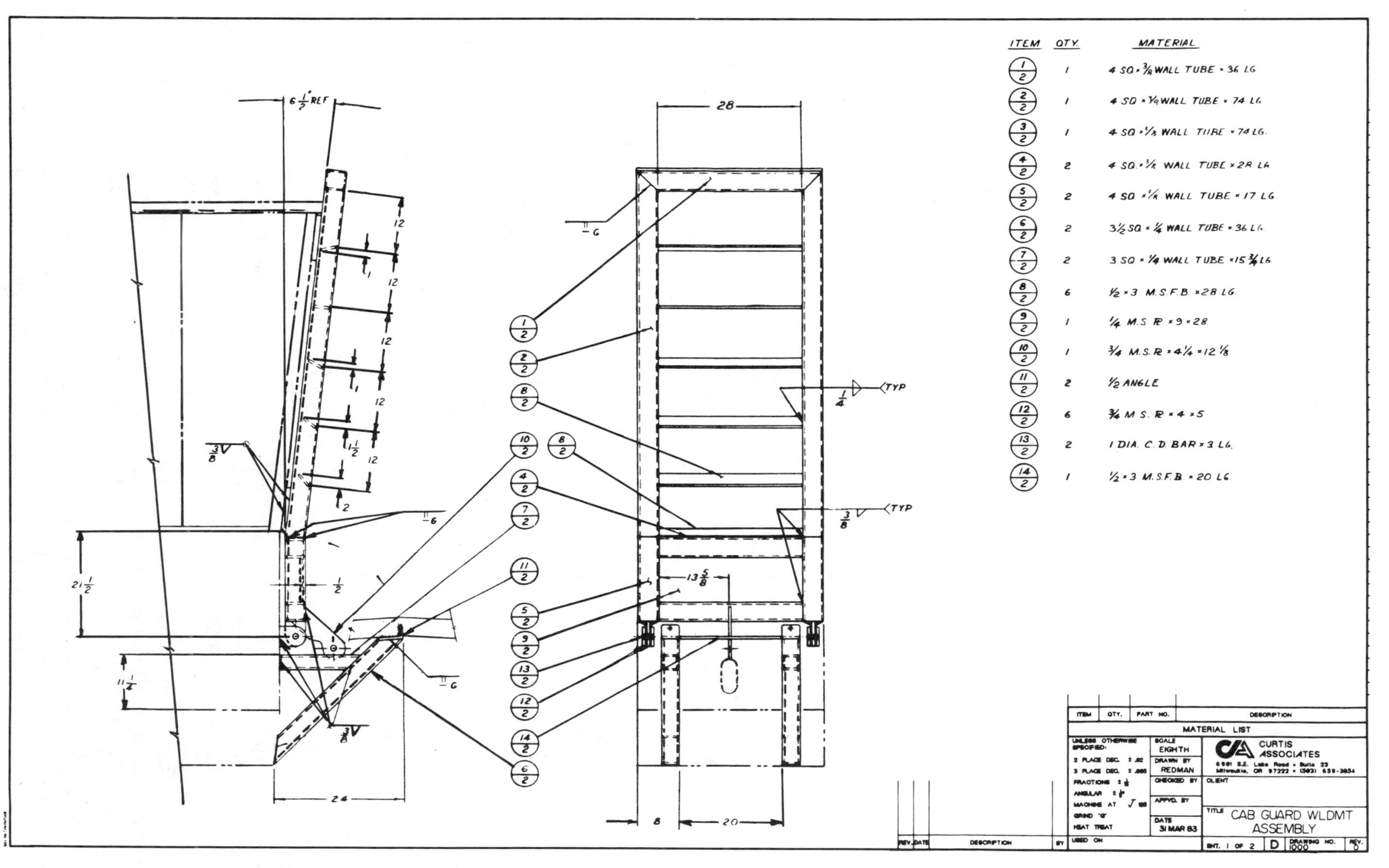

Figure 13–18 Assembly and parts list with page identification balloons. *Courtesy Curtis Associates.*

ASSEMBLY Cross Shaft Assembly USED ON

NUMBER OF UNITS DATE

HAVE	NEED	P/Ø NO. / W/Ø NO.	DET. NO.	PART NO.	DWG.	QTY.	PART NAME	DESCRIPTION	VENDOR
			1		B	1	Bearing Retainer	Ø 3" C.D. Bar	
			2		B	1	Cross Shaft	Ø 3/4" C.D. Bar	
			3		A	1	Spacer	Ø 3/4 O.D. x 11GA Wall Tube x .738 Thick	
			4		A	1	Spacer	Ø 3/4 O.D. x 11GA Wall Tube x .125 Thick	

Figure 13–19 Parts list separate from assembly drawing. *Courtesy Curtis Associates.*

as not to scale with a thick straight line placed under the new dimension.

After the part has been changed on the drawing and the proper R number or letter placed next to the change, the drafter will record the change in the ECN column of the drawing. The location of the ECN column varies with different companies. This ECN identification may be found next to the title block or in a corner of the drawing.

ANSI The ANSI document, *Drawing Sheet Sizes and Format*, ANSI Y14.1, recommends the ECN column be placed in the upper right corner of a drawing. The drafter will record the revision number (number of times revised), ECN number (usually given on the ECR), and the date of the change. Some companies also have a column for a brief description of the change (as recommended by ANSI) and an approval. Figure 13–23 shows an expanded and condensed ECN column format. Notice that changes are added in alphabetic order from top to bottom.

The condensed ECN column in Figure 13–23(b) shows that the drawing has been changed twice. The first change was initiated by ECN number 2604 on September 10, 1983, and the second by ECN 2785 on November 18, 1984. Some companies may also identify the number of times a drawing has been changed by providing the letter of the current change in the title block, as shown in Figure 13–24. The drafter will alter this letter to reflect each change.

When the drafter has made all of the drawing changes as specified on the ECR, then an engineering change notice (ECN) will be filled out. The ECN will complete the process and be filed for future reference. Usually the ECN completely describes the part as it existed before the change and the change that was made as presented on the ECR. Figure 13–25 shows a typical ECN form. With a changed drawing and a filed ECN, anyone can verify what the part was before the change and the reason for the change. The ECN number will be a reference for reviewing the change.

The general engineering change elements, terminology, and techniques are consistent among companies, although the actual format of engineering change documents may be considerably different. Some formats are very simple while others are much more detailed. One of the first tasks of an entry-level drafter is to become familiar with the specific method of preparing engineering changes.

PROFESSIONAL PERSPECTIVE

Many entry-level drafters are often involved in preparing detail drawings or making changes to detail drawings. When a drafter has gained valuable experience with drafting practices, standards, and company products, there is usually an opportunity to advance to a design drafter position. Drafting jobs at these levels can be exciting as the drafters work closely with engineers to create new and updated product designs. An individual drafter in a small company may be teamed with an engineer to implement designs. In larger companies a team of drafters, designers, and engineers may work together to design new products. These are the types of situations where a drafter may have the opportunity to prepare some or all of the drawings for a complete set of working drawings. Generally in these research and development departments of companies the preliminary product drawings are used to build prototypes of the designs for testing. After sufficient tests have been performed on the product, the drawings are revised and released for production. The new product will now become reality.

Aerojet Liquid Rocket Company

ENGINEERING CHANGE REQUEST & ANALYSIS

DATE	ECRA NO	PAGE

ORIGINATOR NAME	DEPT	EXT	DATE	DOCUMENT NEED DATE	PROPOSED EFFECTIVITY

PART DOCUMENT NO	CURRENT REV LTR	PART DOCUMENT NAME

USED ON NEXT ASSY NO	PROGRAM(S) AFFECTED	CEPL(S) AFFECTED

QTSS ITEM YES ____ NO ____

REQUAL REQD YES ____ NO ____

REVISE QTSS FORM YES ____ NO ____

DESCRIPTION OF CHANGE

JUSTIFICATION OF CHANGE

PROJECT ENGINEER SIGNATURE	DEPT	EXT	DATE	CCB REP SIGNATURE	DATE

DESIGN ENGINEERING TECHNICAL EVALUATION

DESIGN ENGINEER SIGNATURE	DEPT	EXT	DATE	CAUSING CONT OR WORK ORDER NO	

ANY AFFECT ON	YES	NO		YES	NO		YES	NO
1 PERFORMANCE			5 WEIGHT			9 OPERATIONAL COMPUTER PROGRAMS		
2 INTERCHANGEABILITY			6 COST SCHEDULE			10 RETROFIT		
3 RELIABILITY			7 OTHER END ITEMS			11 END ITEM IDENT		
4 INTERFACE			8 SAFETY EMI			12 VENDOR CHANGE CRITICAL ITEMS ONLY		

CCB DECISION

SIGNATURES	DEPT	CONCUR	DISSENT
QUALITY ASSURANCE			
MANUFACTURING			
ENGINEERING			
PRODUCT SUPPORT			
MATERIAL			
TEST OPERATION			

CCB CHAIRMAN SIGNATURE	DATE	CLASS I ☐ CLASS II ☐
CUSTOMER SIGNATURE	DATE	EFFECTIVITY

Figure 13–20 Sample engineering change request (ECR) form. *Courtesy Aerojet TechSystems Company.*

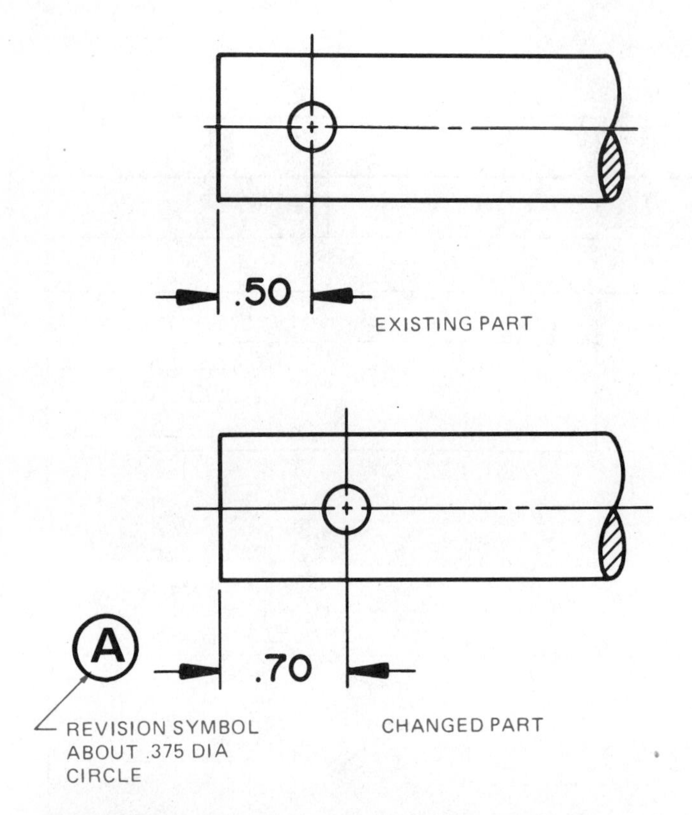

Figure 13–21 An existing part and the same part after a change has been made.

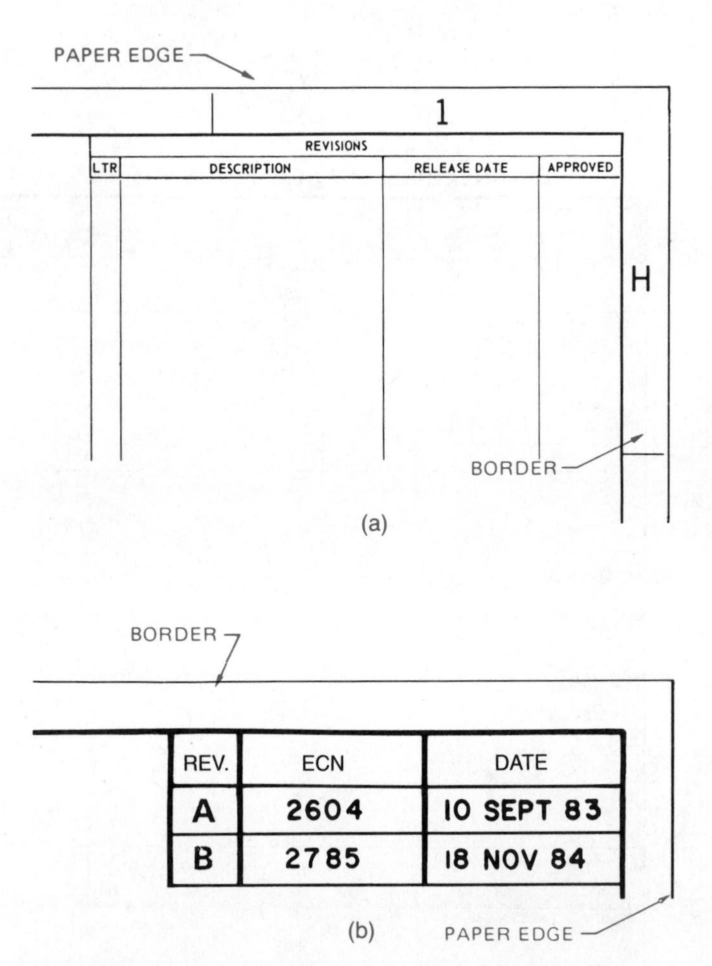

Figure 13–23 (a) Expanded revision columns. *Courtesy Aerojet TechSystems Company;* (b) condensed revision columns.

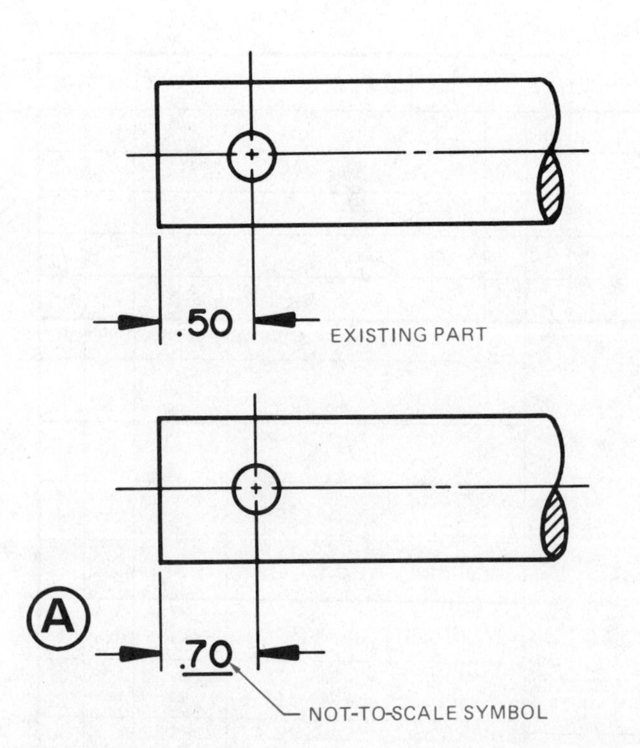

Figure 13–22 An existing part and the same part changed using the not-to-scale symbol.

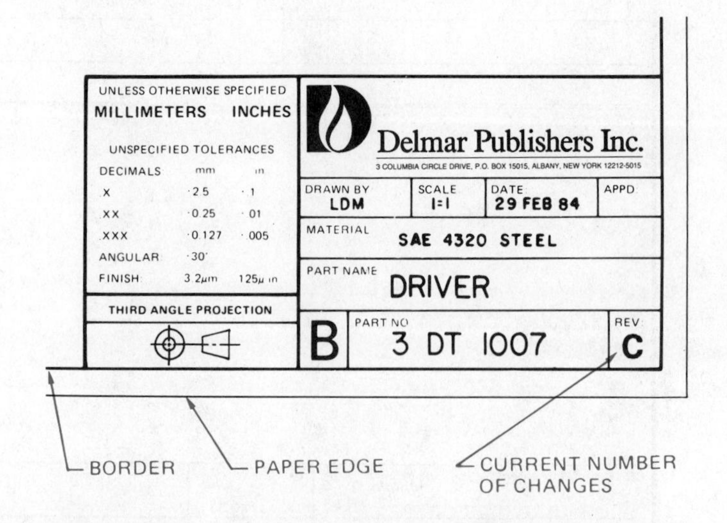

Figure 13–24 Title block displaying the letter indicating the number of times the drawing has been changed.

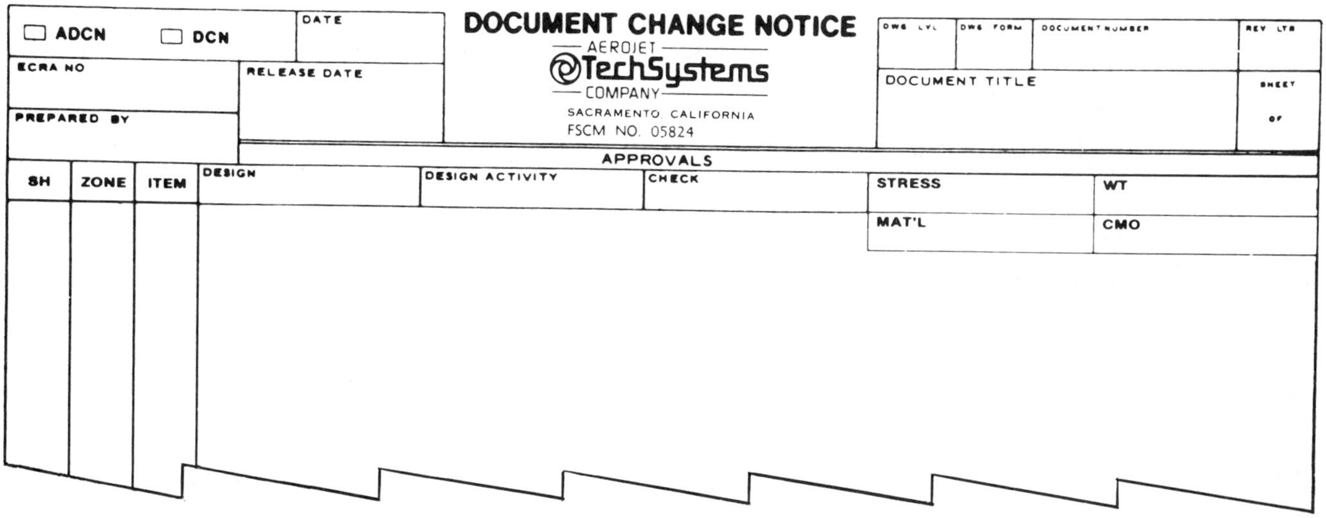

☐ ADCN ☐ DCN
DATE
ECRA NO
RELEASE DATE
PREPARED BY
DOCUMENT CHANGE NOTICE
AEROJET
TechSystems
COMPANY
SACRAMENTO CALIFORNIA
FSCM NO. 05824
DWG LVL
DWG FORM
DOCUMENT NUMBER
REV LTR
DOCUMENT TITLE
SHEET
OF
APPROVALS
SH
ZONE
ITEM
DESIGN
DESIGN ACTIVITY
CHECK
STRESS
WT
MAT'L
CMO

Figure 13–25 Sample engineering change notice. *Courtesy Aerojet TechSystems Company.*

ENGINEERING PROBLEM

Occasionally the engineering drafter has to take an existing product or structure and draw a set of working drawings. This may be a prototype. A prototype is a model or original design that has not been released for production. The prototype is often used for testing and performance evaluations. In most cases there are prototype drawings from which the prototype was built, but during the modification process changes may have been made. Your job, then, is to complete the production drawings, which are established by measuring the prototype, redrawing the prototype drawings with changes, or a combination of these methods. These revised drawings are often referred to as *as-built drawings*. In this type of situation you need to be well versed in the use of measuring equipment such as calipers, micrometers, and surface gages. You need to consult with the engineer, the testing department, and the shop personnel who made any modifications to the prototype. With this research you can fully understand how the new drawings should be created. You need to be careful to make the drawings accurate and take into account the manufacturing capabilities of your company. Then the set of working drawings that you complete is checked by the checker and the engineer before the product is released for production. Once the product is released for production, changes are often costly because tooling and patterns may have to be altered. These types of changes are usually submitted as engineering change requests (ECR). They must be approved by the engineering and manufacturing departments.

WORKING DRAWINGS PROBLEMS

DIRECTIONS

1. From the selected engineering sketches or layouts prepare a complete set of working drawings, including details, assembly, and parts list. Determine which views and dimensions should be used to completely detail each part. Also, determine the views, parts list, dimensions, and notes, if any, for the assembly drawing. Use ANSI standards. Use manual drafting or CADD as required by your course guidelines and objectives.
2. The complete set of working drawings will be prepared with one detail drawing per sheet using multiview projection and with the assembly drawing and parts list on one sheet unless otherwise specified. All purchase (standard) parts will be completely identified in the parts list. Using the sketches as a guide, draw original multiview drawings on adequately sized vellum. Add all necessary dimensions and notes using unidirectional dimensioning. Use computer-aided drafting if required by your course guidelines. Problems 13–1 through 13–10 are designed to be manufactured as projects in the manufacturing (machine) technology shop.

3. Include the following general notes at the lower left corner of the sheet .5 in. each way from the corner border lines for each detail drawing: (NOTE: Number 2 does not apply to the assembly drawing.)

   2. REMOVE ALL BURRS AND SHARP EDGES.

   1. INTERPRET DIMENSIONS AND TOLERANCES PER ANSI Y14.5M–1982.

   NOTES:
   Additional general notes may be required depending on the specifications of each individual assignment.

UNSPECIFIED TOLERANCES:

DECIMALS	mm	IN.
X	±2.5	±.1
XX	±0.25	±.01
XXX	±0.127	±.005
ANGULAR	±30°	
FINISH	3.2μm	125μIN.

Problem 13–1 Working-drawing assembly (metric)

Assembly Name: Plumb Bob
SPECIFIC INSTRUCTIONS:
Prepare a working-drawing assembly that has a detail drawing of each part, an assembly drawing, and a parts list on one sheet.
PARTS LIST:

ITEM	QTY	NAME	MATERIAL
1	1	PLUMB BOB	BRONZE
2	1	CAP	BRONZE

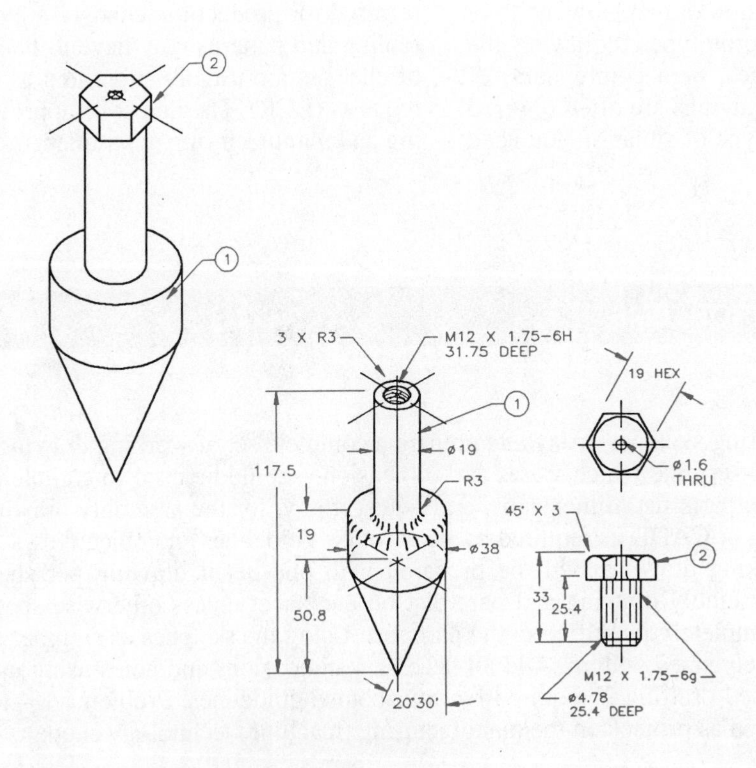

Problem 13–2 Working drawing (in.)

Assembly Name: Hammer
SPECIFIC INSTRUCTIONS:
Prepare a detail drawing for the hammer head and two optional hammer handles on one sheet. Make the assembly drawing and parts list on another sheet.

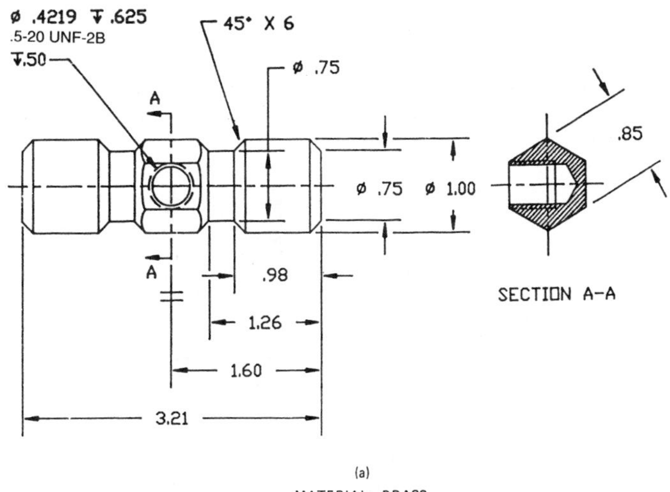

(a)
MATERIAL: BRASS
HAMMERHEAD

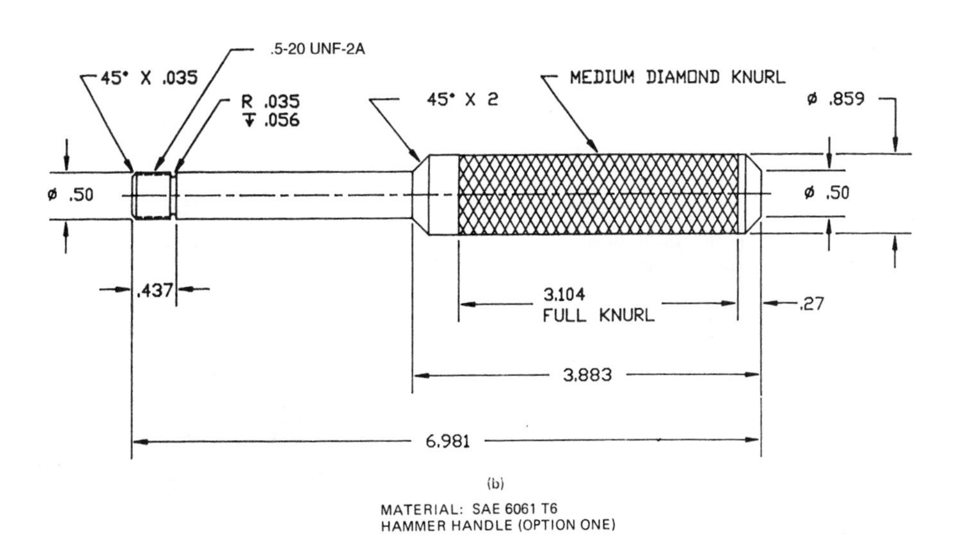

(b)
MATERIAL: SAE 6061 T6
HAMMER HANDLE (OPTION ONE)

(c)
MATERIAL: SAE 6061 T6
HAMMER HANDLE (OPTION TWO)

Problem 13–3 Working drawing (in.)

Assembly Name: Key Holder
SPECIFIC INSTRUCTIONS:
Prepare a detail drawing for each part, an assembly drawing, and a parts list on one sheet, unless otherwise specified by your instructor.

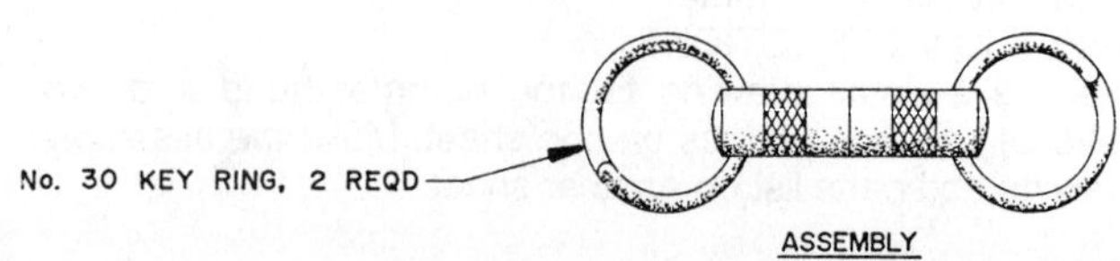

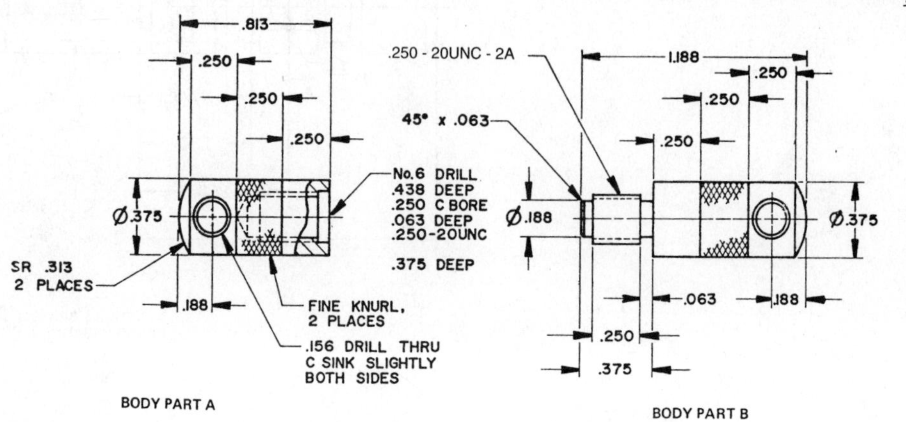

Problem 13–4 Working drawing (metric)

Assembly Name: C-clamp
SPECIFIC INSTRUCTIONS:
Prepare a complete set of working drawings with all of the detail drawings on one sheet and the assembly drawings and parts list on another sheet. Use multiview projection for view layout.

PARTS LIST:

ITEM	QTY	NAME	MATERIAL
1	1	PIN	CRMS
2	1	BODY	SAE 4320
3	1	SCREW	SAE 4320
4	1	SWIVEL	SAE 1020

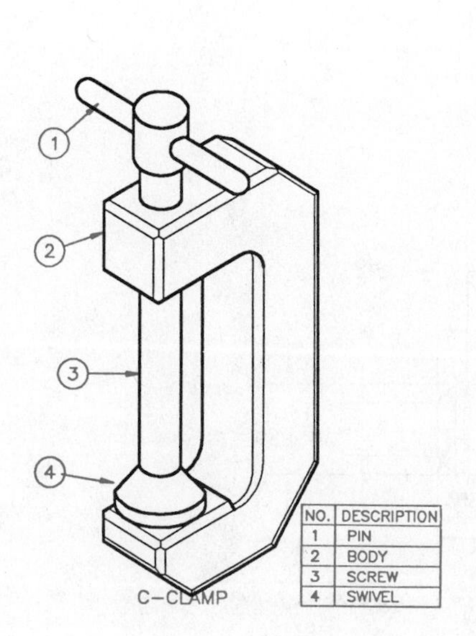

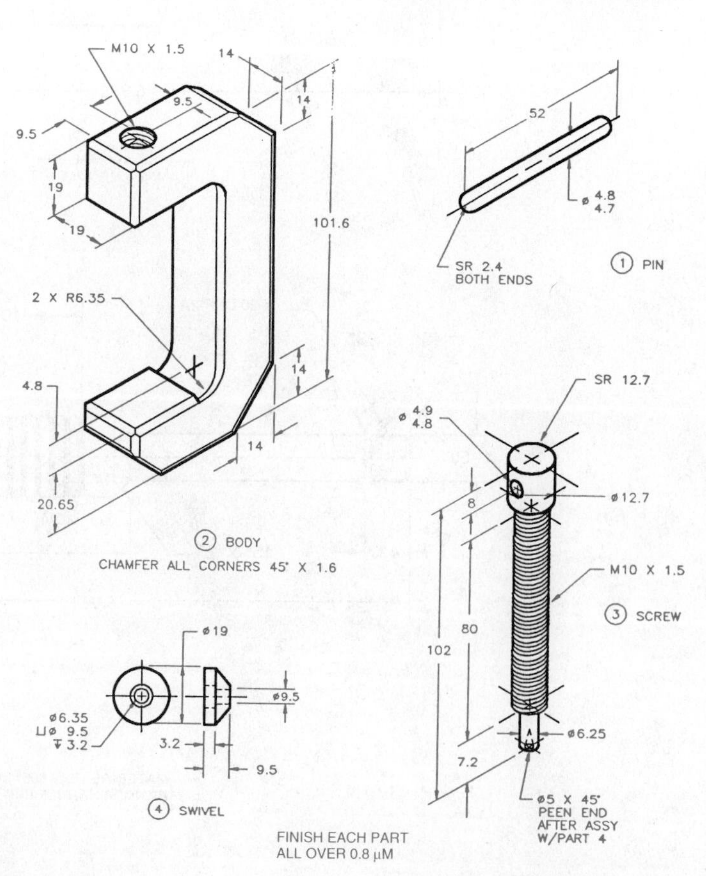

Problem 13–5 Working drawing (in.)

Assembly Name: Mill Work Stop
SPECIFIC INSTRUCTIONS:
Prepare a complete set of working drawings with one detail drawing per sheet and the assembly drawing with parts list on one sheet. Notice: part number 10, TEE STUD, is cut from standard 3/8-16UNC ALL THREAD. Parts 12 and 13, BLANK KNOBS, are purchased as knobs without the threads machined. You will draw the knobs as a "representation" of the purchase part and give the thread note as identified in the parts list. The actual dimensions of the knob to be purchased are not important; however, accurate specifications for the threads to be machined into the purchase part are important.

PARTS LIST:

ITEM	QTY	NAME	DESCRIPTION	MATERIAL
1	1	WING NUT	1/4-28 × 1/2	
2	1	VERTICAL MEMBER	3/4 × 2 × 4	SAE 6061
3	1	BASE MEMBER	1 × 2 × 3.5	SAE 6061
4	2	RIB	1/4 PLATE	SAE 6061
5	1	ARM	3/8 × 1 × 5-1/2	SAE 1018
6	1	CLAMP	1/2 × 1 × 1-5/16	SAE 1018
7	1	TEE PLATE DRILL AND TAP AT CENTER FOR 5/16-18UNC-2 THREAD THRU	5/16 × 1 × 1-1/4	SAE 1018
8	1	STOP ROD	∅ 5/16 × 6	SAE 303 SS
9	1	BUSHING	∅ 1/2 × .31	SAE 1018
10	1	TEE STUD	3/8-16 × 2 LG ALL THREAD	
11	1	CARRIAGE BOLT	5/16–18UNC-2	
12	21	BLANK KNOB	VLIER HK-3 (DRILL AND TAP FOR 3/8-16UNC-2B)	
13	1	BLANK KNOB	VLIER HK-3 (DRILL AND TAP FOR 5/16–18UNC-2B)	

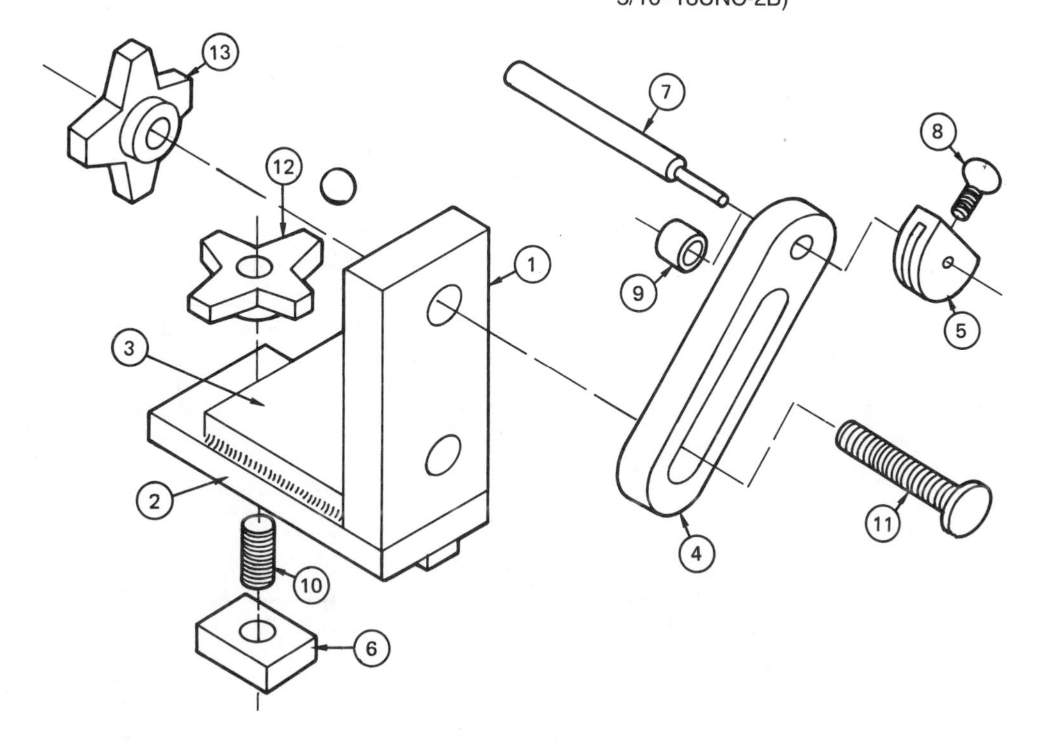

Problem 13–5 (Continued)

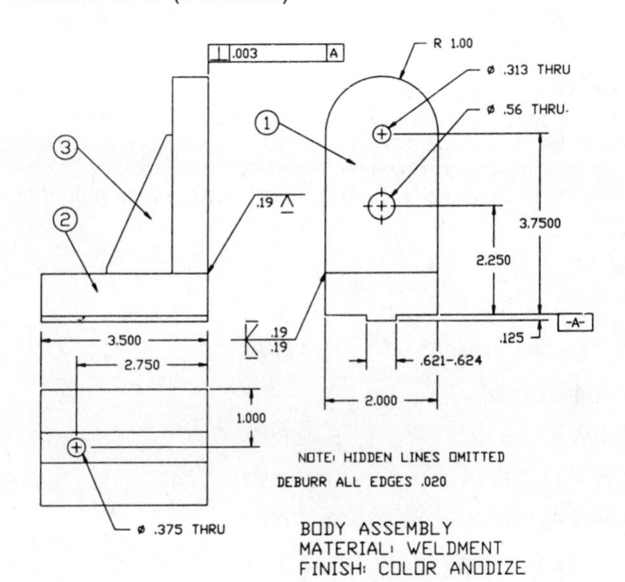

RIB

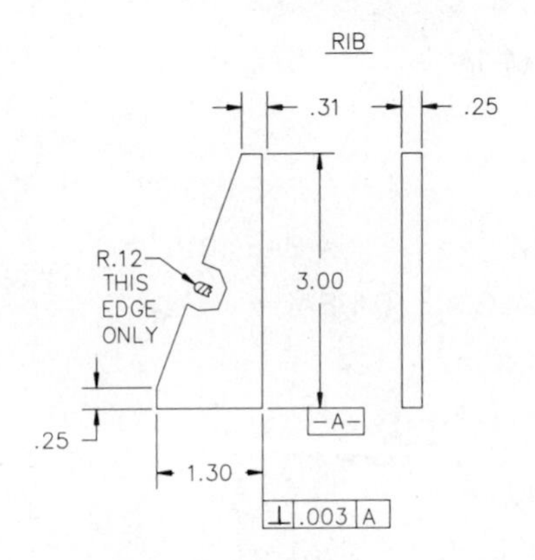

TOLERANCES U.O.S.
ANGLES N/A
.X ±.030
.XX ±.010
.XXX ±.003

MATERIAL 1/4 THICK 6061 ALUM.
FINISH – NONE
2 REQUIRED

ARM

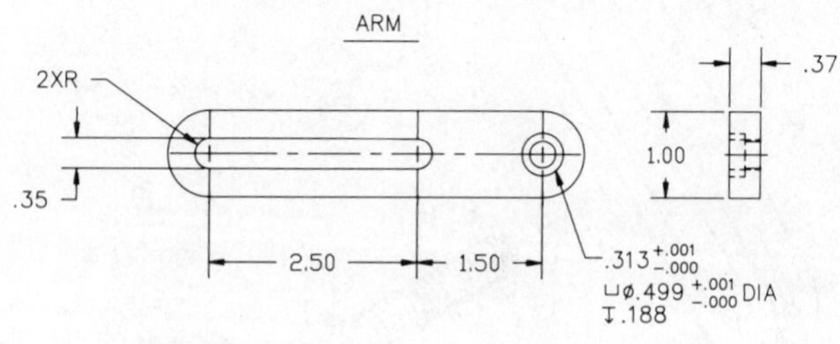

TOLERANCES U.O.S.
ANGLES N/A
.X ±.030
.XX ±.010
.XXX ±.003

MATERIAL – 3/8" X 1" C–1018
FINISH – C'HARDEN .005/.010
BLACK OXIDE

DEBURR ALL EDGES .015

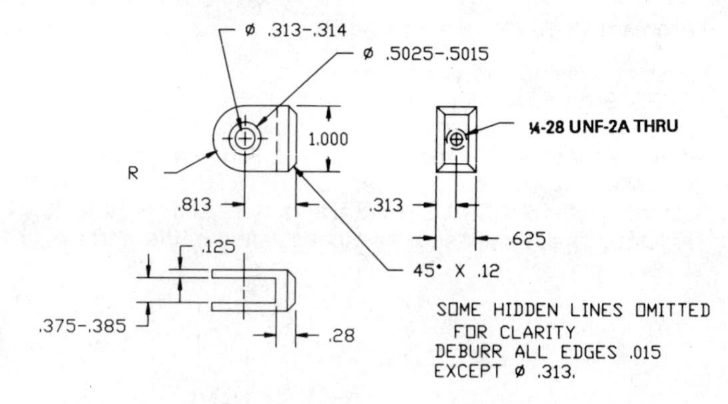

5 CLAMP
MATERIAL: 5/8 X 1 C-1018
FINISH: CASE HARDEN .005/.010
BLACK OXIDE

STOP ROD

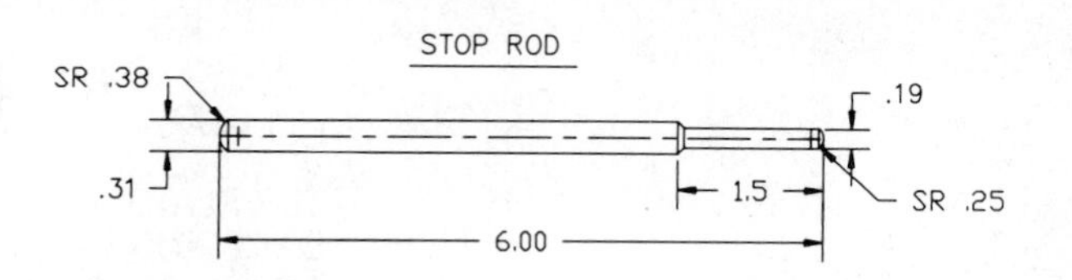

TOLERANCES U.O.S.
ANGLES ±1DEG
.X ±.030
.XX ±.010
.XXX ±.003

5/16" DIA 303 STAINLESS
FINISH – POLISH

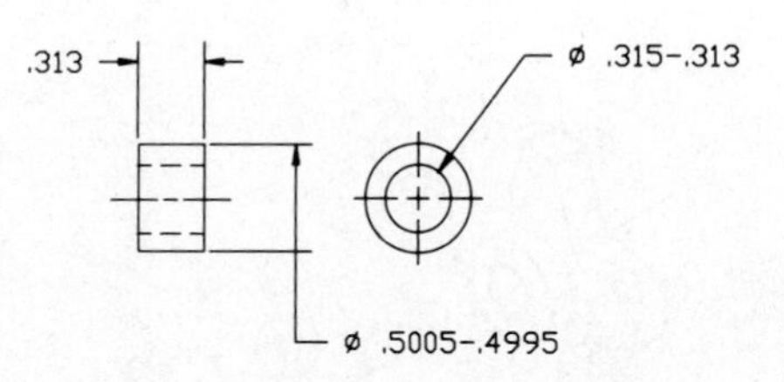

9 BUSHING
MATERIAL: ø 5/8 C-1018
FINISH: CASE HARDEN .005/.010
BLACK OXIDE
DEBURR ALL EDGES .010

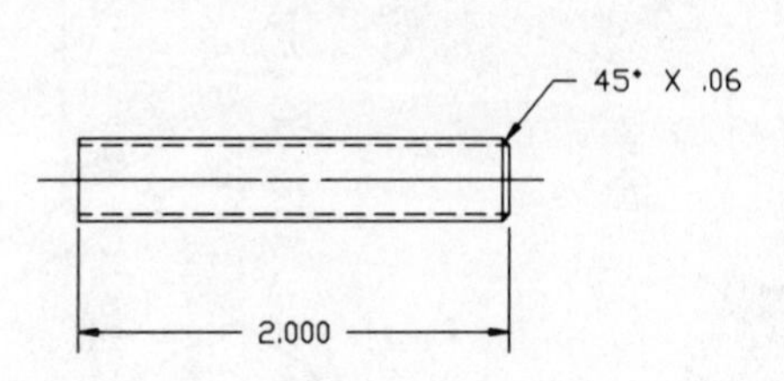

10 TEE STUD
MATERIAL: 3/8-16UNC ALL THREAD C-1020

Problem 13–6 Working drawing (metric)

Assembly Name: Tool Holder
SPECIFIC INSTRUCTIONS:
Prepare a complete set of working drawings with one detail drawing per sheet, and the assembly and parts list on another sheet.

PARTS LIST:

ITEM	QTY	NAME	MATERIAL
1	1	TOOL HOLDER BODY	06 STEEL
2	1	STUD	SAE 1035 STEEL
3	1	KNURL NUT	SAE 3130 STEEL
4	1	WASHER	SAE 1060 STEEL
5	1	SHIM	SAE 4320 STEEL
6	1	ADJUSTMENT SCREW	SAE 1035 STEEL
7	1	PARTING TOOL, 3/32 In. × ½ In. PURCHASE PART	TOOL STEEL
8	1	M 10 × 1.5 HEX NUT PURCHASE PART	

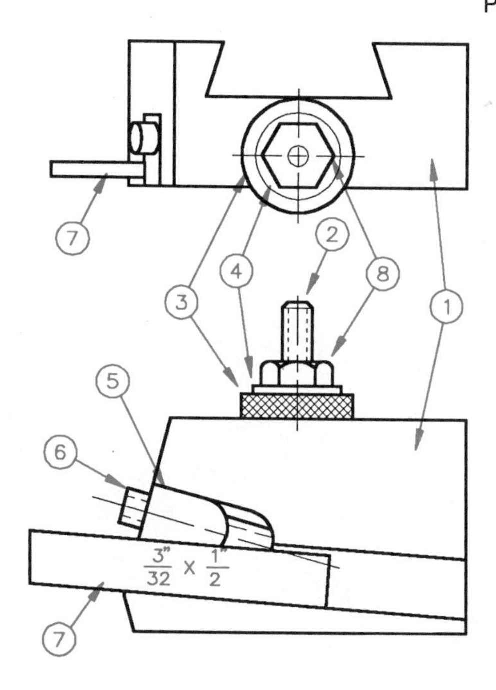

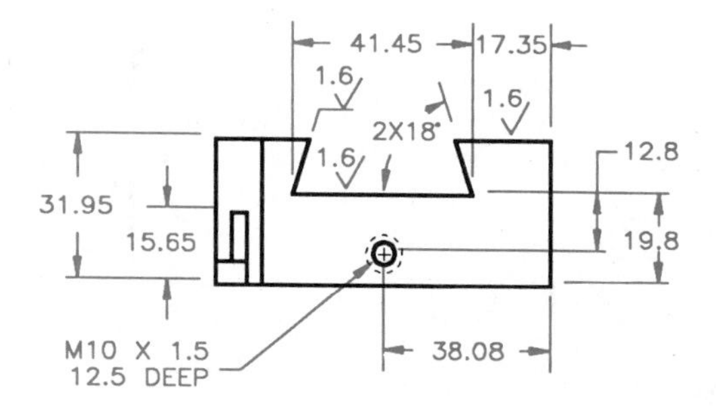

NOTE TO STUDENT: WHEN PREPARING THE ASSEMBLY DRAWINGS, USE SEPARATE BALLOONS FOR EACH PART IN EACH VIEW, OR USE ONLY ONE BALLOON IN THE VIEW THAT MOST CLEARLY IDENTIFIES THE PART.

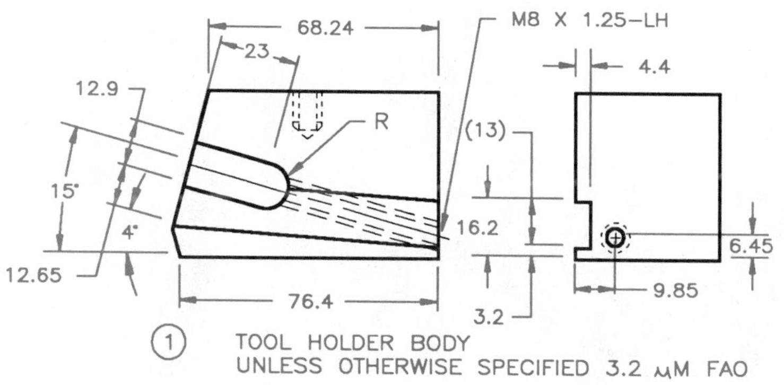

(1) TOOL HOLDER BODY
UNLESS OTHERWISE SPECIFIED 3.2 μM FAO

Problem 13–6 (Continued)

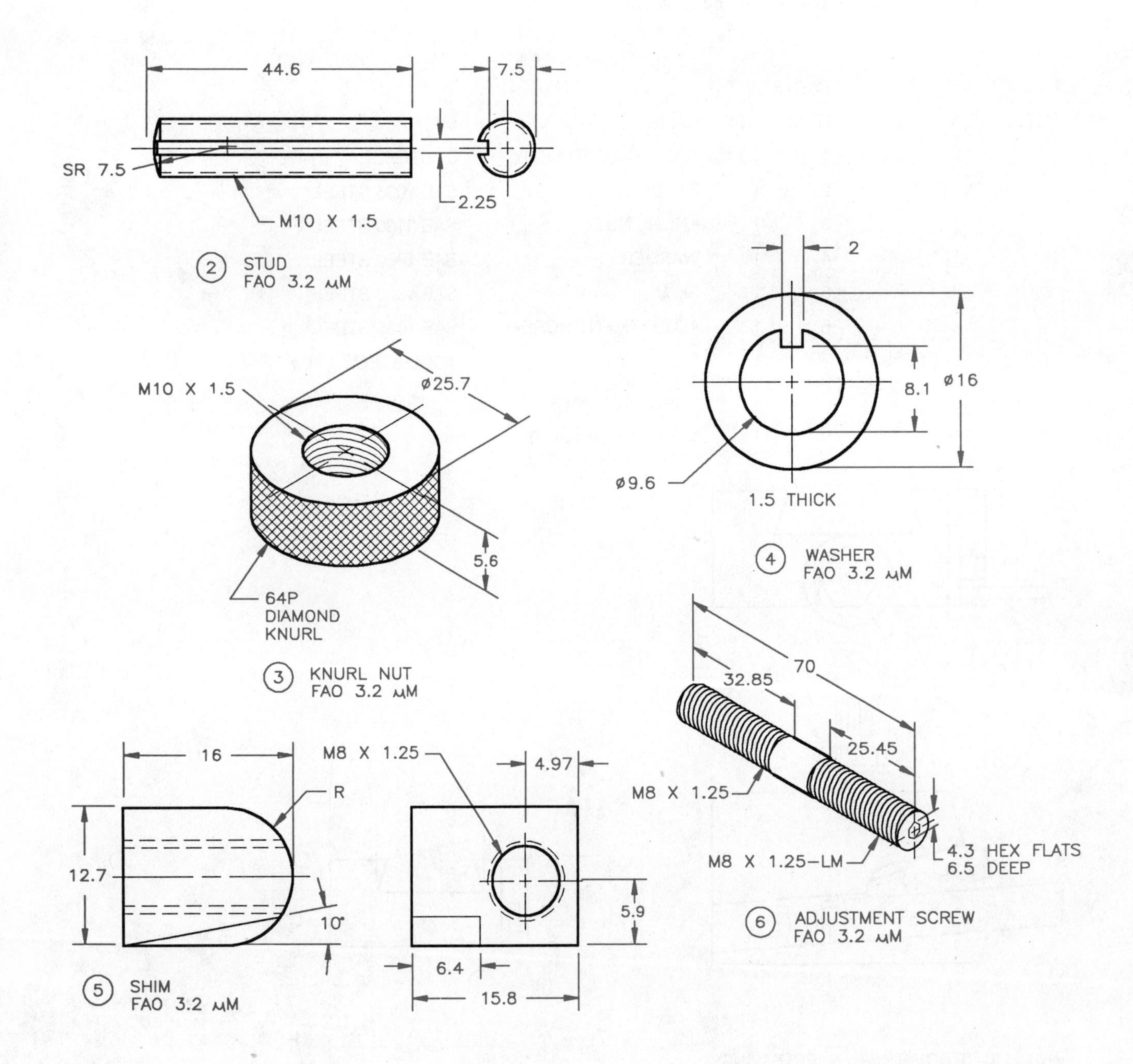

Problem 13–7 Working drawing (in.)

Assembly Name: Adjustable Attachment
SPECIFIC INSTRUCTIONS:
Prepare a complete set of working drawings with detail drawings of individual parts combined on one or more sheets depending on the size of sheet selected. The assembly drawing and parts list will be combined on one sheet.

PARTS LIST:

ITEM	QTY	NAME	DESCRIPTION	MATERIAL
1	1	FRAME		SAE 4340
2	1	ADJUSTING SCREW		SAE 1045
3	1	KNOB		SAE 1024
4	1	CLAMP SCREW		SAE 2330
5	1	PILOT SCREW		SAE 2330
6	1	ADJUSTING NUT		SAE 3130
7	1	CAP SCREW	.25-20UNC-2 × .75 HEX SOC HEAD	STL
8	1	TAPER PIN	0 × .625	STL
9	1	SET SCREW	8-32UNC-2 HEX SOC FLAT POINT	STL

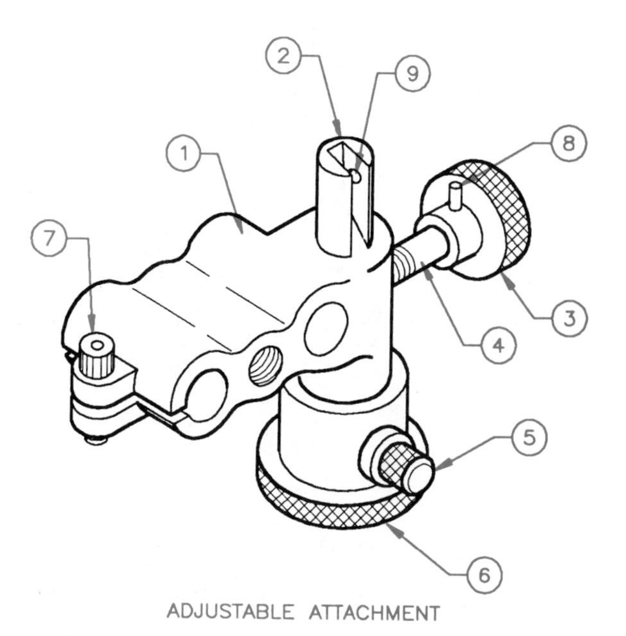

ADJUSTABLE ATTACHMENT

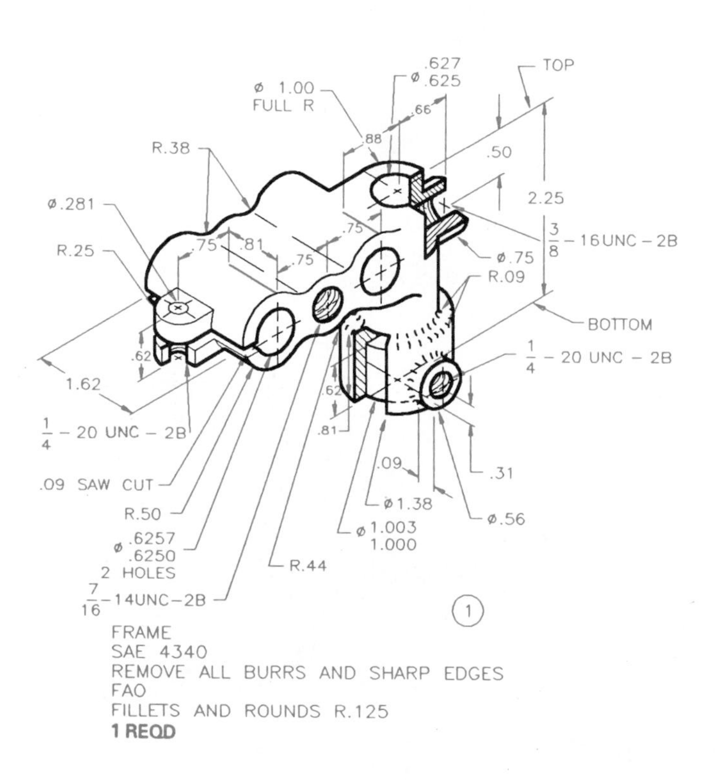

FRAME
SAE 4340
REMOVE ALL BURRS AND SHARP EDGES
FAO
FILLETS AND ROUNDS R.125
1 REQD

.24
.18
#8 – 32 UNC – HEX SOCK
FLAT PT. SET SCREW
1 R'QD
.25
.188
2
2.00
⌀ .623
.621
1/2 – 20UNF – 2A
3.75
.38
1.50
45° X .03

ADJUSTING SCREW
SAE 1045
REMOVE ALL BURRS AND SHARP EDGES
FAO
1 REQD

Problem 13–7 (Continued)

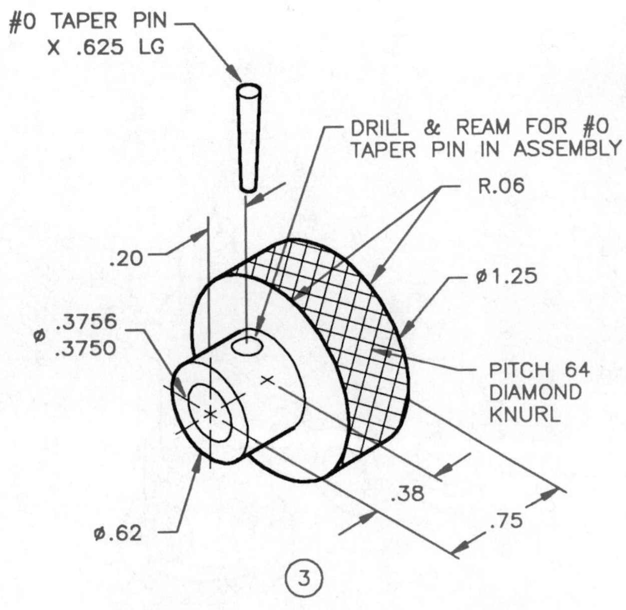

KNURL KNOB
SAE 1024
REMOVE ALL BURRS AND SHARP EDGES
FAO
1 REQD

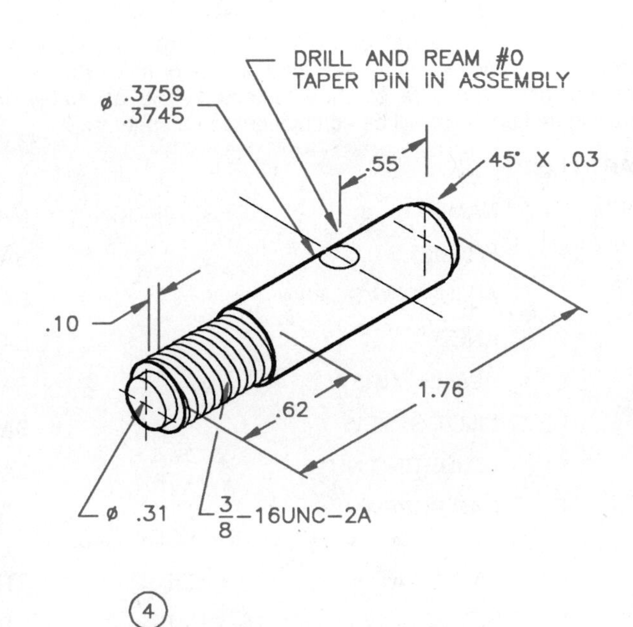

CLAMP SCREW
SAE 2330
REMOVE ALL BURRS AND SHARP EDGES
1 REQD

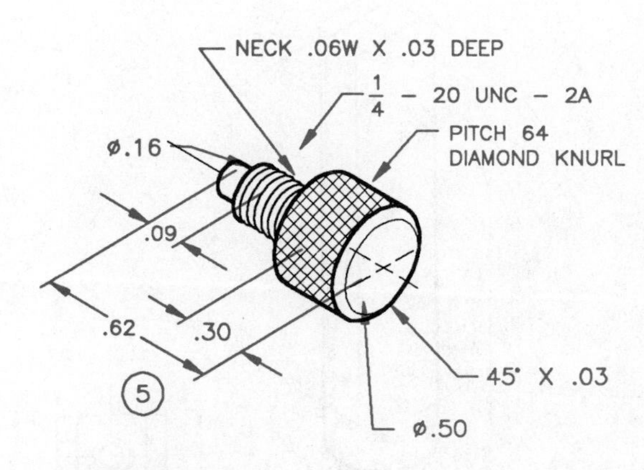

PILOT SCREW
SAE 2330
REMOVE ALL BURRS AND SHARP EDGES
FAO
2 REQD

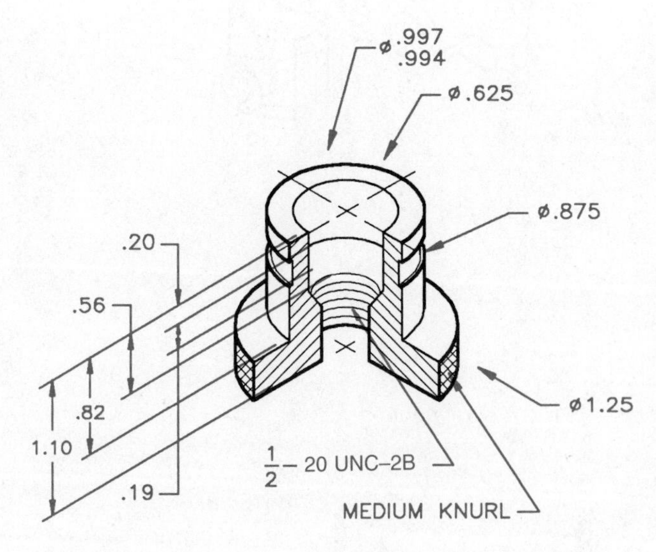

6 ADJUSTING NUT
SAE 3130
REMOVE ALL BURRS AND SHARP EDGES
FAO
1 REQD

Problem 13–8 Working drawing (in.)

Assembly Name: Precision Vice

PARTS LIST:

ITEM	QTY	NAME	DESCRIPTION	MATERIAL
1	1	BASE		SAE 1040
2	2	CAP SCREW	1/4-28NF × 3/4 HEX SOCKET HEAD	STL
3	1	JAW INSERT		SAE 4330
4	1	JAW INSERT		SAE 4330
5	2	MACH SCREW	1/4-28NF × 1/2 FLAT HD	STL
6	1	CAP SCREW	5/16–24NF × 1 1/2 HEX SOCKET HEAD	STL
7	1	BALL WASHER		SAE 1040
8	1	JAW		SAE 1040
9	1	NUT		SAE 1040

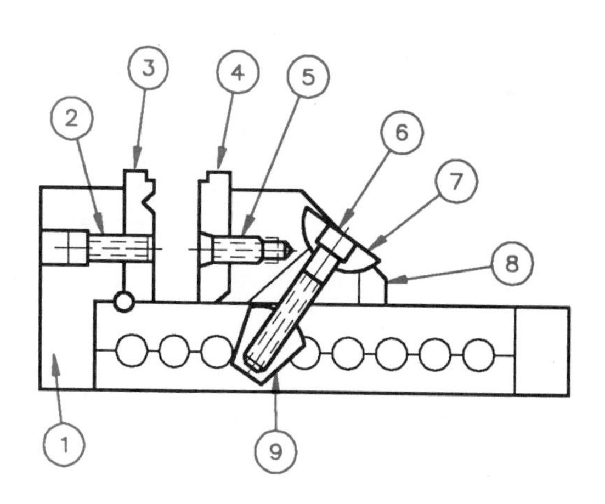

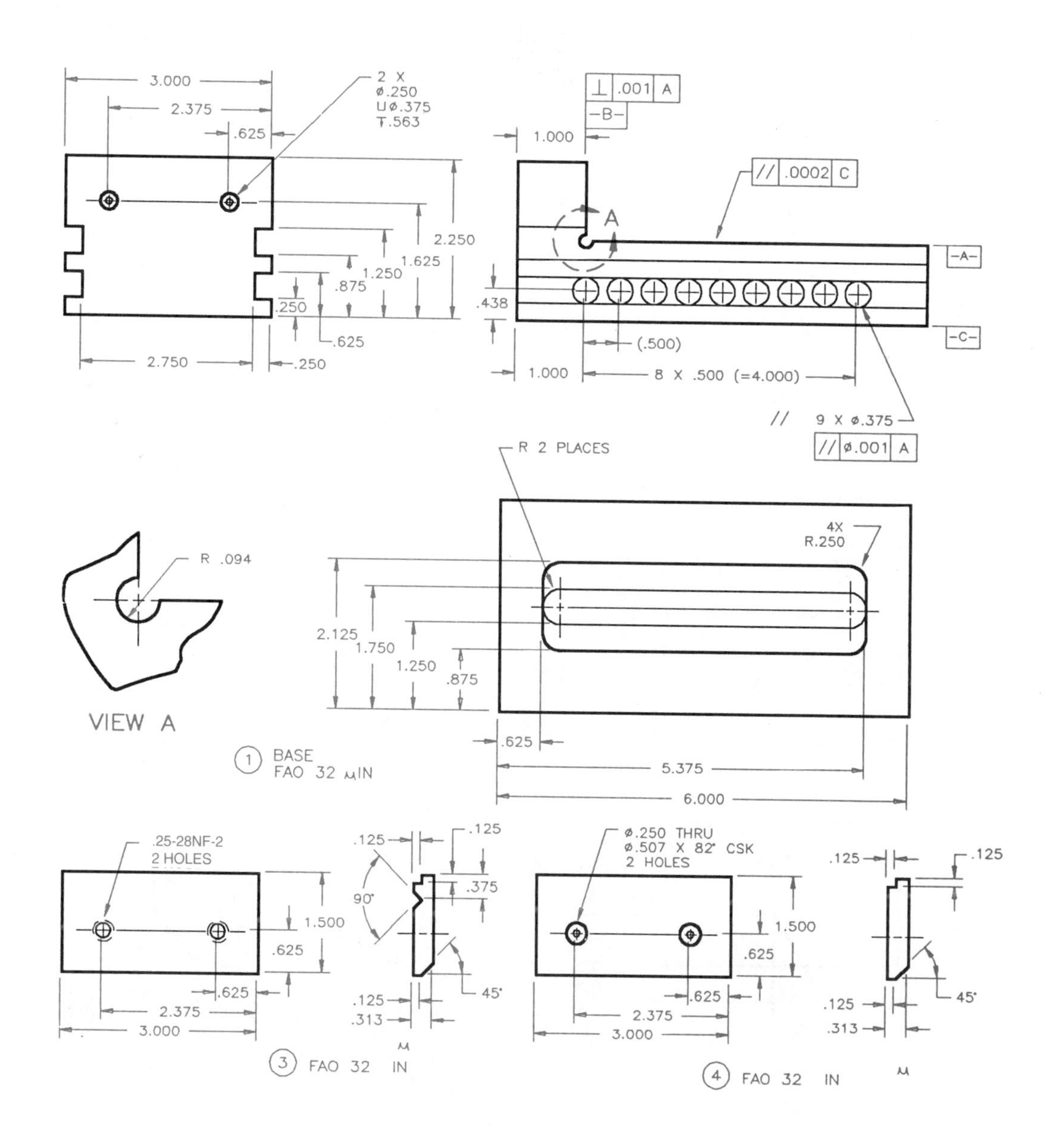

Problem 13–8 (Continued)

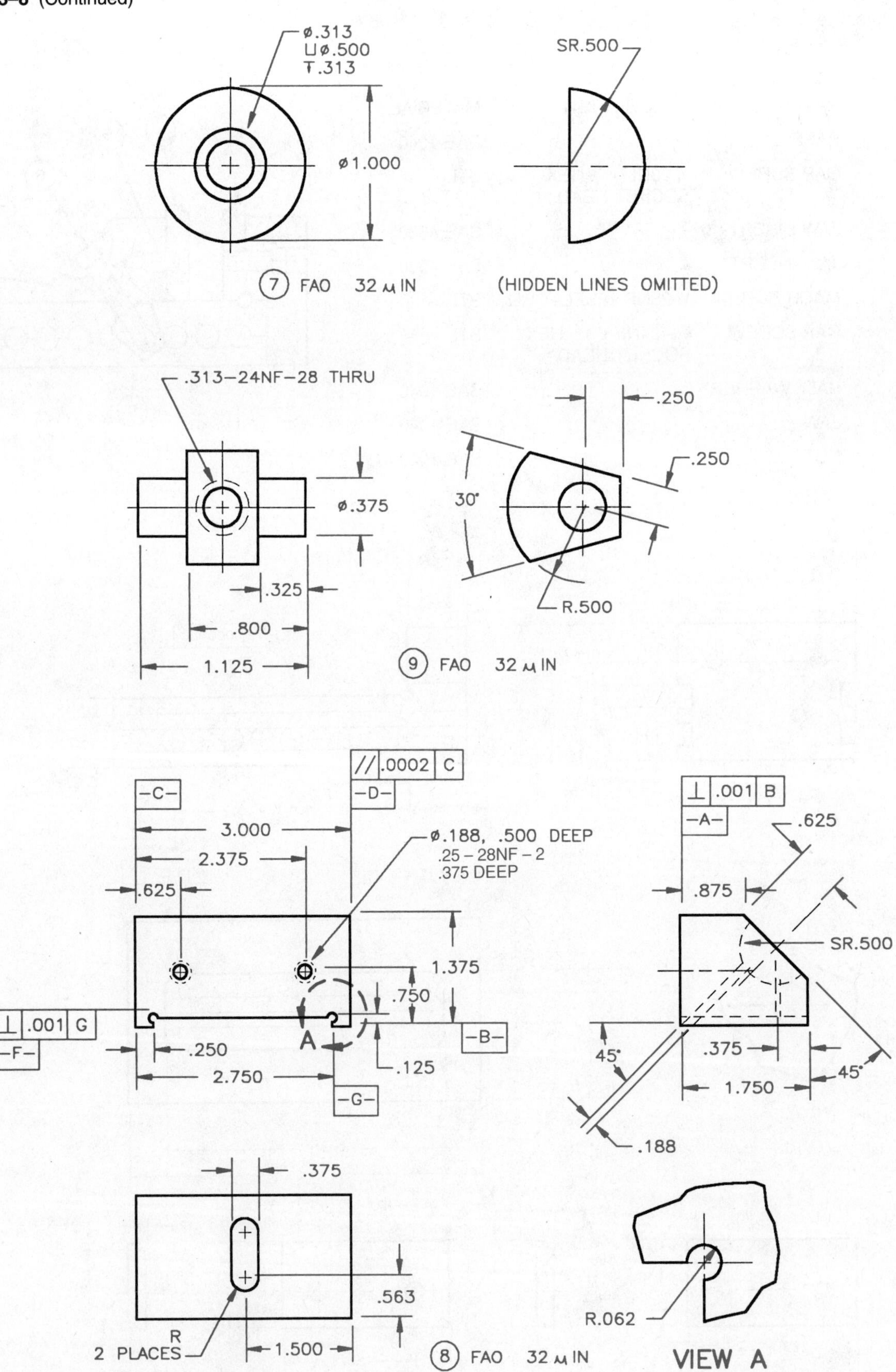

Problem 13–9 Working drawing (in.)

Assembly Name: Machine Vice
SPECIFIC INSTRUCTIONS:
When preparing the assembly drawings, use separate balloons for each part in each view, or use only one balloon in the view that most clearly identifies the part.

PARTS LIST:

ITEM	QTY	NAME	MATERIAL	DESCRIPTION
1	2	HANDLE CAP	MS	
2	1	HANDLE	MS	
3	1	BODY	SAE 4320	
4	1	SCREW	SAE 4320	
5	1	MOVABLE JAW	SAE 1020	
6	1	MOVABLE JAW PLATE	SAE 4320	
7	1	FIXED JAW PLATE	SAE 4320	
8	1	GUIDE	SAE 1020	
9	2	MACHINE SCREW	STL	.25-20UNC-2 × .500 SLOT FIL HD
10	2	MACHINE SCREW	STL	.190-32UNF-2 × .875 SLOT FIL HD
11	1	SET SCREW	STL	.25-20UNC-2 × .250 FULL DOG POINT
12	2	MACHINE SCREW	STL	.190-32UNF-2 × 6 SLOT FIL HD

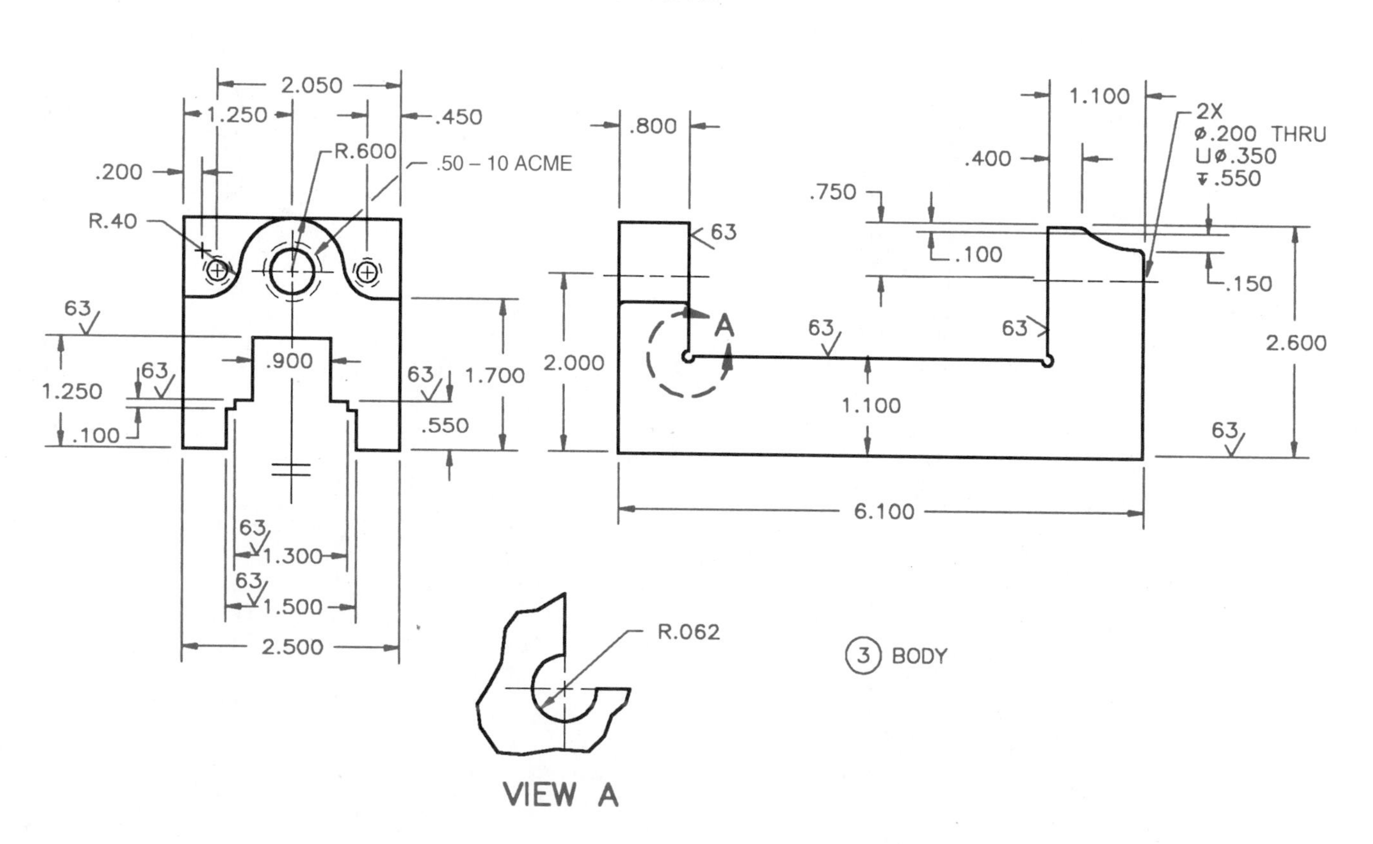

Problem 13–9 (Continued)

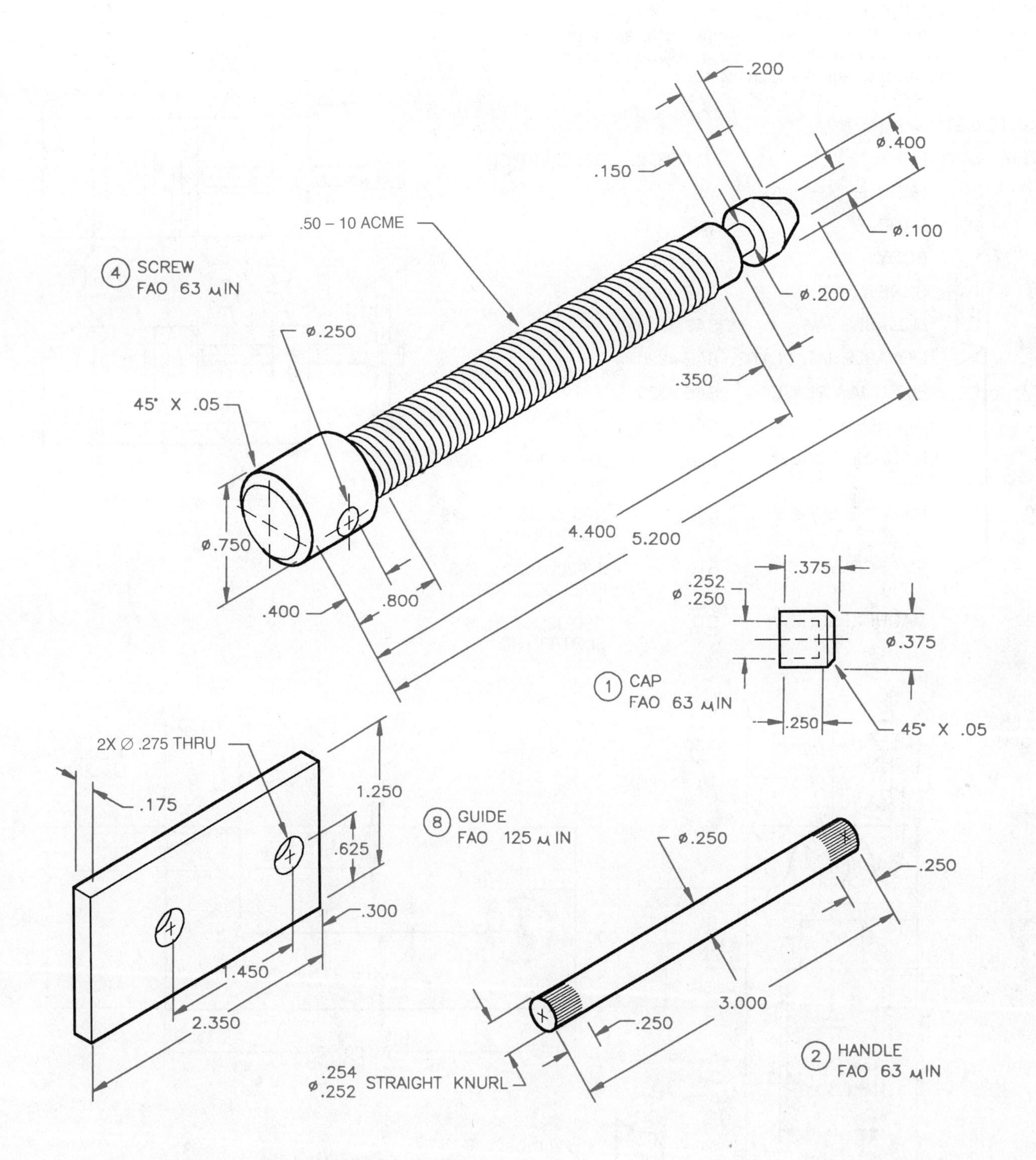

Problem 13–9 (Continued)

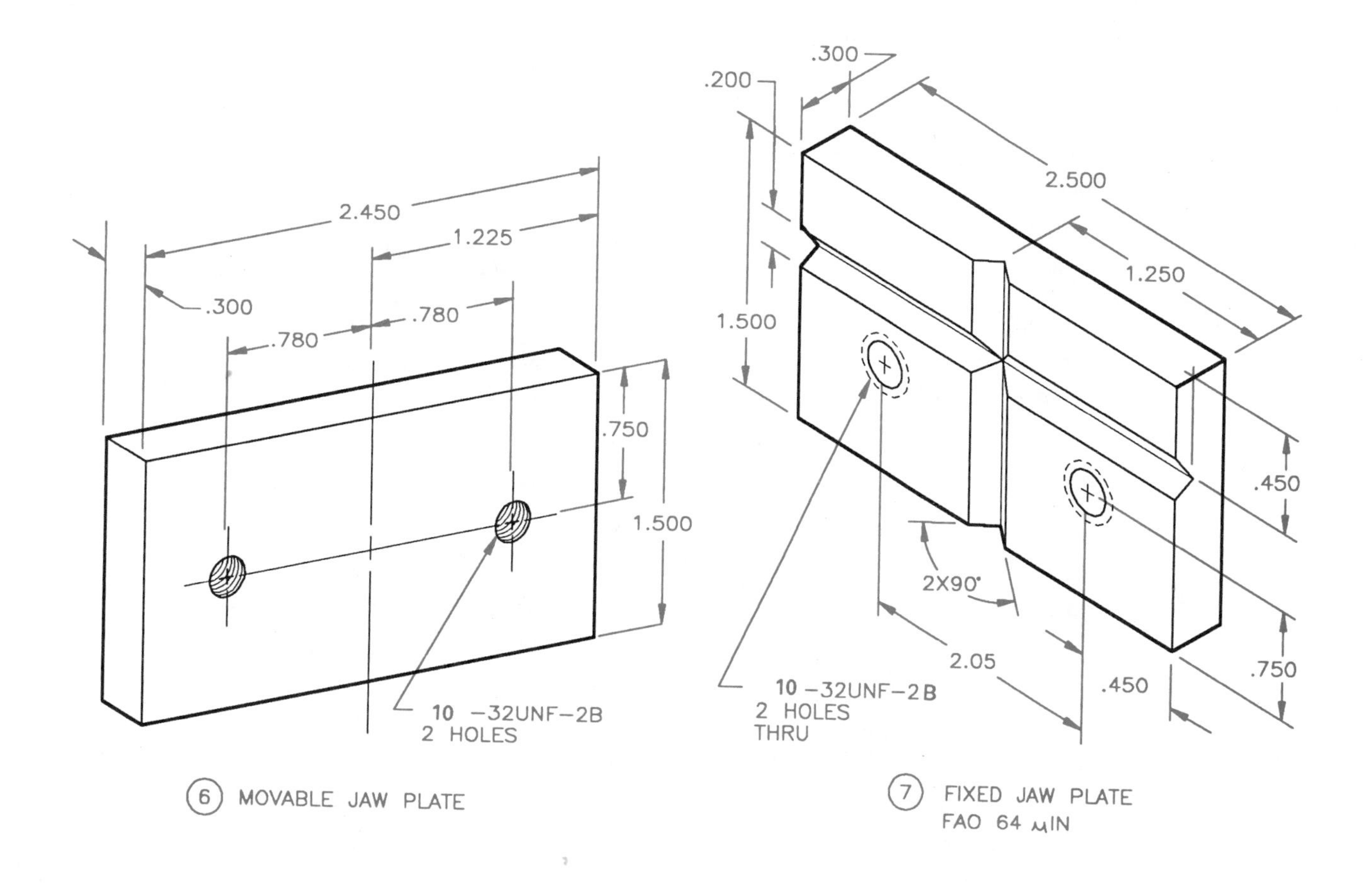

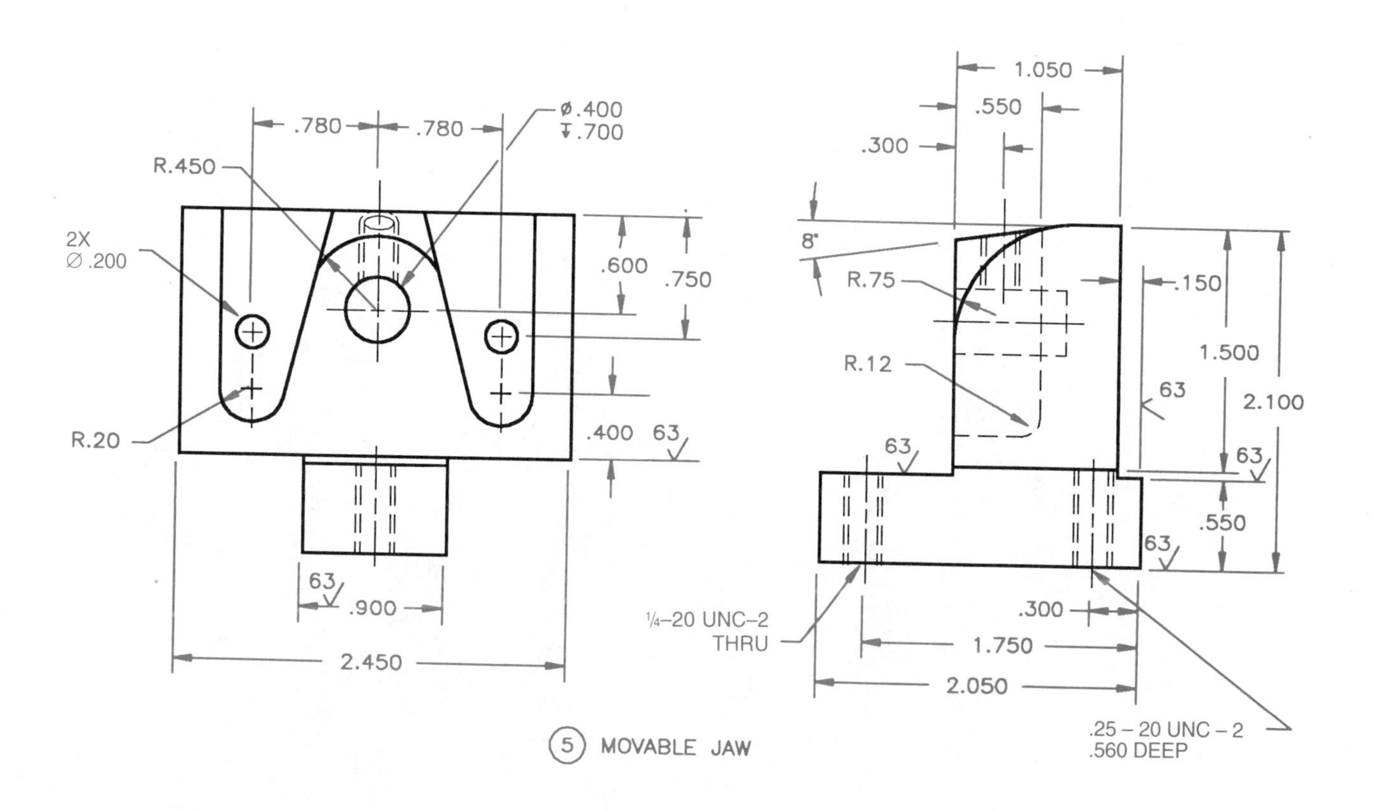

Problem 13–10 Working drawing (in.)

Assembly Name: Arbor Press

PARTS LIST:

ITEM	QTY	NAME	DESCRIPTION	MATERIAL
1	1	BASE		SAE 1020
2	1	COLUMN		SAE 1020
3	1	TABLE		SAE 1020
4	1	TABLE PIN		SAE 1020
5	1	SLEEVE		SAE 1020
6	1	HANDLE		SAE 1020
7	1	GEAR		SAE 4320
8	1	COVER PLATE		SAE 1020
9	1	RACK		SAE 4320
10	1	SCREW		SAE 1040
11	1	RACK PAD		SAE 4320
12	2	BALL END		SAE 1020
13	4	CAP SCREWS	8-32UNC-2 × .50 HEX SOC	STL
14	1	MACHINE SCREW	.375-16UNC-2 × 1.00 HEX HEAD	STL

NOTE TO STUDENT: WHEN PREPARING THE ASSEMBLY DRAWING USE SEPARATE BALLOONS FOR EACH PART IN EACH VIEW OR USE ONLY ONE BALLOON IN THE VIEW THAT MOST CLEARLY IDENTIFIES THE PART.

Problem 13–10 (Continued)

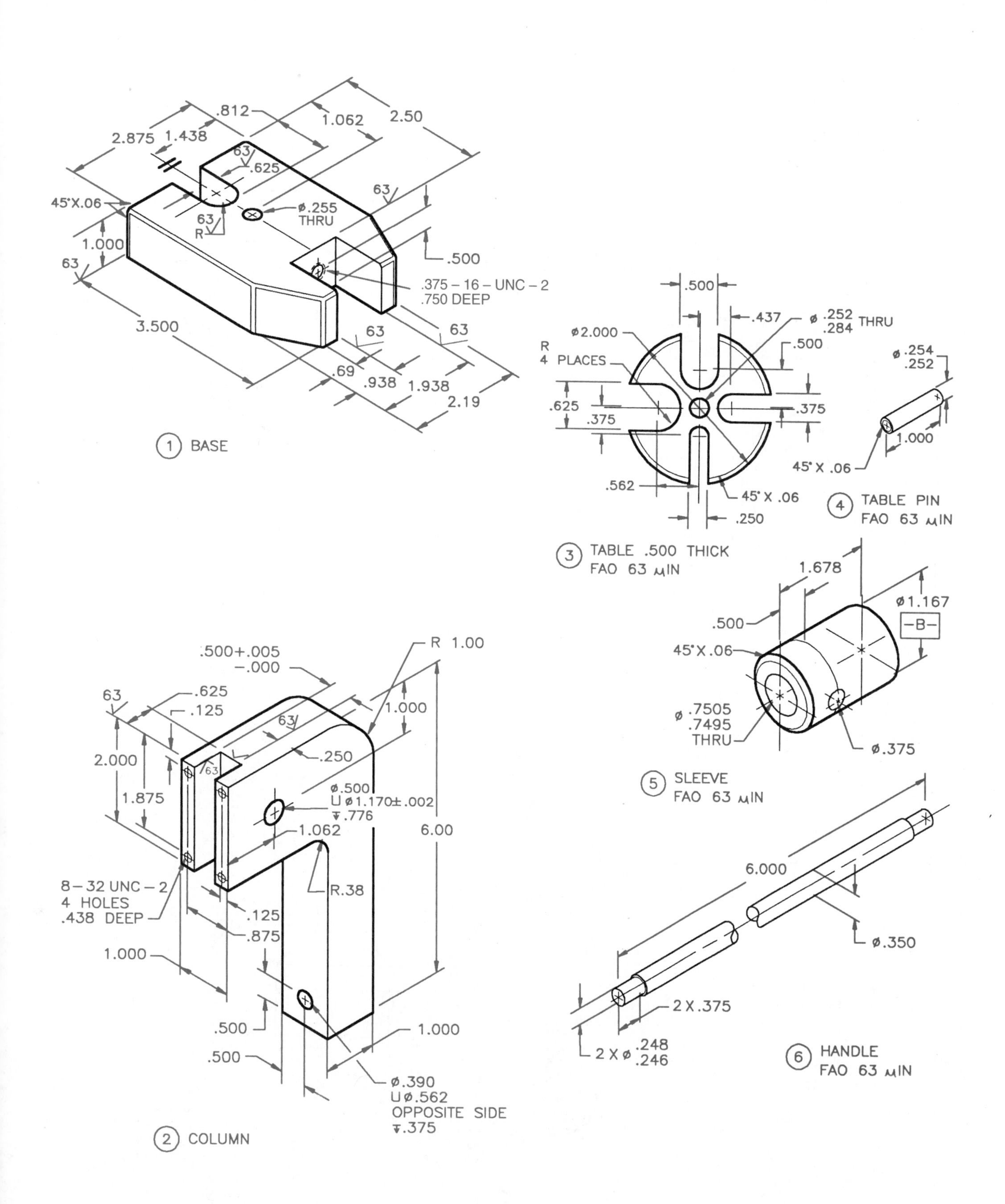

Problem 13–10 (Continued)

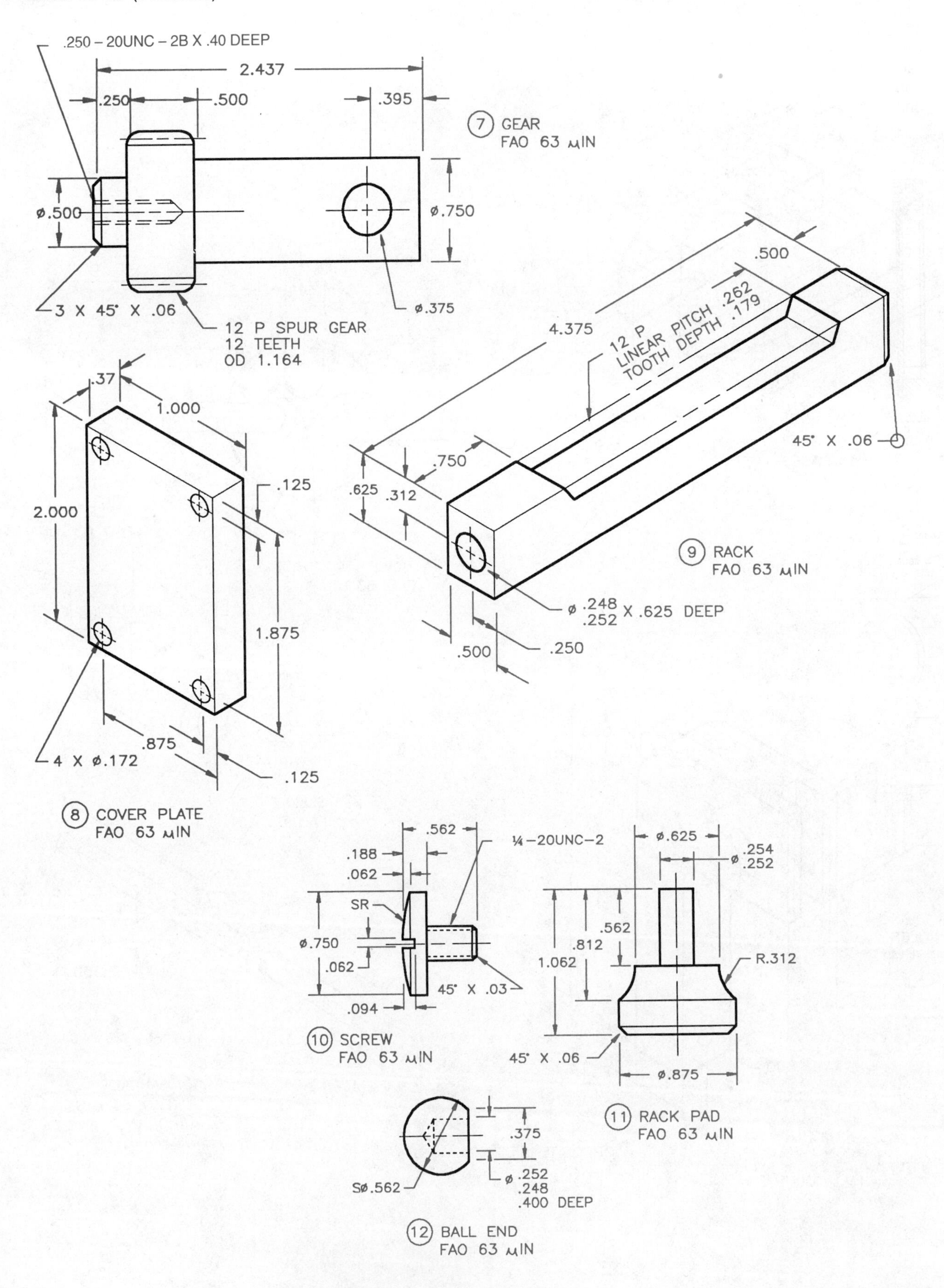

Problem 13–11 Working drawing (in.)

Assembly Name: Hydraulic Jack
SPECIFIC INSTRUCTIONS:
Prepare one detail drawing per sheet, with the assembly drawing and parts list together on one sheet. Determine the sheet sizes by the size and scale of the part, the number of dimensions and notes, and the number of views.

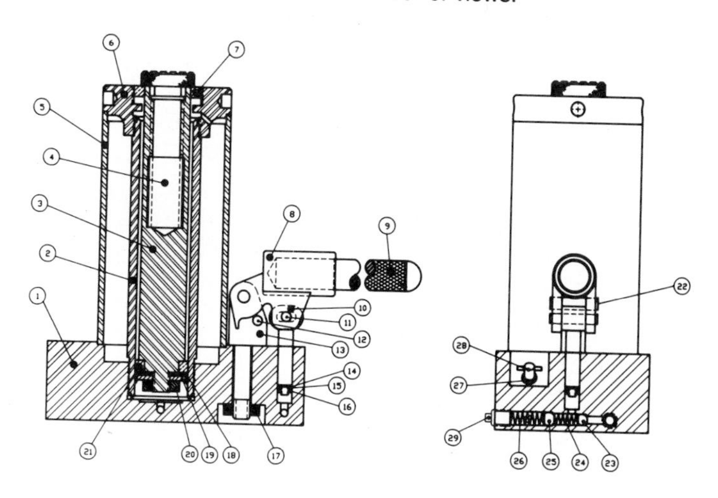

ITEM	QTY.	PART NAME	PART NUMBER	MATERIAL
29	1	1/8-NPT-PLUG	1DT3029	STEEL
28	1	NEEDLE VALVE	1DT3028	M.S.
27	1	NEEDLE VALVE NUT	1DT3027	BRONZE
26	1	.343 SPRING	1DT3026	SPRING STEEL
25	1	.3125 BALL	1DT3025	STEEL
24	1	.281 SPRING	1DT3024	SPRING STEEL
23	1	.25 BALL	1DT3023	STEEL
22	1	PIVOT PIN	1DT3022	.312 DIA. STEEL
21	1	BRONZE GUIDE	1DT3021	BRONZE
20	1	.437-20 UNF NUT	1DT3020	STEEL
19	1	PISTON WASHER	1DT3019	STEEL
18	1	LEATHER CUP	1DT3018	1.5″ DIA. LEATHER
17	1	.437-14 UNC NUT	1DT3017	STEEL
16	1	10-32 NUT	1DT3016	STEEL
15	1	.125 WASHER	1DT3015	STEEL
14	1	CUP SEAL	1DT3014	.445 DIA. NEOPRENE
13	1	PUMP PIVOT SUPPORT	1DT3013	C.R.S.
12	1	STOP PIN	1DT3012	1/4″ DIA. STEEL
11	1	DRIVE PIN	1DT3011	1/4″ DIA. STEEL
10	1	PUMP PLUNGER	1DT3010	H.S.
9	1	PUMP HANDLE	1DT3009	C.R.S.
8	1	PUMP HANDLE SOCKET	1DT3008	H.R.M.S.
7	1	PACKING NUT	1DT3007	BRONZE
6	1	TOP CAP	1DT3006	M.S.
5	1	RESERVOIR TUBE	1DT3005	3″ PIPE STEEL SCH#40
4	1	SCREW	1DT3004	M.S.
3	1	PISTON	1DT3003	H.R.M.S.
2	1	CYLINDER TUBE	1DT3002	FREE MACHINING LEADED / C.R.M.S.
1	1	BASE	1DT3001	M.S.

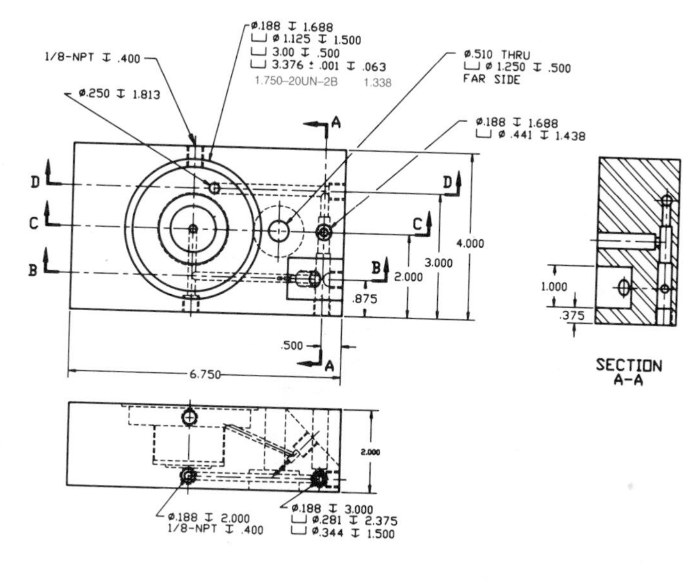

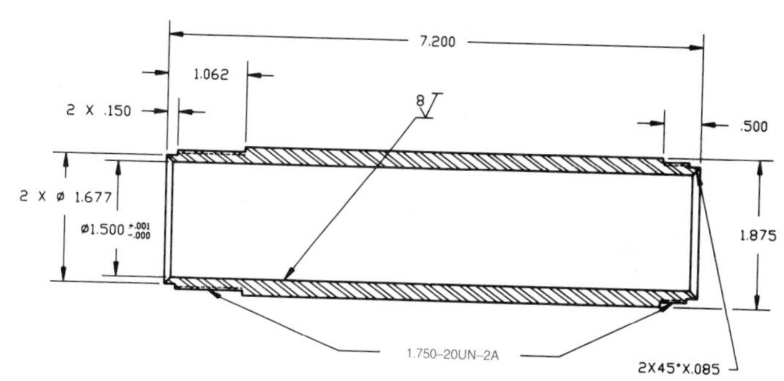

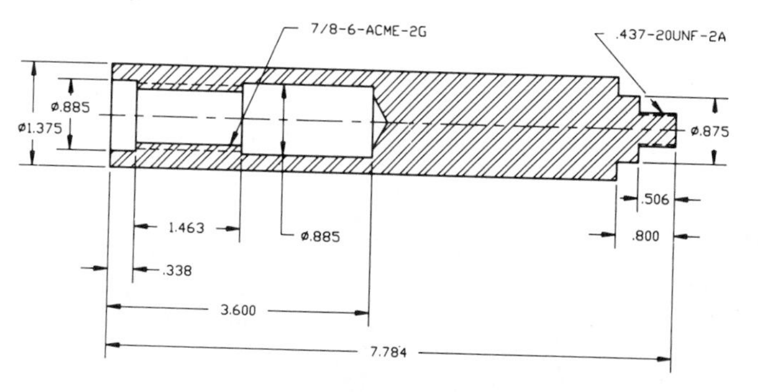

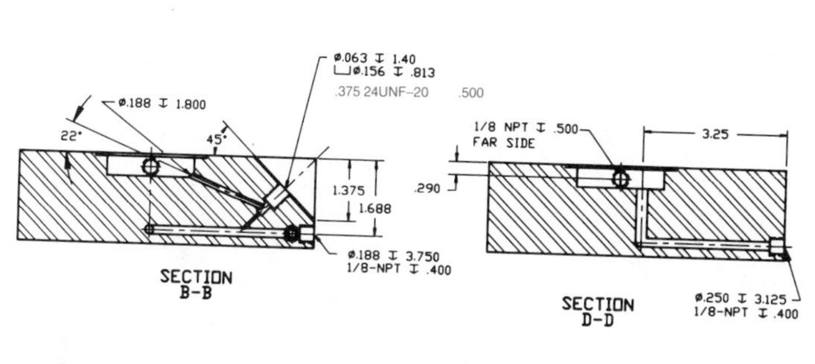

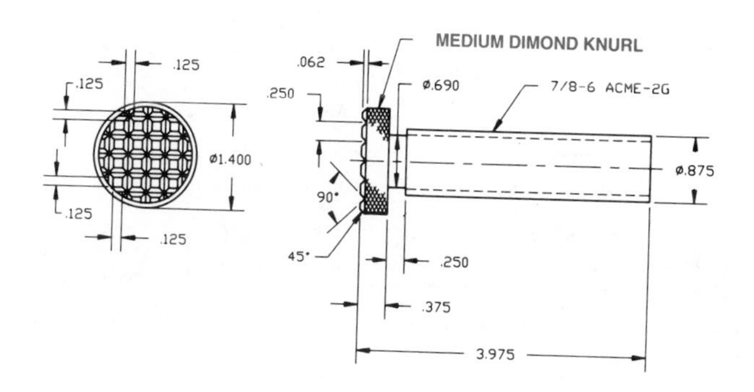

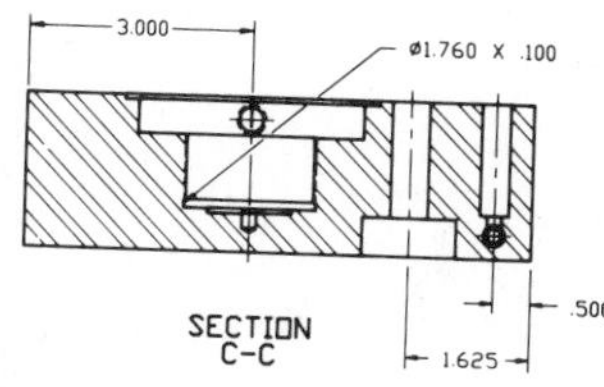

Problem 13–11 (Continued)

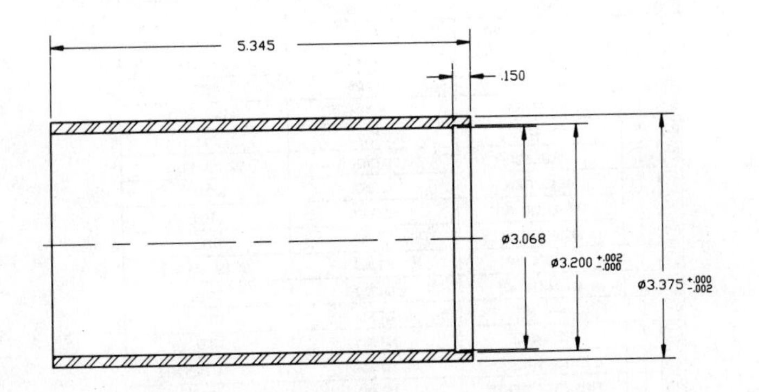

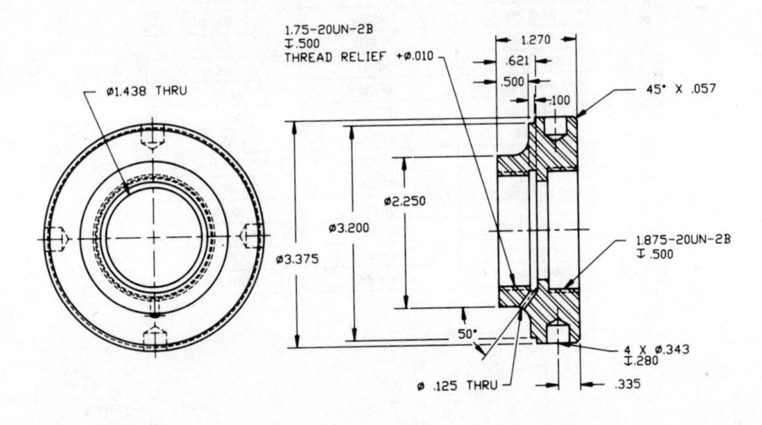

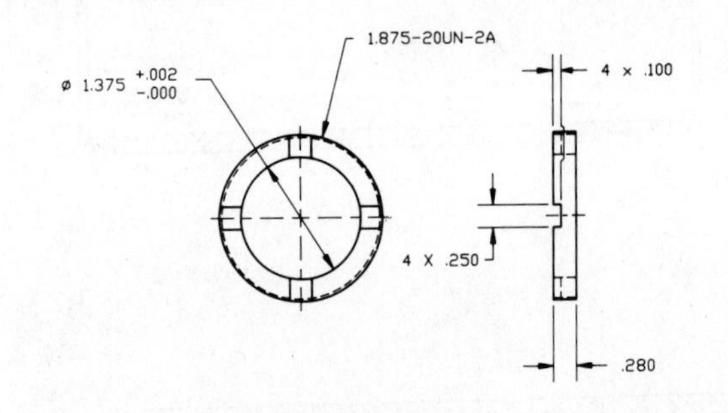

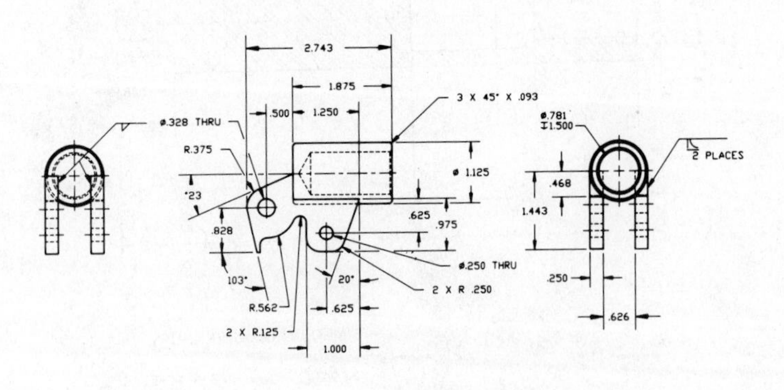

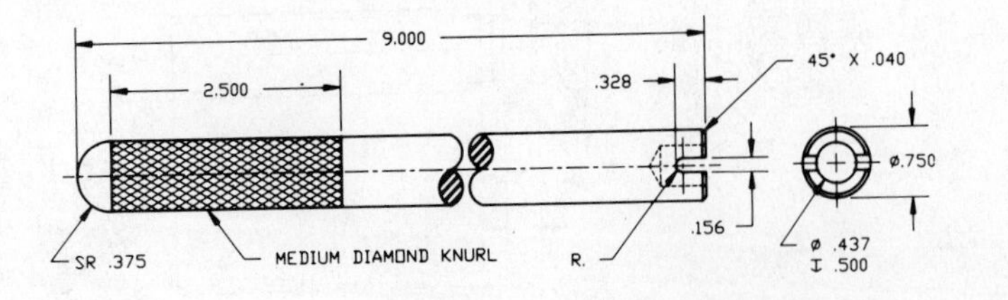

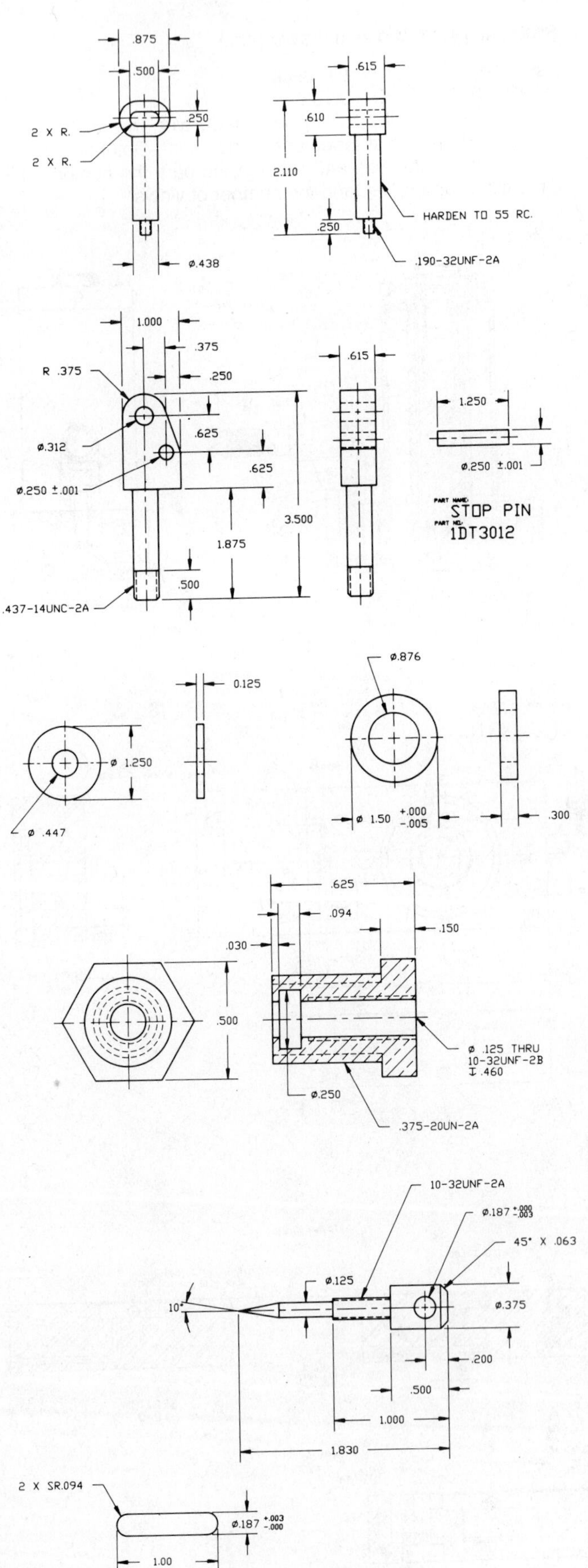

Problem 13–12 Working drawing (in.)

Assembly Name: Worm Gear Reducer
SPECIFIC INSTRUCTIONS:
Prepare one detail drawing per sheet, with the assembly drawing and parts list together on one sheet. Determine the sheet sizes by the size and scale of the part, the number of dimensions and notes, and the number of views.
Courtesy Winsmith.

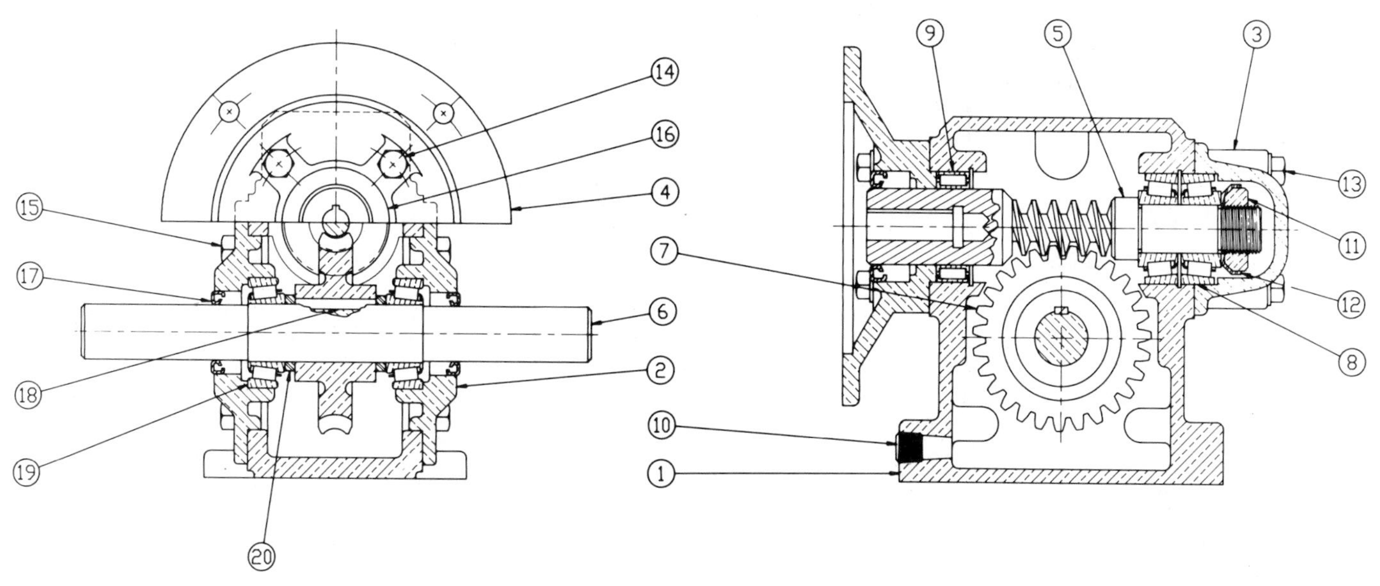

MDT LINE WORM GEAR REDUCER ASSEMBLY
DRAWING NUMBER: 6DT 1000

KEY	QTY	NAME	DESCRIPTION	PART NO
20	2	SLOW SPEED SPACER	TIMKEN TW-506	5DT 1020
19	2	SNGL. ROW TAP. ROLLER BEARING	KOYO 32005J	5DT 1019
18	1	SLOW SPEED KEYWAY	.1875 X .245 X 1.450	5DT 1018
17	2	SLOW SPEED OIL SEAL	PARKER 2-020	5DT 1017
16	1	HIGH SPEED OIL SEAL	PARKER 2-028	5DT 1016
15	8	MACHINE SCREW	.375-16UNC-2A X .625 HEX HEAD	5DT 1015
14	4	MACHINE SCREW	.375-16UNC-2A X 1.875 HEX HEAD	5DT 1014
13	4	MACHINE SCREW	.375-16UNC-2A X 2.250 HEX HEAD	5DT 1013
12	1	HIGH SPEED LOCKWASHER	TIMKEN TW-105	5DT 1012
11	1	HEX NUT	.875-16 UN-2B	5DT 1011
10	1	TAPER PLUG	.500-16NPT PLUG	5DT 1010
9	1	SNGL. ROW CYL. ROLLER BEARING	KOYO CRL11	5DT 1009
8	1	DBL. ROW TAPERED ROLLER BEARING	KOYO46T30305DJ/29.5	5DT 1008
7	1	WORM GEAR	BRONZE	1DT 1005
6	1	SLOW SPEED SHAFT		1DT 1006
5	1	HIGH SPEED SHAFT		1DT 1005
4	1	MOTOR ADAPTOR		1DT 1004
3	1	BEARING CAP		1DT 1003
2	2	RETAINING PLATE		1DT 1002
1	1	HOUSING		1DT 1001

PARTS LIST

Problem 13–12 (Continued)

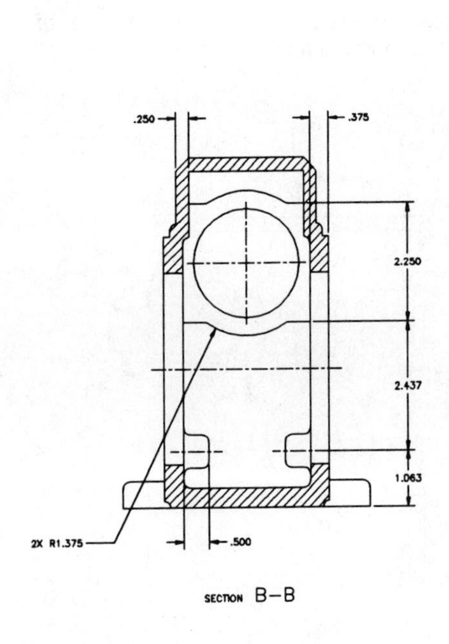

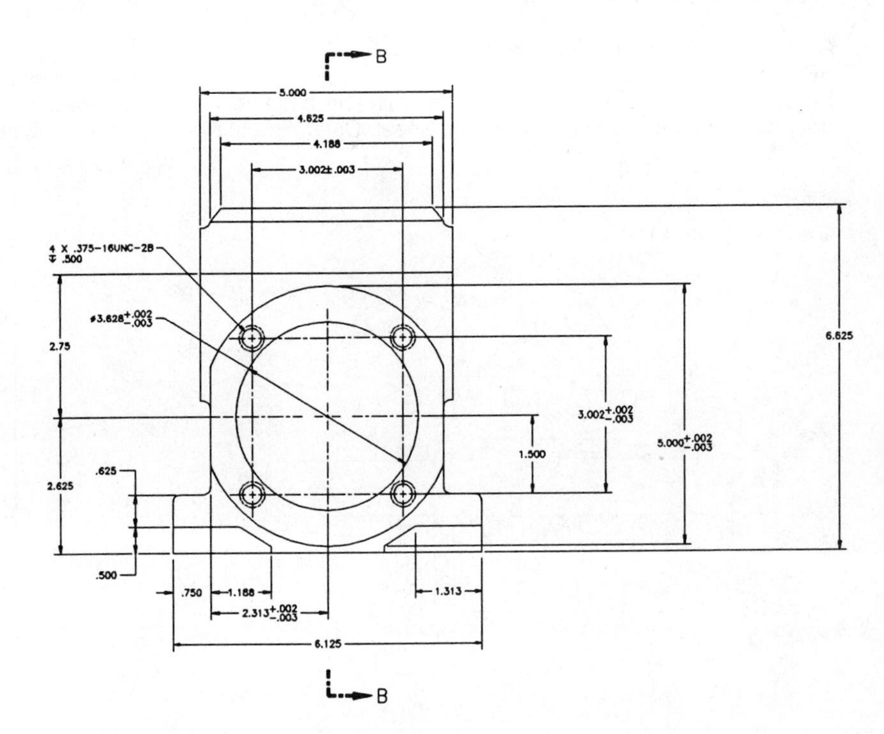

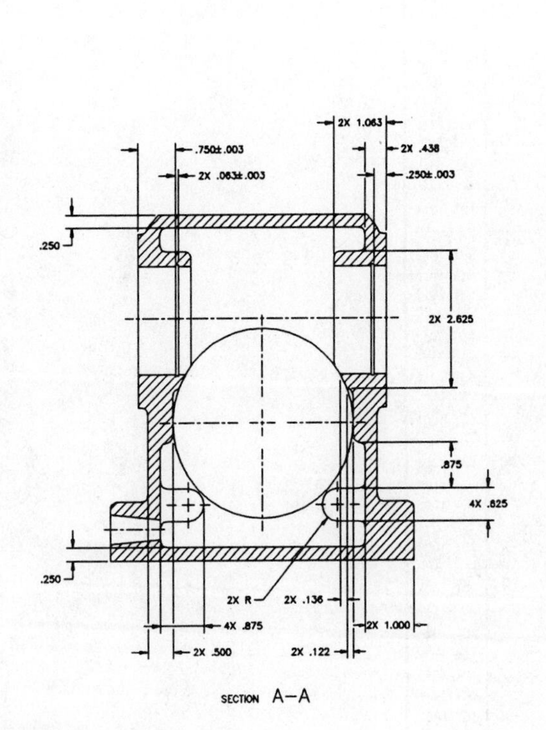

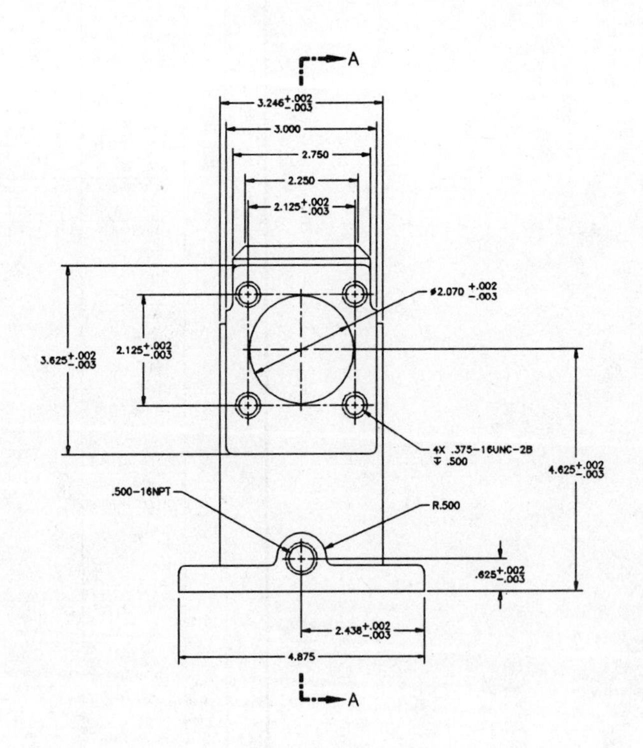

PART NAME: HOUSING
MATERIAL: CI
DRAWING NUMBER: 1DT 1001

Problem 13–12 (Continued)

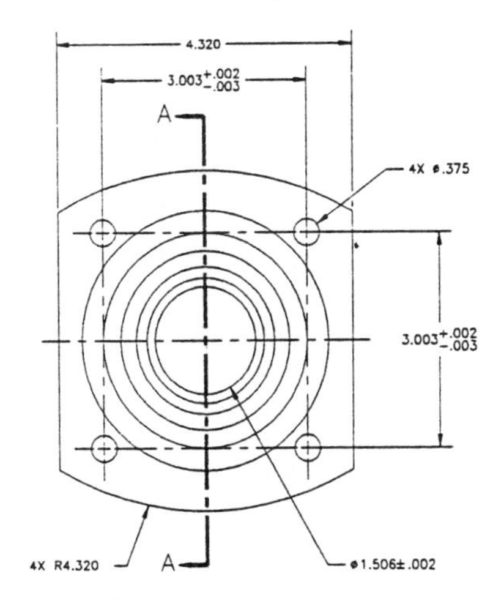

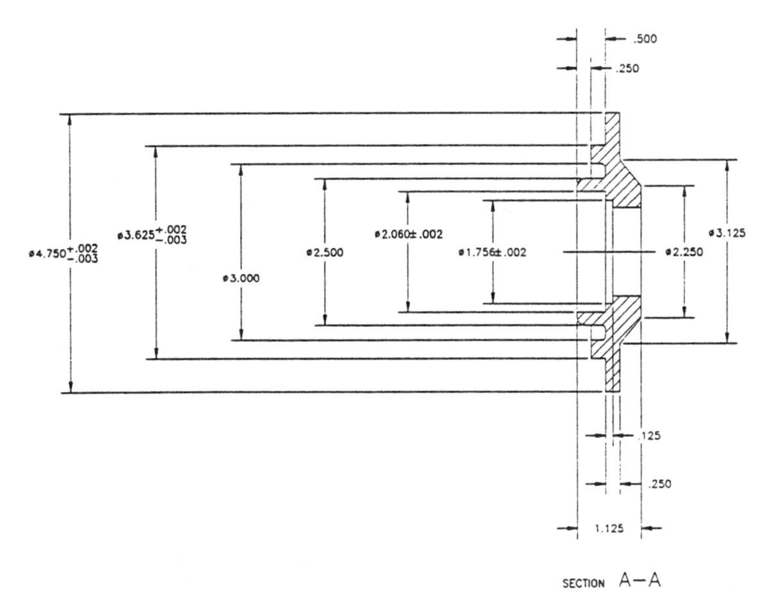

PART NAME: RETAINING PLATE
MATERIAL: CI
DRAWING NUMBER: 1DT 1002

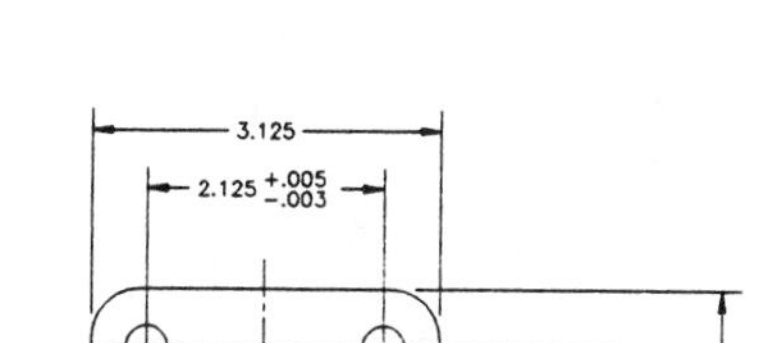

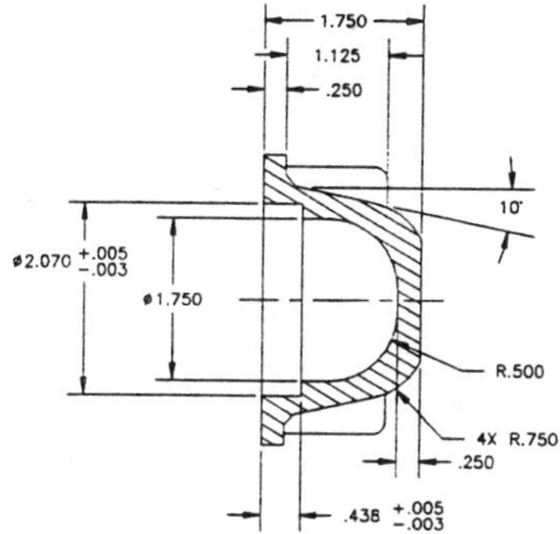

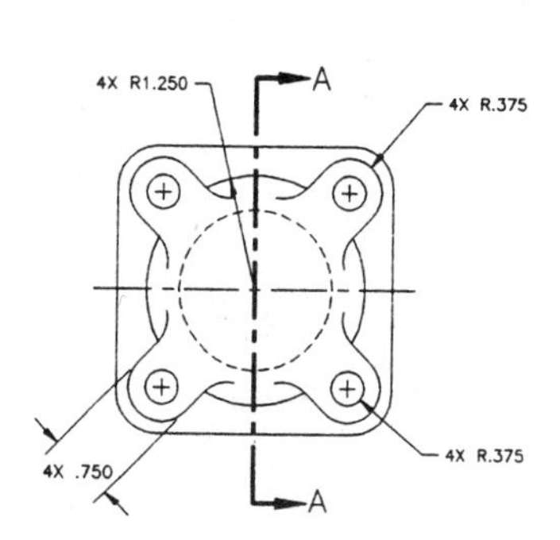

PART NAME: BEARING CAP
MATERIAL: CRS
DRAWING NUMBER: 1DT 1003

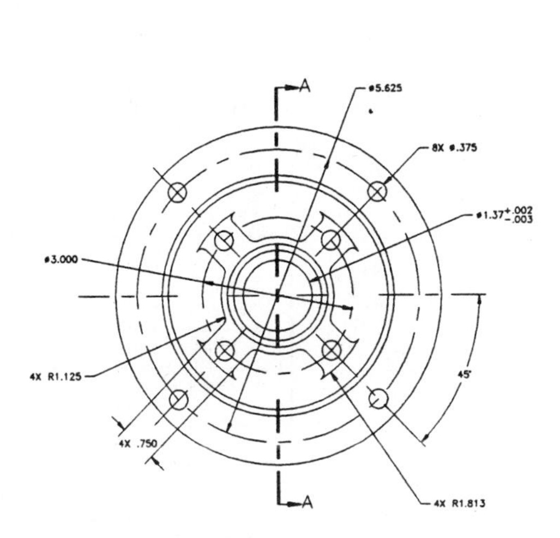

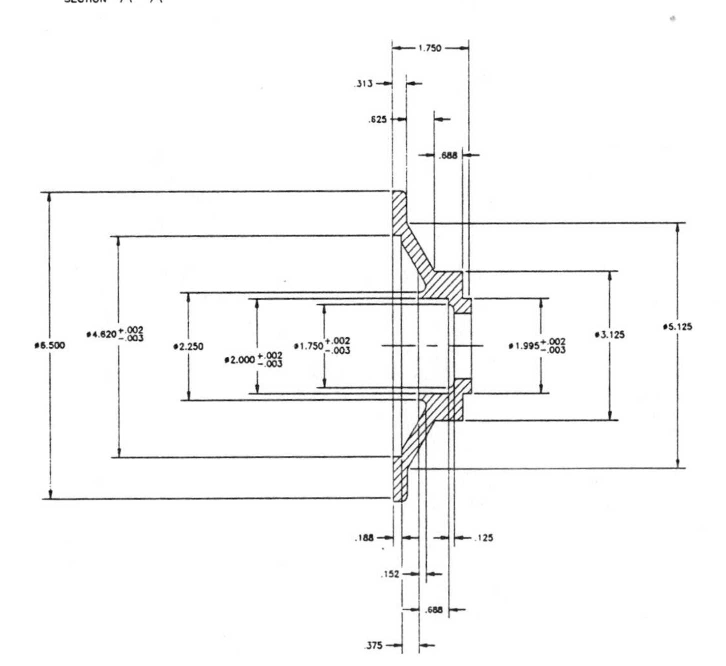

PART NAME: MOTOR ADAPTOR
MATERIAL: CI
DRAWING NUMBER: 1DT 1004

Problem 13–12 (Continued)

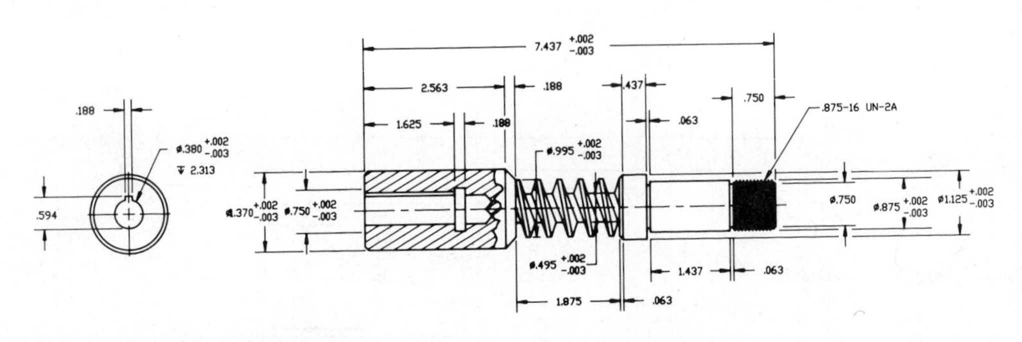

WORM GEAR DATA	
NUMBER OF THREADS	5
AXIAL PITCH	
PRESSURE ANGLE	20°
PITCH DIAMETER	.750
LEAD RIGHT HAND	.300
LEAD ANGLE	169°
ADDENDUM	.125
WHOLE DEPTH	.250
CHORDIAL THICKNESS	.163
WORMGEAR PART NUMBER	1DT 1005

PART NAME: HIGH SPEED SHAFT
MATERIAL: SAE 4320
DRAWING NUMBER: 1DT 1005

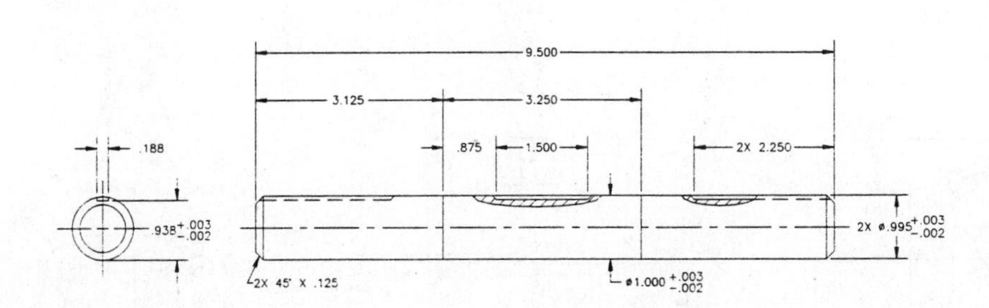

PART NAME: SLOW SPEED SHAFT
MATERIAL: SAE 4320
DRAWING NUMBER: 1DT 1006

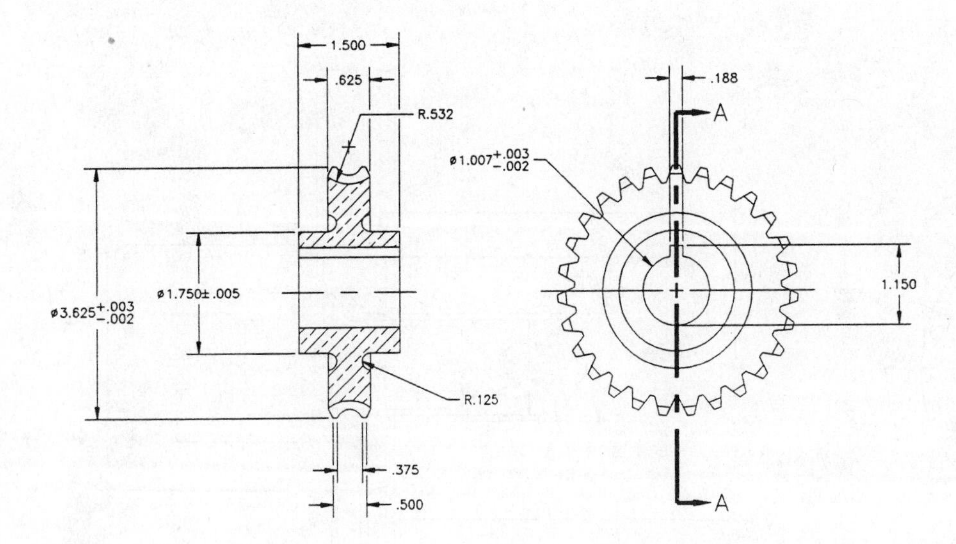

SPUR GEAR DATA	
DIAMETRAL PITCH	8
NUMBER OF TEETH	27
PRESSUR ANGLE	20°
PITCH DIAMETER	3.375
BASE CIRCLE DIAMETER	3.6187
CIRCULAR PITCH	.3927
CIRCULAR THICKNESS	.1964
ROOT DIAMETER	3.125

PART NAME: WORM GEAR
MATERIAL: PHOSPHOR BRONZE
PART NUMBER: 1DT 1007

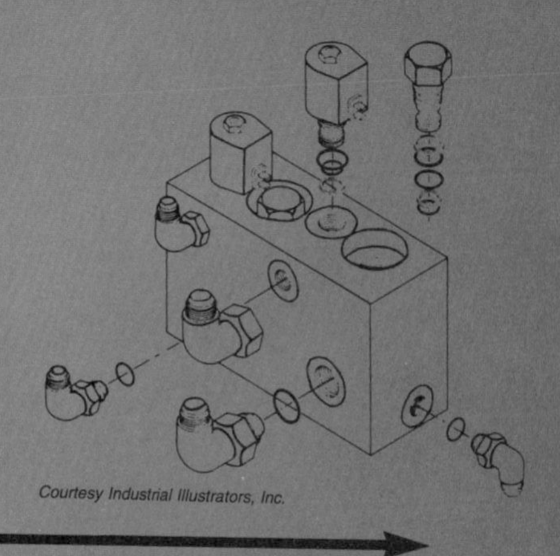

CHAPTER 14

Pictorial Drawings

LEARNING OBJECTIVES

After completing this chapter, you will:

- Draw a three-dimensional object using 3-D coordinates.
- Construct objects using isometric, dimetric, or trimetric methods.
- Construct objects using oblique drawing methods.
- Draw objects using one-, two-, or three-point perspective.
- Apply a variety of shading techniques to pictorial drawings.
- Given an orthographic engineering sketch of a part or assembly, draw it in pictorial form using proper line contrasts and shading techniques.

PICTORIAL DRAWING

Most products are made from orthographic drawings that allow us to view an object with our line of sight perpendicular to the surface we are looking at. The one major shortcoming of this form of drawing is the lack of depth. Certain situations demand a single view of the object that provides a more realistic representation. This realistic single view is achieved with pictorial drawing.

The most common forms of pictorial drawing used in mechanical drafting are isometric and oblique. These two basic forms of pictorial drawing are easy to master, if you can visualize objects in orthographic projection and three dimensions.

Isometric drawing belongs to a family of pictorial representation known as *axonometric* projection. Two other forms of drawing occupy this group, and they are similar to isometric. Dimetric projection involves the use of two different scales as opposed to the single scale used in isometric. Trimetric projection is the most involved of the three and uses three different scales for measurement. Figure 14–1 illustrates the differences in scale between isometric and a common dimetric and trimetric drawing.

The terms drawing and projection should be clarified. A *projection* is an exact representation of an object projected onto a plane from a specific position. The observer's line of sight to the various points on the object passes through a projection plane. The representation on a drawing sheet of the points of the object on the projection plane becomes the *drawing*. Exact projections of objects are time-consuming to make and often involve the use of odd angles and scales. Therefore most drafters and illustrators work with axonometric drawing tech

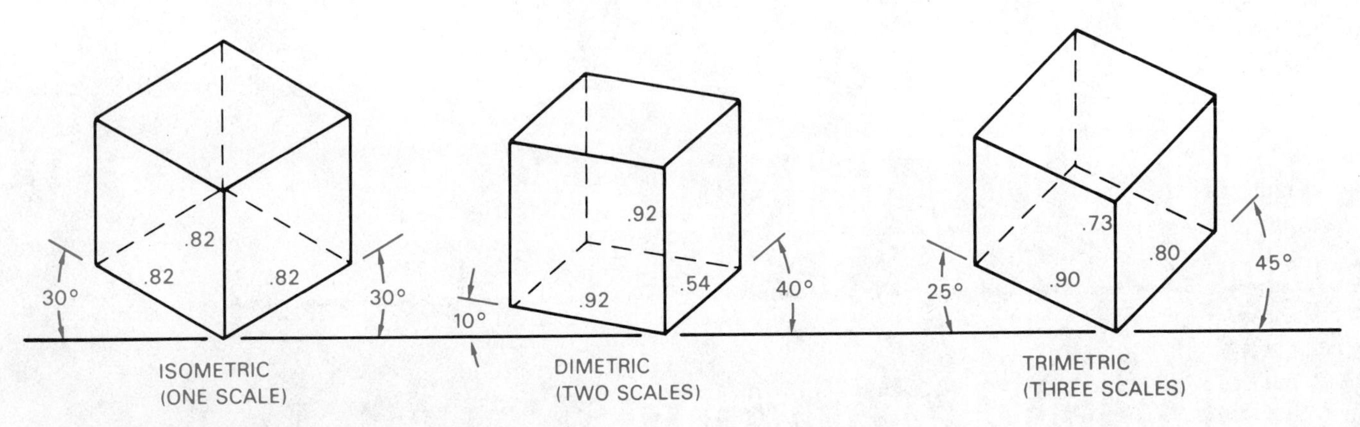

Figure 14–1 The three types of axonometric projections.

niques rather than true projection techniques. Creating an axonometric drawing involves the use of approximate scales and angles that are close enough to the projection scales and angles to be acceptable.

The most realistic type of pictorial illustration is perspective drawing. The use of vanishing points in the projection of these drawings gives them the depth and distortion that we see with our eyes. We will examine each of these types of pictorial drawings in this chapter and discuss step-by-step construction methods for them. In addition we will see how pictorial drawings can be drawn with a computer drafting system.

Technical Illustration

Pictorial drawing is a term that is often used interchangeably with technical illustration. But pictorial drawing includes only line drawings done in one of several three-dimensional methods, whereas technical illustration involves the use of a variety of artistic and graphic arts skills and a wide range of media in addition to pictorial drawing techniques. Figure 14–2 is an example of a pictorial drawing, and Figure 14–3 shows a technical illustration. Pictorial drawings are most often the basis for technical illustrations.

Uses of Pictorial Drawings

Pictorial drawings are excellent aids in the design process, for they allow designers and engineers to view the objects at various stages in their development. Pictorial drawings are used in instruction manuals, parts catalogs, advertising literature, technical reports and presentations, and as aids in the assembly and construction of products.

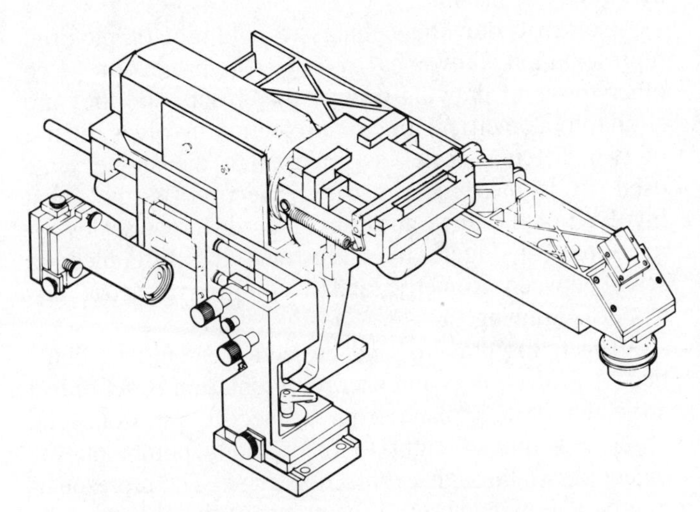

Figure 14–2 Pictorial drawing. *Courtesy Industrial Illustrators, Inc.*

Figure 14–3 Cutaway technical illustration. *Courtesy Industrial Illustrators, Inc.*

ISOMETRIC PROJECTION

The word *isometric* means equal (iso) measure (metric). With this kind of drawing, the three principal planes and edges make equal angles with the plane of projection. An isometric projection is achieved by first revolving the object, in this case a 1-in. cube, 45° in a multiview drawing, Figure 14–4(a), then tilting it forward until the diagonal line AE is perpendicular to the projection plane, as seen in the side view of Figure 14–4(b). This creates an angle of 35° 16' between the vertical axis AD and the plane of projection. When viewed in the isometric or front view, this axis appears vertical. The remaining two principal axes, AB and AC, are at 30° to a horizontal line.

The three principal axes are called isometric lines, and any line parallel to them is also an isometric line. These lines can all be measured. Any lines not parallel to these three axes are nonisometric lines and cannot be measured. The angles between each of these three isometric axes are 120°. The three planes between the isometric axes, and any plane parallel to them, are called isometric planes. (See Figure 14–5.)

Isometric Scale

The isometric projection is a true representation of an object rotated and tilted in the manner just described.

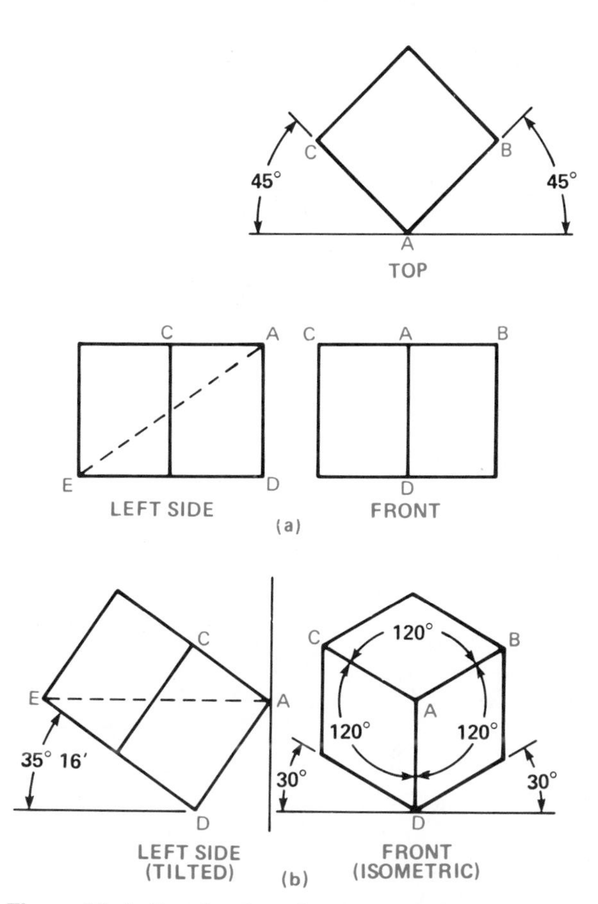

Figure 14–4 Construction of an isometric projection.

An isometric projection must be drawn using an isometric scale. An isometric scale is created by first laying a regular scale at 45° and projecting the increments of that scale vertically down to a blank scale drawn at an angle of 30°. The resulting isometric scale is seen in Figure 14–6. A 1-in. measurement on the regular scale now measures .816 in. on the isometric scale.

An isometric drawing is done using a regular scale. This is most common in industry because it does not involve the creation of special scales. The only difference between an isometric drawing and an isometric projection is the size. The drawing appears slightly larger than the projection. Figure 14–7 illustrates the differences between the drawing and the projection.

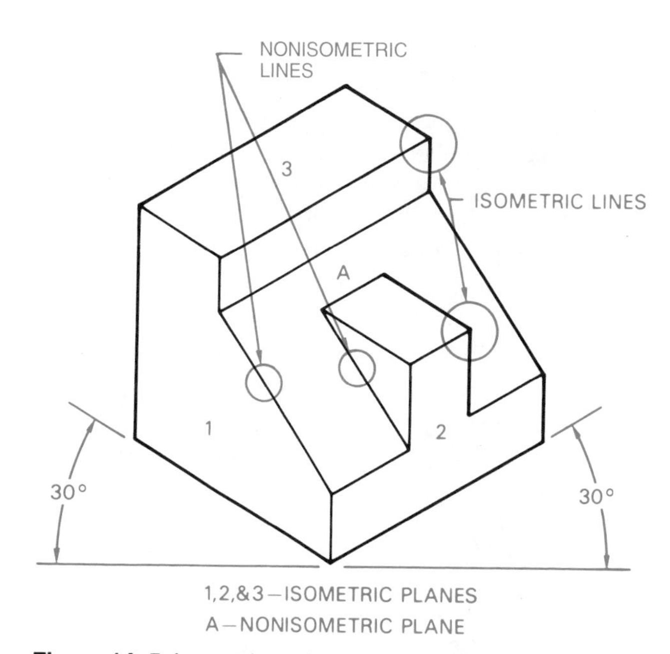

Figure 14–5 Isometric and nonisometric planes.

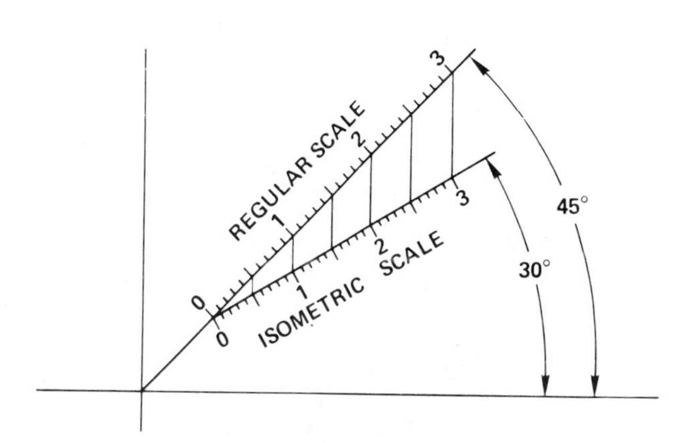

Figure 14–6 Projection of regular scale to isometric scale.

TYPES OF ISOMETRIC DRAWINGS

Isometric drawing is a form of pictorial drawing in which the receding axes are drawn at 30° from the horizontal, as shown in Figure 14–4. There are three basic forms of isometric drawing known as regular, reverse, and long-axis isometric.

Regular Isometric

The top of an object can be seen in the regular isometric form of drawing. An example can be seen in Figure 14–8(a). This is the most common form of isometric drawing, and when using it, the illustrator can choose to view the object from either side.

Reverse Isometric

The only difference between reverse and regular isometric is that you can view the bottom of the part instead of the top. The 30° axis lines are drawn downward from the horizontal line instead of upward. Figure 14–8(b) shows an example of reverse isometric.

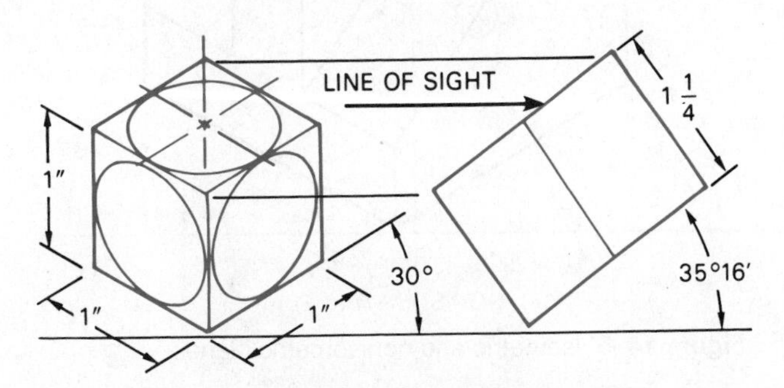

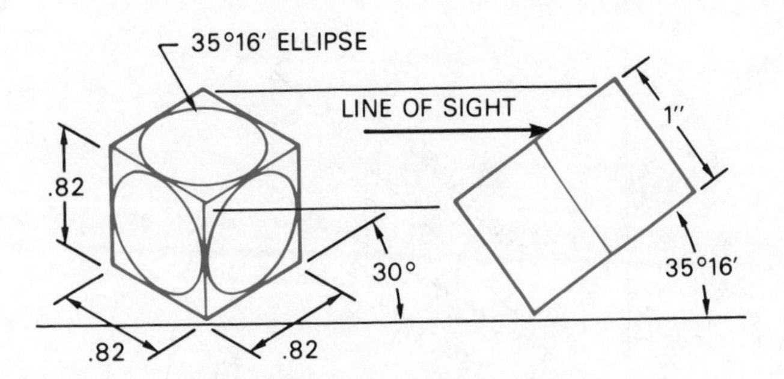

Figure 14–7 The differences between isometric drawing and isometric projection.

Long-axis Isometric

The long-axis isometric drawing is normally used for objects that are long, such as shafts. Figure 14–8(c) shows an example of the long-axis form.

The drafter should choose the view that will give the most realistic presentation of the object. For example, if the object is normally seen from below, then the reverse isometric would be the proper form to use.

ISOMETRIC CONSTRUCTION TECHNIQUES

Just as objects differ in their geometric makeup, so do the construction methods used to draw the object.

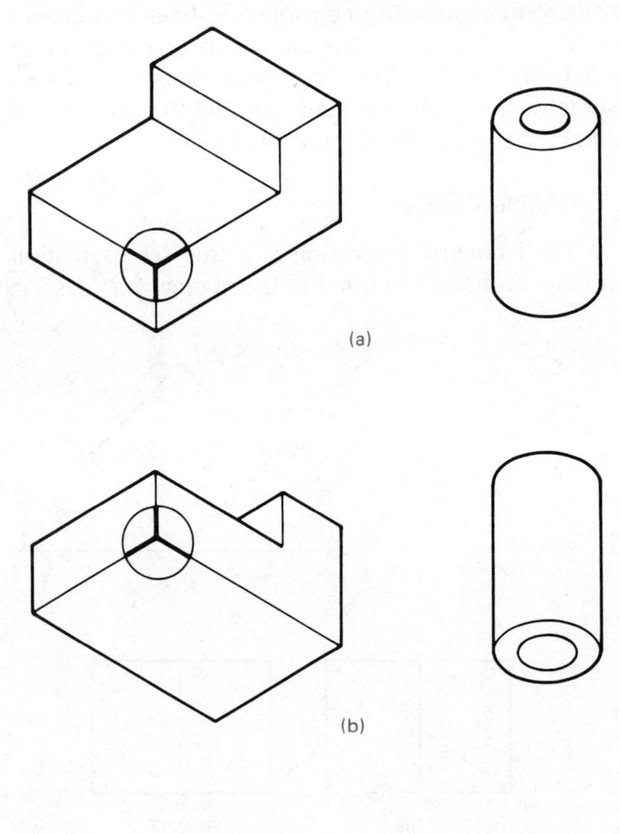

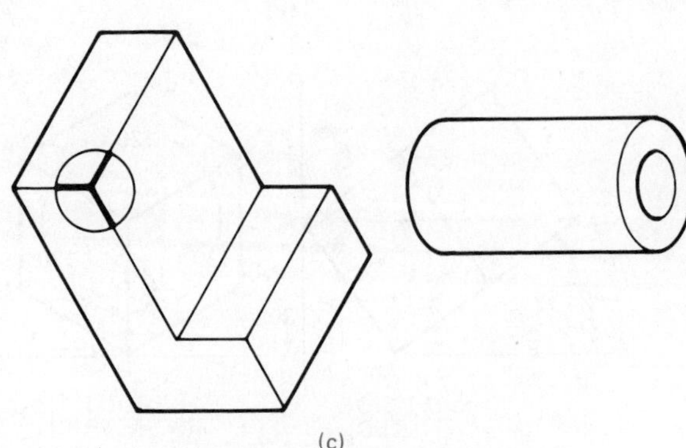

Figure 14–8 Isometric axis variations: (a) regular; (b) reverse; (c) long-axis.

Different techniques exist to assist the drafter in constructing the various shapes. We will look at a few here.

Box, or Coordinate, Method

The most common form of isometric construction is the box, or coordinate, method and is used on objects that have angular or radial features. The orthographic views of the object to be drawn are shown in Figure 14–9(a).

First an isometric box the size of the overall dimensions of the object (X, Y, and Z) must be drawn. (See Figure 14–9(b).) Then the measurements of the features of the object can be transferred to the isometric box. To locate points D and C, just measure dimension U from points F and G. Point E is located at the lower left corner of the box. Draw a line from E to D. Next locate point H by measuring up distance W from the bottom right corner of the box. Draw a light construction line at a 30° angle toward the front vertical axis. Now draw a line from A parallel to line ED until it intersects the line from H. This intersection will be point B. The location of point B can also be found by measuring the horizontal dimension T from point A, then the vertical dimension W.

It may be necessary to draw construction lines on the orthographic views and transfer measurements directly from these views to the isometric view. This method may work well with irregularly shaped objects such as the one shown in Figure 14–10.

Centerline Layout Method

The centerline method begins with the skeleton of the object, the centerlines. This technique is used on objects with many circles and arcs. The use of this method is seen in Figure 14–11(a) through (d). The center points of all of the circles and arcs should be located first. Begin with a good reference point from which you can work, such as the bottom of the object in Figure 14–11(a). Center points A and B are first established,

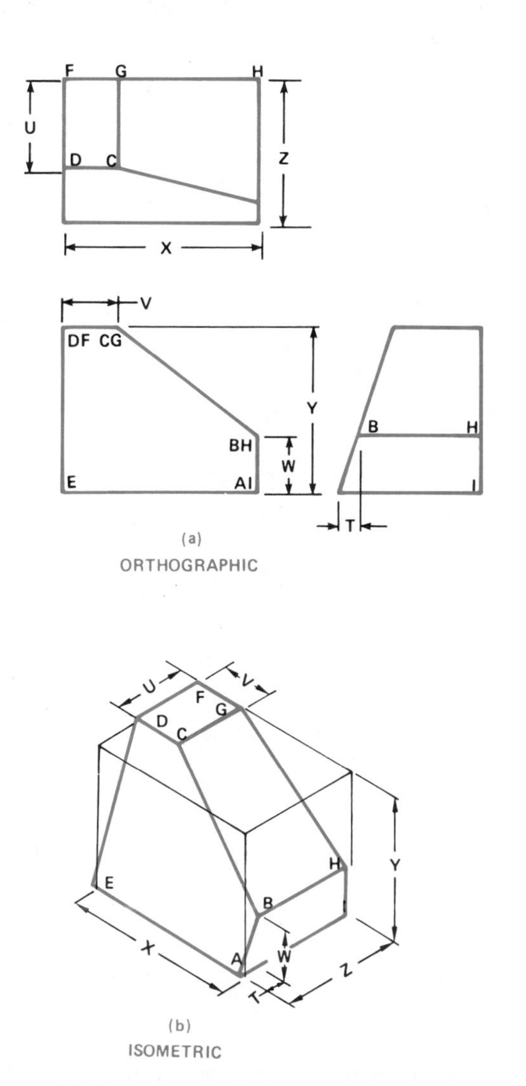

Figure 14–9 Box method of isometric construction: (a) orthographic view; (b) isometric view.

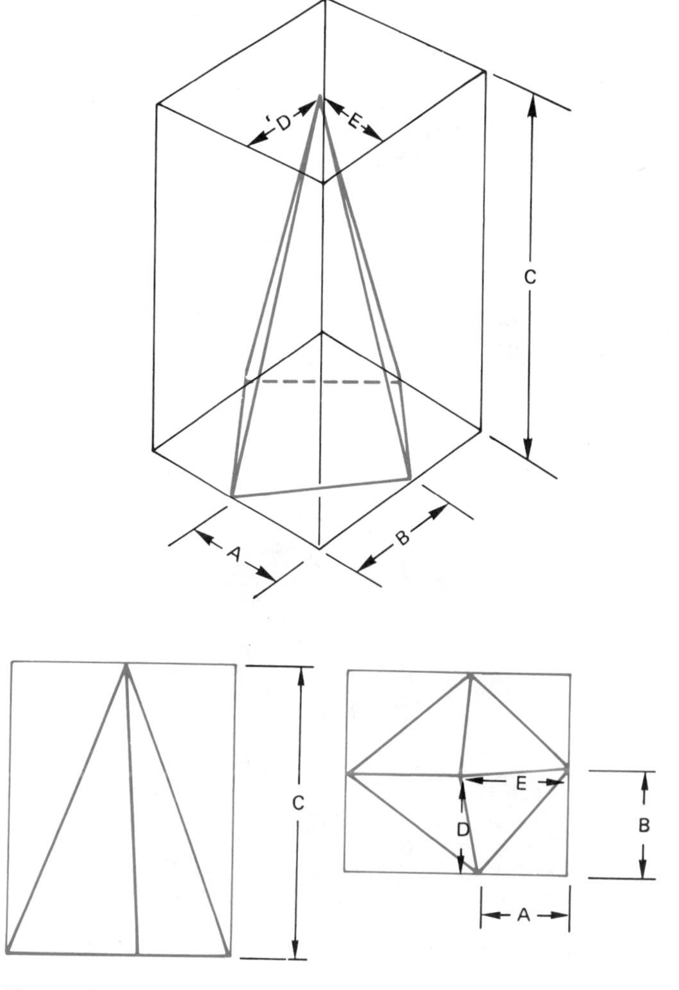

Figure 14–10 Isometric box method for an irregular object.

then vertical axis lines are drawn up from these. Now the vertical locations of center points C, D, and E can be measured as seen in Figure 14–11(b). Then determine the various sizes of the ellipses and draw them at their proper locations as shown in Figure 14–11(c). The finished drawing is shown in Figure 14–11(d).

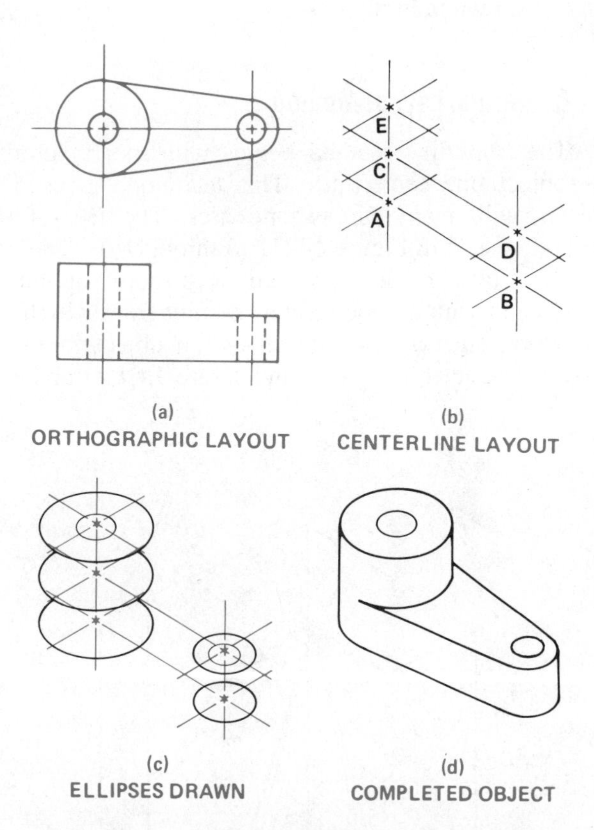

Figure 14–11 Isometric centerline layout method: (a) orthographic layout; (b) centerline layout; (c) ellipses drawn; (d) completed object.

Circle and Arc Construction

After the center points of circles and arcs are located, you need to select the correct size of circle from an isometric ellipse template, such as the one shown in Figure 14–12. Note that each ellipse on the template has several tick marks around it. A description of these tick marks is shown in Figure 14–13. An isometric ellipse is measured on the marks at 30° angles from the large (horizontal) diameter of the ellipse. When aligning an isometric ellipse on the centerlines of your layout, always align the tick marks on the *minor* diameter of the ellipse with the *axis* of the feature, as shown in Figure 14–14. When you do this, two other sets of tick marks on the template will align with the ellipse centerlines.

Arc locations are found in the same way as circles, and the use of the isometric ellipse template is also the same, except just a portion of the circle is drawn. First, find the center point of the arc, then the tangent points. (See Figure 14–15.) Draw the axis line of the arc lightly and again align the minor diameter tick marks with the axis line. Now draw only the portion of the ellipse that is needed for the arc.

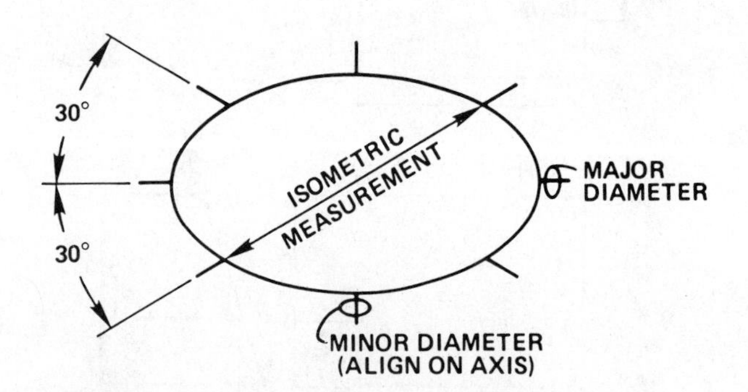

Figure 14–13 Meaning of isometric ellipse template markings.

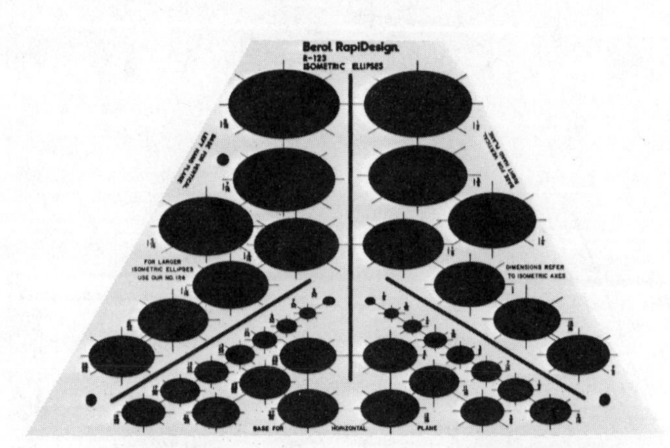

Figure 14–12 Isometric ellipse template. *Courtesy Berol USA RapiDesign.*

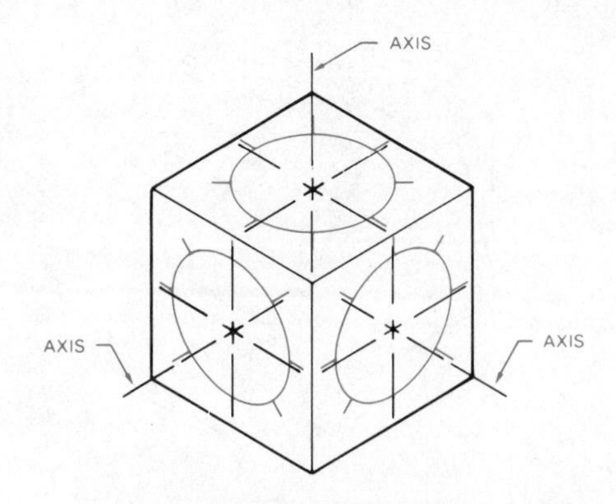

Figure 14–14 Align minor diameter of isometric ellipse template on axis of hole.

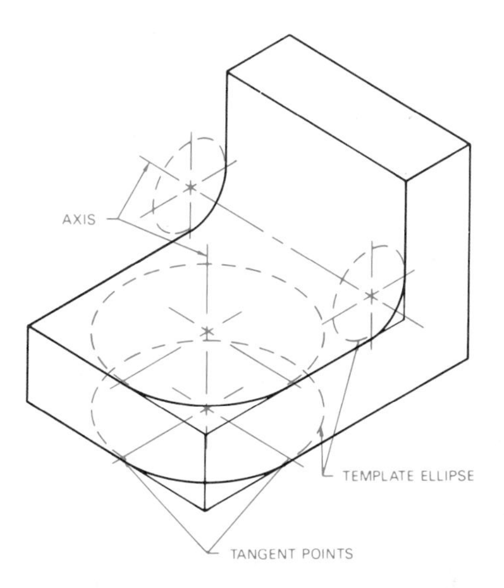

Figure 14–15 Isometric arc layout using ellipse template.

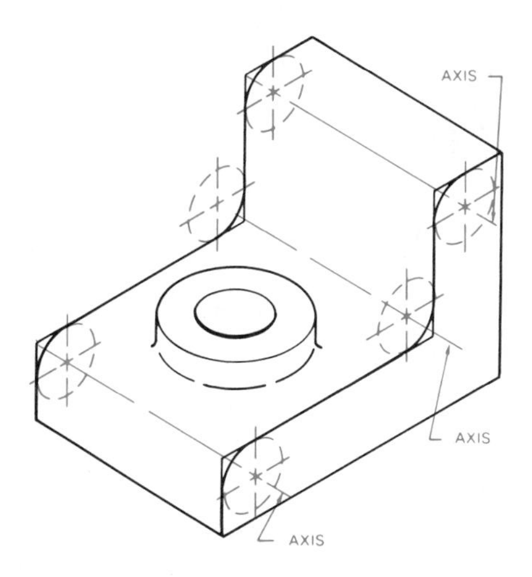

Figure 14–16 Isometric fillet and round layout.

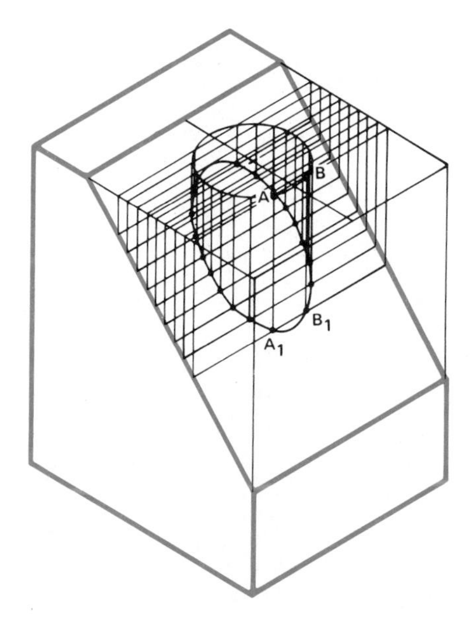

Figure 14–17 Construction of ellipse on isometric oblique plane.

Fillets and Rounds

The key to drawing isometric circles and arcs is getting the ellipse aligned properly. Study the object in Figure 14–16. Determine the axis of the circular feature, then align the minor diameter tick marks of the template with the axis of the fillet or round.

Intersections

A common type of intersection is that of a cylindrical hole passing through an oblique plane. First draw an isometric ellipse at the top of the boxed surface of the object. (See Figure 14–17.) Then draw a series of construction lines parallel to one side of the box that pass through the ellipse. Project each of these lines down the sides of the box and across the oblique surface. Each line forms a trapezoid or parallelogram depending on the shape of the object. The points at which each line intersects the ellipse (A and B, for example) should be projected straight down to the opposite side of the parallelogram. These new points, A_1 and B_1, are now points on the perimeter of the hole in the oblique surface.

Another common form of intersection occurs when two cylinders meet. This construction is similar to the one just described. In Figure 14–18, the center point location of the branch cylinder is determined, and the ellipse is drawn. Next draw as many construction lines passing through the ellipse as needed to produce a smooth curve on the intersection. Project these lines to the end of the main cylinder, down to the edge, then back along the cylinder. Now project points on the ellipse down to the corresponding construction line on the cylinder, A to A_1 and B to B_1, for example. Then connect the points using an irregular curve to produce the intersection.

Sections

Full and half sections are common in technical illustration and should be drawn along the isometric axes. The section lines in full sections should all be drawn in the same direction, while those on a half section should appear in opposite directions. See the section lines illustrated in Figure 14–19(a) and (b). The section lines in offset sections should change directions with each jog in the part, as seen in Figure 14–19(c).

There is no preferred way to draw an isometric section. One technique that you might try is to imagine that the cutting-plane line is a long string, and you are stretching it out just above the axis along which you plan to cut. If you let the string fall, it would come to rest along the axis where the cut is to be made, as illustrated in Figure 14–20.

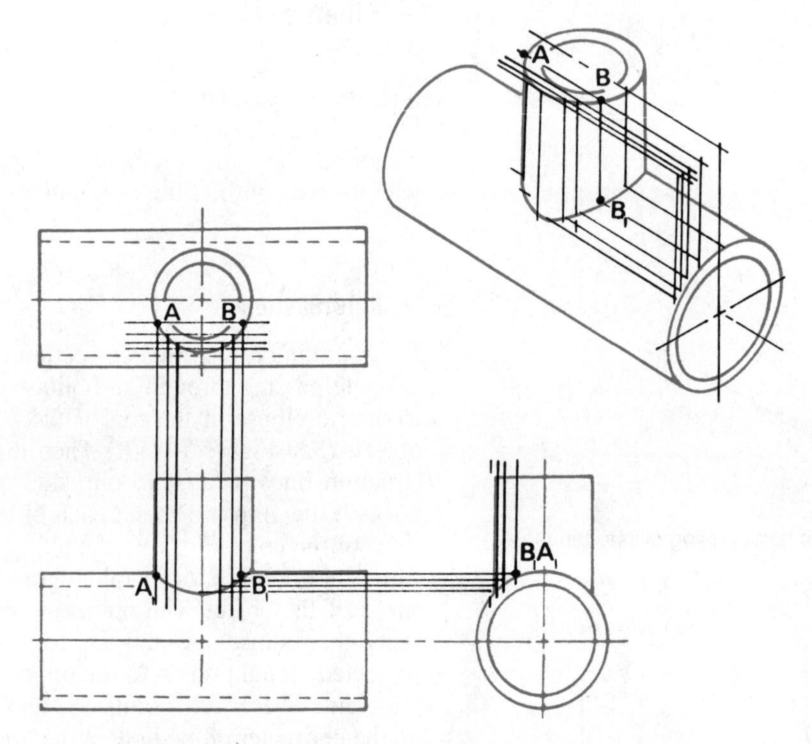

Figure 14–18 Isometric construction of two intersecting cylinders.

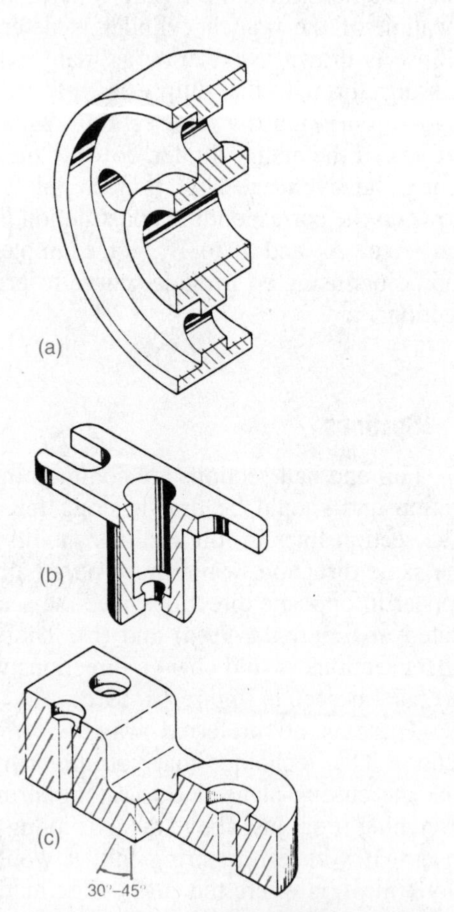

Figure 14–19 Isometric sections: (a) full; (b) half; (c) offset.

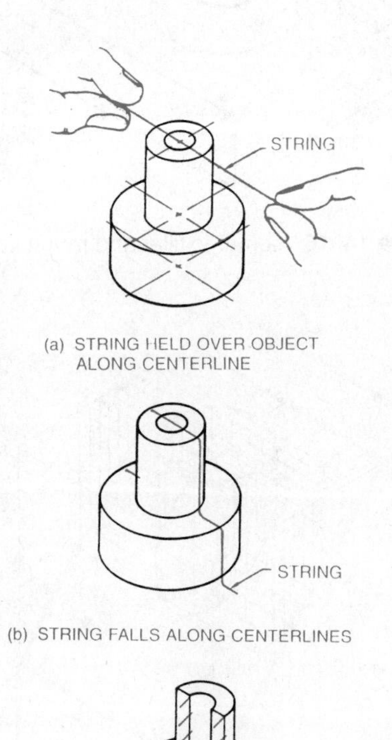

Figure 14–20 Visualize a string dropped along the cutting plane to create an isometric full section.

Threads

ANSI Screw threads can be drawn in isometric by first measuring equal spaces along the shaft or hole to be threaded. Then, using the same size ellipse as the diameter of the shaft or hole, draw a series of parallel ellipses. (See Figure 14–21.) These ellipses represent the crests of the threads. This method achieves a simple isometric representation of threads.

A more detailed thread appearance can be achieved by giving each ellipse rounded ends instead of butting the ellipse into the straight sides of the shaft. (See Figure 14–22.) When using this technique, begin drawing the threads from the head of the shaft. Be sure to lay out guidelines for the shaft diameter so that the ends of the ellipse do not fall short of the edge of the shaft or go beyond it, as seen in Figure 14–23. This can create a phenomenon known as wandering threads.

Spheres

A sphere drawn isometrically is nothing more than a true circle. Figure 14–24 illustrates the construction of a sphere using three isometric ellipses drawn to represent the perpendicular axes of the circle. If you need to draw a sphere, remember to choose a circle that is 1¼ times larger than the actual sphere, because isometric drawings are 1¼ times larger than the actual representation.

Should you need to draw spheres that have been cut in some manner, refer to Figure 14–25(a), (b), and (c). They illustrate a half sphere, three-quarters of a sphere, and a sphere with a flat side.

Isometric Dimensioning

It is not common for isometric drawings to be dimensioned; however, some isometric piping drawings rely heavily on dimensioning to get their message across. One technique that can be used when dimensioning is to letter vertical strokes parallel to extension lines, as shown in Figure 14–26. This gives the appearance that the dimension is lying in the plane of the extension lines. A second simple technique is to draw the heel of the arrowhead parallel with the extension line. This further emphasizes the plane that the dimension is lying in. (See Figure 14–26(a).) Examples of unidirectional, aligned, and one-plane (horizontal) isometric dimensioning are shown in Figure 14–26(b).

DIMETRIC PICTORIAL REPRESENTATION

Dimetric drawing is similar to isometric drawing, but instead of all three axes forming equal angles with the plane of projection, only two axes form equal angles. These can be greater than 90° and less than 180°, but cannot have an angle of 120° because that is the isometric angle. The third axis may have an angle less or greater than the two equal axes, depending on the angles chosen. The two equal angles are said to be foreshortened because they are not measured at full scale. They are foreshortened equally. Since the third axis is projected at a different angle, it is foreshortened at a different scale. Full-size and foreshortened approximate scales are used on the dimetric axes. Some common approximate dimetric angles and scales are shown in Figure 14–27.

Figure 14–21 Isometric thread representation.

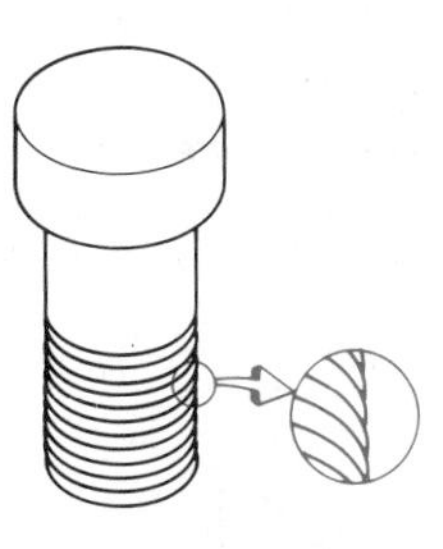
Figure 14–22 Detailed isometric thread representation.

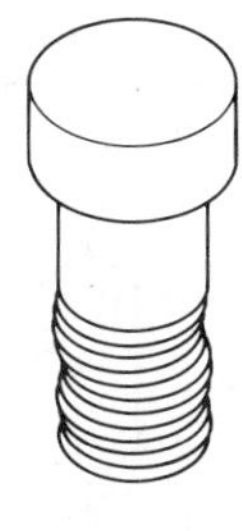
Figure 14–23 Off-center ellipses produce wandering threads.

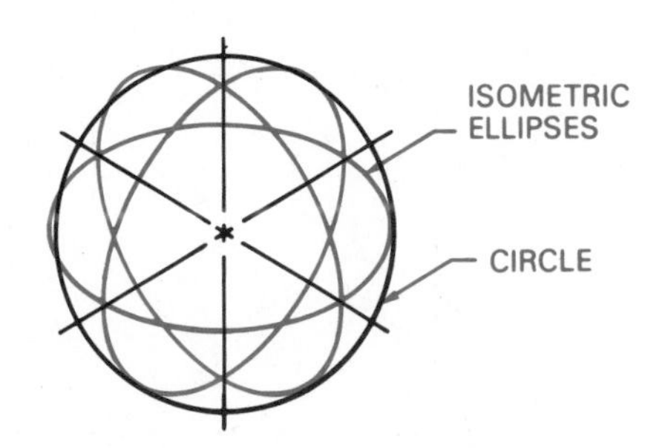

Figure 14–24 Isometric sphere construction.

TRIMETRIC PICTORIAL REPRESENTATION

The term trimetric refers to three measurements. It is a type of pictorial drawing in which none of the three principal axes makes an equal angle with the plane of projection. Since all three angles are unequal, the scales used to measure on the three axes are also unequal. Trimetric projection provides an infinite number of projections.

Trimetric drawing is similar to dimetric drawing. Figure 14–28 shows some common trimetric angles for the width and depth axes, plus the scales to use on each of the three axes. The size of angle ellipse to use on each principal plane is also indicated.

(a) HALF SPHERE (b) THREE QUARTER SPHERE

Figure 14–25 Portions of isometric spheres: (a) half sphere; (b) three-quarter sphere; (c) sphere with flat side.

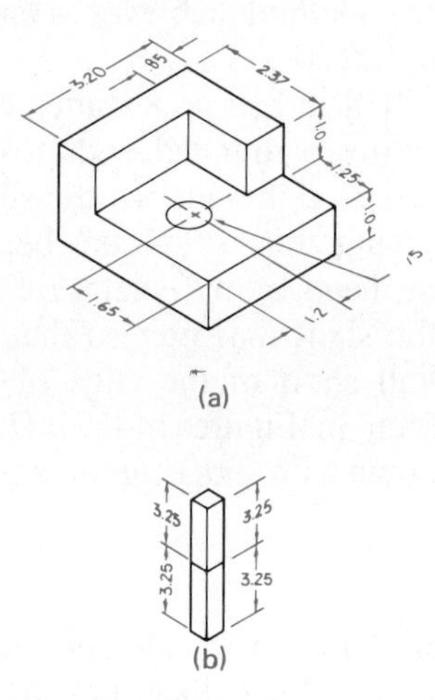

Figure 14–26 Isometric dimensioning: (a) dimensions parallel to extension lines; (b) various styles of dimensioning.

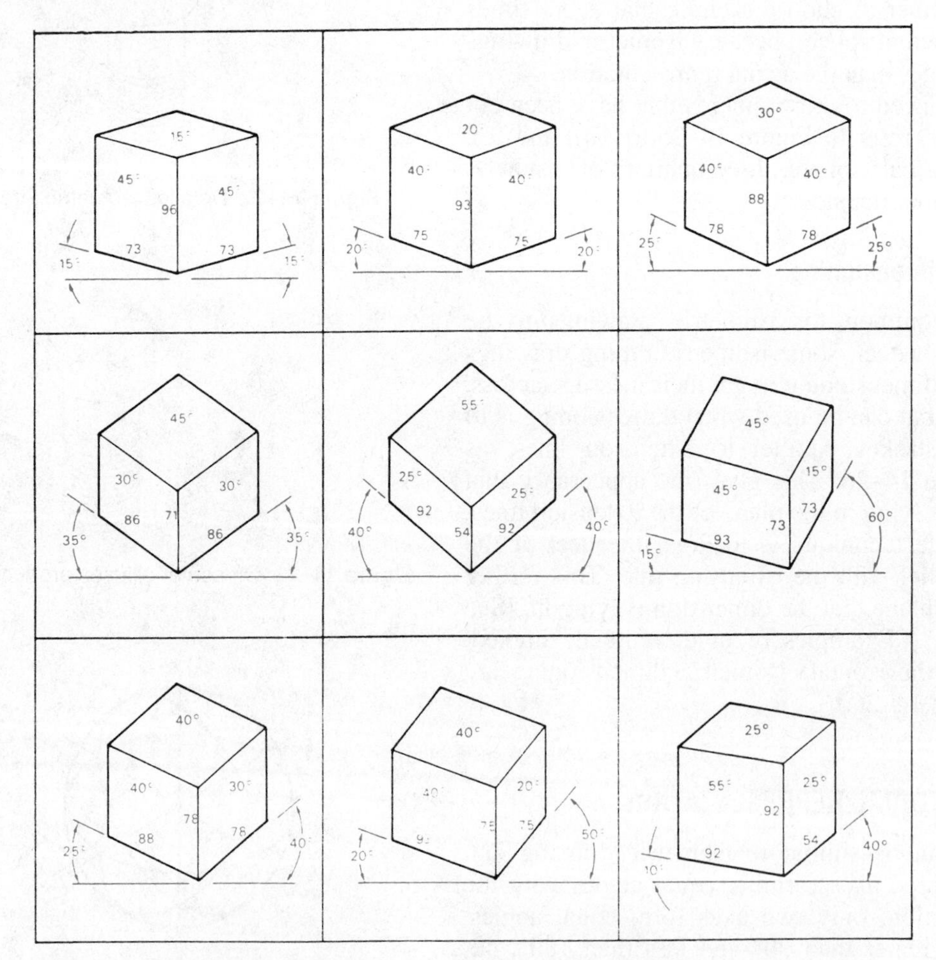

Figure 14–27 Common approximate dimetric angles, scales, and ellipse angles.

Figure 14–28 Common approximate trimetric angles and scales.

EXPLODED PICTORIAL DRAWING

Complicated parts and mechanisms are often illustrated as an exploded assembly in order to show the relationship of the parts in the most realistic manner. (See Figure 14–31(a).)

An exploded assembly is a collection of parts, each drawn in the same axonometric method. Any of the pictorial drawing methods mentioned in this chapter can be used to create exploded assemblies. The most important aspect of this type of drawing is to select the viewing direction that will illustrate as much of the assembly as possible.

Centerlines are used in exploded views to represent lines of explosion. These lines aid the eye in following a part to its position in the assembly. Centerlines should avoid crossing other centerlines and parts. Centerlines should always be drawn parallel to the axis lines of the drawing regardless of the drawing method selected. The neatest presentation is achieved by leaving gaps where the centerline intersects the part to which it applies and the feature it is mating with. This is shown in Figure 14–31(b).

OBLIQUE DRAWING

Oblique drawing is a form of pictorial drawing in which the plane of projection is parallel to the front surface of the object. The lines of sight are at an angle to the plane of projection and are parallel to each other. This allows

3–D CAPABILITIES

Computer-generated pictorial line drawings have given drafters, designers, and engineers the ability to draw an object once and then create as many different displays of that object as they can think of. Three-dimensional wire-form capabilities, seen in Figure 14–29, are available with most commercial CADD systems. They allow the operator to view the object as if it were constructed of wire. All edges can be seen at the same time. This is somewhat limiting because complex parts soon become a maze of lines. The solid form, shown in Figure 14–30 is a step closer to reality because it shows only the surfaces that would actually be seen.

The greatest realism in pictorial presentation can be achieved by CADD systems that possess color solids-modeling capabilities. These systems allow the operator to draw the object, shade or color its different parts or surfaces, and rotate it in any direction desired. The operator can also cut into the object at any location to view internal features and then rotate it to achieve the best view. This screen display can then be sent to a color hardcopy unit, a pen plotter, an electrostatic plotter, a 35-mm slide production unit, or videotape.

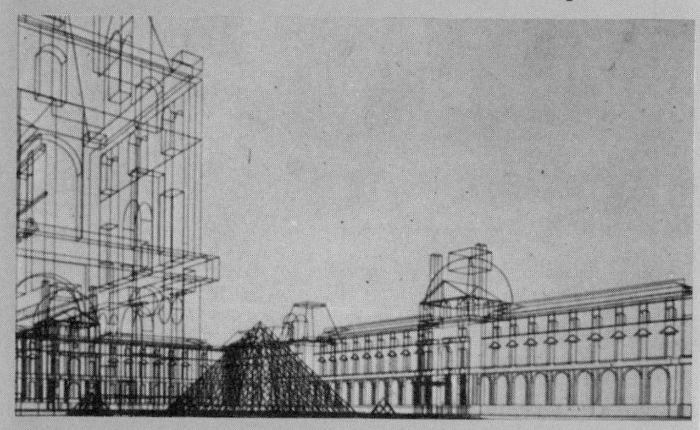

Figure 14–29 The Louvre art museum in Paris, France, drawn as a wire form. *Courtesy Computervision Corporation.*

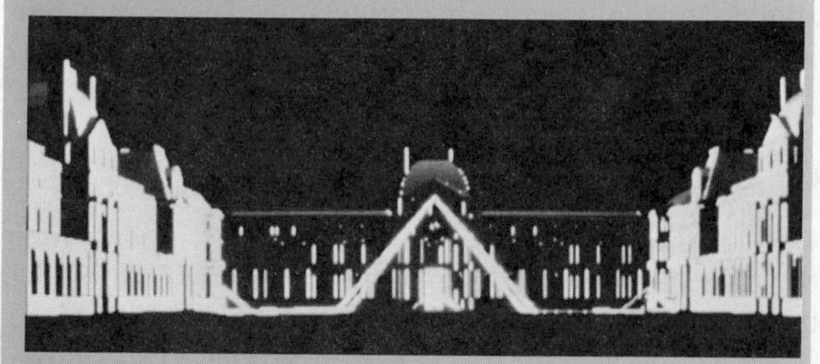

Figure 14–30 Another view of the Louvre, drawn as a solid model. *Courtesy Computervision Corporation.*

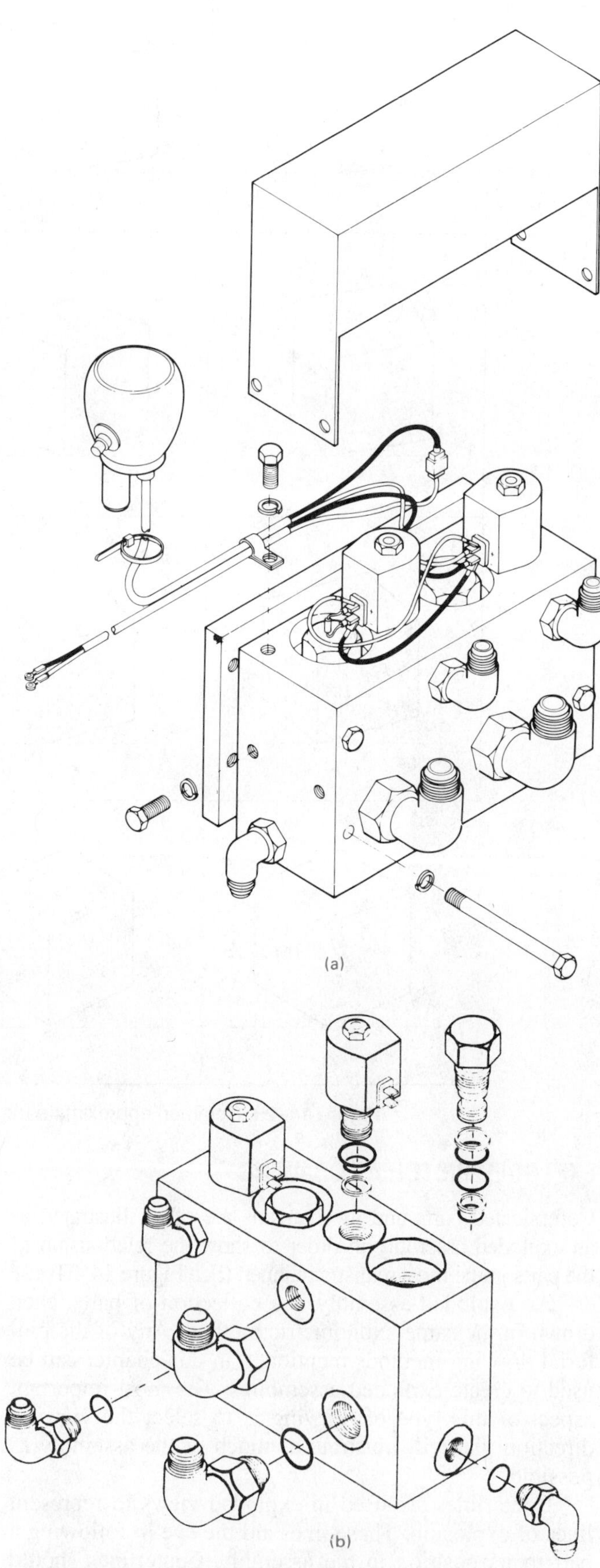

Figure 14–31 Exploded assemblies. *Courtesy Industrial Illustrators, Inc.*

the viewer to see three faces of the object. The front face, and any surface parallel to it, is shown in true shape and size, while the other two faces are distorted in relation to the angle and scale used. Oblique drawing is useful if one face of an object needs to be shown without distortion.

There are three methods of oblique drawing: cavalier, cabinet, and general.

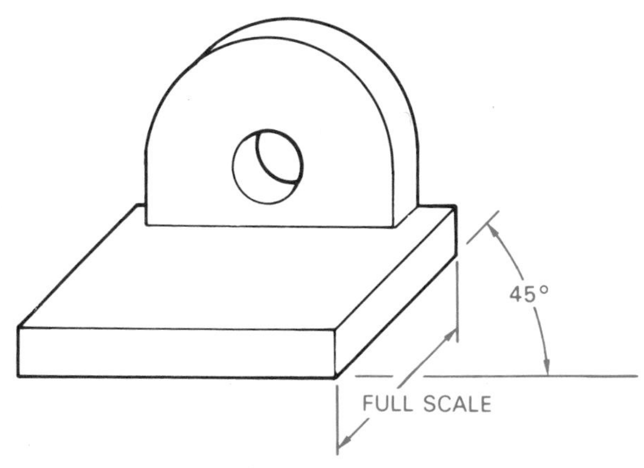

Figure 14–32 Cavalier oblique.

Cavalier Oblique

The cavalier projection is one in which the receding lines are drawn true size, or full scale. This form of oblique drawing is usually drawn at an angle to a horizontal of 45°, which approximates a viewing angle of 45°. Because of the scale used on the receding axis, it is not a good idea to draw long objects with the long axis perpendicular to the front face, or projection plane. Objects that have a depth that is smaller than the width can be drawn in the cavalier form without too much distortion. Figure 14–32 shows an object drawn in cavalier oblique.

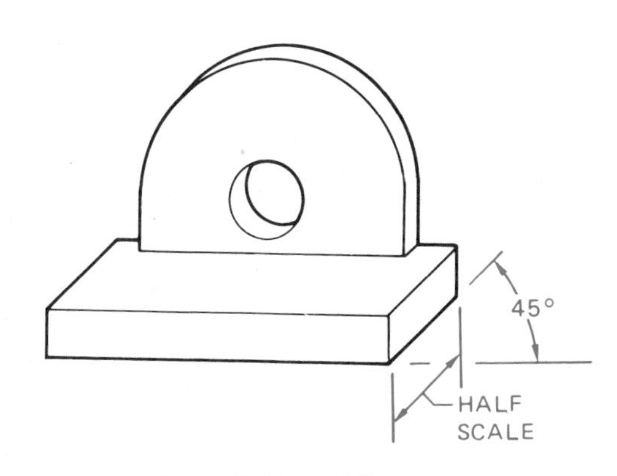

Figure 14–33 Cabinet oblique.

Cabinet Oblique

A cabinet oblique drawing is also drawn with a receding angle of 45°, but the scale along the receding axis is half size. Objects having a greater depth than width can be drawn in this form without the appearance of too much distortion. Cabinet makers often manually draw their cabinet designs using this type of oblique drawing. A cabinet drawing is shown in Figure 14–33.

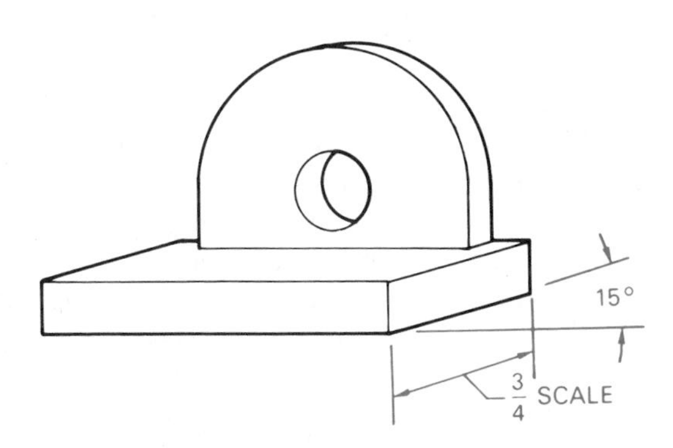

Figure 14–34 General oblique.

General Oblique

The general oblique drawing is normally drawn at an angle other than 45°, and the scale on the receding axis is also different from those used in cavalier and cabinet. The most common angles for a general oblique drawing are 30°, 45°, and 60°, but the drafter can use any angle desired. Any scale from half to full size can be used. A general oblique drawing is shown in Figure 14–34.

PERSPECTIVE DRAWING

Perspective drawing is the most realistic form of pictorial illustration. It reflects the phenomenon of objects appearing smaller the farther away they are until they vanish at a point on the horizon. Lines that are drawn parallel to the principal orthographic planes appear to converge on a vanishing point in perspective drawing.

Computer-aided drafting enables drafters and designers to create pictorial and perspective drawings without much knowledge of the construction techniques involved in either kind of drawing. Architects and drafters have long been using perspective grids to aid in the creation of perspective drawings. But the knowledge of how perspective drawings are made is valuable even for those who use drafting aids and computer systems.

The three types of perspective drawing techniques take their names from the number of vanishing points used in each. One-point, or parallel, perspective has one vanishing point and is used most often when drawing interiors of rooms. Two-point, or angular, perspective is the most popular and is used to illustrate exteriors of houses, small buildings, civil engineering projects, and, occasionally, machine parts. The third type, three-point

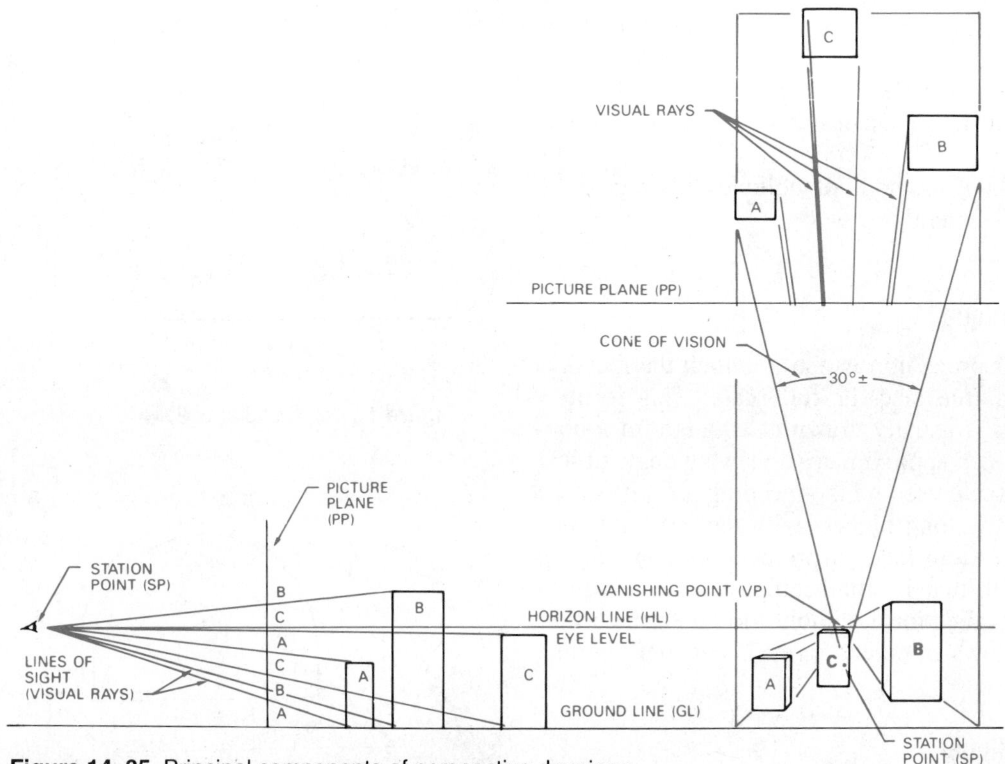

Figure 14–35 Principal components of perspective drawings.

perspective, has three vanishing points and is used to illustrate objects having great vertical measurements, such as tall buildings. It is a lengthy process to draw in three-point perspective; therefore it is not used as often as two-point perspective.

General Concepts

Two principal components of a perspective drawing are the eye of the person viewing the object, and the location of the person in relation to the object. The eye level of the observer is the *horizon line* (HL). This line is established in the elevation (front) view. (See Figure 14–35.) The position of the observer in relation to the object is the *station point* (SP) and is established in relation to the plan (top) view. The location of the station point determines how close the observer is to the object, and the angle at which the observer is viewing the object. The *ground line* (GL) is the line on which the object rests. The *picture plane* (PP), or plane of projection, is the surface (drawing sheet) on which the object is projected. The picture plane can be situated anywhere between the observer and the object. The picture plane can also be located beyond the object. The observer's lines of sight, or visual rays, determine what will show on the picture plane and where. Finally, the three vanishing points discussed in the following sections are referred to as vanishing point right (VPR), vanishing point left (VPL), and vanishing point vertical (VPV).

ONE-POINT PERSPECTIVE

This form of perspective, also known as parallel perspective, has only one vanishing point. The plan view is oriented so that the front surface of the object is parallel to the picture plane. The elevation view is situated below and to the right or left of the plan and rests on the ground line. The following steps will enable you to construct a one-point perspective. Refer to Figure 14–36 as you read the instructions.

Step 1 Locate the station point between the picture plane and the elevation view of the object. The station point can be anywhere, depending on the part of the object you wish to view. The visual rays from the station point to the extreme corners of the object should form an included angle of approximately 30° to provide the most realistic perspective.

Step 2 Determine the eye level of the observer in the elevation view. If you wish to look over the object to see the top surface, the horizon line should be drawn above the elevation view. The horizon line can be drawn at any eye level.

Step 3 The vanishing point is located on the horizon line directly in line with the station point.

Step 4 Project all points in the plan view that touch the picture plane to the corresponding points in the elevation view. These lines are true scale.

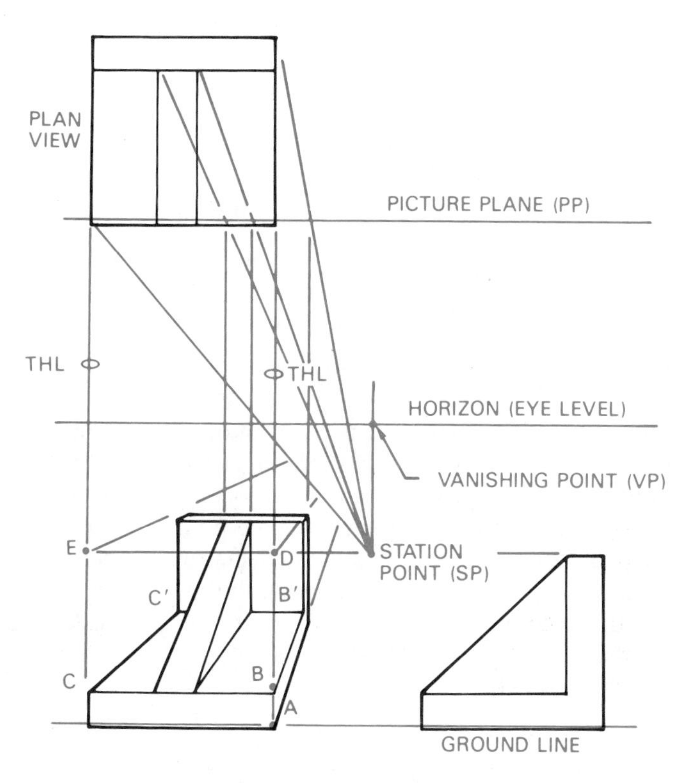

Figure 14–36 Constructing a one-point perspective drawing.

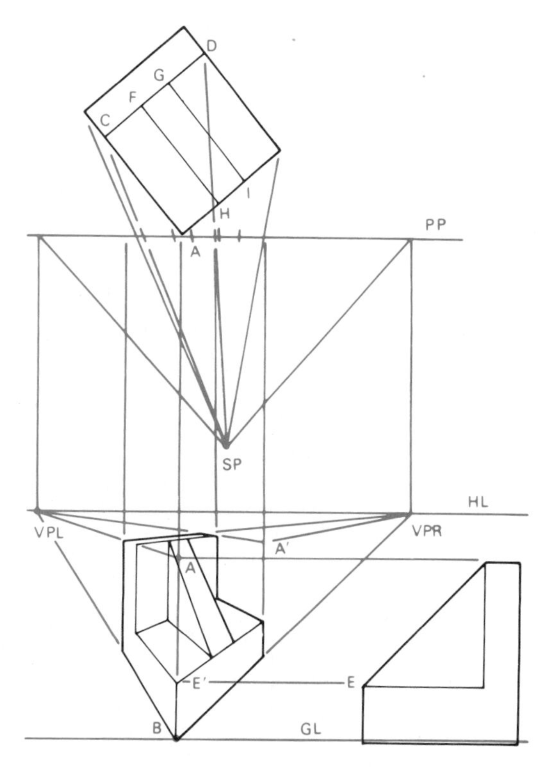

Figure 14–37 Constructing a two-point perspective drawing.

Step 5 Draw visual rays from the station point to the rear corners of the object in the plan view.

Step 6 The points where the visual rays intersect the picture plane are projected vertically down to the perspective view. When drawing complex objects, work with only a portion of these points at one time to avoid confusion.

Step 7 Project points A, B, and C toward the vanishing point to intersect the corresponding projectors from the picture plane. The horizontal line B'C' can then be drawn. Remember, any line that is parallel to the PP in the plan view must be parallel to the GL in the perspective view.

Step 8 Project the height of the object from the elevation view to points D and E on the two *true-height lines* (THL). True-height lines are projected from points that touch the PP.

Step 9 Project the height at points D and E back toward the vanishing point to intersect the projectors from the rear portion of the object in the plan view.

Step 10 Complete the object by connecting the ends of the sloping portion and projecting the base of the sloped feature toward the vanishing point.

TWO-POINT PERSPECTIVE

The two-point perspective method is also termed angular perspective because it is turned so that two of its principal planes are at an angle to the picture plane. This is the most popular form of perspective drawing. Its two vanishing points allow parallel lines on the two principal planes to be projected in two directions, thus giving another dimension to the depth of the perspective.

We will use the same object drawn previously in the discussion of one-point perspective. The object has been positioned at an angle in the plan view, with one corner touching the PP. The elevation view, SP, HL, and GL have all been established. Refer to Figure 14–37 as you read.

Step 1 Draw a line from the station point, parallel to each side of the object, to intersect the PP.

Step 2 Project the points on the PP down to the HL. This establishes vanishing point right (VPR) and vanishing point left (VPL) on the horizon line.

Step 3 Project point A on the PP down to the GL. This becomes the true-height line AB in the perspective view.

Step 4 Begin blocking in the object by projecting points A and B to the two vanishing points. This establishes the two angular, or perspective, sides of the object.

Step 5 Project visual rays from the SP to the extreme corners of the object in the plan view. Remember that this cone of vision formed by the visual rays should be approximately 30°.

Step 6 Project the intersection of these visual rays with the PP down to intersect with the projectors from points A and B to the two vanishing points. This blocks in the two sides of the object.

Step 7 Draw lines from the SP to points C and D. Where these lines intersect the PP, project vertical lines down into the perspective view. The height of the object at points A and A' can now be projected toward the VPL to intersect with the projectors from points C and D.

Step 8 The height at E can be projected to E' on the THL. Project E' toward both vanishing points to create the basic shape of the part.

Step 9 Project points F, G, H, and I to the station point. Where these lines intersect the PP, drop vertical projectors down to the perspective view. These lines will intersect corresponding points on the perspective drawing. Connect the points as shown in Figure 14–37 to complete the perspective view.

THREE-POINT PERSPECTIVE

Three vanishing points are used in three-point perspective. These drawings require more time to construct than do two-point perspectives, and often occupy a considerable area on the drawing sheet. Three-point perspective drawings are used when certain effects are needed for visual stimulation. We will examine the method of constructing a three-point perspective that requires the least amount of drawing space. Refer to Figure 14–38 as you read the instructions.

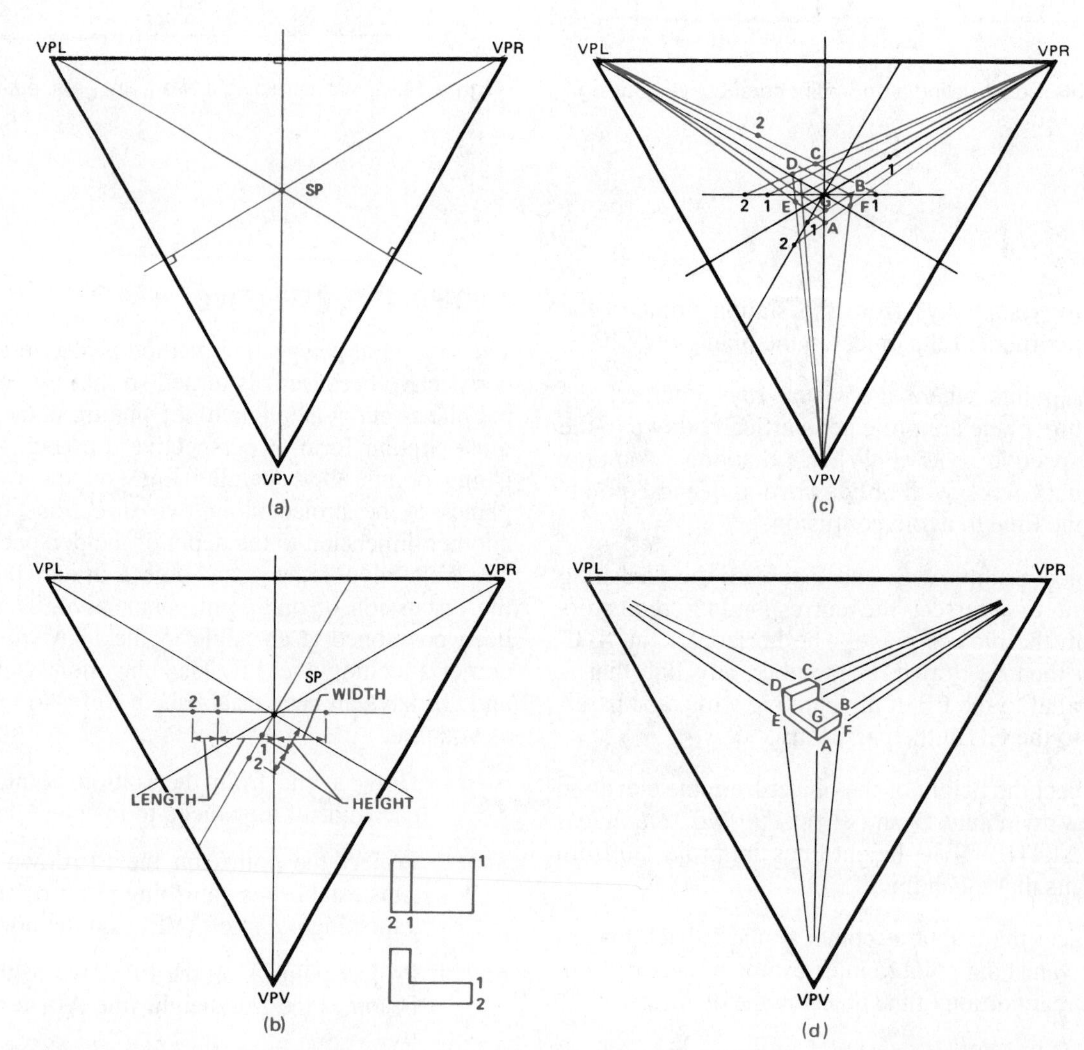

Figure 14–38 Constructing a three-point perspective drawing: (a) establishing vanishing points and the station point; (b) establishing the length, width, and height of the object; (c) establishing the outline of the object; (d) the completed object.

Step 1 Draw an equilateral triangle to occupy as much of the sheet as possible. Label the three corners as vanishing points VPR, VPL, and VPV.

Step 2 Construct perpendicular bisectors of each of the three sides. Their intersection at the center of the triangle should be labeled SP (station point). (See Figure 14–38(a).)

Step 3 Draw a horizontal line through the SP. Measure and mark the length of the object to the left of the SP. Measure and mark the width of the object to the right of the SP. This will allow you to draw an object in perspective as if you had turned the plan view 45° to the picture-plane line. (See Figure 14–38(b).)

Step 4 Draw a line parallel to line VPV, VPR. This becomes the height measuring line. Measure the height of the object from the SP down to the left along this line, as shown in Figure 14–38(b). Note that numbers have been placed at intervals along the measuring lines. These indicate measurements of features on the object.

Step 5 Project lines from points 1 and 2 to the left of the SP to VPR. Next, draw a line from point 1 at the right of the SP to VPL. Draw lines from points 1 and 2 on the height measuring line to VPR. (See Figure 14–38(c).) You have now blocked in the overall measurements of the object. The following points have been located: lower front corner (A), upper right corner (B), upper far corner (C), and upper left corner (D). Note in Figure 14–38(c) that the station point (SP) is the upper front corner of a box that encloses the object.

Step 6 Project points B and D to VPV, then project A to VPR and VPL. The intersections of these lines are points E and F, the lower left and right corners respectively. Point G is the height of the front corner of the object as shown in Figure 14–38(c). Keep in mind that any surface of the object that is parallel to one of the principal planes of the orthographic view must project to one of the vanishing points as you draw the remainder of the object's features. The completed object is shown in Figure 14–38(d).

CIRCLES AND CURVES IN PERSPECTIVE

In most cases, circles in perspective will appear as ellipses. But if a surface of an object is parallel to the picture plane, any circle on that surface will appear as a circle. (See Figure 14–39.) Circles located in planes that are at an angle to the picture plane will appear as ellipses and can be drawn by a method of intersecting lines projected from the elevation and plan view, which is referred to as the coordinate method.

The object having the circles should first be drawn in plan and elevation, and all of the necessary lines and points for your perspective drawing should be determined and placed on the drawing. (See Figure 14–40.) Next divide the circle in the elevation view into a convenient number of pie-shaped sections. Project the intersections of these section lines with the circle to the top and side of the object. The points along the side of the object can then be projected onto the perspective view. Now transfer the distances formed by the intersection of the top of the object and the lines projected from the pie-shaped sections in the elevation view to the plan view.

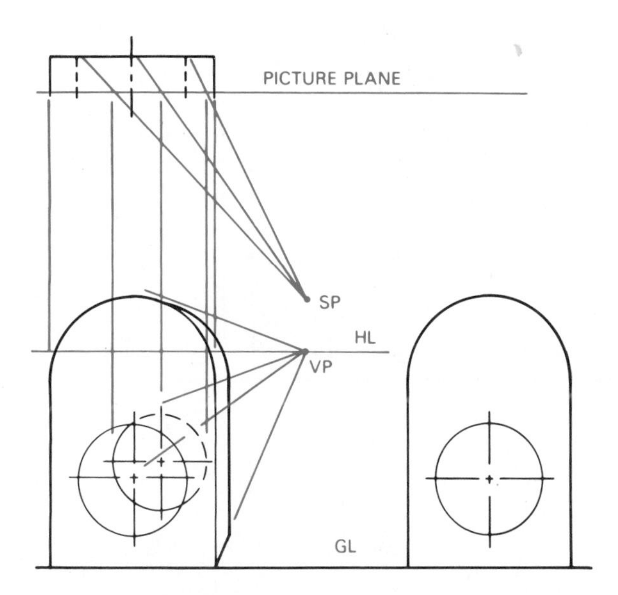

Figure 14–39 Circles in surfaces parallel to the picture plane.

Next draw visual rays from the SP to the points in the plan view. Where the visual rays intersect the picture plane line, project the points straight to the perspective view. These projectors will intersect the ones drawn from the elevation view. Connect these intersection points with an irregular curve or an ellipse template.

The same method can be used to draw irregular curves. Establish a grid, or coordinates, on the curve in the elevation view and place the same divisions on a plan view. Project these two sets of coordinates to the perspective view and then use an irregular curve to connect the points. (See Figure 14–41.)

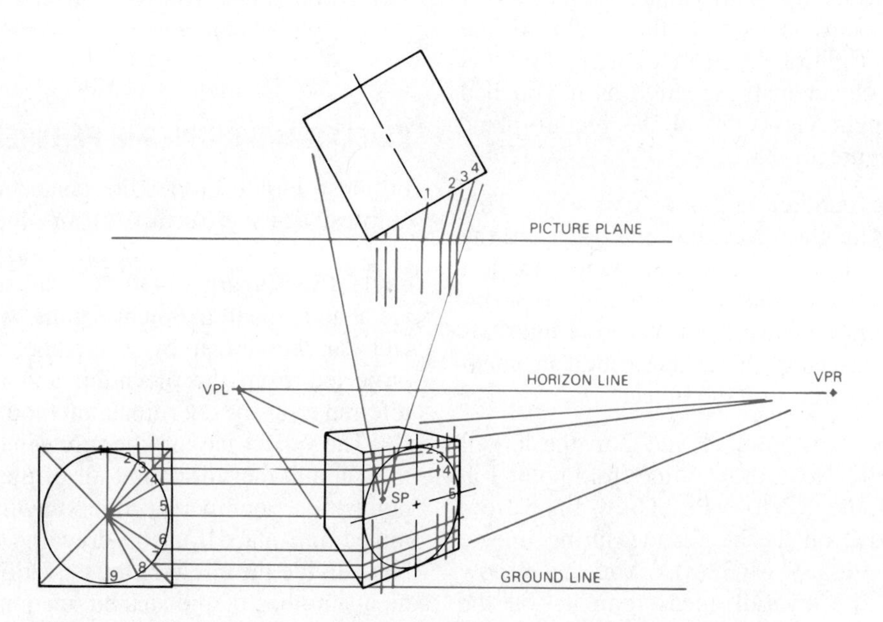

Figure 14–40 Circles plotted in perspective by the coordinate method.

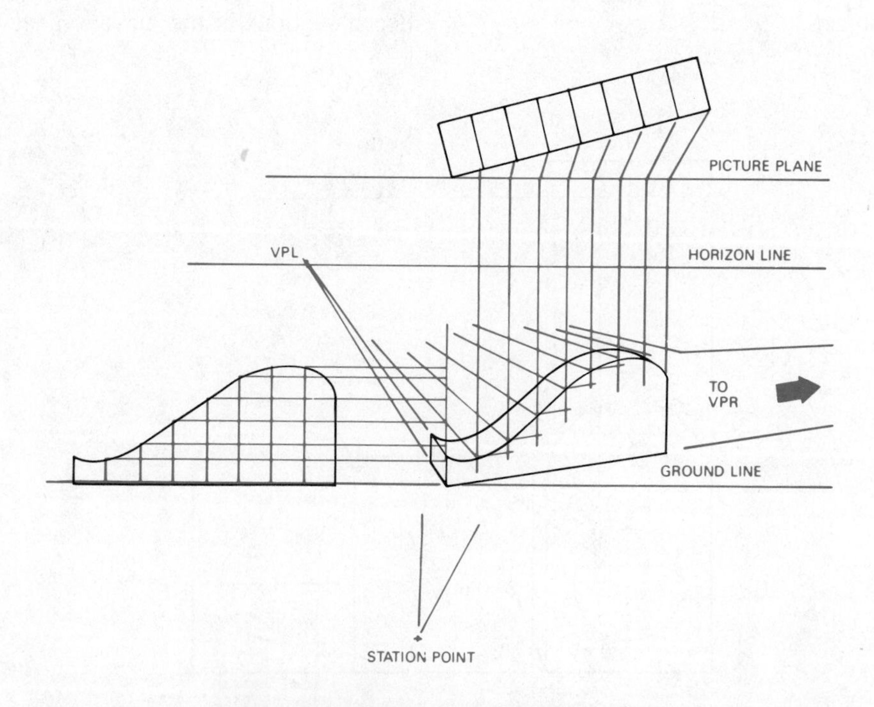

Figure 14–41 Irregular curve plotted in perspective by the coordinate method.

BASIC SHADING TECHNIQUES

Line Shading

Most pictorial drawings are created to illustrate shape description and to portray the relationship of parts in an assembly clearly. Highly artistic renderings are not normal for most industrial purposes; therefore any shading techniques used should be simple while conveying the desired effect.

The objects shown in Figure 14–42 illustrate the most basic form of shading, called line-contrast shading. Vertical lines opposite the light source and bottom edges of the part are drawn with the thick lines. Some illustrators will outline the entire object in a heavy line to make it stand out as shown in Figure 14–42(b).

Straight-line shading is a series of thin straight lines that can be varied to achieve any desired shading. They can be used on flat and curved surfaces. Note in Figure 14–43 the two ways that straight lines can be used on curved surfaces.

Additional emphasis can be given to curved surfaces with the use of block shading. Figure 14–44 shows how block shading on one or both sides of a curved surface can produce a highlight effect. One to three lines of shading are normally used to achieve the block effect. The total amount of block shading should be approximately one-third the width of the object.

Stipple shading is just dots. (See Figure 14–45.) The closer they are, the darker the shading. Although stippling takes longer than line shading, the results can be quite pleasing.

Fillets and Rounds

Fillets and rounds can be depicted in three ways, which are shown in Figure 14–46. Example (a) uses three lines to indicate the two outside edges (tangent points) of the radius and the centerline of the radius. Note the occasional gaps in the outside line.

With the technique in example (b), the same size ellipse is moved along the axis of the radius and repeated at regular intervals. It is important to draw construction lines first for the outside edges of the fillet or round. This ensures that all of the ellipses are aligned. Draw these ellipses with a thin pen or pencil point.

The example at (c) is the simplest and requires the least amount of time to draw. A broken line is drawn along the axis of the radius to indicate the curved surface.

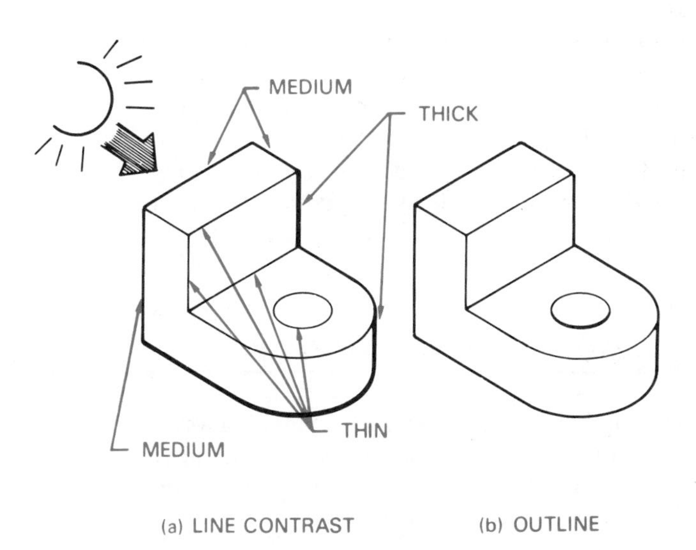

Figure 14–42 Line-contrast shading.

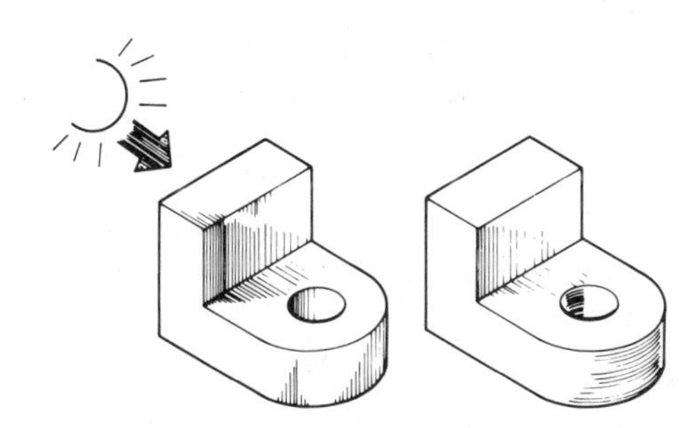

Figure 14–43 Straight-line shading.

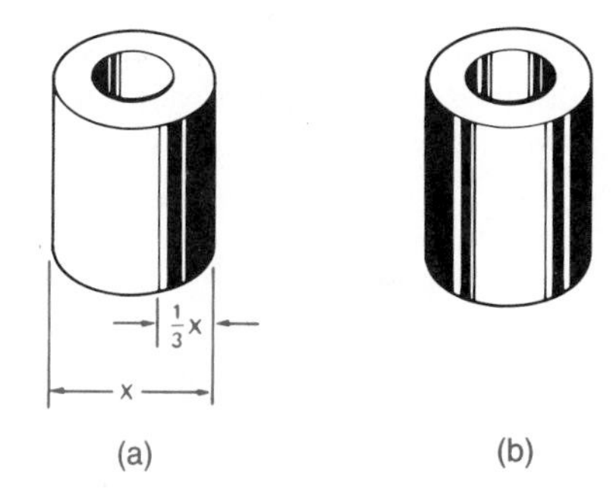

Figure 14–44 Block shading.

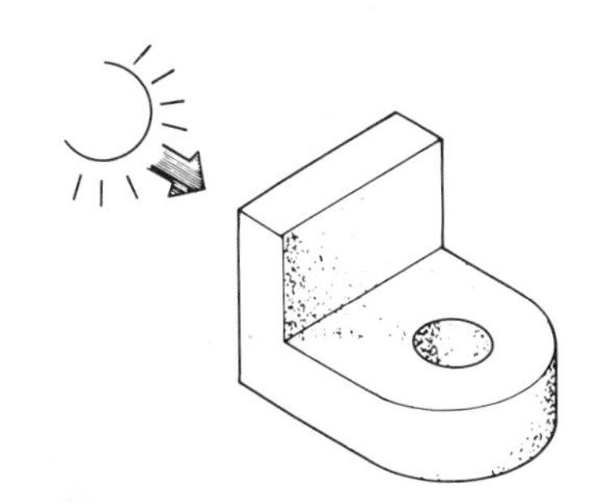

Figure 14–45 Stipple shading.

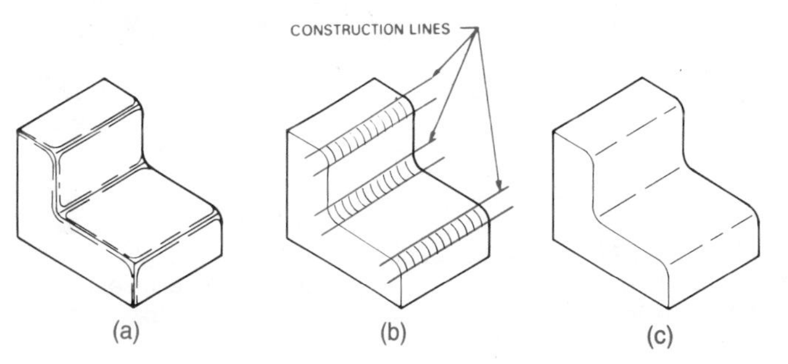

Figure 14–46 Depicting fillets and rounds.

LAYOUT TECHNIQUES

The best way to begin any pictorial drawing, especially if you are working from an orthographic drawing, is to make a quick freehand sketch. This will enable you to see the part in a 3–D form before you try to create a drawing. Always know what the final drawing is going to look like before you begin.

Objects that have boxy or angular shapes should be drawn by first laying out the overall sizes of the part. Box in the shape. Then begin to cut away from the box using measurements from the part or drawing. That is called the coordinate, or "box-in," method. Use construction lines to assist in the layout, be it a manual or a CADD drawing. If the object has circular features, use the following layout techniques.

Step 1 Lay out the centerlines and axis lines of these features first. (See Figure 14–47(a).) They act as the skeleton of the part.

Step 2 Draw the circular features using ellipses (See Figure 14–47(b).)

Step 3 Join ellipses with straight lines (See Figure 14–47(c).)

Step 4 Add additional features such as shading, notes, and item tags. If you are drawing manually, trace the object lines in pencil or ink and add shading as required. Any notes or item tags that are needed can be added last. If you are working with a CADD drawing, you can edit the drawing so only the object lines show. If you have created a special "construction" layer for your CADD drawing, turn this layer off when the drawing is completed. The finished drawing is shown in Figure 14–47(d).

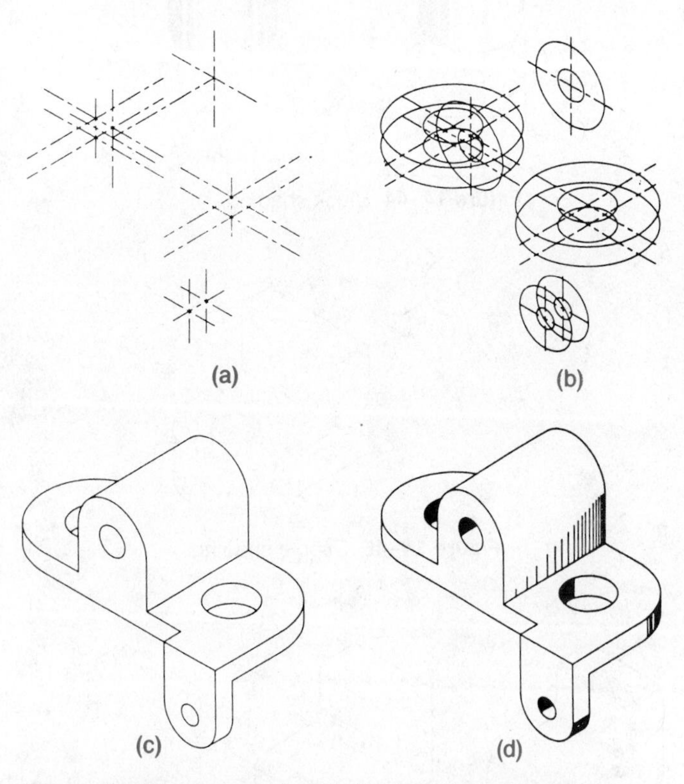

Figure 14–47 (a) Lay out centerlines and axis lines of circular features; (b) draw circular features using ellipses; (c) join ellipses with straight lines; (d) complete the drawing with shading.

PROFESSIONAL PERSPECTIVE

Pictorial drawing requires a good ability to visualize objects in three-dimensional form. Companies that produce parts catalogs, instruction manuals, and presentation drawings require the services of a technical illustrator or someone skilled in pictorial drawing. A part of your professional portfolio should be 8" × 10" photo reductions of your best pictorial drawings. Reductions of large drawings can look good, and small mistakes tend to fade away when reduced.

The field of technical illustration is much more limited than engineering and drafting but is wide open to the free-lancer or student looking for quick jobs. Always keep examples of a variety of pictorial drawing types in your portfolio. Your portfolio should be a three-ring binder with tabbed dividers for the different types of designs and drawings you have created. Manila pocket folders that fit into the binder are excellent for holding folded blueprints and reduced pictorial drawings. Remember that often a small free-lance job can lead to bigger jobs and even to owning your own company.

ENGINEERING PROBLEM

This problem requires you to create a pictorial view of a given object that is in the form of an engineering sketch. (See Figure 14–48.) The following instructions should be used in completing the problem.

1. Choose the view of the object that best shows most of the features of the part. Orient this view facing to the left or right.
2. Use the centerline layout method to locate the axis lines of the circular features.
3. Lay out additional thicknesses and features using the coordinate, or box-in, method.
4. Use proper ellipses to draw circles and arcs.
5. If inking, use different pen widths for line contrast.
6. Apply shading as required.

The completed object is shown in Figure 14–49.

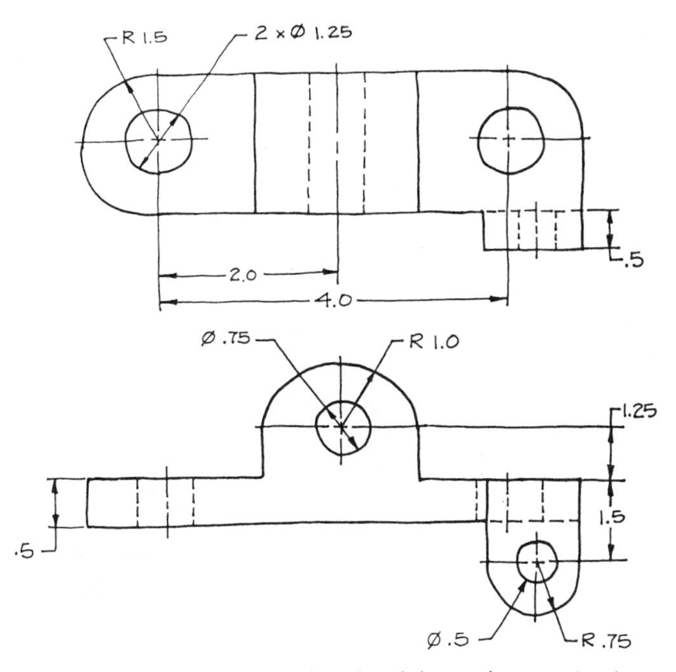

Figure 14–48 This engineering sketch is used to construct an isometric view of the object.

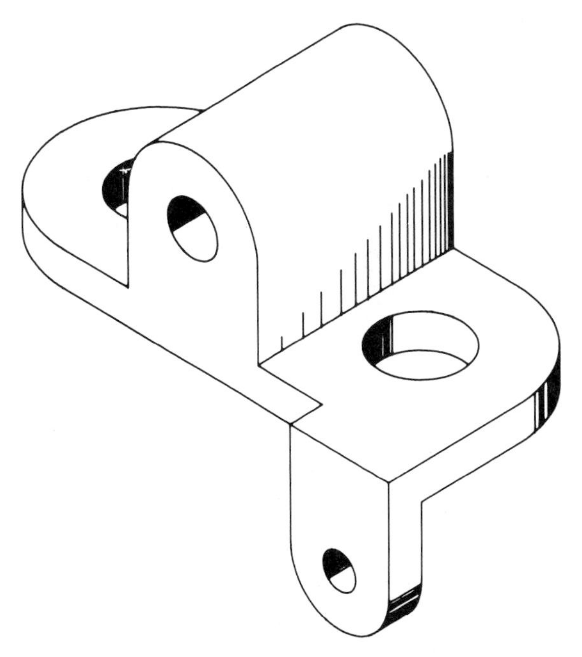

Figure 14–49 The solution to the engineering problem.

PICTORIAL DRAWING PROBLEMS

DIRECTIONS

Choose the best axis to show as many features of the object as possible in your axonometric or oblique drawing.

For axonometric problems — problems numbered 14–1 through 14–11 — draw isometric, dimetric, or trimetric as assigned. For oblique problems — problems numbered 14–12 through 14–17 — draw cavalier, cabinet, or general oblique as assigned. Remember that circular features are best shown in the front plane of the oblique view.

For perspective problems — problems numbered 14–18 through 14–25 — make a one-, two-, or three-point perspective drawing as assigned except for problem 14–18, which should be done as a one-point perspective view. All objects may be turned at any angle on the picture-plane line for viewing from the station point except problem 14–18, which should be drawn in the direction indicated.

1. Make a freehand sketch of the object to assist in visualization and layout of axonometric and oblique problems.
2. For axonometric and oblique problems, select a scale to fit the drawing comfortably on an A- or B-size drawing sheet. Use a C- or D-size drawing sheet for drawing an initial layout of the perspective problems on sketch or butcher paper.
3. Dimension axonometric or oblique problems only if assigned by your instructor. Do not place dimensions on a perspective view.
4. Perspective objects without dimensions can be measured directly and scaled up as indicated or assigned.
5. Trace the perspective view in pencil or ink on vellum or Mylar®.
6. Make your drawings using a CADD system if appropriate with course guidelines.

Problem 14–1 Axonometric projection

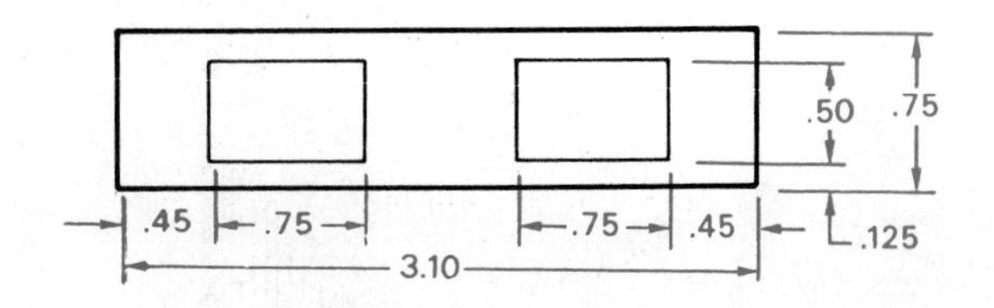

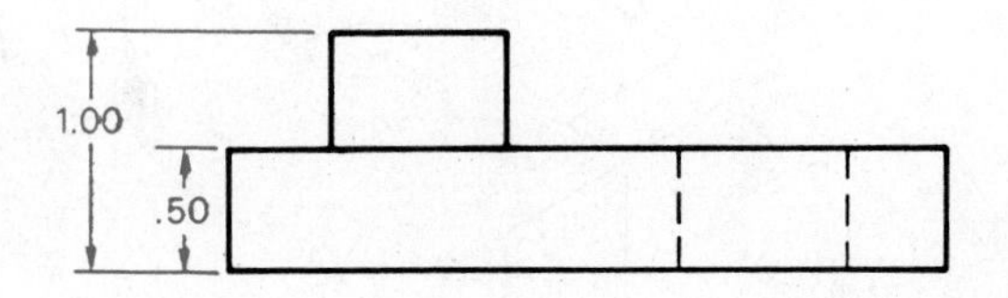

Problem 14–2 Axonometric projection

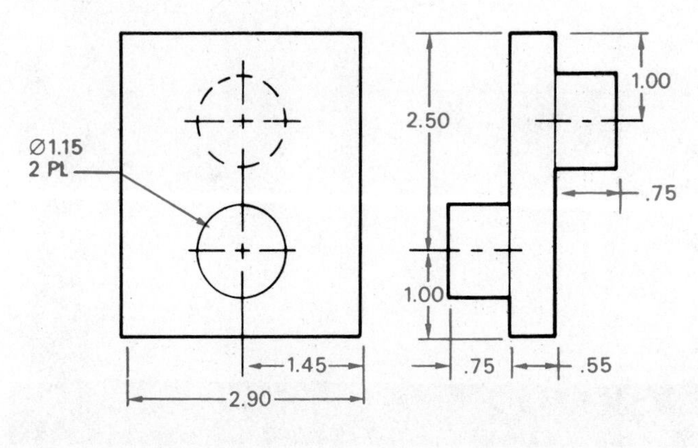

Problem 14–3 Axonometric projection

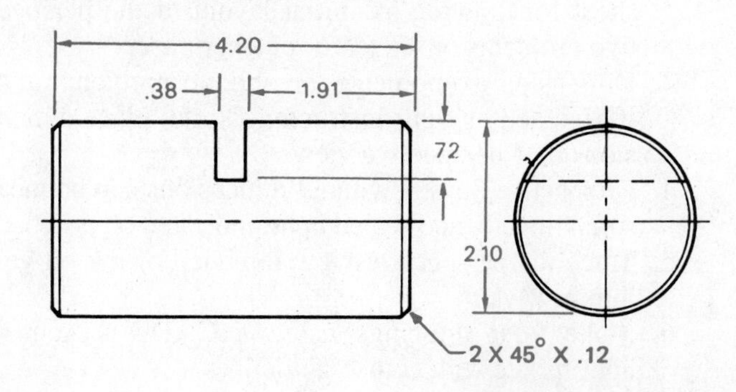

Problem 14–4 Axonometric projection (metric)

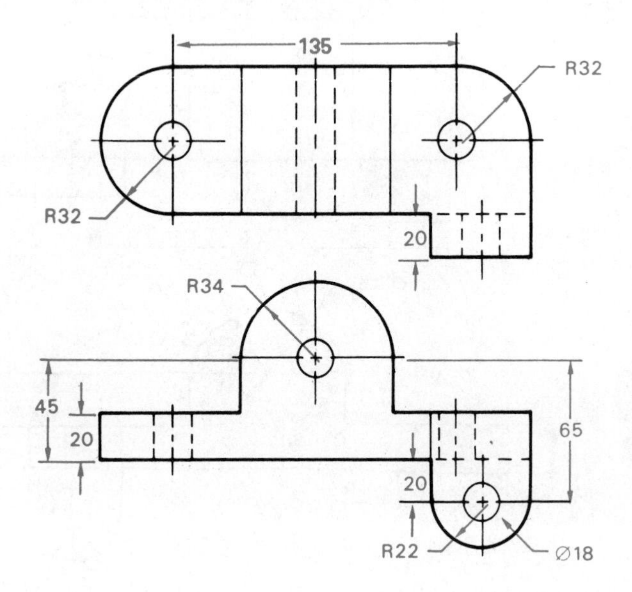

Problem 14–5 Axonometric projection

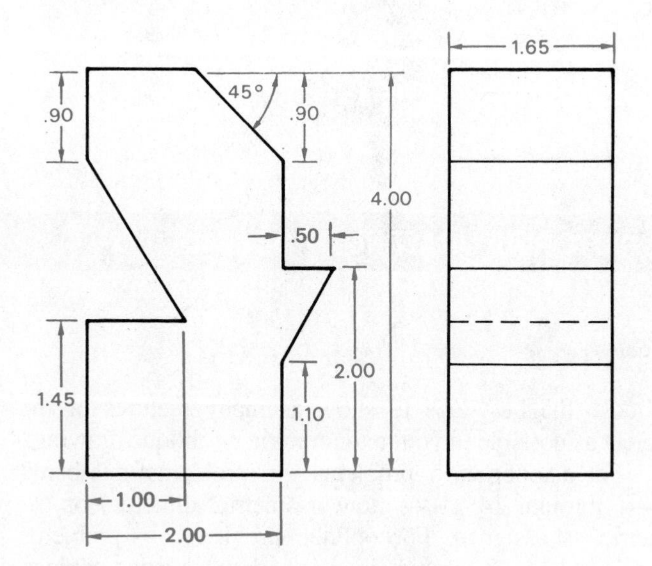

Problem 14–6 Axonometric projection (metric)

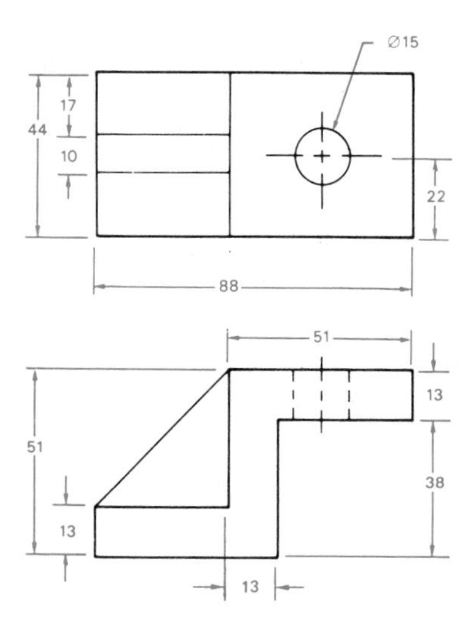

Problem 14–9 Axonometric projection (metric)

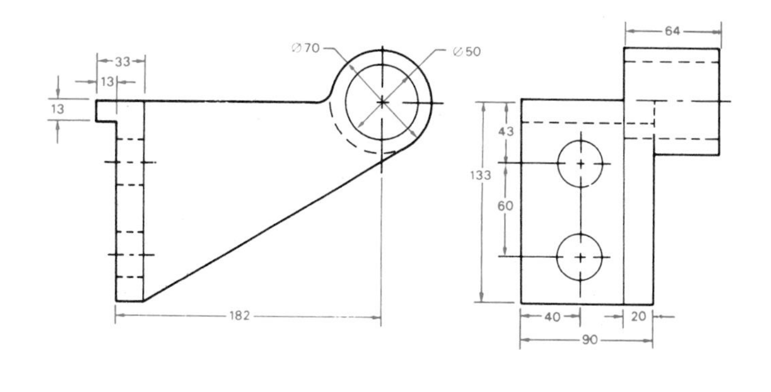

Problem 14–7 Axonometric projection

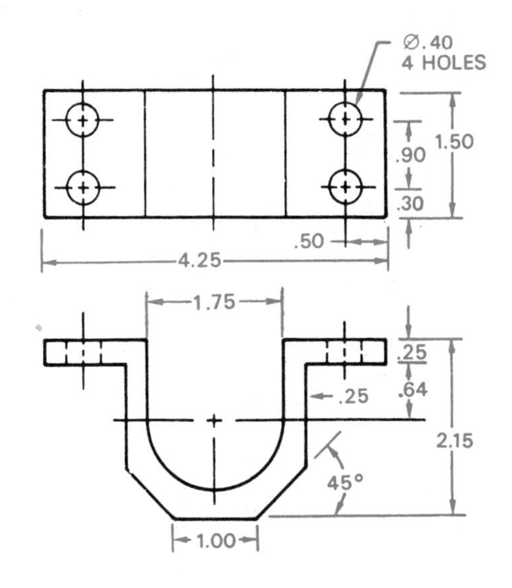

Problem 14–10 Axonometric projection

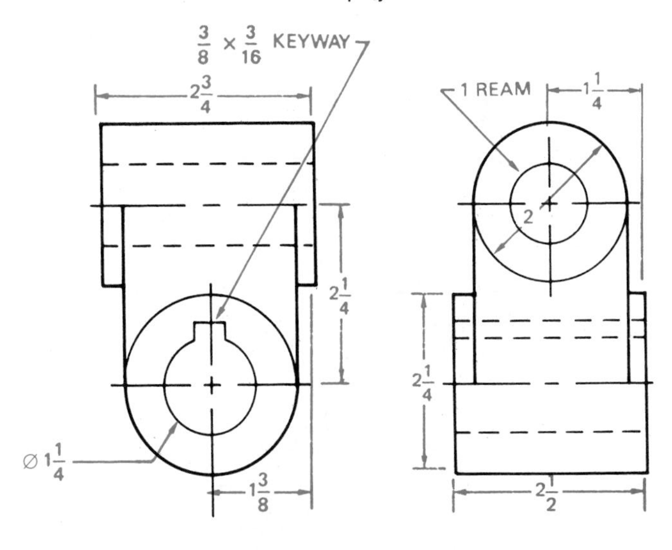

Problem 14–8 Axonometric projection (metric)

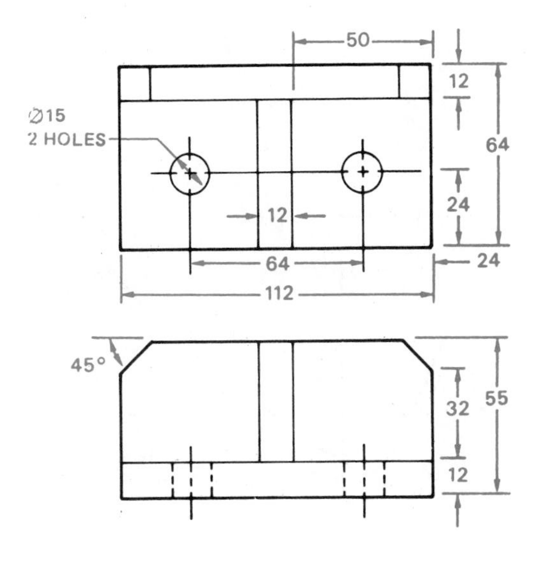

Problem 14–11 Oblique projection

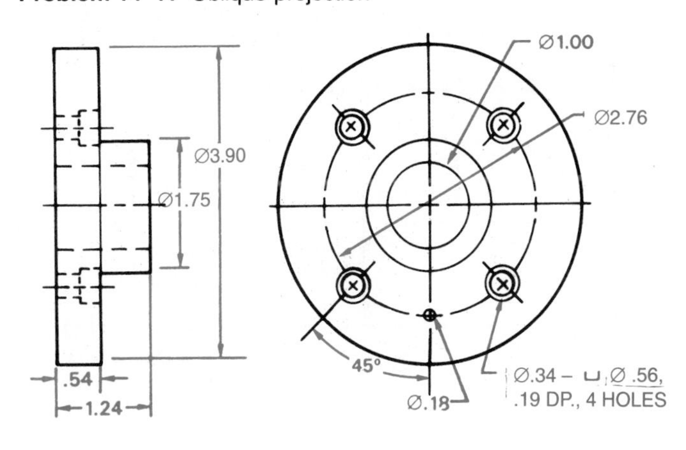

Problem 14–12 Oblique projection

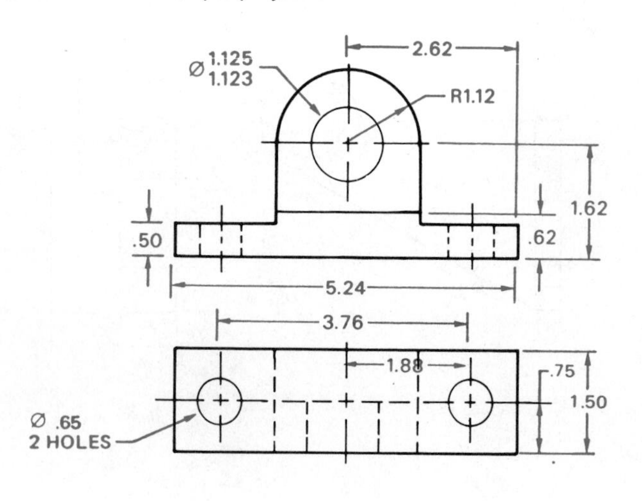

Problem 14–15 Oblique projection

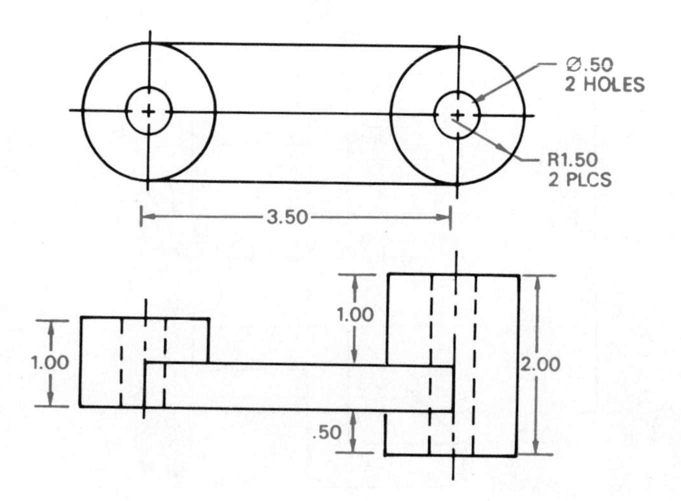

Problem 14–13 Oblique projection

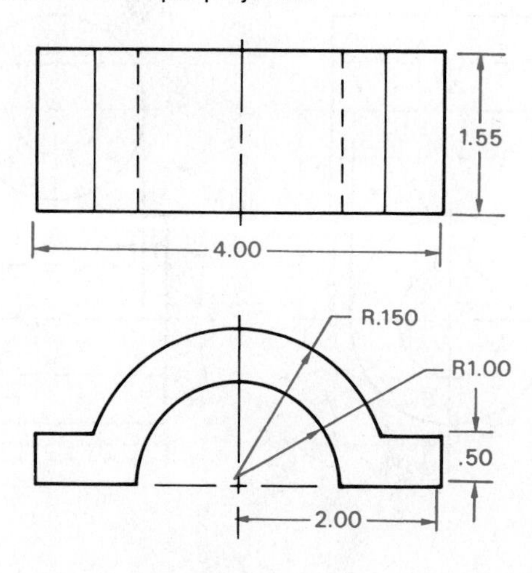

Problem 14–16 Oblique projection

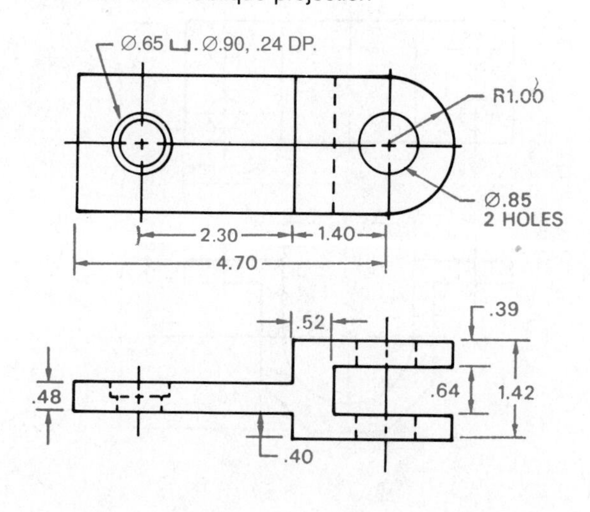

Problem 14–17 Oblique projection

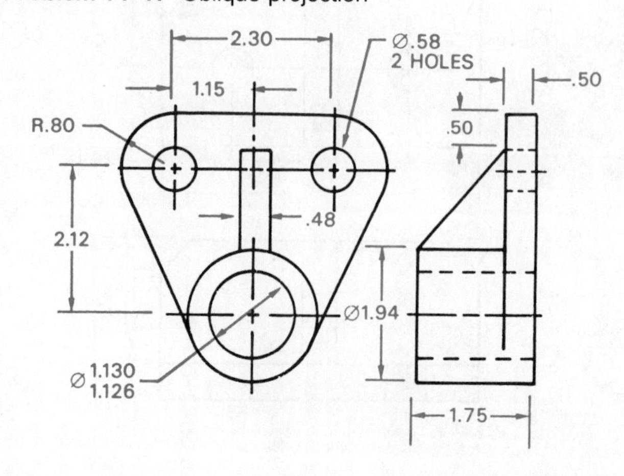

Problem 14–14 Oblique projection (metric)

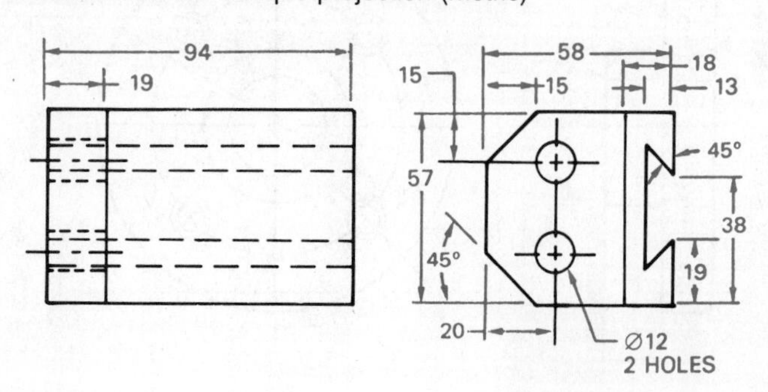

Problem 14–18 Perspective projection

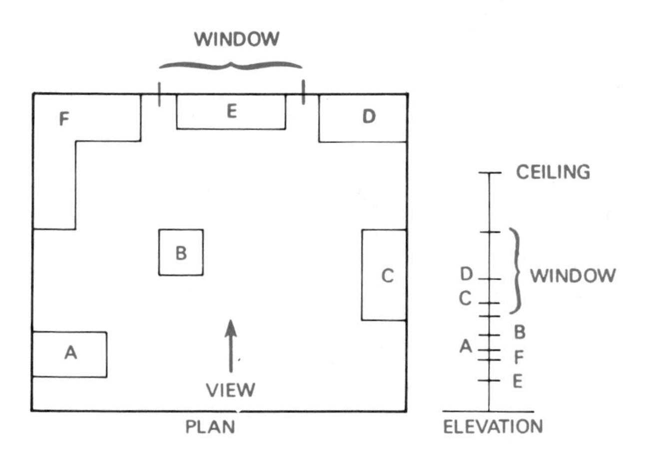

Problem 14–19 Perspective projection

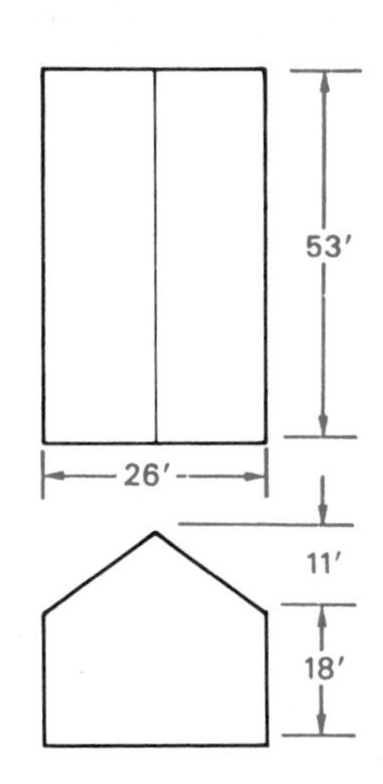

Problem 14–20 Perspective projection

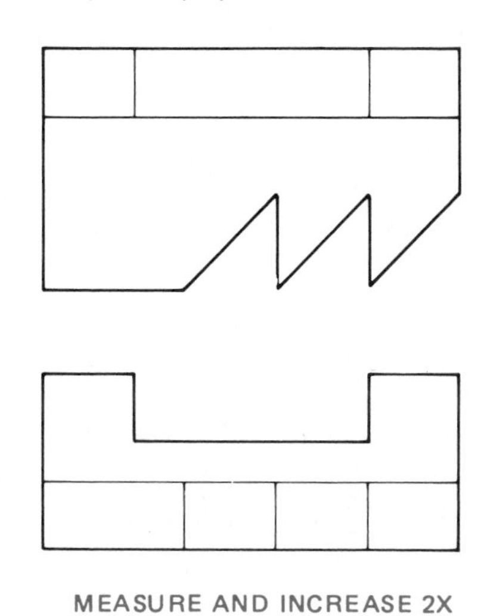

MEASURE AND INCREASE 2X

Problem 14–21 Perspective projection

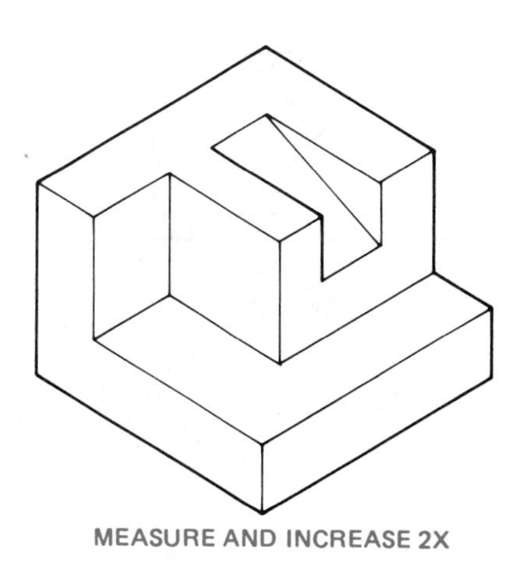

MEASURE AND INCREASE 2X

Problem 14–22 Perspective projection

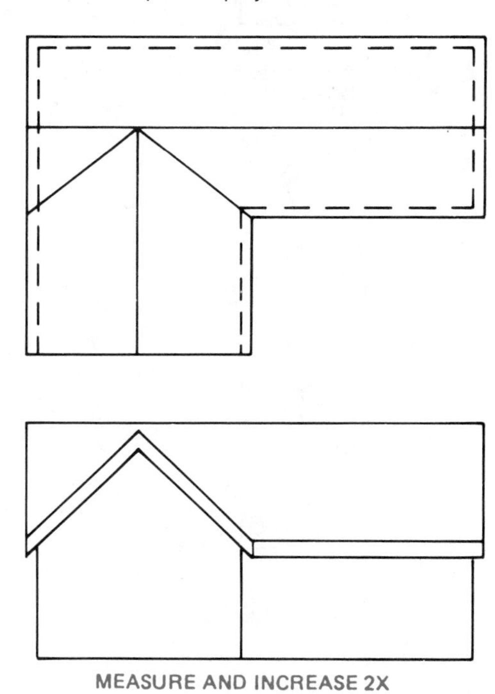

MEASURE AND INCREASE 2X

Problem 14–23 Perspective projection

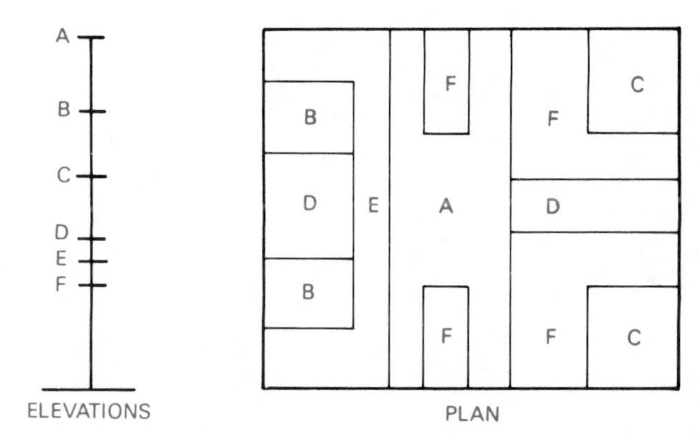

MEASURE AND INCREASE 3X

Problem 14–24 Perspective projection

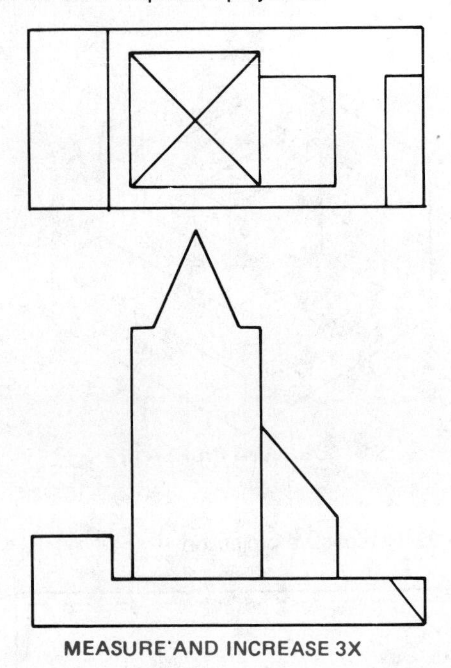

Problem 14–25 Perspective projection

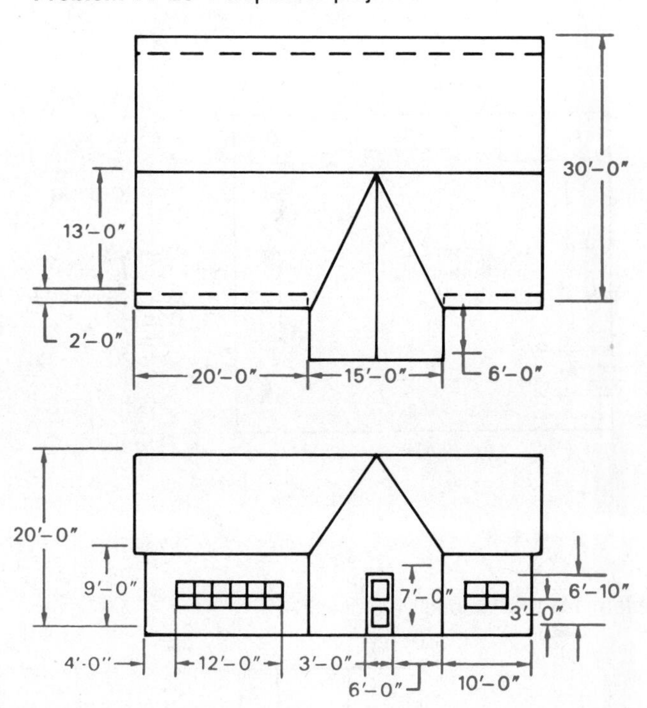

Courtesy Computer Support Corporation.

CHAPTER 15

Engineering Charts and Graphs

LEARNING OBJECTIVES

After completing this chapter, you will:

- Design and draw the following types of charts and graphs: rectilinear, surface, pie, polar, nomographs, trilinear, flow, distribution, and pictorial.
- Use a computer-aided design and drafting system to design and draw a chart from engineering specifications and sketches.

The old saying "a picture is worth a thousand words" goes especially well with charts. Actually it might be better stated, "a drawing is worth a thousand numbers." *Charts* are graphic representations of any measurable data. Charts, graphs, and diagrams are all synonymous with the graphic representation of numerical data. Charts may be defined more specifically as illustrations that give information that would otherwise be arranged in a table (tabular form). Graphs and diagrams are a series of points, a line, a curve, or an area representing the comparison of variables. The advantage of charts and graphs is that technical data may be shown in a manner that quickly and graphically communicates the information.

Professionals in any field can use charts and graphs to graphically explain or demonstrate the results of statistics. *Statistics* is the science of the collection, arrangement, and interpretation of quantifiable information. Statistics has two areas or branches: descriptive statistics and inferential statistics. Descriptive statistics is the summarizing and organization of data. Inferential statistics includes making inferences (drawing conclusions) about a whole population based on information obtained from a sample. Many engineers, quality-control people, and managers are beginning to understand that to really interpret data, they need the statistical data and a picture of the data in the form of a chart or graph. Many people have difficulty understanding and interpreting statistical data unless the quantities are shown on a chart or graph.

The accurate preparation of charts and graphs requires a degree of knowledge of the topic related to the data and an understanding of where, how, and why the chart will be used. There are two fundamental applications for charts: analysis and presentation. Analytical charts are used to analyze, examine, and explore data, or the information is used to calculate specific required values. Presentation charts are generally more artistic in appearance and are used to demonstrate or present information. Presentation charts are used in advertising and may be pictorial for easy communication of data for lay people. While there may be two classifications of charts, there is a crossover. For example, analytical charts may be used for presentation, and presentation charts often contain analytical data.

ASA According to the American Standard ASA Y15.2, the following questions should be answered before a chart design is started:

1. What is the general purpose of the chart?
2. What kind of data are to be presented?
3. What features of the data is the chart to identify?
4. For what audience is the chart intended?
5. What method will be used to show the chart to the audience?

The selection of the specific type of chart is the designer's choice. Not all chart design questions will be applied to each chart. Some charts may be designed to focus on one or more of the related design questions. One of the most important factors in chart design is simplicity. Keep the chart as simple as possible by avoiding the presentation of too much information or detail.

Charts may be designed for any purpose to accurately communicate numerical data from a simple line format to a pictorial presentation. The selection of the chart type depends on the audience, the intent of the presentation, and the type of data. The basic types of charts include rectilinear (line) charts, surface charts, column (bar) charts, pie charts, and pictorial charts.

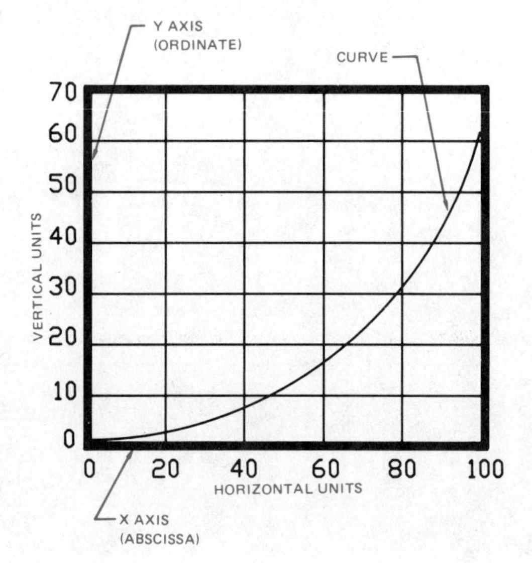

Figure 15–1 Elements of a rectilinear chart.

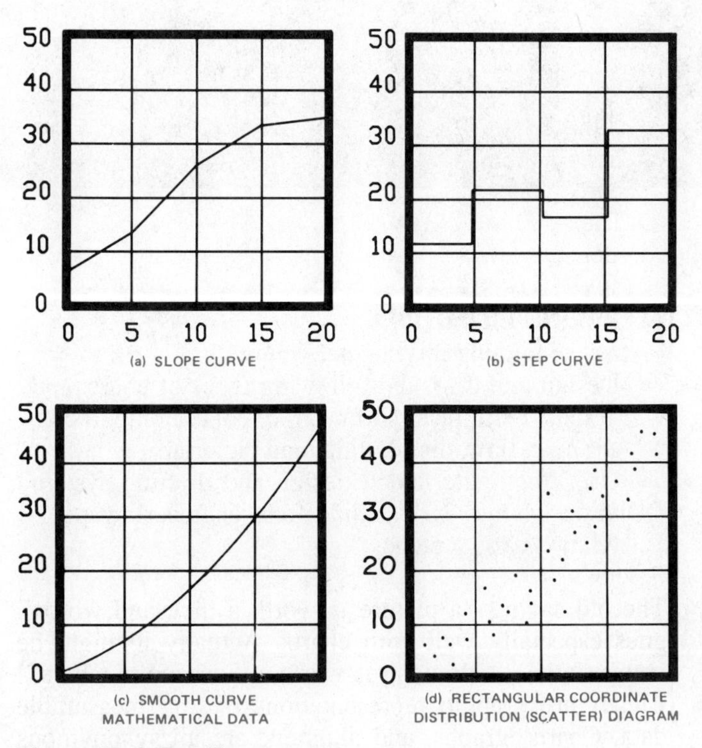

Figure 15–2 Rectilinear curves.

RECTILINEAR CHARTS

Rectilinear charts, also known as line charts, are the most common type of chart used for the presentation of analytical data. The rectilinear chart is commonly set up on a horizontal and vertical grid where the horizontal axis or scale represents amounts of time or other significant independent values. The vertical axis identifies the dependent quantities or values related to the horizontal values. The horizontal axis is called the X axis or *abscissa* and the vertical axis is called the Y axis or *ordinate*. The line formed by connecting the data elements is called the *curve*. (See Figure 15–1.) Rectilinear charts are referred to as a slope curve when the data represent points in time that are not backed by mathematical computations. (See Figure 15–2[a].) Step curves are established by data that changes abruptly. (See Figure 15–2[b].) In some applications a smooth curve is used when the data plotted form a consistent pattern, or when the average of the calculated points is represented. A smooth curve generally acknowledges an empirical relationship between the data and the curve. (See Figure 15–2[c].) A type of rectangular chart where greatly varying data do not fit into a defined curve is called a rectangular coordinate distribution chart or scatter chart. The term *scatter* is used to describe these charts because the plotted data are scattered around on the chart. The purpose of these charts is to observe the distribution of data and to identify any areas of concentration. (See Figure 15–2[d].)

Scales

The selection of horizontal and vertical scales is one of the most important aspects of chart design. A good balance between the scales should be obtained. If one scale is exaggerated out of proportion, then the resulting curve may be misleading. Scale selection may greatly affect chart design. For example, the two charts shown in Figure 15–3 contain the same data; however, the chart in Figure 15–3(a) has a smaller vertical scale than the chart in Figure 15–3(b). Notice the extreme difference in curve representation between the two charts.

When selecting the proper scale, the first consideration is the range of data. The scale range extends from

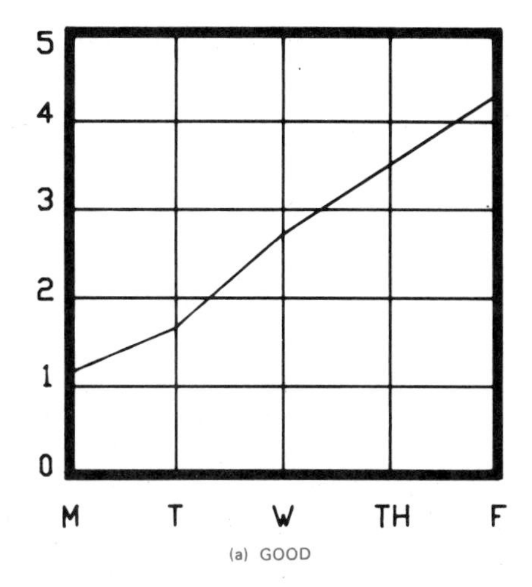

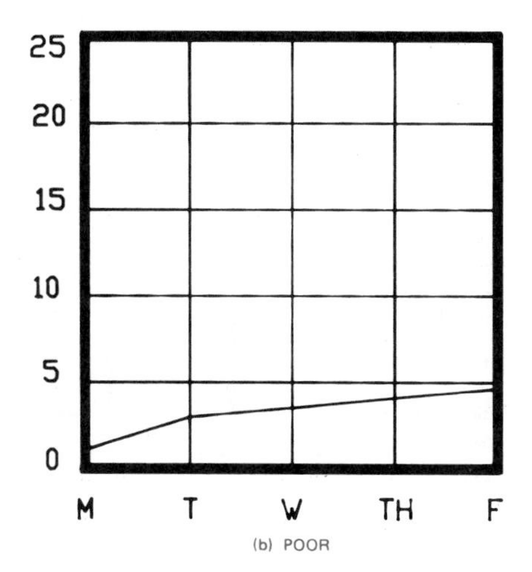

Figure 15–3 Scale selection may greatly affect chart design.

the smallest value to the largest value. Charts often, but not always, begin at zero and extend beyond the largest Y value and are at least as long as the largest X value. The scale should include zero when a comparison of two or more curve magnitudes is made. The scales may not necessarily show zero when the intent of the chart is to show the relationship between one group of units and another, or when the general shape of one curve is compared to another.

The selection of a beginning and ending scale value also depends on the range of values to be displayed. For example, if the display values range from 500 to 1500, then a zero beginning may be too far removed from the first value of 500. When zero does not begin the scale, the first numerical value should be labeled with bold letters or otherwise clearly identified so the reader does not assume a zero beginning. The scale divisions are usually established as convenient units such as 0,2,4 . . . ; or 0,5,10. . . . The scale should also be selected so the curve is well distributed throughout the chart. This requires that the scale range be extended beyond the value limits. (See Figure 15–4.) As previously shown in Figure 15–3(b), too much range is also not desirable because it provides an excessive amount of blank area in the chart. In some situations a certain amount of blank space may be necessary to accommodate notes, details, or other information. When this is appropriate, add enough scale units to provide this space.

When charts are compared, the designer should set the scales of all of the charts the same by starting with the chart that has the greatest range of values. The only potential problem is that some of the curves may be flat; however, this may be necessary to provide a realistic comparison. Figure 15–5 shows two charts placed for comparison. Another method to show the comparison of two or more curves is to place each curve on the same chart. When the curves do not cross, the curve lines may be drawn all the same thickness as shown in Figure 15–6.

When the curves to be compared intersect each other, a technique should be used to differentiate each curve. Figure 15–7 shows curve comparison on the same chart by using different line thicknesses. Another common method used to distinguish curves in a comparative chart is to show the curves using different line representations;

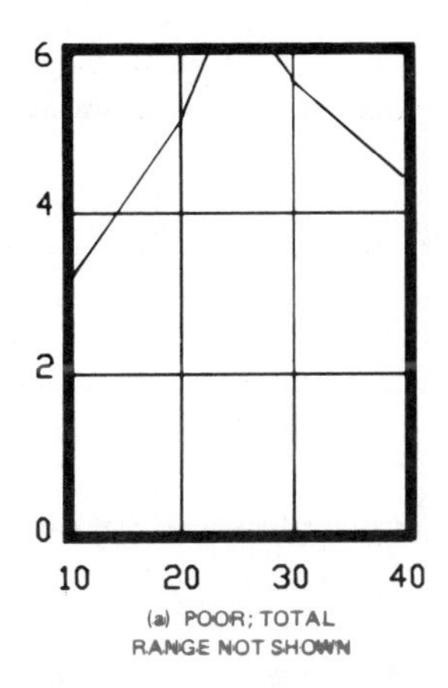

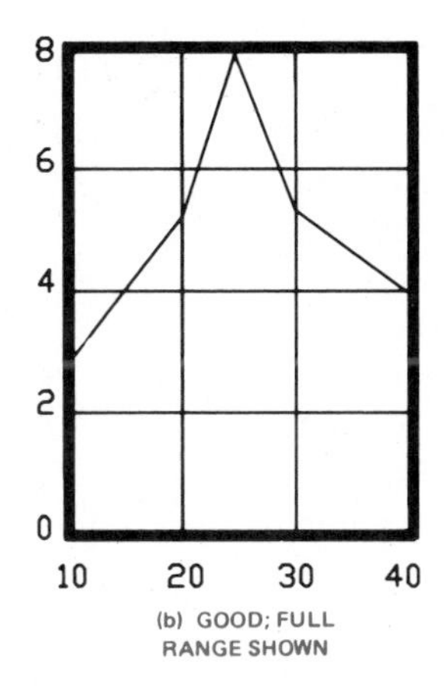

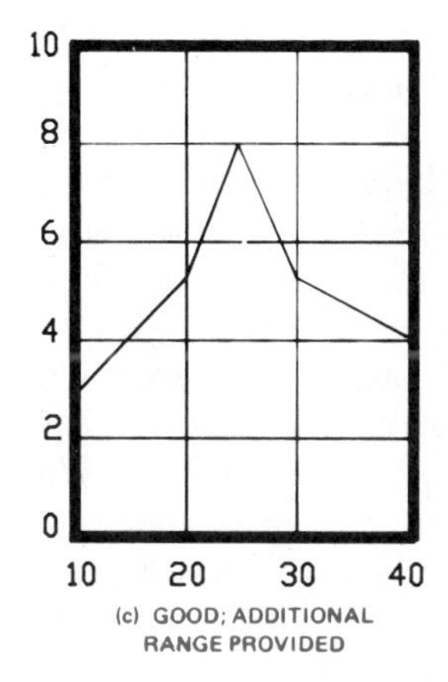

Figure 15–4 Scale range selection affects chart design.

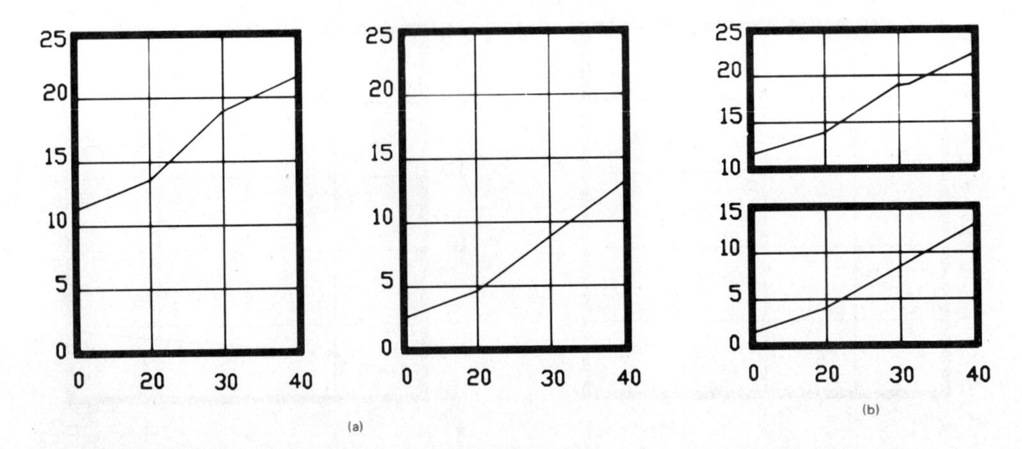

Figure 15–5 (a) Charts placed side by side for comparison — same scale range; (b) comparative charts stacked to save space.

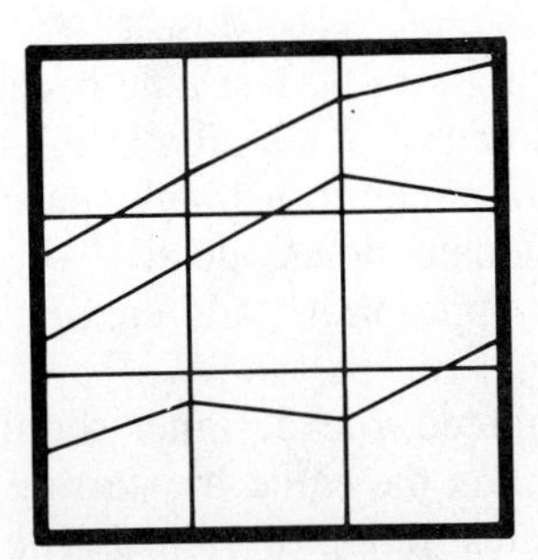

Figure 15–6 Noncrossing curves may be drawn the same thickness for comparison.

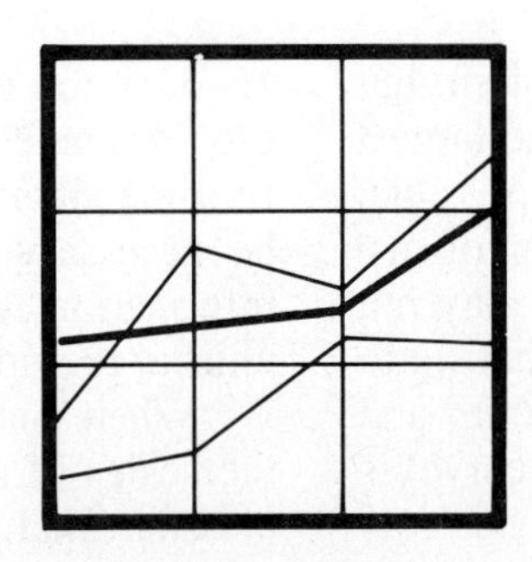

Figure 15–7 Curve comparison using different line thicknesses.

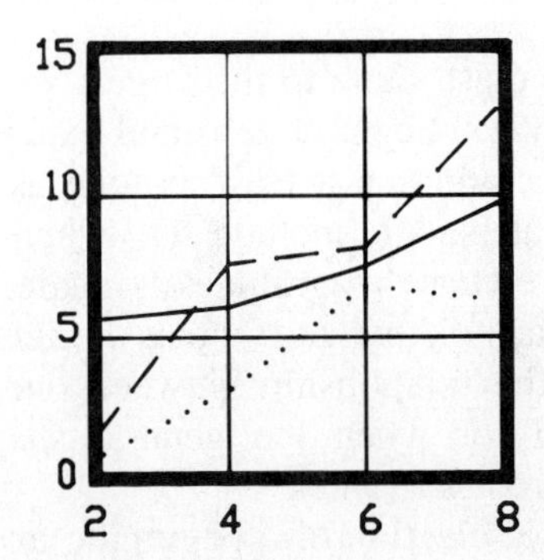

Figure 15–8 A comparative chart using different line representations.

for example, solid line, dash line, and dotted line as shown in Figure 15–8. It is not normally a good idea to design a chart with unusual patterns or shapes. These types of designs often clutter the chart or may be difficult to differentiate without careful analysis. (See Figure 15–9.) Another technique for charts with the same horizontal and vertical scales is to prepare the charts separately on clear polyester film and overlay them for comparison.

The previous discussion and examples have shown charts where numerical values along the abscissa and ordinate have been positive beginning with zero on the X and Y axis. Many charts that are designed from mathematical data do not follow this layout pattern. When numerical values exist as negative quantities, the chart must reflect the quantities below zero as shown in Figure 15–10. Some applications exist where plotted values are negative X and Y or positive X and negative Y, or positive X and positive Y values. When this occurs, the chart will be set up in quadrants as shown in Figure 15–11. Quadrant 1 shows positive X and Y values. Quadrant 2 provides negative X and positive Y values. Quadrant 3 displays negative X and Y values. Quadrant 4 shows positive X and negative Y values.

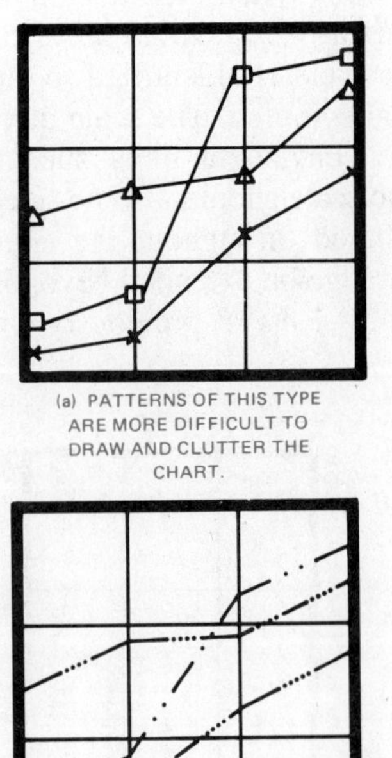

(b) PATTERNS OF THIS TYPE ARE DIFFICULT TO DIFFERENTIATE.

Figure 15–9 Charts designed with unusual patterns or line designations should be avoided.

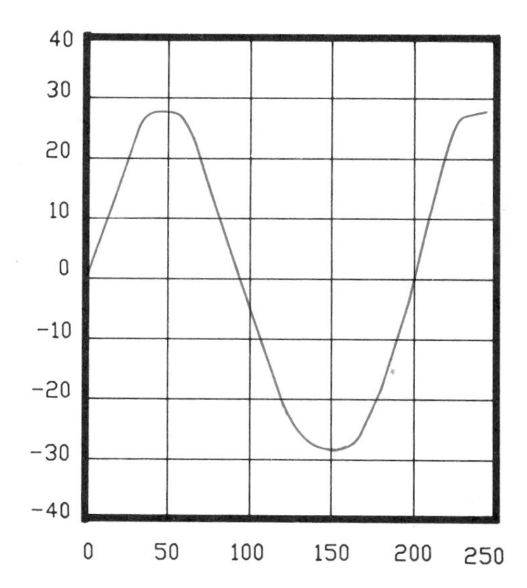

Figure 15–10 Chart showing negative values.

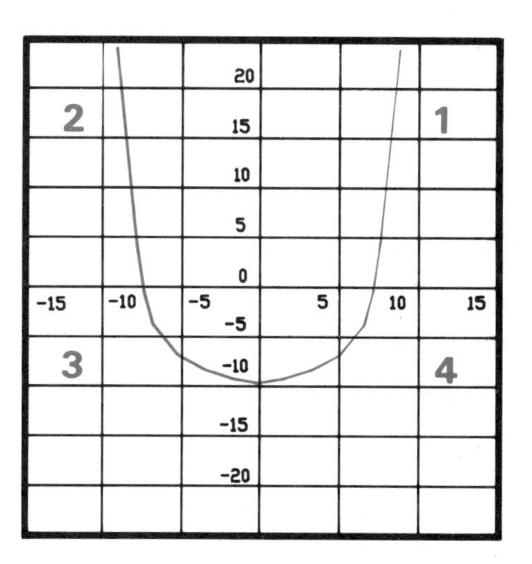

Figure 15–11 Chart set up in quadrants on an X–Y axis.

Ratio Scales

Ratio scales are special scales that are referred to as logarithmic and semilogarithmic scales. The scales that were discussed previously are arithmetic scales, where equal distances represent equal amounts. Arithmetic scales are used to display the absolute dimension of values. Ratio scales represent a different function, where the rate of change of the variables is more important than the absolute values or amount of change. Logarithmic charts use a logarithmic scale for both the X and Y axis. Semilogarithmic charts use a logarithmic scale for the Y axis and an arithmetic scale for the X axis. The advantage of ratio scales exists when it is desirable to display changes at one level either relatively larger or smaller than changes at another level, or for showing a pattern of relative changes. Logarithmic scales are established by making the divisions proportional to the logarithms of the arithmetic scales. A comparison of the arithmetic and logarithmic scales is shown in Figure 15–12.

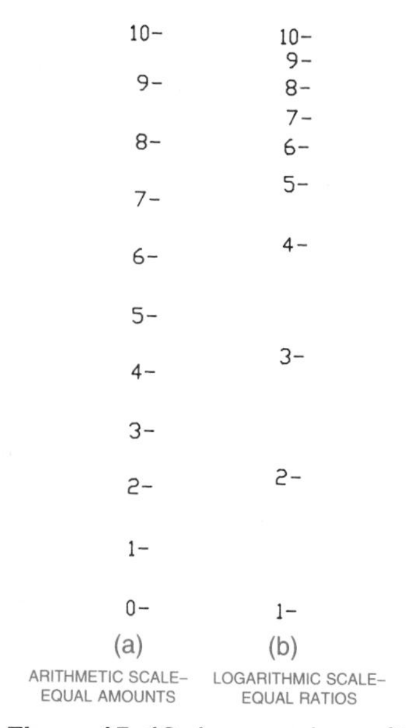

Figure 15–12 A comparison of the arithmetic and logarithmic scales.

There are a variety of logarithmic scales available. The scales most commonly used for chart design are one-cycle, two-cycle, and three-cycle. (See Figure 15–13.)

Logarithmic charts may be prepared directly on preprinted logarithmic paper or designed to fit any required area. Preprinted log scales may be used to divide the given area into proportional divisions by the same technique used when dividing a space into equal parts (Chapter 4), or using scales set up in the CADD program.

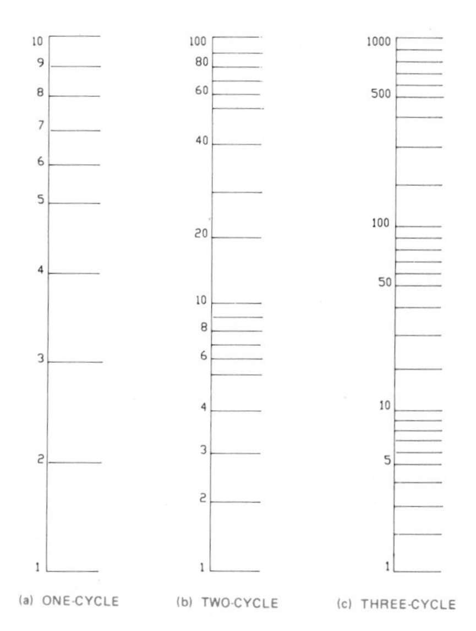

Figure 15–13 Logarithmic scales.

Grids

The grid is the horizontal and vertical lines that form the chart layout. The grid lines are determined by the scales and are used to aid both drawing and reading the chart. Charts and graphs may be prepared on preprinted grid paper or designed to fit specific needs. The actual grid

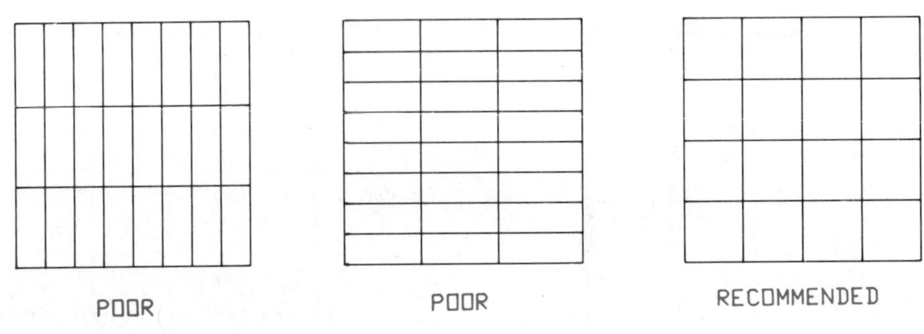

Figure 15–14 Grid spacing.

format depends on the purpose of the chart. In most applications the following guidelines should be considered:

1. A grid with wide horizontal and short vertical increments is used for data that cover a long period of time.
2. A grid with narrow horizontal and high vertical increments is used for data that represent short periods of time or rapid value changes.
3. Most applications work out well if the horizontal and vertical grids are approximately equal. (See Figure 15–14.)

Grid lines are generally thin. The thickness of these lines may increase as the distance between them increases. Close grid lines should usually be drawn very thin.

Certain horizontal or vertical grid lines may be distinguished from other lines by either drawing them thicker or as a dashed line. For example, manufacturing quality control often uses computerized monitoring of dimensional inspections in Statistical Process Control (SPC). When this is done, a chart that shows feature dimensions obtained at inspection intervals is developed. The chart shows the expected limits of sample averages as two dashed, parallel horizontal lines, as shown in Figure 15–15.

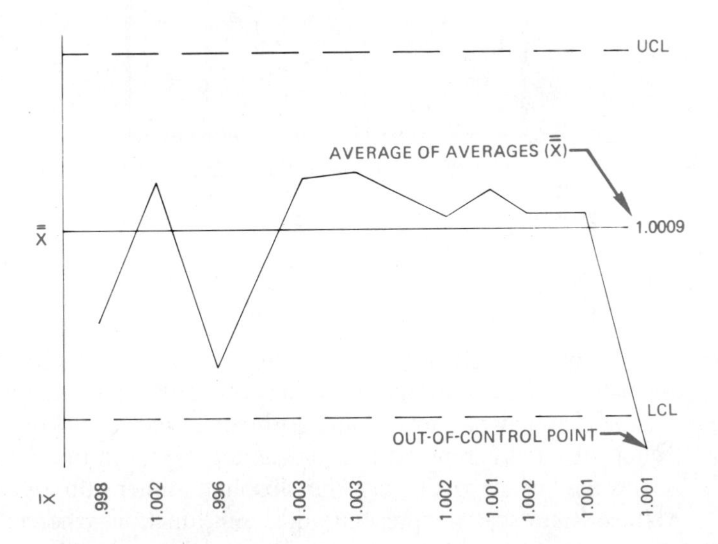

Figure 15–15 A chart showing feature dimensions at inspection intervals.

It is important not to confuse control limits with tolerances—they are not related to each other. The control limits come from the manufacturing process as it is operating. For example, if you set out to make a part 1.000 ± .005 in., and you start the run and periodically take five samples and plot the averages ($\bar{x}$) of the samples, the sample averages will vary less than the individual parts. The control limits represent the expected variation of the sample averages if the process is stable. If the process shifts or a problem occurs, the control limits signal that change.

Notice in Figure 15–15 that the $\bar{x}$ values represent the average of each five samples; $\bar{\bar{x}}$ is the average of averages over a period of sample taking. The upper control limit (UCL) and the lower control limit (LCL) represent the expected variation of the sample averages. A sample average may be "out of control" yet remain within tolerance. During this part-monitoring process, the out-of-control point represents an individual situation that may not be a problem; however, if samples continue to be measured out of the control limits, the situation must be analyzed and the problem rectified. When this process is used in manufacturing, part dimensions remain within tolerance limits and parts will not be scrapped. Statistical Process Control is discussed further in Chapter 7.

Line Chart Labeling

There are standard guidelines for the placement of labels and captions. In general, however, labels should be large enough for easy reading and placed to avoid cluttering the chart. The most important feature of the chart should be emphasized by bold display or special effects such as color. For example, the title or a predominant curve may be emphasized. Avoid overemphasizing a particular feature to the extent that other items are lost.

Chart Titles. Chart titles are generally aligned with the left edge of the paper or centered at the top of the chart. Title lettering is higher and bolder than other labels and captions.

Labeling Scales. The scales are usually identified with numerical values adjacent to the corresponding grid lines. Abscissa scale numerals are placed below the horizontal scale and ordinate scales are labeled to the left or right of the chart. Very long or high charts may have scales labeled at both grid limits. The horizontal and vertical units are identified adjacent to the numerical values or below the X values and directly above the column of Y values. Figure 15–16 shows several examples.

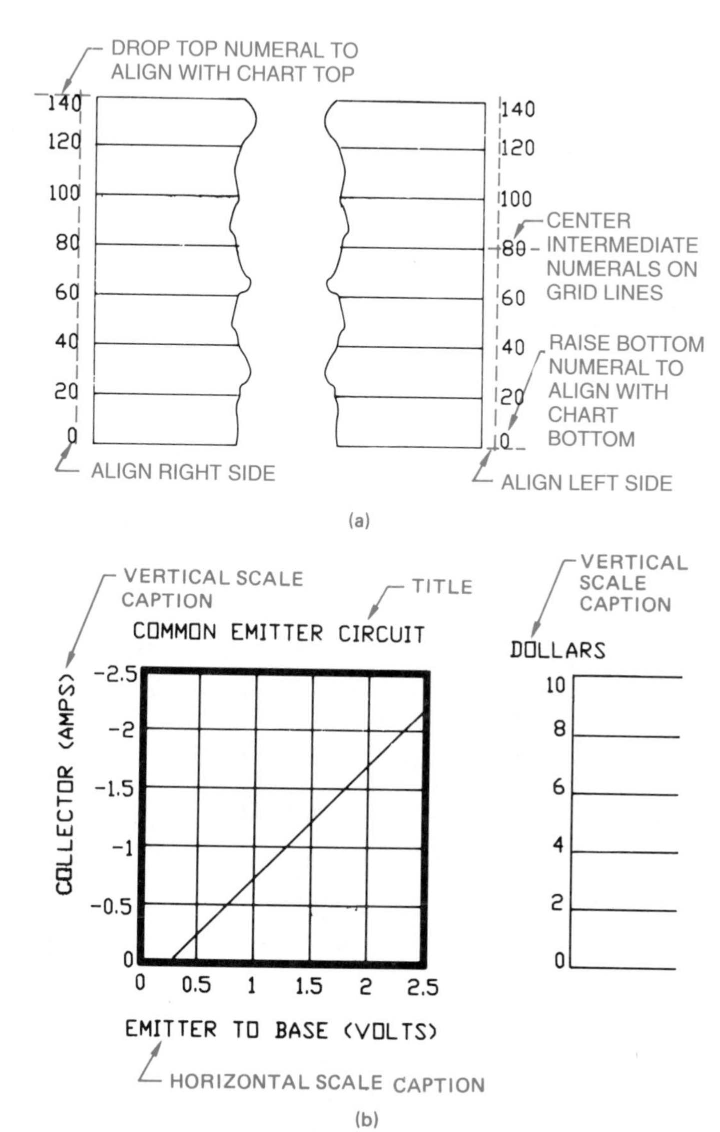

Figure 15–16 Chart labeling.

Scale identification should be kept as simple as possible. For example, long scale numerals should be shortened using the easiest format for the reader to understand, and proper abbreviations should be used when appropriate. (See Figure 15–17.)

Curve Captions. Curve captions are necessary when two or more curves are displayed on the same chart. Single curves are labeled only if identification is not otherwise apparent from other captions and titles. The lettering for curve labels is generally smaller than titles and larger than general notes. Place curve labels on the face of the chart in areas that clearly avoid crowding or confusion. (See Figure 15–18.)

Subtitles and Notes. Subtitles and general notes are subordinate captions that relate to the entire chart. These items may be placed below the title, in the field of the chart, or below the chart, depending on the design influence of the information and the space requirements. Specific notes relate to individual items on the chart and may be placed in the field of the chart or keyed to the chart by identification symbols and placed in the area of general notes. (See Figure 15–19.)

Missing or Projected **Data.** When some data is omitted due to missing or unavailable information, the estimated curve representing the missing data should be shown as demonstrated in Figure 15–20.

There are some applications when the function of the chart shows current data and provides an estimate of projected values. This may be done by continuing the curve into a future interval as shown in Figure 15–20.

In some unique situations one or more values may fall well out of the normal distribution of values. When these "freak" values occur, they may completely alter the appearance of the chart unless intentionally left beyond the grid boundaries. When this is done, the numerical designation of the freak value is labeled as shown in Figure 15–21. Freak values should be evaluated carefully for possible error because in actual practice this approach is uncommon.

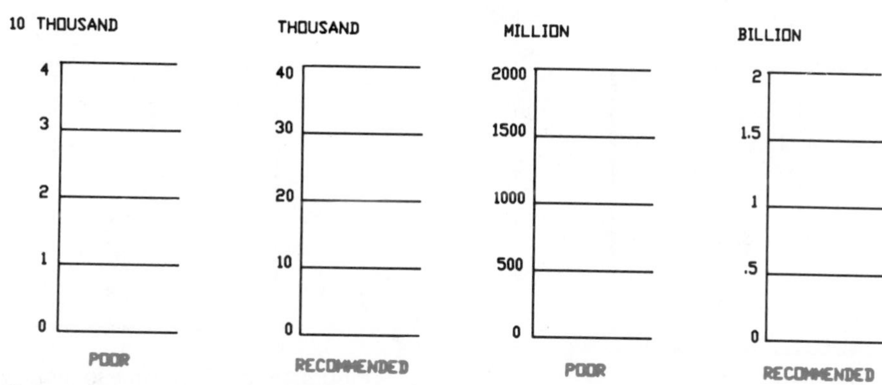

Figure 15–17 Scale identification.

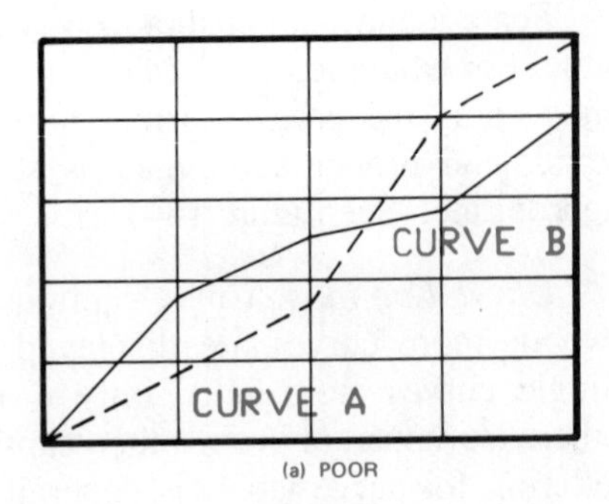

(a) POOR

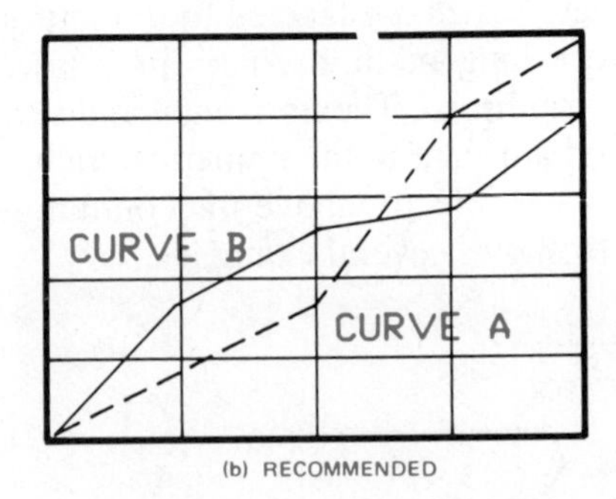

(b) RECOMMENDED

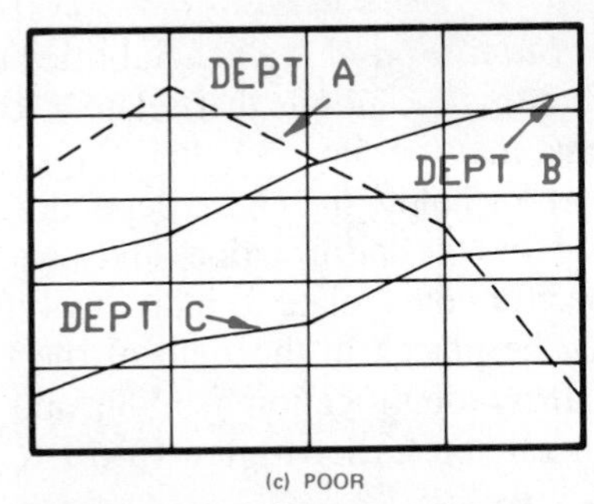

(c) POOR

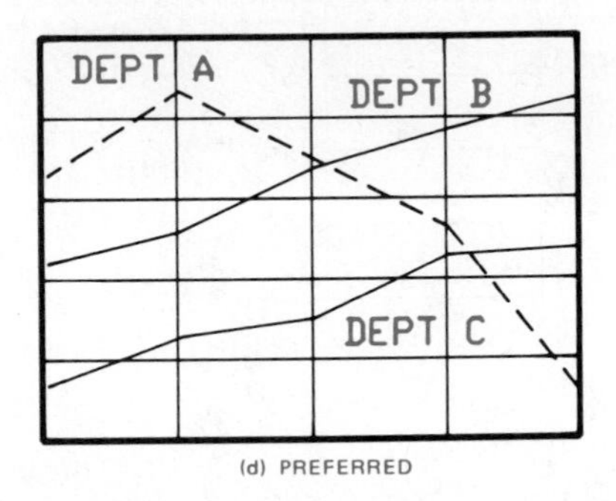

(d) PREFERRED

Figure 15–18 Curve captions.

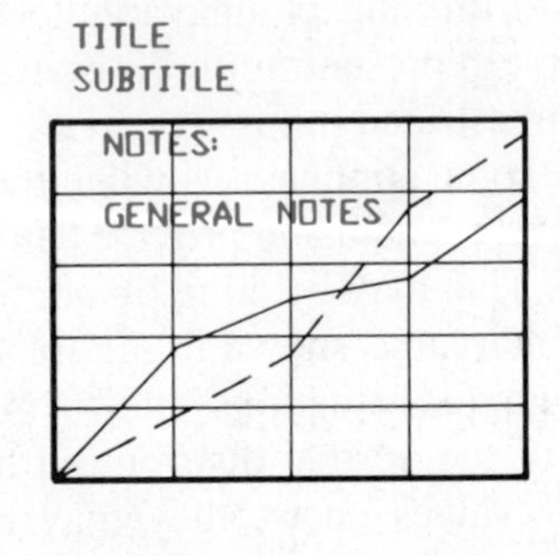

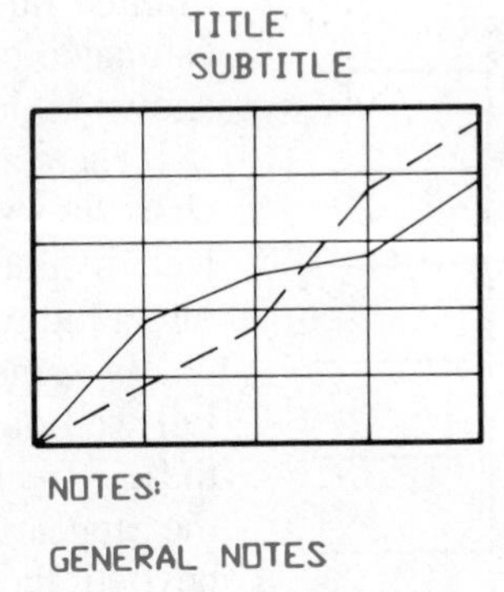

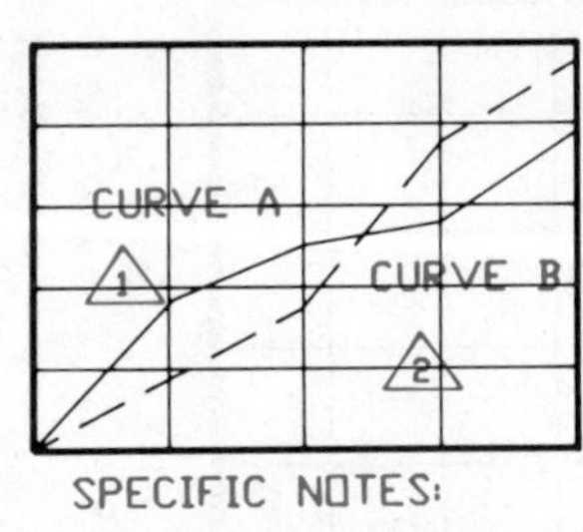

Figure 15–19 Subtitles and notes.

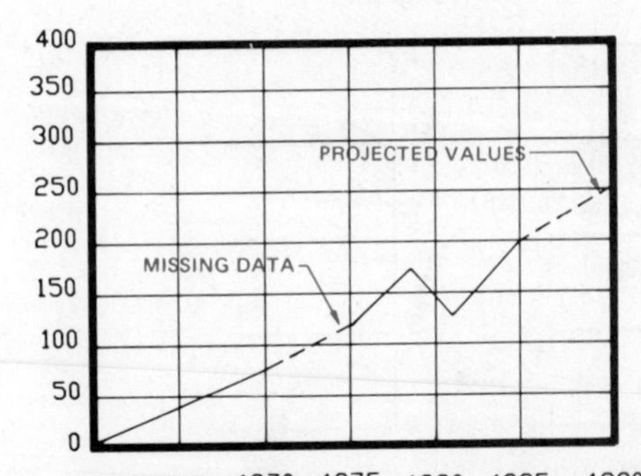

Figure 15–20 Missing and projected data.

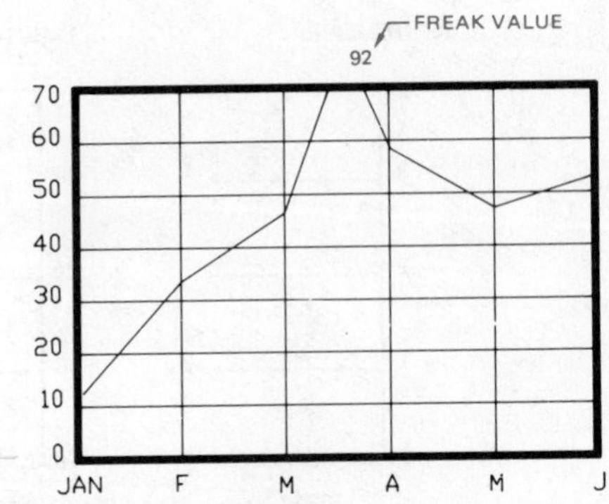

Figure 15–21 Technique for freak values. Avoid this method whenever possible.

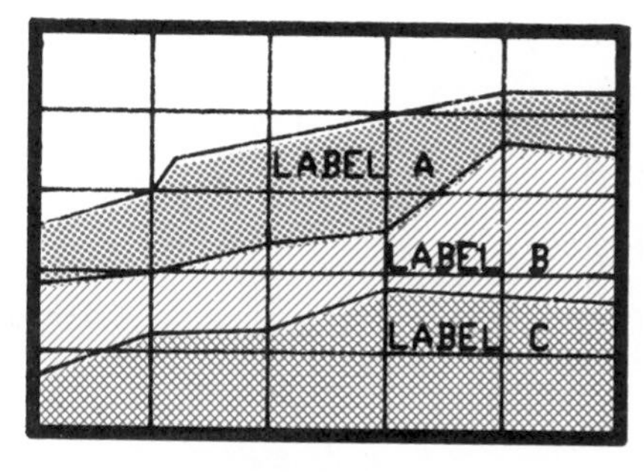

(a) LABELING THE VALUES ON THE CHART FACE

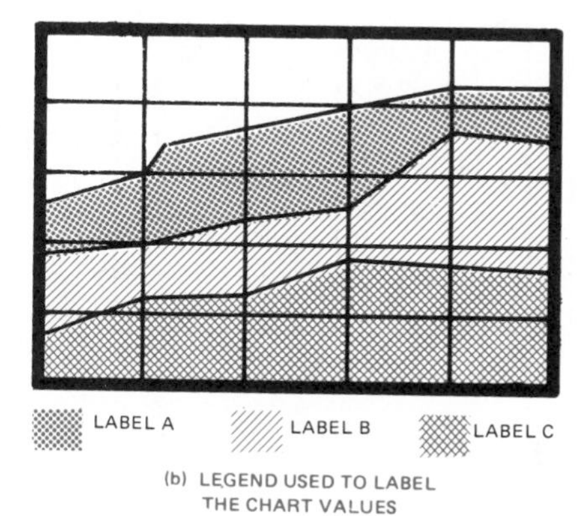

(b) LEGEND USED TO LABEL THE CHART VALUES

Figure 15–22 Surface chart design and layout techniques.

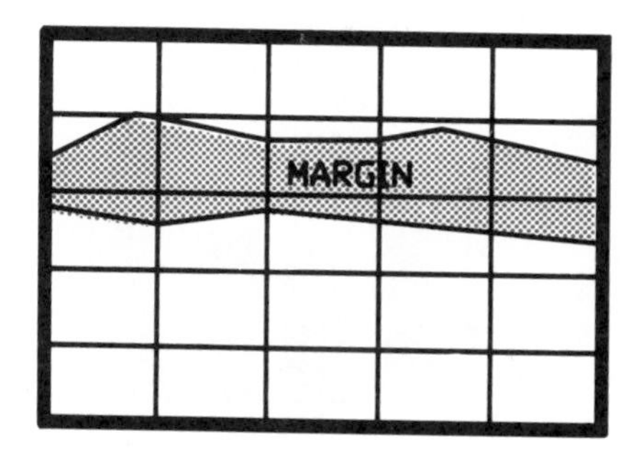

Figure 15–23 Emphasizing the difference between two curves.

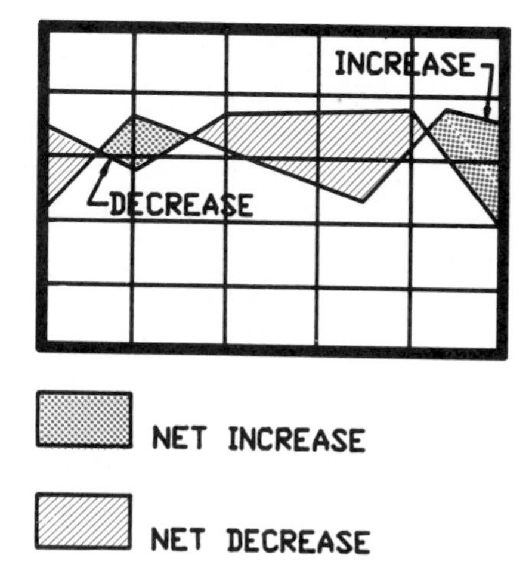

Figure 15–24 Emphasizing the net difference between two curves.

SURFACE CHARTS

Surface charts, also known as area charts, are designed to show values that are represented by the extent of a shaded area. The only difference between a surface chart and a line chart is that the area between the curve and the X axis or the area between curves is shaded for emphasis. The advantage of surface charts is that they define more clearly the difference between curves or the extent of a single curve. Surface charts should be avoided when accurate readings are required, or when greatly irregular layers exist. Surface charts are designed much like line charts, except that transfer films or computer graphics hatching techniques are used to create shading on surface charts. Some of the common design layout and labeling techniques used with surface charts are shown in Figure 15–22.

One of the best design considerations for surface charts exists when the intent of the chart is to show a margin between two curves. (See Figure 15–23.) Another design characteristic is the implementation of a surface chart that shows the net increase or the net decrease between two sets of values. (See Figure 15–24.)

COLUMN CHARTS

Column charts are commonly referred to as bar charts. A bar chart is used to represent numerical values by the height or length of columns. The data presented usually represent total periods of time or total percentages as opposed to various periods of time shown in line charts. Bar charts provide less detail than line charts but provide a more dramatic and easy-to-understand display for nontechnical readers.

The columns of a bar chart may be placed either horizontally or vertically. The vertical application is most common. The design of the bar chart vertical scale is similar to the technique used for line charts. For vertical column charts the horizontal grid is determined by the range of data and vertical grids are replaced by the columns. The design of the columns should assist and not distract from reading the chart. Chart columns should not be too bold or too thin, as shown in Figure 15–25. The column chart shown in Figure 15–25(c) is used a lot in quality control and is referred to as a histogram. The vertical scale is the frequency of occurrence and the horizontal scale is the value measured.

Subdivided column charts are used to show the values to individual components within total quantities of column data. The component values in each column are shaded to stand out as different quantities. The column subdivisions may be labeled on the face of the chart within the designated areas or with leaders pointing to

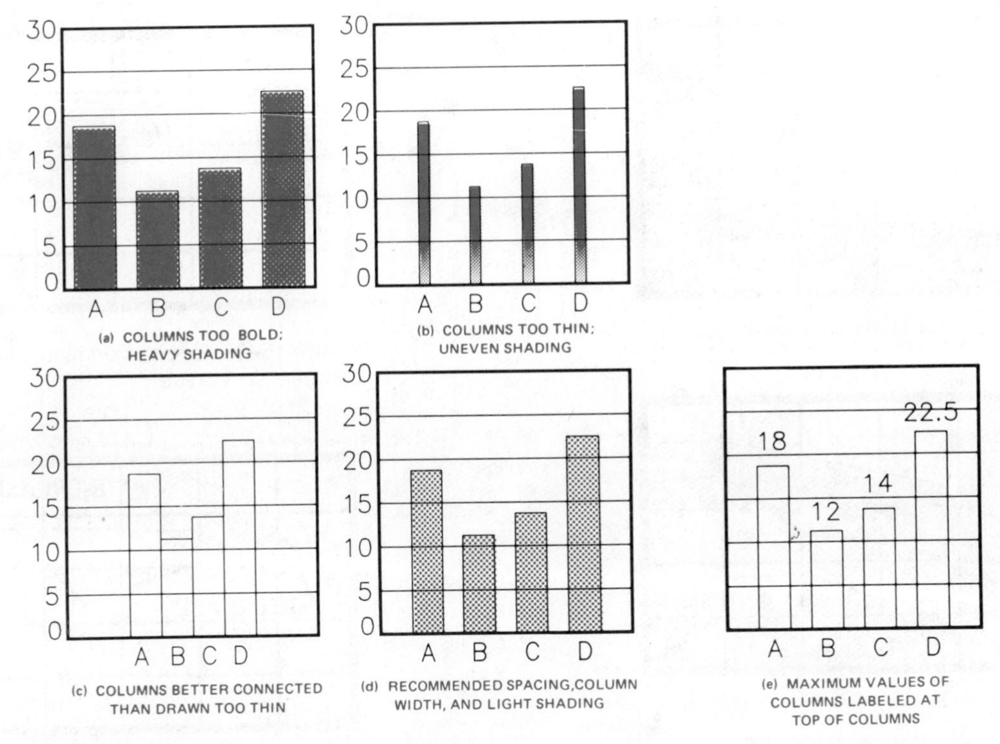

Figure 15–25 Column chart techniques.

the area, or a legend may be used that differentiates the components. Caution should be exercised when using these charts as they may be deceiving, and hard to read and interpret. (See Figure 15–26.)

Grouped column charts are used to compare two quantities at different intervals. Two columns are attached to show the comparison at the given periods, as shown in Figure 15–27.

One-hundred-percent column charts are used to show the relationship between the distribution of components where the totals are always equal values, such as 100 percent of any given unit. (See Figure 15–28.) These charts are effective in some applications but may be hard to read. It is usually better to separate the data.

Over-under column charts are used to emphasize the difference between two variables or the difference

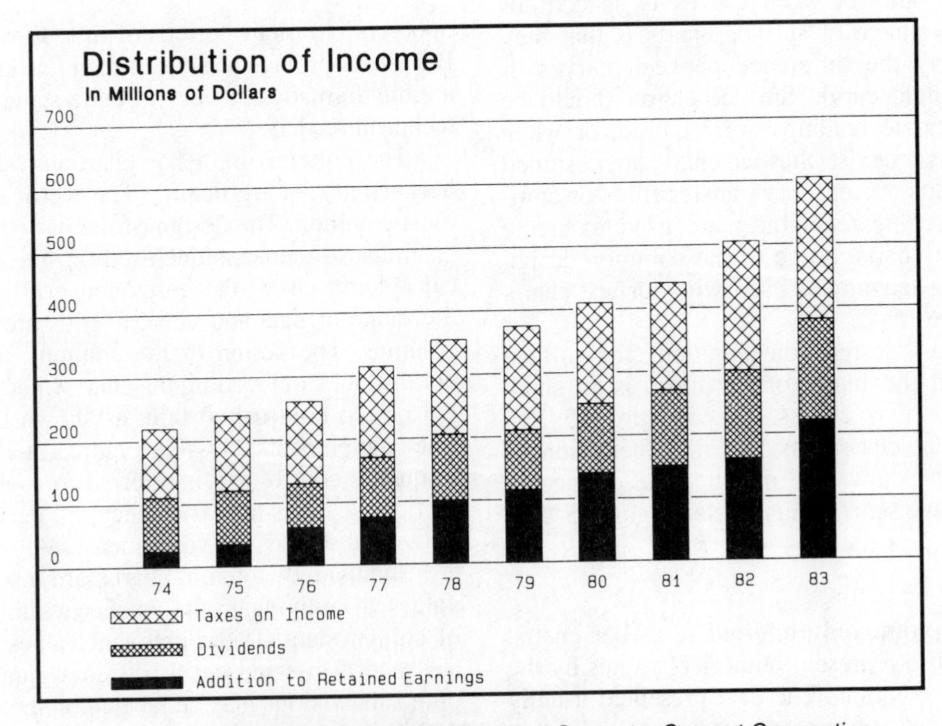

Figure 15–26 Subdivided column chart. *Courtesy Computer Support Corporation.*

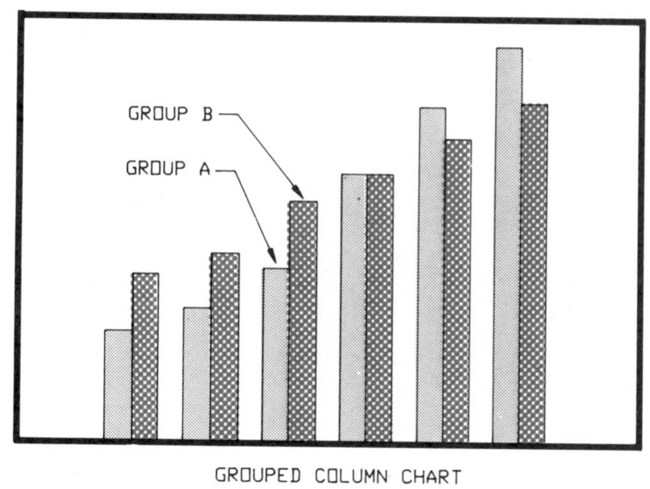

Figure 15–27 Grouped column chart.

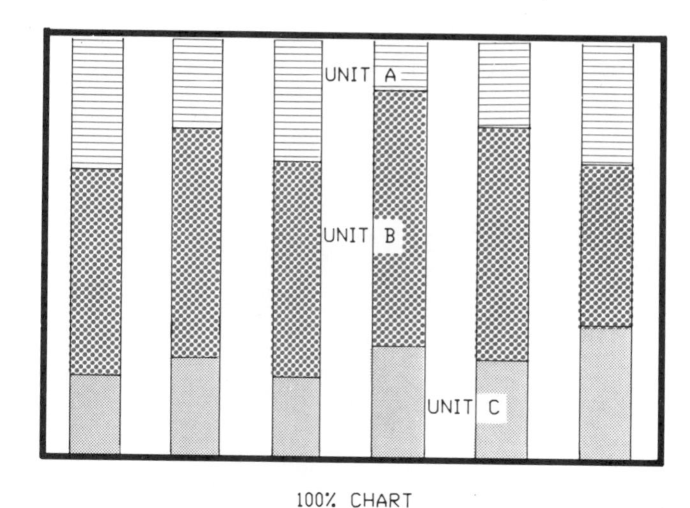

Figure 15–28 One hundred percent chart.

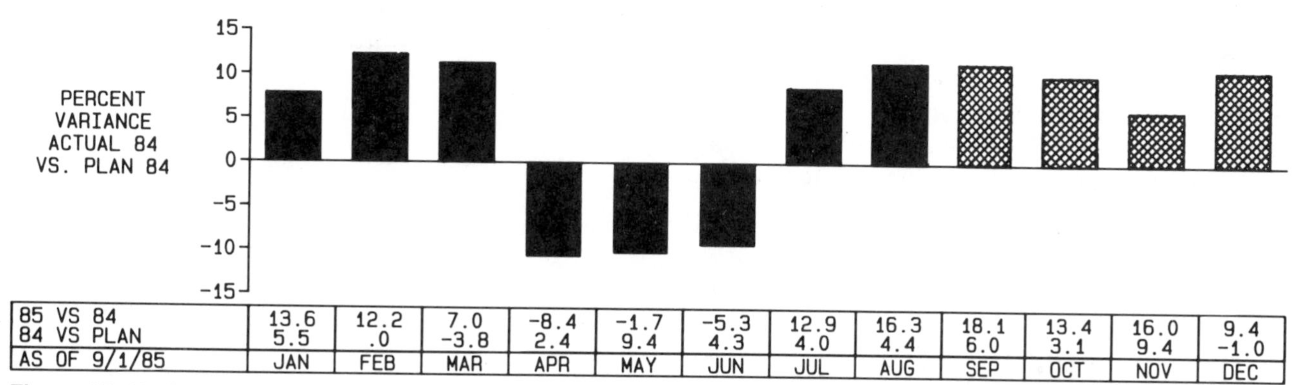

85 VS 84 84 VS PLAN	13.6 5.5	12.2 .0	7.0 -3.8	-8.4 2.4	-1.7 9.4	-5.3 4.3	12.9 4.0	16.3 4.4	18.1 6.0	13.4 3.1	16.0 9.4	9.4 -1.0
AS OF 9/1/85	JAN	FEB	MAR	APR	MAY	JUN	JUL	AUG	SEP	OCT	NOV	DEC

Figure 15–29 Over-under column chart. *Courtesy Computer Support Corporation.*

between a variable and a given standard. The over-under column chart has columns that originate at a common value or standard line, as shown in Figure 15–29.

The previously discussed bar charts have been index charts, where each of the columns originates at a common base. Another type of bar chart is the range bar, where individual columns represent segments of the total applicable range. Range column charts are used to emphasize the difference between values where one value is always higher or larger than the other. Many applications of range bar charts also show the average value for each column. (See Figure 15–30.) These charts show the central tendency (average) of data and the variability (range) of the data. Range bar charts are also used with median values, rather than average values, marked, and with control limits the same as the chart in Figure 15–15.

AVERAGE

Figure 15–30 Range column chart.

PIE CHARTS

Pie charts are used for presentation purposes and are one of the most popular methods of making graphic representations of data for people to understand. It is easy to understand how the portions of a pie represent quantity. Even without reference to the values of the pieces, readers can quickly determine the relationship between large and small portions. Pie charts are also referred to as 100-percent circle charts because the entire pie or circle represents 100 percent of the total, and each piece of pie relates to a percentage of the total. For example, a pie

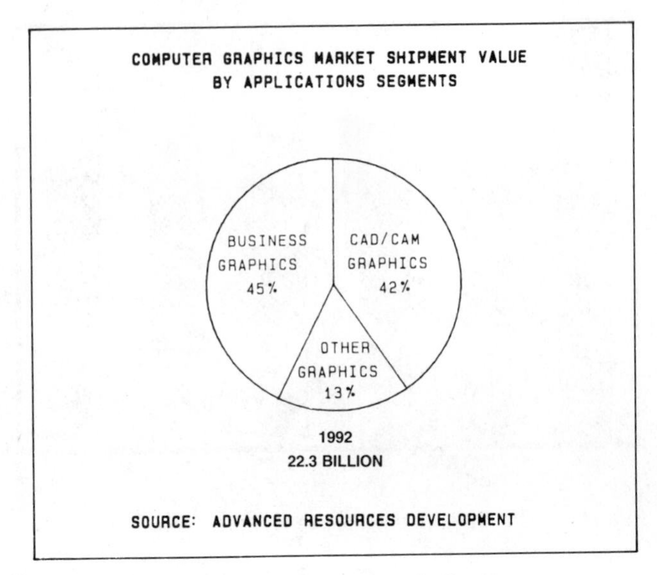

Figure 15–31 Comparative pie charts. *Courtesy MICROGRAFX.*

chart may represent 100 percent of 1992 computer graphics market shipment, where the percentage of each segment's shipment is shown as a piece of the pie. (See Figure 15–31.)

POLAR CHARTS

Polar charts are designed by establishing polar coordinate scales. Polar coordinates are points determined by an angle and distance from a center or pole. There are two scales on the polar chart. The first scale is related to the degrees or radians of a circle, and these radiate from the pole. The second scale represents distances from the center. Each distance is represented by a concentric circle. (See Figure 15–32.) Polar coordinate charts are used to display the effects of data that radiate from a source. The applications may include the range of intensity of a light source or the effective range of communication signals.

NOMOGRAPHS

Nomographs are graphic representations of the relationship between two or more variables of a mathematical equation. Nomographs are used as a quick graphic reference where unknown values of a given equation may be determined by aligning given numerals on a series of scales. There are two general types of nomographs. The most elementary is a concurrency chart, also known as a rectangular Cartesian coordinate chart. The other type of nomograph is called an alignment chart. Nomographs are generally designed only when the chart will get a great deal of use; otherwise the mathematical solution of unknown values would be the easiest and fastest approach. Nomographs, in general, are difficult and time-consuming to design. A nomograph is shown in Figure 15–33.

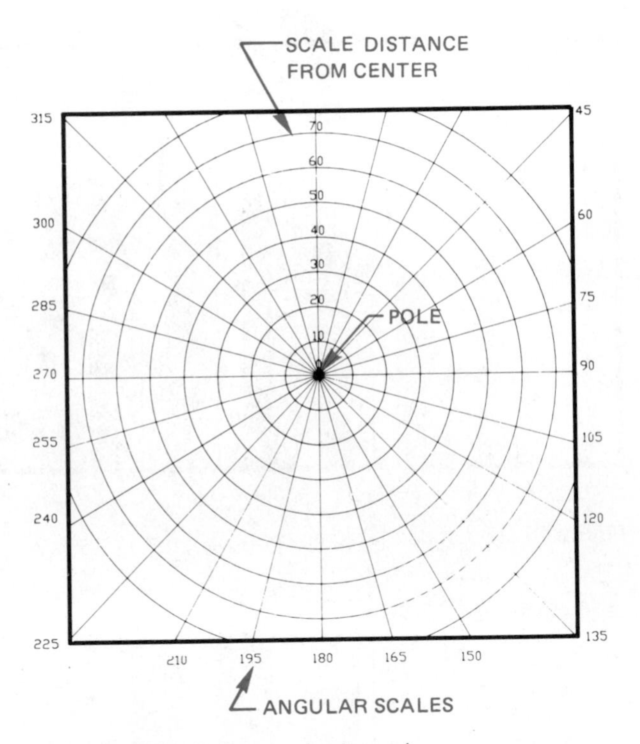

Figure 15–32 Polar coordinate scales.

Nomogram
Horsepower to Torque
Conversion Table

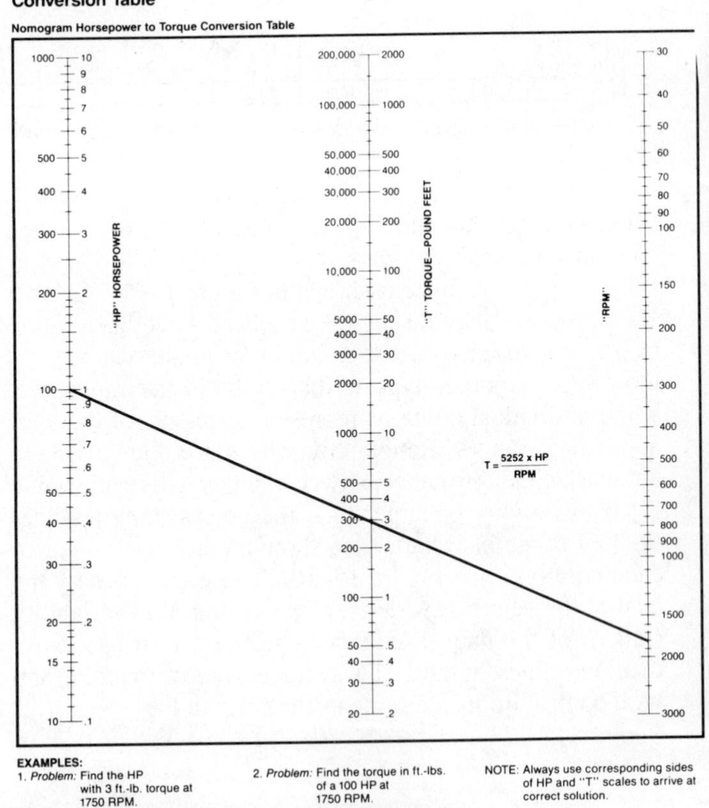

Figure 15–33 Nomograph. *Courtesy Lovejoy, Inc., Downers Grove, Illinois.*

GRAPHICS

The examples displayed in this chapter have been designed and plotted or printed by computer. CADD software is available that transforms numerical data into presentation-quality graphics using any desired format — for example, line, bar, pie, or pictorial charts. Charts may be designed with any grid, scales, line type, shading technique, or text size and style. Programs are available that allow the designer to use a variety of predetermined symbols for two- or three-dimensional applications. Using a CADD system of this type to design charts is the same as having a portfolio of artwork that can be automatically used in any desired configuration for unlimited creativity. With CADD, scales may be automatically reduced, enlarged, or otherwise changed to accommodate any format. (See Figure 15–34.)

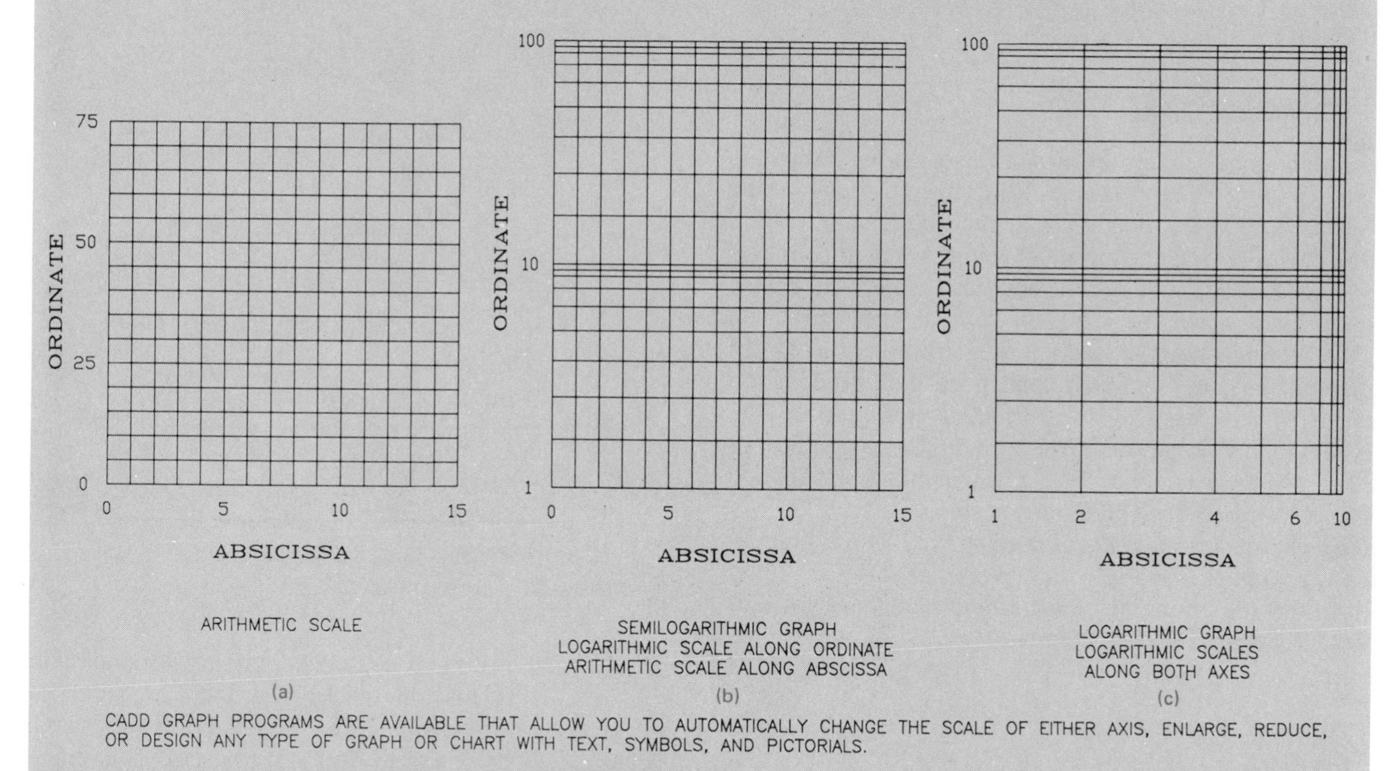

Figure 15–34 Computer-generated charts automatically change the scales by changing the ordinate and abscissa requirements: (a) arithmetic scale; (b) semilogarithmic scale; (c) logarithmic scale.

Concurrency Charts

The concurrency chart uses the X and Y axis, as previously described in the discussion on rectilinear charts, to graphically solve values for given mathematical equations. For example, given the equation a + b = c, a series of curves are established on a rectilinear chart where each line represents one of the equation values. (See Figure 15–35.) The chart is used to solve for an unknown value by selecting any two known values. Use Figure 15–35 to solve the formula a + b = c when a = 5 and b = 15. Find 5 on the X axis. Then project directly up to the curve where b = 15. Project to the left and locate the c value at the Y axis. The answer should be 20. This is an extremely simplified version of the concurrency chart; however, the technique is the same for more complex versions.

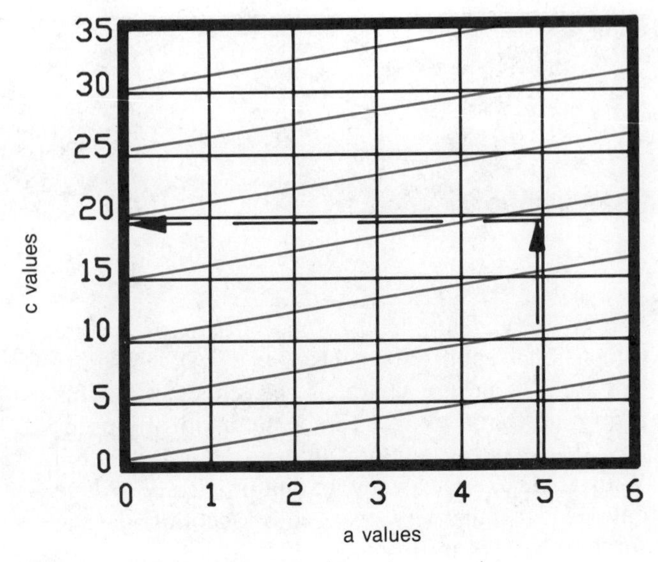

Figure 15–35 Concurrency chart where a + b = c.

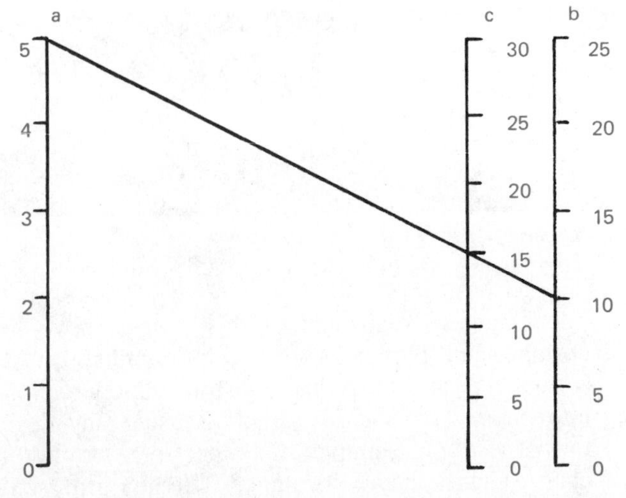

Figure 15–36 Alignment charts where a + b = c.

Alignment Charts

Alignment charts are designed to graphically solve mathematical equation values using three or more scaled lines. The known values are aligned on a combination of scales, and the solution is found when the line is extended or aligned with the additional scale. The alignment chart scales are typically vertical lines; however, other combinations of scale arrangements are used. To demonstrate the alignment chart principle, the equation a + b = c is used again, as shown in Figure 15–36. Notice that if a line is drawn between the 5 on the "a" scale and the 10 on the "b" scale, the unknown value on the "c" scale is 15. Therefore 5 + 10 = 15.

Scales may have equal graduations. These graduations are called *uniform scales*. When the scales resulting from the solution of the equation variables are not uniform, they are called functional scales.

TRILINEAR CHARTS

Trilinear means consisting of three lines. A trilinear chart is designed in the shape of an equilateral triangle. These charts are used to show the interrelationship among three variables on a two-dimensional diagram. The amounts of the three variables, usually presented as percentages, can be represented as shown in Figure 15–37. The corners of the triangle labeled A, B, and C correspond to 100-percent values of A, B, and C. The side of the triangle opposite the corner labeled A represents the absence of A values. Therefore the horizontal lines across the triangle show increasing percentages of A from 0 percent at the base to 100 percent at the A vertex. In a similar interpretation the percentages of B and C, respectively, are given by the distances opposite the two sides of the vertices at either B or C. From the three scales of the triangle the value corresponding to any point can be read, and the total of the three values at any given point is always 100 percent. This is possible because of the geometric result that the sum of the three perpendicular distances from any point to the three sides of the triangle is equal to the height of the triangle.

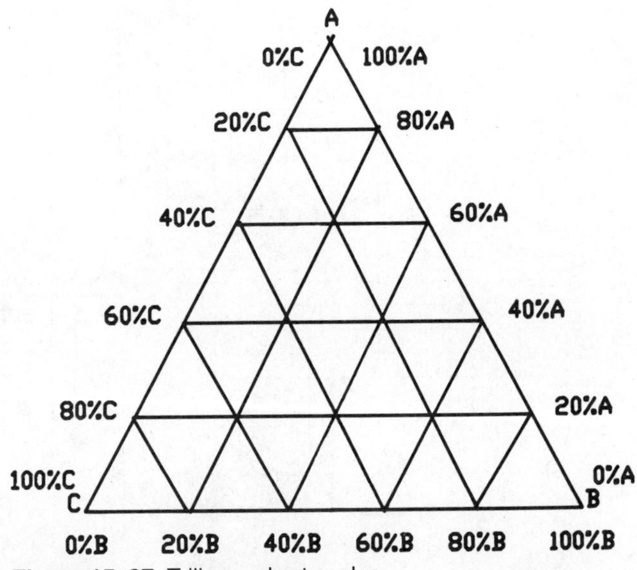

Figure 15–37 Trilinear chart scales.

FLOWCHARTS

Flowcharts are used to show organizational structure, steps in a series of operations, the flow of something through a system, or the progression of materials through manufacturing processes. Organizational charts show the relationship of the different levels of an organization, as shown in Figure 15–38.

Production control or process charts are used for planning and coordinating a series of activities, such as a marketing planning cycle as shown in Figure 15–39.

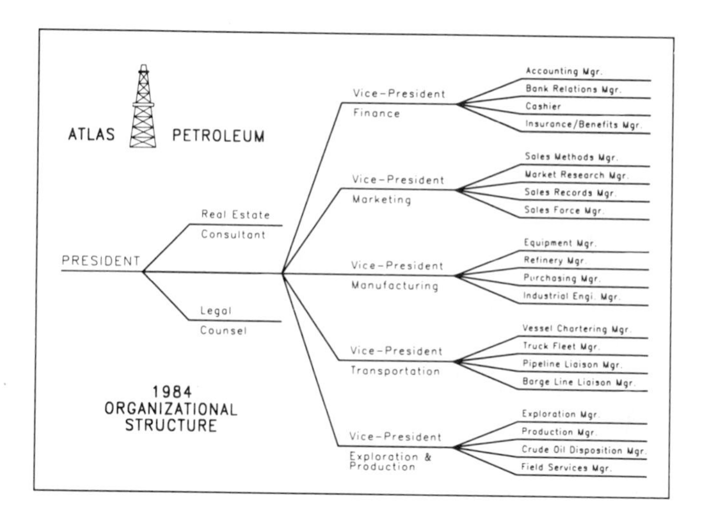

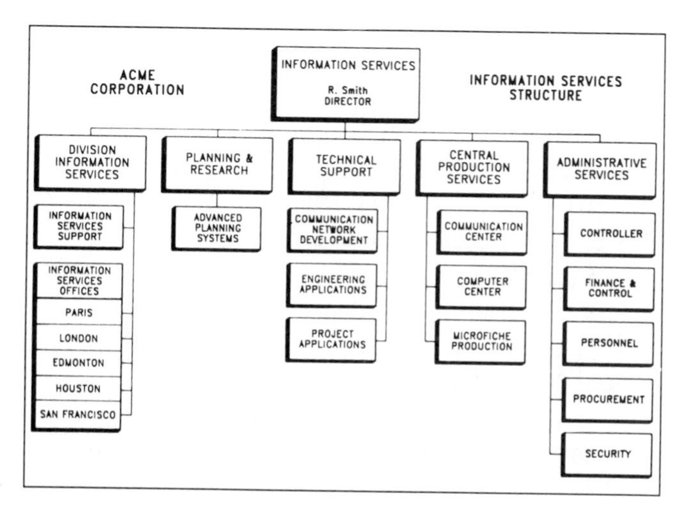

Figure 15–38 Organizational charts. *Courtesy Computer Support Corporation.*

DISTRIBUTION CHARTS

Distribution charts are used to display data based on geographical region. The data may be defined by outlining geographic areas. These charts are often in the form of maps to provide locational information. For example, the average frost depth in the United States is shown in Figure 15–40(a). Another use of the distribution chart is placing dots or other symbols to define specific locations as shown in Figure 15–40(b).

PICTORIAL CHARTS

Any chart type may be designed in a pictorial manner. Pictorial charts are used for advertising promotions where the pictorial representation is more important than the data presented. Pictorial charts are used to enhance the meaning of the data using two- or three-dimensional graphics, photography, or color. Any pictorial representation may be used to help the reader visualize the intent of the chart. Pictorial bar charts may be drawn with the

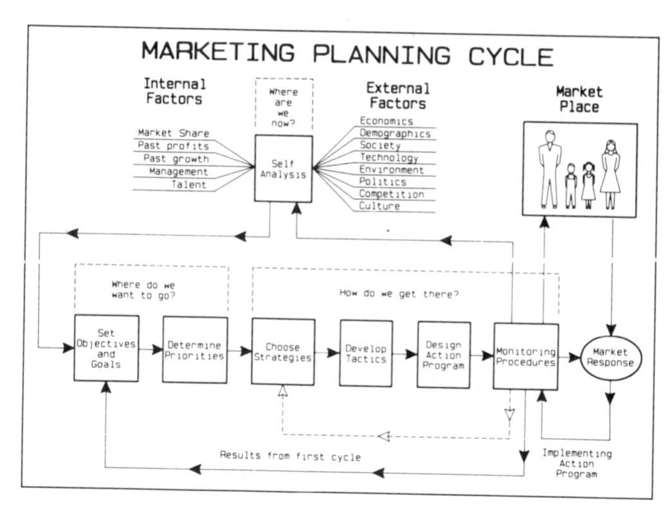

Figure 15–39 Production control or process chart. *Courtesy Computer Support Corporation.*

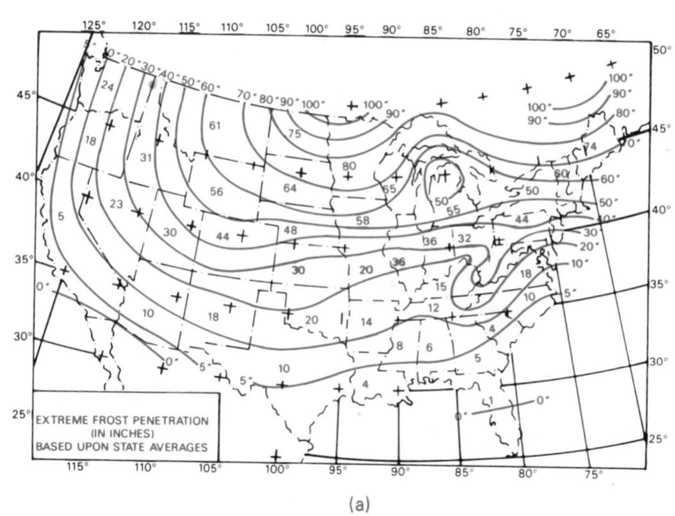

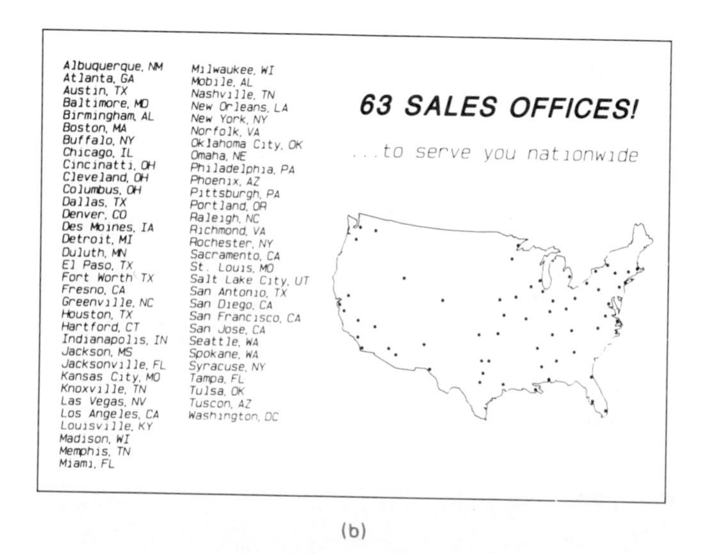

Figure 15–40 Distribution charts: (a) Average frost depth in the United States; (b) national sales office. *Courtesy Computer Support Corporation.*

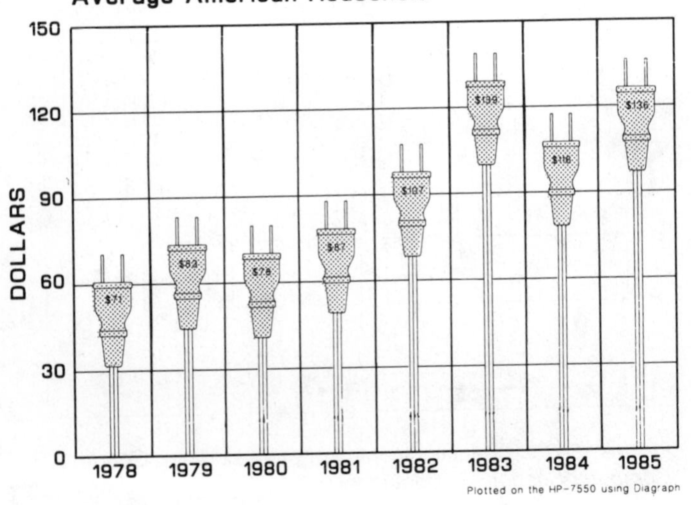

Figure 15–41 Pictorial bar chart. *Courtesy Computer Support Corporation.*

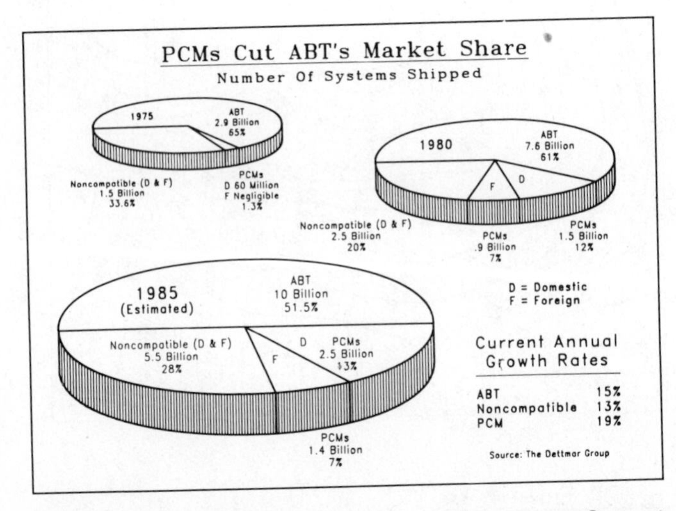

Figure 15–42 Pictorial pie chart. *Courtesy Computer Support Corporation.*

bars shown three-dimensionally or with pictorial representations of the product, as shown in Figure 15–41. Pie charts may be easily shown pictorially by drawing the pie isometrically, as shown in Figure 15–42. The designer may also prepare creative pictorial line charts, as shown in Figure 15–43. The chart design, format, and layout may serve any function using a variety of symbols to pictorially depict recognizable features, as shown in Figure 15–44 (a) and (b).

PROFESSIONAL PERSPECTIVE

Following are some professional tips that can make your charts and graphs accurate for the intended audience.

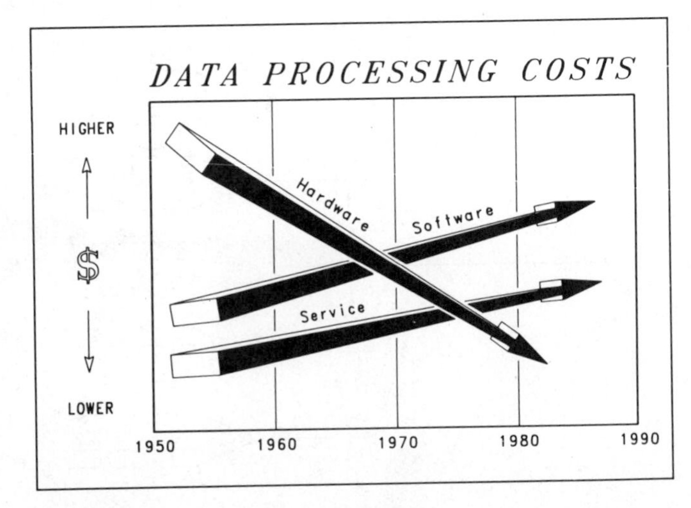

Figure 15–43 Pictorial line chart. *Courtesy Computer Support Corporation.*

- Make your graphs and charts clean and simple, avoiding unnecessary detail and information.
- Do not misrepresent the facts.
- Keep all sets of data proportional.
- Avoid distorting the data.
- Study the background of the data to insure you understand the intended representation.
- Know the audience and know what they want to have the graph or chart represent. If you display one set of information and they want something else, then your work is useless.
- Be sure the numbers you are representing are accurate and appropriate.

You should consider taking some classes in statistics, or have someone who understands statistics available to evaluate your designs to insure the data is represented accurately. If you are unprepared or if the data are misrepresented, someone will question the quality of the presentation.

ENGINEERING PROBLEM

Given the following statistical data, design a rectilinear chart that represents the information.

Projected population of the United States from 1985 to 2165. The source is World Bank data, projected assuming constant fertility and migration. The year is followed by the population in millions: 1985 — 238, 1995 — 255, 2005 — 269, 2015 — 282, 2025 — 290, 2035 — 291, 2045 — 287, 2055 — 283, 2065 — 279, 2075 — 273, 2085 — 269, 2095 — 264, 2105 — 259, 2115 — 255, 2125 — 251, 2135 — 247, 2145 — 243, 2155 — 240, 2165 — 239.

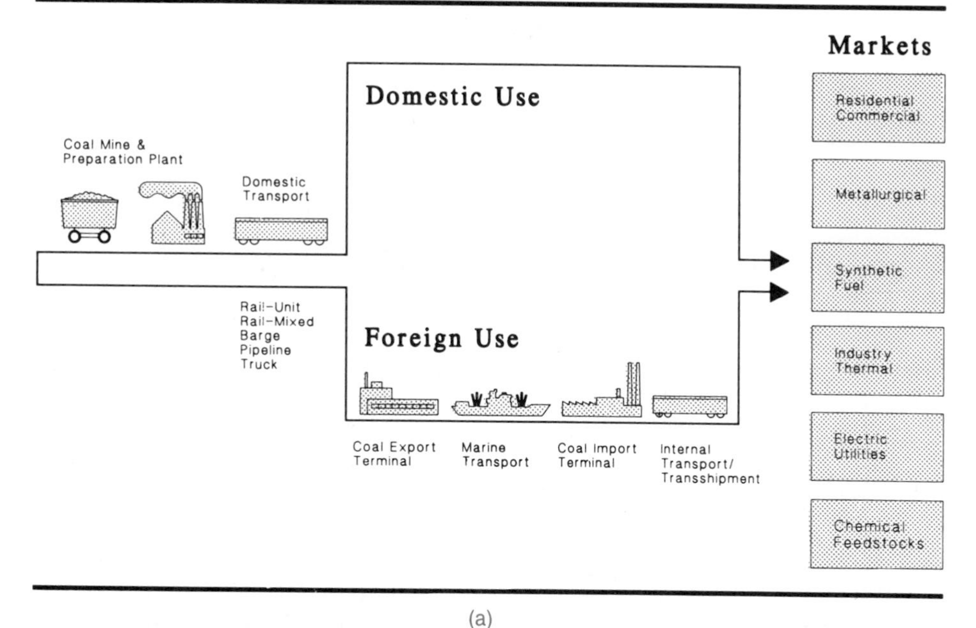

(a)

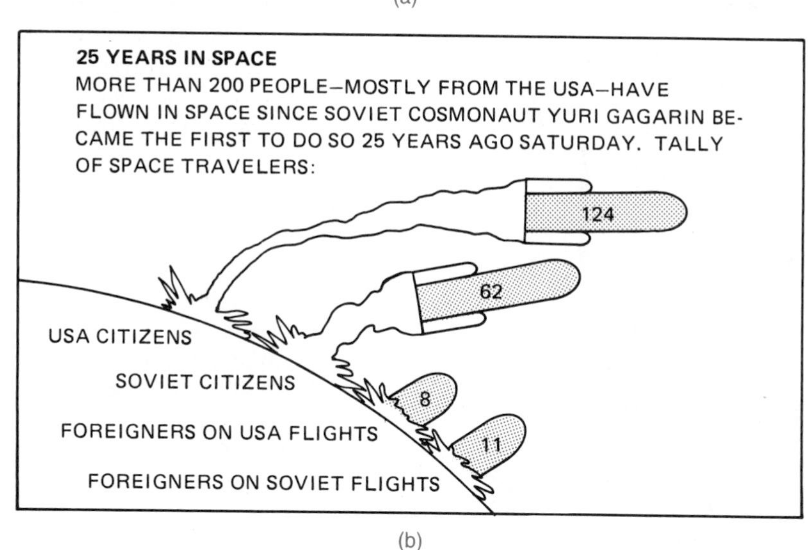

(b)

Figure 15–44 Pictorial flowcharts using symbols. (a) *Courtesy Computer Support Corporation*; (b) *Courtesy* USA Today.

Step 1 Establish the range of units. In this case the lowest unit in years is 1985 years and the highest is 2165 years. The lowest unit in population is 238 million and the highest is 291 million.

Step 2 Determine the vertical scale to accommodate the range. For example, 230 to 300 million as shown in Figure 15–45.

Step 3 Determine the horizontal scale. In this case there is a set of data every 10 years. The horizontal scale will be divided into units representing 10-year modules. It is best to make the horizontal scale approximately equal to the vertical units without crowding. In this case there are more horizontal than vertical units, so the 10-year divisions will be closer than preferred; however, the 30-year divisions will be the main units as shown in Figure 15–46.

Step 4 Plot the points and draw the curve as shown in Figure 15–47.

Step 5 Complete the chart by labeling the title and any required captions. (See Figure 15–48.)

Figure 15–45 Determine the range for the vertical scale units — step 2.

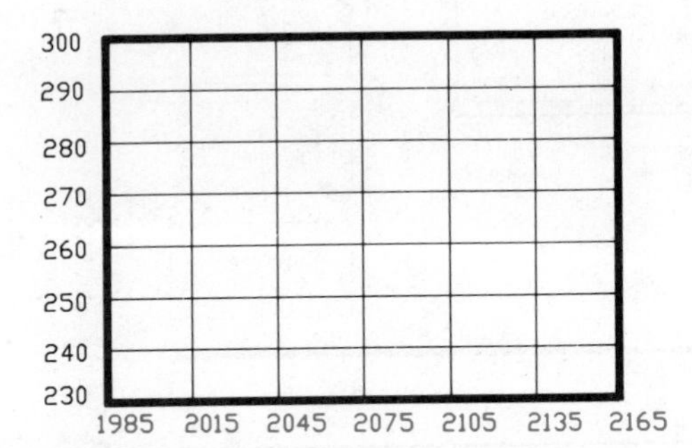

Figure 15–46 Determine the horizontal scale and establish the grid — step 3.

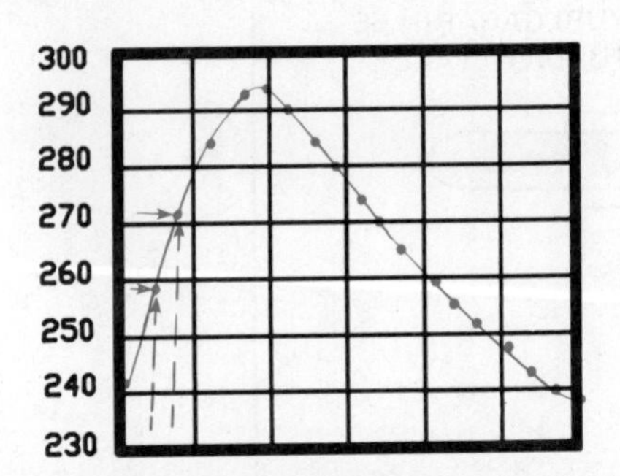

Figure 15–47 Plot points from given data and draw the curve. Note that points are exaggerated for emphasis. The points should be plotted very lightly — step 4.

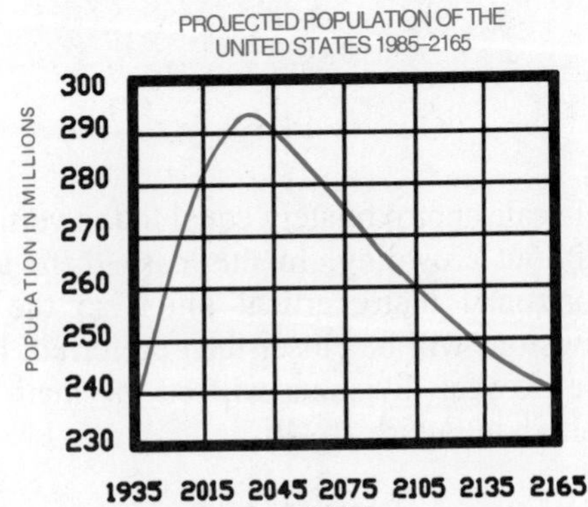

Figure 15–48 Complete the chart by labeling the title and required captions — step 5.

PIE CHART DESIGN

Given the following information, design a pie chart that graphically displays the data:

1994 agricultural sales for Washington County:
Nursery crops = 44%
Vegetable crops = 26%
Fruits and berries = 11%
Grain and hay = 8%
Greenhouse crops = 6%
Miscellaneous crops = 5%

Step 1 Using 8½" × 11" paper, draw a circle with a diameter of approximately 4 in. The diameter of the circle depends on the sheet size and the information to be included. Individual judgment must be used.

Step 2 Convert each category into degrees. There are 360° in the circle, representing 100% of the values. Use the formula 3.6 × % = degrees.
Nursery crops = 3.6 × 44 = 158.4°.
Vegetable crops = 3.6 × 26 = 93.6°.
Fruits and berries = 3.6 × 11 = 39.6°.
Grain and hay = 3.6 × 8 = 28.8°.
Greenhouse crops = 3.6 × 6 = 21.6°.
Miscellaneous crops = 3.6 × 5 = 18°.

Step 3 Using the degrees calculated in Step 2, lay out each piece of pie on the circle, as shown in Figure 15–49. Notice that at least one of the small pie parts is separated from the group of small parts for ease in labeling later.

Step 4 Add the title, notes, and captions, and do any shading to suit the design. (See Figure 15–50.)

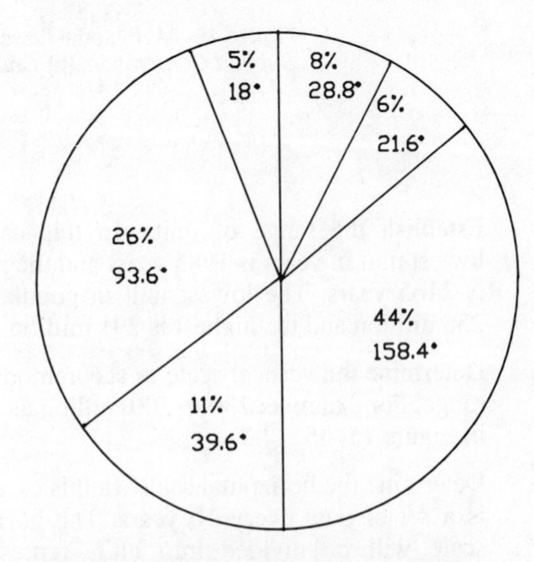

Figure 15–49 Pie chart design and layout.

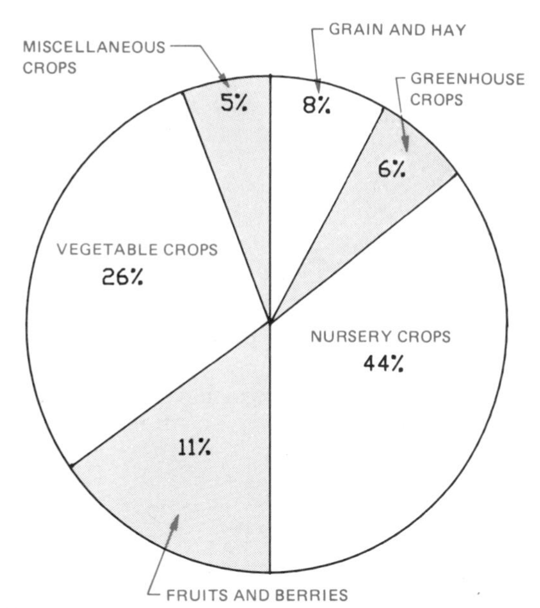

Figure 15–50 Pie chart design shading and labeling.

ENGINEERING CHARTS AND GRAPHS PROBLEMS

DIRECTIONS

Given the data for the following chart problems, design and draw the charts using a CADD system or manual drafting tools.

Problem 15–1 Draw the following line chart. Set up the horizontal scale for the nine years ending last year, and project this year and next year. *Courtesy Computer Support Corporation.*

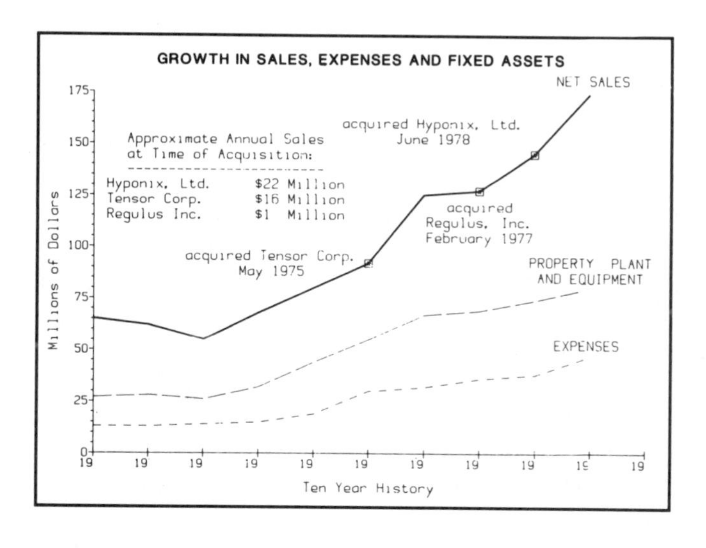

Problem 15–2 Redesign similar to the format shown in Figure 15–26, and draw the following column chart, with modifications projecting to the year 2000. *Courtesy Computer Support Corporation.*

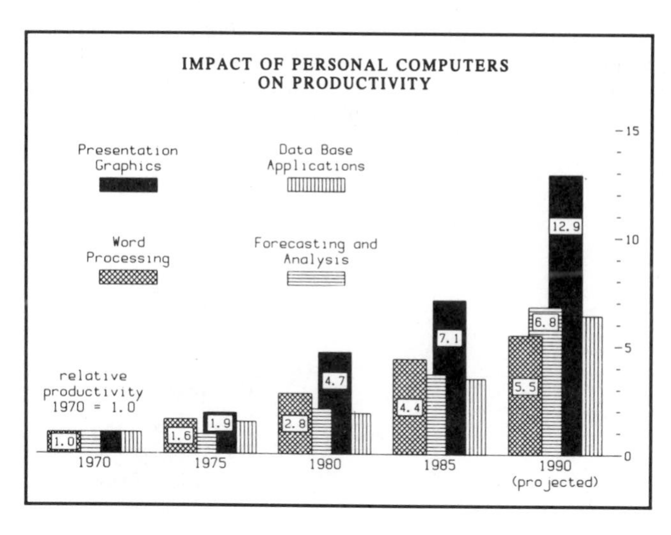

Problem 15–3 Design a rectilinear chart that represents the increase in recommended current amperage with the increase in round solid copper wire gage size.

AGW gage size.	Recommended maximum current in amperes.
28	.4
27	
26	.7
25	
24	1.1
23	
22	1.8
21	
20	3
19	
18	4.5
17	
16	7
15	
14	12
13	
12	18
11	
10	30

Problem 15–4 Design a comparative rectilinear chart showing the five-speed transmission diagram for the Porsche 911 Carrera (establish straight line curve average for each gear).

	SPEED (MPH)				
ENGINE SPEED (RPM)	1ST GEAR	2ND	3RD	4TH	5TH
1000	10	17	25	30	40
2000	20	34	48	62	78
3000	38	50	74	92	116
4000	38	67	97	122	155
5000	47	84	121	152	192
6000	56	100	145	182	230
7000	66	105	168	212	268

Tire size: 185/70-VR15-215/60 VR15

NOTE: This diagram shows guiding values, based on a medium effective rolling radius. Slight deviations due to tire tolerance, variations in the rolling radius, wear, and skidding on the wheels have not been taken into account.

Problem 15–5 Design a chart similar to Figure 15–15 showing feature dimensions of a part obtained at inspection intervals based on the following information:

1. Average of samples taken at given intervals ($\overline{x}$): 2.6248, 2.6252, 2.6250, 2.6250, 2.6244, 2.6248, 2.6251.
2. Upper Control Limit (UCL) = 2.6253.
3. Lower Control Limit (LCL) = 2.6243.
4. Determine the average of averages ($\overline{\overline{x}}$).

Problem 15–6 Design a surface semilogarithmic chart from the following information. Ten years of community college graduates with technical associate of science (AS) degrees vs. associate of arts (AA) degrees.

YEAR	AS	AA
1983–84	266	37
1984–85	214	24
1985–86	225	26
1986–87	225	34
1987–88	269	32
1988–89	265	33
1989–90	232	24
1990–91	233	14
1991–92	242	18
1992–93	263	30
1993–94	270	36

Problem 15–7 Design a column chart based on the following information: Where top companies reside; more than half of the USA's Fortune 500 companies are headquartered in six states. *Information courtesy* USA Today.

STATE	NUMBER OF FORTUNE 500 HEADQUARTERS
NY	75
IL	49
CA	38
OH	37
CT	35
PA	34

Problem 15–8 Design a column chart that shows net sales with values on the left vertical scale and a step curve to show income per share with values on the right vertical scale. The horizontal scale shows the range of years from 1985 to 1994 for Dial Industries, 10-year summary of operations and financial review.

YEAR	1985	1986	1987	1988	1989	1990	1991	1992	1993	1994
Net Sales in Millions	1400	1500	1600	1825	2150	2275	2500	2900	3225	3475
Income/Share $	1.39	1.56	1.85	1.87	2.06	2.00	2.36	2.61	1.55	2.62

Problem 15–9 Design an expense/revenue over-under column chart for Precision Manufacturing, Ltd. by country ($ in millions).

	EXPENSE	REVENUE
United States	343.2	651.7
England	264.8	466.9
Canada	301.6	485.2
Mexico	241.8	413.3
Australia	213.7	339.9
France	239.4	302.7
Taiwan	201.6	224.3
Hong Kong	214.7	102.5
Brazil	225.0	88.2

Problem 15–10 Design a range bar chart that represents values for surface roughness in microinches and micrometers produced by common processing methods.

	RANGE		AVERAGE	
PROCESS	MICROINCH	MICROMETER	MICROINCH	MICROMETER
Flame cutting	2000–250	50–6.3	1000–500	25–12.5
Snagging	2000–125	50–3.2	1000–250	25–6.3
Sawing	2000–32	50–0.80	1000–63	25–1.6
Planing,shaping	1000–16	25–0.40	500–63	12.5–1.6
Drilling, chemical milling, elect. discharge machining	500–32	12.5–0.80	250–63	6.3–1.6
Milling	1000–8	25–0.20	250–32	6.3–0.80
Broaching, reaming	250–16	6.3–0.40	125–32	3.2–0.80
Electron beam, laser	250–8	6.3–0.40	250–32	6.3–0.80
Electrochemical	500–2	12.5–0.05	125–8	3.2–0.20
Boring, turning	1000–1	25–0.025	250–16	6.3–0.40
Barrel finishing	125–2	3.2–0.05	32–8	0.80–0.20
Electrolytic grinding	32–4	0.80–0.10	20–8	0.60–0.20
Roller burnishing	32–4	0.80–0.10	16–8	0.40–0.20
Grinding	250–1	3.2–0.025	63–4	1.6–0.10
Honing	63–1	1.6–0.025	32–4	0.80–0.10
Electropolish	63–0.5	1.6–0.012	32–4	0.80–0.10
Polishing	32–0.5	0.80–0.012	16–4	0.40–0.10
Lapping	32–0.05	0.80–0.012	16–2	0.40–0.05
Superfinish	32–0.05	0.80–0.012	8–1	0.20–0.025

NOTE: The ranges indicated are typical of the process; higher or lower values may be obtained under special conditions.

Problem 15–11 Design a pie chart that represents the agricultural sales for Washington County during 1994.

Nursery crops = 44%
Vegetable crops = 26%
Small fruit and berries = 11%
Other crops = 8%
Miscellaneous animals = 6%
Greenhouse crops = 5%

Problem 15–12 Design a pie chart that represents Product Revenue by category for 1994. *Data courtesy Computer Support Corporation.*

Automated process equipment = 28%
Marine products = 20%
Energy services = 17%
Sports products = 13%
Electronic controls = 10%
Specialty products = 7%
Wheel goods = 5%

Problem 15–13 Design a nomograph that may be used to determine grades for the Engineering Drafting I class drafting projects and term tests based on the following assumptions. Total points for drafting projects equal 500, and total points for tests equal 300. Grades are based on the following percentages: A = 100–91%; B = 90–81%; C = 80–71%; D = 70–61%; F = below 61%.

Problem 15–14 Design a flowchart for the chain of command of your school or company.

Problem 15–15 Design a pictorial chart that provides the following information: 25 years in space from 1960 to 1985. More than 200 people — mostly from the USA — have flown in space since Soviet cosmonaut Yuri Gagarin became the first to do so. Tally of space travelers:

USA citizens = 124
Soviet citizens = 62
Foreigners on USA flights = 8
Foreigners on Soviet flights = 11

Source: Congressional Research Service.
See Figure 15–44(b) for an example.

Problem 15–16 Design and draw an alignment chart for the formula a + b = c.

Problem 15–17 Design and draw a distribution chart showing the cities in the United States with populations over 500,000.

Problem 15–18 Review your family electric bills over the past year and design and draw a pictorial chart similar to the one shown in Figure 15–41.

Problem 15–19 Research the monthly average rainfall in your state over the past year and design and draw a pictorial chart. Consider clouds with rain below, for example, as the bar design.

Appendices

APPENDIX A Abbreviations 552

APPENDIX B Tables 556

Table 1 Inches to Millimeters 556
Table 2 Millimeters to Inches 556
Table 3 Inch/Metric—Equivalents 557
Table 4 Inch/Metric—Conversion 558
Table 5 Rules Relative to the Circle 559
Table 6 Dimensioning Symbols 560
Table 7 ASTM and SAE Grade Markings for Steel Bolts and Screws 561
Table 8 Unified Standard Screw Thread Series 562
Table 9 Dimensions of Hex Cap Screws (Finished Hex Bolts) 563
Table 10 Dimensions of Hexagon and Spline Socket Head Cap Screws (1960 series) 564
Table 11 Dimensions of Hexagon and Spline Socket Flat Countersunk Head Cap Screws 565
Table 12 Dimensions of Hexagon and Spline Socket Set Screws 566
Table 13 Dimensions of Slotted Flat Countersunk Head Cap Screws 568
Table 14 Dimensions of Slotted Round Head Cap Screws 569
Table 15 Dimensions of Slotted Fillister Head Cap Screws 570
Table 16 Dimensions of Slotted Flat Countersunk Head Machine Screws 571
Table 17 Dimensions of Hex Nuts and Hex Jam Nuts 572
Table 18 Woodruff Key Dimensions 573
Table 19 Woodruff Keyseat Dimensions 576
Table 20 Key Size versus Shaft Diameter 578
Table 21 Key Dimensions and Tolerances 579
Table 22 Gib Head Nominal Dimensions 580
Table 23 Class 2 Fit for Parallel and Taper Keys 581
Table 24 Decimal Equivalents and Tap Drill Sizes 582

APPENDIX C American National Standards of Interest to Designers, Architects, and Drafters 583

APPENDIX D Unified Screw Thread Variations 584

APPENDIX E Metric Screw Thread Variations 586

APPENDIX F Geometric Tolerancing Symbol Trees 587

APPENDIX A Abbreviations

The following are standardized abbreviations commonly used on working drawings.

Accessory ... ACCESS.
Accumulate ... ACCUM
Adaption ... ADAPT.
Addendum ... ADD.
Addition ... ADD.
Alteration ... ALT
Alternate ... ALT
Alternating current ... AC
Altitude ... ALT
Aluminum ... AL
American Iron and Steel Institute ... AISI
American Society for Testing Materials ... ASTM
American Society of Mechanical Engineers ... ASME
American Wire Gage ... AWG
Ampere ... AMP
Approved ... APPD
Approximate ... APPROX
Assembly ... ASM
Attach ... ATT
Authorize ... AUTH
Auxiliary ... AUX
Average ... AVG

Balance ... BAL
Battery ... BAT
Bearing ... BRG
Bill of material ... B/M
Bracket ... BRKT
Brass ... BRS
British thermal unit ... BTU
Bronze ... BRZ
Brown and Sharpe ... B&S
Bushing ... BUSH.

Cadmium ... CAD.
Capacity ... CAP.
Carbon steel ... CS
Casting ... CST
Cast iron ... CI
Cast steel ... CS
Center ... CTR
Centerline ... (℄) or CL
Center of gravity ... CG
Centigrade ... C
Centimeter ... CM
Chamfer ... CHAM
Change ... CHG
Check ... CHK
Chief ... CH
Chromium ... CHR
Circular pitch ... CP
Circumference ... CIRC
Clockwise ... CW
Cold-drawn steel ... CDS
Cold-rolled steel ... CRS
Company ... CO
Composition ... COMP
Compression, Compressor ... COMP
Concentric ... CONC
Condition ... COND
Conductor ... COND
Conduit ... CDT
Connector ... CONN
Continue ... CONT
Control ... CONT
Copper ... CU
Corporation ... CORP
Correspond ... CORRES
Corrosion resistant steel ... CRES
Counterbalance ... CBAL
Counterbore ... CBORE
Counterclockwise ... CCW
Counterdrill ... CDRILL
Countersink ... CSK
Counterweight ... CTWT
Cubic ... CU
Cubic centimeter ... CC
Cubic feet per minute ... CFM
Cubic foot ... CU FT
Cubic inch ... CU IN.
Cycle ... CY

Dedendum ... DED
Degree ... (°) or DEG
Department ... DEPT
Detail ... DET
Develop ... DEV
Deviation ... DEV
Diagonal ... DIAG
Diameter ... DIA, ∅
Diametral pitch ... DP
Dimension ... DIM.
Direct current ... DC
Distance ... DIST
Division ... DIV
Drawing ... DWG

Each ... EA
Eccentric ... ECC
Effective ... EFF
Electric ... ELEC
Elevation ... ELEV

Engineer .... ENGR
Engineering .... ENGRG
Equipment .... EQUIP.
Equivalent .... EQUIV
Estimate .... EST
Etcetera .... ETC
Extension .... EXT
External .... EXT
Extrusion .... EXTR

Fahrenheit .... F
Feet per minute .... FPM
Figure .... FIG.
Fillister head .... FIL HD
Finish .... FIN.
Fitting .... FTG
Flat .... FL
Flat head .... FHD
Flexible .... FLEX.
Foot .... (') or FT
Foot pounds .... FT LB
Forging .... FORG
Forward .... FWD
Front .... FRT

Gage (Gauge) .... GA
Gallon .... GAL
Galvanize .... GALV
Gasket .... GSKT
Generator .... GEN
Grain .... GRN
Gravity .... G
Grind .... GRD

Harden .... HDN
Hardware .... HDW
Head .... HD
Heat-treat .... HT TR
Height .... HGT
Hexagon .... HEX
Horizontal .... HORIZ
Hot-rolled steel .... HRS
Hour .... HR
Hydraulic .... HYD

Illustration .... ILLUS
Inch .... (") or IN.
Inch ounce .... IN. OZ
Inch pound .... IN. LB
Inclusive .... INCL
Information .... INFO
Inside diameter .... ID
Inside radius .... IR
Inspection .... INSP
Installation .... INSTL
Instrument .... INST
Insulation .... INSL

Interchangeable .... INTCHG
Intermediate .... INTER.
Internal .... INT

Joint .... JT

Kilometer .... KM
Knock out .... KO

Laboratory .... LAB
Left hand .... LH
Length .... LG
Lock washer .... LWASH
Longitude, Longitudinal .... LONG.
Lower .... LWR
Lubricate .... LUB

Machine .... MACH
Magnesium .... MG
Maintenance .... MAINT
Malleable .... MAL
Manufacture .... MFR
Material .... MATL
Maximum .... MAX
Maximum material condition .... MMC
Mechanical .... MECH
Medium .... MED
Memorandum .... MEMO
Mercury .... HG
Mile .... MI
Miles per hour .... MPH
Millimeter .... MM
Minimum .... MIN
Minute .... (') or MIN
Miscellaneous .... MISC
Modify .... MOD
Molding .... MLDG
Mounting .... MTG

National .... NATL
National Electrical Mfg. Association .... NEMA
National Machine Tool Builders Association ... NMTBA
Negative .... (–) or NEG
Nickel .... NI
No drawing .... ND
Nominal .... NOM
Nonstandard .... NONSTD
Number .... NO.

Obsolete .... OBS
Of true position .... OTP
On center .... OC
Operate .... OPER
Opposite .... OPP
Optional .... OPT
Original .... ORIG
Ounce .... OZ

Outside diameter .... OD
Outside radius .... OR.
Oval head .... OV HD
Overall .... OA
Oxygen .... OXY

Package, Packing .... PKG
Page .... P
Part .... PT
Parting line .... PL
Patent .... PAT.
Pattern .... PATT
Perpendicular .... PERP
Pint .... PT
Pitch .... P
Pitch circle .... PC
Pitch diameter .... PD
Plan view .... PV
Point on line .... POL
Position .... POSN
Positive .... (+) or POS
Pound .... LB
Pounds per square inch .... PSI
Preliminary .... PRELIM
Pressure .... PRESS.
Process .... PROC
Product, Production .... PROD.

Quality .... QUAL
Quantity .... QTY
Quart .... QT
Quarter .... QTR

Rear view .... RV
Rectangular .... RECT
Reduction .... RED.
Reference .... REF
Regardless of feature size .... RFS
Regular .... REG
Reinforce .... REINF
Remove .... REM
Require .... REQ
Required .... REQD
Resistor .... RES
Reverse .... REV
Revision .... REV
Revolution .... REV
Revolutions per minute .... RPM
Right hand .... RH
Root diameter .... RD
Round .... RD
Round head .... RD HD
Rubber .... RUB.

Screw .... SCR
Screw threads
American National coarse .... NC
American National fine .... NF
American National extra fine .... NEF
American National 8 pitch 8N
American National 12 pitch 12N
American National 16 pitch 16N
American Standard taper pipe .... NPT
American Standard straight pipe coupling .... NPSC
American Standard taper .... NPTF
Unified screw thread coarse .... UNC
Unified screw thread fine .... UNF
Unified screw thread extra fine .... UNEF
Unified screw thread 8 thread 8UN
Unified screw thread 12 thread 12UN
Unified screw thread 16 thread 16UN
Unified screw thread special .... UNS
Second .... (") or SEC
Section .... SECT.
Serial, Series .... SER
Sheet .... SH
Side view .... SV
Sketch .... SK
Society of Automotive Engineers .... SAE
Special .... SPL
Specification .... SPEC
Specific gravity .... SP GR
Spherical .... SPHER
Spotface .... SF
Spring .... SPR
Square .... SQ
Standard .... STD
Steel .... STL
Support .... SUPT
Switch .... SW
Symbol .... SYM
Symmetrical .... SYM
Synthetic .... SYN

Tangent .... TAN.
Technical .... TECH
Teeth .... T
Temperature .... TEMP
Tensile strength .... TS
Terminal .... TERM.
Theoretical .... THEO
Thickness .... THK
Thread .... THD
Through .... THRU
Tolerance .... TOL
Tracer .... TCR
Trade mark .... TM
Transformer .... TRANS
Transmission .... TRANS
Transverse .... TRANSV
True length .... TL
True position .... TP
True view .... TV
Turnbuckle .... TRNBKL

Typical .... TYP

Ultimate .... ULT
United States .... US
United States of America Standards Institute ...... USASI
United States gage .... USG
Universal .... UNIV
Unless otherwise specified .... UOS
Upper .... UPR

Vacuum .... VAC
Velocity V
Versus .... VS
Vertical .... VERT
Volt V
Volume .... VOL

Water line .... WL
Watt .... W
Weight .... WT
Wood .... WD
Wood screw .... WD SCR
Wrought iron .... WI

Yard .... YD
Year .... YR

APPENDIX B Tables

TABLE 1 INCHES TO MILLIMETERS

in.	mm	in.	mm	in.	mm	in.	mm
1	25.4	26	660.4	51	1295.4	76	1930.4
2	50.8	27	685.8	52	1320.8	77	1955.8
3	76.2	28	711.2	53	1346.2	78	1981.2
4	101.6	29	736.6	54	1371.6	79	2006.6
5	127.0	30	762.0	55	1397.0	80	2032.0
6	152.4	31	787.4	56	1422.4	81	2057.4
7	177.8	32	812.8	57	1447.8	82	2082.8
8	203.2	33	838.2	58	1473.2	83	2108.2
9	228.6	34	863.6	59	1498.6	84	2133.6
10	254.0	35	889.0	60	1524.0	85	2159.0
11	279.4	36	914.4	61	1549.4	86	2184.4
12	304.8	37	939.8	62	1574.8	87	2209.8
13	330.2	38	965.2	63	1600.2	88	2235.2
14	355.6	39	990.6	64	1625.6	89	2260.6
15	381.0	40	1016.0	65	1651.0	90	2286.0
16	406.4	41	1041.4	66	1676.4	91	2311.4
17	431.8	42	1066.8	67	1701.8	92	2336.8
18	457.2	43	1092.2	68	1727.2	93	2362.2
19	482.6	44	1117.6	69	1752.6	94	2387.6
20	508.0	45	1143.0	70	1778.0	95	2413.0
21	533.4	46	1168.4	71	1803.4	96	2438.4
22	558.8	47	1193.8	72	1828.8	97	2463.8
23	584.2	48	1219.2	73	1854.2	98	2489.2
24	609.6	49	1244.6	74	1879.6	99	2514.6
25	635.0	50	1270.0	75	1905.0	100	2540.0

The above table is exact on the basis: 1 in. = 25.4 mm

TABLE 2 MILLIMETERS TO INCHES

mm	in.	mm	in.	mm	in.	mm	in.
1	0.039370	26	1.023622	51	2.007874	76	2.992126
2	0.078740	27	1.062992	52	2.047244	77	3.031496
3	0.118110	28	1.102362	53	2.086614	78	3.070866
4	0.157480	29	1.141732	54	2 125984	79	3.110236
5	0.196850	30	1.181102	55	2.165354	80	3.149606
6	0.236220	31	1.220472	56	2.204724	81	3.188976
7	0.275591	32	1.259843	57	2.244094	82	3.228346
8	0.314961	33	1.299213	58	2.283465	83	3.267717
9	0.354331	34	1.338583	59	2.322835	84	3.307087
10	0.393701	35	1.377953	60	2.362205	85	3.346457
11	0.433071	36	1.417323	61	2.401575	86	3.385827
12	0.472441	37	1.456693	62	2.440945	87	3.425197
13	0.511811	38	1.496063	63	2.480315	88	3.464567
14	0.551181	39	1.535433	64	2.519685	89	3.503937
15	0.590551	40	1.574803	65	2.559055	90	3.543307
16	0.629921	41	1.614173	66	2.598425	91	3.582677
17	0.669291	42	1.653543	67	2.637795	92	3.622047
18	0.708661	43	1.692913	68	2.677165	93	3.661417
19	0.748031	44	1.732283	69	2.716535	94	3.700787
20	0.787402	45	1.771654	70	2.755906	95	3.740157
21	0.826772	46	1.811024	71	2.795276	96	3.779528
22	0.866142	47	1.850394	72	2.834646	97	3.818898
23	0.905512	48	1.889764	73	2.874016	98	3.858268
24	0.944882	49	1.929134	74	2.913386	99	3.897638
25	0.984252	50	1.968504	75	2.952756	100	3.937008

The above table is approximate on the basis: 1 in. = 25.4 mm, 1/25.4 = 0.039370078740+

TABLE 3 INCH/METRIC — EQUIVALENTS

Fraction	Decimal Equivalent		Fraction	Decimal Equivalent	
	Customary (in.)	Metric (mm)		Customary (in.)	Metric (mm)
1/64	.015625	0.3969	33/64	.515625	13.0969
1/32	.03125	0.7938	17/32	.53125	13.4938
3/64	.046875	1.1906	35/64	.546875	13.8906
1/16	.0625	1.5875	9/16	.5625	14.2875
5/64	.078125	1.9844	37/64	.578125	14.6844
3/32	.09375	2.3813	19/32	.59375	15.0813
7/64	.109375	2.7781	39/64	.609375	15.4781
1/8	.1250	3.1750	5/8	.6250	15.8750
9/64	.140625	3.5719	41/64	.640625	16.2719
5/32	.15625	3.9688	21/32	.65625	16.6688
11/64	.171875	4.3656	43/64	.671875	17.0656
3/16	.1875	4.7625	11/16	.6875	17.4625
13/64	.203125	5.1594	45/64	.703125	17.8594
7/32	.21875	5.5563	23/32	.71875	18.2563
15/64	.234375	5.9531	47/64	.734375	18.6531
1/4	.250	6.3500	3/4	.750	19.0500
17/64	.265625	6.7469	49/64	.765625	19.4469
9/32	.28125	7.1438	25/32	.78125	19.8438
19/64	.296875	7.5406	51/64	.796875	20.2406
5/16	.3125	7.9375	13/16	.8125	20.6375
21/64	.328125	8.3384	53/64	.828125	21.0344
11/32	.34375	8.7313	27/32	.84375	21.4313
23/64	.359375	9.1281	55/64	.859375	21.8281
3/8	.3750	9.5250	7/8	.8750	22.2250
25/64	.390625	9.9219	57/64	.890625	22.6219
13/32	.40625	10.3188	29/32	.90625	23.0188
27/64	.421875	10.7156	59/64	.921875	23.4156
7/16	.4375	11.1125	15/16	.9375	23.8125
29/64	.453125	11.5094	61/64	.953125	24.2094
15/32	.46875	11.9063	31/32	.96875	24.6063
31/64	.484375	12.3031	63/64	.984375	25.0031
1/2	.500	12.7000	1	1.000	25.4000

TABLE 4 INCH/METRIC — CONVERSION

Measures of Length
1 millimeter (mm) = 0.03937 inch
1 centimeter (cm) = 0.39370 inch
1 meter (m) = 39.37008 inches
= 3.2808 feet
= 1.0936 yards
1 kilometer (km) = 0.6214 mile
1 inch = 25.4 millimeters (mm)
= 2.54 centimeters (cm)
1 foot = 304.8 millimeters (mm)
= 0.3048 meter (m)
1 yard = 0.9144 meter (m)
1 mile = 1.609 kilometers (km)
Measures of Area
1 square millimeter = 0.00155 square inch
1 square centimeter = 0.155 square inch
1 square meter = 10.764 square feet
= 1.196 square yards
1 square kilometer = 0.3861 square mile
1 square inch = 645.2 square millimeters
= 6.452 square centimeters
1 square foot = 929 square centimeters
= 0.0929 square meter
1 square yard = 0.836 square meter
1 square mile = 2.5899 square kilometers
Measures of Capacity (Dry)
1 cubic centimeter (cm^3) = 0.061 cubic inch
1 liter = 0.0353 cubic foot
= 61.023 cubic inches
1 cubic meter (m^3) = 35.315 cubic feet
= 1.308 cubic yards
1 cubic inch = 16.38706 cubic centimeters (cm^3)
1 cubic foot = 0.02832 cubic meter (m^3)
= 28.317 liters
1 cubic yard = 0.7646 cubic meter (m^3)
Measures of Capacity (Liquid)
1 liter = 1.0567 U.S. quarts
= 0.2642 U.S. gallon
= 0.2200 Imperial gallon
1 cubic meter (m^3) = 264.2 U.S. gallons
= 219.969 Imperial gallons
1 U.S. quart = 0.946 liter
1 Imperial quart = 1.136 liters
1 U.S. gallon = 3.785 liters
1 Imperial gallon = 4.546 liters
Measures of Weight
1 gram (g) = 15.432 grains
= 0.03215 ounce troy
= 0.03527 ounce avoirdupois
1 kilogram (kg) = 35.274 ounces avoirdupois
= 2.2046 pounds
1000 kilograms (kg) = 1 metric ton (t)
= 1.1023 tons of 2000 pounds
= 0.9842 ton of 2240 pounds
1 ounce avoirdupois = 28.35 grams (g)
1 ounce troy = 31.103 grams (g)
1 pound = 453.6 grams
= 0.4536 kilogram (kg)
1 ton of 2240 pounds = 1016 kilograms (kg)
= 1.016 metric tons
1 grain = 0.0648 gram (g)
1 metric ton = 0.9842 ton of 2240 pounds
= 2204.6 pounds

TABLE 5 RULES RELATIVE TO THE CIRCLE

To Find Circumference—			
Multiply diameter by	3.1416	Or divide diameter by	0.3183
To Find Diameter—			
Multiply circumference by	0.3183	Or divide circumference by	3.1416
To Find Radius—			
Multiply circumference by	0.15915	Or divide circumference by	6.28318
To Find Side of an Inscribed Square—			
Multiply diameter by	0.7071		
Or multiply circumference by	0.2251	Or divide circumference by	4.4428
To Find Side of an Equal Square—			
Multiply diameter by	0.8862	Or divide diameter by	1.1284
Or multiply circumference by	0.2821	Or divide circumference by	3.545
Square—			
A side multiplied by	1.4142	equals diameter of its circumscribing circle.	
A side multiplied by	4.443	equals circumference of its circumscribing circle.	
A side multiplied by	1.128	equals diameter of an equal circle.	
A side multiplied by	3.547	equals circumference of an equal circle.	
To Find the Area of a Circle—			
Multiply circumference by one-quarter of the diameter.			
Or multiply the square of diameter by	0.7854		
Or multiply the square of circumference by	.07958		
Or multiply the square of $^1/_2$ diameter by	3.1416		
To Find the Surface of a Sphere or Globe—			
Multiply the diameter by the circumference.			
Or multiply the square of diameter by	3.1416		
Or multiply four times the square of radius by	3.1416		

TABLE 6 DIMENSIONING SYMBOLS

SYMBOL FOR:	ANSI Y14.5	ISO
DIMENSION ORIGIN	⌖→	NONE
FEATURE CONTROL FRAME	⌖ | Ø0.5Ⓜ | A | B | C	⌖ | Ø0.5Ⓜ | A | B | C
CONICAL TAPER	▷	▷
SLOPE	◺	◺
COUNTERBORE/SPOTFACE	⌴	NONE
COUNTERSINK	⌵	NONE
DEPTH/DEEP	↧	NONE
SQUARE (SHAPE)	□	□
DIMENSION NOT TO SCALE	<u>15</u>	<u>15</u>
NUMBER OF TIMES/PLACES	8X	8X
ARC LENGTH	⌒105	NONE
RADIUS	R	R
SPHERICAL RADIUS	SR	NONE
SPHERICAL DIAMETER	SØ	NONE

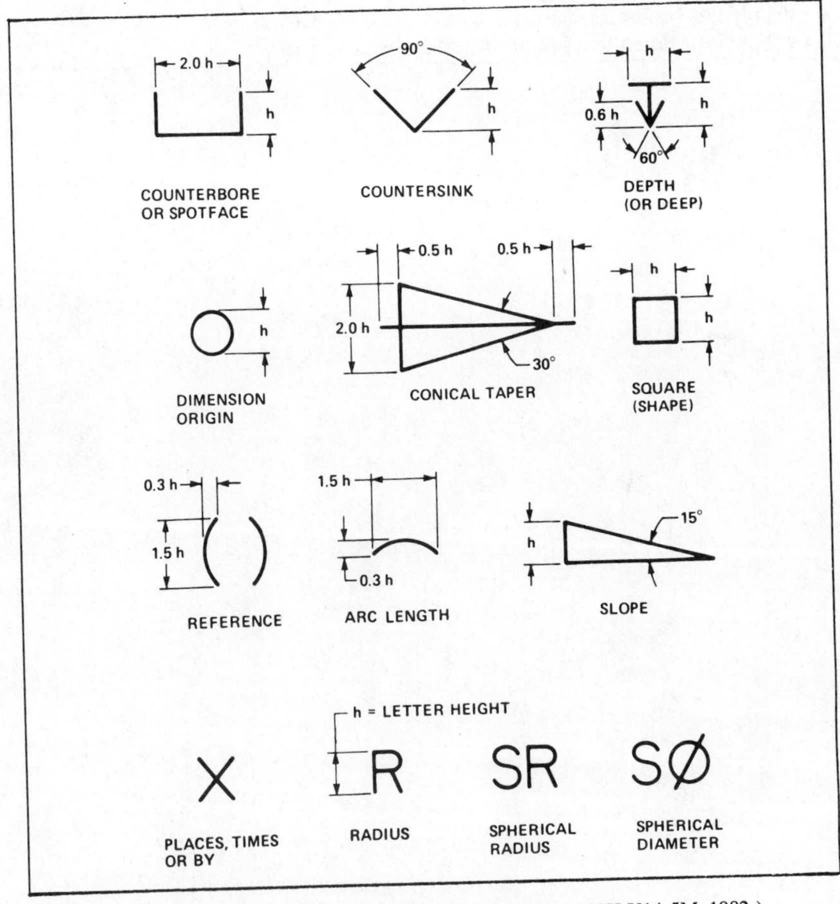

(Reprinted from The American Society of Mechanical Engineers—ANSI Y14.5M-1982.)

TABLE 7 ASTM AND SAE GRADE MARKINGS FOR STEEL BOLTS AND SCREWS

Grade Marking	Specification	Material
 NO MARK	SAE–Grade 1	Low or Medium Carbon Steel
	ASTM–A307	Low Carbon Steel
	SAE–Grade 2	Low or Medium Carbon Steel
	SAE–Grade 5	Medium Carbon Steel, Quenched and Tempered
	ASTM–A 449	
	SAE–Grade 5.2	Low Carbon Martensite Steel, Quenched and Tempered
 A 325	ASTM–A 325 Type 1	Medium Carbon Steel, Quenched and Tempered Radial dashes optional
 A 325	ASTM–A 325 Type 2	Low Carbon Martensite Steel, Quenched and Tempered
 A 325	ASTM–A 325 Type 3	Atmospheric Corrosion (Weathering) Steel, Quenched and Tempered
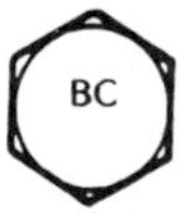 BC	ASTM–A 354 Grade BC	Alloy Steel, Quenched and Tempered
	SAE–Grade 7	Medium Carbon Alloy Steel, Quenched and Tempered, Roll Threaded After Heat Treatment
	SAE–Grade 8	Medium Carbon Alloy Steel, Quenched and Tempered
	ASTM–A 354 Grade BD	Alloy Steel, Quenched and Tempered
	SAE–Grade 8.2	Low Carbon Martensite Steel, Quenched and Tempered
 A 490	ASTM–A 490 Type 1	Alloy Steel, Quenched and Tempered
 A 490	ASTM–A 490 Type 3	Atmospheric Corrosion (Weathering) Steel, Quenched and Tempered

(Reprinted from The American Society of Mechanical Engineers—ANSI B18.2.1-1981.)

TABLE 8 UNIFIED STANDARD SCREW THREAD SERIES

Sizes		Basic Major Diameter	THREADS PER INCH											Sizes
			Series with graded pitches			Series with constant pitches								
Primary	Secondary		Coarse UNC	Fine UNF	Extra fine UNEF	4UN	6UN	8UN	12UN	16UN	20UN	28UN	32UN	
0		0.0600	–	80	–	–	–	–	–	–	–	–	–	0
	1	0.0730	64	72	–	–	–	–	–	–	–	–	–	1
2		0.0860	56	64	–	–	–	–	–	–	–	–	–	2
	3	0.0990	48	56	–	–	–	–	–	–	–	–	–	3
4		0.1120	40	48	–	–	–	–	–	–	–	–	–	4
5		0.1250	40	44	–	–	–	–	–	–	–	–	–	5
6		0.1380	32	40	–	–	–	–	–	–	–	–	UNC	6
8		0.1640	32	36	–	–	–	–	–	–	–	–	UNC	8
10		0.1900	24	32	–	–	–	–	–	–	–	–	UNF	10
	12	0.2160	24	28	32	–	–	–	–	–	–	UNF	UNEF	12
1/4		0.2500	20	28	32	–	–	–	–	–	UNC	UNF	UNEF	1/4
5/16		0.3125	18	24	32	–	–	–	–	–	20	28	UNEF	5/16
3/8		0.3750	16	24	32	–	–	–	–	UNC	20	28	UNEF	3/8
7/16		0.4375	14	20	28	–	–	–	–	16	UNF	UNEF	32	7/16
1/2		0.5000	13	20	28	–	–	–	–	16	UNF	UNEF	32	1/2
9/16		0.5625	12	18	24	–	–	–	UNC	16	20	28	32	9/16
5/8		0.6250	11	18	24	–	–	–	12	16	20	28	32	5/8
	11/16	0.6875	–	–	24	–	–	–	12	16	20	28	32	11/16
3/4		0.7500	10	16	20	–	–	–	12	UNF	UNEF	28	32	3/4
	13/16	0.8125	–	–	20	–	–	–	12	16	UNEF	28	32	13/16
7/8		0.8750	9	14	20	–	–	–	12	16	UNEF	28	32	7/8
	15/16	0.9375	–	–	20	–	–	–	12	16	UNEF	28	32	15/16
1		1.0000	8	12	20	–	–	UNC	UNF	16	UNEF	28	32	1
	1 1/16	1.0625	–	–	18	–	–	8	12	16	20	28	–	1 1/16
1 1/8		1.1250	7	12	18	–	–	8	UNF	16	20	28	–	1 1/8
	1 3/16	1.1875	–	–	18	–	–	8	12	16	20	28	–	1 3/16
1 1/4		1.2500	7	12	18	–	–	8	UNF	16	20	28	–	1 1/4
	1 5/16	1.3125	–	–	18	–	–	8	12	16	20	28	–	1 5/16
1 3/8		1.3750	6	12	18	–	UNC	8	UNF	16	20	28	–	1 3/8
	1 7/16	1.4375	–	–	18	–	6	8	12	16	20	28	–	1 7/16
1 1/2		1.5000	6	12	18	–	UNC	8	UNF	16	20	28	–	1 1/2
	1 9/16	1.5625	–	–	18	–	6	8	12	16	20	–	–	1 9/16
1 5/8		1.6250	–	–	18	–	6	8	12	16	20	–	–	1 5/8
	1 11/16	1.6875	–	–	18	–	6	8	12	16	20	–	–	1 11/16
1 3/4		1.7500	5	–	–	–	6	8	12	16	20	–	–	1 3/4
	1 13/16	1.8125	–	–	–	–	6	8	12	16	20	–	–	1 13/16
1 7/8		1.8750	–	–	–	–	6	8	12	16	20	–	–	1 7/8
	1 15/16	1.9375	–	–	–	–	6	8	12	16	20	–	–	1 15/16
2		2.0000	4 1/2	–	–	–	6	8	12	16	20	–	–	2
	2 1/8	2.1250	–	–	–	–	6	8	12	16	20	–	–	2 1/8
2 1/4		2.2500	4 1/2	–	–	–	6	8	12	16	20	–	–	2 1/4
	2 3/8	2.3750	–	–	–	–	6	8	12	16	20	–	–	2 3/8
2 1/2		2.5000	4	–	–	UNC	6	8	12	16	20	–	–	2 1/2
	2 5/8	2.6250	–	–	–	4	6	8	12	16	20	–	–	2 5/8
2 3/4		2.7500	4	–	–	UNC	6	8	12	16	20	–	–	2 3/4
	2 7/8	2.8750	–	–	–	4	6	8	12	16	20	–	–	2 7/8
3		3.0000	4	–	–	UNC	6	8	12	16	20	–	–	3
	3 1/8	3.1250	–	–	–	4	6	8	12	16	–	–	–	3 1/8
3 1/4		3.2500	4	–	–	UNC	6	8	12	16	–	–	–	3 1/4
	3 3/8	3.3750	–	–	–	4	6	8	12	16	–	–	–	3 3/8
3 1/2		3.5000	4	–	–	UNC	6	8	12	16	–	–	–	3 1/2
	3 5/8	3.6250	–	–	–	4	6	8	12	16	–	–	–	3 5/8
3 3/4		3.7500	4	–	–	UNC	6	8	12	16	–	–	–	3 3/4
	3 7/8	3.8750	–	–	–	4	6	8	12	16	–	–	–	3 7/8
4		4.0000	4	–	–	UNC	6	8	12	16	–	–	–	4
	4 1/8	4.1250	–	–	–	4	6	8	12	16	–	–	–	4 1/8
4 1/4		4.2500	–	–	–	4	6	8	12	16	–	–	–	4 1/4
	4 3/8	4.3750	–	–	–	4	6	8	12	16	–	–	–	4 3/8
4 1/2		4.5000	–	–	–	4	6	8	12	16	–	–	–	4 1/2
	4 5/8	4.6250	–	–	–	4	6	8	12	16	–	–	–	4 5/8
4 3/4		4.7500	–	–	–	4	6	8	12	16	–	–	–	4 3/4
	4 7/8	4.8750	–	–	–	4	6	8	12	16	–	–	–	4 7/8
5		5.0000	–	–	–	4	6	8	12	16	–	–	–	5
	5 1/8	5.1250	–	–	–	4	6	8	12	16	–	–	–	5 1/8
5 1/4		5.2500	–	–	–	4	6	8	12	16	–	–	–	5 1/4
	5 3/8	5.3750	–	–	–	4	6	8	12	16	–	–	–	5 3/8
5 1/2		5.5000	–	–	–	4	6	8	12	16	–	–	–	5 1/2
	5 5/8	5.6250	–	–	–	4	6	8	12	16	–	–	–	5 5/8
5 3/4		5.7500	–	–	–	4	6	8	12	16	–	–	–	5 3/4
	5 7/8	5.8750	–	–	–	4	6	8	12	16	–	–	–	5 7/8
6		6.0000	–	–	–	4	6	8	12	16	–	–	–	6

TABLE 9 DIMENSIONS OF HEX CAP SCREWS (FINISHED HEX BOLTS)

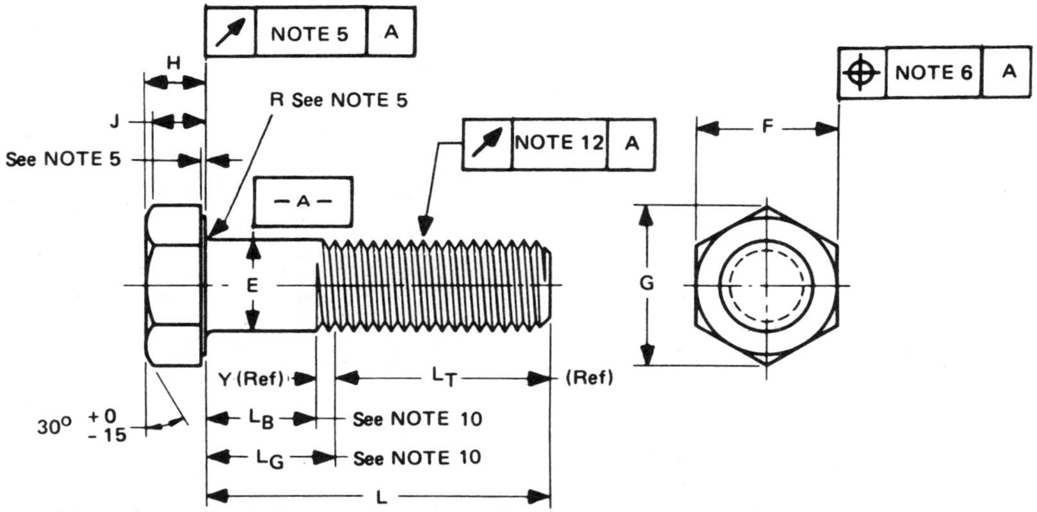

Nominal Size or Basic Product Dia (18)	E		F			G		H			J	L_T		Y	Runout of Bearing Surface FIM (5)
	Body Dia (8)		Width Across Flats			Width Across Corners (4)		Height			Wrenching Height (4)	Thread Length For Screw Lengths (10): 6 in. and Shorter	Thread Length For Screw Lengths (10): Over 6 in.	Transition Thread Length (10)	
	Max	Min	Basic	Max	Min	Max	Min	Basic	Max	Min	Min	Basic	Basic	Max	Max
1/4 0.2500	0.2500	0.2450	7/16	0.438	0.428	0.505	0.488	5/32	0.163	0.150	0.106	0.750	1.000	0.250	0.010
5/16 0.3125	0.3125	0.3065	1/2	0.500	0.489	0.577	0.557	13/64	0.211	0.195	0.140	0.875	1.125	0.278	0.011
3/8 0.3750	0.3750	0.3690	9/16	0.562	0.551	0.650	0.628	15/64	0.243	0.226	0.160	1.000	1.250	0.312	0.012
7/16 0.4375	0.4375	0.4305	5/8	0.625	0.612	0.722	0.698	9/32	0.291	0.272	0.195	1.125	1.375	0.357	0.013
1/2 0.5000	0.5000	0.4930	3/4	0.750	0.736	0.866	0.840	5/16	0.323	0.302	0.215	1.250	1.500	0.385	0.014
9/16 0.5625	0.5625	0.5545	13/16	0.812	0.798	0.938	0.910	23/64	0.371	0.348	0.250	1.375	1.625	0.417	0.015
5/8 0.6250	0.6250	0.6170	15/16	0.938	0.922	1.083	1.051	25/64	0.403	0.378	0.269	1.500	1.750	0.455	0.017
3/4 0.7500	0.7500	0.7410	1 1/8	1.125	1.100	1.299	1.254	15/32	0.483	0.455	0.324	1.750	2.000	0.500	0.020
7/8 0.8750	0.8750	0.8660	1 5/16	1.312	1.285	1.516	1.465	35/64	0.563	0.531	0.378	2.000	2.250	0.556	0.023
1 1.0000	1.0000	0.9900	1 1/2	1.500	1.469	1.732	1.675	39/64	0.627	0.591	0.416	2.250	2.500	0.625	0.026
1 1/8 1.1250	1.1250	1.1140	1 11/16	1.688	1.631	1.949	1.859	11/16	0.718	0.658	0.461	2.500	2.750	0.714	0.029
1 1/4 1.2500	1.2500	1.2390	1 7/8	1.875	1.812	2.165	2.066	25/32	0.813	0.749	0.530	2.750	3.000	0.714	0.033
1 3/8 1.3750	1.3750	1.3630	2 1/16	2.062	1.994	2.382	2.273	27/32	0.878	0.810	0.569	3.000	3.250	0.833	0.036
1 1/2 1.5000	1.5000	1.4880	2 1/4	2.230	2.175	2.598	2.480	1 5/16	0.974	0.902	0.640	3.250	3.500	0.833	0.039
1 3/4 1.7500	1.7500	1.7380	2 5/8	2.625	2.538	3.031	2.893	1 3/32	1.134	1.054	0.748	3.750	4.000	1.000	0.046
2 2.0000	2.0000	1.9880	3	3.000	2.900	3.464	3.306	1 7/32	1.263	1.175	0.825	4.250	4.500	1.111	0.052
2 1/4 2.2500	2.2500	2.2380	3 3/8	3.375	3.262	3.897	3.719	1 3/8	1.423	1.327	0.933	4.750	5.000	1.111	0.059
2 1/2 2.5000	2.5000	2.4880	3 3/4	3.750	3.625	4.330	4.133	1 17/32	1.583	1.479	1.042	5.250	5.500	1.250	0.065
2 3/4 2.7500	2.7500	2.7380	4 1/8	4.125	3.988	4.763	4.546	1 11/16	1.744	1.632	1.151	5.750	6.000	1.250	0.072
3 3.0000	3.0000	2.9880	4 1/2	4.500	4.350	5.196	4.959	1 7/8	1.935	1.815	1.290	6.250	6.500	1.250	0.079

(Reprinted from the American Society of Mechanical Engineers—ANSI B18.2.1–1981. Refer to this document for notes specified in drawings.)

TABLE 10 DIMENSIONS OF HEXAGON AND SPLINE SOCKET HEAD CAP SCREWS (1960 SERIES)

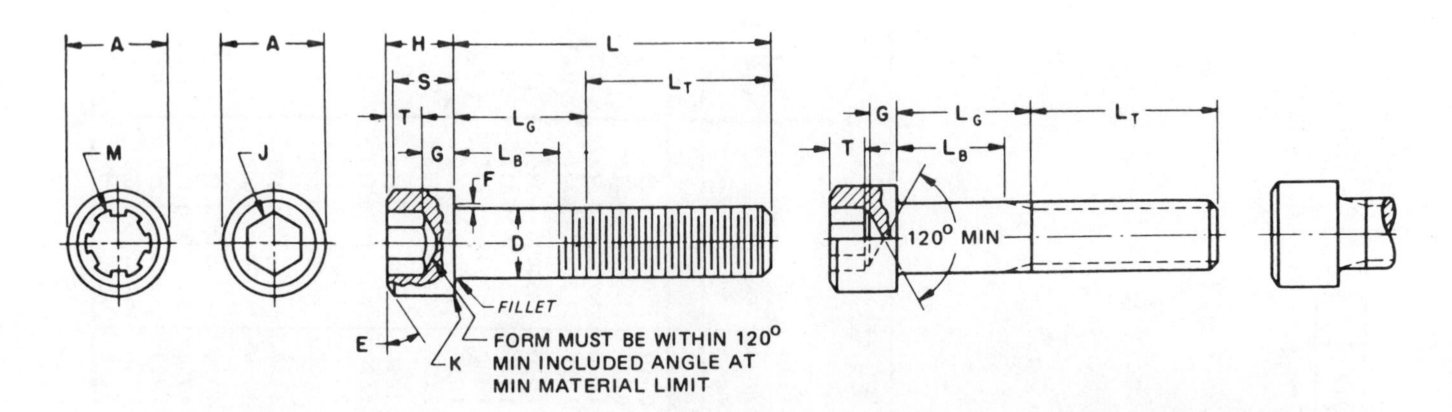

Nominal Size or Basic Screw Diameter		D Body Diameter		A Head Diameter		H Head Height		S Head Side Height	M Spline Socket Size	J Hexagon Socket Size	T Key Engagement	G Wall Thickness	K Chamfer or Radius
		Max	Min	Max	Min	Max	Min	Min	Nom	Nom	Min	Min	Max
0	0.0600	0.0600	0.0568	0.096	0.091	0.060	0.057	0.054	0.060	0.050	0.025	0.020	0.003
1	0.0730	0.0730	0.0695	0.118	0.112	0.073	0.070	0.066	0.072	1/16 0.062	0.031	0.025	0.003
2	0.0860	0.0860	0.0822	0.140	0.134	0.086	0.083	0.077	0.096	5/64 0.078	0.038	0.029	0.003
3	0.0990	0.0990	0.0949	0.161	0.154	0.099	0.095	0.089	0.096	5/64 0.078	0.044	0.034	0.003
4	0.1120	0.1120	0.1075	0.183	0.176	0.112	0.108	0.101	0.111	3/32 0.094	0.051	0.038	0.005
5	0.1250	0.1250	0.1202	0.205	0.198	0.125	0.121	0.112	0.111	3/32 0.094	0.057	0.043	0.005
6	0.1380	0.1380	0.1329	0.226	0.218	0.138	0.134	0.124	0.133	7/64 0.109	0.064	0.047	0.005
8	0.1640	0.1640	0.1585	0.270	0.262	0.164	0.159	0.148	0.168	9/64 0.141	0.077	0.056	0.005
10	0.1900	0.1900	0.1840	0.312	0.303	0.190	0.185	0.171	0.183	5/32 0.156	0.090	0.065	0.005
1/4	0.2500	0.2500	0.2435	0.375	0.365	0.250	0.244	0.225	0.216	3/16 0.188	0.120	0.095	0.008
5/16	0.3125	0.3125	0.3053	0.469	0.457	0.312	0.306	0.281	0.291	1/4 0.250	0.151	0.119	0.008
3/8	0.3750	0.3750	0.3678	0.562	0.550	0.375	0.368	0.337	0.372	5/16 0.312	0.182	0.143	0.008
7/16	0.4375	0.4375	0.4294	0.656	0.642	0.438	0.430	0.394	0.454	3/8 0.375	0.213	0.166	0.010
1/2	0.5000	0.5000	0.4919	0.750	0.735	0.500	0.492	0.450	0.454	3/8 0.375	0.245	0.190	0.010
5/8	0.6250	0.6250	0.6163	0.938	0.921	0.625	0.616	0.562	0.595	1/2 0.500	0.307	0.238	0.010
3/4	0.7500	0.7500	0.7406	1.125	1.107	0.750	0.740	0.675	0.620	5/8 0.625	0.370	0.285	0.010
7/8	0.8750	0.8750	0.8647	1.312	1.293	0.875	0.864	0.787	0.698	3/4 0.750	0.432	0.333	0.015
1	1.0000	1.0000	0.9886	1.500	1.479	1.000	0.988	0.900	0.790	3/4 0.750	0.495	0.380	0.015
1 1/8	1.1250	1.1250	1.1086	1.688	1.665	1.125	1.111	1.012	....	7/8 0.875	0.557	0.428	0.015
1 1/4	1.2500	1.2500	1.2336	1.875	1.852	1.250	1.236	1.125	....	7/8 0.875	0.620	0.475	0.015
1 3/8	1.3750	1.3750	1.3568	2.062	2.038	1.375	1.360	1.237	....	1 1.000	0.682	0.523	0.015
1 1/2	1.5000	1.5000	1.4818	2.250	2.224	1.500	1.485	1.350	....	1 1.000	0.745	0.570	0.015
1 3/4	1.7500	1.7500	1.7295	2.625	2.597	1.750	1.734	1.575	....	1 1/4 1.250	0.870	0.665	0.015
2	2.0000	2.0000	1.9780	3.000	2.970	2.000	1.983	1.800	....	1 1/2 1.500	0.995	0.760	0.015
2 1/4	2.2500	2.2500	2.2280	3.375	3.344	2.250	2.232	2.025	....	1 3/4 1.750	1.120	0.855	0.031
2 1/2	2.5000	2.5000	2.4762	3.750	3.717	2.500	2.481	2.250	....	1 3/4 1.750	1.245	0.950	0.031
2 3/4	2.7500	2.7500	2.7262	4.125	4.090	2.750	2.730	2.475	....	2 2.000	1.370	1.045	0.031
3	3.0000	3.0000	2.9762	4.500	4.464	3.000	2.979	2.700	....	2 1/4 2.250	1.495	1.140	0.031
3 1/4	3.2500	3.2500	3.2262	4.875	4.837	3.250	3.228	2.925	....	2 1/4 2.250	1.620	1.235	0.031
3 1/2	3.5000	3.5000	3.4762	5.250	5.211	3.500	3.478	3.150	....	2 3/4 2.750	1.745	1.330	0.031
3 3/4	3.7500	3.7500	3.7262	5.625	5.584	3.750	3.727	3.375	....	2 3/4 2.750	1.870	1.425	0.031
4	4.0000	4.0000	3.9762	6.000	5.958	4.000	3.976	3.600	....	3 3.000	1.995	1.520	0.031

(Reprinted from The American Society of Mechanical Engineers—ANSI/ASME B18.3-1982.)

TABLE 11 DIMENSIONS OF HEXAGON AND SPLINE SOCKET FLAT COUNTERSUNK HEAD CAP SCREWS

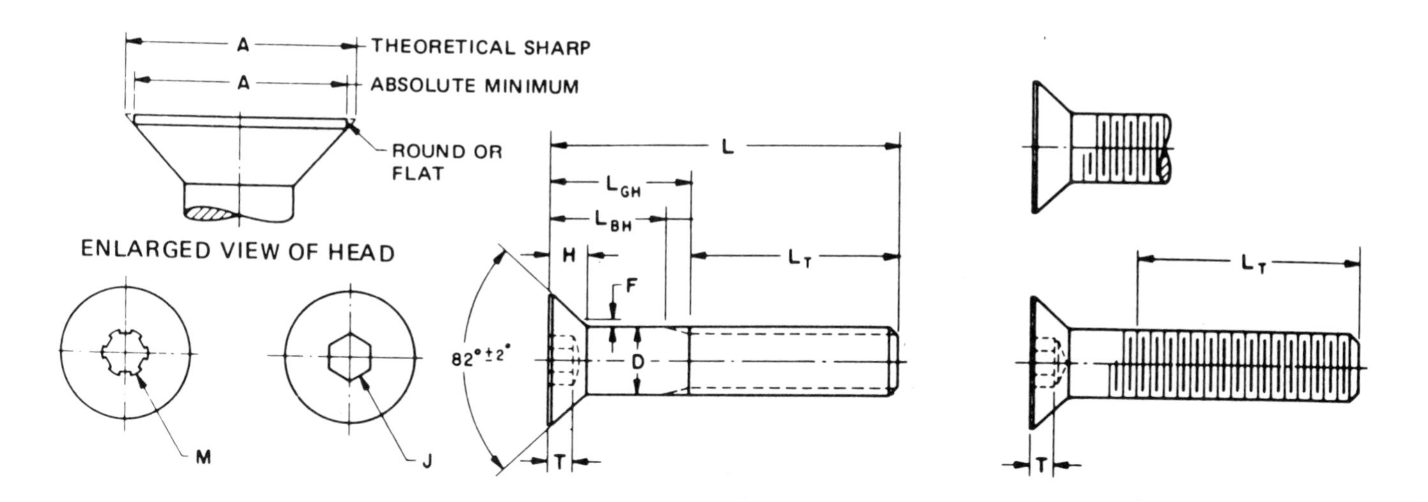

Nominal Size or Basic Screw Diameter		D		A		H		M	J	T	F
		Body Dia		Head Diameter		Head Height		Spline Socket Size	Hexagon Socket Size	Key Engagement	Fillet Extension Above D Max
				Theoretical Sharp Max	Abs. Min	Reference	Flushness Tolerance				
		Max	Min						Nom	Min	Max
0	0.0600	0.0600	0.0568	0.138	0.117	0.044	0.006	0.048	0.035	0.025	0.006
1	0.0730	0.0730	0.0695	0.168	0.143	0.054	0.007	0.060	0.050	0.031	0.008
2	0.0860	0.0860	0.0822	0.197	0.168	0.064	0.008	0.060	0.050	0.038	0.010
3	0.0990	0.0990	0.0949	0.226	0.193	0.073	0.010	0.072	1/16 0.062	0.044	0.010
4	0.1120	0.1120	0.1075	0.255	0.218	0.083	0.011	0.072	1/16 0.062	0.055	0.012
5	0.1250	0.1250	0.1202	0.281	0.240	0.090	0.012	0.096	5/64 0.078	0.061	0.014
6	0.1380	0.1380	0.1329	0.307	0.263	0.097	0.013	0.096	5/64 0.078	0.066	0.015
8	0.1640	0.1640	0.1585	0.359	0.311	0.112	0.014	0.111	3/32 0.094	0.076	0.015
10	0.1900	0.1900	0.1840	0.411	0.359	0.127	0.015	0.145	1/8 0.125	0.087	0.015
1/4	0.2500	0.2500	0.2435	0.531	0.480	0.161	0.016	0.183	5/32 0.156	0.111	0.015
5/16	0.3125	0.3125	0.3053	0.656	0.600	0.198	0.017	0.216	3/16 0.188	0.135	0.015
3/8	0.3750	0.3750	0.3678	0.781	0.720	0.234	0.018	0.251	7/32 0.219	0.159	0.015
7/16	0.4375	0.4375	0.4294	0.844	0.781	0.234	0.018	0.291	1/4 0.250	0.159	0.015
1/2	0.5000	0.5000	0.4919	0.938	0.872	0.251	0.018	0.372	5/16 0.312	0.172	0.015
5/8	0.6250	0.6250	0.6163	1.188	1.112	0.324	0.022	0.454	3/8 0.375	0.220	0.015
3/4	0.7500	0.7500	0.7406	1.438	1.355	0.396	0.024	0.454	1/2 0.500	0.220	0.015
7/8	0.8750	0.8750	0.8647	1.688	1.604	0.468	0.025	. . .	9/16 0.562	0.248	0.015
1	1.0000	1.0000	0.9886	1.938	1.841	0.540	0.028	. . .	5/8 0.625	0.297	0.015
1 1/8	1.1250	1.1250	1.1086	2.188	2.079	0.611	0.031	. . .	3/4 0.750	0.325	0.031
1 1/4	1.2500	1.2500	1.2336	2.438	2.316	0.683	0.035	. . .	7/8 0.875	0.358	0.031
1 3/8	1.3750	1.3750	1.3568	2.688	2.553	0.755	0.038	. . .	7/8 0.875	0.402	0.031
1 1/2	1.5000	1.5000	1.4818	2.938	2.791	0.827	0.042	. . .	1 1.000	0.435	0.031

(Reprinted from The American Society of Mechanical Engineers—ANSI/ASME B18.3-1982.)

TABLE 12 DIMENSIONS OF HEXAGON AND SPLINE SOCKET SET SCREWS

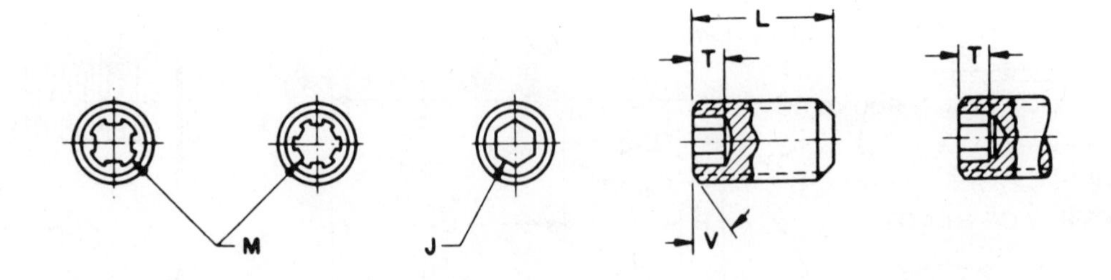

Nominal Size or Basic Screw Diameter		J	M	T		C		R	Y
		Hexagon Socket Size	Spline Socket Size	Min Key Engagement to Develop Functional Capability of Key		Cup and Flat Point Diameters		Oval Point Radius	Cone Point Angle 90° ±2° For These Nominal Lengths or Longer; 118° ±2° For Shorter Nominal Lengths
		Nom	Nom	Hex Socket T_H Min	Spline Socket T_S Min	Max	Min	Basic	
0	0.0600	0.028	0.033	0.050	0.026	0.033	0.027	0.045	5/64
1	0.0730	0.035	0.033	0.060	0.035	0.040	0.033	0.055	3/32
2	0.0860	0.035	0.048	0.060	0.040	0.047	0.039	0.064	7/64
3	0.0990	0.050	0.048	0.070	0.040	0.054	0.045	0.074	1/8
4	0.1120	0.050	0.060	0.070	0.045	0.061	0.051	0.084	5/32
5	0.1250	1/16 0.062	0.072	0.080	0.055	0.067	0.057	0.094	3/16
6	0.1380	1/16 0.062	0.072	0.080	0.055	0.074	0.064	0.104	3/16
8	0.1640	5/64 0.078	0.096	0.090	0.080	0.087	0.076	0.123	1/4
10	0.1900	3/32 0.094	0.111	0.100	0.080	0.102	0.088	0.142	1/4
1/4	0.2500	1/8 0.125	0.145	0.125	0.125	0.132	0.118	0.188	5/16
5/16	0.3125	5/32 0.156	0.183	0.156	0.156	0.172	0.156	0.234	3/8
3/8	0.3750	3/16 0.188	0.216	0.188	0.188	0.212	0.194	0.281	7/16
7/16	0.4375	7/32 0.219	0.251	0.219	0.219	0.252	0.232	0.328	1/2
1/2	0.5000	1/4 0.250	0.291	0.250	0.250	0.291	0.270	0.375	9/16
5/8	0.6250	5/16 0.312	0.372	0.312	0.312	0.371	0.347	0.469	3/4
3/4	0.7500	3/8 0.375	0.454	0.375	0.375	0.450	0.425	0.562	7/8
7/8	0.8750	1/2 0.500	0.595	0.500	0.500	0.530	0.502	0.656	1
1	1.0000	9/16 0.562	...	0.562	...	0.609	0.579	0.750	1 1/8
1 1/8	1.1250	9/16 0.562	...	0.562	...	0.689	0.655	0.844	1 1/4
1 1/4	1.2500	5/8 0.625	...	0.625	...	0.767	0.733	0.938	1 1/2
1 3/8	1.3750	5/8 0.625	...	0.625	...	0.848	0.808	1.031	1 5/8
1 1/2	1.5000	3/4 0.750	...	0.750	...	0.926	0.886	1.125	1 3/4
1 3/4	1.7500	1 1.000	...	1.000	...	1.086	1.039	1.312	2
2	2.0000	1 1.000	...	1.000	...	1.244	1.193	1.500	2 1/4

TABLE 12 (CONTINUED)

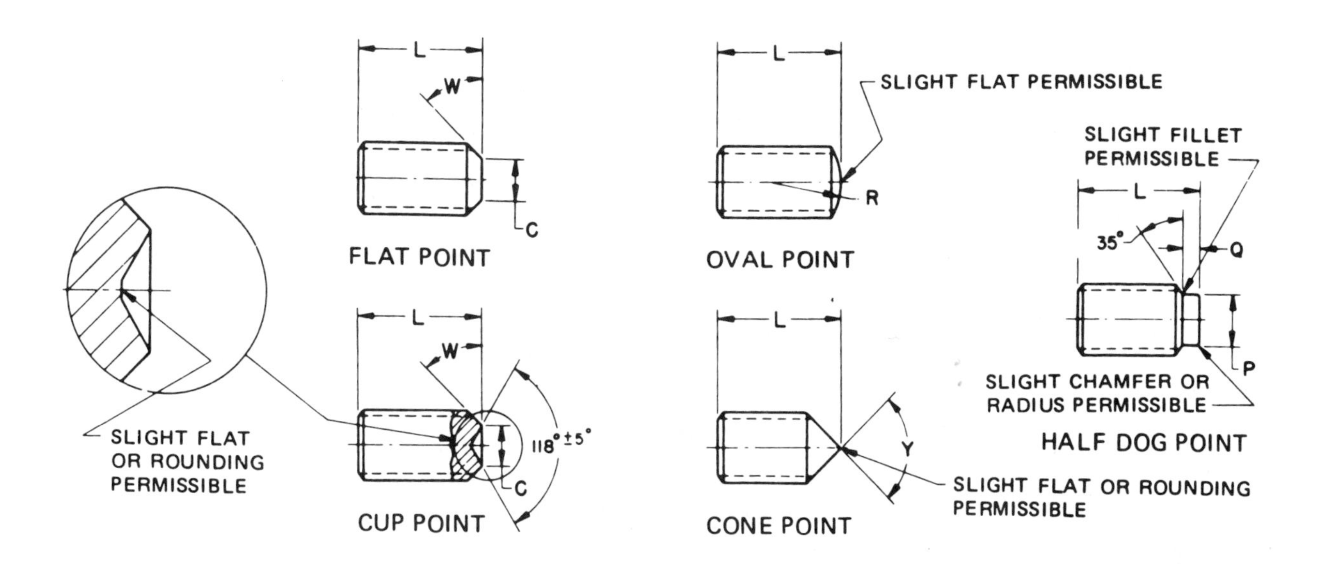

Nominal Size or Basic Screw Diameter		P		Q		B			B₁		
		Half Dog Point				Shortest Optimum Nominal Length To Which Column T_H Applies			Shortest Optimum Nominal Length To Which Column T_S Applies		
		Diameter		Length		Cup and Flat Points	90° Cone and Oval Points	Half Dog Point	Cup and Flat Points	90° Cone and Oval Points	Half Dog Point
		Max	Min	Max	Min						
0	0.0600	0.040	0.037	0.017	0.013	7/64	1/8	7/64	1/16	1/8	7/64
1	0.0730	0.049	0.045	0.021	0.017	1/8	9/64	1/8	3/32	9/64	1/8
2	0.0860	0.057	0.053	0.024	0.020	1/8	9/64	9/64	3/32	9/64	9/64
3	0.0990	0.066	0.062	0.027	0.023	9/64	5/32	5/32	3/32	5/32	5/32
4	0.1120	0.075	0.070	0.030	0.026	9/64	11/64	5/32	3/32	11/64	5/32
5	0.1250	0.083	0.078	0.033	0.027	3/16	3/16	11/64	1/8	3/16	11/64
6	0.1380	0.092	0.087	0.038	0.032	11/64	13/64	3/16	1/8	13/64	3/16
8	0.1640	0.109	0.103	0.043	0.037	3/16	7/32	13/64	3/16	7/32	13/64
10	0.1900	0.127	0.120	0.049	0.041	3/16	1/4	15/64	3/16	1/4	15/64
1/4	0.2500	0.156	0.149	0.067	0.059	1/4	5/16	19/64	1/4	5/16	19/64
5/16	0.3125	0.203	0.195	0.082	0.074	5/16	25/64	23/64	5/16	25/64	23/64
3/8	0.3750	0.250	0.241	0.099	0.089	3/8	7/16	7/16	3/8	7/16	7/16
7/16	0.4375	0.297	0.287	0.114	0.104	7/16	35/64	31/64	7/16	35/64	31/64
1/2	0.5000	0.344	0.334	0.130	0.120	1/2	39/64	35/64	1/2	39/64	35/64
5/8	0.6250	0.469	0.456	0.164	0.148	5/8	49/64	43/64	5/8	49/64	43/64
3/4	0.7500	0.562	0.549	0.196	0.180	3/4	29/32	51/64	3/4	29/32	51/64
7/8	0.8750	0.656	0.642	0.227	0.211	7/8	1 1/8	63/64	7/8	1 1/8	63/64
1	1.0000	0.750	0.734	0.260	0.240	1	1 17/64	1 1/8	...	...	...
1 1/8	1.1250	0.844	0.826	0.291	0.271	1 1/8	1 25/64	1 3/16	...	...	...
1 1/4	1.2500	0.938	0.920	0.323	0.303	1 1/4	1 1/2	1 5/16	...	...	...
1 3/8	1.3750	1.031	1.011	0.354	0.334	1 3/8	1 21/32	1 7/16	...	...	...
1 1/2	1.5000	1.125	1.105	0.385	0.365	1 1/2	1 51/64	1 9/16	...	...	...
1 3/4	1.7500	1.312	1.289	0.448	0.428	1 3/4	2 7/32	1 61/64	...	...	...
2	2.0000	1.500	1.474	0.510	0.490	2	2 25/64	2 5/64	...	...	...

(Reprinted from The American Society of Mechanical Engineers—ANSI/ASME B18.3-1982.)

TABLE 13 DIMENSIONS OF SLOTTED FLAT COUNTERSUNK HEAD CAP SCREWS

CAP SCREWS
FLAT

Type of Head

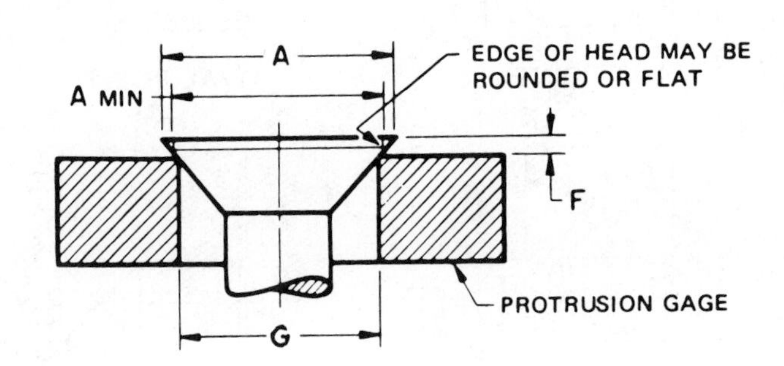

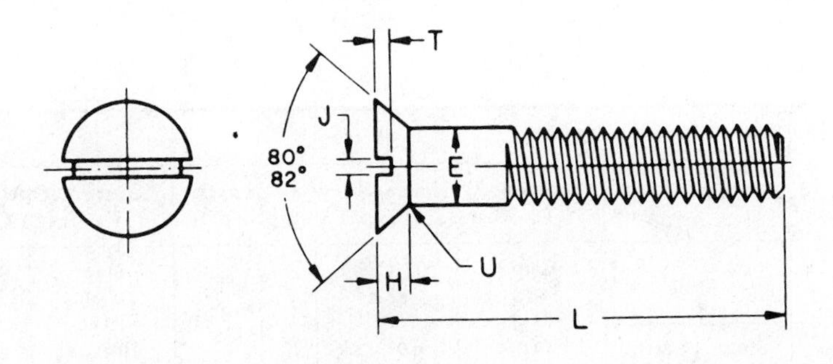

Nominal Size[1] or Basic Screw Diameter	E Body Diameter		A Head Diameter		H[2] Head Height	J Slot Width		T Slot Depth		U Fillet Radius	F[3] Protrusion Above Gaging Diameter		G[3] Gaging Diameter
	Max	Min	Max, Edge Sharp	Min, Edge Rounded or Flat	Ref	Max	Min	Max	Min	Max	Max	Min	
1/4 0.2500	0.2500	0.2450	0.500	0.452	0.140	0.075	0.064	0.068	0.045	0.100	0.046	0.030	0.424
5/16 0.3125	0.3125	0.3070	0.625	0.567	0.177	0.084	0.072	0.086	0.057	0.125	0.053	0.035	0.538
3/8 0.3750	0.3750	0.3690	0.750	0.682	0.210	0.094	0.081	0.103	0.068	0.150	0.060	0.040	0.651
7/16 0.4375	0.4375	0.4310	0.812	0.736	0.210	0.094	0.081	0.103	0.068	0.175	0.065	0.044	0.703
1/2 0.5000	0.5000	0.4930	0.875	0.791	0.210	0.106	0.091	0.103	0.068	0.200	0.071	0.049	0.756
9/16 0.5625	0.5625	0.5550	1.000	0.906	0.244	0.118	0.102	0.120	0.080	0.225	0.078	0.054	0.869
5/8 0.6250	0.6250	0.6170	1.125	1.020	0.281	0.133	0.116	0.137	0.091	0.250	0.085	0.058	0.982
3/4 0.7500	0.7500	0.7420	1.375	1.251	0.352	0.149	0.131	0.171	0.115	0.300	0.099	0.068	1.208
7/8 0.8750	0.8750	0.8660	1.625	1.480	0.423	0.167	0.147	0.206	0.138	0.350	0.113	0.077	1.435
1 1.0000	1.0000	0.9900	1.875	1.711	0.494	0.188	0.166	0.240	0.162	0.400	0.127	0.087	1.661
1 1/8 1.1250	1.1250	1.1140	2.062	1.880	0.529	0.196	0.178	0.257	0.173	0.450	0.141	0.096	1.826
1 1/4 1.2500	1.2500	1.2390	2.312	2.110	0.600	0.211	0.193	0.291	0.197	0.500	0.155	0.105	2.052
1 3/8 1.3750	1.3750	1.3630	2.562	2.340	0.665	0.226	0.208	0.326	0.220	0.550	0.169	0.115	2.279
1 1/2 1.5000	1.5000	1.4880	2.812	2.570	0.742	0.258	0.240	0.360	0.244	0.600	0.183	0.124	2.505

[1]Where specifying nominal size in decimals, zeros preceding decimal and in the fourth decimal place shall be omitted.
[2]Tabulated values determined from formula for maximum H.
[3]No tolerance for gaging diameter is given. If the gaging diameter of the gage used differs from tabulated value, the protrusion will be affected accordingly and the proper protrusion values must be recalculated.

FOOTNOTES REFER TO ANSI B18.6.2 – 1972

(Reprinted from The American Society of Mechanical Engineers—ANSI B18.6.2-1972.)

TABLE 14 DIMENSIONS OF SLOTTED ROUND HEAD CAP SCREWS

CAP SCREWS
ROUND

Type of Head

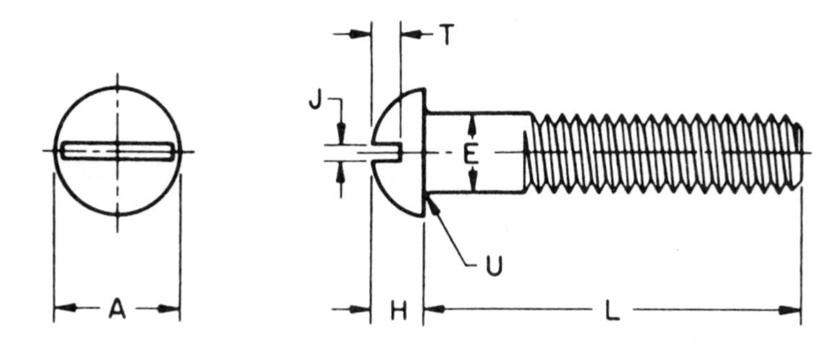

Nominal Size[1] or Basic Screw Diameter	E		A		H		J		T		U	
	Body Diameter		Head Diameter		Head Height		Slot Width		Slot Depth		Fillet Radius	
	Max	Min	Max	Min	Max	Min	Max	Min	Max	Min	Max	Min
1/4 0.2500	0.2500	0.2450	0.437	0.418	0.191	0.175	0.075	0.064	0.117	0.097	0.031	0.016
5/16 0.3125	0.3125	0.3070	0.562	0.540	0.245	0.226	0.084	0.072	0.151	0.126	0.031	0.016
3/8 0.3750	0.3750	0.3690	0.625	0.603	0.273	0.252	0.094	0.081	0.168	0.138	0.031	0.016
7/16 0.4375	0.4375	0.4310	0.750	0.725	0.328	0.302	0.094	0.081	0.202	0.167	0.047	0.016
1/2 0.5000	0.5000	0.4930	0.812	0.786	0.354	0.327	0.106	0.091	0.218	0.178	0.047	0.016
9/16 0.5625	0.5625	0.5550	0.937	0.909	0.409	0.378	0.118	0.102	0.252	0.207	0.047	0.016
5/8 0.6250	0.6250	0.6170	1.000	0.970	0.437	0.405	0.133	0.116	0.270	0.220	0.062	0.031
3/4 0.7500	0.7500	0.7420	1.250	1.215	0.546	0.507	0.149	0.131	0.338	0.278	0.062	0.031

[1] Where specifying nominal size in decimals, zeros preceding decimal and in the fourth decimal place shall be omitted.

(Reprinted from The American Society of Mechanical Engineers—ANSI B18.6.2-1972.)

TABLE 15 DIMENSIONS OF SLOTTED FILLISTER HEAD CAP SCREWS

CAP SCREWS
FILLISTER

Type of Head

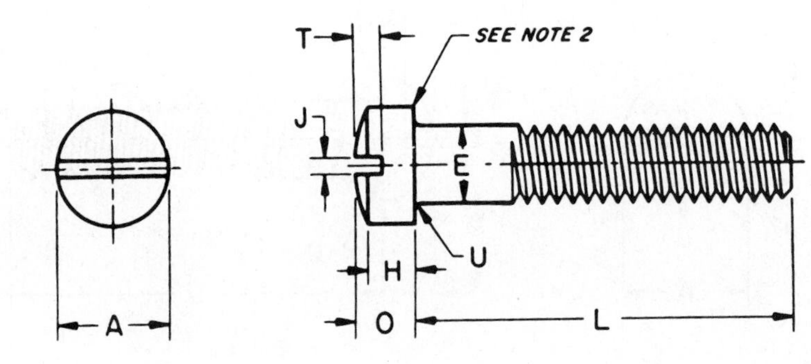

Nominal Size[1] or Basic Screw Diameter	E		A		H		O		J		T		U	
	Body Diameter		Head Diameter		Head Side Height		Total Head Height		Slot Width		Slot Depth		Fillet Radius	
	Max	Min	Max	Min	Max	Min	Max	Min	Max	Min	Max	Min	Max	Min
1/4 0.2500	0.2500	0.2450	0.375	0.363	0.172	0.157	0.216	0.194	0.075	0.064	0.097	0.077	0.031	0.016
5/16 0.3125	0.3125	0.3070	0.437	0.424	0.203	0.186	0.253	0.230	0.084	0.072	0.115	0.090	0.031	0.016
3/8 0.3750	0.3750	0.3690	0.562	0.547	0.250	0.229	0.314	0.284	0.094	0.081	0.142	0.112	0.031	0.016
7/16 0.4375	0.4375	0.4310	0.625	0.608	0.297	0.274	0.368	0.336	0.094	0.081	0.168	0.133	0.047	0.016
1/2 0.5000	0.5000	0.4930	0.750	0.731	0.328	0.301	0.413	0.376	0.106	0.091	0.193	0.153	0.047	0.016
9/16 0.5625	0.5625	0.5550	0.812	0.792	0.375	0.346	0.467	0.427	0.118	0.102	0.213	0.168	0.047	0.016
5/8 0.6250	0.6250	0.6170	0.875	0.853	0.422	0.391	0.521	0.478	0.133	0.116	0.239	0.189	0.062	0.031
3/4 0.7500	0.7500	0.7420	1.000	0.976	0.500	0.466	0.612	0.566	0.149	0.131	0.283	0.223	0.062	0.031
7/8 0.8750	0.8750	0.8660	1.125	1.098	0.594	0.556	0.720	0.668	0.167	0.147	0.334	0.264	0.062	0.031
1 1.0000	1.0000	0.9900	1.312	1.282	0.656	0.612	0.803	0.743	0.188	0.166	0.371	0.291	0.062	0.031

[1]Where specifying nominal size in decimals, zeros preceding decimal and in the fourth decimal place shall be omitted.
[2]A slight rounding of the edges at periphery of head shall be permissible provided the diameter of the bearing circle is equal to no less than 90 percent of the specified minimum head diameter.

(Reprinted from The American Society of Mechanical Engineers—ANSI B18.6.2-1972.)

TABLE 16 DIMENSIONS OF SLOTTED FLAT COUNTERSUNK HEAD MACHINE SCREWS

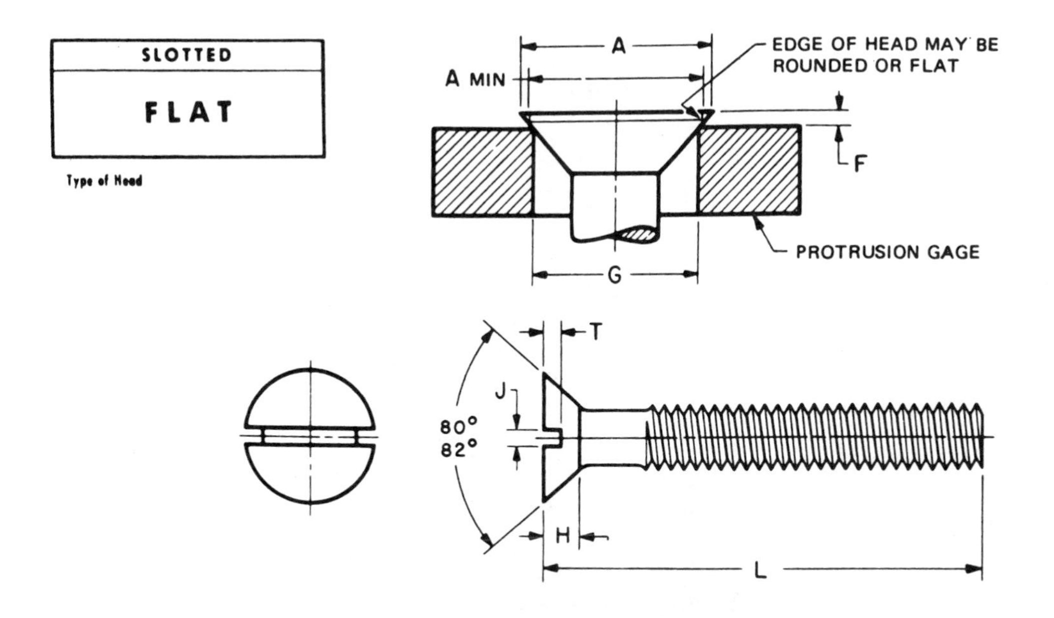

Nominal Size[1] or Basic Screw Diameter	L[2]	A		H[3]	J		T		F[4]		G[4]
	These Lengths or Shorter are Undercut.	Head Diameter		Head Height	Slot Width		Slot Depth		Protrusion Above Gaging Diameter		Gaging Diameter
		Max, Edge Sharp	Min, Edge Rounded or Flat	Ref	Max	Min	Max	Min	Max	Min	
0000 0.0210	—	0.043	0.037	0.011	0.008	0.004	0.007	0.003	*	*	*
000 0.0340	—	0.064	0.058	0.016	0.011	0.007	0.009	0.005	*	*	*
00 0.0470	—	0.093	0.085	0.028	0.017	0.010	0.014	0.009	*	*	*
0 0.0600	1/8	0.119	0.099	0.035	0.023	0.016	0.015	0.010	0.026	0.016	0.078
1 0.0730	1/8	0.146	0.123	0.043	0.026	0.019	0.019	0.012	0.028	0.016	0.101
2 0.0860	1/8	0.172	0.147	0.051	0.031	0.023	0.023	0.015	0.029	0.017	0.124
3 0.0990	1/8	0.199	0.171	0.059	0.035	0.027	0.027	0.017	0.031	0.018	0.148
4 0.1120	3/16	0.225	0.195	0.067	0.039	0.031	0.030	0.020	0.032	0.019	0.172
5 0.1250	3/16	0.252	0.220	0.075	0.043	0.035	0.034	0.022	0.034	0.020	0.196
6 0.1380	3/16	0.279	0.244	0.083	0.048	0.039	0.038	0.024	0.036	0.021	0.220
8 0.1640	1/4	0.332	0.292	0.100	0.054	0.045	0.045	0.029	0.039	0.023	0.267
10 0.1900	5/16	0.385	0.340	0.116	0.060	0.050	0.053	0.034	0.042	0.025	0.313
12 0.2160	3/8	0.438	0.389	0.132	0.067	0.056	0.060	0.039	0.045	0.027	0.362
1/4 0.2500	7/16	0.507	0.452	0.153	0.075	0.064	0.070	0.046	0.050	0.029	0.424
5/16 0.3125	1/2	0.635	0.568	0.191	0.084	0.072	0.088	0.058	0.057	0.034	0.539
3/8 0.3750	9/16	0.762	0.685	0.230	0.094	0.081	0.106	0.070	0.065	0.039	0.653
7/16 0.4375	5/8	0.812	0.723	0.223	0.094	0.081	0.103	0.066	0.073	0.044	0.690
1/2 0.5000	3/4	0.875	0.775	0.223	0.106	0.091	0.103	0.065	0.081	0.049	0.739
9/16 0.5625	—	1.000	0.889	0.260	0.118	0.102	0.120	0.077	0.089	0.053	0.851
5/8 0.6250	—	1.125	1.002	0.298	0.133	0.116	0.137	0.088	0.097	0.058	0.962
3/4 0.7500	—	1.375	1.230	0.372	0.149	0.131	0.171	0.111	0.112	0.067	1.186

[1]Where specifying nominal size in decimals, zeros preceding decimal and in the fourth decimal place shall be omitted.
[2]Screws of these lengths and shorter shall have undercut heads.
[3]Tabulated values determined from formula for maximum H.
[4]No tolerance for gaging diameter is given. If the gaging diameter of the gage used differs from tabulated value, the protrusion will be affected accordingly, and the proper protrusion values must be recalculated.
*Not practical to gage.

For additional requirements refer to General Data on Pages 3, 4, and 5.

FOOTNOTES REFER TO ANSI B18.6.3 – 1972

(Reprinted from The American Society of Mechanical Engineers—ANSI B18.6.3-1972.)

TABLE 17 DIMENSIONS OF HEX NUTS AND HEX JAM NUTS

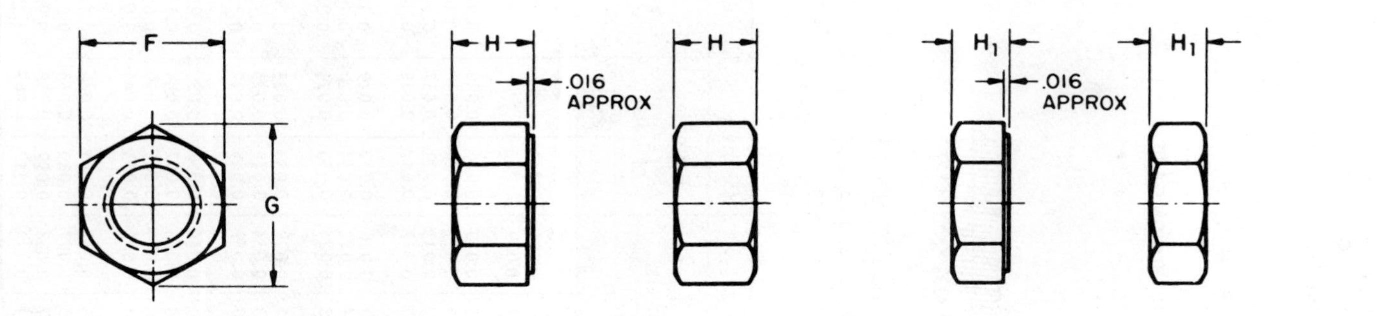

Nominal Size or Basic Major Dia of Thread		F Width Across Flats			G Width Across Corners		H Thickness Hex Nuts			H_1 Thickness Hex Jam Nuts			Hex Nuts Specified Proof Load: Up to 150,000 psi	Hex Nuts Specified Proof Load: 150,000 psi and Greater	Jam Nuts All Strength Levels
		Basic	Max	Min	Max	Min	Basic	Max	Min	Basic	Max	Min	Runout of Bearing Face, FIR Max		
1/4	0.2500	7/16	0.438	0.428	0.505	0.488	7/32	0.226	0.212	5/32	0.163	0.150	0.015	0.010	0.015
5/16	0.3125	1/2	0.500	0.489	0.577	0.557	17/64	0.273	0.258	3/16	0.195	0.180	0.016	0.011	0.016
3/8	0.3750	9/16	0.562	0.551	0.650	0.628	21/64	0.337	0.320	7/32	0.227	0.210	0.017	0.012	0.017
7/16	0.4375	11/16	0.688	0.675	0.794	0.768	3/8	0.385	0.365	1/4	0.260	0.240	0.018	0.013	0.018
1/2	0.5000	3/4	0.750	0.736	0.866	0.840	7/16	0.448	0.427	5/16	0.323	0.302	0.019	0.014	0.019
9/16	0.5625	7/8	0.875	0.861	1.010	0.982	31/64	0.496	0.473	5/16	0.324	0.301	0.020	0.015	0.020
5/8	0.6250	15/16	0.938	0.922	1.083	1.051	35/64	0.559	0.535	3/8	0.387	0.363	0.021	0.016	0.021
3/4	0.7500	1 1/8	1.125	1.088	1.299	1.240	41/64	0.665	0.617	27/64	0.446	0.398	0.023	0.018	0.023
7/8	0.8750	1 5/16	1.312	1.269	1.516	1.447	3/4	0.776	0.724	31/64	0.510	0.458	0.025	0.020	0.025
1	1.0000	1 1/2	1.500	1.450	1.732	1.653	55/64	0.887	0.831	35/64	0.575	0.519	0.027	0.022	0.027
1 1/8	1.1250	1 11/16	1.688	1.631	1.949	1.859	31/32	0.999	0.939	39/64	0.639	0.579	0.030	0.025	0.030
1 1/4	1.2500	1 7/8	1.875	1.812	2.165	2.066	1 1/16	1.094	1.030	23/32	0.751	0.687	0.033	0.028	0.033
1 3/8	1.3750	2 1/16	2.062	1.994	2.382	2.273	1 11/64	1.206	1.138	25/32	0.815	0.747	0.036	0.031	0.036
1 1/2	1.5000	2 1/4	2.250	2.175	2.598	2.480	1 9/32	1.317	1.245	27/32	0.880	0.808	0.039	0.034	0.039

(Reprinted from The American Society of Mechanical Engineers—ANSI B18.2.2-1972.)

TABLE 18 WOODRUFF KEY DIMENSIONS

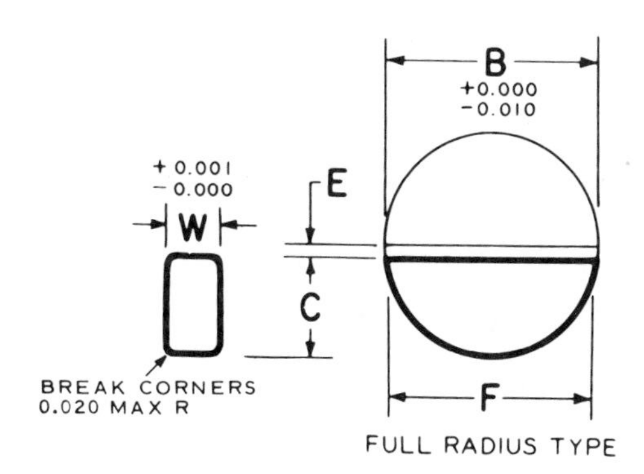

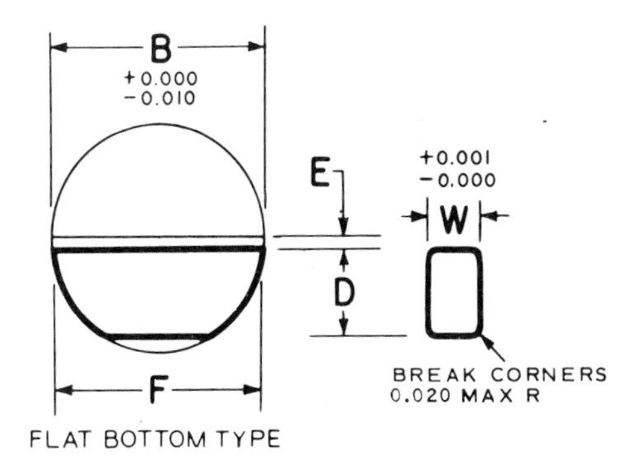

Key No.	Nominal Key Size W × B	Actual Length F +0.000-0.010	Height of Key				Distance Below Center E
			C		D		
			Max	Min	Max	Min	
202	1/16 × 1/4	0.248	0.109	0.104	0.109	0.104	1/64
202.5	1/16 × 5/16	0.311	0.140	0.135	0.140	0.135	1/64
302.5	3/32 × 5/16	0.311	0.140	0.135	0.140	0.135	1/64
203	1/16 × 3/8	0.374	0.172	0.167	0.172	0.167	1/64
303	3/32 × 3/8	0.374	0.172	0.167	0.172	0.167	1/64
403	1/8 × 3/8	0.374	0.172	0.167	0.172	0.167	1/64
204	1/16 × 1/2	0.491	0.203	0.198	0.194	0.188	3/64
304	3/32 × 1/2	0.491	0.203	0.198	0.194	0.188	3/64
404	1/8 × 1/2	0.491	0.203	0.198	0.194	0.188	3/64
305	3/32 × 5/8	0.612	0.250	0.245	0.240	0.234	1/16
405	1/8 × 5/8	0.612	0.250	0.245	0.240	0.234	1/16
505	5/32 × 5/8	0.612	0.250	0.245	0.240	0.234	1/16
605	3/16 × 5/8	0.612	0.250	0.245	0.240	0.234	1/16
406	1/8 × 3/4	0.740	0.313	0.308	0.303	0.297	1/16
506	5/32 × 3/4	0.740	0.313	0.308	0.303	0.297	1/16
606	3/16 × 3/4	0.740	0.313	0.308	0.303	0.297	1/16
806	1/4 × 3/4	0.740	0.313	0.308	0.303	0.297	1/16
507	5/32 × 7/8	0.866	0.375	0.370	0.365	0.359	1/16
607	3/16 × 7/8	0.866	0.375	0.370	0.365	0.359	1/16
707	7/32 × 7/8	0.866	0.375	0.370	0.365	0.359	1/16
807	1/4 × 7/8	0.866	0.375	0.370	0.365	0.359	1/16
608	3/16 × 1	0.992	0.438	0.433	0.428	0.422	1/16
708	7/32 × 1	0.992	0.438	0.433	0.428	0.422	1/16
808	1/4 × 1	0.992	0.438	0.433	0.428	0.422	1/16
1008	5/16 × 1	0.992	0.438	0.433	0.428	0.422	1/16
1208	3/8 × 1	0.992	0.438	0.433	0.428	0.422	1/16
609	3/16 × 1 1/8	1.114	0.484	0.479	0.475	0.469	5/64
709	7/32 × 1 1/8	1.114	0.484	0.479	0.475	0.469	5/64
809	1/4 × 1 1/8	1.114	0.484	0.479	0.475	0.469	5/64
1009	5/16 × 1 1/8	1.114	0.484	0.479	0.475	0.469	5/64

TABLE 18 (CONTINUED)

Key No.	Nominal Key Size W × B	Actual Length F +0.000-0.010	Height of Key				Distance Below Center E
			C		D		
			Max	Min	Max	Min	
610	3/16 × 1¼	1.240	0.547	0.542	0.537	0.531	5/64
710	7/32 × 1¼	1.240	0.547	0.542	0.537	0.531	5/64
810	¼ × 1¼	1.240	0.547	0.542	0.537	0.531	5/64
1010	5/16 × 1¼	1.240	0.547	0.542	0.537	0.531	5/64
1210	⅜ × 1¼	1.240	0.547	0.542	0.537	0.531	5/64
811	¼ × 1⅜	1.362	0.594	0.589	0.584	0.578	3/32
1011	5/16 × 1⅜	1.362	0.594	0.589	0.584	0.578	3/32
1211	⅜ × 1⅜	1.362	0.594	0.589	0.584	0.578	3/32
812	¼ × 1½	1.484	0.641	0.636	0.631	0.625	7/64
1012	5/16 × 1½	1.484	0.641	0.636	0.631	0.625	7/64
1212	⅜ × 1½	1.484	0.641	0.636	0.631	0.625	7/64

All dimensions given are in inches.

The key numbers indicate nominal key dimensions. The last two digits give the nominal diameter B in eighths of an inch, and the digits preceding the last two give the nominal width W in thirty-seconds of an inch.

Example:

No. 204 indicates a key 2/32 × 4/8 or 1/16 × ½.
No. 808 indicates a key 8/32 × 8/8 or ¼ × 1.
No. 1212 indicates a key 12/32 × 12/8 or ⅜ × 1½.

TABLE 18 (CONTINUED)

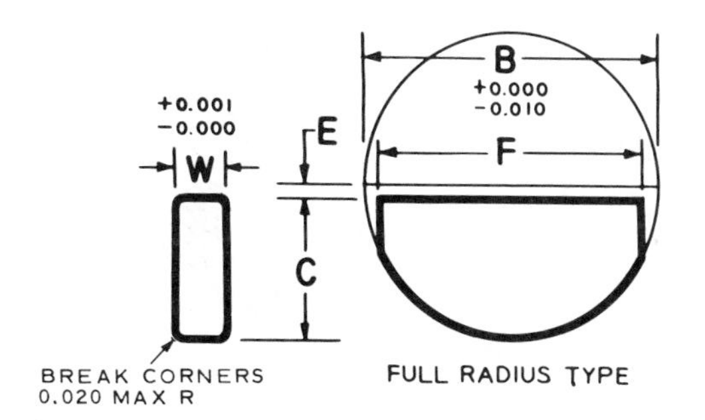

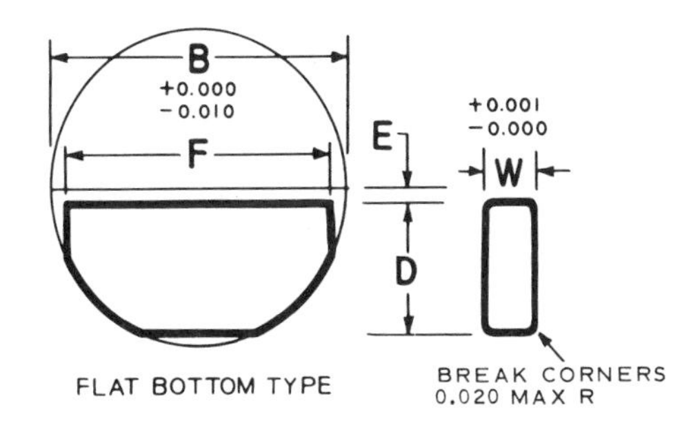

Key No.	Nominal Key Size W × B	Actual Length F +0.000-0.010	Height of Key				Distance Below Center E
			C		D		
			Max	Min	Max	Min	
617-1	3/16 × 2 1/8	1.380	0.406	0.401	0.396	0.390	21/32
817-1	1/4 × 2 1/8	1.380	0.406	0.401	0.396	0.390	21/32
1017-1	5/16 × 2 1/8	1.380	0.406	0.401	0.396	0.390	21/32
1217-1	3/8 × 2 1/8	1.380	0.406	0.401	0.396	0.390	21/32
617	3/16 × 2 1/8	1.723	0.531	0.526	0.521	0.515	17/32
817	1/4 × 2 1/8	1.723	0.531	0.526	0.521	0.515	17/32
1017	5/16 × 2 1/8	1.723	0.531	0.526	0.521	0.515	17/32
1217	3/8 × 2 1/8	1.723	0.531	0.526	0.521	0.515	17/32
822-1	1/4 × 2 3/4	2.000	0.594	0.589	0.584	0.578	25/32
1022-1	5/16 × 2 3/4	2.000	0.594	0.589	0.584	0.578	25/32
1222-1	3/8 × 2 3/4	2.000	0.594	0.589	0.584	0.578	25/32
1422-1	7/16 × 2 3/4	2.000	0.594	0.589	0.584	0.578	25/32
1622-1	1/2 × 2 3/4	2.000	0.594	0.589	0.584	0.578	25/32
822	1/4 × 2 3/4	2.317	0.750	0.745	0.740	0.734	5/8
1022	5/16 × 2 3/4	2.317	0.750	0.745	0.740	0.734	5/8
1222	3/8 × 2 3/4	2.317	0.750	0.745	0.740	0.734	5/8
1422	7/16 × 2 3/4	2.317	0.750	0.745	0.740	0.734	5/8
1622	1/2 × 2 3/4	2.317	0.750	0.745	0.740	0.734	5/8
1228	3/8 × 3 1/2	2.880	0.938	0.933	0.928	0.922	13/16
1428	7/16 × 3 1/2	2.880	0.938	0.933	0.928	0.922	13/16
1628	1/2 × 3 1/2	2.880	0.938	0.933	0.928	0.922	13/16
1828	9/16 × 3 1/2	2.880	0.938	0.933	0.928	0.922	13/16
2028	5/8 × 3 1/2	2.880	0.938	0.933	0.928	0.922	13/16
2228	11/16 × 3 1/2	2.880	0.938	0.933	0.928	0.922	13/16
2428	3/4 × 3 1/2	2.880	0.938	0.933	0.928	0.922	13/16

All dimensions given are in inches.

The key numbers indicate nominal key dimensions. The last two digits give the nominal diameter B in eighths of an inch, and the digits preceding the last two give the nominal width W in thirty-seconds of an inch.

Example:

No. 617 indicates a key 6/32 × 17/8 or 3/16 × 2 1/8
No. 822 indicates a key 8/32 × 22/8 or 1/4 × 2 3/4
No. 1228 indicates a key 12/32 × 28/8 or 3/8 × 3 1/2

The key numbers with the -1 designation, while representing the nominal key size, have a shorter length F and due to a greater distance below center E are less in height than the keys of the same number without the -1 designation.

(Reprinted from The American Society of Mechanical Engineers—ANSI B17.2-1967.)

TABLE 19 WOODRUFF KEYSEAT DIMENSIONS

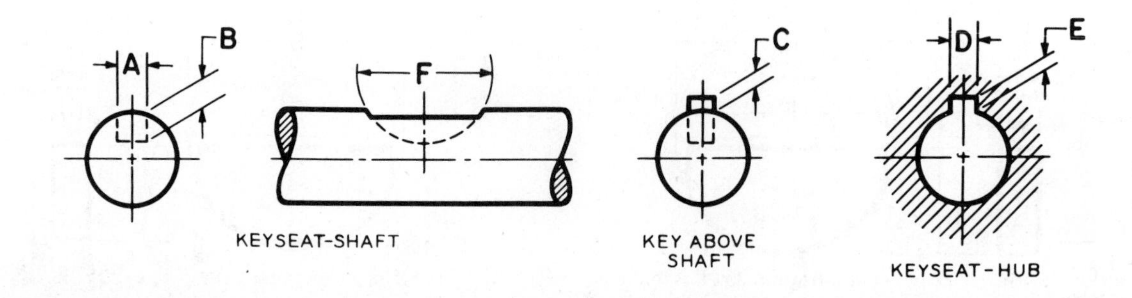

Key Number	Nominal Size Key	Keyseat – Shaft					Key Above Shaft	Keyseat – Hub	
		Width A*		Depth B	Diameter F		Height C	Width D	Depth E
		Min	Max	+0.005 -0.000	Min	Max	+0.005 -0.005	+0.002 -0.000	+0.005 -0.000
202	1/16 × 1/4	0.0615	0.0630	0.0728	0.250	0.268	0.0312	0.0635	0.0372
202.5	1/16 × 5/16	0.0615	0.0630	0.1038	0.312	0.330	0.0312	0.0635	0.0372
302.5	3/32 × 5/16	0.0928	0.0943	0.0882	0.312	0.330	0.0469	0.0948	0.0529
203	1/16 × 3/8	0.0615	0.0630	0.1358	0.375	0.393	0.0312	0.0635	0.0372
303	3/32 × 3/8	0.0928	0.0943	0.1202	0.375	0.393	0.0469	0.0948	0.0529
403	1/8 × 3/8	0.1240	0.1255	0.1045	0.375	0.393	0.0625	0.1260	0.0685
204	1/16 × 1/2	0.0615	0.0630	0.1668	0.500	0.518	0.0312	0.0635	0.0372
304	3/32 × 1/2	0.0928	0.0943	0.1511	0.500	0.518	0.0469	0.0948	0.0529
404	1/8 × 1/2	0.1240	0.1255	0.1355	0.500	0.518	0.0625	0.1260	0.0685
305	3/32 × 5/8	0.0928	0.0943	0.1981	0.625	0.643	0.0469	0.0948	0.0529
405	1/8 × 5/8	0.1240	0.1255	0.1825	0.625	0.643	0.0625	0.1260	0.0685
505	5/32 × 5/8	0.1553	0.1568	0.1669	0.625	0.643	0.0781	0.1573	0.0841
605	3/16 × 5/8	0.1863	0.1880	0.1513	0.625	0.643	0.0937	0.1885	0.0997
406	1/8 × 3/4	0.1240	0.1255	0.2455	0.750	0.768	0.0625	0.1260	0.0685
506	5/32 × 3/4	0.1553	0.1568	0.2299	0.750	0.768	0.0781	0.1573	0.0841
606	3/16 × 3/4	0.1863	0.1880	0.2143	0.750	0.768	0.0937	0.1885	0.0997
806	1/4 × 3/4	0.2487	0.2505	0.1830	0.750	0.768	0.1250	0.2510	0.1310
507	5/32 × 7/8	0.1553	0.1568	0.2919	0.875	0.895	0.0781	0.1573	0.0841
607	3/16 × 7/8	0.1863	0.1880	0.2763	0.875	0.895	0.0937	0.1885	0.0997
707	7/32 × 7/8	0.2175	0.2193	0.2607	0.875	0.895	0.1093	0.2198	0.1153
807	1/4 × 7/8	0.2487	0.2505	0.2450	0.875	0.895	0.1250	0.2510	0.1310
608	3/16 × 1	0.1863	0.1880	0.3393	1.000	1.020	0.0937	0.1885	0.0997
708	7/32 × 1	0.2175	0.2193	0.3237	1.000	1.020	0.1093	0.2198	0.1153
808	1/4 × 1	0.2487	0.2505	0.3080	1.000	1.020	0.1250	0.2510	0.1310
1008	5/16 × 1	0.3111	0.3130	0.2768	1.000	1.020	0.1562	0.3135	0.1622
1208	3/8 × 1	0.3735	0.3755	0.2455	1.000	1.020	0.1875	0.3760	0.1935
609	3/16 × 1 1/8	0.1863	0.1880	0.3853	1.125	1.145	0.0937	0.1885	0.0997
709	7/32 × 1 1/8	0.2175	0.2193	0.3697	1.125	1.145	0.1093	0.2198	0.1153
809	1/4 × 1 1/8	0.2487	0.2505	0.3540	1.125	1.145	0.1250	0.2510	0.1310
1009	5/16 × 1 1/8	0.3111	0.3130	0.3228	1.125	1.145	0.1562	0.3135	0.1622

TABLE 19 (CONTINUED)

Key Number	Nominal Size Key	Keyseat – Shaft					Key Above Shaft	Keyseat – Hub	
		Width A*		Depth B	Diameter F		Height C	Width D	Depth E
		Min	Max	+0.005 -0.000	Min	Max	+0.005 -0.005	+0.002 -0.000	+0.005 -0.000
610	$\frac{3}{16} \times 1\frac{1}{4}$	0.1863	0.1880	0.4483	1.250	1.273	0.0937	0.1885	0.0997
710	$\frac{7}{32} \times 1\frac{1}{4}$	0.2175	0.2193	0.4327	1.250	1.273	0.1093	0.2198	0.1153
810	$\frac{1}{4} \times 1\frac{1}{4}$	0.2487	0.2505	0.4170	1.250	1.273	0.1250	0.2510	0.1310
1010	$\frac{5}{16} \times 1\frac{1}{4}$	0.3111	0.3130	0.3858	1.250	1.273	0.1562	0.3135	0.1622
1210	$\frac{3}{8} \times 1\frac{1}{4}$	0.3735	0.3755	0.3545	1.250	1.273	0.1875	0.3760	0.1935
811	$\frac{1}{4} \times 1\frac{3}{8}$	0.2487	0.2505	0.4640	1.375	1.398	0.1250	0.2510	0.1310
1011	$\frac{5}{16} \times 1\frac{3}{8}$	0.3111	0.3130	0.4328	1.375	1.398	0.1562	0.3135	0.1622
1211	$\frac{3}{8} \times 1\frac{3}{8}$	0.3735	0.3755	0.4015	1.375	1.398	0.1875	0.3760	0.1935
812	$\frac{1}{4} \times 1\frac{1}{2}$	0.2487	0.2505	0.5110	1.500	1.523	0.1250	0.2510	0.1310
1012	$\frac{5}{16} \times 1\frac{1}{2}$	0.3111	0.3130	0.4798	1.500	1.523	0.1562	0.3135	0.1622
1212	$\frac{3}{8} \times 1\frac{1}{2}$	0.3735	0.3755	0.4485	1.500	1.523	0.1875	0.3760	0.1935
617-1	$\frac{3}{16} \times 2\frac{1}{8}$	0.1863	0.1880	0.3073	2.125	2.160	0.0937	0.1885	0.0997
817-1	$\frac{1}{4} \times 2\frac{1}{8}$	0.2487	0.2505	0.2760	2.125	2.160	0.1250	0.2510	0.1310
1017-1	$\frac{5}{16} \times 2\frac{1}{8}$	0.3111	0.3130	0.2448	2.125	2.160	0.1562	0.3135	0.1622
1217-1	$\frac{3}{8} \times 2\frac{1}{8}$	0.3735	0.3755	0.2135	2.125	2.160	0.1875	0.3760	0.1935
617	$\frac{3}{16} \times 2\frac{1}{8}$	0.1863	0.1880	0.4323	2.125	2.160	0.0937	0.1885	0.0997
817	$\frac{1}{4} \times 2\frac{1}{8}$	0.2487	0.2505	0.4010	2.125	2.160	0.1250	0.2510	0.1310
1017	$\frac{5}{16} \times 2\frac{1}{8}$	0.3111	0.3130	0.3698	2.125	2.160	0.1562	0.3135	0.1622
1217	$\frac{3}{8} \times 2\frac{1}{8}$	0.3735	0.3755	0.3385	2.125	2.160	0.1875	0.3760	0.1935
822-1	$\frac{1}{4} \times 2\frac{3}{4}$	0.2487	0.2505	0.4640	2.750	2.785	0.1250	0.2510	0.1310
1022-1	$\frac{5}{16} \times 2\frac{3}{4}$	0.3111	0.3130	0.4328	2.750	2.785	0.1562	0.3135	0.1622
1222-1	$\frac{3}{8} \times 2\frac{3}{4}$	0.3735	0.3755	0.4015	2.750	2.785	0.1875	0.3760	0.1935
1422-1	$\frac{7}{16} \times 2\frac{3}{4}$	0.4360	0.4380	0.3703	2.750	2.785	0.2187	0.4385	0.2247
1622-1	$\frac{1}{2} \times 2\frac{3}{4}$	0.4985	0.5005	0.3390	2.750	2.785	0.2500	0.5010	0.2560
822	$\frac{1}{4} \times 2\frac{3}{4}$	0.2487	0.2505	0.6200	2.750	2.785	0.1250	0.2510	0.1310
1022	$\frac{5}{16} \times 2\frac{3}{4}$	0.3111	0.3130	0.5888	2.750	2.785	0.1562	0.3135	0.1622
1222	$\frac{3}{8} \times 2\frac{3}{4}$	0.3735	0.3755	0.5575	2.750	2.785	0.1875	0.3760	0.1935
1422	$\frac{7}{16} \times 2\frac{3}{4}$	0.4360	0.4380	0.5263	2.750	2.785	0.2187	0.4385	0.2247
1622	$\frac{1}{2} \times 2\frac{3}{4}$	0.4985	0.5005	0.4950	2.750	2.785	0.2500	0.5010	0.2560
1228	$\frac{3}{8} \times 3\frac{1}{2}$	0.3735	0.3755	0.7455	3.500	3.535	0.1875	0.3760	0.1935
1428	$\frac{7}{16} \times 3\frac{1}{2}$	0.4360	0.4380	0.7143	3.500	3.535	0.2187	0.4385	0.2247
1628	$\frac{1}{2} \times 3\frac{1}{2}$	0.4985	0.5005	0.6830	3.500	3.535	0.2500	0.5010	0.2560
1828	$\frac{9}{16} \times 3\frac{1}{2}$	0.5610	0.5630	0.6518	3.500	3.535	0.2812	0.5635	0.2872
2028	$\frac{5}{8} \times 3\frac{1}{2}$	0.6235	0.6255	0.6205	3.500	3.535	0.3125	0.6260	0.3185
2228	$\frac{11}{16} \times 3\frac{1}{2}$	0.6860	0.6880	0.5893	3.500	3.535	0.3437	0.6885	0.3497
2428	$\frac{3}{4} \times 3\frac{1}{2}$	0.7485	0.7505	0.5580	3.500	3.535	0.3750	0.7510	0.3810

Width A values were set with the maximum keyseat (shaft) width as that figure which will receive a key with the greatest amount of looseness consistent with assuring the key's sticking in the keyseat (shaft). Minimum keyseat width is that figure permitting the largest shaft distortion acceptable when assembling maximum key in minimum keyseat.

Dimensions A, B, C, D are taken at side intersection.

(Reprinted from The American Society of Mechanical Engineers—ANSI B17.2-1967.)

TABLE 20 KEY SIZE VERSUS SHAFT DIAMETER

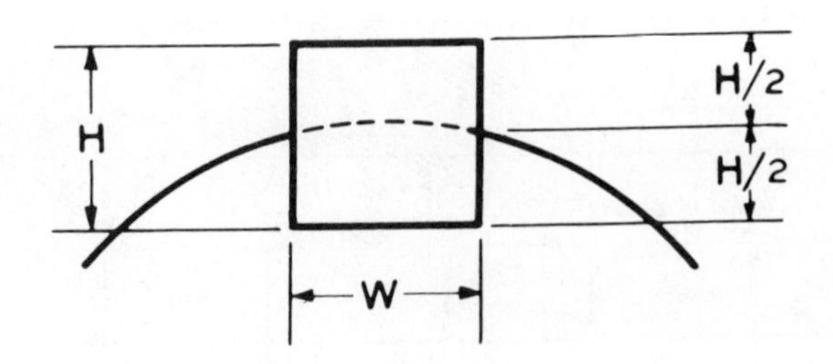

NOMINAL SHAFT DIAMETER		NOMINAL KEY SIZE			NOMINAL KEYSEAT DEPTH	
			Height, H		$H/2$	
Over	To (Incl)	Width, W	Square	Rectangular	Square	Rectangular
5/16	7/16	3/32	3/32	3/32	3/64	3/64
7/16	9/16	1/8	1/8	1/8	1/16	1/16
9/16	7/8	3/16	3/16	3/16	3/32	3/32
7/8	1-1/4	1/4	1/4	1/4	1/8	1/8
1-1/4	1-3/8	5/16	5/16		5/32	
1-3/8	1-3/4	3/8	3/8	1/4	3/16	1/8
1-3/4	2-1/4	1/2	1/2	3/8	1/4	3/16
2-1/4	2-3/4	5/8	5/8	7/16	5/16	7/32
2-3/4	3-1/4	3/4	3/4	1/2	3/8	1/4
3-1/4	3-3/4	7/8	7/8	5/8	7/16	5/16
3-3/4	4-1/2	1	1	3/4	1/2	3/8
4-1/2	5-1/2	1-1/4	1-1/4	7/8	5/8	7/16
5-1/2	6-1/2	1-1/2	1-1/2	1	3/4	1/2
6-1/2	7-1/2	1-3/4	1-3/4	1-1/2*	7/8	3/4
7-1/2	9	2	2	1-1/2	1	3/4
9	11	2-1/2	2-1/2	1-3/4	1-1/4	7/8
11	13	3	3	2	1-1/2	1
13	15	3-1/2	3-1/2	2-1/2	1-3/4	1-1/4
15	18	4		3		1-1/2
18	22	5		3-1/2		1-3/4
22	26	6		4		2
26	30	7		5		2-1/2

*Some key standards show 1-1/4 in. Preferred size is 1-1/2 in.

All dimensions given in inches.

Shaded areas:

For a stepped shaft, the size of a key is determined by the diameter of the shaft at the point of location of the key, regardless of the number of different diameters on the shaft.

Square keys are preferred through 6½-inch diameter shafts and rectangular keys for larger shafts. Sizes and dimensions in unshaded area are preferred.

If special considerations dictate the use of a keyseat in the hub shallower than the preferred nominal depth, it is recommended that the tabulated preferred nominal standard keyseat be used in the shaft in all cases.

(Reprinted from The American Society of Mechanical Engineers—ANSI B17.1-1967.)

TABLE 21 KEY DIMENSIONS AND TOLERANCES

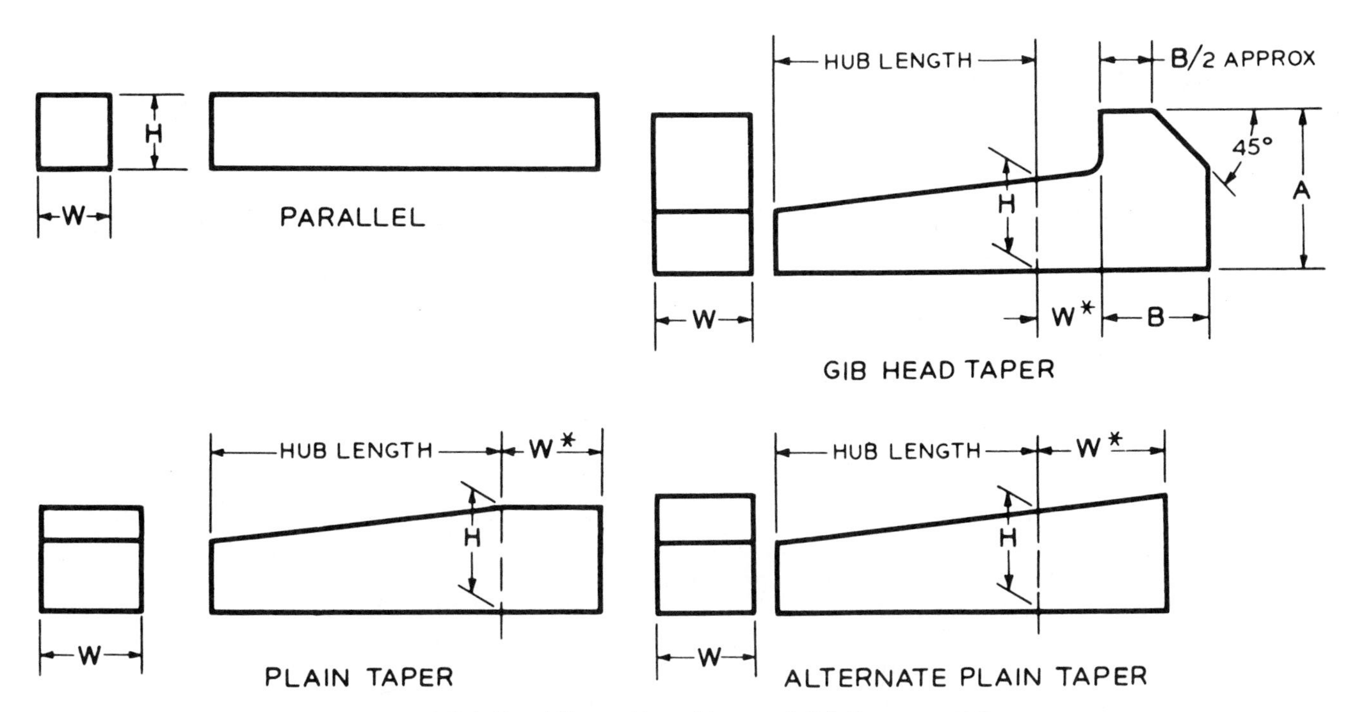

Plain and Gib Head Taper Keys Have a 1/8" Taper in 12"

KEY			NOMINAL KEY SIZE Width, W		TOLERANCE	
			Over	To (Incl)	Width, W	Height, H
Parallel	Square	Bar Stock	–	3/4	+0.000 –0.002	+0.000 –0.002
			3/4	1-1/2	+0.000 –0.003	+0.000 –0.003
			1-1/2	2-1/2	+0.000 –0.004	+0.000 –0.004
			2-1/2	3-1/2	+0.000 –0.006	+0.000 –0.006
		Keystock	–	1-1/4	+0.001 –0.000	+0.001 –0.000
			1-1/4	3	+0.002 –0.000	+0.002 –0.000
			3	3-1/2	+0.003 –0.000	+0.003 –0.000
	Rectangular	Bar Stock	–	3/4	+0.000 –0.003	+0.000 –0.003
			3/4	1-1/2	+0.000 –0.004	+0.000 –0.004
			1-1/2	3	+0.000 –0.005	+0.000 –0.005
			3	4	+0.000 –0.006	+0.000 –0.006
			4	6	+0.000 –0.008	+0.000 –0.008
			6	7	+0.000 –0.013	+0.000 –0.013
		Keystock	–	1-1/4	+0.001 –0.000	+0.005 –0.005
			1-1/4	3	+0.002 –0.000	+0.005 –0.005
			3	7	+0.003 –0.000	+0.005 –0.005
Taper	Plain or Gib Head Square or Rectangular		–	1-1/4	+0.001 –0.000	+0.005 –0.000
			1-1/4	3	+0.002 –0.000	+0.005 –0.000
			3	7	+0.003 –0.000	+0.005 –0.000

*For locating position of dimension H. Tolerance does not apply.
All dimensions given in inches.

TABLE 22 GIB HEAD NOMINAL DIMENSIONS

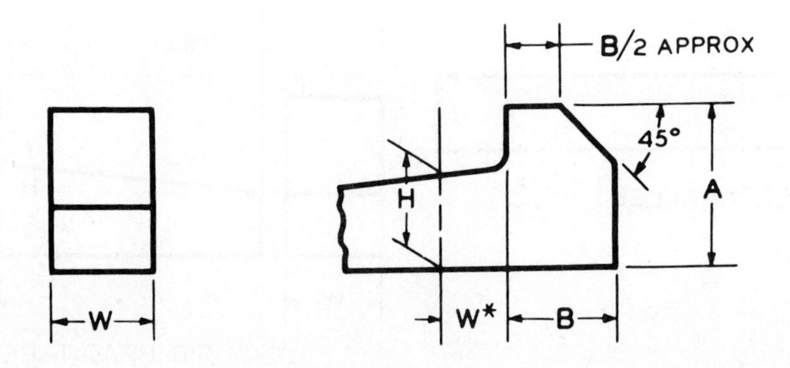

Nominal Key Size Width, *W*	SQUARE			RECTANGULAR		
	H	*A*	*B*	*H*	*A*	*B*
1/8	1/8	1/4	1/4	3/32	3/16	1/8
3/16	3/16	5/16	5/16	1/8	1/4	1/4
1/4	1/4	7/16	3/8	3/16	5/16	5/16
5/16	5/16	1/2	7/16	1/4	7/16	3/8
3/8	3/8	5/8	1/2	1/4	7/16	3/8
1/2	1/2	7/8	5/8	3/8	5/8	1/2
5/8	5/8	1	3/4	7/16	3/4	9/16
3/4	3/4	1-1/4	7/8	1/2	7/8	5/8
7/8	7/8	1-3/8	1	5/8	1	3/4
1	1	1-5/8	1-1/8	3/4	1-1/4	7/8
1-1/4	1-1/4	2	1-7/16	7/8	1-3/8	1
1-1/2	1-1/2	2-3/8	1-3/4	1	1-5/8	1-1/8
1-3/4	1-3/4	2-3/4	2	1-1/2	2-3/8	1-3/4
2	2	3-1/2	2-1/4	1-1/2	2-3/8	1-3/4
2-1/2	2-1/2	4	3	1-3/4	2-3/4	2
3	3	5	3-1/2	2	3-1/2	2-1/4
3-1/2	3-1/2	6	4	2-1/2	4	3

*For locating position of dimension *H*.

For larger sizes the following relationships are suggested as guides for establishing *A* and *B*:

$A = 1.8\,H \qquad B = 1.2\,H$

All dimensions given in inches.

(Reprinted from The American Society of Mechanical Engineers—ANSI B17.1-1967.)

TABLE 23 CLASS 2 FIT FOR PARALLEL AND TAPER KEYS

Type of Key	KEY WIDTH		SIDE FIT			TOP AND BOTTOM FIT			
	Over	To (Incl)	Width Tolerance		Fit Range*	Depth Tolerance			Fit Range*
			Key	Keyset		Key	Shaft Keyseat	Hub Keyseat	
Parallel Square	–	1-1/4	+0.001 –0.000	+0.002 –0.000	0.002 CL 0.001 INT	+0.001 –0.000	+0.000 –0.015	+0.010 –0.000	0.030 CL 0.004 CL
	1-1/4	3	+0.002 –0.000	+0.002 –0.000	0.002 CL 0.002 INT	+0.002 –0.000	+0.000 –0.015	+0.010 –0.000	0.030 CL 0.003 CL
	3	3-1/2	+0.003 –0.000	+0.002 –0.000	0.002 CL 0.003 INT	+0.003 –0.000	+0.000 –0.015	+0.010 –0.000	0.030 CL 0.002 CL
Parallel Rectangular	–	1-1/4	+0.001 –0.000	+0.002 –0.000	0.002 CL 0.001 INT	+0.005 –0.005	+0.000 –0.015	+0.010 –0.000	0.035 CL 0.000 CL
	1-1/4	3	+0.002 –0.000	+0.002 –0.000	0.002 CL 0.002 INT	+0.005 –0.005	+0.000 –0.015	+0.010 –0.000	0.035 CL 0.000 CL
	3	7	+0.003 –0.000	+0.002 –0.000	0.002 CL 0.003 INT	+0.005 –0.005	+0.000 –0.015	+0.010 –0.000	0.035 CL 0.000 CL
Taper	–	1-1/4	+0.001 –0.000	+0.002 –0.000	0.002 CL 0.001 INT	+0.005 –0.000	+0.000 –0.015	+0.010 –0.000	0.005 CL 0.025 INT
	1-1/4	3	+0.002 –0.000	+0.002 –0.000	0.002 CL 0.002 INT	+0.005 –0.000	+0.000 –0.015	+0.010 –0.000	0.005 CL 0.025 INT
	3	Δ	+0.003 –0.000	+0.002 –0.000	0.002 CL 0.003 INT	+0.005 –0.000	+0.000 –0.015	+0.010 –0.000	0.005 CL 0.025 INT

*Limits of variation. CL= Clearance; INT = Interference

Δ To (Incl) 3-1/2 Square and 7 Rectangular key widths.

All dimensions given in inches.

(Reprinted from The American Society of Mechanical Engineers—ANSI B17.1-1967.)

TABLE 24 DECIMAL EQUIVALENTS AND TAP DRILL SIZES

FRACTION OR DRILL SIZE	DECIMAL EQUIVALENT	TAP SIZE
NUMBER SIZE DRILLS 80	.0135	
79	.0145	
1/64	.0156	
78	.0160	
77	.0180	
76	.0200	
75	.0210	
74	.0225	
73	.0240	
72	.0250	
71	.0260	
70	.0280	
69	.0292	
68	.0310	
1/32	.0312	
67	.0320	
66	.0330	
65	.0350	
64	.0360	
63	.0370	
62	.0380	
61	.0390	
60	.0400	
59	.0410	
58	.0420	
57	.0430	
56	.0465	
3/64	.0469	0-80
55	.0520	
54	.0550	1-56
53	.0595	1-64, 72
1/16	.0625	
52	.0635	
51	.0670	
50	.0700	2-56, 64
49	.0730	
48	.0760	
5/64	.0781	
47	.0785	3-48
46	.0810	
45	.0820	3-56, 4-32
44	.0860	4-36
43	.0890	4-40
42	.0935	4-48
3/32	.0938	
41	.0960	
40	.0980	
39	.0995	
38	.1015	5-40
37	.1040	5-44
36	.1065	6-32
7/64	.1094	
35	.1100	
34	.1110	6-36
33	.1130	6-40
32	.1160	
31	.1200	
1/8	.1250	
30	.1285	
29	.1360	8-32, 36
28	.1405	8-40
9/64	.1406	
27	.1440	
26	.1470	
25	.1495	10-24
24	.1520	
23	.1540	
5/32	.1562	
22	.1570	10-30
21	.1590	10-32
20	.1610	
19	.1660	
18	.1695	
11/64	.1719	
17	.1730	
16	.1770	12-24
15	.1800	
14	.1820	12-28
13	.1850	12-32
3/16	.1875	
12	.1890	
11	.1910	
10	.1935	
9	.1960	
8	.1990	
7	.2010	1/4-20
13/64	.2031	
6	.2040	
5	.2055	
4	.2090	
3	.2130	1/4-28
7/32	.2188	
2	.2210	
1	.2280	
LETTER SIZE DRILLS A	.2340	
15/64	.2344	
LETTER SIZE DRILLS B	.2380	
C	.2420	
D	.2460	
1/4 E	.2500	
F	.2570	5/16-18
G	.2610	
17/64	.2656	
H	.2660	
I	.2720	5/16-24
J	.2770	
K	.2810	
9/32	.2812	
L	.2900	
M	.2950	
19/64	.2969	
N	.3020	
5/16	.3125	3/8-16
O	.3160	
P	.3230	
21/64	.3281	
Q	.3320	3/8-24
R	.3390	
11/32	.3438	
S	.3480	
T	.3580	
23/64	.3594	
U	.3680	7/16-14
3/8	.3750	
V	.3770	
W	.3860	
25/64	.3906	7/16-20
X	.3970	
Y	.4040	
13/32	.4062	
Z	.4130	
27/64	.4219	1/2-13
7/16	.4375	
29/64	.4531	1/2-20
15/32	.4688	
31/64	.4844	9/16-12
1/2	.5000	
33/64	.5156	9/16-18
17/32	.5312	5/8-11
35/64	.5469	
9/16	.5625	
37/64	.5781	5/8-18
19/32	.5938	11/16-11
39/64	.6094	
5/8	.6250	11/16-16
41/64	.6406	
21/32	.6562	3/4-10
43/64	.6719	
11/16	.6875	3/4-16
45/64	.7031	
23/32	.7188	
47/64	.7344	
3/4	.7500	
49/64	.7656	7/8-9
25/32	.7812	
51/64	.7969	
13/16	.8125	7/8-14
53/64	.8281	
27/32	.8438	
55/64	.8594	
7/8	.8750	1-8
57/64	.8906	
29/32	.9062	
59/64	.9219	
15/16	.9375	1-12, 14
61/64	.9531	
31/32	.9688	
63/64	.9844	1 1/8-7
1	1.0000	
1 3/64	1.0469	1 1/8-12
1 7/64	1.1094	1 1/4-7
1 1/8	1.1250	
1 11/64	1.1719	1 1/4-12
1 7/32	1.2188	1 3/8-6
1 1/4	1.2500	
1 19/64	1.2969	1 3/8-12
1 11/32	1.3438	1 1/2-6
1 3/8	1.3750	
1 27/64	1.4219	1 1/2-12
1 1/2	1.5000	

PIPE THREAD SIZES

THREAD	DRILL	THREAD	DRILL
1/8-27	R	1 1/2-11 1/2	1 47/64
1/4-18	7/16	2-11 1/2	2 7/32
3/8-18	37/64	2 1/2-8	2 5/8
1/2-14	23/32	3-8	3 1/4
3/4-14	59/64	3 1/2-8	3 3/4
1-11 1/2	1 5/32	4-8	4 1/4
1 1/4-11 1/2	1 1/2		

Courtesy The L. S. Starrett Company, Athol, Massachusetts.

APPENDIX C American National Standards of Interest to Designers, Architects, and Drafters

TITLE OF STANDARD

- Abbreviations . . . Y1.1-1972
- American National Standard Drafting Practices
  - Size and Format . . . Y14.1-1980
  - Line Conventions and Lettering . . . Y14.2M-1992
  - Multi and Sectional View Drawings . . . Y14.3-1975(R1980)
  - Pictorial Drawing . . . Y14.4-1957
  - Dimensioning and Tolerancing (revision in process 1993–1994) . . . Y14.5M-1982
  - Screw Threads . . . Y14.6-1978
  - Screw Threads (Metric Supplement) . . . Y14.6aM-1981
  - Gears and Splines
    - Spur, Helical, and Racks . . . Y14.7.1-1971
    - Bevel and Hyphoid . . . Y14.7.2-1978
  - Forgings . . . Y14.9-1958
  - Springs . . . Y14.13M-1981
  - Electrical and Electronic Diagram . . . Y14.15-1966(R1973)
  - Interconnection Diagrams . . . Y14.15a-1971
  - Information Sheet . . . Y14.15b-1973
  - Fluid Power Diagrams . . . Y14.17-1966(R1980)
  - Digital Representation for Communication of Product Definition Data . . . Y14.26M-1981
  - Computer-Aided Preparation of Product Definition Data Dictionary of Terms . . . Y14.26.3-1975
  - Digital Representation of Physical Object Shapes . . . Y14 Report
  - Guideline – User Instructions . . . Y14 Report No. 2
  - Guideline – Design Requirements . . . Y14 Report No. 3
  - Ground Vehicle Drawing Practices . . . In Preparation
  - Chasis Frames . . . Y14.32.1-1974
  - Parts Lists, Data Lists, and Index Lists . . . Y14.34M-1982
  - Surface Texture Symbols . . . Y14.36-1978
- Illustrations for Publication and Projection . . . Y15.1M-1979
- Time Series Charts . . . Y15.2M-1979
- Process Charts . . . Y15.3M-1979
- Graphic Symbols for:
  - Electrical and Electronics Diagrams . . . Y32.2-1975
  - Plumbing . . . Y32.4-1977
  - Use on Railroad Maps and Profiles . . . Y32.7-1972(R1979)
  - Fluid Power Diagrams . . . Y32.10-1967(R1974)
  - Process Flow Diagrams in Petroleum and Chemical Industries . . . Y32.11-1961
  - Mechanical and Acoustical Element as Used in Schematic Diagrams . . . Y32.18-1972(R1978)
  - Pipe Fittings, Valves, and Piping . . . Z32.2.3-1949(R1953)
  - Heating, Ventilating, and Air Conditioning . . . Z32.2.4-1949(R1953)
  - Heat Power Apparatus . . . Z32.2.6-1950(R1956)
- Letter Symbols for:
  - Glossary of Terms Concerning Letter Symbols . . . Y10.1-1972
  - Hydraulics . . . Y10.2-1958
  - Quantities Used in Mechanics for Solid Bodies . . . Y10.3-1968
  - Heat and Thermodynamics . . . Y10.4-1982
  - Quantities Used in Electrical Science and Electrical Engineering . . . Y10.5-1968
  - Aeronautical Sciences . . . Y10.7-1954
  - Structural Analysis . . . Y10.8-1962
  - Meteorology . . . Y10.10-1953(R1973)
  - Acoustics . . . Y10.11-1953(R1959)
  - Chemical Engineering . . . Y10.12-1955(R1973)
  - Rocket Propulsion . . . Y10.14-1959
  - Petroleum Reservoir Engineering and Electric Logging . . . Y10.15-1958(R1973)
  - Shell Theory . . . Y10.16-1964(R1973)
  - Guide for Selecting Greek Letters Used as Symbols for Engineering Mathematics . . . Y10.17-1961(R1973)
  - Illuminating Engineering . . . Y10.18-1967(R1977)
  - Mathematical Signs and Symbols for Use in Physical Sciences and Technology . . . Y10.20-1975

APPENDIX D Unified Screw Thread Variations

THE FOLLOWING list is a ready reference of available standard series and selected combinations of Unified screw threads. Each thread is given as part of a proper thread note including major diameter, threads per inch, and series identification.

0–80 UNF
1–64 UNC
1–72 UNF
2–56 UNC
2–64 UNF
3–48 UNC
3–56 UNF
4–40 UNC
4–48 UNF
5–40 UNC
5–44 UNF
6–32 UNC
6–40 UNF
8–32 UNC
8–36 UNF
10–24 UNC
10–28 UNS
10–32 UNF
10–36 UNS
10–40 UNS
10–48 UNS
10–56 UNS
12–24 UNC
12–28 UNF
12–32 UNEF
12–36 UNS
12–40 UNS
12–48 UNS
12–56 UNS
1/4–20 UNC
1/4–24 UNS
1/4–27 UNS
1/4–28 UNF
1/4–32 UNEF
1/4–36 UNS
1/4–40 UNS
1/4–48 UNS
1/4–56 UNS
5/16–18 UNC
5/16–20 UN
5/16–24 UNF
5/16–27 UNS
5/16–28 UN
5/16–32 UNEF
5/16–36 UNS
5/16–40 UNS
5/16–48 UNS
3/8–16 UNC
3/18–16 UNC
3/8–18 UNS
3/8–20 UN
3/8–24 UNF
3/8–24 UNF
3/8–27 UNS
3/8–28 UN
3/8–32 UNEF
3/8–36 UNS
3/8–40 UNS
.390–27 UNS
7/16–14 UNC
7/16–16 UN
7/16–18 UNS
7/16–20 UNF
7/16–24 UNS
7/16–27 UNS
7/16–28 UNEF
7/16–32 UN
1/2–12 UNS
1/2–13 UNC
1/2–14 UNS
1/2–16 UN
1/2–18 UNS
1/2–20 UNF
1/2–24 UNS
1/2–27 UNS
1/2–28 UNEF
1/2–32 UN
9/16–12 UNC
9/16–14 UNS
9/16–16 UN
9/16–18 UNF
9/16–20 UN
9/16–24 UNEF
9/16–27 UNS
9/16–28 UN
9/16–32 UN
5/8–11 UNC
5/8–12 UN
5/8–14 UNS
5/8–16 UN
5/8–18 UNF
5/8–20 UN
5/8–24 UNEF
5/8–27 UNS
5/8–28 UN
5/8–32-UN
11/16–12 UN
11/16–16 UN
11/16–20 UN
11/16–24 UNEF
11/16–28 UN
11/16–32 UN
3/4–10 UNC
3/4–12 UN
3/4–14 UNS
3/4–16 UNF
3/4–18 UNS
3/4–20 UNEF
3/4–24 UNS
3/4–27 UNS
3/4–28 UN
3/4–32 UN
13/16–12 UN
13/16–16 UN
13/16–20 UNEF
13/16–28 UN
13/16–32 UN
7/8–9 UNC
7/8–10 UNS
7/8–12 UN
7/8–14 UNF
7/8–16 UN
7/8–18 UNS
7/8–20 UNEF
7/8–24 UNS
7/8–27 UNS
7/8–28 UN
7/8–32 UN
15/16–12 UN
15/16–16 UN
15/16–20 UNEF
15/16–28 UN
15/16–32 UN
1–8 UNC
1–10 UNS
1–12 UNF
1–14 UNS
1–16 UN
1–18 UNS
1–20 UNEF
1–24 UNS
1–27 UNS
1–28 UN
1–32 UN
1-1/16–8 UN
1-1/16–12 UN
1-1/16–16 UN
1-1/16–18 UNEF
1-1/16–20 UN
1-1/16–28 UN
1-1/8–7 UNC
1-1/8–8 UN
1-1/8–10 UNS
1-1/8–12 UNF
1-1/8–14 UNS
1-1/8–16 UN
1-1/8–18 UNEF
1-1/8–20 UN
1-1/8–24 UNS
1-1/8–28 UN
1-3/16–8 UN
1-3/16–12 UN
1-3/16–16 UN
1-3/16–18 UNEF
1-3/16–20 UN
1-3/16–28 UN
1-1/4–7 UNC
1-1/4–8 UN
1-1/4–10 UNS
1-1/4–12 UNF
1-1/4–14 UNS
1-1/4–16 UN
1-1/4–18 UNEF
1-1/4–20 UN
1-1/4–24 UNS
1-1/4–28 UN
1-5/16–8 UN
1-5/16–12 UN
1-5/16–16 UN
1-5/16–18 UNEF
1-5/16–20 UN
1-5/16–28 UN
1-3/8–6 UNC
1-3/8–8 UN
1-3/8–10 UNS
1-3/8–21 UNF
1-3/8–14 UNS
1-3/8–16 UN
1-3/8–18 UNEF
1-3/8–20 UN
1-3/8–28 UN
1-7/16–6 UN
1-7/16–8 UN
1-7/16–12 UN
1-7/16–16 UN
1-7/16–18 UNEF
1-7/16–20 UN
1-7/16–28 UN
1-1/2–6 UNC
1-1/2–8 UN

1-1/2–10 UNS
1-1/2–12 UNF
1-1/2–14 UNS
1-1/2–16 UN
1-1/2–18 UNEF
1-1/2–20 UN
1-1/2–24 UNS
1-1/2–28 UN
1-9/16–6 UN
1-9/16–8 UN
1-9/16–12 UN
1-9/16–16 UN
1-9/16–18 UNEF
1-9/16–20 UN
1-5/8–6 UN
1-5/8–8 UN
1-5/8–10 UNS
1-5/8–12 UN
1-5/8–14 UNS
1-5/8–16 UN
1-5/8–18 UNEF
1-5/8–20 UN
1-5/8–24 UNS
1-11/16–6 UN
1-11/16–8 UN
1-11/16–12 UN
1-11/16–16 UN
1-11/16–18 UNEF
1-11/16–20 UN
1-3/4–5 UNC
1-3/4–6 UN
1-3/4–8 UN
1-3/4–10 UNS
1-3/4–12 UN
1-3/4–14 UNS
1-3/4–16 UN
1-3/4–18 UNS
1-3/4–20 UN
1-13/16–6 UN
1-13/16–8 UN
1-13/16–12 UN
1-13/16–20 UN
1-7/8–6 UN
1-7/8–8 UN
1-7/8–10 UNS
1-7/8–12 UN
1-7/8–14 UNS
1-7/8–16 UN
1-7/8–18 UNS
1-7/8–20 UN
1-15/16–6 UN
1-15/16–8 UN
1-15/16–12 UN
1-15/16–16 UN
1-15/16–20 UN
2–4 1/2 UNC
2–6 UN
2–8 UN
2–10 UNS
2–12 UN
2–14 UNS
2–16 UN
2–18 UNS
2–20 UN
2-1/16–16 UNS
2-1/8–6 UN
2-1/8–8 UN
2-1/8–12 UN
2-1/8–16 UN
2-1/8–20 UN
2-3/16–16 UNS
2-1/4–4-1/2 UNC
2-1/4–6 UN
2-1/4–8 UN
2-1/4–10 UNS
2-1/4–12 UN
2-1/4–14 UNS
2-1/4–16 UN
2-1/4–18 UNS
2-1/4–20 UN
2-5/16–16 UNS
2-3/8–6 UN
2-3/8–8 UN
2-3/8–12 UN
2-3/8–16 UN
2-3/8–20 UN
2-7/16–16 UNS
2-1/2–4 UNC
2-1/2–6 UN
2-1/2–8 UN
2-1/2–10 UNS
2-1/2–12 UN
2-1/2–14 UNS
2-1/2–16 UN
1-1/2–18 UNS
2-1/2–20 UN
2-5/8–6 UN
2-5/8–8 UN
2-5/8–12 UN
2-5/8–16 UN
2-5/8–20 UN
2-3/4–4 UNC
2-3/4–6 UN
2-3/4–8 UN
2-3/4–10 UNS
2-3/4–12 UN
2-3/4–14 UNS
2-3/4–16 UN
2-3/4–18 UNS
2-3/4–20 UN
2-7/8–6 UN
2-7/8–8 UN
2-7/8–12 UN
2-7/8–16 UN
2-7/8–20 UN
3–4 UNC
3–6 UN
3–8 UN
3–10 UNS
3–12 UN
3–14 UNS
3–16 UN
3–18 UNS
3–20 UN
3-1/8–6 UN
3-1/8–8 UN
3-1/8–12 UN
3-1/8–16 UN
3-1/4–4 UNC
3-1/4–6 UN
3-1/4–8 UN
3-1/4–10 UNS
3-1/4–12 UN
3-1/4–14 UNS
3-1/4–16 UN
3-1/4–18 UNS
3-3/8–6 UN
3-3/8–8 UN
3-3/8–12 UN
3-3/8–16 UN
3-1/2–4 UNC
3-1/2–6 UN
3-1/2–8 UN
3-1/2–10 UNS
3-1/2–12 UN
3-1/2–14 UNS
3-1/2–16 UN
3-1/2–18 INS
3-5/8–6 UN
3-5/8–8 UN
3-5/8–12 UN
3-5/8–16 UN
3-3/4–4 UNC
3-3/4–6 UN
3-3/4–8 UN
3-3/4–10 UNS
3-3/4–12 UN
3-3/4–14 UNS
3-3/4–16 UN
3-3/4–18 UNS
3-7/8–6 UN
3-7/8–8 UN
3-7/8–12 UN
3-7/8–16 UN
4–4 UNC
4–6 UN
4–8 UN
4–10 UNS
4–12 UN
4–14 UNS
4–16 UN

APPENDIX E Metric Screw Thread Variations

THE FOLLOWING list is a ready reference of available standard coarse pitch series ISO metric screw threads. Each thread is given as part of a proper thread note, including metric symbol, major diameter, and thread pitch.

M1 × 0.25
M1.1 × 0.25
M1.2 × 0.25
M1.4 × 0.3
M1.6 × 0.35
M1.8 × 0.35
M2 × 0.4
M2.2 × 0.45
M2.5 × 0.45
M3 × 0.5
M3.5 × 0.6
M4 × 0.7
M4.5 × 0.75
M5 × 0.8
M6 × 1
M7 × 1
M8 × 1.25
M9 × 1.25
M10 × 1.5
M11 × 1.5
M12 × 1.75
M14 × 2
M16 × 2
M18 × 2.5
M20 × 2.5
M22 × 2.5
M24 × 3
M27 × 3
M30 × 3.5
M33 × 3.5
M36 × 4
M39 × 4
M42 × 4.5
M45 × 4.5
M48 × 5
M52 × 5
M56 × 5.5
M60 × 5.5
M64 × 6
M68 × 6

APPENDIX F Geometric Tolerancing Symbol Trees

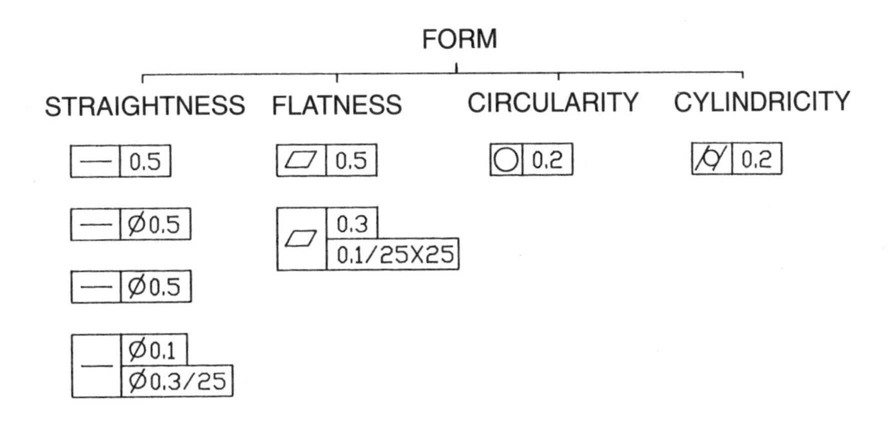

ORIENTATION

PARALLELISM

0.5 A

Ø 0.5 A

0.5 A
EACH ELEMENT

Ø 0.5 A
EACH RADIAL ELEMENT

0.5 A B

PERPENDICULAR

0.5 A

Ø 0.5 A

0.5 A
EACH ELEMENT

Ø 0.5 A

ANGULARITY

0.5 A

0.5 A
EACH ELEMENT

PROFILE

PROFILE OF A LINE

0.0

0.5 A

0.5 A B

0.5 A B C

BETWEEN X & Y

2(NO. OF)SURFACES

PROFILE OF A SURFACE

0.0

0.5 A

0.5 A B

0.5 A B C

BETWEEN X & Y

2(NO. OF)SURFACES

RUNOUT

CIRCULAR

| ↗ | 0.5 | A |

| ↗ | 0.5 | A–B |

| ↗ | 0.5 | A | B |

TOTAL

| ⌰ | 0.5 | A |

| ⌰ | 0.5 | A–B |

| ⌰ | 0.5 | A | B |

LOCATION

POSITION

| ⌖ | ⌀0.5 Ⓜ | A |

| ⌖ | ⌀0.5 Ⓜ | A | B |

| ⌖ | ⌀0.5 Ⓜ | A | B | C |

| ⌖ | ⌀0.5 Ⓢ | A | B |

| ⌖ | ⌀0.5 Ⓜ | A | B Ⓜ |

| ⌖ | ⌀0.5 Ⓢ | A | B | C |

| ⌖ | ⌀0.5 Ⓛ | A | B | C |

| ⌖ | ⌀0.3 Ⓜ | A | B |
| | ⌀0.1 Ⓜ |

| ⌖ | ⌀0.5 Ⓜ | A | B | C |
| | ⌀0.2 Ⓜ | A |

CONCENTRICITY

| ◎ | ⌀0.5 | A |

| ◎ | ⌀0.5 Ⓜ | A |

Glossary

ACME A thread system used especially for feed mechanisms.
ADDENDUM (SPUR GEAR) The radial distance from the pitch circle to the top of the tooth.
ADDENDUM ANGLE (BEVEL GEAR) The angle subtended by the addendum.
ALIGNED SECTION The cutting plane staggered to pass through offset features of an object.
ALIGNMENT CHARTS Designed to graphically solve mathematical equation values using three or more scaled lines.
ALLOWANCE The tightest possible fit between two mating parts.
ALLOYS A mixture of two or more metals.
AMPLIFIER (AMP) A device that allows an input signal to control power; capable of having an output signal greater than the input signal.
ANNEALING Under certain heating and cooling conditions and techniques, steel may be softened.
APPARENT INTERSECTION This is a condition where lines or planes *look* as if they may be intersecting, but in reality they may not be intersecting.
AUXILIARY VIEW A view that is required when a surface is not parallel to one of the principal planes of projection; the auxiliary projection plane is parallel to the inclined surface so that the surface may be viewed in its true size and shape.
AXIS The centerline of a cylindrical feature.

BALL BEARING A friction-reducer where balls roll in two grooved rings.
BASE CIRCLE (CAM) The smallest circle tangent to the CAM follower at the bottom of displacement.
BASE CIRCLE DIAMETER (SPUR GEAR) The diameter of a circle from which the involute tooth is generated.
BASIC DIMENSION A numerical value used to describe the theoretically exact size, profile, orientation, or location of a feature or datum target. It is the basis from which permissible variations are established by tolerances on other dimensions, in notes, or in feature control frames.
BEARING A mechanical device that reduces friction between two surfaces.
BEARING ANGLE The bearing angle of a line is always 90° or less and is identified either from the north or the south.
BEARING OF A LINE The angular relationship of the horizontal projection of the line relative to the compass, expressed in degrees.
BEARING SEAL A rubber, felt, or plastic seal on the outer and inner ring of a bearing. Generally it is filled with a special lubricant by the manufacturer.
BEARING SHIELD A bearing shield is a metal plate on one or both sides of the bearing; serves to retain the lubricant and keep the bearing clean.
BELL AND SPIGOT A pipe connection in which one end of a piece of pipe has a bell-shaped opening and the other end is tapered or notched to fit into the bell.
BELLCRANK A link, pivoted near the center, that oscillates through an angle.
BEND ALLOWANCE The amount of extra material needed for a bend to compensate for compression during the bending process.
BEND RELIEF Cutting away material at a corner to help relieve stress.
BEVEL The term used to denote the slope of beams, as in structural engineering.
BEVEL GEAR Used to transmit power between intersecting shafts; takes the shape of a frustum of a cone.
BIAS The voltage applied to a circuit element to control the mode of operation.
BILATERAL TOLERANCE A tolerance in which variation is permitted in both directions from the specified dimension.
BIT Binary digit.
BOLT CIRCLE Holes located in a circular pattern.
BORE To enlarge a hole with a single-pointed machine tool in a lathe, drill press, or boring mill.
BOSS A cylindrical projection on the surface of a casting for forging.
BOW'S NOTATION A system of notation used to label a vector system. A letter is given to the space on each side of the vector, and each vector is then identified by the two letters on either side of it, read in a clockwise direction.
BROKEN-OUT SECTION A portion of a part is broken away to clarify an interior feature; there is no associated cutting plane line.
BUS An aluminum or copper plate or tubing that carries the electrical current.
BUSHING A replaceable lining or sleeve used as a bearing surface.
BUTT WELD A form of pipe manufacture in which the seam of the pipe is a welded flat-faced joint. Also, a form of welding in which two pieces of material are "butted" against each other and welded.

CABINET OBLIQUE DRAWING A form of oblique drawing in which the receding lines are drawn at half scale and usually at a 45° angle from horizontal.
CAD Computer-Aided Design.

CADD Computer-Aided Design Drafting.

CAD/CAM Computer-Aided Design/Computer-Aided Manufacturing.

CAE Computer-Aided Engineering.

CAM A machine part used to convert constant rotary motion into timed irregular motion.

CAM MOTION The base point from which to begin cam design. There are four basic types of motion: simple harmonic, constant velocity, uniform accelerated, and cycloidal.

CAPACITOR An electronic component that opposes a change in voltage and storage of electronic energy.

CARBURIZATION A process where carbon is introduced into the metal by heating to a specified temperature range while in contact with a solid, liquid, or gas material consisting of carbon.

CARTESIAN COORDINATE SYSTEM A measurement system based on rectangular grids to measure width, height, and depth (X, Y, and Z).

CASTING An object or part produced by pouring molten metal into a mold.

CAVALIER OBLIQUE DRAWING A form of oblique drawing in which the receding lines are drawn true size, or full scale. Usually drawn at an angle of 45 horizontal degrees.

CENTRAL PROCESSING UNIT (CPU) The processor and main memory chips in a computer. Specifically, the CPU is just the processor, but generally it refers to the computer.

CHAIN DIMENSIONING Also known as point-to-point dimensioning, when dimensions are established from one point to the next.

CHAMFER A slight surface angle used to relieve a sharp corner.

CHORDAL ADDENDUM (SPUR GEAR) The height from the top of the tooth to the line of the chordal thickness.

CHORDAL THICKNESS (SPUR GEAR) The straight line thickness of a gear tooth on the pitch circle.

CIM Computer-integrated manufacturing that combines CADD, CAM, and CAE into a controlled system.

CIRCULAR PITCH (SPUR GEAR) The distance from a point on one tooth to the corresponding point on the adjacent tooth, measured on the pitch circle.

CIRCULAR THICKNESS (SPUR GEAR) The length of an arc between the two sides of a gear tooth on the pitch circle.

CLEARANCE (SPUR GEAR) The radial distance between the top of a tooth and the bottom of the mating tooth space.

COIL OR INDUCTOR A conductor wound on a form or in a spiral; contains inductance.

COLD-ROLLED STEEL (CRS) The additional cold forming of steel after initial hot rolling; cleans up hot-formed steel.

COMMAND A specific instruction issued to the computer by the operator. The computer performs a function or task in response to a command.

COMPRESSIVE Pushing toward the point of currency, as in forces that are compressed.

CONCENTRIC Two or more circles sharing the same center.

CONCURRENT FORCES Forces acting on a common point.

CONE DISTANCE (BEVEL GEAR) The slant height of the pitch cone.

CONSTRUCTION LINES Very lightly drawn, nonreproducing lines used for the layout of a drawing.

CONTOUR LINE Denotes a series of connected points at a particular elevation.

COPLANAR FORCES All lie in the same plane.

COUNTERBORE To cylindrically enlarge a hole; generally to allow the head of a screw or bolt to be recessed below the surface of an object.

COUNTERDRILL A machined hole that looks similar to a countersink counterbore combination.

COUNTERSINK Used to recess the tapered head of a fastener below the surface of an object.

CRANK A link, usually a rod or bar, that makes a complete revolution about a fixed point.

CROWN BACKING (BEVEL GEAR) The distance between the cone apex and the outer tip of the gear teeth.

CRT Cathode Ray Tube.

CURSOR A small rectangle, underline, or set of crosshairs that indicate present location on a video display screen. Also, a hand-held input device used in conjunction with a digitizer.

DATUM A theoretically exact point, axis, or plane derived from the true geometric counterpart of a specified datum feature. The origin from which the location or geometric characteristics of features of a part are established.

DATUM DIMENSIONING A dimensioning system where each dimension originates from a common surface, plane, or axis.

DECLINATION A line that goes downward from its origin; assigned negative values.

DEDENDUM (SPUR GEAR) The radial distance from the pitch circle to the bottom of the tooth.

DEDENDUM ANGLE (BEVEL GEAR) The angle subtended by the dedendum.

DEFAULT An action taken by computer software unless the operator specifies differently.

DETAIL A drawing of an individual part that contains all of the views, dimensions, and specifications necessary to manufacture the part.

DIAMETRAL PITCH A ratio equal to the number of teeth on a gear per inch of pitch diameter.

DIAZO A printing process that produces blue, black, or brown lines on various media (other resultant colors are also produced with certain special products). The print process is a combination of exposing an original in contact with a sensitized material exposed to an ultraviolet light and then running the exposed material through an ammonia chamber to activate the remaining sensitized image to form the desired print. This is a fast and economical method of making prints commonly used in drafting.

DIGITIZE The act of locating points and selecting commands using an input device (puck or stylus, used with a digitizer tablet).

DIGITIZER An electronically sensitized flat board or tablet that serves as a drawing surface for the input of graphics data. Images can be drawn or traced, and commands and symbols can be selected from a menu attached to the digitizer.

DIHEDRAL ANGLE The angle that is formed by two intersecting planes.

DIMETRIC DRAWING A pictorial drawing in which two axes form equal angles with the plane of projection. These can be greater than 90° but less than 180° and cannot have an angle of 120 degrees. The third axis may have an angle less or greater than the two equal axes.

DIP The slope of a stratum.

DISPLACEMENT DIAGRAM A graph; the curve on the diagram is a graph of the path of the cam follower. In the case of a drum cam displacement diagram, the diagram is actually the developed cylindrical surface of the cam.
DOCUMENTATION Instruction manuals, guides, and tutorials provided with any computer hardware/software system.
DOWEL PIN A cylindrical fastener used to retain parts in a fixed position or to keep parts aligned.
DRAFT The taper on the surface of a pattern for castings of the die for forgings, designed to help facilitate removal of the pattern from the mold or the part from the die. Draft is often 7–10 degrees but depends on the material and the process.
DRILLING DRAWING Used to provide size and location dimensions for trimming the printed circuit board.
DRUM CAM A drum, or cylindrical, cam is a cylinder with a groove in its surface. As the cam rotates, the follower moves through the groove, producing a reciprocating motion parallel to the axis of the camshaft.
DRUM PLOTTER A graphics pen plotter that can accommodate continuous-feed paper, or in which sheet paper or film is attached to a sheet of flexible material mounted to a drum. The pen moves in one direction and the drum in the other.
DUCT Sheet metal, plastic, or other material pipe designed as the passageway for conveying air from the HVAC equipment to the source.
DUCTILITY The ability to be stretched, drawn, or hammered without breaking.

ECCENTRIC CIRCLE Not having the same center.
ELECTRICAL RELAYS Magnetic switching devices.
ELECTRODISCHARGE MACHINING (EDM) A process where material to be machined and an electrode are submerged in a fluid that does not conduct electricity, forming a barrier between the part and the electrode. A high-current, short-duration electrical charge is then used to remove material.
ELECTROLESS The depositing of metal on another material through the action of an electric current.
ELECTRON BEAM (EB) Generated by a heated tungsten filament used to cut or machine very accurate features in a part.
ELEMENT Any line, group of lines, shape, or group of shapes and text that is so defined by the computer operator.
ELEMENTARY DIAGRAMS Provide the detail necessary for engineering analysis and operation or maintenance of substation equipment.
ENGINEERING CHANGE DOCUMENTS Documents used to initiate and implement a change to a production drawing; engineering change request (ECR) and engineering change notice (ECN) are examples.
ENTITY *See* ELEMENT.
EQUILIBRANT A vector that is equal in magnitude to the resultant and has the opposite direction and sense.
EQUILIBRIUM When a vector system has a resultant of zero, the system is said to be in equilibrium.
EXPLODED ASSEMBLY A pictorial assembly showing all parts removed from each other and aligned along axis lines.

FACE ANGLE (BEVEL ANGLE) The angle between the top of the teeth and the gear axis.
FAULT CONDITION A short circuit that is a zero resistance path for electrical current flow.
FILLET A curve formed at the interior intersection between two or more surfaces.
FIXTURE A device for holding work in a machine tool.
FLANGE A thin rim around a part.
FLATBED PLOTTER A pen plotter where the drawing surface (bed) is oriented horizontally, and paper or film is attached to the surface by a vacuum or an electrostatic charge. The pen moves in both the x and y directions.
FLOPPY DISK A thin, circular, magnetic storage medium encased in a cover. It comes in 8", 5¼", and 3½" sizes.
FLOWCHARTS Used to show organizational structure, steps, or progression in a process or system.
FLOW DIAGRAM A chart-type drawing that illustrates the organization of a system in a symbolic format.
FOLD LINES The reference line of intersection between two reference planes in orthographic projection.
FOLLOWER The cam follower is a reciprocating device whose motion is produced by contact with the cam surface.
FONT A specific type face, such as Helvetica or Gothic.
FORESHORTENED LINE A line that appears shorter than its actual length, because it is at an angle to the line of sight.
FOUR-BAR LINKAGE The most commonly used linkage mechanism. It contains four links: a fixed link called the ground link, a pivoting link called a driver, another pivoted link called a follower, and a link between the driver and follower, called a coupler.
FREE-BODY DIAGRAM A diagram that isolates and studies a part of the system of forces in an entire structure.
FULL SECTION The cutting plane extends completely through the object.
FUNCTION KEYS Extra keys on an alphanumeric keyboard that can be used in a computer program to represent different commands and functions. The active commands for function keys may change several times in a program.

GATE The part of an electronic system that makes the electronic circuit operate; permits an output only when a predetermined set of input conditions are met.
GEAR A cylinder or cone with teeth on its contact surface; used to transmit motion and power from one shaft to another.
GEAR RATIO Any two mating gears have a relationship to each other called a gear ratio. This relationship is the same between any of the following: RPMs, number of teeth, and pitch diameters of the gears.
GEAR TRAIN Formed when two or more gears are in contact.
GRADE OF A LINE A way to describe the inclination of a line in relation to the horizontal plane. The percent grade is the vertical rise divided by the horizontal run multiplied by 100.
GRAPHICAL KINEMATIC ANALYSIS The process of drawing a particular mechanism in several phases of a full cycle to determine various characteristics of the mechanism.
GRAPHICS TABLET *See* DIGITIZER.

HALF SECTION Used typically for symmetrical objects; the cutting-plane line actually cuts through one quarter of the part. The sectional view shows half of the interior and half of the exterior at the same time.
HARDCOPY A paper copy.
HARDWARE The physical computer equipment.
HIGHWAY DIAGRAM A simplified or condensed representation of a point-to-point, interconnecting wiring diagram for an electrical circuit.

HONE A method of finishing a hole or other surface to a desired close tolerance and fine surface finish using an abrasive.

INCLINATION A line that goes upward from its origin; assigned positive values.
INDUCTANCE The property in an electronic circuit that opposes a change in current flow or where energy may be stored in a magnetic field, as in a transformer.
INTEGRATED CIRCUIT (IC) All of the components in a schematic are made up of one piece of semiconductor material.
INTERSECTING LINES When lines are intersecting, the point of intersection is a point that lies on both lines.
ISOMETRIC DRAWING A form of pictorial drawing in which all three drawing axes form equal angles (120°) with the plane of projection.

JIG A device used for guiding a machine tool in the machining of a part or feature.
JOINT The connection point between two links.
JOYSTICK A graphics input device composed of a lever mounted in a small box that allows the user to control the movement of the cursor on the video display screen.

KERF A groove created by the cut of a saw.
KINEMATICS The study of motion without regard to the forces causing the motion.

LAP WELD A form of pipe manufacture in which the seam of the pipe is an angular "lap."
LARGE SCALE INTEGRATION (LSI) More circuits on a single small IC chip.
LASER (LIGHT AMPLIFICATION BY STIMULATED EMISSION OF RADIATION) A device that amplifies focused light waves and concentrates them in a narrow, very intense beam.
LAY Describes the basic direction or configuration of the predominant surface pattern in a surface finish.
LAYER An individual aspect of a CADD drawing that makes a complete drawing when combined.
LEAD (WORM THREAD) The distance that the thread advances axially in one revolution of the worm or thread.
LEVER A link that moves back and forth through an angle; also known as a rocker.
LIGHT PEN A video display screen input device. It is a light-sensitive stylus connected to the terminal by a wire; enables the user to draw or select menu options directly on the screen.
LINE OF SIGHT An imaginary straight line from the eye of the observer to a point on the object being observed. All lines of sight for a particular view are assumed to be parallel and are perpendicular to the projection plane involved.
LOGIC DIAGRAMS A type of schematic that is used to show the logical sequence in an electronic system.

MALLEABLE The ability to be hammered or pressed into shape without breaking.
MASTER PATTERN A one-to-one scale circuit pattern that is used to produce a printed circuit board.
MAXWELL DIAGRAM A combination vector diagram used to analyze the forces acting in a truss.
MECHANICAL JOINT A pipe connection that is a modification of the "bell and spigot" in which flanges and bolts are used with gaskets, packing rings, or grooved pipe ends providing a seal.
MECHANISM A combination of two or more machine members that work together to perform a specific motion.
MOUSE A hand-held input device connected to the terminal by a wire. It is moved across a flat surface to control the movement of the cursor on the screen. It rolls on a small ball that sends directional signals to the computer and may have one or more buttons that serve as function keys.
MULTIVIEW PROJECTION The views of an object as projected upon two or more picture planes in orthographic projection.

NECK A groove around a cylindrical part.
NOMINAL SIZE The designation of the size for a commercial product.
NOMOGRAPH A graphic representation of the relationship between two or more variables of a mathematical equation.
NONDESTRUCTIVE TESTING (NDS) Tests for potential defects in welds; they do not destroy or damage the weld or the part.
NORMAL PLANE A plane surface that is parallel to any of the primary projection planes.
NORMALIZING A process of heating steel to a specific temperature and then allowing the material to cool slowly by air, bringing the steel to a normal state.
NUMERICAL CONTROL (NC) A system of controlling a machine tool by means of numeric codes that direct the commands for the machine movements; computer numerical control (CNC) is a computer command control of the machine movement.

OBLIQUE DRAWING A form of pictorial drawing in which the plane of projection is parallel to the front surface of the object and the receding angle is normally 45°.
OBLIQUE LINE A straight line that is not parallel to any of the six principal planes.
OBLIQUE PLANE Inclined to all of the principal projection planes.
OFFSET SECTION The cutting plane is offset through staggered interior features of an object to show those features in section as if they were in the same plane.
OPERATIONAL AMPLIFIER (OPAMP) A high-gain amplifier created from an integrated circuit.
OUTSIDE DIAMETER (SPUR GEAR) The overall diameter of the gear; equal to the pitch diameter plus two addenda.

PADS Or lands, are the circuit-termination locations where the electronic devices are attached.
PATTERN DEVELOPMENT Based on laying out geometric forms in true size and shape flat patterns.
PERSPECTIVE DRAWING A form of pictorial drawing in which vanishing points are used to provide the depth and distortion that are seen with the human eye. Perspective drawings can be drawn using one, two, and three vanishing points.
PHOTODRAFTING A combination of a photograph(s) with line work and lettering on a drawing.
PICTORIAL DRAWING A form of drawing that shows an object's depth. Three sides of the object can be seen in one view.

PIE CHARTS Used for presentation purposes where portions of a circle represent quantity.
PIERCING POINT A piercing point is a point where a particular line intersects a plane.
PINION GEAR When two gears are mating, the pinion gear is the smaller — usually the driving gear.
PITCH A distance of uniform measure determined at a point on one unit to the same corresponding point on the next unit; used in threads, springs, and other machine parts.
PITCH (WORM) The distance from one tooth to the corresponding point on the next tooth measured parallel to the worm axis; equal to the circular pitch on the worm gear.
PITCH ANGLE (BEVEL GEAR) The angle between an element of a pitch cone and its axis.
PITCH DIAMETER (BEVEL GEAR) The diameter of the base of the pitch cone.
PITCH DIAMETER (SPUR GEAR) The diameter of an imaginary pitch circle on which a gear tooth is designed. Pitch circles of two spur gears are tangent.
PLAIN BEARING Based on a sliding action between the mating parts; also called sleeve or journal bearings.
PLANE A surface that is not curved or warped. It is a surface in which any two points may be connected by a straight line, and the straight line will always lie completely within the surface.
PLATE CAM A cam in the shape of a plate or disk. The motion of the follower is in a plane perpendicular to the axis of the camshaft.
POLAR CHARTS Designed by establishing polar coordinate scales, where points are determined by an angle and distance from a center or pole.
POLYESTER FILM A high-quality drafting material with excellent reproduction, durability, and dimensional stability; also known by the trade name Mylar®.
POLYGONS Polygons are enclosed figures such as triangles, squares, rectangles, parallelograms, and hexagons.
PRESSURE ANGLE The direction of pressure between contacting gear teeth. It determines the size of the base circle and the shape of the involute spur gear tooth, commonly 20°.
PRIME CIRCLE (CAM) A circle with a radius equal to the sum of the base circle radius and the roller follower radius.
PRINTED CIRCUITS (PC) Electronic circuits printed on a board that form the interconnection between electronic devices.
PRINTER A device that receives data from the computer and converts it into alphanumeric or graphic printed images.
PROFILE LINE A profile line is one that is parallel to the profile projection plane; its projection appears in true length in the profile view.
PROJECTION LINE A projection line is a straight line at 90° to the fold line, which connects the projection of a point in a view to the projection of the same point in the adjacent view.
PROJECTION PLANE A projection plane is an imaginary surface on which the view of the object is projected and drawn. This surface is imagined to exist between the object and the observer.

QUENCH To cool suddenly by plunging into water, oil, or other liquid.

RACK Basically a straight bar with teeth on it. Theoretically it is a spur gear with an infinite pitch diameter.
RADIAL MOTION Exists when the path of the motion forms a circle, the diameter of which is perpendicular to the center of the shaft; also known as rotational motion.
RATIO SCALES Special scales that are referred to as logarithmic and semilogarithmic.
REAM To enlarge a hole slightly with a machine tool called a reamer to produce greater accuracy.
RECTILINEAR CHARTS Set up on a horizontal and vertical grid, where the vertical axis identifies the quantities or values related to the horizontal values; also known as line charts.
RELIEF A slight groove between perpendicular surfaces to provide clearance between the surfaces for machining.
REMOVED SECTION A sectional view taken from the location of the section cutting plane and placed in any convenient location of the drawing, generally labeled in relation to the cutting plane.
RESISTORS Components that contain resistance to the flow of electric current.
REVOLUTION An alternate method for solving descriptive geometry problems in which the observer remains stationary and the object is rotated to obtain various views.
REVOLVED SECTION A sectional view established by revolving 90°, in place, into a plane perpendicular to the line of sight; generally used to show the cross section of a part or feature that has consistent shape throughout the length.
RIB A thin metal section between parts to reinforce while reducing weight in a part.
RIGHT ANGLE A right angle is an angle of 90 degrees.
ROCKER A link that moves back and forth through an angle; also known as a lever.
ROLLER BEARINGS A bearing composed of two grooved rings and a set of rollers. The rollers are the friction-reducing element.
ROOT DIAMETER (SPUR GEAR) The diameter of a circle coinciding with the bottom of the tooth spaces.
ROUND Two or more exterior surfaces rounded at their intersection.
RPM Represents revolutions per minute.
RUNOUTS Characteristics of intersecting features, determined by locating the line of intersection between the mating parts.

SCHEMATIC DIAGRAMS Drawn as a series of lines and symbols that represent the electrical current path and the components of the circuit. Provides the basic circuit connection information for electronic products.
SEMICONDUCTORS Devices that provide a degree of resistance in an electronic circuit; types include diodes and transistors.
SKEW LINES Lines that are neither parallel nor intersecting.
SLIDER A link that moves back and forth in a straight line.
SLOPE ANGLE The angle in degrees that the line makes with the horizontal plane.
SOCKET WELD A form of pipe connection in which a plain-end pipe is slipped into a larger opening or "socket" of a fitting. One exterior weld is required; thus no weld material protrudes into the pipe.
SOFTWARE Computer programs stored on magnetic tape or disk that enable a computer to perform specific functions to accomplish a task.
SOLDER An alloy of tin and lead.

SOLDER MASK A polymer coating to prevent the bridging of solder between pads or conductor traces on a printed circuit board.
SOLIDS MODELING A design and engineering process in which a 3-D model of the actual part is created on the screen as a solid part showing no hidden features.
SPACE DIAGRAM A drawing of a vector system showing the correct direction and sense but not drawn to scale.
SPLINE One of a series of keyways cut around a shaft and mating hole; generally used to transfer power from a shaft to a hub while allowing a sliding action between the parts.
SPUR GEAR The simplest, most common type of gear used for transmitting motion between parallel shafts. Its teeth are straight and parallel to the shaft axis.
STRETCHOUT LINE Typically, the beginning line upon which measurements are made and the pattern development is established.
SURFACE CHARTS Designed to show values represented by the extent of a shaded area; also known as area charts.
SURFACE FINISH Refers to the roughness, waviness, lay, and flaws of a machine surface.
SURFACE MOUNT TECHNOLOGY (SMT) The traditional component lead through is replaced with a solder paste to hold the electronic components in place on the surface of the printed circuit board and take up to less than one-third of the space of conventional PC boards.

TANGENT A straight or curved line that intersects a circle or arc at one point only; is always 90° relative to the center.
TAPER A conical shape on a shaft or hole, or the slope of a plane surface.
TEMPERING A process of reheating normalized or hardened steel to a specified temperature, followed by cooling at a predetermined rate to achieve certain hardening characteristics.
TENSILE FORCES Forces that pull away.
TENSILE STRENGTH Ability to be stretched.
THERMOPLASTIC Plastic material may be heated and formed by pressure. Upon reheating, the shape can be changed.
THERMOSET Plastics are formed into permanent shape by heat and pressure and may not be altered after curing.
THERMOSTAT An automatic mechanism for controlling the amount of heating or cooling given by a central or zoned heating or cooling system.
TILT-UP CONSTRUCTION Method using formed wall panels that are lifted or tilted into place.
TOLERANCE The total permissible variation in a size or location dimension.
TRACKBALL An input device consisting of a smooth ball mounted in a small box. A portion of the ball protrudes above the top of the box and is rotated with the hand to move the cursor on the computer screen.
TRANSISTORS Semiconductor devices in that they are conductors of electricity with resistance to electron flow applied and are used to transfer or amplify an electronic signal.
TRANSITION PIECE A duct component that provides a change from square or rectangular to round; also known as a square to round.
TRANSLATIONAL MOTION Linear motion.
TRIANGULATION A technique used to lay out the true size and shape of a triangle with the true lengths of the sides; used in pattern development on objects such as the transition piece.
TRILINEAR CHART Designed in the shape of an equilateral triangle; used to show the interrelationship between three variables on a three-dimensional diagram.
TRIMETRIC DRAWING A type of pictorial drawing in which all three of the principal axes do not make equal angles with the plane of projection.
TRUE LENGTH OR TRUE SIZE AND SHAPE When the line of sight is perpendicular to a line, surface, or feature.
TRUE POSITION The theoretically exact location of a feature established by basic dimensions.

ULTRASONIC MACHINING A process where a high-frequency mechanical vibration is maintained in a tool designed to a desired shape; also known as impact grinding.
UNDERCUT A groove cut on the inside of a cylindrical hole.
UNILATERAL TOLERANCE A tolerance in which variation is permitted in only one direction from the specified dimension.
UPSET A forging metal used to form a head or enlarged end on a shaft by pressure or hammering between dies.

VALVE Any mechanism, such as a gate, ball, flapper, or diaphragm, used to regulate the flow of fluids through a pipe.
VECTOR ANALYSIS A branch of mathematics that includes the manipulation of vectors.
VECTOR DIAGRAM A drawing of the vector system in which the vectors are drawn with the correct magnitude and sense, and to scale.
VECTOR QUANTITY A quantity that requires both magnitude and direction for its complete description.
VELLUM A drafting paper with translucent properties.
VIEWING PLANE LINE Represents the location of where a view is established.
VISUALIZATION The process of recreating a three-dimensional image of an object in a person's mind.

WEB *See* RIB.
WHOLE DEPTH (SPUR GEAR) The full height of the tooth. It is equal to the sum of the addendum and the dedendum.
WIRE FORM A three-dimensional form in which all edges and features show as lines, thus appearing to be constructed of wire.
WIRELESS DIAGRAM Similar to highway diagrams except that interconnecting lines are omitted. The interconnection of terminals is provided by coding.
WIRING DIAGRAM A type of schematic that shows all of the interconnections of the system components, also referred to as a point-to-point interconnecting wiring diagram.
WORKING DEPTH (SPUR GEAR) The distance that a tooth occupies in the mating space. It is equal to two times the addendum.
WORM GEARS Used to transmit power between nonintersecting shafts. The worm is like a screw and has teeth similar to the teeth on a rack. The teeth on the worm gear are similar to the spur gear teeth, but they are curved to form the teeth on the worm.

ZONING A system of numbers along the top and bottom and letters along the left and right margins of a drawing used for ease of reading and locating items.

Index

A

Abbreviations, 552–55 table
Abrasive saw, 226
Abscissa, 530
Absolute coordinates, 58–59
Acme thread, 304
Actual size, 243
Adhesion, 212
Adjacent views, 193
Aeronautical drafter, 3
AGMA, 434
Aligned dimensioning, 240–41
Aligned section, 337
Alignment chart, 542
Allowance, 243, 266
  machining, 270
  shrinkage, 269
Alloy cast iron, 212
Alloys, 211
Aluminum, 214, 216
American Design Drafting Association, 9
American Gear Manufacturers Association. *See* AGMA
American Iron and Steel Institute. *See* AISI
American National Standards Institute. *See* ANSI
American National Standard taper pipe thread, 304
American National thread, 303, 309–10
American Society for Testing Materials. *See* ASTM
Ames Lettering Guide, 102–4
Ammonia, 42–43
Angles, 129–30
  bisector of, 133
  dimensioning, 248
  transferring, 134
Angularity, 387
Annealing, 214
ANSI, 31
  titles of standards, 583 table
Architect's scale, 33–34
Architectural drafter, 1
Architectural lettering, 98
Arcs
  dimensioning, 249–52
  isometric, 97, 508
  projecting on inclined planes, 160
  tangent to a line and circle, 142–44
Area Fill, 80
Arm drafting machine, 25–26
Arrowheads, 114
  dimension line and leader termination with, 245
Arrowless dimensioning, 241
Assembly drawings, 462–74
  detail, 467
  erection, 467, 470
  general, 467
  layout, 465
  pictorial, 470
  subassembly, 470
ASTM, 212
Automatic pencils, 14
Automotive design drafter, 7
Auxiliary section, 340
Auxiliary views, 168–75
  dimensioning, 261
  enlargements of, 172
  plotting curves in, 172
  secondary views, 172–75
Axonometric drawing, 503–4

B

Backspace key, 57
Ball bearings, 445–46
Balloons, 470–71
Band saw, 225
BASIC, 56
Basic dimension, 243, 378
Bearings, 443–52
  codes for, 447
  dimensioning, 448–49
  linear, 443
  lubrication of, 450
  mountings for, 452
  oil grooving of, 450
  plain, 445
  rolling element, 445–46
  rotational, 443
  sealing methods, 450–52
  selection of, 448
  shaft and housing fits, 449
  shaft shoulder and housing shoulder fits, 449–50
  surface finish of shaft and housing, 450
  symbols for, 446–47
Bending, 221
Bevel gears, 435, 443
Bilateral tolerance, 243, 264
Bit, 63, 76
Blacksmithing, 220
Blind hole, 228
Blowback, 43
Blueprint, 41
Bolts, 312
  ASTM and SAE grade markings, 561 table
Bore, 228
Boss, 230
Brass, 216
Break lines, 116
Brinell hardness test, 214
Broken-out section, 339
Bronze, 216
Buttress thread, 304
Byte, 64

C

C, 56
CAD/CAM, 53
CADD, 10, 45
  educational preparation for, 55–56
  future hardware and software for, 73–74

getting started with, 78-82
job market for, 54-55
users of, 53-54
CADD materials, 74-78
Calculator keypad, 58
Cams, 425-33
cam displacement diagrams
constant velocity motion, 428
cycloidal motion, 429-30
simple harmonic motion, 427-28
uniform accelerated motion, 428-29
using different motions, 430
cam followers, 426
in-line, construction of, 430-32
offset, construction of, 432
drum cams, 432-33
types of, 425
Cap screws, 313
Carat, 216
Carburization, 213
Careers in drafting, 1-10
Cartesian coordinate system, 58-59
Cartographic drafter, 8
Case hardened, 213
Casting drawing, 269
Castings, 216-19
centrifugal, 218
cores, 217-18
die, 218
fillets and rounds for, 269-70
investment, 219
permanent, 218-19
sand, 217
Castings drafter, 5
Cast iron, 212
CD-ROM, 78
Centerlines, 112-13
Centers, 222
Central processing unit. *See* CPU
Centrifugal casting, 218
Chain dimensioning, 246
Chain lines, 116-17
Chairs, drafting, 14
Chamfers, 230-31
dimensioning, 249
Chart drawings, 241-42
Charts, 529-44
alignment, 542
column, 537-39
concurrency, 541
distribution, 543
flowcharts, 542
grids, 533-34
labeling, 534-35
nomographs, 540
pictorial, 543-44
pie, 539-40
polar, 540
rectilinear, 530
scale selection in, 530-32
ratio, 533
surface, 537
trilinear, 542
Chemical machining, 226-27
Chilled cast iron, 212
Chuck, 222
CIM, 53-54
Circles, 131
isometric, 96, 508
in perspective, 519-20
projection of, 159-60
rules for, 559 table
tangencies to, 141-45
Circle template, 21-22
Circularity, 383
Circular lines
hand-compass method for sketching, 89
trammel method for sketching, 88-89
Cire perdue, 219
Civil drafter, 3, 5
Civil engineer's scale, 32-33
Cleaning
drafting equipment, 34
of pens, 16-17
Cleaning agents, 18
Clearance fit, 266
Cold-rolled steel, 213
Cold saw, 226
Collet, 222
Color monitor, 63-64
Columbium, 216
Column chart, 537-39
Commands, 57, 80-82
Compasses
beam, 18
bow, 18
circuit-scribing, 18
drop-bow, 18
friction-head, 18
use of, 19-20
Composites, 212
Computer-aided design/computer-aided manufacturing. *See* CAD/CAM
Computer-aided design drafting. *See* CADD
Computer drafting equipment, 62-73
Computer graphics, 47
Computer-integrated manufacturing. *See* CIM
Computer languages, 56-57
Computers, 10, 44-45, 50-53
Concentricity, 397
Concurrency chart, 541
Conical shapes
dimensioning, 249
Construction lines, 110
Contours
dimensioning, 252
visualizing, 158
Control key, 58
Conventional breaks, 340-41
Conversion
inches/metric, 558 table
inches/millimeters, 556 table

Coordinate systems, 58–62
Cope, 217
Copper alloys, 216
Copy, 80
Cores, 217–18
Counterbore, 229, 256
Counterdrill, 229, 256
Countersink, 229, 256
CPU, 64
Crossed helical gears, 436
Cursor keys, 57
Curves
  in perspective, 519–20
  plotting in auxiliary views, 172
Cutting, 221
Cutting-plane lines, 114, 332, 334
Cutting-wheel pointer, 15–16
Cycle, 427
Cylindricity, 383–84

D

Dardelet thread, 304
Data glove, 74
Datum, 243
  axis, 374, 377
  coplanar surface, 372
  features, 243, 370, 372
  feature symbols, 368–69
  partial surface, 372
  precedence and material condition, 400
  reference frame, 370
  target symbols, 374
Datum dimensioning, 247–48
Decimal points, 244
Definitions. *See* Terminology
Descriptive geometry, 191–204
  lines in space, 195–200
  planes, 200–204
  projection of a point, 193–94
  terminology in, 193
Detail drawings, 460–62
Diametral pitch, 438
Diazo prints, 41–43
  making, 41
  safety precautions for, 42–43
  sepias, 42
  storing materials for, 42
Die, 232, 302
Die casting, 218
Digitizer cursor, 48, 67–68, 79–80
Dimension, 243
Dimensional stability, 34
Dimensioning, 240–79
  aligned, 240–41
  angles, 248
  ANSI standard for, 242–43
  arcs, 249–52
  arrowheads, 245
  arrowless, 241
  auxiliary views, 261
  bearings, 448–50
  chain, 246
  chamfers, 249
  chart drawing, 241–42
  conical shapes, 249
  contours, 252
  datum, 247–48
  decimal points, 244
  dual, 244
  fractions, 244–45
  general notes, 262, 264
  hexagons and other polygons, 249
  isometric, 511
  lines, 245
  locations, 259
  machined surfaces, 273, 275
  maximum and minimum dimensions, 269–73
  notes for size features, 255–59
  numerals and dimension lines, 245–46
  origin feature, 261
  for platings and coatings, 268–69
  symbols for, 560 table
  tabular, 241
  terminology for, 243–44
  tolerancing. *See* Tolerancing
  unidirectional, 240
Dimensioning and Tolerancing (ANSI Y14.5M-1982), 242–43, 367
Dimension lines, 113
  numerals and, 245–46
  spacing of, 245
Dimetric drawings, 511
Directional survey drafter, 6
Disk pack, 76
Displacement, 427
Distribution chart, 543
Dividers, 20
Dovetail, 231
Dowel pins, 317
Draft, 269–70, 273
Drafter
  advancement for, 9
  how to become, 9
  job opportunities for, 9
  salaries and working conditions, 10
Drafting equipment, 13–14
Drafting instruments, 18–24
Drafting machines
  arm drafting, 25–26
  controls of and machine head operation, 27–29
  coordinate reading and processing machine, 27
  digital display, 26
  scales of, 31–34
  setup of, 29–30
  sizes of, 26
  track drafting, 26
Drafting media
  factors influencing choice of, 34
Drag, 217
Drawing media, 74
Drawing mode, 78
Drawing pens, 16–17, 74–76

Drawings
  layout of, 164–67
  manipulating, 80–82
  pictorial. *See* Pictorial drawings
  storing, 83–84
  symbols for, 82–83
  working. *See* Working drawings
Drawing Sheet Sizes and Format (ANSI Y14.1), 476
Drill, 227–28
Drilling machine, 222
Drum cams, 432–33
Drum plotter, 70–71
Dry sand, 217
Dual dimensioning, 244
Ductility, 212
Durability, 34
Dusting brush, 18
Dwell, 427

E

Electrical drafter, 3
Electric erasers, 17
Electric lead pointer, 16
Electrochemical machining (ECM), 227
Electrodischarge machining (EDM), 227
Electron beam (EB) cutting and machining, 227
Electronic drafter, 3
Electrostatic plotter, 71
Ellipse, 145, 147
Ellipse template, 23–24, 159, 508
End thrust, 435
Engineering change notice (ECN), 474
  ANSI recommendations for, 476
Engineering change request (ECR), 474
Enlargement
  of auxiliary view, 172
  of multiview, 159
Equivalents (inch/metric), 557 table
Eradicating fluid, 17–18
Erasability, 34
Erasers, 17
Erasing shield, 17
Escape key, 57
Exploded assembly, 513
Extension lines, 113
External threads, 232
Extrusion, 214

F

Face gears, 436
Fall, 427
Fasteners, 300–319
  dowel pins, 317
  keys, keyways, and keyseats, 318
  retaining rings, 318
  rivets, 318–19
  screw-threads, 300–311
  taper pins, 317–18
  threaded, 312–16
  washers, 316
Feature control frame, 377–78
Features, 243
  multiple, position of, 400, 402
  unsectioned, 337
Ferrous metals, 211
Fillets, 161, 231, 269–70, 521
Film, 35–36
  gouging of, 120
First-angle projection, 154, 163–64
Fits
  ANSI/ISO standard for, 268
  clearance, 266, 268
  establishing dimensions for, 267
  force, 267
  interference, 266
  locational, 267
  running and sliding, 267
  transition, 268
Fixed hard disk, 76
Flask, 217
Flatbed plotter, 70
Flatness, 382
Flat-panel display, 64
Flat springs, 320
Floppy disk, 76
Flowchart, 542
Fold lines, 154, 193
Force fit, 267
Forging
  draft for, 270, 273
  hand, 220
  machine, 220–21
Forging drawing, 273
Form tolerance, 378, 381–84
FORTRAN, 56
Founding, 216
Fractions, 244–45
Freak values, 535
Free state variation, 383
French curves, 24
Front projection plane, 193
Front view, 156
Full section, 335
Function board, 67
Function keys, 58

G

Gear and bearing assemblies, 452, 454
Gear Drawing Standards-Part I (ANSI Y14.7.1-1971), 439
Gears, 434–43
  bevel, 443
  intersecting shafting, 435–36
  nonintersecting shafting, 436
  parallel shafting, 435
  rack and pinion, 436, 441
  spur, 436–41
  worm, 443
Gear trains, 439–41
Generation, 43
Geological drafter, 6
Geometric construction, 129–47
  of circles and tangencies, 141–45
  common constructions, 132–35

of ellipse, 145, 147
geometric shapes, 129–31
line characteristics, 129
of polygons, 135–41
Geometric tolerancing, 367–408
basic dimensions, 378
combination controls, 408
composite positional, 402–3
datums. *See* Datums
feature control frame, 377–78
of form, 378, 381–84
of location, 396–400
material condition
datum precedence and, 400
symbols, 393–96
multiple features location, 400, 402
of orientation, 386–87
of profile, 384–86
projected tolerance zone, 403, 405
of runout, 387, 389–90, 393
symbols for, 368, 587–88 table
virtual condition, 407–8
Geophysical drafter, 6
Ghosting, 34
Gib head dimensions, 580 table
Glossary, 589–94
Gold, 216
Gray cast iron, 212
Green sand, 217
Grid, 81
Grids, 533–34
Grinding machine, 222
Grooves
dimensioning, 259
Ground line, 516
Guidelines, 110
Ames Lettering Guide
column of holes on the frame, 104
column with equally speced holes, 102–3
cross-sectioning, 103
four guidelines, 103
metric, 103–4
slanted or vertical, 103
symbol template, 104
three-fifths column of holes, 104
two-thirds ratio column of holes, 104

H

Half section, 336
Hand forging, 220
HATCH, 80
Head cap screws
hexagon and spline socket dimensions, 564 table
hexagon and spline socket flat countersunk dimensions, 565 table
slotted fillister dimensions, 570 table
slotted flat countersunk dimensions, 568 table
slotted round dimensions, 569 table
Heating and ventilating drafter, 6–7
Heat treating, 213
Helical gears, 435
Hexagons, 138
across the corners method of construction, 139–40
across the flats method of construction, 139
dimensioning, 249
Hex cap screw dimensions, 563 table
Hex nut dimensions, 572 table
Hidden lines, 112
High carbon steel, 212
High-speed steel, 213
Holes
dimensioning, 255–56
Home key, 57
Honing, 222
Horizon line, 516
Horizontal mill, 224
Horizontal projection plane, 193
Hot-rolled steel, 212
Housing fits, 449
HVAC drafter, 6–7
Hypoid gears, 436

I

Impact printer, 71–72
Inclined lettering, 98
Inclined plane
projection of circles and arcs on, 159–60
In control, 237
Incremental coordinates, 59–60
Inks, 17
erasing, 120
Input devices, 66–68
Interference fit, 266
Internal threads, 233
International Organization for Standardization. *See* ISO
Investment casting, 219
Involute curve, 437
Irregular curves, 24
Irregular shapes, 91–92
ISO, 303
Isometric drawings, 505–11
construction techniques
box or coordinate method, 507
centerline layout method, 507–8
circles and arcs, 508
fillets and rounds, 509
intersections, 509
sections, 509
spheres, 511
threads, 511
dimensioning, 511
isometric scale, 505
long-axis, 506
regular, 506
reverse, 506
Isometric lines, 95
Isometric sketches, 93–97
arcs, 97
circles, 96
nonisometric lines, 95
setting up an axis, 93–94

J

Jobs in drafting, 1-10
Joystick, 68

K

Kerf, 225, 232
Keyboard, 57-58, 66-67
Keys, 232, 318
  class 2 fit for parallel and taper, 581 table
  dimensions and tolerances, 579 table
  size of vs. shaft diameter, 578 table
  Woodruff dimensions, 573-75 table
Keyseats, 232, 318
  dimensioning, 258
  Woodruff dimensions, 576-77 table
Keyways, 232, 318
Killed steel, 213
Knurls, 234
  dimensioning, 258-59

L

Lag screws, 314
Landscape drafter, 1, 3
Lapping, 222
Laser machining, 227
Lathe, 222, 302
Lay, 275
Layout techniques, 164-67, 522
Leader lines, 114
Lead grades, 14-15
Least material condition. *See* LMC
Lettering, 97-108
  architectural, 98
  composition and, 102
  guidelines for, making, 102-5
  inclined, 98
  in industry, 100
  legibility of, 100
  lettering guide templates, 105
  lower-case, 98
  machine lettering, 106-7
  mechanical lettering equipment, 105-6
  microfont, 97
  single-stroke Gothic, 97
  techniques for, 102
  transfer lettering, 107
  using lettering equipment, 106
  vertical freehand, 100-102
Lettering guide templates, 105
Lettering pens, 106
Limits, 264
Line Conventions and Lettering (ASME Y14.2M-1992), 97, 109
Line of sight, 193
Lines, 109-17
  arrowheads, 114
  break, 116
  centerlines, 112-13
  chain, 116-17
  characteristics of, 129
  classification of, 195
  construction and guidelines, 110
  cutting-plane and viewing-plane, 114-15, 332, 334
  dimension and leader, 113-14, 245
  dividing into equal parts, 134-35
  end view of, 200
  extension, 113
  fold, 154, 193
  hidden, 112
  object, 110
  oblique, 195
  parallel, 132-33, 195, 198
  pencil and ink techniques for drawing, 117-21
  perpendicular, 133
  perpendicular bisector of, 133
  phantom, 116
  projection, 193
  section, 115-16, 334-35
  stitch, 117
  true length of, 198-200
Line shading, 521
LISP, 56
LMC, 243, 266, 396
  positional tolerance at, 400
Locational fit, 267
Location dimensions, 240, 259
Location tolerance, 396-400
Lost-wax casting, 219
Lower-case lettering, 98
Lower control limit (LCL), 238
Lubrication of bearings, 450
Lug, 230

M

Machined features, 227-35
  design and drafting of, 234-35, 278-79
  terminology for, 227-34
Machined surfaces, 273, 275-76
Machine forging, 220-21
Machine lettering, 106-7
Machine processes, 221-35
Machine screws, 312
  slotted flat countersunk head dimensions, 571 table
Machine tools, 222-26
Machining allowance, 270
Magnetic disks, 76-77
Magnetic tape, 77-78
Mainframe computer, 65
Malleable cast iron, 212
Manufacturing materials, 211-16
  cast iron, 212
  nonferrous metals, 214, 216
  plastics, 212
  precious and specialty metals, 216
  steel, 212-14
Manufacturing processes, 216-21. *See also* Machine processes; Manufacturing materials
  castings, 216-19
  forgings, 220-21
Mass storage, 76
Matrix display, 64
Matte, 35
Maximum material condition. *See* MMC

Measurement lines, 90
Mechanical drafter, 6
Mechanical lettering equipment, 105–6
Mechanical pencils, 14
Mechanical Spring Representation (ANSI Y14.13M-1981), 300
Mechanism, 425
Medium carbon steel, 212
Mental skills, 49
Menu, 78–80
Metric scale, 31–32
Metric thread, 303, 309
  screw thread variations, 586 table
Microcomputer, 65–66
Microfilm, 43–45
  aperture card and roll film, 44
  computerized filing and retrieval, 44–45
  reader-printers, 44
Microfont lettering, 97
Mild steel, 212
Millimeter, 31
Milling machine, 224
Minicomputer, 65
MIRROR, 80
MMC, 243–44, 266
  effect on form tolerance, 395–96
  positional tolerance at, 398–99
  zero positional tolerance at, 399
Model mode, 78
Monitors, 63–64
Monochrome monitor, 63–64
Motor skills, 48–49
Mouse, 68, 80
Multi and Sectional View Drawings (ANSI Y14.3-1975), 153, 332
Multiviews, 153–67
  first-angle projection, 163–64
  layout techniques, 164–67
  projection of circles and arcs, 159–62
  sketching, 92
  third-angle projection, 162
  view selection, 156–59
Mylar, 35–36, 74

N

Necks, 232
  dimensioning, 259
Negative angles, 29
Nodular cast iron, 212
Nomograph, 540
Nonferrous metals, 211, 214, 216
Nonimpact printer, 72
Nonisometric lines, 95
Normalizing, 213
Normal planes, 201
Notes
  general, 262, 264
  for size features, 255–59
  for thread, 309–11
Nuts, 312, 316

O

Object lines, 110
Oblique drawings, 513, 515
  cabinet, 515
  cavalier, 515
  general, 515
Oblique lines, 195
Oblique planes, 201
Occupations in drafting, 1–10
Octagons, 140
Offset section, 336–37
Ogee curve, 144–45
Oil and gas drafter, 7
One-point perspective, 516–17
Optical disks, 78
Ordinate, 530
Orientation tolerance, 386–87
Origin, 58, 261
Orthographic projection, 92, 153
Out of control, 237
Output devices, 68, 70–72

P

Pad, 230
Papers, 34–35
Parallel bar, 20
Parallelism, 386–87
Parallel lines, 132–33, 195, 198
Partial views, 158–59
Pascal, 56
Patent drafter, 5
Pencils, 14
Pens, 16–17, 74–76
Perfect form envelope, 393
Period, 427
Permanent casting, 218–19
Perpendicularity, 387
Perpendicular lines, 133
  bisector of, 133
Perspective drawings, 515–20
  circles and curves in, 519–20
  general concepts of, 516
  one-point, 516–17
  three-point, 518–19
  two-point, 517–18
Phantom lines, 116
Photocopy printers, 43
Photogrammetrist, 8–9
Pictorial chart, 543–44
Pictorial drawings, 153, 503–22
  dimetric, 511
  exploded, 513
  isometric. *See* Isometric drawings
  layout techniques, 522
  oblique, 513, 515
  perspective. *See* Perspective drawings
  shading techniques, 521
  technical illustration and, 504
  trimetric, 512
  uses of, 504
Picture plane, 516
Pie chart, 539–40
Piping drafter, 7

Pixel, 63
Plain bearings, 445
Plane of projection, 153
Planes, 200–204
  classification of surfaces, 201
  edge view of, 202–3
  locating in space, 201
  true shape of, 203–4
Plastic leads, 15
Plastics, 212
Platinum, 216
Plotters, 70–71
Plumbing drafter, 7
Pocket pointer, 15
Point
  projection of a, 193–94
Pointing devices, 79–80
Polar chart, 540
Polar coordinates, 60
Polyester film, 35–36
Polyester leads, 15, 120–21
Polygons, 131
  constructing, 135–41
  dimensioning, 249
Polymers, 212
Portables, 65
Positive angles, 28–29
Power hacksaw, 225
Precious metals, 216
Printers, 71–72
Profile projection plane, 193
Profile tolerance, 384–86
Projection
  of circles and arcs, 159–62
  definition of, 503
  first-angle, 154, 163–64
  of point, 193–94
  third-angle, 154, 162
Projection lines, 193
Projection plane, 193
Prompts, 57
Proportional dividers, 20
Puck, 67–68, 80
Punching, 221
Purchase parts, 474

Q

Quadrilaterals, 130–31
Quality control, 236–38
Quenching, 213
QWERTY, 57

R

Rack and pinion, 436, 441
Radial load, 443
RAM, 64
Random access memory. *See* RAM
Raster display, 63
Ratio scales, 533
Read only memory. *See* ROM
Ream, 228
Reamer, 228
Rectilinear chart, 530
Refractories, 212
Regardless of feature size. *See* RFS
Regular polygons, 131
Relative coordinates, 59–60
Removed sections, 342–43
Rendering, 54
Repetitive features, 259
Reproduction, 36
Retaining rings, 318
Return key, 57
Revision block, 39, 41
Revolution
  conventional, 339
  of sections, 341
RFS
  datum feature, 395
  effect on axis straightness, 393
  effect on surface straightness, 393
  positional tolerance at, 400
Rimmed steel, 213
Rise, 427
Rivets, 318–19
Rockwell hardness test, 214
Rolled thread, 304
Roller bearings, 446
ROM, 64
Rounded corners, 162
Rounds, 161, 231, 269–70, 521
Runner and sprue, 217
Running and sliding fits, 267
Runouts, 162
Runout tolerance, 387, 389–90, 393
  circular runout, 390
  total runout, 390, 393

S

Sand casting, 217
Sanding block, 15
Saw machines, 225–26
Scales
  aligning, 29–30
  architect's, 33–34
  arithmetic, 530–32
  civil engineer's, 32–33
  labeling, 534–35
  metric, 31–32
  notation for, 31
  ratio, 533
  shapes of, 31
  uniform, 542
Scatter, 530
Screws. *See also* Screw-thread fasteners; Threaded fasteners
  ASTM and SAE grade markings, 561 table
Screw-thread fasteners, 232, 300–311
  measuring, 311
  terminology for, 301–2
  thread forms, 303–4
  thread notes, 309–11
  thread representations, 305–9

detailed, 305, 307–9
schematic, 305–7
simplified, 305–6
tools for, 302
Screw thread inserts, 314–15
Screw Thread Representation (ANSI Y14.6-1978), 300
Screw Thread Representation (Metric Supplement) (ANSI Y14.6aM-1981), 300
Scriber, 105
Sealing methods, 450–52
Secondary auxiliary views, 172–75
Second generation drawing, 43
Section lines, 115–16, 334–35
Sections, 332–43
aligned, 337
auxiliary, 340
broken-out, 339
conventional breaks in, 340–41
conventional revolutions in, 339
cutting-plane lines, 332, 334
full, 335
half, 336
isometric, 509
offset, 336–37
removed, 342–43
revolved, 341
section lines, 334–35
unsectioned features, 337
Self-tapping screws, 314
Sepias, 42
Set screws, 314
hexagon and spline socket dimensions, 566–67 table
Shading techniques, 521
Shaft fits, 449
Shaper, 226
Sharp-V thread, 303
Sheet sizes, 36
Shrinkage allowance, 269
Silver, 216
Single-stroke Gothic lettering, 97
Size dimensions, 240
Sketching, 87–97
circular lines, 88–89
irregular shapes, 91–92
isometric drawings, 93–97
measurement lines and proportions, 90
multiviews, 92
straight lines, 88
tools and materials for, 87–88
Slots
dimensioning, 258
Smithing, 220
Smoothness, 34
Society of Automotive Engineers. *See* SAE
Software, 73
Solids modeling, 54
Specified dimension, 244
Spheres
isometric, 511
Spline, 232
Spotface, 229–30, 256
Springs, 319–23
flat, 320
representations of, 321–23
detailed, 322
schematic, 322–23
simplified, 323
terminology for, 320
torsion, 320
Spur gears, 435–39
Squares, 137–38
Square thread, 304
Station point, 516
Statistical process control (SPC), 237–38
Statistics, 529
Steel, 212–14
cold-rolled, 213
hardening of, 213–14
high-speed, 213
hot-rolled, 212
numbering systems for, 213
properties of, 212
Stitch lines, 117
Storage, 72–73
Storage media, 76–78
Straight lines, 88
Straightness, 381–82
Structural drafter, 5
Stylus, 67–68, 80
Surface chart, 537
Surface finish, 273
for bearings, 450
Surface roughness, 273, 275
Surface texture, 234
Surface waviness, 275
Swaging, 221
Symbols, 82–83, 279
bearings, 446–47
datum
features, 368–69
targets, 374
dimensioning, 560 table
geometric tolerancing, 368
material condition, 393–96
surface finish, 275–76
Symbols for Welding and Nondestructive Testing Including Brazing (ANSI/AWS A2.4-1979), 300
Symmetry, 398

T

Tables, drafting, 14
Tabular dimensioning, 241
Tangents, 131
constructing, 141–45
Tap, 233, 302
Tap drill
sizes and decimal equivalents, 582 table
Taper pins, 317
Technical illustration, 504
Technical illustrator, 7–8
Technical pens, 16

Tempering, 214
Templates
  bolt or screw head, 315–16
  circle, 21–22
  ellipse, 23–24
  lettering, 105
Tensile strength, 212
Terminology
  for cam displacement diagrams, 427
  for descriptive geometry, 193
  for dimensioning, 243–44
  for machined features, 227–34
  for machined surfaces, 273, 275
  for screw-threads, 301–2
  for springs, 320
  spur gears, 437
  for title blocks, 39
  for tolerancing, 264–66
Thermoplastics, 212
Thermosets, 212
Third-angle projection, 154, 162
Threaded fasteners, 312–16
  bolts and nuts, 312
  cap screws, 313
  drawing, 315
    templates for, 315–16
  lag and wood screws, 314
  machine screws, 312
  nuts, 316
  self-tapping screws, 314
  set screws, 314
  thread inserts, 314–15
Threads, 232–33
  forms of, 303–4
  isometric, 511
  notes for, 309–11
  representations of, 305–9
3-D coordinates, 60–62
Three-point perspective, 518–19
Thrust load, 443
Titanium, 216
Title blocks, 37–39, 41
  revision block instructions, 39, 41
  terminology for, 39
Tolerance, 244
Tolerancing, 264–68. *See also* Dimensioning; Geometric tolerancing
  fits
    ANSI/ISO metric limits and, 268
    dimensions for, 267–68
    force, 267
    locational, 267
    running and sliding, 267
    selecting, 266
    standard ANSI designation, 266
  terminology for, 264–66
Tool design, 235
Tool design drafter, 5
Top projection plane, 193
Torsion springs, 320
Tracing pins, 106
Trackball, 68
Track drafting machine, 26
Transfer lettering, 107
Transition fit, 268
Transparency, 34
Triangles, 20–21, 130
  constructing by triangulation, 135–37
Trilinear chart, 542
Trimetric drawings, 512
True-height lines, 517
True shape of a plane, 201
T-slot, 233
Tungsten, 216
Turret lathe, 222
Two-point perspective, 517–18

U

Ultrasonic machining, 227
Ultraviolet light, 43
Unidirectional dimensioning, 240
Unified Thread Series, 300, 303, 309–10, 562 table
  screw thread variations, 584–85 table
Uniform scales, 542
Unilateral tolerance, 244, 264
Universal milling machine, 224–25
Upper control limit (UCL), 238
Upset forging, 220–21

V

Vector monitor, 63
Vellum, 34–35
Vernier scale, 28–29
Vertical freehand lettering, 100–102
  curved elements, 101
  decimal points, 102
  fractions, 101–2
  numerals, 101
  straight elements, 101
Vertical mill, 224
Viewing-plane lines, 114–15
Viewpoint, 81–82
Virtual condition, 407–8
Virtual reality, 74

W

Washers, 316
Welding, 221
White cast iron, 212
Whitworth thread, 304
Wire form, 54
Wood screws, 314
Working drawings, 460–76
  assembly drawings, 462–70
  detail drawings, 460–62
  engineering change notice, 474, 476
  engineering change request, 474
  identification numbers, 470–71
  parts list, 472, 474
  purchase parts, 474
Worm gears, 436, 443

X

X axis, 58, 530

Y

Y axis, 58, 530

Z

Zoning, 36–37
ZOOM, 80